International Conference on Optoelectronic Science and Materials 2019

IOP Conference Series: Materials Science and Engineering
Volume 711

Hefei, China
20 - 22 September 2019

ISBN: 978-1-7138-2570-8
ISSN: 1757-8981

Printed from e-media with permission by:

Curran Associates, Inc.
57 Morehouse Lane
Red Hook, NY 12571

Some format issues inherent in the e-media version may also appear in this print version.

Institute of Physics
Dirac House, Temple Back
Bristol BS1 6BE UK

Phone: 44 1 17 929 7481
Fax: 44 1 17 920 0979

techtracking@iop.org

Additional copies of this publication are available from:

Curran Associates, Inc.
57 Morehouse Lane
Red Hook, NY 12571 USA
Phone: 845-758-0400
Fax: 845-758-2633
Email: curran@proceedings.com
Web: www.proceedings.com

TABLE OF CONTENTS

IOP Conf. Series: Materials Science and Engineering **711** (2020) 011001 doi:10.1088/1757-899X/711/1/011001

Preface

The 2019 International Conference on Optoelectronic Science and Materials (ICOSM2019), co-sponsored by the Key Laboratory of Infrared and Low Temperature Plasma in Anhui Province and Guangdong Academy of International Academic Exchange, was held on September 20-22, 2019 in Hefei, China. As an International Symposium on optoelectronics and electronic engineering, the conference aims to provide a communication platform for experts, scholars, engineers and technicians from home and abroad, which covers broad research fields of "Optoelectronic Technology Application", "Photovoltaic Science Material", "Photoelectric Signal Processing", "Low Temperature Plasma Technology and Application", and "Laser Technology and Application". The symposium is therefore expected to spread frontier technologies in these fields, broaden the research ideas, strengthen the academic research and discussion, as well as promote the industrialization of academic achievements.

We were very honored to have Prof. Yongshun Ling representing Chinese Academy of Science as honorary chairman, and Prof. Yihua Hu from National University of Defense Technology, as well as Prof. Pei Wang from University of Science and Technology of China as other two chairmen. It's wonderful to invite four outstanding keynote speakers with a variety expertise as follows: 1) Prof. Chongwen Zhou from University of Science and Technology of China; 2) Prof. Ming Li from Substance research institute, China; 3) Prof. Yujun Zhang from Electronic Engineering Institute of PLA, China and State Key Laboratory of Pulsed Power Laser Technology; 4) Prof. Liang Li from Anhui University, China.

We are pleased to introduce the Proceedings of ICOSM2019 which contain a selection of papers from contributed presentation in the conference. The Proceedings features recent result and contemporary research in following topics: 1) Infrared technology and infrared information processing; 2) Optical information processing technology and application; 3) Photoelectric functional materials and devices; 4) Plasma technology; 5) Laser technology and applications. All the papers have been through rigorous review and process to meet the requirements of International publication standard.

On behalf of the committee, we would like to express our sincere gratitude to the distinguished speakers, to the chairman and all participants for their great support. We would also like to thank the contributors for submitting their papers and patiently revised them according to the suggestions of the designated reviewers. We hope that those volumes will not only provide the readers a broad overview of the latest advances but also provide the researchers a valuable summary and reference in this field.

The Organizing Committee of ICOSM2019

IOP Conf. Series: Materials Science and Engineering **711** (2020) 011001 doi:10.1088/1757-899X/711/1/011001

Committee member

Honorary Chairman

Prof.Yongshun Ling, Chinese Academy of Science (Academician)

Chairman

Prof.Yihua Hu
Prof. Pei Wang

Secretary-general

Prof. Lu yuan (State Key Laboratory of Pulsed Power Laser Technology)

Committee member

Prof. Ming-Yang Chen, Jiangsu University, Department of Opto-Electronic Information Science and Engineering, China
Dr. Jinhua Hu, School of Information & Electrical Engineering, Hebei University of Engineering, China
Prof. Chongwen Zhou, University of Science and Technology of China
Prof. Dabin Yu, China State Key Lab of Pulsed Power Laser
Prof. Hua Yang, Shanghai Jiaotong University, China
Prof. Jiaming Shi, China State Key Lab of Pulsed Power Laser
Prof. Karima HORCHANI-NAIFER, Centre National des Recherches en Sciences des Matériaux, Technopole Borj Cedria, Laboratoire de physico-Chimie des Matériaux Minéraux et leurs Applications, TUNISIA
Prof. Liang Li, Anhui University/ Institutes of Physical Science and Information Technology
Prof. Lu yuan , State Key Laboratory of Pulsed Power Laser Technology
Prof. Marko Topič, University of Ljubljana, Slovenia
Prof. Ming Li, Substance research institute, China
Prof. Ranjana Jha, Netaji Subhas University of Technology, India
Prof. Xiaoning Fu, University of Electronic Science and Technology, China
Prof. Xiaoxia Li, State Key Laboratory of Pulsed Power Laser Technology
Prof. Yanlin Liao, Anhui University, China
Prof. Yujun Zhang, China State Key Lab of Pulsed Power Laser, China
Asso. Prof. Ming Li, Chinese Academy of Science/ Solid State Physics
Asst. Prof. Reena Ittyachan, Department of Physics, Sacred Heart College, Chalakudy, Affiliated to University of Calicut, India
Dr. Beddiaf ZAIDI, Department of Physics, Faculty of Material Sciences, University of Batna 1, Algeria
Dr. C. Shivakumara, Indian Institute of Science / Solid State and Structural Chemistry Unit., India
Dr. D.Sajan, Department of Physics, Bishop Moore College, Mavelikara, Affiliated to University of Kerala, India
Dr. Zaineb Baccouch, National School of Engineers of Sfax/Mechanical Department, Tunisia
Dr. Zhaohui Hu, Beihang University / School of Instrumentation and Optoelectronic Engineering, China
Dr. Ruixue Liu, Retina Health Center/ Research Department

Peer review statement

All papers published in this volume of *IOP Conference Series: Materials Science and Engineering* have been peer reviewed through processes administered by the proceedings Editors. Reviews were conducted by expert referees to the professional and scientific standards expected of a proceedings journal published by IOP Publishing.

Evolution of Microstructure and Properties of ZYK530 Magnesium Alloy during Extrusion-Forging compound Forming

Fenghong Cao[1*], Chang Chen[1] and Yaohui Xu[1]

[1] School of Physics and Electronic Engineering, Laboratory for Functional Materials, Leshan Normal University, Leshan, Sichuan 614004, China

[*]Corresponding author's e-mail: lscaofh2004@sina.com

Abstract. The evolution of microsturcture and mechanical properties of ZYK530 Magnesium alloy during extrusion and die forging compound forming process were studied and analyzed. The results show that the as-cast ZYK530 alloy mainly consists of a-Mg matrix and I-Mg_3Zn_6Y and a small amount of W-$Mg_3Y_2Zn_3$ and Mg_2Zn_3 phases. During extrusion to die forging forming, the grain size of as-cast ZYK530 Magnesium alloy is refined and uniform under the action of three-dimensional compressive stress and high temperature deformation. The I-Mg_3Zn_6Y phase of ZYK530 magnesium alloy decrease relatively, while the distribution of W-$Mg_3Y_2Zn_3$ is more obvious, a small amount of Z-$Mg_{12}ZnY$ and Mg-Zn phases and undetermined phases can be observed in the matrix. Tensile test results show that the tensile strength, yield strength and elongation of extruded preformed specimens are increase by 32.6%, 18% and 18% respectively compared with as-cast specimens. The mechanical properties of final forged specimens are improved by 9%, 38.6% and 23% respectively compared with those of extruded preformed specimens. The fracture mechanism changes form as-cast cleavage fracture to mixed fracture mechanism of toughness and brittleness.

1. Introduction

As the lightest and environmentally friendly metal structure material, magnesium alloy has received more and more attention in energy saving and emission reduction and product weight reduction, and it is becoming more and more widely used in automobiles, ships, military industry, 3c and other fields[1~4]. Y, Zn and Zr are important alloying elements for improving the mechanical properties of magnesium alloys[5~9]. According to the related literature[10~13], the Zn/Y mass ratio affects the number and distribution of the I-Mg_3Zn_6Y phase (the icosahedral structure), W-$Mg_3Zn_3Y_2$ phase (cubic structure) and Z-$Mg_{12}ZnY$ phase (18R modulation structure) in the magnesium alloy matrix, wherein the plastic properties of the three phases are I-Mg_3Zn_6Y phase>W-$Mg_3Zn_3Y_2$ phase>Z-$Mg_{12}ZnY$ phase, therefore, the selection of a suitable Zn/Y is an important way to control the mechanical properties of the alloy.

Deformation is an important way to change the properties of magnesium alloys[13], by refining alloy grain shape deformation, changing the grain orientation, eliminating the internal stress of formation in the process of casting and extrusion forming. However, at present, there is not much research on the microstructure and properties of semi-continuous casting ZYK530 alloy during the extrusion forming process. It is reported[8] that the mechanical properties at room temperature are excellent when the mass ratio of Zn/Y is 1.33. Therefore, in this paper, based on the research of alternative materials for automobile steering arm, the as-cast ZYK530 Mg alloy is taken as the research object, and the evolution of microstructure and mechanical properties of ZYK530 Magnesium alloy during extrusion and forging compound forming is discussed, which provides a certain experimental theoretical basis for further improvement in strength and ductility of Mg-Zn-Zr series alloy, and also provides a certain experimental basis for the research on the lightweight of Magnesium alloy in automotive materials.

2. Experimental methods and materials

The experimental alloys were prepared by semi-continuous casting method, and the raw materials were ZK60 magnesium alloy, pure Mg ingots, pure Zn ingots, Mg-Zr (containing Zr30%) and Mg-Y (containing Y30%) intermediate alloys. The experimental alloys were smelted in a crucible, and the ZK60 magnesium alloy was used as a base material. 5wt%Zn, 3wt%Y and 0.6wt%Zr were respectively added to Zn, Mg-Y and Mg-Zr intermediate alloy when the melt temperature reached 740℃, the melt was refined by Ar gas and then continuously cast into Φ112mm×125mm ingots at 730℃. The composition of the alloys was analyzed by XRF(X-Ray Fluorescence) as shown in Table1. The sample of extrusion preformed billet was prepared on XJ-800T extruder after the ingot was solution treated at 400℃×12h, as shown in Fig.1. The extrusion temperature is 420℃,the extrusion ratio and speed are 16:1 and 17mm/s respectively, and then the extrusion pre-formed sample is put into the die cavity for final forging. The forging direction is consistent with the extrusion direction. The forging temperature is 440℃, the degree of deformation is 50%, and the forged speed is 5.6mm/s, as shown in Fig.1. The ASTM(B557M-06) gauge size is 25mm×6mm×3.5mm.

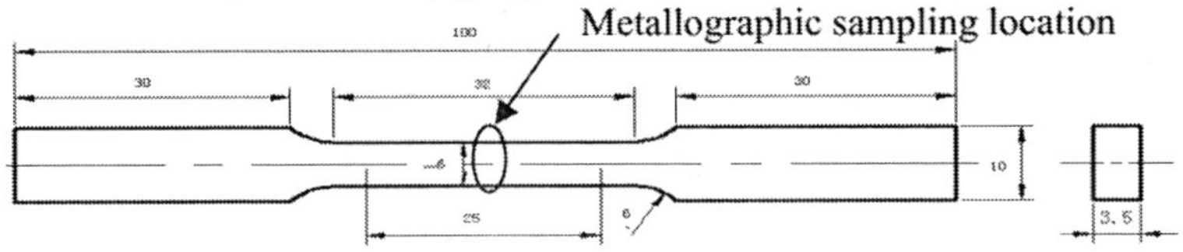

Fig.1 Tensile Samples

Table 1 Chemistry composition of alloy (mass fraction: %)

Alloy	Chemistry composition (mass fraction: %)								
	Zn	Zr	Y	Mn	Fe	Si	Ni	Cu	Mg
ZYK530	4.8918	0.5187	2.8956	0.0063	0.0018	0.0029	0.0009	0.0026	Bal.

After pre-grinding and polishing the cast, extruded and forged samples, the metallographic samples were corroded by picric acid to prepare metallographic samples. The corroded time was for 4~10s. The metallographic structure of the samples was observed and analyzed by OM(Optical Microscope), XRD(X-ray Diffraction), SEM(Scanning Electron Microscope), EDS(Energy Dispersive Spectrometer). The mechanical properties of the samples were tested by universal tensile testing machine, and the fracture morphology of the samples was scanned by SEM, and the fracture mechanism was analyzed.

3. Experimental results and discussion

3.1. Microstructure Evolution of ZYK530 Magnesium Alloy

It can be seen from the Fig.2(a) that the grain size is uniform, the grain boundary bends like plum blossom, the average grain size is about 27μm, the black second phase particles are dispersed in the crystal; at the same time, there are lots of lamellar and fish-bone eutectic microstructures at the triangular grain boundary as show in Fig.3(a), a, b and c, and a large number of fine line microstructures are pointed from the gain boundary to the grain interior.

Fig.2(b) shows the microstructure of as-extruded ZYK530 Magnesium alloy. It can be seen from the picture that the large dendrites in the as-cast state have disappeared completely, but the large grains distributed along the extrusion flow line can be seen, as shown in Fig.2(b). The point “A” indicates that incomplete dynamic recrystallization is incomplete during the cast-extrusion preforming, which may be due to uneven deformation during the extrusion preforming. The flow rate of the metal near the center of the mold is faster than that of the metal near the mold, and supercooling degree of alloy liquid near die wall is high. Therefore, during the extrusion preforming, the metals in the center of the mold are too fast to form dynamic recrystallized grains, while dynamic recrystallization of the metal near inner wall of the mold is relatively complete, forming a distinct equiaxed crystal. According to the related literature[10], the deformation of magnesium alloy above 350℃ is high temperature deformation, and temperature is an important factor affecting the deformation of the alloy. During the high temperature deformation process, the nucleation rate of dynamic recrystallization of the alloy is controlled by thermal activation. The temperature has a great effect on improving the plasticity and deformation speed of the alloy, so the dynamic recrystallization of the alloy is obvious and the grain of the alloy is remarkably refined, and the average grain size is 2.87μm. Compared with the extruded state, the incomplete dynamic recrystallized grains in the alloy along the extrusion direction disappear completely after die forging, as shown in Fig.2(c), and the microstructure presented obvious distribution of drum streamline along the die forging direction.

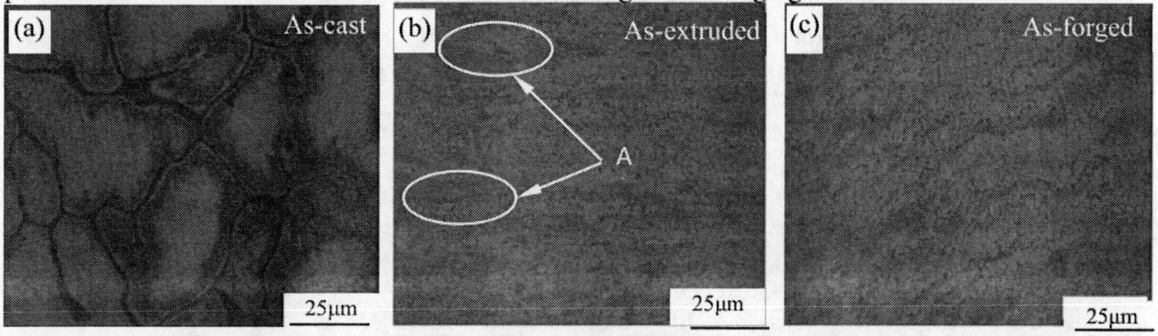

Fig.2 Changes of microstructure of ZYK530 Mg alloy extruded-forged forming

In the process of die forging forming, the length direction of extruded billet remains unchanged, so the metal flow direction is perpendicular to the die forging direction, and the alloy grains turn along the direction of die forging deformation occurred at the same time and the average gain size is 2.87μm which indicates the improvement of the grain size of the alloy is limited under high temperature and large deformation conditions.

From the analysis of SEM and EDS in Fig.3(a)~(b), it can be seen that the mass ratio of Zn/Y is about 3.8. Combined with the XRD diagram of as-cast ZYK530 alloy in Fig.3(c), the alloy is mainly composed of I-Mg_3Zn_6Y phase, a-Mg matrix, a small amount of W-$Mg_3Y_2Zn_3$ phase and Mg_2Zn_3 phase.

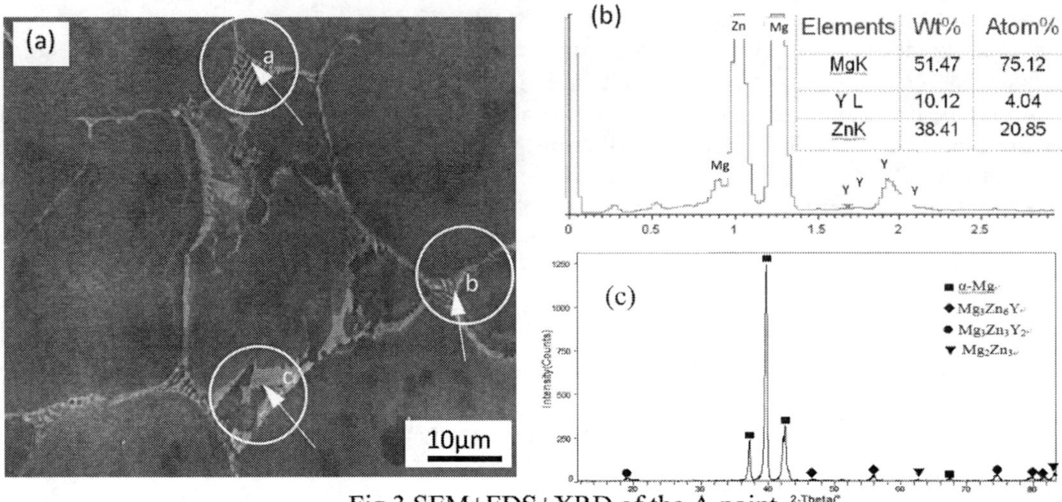

Fig.3 SEM+EDS+XRD of the A point
(a) A point SEM; (b) A point EDS; (c) XRD of as-cast ZYK530

Fig.4(a) is the SEM morphology and EDS of as-extruded ZYK530 alloy. It can be seen from the diagram that a large number of precipitated second phases are precipitated along the extrusion direction in the alloy. The mass ratio of Zn/Y is 1.22. Combining with the XRD pattern of Fig.5(a), it can be seen that the main alloying phases are mainly W-$Mg_3Zn_3Y_2$ and Z-$Mg_{12}ZnY$. No peak phases of I-Mg_3YZn_6 and Mg_2Zn_3 phases have been observed. Fig.4(c) is an enlarged view of the microsturcture of the die-forged ZYK530 Magnesium alloy of Fig.4(b). During high temperature die forging, the number of second phases precipitated along the direction of forging deformation increases obviously under the action of three-dimensional compressive stress. According to EDS in Fig.4(c), the mass ratio of Zn /Y is about 1.43. In addition to I-Mg3Zn6Y, W-$Mg_3Y_2Zn_3$, there are a small number of Mg_2Zn_3 and Z-$Mg_{12}ZnY$ phases and uncertain phases in the XRD, as shown in Fig.5(b).

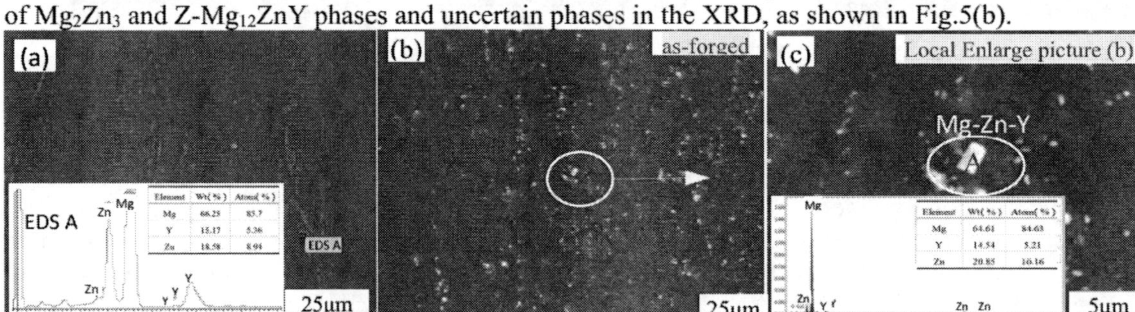

Fig.4 SEM of as-extruded and as-forged ZYK530

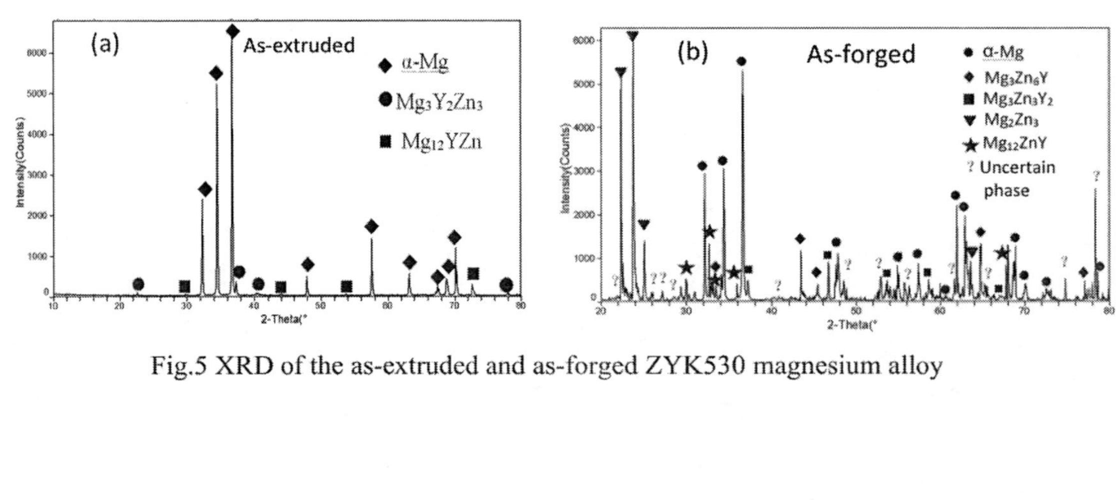

Fig.5 XRD of the as-extruded and as-forged ZYK530 magnesium alloy

3.2. Discussion and analysis

According to the literature, when the mass ratio of Zn/Y is greater than 4.38, the quasicrystalline phase I-Mg_3Zn_6Y is easily formed by conventional casting; when the mass ratio of Zn/Y is less than 1.10, the main phase is W-Mg_3Zn_3Y; while when the mass ratio of Zn/Y is 1.10~4.38, the main phases in the alloy are I phase and W-$Mg_3Zn_3Y_2$ phase[14]. From the experimental results, it can be seen that the phase composition of as-cast to as-extruded forming is inconsistent with that reported.

In the process of semi-continuous casting, according to the phase diagrams of Mg-Zn and Mg-Y binary alloy[15], the solid solubility of Y in Mg is higher than that of Zn, and it is easy to form segregation at the grain boundary to form compound phase. That is to say, during the semi-continuous casting process, α-Mg precipitates first, then α-Mg+W-$Mg_3Zn_3Y_2$ is formed under eutectic condition. With the continuation of solidification, the alloy undergoes peritectic reaction and forms α-Mg+W-Mg_3Zn_6Y, because of uneven crystallization in the casting process. In the presence of equilibrium segregation, there may be a small amount of residual peritectic reaction, namely W-$Mg_3Y_2Zn_3$ phase and Mg_2Zn_3 phase. The I-Mg_3Zn_6Y phase has a good coherence with Mg matrix, but I-Mg_3Zn_6Y phase has a larger volume and a certain distribution of W-$Mg_3Y_2Zn_3$ phase, so the mechanical properties of as-cast ZYK530 alloy is poor.

The I-Mg_3Zn_6Y and Mg_2Zn_3 peaks were not observed in the extruded XRD pattern. The reason may be that the large block I-Mg_3Zn_6Y is crushed into small particles under the action of the triaxial compressive stress, and forms the core of recrystallization, the I-Mg_3Zn_6Y phase is completely dissolved in the matrix, and the Y supersaturation diffused into the matrix tends to form a supersaturated solid solution, so that excess Y precipitates, forming a W-$Mg_3Zn_3Y_2$ phase with Mg and Zn according to an atomic ratio, and the Z-$Mg_{12}ZnY$ phase is easily formed when the mass ratio of Zn/Y is very low, therefore, it consists of W-$Mg_3Zn_3Y_2$ phase in as-extruded alloy and distributes a small amount of Z-$Mg_{12}ZnY$ phase. In the process of die forging, the phase of the alloy is decomposed during the high temperature and three-direction stress deformation, and I-Mg_3ZnY_6 phase, W-Mg_3Zn_3Y phase and the Z-$Mg_{12}ZnY$ phase are decomposed, while the I-Mg_3ZnY_6 phase reinforcing phase is retained, indicating that the alloy is in the process of high temperature deformation and cooling. The Zn and Y atoms are redistributed, and the crushed grains are also used as the core of the recrystallized nucleus while forming a small amount of W-$Mg_3Y_2Zn_3$ phase, Z-$Mg_{12}ZnY$ phase, Mg_2Zn_3 phase and a small amount of undetermined phase.

In summary, dynamic recrystallization occrus during deformation when the deformation temperature of the alloy exceeds 1/3 of the melting point of the alloy[16]. During the process from casting to forging, besides obvious dynamic recrystallization, there are also obvious rearrangements of Zn and Y atoms, which makes the volume fraction and distribution of I-Mg_3Zn_6Y, W-$Mg_3Y_2Zn_3$ and Z-$Mg_{12}ZnY$ change obviously. The mechanical properties of the alloy will be affected by melting, and the grain size does not change significantly from extrusion to forging, which indicates that the effect of high temperature and large deformation on grain size is limited.

3.3. Mechanical properties analysis

Fig.6 shows the tensile properties under different forming conditions. The results show that the tensile strength σ_b and yield strength σs of ZYK530 magnesium alloy are increased by 44.5%, 63.5% and elongation δ by 8.8%, respectively.

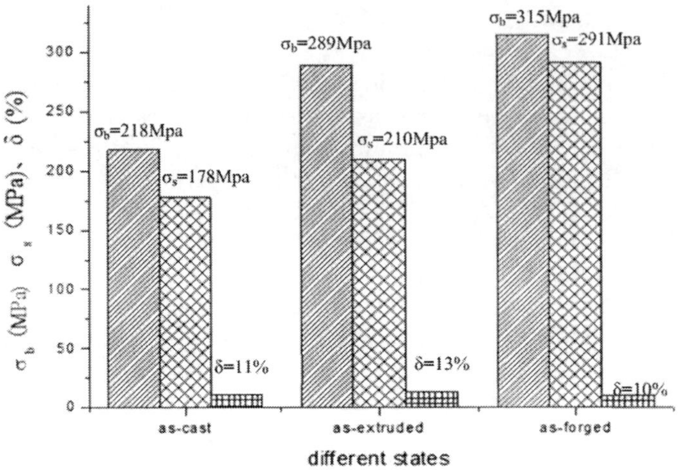

Fig.6 Mechanical properties of ZYK530 Alloy under Different Conditions

With the change of deformation mode, the yield strength of the alloy increases significantly, mainly due to the change of grain size and the change of deformation strengthening mechanism. ZYK530 magnesium alloy undergoes obvious continuous dynamic recrystallization during extrusion preforming and die forging, and the grain size is difficult to grow. At the same time, the second phase particles also hinder the dynamic crystal growth of the alloy. Therefore, compared with as-cast ZYK530, the phase of the extrusion preformed alloy is mainly composed of W-$Mg_3Zn_3Y_2$ and Z-$Mg_{12}ZnY$, in which W-$Mg_3Zn_3Y_2$ has weaker bonding ability with magnesium matrix and deteriorates the mechanical properties of the alloy; Z-$Mg_{12}ZnY$ phase is an unstable phase, but it helps to improve the properties of the alloy, and its mechanical properties are better than those of as-cast alloy. The results show that grain refinement plays a decisive role in the mechanical properties of the alloy. In addition to W-$Mg_3Y_2Zn_3$ phase and Z-$Mg_{12}ZnY$ phase, I-Mg_3Zn_6Y phase and a small amount of Mg_2Zn_3 phase in the alloy which have strong bonding force with magnesium matrix are added to the alloy during the final forging process, which in conducive to further improving the properties of alloy. Although the grain size of the die-forged specimen does not change significantly compared with that of the extruded specimen, the mechanical properties of the die-forged specimen are also improved significantly compared with that of the extruded specimen. The Mg-Zn-Y phase transformation of ZYK530 alloy is complex in the process of casting-extrusion-forging compound forming, and the I-Mg_3Zn_6Y phase is not easy to move, which can effectively hinder the movement of dislocations and improve the mechanical properties of the alloy. Grain refinement is the decisive factor for the improment of mechanical properties of ZYK530 alloy. The mechanical properties of ZYK530 magnesium alloy are obviously improved after extrusion-forging compound forming.

3.4. Fracture analysis

Fig.7 shows the tensile fracture morphology of the ZYK530 magnesium alloy in different states. The ZYK530 magnesium alloy has a jagged fracture with no obvious necking, and the tearing edge is the main way to connect the cleavage facets and form the cleavage river. At the same time, white precipitates were observed on the grain boundaries, indicating that the fracture mechanism was mainly brittle fracture, as shown in Fig.7(a). After extrusion-forging compound forming, the macro-fractures have obvious necking, and there are obvious dimples on the fractures. The dimples on the die forged sample fracture surface increase obviously and are uniform and small, and distribute in the streamline along the forging direction, as shown in Fig.7(b)~(c), which shows that the dimples on the extrusion fracture surface distribute along the extrusion direction. In addition, fine white particles precipitate on the fracture surface of as-extruded and die forged specimens, and there are a few cleavage steps, so the fracture mechanism is ductile-brittle mixed fracture. The results show that the effect of deformation on grain size is limited during extrusion forging process, but it has a great influence on the volume fraction and phase distribution of the alloy, which affects the mechanical properties of the alloy.

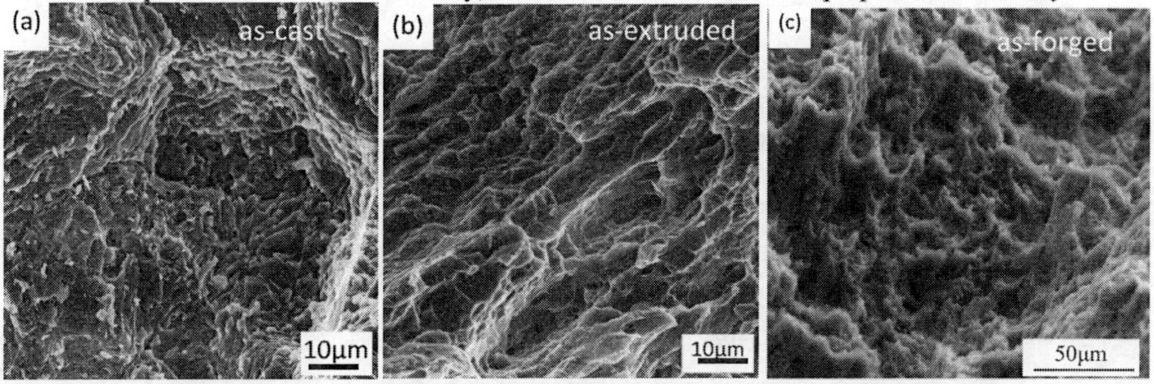

Fig.7 Fracture morphology of ZYK530 magnesium alloys in different states

4. Conclusion

1) The I-Mg_3Zn_6Y and W-$Mg_3Y_2Zn_3$ phases and a small amount of Mg_2Zn_3 phase in the as-cast ZYK530 alloy are mainly distributed at triangular grain boundaries and the grain size is relatively coarse. After the extrusion-forging compound forming, the dynamic recrystallization of ZYK530 alloy under three-dimensional stress makes the grain refined obviously, and the number and distribution of precipitated I-Mg_3Zn_6Y, Z-$Mg_{12}ZnY$ phase, W-$Mg_3Y_2Zn_3$ phase and Mg_2Zn_3 phase undergo complex changes, which play an important role in the mechanical properties of the alloy, but the mechanical properties a key to the role is grain refinement.

2) The as-cast ZYK530 magnesium alloy is obviously a brittle fracture mechanism; there are more and more fine dimple fracture characteristics as-forged state than as-extruded state. In the former, in the former, equiaxed dimples are mainly distributed with small quasi-cleavage surface and tearing edges, which belong to the mixed fracture characteriscs of toughness and brittleness, while in the latter, equiaxed dimples are mainly distributed. In the latter, the fracture is an equiaxed dimple morphology, which belongs to the ductile fracture mechanism.

Acknowledgments

This research was financially supported by Leshan Normal University talent introduction project (Grant No.: XJR17003) and Leshan Normal University Youth Project (Grant No.:Z16005)

Reference

[1] Koichi I, Hiroyuki W, Toshiji M. High strain rate deformation behavior of an AZ91 magnesium alloy at elevated temperatures[J]. Materials Letters.2005, 59: 1511-1515.

[2] Xi Junjie, Liu Jie. Research and Application of High performance Rare Earth Magnesium Alloy[J]. Hot working technology. 2014, 43(10): 6-11

[3] Yang Lianfu. Analysis on Effects of Alloying on Magnesium Alloy Corrosion Resistance for Auto mobile Hub[J]. Hot working technology. 2015, 44(2):73-75

[4] Luo A. Recent magnesium alloy development for automotive power train applications[J]. Materials Science Forum, 2003, 419-422: 57-56.

[5] Li Shubo, Du Wenbo, Wang Xudong, et. Effect of Zr Addition on the Grain Refinement Mechanism of Mg-Gd-Er alloys[J]. Acta Metallurgica Sinica. 2018, 54(6): 911-917.

[6] Xu Daokui. Study on Mechanical Behavior and Microstructure Characteristics of Mg-Zn-Y-Zr Magnesium Alloy[D].2008. 05.01

[7] Zhang Qing, Li Quanan, Jing Xiaotian, et. Application of Yttrium in Heat Resistant Magnesium Alloy[J]. Chinese Rare Earths, 2011, 32(2): 71-75.

[8] Cao Fenghong, Long Siyuan. Effects of Mass Ratio of Y to Zn on Microstructures and Mechanical Properties of ZK Magnesium Alloys[J]. Special Casting & Nonferrous Alloys, 2018, 34(2): 26-29
[9] Suzkui M, Kimura T, Koike.J, et al. Strengthening effect of Zr in heat resistant Mg-Y-Zn Solid alloys[J]. Scripta Materialia, 2003, 48: 997.
[10] Singh Alok, Tsai A. P. On the cubic W phase and its relationship to the icosahedral phase in Mg-Zn-Y alloys[J]. Scripta Materialia. 2003. 49: 143
[11] Singh Alok, Watanabe M, Kato A.etal. Formation of iscosahedral-hexagonal H phase nano-composition Mg-Zn-Y alloys[J]. Scripta Materialia, 2004, 51: 955
[12] Zhang Ya. Study on Microstructure, Properties and Deformation Behavior of Mg-Zn-Y-Zr Wrought Magnesium Alloy[D]. 2005.12.
[13] Yu Kun. Study on Microstructure, Properties and Processing Technology of Rare Earth Wrought Magnesium Alloy[D]. Central South University, 2011.08.
[14] Fu Binguo. Microstructure and Mechanical Properties of Mg-Zn-Y-Zr Alloys[D]. 2010.07.01
[15] Feng Juanni, Feng Juanni， Guo Xuefeng, et.. Microstructure and phase Transition Points of As-Cast Mg-Zn-Y System Alloys[J]. Journal of China Rare Earth Science.2006. 24(1): 86-90.

ROS-based robot offline planning simulation system

Liu Zhiheng*, **Chen Jian, Mei Zhen and Li Chao**

Wuhu HIT Robot Technology Research Institute Co., Ltd.

*Corresponding author's e-mail: liuzhiheng@hitrobot.com.cn

Abstract. In order to realize the rapid and automatic generation of robot motion trajectory, a robot offline planning simulation system was developed based on ROS platform. This paper analyzes the model STL file, reconstructs the topology data structure, generates the model minimum bounding box, and obtains the model surface path through the slicing algorithm to generate the processing path. The system loads the Moveit motion planner, applies different motion planning methods, generates a collision-free, continuous motion trajectory, and visualizes the offline planning process. The simulation verification system can realize the functions of rapid path generation on the workpiece surface and obstacle avoidance motion planning of the robot arm. It can realize offline programming of robots and improve the efficiency of robot teaching.

1. Introduction

Robots are used more and more in modern industrial manufacturing. The traditional online teaching method is reliable, efficient, low-cost and easy to implement for low precision and simple operation objects. However, with the increase in product complexity and manufacturing accuracy requirements, there are problems such as cumbersome programming, low efficiency, and low precision. The method of teaching the robot to generate the robot control program can not meet the production needs, so it is urgent to develop the robot offline programming technology[1].

Since the application of off-line planning technology improves the efficiency of robot programming, the manufacturing flexibility, production efficiency and product quality of the system are improved[2-3]. Foreign offline programming technology has developed earlier and has mature commercial software, but most of them are expensive and have certain limitations. e.g. ABB Robot's RobotStudio can provide a complete offline planning system[4], but it can only be applied to its own brand of robots, and it is not universal; Robotmaster is mainly used for 3D milling, and it is not suitable for other processing fields. DELMIA is mainly used for processing in the automotive industry, and its expert software is difficult to operate. Robcad is mainly used for production line simulation, and its offline function is weak.Domestic offline programming technology started relatively late, mostly research in university laboratories and research institutes. e.g. Guo Jichang and others used the UG/Open function library to develop the offline programming system of KUKA KR30L16 robot, and realized key technologies such as motion simulation and detection[5]. Song Pengfei based on SolidworksAPI secondary development function, designed offline programming simulation system to achieve automatic generation of

FANUC robot operating program in Solidworks environment[6]. Li Xiang proposed an interactive 3D offline programming and simulation system based on OpenGL, which can detect collisions or singular points during robot operation[7]. Xu Xiang and other Qt for the software development framework combined with the functions of OpenGL and VRML, established a modular robot offline programming system[8]. With the Unity3D software as the system development platform, Tao Tao designed a cross-platform offline programming system to realize the motion simulation of the mobile client-to-system[9].

In summary, the offline programming system based on third-party software development has realized basic functions, but does not provide robot obstacle avoidance planning functions for complex working conditions. The ROS system is a robot software platform that provides complete obstacle avoidance planning. At present, the offline programming system based on ROS platform has no process of model processing, so it is not a complete offline programming system[10-11].

This paper builds a complete offline planning system based on the open source ROS platform combined with the processing technology to explore the offline planning method. Through the secondary development of the scalable view tool Rviz, the machining path planning of the workpiece surface is carried out. The integrated motion planning library Moveit plans the motion trajectory with obstacle avoidance function, and builds a visual simulation platform to realize the motion simulation function.

2. Model surface path planning

The STL file is imported from the outside through the Rviz visualization function. Since the shape of the geometric model is composed of a triangular mesh, the surface path of the model can be quickly planned by the slice processing method. The planning process is:

- Reading, topology reconstruction and storage of triangular mesh information of the STL model.
- Obtain the minimum bounding box OBB of the model component.
- Generating a set of plane clusters parallel to the calibration surface with any section of the minimum bounding box as the initial plane.
- Obtaining the intersection of each plane and the triangular grid of the spraying component in turn, and connecting the intersection points in sequence to form a strip path stroke.
- The path point at the end of the previous stroke is connected with the path point of the start point of the latter stroke, and finally the surface path of the model is formed.

2.1. Topological reconstruction of data structures

The topology between the meshes is reconstructed by reading the triangular mesh information[12]. The topology is built according to the following criteria: a triangular mesh has at most three co-edge triangles and at least one. All co-edge triangles are obtained by traversing the vertex information of the triangular mesh to establish a topological data structure of faces, edges, and vertices. It is possible to quickly retrieve the three sides and three vertex numbers of the face, the two vertices of the edge and the face number, the coordinates of the vertex and the face number. In order to save storage time and space, the coordinate information of the points only needs to be recorded once and indexed by number. The specific data structure is as follows:

```
    Struct Facet  // Data structure
{
int f_inx_; //Face number
int vs_inx_[3]; // Number of three vertices
int es_inx_[3]; // Number of three sides
Ogre::Vector3 normal_; // Normal vector
double area_; // Area
}
```

IOP Conf. Series: Materials Science and Engineering **711** (2020) 012002 doi:10.1088/1757-899X/711/1/012002

```
    Struct Edge // Edge data structure
{
int e_inx_; // Side number
int vs_inx_[2]; // Number of two vertices
int fs_inx_[2]; // Number of two faces
double length; // Length
}
    Struct Vertex // Vertex data structure
{
int v_inx_; // Vertex number
Ogre::Vector3 v_data_; // Coordinate information
std::vector<int> face_inx; // Face number
Ogre::Vector3 v_nor; // Normal vector
}
```

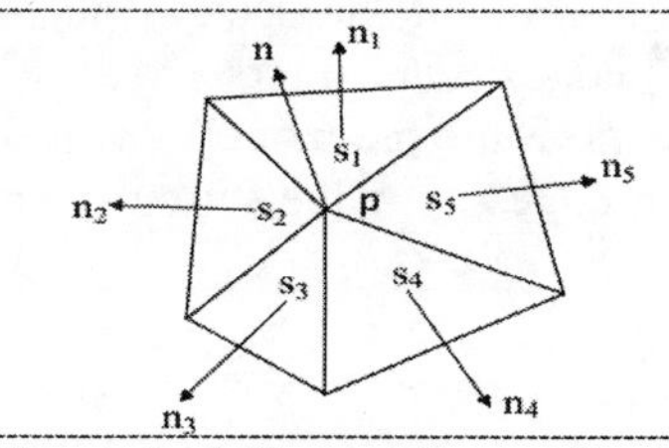

Figure 1. Adjacent faces of point

Since the STL model only provides the normal vector of the face, the normal vector of the intersection can be obtained indirectly from the normal vector of the face. Here we use the method of calculating the normal vector based on the area weight[13]. Analyze the geometric properties of the adjacent faces of the vertex p in figure 1, and retrieve its normal vector and area, respectively , and the normal vector of p point is:

$$n = \sum_{i=1}^{m} s_i n_i \left(\sum_{i=1}^{m} s_i \right)^{-1} \tag{1}$$

2.2. Generation of truncated plane clusters

In this paper, a plane of the model minimum bounding box (OBB) is used as the initial calibration plane of the truncated plane cluster, and a truncated plane cluster parallel to the calibration plane is generated according to the set stroke spacing parameter, as shown in figure 2. The OBB bounding box is obtained according to the reconstructed topological data structure, and principal component analysis (PCA) is performed on all the vertex information to obtain three feature vectors and serve as three main axes of the OBB.

The specific steps are: First, index the area of the adjacent triangle patch $\Delta S_1, \Delta S_2, \cdots, \Delta S_k$ of each vertex p, calculate its weight ω_p , the mean of the weights $\overline{p}$:

$$\omega_p = sum(\Delta S_1, \Delta S_2, \cdots, \Delta S_k)/k \tag{2}$$

$$\overline{p} = \frac{1}{n} \sum_{i=1}^{n} \omega_i p_i \tag{3}$$

where k is the number of patches, n is the number of vertices. Therefore, construct the covariance matrix as:

$$C = \frac{1}{n} \sum_{i=1}^{n} (\omega_i p_i - \overline{p})(\omega_i p_i - \overline{p})^T \tag{4}$$

Then calculate its eigenvector v_1, v_2, v_3, and set it to the direction of the major axis of the bounding box. Finally, all the vertices are projected on three main axes to obtain the bounding box center point and three axis lengths.

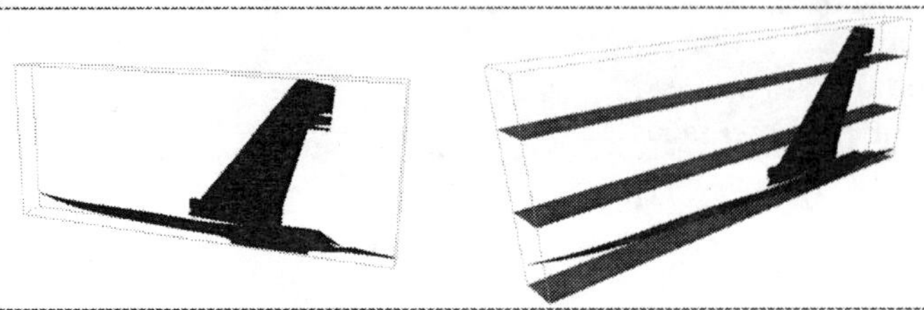

Figure 2. Bounding box and Section plane cluster

2.3. Generation of truncated plane clusters

Here, a slice algorithm is used to obtain a path point. Traverse the edges in the model to determine whether it intersects the section plane. If it intersects, it terminates the loop and calculates the intersection point, which is the path point of the section plane.Index the two adjacent faces of the intersecting edge, find the other edges that intersect the intersecting plane and calculate the intersection. By analogy, the intersections are recursively reciprocated to the two sides until there is no intersecting edge or the intersecting edges coincide (circular path). Finally, a path point intersecting the section can be obtained.

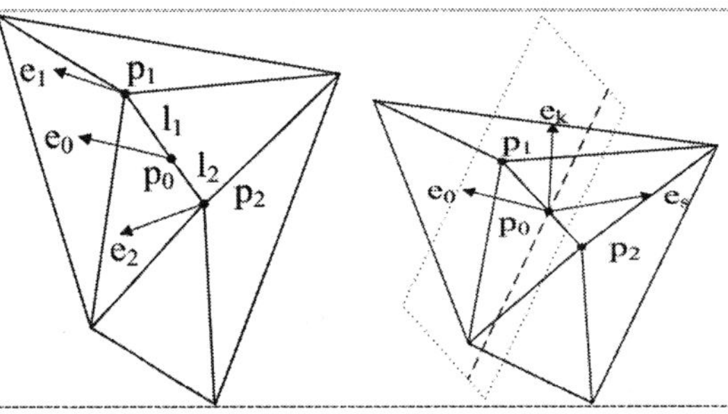

Figure 3. Point normal vector and Point pose

Analysis of the geometrical characteristics of the intersecting plane intersection point shows that the p_0 point pose can be obtained from its normal vector and machining direction, as shown in figure 3. The normal vector of point p_0 is determined by the characteristics of the surface. By indexing the data structures of the vertices p_1 and p_2 of the intersecting edges, the normal vectors e_1 and e_2 of the two points can be obtained respectively. Here, the position of p_0 point between point p_1 and point p_2 is used as the weight, and the normal vector e_0 of point p_0 is calculated:

$$e_0 = k_1 e_1 + k_2 e_2 \tag{5}$$

$$k_1 = l_1 / (l_1 + l_2) \qquad k_2 = l_2 / (l_1 + l_2) \tag{6}$$

Where k_1 is the weight of point p_1, k_2 is the weight of point p_2, l_1 is the distance between point p_1 and point p_0, and l_2 is the distance between point p_2 and point p_0.

The machining direction on the same section plane is in the section plane, so it is perpendicular to the normal vector es of the section plane. And the machining direction is along the tangential direction of the curved surface, so it is perpendicular to the surface normal vector and is approximated by the normal vector e0 of the point p_0. Therefore, the machining direction vector e_k is:

$$e_k = e_0 \times e_s \tag{7}$$

According to the general situation established by the robot tool coordinate system, the normal vector e_0 is taken as the z-vector RZ of the pose, the direction vector e_k is taken as the x-vector RX of the pose, and the y-vector RY of the pose is obtained by RZ×RX. Then get the spatial pose of p_0.

At this point, a path point position and posture of the section plane can be obtained. Generate path points for all truncated planes, and connect the end points of each leg to form an offline path for the entire model. The offline path is a series of discrete pose points, although it reflects the actual feature information of the workpiece surface, but cannot be directly sent to the arm movement. Since the offline path requires smooth processing, robot arm collision detection, trajectory planning, etc. Moveit is required to optimize the path.

3. Motion planning of the manipulator

The Moveit motion planning library includes forward and inverse kinematics calculations, obstacle introduction, collision detection, and motion planning. Based on the Rviz plugin mechanism, a robotic arm scene can be generated for visual simulation. Load the motion planner specifically through the following modules:

- Robot Model, the configuration assistant loads the arm URDF model and configures the joint parameters, the robot arm configuration, the end operator, etc.
- Planning Scene, by adding a restraining object in the movement space of the robot arm, the movement of the mechanical arm avoids the obstacle and ensures the normal operation of the mechanical arm.
- Motion Plan, instantiate the object Robot Model and Planning Scene, load the motion planner implementation to call the motion planning function.

Complete the loading of the motion planner to plan the motion of the path point. Specific exercise planning process:

(1) Set the initial state of the manipulator motion planning according to the robot processing technology requirements. Here, the setStartState function is used to reset the initial state of the arm.

(2) Since the planned offline path is the position that the mechanical arm needs to reach in Cartesian space, the motion path can be obtained by planning the mechanical arm motion using the Cartesian coordinate path (computeCartesianPath function) provided in Moveit. However, this function only has collision detection and cannot perform obstacle avoidance planning. Output failure results for a path point with collisions.

(3) The success rate fraction of the plan is returned by the computeCartesianPath function to determine whether the current plan is successful. Here, the fraction threshold is set to 0.1 (10%). When the planned success rate is less than 10%, it is regarded as a planning failure and the planning process is interrupted. Save the planned path points and generated motion points, and set the current path point to the initial state of the next stage plan.

(4) The current initial state point does not satisfy the condition of the Cartesian coordinate path planning, and the plan function with obstacle avoidance planning is used. Set the position of the planning point by the setPoseTarget function, which can adjust the attitude of the arm in a wide range and be used for the collision point in the transition path. This function is based on sampling-based obstacle avoidance motion planning, which has a large amount of calculation and a long planning time, and is not suitable for planning of large-scale points. Therefore, after the transition point adjusts the posture, continue to use the computeCartesianPath function for motion planning.

(5) Repeat steps 2-4 until the robot motion planning for all path points is completed.

After Moveit's robotic arm motion planning, a series of motion points that the robot can perform are obtained. The post-processing module of the robot controller generates a command recognized by the robot to control the movement of the robot arm along the planned trajectory.

4. Simulation test and result analysis

The offline path planning method proposed in this paper is mainly for the path planning of the workpiece surface, and is suitable for processing fields such as spraying and cleaning. The effectiveness, completeness and accuracy of offline path planning are verified by planning different workpiece surfaces. Here, the surface path planning is performed on the plane, the curved surface and the annular surface respectively, and the line spacing is set to 50 mm, and the Ubuntun 16.04 system and the ROS Kinetic version are adopted. Obtain the path point pose, as shown in figure 4, where

IOP Conf. Series: Materials Science and Engineering **711** (2020) 012002 doi:10.1088/1757-899X/711/1/012002

yellow represents the machining path, red represents the z-direction, blue represents the x-direction, and green represents the y-direction.

Analyze the simulation results:

- The path point position is obtained by the intersection of the section plane and the patch, and it can be visually seen that the plane path points are regularly distributed according to the shape of the triangle mesh.
- The z-direction of the path point is always perpendicular to the patch, and the x-direction is along the direction of motion. It can be seen that the attitude of the curved path point can change as the shape of the patch changes.
- The path of the toroidal plane is planned along the surface of the patch, and there is no overlap or missing.

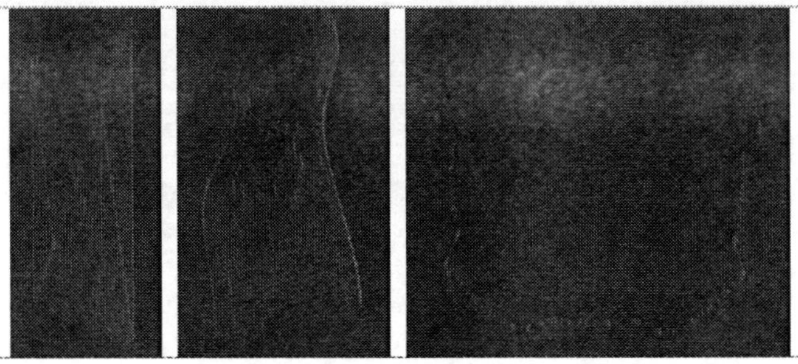

Figure 4. Path planning for different faces

The above path planning for different surfaces verifies that it has offline planning performance. However, for a workpiece, it is sometimes necessary to process a plurality of faces, so that a combined path planning of a plurality of patches is required.According to the model features, the 3D modeling software divides into different planes or surfaces for surface path planning. Here, the cuboid is taken as an example for path planning, and the path point pose is obtained as shown in figure 5.

Figure 5. Cuboid path planning

The simulation platform of the offline planning system was built, and the six-degree-of-freedom industrial robot was used to simulate the surface contour of the die casting.The offline planning simulation system interface is written based on QT, as shown in figure 6. It can realize the modification of the imported model pose, select different patches for path planning and motion planning, import and delete obstacles, save path points and motion points. Figure 7 shows the motion simulation of the robot along the contour of the die casting, which is convenient for the user to observe the process of the robot. The robot runs from the initial state to the infeed point, moves along the blue contour to the retraction point, and returns to the initial state. After confirming that there is no problem during the exercise, save the robot movement point. The robot is processed and sent to the robot controller to complete the offline planning and simulation process of the robot.

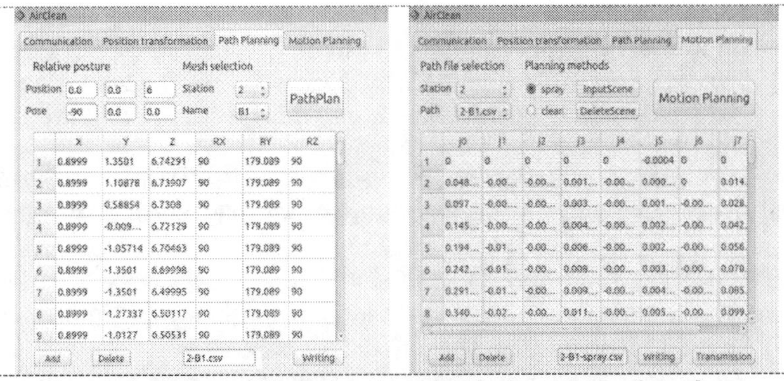

Figure 6. Robot offline planning simulation system interface

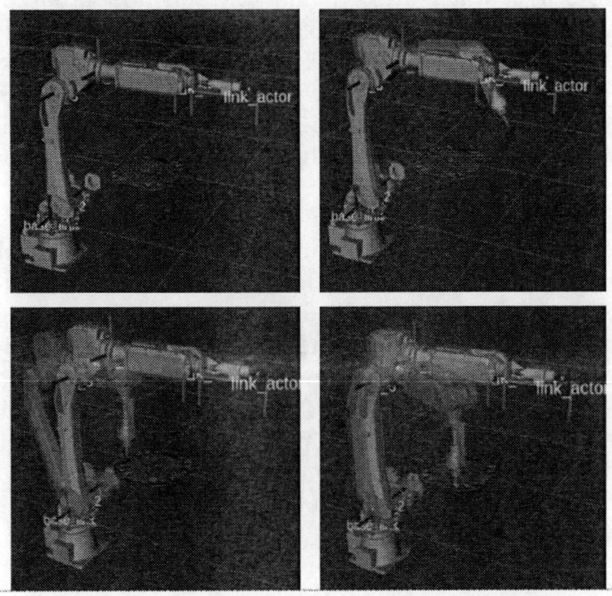

Figure 7. Robot motion simulation process

5. Conclusion

Based on the open source ROS platform, this paper develops an off-line simulation system for surface path planning and robot arm motion planning. The model layered slicing algorithm is used to solve the problem that the model can't handle in the ROS system. Reconstruct the mesh information topology data structure, generate the model minimum bounding box, quickly slice the path point position and posture, and complete the path planning of the workpiece surface. Based on the Moveit motion planning open source library, the motion planning process is executed to simulate robot motion and generate continuous, collision-free motion trajectories. Through simulation experiments on different workpieces, it is verified that the simulation system has the offline planning function of the robot, which can improve the efficiency and accuracy of robot programming. The system has great application potential for offline planning in the fields of industrial robot cleaning and spraying.

Acknowledgement

This work is supported by National Key Research and Development Program of China under Grant 2017YFB1303900.

References

[1] Lin, Y.Z., Liu, Q.Q., Xu J., Liao, J.F. (2015) Research status and development trend of industrial robot off-line programming system. Mechatronics, 21(7): 8-10.

[2] Polden, J., Pan, Z., Larkin, N., Duin, S.V., Norrish, J. Offline programming for a complex welding system using DELMIA automation. (2011) Lecture Notes in Electrical Engineering, 88: 341-349.

[3] Wylie, N., Wylie, S.R., Cullen, J.D., Al-Jader, M. (2010) NDE system for the quality control of spot welding in the automotive industry. Sensors Applications Symposium. IEEE.

[4] Holubek, R., Delgado Sobrino, D.R., Košťál Peter, Ružarovský, R. Offline programming of an ABB robot using imported CAD models in the RobotStudio software environment. (2014) Applied Mechanics and Materials, 693: 62-67.

[5] Guo, J.C., Zhu, Z.M., Yan, G.R., Ma G.R. Development of off-line programming system for arc welding robot based on UG. (2016) Electric Welding Machine, 46(10).

[6] Song, P.F., He, R.L., Miao, J.Z., Shan, D.R. Offline programming system based on Solidworks for industrial robot. (2013) Manufacturing automation, 35(9): 1-4.

[7] Li, X., Qian, H., Fu, Z., Zhao, Y.Z. The off-line programming and simulation system design of robot based on QT and OpenGL. (2013) Mechatronics, 19(4): 56-59.

[8] Xu, X., Gan, Y.H., Dai, X.Z., Duan, J.J. Extensible robot off-line programming system based on QT. (2015) J. Huazhong Univ. of Sci.&Tech. (Nature Science Edition). (s1): 513-516.

[9] Yan, T., Zhang, P., He, C.J., Zhang, Z.P. Research and development of the cross-platform industrial robot off-line programming system based on Unity3D. (2016) Modular Machine Tool & Automatic Manufacturing Technique, (10): 99-104.

[10] Lu, Z., Han, Y.Y., Ji, L., Guo, W. Development of an off-line programming system for UR robots based on ROS. (2017) Computer Simulation, (09): 322-326+331.

[11] Wen, K.C. ROS-based motion simulation and trajectory planning of palletizing robot. 2016.

[12] Yang, G., Liu, W.J., Wang, W., Tian, F.J. Research on the rapid slicing algorithm based on STL topology construction. (2009) Modern Manufacturing Engineering, (10): 32-35.

[13] Sun, Y.W., Liu, W.J., Wang, Y.C. Research on the algorithm of NC tool path calculation for triangular surface machining. (2002) Chinese Journal Of Mechanical Engineering, 38(10): 50-53.

Research on productivity effect of manufacturing capital accumulation——an empirical study based on data of manufacturing enterprises in Beijing, Tianjin and Hebei

Xiaonan Fan[1], Qiyue Li[1*] and Chong Feng[2]

[1]College of Management, Dalian Polytechnic University, Dalian, Liaoning Province, 116034, China

[2]College of Management, Xiamen University, Xiamen, Fujian Province,361005, China

*Corresponding author' s e-mail:1115366484@qq.com

Abstract: Based on the panel data quantile regression, the horizontal and vertical comparisons reveal the marginal effect of the capital accumulation of the Beijing-Tianjin-Hebei region on the growth of total factor productivity. Combined with the heterogeneity of regions and industries, this paper provides a theoretical basis for how to effectively invest and how the government solves the mismatch of capital. The study found that although the increase in investment has a positive effect on the technological progress of enterprises, the dividend of capital investment has gradually disappeared, and the overall trend is inverted U-shaped. In terms of sub-regions, increasing investment in quality is still the main problem that Hebei Province facing in the future. Different industries also present different characteristics. A few industries have shown an increase in investment increased the growth rate of total factor productivity.

1. Introduction

Neoclassical economics believes that factor input and the improvement of total factor productivity jointly promote economic growth, while the scarcity of resources determines that factor-driven growth is difficult to sustain, and total factor productivity will play an increasingly important role. Since China entered the new normal, the government actively encourages enterprises and regions to explore new development models. In order to achieve sustainable economic growth and supply-side reform, it is necessary to clarify the following question: Changing the growth mode that depends on investment is the key to China's economic transformation. Does this transformation apply to all enterprises? Does the existence of heterogeneity in enterprises and industries affect the contribution of capital investment to the growth of total factor productivity?

Through horizontal and vertical comparisons, enterprises should increase effective investment, improve input-output relationship, and promote sustainable and healthy development of enterprises. In April 2017, the CPC Central Committee and the State Council decided to establish a national-level new district in Xiong'an New District as a major historic strategic project. Therefore, understanding the characteristics of manufacturing capital stock and total factor productivity in the Beijing-Tianjin-Hebei region is a prerequisite for effective investment and further research.

Based on the above analysis, the full text is divided into three parts. The first part is literature review, which summarizes the relationship between capital investment and total factor productivity and related literature, and briefly summarizes the quantile regression. The second part is to measure the total factor productivity of the manufacturing industry in the Beijing-Tianjin-Hebei region using LP semi-parameter method, use capital stock to represent the amount of capital accumulation and adopts panel quantile regression to empirically analyze the characteristics of capital stock of enterprises from the perspective of the marginal effect of capital on total factor productivity. The marginal effects of different locations and time periods are further explored. The third part is the conclusions and recommendations of the article.

2. Literature review

Total Factor Productivity (TFP) is an indicator of the relationship between input and output. The measurement of total factor productivity generally refers to the measurement of the growth rate of total factor productivity. Meeusen & Brocck, Battese & Coelli improved SFA, using the measurement method to estimate the frontier of the production function, and divided the total production function into the front-end production function and the non-efficiency part. Relatively speaking, there is less literature on enterprise TFP. In order to solve the problem of an inevitable measurement technology when enterprise TFP estimation, that is, simultaneity deviation. Olley and Pakes (1996) developed a method based on the consistent semi-parametric estimation method, OP method. Levinsohn and Petrin (2003) further proposed the LP method to reduce the amount of sample loss, but this method is still used less in China.

Capital is the most important input factor in the process of economic growth. The efficiency of capital use and its mechanism of action on total factor productivity have always been the focus of research. As capital density increases, the marginal productivity of capital begins to decline, and the rate of economic growth will inevitably decline (Kim, Lau, 1996). At the same time, China's economy has accumulated relatively excess capital too early and too quickly because of excessive competition by local governments, excessive investment has led to a large amount of capital deposited in the field of overcapacity, and weakened the ability of the economy to absorb surplus labor, thereby restricting the continued growth of economy (Zhigang Yuan, 2002).

According to China's data from 1978 to 1998, Chow and Lin (2002) found that the contribution of physical capital investment to China's economic growth was 62%. Wang and Yao (2003) found that the contribution rate of physical capital investment to China's economic growth was 47.7% between 1978 and 1999. Wang Dihai et al. (2008) found that the healthy investment rate is positively correlated with the economic growth rate. However, the current research on the relationship between capital and total factor productivity is mainly concentrated in the macroscopic field, and few studies have analyzed it based on micro data.

The quantile regression can more fully describe the relationship between the independent variable and the dependent variable. Koenker and Hallock (2001) proposed a quantile regression method for panel data, and Karlssom (2005) introduced nonlinear data into panel quantile regression. Zhu Jianping (2012) proposed a two-stage panel quantile regression method, and compared with the fixed effect transform quantile regression method proposed by Luo Youxi et al.

Capital accumulation is the dynamic process of turning profit into capital, capital stock refers to the total amount of capital in a society at a certain moment. We use capital stock to represent the amount of capital accumulation in this paper. Next, we will compare the capital stock characteristics of enterprises with different developments through the manufacturing enterprises above designated size in the Beijing-Tianjin-Hebei region, and compare the trends of different regions and different time periods.

3. Research design

3.1 Samples and data sources

The purpose of this paper is to study the marginal role of corporate capital stock for total factor productivity growth in the Beijing-Tianjin-Hebei region. The data used in this paper is the industrial enterprise database data from 2004 to 2013. Due to the large number of industrial enterprise database samples and some enterprise information errors or missing, this paper filters and adjusts the enterprise samples before use.

First of all, the paper excludes the sample of enterprises with zero industrial output value or zero number of employees, and limits the total output value of fixed price, total fixed assets, management expenses, payable wages, etc., which must be positive, depreciation, paid-in capital, etc. cannot be negative. Secondly, according to the matching process similar to Brandt (2012), the matching and unmatched data were combined to obtain the unbalanced panel data from 2004 to 2013. Finally, the price index of the enterprise's industrial output value, basic input and intermediate input is deflated in the time dimension. Based on 2004, the industrial product price index, the fixed asset investment price index, the raw materials, fuel, and power purchase price indices were deflated. The capital stock is measured by the perpetual deposit method.

3.2 Measurement of total factor productivity growth

In this part, the LP method is used to obtain the level of total factor productivity at the enterprise level. In order to more clearly understand the level and characteristics of the total factor productivity growth of Beijing-Tianjin-Hebei, the results of the calculation are divided into time and region.

This paper uses the proportion of total industrial output value as the weight of the enterprise, focusing on the trends of total factor productivity in different years and regions and industries. Figure 1 shows the total factor productivity from 2004 to 2013 in China.

Figure 1 shows that from 2004 to 2009, the growth rate of total factor productivity showed a steady increase. After 2009, it increased sharply to 2.48, and after 2010, there was a downward trend. Due to the financial crisis, some enterprises stopped production and exited the market in 2009. With the government's bailout policy and the partial withdrawal of capital dividends and factor inputs, the average total factor productivity growth rate in 2009 increased significantly. After 2010, China's financial industry has become a new source of economic development. Housing prices and stock prices have risen sharply, attracting a large amount of capital investment.

The growth rates of TFP in Beijing, Tianjin and Hebei are basically consistent with the overall trend, but the technological progress and innovation of enterprises in Hebei Province lags behind that of Beijing and Tianjin. The innovation capacity of Hebei Province does not match its geographic area, population and economic aggregate. Its economic development relies heavily on resource input. Paradoxically, the mineral resources of Hebei Province are not very rich, which affects the effective allocation of resources to a certain extent, and has a crowding-out effect on high-tech industries.

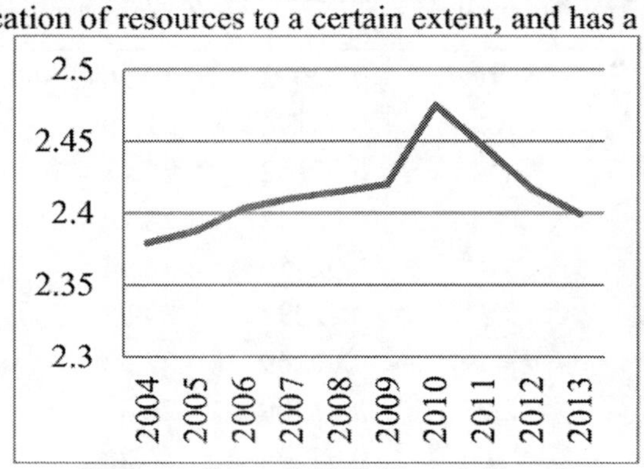

Figure 1 Overall TFP Growth Trend.

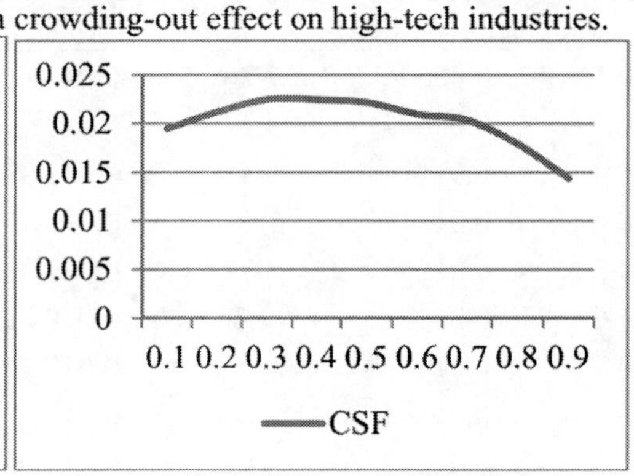

Figure 2 Overall Quantile Coefficient.

4. The analysis of empirical results

4.1. Overall research

With the advancement of technology and the improvement of overall efficiency, the marginal utility of capital for total factor productivity growth is not static. Studying the investment behavior of economic entities at different stages of development, the quantile regression of panel data can study the relationship between the two more comprehensively at different grading levels. Table 1 and Figure 2 shows the results of the quantile regression of the entire Beijing-Tianjin-Hebei region.

Table 1 Overall Quantile Regression.

TFP Quantile	Capital stock factor	P	T
0.1	0.01949	0	12.7
0.2	0.02119	0	19.46
0.3	0.02248	0	25.66
0.4	0.02247	0	27.59
0.5	0.02216	0	31.1
0.6	0.02096	0	32.88
0.7	0.02037	0	35.23
0.8	0.01791	0	28.07
0.9	0.01439	0	18.7

The regression results in Table 1 show that as the technological level of the enterprise increases, the marginal role of the capital stock gradually increases, and reaches a relatively stable peak between 0.3 and 0.5. For companies with faster total factor productivity growth, the role of capital in technology growth is significantly lower. For companies with a TFP of 0.9, the marginal value of capital stock is only 0.014, which is 64% of the median.

The increasing trend in Figure 2 shows that increasing investment can bring greater economic benefits and promote the growth of total factor productivity. Between 0.1 and 0.2 points, the increase in capital investment is most significant. At this time, the capital stock of enterprises does not reach the optimal ratio of capital and labor input. Increasing investment can not only achieve higher profits, but also significantly promote technological progress. However, for enterprises after the median quintile, their profit status is relatively good, and there is a certain degree of scale advantage. Although the investment increase has a positive effect on the technological progress of the enterprise, the dividend of capital investment gradually disappears.

4.2. Region-divided study

Table 2 Region-divided Regression.

TFP Quantile	Capital stock factor(Beijing)	T(Beijing)	Capital stock factor(Tianjin)	T(Tianjin)	Capital stock factor(Hebei)	T(Hebei)	P
0.1	0.00606	2.26	0.01097	4.16	0.03093	10.96	0.024/0/0
0.2	0.00808	5.24	0.01311	7.2	0.03194	21.28	0
0.3	0.01214	8.46	0.01366	7.14	0.03152	26.03	0
0.4	0.01249	7.05	0.01525	9.44	0.03042	28.26	0
0.5	0.01198	8.97	0.01776	13.46	0.02756	33.94	0
0.6	0.01208	11.31	0.01994	13.31	0.02497	30.67	0
0.7	0.01326	11.59	0.01721	13.49	0.02249	27.52	0
0.8	0.01516	9.89	0.01802	12.27	0.01909	26.31	0
0.9	0.01359	8.85	0.01558	7.93	0.01544	16.26	0

Table 2 and Figure 3 shows the results of the regression of the region-divided of Beijing, Tianjin, and Hebei. The economic development level of Beijing and Tianjin is relatively high. This promotes the development of specialized inputs and services, provides a centralized market for workers with specialized skills and enables companies to benefit from technology spillovers, further enhancing the utilization of capital by firms. Therefore, the role of the capital stocks in these two regions shows a steady rise. The capital stock factor is between 0.006 and 0.019.

By contrast, the economic development level of Hebei Province is relatively backward, and most enterprises have not reached the optimal production scale. The 0.1 to 0.4 quantile interval increases the marginal output value of unit capital input by as much as 0.032. It can be foreseen that increasing quality investment is still the main problem facing Hebei Province in the future. Hebei provincial government should make full use of the geographical advantages of the adjacent Beijing-Tianjin-Second City, and use inter-regional links to fully absorb advanced knowledge and technology spillovers, further deepen structural reforms through the absorption of coding and non-coding knowledge.

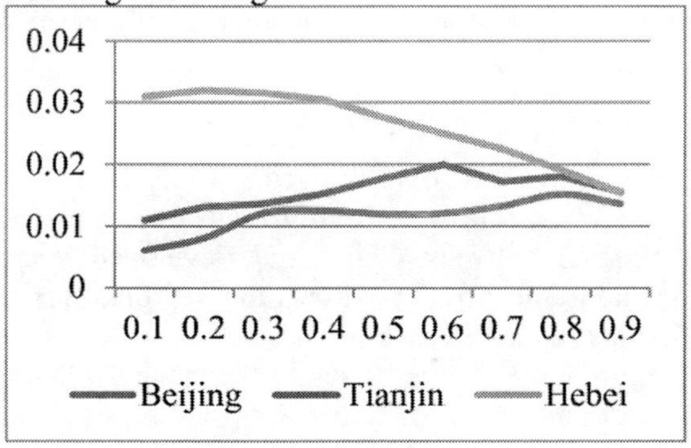

Figure 3 Region-divided Regression Results.

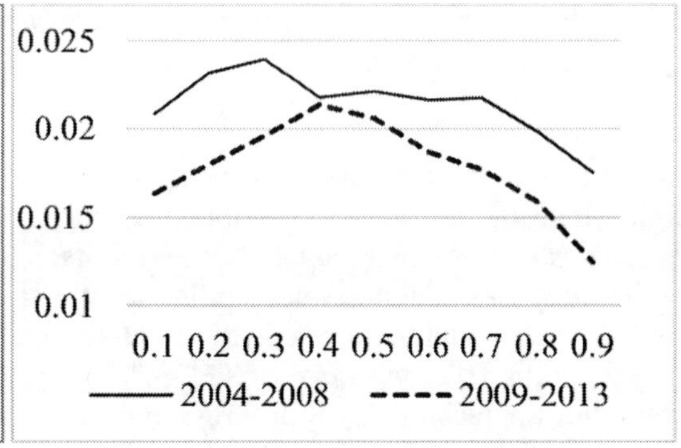

Figure 4 Time-divided Regression.

4.3. Time-phase research

After 2008, in order to cope with the impact of the world financial crisis on China's economic operation, the Chinese government has implemented stimulus investment policies including RRR cuts and interest rate cuts. Therefore, this paper divides the time period into two periods: 2004-2008, 2009-2013,Table 3 and Figure 4 shows the regression results for two periods.

Table 3 Time-divided Regression.

TFP Quantile	Capital stock factor(2004-2008)	T(2004-2008)	Capital stock factor(2009-2013)	T(2009-2013)	P[a]
0.1	0.02085	9.92	0.01636	6.91	0
0.2	0.02318	10.37	0.018	12.04	0
0.3	0.02394	19.13	0.01961	15.2	0
0.4	0.02177	19.93	0.02136	19.06	0
0.5	0.02211	24.08	0.0206	21.33	0
0.6	0.02162	24.45	0.01868	19.87	0
0.7	0.02175	24.03	0.01771	22.78	0
0.8	0.01985	18.82	0.01589	14.41	0
0.9	0.01753	18.99	0.01246	11.06	0

[a] represents the P value of capital stock in the two periods of 2004-2008 and 2009-2013.

In the two periods from 2004 to 2008 and 2009 to 2013, the marginal output of capital stock to total factor productivity also showed the characteristics of inverted U-shaped. From 2004 to 2008, in the interval of [0.1, 0.3], the marginal output value of capital stock increased significantly, and began to decrease after reaching the peak. It stabilized at 0.022 in [0.4, 0.7], and then decreased rapidly after 0.7 points. The overall characteristics were not much different from the previous period 2009 to 2013, and there was a rapid upward trend between 0.1 and 0.4 points, followed by a sharp decline.

4.4. Industry-divided research

By analyzing the data, we can find the regression results for the 0.25, 0.5, and 0.75 quantiles for each industry in the manufacturing industry. Most industries show a trend of increasing first and then decreasing, which is consistent with the overall trend. In addition, with the improvement of the technical level, the role of capital stock has gradually decreased in nine industries. This part of the industry is mainly concentrated in the textile industry, shoes and hats, leather, wood processing and other light industries. Since the technology required by enterprises is relatively simple, increasing capital investment does not significantly promote the improvement of corporate technology.

In the beverage, stationery, chemical fiber and plastics industries, the role of capital stock is increasing rapidly. By introducing investment and expanding the scale of the enterprise and the management of standardization, it is possible to promote the technological level of the enterprise extremely effectively. Appropriate policy tilt can more fully play the role of this part of enterprises in promoting the improvement of regional total factor productivity.

5. Summary

First, the average growth rate of total factor productivity in China over the years was 2.42, which was higher than the average of developed countries by 2%. The relatively high growth rate indicates that there is still a certain gap between the technical level between China and developed countries. Technology introduction and spillover are still important driving forces for regional economic development. Total factor productivity increases first and then decreases.

Secondly, the growth rate of TFP in Beijing, Tianjin and Hebei is basically the same as the overall trend, but the technological progress and innovation of enterprises in Hebei Province lags behind that of Beijing and Tianjin. In order to solve the problem of uneven development in the Beijing-Tianjin-Hebei region, the geographical advantage provides the basis for Hebei Province to absorb advanced professional knowledge.

In addition, although the increase in investment volume still has a positive effect on the technological progress of enterprises, the dividend of capital investment gradually disappears, and the overall trend is inverted U-shaped. After 2009, relying on the original capital investment is obviously unable to achieve long-term growth in total factor productivity.

Finally, Increase quality investment is still the main problem Hebei Province facing in the future. Different industries also present different characteristics, according to the characteristics of different industries, different incentive mechanisms should be implemented to encourage effective free investment. By deepening the reform of investment and financing systems, the key role of investment in optimizing supply structure should be exerted.

In general, as technology advances and overall efficiency increase, the marginal utility of capital for total factor productivity growth is not static. Enterprises should use horizontal and vertical comparisons to identify their own nodes and increase effective investment.

Acknowledgments

This paper is grateful for the support of the following items:National Natural Science Foundation of China Youth Fund Project “Research on Technology Innovation and Productivity Improvement of Real Economy from the Perspective of Financialization” (No. 71703012); Key Project of Natural Science Foundation of Liaoning Province “Design and Development of Manufacturing Database Based on Innovation Needs of SMEs” (No. 20170520264); Liaoning Provincial Science and Technology Association Science and Technology Innovation Think Tank Project Class A Project “Research on the Dynamic Change and Policy Choice of High-Quality Development of Liaoning Province Equipment Manufacturing Industry in New Era” (No. LNKX2018-2019A06); Liaoning Provincial Social Science Planning Fund Project “Building a virtuous cycle of Liaoning's real economy and financial innovation from the perspective of symbiosis theory” (No. L18BJY007); Liaoning Provincial Social Science Association Project “Research on the High-quality Development Strategy of Liaoning Province Equipment Manufacturing Industry under the Background of Supply-side Structural Reform” (No. 2020lslktyb-006); Dalian Academy of Social Sciences think tank major research topic "Dynamic change and policy choice research on the high-quality development of Dalian advanced manufacturing industry" (No. 2019dlsky047).

References

[1] Quo,Q.,Zhao,Z.,andJia,J.(2006).Analysis of Total Factor Productivity of Chinese Provincial Economy[J]. Frontiers of Economics in China, 3(1).

[2] Coelli T.J,Rao D.S.P,Christopher J,Donnell O,Battese G.E.(2006).An Introduction to efficiency and productivity analysis[M].Springer Publishers.

[3] Felipe J. (1999).Total factor productivity growth in east Asia:a critical survey[J].Journal of Development Studies,35(4):1-41.

[4] Zheng Y.(2007).Re-recognition of Total Factor Productivity——Analysis of Some Limitations of Economic Growth Quality by TFP[J].Quantities Economics & Technology Research,(9):3-11.

[5] Meeusen W,Broeck J.V.D.(1977).Efficiency estimation from Cobb-Douglas production functions with composed error[J].International Economic Review,18:435-444.

[6] Battese G.E,Coelli T.J.(1992).Frontier production functions, technical efficiency and panel data:with application to paddy farmers in India[J].Journal of Productivity Analysis,3:153-169.

[7] Jiang C, Wu L,Wei T.(2010). Changes in China's Total Factor Productivity: 2000-2008 [J]. Finance and Economics, (7): 55-62.

[8] Olley G.S, Pakes A.(1992).The dynamics of productivity in the telecommunications equipment industry[R]. National Bureau of Economic Research.

[9] Levinsohn J, Petrin A.(2003).Estimating production functions using inputs to control for unobservables[J]. The Review of Economic Studies,70(2): 317-341.

[10] Zhang J.(2010).Reform, Transformation and Growth: Observation and Interpretation [M]. Beijing Normal University Press.

[11] Kim J.I, Lau L.J.(1996).The sources of Asian Pacific economic growth[J]. The Canadian Journal of Economics/Revue canadienned'Economique,29: S448-S454.

[12] Yuan Z.G.(2002).China Employment Report [M]. Economic Science Press.

[13] Chow G, Lin A.(2002).Accounting for economic growth in Taiwan and Mainland China: a comparative analysis[J]. Journal of Comparative Economics,30(3): 507-530.

Comparison and Analysis of Electromagnetic Characteristics of New Quadrupole Track Electromagnetic Launcher

Tao Shu[1], Qingrong Chen[2], Gang Feng[1], Shaowei Liu[1], and Mengyang Cui[2]

[1] Air Defense and Missile Defense College, Air Force Engineering University, Xi'an, Shan xi, 710051, China

[2] Graduate College, Air Force Engineering University, Xi'an, Shan xi, 710051, China

*Corresponding author's e-mail: 1034021926@qq.com

Abstract. Electromagnetic railgun is the main research type of electromagnetic launcher. Researchers have found that the electromagnetic launcher with multipole track structure is capable of greatly improving the comprehensive performance of simple railgun. In this paper, a new quadrupole track electromagnetic launcher model is designed on the basis of simple railgun and quadrupole railgun. In order to fully analyze the electromagnetic characteristics of the model, the theoretical analysis of the structure is carried out at first. Then, under the same conditions as possible, the electromagnetic characteristics of it and the quadrupole railgun model are simulated and compared by using Ansys Maxwell simulation software. The results reveal that the new quadrupole track electromagnetic launcher model has the advantages of more concentrated magnetic field, more uniform armature current distribution and greater electromagnetic thrust.

1. Introduction

Eectromagnetic railgun, referred to as railgun, is the main research type of electromagnetic launcher. It relies on the electromagnetic force between the track and armature to accelerate the projectile. It has many outstanding advantages, such as large kinetic energy, high speed, high precision, wide mass range of payload, good controllability and concealment, simple energy and low cost, safety and reliability. However, it also encounters various problems like high thrust and high current supply, extrusion, wear, planer, ablation and transition caused by assembly size and thermal expansion of armature and rail, sliding electrical contact between armature and rail, etc [1-3].

The structure of electromagnetic launcher plays a fundamental and decisive role in its performance. Therefore, the innovation of the structure of electromagnetic launcher is of great significance to promoting the practical application of electromagnetic launching technology. In order to achieve better comprehensive performance, researchers at home and abroad have done a lot of research on multipolar track electromagnetic launcher, among which plane augmented railgun and stacked augmented railgun are typical representatives. It has been discovered that the electromagnetic launcher with multipole track can provide greater electromagnetic thrust for the projectile and alleviate the obstruction of material and power supply technology. However, with the increase in the number of tracks, the structure of the model becomes very complex, which requires high processing technology, and the skin effect of current is more obvious. At the same time, a lot of Joule heat is generated in multiple tracks, resulting in a large amount of energy waste. Therefore, for the electromagnetic launcher, the relationship between the number of tracks and the performance of the launcher is not linear [4-6].

As early as the 1990s, some scholars had suggested the concept of small caliber plasma armature quadrupole railgun [7]. In recent years, more and more researchers have conducted a lot of research into the electromagnetic launcher model of quadrupole track structure. The static performance and transient dynamics of quadrupole track electromagnetic launcher are analyzed in reference [8-9]. The structural design and thrust simulation analysis of the enhanced quadrupole track electromagnetic launcher are carried out in reference [10]. The contact pressure between track and armature of quadrupole track electromagnetic launcher is simulated and analyzed, while the interference fit parameters between armature and track are optimized in reference [11-13]. In this paper, a new quadrupole track electromagnetic launcher model is designed on the basis of simple railgun and quadrupole railgun. The electromagnetic characteristics of the model are compared and analyzed by combining theory with simulation.

2. Modeling of the new quadrupole track electromagnetic launcher

2.1. Model structural analysis

The basic structure of the new quadrupole track electromagnetic launcher model is shown in figure 1. The model is symmetrical in structure, and the armature moves rapidly towards the outlet under the combined action of magnetic fields generated by four tracks of identical size and material. Current flows from the two tracks marked blue on the left, through the armature clamped in the upper and lower tracks, and then from the two tracks marked red on the right to form a closed loop. By applying pre-tightening force from four armature-rail contact surfaces, the structure can better maintain the high-speed sliding electrical contact between armature and rail, thus reducing the vibration during motion. By adding insulation materials, supporting materials and other auxiliary materials around the track form the launcher barrel, the left and right armature displacement can be prevented, and the overall stability of the model can be enhanced. Obviously, the model has various advantages, such as simple and stable structure, easy maintenance, convenient replacement of track and armature, etc.

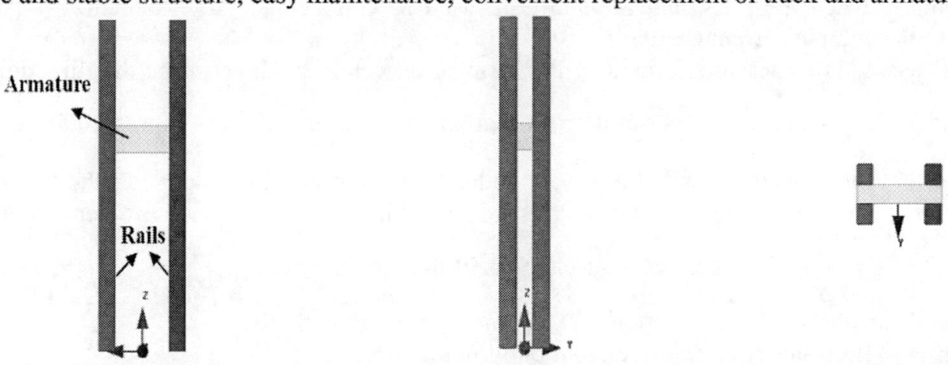

(a) Main view (b) Left view (c) Top view

Figure1. Structural schematic diagram of the new quadrupole track electromagnetic launcher

2.2. Theoretical analysis of electromagnetic thrust

Prior to the analysis of the electromagnetic thrust of the model, it is necessary to construct the current and magnetic field coordinate system of the new quadrupole track electromagnetic launcher as shown in figure 2. In this coordinate system, the tail end faces of four rails are all in the XOY plane, the upper and lower rails are X-axisymmetric, and the left and right rails are Z-axisymmetric. The sizes of the four rails are the same. The distance between left and right tracks is b, the distance between upper and lower track is d, the width of the track is a, and the height of the track is h. $l(t)$ represents the length of the armature tail end face from the rail tail end face at t time, and m indicates the length of the armature. It is supposed that the source point of orbital current is $S(x, y, z)$, and that the field point of armature is $P(x',y',z')$. The current I is put into the tail end faces of the rail 1 and 3. Then the current in the end faces of the rail 2 and 4 is $-I$.

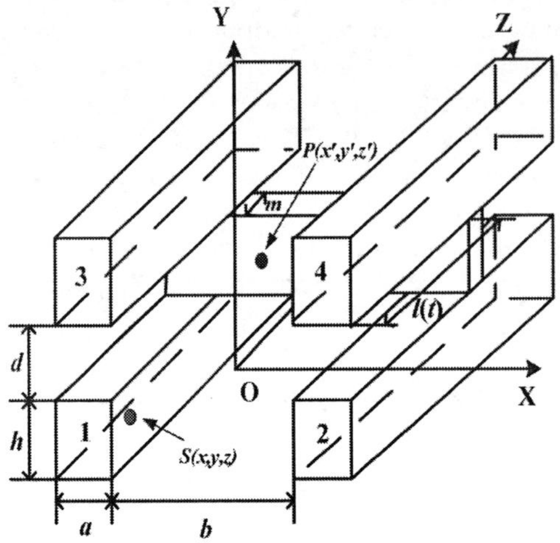

Figure2. Current and magnetic field coordinates of the new quadrupole track electromagnetic launcher

For current-carrying conductors, the line element vector is $\boldsymbol{dl}$, and the product of current I and $\boldsymbol{dl}$ flowing through a line element vector is called current element. Meanwhile, the direction of the current in the current element is taken as the direction of the line element vector. Then, we can regard a current-carrying wire as being connected by many current elements $I\boldsymbol{dl}$. Thus, the magnetic inductance $\boldsymbol{B}$ excited by a current-carrying conductor at a certain point in the magnetic field is the superposition of the magnetic inductance $\boldsymbol{dB}$ at that point by all the current elements of the conductor [10]. Therefore, when the track and armature are used as conductors, they ought to be calculated according to the volume current element.

Current density is a measure defined in the form of a vector. Its direction is the direction of the current, and its size is the current per unit cross-section area. The equation is expressed as $\boldsymbol{J} = \frac{I}{\boldsymbol{S}}$. In the formula, $\boldsymbol{J}$ represents the current density, I indicates the current and $\boldsymbol{S}$ denotes the cross-section vector. Without considering the skin effect, under ideal conditions, the body current element in orbit is $J_V dxdydz\boldsymbol{k}$, $J_V = \frac{I}{ah}$. The body current element in the armature is $J'_V = \frac{I}{dm}$, $J'_V = \frac{I}{dm}$. $\boldsymbol{i}, \boldsymbol{j}, \boldsymbol{k}$ represent the unit vectors in the direction of X, Y and Z axes respectively.

According to Biot-Savart's law, there are the following formulas

$$d\boldsymbol{B} = \frac{\mu_0 (I\boldsymbol{dl} \times \boldsymbol{R})}{4\pi R^3} \tag{1}$$

Among them, μ_0 represents vacuum permeability, and its value is $\mu_0 = 4\pi \times 10^{-7}\ \mathrm{N \cdot A^{-1}}$. If the coordinates of track 1 current source point are $S_1(x,y,z)$ and armature field point are $P(x',y',z')$, then there are

$$d\boldsymbol{B}_1 = \frac{\mu_0 (J_V dxdydz\boldsymbol{k} \times \boldsymbol{R}_1)}{4\pi R_1^3} \tag{2}$$

$$\begin{aligned} \boldsymbol{R}_1 &= (x'-x)\boldsymbol{i} + (y'-y)\boldsymbol{j} + (z'-z)\boldsymbol{k} \\ R_1 &= \sqrt{(x'-x)^2 + (y'-y)^2 + (z'-z)^2} \end{aligned} \tag{3}$$

$$\begin{aligned} d\boldsymbol{B}_1 &= \frac{\mu_0}{4\pi R_1^3} J_V dxdydz\boldsymbol{k} \times \left[(x'-x)\boldsymbol{i} + (y'-y)\boldsymbol{j} + (z'-z)\boldsymbol{k}\right] \\ &= \frac{\mu_0 J_V dxdydz}{4\pi R_1^3}[(y-y')\boldsymbol{i} + (x'-x)\boldsymbol{j}] \end{aligned} \tag{4}$$

Therefore, the electromagnetic induction intensity produced by orbital 1 to field point P is as follows

$$\begin{aligned} \boldsymbol{B}_1 &= \int_0^{l(t)+m} \int_{-h-\frac{d}{2}}^{-\frac{d}{2}} \int_{-a-\frac{b}{2}}^{-\frac{b}{2}} \frac{\mu_0 J_V}{4\pi R_1^3}[(y-y')\boldsymbol{i} + (x'-x)\boldsymbol{j}]dxdydz \\ &= \frac{\mu_0 J_V}{4\pi} \int_0^{l(t)+m} \int_{-h-\frac{d}{2}}^{-\frac{d}{2}} \int_{-a-\frac{b}{2}}^{-\frac{b}{2}} \frac{[(y-y')\boldsymbol{i} + (x'-x)\boldsymbol{j}]dxdydz}{\sqrt{(x'-x)^2 + (y'-y)^2 + (z'-z)^2}^3} \end{aligned} \tag{5}$$

The electromagnetic force at any point on the armature can be calculated by combining the current intensity of the armature. The electromagnetic force produced by track 1 on the armature field point is as follows

$$d\boldsymbol{F}_1 = J'_V dx'dy'dz'\boldsymbol{i} \times \boldsymbol{B}_1 \tag{6}$$

When the armature acts as a current-carrying conductor, the ampere force acts on the whole armature area. Therefore, by integrating the whole armature volume, the electromagnetic propulsion force produced by track 1 on the armature as a whole can be derived as follows

$$\begin{aligned} \boldsymbol{F}_1 &= \int_{-a-\frac{b}{2}}^{a+\frac{b}{2}} \int_{-\frac{d}{2}}^{\frac{d}{2}} \int_{l(t)}^{l(t)+m} J'_V dx'dy'dz'\boldsymbol{i} \times \boldsymbol{B}_1 \\ &= \frac{\mu_0 J_V J'_V}{4\pi} \int_{-a-\frac{b}{2}}^{a+\frac{b}{2}} \int_{-\frac{d}{2}}^{\frac{d}{2}} \int_{l(t)}^{l(t)+m} [\int_0^{l(t)+m} \int_{-h-\frac{d}{2}}^{-\frac{d}{2}} \int_{-a-\frac{b}{2}}^{-\frac{b}{2}} \frac{x'-x}{R_1^3} dxdydz]dx'dy'dz'\boldsymbol{k} \end{aligned} \tag{7}$$

Due to the symmetry of the structure, the electromagnetic thrust of tracks 2, 3 and 4 is analyzed using the same method. The results are as follows.

$$\boldsymbol{F}_2 = \boldsymbol{F}_3 = \boldsymbol{F}_4 = \boldsymbol{F}_1 \tag{8}$$

Therefore, the theoretical value of electromagnetic thrust generated by the launch model is

$$\boldsymbol{F} = 4\boldsymbol{F}_1 \tag{9}$$

Its direction is along the Z axis.

3. Comparison and analysis of simulation

In order to perform analysis of the electromagnetic performance of the new quadrupole track electromagnetic launcher more intuitively and deeply, the electromagnetic performance of the structure and the quadrupole railgun model are simulated and analyzed by using Ansys Maxwell simulation software under the same conditions to the greatest extent. Although the two models are quadrupole track structures, the model track layout, armature pre-tightening and current mode are completely different.

For convenience, the new quadrupole track electromagnetic launcher model shown in figure 1 is called configuration 1, and the quadrupole railgun model shown in figure 3 is called configuration 2.

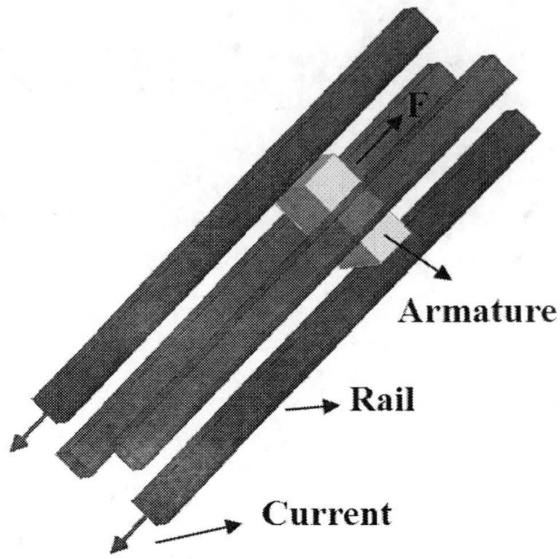

Figure3. Basic model of configuration 2

Configuration 2 model is shown in figure 3. In this model, the structure, size and material of the four rails are the same, the armature is installed in the middle of the four rails, and the whole model is symmetrical about the geometric center of the armature. In the two relative rails, the same current of equal magnitude is introduced, which flows through the armature and finally flows out of the other two relative rails. The current in the guideway generates a quadrupole magnetic field area in the launching pipeline. The interaction between the magnetic field and the current in the armature generates thrust, which drives the armature forward rapidly, thus completing the launching of the projectile.

From the perspective of model structure, these two structures exhibit the characteristics of simple and stable structure, easy processing and replacement. From the point of view of model structure, these two structures have the characteristics of simple and stable structure, easy processing and replacement. Next, the electromagnetic characteristics of the model are analyzed in more depth from the perspective of simulation. The basic parameters of the simulation models of two quadrupole track electromagnetic launchers are shown in Table 1. Both models adopt copper track and aluminum armature. The armature is located 300 mm away from the rear end of the track. 100 kA impulse current excitation is applied to each model. Ansys Maxwell finite element simulation software is used. The two configurations are analyzed by appropriate solution domain and mesh size. The following simulation results are obtained:

Table 1. Basic parameters of simulation model.

Parameter type	Configuration 1	Configuration 2
Track length (mm)	500	500
Track width (mm)	30	30
Track height (mm)	30	30
Upper and lower track spacing (mm)	30	100
Left and right track spacing (mm)	100	100
Armature length (mm)	40	40
Armature width (mm)	160	100
Armature height (mm)	30	100

3.1. Distribution of magnetic field between tracks

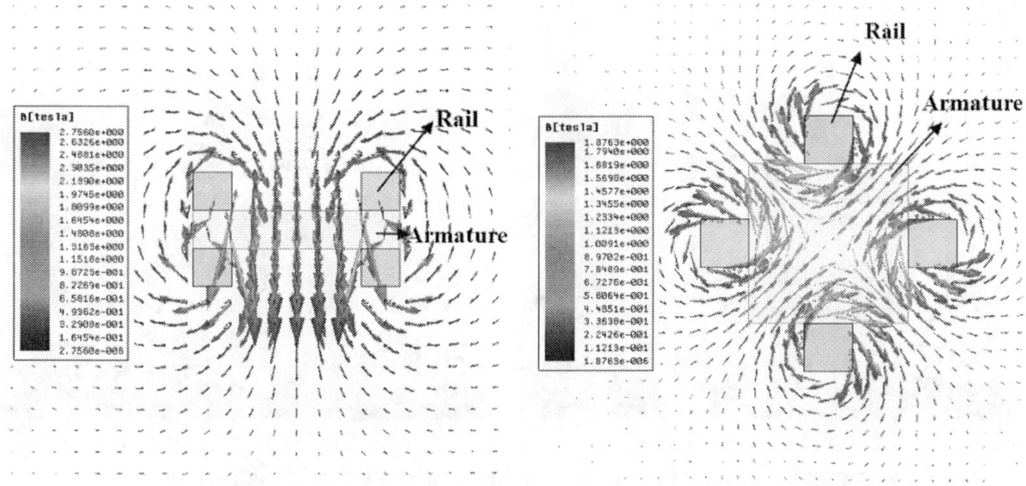

(a) Configuration 1 (b) Configuration 2

Figure4. Distribution vector diagram of magnetic induction lines between tracks

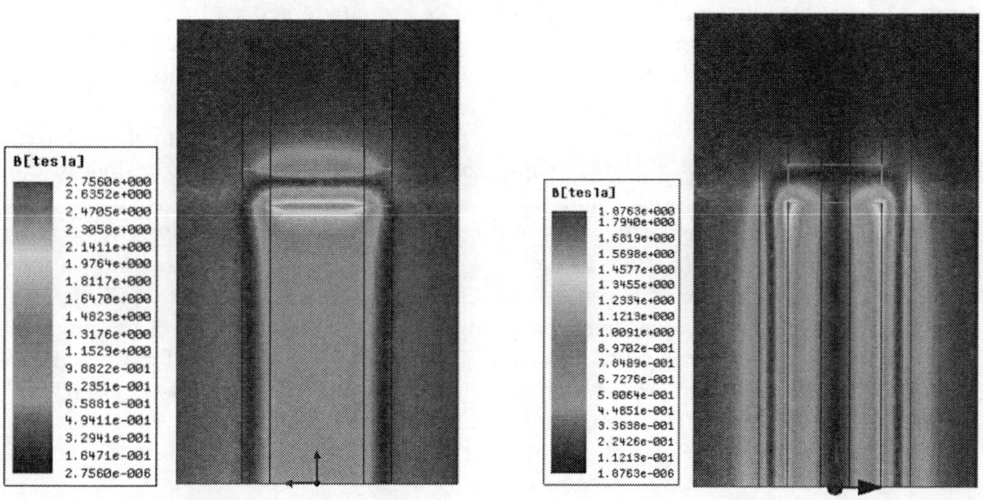

(a) Main view of configuration 1 (b) Main view of configuration 2

IOP Conf. Series: Materials Science and Engineering **711** (2020) 012004 doi:10.1088/1757-899X/711/1/012004

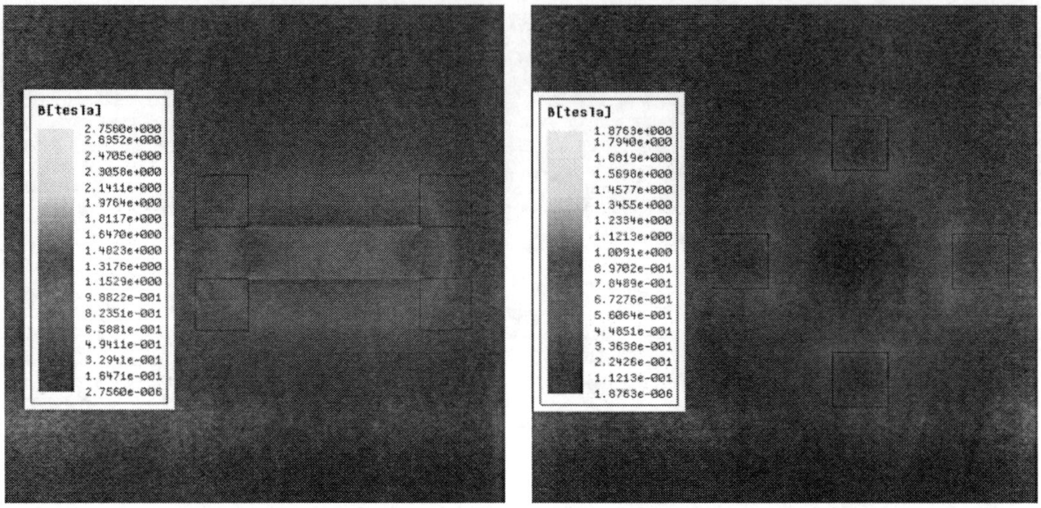

（c）Top view of configuration 1 （d）Top view of configuration 2

Figure5. Clouds of track and armature magnetic field distribution

As revealed by the figure above, the maximum magnetic induction intensity of configuration 1 is 2.756T and that of configuration 2 is 1.8763T. From figure 4, it can be seen that in configuration 1, the magnetic induction lines generated by the polar tracks are converged and strengthened between the tracks, which is due to the way that the structure passes through the same direction current at the same side of the tracks. Meanwhile, the magnetic field generated by the polar tracks cancels each other in the armature center region, and the distribution of magnetic inductance lines near the armature is relatively sparse because the current in the same direction is introduced into the relative tracks in configuration 2. Figure 5 shows that the magnetic field of configuration 1 is mainly concentrated between tracks and near armature, especially at the end of armature, and there is almost no divergence of magnetic energy around tracks. In configuration 2, magnetic field mainly concentrates between orbits. However, due to the offset effect of magnetic field, the magnetic field radiates to the outer side of the orbits, causing a large amount of magnetic energy dissipation. At the same time, the magnetic field in the armature region is weak, especially in the central region of the armature, which is not conducive to improving the force on the armature. Thus, the utilization rate of magnetic energy is not high. Therefore, configure 1 has a more concentrated magnetic field and less energy loss due to its unique structure and the way in which current is fed.

3.2. Armature current distribution

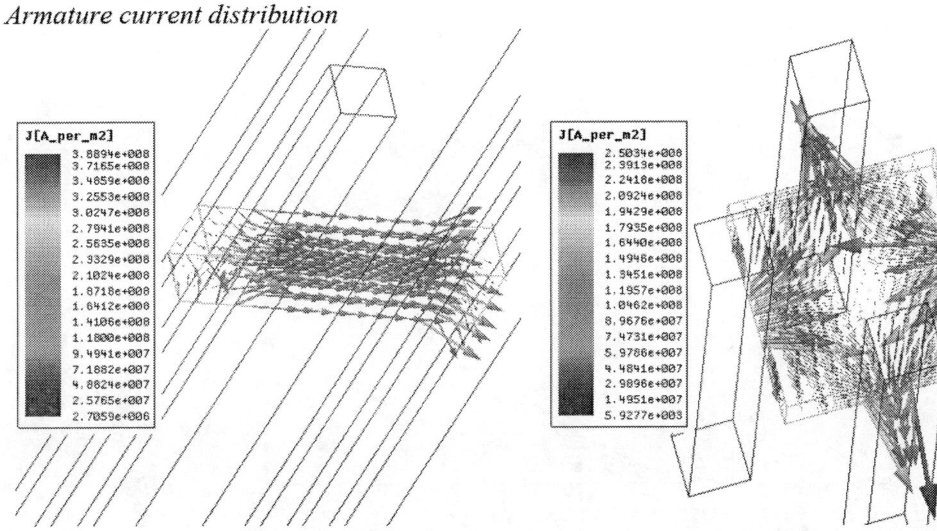

(a) Configuration 1 (b) Configuration 2

Figure6. Vector diagram of current distribution in armature

As indicated by the figure above, the maximum current density of armature in configuration 1 is 3.8894×10^{8} A/m^2. Current is injected into armature from two tracks on the left side, which makes the current density in armature increase further without excessive current concentration. Meanwhile, the current distribution inside armature is quite uniform, which greatly inhibits the skin effect of current and is beneficial to the uniform force of armature in the process of high-speed movement. In configuration 2, the maximum current density of armature is 2.5034×10^{8} A/m^2, which is small. However, due to the obvious tendency of choosing the shortest path to pass through the conductor, the phenomenon of current concentration occurs. In the mean time, the distribution of current in armature is dispersed, especially in the central region of armature, and there is almost no current. To a large extent, this results in the waste of energy and materials, and the consistency of armature force is also affected to a certain extent. Therefore, the armature current distribution of configuration 1 is more uniform, and it has certain advantages in armature stable force and restraining current concentration effect.

3.3. Electromagnetic thrust

The electromagnetic thrust produced by the two types of quadrupole track electromagnetic launchers at a time when 100kA pulse current excitation is applied to each model is shown in Table 2 below.

Table 2. Simulation results of electromagnetic thrust.

structure type	F(x) /N	F(y) /N	F(z) /N	Mag(F) /N
Configuration 1	7.013	0.1015	11240	11240
Configuration 2	3.7239	10.106	4747.2	4747.2

In the actual process of movement, the force on armature is unlikely to be completely uniform, which leads to the existence of X-axis and Y-axis transverse forces in both configurations, but the numerical values are very small. As compared to configuration 1, the transverse force of configuration 1 is smaller, for which the impact on track is smaller. It is confirmed from another point of view that the force on configuration 1 armature is more consistent. The axial force produced by configuration 1 is 11240N, and that produced by configuration 2 is 4747.2N. Under the same current excitation, the electromagnetic thrust produced by configuration 1 is 2.37 times that in configuration 2.

In order to compare and analyze the change of thrust with current of each model more comprehensively, the scanning settings of current parameters are created by using simulation software

[11], and the curve of thrust with current of each model in the range of $\left[0,300\right]$ kA is obtained as follows:

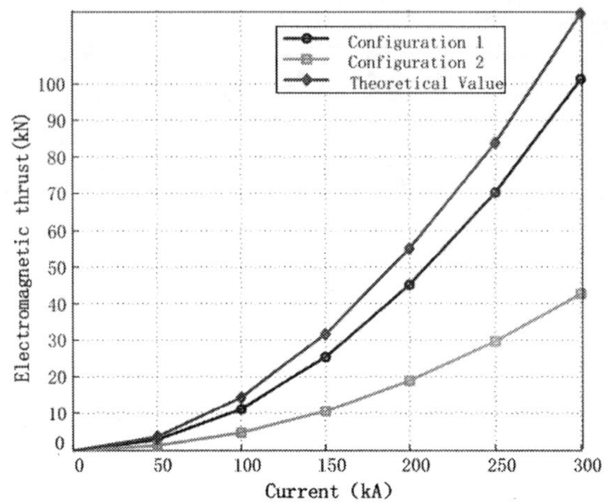

Figure7. The relationship between electromagnetic thrust and current

The basic parameters of configuration 1 simulation model and related modeling data are substituted by formulas (7), (8). After calculation, the red curve shown in figure 7 is obtained, which represents the curve of theoretical value of electromagnetic thrust varying with current. From figure 7, it can be seen that the electromagnetic thrust of each model increases with the increase of current, and the magnitude of the increase of configuration 1 is larger than that of configuration 2, which indicates the higher energy utilization of configuration 1. In addition, without considering the actual losses such as Joule heat and friction heat, the simulation results of configuration 1 electromagnetic thrust are similar to the theoretical values, which highlight the advantages of configuration 1 in electromagnetic thrust and energy utilization.

The simulation results of electromagnetic thrust are consistent with the analysis results of magnetic field distribution between tracks and armature current distribution, which further validates the excellent electromagnetic comprehensive characteristics of configuration 1.

4. Conclusion

Based on the simple rail gun and the four-rail gun, a new type of quadrupole rail electromagnetic launcher model is designed in this paper. The theoretical and simulation analysis shows that the configuration has the following advantages: firstly, the structure is simple and stable, and it is easy to realize in engineering; secondly, the magnetic field distribution is concentrated, and the magnetic energy utilization rate is high; thirdly, the uniform distribution of armature current can better keep armature moving smoothly and restrain the current concentration effect; fourthly, it can generate greater electromagnetic thrust under smaller current excitation, which reduces the dependence of electromagnetic launcher on power supply technology.

References

[1] Marshall R A,WANG Ying, “The science and technology of electromagnetic railgun,” Beijing: The Press of Ordnance Industry,2006, pp.23-25.

[2] J. Li, P. Yan, W. Q. Yuan, “Electromagnetic gun technology and its development,” High Voltage Engineering, vol. 40, pp. 1052-1064, April .2014.

[3] Shen Max Chung,Yu-Chou Chuang, “Characteristics of electromagnetic radiation of a railgun at the final firing stage,” IEEE Transactions on Plasma Science,2016,44(1): 49-58.

[4] X. Yu and Z. Fan, "Simulation and two-objective optimization of the electromagnetic-railgun model considering VSEC resistance and contact resistance," IEEE Trans. Plasma Sci., vol. 39, no. 1,pp. 405 - 410, Jan. 2011.

[5] QIAO Zhiming, LEI Bin, XIANG Hongjun,etc, " The current state of development of reinforced electromagnetic railgun and its technical characteristics, " Cruise Missile, 2016(10):28-31.

[6] Ferrero R, Marracci M, Tellini B, " Characterization of inductance gradient and current distribution in electromagnetic launchers," IEEE Trans Instrumentation and Measurement, 2011, 60(5): 1795-1801.

[7] E. Igenbergs, " A symmetrical rail accelerator, " IEEE Transactions on magnetics, 1991, 27(1):650-653.

[8] YANG Z Y,FENG G,XUE X P, "An Electromagnetic Rail Launcher by Quadrupole Magnetic Field for Heavy Intelligent Projectiles," IEEE Transactions on Plasma Science,2017,45(1): 1095-4099.

[9] MIAO Haiyu, LIU Shaowei, LIU Ming, YANG Zhiyong, "The simulation of electromagnetic thrust of series reinforced quadrupole rail launcher," The Academic Journal of Air Force Engineering University (Science of Nature),2018,19(03):71-76.

[10] TONG Siyuan, FENG Gang, YANG Zhiyong,etc, "The parameter optimization of interference fit for armature track of missile quadrupole magnetic field," High Power Laser and Particle Beams, 2019,31(01):38-44.

[11] BAI Xiangzhong, ZHAO Jianbo, TIAN Zhenguo, "The mechanical analysis of launching components of electromagnetic track, " Beijing : The Press of National Defense Industry,2015.6,pp.24-27.

[12] ZHAO Bo, ZHANG Hongliang, "The application of Ansoft 12 in Engineering electromagnetic field," Beijing: China Water & Power Press,2010.1,pp.236-243.

[13] R. Ferrero, M. Marracci, B. Tellini, " Characterization of inductance gradient and current distribution in electromagnetic launchers," IEEE Trans Instrumentation and Measurement, vol. 60, pp. 1795-1801, May. 2011.

Dynamic Performance Analysis of the micro-hole EDM Machine Tool

Liu Jian-yong [1,2], Wang Min [3], Hu Jian-fei [1,2], Qin Peng [3], Shi Long [1,2*], Cai Yan-hua[1,2]

[1] Beijing Institute of Electro-Machining, Beijing 100191, China

[2] Beijing Key Laboratory of Electrical Discharge Machining Technology, Beijing 100191, China

[3] Advanced Manufacturing technology of the Key Laboratory of Beijing Municipality, College of Mechanical Engineering and Applied Electronics Technology, Beijing University of Technology, Beijing 100124, China

[*]Corresponding author's e-mail: 875598674@qq.com

Abstract. The EDM machine is special processing machine using the principle of EDM process of conductive materials. By means of modal analysis, ODS (Operational, Deflection, Shapes) analysis and order analysis, the static and dynamic performance parameters of the EDM machine tool are studied. According to the processing requirements, the structure of the machine tool is optimized, and the weak links are strengthened so as to reduce the aperture dispersion of the machined holes. Through the modal test this experiment firstly analyzes the natural frequency, damping ratio and modal shape in each mode of machine tool structure to understanding the characteristics of each order machine main mode. Then, through ODS (Operational Deflection Shapes) test, it analyzes the vibration characteristics of machine tool structure in operation condition. And it also analyzes the relation-shape between the vibration of rotating machinery and the speed of the machine through the order analysis to predict the vibration of actual machine tool structure in a certain band external or internal variety of vibration response. Finally, it makes a comprehensive study on the static and dynamic performance of machine tools, and provides a reference for the design and optimization of structure dynamic strength.

1. INTRODUCTION

EDM machine tool is a special machine tool for processing conductive material by using EDM principle [1]. Because it has the characteristics of non-contact processing, no macroscopic cutting force, killing with kindness, precision micro etc, precision micro EDM technology has been widely used in aviation, aerospace, computer technology, instrumentation, precision machinery, automotive and other industries in some of the cutting-edge products [2,3].

The research object of this paper is the seven axis precision NC EDM machine tool developed by Beijing Institute of electrical processing [4,5]. The three dimensional diagram is shown in Figure 1. It is found that when the spindle speed reaches 1000 revolutions per minute, the machine tool will vibrate violently, resulting in a larger diameter difference of the machined holes, which will affect the quality of the machining. Therefore, through the experimental analysis, the structure of the machine tool is optimized, and the performance is improved, and the aperture difference of the hole is reduced.

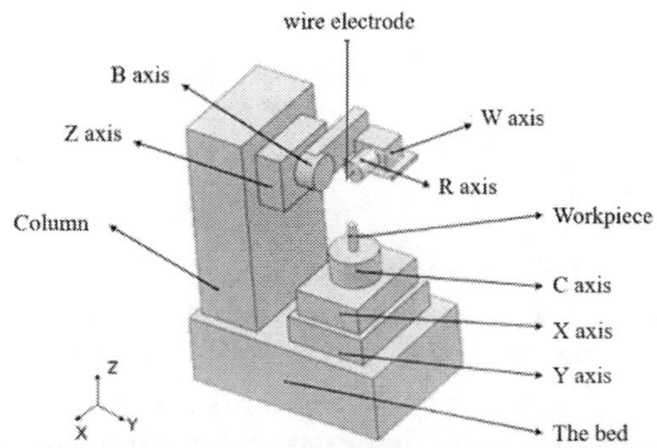

Figure1. Three-dimensional diagram of seven axis EDM machine tool

For example, 0.26mm tungsten wire is used to process metal plates of 1.25mm thickness. The ideal machining aperture is 0.3mm, and the diameter of the actual machined hole is shown in Figure 2. A total of 32 holes were machined, and the bore number was 1-32.

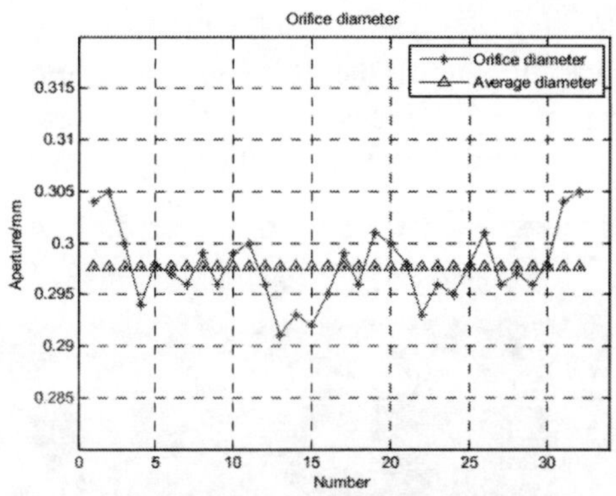

Figure2. Orifice diameter

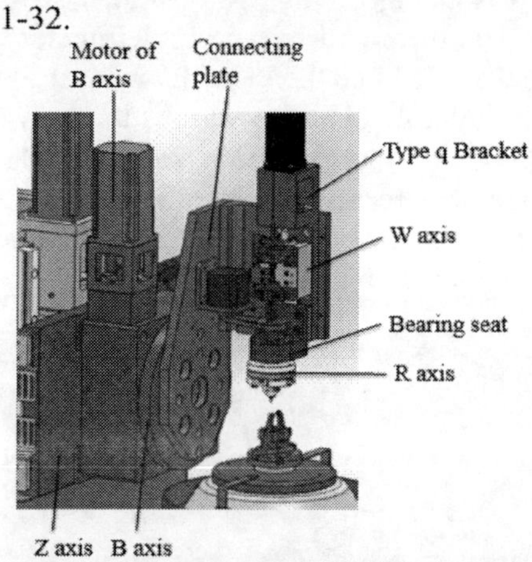

Figure3. Partial structure three-dimensional graph

The maximum aperture difference of these holes is 0.014mm. Modal test and ODS analysis can be used to understand the mode of machine tool. According to the modal shape, the reference is put forward to optimize the structure and reduce the vibration of machine tool, so as to reduce the aperture difference.

The structure near the R axis of this machine tool needs to be analyzed in detail, therefore, a three-dimensional drawing of this part of the structure is drawn, as shown in figure 3.

2. Modal test

Mode is the inherent vibration characteristic of mechanical structure. Modal analysis is an important method for structural dynamic design and equipment fault diagnosis, and provides an effective way to study vibration of various practical structures. First of all, the machine tool is subjected to artificial vibration in the static state, and the transfer function between any two points is obtained by analyzing the two channel signal of the vibration force of the exciter and the vibration acceleration of the machine tool. Then, the modal parameters of the machine tool are identified by the curve fitting of the test transfer function. Finally, modal analysis theory is applied to solve the modal shapes of machine tools [6].

2.1 Test equipment Formatting the title

The modal experiment was carried out by the modal testing and analysis system of the B&K company of Denmark. The main test equipment include three direction acceleration sensor, 5320-50 type hammer, pulse system, 3560B data acquisition front end and so on. The measured frequency response function data is imported into ME'scopeVES software to identify the modal parameters. In the experiment, the impact hammer is used to shock the fixed excitation point. Because the striking force is grasped by people so it is difficult to control each time. In order to reduce the random error, each set of points should record 3 sets of data, which makes the frequency response function smoother. According to the correlation coefficient of frequency response function is good or bad judgment. In general, the closer the coherence coefficient is to 1, the smaller the experimental interference is, and the more reliable the experimental result is.

2.2 Selection of excitation mode and measuring point

Due to the smaller space of the tested machine tool,and the three direction force acceleration sensor requires the angle of the hammer striking direction is 45 degrees with the angle of the X, y and Z direction（the directions of X, y, and Z are consistent with figure 3）. Because using a mobile sensor for multi-point pickup is more convenient and easier to guarantee the same angle，so the excitation mode of this modal experiment is single point excitation and multi-point vibration pickup. That is, in the experimental process, the excitation point is not changed and the position of the vibration pickup point is changed only. The hammer strikes the test specimen and gives it an impulse. The vibration signal is measured by an accelerometer mounted on a machine tool. The force signal and response signal are fed into the PULSE analyzer to calculate the frequency response function.

In order to stimulate various modes which affect the performance of machine tools under practical conditions，the excitation point is selected at the upper end of the R axis. The angle between excitation direction and the the x y and z directions are all 45 degrees. The sensors are respectively arranged on the bed, Y axis, C axis, column, Z axis, W axis and R axis. A total of 31 measuring points were selected in this experiment.

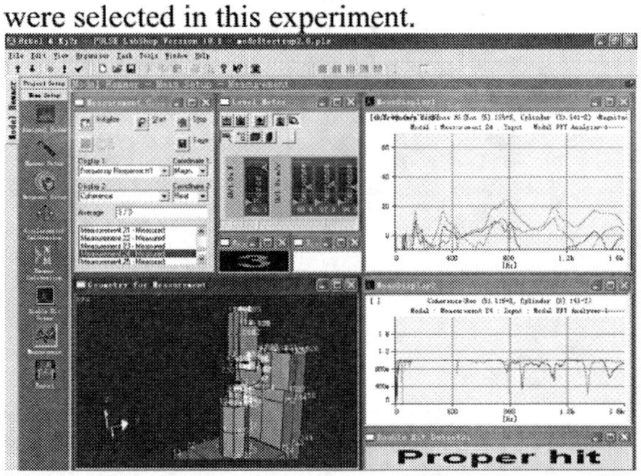

Figure4. Pick up points and exciting points position

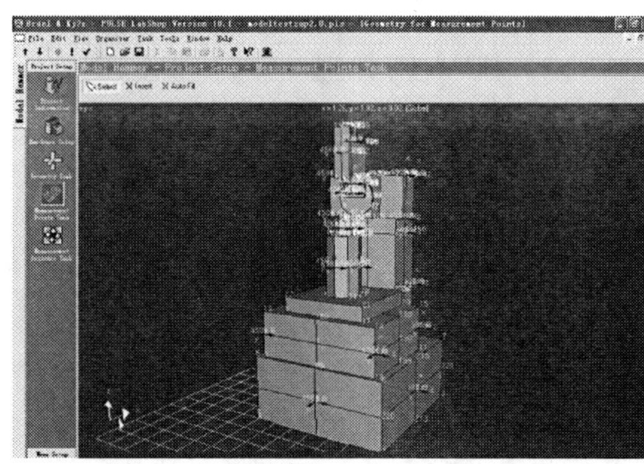

Figure5. PULSE system test panel

2.3 Test procedure and results

According to the size of the machine tools, a wireframe model built for machine tool entities inside the Geometry. Because those points need to collect data, the direction of the points should be in accordance with the three direction force sensor.

Test the mode of the machine tool and the test system panel as shown in figure 5.

In the case of good coherence, save the results and import the results into the post-processing software ME'scopeVES for analysis. The modal parameters of the machine tool can be obtained, as shown in Table 1. The modal shapes are shown in Table 2, and the mode of operation is shown in figure 6. Because of the limited length of the article, only the first order vibration modes and the first five modal formation characteristics are listed.

Table1. Modal parameters

Modal order	Natural frequency (Hz)	Damping ratio (%)	Modal stiffness (N/m)
1	35.4	1.88	355e6
2	54.7	3.57	114e6
3	142	3.28	3.27e9
4	191	2.83	303e6
5	333	1.42	529e6

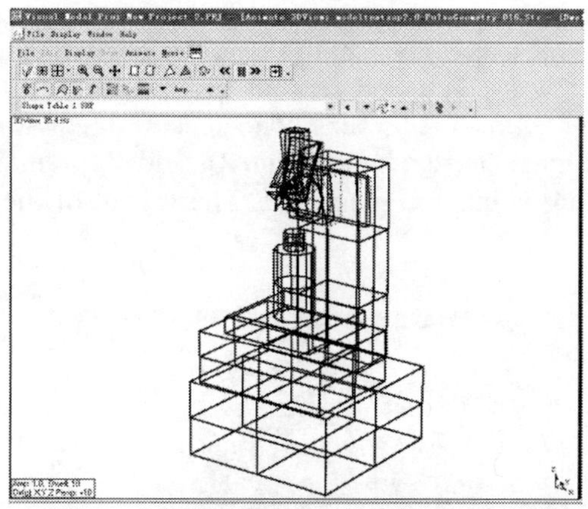

Figure6. Working mode of first order

Through the analysis of the vibration modes obtained by modal testing, it is found that when R axis is used the low dynamic stiffness of EDM machine tools is caused by the two order modes of natural frequency 54.7Hz, which is exactly 3 times the frequency of 1000rpm. By analyzing the working vibration modes of the machine tools, it is found that the low stiffness is caused by that the stiffness of the joint of the connecting plate and the bearing seat is poor. At the same time, through analysis of other main modal formation, it is found that the Q type stent is easy to produce Z direction vibration, and the joint surface of connecting plate and B axis also has the problem of unreliable connection. These are likely to restrict the improvement of the anti-vibration performance of the spindle of EDM machine tools.

Table2. Modal formation and characteristics

Modal order	Natural frequency (Hz)	Mode characteristic
1	35.4	The connecting plate and the R axis vibrate in the Y direction.
2	54.7	R axis and the connecting plate vibrated torsionally in the X-Y plane.
3	142	The connecting plate and the W axis vibrate violently in the x-z plane.
4	191	The B shaft motor is vibrating in the Z direction and the connecting plate drives the W shaft and the R shaft to vibrate violently in the x-z plane.
5	333	The connecting plates vibrate violently in both the x-z plane and the Y-Z plane.

3. ODS test

ODS is defined as the analysis of vibration and deformation of a structure at a certain frequency, which indicates the vibration state of the structure under certain working conditions.

3.1 Test equipment

The test was conducted using the Danish B&K company's ODS test and analysis system. The main test equipment include 3 unidirectional acceleration sensors, 5320-50 force hammer, pulse system, 3560B data acquisition front end, etc. The measured frequency response function data is imported into ME'scopeVES software to identify the modal parameter identification. During the experiment, a sensor is fixed to a certain position, (the reference sensor is chosen to be fixed on the flange of the clamping electrode wire) , The rest of the sensors are moving at the point of interest to obtain the signal. The ODS test was carried out under the condition of reciprocating feeding of the W axis.

3.2 Excitation mode and selection of measuring points

According to the shape, structure and mode shapes of machine tools, in the machine outside choose 8 pick-up points and reference pick-up points set in the upper edge of R axis. Due to the limited number of sensors (three), each test was conducted in four groups. Due to the limited number of sensors (three), each test was conducted in four groups. The sensor signal at the upper end of the R axis is determined as a reference signal (an excitation signal), and the remaining two sensor positions are moved until all the measuring points are measured. The layout of the measuring points is shown in figure 7.

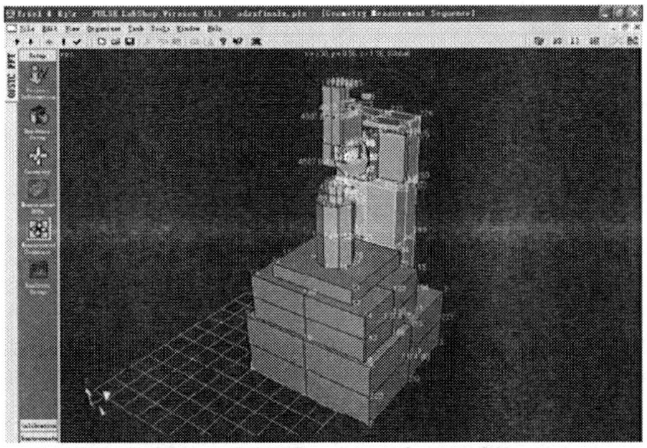

Figure7. Sketch map of ODS measuring points arrangement

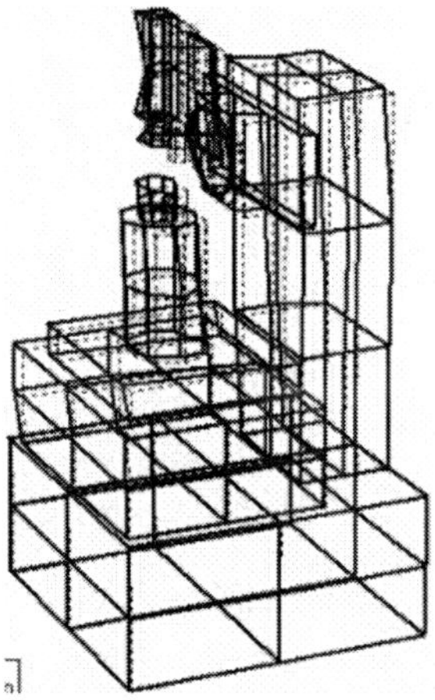

Figure8. Working mode of first order

3.3 Test procedure and results

This ODS carries on the experiment to the machine tool using the W axis feed processing. The sensor is inserted into the model and the direction of the sensor is adjusted. The machine tools are measured in three directions: X, Y and Z. The sensor position is shown in figure 7. Record the vibration and deformation of the machine tool in the three directions of X, Y and Z when feeding the W axis. After analyzing the experimental data into the post-processing software ME'scopeVES, the modal shapes of the machine tools can be obtained, as shown in figure 8. Because of the length of the article, only the first order modes of vibration are listed here. The natural frequencies in the three directions of X, Y and Z are shown in Table 3.

Table3. Natural frequency

X direction	49.1	139	195	356	443	588
Y direction	50.3	115	207	328	485	615
Z direction	-	150	210	371	474	685

Compared the experimental modal test results with the ODS test results it can be found that in the static state of the spindle, the excitation frequency has six orders, which is similar to the natural frequency. At this speed, chatter may occur near the six frequency ranges, which means that the six frequency components are more likely to be excited in the actual running state of the machine.

By comparing the vibration modes of ODS test with the modal test, it can be found that in the two experiments the larger deformation structures are the union surface of the B axis and the connecting plate, the connecting plate and the Q type bracket, including the bearing seat. Therefore, the results of the two experiments are consistent, indicating that the experimental results are credible and reliable.

4. Vibration order analysis

The vibration of rotating machinery is closely related to the speed of the machine. The mechanical axis of the mass imbalance and eccentric will cause vibration. Its vibration frequency is an integer multiple of the speed of revolution frequency. In addition, there is a transmission ratio between the active and the driven parts in the mechanical drive. Therefore, the vibration signal analysis is often based on the basic axis speed. The multiple relative to the base axis speed is called order. It is called order analysis that order speed is used as independent variable.

4.1 Test equipment

This order analysis experiment is conducted by the test and analysis system of the Danish B&K company. The main test equipment include 1 unidirectional acceleration sensors, 1 MM0024 infrared tachometer, pulse system, 3560B data acquisition front end, etc..

4.2 Test points selection

In actual conditions, the machine will cause noise and vibration when the R axis speed is known to be 1000rpm, Therefore, this vibration order analysis will set R axis speed as reference speed and an insulating black tape is coated on the flange surface at the lower part of the R axis, and put a reflective paper on the adhesive tape. Align the MM0024 infrared tachometer which is placed at about 30mm from the R axis to the lower flange surface of the R shaft. It fixed with a magnetic mount. The unidirectional acceleration sensor is arranged on the bearing seat, and since the vibration in the X direction is much larger than that in the Y direction, the sensor selects the direction of the X to be measured.

4.3 Test procedure and results

The reference speed is R axis speed, which is increased from 500rpm to 1800rpm by manual speed regulation. The pulse system will automatically draw the three-dimensional spectrum of vibration nephogram and order vibration nephogram.

The order spectrum of the vibration signal in the R axis X direction is shown in Figure 9, The X direction here is consistent with the X direction in ODS.

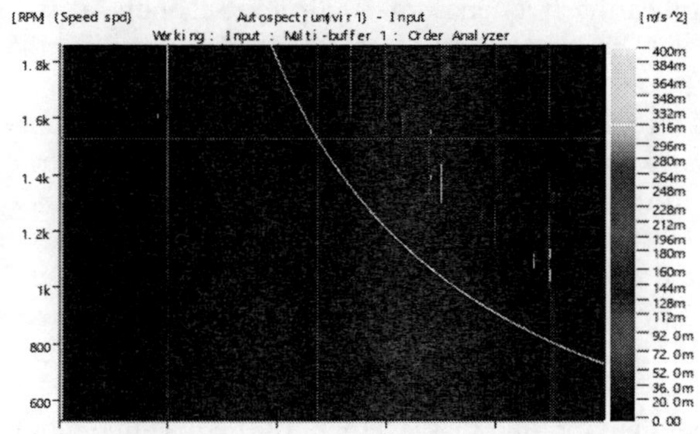

Figure 9. vibration acceleration of X direction

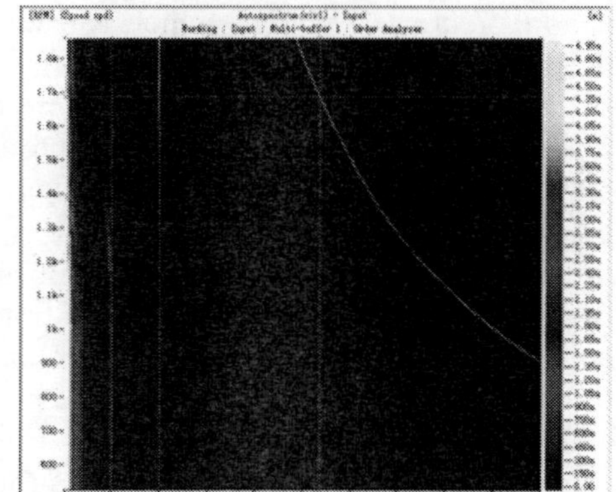

Figure 10. Vibration displacement of X direction

In Fig. 9, the abscissa is the order of the reference speed, the ordinate is the reference axis speed, and the color depth indicates the severity of the vibration acceleration. A number of vertical bright lines can be seen from Fig. 9, indicating that there exists a strong forced oscillation on a number of orders with the spindle speed as the fundamental frequency. The order represented by these bright lines is shown in table 4.

Table 4. X direction vibration order

Order						
1	2	4	6	7	8	9

The order of the maximum amplitude forced vibration is 2 order. The strength of the 1 order vibration is weaker. The vibrations of the 7, 8 and 9 order are low at low speed but when the speed increases to 1100rpm, the vibration intensity will increase by two to three times. The vibration intensity of the 2 order is the largest, and the main reason is that the R axis is driven by the motor through the belt, which drive ratio is 2. Therefore, the 2 order represents the forced vibration caused by the rotation of the motor.

The two order integral of Fig. 9 is used to obtain the order-displacement spectrum of vibration signal, as shown in Figure 10. The abscissa is the order of the reference speed, the ordinate is the reference axis speed, and the color depth indicates severity of the vibration displacement. As you can see from the diagram, the bright lights of the 1 and 2 order are still very large. Especially in the 2 order, the maximum value can reach 4.09um and the maximum speed is 1741rpm, and the frequency is 58.05Hz. This kind of larger displacement will result in a certain degree of accuracy loss for precision edm.

Through the order analysis, we can draw the following conclusions:

(1) The maximum order of vibration in the X direction is the first order and the two order, and the vibration of the two order is the most violent. The forced vibration at the first order shows that the forced vibration whose frequency is equal to that of the main shaft exists in the process of machine tool work. The main reason is that the dynamic balance of the bearing seat is not good. Forced oscillation at the two order is caused by the rotation of the motor, because the R shaft is driven by the motor through the belt which drive ratio is 2:1.

(2) When the speed reaches 1400rpm, the vibration intensity begins to increase and when the speed reaches 1600rpm, the vibration is the most violent. Therefore, the speed higher than 1400rpm should be avoided in actual processing. This is also explained that the improvement of the structure, the structural rigidity of the X direction and the increase of the stable range of the speed can improve the production efficiency of the machine tool.

5. Conclusion

Through the above series of test and analysis, we can draw the following conclusions:

(1) By analyzing the vibration modes obtained by modal testing, first of all, it is found that the dynamic stiffness of EDM machine tools is low, which is mainly caused by the second order modal of natural frequency 54.7Hz, which is exactly 3 times the frequency of 1000rpm. Secondly, it is found that the weak stiffness between the connecting plate and the bearing seat is the main reason for the low stiffness of machine tools. At the same time, through analysis of other main modal formation, it is found that the Q type stent is easy to produce Z direction vibration, and the joint surface of connecting plate and B axis also has the problem of unreliable connection. These are likely to restrict the improvement of the anti-vibration performance of the spindle of EDM machine tools.

(2) By comparing the vibration modes of ODS test with the modal test, it can be found that in the two experiments the larger deformation structures are the union surface of the B axis and the connecting plate, the connecting plate and the Q type bracket, including the bearing seat. Therefore, the results of the two experiments are consistent, indicating that the experimental results are credible and reliable.

(3) Through the order analysis, it is found that the vibration displacement of the machine tool in the X direction is relatively large. Therefore, it is necessary to strengthen the structural rigidity of the X direction and increase the stability interval of the speed, so as to improve the production efficiency of the machine tool. Specific measures include increasing the thickness of the board and designing stiffeners or dampers on the surface of the board.

Through a series of tests on EDM machine tools, the weak links of the machine tools are found, which provides a good reference for the structural optimization of machine tools.

Acknowlegements
This project is supported by 2018 Nova Program(Z181100006218078).

REFERENCE

[1] Wang Zhen-long, Zhao Wan-sheng. Micro EDM Technology in micro manufacturing system, 2003(9): 23- 27.
[2] Cao Feng-guo. Special Processing Manual [M]. Beijing: China Machine Press, 2010.
[3] Chen Yun-xin, Wu Fu-ming, Deng Cheng-wen. Deliberated Wire Electrode Vibration Based on ANSYS [J].Electromachining and Mould, 2006,(4), 26-28.
[4] Yang Da-yong. The development of EDM Technology [J]. Electromachining and Mould (Supplement), 2010:45-52.
[5] Yang Da-yong, Cao Feng-guo, Wang Shu-feng, et al. Development of multi axis precision CNC EDM machine [J]. Electromachining and Mould (Supplement), 2007:65-69.
[6] An Li-bao, Zhao Guo-guang. Spindle vibration test of EDM machine tool [J]. Aviation measurement technology, 1994 17-20

IOP Conf. Series: Materials Science and Engineering **711** (2020) 012006 doi:10.1088/1757-899X/711/1/012006

Design and Application of Intelligent Live Cleaning Device for Substation Pillar Insulator

Qiang Chen, Jian Li, Jianxiang Li, Zhenli Wang, Xu Dong, Jizhi Liu, Tao Li

State Grid Intelligence Technology Co., Ltd, Jinan, Shandong, China

e-mail: chq0531@126.com

Abstract. As the core and key hub of a strong smart grid, substation plays an important role in the stable operation of power system. According to the technical specifications of live operation in substation, an intelligent live cleaning device for substation pillar insulators is designed and applied in substation field.

1. Introduction

Substation pillar insulators have been placed outdoors for a long time. Industrial pollution and natural pollution in the air deposit on the surface of insulators to form a pollution layer. Pollution flashover accidents are easily caused by the pollution layer under humid conditions, which seriously affects the safe operation of the power system.

Therefore, in order to prevent insulator pollution flashover caused blackouts and other accidents, it is necessary to use external forces to clean up the polluted layer accumulated on the insulator surface. It is necessary to carry out power outage cleaning work regularly every year in areas with serious contamination. With the continuous development of society, the development of power grid is facing the demand of high power supply reliability, high load density and high power quality. Users are more and more strict about outage time. It is difficult to arrange outage for important key lines, and the opportunity of outage maintenance is becoming less and less.

At present, there are the following shortcomings in the live cleaning device of substation pillar insulators: The existing live cleaning device for insulators is mainly used to lift two brushes to the contact position of insulators manually through the insulating rod, and the motor installed at the bottom of the insulating rod drives the brush to rotate through the insulating transmission rod. This kind of structure and operation mode causes great physical and psychological burden to the operator, and the cleaning effect is only observed manually from the ground, so the evaluation error is large.

Substation pillar insulator live intelligent cleaning device can use mobile operating platform (insulated bucket arm truck) to enter the equipment area of substation to complete live cleaning of pillar insulator, which improves the efficiency and automation level of substation pillar insulator cleaning.

2. Mechanical structure

The live intelligent cleaning device for substation support insulators is mainly composed of vertical lifting mechanism, encircling rotating mechanism, cleaning actuator and intelligent control system. The encircling rotating mechanism is installed in the vertical lifting mechanism and can move up and down along the lifting mechanism. The left and right ends of the encircling rotating mechanism are equipped with a cleaning actuator, which can drive the cleaning actuator to rotate horizontally around

IOP Conf. Series: Materials Science and Engineering **711** (2020) 012006 doi:10.1088/1757-899X/711/1/012006

the axis of the substation pillar insulator. The cleaning mechanism sweeps the substation pillar insulator. Intelligent control system can control the cleaning device to clean the target insulator independently and intelligently.

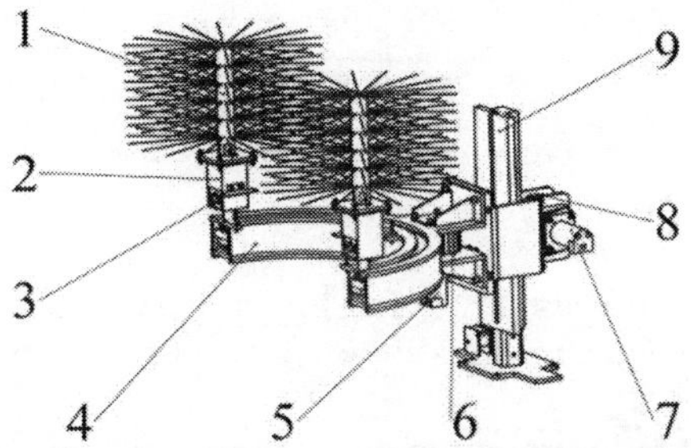

Figure 1. Structural schematic diagram of live intelligent cleaning device for pillar insulators.
1. Brush; 2. Cleaning motor bracket; 3. Sweeping motor; 4. Embracing device; 5. Embracing motor
6. Enclosure drive gear; 7. Lift drive motor; 8. Lift drive gear; 9. Lifting rack

The vertical lifting mechanism includes lifting rack, connecting bracket, lifting hydraulic motor and lifting gear. The connecting bracket is placed on the lifting rack. The lifting hydraulic motor is connected with the lifting gear through the worm gear reducer. The lifting gear and the lifting rack constitute the transmission pair. The lifting hydraulic motor is used to drive the lifting gear along the lifting gear. The rack moves up and down, thus driving the connecting bracket to move up and down.

The embracing rotating mechanism is composed of embracing bracket, gear ring, gear ring bracket and guide groove, etc. The embracing bracket is welded by engineering plastic sheet, and a guide groove is arranged on the bottom of the inner cavity of the bracket. The ring and gear ring bracket are confined in the guide groove and moved along the guide groove.

The hydraulic motor of the embracing mechanism is directly connected with the gear and exposed through an opening on one side of the embracing support to match the gear ring. The hydraulic motor of the embracing mechanism is used to drive the gear ring to rotate horizontally around the axis of the embracing support in the inner cavity of the embracing support.

The circumferential rotating mechanism also includes a rotating control device, which includes two circumferential limit position proximity switches and a retaining iron, the retaining iron is set on the cleaning mechanism respectively, and the circumferential limit position proximity switch is set on the embracing bracket respectively, so as to send the electric signal to the ratio of the circumferential hydraulic motor when the position of the retaining iron is sensed. The directional valve reverses the proportional directional valve to drive the rotary gear to turn.

The cleaning actuator includes a brush, a rotary motor and a cleaning motor bracket. The cleaning motor bracket is fixed on the encircling rotating mechanism. The rotary motor is installed on the cleaning motor bracket. The brush is connected to the output shaft of the rotary motor and driven by the rotating motor.

3. Hydraulic system

Considering the special working conditions of live work, the device uses hydraulic transmission as the main power source, and its hydraulic schematic diagram is shown in the following figure. The hydraulic system mainly includes quick change joint, high precision filter, electro-hydraulic proportional directional valve and hydraulic motor.

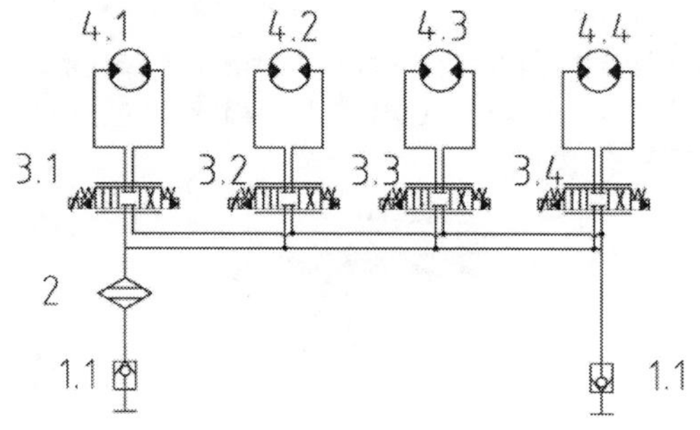

Figure 2. Hydraulic Principle Diagram of Intelligent Live Cleaning Device for Pillar Insulator
1. Quick Change Connector; 2. Filter; 3. Electromagnetic directional valve;

According to the speed regulation requirements of intelligent cleaning device, the hydraulic system adopts electro-hydraulic proportional directional valve, which is a hydraulic load-sensitive multi-way valve, mainly composed of pump side module, basic module and end cover. Load sensing system is a closed-loop system with differential pressure feedback, which can realize the following control of the load pressure of the pump. That is to say, the load pressure is fed back to the load sensing pump, so that the load sensing pump can output a slightly larger pressure than the load, and the difference is set by the spring of the load sensing valve.

Table 1. Main Technical Parameters of Electro-hydraulic Proportional Directional Valve.

Type	Parameter
Pressure(MPa)	37.5
flow rate(L/min)	10
voltage (V)	0～12
response time(S)	0.45
power (W)	8
temperature (℃)	-30～60

When the load decreases, the output pressure of the load sensitive pump decreases, which reduces the energy consumption. The system can not only ensure the automatic matching of the pressure and load of the hydraulic pump, but also control the load speed accurately. Because the pressure compensation valve can ensure that the pressure difference between the front and back of the reversing valve (i.e. the difference between the pump outlet pressure and the load pressure) is constant, the flow rate of the pump is only determined by the opening size of the reversing valve, and has nothing to do with the load pressure.

The reliability of the hydraulic system of the live intelligent sweeping device for substation pillar insulators has an extremely important influence on the overall reliability of the whole device in the operation process. It is required that the hydraulic system should be in normal operation state and be able to realize its pre-operation without failures or fewer failures during long-term storage and repeated use under specific conditions. Periodic effectiveness. Therefore, it is necessary to analyze the reliability of hydraulic system.

Table 2. Basic Failure Probability Table for Hydraulic Components

Components	failure probability
tank	1.3
Oil filter	0.4
Hydraulic pump	12.5
Unidirectional hydraulic valve	4.5

Hydraulic relief valve	6.7
Hydraulic motor	6.7
Proportional reversing valve	4.6

Hydraulic system is composed of several hydraulic components in series or in parallel. According to the principle of reliability, the reliability of n components in series is as follows:

$$R_S(t) = R_1(t) \times R_2(t) \cdots R_n(t) = \prod_{i=1}^{n} R_i(t) = e^{-\sum_{i=1}^{n} \lambda i t} \quad (1)$$

λ – Real failure probability of components; $\lambda = \kappa \times \lambda_0$

κ – Environmental coefficient , Usually take 10

$R_i(t)$ –Reliability of i components, $R_i(t) = e^{-\lambda t}$, $i = 1, 2 \cdots n$

The reliability of components in parallel is as follows:

$$R_s(t) = 1 - [1 - R_1(t)] \times [1 - R_2(t)] \cdots [1 - R_n(t)] = 1 - \prod_{i=1}^{n} [1 - R_i(t)] \quad (2)$$

During its operation, the walking motor is shut down, four leg cylinders are in parallel, and the rest hydraulic components are in series. So its reliability is:

$$R(t) = \left[1 - \left(1 - e^{-3.5\times10^{-6} t}\right)\right]^7 \times \left[1 - \left(1 - e^{-4.6\times10^{-6} t}\right)\right]^7 \times e^{-1.3\times10^{-6} t} \times e^{-6.7\times10^{-6} t} \quad (3)$$

The trouble-free running time of the hydraulic system during operation is as follows:

$$MTVF = \int_0^{\infty} R(t) dt = 2387h \quad (4)$$

4. Test and Application

According to the "Basic Technical Requirements and Design Guidelines for Live Work Tools", live work tools need to be tested before acceptance. Under the test voltage, the electrical insulation performance experiments of 110-220 kV open substation live overhaul operation and robot-mounted post insulator operation are carried out. The experimental pictures are shown in the following figure. The experimental verification shows that the robot platform is equipped with insulator cleaning tool. Under the test voltage of 220 kV, the test current is 329.7 μA, which is small. 500 μA stipulated in "Basic Technical Requirements and Design Guidelines for Live Working Tools".

Table 3.Test data of live operation of robot with insulating cleaning tool

Voltage /kV	leakage current/ μA
50	64.94
110	173.4
160	262.2
220	329.7

Figure3. Field Diagram of Electrical Insulation Performance Test of Cleaning Device

The live cleaning device of substation pillar insulator has been tested in a 220 kV substation of Southern Power Grid Company. The test results show that the live cleaning device of substation pillar insulator can meet the live operation demand of substation under 220 kV voltage level. Compared with manual live cleaning, it can reduce labor intensity and enhance work. Industry efficiency, reduce blackout time.

Figure4. On-site live working diagram of Substation

5. Conclusion and Prospect

Successful development of live cleaning device for substation pillar insulators has opened up a new era of live operation in substations, and achieved breakthroughs in insulation protection, mobile platform and self-positioning of operation targets. Next, on the basis of successful research and development of the live cleaning device for the pillar insulators, dry ice cleaning, wire repair and foreign body removal of the pillar insulators in substations will be carried out.

ACKNOWLEDGE

This paper from the project of Technology Research and Demonstration Application of Live Working Robot for Substation Equipment is supported by the Key Research and Development Plan of Shandong Province in 2017(2017CXGC0918).

References

[1] G. W. Jointt Type Ultra-high Voltage Hot-line Sweeping Robot. HVCR-II[J]. Journal of Shanghai Jiaotong University, 2006,40(12):2173-2176,2181.

[2] ZHAO P Y,CHEN Y L,ZHOU H. Overview of hydraulic hybrid engineering machinery system and control strategy[J]. Journal of Zhejiang University(Engineering Science), 2016(3): 449—459.

[3] LI F,WANG X,WANG P P,et al. Energy-saving research on c1osed—loop hydraulic syStem of hoist mecIlallism[J]. Construction Machinery, 2015(7):93—97.
[4] SAKUAI Y,NAKADA T,TANAKA K. Design method of an intelligent oil-hydraulic system(1oad sensiIlg oil-hydraulic system)[C]//Intelligent control, 2002 Proceedings of the 2002 IEEE Intemational Symposium on, lEEE. 2002: 626—630.

IOP Conf. Series: Materials Science and Engineering **711** (2020) 012007 doi:10.1088/1757-899X/711/1/012007

Flexible Connection of Cockpit Canopy Defect Detection Device Based on Machine Vision

Pengfei Wang[1], Xiaojia Lu[2], Jian Chen[2*] and Ping Zhang[1*]

[1] State Wuhu Machine Factory

[2] Wuhu HIT Robot Technology Research Institute Co., Ltd

[*]Corresponding author's e-mail: 1071788036@qq.com

Abstract. In this paper, a detection device for air bubbles, impurities and other defects inside the flexible connection of aviation aircraft canopy is designed. This device is mainly composed of prism, industrial camera, lighting and analysis software. The prism is repeatedly optimized using light tools optical simulation software to achieve the best imaging results. The internal situation of the glued surface of the entire cockpit canopy is realized by the stitching algorithm based on SURF algorithm with additional feature points on images. At the same time, the image processing method based on OpenCV is used to realize the intelligent detection of the bonding surface defects. Compared with manual visual inspection, the device can complete the image capture, quantitative analysis and health monitoring of the surface of the flexible connection, effectively ensuring flight safety. This detection device improves the detection efficiency by about 50%.

1. Introduction

The aeronautical transparent member forms a sealed structure with the metal skeleton through the flexible connection, which is a common connection form of the fighter cockpit canopies. The edge of the transparent member is bonded to the body through the flexible polyester ribbon. Therefore, the quality of bonding has an important impact on flight safety[1]~[4]. In recent years, edge degumming in cockpit canopies with flexible connection between plexiglass and frame occurs frequently, which seriously affects the daily flight and training of the army, and attracts great attention from industrial departments, colleges and military. In addition to the perfect bonding process, the flexible connection technology must also check the quality through reliable testing methods[5]. After the transparent member is combined with the body, the flexible connection area is in the concealed area of the structure. Prism is generally used at home and abroad, and the edge connection area is visually inspected by optical path refraction and total reflection [6]. Peng Chun proposed to use the camera device to improve the clarity and defect characteristics of the observation based on the prism detection. Zhang et al[7]. analyzed the influence of bonding defect area on bonding strength by artificially manufacturing quantitative defect area, and believed that when the defect area does not exceed 35%, flight safety can be guaranteed. But the defects in the process of bonding are often random and not quantitative, the previous detection methods cannot determine whether the defects are within the tolerance range, and whether the defects change during the use of cockpit canopy cannot be effectively identified. Li Zhisheng put forward the use of plastic film instead of paste gel brushing to reduce the connection defects caused by glue shortage, Yu Guoliang et al[8]. improved the identification of defect features by adding pigments[9]. All domestic institutions are exploring methods to reduce the defect rate of adhesion or improving detection methods,

[1] Science and Technology Planning Project - Key Research and Development Plan in Wuhu City (2017fy11)

but they have not achieved quantitative analysis of bonding defect and judged whether the defects are within tolerance.

Compared to visual inspection, machine vision can quantitatively and qualitatively analyze target object defects, and can also classify defect types. Machine vision is widely used in various industries, such as auto parts, food packaging, medicine detect and so on. Bao Ting, Tang Xia, et al. [10] proposed the use of machine vision to detect the defects of automobile shock absorbing discs. The main test is whether the parts on the damper disc are missing. The detection system has high real-time performance and the efficiency meets the needs of enterprises. In the pharmaceutical industry, the traditional manual inspection has always been used in drug detection, which is inefficient and the rate of false detection increases over a long period of time. Machine vision is used to detect drugs with specific shapes [11], so that the detection rate can be guaranteed and the batch production demand can be met at the same time. However, in the field of machine vision, there is currently no report on the analysis about transparent material. In this paper, a set of highly integrated visual inspection device is used to study the optimal imaging scheme of the image of the flexible connection area, the device will finally store and quantify the capture images.

2. Mechanical Structure Design

2.1. Integrated Design

The mechanical structure of the device mainly includes a camera fixing device, a holding structure and a prism fixing structure. In order to adjust the lens to the best angle, the camera fixture and the grip structure are connected in a way that can rotate at a certain angle. The prism is placed in the fixing device, which is connected with the base through a U-groove. This design allows the prism to move backwards and forwards, and the two-way adjustment increases the flexibility of the whole device, Then the prism fixing device and the base can be fixed by the fixing screw. The overall mechanical structure is shown in Figure 1.

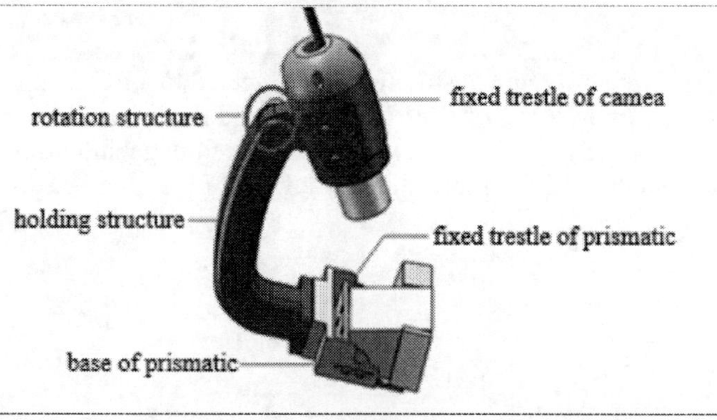

Figure 1. Mechanical mechanism diagram of the testing device

2.2. Prism Design

Since the internal adhesion of the glued surface is observed, it is necessary to use a prism. According to the principle of refraction, one of its surfaces is bonded to 10mm of the glued surface through a coupling agent (the purpose is to remove air), Then refract the internal adhesion of the glued surface to the other side of the prism. Figure 2 is a road diagram of light passing through the prism[12] and glued surface simulated by Light Tools simulation software.

IOP Conf. Series: Materials Science and Engineering **711** (2020) 012007 doi:10.1088/1757-899X/711/1/012007

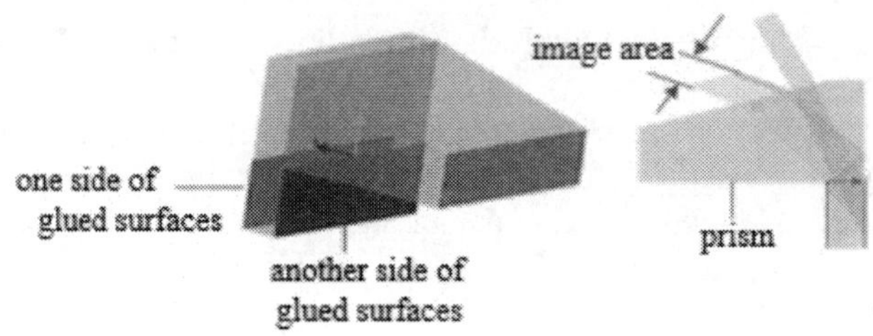

Figure 2. Simulated path diagram about prism

As shown in Figure 3, the measured surface close to the prism needs to be totally reflected to reach the field of view, while the surface away from the prism can be observed only by refraction, the depth of observation is affected, so the design of the prism structure needs to be optimized. Using the principle of reverse ray tracing[13], you can find the best imaging field of view and imaging position.

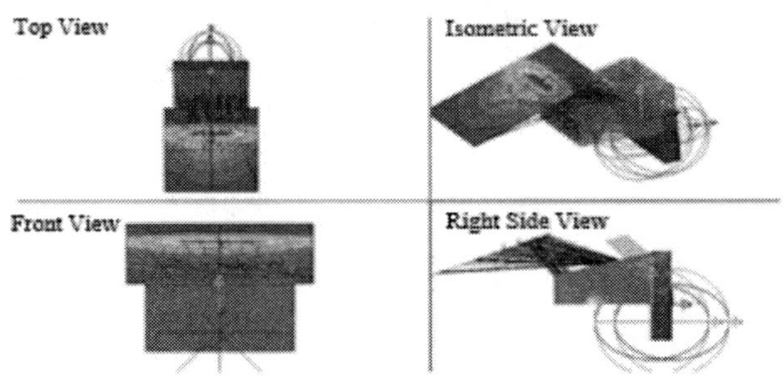

Figure 3. Simulation of the best imaging field of view and position map of the prism surface

3. Design of the software algorithms

3.1. Integrated Design

The camera in the above mechanical structure is connected to a PC with high performance that contains image processing program. One of the functional modules is used to control the camera to capture images; the other functional module is used for multiple image stitching; the last functional module is used to detect and display the internal adhesive defects of the glued surface.

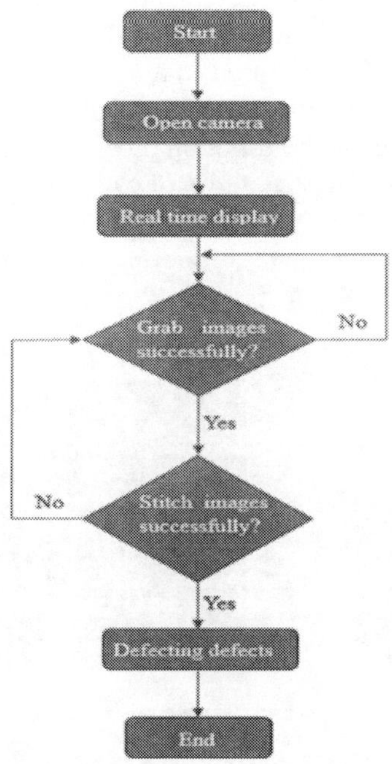

Figure 4. Program flow chart

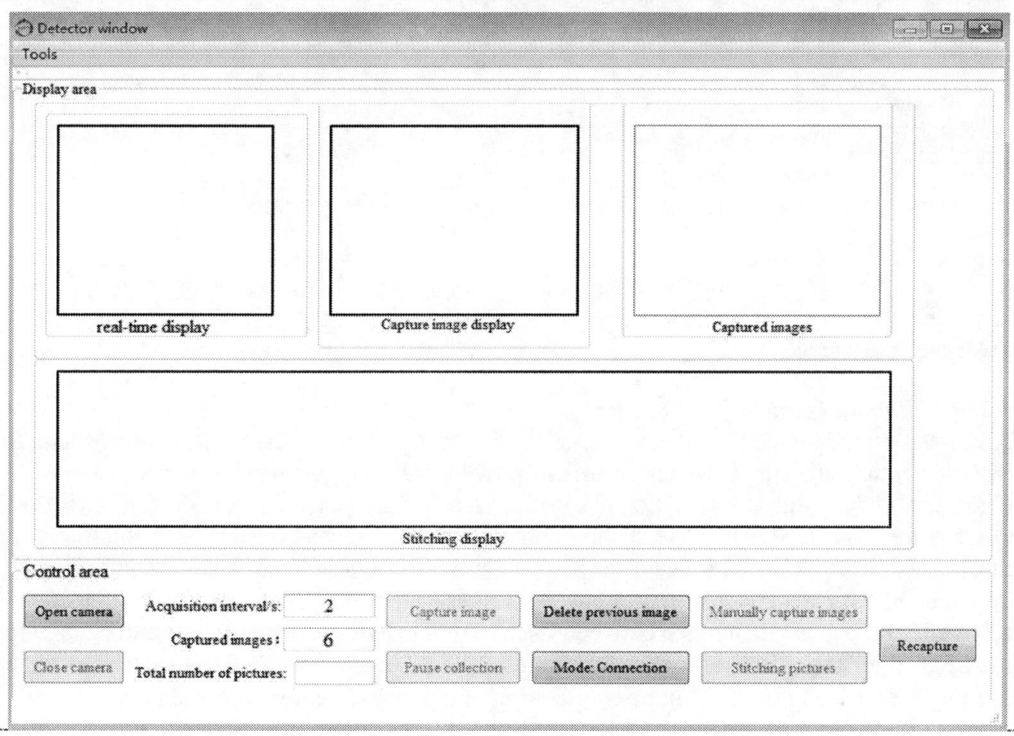

Figure 5. The main interface of the detection software

A flow chart of the entire software operation is shown in Figure 4, the main part of which is the image stitching and defect detection. we can see the main window of the entire software from Figure 5, The display area is used to match the camera to capture the internal images of the glue surface, which ensures that the prism fits the cockpit canopy transparently. The "real-time display" window displays a frame-by-frame image captured by the camera, which greatly facilitates the operator to find the right location for the captured image and improves the accuracy of capturing images. A thumbnail of the total captured images is shown in "Captured images" window, which is convenient for the operator to observe. The middle of the "Stitching display" window is used to display the effect of the images stitching. The bottom control area is the function block of camera communication, such as capture mode setting and picture stitching. A display diagram of capture images and stitching images during operation of the device is shown in Figure 6.

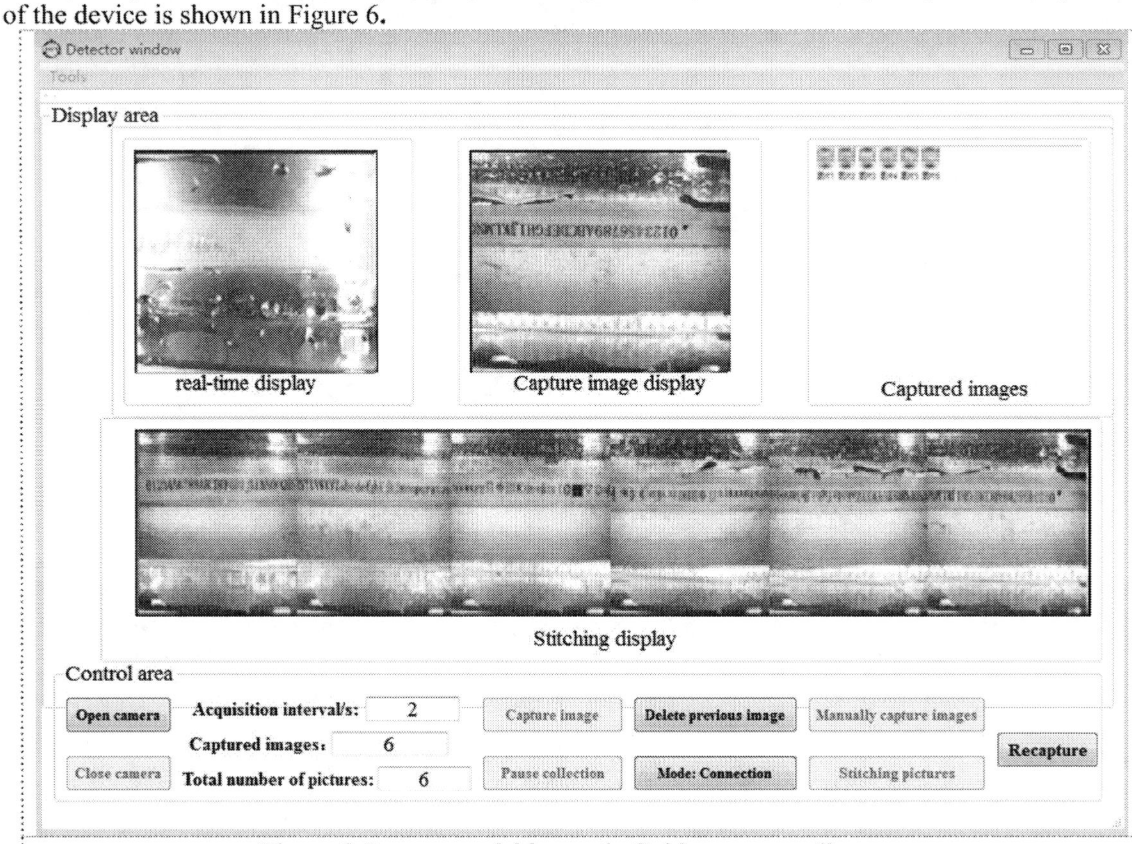

Figure 6. Image acquisition and stitching process diagram

3.2. Image stitching

3.2.1 Additional Feature Points Detection

A key step in the image stitching process[14]~[15] is feature points detection. Among them, the traditional SIFT detection algorithm, based on this improved SURF algorithm[16], as well as the ORB detection algorithm[17]with high computational speed and real-time performance. In terms of detection speed, ORB is the fastest, SURF is the second and SIFT is the slowest, in terms of rotational robustness, all three are similar, SURF is superior to SIFT in terms of invariance of scale transformation, while ORB does not have scale transformation characteristics. To summarize, the SURF detection algorithm is selected. The biggest difference between the SURF and SIFT is when looking for image feature points, the SIFT algorithm searches for feature points in DOG images, while SURF searches for feature points in images consisting of Hessian matrix determinant approximations, formula (1) is calculation formula:

$$det\left(H_{approx}\right)=L_{xx}L_{yy}-\left(0.9L_{xy}\right)^{2} \tag{1}$$

The Huessian matrix formula for calculating a pixel in an image is as follows:

$$H(x,\sigma)=\begin{bmatrix} L_{xx}(x,\sigma)L_{xy}(x,\sigma) \\ L_{xy}(x,\sigma)L_{yy}(x,\sigma) \end{bmatrix} \tag{2}$$

Where L_{xx} , L_{xy} ,and L_{yy} are the second derivatives of the Gaussian filtered image in each direction respectively. There is no SIFT drop step sampling process in the SURF algorithm, so the processing speed is increased. Moreover, when constructing feature descriptors, SIFT uses 128-dimensional, while SURF uses 64-dimensional, so the SURF algorithm workload is reduced by half.

Figure 7. Capture of glued surface image-1

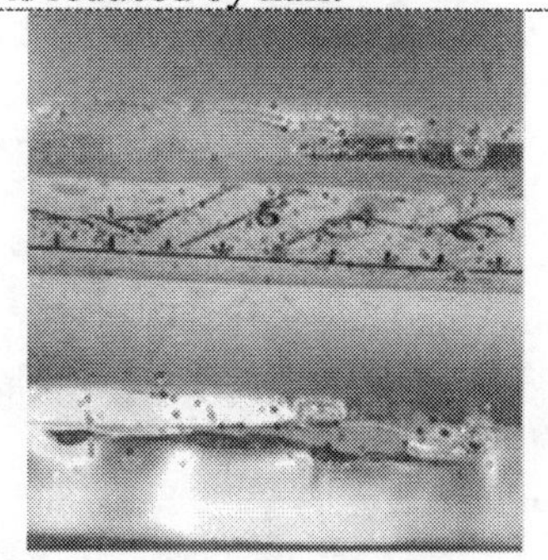

Figure 8. Capture of glued surface image-2

The two images adjacent to the inside of the glued surface captured by the camera are shown in Figure 7 and Figure 8 respectively. The internal features of the glued surface are not obvious, so the feature points are added artificially on images. The feature points detected by SURF in two of the images have been marked.

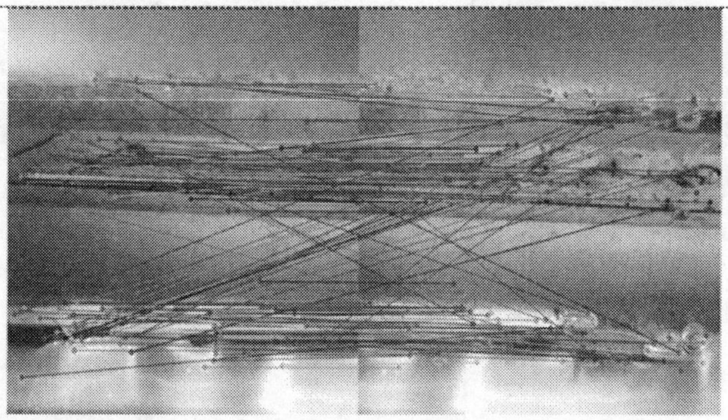

Figure 9. Feature points matching image

Figure 9 is a feature matching image of figure.7 and figure 8, in which the matching points are 354 and the mismatching rate is about 5%.

3.3. Defect detection

After optimizing the prism, the field of view is optimized, but the maximum length of a single image is still only about 60 mm, while the total length of the whole flexible connection surface is about 1 000 mm. If the image stitching cannot be realized, the image rendering effect and detection efficiency will be greatly affected.

After the image captured is completed, the final stitching effect can be achieved through image registration, image copying and image fusion. The middle part of Figure 6 is the stitching effect of the image after continuously capture some glued surface images. The result shows that multiple images can be successfully stitched together, but the fusion effect of images needs to be improved. After the images are stitched, the entire image can be detected for defects. The first step is filtering. Median filtering[18]~[19] has a good filtering effect on the noise caused by outliers and a small number of pixels. The purpose of this filtering is to eliminate some additive noise in the image and prevent the influence of calculating the feature in the subsequent algorithm on the result parameters. The original images and median filtered images are shown in Figure 10 and Figure 11 respectively. There is no difference on the surface, and they have little influence on background images and defect areas on the glue surface.

Figure 10. Original image of the glued surface

Figure 11. Median filtered image

After transforming the collected image into gray scale image, a contrast stretching method is adopted to enhance the contrast of defect area and make the defect more obvious, as shown in Figure 13.

Figure 12. Original image of the glued surface

Figure 13. Contrast stretch gray scale image

In order to solve the problem of uneven illumination of the glued surface due to capture, an improved OpenCV local adaptive fast binarization[20]~[21] method is adopted, and then subjected to morphological processing[22], and finally the target contour is drawn, as shown in Figure 14 and Figure 15.

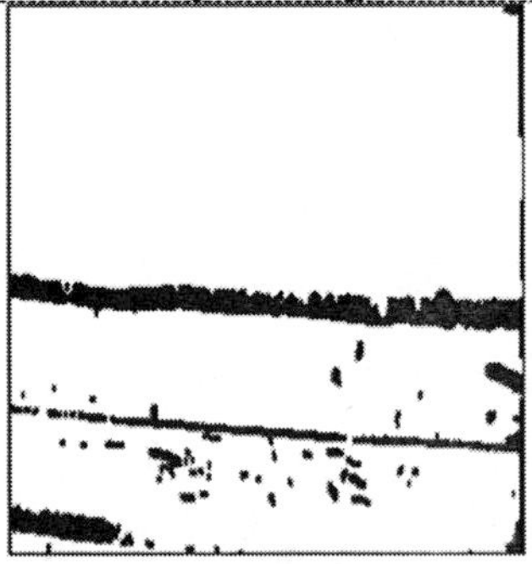

Figure 14. The image after morphological processing

Figure 15. The picture of defect contours

According to the threshold setting for defect area, the program will automatically eliminate some contours that are not within the threshold range, and eventually mark the selected defect contours. Considering the complexity of defects, the program also makes manual circle selection of defects, and then automatically calculates the defect area.

In order to facilitate statistical analysis, the program shows the number and area of defects in tabular form, which are shown Figure 16 and Figure 17 respectively.

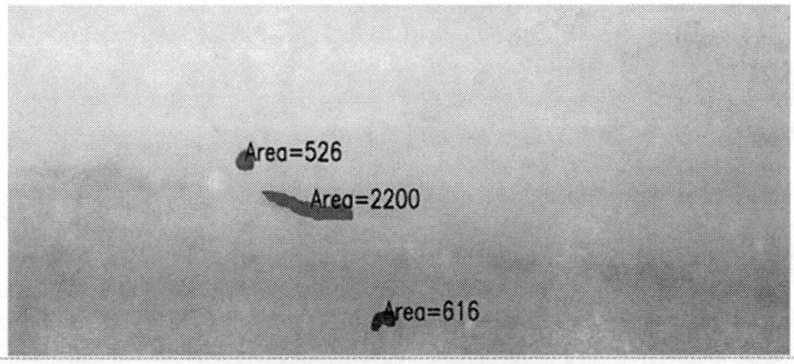

Figure 16. Defect area labeling

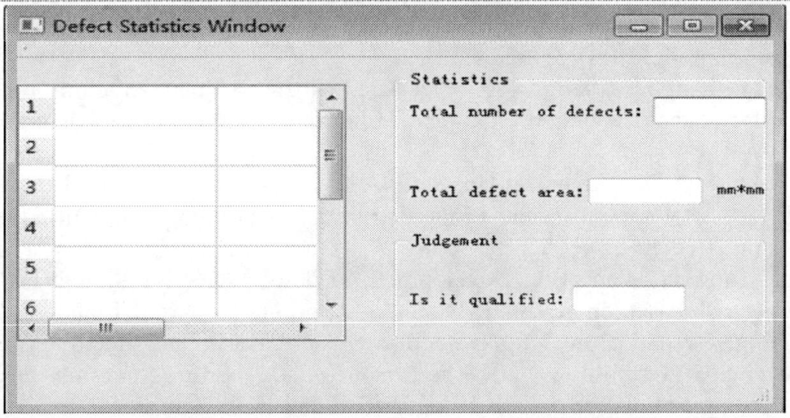

Figure 17. Defect statistics interface diagram

The unit of the defect area is square millimeter. The camera needs to be calibrated by checkerboard before use. The program finally gives the physical size of the defect and will be displayed in the "defect statistics" window. According to the actual set judgment standard, the "result judgment" will give "qualified" or "unqualified" prompt.

4. Conclusion

Through various tests and trials, the overall efficiency of the device based on machine vision for detecting flexible connection defects of transparent parts of cockpit canopy has been improved by about 50% compared with the prism-based equipment. At the same time, it has the following advantages:

(1) The light Tools optical simulation software optimizes the design of the prism to obtain the best viewing field of view for the flexible connection.

(2) The stitching algorithm based on SURF algorithm with additional feature points is used to get the whole image of transparent glued surface. Image acquisition and archiving function can be used as the basis of whether the defects of plywood of cockpit canopy have expanded or not during flight, and the health monitoring of key bearing areas can be realized.

(3) The image processing method based on OpenCV can realize the quantitative analysis of the glued surface defects with an accuracy of 0.1mm, which can accurately determine whether the defect area is within the safe tolerance range.

The device also has certain limitations. Due to factors such as small color difference between the white rubber layer and the bubble, interference of the couplant bubble, optical distortion through the prism, etc., the defect recognition and stitching effect have certain deviations. In special cases, the operator is also required to circle or filter the defect area for quantitative analysis. The algorithm needs to be further optimized to accelerate the detection efficiency on the basis of the guaranteed function.

References

[1] Huang Baochen, Yang Xu. Design of the Edge Joints for Wind Shield and Canopy Transparency [J].Aircraft Design,2014,34(04):37-41.
[2] Wang Jiaying, Pei Xiangfa. Analyses& Validation of an Aircraft Cockpit Canopy Finite Element Model [J]. Aircraft Design, 2006(04): 43-45.
[3] Liu Ye,Zhao ,Yilei,Yu Guoliang,Wang Jieying. Study on a Polyurethane Adhesive for the Aeronautical Perspex Bonding [J].Chemistry and Adhesion,2016,38(03):197-198+210.
[4] Zhang Hongfeng, Shang Lei,Zhong Yanli. Study on Properties of Modified Acrylate Adhesive for Cabin Edge Bonding [J].Chemistry and Adhesion,2006(04):242-245.
[5] Qi Xiangfa,Wang Jiaying. Failure Detection and Analysis for Cockpit Canopies withFlexible Connection Between Plexiglass and Frame [J].Aircraft Design,2006(03):35-39+47.
[6] Pengchun Transparent Edge Hidden Structure Visualization Inspection Device: CN, 102221564 2011 [P]. 2011-10-19
[7] Shang Lei,Dai Wenming,Kong Xianzhi,Sun Dongzhou,Sun Wei. Study on Glue Spreading Technics and Natural Aging Properties of J-189 Polyurethane Adhesive [J].Chemistry and Adhesion,2014,36(03):218-221.
[8] Li Zhisheng,Zhang Hongfeng,Li Lei,Yan Yue.Influence of new structural bonding materials on the performance of aviation plexiglass[J].Journal of Aeronautical Materials,2017,37(03):78-83.
[9] Yu Guoliang, Sun Dongzhou, Lu Hu, Sun Wei, Li Yue, Zhang Liying, Kong Xianzhi. Study on the Effects of Pigment on the Properties of J-189 Polyurethane Adhesive [J].Chemistry & Adhesion,2018,40(02):105-110.
[10] Bao Ting, Tangxia, Wang Lili. Development of defect detection system for automobile shock absorber disk based on machine vision [J]. Shandong Industrial Technology, 2018 (21): 1-2.
[11] Hu Xiubing. Research on Drug Detection System Based on Machine Vision [D]. Tianjin University of Technology, 2018. le Prism Coded Aperture Snapshot Spectral Imager using Ray Tracing Simulation," 2018 IEEE British and Irish Conference on Optics and Photonics (BICOP), London, United Kingdom, 2018, pp. 1-4.
[12] K. Fanzeng, C. Guojun, Z. Nae and C. Song, "A reverse ray tracing algorithm based on cluster analysis," 2016 IEEE Information Technology, Networking, Electronic and Automation Control Conference, Chongqing, 2016, pp. 529-537.
[13] S. Yeh and S. Lai, "Real-time video stitching," 2017 IEEE International Conference on Image Processing (ICIP), Beijing, 2017, pp. 1482-1486.
[14] S. Yeh and S. Lai, "Real-time video stitching," 2017 IEEE International Conference on Image Processing (ICIP), Beijing, 2017, pp. 1482-1486.
[15] C. Huang, S. Lin and J. Chen, "Efficient Image Stitching of Continuous Image Sequence With Image and Seam Selections," in IEEE Sensors Journal, vol. 15, no. 10, pp. 5910-5918, Oct. 2015.
[16] L. Zhu, Y. Wang, B. Zhao and X. Zhang, "A Fast Image Stitching Algorithm Based on Improved SURF," 2014 Tenth International Conference on Computational Intelligence and Security, Kunming, 2014, pp. 171-175.
[17] C. Ma, X. Hu, L. Fu and G. Zhang, "An Improved ORB Algorithm Based on Multi-Feature Fusion," 2018 IEEE 27th International Symposium on Industrial Electronics (ISIE), Cairns, QLD, 2018, pp. 729-734.
[18] C. Chang, J. Hsiao and C. Hsieh, "An Adaptive Median Filter for Image Denoising," 2008 Second International Symposium on Intelligent Information Technology Application, Shanghai, 2008, pp. 346-350.

[19] B. Sravani and M. V. Nageswara Rao, "Removing of high density salt and pepper noise using fuzzy median filter," 2014 International Conference on High Performance Computing and Applications (ICHPCA), Bhubaneswar, 2014, pp. 1-6.

[20] X. Zheng, W. Tan and J. Du, "A Fast Adaptive Binarization Method Based on Sub Block OSTU and Improved Sauvola," 2011 7th International Conference on Wireless Communications, Networking and Mobile Computing, Wuhan, 2011, pp. 1-5.

[21] D. Gaceb, F. Lebourgeois and J. Duong, "Adaptative Smart-Binarization Method: For Images of Business Documents," 2013 12th International Conference on Document Analysis and Recognition, Washington, DC, 2013, pp. 118-122.

[22] L. Yuan and X. Xu, "Adaptive Image Edge Detection Algorithm Based on Canny Operator," 2015 4th International Conference on Advanced Information Technology and Sensor Application (AITS), Harbin, 2015, pp. 28-31.

IOP Conf. Series: Materials Science and Engineering **711** (2020) 012008 doi:10.1088/1757-899X/711/1/012008

Study on the prediction method of supporting force distribution and experiment

Hongfeng Ma[1], **Gaojie Liu**[2]

[1]School of Mechanical and Aerospace Engineering, Jilin University, Changchun, 130022, China

[2]Taizhou First Technician College, Taizhou, 318020, China

[*]Corresponding author's e-mail: 290297738@qq.com

Abstract. The supporting force distribution is very important for the vehicle to maintain safety and stability, especially for the heavy-duty refitting vehicle with eccentric load. A simple and generic model to compute the supporting force distribution is essential in the design stage of refitting vehicle, whereas there is little research involving in solving this problem by now. This article presents an efficient and simple method to predict the supporting force of tire (supporting point). Firstly, the model for the vehicle with three or four axles is simplified. The mathematical matrix is established by force equilibrium equation, moment equilibrium equation and deformation compatibility equation. With the known parameters of load, position and dimension, the force of every supporting point can be calculated, and so the supporting force distribution is acquired. Then, a corresponding experiment is designed and implemented. According to the results of the comparison, the results indicate that the general prediction method to calculate supporting force distribution presented in this paper is accurate and validated in the pre-design stage of the refitting vehicle.

1. Introduction

The refitting vehicles are very popular in engineering applications, majority of them have four or three axles, multi tires (supporting points) are supporting on the ground. It is the statically indeterminate system, the force distribution of supporting points is very important to ensure the safety and stability of the whole vehicle. Though the FEA for the whole vehicle can solve this problem, whereas, in the initial stage of design, the frequent improvement of design needs the FE model changes frequently, resulting in wasting a lot of time. So a simple and generic mathematical model to compute the force distribution is essential in the design stage of refitting vehicle. This paper presents an efficient and simple method to predict the supporting force of tire (supporting point).

There is little research involving in proposing simple and generic model to compute the force distribution for the vehicle by now. Some researchers studied the axle load and solved engineering problems in the recent years [1-4]. Timm et al. proposed a mixed distribution model of two or more theoretical distributions to accurately characterize axle load spectra [5]. Battini et al. used a plane finite element analysis to study the axle load distribution, and based on this, a triangular load distribution was proposed [6]. Law et al. presented a new moving load identification method to identify a system of general moving loads or interaction forces between the vehicle and the bridge deck [7]. Gerbert et al. presented a theory for determining the distribution of the belt tension and the tooth load in timing belts [8]. Zhang et al. showed that the distribution of the belt tension and the tooth load in metal timing belts was theoretically analyzed [9]. Nirut and Kitjapat presented an analytical method for determining the load distribution of single-column multibolt connection [10].

This article presents an efficient and simple method to predict the supporting force of tire (supporting point). The subsequent sections are organized as follows. In section 2, the model for the vehicle with three or four axles is simplified. The mathematical matrix is established by force equilibrium equation, moment equilibrium equation and deformation compatibility equation. In section 3, an experiment is designed and carried out, and the FEA for the experimental model is also implemented. Finally, the work is summarized in section 4.

2. Mathematical model

For the vehicle with four axles, a balanced suspension is typically designed between the rear two axles to homogenize the supporting force, as shown in Figure 1. According to the force analysis, the vehicle with four axles can be simplified to the force state of the vehicle with three axles. Each axle has two supporting points of tire, so there are 8 supporting points for the vehicle with four axles, there are 6 supporting points for the vehicle with three axles. In the following study, the calculation formula of the force distribution will be deduced based on the force model of the vehicle with three axles.

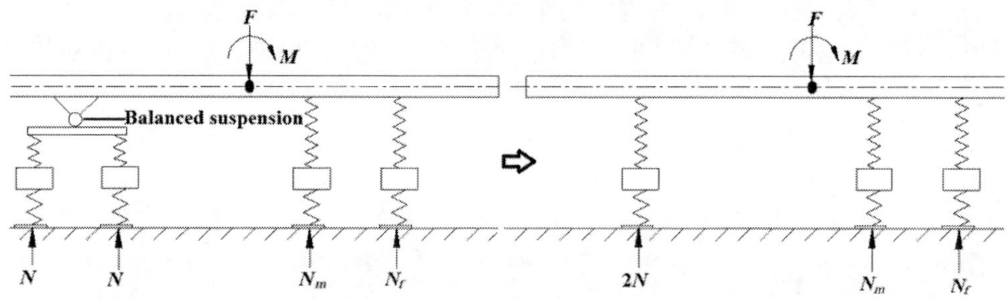

a) Vehicle with four axles b) Vehicle with three axles

Figure 1: Force analysis diagram in the front view

The main factor affecting the supporting force is the vertical stiffness of the supporting point, so in this study, the vertical stiffness of leaf spring and tire is considered, and the stiffness in other directions is ignored. As the stiffness of the chassis is much higher than leaf spring and tire, the structure of the chassis is assumed to be non-deformed. The vehicle is in the static and stable state. The supporting force distribution is as follows.

$$G+F=\sum_{i=1}^{6} N_i \quad (i=1,2,\ \ldots\ldots,\ 6) \tag{1}$$

where G is the gravity of the whole vehicle, F is the load, and N_i represents the force of the i-th supporting point. N_1 and N_2 are the supporting forces of the front axle, respectively. N_3 and N_4 are the supporting forces of the mid axle, respectively. N_5 and N_6 are the supporting force of the rear axle, respectively.

$$N_f = N_1 + N_2 \tag{2}$$

$$N_m = N_3 + N_4 \tag{3}$$

$$N_r = N_5 + N_6 \tag{4}$$

where N_f is the supporting force of the front axle, N_m is the supporting for of mid axle, and N_r is the supporting force of the rear axle. The calculation model is simplified, the deformation form of the vehicle is assumed to be the configuration shown in Figure 2, and the stiffness of the leaf spring and the tire is assumed linear elasticity. The point o represents the center of gravity, M represents the moment caused by F, δ_f is the deformation of front axle, δ_m is the deformation of mid axle, and δ_r is the deformation of rear axle. L means the distance from the center of gravity to the front axle, L_m means the distance from the mid axle to the front axle, and L_r means the distance from the rear axle to the front axle.

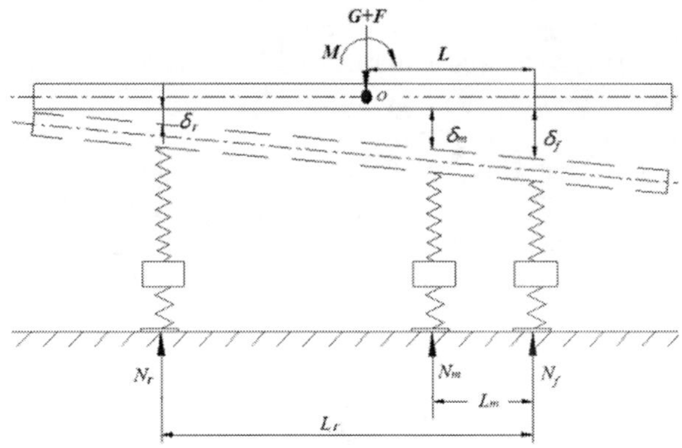

Figure 2: The calculation model diagram of multi-spring

The static force equilibrium equation can be defined as

$$N_f + N_m + N_r = G + F \tag{5}$$

The moment equilibrium equation can be defined as

$$N_r L_r + N_m L_m = (G+F)L - M \tag{6}$$

According to the geometric relation, the deformation compatibility equation can be defined as

$$\frac{\delta_m - \delta_r}{\delta_f - \delta_r} = \frac{L_r - L_m}{L_r} \tag{7}$$

The stiffness of the leaf spring and the tire is assumed linear elasticity, the elastic equation can be defined as

$$N_f = k_f \delta_f \tag{8}$$

$$N_m = k_m \delta_m \tag{9}$$

$$N_r = k_r \delta_r \tag{10}$$

where k_f, k_m and k_b represent the stiffness of the front axle, the stiffness of the mid axle and the stiffness of the rear axle, respresently.

The mathematical matrix can be acquired based on the Eq. (5-10), as shown in Eq. 11. For the structure designed in the initial stage, the parameters of G, F, L, M, k_f, k_m and k_b can be acquired from geometric model and the stiffness of the leaf spring and the tire. So δ_f, δ_m and δ_r are obtained by calculating the mathematical matrix, then the supporting force of each axle is acquired by the Eq. (8-10).

$$\begin{bmatrix} k_r & k_m & k_f \\ k_r L_r & k_m L_m & 0 \\ L_m & -L_r & L_r - L_m \end{bmatrix} \begin{bmatrix} \delta_r \\ \delta_m \\ \delta_f \end{bmatrix} = \begin{bmatrix} G+F \\ (G+F)L - M \\ 0 \end{bmatrix} \tag{11}$$

The force state in left view of the vehicle is presented in Figure 3. The point o is the gravity center of the chassis of the vehicle. The point o' is the equivalent loading position. When the load is on the symmetric plane, the forces of the two supporting points in every axle are the same. While the eccentric load occurs, the additional moment is formed, the forces of the two supporting points in every axle are different, resulting in every force of supporting points of the vehicle is different.

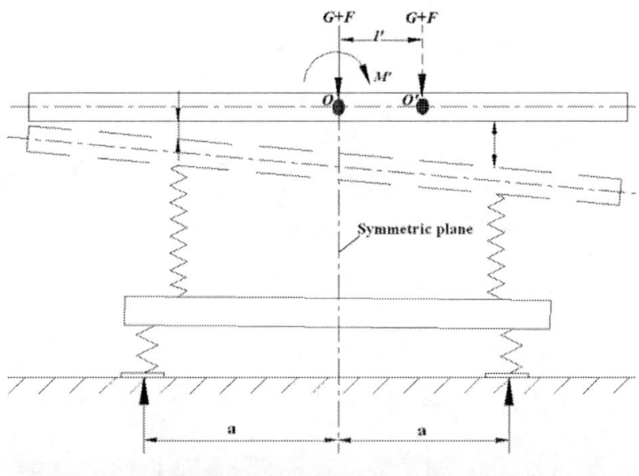

Figure 3: Force analysis diagram in the left view

As shown in Figure 3, *G* is the gravity of the chassis, *F* is the load, and *a* is the distance from the symmetric plane to the supporting point. Based on the force translation theorem, the additional moment can be equivalent from point *o* to point *o'*, so

$$l' = M' / (G+F) \tag{12}$$

where *l'* is the equivalent distance from the point *o* to the point *o'*, *M'* is the additional moment caused by the eccentric load. (*G+F*) is the total load applied to the vehicle. As the structure of the chassis is assumed to be non-deformed, so the distance *l'* is also equivalent and applicable to the force state of each axle, as shown in Figure 4.

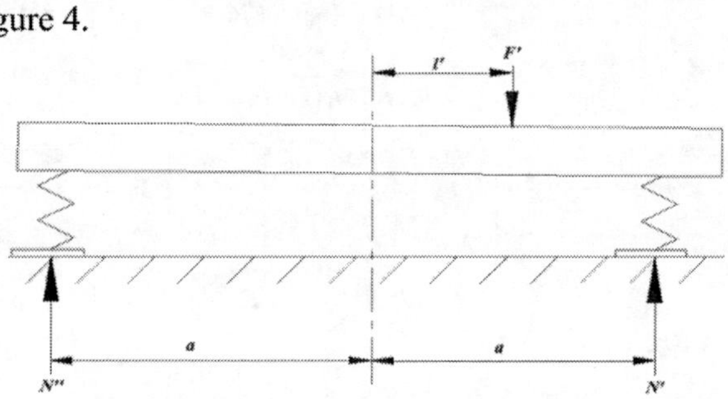

Figure 4: The force analysis for each axle

where *F'* is the load applied on the axle, *N'* and *N"* are the forces of the two supporting points of each axle, so the static force equilibrium equation can be defined as

$$N = N' + N'' \tag{13}$$

$$N = F' \tag{14}$$

The moment equilibrium equation can be defined as

$$N'' * 2a = F'(a - l') \tag{15}$$

N' and *N"* can be calculated by Eq. (12), Eq. (13) and Eq. (15)

$$N' = \frac{N}{2} + \frac{M'}{2(G+F)a} N \tag{16}$$

$$N'' = \frac{N}{2} - \frac{M'}{2(G+F)a} N \tag{17}$$

where *N* is the supporting force of each axle, so the force distribution of the vehicle is as follows

$$\begin{bmatrix} N_1 = \dfrac{N_f}{2} + \dfrac{M'}{2(G+F)a} N_f \\ N_2 = \dfrac{N_f}{2} - \dfrac{M'}{2(G+F)a} N_f \\ N_3 = \dfrac{N_m}{2} + \dfrac{M'}{2(G+F)a} N_m \\ N_4 = \dfrac{N_m}{2} - \dfrac{M'}{2(G+F)a} N_m \\ N_5 = \dfrac{N_r}{2} + \dfrac{M'}{2(G+F)a} N_r \\ N_6 = \dfrac{N_r}{2} - \dfrac{M'}{2(G+F)a} N_r \end{bmatrix} \tag{18}$$

For the vehicle with four axles, the force distribution of the vehicle is as follows

$$\begin{bmatrix} N_1 = \dfrac{N_f}{2} + \dfrac{M'}{2(G+F)a} N_f \\ N_2 = \dfrac{N_f}{2} - \dfrac{M'}{2(G+F)a} N_f \\ N_3 = \dfrac{N_m}{2} + \dfrac{M'}{2(G+F)a} N_m \\ N_4 = \dfrac{N_m}{2} - \dfrac{M'}{2(G+F)a} N_m \\ N_5 = \dfrac{N_r}{4} + \dfrac{M'}{4(G+F)a} N_r \\ N_6 = \dfrac{N_r}{4} - \dfrac{M'}{4(G+F)a} N_r \\ N_7 = \dfrac{N_r}{4} + \dfrac{M'}{4(G+F)a} N_r \\ N_8 = \dfrac{N_r}{4} - \dfrac{M'}{4(G+F)a} N_r \end{bmatrix} \tag{19}$$

3. Experiment and comparison

3.1. Experiment in the structural laboratory

In order to verify the validity and accuracy of the above mathematical model, an experiment is designed and implemented in the structural laboratory. As the main factors influenced on the supporting force of the vehicle are the leaf spring and tire, and their stiffness are assumed to be linear, so in the laboratory, the springs are utilized to simulate the leaf spring and tire of the vehicle. As shown in Figure 5, the experimental model has 3 axles, and 6 supporting points are laid on 6 weighing sensors. The rectangular beam are simulated as axle, the welded structure is simulated as the chassis. Two actuators are set in the two side of experimental model, which are used to apply symmetrical load and eccentric load. The spring near the supporting point is simulated as the tire, whose stiffness is 700 N/mm. The spring between the axle and the chassis is simulated as the leaf spring, whose stiffness is 500 N/mm. The weighing sensors connecting to data acquisition equipment are set under every supporting point. The precision value of the weighing sensor is 3‰. The test process is as following, six weighing sensors are reset and placed on the ground in the laboratory. Then, the whole experimental model is laid on the weighing sensors by the staff utilizing lifting equipment. In the end, the symmetrical load or eccentric load is applied on the experimental model by the actuators, and there are 3 load cases in all, listed in Table 1.

IOP Conf. Series: Materials Science and Engineering **711** (2020) 012008 doi:10.1088/1757-899X/711/1/012008

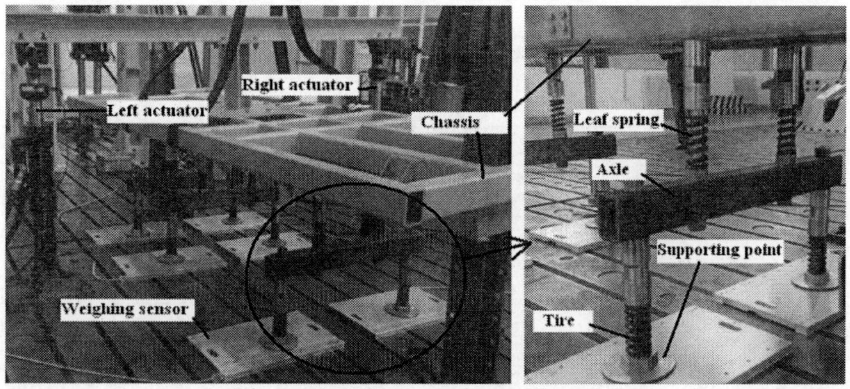

Figure 5: The experimental model

Table 1: Load cases

Load case	Description	Remarks
Load case 1	Gravity 9.22 kN, left load 10kN, right load 10kN	With symmetrical load
Load case 2	Gravity 9.22 kN, left load 20kN, right load 15kN	With eccentric load
Load case 3	Gravity 9.22 kN, left load 25kN, right load 15kN	With eccentric load

3.2. Calculated result and comparison

The geometric parameters can be acquired from the geometric model, as shown in Figure 6. L_f=2074 mm, L_m=2074 mm, L=2074 mm, M= -1862240 N*mm, G=9220 N, a=435 mm. For the load case 1, F=20000 N, for the load case 2, F=35000 N, for the load case 3, F=40000 N. As the stiffness of the spring and the tire are 500 N/mm and 700 N/mm respectively, so

$$k_f = k_m = k_b = 2*\frac{770*500}{(770+500)} = 583N/mm \quad (20)$$

Based on the Eq. 10, the supporting force of each axle (δ_f, δ_m and δ_r) can be acquired. Then, the force distribution of the vehicle can be calculated by the Eq. 18. For the load case 1, M'=0, for the load case 2, M'=6465000 N*mm, and for the load case 3, M'=12930000 N*mm.

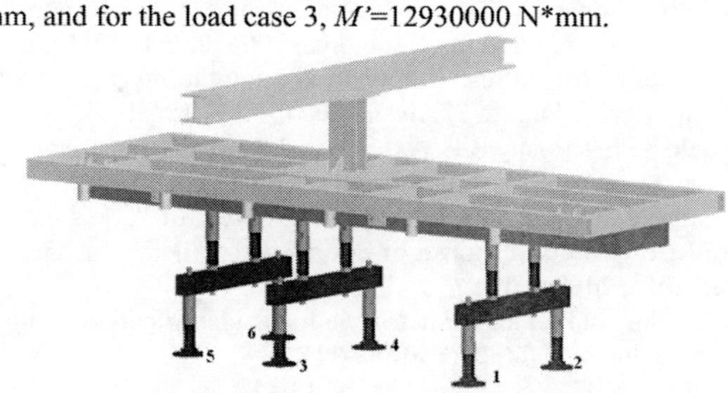

Figure 6: The geometric model of experiment

The calculated and experimental results are presented in Table 2, the calculated result means the supporting force calculated by the mathematical model, and the error means the calculated accuracy of supporting force relative to the experimental result. According to the data and the comparison, the maximum error is 5.9%, while the minimum error is 1.5%. And the average error is only 3.9%, which is not high for the engineering design. In brief, the experimental results are close to the calculated results in all the load cases, and this verifies the validity of the mathematical model proposed in this paper.

Table 2: The comparison between calculation and experiment

Supporting points		1	2	3	4	5	6
	Calculation / kN	4.840	4.840	4.908	4.908	4.862	4.862
Load case 1	Experiment / kN	4.735	4.665	4.755	4.835	4.995	5.125
	Error	2.2%	3.8%	3.2%	1.5%	2.7%	5.1%
	Calculation / kN	10.053	4.995	9.585	4.763	9.901	4.920
Load case 2	Experiment / kN	9.565	4.720	9.815	5.010	9.755	5.215
	Error	5.1%	5.8%	2.3%	4.9%	1.5%	5.6%
	Calculation / kN	13.501	3.334	12.720	3.141	13.247	3.272
Load case 3	Experiment / kN	12.745	3.165	13.25	3.285	13.465	3.455
	Error	5.9%	5.3%	4.0%	4.4%	1.6%	5.3%

4. Conclusion

The main factor affecting the supporting force is the vertical stiffness of the supporting point, so in this study, the vertical stiffness of leaf spring and tire is considered and assumed linear. Simultaneously, the structure of the chassis is assumed to be non-deformed. When the vehicle is in the static and stable state the mathematical matrix is established by force equilibrium equation, moment equilibrium equation and deformation compatibility equation. The force distributions of the vehicle with 3 axles and 4 axles are presented in Eq. 18 and Eq. 19. To verify the validity of the mathematical model, the experiment is designed and implemented in the laboratory.

The mathematical model proposed in this paper can solve the problem of force distribution for the refitting vehicle, however, there are some assumptions in this prediction method, and they are also the limiting factors for the application of this model. Firstly, the chassis or the frame should have enough stiffness, so that the structural deformation is very small. Secondly, the nonlinear characteristic of stiffness for the leaf spring and tire is not big. To further improve the calculation model proposed in this paper, nonlinear factors need to be further considered and studied.

References

[1] Wu S Q, Law S S. Vehicle axle load identification on bridge deck with irregular road surface profile. Engineering Structures, 2011, 33(2):591-601.

[2] Mutton P J, Epp C J, Dudek J. Rolling contact fatigue in railway wheels under high axle loads. Wear, 1991, 144(1–2):139-152.

[3] Pinkaew T. Identification of vehicle axle loads from bridge responses using updated static component technique. Engineering Structures, 2006, 28(11):1599-1608.

[4] Mutton P J , Alvarez E F . Failure modes in aluminothermy rail welds under high axle load conditions. Engineering Failure Analysis, 2004, 11(2):151-166.

[5] Timm D H, Tisdale S M, Turochy R E. Axle Load Spectra Characterization by Mixed Distribution Modelling. Journal of Transportation Engineering, 2005, 131(2):83-88.

[6] Battini J M, Mahir Ãlker-Kaustell, Syk A , et al. Effect of Axle Load Spreading and Support Stiffness on the Dynamic Response of Short Span Railway Bridges. Structural Engineering International, 2014, 24(4):1-10.

[7] Law S S, Bu J Q, Zhu X Q, et al. Vehicle axle loads identification using finite element method. Engineering Structures, 2004, 26(8):1143-1153.

[8] Gerbert G, H. Jönsson H, Persson U, et al. Load Distribution in Timing Belts. Journal of Mechanical Design, 1978, 100(2):208.

[9] Zhang W, Koyama T. Load Distribution of Metal Timing Belts. ASME. J. Mech. Des. 1999, 123(1):98-103.

[10] Nirut K, Kitjapat P. An Analytical Method for Determining the Load Distribution of Single-Column Multibolt Connection. Advances in Civil Engineering, 2017, 2017:1-19.

Effect of Starting Microstructure in High Speed Steel on the Grain Refinement Potency of Electropulsing Treatment

Jiatao Zhang*, Penglei Hu, Xiongqi Li

College of Materials Engineering, Hubei University of Automotive Technology, Shiyan 442002, Hubei, PR China

*Corresponding author E-mail: jiatao_zhang@qq.com

Abstract: High speed steels (HSSs) are widely used materials for tools production, which grab a third part of the global cutting tool market. Heat treatment is an indespensable process for the fabrication of HSS tools. Modifications of carbides and grain size of the matrix are two main aspects for the heat treatment of HSSs. In order to obtain better wear property, hardness or red hardness, modification of carbides attracts more researchers' attention, but grain refinement benefits to the toughness and tool life. Compared with the grain refinement methods used for other steels, such as controlled rolling, available method for grain refinement of HSSs seems deficient. Our previous study provided a new method to produce ultrafine grained M2 steel, but the effect of the starting structures on the grain refinement potential of electropulsing treatment (EPT) was not involved. Results in this work show that the grain size can both be refined or even ultra-refined by EPT, whatever the difference in the starting structures. But the austenite transformation of the matrix was completed at lower peak temperature when using starting tempered structure. And the pre-tempered M2 samples have higher hardness than that obtained after heat treatment or EPT when using the starting annealed structures.

1. Introduction

High speed steels are alloys of complex Fe-Cr-C-X system, where X represents different strong carbide forming elements, such as W, Mo, Cr, V, or Co[1, 2]. Due to their excellent wear resistance, high red hardness and good toughness, HSSs are important industry materials and widely used for cutting tools, molds and rollers[1-3]. The mechanical properties of HSSs basically depend on the conditions of carbides, as well as the grain size of the matrix. Modification of carbides in HSSs has been intensively studied[1, 2, 4]. But the grain refinement method for HSSs seems scarce.

Grain refinement is a popular subject in modification of microstructure and mechanical properties in steels. But researches on modification of carbides in HSSs attract more attentions than that on grain refinement. This is because morphologies of carbides contribute the red hardness and wear property of HSSs, which is very important for this material used for cutting tools. Nevertheless, grain refinement profits for the improvement of the toughness and tool life[5].

Heat treatment is a vital process to control the grain size in HSSs. Grain size of commercial HSSs ranges from 5 ~ 60μm [6-9], depending on detailed chemical compositions and austenitization temperature. Grain size obtained at low temperature austenitization is finer than that at high temperature austenitization. But in order to obtain sufficient secondary hardening effect during triple tempering, the austenitization temperature must be high enough to obtain higher supersaturated carbon and alloying elements in solution in the matrix before quenching. Therefore, assurance of sufficient carbides dissolution in HSSs is the first important thing during heat treatment. Hence, it seems hard to refine the grain size and simultaneously promote the dissolution carbides during heat treatment.

Our previous study[10] reported a new method by using EPT to produce ultrafine grained M2 HSS. But it is not clear whether the starting structures have effect on the grain refinement of M2 steel after EPT. In this work, the effect of starting structures on the grain size in M2 steel after EPT has been studied.

2. Experimental Procedure

Commercial M2 HSS was used as the tested material in this study. The two starting structure types used for EPT in this work were annealed and tempered, respectively. The as-received M2 steel was annealed, which was used as one type of the starting structures. In order to obtain tempered starting structures, the as-received M2 steel was austenitized at 1180°C for 10min in the vacuum tube furnace, followed by oil quenching. Then these quenched samples were tempered at 550°C for 1 hour.

Electropulsing was performed under ambient condition by self-made electropulsing generator. The discharging duration and pulse current density were controlled by the program on a computer. The temperatures during discharging were measured infrared thermometer. Table 1 and Table 2 list the used pulse current parameters and corresponding peak temperatures of samples. At the end of the discharging duration, the sample was quenched by oil immediately. The schematic illustration of the EPT system can be found in our previous work [10]. For comparison, the as-received M2 steel was also treated by conventional heat treatment. The grain size was measured by Image-pro plus 6 software. The final result of each sample was taken from more than 20 images observed at different fields with a magnification of 1000.

Table 1. Pulse current parameters used for the starting annealed structure group.

Peak current density (A/mm^2)	1700					
Discharging duration (ms)	180	190	200	220	230	240
Peak temperature (°C)	1190	1217	1251	1276	1286	1308

Table 2. Pulse current parameters used for the starting tempered structure group.

Peak current density (A/mm^2)	1700								
Discharging duration (ms)	170	174	176	182	185	190	193	195	210
Peak temperature (°C)	1034	1075	1095	1138	1188	1202	1218	1255	1278

3. Results

The as-received M2 steel consisted of coarse eutectic carbides and spheroidal pearlite (Fig. 1a). In order to obtain the tempered starting structures, some M2 samples were pre-heat treated. During conventional austenitization at 1180°C for 10min, most of carbides were dissolved into the matrix (Fig. 1b). But these primary carbides were not dissolved completely. These primary carbides are MC and M_6C types, which precipitate at 1294 and 1288°C in equilibrium, respectively, much higher than the austenitizing temperature in most cases[11]. So, coarse primary carbides are hardly dissolved completely during heat treatment. The hardness was improved to 64 in HRC after quenching. The average grain size was 18μm. When the quenched sample was tempered at 550°C for 1 hour, numerous secondary carbides precipitated (Fig. 1c). And the hardness was maintained at 59.3 in HRC.

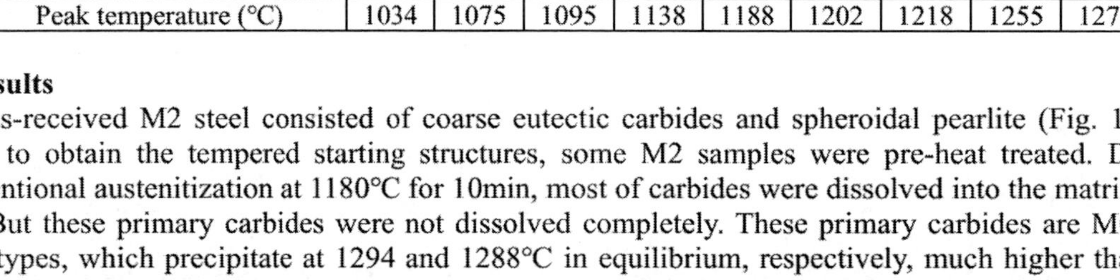

Figure 1. Microstructures of M2 samples: (a) annealed (as-received), (b) conventional quenched from 1180°C, (c) tempered at 550°C for 1 hour.

IOP Conf. Series: Materials Science and Engineering **711** (2020) 012009 doi:10.1088/1757-899X/711/1/012009

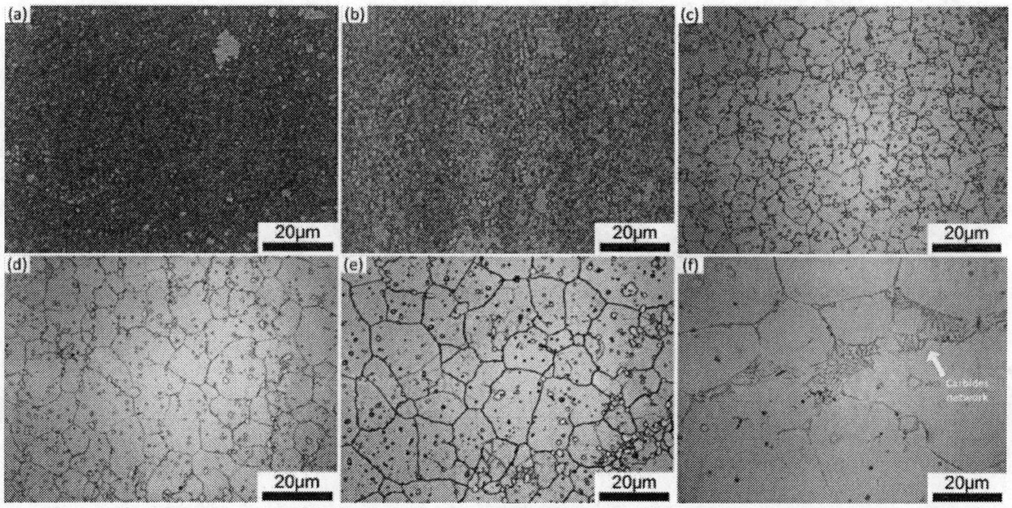

Figure 2. Microstructures of M2 samples quenched from different austenitization temperatures during conventional heat treatment:
(a)1050°C, (b)1100°C, (c)1150°C, (d)1200°C, (e)1250°C, (f)1286°C.

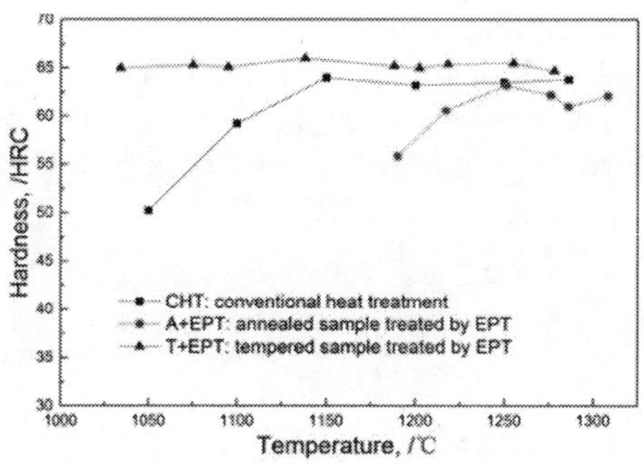

Figure 3. Hardness of M2 samples

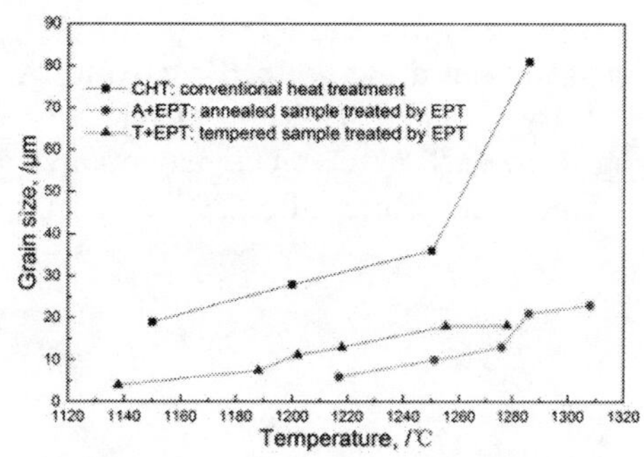

Figure 4. Grain size of M2 samples

For comparison, the microstructure with the austenitization temperatures during heat treatment are shown in Figure 2. A_1 and A_{cm} temperatures of M2 steel are 795°C and 821°C respectively [12]. When the steel was heated to 1050°C during heat treatment, the matrix transformed into austenite. But a majority of carbides in the starting annealed structures was not dissolved. The supersaturated carbon and alloying elements in solution in the matrix was not high enough to obtain full martensitic matrix during quenching (Fig. 2a). When the austenitizing temperature was increased to 1100°C, most spheroidal pearlitic carbides were dissolved. After quenching the matrix transformed into martensite, and the hardness was improved to 59.4 in HRC (Fig. 3). While quenching from 1150°C, more primary carbides were dissolved, and the newly formed austenite grain boundaries were clearly revealed. With further increasing the austenitization temperature, more primary carbides were dissolved. But the grain size grew coarse further. When quenching from 1286°C, fish born-like carbides network was formed at the grain boundaries (Fig. 2f). The grain size was increased to 81μm (Fig. 4).

Microstructures of M2 samples with starting annealed structures after EPT are shown in Fig. 5. The hardness of the sample quenched from 1190°C during EPT was improved to 56.7HRC (Fig. 3). This indicated the starting structures had undergone austenite transition upon heating during discharging. The A+EPT-1190°C sample consists coarse primary carbides, undissolved pearlite carbides and partial martensitic matrix (Fig. 5a). With increasing of peak temperature during EPT, more and more carbides were dissolved, and the matrix transformed into martensite. When the peak temperature was increased to 1286°C, fish born-like carbides network was formed at the grain boundaries (Fig. 5f).

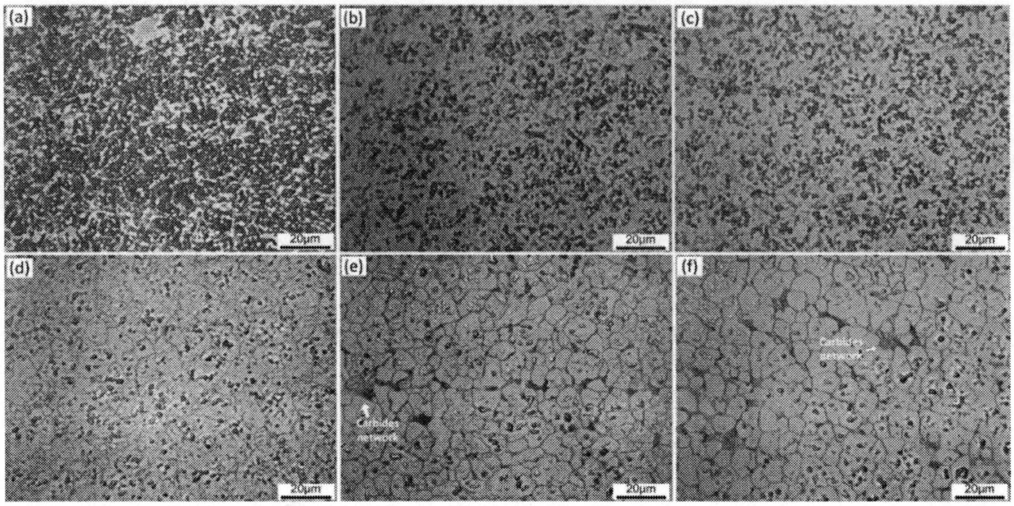

Figure 5. Microstructures of M2 samples with starting annealed structures after EPT and quenching from peak temperatures: (a) 1190°C, (b) 1217°C, (c)1251°C, (d) 1276°C, (e)1286°C, (f)1308°C.

Microstructures of M2 samples with starting tempered structures after EPT are shown in Fig. 6. It can be found that most of secondary carbides in the starting tempered structures were dissolved when quenched from 1034°C (Fig. 6a). With the peak temperature increasing to 1095°C, some very fine grained austenite were formed (Fig. 6c). When quenching from 1138°C, the starting matrix had transformed into ultrafine grained austenite. And all carbides were dissolved into the matrix except coarse primary carbides. With the increasing of peak temperature, grain size grew coarse. But when quenching from 1278°C, carbides networks were formed (Fig. 6i). The hardness of M2 samples with starting tempered structures after EPT was stably maintained to ~ 65HRC.

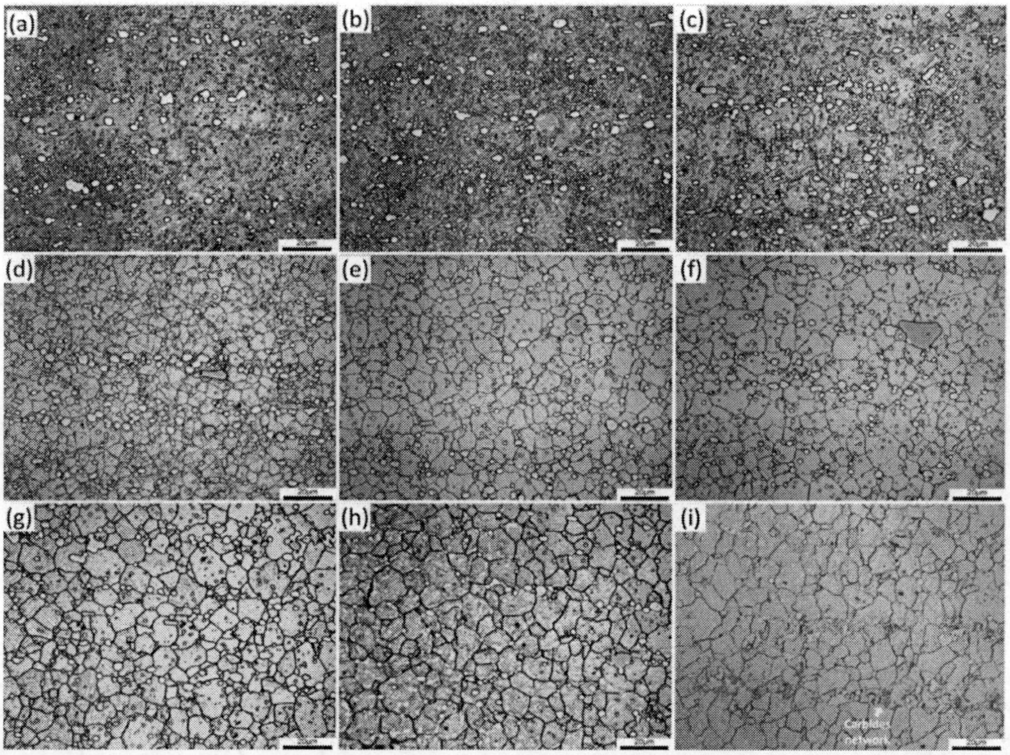

Figure 6. Microstructures of M2 samples with starting tempered structures after quenching from various peak temperatures during EPT: (a) 1034°C, (b) 1075°C, (c)1095°C, (d) 1138°C, (e)1188°C, (f)1202°C, (g)1218°C, (h)1255°C, (i)1278°C.

4. Discussion

Heat treatment is an imperative process for the fabrication of HSS tools. And it was a basic way to control the grain size for a specified HSS. In order to obtain finer grain size during conventional heat treatment, HSSs should be austenitized at lower temperature. But this could decrease the dissolved volumetric fraction of primary carbides. For example, the grain size of M2 steel can be refined to 6μm when quenched from temperature less than 1140°C, but the dissolution of primary M_6C carbide can be increased 50% if quenching from 1180°C[9]. So grain refinement in HSSs by lowing austenitization temperature is not a sensible choice.

Previous study[10] and results in this work show that EPT demonstrates remarkable grain refinement ability in M2 steel. The grain size of M2 steel after EPT was much smaller than that after heat treatment (Fig. 4), regardless of the starting microstructure used during EPT. But the ultrafine grained microstructure was obtained at lower peak temperature when using the starting tempered structure. This is probably because of the difference related to the austenite formation mechanisms. The formation of austenite in the starting annealed structures is a diffusion controlled process. This process involves formation of austenitic nucleus and the dissolution of carbides, controlled by the diffusion around the austenite nucleus, matrix and carbide. A_{cm} temperature of the matrix is 821°C, much lower than the peak temperature needed for the actual completion of austenite transformation in the matrix. But the dissolution temperature of carbides, formed during annealing, is higher than the A_{cm} temperature. The temperature needed for the completion of austenite transformation of the matrix during heat treatment is about 1100°C (Fig. 2b). While the austenite transformation of the matrix was completed at about 1217°C. This is because the heating rate during EPT is much higher than that during heat treatment. High heating rate increases the austenite transformation finishing temperature.

However, the microstructure transformation process in the starting tempered samples was different. Even though this process was also relevant to some diffusion controlled transition, such as the dissolution of secondary carbides in the starting tempered structures, the austenite transformation of the matrix was completed at lower peak temperature, compared with that when using the starting annealed structures. This phenomenon resulted from two reasons. The first reason was related to the formation mechanism of revised austenite. The starting tempered structures could contain residual austenite and un-decomposed martensite. Even new martensite probably formed during cooling from the temperature of tempering, due to the decrease of stability of residual austenite after decomposing during tempering. Reverse transformation from martensite during rapid heating is not a diffusion process[13]. So, formation of austenite from martensite during EPT was a time independent transition, and probably not influenced by the dissolution of carbide. The other reason was less time needed for dissolution of secondary carbide. For the starting tempered sample, most of the carbides except the coarse primary carbides in the as-received steel were fully dissolved during the pre-heat treatment. Although lots of secondary carbides precipitated during tempering, the size of secondary carbides was much smaller than that of the spheroidal pearlite carbides in the annealed sample. Meanwhile, spheroidal pearlite carbides form during annealing, which have higher dissolution temperature than that for secondary carbides, precipitated during tempering. So coarse spheroidal pearlite carbide is more stable thermodynamically, which needs longer time or higher temperature for the completion of dissolution.

In the view of the hardness after EPT or heat treatment, samples after EPT from the starting tempered structure have higher hardness (Fig. 3). And the hardness was stably maintained at about 65HRC around the whole austenitization temperature range during EPT. Although the grain size in M2 samples with two different starting structures can both be ultra-refined by EPT, the hardness of ultrafine grained M2 sample with the starting tempered structure is higher than that of ultrafine grained sample with the starting annealed structure, because more carbides were dissolved during EPT when using the starting tempered structure.

5. Conclusion

The grain size of M2 steel can be refined by EPT, regardless of the difference in starting microstructure used. And the ultrafine grained matrix can both be obtained after EPT. But the formation of ultrafine grained microstructure is at lower peak temperature when using the starting tempered structure. And the pre-tempered M2 samples after EPT have the highest hardness, compared with that obtained after heat treatment or EPT when using the starting annealed structures. Consequently, the method combined the pre-tempering with EPT is a better choice to refine the microstructure of HSS.

Acknowledgements

Fund supports: The Science Foundation of Hubei Province (No. 2017CFB293); The PhD Start-up Fund of HUAT (No. BK201605).

References

[1] Jardin R. T., Tchoufang Tchuindjang J., Duchêne L., *et al.* (2019) Thermal histories and microstructures in Direct Energy Deposition of a High Speed Steel thick deposit, *Mater. Lett.*, **236**: 42-45.

[2] Peng H., Hu L., Ngai T., *et al.* (2018) Effects of austenitizing temperature on microstructure and mechanical property of a 4-GPa-grade PM high-speed steel, *Mater. Sci. Eng., A*, **719**: 21-26.

[3] Vasilko K. and Murčinková Z. (2016) Tool Life Extension Methods for Cut-off Tools Made of High-speed Steel, *Procedia Engineering*, **149**: 520-525.

[4] Qu H., Liao B., Liu L., *et al.* (2012) Precipitation rule of carbides in a new high speed steel for rollers, *Calphad*, **36**: 144-150.

[5] Bhattacharyya D., Hajra A., Basu A., *et al.* (1977) The effect of grain size on the wear

characteristics of high speed steel tools, *Wear,* **42**: 63-69.

[6] Yu T. H. and Yang J. R. (2007) Effect of Retained Austenite on GPM A30 High-Speed Steel, *J. Mater. Eng. Perform.,* **16**: 500-507.

[7] Lu L., Hou L. G., Zhang J. X., *et al.* (2016) Improved the microstructures and properties of M3:2 high-speed steel by spray forming and niobium alloying, *Mater. Charact.,* **117**: 1-8.

[8] Wu Q., Sun D. and Liu C. (2008) Effect of Austenitizing Temperature on Microstructure and Mechanical Properties of Semi-High-Speed Steel Cold-Forged Rolls, *J. Mater. Eng. Perform.,* **18**: 952-958.

[9] Nogueira R. A., Filho O. O. A., Souza L. F. M., *et al.* (2006) Grain Size of Commercial High Speed Steel, *Mater. Sci. Forum,* **530-531**: 16-21.

[10] Zhang J., Zhao H., Shi Q., *et al.* (2019) Fabrication of Ultrafine Grained High Speed Steel with Satisfactory Carbide Dissolution by Electropulsing Treatment, *ISIJ Int.*: ISIJINT-2018-2774.

[11] Wang Q., Cheng G. and Huang Y. (2018) Morphology and precipitation mechanism of large carbides in M2 high speed steel, *Iron and Steel* **53**: 65-71.

[12] Liu B. L., Lü Z. Q., Feng W. W., *et al.* (2017) Precipitation and decomposition behaviors of carbides in AISI M2 high-speed steel with nitrogen and mischmetal, *J. Cent. South Univ.,* **24**: 782-788.

[13] Zhang J. T., Zhao Y. G., Tan J., *et al.* (2015) Austenite Grain Refinement by Reverse $\alpha' \rightarrow \gamma$ Transformation in Metastable Austenitic Manganese Steel, *J. Iron Steel Res. Int.,* **22**: 157-162.

IOP Conf. Series: Materials Science and Engineering **711** (2020) 012010 doi:10.1088/1757-899X/711/1/012010

Finite Element Analysis of the Maintenance Robot with Charged Used in Substation

Xu Dong*, Jianxiang Li, Jian Li, Qiang Chen, Jizhi Liu and Tao Li

State Grid Intelligence Technology Co., Ltd., Jinan, Shandong, 250101, China

*dxengineer@163.com

Abstract. The Maintenance Robot with Charged Used in Substation is mainly used for the maintenance tasks with hot-line in 220kV substation. It is mainly composed of the mobile chassis and insulated lifting arm. The end of insulated lifting arm is matched with different working tools to achieve different tasks. The robot needs to maintain the stability of the mechanical body during working, so the dynamic characteristics of the robot cannot be ignored. The finite element structure analysis of the robot body under working conditions is carried out, and the stress distribution and structural deformation diagram of the robot mechanical body under working conditions are obtained in this paper. The qualitative analysis of the material concentration in the stress concentration part is carried out due to the calculation results. The low-order vibration frequency of the robot and the corresponding mode shape are obtained by preload modal analysis. The dynamic characteristics of the robot during the operation are analyzed based on the preload modal analysis results.

1. Introduction

The maintenance robot with charged used in substation mainly carries out the post-insulator cleaning operation in the equipment area of the substation, and the foreign body removal work in the substation can be carried out when the hydraulic mechanical arm is equipped at the end of the insulated lifting arm [1]. In the process of carrying out the robot's live working, the robot body needs to maintain a certain stability and provide support for live working. The vibration characteristics of the robot are related to its own structure. Under different working conditions, the boundary conditions are different, and the vibration characteristics are also different. The moving chassis of the robot is driven by a motor. The vibration frequency of the rated speed of the motor is about 93.3 Hz. The structure of the robot should be designed to avoid the various operating conditions of the robot, especially under the working conditions. The structural frequency of the robot should be avoided at the same or similar to the vibration frequency of the drive motor to avoid resonance and then affects the safety of the robot.

The legs of the robot are opened, the insulation lifting arms are raised, and the end of working platform is equipped with different operational tools when the robot is working. The model of the robot under the working conditions is shown in Figure 1. The finite element analysis of the robot is carried out under the load of the robot chassis and the insulating lifting arm and the tool load of working platform, and the stress distribution and stress concentration part of the robot are obtained to conduct qualitative analysis. The deformation of the robot under external load conditions can also be obtained at the same time. The low-order vibration frequency and mode shape of the robot are obtained through the finite element preload modal analysis, and the modal analysis results are compared with the motor vibration frequency to analyze the stability of the robot under working conditions.

Figure 1. The model of the robot under the working conditions

2. Theoretical Basis

The theoretical basis of the finite element modal analysis of the robot is structural dynamics. The general structural dynamics equation is as follows:

$$[M]\{\ddot{u}\}+[C]\{\dot{u}\}+[K]\{u\}=\{F(t)\} \quad (1)$$

In the equation, [M] is the mass matrix of the structure, [C] is the damping matrix of the structure, and [K] is the stiffness matrix of the structure, $\{\ddot{u}\},\{\dot{u}\},\{u\}$ are the acceleration vector, the velocity vector and the displacement vector of the structure, {F(t)} is the vector of load.[2]

The robot's system is mainly based on rigid connections, so the influence of damping characteristics on the structure is not considered at present. Therefore, the finite element analysis equation under preload is:

$$[M]\{\ddot{u}\}+[K]\{u\}=\{F(t)\} \quad (2)$$

3. Establishment of the finite element model

The robot's finite element model is discretized on the basis of the 3D model. The model of the insulation lifting arm is mainly composed of the rotary platform, the main arm, the upper arm and the telescopic arm, which simplifies the structure of the rotary support and integrates the bolts of the connecting part. The finite element model of the robot is shown in Fig. 2.

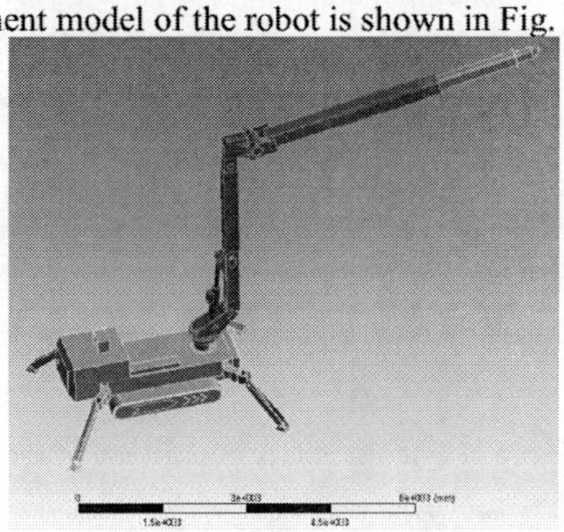

Figure 2. The finite element model of the robot

The material properties of the robot are mainly composed of three types: the Q345 steel, the rubber, and an insulating arm with FRP material properties. The material properties of Q345 have an elastic modulus of 206 GPa, a Poisson's ratio of 0.28, a density of 7.85×10^3kg/cm^3. The material properties of rubber have an elastic modulus of 7.84MPa, a Poisson's ratio of 0.47, a density of 1.3×10^3kg/cm^3. The material properties of FRP have an elastic modulus of 20 GPa, a Poisson's ratio of 0.22, a density of 1.7×10^3kg/cm^3.

4. Meshing

The meshing is divided into tetrahedron and hexahedron. The division method adopts free division. The grid correlation is selected as Fine+0 according to the size of the model and the accuracy of the calculation. Local adjustment is performed on the mesh with large distortion, and localized encryption processing is performed to reduce the distortion. The number of nodes is 146,414, and the number of cells is 69,341 after meshing. The meshing of the robot is shown in Figure 3.

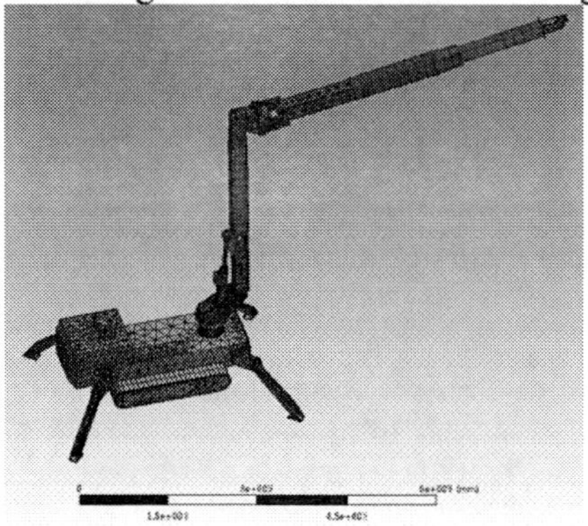

Figure 3. The meshing of the robot

5. Boundary Conditions

The boundary conditions of the robot are divided into mechanical boundary conditions and displacement boundary conditions. The mechanical boundary conditions are mainly composed of three parts. The mounting plate of motor applies 3000N equal force to simulate the force of the motor and hydraulic tank. An equivalent force of 3000N is applied to position of the battery package to simulate the force of the battery package. An equivalent force of 2000N is applied to the end of the telescopic arm of the insulated lift arm to simulate the force of the load. The displacement boundary condition is that the contact surface between the leg and the ground is fully constrained by using Fixed, and then the DOFs of displacement is limited to simulate the actual working condition. The Bonded constraint is adopted at the contact part of the component, and the Bonded constraint is a linear constraint, so that the contact surface does not have tangential sliding and normal separation, and then no large separation deformation occurs between the components during the calculation process. The boundary conditions of the robot are shown in Figure 4.

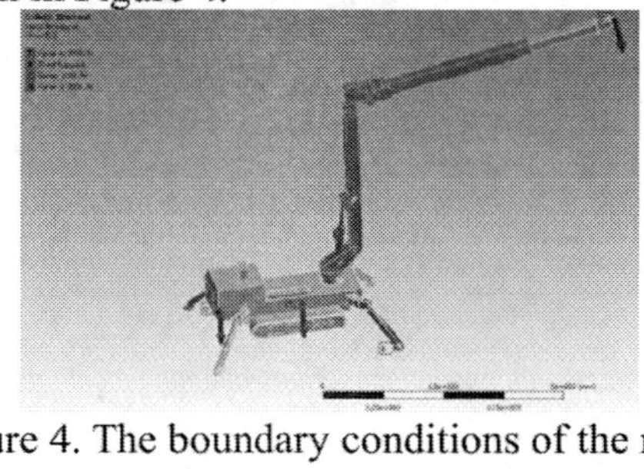

Figure 4. The boundary conditions of the robot

6. Static Analysis

After the model processing, meshing and boundary constraints are applied to the finite element model of the robot, the solution calculation and post-processing operations are performed to obtain the static simulation deformation of the robot (Fig. 5) and the stress distribution of the robot(Fig. 6).

IOP Conf. Series: Materials Science and Engineering **711** (2020) 012010 doi:10.1088/1757-899X/711/1/012010

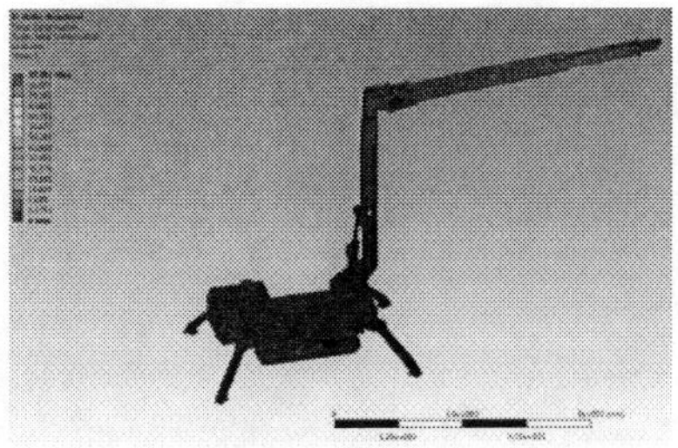

Figure 5. The static simulation deformation of the robot

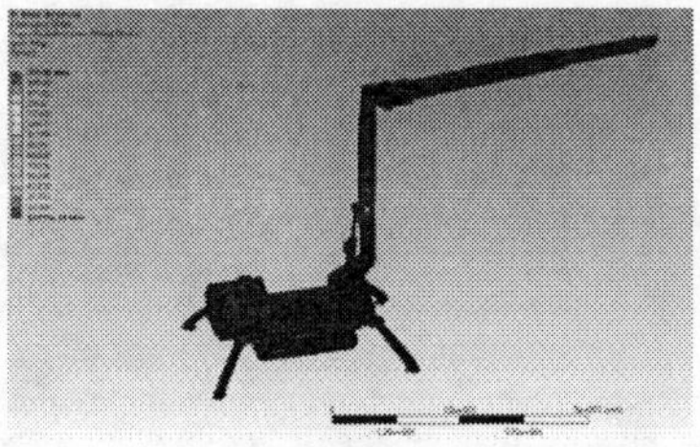

Figure 6. The stress distribution of the robot

It can be seen from the static simulation deformation of the robot under working condition that the maximum deformation occurs on the end of the telescopic arm of the insulating lifting arm, and the maximum deformation is about 87.85 mm. It can be seen from the stress distribution of the robot that the overall stress distribution of the robot is uniform, and the material structure generally meets the design requirements. The stress concentration occurs at the joint between the main arm and the upper arm, and the maximum stress is 203.33 MPa, as shown in Fig. 7. The material of the joint between the main arm and the upper arm is Q345, and its yield strength is 345 MPa. The calculated maximum stress is much smaller than the yield strength. And then the structural design and material properties are all meet the working requirements.

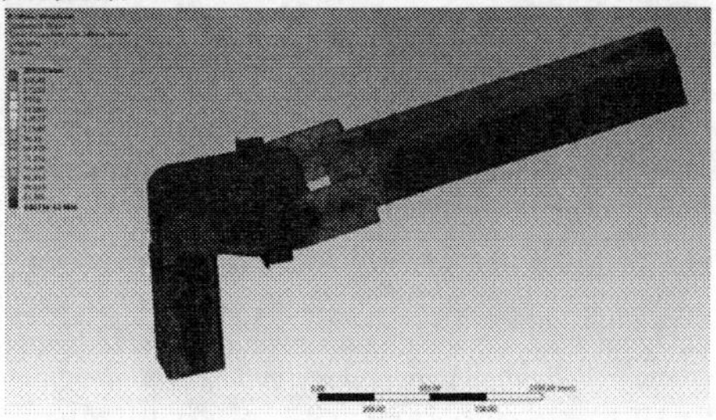

Figure 7. The stress concentration part

7. Modal Analysis with Prestress

Prestressed modal analysis of the robot based on static analysis. The finite element model, mesh and boundary conditions of the robot are unchanged. The robot is solved and post-processed after proper pre-processing. The first mode to sixth mode are extracted in the solution process. The low-order vibration frequency is shown in Table 1, and the corresponding vibration modes are shown in Fig. 8.

Table 1. The low-order vibration frequency of the robot

Modal Order	Vibration Frequency (Hz)
1	5.96
2	10.57
3	13.06
4	15.21
5	24.25
6	29.33

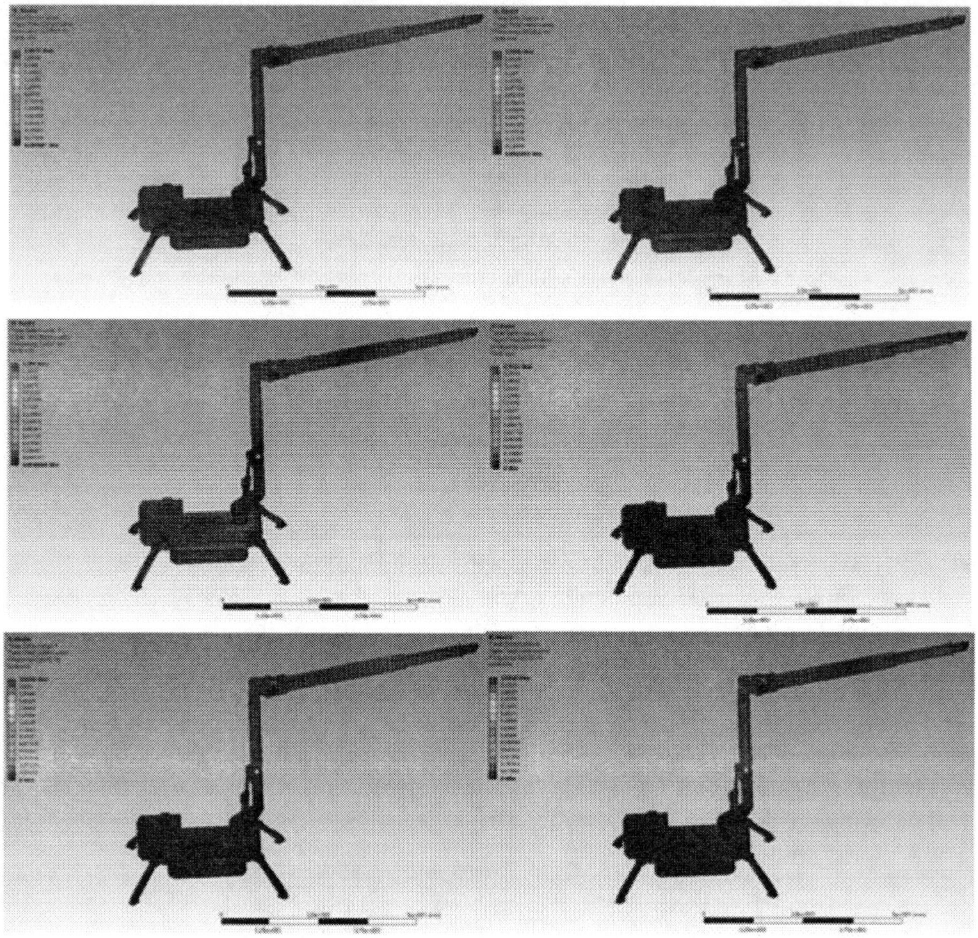

Figure 8. The corresponding vibration modes

It can be seen from the Table 1 and Figure 8:

The first order vibration frequency is 0.51 Hz. The vibration performance is rotating vibration and pitching vibration before and after around the horizontal axis of the robot. The telescopic arm end of the insulated lifting arm is the largest vibration part.

The second-order vibration frequency is 0.61 Hz. The vibration performance is rotating vibration left and right around the horizontal axis of the robot. The telescopic arm end of the insulated lifting arm is the largest vibration part.

The third-order vibration frequency is 1.49 Hz. The vibration performance is rotating vibration left and right around the vertical axis of the robot. The front leg contacted with the ground is the largest vibration part.

The fourth-order vibration frequency is 1.73 Hz. The vibration performance is the insulating lifting arm swings to the left and right around the vertical axis of the rotating platform. The telescopic arm end of the insulated lifting arm has the maximum vibration displacement.

The fifth-order vibration frequency is 1.84 Hz. The vibration performance is the upper arm of the insulating lift arm and the telescopic arm pitching vibration up and down. The end of the telescopic arm of the insulating lifting arm has the maximum vibration displacement.

The sixth-order vibration frequency is 4.83 Hz. The vibration performance is the insulating lifting arm swings to the left and right around the vertical axis of the rotating platform. The ear plate of the telescopic arm of insulated lifting arm is the largest vibration part.

It can be seen from the preload modal analysis that the low-order vibration frequency of the robot is between 0.51 Hz and 4.83 Hz. The main vibration is the vibration of the insulated lifting arm. The vibration frequency of the driving motor is 93.3 Hz, which is much larger than the low-order vibration frequency of the robot. The vibration of the drive motor has little effect on the dynamic characteristics of the robot under working conditions, and the robot has good stability during operation.

8. Conclusions

The finite element analysis of the robot under working conditions is conducted to obtain the stress distribution and stress concentration part of the robot in this paper. By comparing the maximum stress part of the robot with the material properties, the conclusion that the material performance meets the requirements is obtained. And the rationality of the robot structure design is verified. The vibration of the robot under the working condition shows the vibration performances are most the main body of the robot. The maximum position of the vibration is mostly the end of the telescopic arm of the insulated lifting arm. The comparison of the low-order vibration frequency of the robot and the vibration of the driving motor shows that the robot has good dynamic characteristics under the action of the drive motor.

Acknowledgments

This work from the project of Technology Research and Demonstration Application of Live Working Robot for Substation Equipment is supported by the Key Research and Development Plan of Shandong Province in 2017(2017CXGC0918).

References

[1] X Zhang, Rh Huang, etc.(2017). Development of the Maintenance Robot with Electrification Used in Substation. In International Conference on Mechanical, Control and Computer Engineering. Harbin.pp.30-33.
[2] Z Pan(2004). Finite Element Analysis and Applications. Tsinghua University Press, Beijing.
[3] X Dong, Jj Su, etc(2016). Finite element analysis and vibration control of the Substation Equipment Water Washing Robot with Hot-line. International Conference on Applied Robotics for the Power Industry. Jinan. pp:1-4.
[4] X Dong, Zc Zheng, etc(2014). Modal Analysis of the Line Inspection Robot's Guide Platform, Applied Mechanics and Materials,Vol.685,pp265-270,2014
[5] X Zhang, Rh Huang,etc(2016). Finite Element Analysis and Vibration Control of the Substation Charged Maintenance Robot. International Conference on Applied Robotics for the Power Industry,Jinan,pp:13-16.
[6] Sg L,X Zhang, R Huang etc(2018).Optimization Design of the Maintenance Robot with Charged Used in Substation,Huhhot.pp:141-144.
[7] X Dong, Jx Li, etc(2019). Finite Element Analysis of Mobile Chassis for the Maintenance Robot with Charged Used in Substation under Idle Condition, IOP Conference Series Earth and Environmental Science, Taiwan, Volume 233.pp:1-5

The study of Industrial design: The transformation of welder based on Tandem double-wire MIG welding

Xue Lin[1], Zhisheng Zhang[2]

[1] Faculty of Mechanical Engineering, University of Southeast, Nanjing, Jiangsu, 211189, China.

[2] Faculty of Mechanical Engineering, University of Southeast, Nanjing, Jiangsu, 211189, China

*Corresponding author's e-mail: oldbc@seu.edu.cn

Abstract. With the advent of industry 4.0, the new demands of welding industry are to adopt advanced welding technology and to realize automation. As a typical representative of welding technology, Tandem double-wire Metal-Inert Gas (MIG) welding could be used in precision welding and high-quality welding occasion. It has some superiority in higher welding efficiency and achieves automatic control easily. The paper presents a possible solution with transforming the common MIG welding tractor into special welding machine based on Tandem double-wire MIG welding. On the basis of product design and improvement methods, this paper integrates systems science and ergonomics to deeply discuss the application of Tandem MIG welding. The aim of transformation design is available to enhance the welding quality as well as to lower the production cost. And the research has consulting value on the design improvement of welding equipment for the manufacturing industries.

1. Introduction

Modern manufacturing industry reflects the level of a national integrated productivity. Welding, regarded as joining technology with low cost, is one of the most important synthesis techniques in the manufacturing. Therefore, welding equipment sales increase greatly to meet demand growth of welding work with the rapid development of manufacture. Under the background of market globalization and production modernization [1], the welding machine has been the development trend in the following areas. They are adopting advanced welding technology, increasing productivity and achieving automation in production operations.

Metal-Inert Gas (MIG) welding occupies the most important position among the fusion welding processes, due to its flexibility and cost effectiveness. Significantly higher travel speeds and deposition rates were achieved by Tandem double wire welding(Fig.1) compared to a common double wire welding(Fig.2). The Tandem MIG equipment comprises of two power sources, two wire feeders and a special torch for accommodating the two wires. The torch houses two contact tips which are electrically isolated from each other [2].

IOP Conf. Series: Materials Science and Engineering **711** (2020) 012011 doi:10.1088/1757-899X/711/1/012011

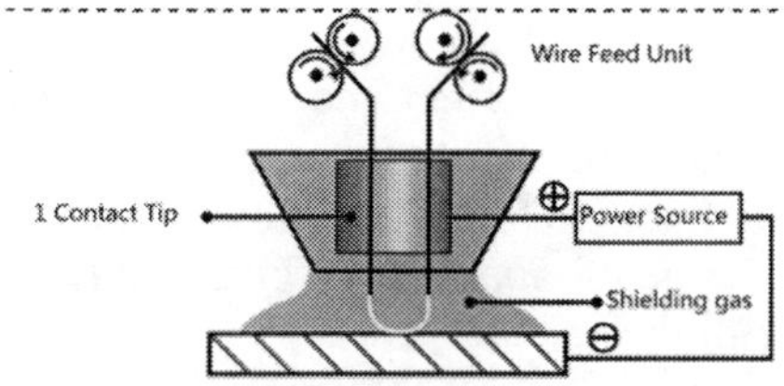

Figure 1. Schematic of double wire welding process

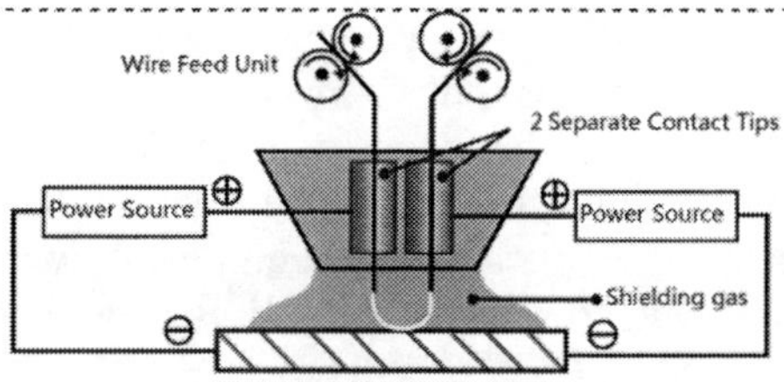

Figure 2. Schematic of Tandem double wire welding process

Tandem MIG welding can only be employed as a robot integrated system or integrated equipment, such as the special welding machine [4]. Special welding machine, as carrier of automatic welding technology, plays an important role on the welding of large equipment. It could increase welding speeds and enhance productivity , and the main advantages of the Tandem welding process are claimed to be faster travel speeds, higher deposition rate, and better comprehensive mechanical properties. Hence, special welding machine with Tandem welding could improve welding quality and reliability of engineering machinery. Compared with robot welding, it is easier to maintain or operate and has high performance cost ratio.

2. Background

Welding operators apply gas-shielded welding technology and use simple and crude single-wire welder manually in the mechanical manufacturing industry. And welding machine focuses on the realization of function rather than the Man-Machine-Environment System. In conclusion, welding machine is backward with low degree of automation in production operations. Aiming at the current situation, we carry on the study and transform MIG welding tractor (Fig.3) to special welding machine based on Tandem double-wire MIG welding. Therefore, special welder can improve not only automation degree but also welding efficiency, and ensure welding quality through systematic design theory.

Figure 3. MIG welding tractor

Figure 4. Wire feeder

3. Study approach

The system is the combination or interrelation of hardware, software, people and the operating environment. This paper takes the man-machine-environment system as the research object and discusses several influence factors from the angle of systematic design and ergonomics. And it emphasizes how to achieve top performance by uniting and coordinating the three elements: man, machine and environment [5]. According to the systematic design thought, human factor engineering introduces the method to study users and the operating environment to design the most human-friendly welding equipment.

IOP Conf. Series: Materials Science and Engineering **711** (2020) 012011 doi:10.1088/1757-899X/711/1/012011

3.1. Welder

People engaged in welding shall be composed of manual welders with work permits, relevant technical management fellows and occupations (boilermaker, pipe fitter and so on) [6]. After the transformation, the whole welding process depends on the motion of the tractor on the track to complete automatically. The welder has a high degree of automation and low labor intensity. Furthermore, this welding equipment will have wider range of application because there are no requirements in welding work license or experience on welding.

3.2. Special welding machine

Industrial design concerning user, an important tool of realizing product diversification, refers to the appearance and interface design of product. This paper elaborates the design of special welding machine based on the purpose and direction.

3.2.1. Appearance design. The appearance design of welder was illustrated through explaining the working principle, function division and the appearance design style. Firstly, there is a welding power, a wire feeder, a welding torch, a gas cylinder, a travelling mechanism and a controller in welding tractor. Special welding machine includes two welding powers, two wire feeders, a welding torch, a gas cylinder, a travelling mechanism and a controller [8]. Comparison between special welding machine and MIG welding tractor in structure (Tab.1) must be the basic premise of transformation.

Table 1. Comparison between Special welding machine and MIG welding tractor

Structure	**MIG welding tractor**	**Special welding machine**
Welding power	1	2
Wire feeder	1	2
Welding torch	1(single wire)	1(double wire)
Gas cylinder	1	1
Travelling mechanism	1	1
Controller	1	1

This paper analyzes theoretical points how to transform the welding equipment from welding tractor into special welding machine. The conversion was done by combining welding tractor with additional wire reel from another wire feeder (Fig.4). The preliminary result of special welding machine is as shown in the figure 5. Two power supplies were adopted, and two wires are fed into a single molten pool through the same torch, but they are electrically isolated from each other in the tandem MIG welding process.

Figure 5. Preliminary result of special welding machine

Then, the confinements and influences of function factors of welding equipment on appearance design are studied. Given structure and principle, distributing the function between human and welder need to research in depth. For instance, operators are responsible for adjusting welding parameters of controller rather than welding manually with the application of special welding machine.

IOP Conf. Series: Materials Science and Engineering **711** (2020) 012011 doi:10.1088/1757-899X/711/1/012011

Finally, frequently-used representative welding equipment looks rather crude in appearance and lack of variation or change. Nowadays, the market considers not only the function of products, but also pay special attention to the appearance design. Therefore, welder machine need change the low-quality image and completely meet the demand of users with emphasis on industrial design.

3.2.2. Interface design. The man-machine interface of welding machine exhibits the following characteristics: ① Higher efficiency of the operation interface. The accurate information transfer and human-machine interaction contribute to the steady operation of the system [9]. ② Friendlier operation interface. It should reduce skill requirements for operators and costs of study.

Figure 6. Control panel

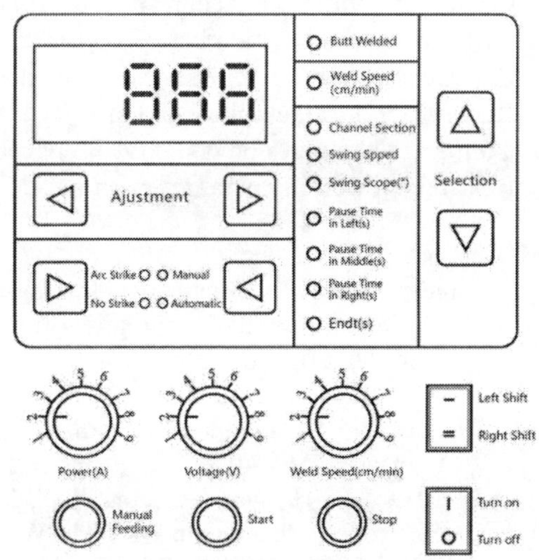

Figure 7. Interface prototype

Firstly, how to optimize so many welding parameters in human interface is a problem to be solved imperatively. We need reconsider the availability and priority of parameters in control panel (Fig.6) because some of them cannot adapt to the demand of special welding equipment. For instance, MIG welding tractor has two modes to choose from manual mode and automatic mode. We take the actual situation of transformation into account and then eliminate the manual welding button in the special welder.

Then, the study targets the welding operator group and focuses on mental structures, emotional features and behavioral properties in the view of man-machine engineering. The parameters of button in control panel (such as type, distribution and precision) need conform to human's physiology and psychic mechanism completely. Except the welding current and voltage and weld speed system use knob switch, other switches use touch button. It is worth considering what we can greatly simplify the operation of the welder in the interface prototype (Fig.7).

Finally, the interface design asks the request of keeping the complexities of operation to a minimum and achieving optimal functionality. Reasonable division of welding parameter is a critical contributor to improve the operation interface. Therefore, welding parameter need be rearranged and divided into different control areas [11], such as the main menu, welder control area, traveling control area and torch control area. Obviously, when users find a button or knob accurately without thinking, that means the control system has good and easy operation interface.

3.3. Welding workshop

During the welding process, the bad operating environment includes electrical hazards, fume, arc radiation and noise. Due to its severe application environment [12], more factors should be taken into account in design. Welding smoke and dust dispersing in workshop influence welders directly. Designing dust cover can improve the severe condition effectively. And the welding equipment is required to provide good mechanical and environmental performance.

4. Main conclusion

The research of the transformation design focuses on the integrated effect of man, machine and environment. The paper thoroughly studies the advanced weld technology of Tandem double-wire welding. And then MIG welding tractor is transformed into special welding machine with good operability and high degree automation. Additionally, it highlights the combination of human and environmental factors with welding machine design based on the man-machine-environment principles. In short, the appearance and interface design points in industrial design concerning user are as follows.

- Combining with wire feeder, two power supplies, two wires through the same torch
- Distributing the function between human and welder
- Pay special attention to the appearance design
- Reconsider the availability and priority of welding parameters
- Conform to human's physiology and psychic mechanism completely
- Rearranged welding parameters and divided into different control areas

References

[1] Kagermann H., Helbig J., Hellinger A., Wahlster W. (2013) Recommendations for implementing the strategic initiative industrie 4.0: securing the future of german manufacturing industry; fifinal report of the industrie 4.0 working group. Technical Report. Acatech – National Academy of Science and Engineering.

[2] Goecke S., Hedegard J., Lundin M., Kaufmann H. (2001) Tandem MIG/MAG welding. Svetsaren, 56: 24–8.

[3] Wei H.L., Li H., Gao Y., Ding X.P., Yang L.J. (2015) Welding process of consumable double electrode with a single arc GMAW. Int J Adv Manuf Technol, 76: 435–46.

[4] Klas W. (2012) Welding Processes Handbook (Second edition). Woodhead Publishing Series in Welding and Other Joining Technologies. pp. 75-97.

[5] Jan D., Henk de V., Sandra V., Wietske E., Albert F. (2004) Combining economic and social goals in the design of production systems by using ergonomics standards[J]. Computers &Industrial Engineering, 47: 207-222.

[6] Michael R.F., Pam S. (2009) Neurological risks associated with manganese exposure from welding operations–A literature review. International Journal of Hygiene and Environmental Health, 212: 459-469.

[7] David O.N., Leila A.G. (2015) The complexity of ergonomic in product design requirements. Procedia Manufacturing, 3: 6169-6174.

[8] Luis F.R., Rodrigo A.R., José E.S. (2019) Double-wire tandem GMAW welding process of HSLA50 steel. Journal of Manufacturing Processes, 45: 227-233.

[9] Valeria V., Fabio P., Francesco L.,Cristian S. (2018) Survey on human－robot collaboration in industrial settings: Safety, intuitive interfaces and applications. Mechatronics, 55: 248-266.

[10] Julian L., Tillmann F., Jan W., Hermann K. (2015) Principles to develop size ranges of products with ergonomic requirements, using a robust design approach. Procedia Manufacturing, 3: 6305-6312.

[11] Giles C. (2018) Simple and usable Web, Mobile, and Interaction Design, 2nd Edition.

[12] Olivera P., Radica P., Meri B., Uroš L., Biljana B. (2014) Fume and gas emission during arc welding: Hazards and recommendation. Renewable and Sustainable Energy Reviews,37: 509-516.

IOP Conf. Series: Materials Science and Engineering **711** (2020) 012012 doi:10.1088/1757-899X/711/1/012012

Experimental research on explosion resistance of masonry structures reinforced with modified polyurea elastomer

TAO Xigui, JIA Yifan*, FAN Chengfei

Institute of Defense Engineering, AMS, PLA, Beijing 100850, China

*Corresponding author's e-mail:tonytxg@126.com

Abstract. In order to improve the explosion resistance performance of masonry structures during the terrorist attacks and accidental disasters, clay brick and concrete block walls are reinforced by modified polyurea materials. Explosion resistance tests of these two kinds of masonry wall structures are carried out to analyze the failure modes and anti-explosion effects of reinforced masonry walls under different thickness and reinforcement modes. The results show that modified polyurea elastomer not only has good spraying adhesion and compatibility with the base material surface, but also can effectively encapsulate the explosive debris generated by the back blasting surface of the masonry wall under explosive load, which reduces the deformation and displacement of the wall.

1. Introduction

The explosion caused by terrorist attacks and accident disasters forms a huge pressure shock wave, which leads to great damage to nearby buildings and objects, and even causes continuous collapse and casualties due to local damage. Therefore, it is urgent to study and analyze how to improve the explosion-proof and explosion-resistant ability of buildings.

Polyurea is an elastic block copolymer formed by the reaction of isocyanate and ammonia. Because of its excellent mechanical properties[1,2], high elastic modulus[3], high tensile strength, high ductility[4] and good impact resistance[5,6], polyurea has been extensively studied and applied in the materials mitigating explosive shock in recent years. In this paper, the brick wall and concrete block walls are reinforced with modified polyurea elastomer. Then explosion tests are carried out on these masonry walls with different thickness of polyurea elastomer, aiming to study and analyze the failure mode and anti-explosion effect of the wall reinforced with polyurea elastomer under different thickness and reinforcement modes, which provides a reasonable basis for practical application.

2. Modification and properties of polyurea elastomer

Polyurea elastomer materials with amino-terminated polyoxypropylene ether as the main component have poor adhesion and compatibility with the substrate surface, which limits the application of polyurea elastomer materials in engineering.

For amino-terminated polyoxypropylene ether polyurea elastomer materials, the research on their properties is mainly aimed at resin chain extender. It is prepared by the reaction of R component consisting of high active amino-terminated polyether and polyamine chain extender with A component consisting of polyisocyanate prepolymer. No catalyst is needed because the reaction speed between amino group and NCO group is very fast. In general, too fast film formation produces some fatal surface defects, sometimes even discontinuous coatings. In order to overcome these shortcomings, the chain extender of primary amine with high activity is modified to reduce its reaction rate with isocyanate to achieve the protective effect of polyurea elastomer.

The modification methods of primary amine chain extender mainly include cyanoethylation of amine chain extender, addition with acrylate, ketoimidation of amine chain extender, etc. In this paper, primary amine chain extender is modified with hexamethylenediamine chain extender and glacial acetic acid to form amide chain extender, which reacts with polyoxypropylene ether of different concentration to form amino-terminated polyether R component, and then reacts with polyisocyanate prepolymer to form high performance polyurea elastomer materials with good adhesion and compatibility with the surface of the base material. According to ASTM D412 standard, the static tensile mechanical properties of the modified polyurea elastomer are tested, the physical and quasi-static mechanical properties of which are shown in Table 1.

Table 1. Physical and Quasi-static Mechanical Properties of Polyurea.

Density g/cm3	Poisson's rate	Elastic modulus MPa	Shear modulus MPa	Yield stress MPa	Fracture stress MPa	Fracture strain
1.12	0.4	161	7.5	6.2	15.8	1.6

3. Explosion resistance test scheme and specimen making of masonry structures reinforced by modified polyurea elastomer

The common filling parts in building structures are clay brick wall, concrete block wall and reinforced concrete, which are divided into load-bearing wall and non-load-bearing wall according to different structural forces. The projection of the damaged part of the wall and high-speed debris caused by explosive load are the main causes for indoor casualties. In this experimental research, clay brick wall and concrete block wall are chosen as the objects of this modified polyurea elastic reinforcement.

3.1. Test Scheme

The failure modes of walls under external loads are different with different failure pressure and resistance limit of walls. The main purpose of the test is to investigate the failure mode and explosion resistance limit of the wall reinforced with polyurea elastomer under different thickness and reinforcement methods. To achieve this purpose, the test plan is drawn up as shown in Table 2. The shock wave load is obtained by the cylindrical TNT explosion in the experiment.

Table 2. Test Scheme.

Wall material	Number	Dimension (cm)	Spraying method	Coating thickness (mm)	Explosion mode
Clay brick	4	250×310×24	Single side	3、5、8	Contact explosion
Concrete block	4	250×310×24	Single side	4、6	Contact explosion

3.2 Specimen making

Brick walls and concrete block walls are constructed on the test site. The process of spraying polyurea elastomer is strictly followed to ensure the quality of sprayed polyurea elastomer coating. Single-wall masonry is completed in 1-2 days as soon as possible. All walls are maintained for 28 days at the same time and under the same conditions. The process of specimen preparation is shown in Figs. 1 and Figs 2.

Fig 1. specimen making of clay brick wall.

Fig 2. specimen making of concrete block wall.

4. Experiment results and analysis of explosion resistance of masonry structures reinforced with modified polyurea elastomer

The influence of the thickness of polyurea elastomer on the explosion resistance of the wall is explored through the explosion test of brick wall and concrete block wall specimens reinforced with elastomer under different working conditions.

4.1. Brick wall

The explosion tests are carried out on the brick wall without sprayed polyurea elastomer (Z-0-0) and those with spayed thickness of 3mm（Z-0-3）, 5mm（Z-0-5） and 8mm（Z-0-8） respectively. The phenomena of the clay brick masonry walls after the test are shown in Figure 3.

(a) Z-0-0 (b) Z-0-3 (c) Z-0-5 (d) Z-0-8

Fig 3. Phenomena of clay brick masonry wall specimens after explosion test.

The brick wall unsprayed with polyurea elastomer collapses and spreads under the action of TNT explosive load with an equivalent amount of 2.17 kg and a detonation distance of 10.5 cm, and the damaged parts splash within the radius of 8m on the ground of the back blasting surface of the wall.

When the thickness of sparyed polyurea elastomer is 3mm, under the action of TNT explosive with an equivalent amount of 2.16 kg and a detonation distance of 10.5 cm, there are dislocation cracks in the central area of the wall, with ruptures in the reinforcement layer and a horizontal displacement of 12 cm. Though slight concave deformation appears, it is relatively intact and does not collapse compared with the non-sprayed wall.

When the thickness of sprayed polyurea elastomer is 5mm, under the action of TNT explosive with an equivalent amount of 2.24 kg and a detonation distance of 10.5 cm, the horizontal displacement of the central area of the wall is 21cm, without obvious rupture in the reinforced layer. The tension strain of the back blasting surface of the wall increases, which leads to the increase of the central displacement of the wall compared with those with polyurea elastomers 3 mm thick (Fig.5).

Under the action of TNT explosive with an equivalent amount of 2.06 kg and a detonation distance of 10.5 cm, the brick wall with sprayed polyurea elastomer 8 mm thick has obvious cracks in the central area, with a horizontal displacement of 33 cm, and obvious ruptures in the reinforcement layer. Compared with other thickness of polyurea elastomer, there is increased response of concave deformation of back blasting surface of brick wall.

Compared with brick wall without sprayed polyurea elastomer, the explosion resistance of the brick wall sprayed with polyurea elastomer is obviously improved, which does not collapse entirely, only with cracks on the back blasting surface of the wall, the surface complete and relatively smooth. The encapsulation of polyurea elastomer can effectively reduce the spattering debris of the brick caused by blast wave and slow down the secondary damage.

The test results show that (Table 3) the brick wall sprayed with different thickness of polyurea elastomer has concave deformation response under explosive load, but polyurea elastomer can effectively reduce the displacement of the center of the wall. With the increase of the thickness of polyurea elastomer, the central displacement of brick wall becomes larger. Polyurea elastomer can effectively restrain the displacement of brick wall and reduce the concave deformation of the wall.

Table 3. Statistics of explosion test results of brick wall specimens.

Specimen number	Coating thickness mm	Booster charge kg	Detonation distance cm	Distance from the ground cm	Result
Z-0-0	0	2.17	10.5	90	building blocks spatter
Z-0-3	3	2.16	10.5	90	horizontal displacement of the center of brick wall 12cm
Z-0-5	5	2.24	10.5	90	horizontal displacement of the center of brick wall 21cm
Z-0-8	8	2.06	10.5	90	horizontal displacement of the center of brick wall 33cm

The brick wall, which is sprayed with 3- mm- thick polyurea elastomer on the back blasting surface, remains relatively intact without damage or debris throwing under explosive load. The slight concave deformation indicates that the middle area is the plastic area under strong dynamic loads, and the boundary of the plastic area varies in the course of plate deformation. But under explosive load, cracks in the central area of the brick wall sprayed with polyurea elastomer almost coincide, the specific magnitude of impact of incident explosive load and magnitude of impact of back blasting surface on unloading wave can not be quantitatively analyzed, but it can be explained that spaying polyurea elastomer on the back blasting surface of brick wall significantly reduces the damage of explosive load on wall structure.

4.2. Concrete block wall

The explosion tests are carried out on concrete block walls unsprayed with polyurea elastomer (Q-0-0) and sprayed with thickness of 4 mm （Q-0-4） and 6mm （Q-0-6） respectively. The phenomena of concrete block walls after explosion test are shown in Fig. 4.

（a） Q-0-0 （b） Q-0-4 （c） Q-0-6

Fig 4. Phenomena of concrete block masonry wall specimens after explosion test.

The concrete block wall without sprayed polyurea elastomer, collapses and disperses under the action of TNT explosive load. The damaged parts spatter within the radius of 16m on the ground of the back blasting surface of the wall.

When the thickness of sparyed polyurea elastomer is 4mm, under the action of TNT explosive with an equivalent amount of 1.2kg and a detonation distance of 20.5cm, the back blasting surface of the concrete block wall is smooth and complete, without rupture or tension damage in the reinforcement layer. The surface of some blocks of the front blasting face is broken. Compared with the wall without polyurea elastomer, the concrete block wall is intact and does not collapse.

The concrete block wall with sprayed polyurea elastomer 6 mm thick collapses, and the middle of the reinforcement layer cracks under the action of TNT explosive with an equivalent amount of 1.25 kg and a detonation distance of 10.5 cm. But there is no spattered damaged blocks.

Test results show that (Table 4) the concrete block wall without sprayed polyurea elastomer collapses and spatters under explosive load and the damaged parts of the wall are thrown into the ground of the back blasting surface, while the concrete block wall with sprayed polyurea elastomer is effectively encapsulated. Even if the wall collapses after being damaged, the debris of the back-burst surface of the wall is obviously decreased, restraining the concave deformation and failure of walls.

The explosion resistance performance of the concrete block wall with sprayed polyurea elastomer 4mm thick is obviously improved, and there is no concave deformation in the wall. Under the explosive load, the center of concrete block wall with sprayed polyurea elastomer with thickness of 6mm is broken and dislocates. The block wall collapses after being torn because of the strong adhesion of polyurea elastomer.

Table 4. Statistics of explosion test results of concrete block wall specimens.

Specimen number	Coating thickness mm	Booster charge kg	Detonation distance cm	Distance from the ground cm	Result
Q-0-0	0	1.22	20.5	70	block throwing, the longest distance from the wall 16.3m
Q-0-4	4	1.20	20.5	70	back blasting face smooth and complete, cracks on the front blasting face of the block
Q-0-0	0	1.18	10.5	90	block throwing, the longest distance from the wall 16m
Q-0-6	6	1.25	10.5	90	middle of the wall tore and collapsed ,without spattered blocks

It can be seen that wall response at the early stage under explosive load experiences the impedance matching of blasting face and the dispersion of shock wave. In the dynamic response of wall, the strain process of polyurea elastomer occurs, which leads to the dispersion of wall strain, and ultimately leads to the transformation of failure mode of the reinforced wall.

5. Conclusion

Through the explosion tests on brick walls and concrete block walls reinforced with polyurea elastomer under different working conditions, the influence of thickness of polyurea elastomer on the explosion resistance of the wall is analyzed. The main conclusions are as follows:

The brick wall sprayed with polyurea elastomer 3mm thick on the back blasting surface remains relatively intact under the action of explosive load, without damage or debris throwing, only with slight concave deformation, which can be considered for application in practical engineering.

The concrete block wall, which is sprayed with polyurea elastomer with the thickness of 4mm on the back blasting surface, shows good explosion resistance.

Polyurea elastomer can effectively reduce the debris generation of the back blasting surface of the wall, and decrease the deformation and displacement of masonry wall under explosive load, which, therefore, improves the explosive resistance performance of the masonry wall.

References

[1] Cai Guijie. (2015)The design method of the reinforced concrete plate coated polyurea under the action of explosion. North University of China, Taiyuan.

[2] Remennikov AM, Ngo TD, Mohotti D, et al. (2017)Experimental in vestigation and simplified modeling of response of steel plates subjected to close-in blast loading from spherical liquid explosive charges. International Journal of Impact Engineering,101: 78-89.

[3] Song Bin. (2016)Anti-detonation properties of explosion-proof pots made of sandwich structures with polyurea elastomer. Nanjing University of Science &Technology, Nanjing.

[4] Sarva SS, Deschanel S, Boyce MC, et al. (2007)Stress strain behavior of a polyurea and a polyurethane from low to high strain rates[J]. Polymer, 48:2208-2213.
[5] Chattopadhyay DK, Raju KVSN. (2007)Structural engineering of polyurethane coatings for high performance applications. Progress in Polymer Science, 32: 352-418.
[6] Parniani S, Toutanji H.(2015) Monotonic and fatigue performance of RC beams strengthened with a polyurea coating system. Construction and Building Materials, 101: 22-29.

Design of Full Automatic Inspection Robot and Optimized Modification of Overhead Ground Transmission Line

Lei Cao, Rui Guo, Jianxiang Li, Feng Zhang, Dazhou Zhou, Juan Jia and Shibin Lu

State Grid Intelligence Technology Co., Ltd., Jinan, Shandong, 250001, China

e-mail: cl_clei@qq.com

Abstract. In order to achieve the full autonomous inspection of the robot along the overhead ground line, a new type of robot operation mode is proposed by considering the overhead ground line and the robot as a whole. According to the characteristics of obstacles encountered by the robot, the optimization design of the overhead line and the structure design of the robot body are carried out. By transforming the line, a obstacle-free path suitable for the robot was build to pass through, which simplifies the obstacle crossing action and improves the obstacle crossing efficiency.

1. Introduction

Overhead transmission line is one of the main power transmission equipment, its safe and stable operation directly affects the reliability of power supply system.

Because the overhead lines are exposed to the wild for a long time, broken strand, foreign object and corrosion often happen. If not timely repair and replacement, small breakages and defects can expand and eventually lead to serious accidents that cause widespread blackouts and huge economic losses. Therefore, the transmission lines must be regularly inspected to detect the defects, taking scientific measures to eliminate potential risks timely and ensuring the safety of power supply.

Traditionally, workers use telescopes to detect the lines. In this way, workers often face bad weather, which leads to time-consuming, inefficient and poor accuracy. With the development of science and technology, robots have been used in the field of high voltage power line detection. The research on the inspection robots of transmission lines abroad has a long history. For example, “LineScout” of Hydro-Quebec research institute in Canada[1-2], “Expliner” of KEPCO in Japan[3] and “Ti” of EPRI in America[3] are most representative. The domestic research is late, but a series of technical achievements have been achieved. The most type robots are “LineBot” of Wuhan University[4-5] and “Ground inspection robot” of the National Academy of Sciences[6]. However, these robots still use multiple arms to overcome the obstacle alternately. This solution of the robot leads to complex mechanical mechanism, poor automation and low safety. Because of these reasons, the application of the transmission line inspection robots has been greatly hindered.

This paper presents the Full Automatic Inspection Robot that can cross any obstacles(dampers, straight towers, resisting-tensile towers, etc.) of overhead ground transmission lines without any obstacle striding movement. The application of barrier-free accessibility technology of this robot has realized the automatic inspection from substation to substation, which benefits for application generalization and improves the automation of transmission lines.

2. Total solution

For the robot, the obstacles on the overhead ground wire mainly include press-connect-tubes, dampers, straight towers and resisting-tensile towers. After analysis, we found that the press-connect-tubes and dampers are small in size and can be directly pressed by the robot. But the other two are obstructions, especially resisting-tensile towers have large spans and intricate lines, the robot cannot pass directly. Traditionally, the robot usually adapts to the lines by designing its own complex mechanical structures and overcomes obstacles in a complicated way, resulting in large size, heavy mass, difficult control, low efficiency and inability to be autonomous.

Therefore, we take the research method of overhead transmission line and inspection robot as a whole system. By simplifying the mechanical structure of the robot and building a obstacle-free path to improve its mobility. This solution not only does not change the original characteristics of the transmission line, but also can achieve full and automatic operation and inspection of the robot.

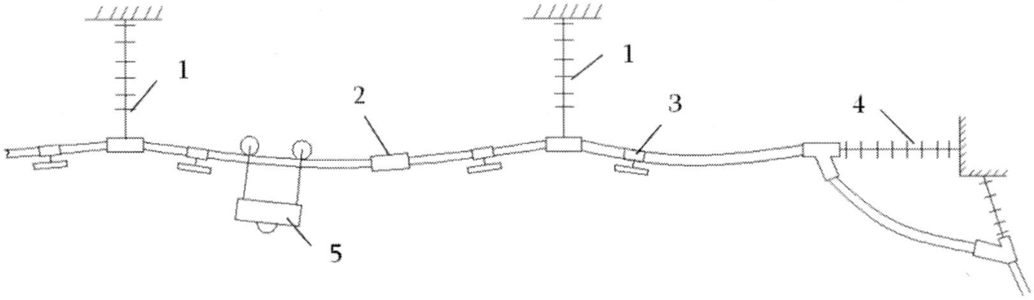

Figure 1. Major obstacles of overhead ground transmission lines.
1. Straight tower; 2. Press-connect-tube; 3. Damper;
4. Resisting-tensile tower; 5. Transmission line inspection robot

3. Optimized design of overhead ground wires

The straight tower and resisting-tensile tower are connected with ground line in different ways. So, it is necessary to optimize the design of specific schemes.

3.1. Modification of straight tower

Ground transmission line and straight tower are connected vertically with suspension clamp, U-shackle and suspension insulator string, etc. These objects block the robot's forward path. In order to reduce the complexity of the mechanical structure of inspection robot and improve the passing efficiency, it is proposed that the structure of the above blocking fittings be transformed into a non-blocking structure. The modified ground suspension clamp is shown in figure 2. The suspension clamp is designed as "C" type, so that the robot can pass through its center directly without any obstacle-crossing action, which greatly improves the efficiency of the robot rolling over the straight tower.

Figure 2. Unmodified & Modified Suspension clamp

3.2. Modification of Resisting-tensile tower

The resisting-tensile tower of transmission line has different shapes, sizes and angles of rotation, which makes it very difficult for the robot to cross. According to the overall project plan, setting up auxiliary guide rails on the outside of the tower to guide the robot to pass smoothly as shown in Figure 3. The guide rail adopts two circular rails parallel to each other, which consists of guide device, rigid rail, flexible rail and tower connectors.

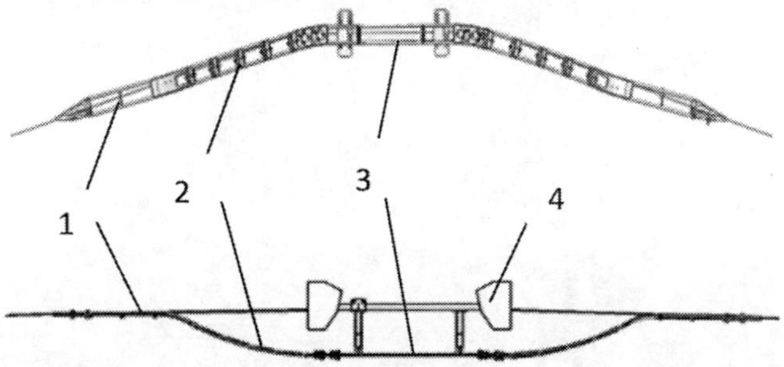

Figure 3. 3D graphical of modified guide rail
1. guide device; 2. flexible rail; 3. rigid rail; 4. resisting-tensile tower

The guide device is set at the front end of the whole rail to guide the robot to get off the ground line. As shown in Figure 4.The slope at the front and the arc at the back of the device gradually lead the robot away from the line and tower, so that the robot can run along the rail.

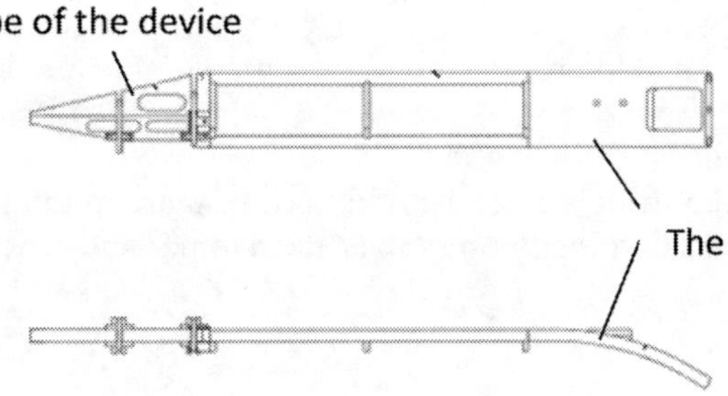

Figure 4. The guide device

Flexible guide rail including two types of flexible joints can be adjusted adaptively in both horizontal and vertical directions as shown in Figure 5. Specifically, the horizontal flexible joint can adapt to different sizes of pole towers, and the vertical flexible joint can adapt to all kinds of line sag. The number of these two joints can be adjusted according to different types and sizes of towers, thus forming a whole smooth rail.

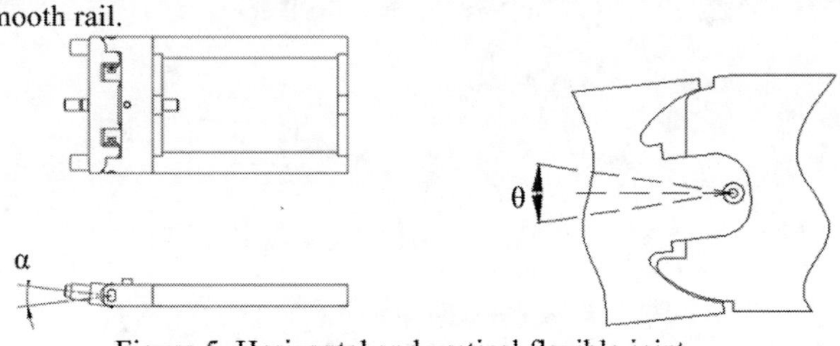

Figure 5. Horizontal and vertical flexible joint

4. Mechanical structure of the robot

The ground wire has been modified to form a obstacle-free passageway. Because of the arch connection between the guide rail and the ground wire and the slope of the line, the robot is required to have the ability of varying curvature. This robot adopts two wheel-arm composite mechanisms with the same structure, which is mainly composed of driving arms, grippers, locking joints, battery control box and monitoring devices. Among them, the front and back two driving arms of the robot have the same structure, and each driving arm is composed of driving wheel device, lifting device and guide device.

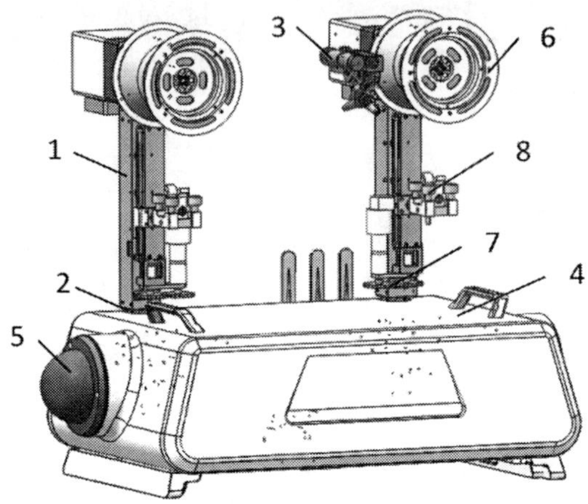

Figure 6. 1. Driving arms; 2. Locking joints; 3. Grippers; 4. Battery control box;
5. Monitoring devices; 6. Driving wheel device; 7. Lifting device; 8. Guide device.

Locking joints act as locking and loosening the driving arm. When the robot runs along the straight line, the locking of the rotating joint ensures that the driving wheel runs along the straight line. When the robot crosses the tension tower, the rotating joint is released. At this time, the driving arm can rotate freely in a certain range to provide the robot with variable curvature profiling motion.

5. Obstacle-crossing process of the robot

When the robot runs in the straight line, the robot grippers are in the open state and the locking joint is in the locked state, which can directly press over the damper and cross the straight tower. As shown in Figure 7.

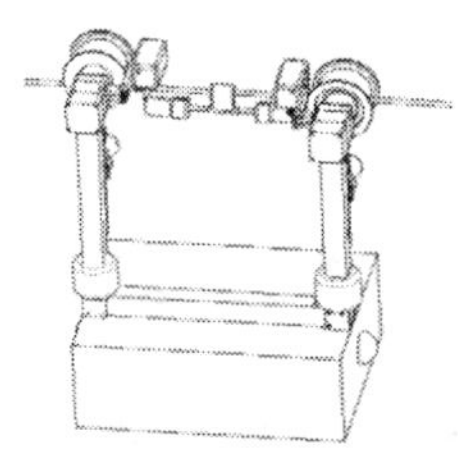

a: Pressing over the damper

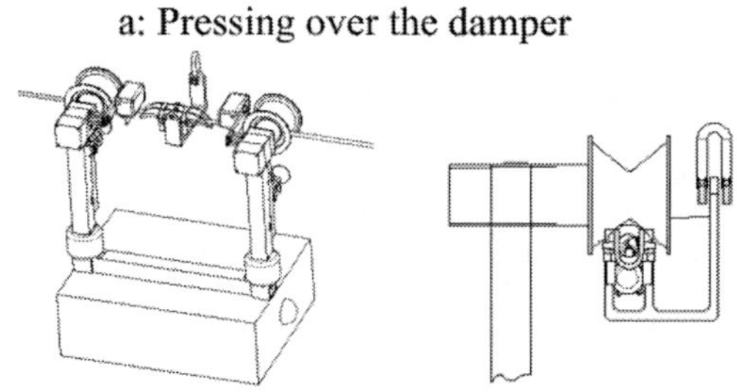

b: Crossing the straight tower

Figure 7. Running in the straight line

When the robot crosses the tension tower, it needs multi-joints cooperation. The obstacle crossing process is as follows:

- When the whole robot rolls to the front end of the bridge rail, the guide wheels of the robot rise and contact with the bottom of the rail, and then the locking joint is unlocked. At this time, under the combined action of the driving wheel and the guide wheel, the robot can follow the arc of the bridge rail to gradually get off the ground line.As shown in Figure 8a.

- When the robot is running on the bridge rail, in order to ensure safety, the robot rolls slowly.As shown in Figure 8b.
- When the robot reaches the other end of the rail, the above joints do the opposite action to 8a. As shown in Figure 8c.

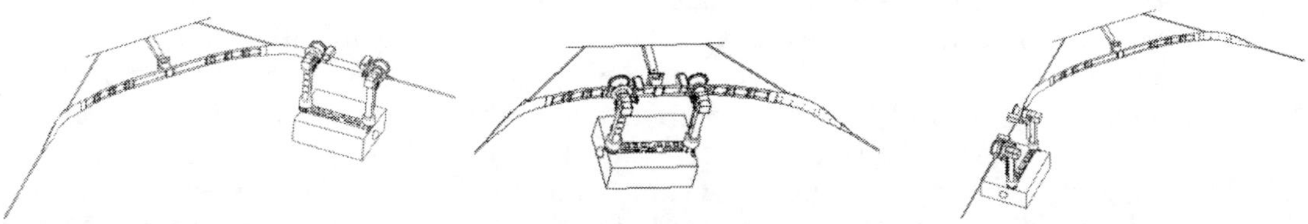

a: Touching the bridge b: Rolling on the rail c: Reaching the end of the rail

Figure 8. Crossing the tension tower

6. Application testing

In order to verify the function and performance of the automatic inspection robot for overhead ground transmission line, a large number of experimental research and application tests have been carried out on outdoor simulated training line and 500kV actual line. Figure 9 shows the operation test of the robot on the 500kV overhead transmission line in the training sub-center of State Grid Shandong Electric Power Company Jining Power Supply Company. After many tests, the obstacle-crossing rail matches well with the overhead ground wire, and realizes its self-adaptation to the sag and horizontal steering of the ground wire. The robot can run steadily along the rail, smoothly across the dampers, straight towers and tension towers, etc., completing the detection image acquisition of the line and its hardware, achieving the expected effect.

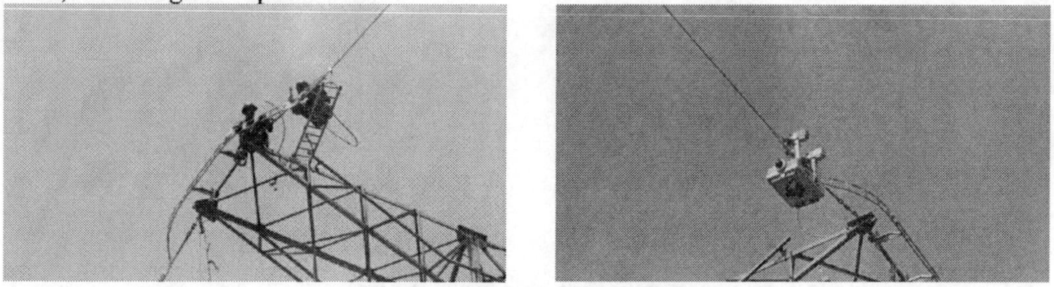

Figure 9. Robot online testing

7. Conclusion and Prospect

According to the general idea of optimizing and improving the overhead ground line, an optimizing and improving scheme is put forward. The blocking obstacle of the line is transformed into a non-blocking obstacle suitable for robot to cross. The structure form and the principle of the mobile platform of overhead ground transmission line inspection robot are put forward pertinently, which realizes the appropriateness. Realizing the function of profiling motion of the rail with variable curvature reduces the complexity of the mechanical structure as much as possible, reduces the obstacle-crossing action and improves the traveling efficiency of the robot.

Acknowledgments

The research and development work of this project has been strongly supported by the National Key R&D Program of China, 2018YFB1307400.

References

[1] Pouliot, N., Montambault, S.. (2009) LineScout Technology: From Inspection to Robotic Maintenance on Live Transmission Power Lines. In: IEEE International Conference on Robotics and Automation. IEEE. pp. 1034-1040.

[2] Montambault, S., Pouliot, N.. (2006) LineScout Technology: Development of an inspection robot capable of clearing obstacles while operation on a live line, In: Proc. IEEE 11th International Conference on Transmission and Distribution Construction, Operation and Live-Line Maintenance. pp. 15-19.

[3] Debenest, P., Guarnieri, M., Takita K.. (2008) Expliner-Robot for Inspection of Transmission Lines. In: IEEE International Conference on Robotics and Automation. IEEE. pp : 469-472.

[4] Gongping, W., Jinchun D., et al. (1999) A high voltage wire trolley with automatic barrier crossing function. Hydraulic and Electric Machinery, 1: 46-49.

[5] Gongping, W., Xiaohui, X., Yinglong, G., et al. (2006) Development of automatic crawler robot for overhead high voltage transmission line. China Mechanical Engineering, 3 : 237-240.

[6] Zhou F, Y., Wang J, D., Li Y, B., et al. (2005) Control of an Inspection Robot for 110kV Power Transmission Lines Based on Expert System Design Methods. In: Control Applications. IEEE. pp : 1563-1568.

Research Status and Development Trend of 3D Printing Technology

Yiran Yuan[1*]

[1] International department of shude high school, Chengdu 610066, China

*Corresponding author's e-mail: 13408093577m@sina.cn

Abstract. As the product of Industry 4.0, Intelligent Manufacturing (IM for short), and the enabling technology, i.e. 3D printing technology, has made rapid development in recent years and widely applied in many fields of social life. Moreover, 3D printing technology is a kind of high-tech manufacturing, which is playing a revolutionary role in the industrial development. Therefore, in order to know the study status and development trend of 3D printing technology, this paper systematically analyzes some typical methods and applications in some fields, as well as predicts its development trend in the future based on the brief introduction to the working principle of 3D printing technology.

1. Introduction

The 3D printing technology (also called additive manufacturing (AM for short) or rapid prototyping technology) is the emerging manufacturing technology, which is well promised by the worldwide. The 3D printing technology is based on the digital model file, bonding some adhesive materials together in powder form such as metals or plastics. And then through the layer-by-layer printing method, it is possible for two-dimensional data to transform into 3D materials. The 3D printing technology shows the characteristics of zero-release and zero-waste, because of its various construction forms. Additionally, the 3D printing technology leads to the redesign of some equipment and products, and because of the concept of manufacturing, it gets rid of the constraints of the traditional processing method as well as brings a large space for innovative design [1]. At the same time, 3D printing creates a new production mode, referring to changing the batch manufacturing into personalized customization without much increase of cost and cycle. The technology of 3D printing is entering various fields and plays a key role [2].

2. Overview of 3D printing technology

As mentioned above, the technology of 3D printing is called AM by CAD designing data and adopting the method of material increase step by step to make parts of entities. Compared with the traditional technology of material removal (like machining), it is a manufacturing method of adding materials from bottom to top.

In the late 1980s, with the development, AM was also called material increase manufacturing, rapid prototyping, layered manufacturing, solid free-form fabrication and 3D printing. Different names respectively show the characteristics of the manufacturing technology.

Figure 1. A new type of 3D printer

International committee, i.e. ASTM F42, makes a clear definition of AM and 3D printing. AM is the process of connecting materials to make objects according to 3D CAD data. Compared with the subtraction manufacturing, its material is increased in the process of manufacturing. 3D printing is defined as the technology adopting printing head, nozzle or other deposition materials to produce some products, which is commonly used to the AM technology. When it is applied to the particular device, the 3D printer is the additive manufacturing machine with relatively low price or low overall function.

Without using traditional cutters, fixtures and several processing procedures, the 3D printer can adopt 3D design data to make parts into any complex shapes rapidly, which can effectively solve the problem of making parts in the complex structure. With the rapid development of the technology, the 3D printing technology has been widely applied in many fields, such as consumer electronic, automobile, aerospace, medical treatment, military industry, geographic information and artistic design, etc.

3. Research Status of 3D printing technology

The 3D printing technology is firstly appeared in America in the late 19th century, and developed in the 1980s. With the development of intelligent manufacturing, the 3D printing technology has been widely served in the manufacturing of the intelligent vehicle and other areas. Recently, the 3D printing technology has been widely used by the overseas, which mainly contains aerospace, automobile manufacturing, medical treatment, industrial product design, architectural design, entertainment products, and biotechnology. Besides, it also accounts for a large proportion in intelligent manufacturing of various products [4].

Since the first 3D printer was invented by Charles Hull in 1986. After that, a great achievement has been made on the 3D printing. In 2000, Zcorp, an American enterprise, cooperating with Rilecen Institute, developed a kind of color 3D printer with the principle of ink-jet printing technology. At the end of the year, an Israeli company, Object Geometries launched the 3D printer Quadra [5] based on the combination of stereolithography apparatus (SLA) and 3D Ink-Jet; In 2010, an American company (i.e. Organoxo) and an Australian company (i.e. Invetech) printed human tissues and organs by using living cells, which was the breakthrough of 3D printing technology in the field of medical.

At present, European countries and America has preliminarily set up a successful business mode of 3D printing technology in the consumer electronics (CE), automobile industry, aircraft industry, medical treatment and other areas. Statistically, among more than 700,000 3D products produced by America in 2012, products in electronic area, automobile manufacturing, medical treatment and industrial machinery industry occupy a large share [7]. Additionally, the 3D printing has been widely used in the customization of personal product gradually. For instance, a lot of foreign companies (like American Quirky) can collect design schemes towards customers by the online platform, and utilize the 3D printing technology to make materials and products for sales, which brings a good benefit.

Compared with the foreign study on the 3D printing technology, the domestic study on 3D printing technology starts later, but due to its huge potential of application in various areas, domestic scientific researchers and industrial experts in bulk pay more attention to it, which leads to the rapid development of 3D printing technology in China. A large amount of research is made, such as the largest 3D printer in the world, which is invented by Huazhong University of Science and Technology through more than a decade. The largest length and width of the machinable components are both up to 1.2 meters. Theoretically, as long as the length and width of the component is less than 1.2 meters, the objects can be printed by this machine. In addition, Xi'an Jiaotong University autonomously developed a nozzle of 3D printer and made great achievements on SLA system and molding materials, which helps its molding accuracy reach 0.2 mm. However, with the overall development, the domestic 3D printing technology still has some disadvantages compared with foreign status, mainly showing in the following aspects:

Firstly, in the aspect of basic study on material science, domestic 3D printing products mainly belong to plastic products and the core manufacturing technology of powdered metals and liquid material is not totally grasped yet.

Secondly, in the aspect of precision machinery and information technology, as for the core parts of 3D printers at the current stage (i.e. laser scanner), because of the limitation of domestic industrial design ability and low automatic controlling system level, there is a large gap of the printer developed independently in terms of process stability and printing accuracy between China and foreign countries, so we need to make further efforts on it [8].

Thirdly, in the aspect of technological development, the level of domestic additive manufacturing technology is equal to that of foreign advanced technology, but in terms of key parts, molding materials, intelligent controlling and application range, domestic level fells behind. The additive manufacturing is mainly applied in modelling, and there is still a large space for improvement in the field of directly making high-performance terminal components.

Fourthly, in the aspect of study on process technology, foreign countries focus more on process controlling, also the theoretical basis of the technology while China depends more on experience and repeated testing to verify the result, which results in Chinese lower level on the key additive manufacturing technology.

4. Typical Application of 3D Printing Technology

4.1 Automobile Industrial Field

With the rapid development of automobile industry, people put forward a higher standard of lightweight design of automobiles, short cycle of design and less manufacturing cost, and the appearance of 3D printing technology makes it possible to meet the demand. On 1st May in 2013, the first automobile made by 3D printing, Urbee 2 was launched, which is opening the door of the application of 3D printing technology in the field of automobile manufacturing.

During the production process of the automobile, thermoplastic high-polymer materials are largely used to produce decorative parts and some structural parts. Compared with the traditional method, the 3D printing technology can shorten the manufacturing time and well perform in making parts with complex structure. At the same time, with the technology of one-step molding, 3D printing can omit a lot of traditional connecting parts and obviously lighten the weight of vehicles for it makes good use of engineering thermoplastic with low density.

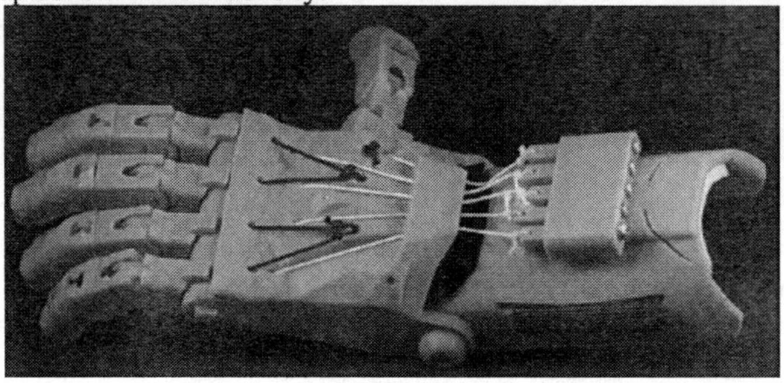

Figure 2. 3D Printing Manipulator

4.2 Aerospace filed

With the development of 3D printing technology, many countries invest plenty of human and material resources in 3D printing technology. At present, the study in the field of aerospace includes two aspects. The first one is the rapid manufacturing of part model and part verification. Aiming at non-metallic materials, using Fused Deposition Modeling, Stereoscopic Lithography, Laminated Object Manufacturing and 3D Printing method. The second one is aiming at the manufacturing of parts of metallic materials and overall structure in a large size, using some techniques including Metal Sintering, Selective Laser Sintering, Laser Rapid Prototyping and Electron-Beam Melting [9].

Different molding techniques have different characteristics [10]. For example, Stereoscopic Lithography shows high accuracy, intensity and hardness, which can also realize color molding, but after molding, its surface is tough. Fused Deposition Modeling has fast molding speed with lower cost, but its accuracy is not high. Selective Laser Sintering is simple with high flexibility and accuracy, which can use various materials (including metal, ceramics and plastics). Laminated Object Manufacturing performs well in model support, low cost and high efficiency. However, the techniques of 3D printing based on adding and accumulation characterize in rapidity, low cost, wide application of materials, high flexibility and high integrated, which shows great advantages compared with traditional techniques [13].

Figure 3. 3D Printing of Aircraft Engine Components

4.3 Civil Engineering Field
During the construction of civil engineering, the applying of 3D printing technology can largely reduce the cost of engineering construction and it is superior in low carbon and environmental protection, the advantages that the traditional construction does not have. At present, after a period of time, 3D printing technology has been maturing. Additionally, 3D printing can help solve the problem of printing the modeling of curved buildings which can not be solved by other production modes, and it also can eventually print the buildings in light weight and high intensity. According to the development situation of 3D printing technology, it cannot print buildings in large volume, but it can adopt the printing method of components and then assemble them together [14]. If we need to print a building in dozens of layers, a large 3D printer is required and its problems of intensity should be solved at the same time [15].

Figure 4. 3D Printing Architectural Model

5. The development trend of 3D printing technology
The 3D printing technology develops rapidly abroad. Although it starts later, it draws wide attention by all sectors of society with the concept of 3D printing appearing.

In the future, the 3D printing technology will be developed with the following aspects [16]:

5.1 Normalization
Normalization is the basic guarantee to select3D printing materials. Due to the expensive materials for 3D printing, mixed materials are used largely. If normalized materials with low waste and low emission can be used for 3D printing, the variety of the available 3D printing materials will be extended [17].

5.2 Scientific
Scientific is also an important part in the technological process. By internal process monitoring and giving feedback regularly, connection, repeatability and unity between machines can be improved. The good process controlling can reduce the down time of devices [18].

5.3 Convenience
Convenience makes the operation of 3D printing simpler. At present, CAD tools should be used in 3D printing. If the operation of 3D printing is simpler and more convenient, the technology will be popularized [19].

5.4 Intelligence
Intelligence simplifies the requirements on software functions and processing, such as the material conversion during the forming process and the removal of waste powder after processing. Simplified requirements of 3D printing technology can better popularize it.

Moreover, some 3D printing materials are harmful and environmentally polluted, which are bad for human health. 3D printing will also give rise to information or technology disclosure, so some printing products will threaten national security (like handgun). In the future, 3D printing will develop for environmental security, intellectual property protection and technological protection [20].

6. Conclusion
The technology of 3D printing plays an important role in the development of the industrial. Due to high efficiency and low cost, the 3D printing technology brings a great reform on traditional manufacturing, which can be regarded as a driving force of economic and social development in China. However, there are also some problems of 3D printing to be coped up, such as without guarantee of production quality, high requirement on production technology, hard personnel training and pollution of waste powder, etc. Although the technical bottleneck still exists currently, there is a bright future for the technology, so the whole world attaches great attention to it. In short, the 3D printing technology will create a new era of the national manufacture.

References
[1] Shao Zhuqiang, Hu Zengrong, Guo Shaoxiong et al. Analysis on advantages and Problems of Metallic 3D Printing Technology [J]. China Southern Agriculture Machinery, 2018, 49(7): 13-13.
[2] Fu Xin. Domestic Development Status and Study Trend of Additive Manufacturing (3D Printing) [J]. China High-Tech Enterprises, 2016(24):27-28.
[3] Tan Lizhong, Fang Fang. Application of 3D Printing Technology in the Aerospace Field [J]. Tactical Missile Technology, 2016(4):1-7.
[4] Li Ka. Analysis on the Development of 3D Printing in the Future Industry [J]. Science & Wealth, 2014(12):4-4.
[5] Guo Libo. Analysis on Principle and Advantages of 3D Printing Technology [J]. Industrial & Science Tribune, 2018, 17(22):61-62.
[6] Wang Wei. Study on Development Status of 3D Printing Materials [J]. New Material Industry, 2019, 303(02):15-19.
[7] Wang Xiaoyan. Six Entry Points of 3D Printing Technology in the Field of Aerospace [J]. World Manufacturing Engineering & Market, 2018(1).
[8] Ben Yue, Zhang Le, Wei Shuai et, al. Study Progress of 3D Printing Ceramic Material [J]. Materials Reports, 2016(21):112-121.
[9] Zhang Hairong, Yu Yong. Application of 3D Printing Technology in the Field of Medical Treatment [J]. Chinese Medical Equipment Journal, 2015, 17(3):174-175.
[10] Han Shoubo, Zhang Yiwen, Tian Xiangjun, et, al. Study and Application of Printing Metallic Powder by High-Quality 3D Printing Technology in the Field of Aerospace[J]. Powder Metallurgy Industry, 2017(27):51.
[11] Wang Fan. Study Status and Key Techniques of 3D Printing Technology [J]. Modern Technical Research, 2017(8):137-138.
[12] Tang Yanyu. Analysis on Application of 3D Printing in the Field of Aerospace [J]. Zhongguo Shebei Gongcheng, 2016(16):123-123.
[13] Wu Xuewei. Development Advantages of 3D Printing at the Present Stage [J]. Wireless Internet Technology, 2014(7):191-191.
[14] Wang Weichen. Study on Application and Outlook of 3D Printing Technology in the Field of Civil Engineering [J]. Value Engineering, 2017, 36(5):85-86.

[15] Chu Min, Jiang Jing. Development Trend and Suggestions of 3D Printing Architecture Technology [J]. Management and Research on Scientific & Technological Achievements, 2017(9).
[16] Zhang Jiao. Study on Influence and Prospect of 3D Printing Technology on Industrial Development [J]. Guide to Business, 2013(17).
[17] Li Xuan, Mo Hong, Li Shuangshuang et, al. Study Progress of Process Controlling of 3D Printing [J]. Acta Automatica Sinica, 2016, 42(7).
[18] Lam C X F, Mo X M, Teoh S H, et al. Scaffold development using 3D printing with a starch-based polymer [J]. Materials Science & Engineering C, 2002, 20(1):49-56.
[19] Wang Hao. Study on the Development Trend of 3D Printing Technology [J]. China Science and Technology Exp, 2015(35):387-387.
[20] Du Xiang. Application and Development Trend of 3D Printing Technology [J]. Powder Industry, 2017:14.

IOP Conf. Series: Materials Science and Engineering **711** (2020) 012015 doi:10.1088/1757-899X/711/1/012015

Cause Analysis and Suggestion on Corrosion Leakage of Pipe in Atmospheric and Vacuum Pressure Unit

Liang DU[1,*], Qingjin WANG[2] , Xuan LI[1], He YAN[1]

[1] China Special Equipment Inspection and Research Institute, Beijing, 100029, China

[2] PetroChina Jiangsu Fuel Asphalt Co., Ltd, Jiangyin, Jiangsu, 214400, China

[*]Corresponding author's e-mail: duliang.sme@163.com

Abstract. An 800,000 tons/year atmospheric and vacuum distillation unit suffered from corrosion and leakage of the atmospheric top volatile pipeline, which resulted in unplanned shutdown, seriously affecting the safe and stable operation of the unit. In this paper, the corrosion mechanism, raw materials, anti-corrosion measures and operation management are analyzed. The causes of corrosion leakage accidents are given and the relevant suggestions for corrosion control and protection are put forward.

1. Introduction

Atmospheric and vacuum distillation unit, as the leader of petrochemical industry, takes advantage of the different physical characteristics of each component of crude oil, carries out pre-treatment, component separation, extraction and other technological processes, and finally obtains representative units of petroleum products such as gasoline, kerosene, diesel oil, lubricating oil and raw materials of downstream units. As an important process of oil refining, the safety status of the device plays a vital role in the long-term safe operation of the refinery. Because crude oil contains a lot of corrosive substances such as water, salt, sulphide and acid, corrosion is the most common problem in atmospheric and vacuum distillation units, especially in low temperature parts, which has always been the focus of equipment managers. Once corrosion leakage occurs, equipment will pose a major threat to the safety of personnel and equipment [1-3].Based on a case of leakage of normal pipe in atmospheric and vacuum distillation unit, the direct causes of corrosion leakage at the top of the tower are proposed from the aspects of medium composition, corrosion mechanism, process parameters and process flow, the subsequent measures and suggestions for corrosion prevention and control are given for specific reasons.

2. Unit Overview

A 800,000-ton/year atmospheric and vacuum distillation unit has been in operation since 1997. It is mainly used to process Venezuelan MEREY16, BOSCAN heavy oil and a small amount of Liaohe crude oil. The acid value and sulfur content of heavy oil in MEREY16 and BOSCAN are shown in Table 1. The main equipment of the device includes primary distillation column, distillation column, atmospheric column, heating furnace and related process pipelines. The technical parameters of atmospheric column with serious corrosion and volatile pipeline at the top of corrosion leak tower are shown in Table 2.

IOP Conf. Series: Materials Science and Engineering **711** (2020) 012015 doi:10.1088/1757-899X/711/1/012015

Table 1. Ingredient content of raw oil

Name of raw oil	Acid value of heavy oil (mg KOH/g)	Sulphur content(%)
MEREY16	1.24	2.60
BOSCAN	1.39	4.9

Table 2. Main technical parameters of corrosion leakage pipeline and equipment

Name of device	Atmospheric tower	Top volatile line
Date of service	November 1997	January 2010
Design pressure (MPa)	0.4	0.24
Design temperature(°C)	250	131
Operating pressure (MPa)	0.3	0.06
Operating temperature(°C)	250	111
Main body material	Head: 16MnR+316L Cylinder: 16MnR+316L	20#
Equipment specification	Diameter:1800 Head: （12mm+3mm） Cylinder: （10mm+3mm）	Φ273×9
Operating medium	Crude oil	Naphtha

In April 2018, the atmospheric and vacuum distillation unit was repaired and started. In October 2018, inspectors found that the leakage occurred at the three-way position 400 mm downward from the injection point of the atmospheric roof volatilization line. The leakage location is shown in Figure 1. The preliminary analysis shows that the reason is the leakage of the pipe wall caused by the corrosion thinning, and it continues to operate after the temporary pipe card treatment[4]. Fifteen days later, the leakage occurred again and the atmospheric and vacuum distillation unit temporarily stopped. Macroscopic inspection showed that the top of the tower was corroded and perforated seriously, as shown in Figure 2. Through manhole inspection, it is found that the inner parts of atmospheric tower are seriously corroded, the valve is corroded and cracked, and the inner wall of the top head of tower is densely covered with surface cracks, as shown in Figure 3 and 4.

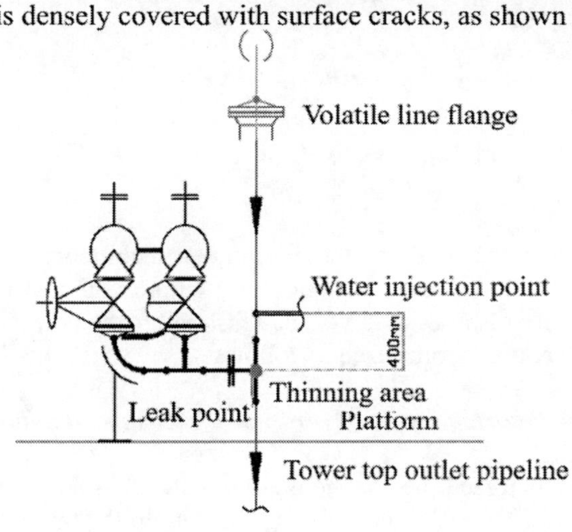

Figure 1. Diagram of corrosion perforation site

Figure 2. Evaporative line corrosion perforation

Figure 3. Corrosion crack on tower wall

Figure 4. Corrosion of floating valve

3. Equipment Damage Pattern Recognition

Based on the comprehensive analysis of the medium characteristics, process parameters and equipment materials, it can be seen that internal corrosion thinning (including uniform corrosion thinning and local corrosion thinning), wet hydrogen sulfide damage and external corrosion are the main failure mechanisms of the atmospheric and vacuum distillation unit[5]. Figure 5 shows the distribution chart of main failure mechanism of atmospheric pressure unit using risk-based inspection technology. From the distribution chart, it can be seen that the proportion of corrosion thinning of equipment is higher, followed by stress corrosion cracking. For the atmospheric tower and pipe jacking in this analysis, internal corrosion is dominant.

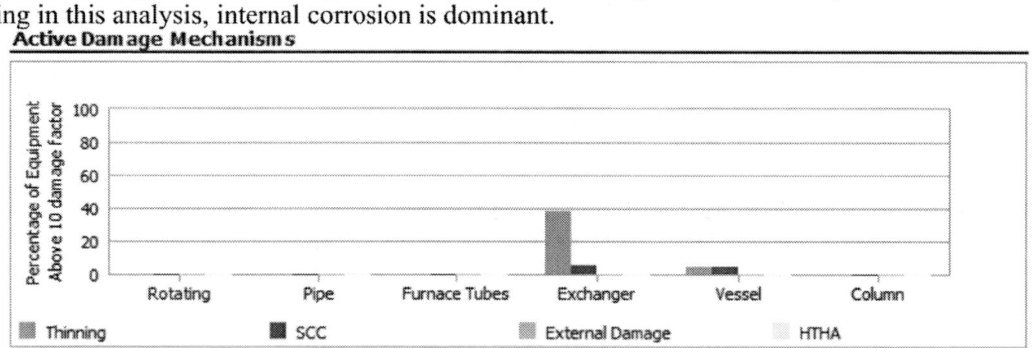

Figure 5. Distribution map of main failure mechanisms.

4. Cause analysis of corrosion leakage

After the leakage accident, based on the potential damage mode of the equipment, technicians start with the possible causes of the corrosion intensification, such as the change of medium composition, the fluctuation of process parameters and the rationality of process flow. The reasons for the corrosion leakage of the pipeline on the top of atmospheric tower are summarized as follows.

4.1. The salt content of crude oil is abnormal and the chloride ion concentration exceeds the standard seriously.

After the corrosion leakage accident, the technicians analyzed the Liaohe oil and water samples of the raw material tank, and the analysis results are shown in Table 3. According to the data in the table, the inorganic salt content in crude oil increased from 10 mg/L to 100 mg/L, and the chloride ion concentration was abnormal. Because the current electric desalination efficiency can not adapt to the abnormal increase of crude oil salt content, the crude oil salt content entering the downstream process exceeds the standard, which causes the rapid decrease of the PH value of condensate at the top of the tower and greatly aggravates the hydrochloric acid corrosion of the pipeline.

Table 3. Analysis results of Liaohe oil and water samples in raw material tank

Sample source	Sample	pH	Chloride ion concentration (mg/L)	Inorganic salt content (mg/L)	Organochlorine content (mg/L)
Oil tank of Liaohe	Crude Oil	2～3	7000	100	5

4.2. The existing electric desalting unit can not remove the organic chlorine in crude oil, which makes the total chlorine content of crude oil exceed the standard.

From the data in Table 1, it can be seen that crude oil from Liaohe Oilfield contains 5 mg/L organic chlorine, which is higher than the process control value of 1-3 mg/L. The main sources of organic chlorine in crude oil are natural existence, oil recovery and transportation aids. Demulsifiers and desalting agents are also the main sources of organochlorine in the process of electric desalting. The electro-desalination process adopted in this unit is only effective for inorganic chlorine in crude oil, and the removal rate of organic chlorine is relatively low or even ineffective. A large amount of organochlorine is hydrolyzed into HCl in the process, which leads to the increase of HCl concentration, thus aggravating the corrosion of hydrochloric acid and acidic water on top equipment and pipelines. After the pipeline leakage accident, the technicians sampled and analyzed the naphtha distillation outlet tank. The composition and content of organic chloride chloromethane and 3-chloropropylene in naphtha samples were analyzed by GC-MS. The test results were shown in Figures 6 and 7.

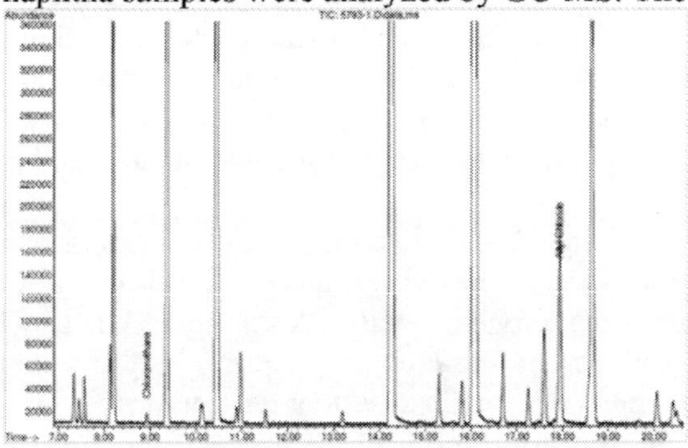

Figure 6. Gas chromatographic analysis of naphtha samples

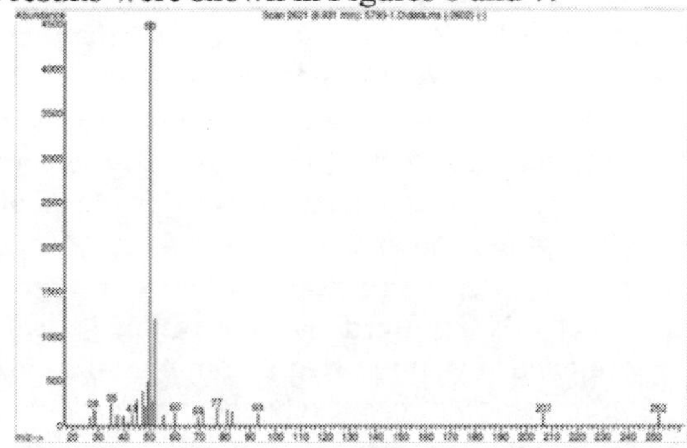

Figure 7. Mass spectrometric analysis of naphtha samples

After further analysis of the data in the diagram, the component data in naphtha samples are given in Table 4. The total chlorine content in naphtha is 1180 mg/kg, of which chloromethane content is 36 mg/kg and 3-chloropropylene content is 491 mg/kg. Higher organic chlorine content not only seriously affects the corrosion of atmospheric tower and pipe jacking line, but also harms the follow-up units such as naphtha hydrogenation and heavy oil processing [6,7].

Table 4. Analysis of water sample in raw material tank

Sample source	Sample	Total chlorine content (mg/kg)	Chloromethane content (mg/kg)	3-chloropropylene content (mg/kg)	Sample source
Naphtha distillation tank	Naphtha	1180	36	491	Naphtha distillation tank

4.3. Dew point corrosion environment formed, which aggravated the degree of local corrosion.

In the process of production and operation, in order to control the final distillation point (> 175 °C) and density of naphtha (between 650 kg/m^3 and 750 kg/m^3), the top temperature of naphtha is usually controlled at about 100 °C, which is very easy to form dew point corrosion environment of hydrochloric acid. The local corrosion environment is more destructive than uniform corrosion, which accelerates the occurrence of corrosion leakage. Macroscopic inspection showed that the serious corrosion occurred mainly in the lower part of the top reflux, the nozzle of the top volatile line safety valve and the tray float valve, and the above positions were in the dew point.

4.4. The design of process flow is unreasonable.

The typical corrosion environment of hydrochloric acid liquid phase zone is formed in the three-way area after the injection point shown in Figure. 1. Irrational setting of water injection point and mode, no sprinkler at water injection point, incomplete atomization of water injection, dissolving HCL directly in liquid form, forming a hydrochloric acid corrosion environment. The design distance between the injection point and the safety valve nozzle is too small (400mm). It is easier to accumulate fluid in the three-way section after the injection point, resulting in corrosion leakage in this section.

In summary, the abnormal medium composition and unreasonable process design caused the serious corrosion of atmospheric tower and the leakage of pipeline at the top of the tower.

5. Corrosion Countermeasures

Internal corrosion of atmospheric and vacuum distillation unit, especially the corrosion at low temperature on the top of tower, is an important problem that perplexes the long-term operation of the unit. It can not be completely avoided by relying solely on material upgrading. In order to prevent corrosion leakage accidents to the greatest extent, it is necessary to coordinate management from process flow, equipment management, corrosion monitoring and other aspects [8-10].

- Strengthen the inspection and evaluation of crude oil quality. Regular inspection and evaluation of crude oil composition, timely understanding of changes in raw material properties, focusing on prevention and control of the impact of organic chlorine concentration changes on the corrosion damage of the unit.
- Strict technological discipline. Strictly control according to the process parameters of the device, strictly observe the device's defensive value and operation boundary, avoid condensation of atmospheric roof oil and gas in non-corrosion-resistant parts to form a corrosive environment.
- Optimize the process flow, rationally select injection points, add special injection nozzles, and pay attention to the blockage of nozzles in actual operation.
- Strengthen the management of electric desalination of crude oil and strictly control the salt content after desalination. Because the raw material is super heavy oil, inorganic salt can not be effectively removed by conventional electric desalting process. Better desalting effect can be achieved by three-stage combination process of vertical-horizontal electric desalting tank, optimization of heat exchange process and improvement of electric desalting temperature.
- Strengthen the management of top injection agent and do a good job of process anticorrosion. Optimizing the top injection agent and strengthening the injection management to ensure that the acid-base value and iron ion concentration of the top condensate are within the reasonable index range.
- Strengthen the quality control of naphtha products. Strengthening the monitoring of side-line products can detect corrosion leakage accidents in time, so as to make timely decisions to avoid unplanned shutdown and other problems, and at the same time to avoid the impact of corrosion on the follow-up units in the combined plant.
- Material upgrading. Because the raw material of the unit is high sulphur and high acid crude oil, the low temperature corrosion on the top of the tower is particularly serious. The practical engineering case shows that the corrosion problem at the top of the tower can be largely solved by replacing the water cooler material with Ti plate wet air cooling and controlling the condensation area in the air-cooled place.
- Strengthen corrosion monitoring and implement fixed-point thickness measurement of pipelines. Fixed-point thickness measurement of pipeline before and after injection point is carried out, and comprehensive inspection is carried out during each overhaul period to ensure timely detection of corrosion phenomena and avoid the occurrence of corrosion accidents.

6. Conclusion

Aiming at a pipeline corrosion leakage accident, the possible failure mechanism of pipeline is given based on the analysis such as medium composition, process temperature, pH value and equipment

IOP Conf. Series: Materials Science and Engineering **711** (2020) 012015 doi:10.1088/1757-899X/711/1/012015

material, The effect of failure mechanism on pipeline leakage was verified by means of analysis. The causes of corrosion perforation and leakage of the atmospheric top volatile pipeline are given. Finally, reasonable suggestions for subsequent corrosion protection and control are put forward.

Acknowledgments

This research is supported by Science and Technology Planning Projects(2017QK177) found by State Administration for Market Regulation, also supported by National Key R&D Program of China under grand of 2018YFF0214705.

References

[1] Liu Xianglan, Wang Ying, Wang Shiying, Shenxin, He Bing. (2009)Corrosion analysis and protection measures of atmospheric and vacuum distillation unit. J. Corrosion and protection, 30 ,02: 142-144.
[2] Zhou Yang, Lu Yunrong, Cheng Sixiang, Wu Xiang. (2014)Corrosion analysis and control suggestions for atmospheric tower top of a large domestic refinery. J. Corrosion and protection, 35,07: 734-736.
[3] Pei Lijun, Li Maodong, Wu Zhihua. (2003)Corrosion and protection of atmospheric and vacuum distillation units for processing low pour point naphthenic crude oil. J. Corrosion and protection, 08: 358-360.
[4] SY-T 6553-2003. Pipeline Inspection Code for Inspection, Repair, Renovation and Re-classification of Pipeline Systems in Use [S].
[5] GB/T 30579-2014. Damage pattern recognition of pressure equipment [S].
[6] Zhang Zhanjun, Wu Shikui, Yu Hongying, Mo Guidi. (2013)Sources, hazards and protective measures of chlorine in crude oil. J. Journal of Guangdong Institute of Petroleum and Chemical Industry, 23,04: 1-4.
[7] Lu Qingxin, Cui Shiwei. (2018)Analysis results of organic chlorine content in Shengli high sulfur and high acid crude oil. J. Qilu Petrochemical Industry, 46, 03: 233-236.
[8] Jiang Hongwei, Yuan Yaoru. Corrosion analysis and protection of atmospheric and vacuum distillation unit [J]. Corrosion and protection of petrochemical industry, 2018, 35 (05): 21-25.
[9] Fan Xiuju, Zhu Jianhua. (2009)Source distribution and removal technology of chloride in crude oil. J. Refining and Chemical Industry, 20, 01: 8-11.
[10] Youxinhua, Wang Yicheng, Li Xiaodong, Yang Liyan. (2015)Effects of crude oil deterioration on atmospheric and vacuum distillation units and Countermeasures. J. Corrosion and protection, 36 ,08: 788-791.

IOP Conf. Series: Materials Science and Engineering **711** (2020) 012016 doi:10.1088/1757-899X/711/1/012016

Analysis of Centroid Variation of Three-floating Gyroscope Based on Error Sensitivity

Zhang Jialun[1], Li Yi[1] and Zhang Faping[1*]

[1] School of Mechanical Engineering, Beijing Institute of Technology, Beijing, 100089, China

[*]Corresponding author's e-mail: zfpnew@163.com

Abstract. The surface error of the three-floating gyroscope caused the centroid variation of the three-floating gyroscope, which caused the navigation accuracy to decrease. In this paper, the error sensitivity model of three-floating gyroscope was established by using the geometric error transfer theory, the influence of each surface error to the three-floating gyroscope's centroid was calculated, and the results were verified using modeling software. Finally, this paper evaluated the assembly capabilities of the assembly surface. This paper has guiding significance for the three-floating gyroscope to reduce production cost under the premise of ensuring navigation accuracy.

1. Introduction

The three-floating gyroscope has been widely used in industrial control, aerospace, automotive and other fields. With the improvement of navigation accuracy requirements in these industries, the accuracy performance of the three-floating gyroscope is particularly important [1]. Accuracy design is an important part of ensuring the accuracy of the gyroscope. In the traditional gyroscope design, the tolerance level of each part of the gyroscope is mainly determined by the empirical method, but due to the size of the part and its position in the assembly, each mating surface has a different degree of influence on the accuracy of the assembly. If the influence of the surface error on the assembly accuracy cannot be quantified, the precision of the machining will not be reasonably determined, resulting in excessive production cost or low system assembly accuracy. Therefore, it is particularly important to quantitatively calculate the influence of these surface errors on the assembly accuracy of the system.

The three-floating gyroscope is used as an inertial device, and the assembly accuracy is reflected in the variation of the assembly's centroid. By establishing the accuracy model of the three-floating gyroscope, the relationship between the error of mating surface and the error of the assembly's centroid can be reflected. In the analysis of mating surface error sensitivity based on error transfer, many scholars have done a lot of research. He Boxia et al. proposed the concept of part basis deviation and matching reference deviation [2]; Lu Cheng et al. analyzed the problem of assembly joint surface error modeling and tolerance optimization design between parts under various tolerance coupling conditions [3]; Wang Ge et al. analyzed the geometric error transfer model problem [4]. Based on the above research, the paper accumulates the process of geometric error transfer caused by the mating surface error of the three-floating gyroscope during the assembly process, and calculates the sensitivity of the mating surface error to the centroid of the assembly.

IOP Conf. Series: Materials Science and Engineering **711** (2020) 012016 doi:10.1088/1757-899X/711/1/012016

2. Structure description and establishment of coordinate system

2.1. Structure Description

There are many kinds of components in the three-floating gyroscope, which are mainly composed of four parts: the housing component, the float component, the torque device/sensor base component and the bellows component. The center-of-gravity offset of the three-floating gyroscope is mainly caused by the misalignment of the center of the float component and the center of the buoyancy support [5]. Since the float component has the greatest influence on the working accuracy of the three-floating gyroscope, this paper mainly studies the centroid variation of the float component. The float component consists of a frame, 2 magnetic suspension rotors (axial),2 magnetic suspension rotors (radial), a torque rotor, a sensor rotor, a motor assembly, and a pontoon, as shown in Figure 1.

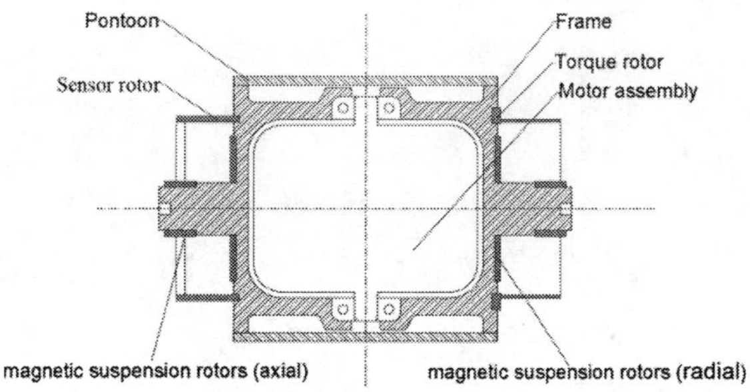

Fig 1. Schematic diagram of the three-floating gyroscope

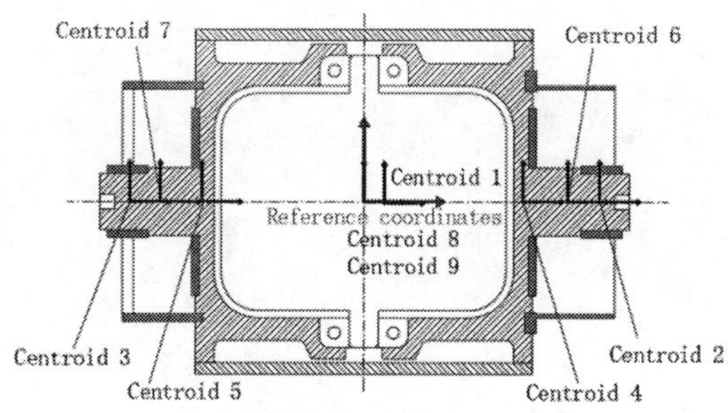

Fig 2. Three-floating gyroscope coordinate system diagram

2.2. Coordinate system establishment

Taking the ideal centroid point of the pontoon as the origin, the axial direction of the frame is the X axis, and the axial direction of the motor is the Y axis, the Z axis direction is determined according to the Cartesian coordinate system, and the reference coordinate system is established, as shown in Fig. 2. It can be seen from the assembly relationship of the float component that all the parts except the frame are directly matched with the frame, it belongs to a single process. After calculation, the ideal position of each part's centroid is shown in Table 1.

Table 1. Ideal relative coordinate values of each part of the float component.

NO.	Part Name	X/mm	Y/mm	Z/mm
1	Frame	-0.1076	0	-0.1382
2	X+ magnetic suspension rotors (radial)	2.5	0	0
3	X- magnetic suspension rotors (radial)	-2.5	0	0
4	X+ magnetic suspension rotors (axial)	0.5	0	0
5	X- magnetic suspension rotors (axial)	-0.5	0	0
6	Torque rotor	3.0403	0	0
7	Sensor rotor	-4.6263	0	0
8	Pontoon	0	0	0
9	Motor assembly	0	0	0

3. Modeling of the error sensitivity

3.1. Calculating of the centroid variation

The most important reason for the variation of the assembly's centroid is that the geometric error of the part mating surface causes the actual mating surface to be inconsistent with the ideal mating surface. If the mating surface is flat, the positional variation of the actual mating surface and the ideal mating surface can be expressed by the six degrees of freedom variation; If the mating surface is a cylindrical surface, the positional variation of the actual mating surface and the ideal mating surface can be expressed by the six degrees of freedom variation of the position and direction of the cylindrical surface axis. In summary, the change between the actual mating surface and the ideal mating surface can be expressed as [4]:

$$U = [\delta_x \quad \delta_y \quad \delta_z \quad d_x \quad d_y \quad d_z]^T \tag{1}$$

where $\delta_x, \delta_y, \delta_z$ is the three tiny rotational components of the pose variation; d_x, d_y, d_z is the three tiny translational components of the pose variation.

The homogeneous transformation matrix of the ideal mating surface coordinate system to the mating assembly surface coordinate system can be expressed as

$$T = \begin{bmatrix} 1 & -\delta_Z & \delta_y & d_x \\ \delta_Z & 1 & -\delta_x & d_x \\ -\delta_y & \delta_x & 1 & d_x \\ 0 & 0 & 0 & 1 \end{bmatrix} \tag{2}$$

Set the ideal centroid coordinate to X (x, y, z), the homogeneous transformation matrix is

$$X_1 = [x \quad y \quad z \quad 1]^T \tag{3}$$

The homogeneous transformation matrix of the actual centroid can be expressed as

$$X_2 = T \times X_1 = \begin{bmatrix} x - y\delta_z + z\delta_y + d_x \\ y - x\delta_z - z\delta_x + d_y \\ z - x\delta_y + y\delta_x + d_z \\ 1 \end{bmatrix} \tag{4}$$

The matrix of centroid variation is

$$\Delta = X_2 - X_1 = \begin{bmatrix} \Delta x \\ \Delta y \\ \Delta z \\ 0 \end{bmatrix} = \begin{bmatrix} -y\delta_z + z\delta_y + d_x \\ -x\delta_z - z\delta_y + d_y \\ -x\delta_y + y\delta_x + d_z \\ 0 \end{bmatrix} \tag{5}$$

Combined centroid drift formula

$$\begin{cases} \Delta_c = [\Delta x_c \quad \Delta y_c \quad \Delta z_c] \\ \Delta x_c = \frac{\Sigma m_i \Delta x_i}{\Sigma m_i}, \Delta y_c = \frac{\Sigma m_i \Delta y_i}{\Sigma m_i}, \Delta z_c = \frac{\Sigma m_i \Delta z_i}{\Sigma m_i} \end{cases} \tag{6}$$

the assembly centroid variation can be obtained.

3.2. Error sensitivity

The influence of each mating surface error on the assembly's centroid can be expressed by error sensitivity. Define the absolute value of the partial derivative of Δ_c for U as the error sensitivity.

$$S_k = \left|\frac{\partial \Delta w_c}{\partial U_k}\right| = \left[\left|\frac{\partial \Delta w_c}{\partial \delta_{xk}}\right| \quad \left|\frac{\partial \Delta w_c}{\partial \delta_{yk}}\right| \quad \left|\frac{\partial \Delta w_c}{\partial \delta_{zk}}\right| \quad \left|\frac{\partial \Delta w_c}{\partial d_{xk}}\right| \quad \left|\frac{\partial \Delta w_c}{\partial d_{yk}}\right| \quad \left|\frac{\partial \Delta w_c}{\partial d_{zk}}\right|\right] \tag{7}$$

where $\mathrm{w} = \mathrm{x}, \mathrm{y}, \mathrm{z}$; k is the number of the mounting surface.

3.3. Sensitivity uncertainty analysis

In order to verify the correctness of the error sensitivity, the uncertainty of the sensitivity is analyzed. Using the point cloud data measured by the coordinate measuring instrument to obtain the fitting

IOP Conf. Series: Materials Science and Engineering **711** (2020) 012016 doi:10.1088/1757-899X/711/1/012016

surface of each mating surface, and analyzed the centroid changes of the assembly using CREO software modeling. The specific steps are as follows:

- The part is ideally modeled and assembled to measure the centroid position $O_1(X_1, Y_1, Z_1)$ of the assembly.
- The part mating faces are modified to fit the fitting surface and assembled to measure the centroid position O_2 (X_2, Y_2, Z_2) of the assembly.
- Calculation error sensitivity uncertainty.

The error sensitivity is measured as

$$S_k' = \frac{\Delta P}{d} \text{ or } S_k' = \frac{\Delta P}{\alpha \times 0.017} \tag{8}$$

where d is the distance from the fitting surface to ideal surface; α is the angle from fitting surface to the ideal surface; ΔP is the distance the centroid moves.

$$\Delta P = \sqrt{(X_1 - X_2)^2 + (Y_1 - Y_2)^2 + (Z_1 - Z_2)^2} \tag{9}$$

Uncertainty can be expressed as

$$u_{rel} = \frac{|S_k - S_k'|}{S_k} \times 100\% \tag{10}$$

3.4. Surface assembly ability evaluation

To reflect the influence of surface error on assembly's centroid, the assembly capability index C_{AP} can be defined as

$$C_{AP} = \frac{\Delta P_{max}}{\Delta P} \tag{11}$$

where ΔP_{max} is the position change of the assembly's centroid caused by the tolerance of mating surface; ΔP is the distance from the actual assembly's centroid to the ideal assembly's centroid.

4. Example

4.1. Sensitivity calculation

The error of each assembly surface is shown in Figure 3:

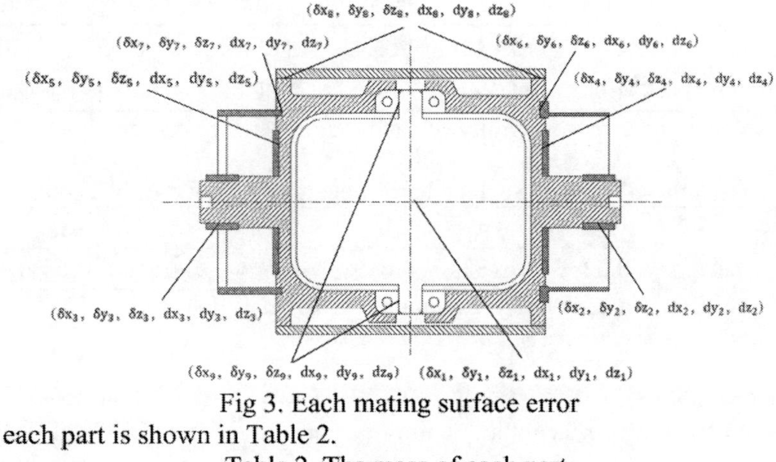

Fig 3. Each mating surface error

The mass of each part is shown in Table 2.

Table 2. The mass of each part.

Part number	1	2	3	4	5
Mass/g	68.4975	0.8239	0.8239	2.6496	2.6496
Part number	6	7	8	9	
Mass/g	4.9059	2.6403	48.0479	119.2581	

Through the above data, combined with the formula (5), (6), (7) and Table 1, the sensitivity of each error can be obtained. After the normalization process, the sensitivity analysis histogram can be drawn, as shown in Figure 4~6.

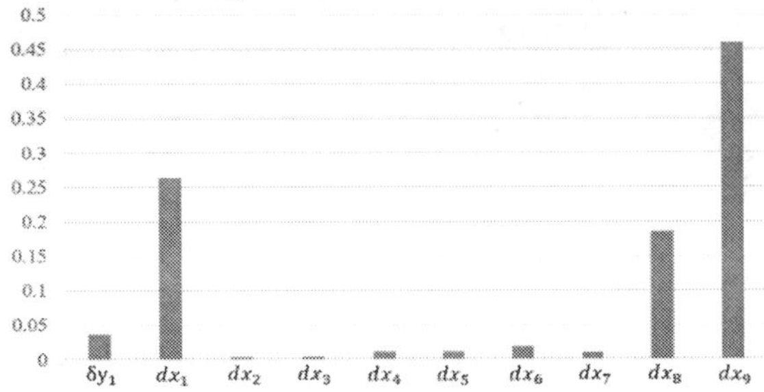

Fig 4. Sensitivity coefficients for each error associated with Δx_c

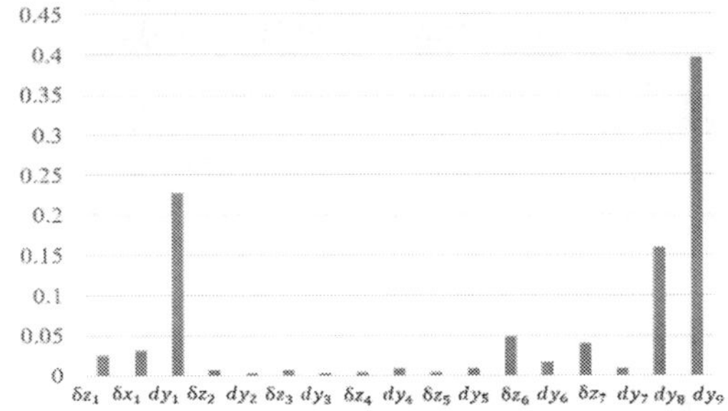

Fig 5. Sensitivity coefficients for each error associated with Δy_c

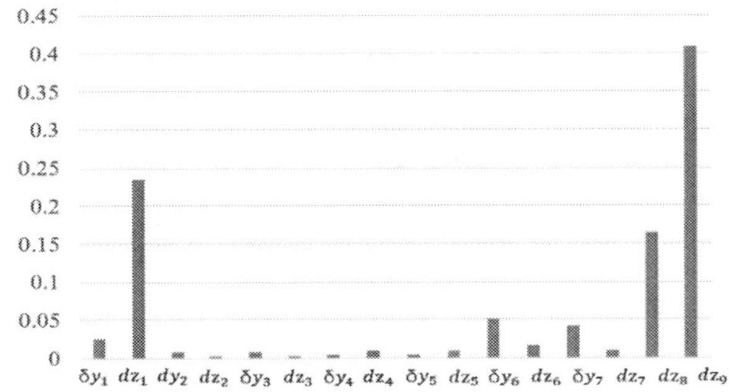

Fig 6. Sensitivity coefficients for each error associated with Δz_c

4.2. Calculation of sensitivity uncertainty

From the above results, it is understood that the error sensitivity of the mating surface 8 and 9 is large, so calculate the sensitivity uncertainty of the mating surface 8 and 9. The data measured by the coordinate measuring instrument can get the error of the matching surface 8 and 9. The centroid variation of the float component can be obtained by CREO software modeling, as shown in Table 3.

Table 3. Table of Variations of Mating Surface Errors.

	dx_8	dy_8	dz_8	dx_9	dy_9	dz_9
U/mm	0.21	0.13	0.16	0.09	0.05	0.03
ΔP /mm	0.0403	0.0250	0.0307	0.0428	0.0238	0.0143

From the above data, the error sensitivity of each mating surface can be obtained as shown in Table 4:

Table 4. Table of Surface Error Sensitivity Uncertainty Table.

	dx_8	dy_8	dz_8	dx_9	dy_9	dz_9
u_{rel}	0.02%	0.18%	0.04%	0.19%	0.05%	0.04%

The results show that the sensitivity calculation method is feasible and the calculation level is reliable (the uncertainty is less than 1%).

4.3. Evaluation of assembly capability

From the tolerance of mating surface, the maximum centroid variation of the float component can be obtained by CREO software modeling. Calculated assembly ability of each assembly surface as shown in Table 5.

Table 5. Table of assembly surface assembly capabilities.

	dx_8	dy_8	dz_8	dx_9	dy_9	dz_9
ΔP_{max} /mm	0.0640	0.0640	0.0640	0.0450	0.0450	0.0450
C_{AP}	1.588	2.56	2.08	1.05	1.89	3.15

The results show that the error of the mating surface of pontoon, motor assembly with frame is within the tolerance range, but the assembly capability of the mating surface of motor assembly with frame is poor, and it is necessary to focus on the translation error in the X direction.

4.4. Analysis of results

The centroid position of float component is influenced by the rotation error and the translational error, it is mainly influenced by the translational error. The translational error of the mating surface of pontoon, motor assembly with frame have a significant influence on the centroid position. Therefore, the values of these three errors should be controlled in the precision design.

5. Summary

This study proposed a method for calculating the influence of mating surface error on assembly's centroid variation based on the single-process error transfer principle. The main contribution of the paper is as follows:

- An error sensitivity analysis model based on the principle of single-process error transfer has been established.
- A method for verifying the correctness of the error sensitivity analysis model has been proposed.
- An evaluation method for the mating surface assembly capability has been proposed.

Acknowledgments

This research is supported by the National Natural Science Foundation of China (Grant No. 51805502), the Project of State department of China (Grant No. JSZL2016204B102) and Equipment Pre-Research Foundation of China (Grant No. JZX7Y20190242012201). The authors gratefully acknowledge the facilities provided by the Industrial Engineering Laboratory (IEL) in Beijing Institute of Technology.

References

[1] An, X., Wan, J. (2009) Research on Random Drift Model of Three-Floated Gyro. Piezoelectectrics & Acoustooptics, 31: 643-645.

[2] He, B., Zhang, Z., Dai, M., Shi, J. (2008) Theory of Modeling Variation Propagation of Mechanical Assembly Processes. Chinese journal of mechanical engineering, 44:62-68.

[3] Lv, C., Liu, Z., Ai, Y., Yu, Z. (2015) Assembly Joint Surface Error Modeling and Tolerance Optimization in the case of Coupled Tolerance. Journal of Mechanical Engineering, 51:108-118.

[4] Wang, G., Zhang, F., Yan, Y., Zhang, T., Guo, S. (2017) Precision Assembly Geometric Error Sensitivity Analysis Based on the Error Transformation Model for Precision Assembly. Transactions of Beijing Institute of Technology, 37: 682-686.

[5] Yu, J., Xu, J. (2014) The Technology for Measuring and Controlling the Disturbance Torque for Magnetic Bearing of Three-floated Gyroscope. Navigation and Control, 13:15-21.

IOP Conf. Series: Materials Science and Engineering **711** (2020) 012017 doi:10.1088/1757-899X/711/1/012017

Understanding of the Modeling Method in Additive Manufacturing

Zhuojun Chen [1*]

[1]School of Mechanical Engineering, Xi'an Jiaotong University, Xi'an, Shaanxi, 710049, China

[*]Corresponding author's e-mail: chenzhuojun@stu.xjtu.edu.cn

Abstract. With the development of additive manufacturing, how to improve the efficiency and accuracy of manufacturing and prediction has attracted more and more attention. This paper focuses on three basic modeling methods, which are summarized according to the issue: empirical method, analytical method and numerical method. These methods are used differently based on practical circumstances. Besides, due to the improvement of computer computing power, machine learning and digital twin have also been applied to the study of additive manufacturing. Machine learning has a good performance in the prediction and optimization of process parameters, but the characteristic of machine learning that requires a lot of data leads to the increase of experimental cost. Digital twin does a good job in monitoring the condition of equipment. In addition, it can replace expensive and time-consuming physical experiments with inexpensive and efficient digital experiments, which can provide data for analysis. However, because of insufficient research, its application is still limited.

1. Introduction

As a new manufacturing technology, additive manufacturing has developed rapidly in the past few years. How to improve the accuracy and manufacturing efficiency along with how to accurately predict the manufacturing process has become focused concerns. In this paper, the method of survey is used to introduce the modeling method in the process of additive manufacturing.

Additive Manufacturing is a new manufacturing technology, which is defined as "the process of joining materials to make parts from 3D model data, usually layer upon layer, as opposed to subtractive manufacturing and formative". Unlike traditional manufacturing technologies, such as subtractive manufacturing and equivalent manufacturing, additive manufacturing adds materials to create the final shape [1]. It can utilize raw materials effectively, produce the least waste and achieve satisfactory geometric accuracy. Additive manufacturing has revolutionized the way prototyping and small batch manufacturing is done [2]. Nowadays, additive manufacturing is not only used in aerospace, medical and automotive industries, but also in fashion, food industry, jewelry production and construction [3]. It has many advantages: (1) customized healthcare products to improve population health and quality of life; (2) reduced environmental impact for manufacturing sustainability; (3) simplified supply chain to increase efficiency and responsiveness in demand fulfillment [4].

This paper will introduce three kinds of modeling methods to solve the problems of additive manufacturing. The first one is the empirical method, which mainly analyses a series of experimental results, then establishes models, draws conclusions and verifies them through further experiments. The second one is the analytical method, which builds models based on physical laws and uses mathematical methods to analyze them. The third one is the numerical method, which is solved by numerical methods on the basis of physical laws in order to simplify and obtain useful results. Finally, some examples are given to illustrate the application of artificial neural network in additive manufacturing in recent years to promote and broaden the development of model methods.

2. Empirical Method

2.1. Introduction

Empirical method is a method which conducts an investigation relying on experiments and not theories. Empirical method focuses on the observations and measurements rather than understanding of the principles themselves. Empirical method consists of a series of experiments, and the results are summarized according to the experiment data. Then, further experiments are also required for verifying and improving previous findings.

The main advantage of empirical method, compared with other modeling methods, is that it requires the least effort to explore the physical laws in the process of additive manufacturing. Empirical method is a good method for some processes that are difficult to explore their specific physical laws, or those whose physical laws are too complex. However, its disadvantages are also obvious: In most cases, the results only apply to the specific case of a particular process. Besides, a large number of experiments are expensive and time consuming.

2.2. Examples

Empirical methods are usually used to find out the relationship between two or several parameters in order to optimize the process.

In vat photo polymerization, Wang et al. [5] use the least-squares method to find out the relationship between the post-cure shrinkage and process parameters. Lan et al. [6] do experimental research on dimensional accuracy of parts and the associated parameters. Karalekas et al. [7] study the shrinkage characteristics of stereolithography built square laminate plates using an acrylic-based photopolymer.

In powder bed fusion, Raghunath et al. [8] investigate the relationship between shrinkage and the various process parameters namely laser power, beam speed, hatch spacing, part bed temperature and scan length.

In material extrusion, Anitha et al. [9] explore the effect of process parameter like layer thickness on design quality. Sood et al. [10] focus on the effect of some important parameters such as layer thickness, part build orientation, raster angle, raster width and air gap on the compressive stress of test specimen.

In sheet lamination, Kechagias and John [11] use typical test part and carry out matrix experiments based on Taguchi design to find out the influence of different process parameters (layer thickness, heater temperature, platform retract, heater speed, laser speed, feeder speed and platform speed) on the roughness of vertical surfaces along Z-axis on ZX-plane of parts.

In previous studies, we can find that most of the researches focus on exploring the relationship between two or several different parameters in order to seek the optimization of process or part quality.

3. Analytical Method

3.1. Introduction

Analytical method is a method which gets exact function relation or forms a theorem from existed theories to establish mathematical model. Its principle is clear and every process has theoretical support. The analytical model is the output of a mathematical analysis of the process that considers the laws of physics and related physical processes.

The main advantage of analytical method is that the result obtained can be easily transferred to other related processes. Besides, it does not need to spend too much time and money on experiments. However, sometimes the physical laws of some processes are too complex or unclear, which will make it difficult to establish analytical models.

3.2. Examples

Analytical methods usually use formulas and models derived from theoretical calculations to control and predict manufacturing process.

Ramanath et al. [12] focus on the study of the melt flow behavior of Poly-ε-caprolactone (PCL) as a representative biomaterial. Mathematical model is established, and the melt flow behavior is studied by changing the inlet filament velocity and the outlet nozzle diameter and angle. Muller [13] introduces a process modeling and a system control to manufacture Functionally Graded Materials parts with a direct laser deposition system. This work enables to choose an adapted manufacturing strategy and control process parameters to obtain the required material distribution and the required geometry. Strano et al. [14] propose a mathematical model to accurately predict surface roughness, which takes into account the staircase effect and the existence of surface particles. This model is conducive to improving the surface quality of parts.

4. Numerical method

4.1. Introduction

The numerical method is a method based on physical laws, which uses numerical step-by-step method to obtain useful results. It processes data and solves problems by means of computer or computational model. The purpose of numerical analysis is to design and analyze some computational methods to obtain approximate but accurate results for the problem. It is often used in situations where the process is complex and difficult to get exact solutions.

The main advantage of numerical method is that it has a wide range of applications. There are many kinds of analytical software based on numerical methods, which can be used to solve problems faster and more accurately.

4.2. Examples

Numerical method is usually a method for solving problems after modeling. Tiebing Chen and Yuwen Zhang [15] use numerical simulation to study the effect of process parameters on the process of stratified sintering in selective laser sintering. Sachs et al. [16] use numerical simulation to develop a mathematical model for binder flight trajectory. Podshivalov et al. [17] apply numerical methods to the generation of microscale scaffolds.

One of the widely used numerical method is finite element method. Finite element analysis uses a mathematical approximation to simulate real physical systems (geometry and load cases). With simple and interacting elements, a finite number of unknowns can be used to approximate an infinitely unknown real system. Since most practical engineering problems are difficult to obtain accurate solutions, the finite element analysis which has high computational accuracy and adapts to various complex shapes makes it an effective engineering analysis tool.

Finite element analysis can be used to solve one-dimensional, two-dimensional and three-dimensional engineering problems. Nelson et al. [18] use one-dimensional finite element model to describe heat transfer of powder bed fusion process. Singh et al. [19] use two-dimensional finite element model to measure the evolution temperature distribution and density of bisphenol-A polycarbonate in selective laser sintering process with time, and determine the important technological parameters affecting the final density of laser sintered parts and their relationship. Bugeda et al. [20] establish a three-dimensional finite element model of single-track sintering in selective laser sintering process, which considers both the thermal phenomena and the sintering phenomena involved in the process. Dong et al. [21] establish a transient three-dimensional finite element model to study the phase transition in selective laser sintering process, which also considers the thermal phenomena and sintering phenomena during sintering. António and Vilar [22] propose a model coupling finite element heat transfer calculations, phase transformation kinetics and microstructure-property relations in Ti-6Al-4V, and use the model to obtain the processing maps of deposition parameters related to the microstructure and properties of the parts.

It can be seen from the above studies that the applications of finite element analysis are mainly focused on the thermodynamic study in the powder bed fusion process.

5. Machine learning and digital twin in additive manufacturing

In addition to the traditional methods mentioned above, due to the improvement of computer computing power, the modeling method of additive manufacturing based on machine learning and digital twin provides a direction for future research.

5.1. Machine Learning

Because machine learning can optimize and predict the parameters only through the relationship between input and output without knowing the internal rules of the system, it is suitable for predicting and optimizing the process parameters of additive manufacturing in the case of difficult to explore the internal physical laws of the process. In addition, compared with traditional methods, the application of machine learning in additive manufacturing is conducive to improving efficiency and accuracy. Therefore, in recent years, machine learning has been widely used in the optimization and prediction of the process parameters in additive manufacturing. Lee et al. [23] build a neural network model to find out the influence of process input parameters on the dimensional accuracy of parts and predict the dimensional accuracy of parts. Rong-Ji et al. [24] use genetic algorithm and back propagation neural network algorithm to determine the optimal process parameters of parts to produce parts with higher accuracy. Munguía et al. [25] propose an estimator based on artificial neural network to estimate the build-time and related costs (labor, machine costs and daily expenses) of selective laser sintering process. Padhye and Deb [26] consider the minimization of surface roughness and construction time in the process of selective laser sintering for multi-objective optimization, which use non-dominated sorting genetic algorithm (NSGA-II) and multi-objective particle swarm optimizer (MOPSO). Garg and Lam [27] apply genetic programming, support vector regression and artificial neural network in formulating the laser power-based-open porosity models. Garg and Lam [28] also use the computational intelligence (CI) approach of multi-gene genetic programming (MGGP) to formulate the model for predicting open porosity in selective laser sintering process. Stoyanov and Bailey [29] put forward State-Space Model Identification-minimization of the prediction error in the space of the model structure to predict and monitor trends and quality online. Machine learning can also be combined with model methods. Salonitis et al. [30] propose a method for the design optimization of lattice components towards weight minimization, which combines finite element analysis and evolutionary computation.

However, machine learning still has limitations. It is expensive to carry out a large number of experiments in additive manufacturing, while machine learning needs a lot of data for training. Therefore, the sample size in most machine learning experiments is insufficient. This problem needs to be solved in the future to further improve accuracy.

5.2. Digital Twin

Digital twin is a simulation process that makes full use of physical model, sensor update, operation history and other data. The purpose of digital twin is to evaluate and predict the condition of equipment.

In additive manufacturing, digital twin can be used to assess the health of the system and to evaluate and predict process parameters. After full experimental verification, this efficient and cheap digital experiment, which has high accuracy and stability, can replace expensive and time-consuming physical experiment. Some attempts have been made in this field. Knapp et al. [31] design a digital twin model to predict cooling rates, temperature gradients, solidification rates, SDAS and micro-hardness values. Debroy et al. [32] study the current status and needs of the first generation digital twin of additive manufacturing. In the above research, we can find that although the digital twin has a good performance in prediction and evaluation, its application is still limited because the current research on it is insufficient.

6. Conclusion

In this paper, three basic modeling methods for additive manufacturing are introduced: empirical method, analytical method, and numerical method. We introduce the main idea and significant applications of these three methods respectively. These methods are used in different situations. Besides, due to the improvement of computer computing power, recently machine learning and digital twin have also been applied to the study of additive manufacturing. With the development of modeling methods for additive manufacturing, the technology of additive manufacturing will surely make rapid progress.

References

[1] ISO/ASTM52900-15: Standard Terminology for Additive Manufacturing – General Principles – Terminology. ASTM International, West Conshohocken, PA; 2015.
[2] Chua C K, Leong K F. (2015) 3D Printing and Additive Manufacturing: Principles and Applications (with Companion Media Pack) Fourth Edition ofRapid Prototyping (4th Edition). World Scientific, 2015.
[3] Mani, M., Lane, B. M., Donmez, M. A., Feng, S. C. and Moylan, S. P. (2016) A Review on Measurement Science Needs for Real-Time Control of Additive Manufacturing Metal Powder Bed Fusion Processes. International Journal of Production Research; 1-19.
[4] Huang S H, Liu P, Abhiram Mokasdar, et al (2013) Additive manufacturing and its societal impact: a literature review. International Journal of Advanced Manufacturing Technology, 67(5-8):1191-1203.
[5] Wang WL, Cheah CM, Fuh JYH, Lu L (1996) Influence of process parameters on stereolithography part shrinkage. J Mater Des 17: 205–213.
[6] Lan PT, Chou SY, Chen LL, Gemmill D (1997) Determining fabrication orientations for rapid prototyping with stereolithography apparatus. J Comput-Aided Des 29(1):53–62.
[7] Karalekas D, Aggelopoulos A (2003) Study of shrinkage strains in a stereolithography cured acrylic photopolymer resin. J Mater Process Technol 6590:1–5.
[8] Raghunath N, Pandey PM (2007) Improving accuracy through shrinkage modelling by using Taguchi method in selective laser sintering. Int J Mach Tools Manuf 47:985–995.
[9] Anitha R, Arunachalam S, Radhakrishan P (2001) Critical parameters influencing the quality if prototypes in fused deposition modeling. J Mater Process Technol 118(1–3):385–388.
[10] Sood AK, Ohdar RK, Mahapatra SS (2012) Experimental investigation and empirical modelling of FDM process for compressive strength improvement. J Adv Res 3(1):81–90.
[11] Kechagias, John. (2007) An experimental investigation of the surface roughness of parts produced by LOM process. Rapid Prototyping Journal, 13(1):17-22.
[12] Ramanath H S, Chua C K, Leong K F, et al. (2008) Melt flow behaviour of poly-ε-caprolactone in fused deposition modelling. J Mater Sci Mater Med, 19(7):2541-2550.
[13] Muller P. (2013) Modeling and control of a direct laser powder deposition process for Functionally Graded Materials (FGM) parts manufacturing. Journal of Materials Processing Technology, 213(5):685-692.
[14] Strano G, Hao L, Everson R M, et al. (2013) Surface roughness analysis, modelling and prediction in selective laser melting. Journal of Materials Processing Technology, 213(4):589-597.
[15] TiebingChen, YuwenZhang. (2004) NUMERICAL SIMULATION OF TWO-DIMENSIONAL MELTING AND RESOLIDIFICATION OF A TWO-COMPONENT METAL POWDER LAYER IN SELECTIVE LASER SINTERING PROCESS. Numerical Heat Transfer, 46(7):17.
[16] Sachs E, Vezzetti E. (2005) Numerical simulation of deposition process for a new 3DP printhead design[J]. Journal of Materials Processing Technology, 161(3):509-515.
[17] Podshivalov L, Gomes C M, Zocca A, et al. (2013) Design, Analysis and Additive Manufacturing of Porous Structures for Biocompatible Micro-Scale Scaffolds.
[18] Nelson, J. C., Xue, S., Barlow, J. W., Beaman, J. J., Marcus, H. L. and Bourell, D. L., (1993) Model of the Selective Laser Sintering of Bisphenol-a Polycarbonate. Industrial & Engineering

Chemistry Research 32.

[19] Singh A K, Prakash R S. (2009) Simulation based two level factorial designed experiments for optimisation of selective laser sintering process[J]. International Journal of Computer Aided Engineering and Technology, 1(3):322.

[20] Bugeda, G., Cervera, M. and Lombera, G., (1999) Numerical Prediction of Temperature and Density Distributions in Selective Laser Sintering Processes. Rapid Prototyping Journal 5; 21-26.

[21] Dong L, Makradi A, Ahzi S, Remond Y (2009) Three-dimensional transient finite element analysis of the selective laser sintering process. J Mater Process Technol 209(2):700–706

[22] António Crespo, Vilar R. (2010) Finite element analysis of the rapid manufacturing of Ti–6Al–4V parts by laser powder deposition. Scripta Materialia, 63(1):140-143.

[23] Lee S H, Park W S, Cho H S, et al. (2001) A neural network approach to the modelling and analysis of stereolithography processes. Proceedings of the Institution of Mechanical Engineers, Part B: Journal of Engineering Manufacture, 215(12):1719-1733.

[24] Rong-Ji W, Xin-Hua L, Qing-Ding W, et al. (2009) Optimizing process parameters for selective laser sintering based on neural network and genetic algorithm. The International Journal of Advanced Manufacturing Technology, 42(11-12):1035-1042.

[25] Munguía, J, Ciurana J, Riba C. (2009) Neural-network-based model for build-time estimation in selective laser sintering. Proceedings of the Institution of Mechanical Engineers, Part B: Journal of Engineering Manufacture, 223(8):995-1003.

[26] Padhye N, Deb K. (2011) Multi-objective Optimisation and Multi-criteria Decision Making for FDM Using Evolutionary Approaches. Rapid Prototyping Journal, volume 17(6):458-478(21).

[27] Garg A, Lam J S L. (2015) Measurement of Environmental Aspect of 3-D Printing Process using Soft Computing Methods. Measurement, 2015: S0263224115002195.

[28] Garg A, Lam J S L, Savalani M M. (2015) A new computational intelligence approach in formulation of functional relationship of open porosity of the additive manufacturing process[J]. The International Journal of Advanced Manufacturing Technology, 80(1-4):555-565.

[29] Stoyanov S, Bailey C. (2017) Machine learning for additive manufacturing of electronics. International Spring Seminar on Electronics Technology.

[30] Salonitis K, Chantzis D, Kappatos V. (2017) A hybrid finite element analysis and evolutionary computation method for the design of lightweight lattice components with optimized strut diameter. The International Journal of Advanced Manufacturing Technology, 90(9-12):2689-2701.

[31] Knapp G L, Mukherjee T, Zuback J S, et al. (2017) Building blocks for a digital twin of additive manufacturing. Acta Materialia, 135:390-399.

[32] Debroy T, Zhang W, Turner J, et al. (2016) Building digital twins of 3D printing machines. Scripta Materialia, 2016: S1359646216305991.

IOP Conf. Series: Materials Science and Engineering **711** (2020) 012018 doi:10.1088/1757-899X/711/1/012018

Integrated preparation of long-life hydrogen storage alloy film/Ni foil composite electrode material

Ying Chen and Zi Wen*

Key Laboratory of Automobile Materials (Jilin University), Ministry of Education, and School of Materials Science and Engineering, Jilin University, Changchun, Jilin, 130022, China

*Corresponding author's e-mail: wenzi@jlu.edu.cn

Abstract. An integrated preparation method of a long-life hydrogen storage alloy film/Ni foil composite electrode material has been proposed in this paper. The hydrogen storage alloy film/Ni foil composite electrode material was prepared by magnetron sputtering. The prepared film composite electrode effectively reduces the electron/ion diffusion distance, increases the specific surface area of alloy, and reduces the contact resistance between the active material and the current collector. Furthermore, the integrated preparation of long-life hydrogen storage alloy film/Ni foil composite electrode material is realized by depositing method. This paper provides a simple and effective method for reducing the internal resistance and increasing the electrochemical reaction rate of the hydrogen storage alloy, and the prepared film composite could be a promising candidate for Ni-MH batteries.

1. Introduction

The nickel-metal hydride (Ni-MH) batteries are used in a variety of electronic devices and equipment due to their high energy density, high power density, environmental friendliness, safety and a wide range of temperature adaptability[1,2]. Ni-MH batteries are generally obtained by cold-pressing hydrogen storage alloy powder (as active material) and nickel carbonyl powder (as current collector and binder). However, this method will create a large contact resistance between the active material and the current collector. Noted that a large contact resistance could slow the electrochemical reaction kinetics, which hinders the performance improvement of the Ni-MH batteries[3-5]. Therefore, the long-life hydrogen storage alloy film/Ni foil composite electrode material could be prepared in order to reduce the contact resistance between the alloy powder and nickel carbonyl powder[6-8].

Magnetron sputtering, as a Physical Vapor Deposition (PVD), has been applied to prepare metal, semiconductor, insulator, etc. Due to the advantages of control easily, large coating area and strong adhesion[9-11]. In this work, an integrated preparation of long-life hydrogen storage alloy film/Ni foil composite electrode material has been presented. With the help of magnetron sputtering coating method, a thin film of hydrogen storage alloy with a certain thickness was deposited on a Ni foil substrate. The long-life hydrogen storage alloy films with different thicknesses and surface morphology were prepared by adjusting the background pressure, sputtering power, sputtering time and the temperature applied on substrate. This method has the following advantages: (1) increase the specific surface area of hydrogen storage alloys; (2) reduce the transmission distance of electrons and ions; (3) reduce the contact resistance between the active material and the current collector; (4) integrated production of hydrogen storage alloy film/Ni foil is realized. The integrated film composite material exhibits great potential to be candidates as the electrode for reducing the internal resistance in Ni-MH battery and increasing the electrochemical reaction rate[12-16].

2. Experiment

2.1. Raw Materials
The rare earth elements La, Ce, Y(99.5%) were purchased from Baotou Xinye New Material Co., Ltd., metallic elements Ni and Co(99.9%) were purchased from Jinchuan Group Co., Ltd., element Mn (99.9%) was purchased from Yiyang Jinneng New Material Co., Ltd., and element Al(99.9%) was purchased from Boyu Metal Co., Ltd.

2.2. Specimen preparation
Alloy preparation: the ingot of alloy $La_{0.6}Ce_{0.3}Y_{0.1}Ni_{3.7}Co_{0.75}Mn_{0.3}Al_{0.35}$ was prepared by inductive melting elements La, Ce, Y, Ni, Co, Mn, Al under high-purity argon atmosphere at 1300 °C. Then the ingot was cooled at a rapidly rate to ensure the homogeneity of the alloy. Moreover, the ingot was annealed under an Ar flow at 1000 °C for 5 h, and mechanically ground to powders[17,18]; Target preparation: the prepared hydrogen storage alloy powders were pressed into a target (the diameter of target is 50 mm, and the thickness is 3 mm), and the target is bound with a copper backing plate by using element In. The obtained object is used for magnetron sputtering; Substrate preparation: cut the Ni foil into proper size (3.5 cm*3.5 cm), and the Ni foil was washed and dried in the following order: acetone–deionized water–3 mol/L HCl–deionized water–ethanol–vacuum drying. Subsequently, a substrate using for sputtering could be obtained; Film preparation: the atmosphere used for deportation was 99.99% purity argon. The sputtering power was controlled as the value of 100 W. The background pressure is 1 Pa. By adjusting the sputtering time X (X = 120, 240 and 360 min) and the temperature applied on substrate Y (Y = 0 and 150 °C), the long-life hydrogen storage alloy film/Ni foil composite electrode material with different thicknesses and surface morphology were obtained (see Table 1).

Table 1. List of prepared film composite samples

	S1	S2	S3	CS1	CS2	CS3
X (min)	120	240	360	120	240	360
Y (°C)	150	150	150	0	0	0

2.3. Characteriation
The morphology properties of the specimens were investigated by a field-emission scanning electron microscopy (FESEM, JSM-6700F, JEOL, 15keV).

3. Result and discussion

3.1. Characteristics of film composite electrode
The specimen shows good elasticity since the shape could recover after bending into an arbitrary angle. It should be noted that the alloy film/Ni foil composite electrodes prepared by magnetron sputtering all have the above characteristics.

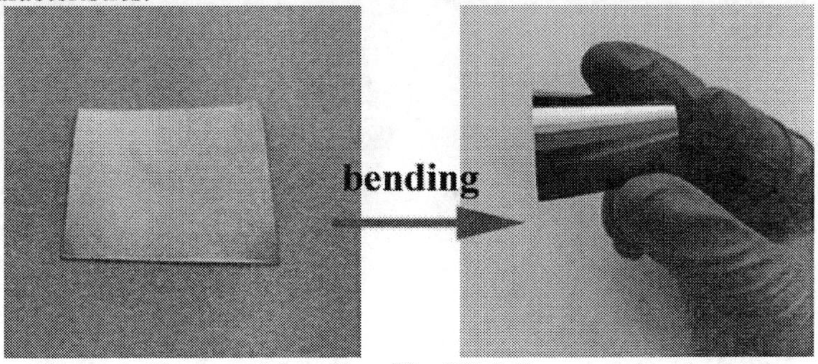

Figure 1. Photos of the alloy film/Ni foil composite electrode.

IOP Conf. Series: Materials Science and Engineering **711** (2020) 012018 doi:10.1088/1757-899X/711/1/012018

3.2. Morphological characterizations of sectional view of film composite electrodes

Figure 2a-c show the SEM images of the sectional views of CS1, CS2 and CS3, respectively. It is clear that the film thicknesses of CS1, CS2 and CS3 are 360, 640 and 800 nm, respectively, which are consistent with the increase of sputtering time. Mentioned that an obvious strip-like traces could be observed in the figure, and this is caused by alloy growth during the deposition process[19,20].

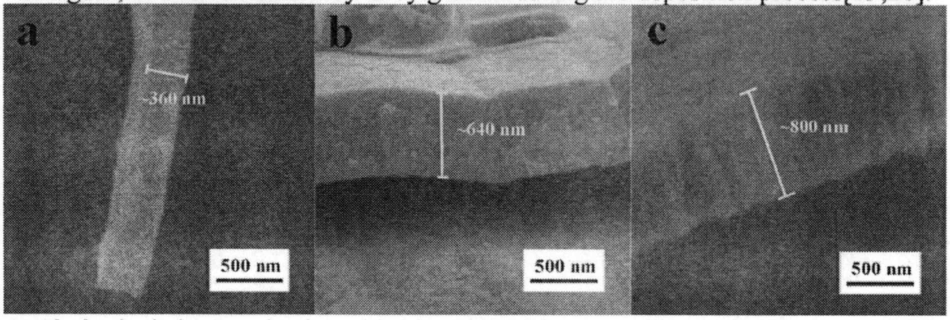

Figure 2. Morphological characterizations (a)-(c) are SEM images ofthe sectional views of CS1, CS2 and CS3, respectively.

3.3. Morphological characterizations of the surface of film composite electrode

Figure 3a-c shows the SEM images of the surface of CS1, CS2 and CS3, respectively. From the figure, there is no significant difference among samples, which are all smooth and flat. Figure 3d-f present the SEM images of the surface of S1, S2 and S3, respectively. It is clear that the film surfaces are covered with oxidized particles, which is distinct from the clean surfaces before applying a temperature of 150 °C on the Ni foil substitute. As illustrated in Figure 3d, the oxidized particles on the surface of S3 is less noticeable, while the oxidized particles became quite obvious when sputtering time is 360 min (see Figure 3f).

The film composite prepared by magnetron sputtering has a larger specific surface area and a smaller ion diffusion distance comparing with the commercial hydrogen storage alloy powder (50 ± 10 μm in diameter). Therefore, the film composite has a large charge/ion transfer rate, which is of great significance for the preparation process and future development of Ni-MH batteries.

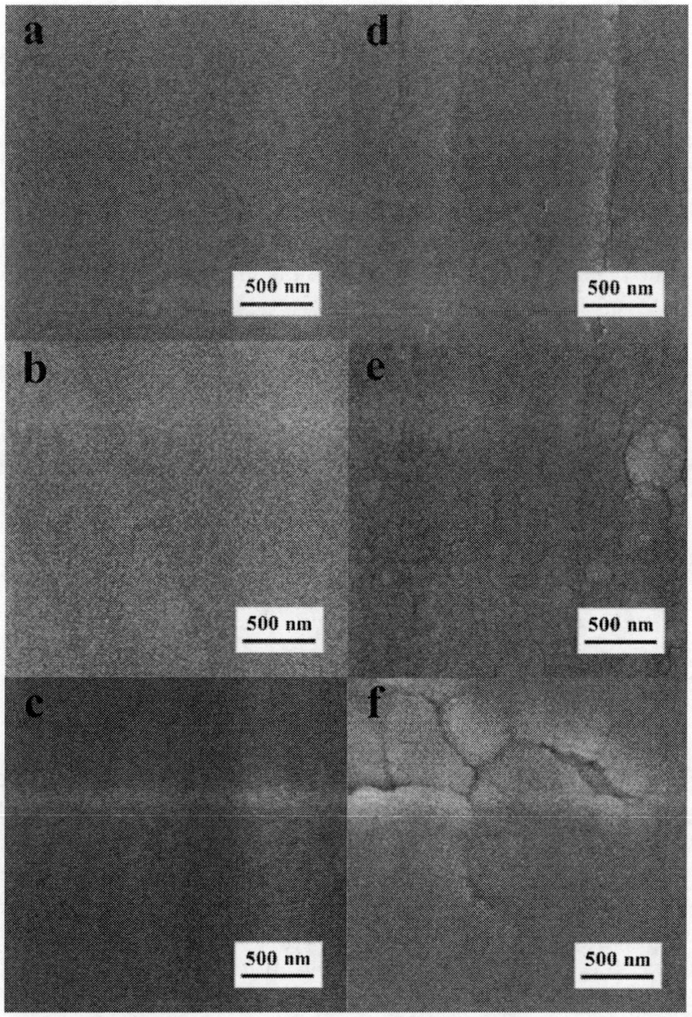

Figure 3. Morphological characterizations. (a)-(c) are the SEM images of surface images of CS1, CS2 and CS3, respectively. (d)-(f) are the surface images of S1, S2 and S3, respectively.

4. Conclusion

In summary, a simple and effective strategy is reported to fabricate film composite for the application of Ni-MH batteries. The film composite material takes the advantages as following: (1) large specific surface area; (2) small transmission distance of electrons and ions; (3) small contact resistance; (4) good elasticity. The prepared long-life hydrogen storage alloy film composite electrode material accelerates the charge transfer rate, and successfully realizes the integrated preparation of the Ni-MH battery. This paper is of great significance for the integrated preparation of Ni-MH battery, which has a promising prospect.

Acknowledgments

This project is financially supported by National Natural Science Foundation of China (Nos. 51671092 and 51631004), Program for JLU Science and Technology Innovative Research Team (No. 2017TD-09) and the Fundamental Research Funds for the Central Universities.

References

[1] Ouyang, L.Z., Huang, J.L., Wang, H., Liu, J.W., Zhu, M. (2017) Progress of hydrogen storage

alloys for Ni-MH rechargeable power batteries in electric vehicles: A review. Mater. Chem. Phys., 200: 164-178.

[2] Young, K. (2016) Compendium of Hydrogen Energy. BASF-Ovonic, the Miami.

[3] Cui, R.C., Yang, C.C., Li, M.M., Jin, B., Ding, X.D., Jiang, Q. (2017) Enhanced high-rate performance of ball-milled $MmNi_{3.55}Co_{0.75}Mn_{0.4}Al_{0.3}$ hydrogen storage alloys with graphene nanoplatelets. J. Alloys Compd., 693: 126-131.

[4] Sakai, T., Yoshinaga, H., Miyamura, H., Kuriyama, N., Ishikawa, H., Uehara, I. (1993) Metal hydride thin film electrodes prepared by r.f. sputtering. J. Alloys Compd., 8: 77-79.

[5] Sakai, T., Ishikawa, H., Miyamura, H., Kuriyama, N. (1991) Thin film preparation of hydrogen storage alloys and their characteristics as metal hydride electrodes. J. Electrochem. Soc., 138: 908-915.

[6] Yang, C.C., Wang, C.C., Li, M.M., Jiang, Q. (2017) A start of the renaissance for nickel metal hydride batteries: A hydrogen storage alloy series with ultra-long cycle life. J. Mater. Chem. A, 5: 1145-1152.

[7] Wang, C.C., Zhou, Y.T., Yang, C.C., Jiang, Q. (2018) Clarifying the capacity deterioration mechanism sheds light on the design of ultra-long-life hydrogen storage alloys. Chem. Eng. J., 352: 325-332.

[8] Wang, C.C., Zhou, Y.T., Yang, C.C., Jiang, Q. (2018) A strategy for designing new AB4.5-type hydrogen storage alloys with high capacity and long cycling life. J. Power Sources, 398: 42-48.

[9] Zhao, Y., Wang, H., Lia, X.Z., Li, J.J., Shi, Z.F., Wu, G.G., Zhuang, S.W., Yin, C.L., Yang, F. (2019) Parametric study on the well-oriented growth of $In_xAl_{1-x}N$ nanodots by magnetron sputtering. Mater. Sci. Semicond. Process., 102: 1-7.

[10] Guo, C.Y., Qi, X.D. (2019) RF magnetron sputter deposition and electrical properties of La and Y doped $SrTiO_3$ epitaxial films. Mater. Des., 179.

[11] Zhang, H.L., Su, R.R., Chen, D.M., Shi, L.Q. (2016) Thermal desorption behaviors of helium in Zr-Co films prepared by sputtering deposition method. Vacuum, 130: 174-178.

[12] Wang, H., Ouyang, L.Z., Peng, C.H., Zeng, M.Q., Chung, C.Y., Zhu, M. (2004) MmM_5/Mg multi-layer hydrogen storage thin films prepared by dc magnetron sputtering. J. Alloys Compd., 370: 4-6.

[13] Wang, H., Ouyang, L.Z., Zeng, M.Q., Zhu, M. (2004) Microstructure and hydrogen sorption properties of Mg–Ni/MmM_5 multi-layer film by magnetron sputtering. Int. J. Hydrogen Energy, 29: 1389-1392.

[14] Li, M.M., Yang, C.C., Jing, W.T., Jin, B., Lang, X.Y., Jiang, Q. (2016) In situ grown Co_3O_4 on hydrogen storage alloys for enhanced electrochemical performance. Int. J. Hydrogen Energy, 41: 8946-8953.

[15] Li, M.M., Wang, Y., Yang, C.C., Jiang, Q. (2018) In situ grown Co_3O_4 nanocubes on N-doped graphene as a synergistic hybrid for applications in nickel metal hydride batteries. Int. J. Hydrogen Energy, 43: 18421-18435.

[16] Li, M.M., Yang, C.C., Chen, L.X., Jiang, Q. (2016) Hydrogen storage alloys/reduced graphite oxide: an efficient hybrid electrode with enhanced high-rate dischargeability. Electrochim. Acta, 200: 59-65.

[17] Li, M.M., Yang, C.C., Wang, C.C., Wen, Z., Zhu, Y.F., Zhao, M., Li, J.C., Zheng, W.T., Lian, J.S., Jiang, Q. (2016) Design of hydrogen storage alloys/nanoporous metals hybrid electrodes for nickel-metal hydride batteries. Sci. Rep., 6: 27601-27610.

[18] Chen, L.X., Zhu, Y.F., Yang, C.C., Chen, Z.W., Zhang, D.M., Jiang, Q. (2016) A new strategy to improve the high-rate performance of hydrogen storage alloys with MoS_2 nanosheets. J. Power Sources, 333: 17-23.

[19] Ouyang, L.Z., Wang, H., Zhu, M., Zou, J., Chung, J. (2004) Microstructure of MmM_5/Mg multi-layer hydrogen storage films prepared by magnetron sputtering. Microsc. Res. Tech., 64: 323-329.

[20] Tsai, D.C., Chang, Z.C., Kuo, B.H., Deng, Y.S., Chen, E.C., Shieu, F.S. (2016) Effects of sputtering power on microstructure and mechanical properties of TiVCr films. Vacuum, 125: 227-233.

IOP Conf. Series: Materials Science and Engineering **711** (2020) 012019 doi:10.1088/1757-899X/711/1/012019

Research on location detection of railway track spike based on neural network

Chen Lu, Yong Shen*

School of Electrical Information Engineering, Yunnan Minzu University, Kunming, Yunnan, 650000, China

*Corresponding author's e-mail: 740568005@qq.com

Abstract. The spike is an important link between the sleeper and the track. It plays a vital role in the quality of the railway and the safe operation of the train. At present, in railway engineering, the position of spike can be obtained by adjusting the parameters of magnetic sensor and repeated measurement tests. This method is simple but requires a large amount of labour cost, so this research attempts to make the "automatic guidance computer system" of the tamping vehicle automatically detect the distance from the iron rail nail according to the driving position of the tamping vehicle by using the BP neural network and the VGG19 neural network, thereby improving the maintenance rate of railway lines and reduced labour cost have laid the foundation for a grand blueprint for unmanned tamping vehicle.

1. Introduction

In recent years, with the advancement of science and technology and the rapid development of social economy, China's railway transportation is moving towards high speed, which puts forward higher requirements on the maintenance speed of railway lines. The spike is an important part of the fixed rail. When using the tamping vehicle to carry out maintenance work on the railway line, it is necessary to judge the position of the pick operation according to the detected position of the spike, so as to maintain and repair railway lines[1]. In railway engineering, the position of spike is determined by magnetic force line data measured by magnetic sensor. Specific rail spike position is obtained by adjusting the parameters of the magnetic sensor and repeated measurement tests. This method requires a large amount of labour cost, so it is envisaged that by using the algorithm of the neural network, the "automatic guiding computer system" of the tamping vehicle can automatically detect the distance between tamping vehicle and spike according to its driving position, thus according to the distance to determine the position of tamping pickaxe[2].

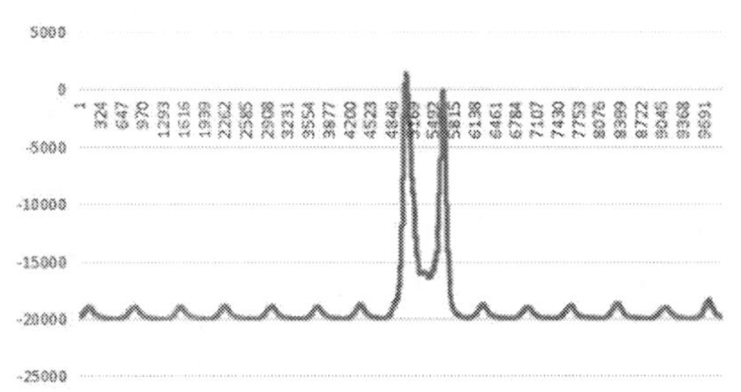

Figure 1. Magnetic signal distribution curve.

Figure 1 shows the distribution curve of magnetic signals measured by the magnetic sensor during a certain operation of the tamping vehicle. The position of each "small protrusion" in the figure is the position of the spike, and the peak of the "M" shape is an abnormal value. This research does not consider the outlier for the time being.

2. BP neural network algorithm

The BP (Back Propagation) network is a multi-layer feedforward network trained by error back propagation algorithm, which is one of the most widely used neural network models[3]. Its learning rule is to use the steepest descent method to continuously adjust the weights and thresholds of the network through back propagation to minimize the sum of squared errors of the network [3-4]. The topological structure of BP neural network model includes input layer, hide layer and output layer.

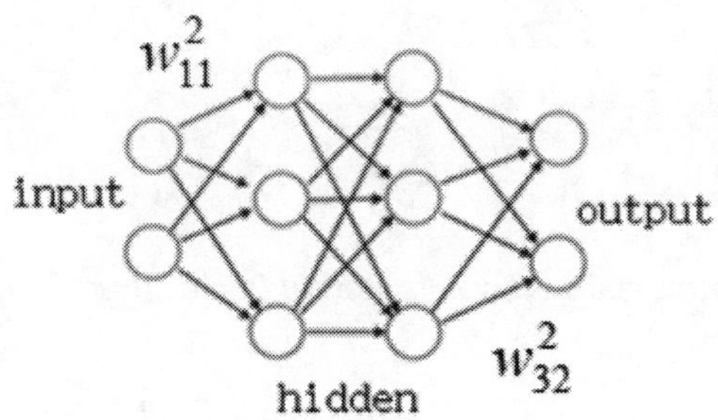

Figure 2. Simple BP neural network structure.

Figure 2 is a simple BP neural network structure, consisting of an input layer, two hidden layers, and an output layer, in which the circle is a neuron and the connecting line represents the transmission of values. The input layer and the output layer have two neurons respectively, which means that a two-dimensional vector can be passed in and a two-dimensional vector can be finally output. Taking the second neuron of hidden layer 1 as an example, the calculation principle process of BP neural network is shown in figure 3:

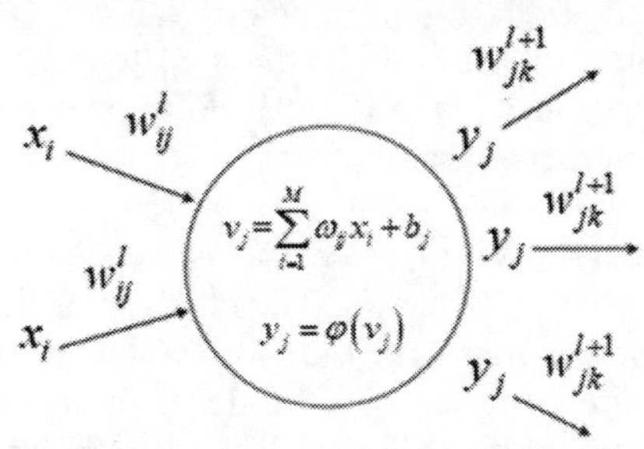

Figure 3. BP calculation schematic diagram.

Looking from left to right, x_i represents the number of incoming values from the previous neuron. In the weight w, i represents the sequence number of the neuron in the previous layer, j represents the sequence number of pointed neuron, and the superscript l is the level, j represents the sequence number of this neuron. The weighted sum gives v_j, where M is the total number of neurons in the previous layer. The results are substituted into the activation function φ, y_i is obtained as the output of this neuron[5].

In order to explain the formula more clearly, the superscript l of the weight w is omitted, and the following formula is used with respect to the neurons of the output layer:

Enter:

$$v_j=\sum_{i=1}^{M}\omega_{ij}x_i+b_j \tag{1}$$

Output:

IOP Conf. Series: Materials Science and Engineering **711** (2020) 012019 doi:10.1088/1757-899X/711/1/012019

$$y_j = \varphi(v_j) \tag{2}$$

Our goal is to find the right w and b, so that the output of the network is as close as possible to the expected output, then when we enter an untrained sample point, we can make an accurate classification[5]. In order to solve the problem, w and b satisfying uniform distribution or normal distribution were generated randomly at the beginning of training, and then the back propagation and gradient descent algorithm were applied to continuously modify w and b to minimize the total error. Now, let's set the output layer neuron code to k, hidden layer 2 code to j, and hide layer 1 code to i. The gradient formula is deduced as follows:

The formula for deriving the gradient is as follows:

For the output layer neuron, the gradient of the total error with respect to the weight w_{jk} is obtained by the gradient descent algorithm:

$$\frac{\partial E}{\partial \omega_{jk}} = \frac{\partial E}{\partial y_k}\frac{\partial y_k}{\partial v_k}\frac{\partial v_k}{\partial \omega_k} = \frac{\partial E}{\partial v_k} x_j \tag{3}$$

Similarly, the gradient of bias b_k can be obtained:

$$\frac{\partial E}{\partial b_k} = \frac{\partial E}{\partial y_k}\frac{\partial y_k}{\partial v_k}\frac{\partial v_k}{\partial b_k} = \frac{\partial E}{\partial v_k} \tag{4}$$

$\frac{\partial E}{\partial v_k}$ is reserved in the formula derivation to derive the local gradient:

$$\delta_k = \frac{\partial E}{\partial v_k} = \frac{\partial E}{\partial y_k}\frac{\partial y_k}{\partial v_k} = -(d_k - y_k)\varphi'(v_k) \tag{5}$$

With regard to local gradients, the weighted sum of all inputs to a neuron is regarded as a whole, weights w and b only affect other parts of the neural network through this whole[6]. Then the weight w adjustment and the bias b adjustment are:

$$\Delta\omega_{jk} = -\eta\frac{\partial E}{\partial \omega_{jk}} = -\eta\delta_k x_j \tag{6}$$

$$\Delta b_k = -\eta\frac{\partial E}{\partial b_k} = -\eta\delta_k \tag{7}$$

Among them, η is the learning rate, or called the step length. According to the actual situation, generally take a smaller number, such as 0.1, 0.01. Putting a negative sign in front of it means searching in the direction of the negative gradient.

3. VGG convolutional neural network

The VGG network uses 3*3 convolution kernels. The changes of the convolution kernel will not only affect the calculation amount of the network, but also affect the size of the network to obtain the receptive field, and ultimately affect the learning mode of the network. Smaller convolution kernels are easier to capture changes in image feature details, and at the same time have a greater impact on the amount of computation of the network[7]. In the network, a smaller convolution kernel (3*3) is used, the stride is 1, and 2-4 convolution layers constitute a "coil unit" of 3*3. Some layers also have a 1*1 convolution layer. The 1*1 convolution kernel can be regarded as a linear transformation and the stride is 1, so the network is deeper. The pooling layers use the maximum pooling layer of 2*2. There are five layers and the stride is 2. In order to obtain more detailed information, ReLU activation function is used for all hidden layers[8]. In the full-image test, the last fully-connected layer is changed to the fully convolutional net, and the parameters during training are reused. Because there is no full connection limit, the network can receive input of any width or height, and it has a remarkable feature: the spatial resolution of the feature map is monotonically decreasing, and the number of channels of the feature graph is monotonically increasing, in order to better convert the image of H*W*3 into the output of 1*1*C.

IOP Conf. Series: Materials Science and Engineering **711** (2020) 012019 doi:10.1088/1757-899X/711/1/012019

4. Analysis of spike detection based on BP and VGG networks

In the data preparation stage, the distribution curve of magnetic signals measured by magnetic sensors is divided into many magnetic signals of the same length according to a step length every 50 meters and a segment length of 784 meters, which were respectively stored in the CSV files. Then, the cut CSV files are randomly divided into training data sets, validation data sets and test data sets. Set a threshold according to the experience, in every section of the magnetic signal, if there is greater than the threshold value is that there is a spike in this signal, and output spike position and the location of tamping car distance value of sample label, if not is greater than the threshold value of the signal, this signal does not exist spike, output sample label value of -1. Finally, the training data and sample labels are input into the established BP neural network model, and the data is trained so that the output result is the distance between the position of the spike and the position of the tamper. And the results are show in figure 4. Then the data is reconstructed to a size of 224*224 and input into the neural network model of VGG19 for train, and the results are show in figure 5.

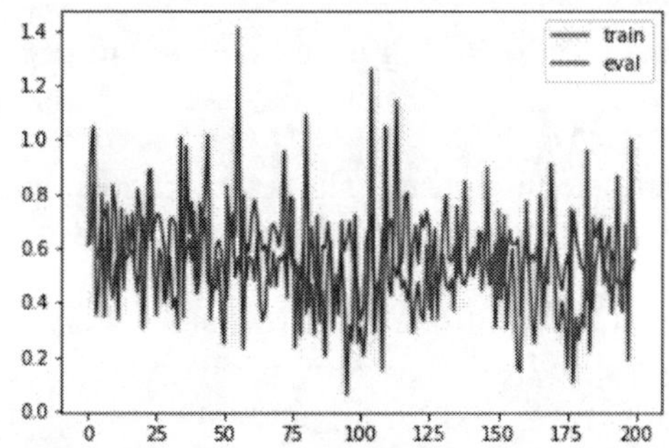

Figure 4. BP neural network training results.

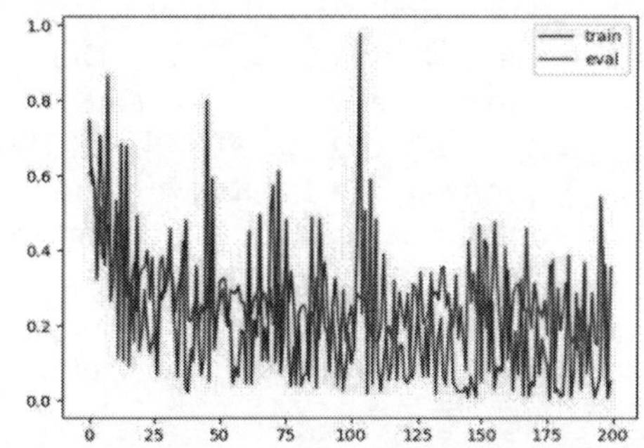

Figure 5. VGG neural network training results.

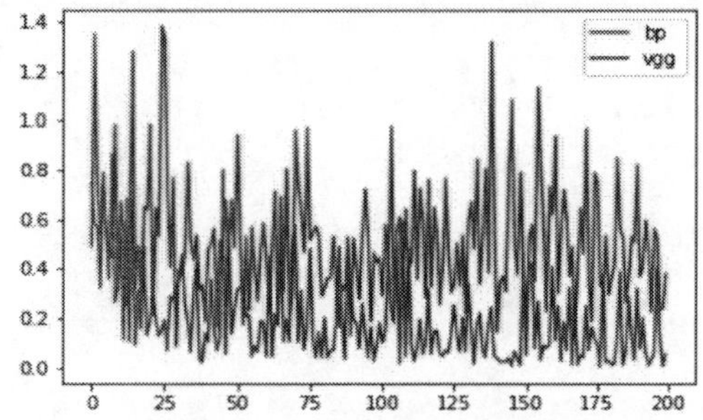

Figure 6. Comparison of BP and VGG loss values.

Figure 4 shows the training results using BP neural network. It can be seen that the effects on the training set and test set cannot converge well, and the loss value has no obvious downward trend. Figure 5 shows the training results using the VGG19 neural network. It can be seen that the loss value decreased from about 0.7 at the beginning of the training to about 0.02, and the error rate is about 0.032. Although there is a certain fluctuation, the effect is relatively ideal. Figure 6 is the comparison diagram of loss value between BP neural network and VGG neural network. It is obvious that at the end of training, the loss value of using VGG model is much smaller than that of BP neural network and converges faster. The conclusion can be drawn that the distance between the rail spike and the tamping vehicle can be determined by the neural network method, and the VGG neural network has a good effect.

5. Conclusion

In this research, BP neural network and VGG neural network are adopted to detect railway track nails, and the advantages and disadvantages of the two methods are compared, and the conclusion is drawn that the railway track nails can be identified by using neural network, which provides a certain theoretical basis for the realization of unmanned tamping vehicle and has certain research significance.

References

[1] Xiaoli Kang. (2018) Application and Research of Large-scale Road Maintenance Machinery for Railway Lines. J. China Strategic Emerging Industry., 40: 186-186.
[2] Lin.J.M. (2013) BP neural network analysis. https://blog.csdn.net/linj_m/article/details/9897839.
[3] Zuoyong Li, Jiayang Wang, Chun Guo. (2008) New method and simulation experiment of PSO algorithm to optimize BP network. J. Acta electronica sinica., 11: 2224-2228.
[4] Jing Jin, Zonghao Li, Liang Zhu, Xinhao Tong, Changwei Yang. (2019) Application of BP neural network in railway construction risk assessment. J. Journal of railway engineering., 36: 103-109.
[5] Rui Fang, Junyang Yu, Lifeng Dong. (2019) Spam text filtering based on B-Feature feature construction and BP neural network. https://doi .org/10.19678/j.issn.1000-3428.0055414.
[6] Shichun Peng. (2019) Research on image semantic segmentation algorithm in deep learning computer network. J. Journal of guiyang university (natural science edition)., 14: 22-26+61.
[7] Simonyan K, Zisserman A. (2014) Very Deep Convolutional Networks for Large-Scale Image Recognition. J. Computer Science., 34: 64-67.
[8] Mengzhuo Yang, Mengjie Guo, Liang Fang. (2019) Research on image classification algorithm based on keras convolutional neural network. J. Science and technology wind., 23:117-118.

Research on the development mode of general exclusive parts for spacecraft integration

Chen Zou[1], Lei Zhang[1], Jiandong Zheng[1], Shuang Zou[1]

[1] Institute of Telecommunication Satellite, China Academy of Space Technology, Peking, 100094, China

Email: zouchen2019@sina.com

Abstract: As an important part of spacecraft, the research on general exclusive parts for spacecraft integration is still in its infancy. In this paper, based on the characteristic of spacecraft and exclusive parts, the development mode of general exclusive parts and the principles of configuration management for general exclusive parts are established. This paper explores the working mode of general exclusive parts adding, modification, abolition and deviation. Furthermore, this paper also suggests the thoughts on the construction of relevant databases.

1. Introduction

Exclusive parts for integration as one of the most important parts of spacecraft indicate connectors for satellite integration (including parts, assembly, and secondary structure). Recently because of the increasingly strict requirement on the cost of spacecraft development, a more efficient and economical working mode for the exclusive parts for integration is urgently necessary. Generalization is an efficient method to reduce management and manufactured cost and improve design efficiency. It is indicated that when general parts account for 13.7%, the manufactured cost can be reduced by 21.25%, while the management cost can be reduced by 20%.[1]

General exclusive parts for integration indicate the exclusive parts that are interchangeably used between satellites. The research on the development mode of general exclusive parts for spacecraft integration is still in its infancy. This paper explores the development mode of general exclusive parts for spacecraft integration.

2. Development mode of general exclusive parts for integration

General parts are wildly used in industry, and the research on them is quite mature, most of which focus on mass production. It is involved that as the diversity of demand increases, using general parts is necessary to adapt to the demand change.[2] With the rapid growth of the number of spacecrafts, the demand for exclusive parts for integration presents a trend of increasing in both quantity and diversity. Therefore, in order to cope with demand growth, generalization is necessary. However, it is also indicated that facing the huge development cost, manufacturers need to consider generalization strategy cautiously.[2] In conclusion, the premise of generalization is to reduce the development cost.

In order to reduce the cost of development, this paper introduces a development mode of "from the project to the project". "From the project" indicates that instead of re-development, the general exclusive parts for spacecraft integration are selected from the parts that have experienced several successful space flights. The specific team is founded to audit the generality of the exclusive parts, separate the suitable ones from the management system for the specific project, and incorporate the suitable ones into an independent management system for general exclusive parts for spacecraft integration. "To the project" indicates the developers of general exclusive parts for spacecraft integration need to focus on serving the whole-spacecraft integration, and the configuration management for the general exclusive parts is based on the whole-spacecraft project management.

3. Configuration management

An independent management system with its own characteristic for the general exclusive parts for spacecraft integration is necessary, that is based on the general principles of spacecraft configuration management.

- It is suggested that the unified functional baseline, development baseline and production baseline are established, indicating stable configuration.
- The codes of ordinary exclusive parts for integration belong to the code system of a specific spacecraft, while the code system of the general exclusive parts is independent, which is significantly different from the ordinary one. Easy use and good expansibility are also necessary for the code system.
- The atlas of general parts for selecting is necessary and the records of characteristic information such as physical characteristics, functional characteristics, and history information are also essential.
- A specific team for configuration management is founded to manage configuration of general parts, whose responsibility is to the audit new parts, update configuration documents, summarize the problems regularly arising during manufacturing process and whole-spacecraft integration, evaluate the existing products, and audit the necessity of the modification and abolition of general exclusive parts.
- It is efficient to use information tools recording configuration of general parts.
- It is necessary to record modification of configuration and catalogue.
- Based on the actual use of the general exclusive parts for integration, regular addition, modification and abolition are suggested.
- Configuration documents need to be approved formally as required.

3.1The addition of the general exclusive parts for integration

According to the development mode of "from the project to the project", the specific team for configuration management audits the generality of the parts having multiple successful experiences of space flight, and the suitable ones are included in the catalogue of general parts. The process is shown in Figure 1.

M1 and M2 are the parts of the specific project development system and they are the prerequisites for the generalization.

M3: Qualification audit is the first step for generalization and the data submitted includes the property and performance, physical characteristics, drawings/models, resumes and other relevant characteristics of the exclusive parts.

M4: The configuration management team audits the exclusive parts according to the submissions, including the generality qualification and the completeness of the data. Due to the strict requirement of reliable and the characteristics of the exclusive parts, the general exclusive parts need to have the following characteristics:

- Normative requirements: General parts should have perfect product specifications, including detailed drawings/models, performance reports, resume reports, and the whole process specifications that indicates the process of design, manufacturing, acceptance and integration.
- Production assurance requirements: General parts should have been verified by batch production and multi-satellite (>=2) flight verification to ensure the consistency of production and quality stability.
- Product performance requirements: The physical properties of the products are well adapted

to the different project demands.
- Product interface requirements: The product interface is well generalized, that is suitable for the shelf products.
- Product demand requirements: In the foreseeable future, there will be multiple demands.
- Product storage requirements: The physical performance of the parts is stable during ground storage, which has long-term storage conditions.

For the general exclusive parts with incomplete information, the applicant needs to resubmit the application after completing the information.

M5: The team for configuration management issues generic codes for the new general exclusive parts.

M6: The team for configuration management approvals the new general exclusive part drawings/models formally.

M7a1: The team for configuration management updates the atlas according to the drawings/models approved.

M7a2: The team for configuration management updates the configuration documents of general exclusive parts according to the information submitted.

M8: After all the process above, the team for configuration management publishes upgraded notice of the general exclusive parts.

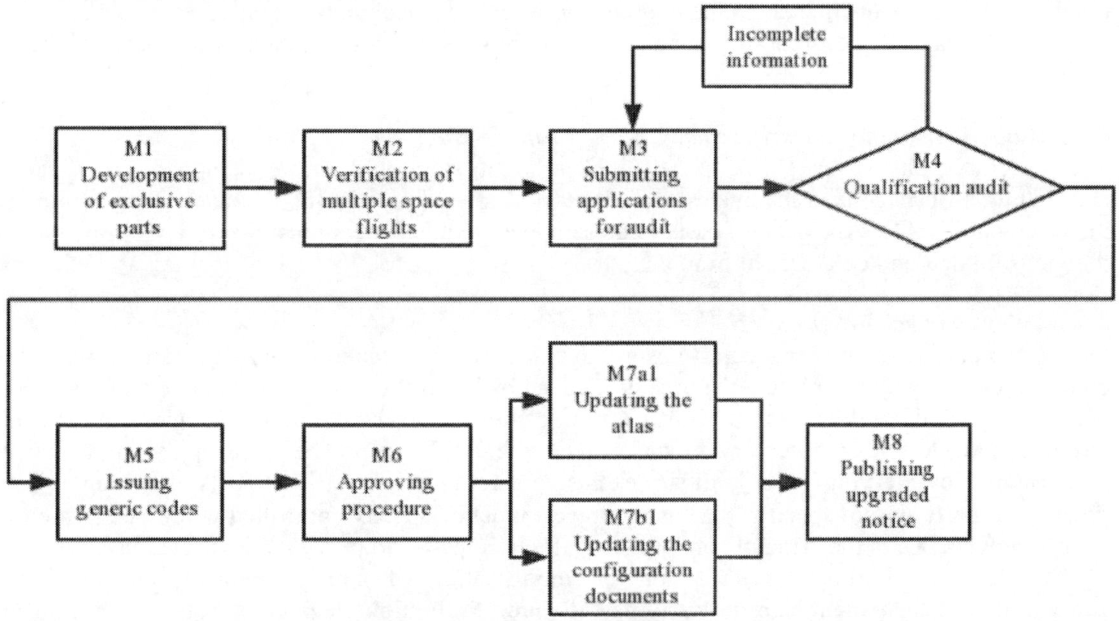

Figure 1. Flow chart for the addition of the general exclusive parts

3.2 The modification of the general exclusive parts for integration

The modification in this section refers to the non-temporary one. Because the configuration of general products usually involves multiple projects, for the general products, subversive modification of critical dimensions (such as interface dimensions) are not acceptable generally, and the modification that induces significant property degradation is prohibited.

When it is necessary to modify general exclusive parts for integration, the configuration management team comprehensively analyzes the impact on the functions, performance and interfaces of relevant projects. After the unified demonstration, collaborative verification and approval accepted by all partners concerned, to reduce the risks induced by the configuration change and realize the unified control of the configuration of general products, the designers implement the modification. The process of the modification of general parts is shown in Figure 2.

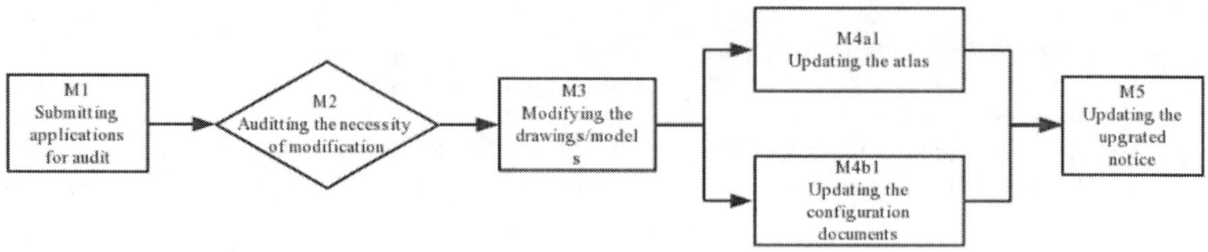

Figure 2. Flow chart for the modification of the general exclusive parts

M1: The designers in charge of general exclusive parts for integration submit a modification application to the configuration management team.

M2: The configuration management team reaches an agreement on the necessity of the submitted modification with the partners who are involved with the modified item.

M3: The designers in charge of general exclusive parts for integration modify the drawings/models according to the items approved, record the modified items completely and execute permitting procedure.

M4a1: The team for configuration management updates the atlas according to the content approved.

M4b1: The team for configuration management records the modified items and updates the version number of the drawings/models in the relative configuration documents.

M5: The team for configuration management publishes upgraded notice of the general exclusive parts.

3.3 The abolition of the general exclusive parts for integration

The general exclusive parts in space flight are not suggested to be scrapped. The team for configuration management evaluates the impact of the abolition. If the team for configuration management identifies the general parts to be scrapped, the designers must make obvious mark on the drawings/models and delete them from the atlas.

3.4. Deviation of general products

Due to the particularity of spacecraft design, the cost of using general exclusive parts in the original configuration may be larger than the benefit induced by general exclusive parts in certain condition. It is necessary to allow deviating the configuration of general exclusive part temporarily. When the designer considers it is necessary to make a temporary deviation, he need apply for a deviation permission. However, unlike the ordinary exclusive parts facing specific projects, the using objects of the general parts are not specific. The using objects must be strictly controlled within the scope of the ones approved, that is the principle of "who deviates, who uses" to prevent unexpected use.

The deviation is not the basis for the modification of configuration documents, but the configuration management team needs to pay attention to multiple deviations of the same product on the same issue, analyze and evaluate whether to modify the current configuration documents of general exclusive parts to form a continuous working mode for improvement.

4. The database for the general exclusive parts for integration

In order to effectively utilize data resources to guide design, it is necessary to construct a database for the general products systemically.[3] It is suggested to establish the database for the general exclusive parts for integration, including all kinds of characteristic, model/drawings and description information necessary. The database has a high utilization rate and does not fail at the end of a specific project. The general product database is defined as two parts: the database for design and the database of drawings/model. The two parts cover the complete information for design.

The database for design includes the information as follows:

- The mechanical properties and electrical properties of general exclusive parts;

- Information of physical properties such as surface characteristics, material, heat treatment status, weight, etc.;
- History such as the source of general parts, project usage, historical change, etc.;
- Requirements for acceptance management;
- Product list;
- The commissioning requirements;
- Other information necessary.

The model database includes both 2D-drawings and 3D-models. The database of 3D-models has appropriate interface with the model base for the whole-craft integration to lay the foundation for the digital integration based on the models.

The database fitted for the actual condition needs to be updated regularly. It is necessary to emphasize the consistency between 2D-drawings and 3D-models and update the data regularly.

5. Conclusion

Because of the small batch production pattern for spacecraft, it is imperative to reduce the development cost. It suggests the development mode of "from project to project" for the general exclusive parts for spacecraft integration to reduce the cost, which indicates the general exclusive parts are selected from the ones verified during a specific space project and separated from the specific project management system. They are constantly expanded and improved to meet project requirements. Furthermore, this paper suggests a database with integrated information improved synchronously to support the whole-spacecraft development.

References

[1] Shen, C., Ye, J.(2009)Discuss on management of modular units. Aerospace Standardization,4:27-30.

[2] Chen, G., Zhang, X., JIN, S.(2013) Product line design based on commonality of development-sensitivity. Forecasting,32:76-80.

[3] Li, S., Huang N., Dong, H., Liu, J.(2019) Research on Construction Proposals of Aero-engine General Parts Design System. Aeronautical Science and Technology,30:21-25.

IOP Conf. Series: Materials Science and Engineering **711** (2020) 012021 doi:10.1088/1757-899X/711/1/012021

Tensile constitutive model of reactive powder concrete based on flexural experiment

Wangwang Li[1*], Wenyu Ji[2], Yonghua Su[1] and Xinlin Ban[1]

[1] Railway Engineering Research Institute, China Academy of Railway Sciences Corporation Limited, Beijing, 100081, China

[2] School of Civil Engineering, Beijing Jiaotong University, Beijing, 100044, China

[*]Corresponding author's e-mail: wangwangli0128@163.com

Abstract. Reactive powder concrete(RPC) is a kind of ultra high performance concrete material, which has ultra high compressive strength and durability as well as good tensile property, elastic modulus, energy absorption and ductility. All these properties make it to be a better choice for the construction of bridges, buildings, military architectures, nuclear structures and the buildings bearing severe environment. The good tensile property, durability and ductility of RPC depend on the steel fibres mixed in the concrete. The bridge effect of steel fibres provides high tensile strength and crack resistant property. The tensile constitutive model of RPC differs from the normal concrete. This study tried to obtain the tensile constitutive model of RPC based on the flexural experiments of RPC prism. On the basis of test results, the tensile constitutive model of RPC is given in this study.

1. Introduction

Reactive powder concrete (RPC) is a kind of ultra high performance concrete. Since its appearance in the Bouyues' laboratory in France[1-2], it has been explored and used for several decades. RPC has ultrahigh performance, including ultra high compressive strength, high tensile strength, high elastic modulus, high energy absorbing, high ductility and good durability. The cement matrix is enhanced by the steel fibres, which develops the high tensile properties and crack resistance of RPC.

These good mechanical properties of RPC make it to be a better choice in the construction of buildings. RPC has been applied in many structures worldwide[3-4]. The first application of RPC was the Sherbrooke Bridge in Canada. The single span this bridge is 60m with RPC truss. Then the Seonyu Bridge was designed and constructed in South Korea with a single span of 120m, in which the RPC slab was used to support the beam. The first application RPC in highway bridges is the Mars Hill Bridge in USA. This is a precast and prestressed high way bridge with a span of 33.5m, RPC was used to improve the durability of bridge structure. The first application of RPC in railway bridges is the RPC girder used in the Caoqian Railway line, which has a lower height of cross section than the normal concrete(NC) bridge for the application of RPC in China.

The RPC material has been widely applied in bridge structures due to its high mechanical properties and durability. RPC material has been explored by the researchers in many countries. The studies at first focus on the composition and preparation of RPC material[5-13]. These studies explored how to optimize the particle size distribution by changing the composition of RPC and how to improve the microstructure of matrix by changing the preparation procedures. These studies make a great contribution to RPC material for behaving ultra high mechanical performance and durability.

IOP Conf. Series: Materials Science and Engineering **711** (2020) 012021 doi:10.1088/1757-899X/711/1/012021

Then the studies of RPC focus the structural behaviours after the application of RPC[14-19]. The flexural properties of RPC girders were explored. The RPC girders have higher rigidity, cracking strength and ultimate strength than NC girder. The bridging effect of steel fibres in RPC can obviously improve the flexural properties of girders. Also the number of cracks on the surface of RPC girders is much higher than NC girders. The smeared cracks property of RPC is utilized when bearing bending moment, a large number of small cracks appeared instead of some large cracks. Then the durability of RPC structures was improved by the appearance of these small cracks. The shear properties of RPC girders were also studied by the explorers. The addition of RPC has a great effect on the shear failure mode and strength. RPC girders showed better shear behaviours than NC girders. But the studies in this scope are not sufficient to fully address the shear mechanism of RPC girders. All the test results show that RPC is a better material when applied in bridge girders compared with NC.

The higher tensile strength of RPC plays an important role in improving the mechanical properties of RPC structures. It is critical to consider the tensile strength of RPC in the calculation of RPC structures, and then the tensile constitutive model of RPC becomes important. However, the tensile constitutive model has not been fully explored, and only a few papers represented the explorations on tensile constitutive model of RPC. Most of these studies explored the tensile property of RPC structures bearing uniaxial tensile force, which react lower tensile strength when applied in RPC girders bearing moment. This study tries to explore the tensile constitutive model of RPC through flexural experiments of RPC prism.

2. Experimental programme

2.1. Mixture proportions of RPC

The mixture proportions of RPC are most important for achieving ultra high performance. The mix proportions of RPC explored in this study are given in Table 1. Silica fume and cement are used to improve the micro structure of RPC mixture. The fine aggregate added in the mixture proportions of RPC included course sand, medium sand and fine sand. The coarse aggregate added in the mixture proportions of RPC is gravel, in order to reduce the shrinkage effect. The steel fibres are added to increase the tensile properties and cracking resistance.

Table 1. Components of the RPC mixture designs(kg/m^3).

Cement	Silica fume	Gravel	Coarse sand	Medium sand	Fine sand	Steel fibre	Water	Water-reducing agent
624	178	547	225	448	112	120	150	18

2.2. Fabrication and specimen dimensions

The fabrication procedure is also important to the mechanical properties of RPC. In order to achieve high performance, the fabrication procedure must include the steam curing process. The mixing and curing procedure of RPC is provided in Fig. 1. Then the mixtures are molded in to 100*100*400mm bars for the flexural tests, the specimens are presented in Fig. 2.

IOP Conf. Series: Materials Science and Engineering **711** (2020) 012021 doi:10.1088/1757-899X/711/1/012021

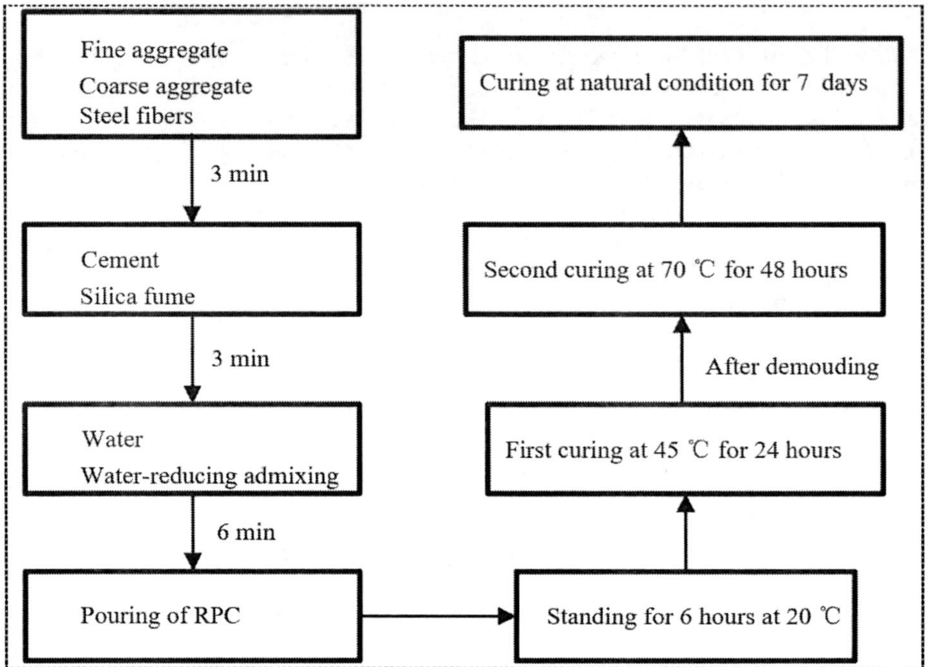

Figure 1. Mixing and curing procedure of RPC.

Figure 2. RPC specimens.

2.3. Test setup and instrumentation

The test setup is shown in Fig. 3. Each specimen was loaded in the compression testing machine, and the load and tensile strains were measured during the loading process. In order to get the accurate loading process, the load and tensile strains were measured by dynamic acquisition instrument (DH5923) with a sampling frequency of 500Hz.

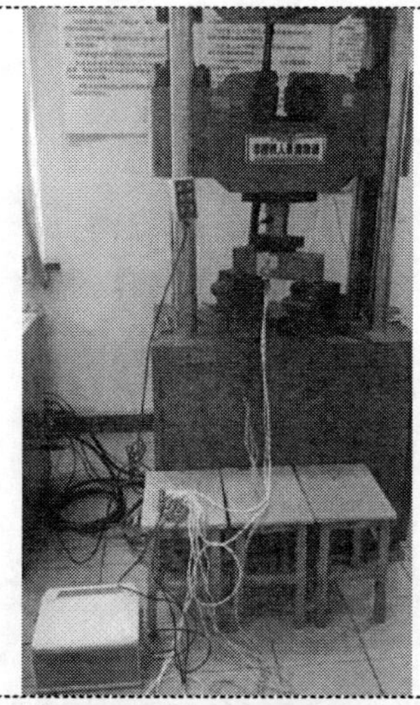

Figure 3. Test setup.

3. Experimental results

The tensile strains at the bottom of the RPC specimen were measured as well as the load applied on the specimen. Then the stresses can be determined through the load. The stress and strain relations are presented in Fig. 4. The descending stage cannot be measured for the insufficient stiffness of testing machine.

The results show that the tensile stress-strain relationship of RPC can be divided into two stages: (1)the stage before cracking, in this stage, stresses increase linearly with strains until the appearance of cracking; (2)the stage after cracking, in this stage, the stresses increase linearly with strain until the broken of specimen. This stage is also called strain hardening.

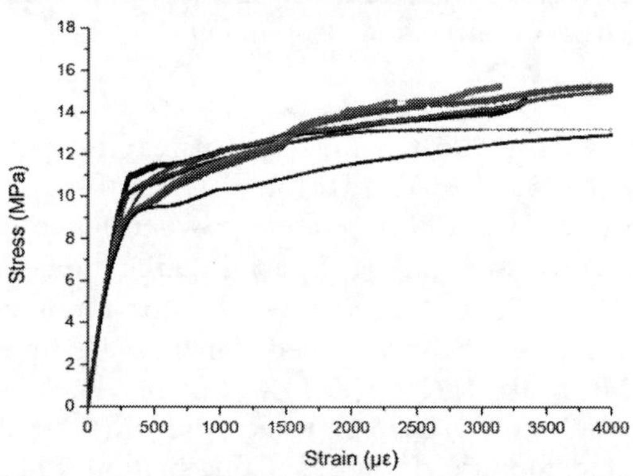

Figure 4. Stress and strain relations of RPC.

4. Tensile constitutive model

Based on the experimental results, the average first cracking strength and the peak strength can be obtained. Also the tensile constitutive model of RPC contains two stages, including the first increasing stage and the second strain hardening stage, in which the first cracking point and the ultimate point are determined from the test results. The fitting curve of RPC tensile constitutive model are presented with the test curves in Fig. 5. The comparison between the fitting curve and the test results indicates that the fitting constitutive model can represent the tensile property of RPC. Then the calculation equations of RPC tensile constitutive model are given in Eq. (1).

$$f_{Rt}=\begin{cases}\varepsilon_{Rt}E_{Rt1} & 0\le\varepsilon_{Rt}\le\varepsilon_{Rt,cr}\\ f_{Rt,cr}+(\varepsilon_{Rt}-\varepsilon_{Rt,cr})\ E_{Rt2} & \varepsilon_{Rt,cr}\le\varepsilon_{Rt}\le\varepsilon_{Rt,\max}\end{cases} \tag{1}$$

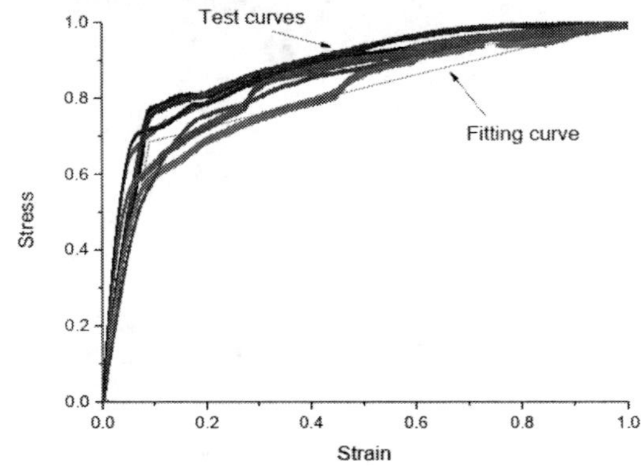

Figure 5. The fitting curve of RPC tensile constitutive model.

5. Conclusions

(1)The specimens 100*100*400mm bars made of RPC were designed and fabricated in this study.

(2)The tensile stress-strain relationship of RPC were tested and measured by dynamic acquisition instrument with a sampling frequency of 500Hz

(3)Based on the experimental results, the tensile constitutive model of RPC was obtained. This can be used in the calculation of flexural properties of RPC girders.

Acknowledgments

This work was supported by the National Natural Science Foundation of China (Project No. 51278040), the Major Project of China Railway (Project No. K2019G005), and the Science and Technology Projects of Hebei Transportation Department (No. C18L00580).

References

[1] Richard P., Cheyrezy M. (1995)Composition of reactive powder concretes[J]. Cement and Concrete Research, 25(7): 1501-1511.

[2] Richard P., Cheyrezy M. H. (1994)Reactive powder concretes with high ductility and 200-800 MPa compressive strength[J]. Special Publication, 144: 507-518.

[3] Yoo D. Y., Yoon Y. S. (2016)A review on structural behavior, design, and application of ultra-high-performance fiber-reinforced concrete[J]. International Journal of Concrete Structures and Materials, 10(2): 125-142.

[4] Shi C., Wu Z., Xiao J. et al. (2015)A review on ultra high performance concrete: Part I. Raw materials and mixture design[J]. Construction and Building Materials, 101: 741-751.

[5] Wille K., Kim D. J., Naaman A. E. (2011)Strain-hardening UHP-FRC with low fiber contents[J]. Materials and Structures, 44(3): 583-598.

[6] Wille K., El-Tawil S., Naaman A. E. (2014)Properties of strain hardening ultra high performance fiber reinforced concrete (UHP-FRC) under direct tensile loading[J]. Cement and Concrete Composites, 48: 53-66.

[7] Wille K., Naaman A. E., El-Tawil S. et al. (2012)Ultra-high performance concrete and fiber reinforced concrete: achieving strength and ductility without heat curing[J]. Materials and structures, 45(3): 309-324.

[8] Yoo D. Y. (2014)Performance enhancement of ultra-high-performance fiber-reinforced concrete and model development for practical utilization[D]. PhD thesis, Korea University.
[9] Hassan A .M. T., Jones S. W., Mahmud G. H. (2012)Experimental test methods to determine the uniaxial tensile and compressive behaviour of ultra high performance fibre reinforced concrete (UHPFRC)[J]. Construction and building materials, 37: 874-882.
[10] Yoo D. Y., Shin H O., Yang J. M., Yoon Y. S. (2014)Material and bond properties of ultra high performance fiber reinforced concrete with micro steel fibers[J]. Composites Part B: Engineering, (58): 122-133.
[11] Barnett S. J., Lataste J F., Parry T., Millard S G., (2010)Soutsos M N., Assessment of fibre orientation in ultra high performance fibre reinforced concrete and its effect on flexural strength[J]. Materials and Structures, 43(7), 1009-1023.
[12] Yoo D. Y., Banthia N., Kang S. T., Yoon Y. S. (2016)Effect of fiber orientation on the rate-dependent flexural behavior of ultra-high-performance fiber-reinforced concrete[J]. Composite Structures, 157: 62-70.
[13] Chan Y. W., Chu S. H. (2004)Effect of silica fume on steel fiber bond characteristics in reactive powder concrete[J]. Cement and Concrete Research, 34(7): 1167-1172.
[14] Yardımcı M. Y., Aydın S., Karabulut A. Ş. (2009)Mechanical properties of reactive powder concrete containing mineral admixtures under different curing regimes[J]. Construction and building materials, 23(3): 1223-1231.
[15] Yoo D. Y., Banthia N. (2016)Mechanical properties of ultra-high-performance fiber-reinforced concrete: A review[J]. Cement and Concrete Composites, 73: 267-280.
[16] Al-Hassani H M, Khalil W I, Danha L S. (2014)Mechanical properties of reactive powder concrete with various steel fiber and silica fume contents[J]. Acta Technica Corviniensis-Bulletin of Engineering, 7(1): 47.
[17] Habel K., Denarie E., Bruhwiler E. (2006)Structural behavior of elements combining ultra-high performance fiber reinforced concretes and reinforced concrete[J]. Structure Engineering, a132: 1793-1800.
[18] Wuest J., Brühwiler E. (2008)Model for predicting the UHPFRC tensile hardening[C]//Ultra High Performance Concrete (UHPC): Proceedings of the Second International Symposium on Ultra High Performance Concrete, Kassel, Germany, (10): 153.
[19] Safdar M., Matsumoto T., Kakuma K. (2016)Flexural behavior of reinforced concrete beams repaired with ultra-high performance fiber reinforced concrete (UHPFRC)[J]. Composite Structures, 157: 448-460.

Design of the remote-control system for PM2.5 capture device

Wan Wang*, Yanlin Lu, Zuo Zhang, Tianzhi Wang

College of Chemistry and Chemical Engineering,
Guizhou University,
Guiyang, Guizhou, 550025, China

wwang7@gzu.edu.cn

Abstract: PM2.5 is the main factor of haze formation. This paper designed a remote control system for PM2.5 capture device by using combined strategies, which were redundant structure, acoustic vibration and reverse blowing. The remote control system is composed of Siemens PLC s7-200 series CPU, software and hardware. The system was simulated by Siemens S7200 SIM programming. The combined strategies realized the continues work. In addition, remote monitoring of the PLC control system implemented to make remote operation more convenient.

1. Introduction

The "grey haze" atmospheric phenomenon occurred in most cities in northern China in 2013. It was caused by the massive emission of PM2.5. PM2.5 refers to the equivalent diameter of less than or equal to 2.5 microns of particulate matter, also known as inhalable lung particles. Its less than 1/20 of the diameter of a human hair, permeate the air, scattering and absorbing visible light, reducing atmospheric visibility and affecting traffic safety [1,2]. In addition, PM2.5 has a small particle size and strong adsorption ability, which can be used as a "carrier" for many pollutants, viruses and bacteria, and stay in the atmosphere for a long time and transport over a long distance, seriously affecting the air quality. The problem of existing capture devices for PM2.5 is that the flux would decrease after working for a period. The reason is that partly PM2.5 adheres to the separated components [3,4]. In this paper, the s7-200 programmable controller from Siemens was selected as the controller of the capture device. The combined strategies, the redundant structure of PM2.5 capture device, acoustic vibration and reverse blowing, were used to realize the continuous operation of the capture device.

2. Working principle and control requirements of PM2.5 capture device

2.1Working principle of PM2.5 capture device

PM2.5 were captured on the surface of the separated components through the delivery of the blower. Clean gas passed through the separated components, and PM2.5 was discharged with the gravity when the capture device stopped. Part of the PM2.5 adhesived on the surface of the component layer, the internal channel was blocked with the increase of thickness, the pressured drop of components on both sides increased; The rate of passing clean gas were reduced, the pressure drop increases. It is difficult to make the adhesion layer fall off by using reverse blowing alone. With the increasing frequency of repeated reverse blowing, the initial flux was down. The paper proposed a combined strategies, redundant device structure; acoustic vibration and reverse blowing. First, the attached layers were destroyed by acoustic vibration, then reverse blowing is carried out. When one device is cleaned, the other device still works. As shown in Fig. 1.1 The redundant structure of PM2.5capture device. It consists of two PM2.5 capture devices. Each device contains the inlet valve (11), the outlet valve (13), draught pan (1), discharging valve (17), differential gauge (9), separation components (24), back-flushing valve (5), acoustic vibration valve (20), acoustic vibration generator (19), container of high pressure gas (39).

This process included 3 steps: first, the inlet valve (11) and outlet valve (13) were closed, when the differential pressure reached the specified value; second, the acoustic vibration valve (20) was opened, which drove the acoustic vibration generator (19); then the back-flushing inlet valve (5) and back-flushing outlet valve (7) were opened, after the structure of attachment layer was broken. The differential pressure is difficult to reach90 the specified value simultaneously, the redundant structure can carry out PM2.5 removal continuously.

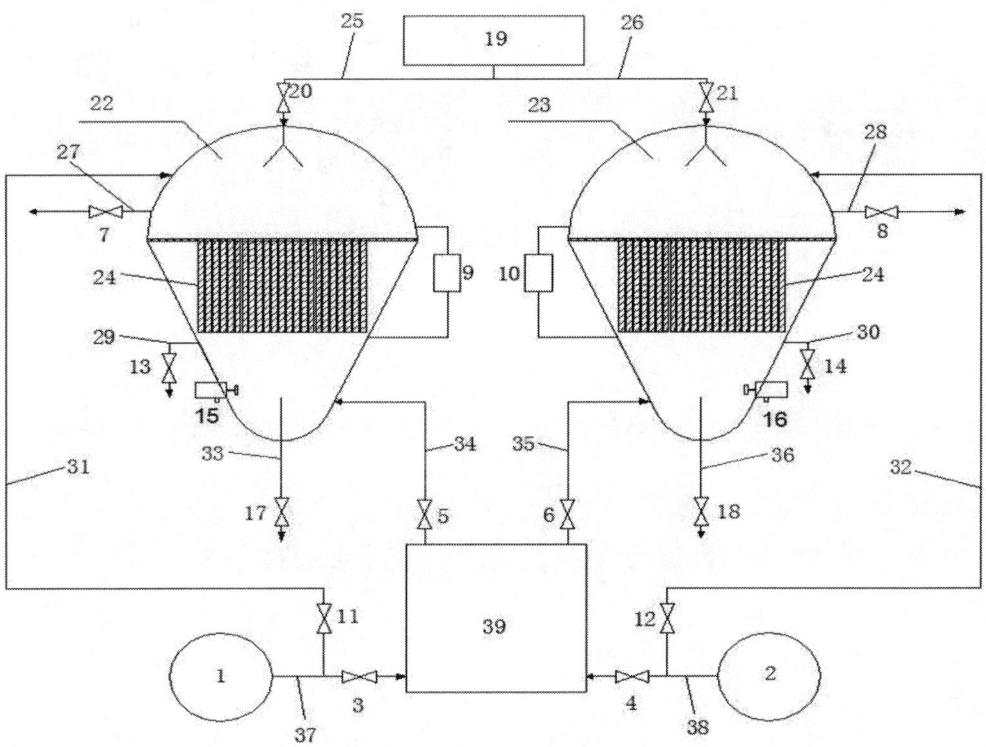

Fig. 1 The redundant structure for PM2.5capture device

(1) (2)draught pan, (5) (6)the back-flushing inlet valve, (7) (8)back-flushing outlet valve, (9) (10)differential gauge, (11) (12)inlet valve, (13) (14) the outlet valve, (15) (16) level switch, (17) (18)discharging valve, (19)acoustic vibration generator, (20) (21)acoustic vibration valve, (24)separation components, (39)container of high pressure gas.

2.2 Control requirements

(1) Normal: We set different working time.10 min for device 1 and 11 min for device 2. That's to say, draught fan (1), inlet valve (11) and outlet valve (13) of device 1 were opened and work for 10 minutes. draught fan (2), inlet valve (12) and outlet valve (14) of device 2 were opened and work for 11 minutes.

(2) Acoustic vibration: At the end of work, inlet valves (11) (12) and outlet valve(13) (14) were closed. Acoustic vibration valve (20) (21) were opened for 15 seconds[5].

(3) Back-flushing: (5) (6)the back-flushing inlet valve, (7) (8)back-flushing outlet valve were opened for 14 seconds[6].

(4) Discharging: (17) or (18) discharging valve were opened for 5 sec seperately, when the level switch (15) or (16) got the set point: if the device is normal, inlet valves(11) or (12) and outlet valve (13)

or (14); if the device is back-flushing, (5) or (6) the back-flushing inlet valve and (7) or (8) back-flushing outlet valve were closed. The program flow chart was shown in Fig. 2.

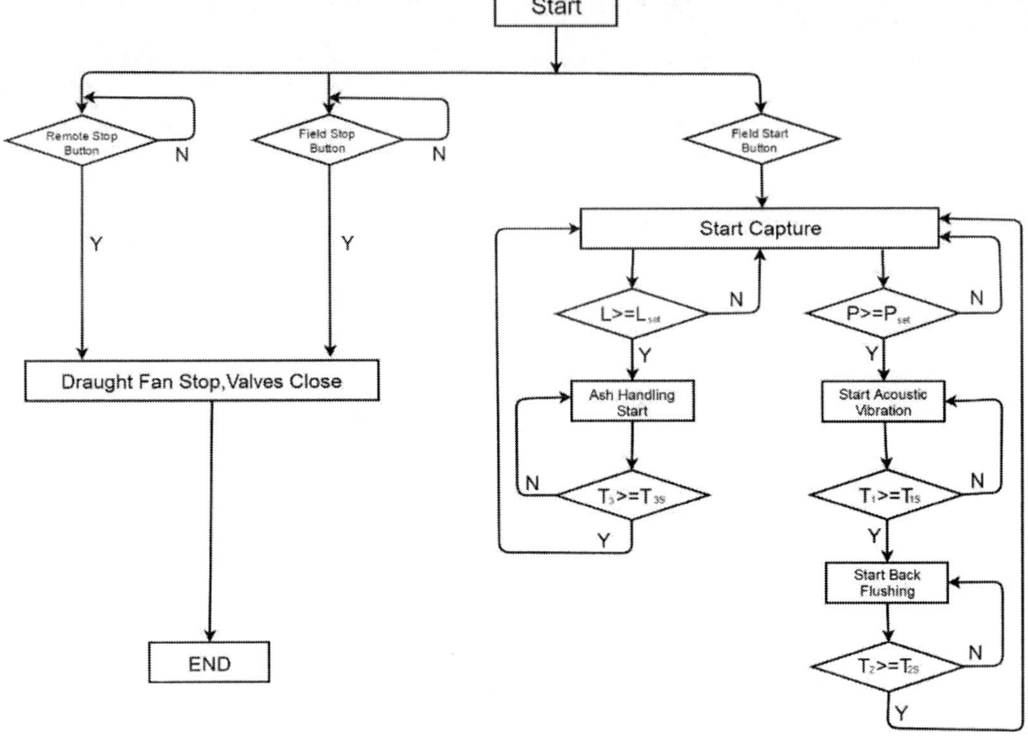

Fig. 2 The program flow chart

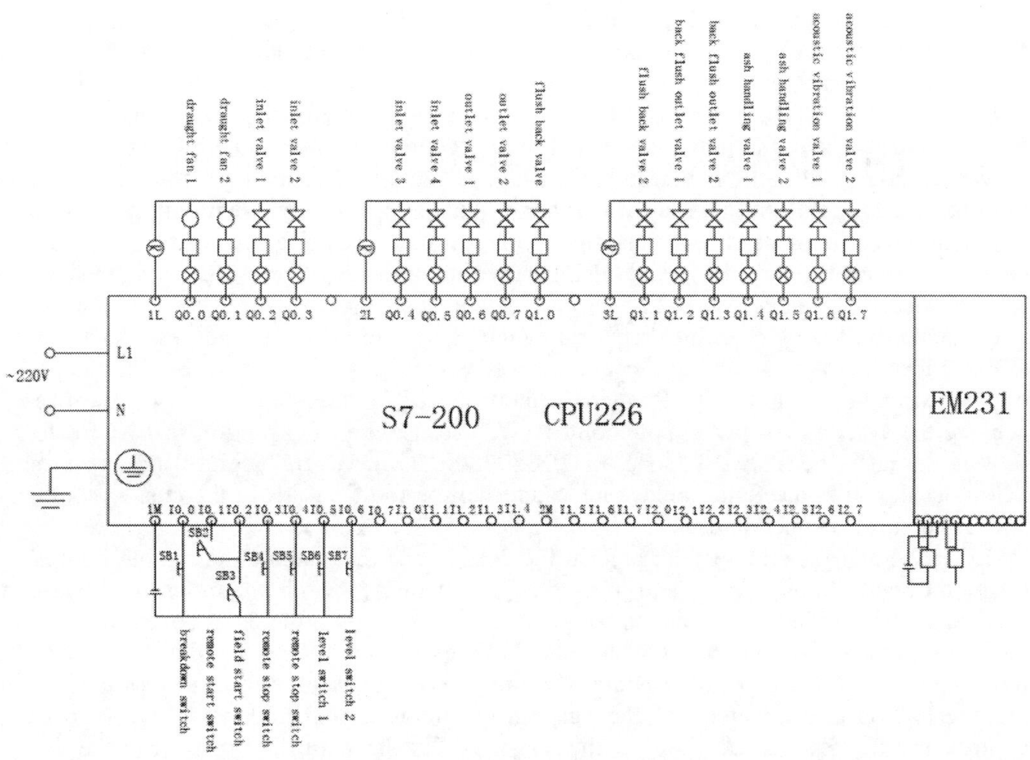

Fig. 3 The connection of CPU226

3. PLC control system design

3.1 Hardware Selectin

The hardware includes solenoid valve, differential pressure meter, draught fan, acoustic vibration generator, level switch, PLC s7-200 series CPU. CPU226 with 24 input points, 16 output points and expanding module EM231, as shown in Fig.3, I0.0 ~ I0.6 as input, which is suitable for fault switch, I0.1 I0.0 for remote start switch, I0.2 start switch at site, I0.3 for remote stop switch I0.4 stop switch, I0.5 at site for level switch (15), I0.6 for level switch (16). Q0.0~Q0.7 and Q1.0~Q1.7 are outputs. Including Q0.0 for draught fan (1), Q0.1 for draught fan (2), Q0.2 for (11), Q0.3 for (12), Q0.4 for (5), Q0.5 for (6), Q0.6 for outlet valve (13), Q0.7 for (14), Q1.0 for the back-flushing inlet valve (5), Q1.1 for (6), Q1.2 for back-flushing outlet valve (7), Q1.3 for back-flushing outlet valve (8), Q1.4 for discharging valve (17), Q1.5 for discharging valve (18), Q1.6 for valve (20) Q1.7 for valve (20).I0.0 and I0.3~I0.6 are normally closed(NC) switches, and I0.1 and I0.2 are normally open(NO) switches.

3.2 Software Design and Simulation

This system adopts PLC s7-200 series of special step-7 MICRO/WIN software for programming. First, the ladder diagram is used to write the logical sequence of the automatic control system. Run s7-200 series of special simulation software "s7-200 Chinese version" for simulation, enter the simulation software, configure CPU226 and two EM231 analog input modules, and select the analog input current value of 0~20mA.Manually adjust the input parameters of the analog quantity in the normal state and load the programmed ladder diagram into the simulator.

The remote control system was realized through 2 steps: communication; PLC remote control. After connecting the computer and PLC with PPI line, run STEP7 programming software, select "communication" and then select "set PG/PC interface", and then set baud rate (generally 9.6kbit/s), and the point-to-point connection between the computer and CPU is completed. Run PC Access software, right mouse click on PLC to set "properties", to ensure that the IP address and TSAP address in PC Access should correspond to the figure above, for example, here we will set the IP address in PC Access to 192.168.0.8, local TSAP address to 11.00, remote TSAP address to 10.00, can complete the Ethernet Settings. Due to the limitation of distance, Ethernet cannot reach a relatively long distance. At this time, we can set up the CPU Internet through CP 243-1 IT module. The setting method is similar to the above Ethernet, but the connection through the Internet can realize ultra-remote monitoring and control of PLC anywhere on earth. After PLC communication is set up, the running state of PLC can be monitored and controlled remotely. For monitoring, OPC Server software -- PC Access as described above -- is needed. Run PC Access, create a new PM2.5 capture device, create a new folder "analog quantity monitoring" to create the six analog quantities required by the device, and make the address (VD0, etc.) correspond to the ladder diagram one to one. Access select RW, that is, read and write, but PC Access software can only read parameters, not write. So there is another need for a reading and writing software, in order to achieve the system of super remote control. We can use the PLC plug-in in EXCEL for reading and writing editing. Open EXCEL, load a "OPCS7200EXCELADDIN" control in the development tools editor, as shown in the figure below, and there will be four tool ids in the custom toolbar. This is the EXCEL control we need to use.

Enter the address and switch name in the cell. Reads the state of the remote boot switch into the location of cell C2 through the add-in's tool. The remaining switch quantities are also filled in the cell of state column in the same order as above. Since the running tool has not been opened yet, the status bar displays the #N/A flag. When the third tool of the load item is pressed, the status bar will display the parameters of the corresponding address in the working state of PLC. At this time, the operation of reading in data is completed and the purpose of remote monitoring is achieved. To achieve remote control, use the second write tool of the add-in. Press the write tool to rewrite the data VD0 of the corresponding cell into a value of 16.9.

4. Results

Start the PLC, open remote start I0.1 and on-site start I0.2, and the system starts to run with electricity, and the draught fan works. The inlet and outlet valves of the left and right two dust removal devices are opened, and then disconnect the I0.1 and I0.2 systems without power off, because a self-locking circuit is formed. When PM2.5 was collected for 10 minutes, the device started to vibrate and device 2 continues to dust. After 15 seconds of acoustic vibration and 14 seconds of back flushing, the device started again. When the pressure differential meter reaches the specified value (manually adjust the AIW2 analog value), device 2 is performing 15-second acoustic vibration, while device 1 continues to dust. At this time, if the pressure difference still fails to return to a normal level after one round of acoustic vibration and back flushing, the system will not carry out normal dust removal, but carry out the second or even the third round of acoustic vibration and blowing, until the membrane flux returns to make the pressure difference within the separation component reach a normal level.

when the ash particle reaches the position of material level switch (1), the material level switch I0.5 is energized, the ash discharge valve (1) is opened for ash discharge, and the ash discharge valve (1) is closed after 5 seconds, and the equipment starts to dust normally once again. When the ash particle reaches the position of the material level switch (ii), the material level switch I0.6 is energized, the ash discharge valve (ii) is opened for ash discharge, and the ash discharge valve (ii) is closed after 5 seconds, and the equipment ii starts to dust normally.

when the circuit fails, the whole system stops running.Or press the remote stop button or the field stop button, and the system will stop running.

At this point, the simulation is finished and the logic of the programming ladder diagram run normally.

5. conclusion

The remote control system was designed for PM2.5 capture device. The redundancy structure of device design can realize continuous uninterrupted operation, with Siemens S7-200 series, 226 type of CPU a shaker automatic control system, and according to the process of flow chart, the ladder diagram, the simulation process can be complete and correct. Through TCP/IP Internet protocol, remote monitoring and control of the whole system can be realized. Due to the limited experimental conditions, many industrial actual production conditions cannot be simulated, so if it is used in actual production, it needs to be adjusted according to the actual situation. This system realizes more intelligent and humanized design, reduces the labor intensity of workers, and makes the control of the whole system more convenient and faster.

Acknowledgment

The paper was fund by Youth science and Technology Talent Growth Program for Education Department of GuiZhou province, Study on sustainable optimization design and application of C-H-O element integrated network in industrial park, Qian jiao he KY zi [2018] 109.

About the Author

Wan Wang*,(1981-), female, Doctor of Engineering, Associate Professor, research field, Process equipment design and integration.

References

[1] Yang X. X., Feng I. H., Wei P. (2012) Atmospheric particulate matter PM2.5 and its hazards [J]. Frontiers of science., 6(01): 22-31.

[2] Zhang H.J., Lu X.S. (2012) Overview of dust removal technology for controlling PM (2.5) [J]. China environmental protection industry. ,(03): 29-33.

[3]Zhu Q. Q., Zhou L. H., Yu C. C., et al. (2012) PM_(2.5) and its management technology for coal-fired power plants [J]. Power station systems engineering., 28(04): 19-20.

[4] Li L. (2007) Design of automatic control system for inorganic membrane separation dust remover [J]. Journal of guizhou university of technology (natural science edition). ,(01): 5-7.

[5] Yuan J. (2012) Application of low-frequency acoustic wave ash cleaner in an aluminum plant [J]. Heilongjiang science and technology information. ,(11): 37.

[6] Wang J., Zou S. S., Tian M. K., et al. Experimental study of collecting performance of silicon carbide ceramic film tube for PM(2.5) [J]. Environmental engineering. 2015, 33(06): 80-83.

On the Influence of Platform Economy on Transformation and Upgrading of Chinese Traditional Manufacturing Industry
- A Case Study of Red Collar Group

Hanyu Song[1] Mengyun Liu[2]

[1]China University of Petroleum. School of Economics and Management. Economics 266580

[2]China University of Petroleum. School of Economics and Management. Marketing 266580

Abstract: The profitability of the traditional manufacturing industry is sluggish and fails to well adapt to the current trend of economic globalization. In the context of industry 4.0, it is imperative for the traditional manufacturing industry to transform and upgrade in an all-round way in the market-oriented and government-led environment. At this time, platform economy furnishes a good impetus for the transformation and upgrading of traditional manufacturing industry. Traditional manufacturing industry adapts to the needs of the external market environment through platform structure, adjusts factor input premised on dynamic changes in factor endowment, integrates scientific and technological revolution, and gradually cultivates competitiveness through continuous technological innovation, thereby achieving the transformation and upgrading of industry. In the background of conversion of era, this paper explores the innovation of traditional enterprises in the current historical background from the angle of evolution, attempting to reveal that platform economy helps traditional manufacturing enterprises to update their production methods, restructure their management structures and set up supply chain thinking, thereby pushing forward their transformation and upgrading. Taking the transformation and upgrading of Red Collar Group as an example, this paper analyzes the process of rapid learning, resource integration, innovation iteration, self-fission, and continuous evolution of the enterprise premised on platform economy, thereby proving the positive impetus of platform economy to the transformation of traditional manufacturing industry and testing its logical rationality. The research concludes that platform economy will become one of the significant strategic modes of transformation and upgrading of traditional manufacturing industry in China.

1. Introduction

China's traditional manufacturing industry makes up the highest proportion of national economy, creates the most tax revenue and offers the most jobs. Its main business income makes up approximately 88% of total revenue of China's industries (data source: China Statistical Yearbook 2017). However, since China's traditional manufacturing industry is at the low end of the global value chain, its extensive economic growth mode fails to well adapt to the new features and requirements of economic globalization and high technicalization. Resulting from the continuous increase of manufacturing costs and the decrease of consumption in the international market, the international competitive advantage of enterprises is declining, and China's domestic demand is insufficient, resulting in that the traditional manufacturing industry is faced with difficulties. To better adapt to international competition, China's traditional manufacturing industry needs to extend from the manufacturing link at the low end of the value chain to the two ends of the "smile curve" (research and development and manufacturing servitization) to realize the strategic transformation of transformation and upgrading of traditional manufacturing industry.

In the context of deepening social division of labor, speeding up technological progress and intensifying market competition, manufacturing industry is confronted with transformation and upgrading mechanisms and paths different from those in the past. As an emerging economic theory, platform economy is rising in a rapid manner, which has played a huge role in advancing the transformation and upgrading of traditional manufacturing industry. Platform economy is a new form of industrial organization, and it is a new format aiming at improving efficiency and decreasing cost premised on digital technology, data drive and platform support and by means of business process change, industry chain consolidation and multi-business combination. At the theoretical level, in the past 30 years, the platform economy theory has been continuously developed, the connotation of platform economy has been continuously extended, and the platform economy thought has been continuously applied in the commercial and industrial fields. Especially in the new economic environment contributed by globalization, informatization and networking, it offers an opportunity to push forward the transformation of traditional manufacturing development strategy. In the meanwhile, platform economy is also of vital practical significance. Data show that all 15 major Internet companies on a global scale operate with the platform mode, and 60 of the world's largest 100 companies also derive their major income from platform mode. While in the 2016 list of domestic unicorn enterprises, 11 of the companies with the highest valuation among the top 15 have all been operated in the platform mode. To sum up, we have entered the era of platform economy. In the national strategic background of "Internet+", platform economy has already become a significant tool for China's economic transformation and a vital part of "mass entrepreneurship and innovation". Platform economic activities can realize the concentration of information and transactions, promote the integration of production factors, improve the efficiency of resource allocation, reconstruct the value chain of the original industry, accelerate the innovation of business models, and enable stakeholders to benefit from the platform.

Taking Red Collar Group as an example, it is the focus of this paper to study how the traditional manufacturing industry can realize transformation and upgrading with this platform economy. Based on the research in related fields at home and abroad and combined with the case analysis of Red Collar Group's development, this paper expounds the influence of platform economy on traditional manufacturing industry in three aspects: changing production concept, innovating business management mode and reshaping industrial supply chain. It clarifies the relationship between the construction of platform economy and the transformation and upgrading of traditional manufacturing industry, offering strategic support for the transformation and upgrading of traditional manufacturing industry under the new economic form.

2. Platform Economy has Triggered Changes in the Economic Operation Mechanism.

The booming development of platform practice calls for related theoretical research. Over the past ten years, the platform has become a topic of common concern in many fields such as economics, management, law, sociology and political science. McAfee and Brynjolfsson1 (2017) regard the rise of the platform as one of the three landmark events of the "digital revolution". It can be said that the development of the platform economy has profoundly changed the face of the global economy. Understanding the characteristics of the platform economy itself and how enterprises use the platform economy is crucial to the transformation and upgrading of traditional manufacturing industries.

First of all, the platform economy is provided with strong core resources and service capabilities, and can create an open and win-win business model/ecosystem (Parker & Van1, 2012, Ceccagnoli2,

[2] Parker G, Van A M, A digital postal platform: Definitions and a roadmap [R], America; The MIT Center of Digital

2014). The emergence of the platform exerts an impact on the change of business model, information transmission and transaction mode: that is, the structure of transaction mode and marginal cost is changed into a charging mode. In the meanwhile, platform operators gather factor resources to the maximum extent and develop a new economic mode: a business mode that allows them to continuously add value while benefiting platform participants (Shi Jianyong3, 2013). Thanks to the ever-accelerated development of business models, the integration elements of the platform have gradually developed into an ecosystem. Traditional industries rely on the ecosystem built by the platform, or rely on network platforms to build their own ecosystem, in a bid to achieve the purpose of industrial transformation and upgrading through the formation of ecosystems. The emergence of platform ecosystems has covered the original industrial pattern and blurred the industrial boundaries, because platform-based enterprises can continuously carry out cross-border operations premised on their core products or services and integrate various businesses to boost industrial transformation and upgrading.

Next, the platform is in possession of a modular division of labor mechanism, an open innovation mechanism, a competition mechanism, a trust mechanism and a network marketing mechanism, which can reshape the industrial chain and boost the transformation and upgrading of the traditional manufacturing industry as a whole. The platform economy reduces the cooperation cost of division of labor, enhances the technological capability of manufacturing enterprises and facilitates the upgrading of the technological process of manufacturing industry; the open innovation mechanism of the platform reduces the cost of product research and development, shortens the research and development cycle, improves the innovation capability of manufacturing industry, and facilitates product upgrading. Moreover, market competition on the platform forces manufacturing enterprises to innovate in research and development, providing development opportunities for small-scale enterprises with market segments and advancing the upgrading of industrial chain functions; the trust and cooperation mechanism of the platform strengthens the cooperation relationship between platform enterprises, reduces the cooperation cost, makes the platform network more extensible, and improves the outward association capability of the platform network (Peng Shiyan[4], 2017). The platform theory of the precise marketing mode based on the platform blurs the concepts of upstream and downstream suppliers and consumers (Ali Research[5], 2016). Yao Kai et al.[6] (2009) expounded that in the network-like industrial chain, platform leaders have realized value creation coordination and standard upgrading among participating entities through a series of strategies.

After the research of various scholars and the analysis of individual enterprises, this paper holds that the impact of platform economy on the transformation and upgrading of traditional industries is principally reflected in three strategic ideas. The first strategic idea for the traditional manufacturing industry to participate in the construction of the platform mode to realize transformation and upgrading is to change the previous cost-leading production mode, abandon the concept of mass production and scale production, and develop personalized and modular flexible production mode by using the platform business mode. On this basis, Red Collar has basically addressed the issues of high inventory, high cost and insufficient innovation capability faced by the traditional manufacturing industry, offering a reference for the successful transformation of the traditional manufacturing industry. The second strategic idea is to change the previous hierarchy-driven management mode to a flat management mode driven by orders placed under the platform economy. The platform makes

Business, 2012.

[3] Ceccagnoli M, Forman C, Huang P, et al. Digital platforms: when is participation valuable?[J]. Communications of the Acm, 2014, 57(2):38-39.

[4] Shi Jianyong. A New Economic Form to Optimize Industrial Structure-Research on Micro-operation Mechanism of Platform Economy[J]. Shanghai Journal of Economics, 2013(8):85-89.

[5] Peng Shiyan. Research on Transformation and Upgrading Mechanism of Manufacturing Industry from the Perspective of Platform Economy[J]. Social Science Front, 2017(7):258-261.

[6] Ali Research. Platform Economy[M]. China Machine Press, 2016.

[7] Yao Kai, Liu Mingyu, Rui Mingjie. Value Innovation Synergy of Network Industry Chain and Platform Leadership [J]. China Industrial Economics, 2009(12):86-95.

information exchange and communication more effective, and the point-to-point communication hierarchy renders the production and sales of traditional manufacturing more efficient. The third strategic thinking is to play the role of supply chain in industrial transformation and upgrading by using information system and four-flow coordinated development premised on platform economy, supply chain thinking, and on the basis of platform economy, to become a value-added chain covering the whole product movement process, and to further build an open supply chain that is mutually beneficial and win-win with suppliers.

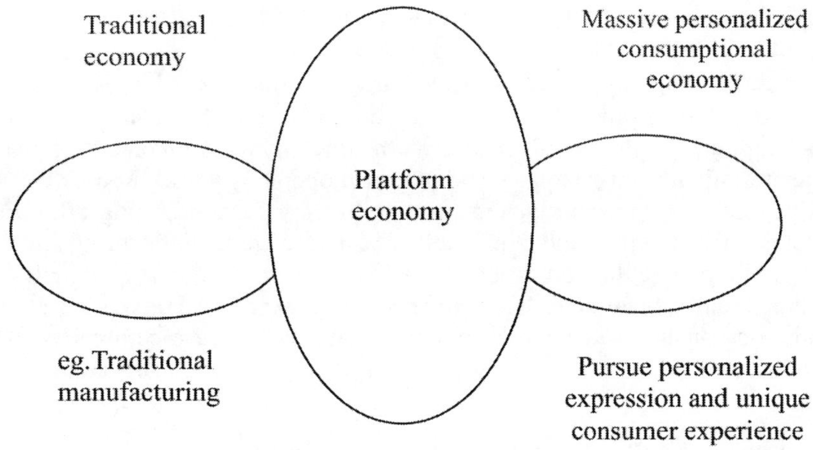

Source: Author drawing
Figure 1 . The connecting function of platform economy

3. Impact of Platform Economy on Transformation and Upgrading of Traditional Manufacturing Industry

Qingdao Red Collar Group Co., Ltd. (hereinafter referred to as Red Collar) has successfully launched a personalized customization platform in the global Internet era, the Global Clothing Customization Supplier Platform with hundreds of millions of capital investment after over 10 years of verification. Red Collar has created a complete solution of full customization, full life cycle and full industrial chain customization of clothing, thoroughly subverting the traditional concept of "workshop tailor" and initially boasting a certain competitive advantage of "speed". After 11 years, Red Collar successfully shaped a new business model (C2M) and a new marketing model (F2C) for garment enterprises in virtue of platform economy and the concept of the Internet. It has built a direct interactive platform between consumers and manufacturers, eliminating the intermediate links. The whole process, by relying on data-driven and network operation, makes use of the lean management of platform economy to realize process and organization reengineering, fast learning, integration of resources, innovation iteration, self-fission, and continuous evolution.

3.1. Platform Economy Contributes to Flexible Production

From the point of view of product production, the platforms are classified into internal (company or product) platform and industrial (external) platform. Internal platform refers to a company's pooling of a series of assets to availably develop and produce a series of products. While making centralized utilization of resources and modular design and production, and enhancing the output efficiency, increase complementary and alternative products to meet the diversified needs of consumers.

Red Collar's Cotte Internet platform (inside) makes full use of information technology, centralizes resources, availably realizes direct interaction between consumers and manufacturers, eliminates information asymmetry and various agency costs resulting from intermediate links, carries out

IOP Conf. Series: Materials Science and Engineering **711** (2020) 012023 doi:10.1088/1757-899X/711/1/012023

industrial assembly line production of personalized products, and realizes personalized customization under large-scale production. Red Collar platform organically combines real economy with virtual economy and initially explores a new path for the transformation and upgrading of traditional manufacturing industry.

(1) LP production mode: Red Collar has fully mobilized the four platform eco-circle entities of platform founder, platform provider (platform leader), user (demander) and user (provider) to develop a personalized production concept. Red Collar has studied and accumulated a database for over 10 years to meet over 99% of customers' personalized suit design needs, which not merely accurately satisfies customers' needs but reduces the enterprise's inventory and completes the effective allocation of resources. The platform employs big data technology to realize customized mass production and meet large-scale differentiated needs.

(2) Agile manufacturing: in the era of platform economy, everything is modular, and every enterprise will specialize in its products. Hence, on the basis of the personalized platform, Red Collar will further launch modular production. Premised on the three major components of platform economy (i.e. platform operator, platform participants and platform operating rules), Red Collar will respond to users' needs in an effective and coordinated manner by rapidly configuring all sorts of resources (including technology, management and personnel) so as to realize manufacturing agility. It delays the difference points on the production line, responds to customer orders quickly, and allows customers to receive suits within 7 days. Meantime, Red Collar actively explores the creation and application of cross-border trade e-commerce retail mode, builds a multi-language e-commerce trading platform, enriches the platform ecosystem, and realizes seamless connection of cross-border electronic commerce.

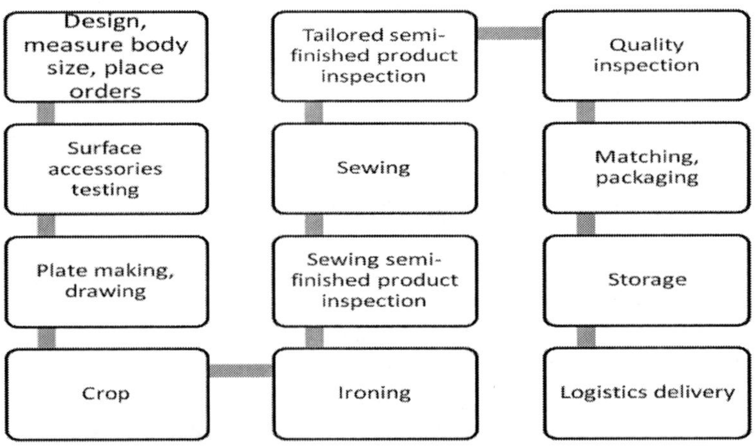

Source: Author drawing

Figure 2. Fully Customized Process - 7 Business Days Process Show

3.2. Change Management Model of Platform Economy

The platform-based business ecosystem breaks through the traditional linear transaction logic, regards the user base as a resource, and puts forward a brand-new network marketing mode, which offers the possibility to expand the traditional strategic management theory. This platform theory is different from the traditional linear marketing theory. It blurs the concepts of upstream and downstream, focuses on orders, and innovates the management mode. Red Collar reversely integrates production factors, comprehensively integrates and eliminates redundant departments, changes hierarchical management into platform-based management, and creates a platform-based and point-to-point flat management mode to realize barrier-free point-to-point communication among departments. Apart from that, it establishes a management mode with the customer service center as the nerve center. All the customers' demands are collected to the customer service center, and the customer service center directly issues instructions point to point, mobilizing the company's resources to complete the order.

IOP Conf. Series: Materials Science and Engineering **711** (2020) 012023 doi:10.1088/1757-899X/711/1/012023

Based on the node management of the platform, Red Collar finally realizes a complete and customized business operation mode and realizes the network-based management of information flow, capital flow and logistics in an all-round way.

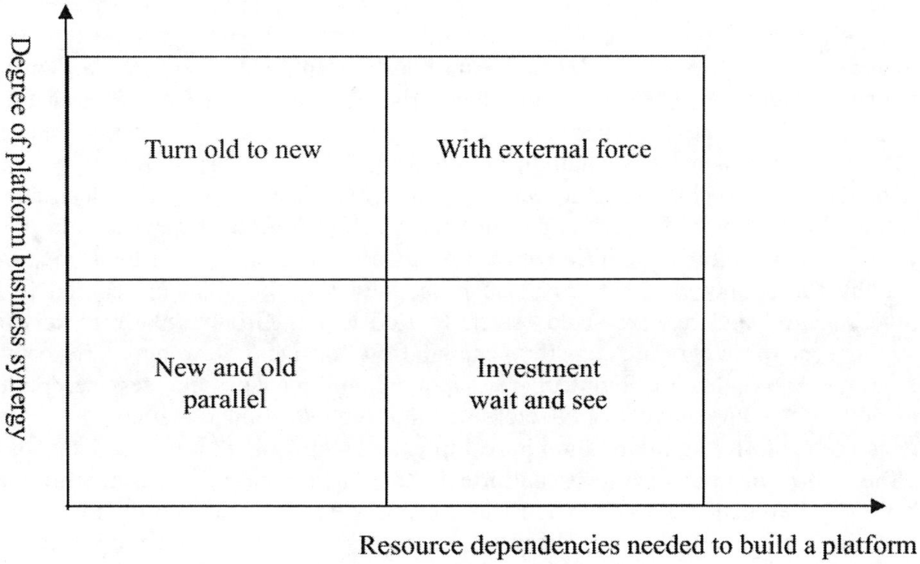

Source: Author drawing
Figure 3. Four platform organization architecture design

3.3. Platform Economy Reconstructs Supply Chain

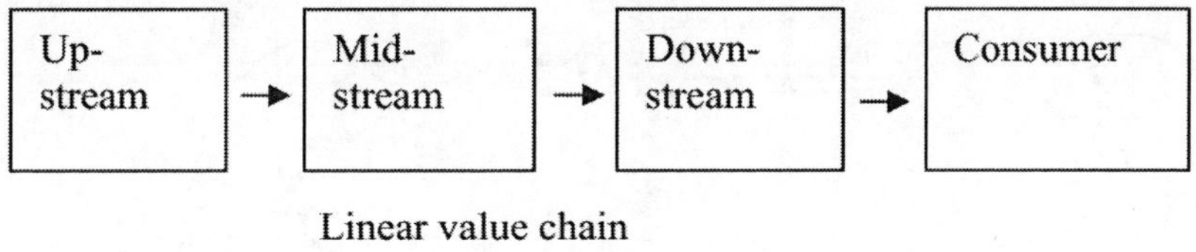

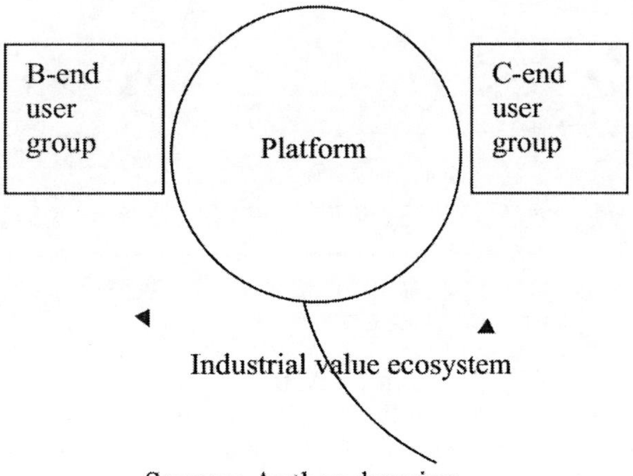

Source: Author drawing
Figure 4. Industrial chain restructuring

One of the driving factors for the transformation and upgrading of China's traditional manufacturing industry is that industrial efficiency in supply chain, organization and management drive the transformation and upgrading of traditional industries. Following the popularization of the platform business model facilitated by Internet economy, a large number of inefficient and redundant links in the traditional supply chain are gradually disappearing and new high-efficiency links are emerging. The traditional internal supply chain operation mode is classified into a pushing-type internal supply chain that uses inventory to drive sales and a pulling-type internal supply chain that uses customer orders to drive production. Currently, most enterprises are in the process of transformation from pushing-type to pulling-type. However, even it is the pulling-type internal supply chain, they are still oriented to industrialized mass production, while Red Collar Group's pulling-type internal supply chain management focusing on "source point theory" has realized industrialized personal customization. "Source point theory" is that the manufacturer produces suits with customer needs as the source point, so as to meet the differentiated needs of each customer. At the same time, compared with the traditional supply chain, it is also in possession of more abundant capital flow, extremely short delivery period and realizes scale effect. In Red Collar Group's platform service mode, the innovative concept of integrating logistics, capital flow, business flow and information flow runs through the whole process of Red Collar's supply chain management service, putting a higher premium on core competitiveness, decreasing intermediate links, saving logistics costs and management costs, mastering information in real time and responding to market demands quickly. Red Collar's "four flows in one" operation platform is shown in Figure 2. This platform can meet the personalized and fragmented needs of industrial manufacturing to return to the high end of the value chain without being "squeezed" by middlemen, distributors and agents, thus overturning the traditional smile curve.

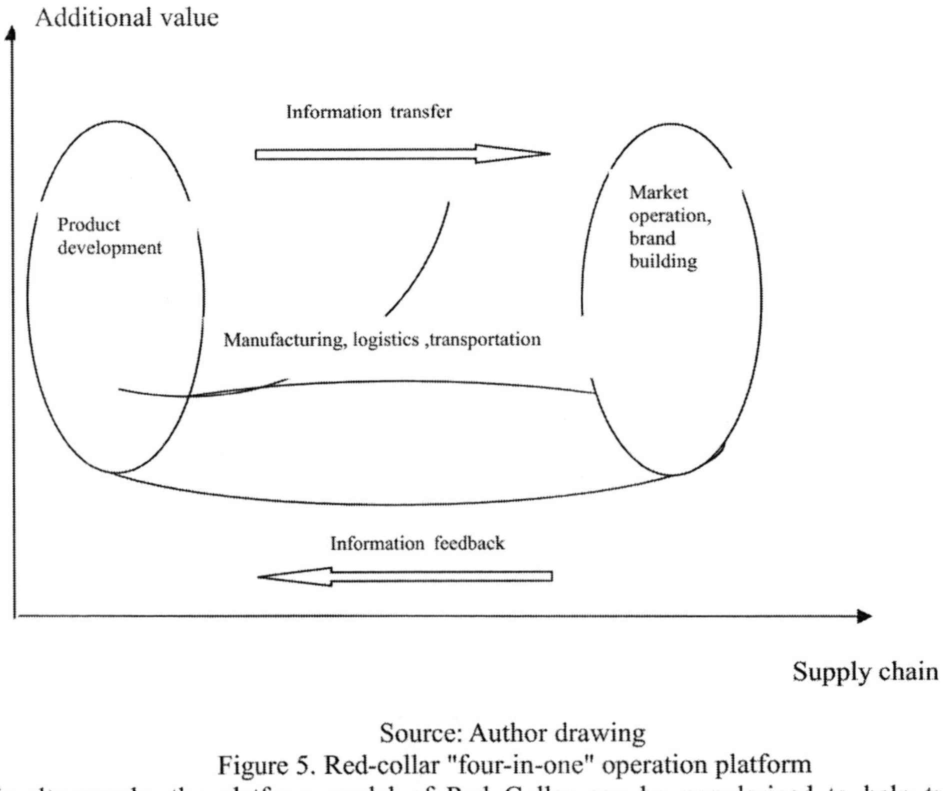

Source: Author drawing

Figure 5. Red-collar "four-in-one" operation platform

Simultaneously, the platform model of Red Collar can be popularized to help transform and upgrade China's traditional manufacturing industry.

4. Conclusions and Prospects

The transformation and upgrading of traditional industries is an old research problem, but on account of the significant position of traditional manufacturing industry in China's economic structure and the arduous difficulties it faces during economic structure transformation, researchers are still keen on studying this issue as ever. Based on a single case analysis of Red Collar Group's use of platform economy for transformation and upgrading, this paper seeks the positive impact of platform economy on the transformation and upgrading of traditional manufacturing industry and offers strategic support for the transformation and upgrading of traditional manufacturing industry in the new economic form. Proceeding from Red Collar Group, this paper analyzes the role of platform economy as a significant means of transformation and upgrading of traditional manufacturing industry, and draws the conclusions as below: in the first place, platform economy has changed the production mode of Red Collar, and enhanced the production efficiency and product diversity via the internal platform; furthermore, it helps Red Collar to build a flat organizational structure, eliminating redundant links of internal communication; finally, platform economy reconstructs Red Collar's supply chain, plays the role of the supply chain as one of the three pillars of the enterprise, and optimizes the traditional channels, operations, costs and other links. The success of Red Collar reveals that platform economy can promote industrial clusters, integration and ecologization with a brand-new mechanism and raise the efficiency of transformation and upgrading of traditional industries. Platform economy will turn into one of the significant strategic modes of transformation and upgrading of traditional industries in China. Platform economy exerts a positive impact on traditional manufacturing and even the entire traditional industry.

As an exploratory and interpretive research, there still exist some deficiencies in this paper: firstly, this paper is merely a single case study, which has its limitations. Thus, it is required to do a comparative study of multiple cases under different scenarios to intensify the test of several conclusions proposed in this paper and expand their applicability. Next, the elaboration and demonstration of the conclusions in this paper need to be further deepened from a large number of researches from multiple perspectives, at multiple layers and using multiple methods, such as empirical research from the point of view of supply chain, or follow-up research, so as to test and expand the conclusions.

References

[1] Wheelwright, S.C., Clark K.B. Creating Project Plans to Focus Product Development [J]. Harvard Business Review, 1992, 70(2).
[2] Parker G, Van A M, A digital postal platform: Definitions and a roadmap [R],America; The MIT Center of Digital Business, 2012.
[3] Ceccagnoli M, Forman C, Huang P, et al. Digital platforms: when is participation valuable?[J]. Communications of the Acm, 2014, 57(2):38-39.
[4] Shi Jianyong. A New Economic Form to Optimize Industrial Structure-Research on Micro-operation Mechanism of Platform Economy [J]. Shanghai Journal of Economics, 2013(8):85-89.
[5] Peng Shiyan. Research on Transformation and Upgrading Mechanism of Manufacturing Industry from the Perspective of Platform Economy [J]. Social Science Front, 2017(7):258-261.
[6] Ali Research. Platform Economy [M]. China Machine Press, 2016.
[7] Yao Kai, Liu Mingyu, Rui Mingjie. Value Innovation Synergy of Network Industry Chain and Platform Leadership [J]. China Industrial Economics, 2009(12):86-95.

IOP Conf. Series: Materials Science and Engineering **711** (2020) 012024 doi:10.1088/1757-899X/711/1/012024

CHAMU: An effective approach for improving the recycling of tea waste

Pin Gao[1]*, Yoshito Ogata[2]

[1] Graduate School of Design, Kyushu University, Fukuoka 815-8540, Japan.

[2] Department of Design Strategy, Faculty of Design, Kyushu University, Fukuoka 815-8540, Japan.

*Corresponding author's e-mail: 418480928@qq.com

Abstract. With the development of tea production, treating tea waste has become a problem that cannot be ignored. Here, we reprocessed tea waste to create a new kind of biomass material and named it "CHAMU". We found that tea waste is a good fiber raw material and can improve the degradation rate of PLA composites in the soil. Based on the above findings, we established a tea waste recycling system. During the processing, it was found that tea stems, tea residue had better plasticity than wood-shaving materials. According to this characteristic, we designed a series of "CHAMU products" to explore "CHAMU" materials in product design.

1. Introduction

Tea, coffee and cocoa are the largest three nonalcoholic drinks in the world [1]. Since tea was distributed over the world, tea demand has increased. The 22nd Intergovernmental Tea Working Group (FAO-IGG/Tea) meeting was held in Naivasha, Kenya, in May 2016. At the meeting, the current tea production and the growth trend of tea production in the next decade were analyzed. According to the analysis, the total world tea production (black, green, instant and other) in 2014 reached 5.13 million tons and will increase at a rate of 3.7% (black tea) and 9.1% (green tea) in the next decade [2]. At the 23rd (FAO-IGG/Tea) meeting in 2018, the meeting confirmed that tea production will continue to grow in the next ten years. The meeting analyzed the growth rate of green tea and black tea and expected world black tea production to grow at an annual rate of 2.2% over the next decade to reach 4.4 million tons in 2027. Global green tea production is expected to grow at a faster annual rate of 7.5% to reach 3.6 million tons in 2027 [3]. The development of tea industry has an important impact on the agricultural development of tea-producing countries. Taking China's tea industry as an example, the output value of China's tea industry exceeds 150 billion yuan (CNY). It has directly affected the livelihood of 80 million tea farmers, and has become one of the core interests of Chinese agriculture [2]. Although the development of tea production and tea economy has many advantages for the tea-producing countries, treating tea waste in the processing has become a problem that cannot be ignored. Tea waste in production can divided into two categories according to production and processing methods. As shown in Figure 1, the first category is the picking up tea stems during processing. Tea stem makes up about 20% of the total weight tea [4]. The other is the tea residue discharged from the tea beverage factory after processing.

Figure 1. Tea Waste (Tea residue and Tea stem)

In the past, producers used in-site incineration for treating tea waste. In recent years, because of the increased awareness of tea waste, the treatment methods have also changed. The treatment method of the tea stem was to collect the tea stems into a bag and use them to absorb the odor to replace the activated carbon. People also can place it in a new car or a newly renovated room to remove odor. Treatment of tea residue was more difficult than tea stems. The watery tea residue carried out from the tea drink factory is prone to rot if not treated in time.

In 2012, the famous Japanese tea drink company "ITONE" produced 51,000 tons of tea residues from tea drinks processing. Of the treatments, 65% was used for compost, 17% was used for heat recovery, and the rest was used as feed or additives [5]. Another Japanese supermarket chain company "TRIAL" also produces tea drinks, and the tea residue each year reaches 1.3 million tons (water-containing tea residue). To this end, the company needs to pay about 12 million yen (JPY) each year to treat tea residue [6].

After realizing the problem of tea waste, the researchers also conducted various studies on it.

Tea waste was rich in cellulose, hemicellulose, lignin, theanine, tea protein, tea polyphenols, tea polysaccharides, caffeine, tea pigments, tea saponins, vitamins [7]. In the fields of biomass materials, food processing, agriculture, medicine, and health care, scholars in fields such as Jinxue Jiang et al. used tea residue as a filler to join the MDF to reduce the formaldehyde released. They found that the protein, as one of the main ingredients of tea, was difficult to dissolve in water, which allowed the tea waste residue to maintain its protein. During MDF pressing, the amino group of the protein peptide chain skeleton in the tea residue can react with formaldehyde, lessening the formaldehyde emission of the composite plate [8]. In addition, polyphenols in tea waste can also react with formaldehyde, and has physical adsorption to formaldehyde. Tuan Su pointed out that tea wastes itself are a porous material, which can be effectively to release formaldehyde or promote the reaction to formaldehyde with tea polyphenols when pressed by hot-press machine [4]. In the adsorption studies on formaldehyde, Wan Shunli et al. analyzed the ability of discarded tea residue to purify heavy metals in water. He pointed out the porous characteristics of tea determine the effectiveness of purifying heavy metal ions in water [9]. Bajpai defined tea as an absorbent material with porosity and large specific surface area. It can be seen that tea waste is a good adsorbing material [10].

Although various studies on tea waste have some results, most of them aimed to a single kind of tea waste. They did not give an effective, systematic solution. Tea residue and tea stem were different from each other as waste. If we want to make more effective use of the characteristics of tea waste, we should adopt different processing methods according to its material composition. In addition, current research on tea stems and tea residues mainly concentrated in the chemical industry and agriculture and there has been little research on product development and design related to people's lives.

Currently, fiber-reinforced plastic (FRP) has been widely used, but most FRPs still cause problems when discarded. For example, when burning waste, the carbon dioxide content in the atmosphere will be increased, and when it is buried in the soil, the waste remains in the soil due to difficulty in degradation [11]. Biomass waste is often mixed as additives with plastics into composite materials in research related to waste degradation. For example, rice shells and bran as additives have been mixed with plastics to improve the degradation of composite materials [12, 13].

IOP Conf. Series: Materials Science and Engineering **711** (2020) 012024 doi:10.1088/1757-899X/711/1/012024

The compostability of tea residue has been proven and has been used to produce tea residue organic fertilizer that promotes plant growth [14]. However, there are few studies on the degradability of wastes such as tea stem or composites made from tea waste.

In order to establish a reliable and environmentally friendly tea waste recycling system, tea waste should have good degradation properties in addition to being processable. Therefore, the degradation performance of tea waste still needs to be proved by experiments.

So far, besides tea waste, many wastes have been reused to make artificial materials. Researchers recycled materials such as crop straw, coffee grounds, and coconut shell fibers, and processed them into furniture and other products for people's daily use [15]. But no research related to improving the reusing of tea waste.

2. What is "CHAMU"

To improve the use efficiency of tea waste, we put forward "CHAMU". The term of "CHAMU" comes from the Chinese Pinyin "cha" and "mu", which means "tea" and "wood". This study's goal was to reprocess various tea wastes like tea stems and tea residues, and use the wastes to create new material "CHAMU". As shown in Figure 2, this means "CHAMU" can be instead of wood or other materials.

Figure 2. "CHAMU" artificial board

3. Material and methods

3.1. Tea stem and tea residue hot pressing

As mentioned above, the main components of tea waste were tea residue (water-containing tea) and tea stems. By measuring the tea, the main ingredients of tea were organic acids, tea polyphenols, amino acids, proteins, alkaloids, sugar, pectin and ash [4]. The tea residue discharged from the tea beverage factory was soaked tea leaves, and water-insoluble components in leaves allowed it still keeps such as protein and tea polysaccharide as well as some tea polyphenols, caffeine and amino acids. Tuan Su tested the tea stems by nitric acid-ethanol test. The test results showed the chemical composition of tea stems mainly included cell wall materials such as cellulose, hemicellulose, and lignin, as well as starch, pectin, and other extracts and ash [4]. After measuring the morphology of the tea stems, the fiber morphology of the tea stems are similar to that of the coniferous wood, which shows the tea stems are a good fiber raw material. According to research on the composition of tea waste, the waste can be divided into two types: fiber (tea stem) and non-fiber (tea residue).

In this study, in order to prove the difference in mechanical strength between fiber and non-fiber, we carried out experiments on forming two materials. In the experiment, tea residue and tea stem used as raw materials, and the same hot pressing method was used for processing (Figure 3). The mechanical properties of the formed sheet were tested under the same conditions of density, adhesive ratio, temperature and time.

Our results show that there is a significant difference in the strength of the two materials. The tea residue artificial boards were not sturdy. The modulus of rupture (MOR) of tea residue artificial board was only 1.23Mpa as determined by the universal testing machine (UTM). The tea residue carbonized during processing because of high temperature, which also reduced the strength of the material. The experiments proved that tea residue is not suitable for producing artificial boards because of its material composition. In contrast to the tea residue material, the tea stems forming a good mechanical strength after hot pressing, and the MOR is 13.2Mpa. Tea stem as a fiber material was suitable for processing artificial board. The formed tea stem artificial board meets the national standard (China) in various mechanical strengths. In addition, artificial boards made from tea stems can effectively control formaldehyde release [6].

Figure 3. Tea stem was formed by hot pressing method

3.2. Study on the relationship of tea stem morphology

Different types of tea stem and even different kinds of tea trees have different fiber shapes. The tea stem has different forms depending on its origin. Taking China's four major tea-growing regions as an example, tea trees were divided into two types, shrub type, and the arbor type, according to the climate, soil, and precipitation of the producing area. In these two categories, they were subdivided according to the size of the leaves.

To prove the relationship between the types of tea stems and their mechanical strength, we selected Oolong tea stems, green tea stems and black tea stems from the different tea areas. First, we crushed tea stems by using a grinder, and then put it into a vibrating screen. As shown in Figure 4, we placed seven different mesh screens on the vibrating screen to find out the distribution of different types of tea stem pieces on the screen. After 5 minutes of vibrating screening, we measured the fragments of tea stems in each sieve to find out the distribution of the tea stems.

Then, the three kinds of tea stems were made into artificial boards by hot pressing. The processing time, temperature and pressure were all the same. The mechanical strength of the three artificial boards was tested using Universal Testing Machine (UTS).

A strong fiber should be long and thick [4]. In Fig.5, the tea stem fragments having a mesh size of 24-mesh or less were powdery, and fiber morphology was not interwoven to form strength during hot press processing. In fragments above 24-mesh, the tea stems of green tea were similar to the tea stems of black tea. It can be seen in Fig.5, the pieces of the Oolong tea stems were fairly average and a large number of long fragments remained on the 4-mesh screen. Oolong tea stems were thicker than the

IOP Conf. Series: Materials Science and Engineering **711** (2020) 012024 doi:10.1088/1757-899X/711/1/012024

stems of black and green tea. This is because the Oolong tea stems used were from tall arbor-type tea trees and the stems of black and green tea came from the dwarf shrub-type tea trees.

It can be seen from the mechanical test that Oolong tea stem from the tall arbor-type tea tree was most suitable for artificial board processing. The tea stem of black and green tea has a certain mechanical strength after molding, but its strength did not reach the Chinese national standard.

Figure 4. Oolong tea stems were placed in vibrating screen machine

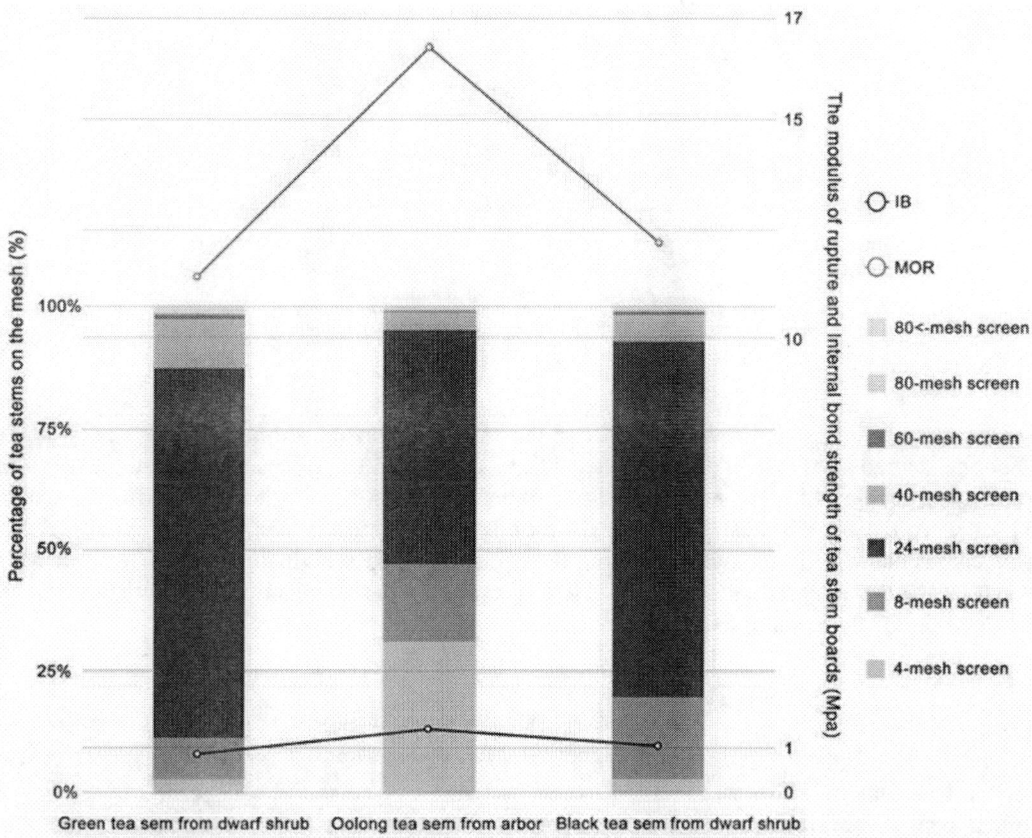

Figure 5. Distribution and mechanical strength of different kinds of tea stem fragments

3.3. Degradation of tea waste

In the experiment of the biodegradability test of samples, we used the soil burial method. The soil contains various microorganisms that can degrade plant material [16-18].

Three different kinds of wastes were used for experiment: the tea stems powder, tea stem fiber and tea residue powder. Tea waste was mixed as a filler with Polylactic Acid（PLA）particles in three different proportions: Sample 1 (50% tea stem powder 50% PLA), sample 2 (50% tea stem fiber 50% PLA), and sample 3 (10% tea residue powder 90% PLA). We pour the material when the extruder temperature is heated to 165 degrees. After high-temperature extrusion molding, the semi-finished product was obtained. The three kinds of semi-finished product obtained from the experiment were further subjected to pulverization, drying, and injection molding to obtain the final experimental sample (Figure 6). Then the shear strength of the three samples was determined by the UTM.

Figure 6. PLA and tea waste composite samples

For the degradation test of PLA and tea waste's composites, we used the method of outdoor soil burial. Outdoor, garden soil was used as an experimental soil and the three different experimental samples were buried. The test period was 4 months. Every month, we took the experimental sample out, washed, and dried, and weighed sample to measure the weight loss rate. The loss weight was calculated with a simple equation [19]. After 4 months, the mechanical strength tests were again performed on the three samples.

$$\%\text{Wt. Loss} = \frac{Wi - Wf}{Wi} \times 100\% \tag{1}$$

Wi= Initial wt. at the beginning

Wf= Final wt. after every month

After 4 months of soil burial method experimental, all three samples degraded to varying degrees. The loss weight rate of sample 1 is 55.2%, sample 2 is 44.4%, and the sample 3 is 18%. In addition, the specific analysis of the experimental data per month shows that the degradation properties of the material changed with the extension of the experimental time. After 4 months of soil burial experiments, the mechanical strength was again tested on the three samples. The test results showed that the mechanical strength of the three samples had different degrees of decrease, sample 1 declined 86%, sample 2 declined 81.5%, and sample 3 declined 40% [20].

It can be seen from experiments that biodegradation is accompanied by a decrease in mechanical strength. Yoshito Otake showed in his study that, it is mentioned that microbial active soil has a greater impact on the degradation of materials [21]. Ogawa mixed 20% of bran with PLA and observed that the mechanical strength for the material dropped drastically in a 30-day soil burial experiment. By measuring the amount of microorganisms in the soil, it is found that these plant wastes have caused the microorganisms in the soil to increase by 100 to 1000 times [12]. Therefore, it can be known that the decrease in the mechanical strength of the material is related to the reproduction of microorganisms.This study proves that using tea waste as a filler to make PLA and tea waste composites can improve the degradation in natural environments.

As shown in Figure 7, we established a tea waste recycling system based on this research results and characteristics of tea waste. Through the cooperation of farms, factories, and shopping malls, the "natural power (the microbes in nature)" will make the "CHAMU" products biodegradable, turning them into fertilizer to return to the land to solve the problem of tea waste disposal and improve the value of waste utilization.

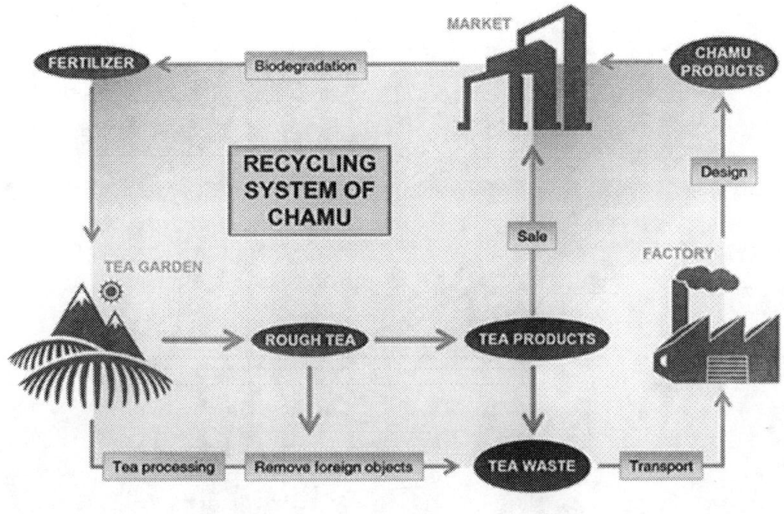

Figure 7. Tea waste recycling system

3.4. The "CHAMU" Product

The processing properties and degradation properties of tea waste studied above. But the production tea waste was far less than wood waste such as wood particles. If only used as an artificial board, the value of tea waste can't fully realize. During the processing, it was found that tea stems, tea residue had better plasticity than wood shaving materials. According to this characteristic, we designed a series of "CHAMU products" to explore "CHAMU" materials in product design. The hot-pressed "CHAMU" material was processed again by CNC (computer numerically controlled) machine, it gives "CHAMU" various shapes to meet the different needs (Figure 8).

It is true that tea waste is not as productive as wood or other agricultural waste, but tea waste has characteristics that can be distinguished from other materials, the tea aroma and the tea culture.

Tea waste contains a large amount of aromatic compounds, especially tea stems, which are rich in aromatic compounds and are the main source of aroma in tea processing. This unique tea aroma is the greatest feature different from other materials. For example, Figure 9 is the design work based on the concept of "CHAMU". This is the tea packaging made by tea stems. Tea stems of the packaging and tea in the packaging come from the same kind of tea, which means uses can judge the type of tea by just smelling the aroma of the CHAMU tea packaging. After the forming of the tea package, the aroma will be slowly released during a whole year and the concentration of aroma will decrease gradually over time. In this situation, users can also judge the approximate storage time of tea by smelling the aroma concentration of CHAMU tea packaging.

Tea culture, which originated in China, has a history of thousands of years. At present, the research on tea waste has not been considered from the cultural point of view. But the tea waste is also a form of "tea". Based on the "tea waste recycling system", the visual performance of "CHAMU" will be redesigned so that "CHAMU" will has the cultural image of "tea". In the past research on cultural imagery, products that effectively express cultural imagery can improve user satisfaction. Therefore, the research on the "connotation characteristics" of tea waste based on the "tea waste recycling system" can expand the design potential of "CHAMU" and to be effectively distinguished from other materials to achieve high value-added reuse of tea waste.

Figure 8. "CHAMU" reprocessed products.

Figure 9. CHAMU Tea Packaging

4. Conclusions

In this study, several methods were used to understand tea waste formation and degradiblity, and the design potential of tea wastes. The research results of various aspects are as follows.

1. Tea waste was processed by the hot-press method. The mechanical properties of tea waste artificial board were tested by a Universal testing machine, which conformed to the national standard of the artificial board (China).
2. The distribution of the tea stem pieces of different tea types was measured, and tea stems of the arbor-type tea trees were more suitable for the processing of artificial board.
3. Over a period of four months, the experiment proved that tea waste as filler could improve the degradation performance of the material.
4. A tea waste recycling system was proposed.
5. We designed a series of "CHAMU products" to explore "CHAMU" materials in product design and discussed the design potential of this material.

Acknowledgments

This work was supported by JSPS KAKENHI Grant Number JP19H04413

References

[1] X.L. Fang. World's three major non-alcoholic beverages (2014). Avaliable at: https://wiki.mbalib.com/wiki/Non-Alcoholic Beverage (Accessed: 15 May 2019).

[2] Projected Production: The Twenty-second Session of the Intergovernmental Group on Tea (2016). Avaliable at:http://www.fao.org/economic/est/est-commodities/tea/tea-meetings/tea22/en/ Current situation and medium term outlook (accessed 12 February 2019).

[3] Production. The Twenty-third Session of the Intergovernmental Group on Tea (2018). Avaliable at: www.indiaenvironmentportal.org.in/.../Current%20Market%20Situation... (accessed 12 February 2019).

[4] T. Su. (2012) Research on the Technology and Control Mechanism of Formaldehyde Emission from Tea Stalk particleboard (In Chinese). Fujian Agriculture and Forestry University.

[5] T. Sato. (2018) About tea husk recycling system (2014). Available at: chrome-extension://oemmndcbldboiebfnladdacbdfmadadm/http://www.3r-suishinkyogikai.jp/data/event/H26S2.pdf (accessed 22 September 2018).

[6] P. Gao, Y. Ogata, C. Song, J. Liu. (2017) The Methods of Tea Waste Reutilization and EconomicBenefits Analysis. *Atlantis Press*. **141**, 418-423.

[7] L.G. Zhu, X.X. Cheng, W.J. Zhang. (2013) Research Progress on the Reuse of Solid Wastes in Tea Industry and Its Application in Environmental Improvement. Fujian Journal of Agricultural Sciences. **28**, 1310.

[8] J.Z. Li, W.R. Zhou, J.B. Dong, J. Mu, D.R. Zhang, D.B. Fan, J.X. Jiang, Faming Zhuanli Shenqing Gongkai Shuomingshu, CN101245172A (2008) (in Chinese)

[9] S.L. Wan, S.Y. Xu, Q.R. Zhang, L.P. Qian, M.H. Ma. (2013) Removal of Heavy Metal From Wastewater By Tea Waste: A Review, Technology of Water Treatment. **39**, 5-9.

[10] Bajpai, Kumar S, Arti J. (2010) Removal of copper (II) from aqueous solution using spent tea leaves (STL) as a potential sorbent. Water SA. **36**, 221-228.

[11] Y. Kihara, G.Ben. (2019) Development and Mechanical Properties of Green Composites Composed of Biodegradable Resin and Kenaf fiber. Avaliable at: www.cit.nihon-u.ac.jp/kouendata/No.38/1_kikai/1-013.pdf (Accessed: 15 May 2019).

[12] T. Ogawa, T.Tsukegi, S. Osawa, K.Kanzaki. (2002) Biodegradation Behavior of Poly(l-lactic acid)/Bran Blend in a Land-fill Test. Materiaru Raifu Gakkaishi. **14**, 81-87.

[13] M. Tanaka, A. Funaki. (2007) Biodegradability of Green Composites of Filled with Chaff Charcoal. Reports of the Mie Industrial Research Institute. **31**,137-139.

[14] L.G. Zhu, X.X Chen, W.J. Zhang, Z.D Wu, and F.Y. Jiang. (2013) Research Progress on the Reuse of Solid Wastes in Tea Industry and Its Application in Environmental Improvement. Fujian Journal of Agricultural Sciences. **28**, 1310-1315.

[15] Materialdistrict: Coquim Coconut Fibre (2018). Available at: https://materialdistrict.com/material/coquim-coconut-fibre/ (accessed 14 February 2019).

[16] H.M. Zhang. (2006) Ecological characteristics of cellulose-decomposing microorganisms and function on the transformation organic matter in Jilin black soil (In Chinese). Jilin Agricultural University.

[17] M. Allgaier, A. Reddy, J. I. Park, N. Ivanova, P. D'haeseleer, S. Lowry, R. Sapra, T. C. Hazen, B. A. Simmons, J. S. VanderGheynst, and P. Hugenholtz. (2001)Targeted discovery of glycoside hydrolases from a switchgrass-adapted compost community. PLoS ONE. **5**, 1-9.

[18] H. Deng, C. Wang, H. Lu, F. Wang, Q. Tu, and W. Wu: Research Progress in Succession of Actinomycetal Communities and Their Capacity of Degrading Lignocellulose During Composting Process. (2013) Chinese Journal of Applied and Environmental Biology. **19**, 581-586.

[19] M. Shovitri, R. Nafi'ah, T.R. Antika, N.H. Alami, N.D. Kuswytasari, and E. Zulaikha. (2017) Soil Burial Method for Plastic Degradation Performed By Pseudomonas PL-01, Bacillus PL-01, and Indigenous Bacteria. AIP Conference Proceedings. **1854**, 020035.

[20] P. Gao, Y. Ogata. (2019) Biodegradability of PLA and Tea Waste Composites Based on "CHAMU"and the "Tea Waste Recycling System". IOP Conference Series: Materials Science and Engineering. **563**.

[21] Y. Otake, Y. Gomi, T. Kobayashi, S. Ito, and M. Yabuki. (1991) Evaluation of Biodegradable Plastics Added Starch by Lay-in-Soil Test I. The Society of Rubber Science and Technology, Japan.**64**, 677-687.

Study on performance of integrated pipeline vibration isolator

Lao Xingsheng*, Liu Yong, Dai Chunhui, Wu Jun and Lyu Weijian

Science and Technology on Thermal Energy and Power Laboratory, Wuhan 2nd Ship Design and Research Institute, Hubei, 430205, P.R. China

Email of corresponding author: 460079620@qq.com

Abstract. For power machine pipeline, a pipeline isolator with integrated pipeline surrounding rubber and base vibration isolating rubber is designed. Adjusting the rubber parameters of the pipeline mainly changes the impedance of the isolator and adjusting the base rubber mainly changes the force transfer rate. Thus the optimal vibration isolation effect can be achieved by increasing the rubber hardness of the pipe surrouding rubber and reducing the rubber hardness of the base rubber. The analysis also shows that the base rubber structure has a decisive influence on the low-frequency vibration isolation performance of the integrated pipeline isolator compared to the pipeline rubber.

1. Introduction

Noise of the power machine includes the mechanical noise generated by the moving parts and the noise generated by the internal fluid motion excitation[1]. Generally, the vibration of the power machine body can be weakened or eliminated by installing isolator arranged centrally, and the power pipeline in power system has a large distribution range and a long distance[2].It is not suitable to reduce vibration through the centralized vibration isolation device. By installing flexible joints and clamps, the vibration transmitted by the power machinery through the pipeline structure and the vibration of the fluid excitation pipeline in the pipeline are weakened, as the clamp is a single-layer rubber-pair, the isolation of vibration is limited, especially in low frequecy condition[3-5].

In order to improve the vibration isolation effect of the power pipeline, an integrated pipeline isolator is proposed, which includes a vibration-insulating rubber tightly surrounding pipeline and a diagonal vibration-isolating rubber coupled with the base. Its vibration isolation performance is analyzed by a finite-point method.

2. Modal analysis of isolator

The structure of the integrated vibration isolator is shown in Figure 1. The vibration isolator has a inner diameter of 250 mm and an axial length of 100 mm. In the finite element model, the unit type is soild185, the unit size is 10mm, total nodes is 9488 with 6730 units. The boundary condition is set as that the degree of freedom in the three directions is constrained at the mounting hole to keep it simple. The material parameters of the isolator are shown in Table 1. The first 10 modal results are calculated as shown in Table 2.

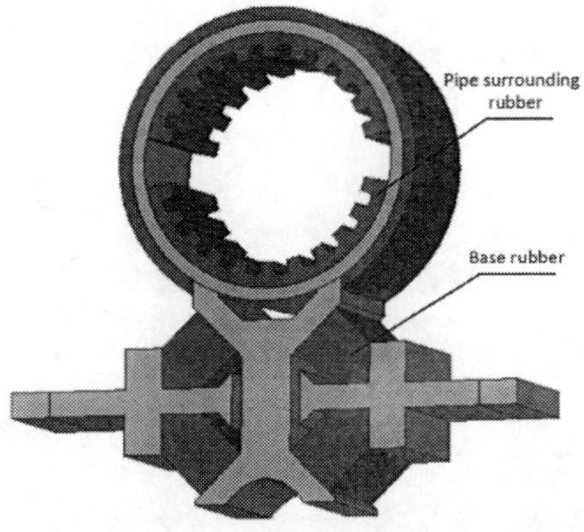

Figure 1. Integrated pipeline vibration isolator.

Table 1. Material settings in the finite element model.

	Density (Kg/m3)	Damping	Elastic modulus	Poisson's ratio
Steel	7800	0	200Gpa	0.3
Rubber	920	0.2	4Mpa	0.49

Table 2. The first 10 modal natural frequencies of the structure

Order	1	2	3	4	5	6	7	8	9	10
Natural frequency (Hz)	11.798	27.12	34.3	38.217	50.175	65.026	366.78	368.91	385.35	385.72

3. Vibration isolation response calculation results and discussion

The origin impedance and the force transmission rate are used as evaluation parameters of the performance of the isolator. The effects of rubber damping, rubber hardness and vibration isolator structure on the performance of the isolator were analyzed.

3.1. Effect of rubber hardness on the performance of the isolator

In general, the hardness of rubber is proportional to the elastic modulus. Therefore, the rubber hardness (Hs) is used as a variable parameter to analyze the rubber hardness of 2MPa (Hs=42), 3Mpa (Hs=51), and 4Mpa (Hs=58). The vibration isolation effect below, the result is shown in Figure 2 and Figure 3. It can be seen that the lower the rubber hardness, the greater the force transmission rate of the isolator, especially at low frequencies, but in opposite, the greater the rubber hardness, the greater the impedance of the origin. The pipeline rubber is close to the excitation source and is the main component that affects the impedance of the origin. The base rubber is the main component that affects the force transmission rate. Therefore, the rubber hardness of the rubber coated in the pipeline can be appropriately increased, and the rubber hardness of the base rubber can be reduced to maximize the vibration isolation effect of the vibration isolator.

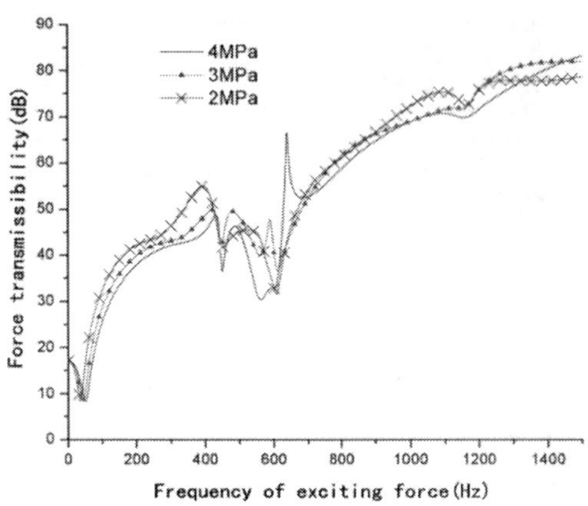

Figure 2. Effect of rubber hardness on force transmissiblity

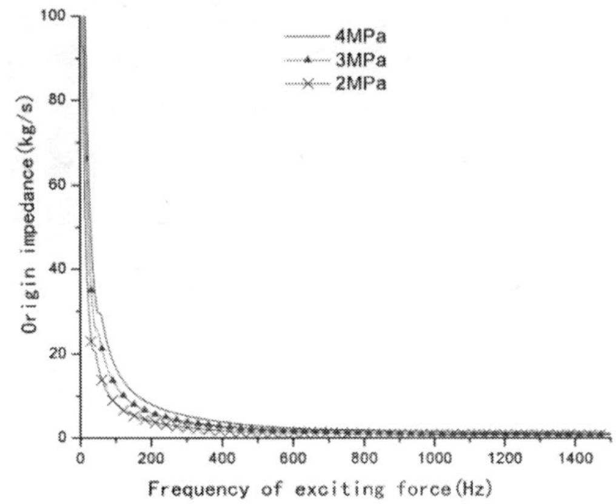

Figure 3. Effect of rubber hardness on origin impedance

3.2. Effect of pipeline surrounding rubber configuration on performance of the isolator

The pipeline rubber is close to the excitation source and is the main component that affects the impedance of the origin. The base rubber is the main component that affects the force transmission rate. In order to more accurately analyze the vibration isolation effect of the isolator, the results of the two main damping components can be changed for comparative analysis. Figure 4 and figure 5 show two structural types of pipeline surrounding rubber, which are sawtooth and continuous rings seperately.

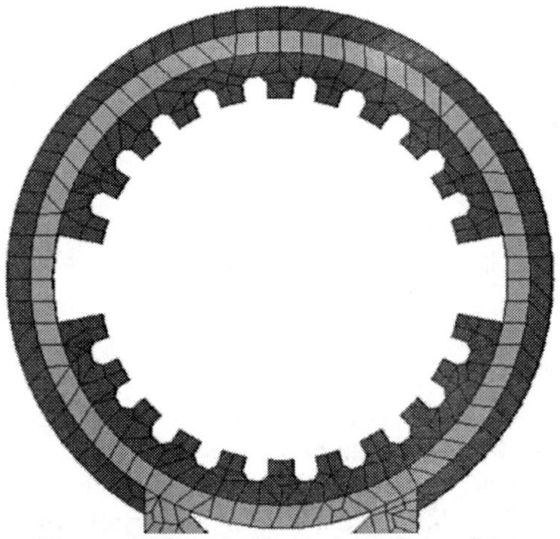

Figure 4. Sawtooth surrounding rubber

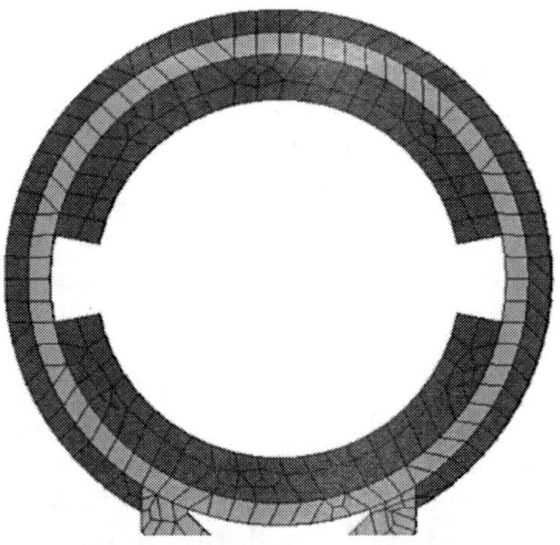

Figure 5. Continuous surrounding rubber

Figures 6 and figure 7 show the change of force transmission rate and origin resistance after the change of the pipe surrounding rubber. It can be seen that as the sawtooth shape is changed to a ring shape, the rubber volume is increased, the impedance of the vibration isolator is significantly improved, and the force transmission rate is slightly increased in the high frequency region.

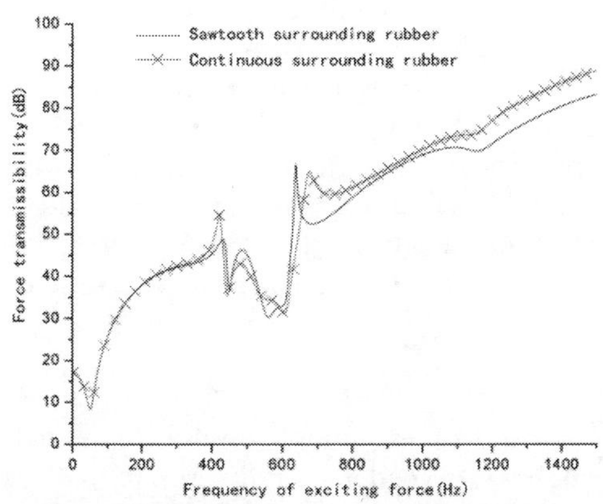

Figure 6. Effect of pipe surrounding rubber configuration on force transmissiblity

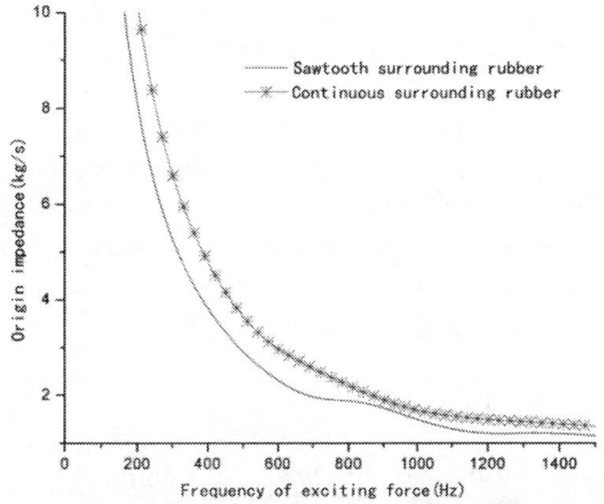

Figure 7. Effect of pipe surrounding rubber configuration on origin impedance

3.3. *Effect of base rubber configuration on performance of the isolator*

Figure 8 shows two configurations of base rubber. One type of vibration isolator has a base rubber with horizontal limit structure and the other has not.

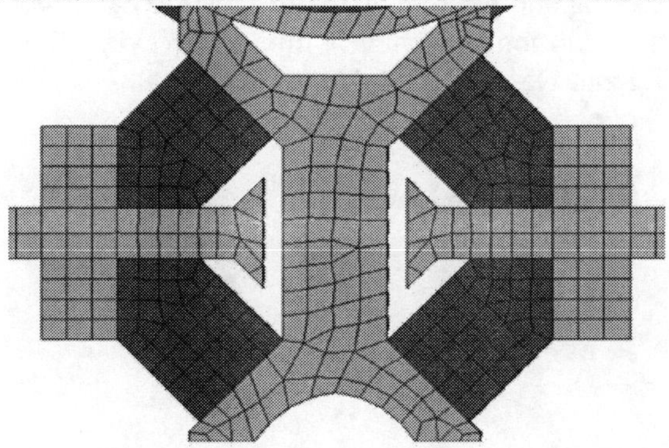

Figure 8. Base rubber with horizontal limit

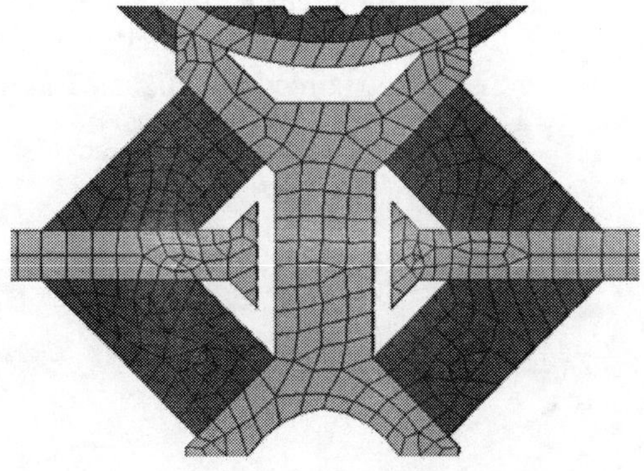

Figure 9. Base rubber without horizontal limit

The change of force transmission rate and orgin impedance is shown in Fig. 10 and Fig. 11. It can be seen that the volume base rubber of the vibration isolator without horizontal limit is large, and the transmission rate of low frequency and high frequency force is somewhat improved.

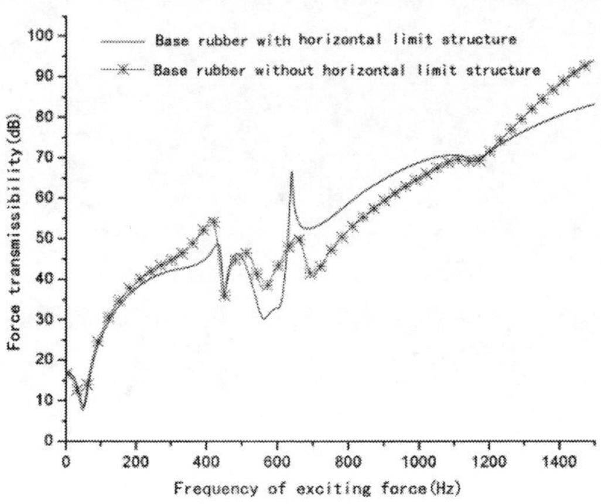

Figure 10. Effect of base rubber configuration on force transmissiblity

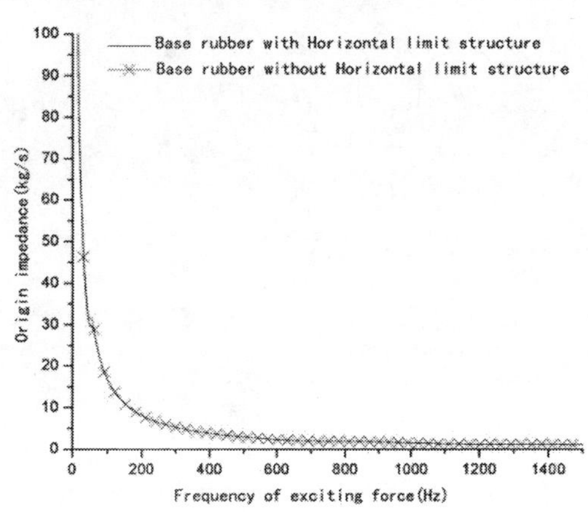

Figure 11. Effect of base rubber configuration on origin impedance

4. Conclusions

An integrated vibration isolator is designed for power machine pipeline, which includes the pipe sourrounding rubber and the obliquely arranged base rubber. The pipeline surrounding rubber is close to the excitation source and is the main component that affects the origin impedance. The base rubber is connected to the foundation and is the main component that affects the force transmission rate. The rubber hardness of the pipe surrouding rubber can be appropriately increased, and the rubber hardness of the base rubber can be reduced to achieve the most suitable vibration isolation effect. The analysis also shows that the base rubber structure has a decisive influence on the low-frequency vibration isolation performance of the pipeline isolator compared to the pipe surrounding rubber, which is very important in power machine pipeline.

References

[1] Anatolii Batura, Andrii Novikov, et.al. An application of the transfer matrix approach for a dynamic analysis of complex spatial pipelines, Nuclear Engineering and Design, 2019,Vol.349(8):174-182

[2] T. Lato, A. Mohany. Passive damping of pressure pulsations in pipelines using Herschel-Quincke tubes, Journal of Sound and Vibration, 2019, Vol.448(5):160-177

[3] R. T. Faal, D. Derakhshan. Flow-Induced Vibration of Pipeline on Elastic Support, Procedia Engineering, 2011, Vol.14(1):2986-2993

[4] Zhenzhou Lu, Dawei Li, Feng Zhang. Optimal design of the positions of the hoops for a hydraulic pipelines system, Nuclear Engineering and Design, 2011, Vol.241(12):4840-4855

[5] Shuai-jun Li, Gong-min Liu, Wei-tao Kong. Vibration analysis of pipes conveying fluid by transfer matrix method,Nuclear Engineering and Design, 2014, Vol.266(1):78-88

Preparation of pesticide microspheres based on polylactic acid: optimized by response surface methodology

Yan Li[1], Jiakun Dai[2*], Duo Cao[3,4*], Yu Ma[1], Lisha Zhen[1] and Fan Chang[1]

[1]Microbiology institute of Shaanxi, Xi'an, Shaanxi, SN 29, China.

[2]Bio-Agriculture institute of Shaanxi, Xi'an, SN 29, China.

[3]The College of Life Science, Northwest University, Xi'an, Shaanxi, SN 29, China

[4]The College of Life Science, Yan'an University, Yan'an, SN 29, China

[*]Authors to whom any correspondence should be addressed

E-mail: djkxa@163.com (Jiakun Dai); E-mail: caoduo2013@163.com (Duo Cao)

Abstract. Polylactic acid, as a kind of environment-friendly material, has excellent biodegradability, biocompatibility and plasticity. In recent years, it has been widely used in medicine, packaging and other fields, but rarely in agriculture. In this study, the process of buprofezin - polylactic acid microsphere was optimized by response surface method. Changing parameters, especially the concentration of PLA and the PLA/drug ratio, can increase the encapsulation efficiency of buprofezin - PLA microspheres. Ultimately, the highest encapsulation efficiency reached to 96.47%. The result of FESEM and laser particle size distribution analyzer showed that the microspheres have smooth surface and good dispersion, with particle size of 34.2 μm.

1. Introduction

Due to the importance of sustainable development, non-petroleum based biodegradable materials have gained more and more attention. Polylactic acid (PLA) is a polyester obtained by polymerization of lactic acid, which mainly derived from the fermentation of starch (such as corn and rice) [1]. PLA is well known for its complete degradation into CO_2 and H_2O_2, and its good performance in biocompatibility, moldability, mechanical strength and thermoplastic [2,3]. As one of the most widely used biodegradable polymer, PLA has been applied in many fields such as medicine, manufacturing and food packaging [4,5,6]. Currently, PLA source products began to appear in agriculture, such as biodegradable plastic film and nursery tray like [7]. The application of non-petroleum based biodegradable materials (e.g. PLA), can help to protect the environment and alleviate the energy shortage.

Because of the photolysis, hydrolysis and the loss by rain/wind, conventional pesticide formulations exhibit very low efficiency on target [8]. To maintain the effect on pests and diseases, people have to increase the amount and frequency of the pesticide. Over-use of pesticides leads to serious environmental pollution and excessive pesticide residues [9]. Since the controlled release microsphere can reduce the total amount of pesticides and the frequency, the new formulation can be an effective solution. In this study, the PLA microspheres of buprofezin (a kind of easily degradable pesticide) was prepared by single emulsion solvent evaporation method.

For there're several factors affecting the reaction, the experimental design is very important. The response surface method (RSM) is a classic statistical method to evaluating the role of parameters and searching optimum process for the response. Recently, the method has been used in optimizing the preparation of sustained release microspheres [10,11,12]. In order to reduce costs and improve the efficiency, we optimized the process with RSM to obtain conditions for a high encapsulation efficiency (E.E.) of the microspheres.

2. Materials and Methods

2.1. Materials

Polylactic acid was purchased from eSUN Industrial Corporation (Shenzhen, China), with a molecular weight (Mw) of 100,000. Buprofezin was purchased from Yinong agricultural Corporation (Changshu, China). Dichloromethane (DCM, AR) and polyvinyl alcoholwa (PVA, 87-89% hydrolyzed) was purchased from Aladdin Industrial Corporation (Shanghai, China).

2.2. Preparation of Buprofezin - PLA microspheres

PLA was dissolved in DCM to prepare the polymer solutions with different concentrations. Buprofezin was added into the polymer solution, and sonicated 3 min (oil phase). PVA was dissolved in distilled water by heating (water phase). Then the oil phase was slowly added to the water phase with rapid stirring to emulsify into small droplets. The emulsion was maintained stirring and heated to 45 °C to remove DCM from the droplet, so that the PLA can be solidified into microspheres. Then collected the microspheres by vacuum filtration. Washed with distilled water for three times to remove residual PVA. The microspheres were harvested after drying and stored at room temperature.

2.3. Encapsulation efficiency determination

The E.E.% was measured at a wavelength of 245 nm with an ultraviolet (UV) spectrometer. The standard curve was prepared using drug concentration ranging from 0.078 to 0.431 μg/mL and had a regression equation of $y = 28.719x + 0.1664$, with $R^2 = 0.9994$. The E.E.% of the drug loading was calculated using the equation (1) as follow:

$$\text{E.E.}(\%) = \frac{\text{The actual drug amount}}{\text{The theoretical drug amount}} \times 100\% \tag{1}$$

2.4. Box-Behnken experimental design

Before design, a lot of preliminary experiments had been performed to choose the parameters and their levels. Four parameters were selected, including PLA concentration, PVA concentration, PLA/drug ratio and the stirring speed. In order to optimize the preparation parameters, 29 experiments were designed according to Design Expert® software (version 10.0.7), and the effects of four parameters on the E.E.% were evaluated. The four independent variables and their coded factors are listed in **Table 1**.

Table 1. Variables and their levels used for the Box-Behnken design

Independent variables	Levels	
	Low, (-1)	High, (+1)
A = concentration of PLA (mg/mL)	40	80
B = concentration of PVA (%)	0.5	1.5
C = ratio of PLA to buprofezin (m/m)	3	5
D = stirring speed (r/min)	800	1200

3. Result and discussion

3.1. Design of experiments and model fitting

The results of actual and predicted response values as well as residuals of all the 29 experiments carried out were shown in **Table 2**.

Table 2. Response values and the residuals of encapsulation efficiency

Run	Independent variables				E.E.%		Residual
	A	B	C	D	Actual	Predicted	
1	1	0	0	1	87.78	89.74	-1.96
2	1	0	1	0	93.45	93.69	-0.24
3	0	0	0	0	94.90	94.30	0.60
4	0	0	1	1	89.81	89.23	0.58
5	-1	0	0	1	79.43	82.11	-2.68
6	0	0	-1	1	83.19	81.05	2.14
7	0	0	1	-1	88.46	89.38	-0.92
8	0	1	0	1	87.35	86.92	0.43
9	0	-1	-1	0	86.33	87.72	-1.39
10	0	0	0	0	93.24	94.30	-1.06
11	1	0	-1	0	92.74	93.14	-0.40
12	-1	1	0	0	83.96	82.86	1.10
13	-1	0	0	-1	83.60	83.66	-0.06
14	0	-1	1	0	88.63	88.62	-0.01
15	1	-1	0	0	91.11	90.99	0.12
16	0	0	0	0	94.71	94.30	0.41
17	0	1	1	0	93.36	93.99	-0.63
18	0	0	0	0	95.88	94.30	1.58
19	-1	0	1	0	93.29	92.10	1.19
20	0	1	-1	0	79.50	81.53	-2.03
21	0	0	-1	-1	84.83	84.19	0.64
22	1	1	0	0	96.04	94.22	1.82
23	-1	-1	0	0	86.31	86.91	-0.60
24	-1	0	-1	0	80.31	79.28	1.03
25	1	0	0	-1	92.14	91.48	0.66
26	0	-1	0	1	83.23	81.74	1.49
27	0	-1	0	-1	89.34	88.98	0.36
28	0	0	0	0	92.79	94.30	-1.51
29	0	1	0	-1	82.28	82.98	-0.70

By analysing the data of the 29 runs in **Table 2**, the final empirical formula for E.E.% was shown in equation (2) as follow:

$$Y = 94.30 + 3.86A - 0.20B + 3.34C - 0.82D + 1.82AB - 3.07AC - 0.048AD + 2.89BC + 2.79BD + 0.75CD - 1.98A^2 - 3.57B^2 - 2.77C^2 - 5.57D^2 \quad (2)$$

where Y is the predicted response of E.E.%, the variables of A to D are the coded values of PLA concentration, PVA concentration, PLA/drug ratio and stirring speed, respectively.

3.2. Analysis of variance (ANOVA)

The significance and validity of the model were tested by ANOVA, which were showed in **Table 3**. The p-value less than 0.05 means significant. The results of ANOVA showed the model's p value <0.0001, indicating that the model is highly significant and reasonable. The lack of fit is not significant, for the p-value = 0.2390. The correlation coefficient $R^2 = 0.9443$ and Adj. $R^2 = 0.8887$, indicating that the equation fitted well. Adeq Precision = 12.199 > 4, indicating the response signal was strong enough. So the regression equation mentioned (equation 2) can be used to predict and analyze the E.E.% of microspheres.

Table 3. ANOVA for encapsulation efficiency

Source	Sum of Squares	Degree of Freedom	Mean Square	F-Value	p-Value Prob.> F
Model	696.56	14	49.75	16.97	< 0.0001*
A	179.10	1	179.10	61.07	< 0.0001*
B	0.50	1	0.50	0.17	0.6847
C	134.00	1	134.00	45.69	< 0.0001*
D	8.10	1	8.10	2.76	0.1187
AB	13.25	1	13.25	4.52	0.0518
AC	37.64	1	37.64	12.83	0.0030*
AD	9.025×10^{-3}	1	9.025×10^{-3}	3.077×10^{-3}	0.9565
BC	33.41	1	33.41	11.39	0.0045*
BD	31.25	1	31.25	10.66	0.0057*
CD	2.24	1	2.24	0.76	0.3974
A^2	25.56	1	25.56	8.71	0.0105*
B^2	82.90	1	82.90	28.27	0.0001*
C^2	49.68	1	49.68	16.94	0.0010*
D^2	201.60	1	201.60	68.74	< 0.0001*
Residual	41.06	14	2.93		
Lack of Fit	34.63	10	3.46	2.15	0.2390[N]
Pure Error	6.43	4	1.61		
Cor Total	737.62	28			

$R^2 = 0.9443$, Adj. $R^2 = 0.8887$, Pred. $R^2 = 0.7160$, Adeq Precision = 12.199

* Significant ("p-Value Prob. > F" less than 0.05).
[N] Non-significant.

The analysis demonstrated that A, C, AC, BC, BD, A^2, B^2, C^2, D^2 are significant model terms. The PLA concentration (A), PLA/drug ratio (C) and the quadratic term (D^2) had a highly significant effect ($P < 0.001$) at the E.E.%. The PLA concentration - PLA/drug ratio interaction term (AC), PVA concentration - PLA/drug ratio (BC) and PVA concentration - stirring speed (BD) had a significant effect ($P < 0.05$), and also the quadratic terms (A^2, B^2 and C^2).

3.3. Process optimization

The 3D response surfaces and the 2D contours for the effects of the variables on the E.E.% of the microspheres are shown in **Figure 1**.

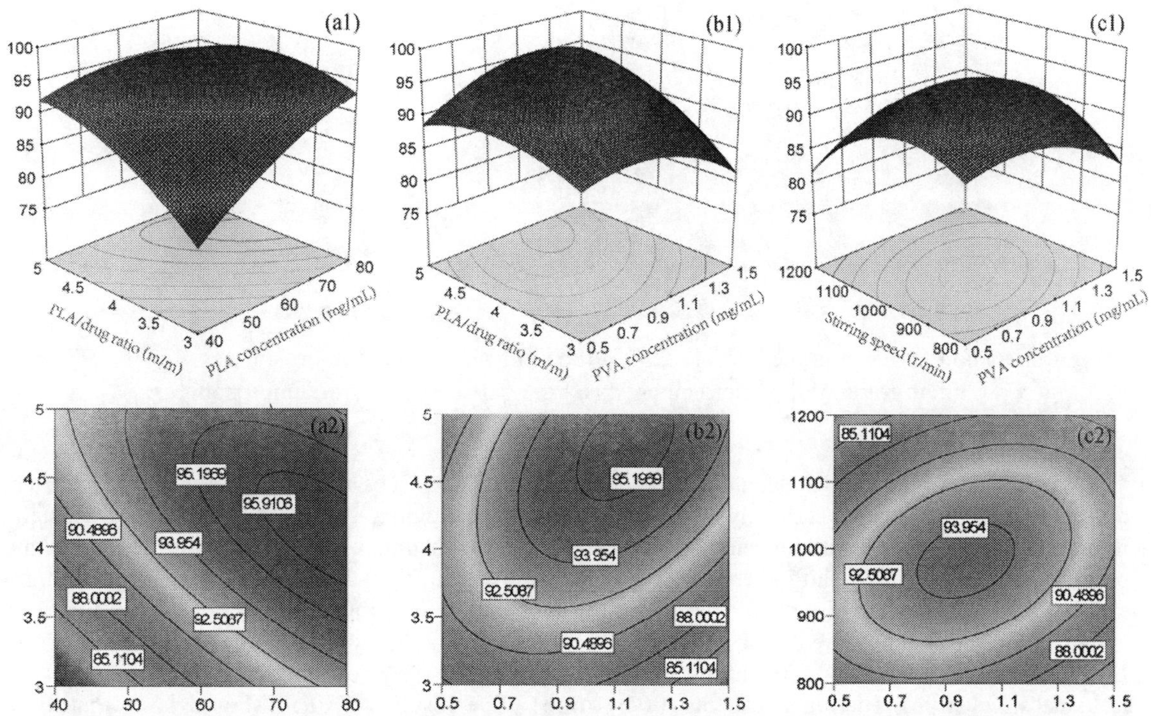

Figure 1. The 3D response surfaces and 2D contours for the interactions of variables on E.E.%: (a) PLA concentration - PLA/drug ratio (AC); (b) PVA concentration - PLA/drug ratio (BC); (c) PVA concentration - stirring speed (BD).

The results depicted the interaction between PLA concentration (A) and PLA/drug ratio (C), PVA concentration (B) and PLA/drug ratio (C), PVA concentration (B) and stirring speed (D), respectively. Consistent with ANOVA analysis, the elliptical contours demonstrate that there are significant interaction between variables.

The maximum E.E.% (96.47%) is caculated by Design Expert Software with the optimal condition as: PLA concentration = 78.24 (mg/mL), PVA concentration = 1.16%, PLA/drug ratio = 4.27 (m/m), stirring speed = 1004 (r/min). Under the condition, the experimentally result was found to be 96.73±0.52%, which is highly consistent with the model prediction.

3.4. Appearance and particle size of the microspheres

The field emission scanning electron microscopy (FESEM) was used to show the appearance of buprofezin - PLA microspheres. The micrograph demonstrated that the microparticals were smooth spherical, with a good dispersion. The particle size was measured by a laser particle size distribution analyzer. Both of the results were shown in **Figure 2**.

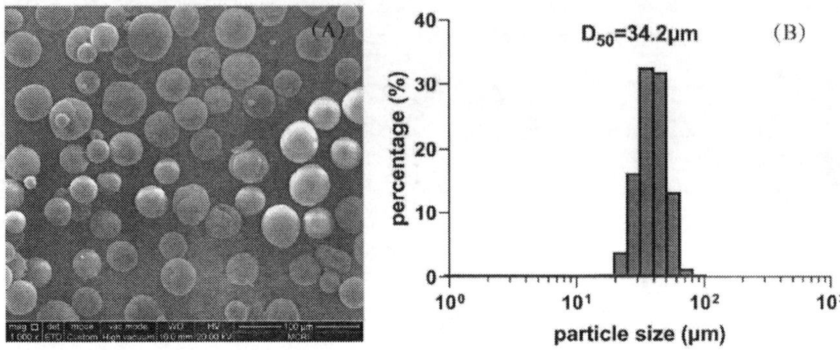

Figure 2. The FESEM and the particle size of the buprofezin - PLA microspheres. (A) the FESEM micrograph of the microspheres; (B) the particle size of the microspheres.

4. Conclusions

The utilization of pesticides is about 10%, which lead to excessive application. The over-using polluted the environment seriously and increased the pesticide residues. Application of sustained release formulations of pesticide can play a positive role on sustainable development, the environment protection and human health. We prepared the buprofezin - PLA microspheres by evaporation emulsification method. Since cost is an important factor limiting application, we optimize the encapsulation efficiency by RSM. After optimization, the conditions for obtaining the maximum E.E.% (96.47%) were PLA concentration at 78.24 (mg/mL), PVA concentration at 1.16%, PLA/drug ratio at 4.27 (m/m), stirring speed at 1004 (r/min). The results of FESEM and laser distribution analyzer showed that the size of the microspheres is about of 34μm, with a smooth surface and good dispersion.

Acknowledgement

This work was supported by the Key Research and Development Program of Shaanxi Province [No. 2017NY-158] and the Science and Technology Plan Project of Shaanxi Academy of Science [No. 2017k-07].

References

[1] John R.P., Nampoothiri K.M., Pandey A. (2006a) Solid-state fermentation for L-lactic acid production from agro wastes using Lactobacillus delbrueckii. Proc. Biochem., 41: 759-763.
[2] Ghorpade V.M., Gennadios A., Hanna M.A. (2001) Laboratory composting of extruded poly (lactic acid) sheets. Bioresour. Technol., 76: 57-61.
[3] Lim L.T., Auras R., Rubino M. (2008) Processing technologies for poly (lactic acid). Prog. Polym. Sci., 33:820-852.
[4] Aguirre E.C., Franco F.I., Samsudin H., Fang X., Auras R. (2016) Poly(lactic acid)-Mass production, processing, industrial applications, and end of life. Adv. Drug Deliver. Rev., 107: 333-366.
[5] Magdalena W., Marlene J.C., Cristina N., Stephen W.B. (2017) Development and characterisation of HPMC films containing PLA nanoparticles loaded with green tea extract for food packaging applications. Carbohyd. Polym.,156: 108-117.
[6] Ghosh S.K. (2007) Application of poly(lactic acid) fibres in automotive interior, Indian J. Fibre Text., 32: 119-121.
[7] Finkenstadt V.L., Tisserat B. (2010) Poly(lactic acid) and Osage Orange wood fiber composites for agricultural mulch films. Ind. Crop. and Prod., 31: 316-320.
[8] Margni M., Rossier D., Crettaz P., Jolliet O. (2002) Life cycle impact assessment of pesticides on human health and ecosystems. Agr., Ecosyst. Environ., 932: 379-392.

[9] Enserink M., Hines P.J., Vignieri S.N., Wigginton N.S., Yeston J.S. (2013) The Pesticide Paradox., Science 341: 728-729.
[10] Deshmukh R.K., Naik J.B. (2015) Optimization of sustained release aceclofenac microspheres using response surface methodology. Mat. Sci. Eng., C 48: 197-204.
[11] Paulo F., Santos L. (2018) Double emulsion solvent evaporation approach as a novel eugenol delivery System - Optimization by response surface methodology. Ind. Crop. Prod., 126: 287-301.
[12] Bouriche S., Cózar-Bernal M.J., Rezgui F., Rabasco Álvarez A.M., González-Rodríguez M.L. (2019) Optimization of preparation method by W/O/W emulsion for entrapping metformin hydrochloride into poly (lactic acid) microparticles using Box-Behnken design. J. Drug Deliv. Sci. Tec., 51: 419-429.

IOP Conf. Series: Materials Science and Engineering **711** (2020) 012027 doi:10.1088/1757-899X/711/1/012027

Research on Laser Welding Organization Performance of Fe-Mn-Si Alloy

Linlin Liu[1], Huajun Dong[1], Xiaochi Wang[1], Chaoyu Zhou[2],

[1]School of Mechanical Engineering, Dalian Jiaotong University, Dalian, Liaoning,116028, China

[2]Department of Mechanical and electrical, Weihai, Shandong, 264210, China

[*]Corresponding author's e-mail: good_linlin@163.com

Abstract. The crystal morphology and formation of the microstructure of the laser weld of Fe-Mn-Si memory alloy were analyzed by metallographic microscope and scanning electron microscope(SEM) and X-ray diffractometer (XRD). The results show that the welding seam microstructure of Fe-Mn-Si memory alloy crystallizes symmetrically from the fusion zone on both sides to the weld center. The $\gamma \rightarrow \varepsilon$ martensite transformation occurs in the welding seam and its adjacent tissues due to laser welding. This is because of the residual stress generated in the welding process, and this kind of ε martensite transformation and its deformation relaxes the residual stress of the weld, thus affecting the mechanical properties of the weld.

1. Introduction

Laser welding has recently attracted research attention because of its unique properties,which include good directivity,precision machining, small heat affected zone of weld,small welding deformation,etc.[1-4].The repetitive rate of excellent weld formation can be realized through proper adjustment of laser process parameters;therefore,laser welding is one of the most important methods of structural parts welding.However,at present,there are few reports on laser welding shape memory alloy[5-6].In this paper,the crystal morphology and formation of the microstructure of the laser weld of Fe-Mn-Si memory alloy were analyzed by metallographic microscope and scanning electron microscope,and the phase transformation of the microstructure during laser welding was discussed.

2. Experiment Material and Method

In this study,backing materials for Fe-17Mn-5Si-10Cr-5Ni (wt%) (shortened form A alloy) were prepared by vacuum median frequency furnace;the cast ingot was first homogenized at 1200℃,reheated to 1100℃ for 1 h.Specimens with dimensions of 5 mm×2 mm×1 mm were wire cut by electrical discharge machine.The solution treatment was carried out by austenitizing at 1000℃ for 1h in the presence of argon atmosphere.

A continuous wave (CW) CO_2 laser of 10 μ m wavelength (5 kW maximum power) was utilized for material processing,and the optimal welding process parameters were obtained through orthogonal test:P=2000 W, V =250 mm/min, z=-0.5 mm.

Metallographical specimens were observed by OLYMPUS G×51microscope.The sample was polished with No. 1200 water sandpaper.In order to further analyze the microstructure of the weld,etched with a solution of 3 $gCuSO_4$+10 mLHCl+30 mLH_2O and followed by etching with a solution of volume ratio of 3:1 HCL+HNO_3.XRD specimens was complemented by the D/max—3B

IOP Conf. Series: Materials Science and Engineering **711** (2020) 012027 doi:10.1088/1757-899X/711/1/012027

X-ray diffraction apparatus of Japan electrical machine,using CuKα radiation,35 kV work voltage,100 mA electric current,2 °/min scanning velocity.Microstructure morphology of weld was analyzed by scanning electron microscope (SEM).

3. Result and Discussion

Figure 1 shows microstructure of cross section of A alloy welding seam.It can be seen that there are no obvious defects in the welding seam,and the overall appearance is T-shaped.The upper surface of the welding seam is wider than the lower surface,and there is a long distance from the upper surface to the lower surface with equal width,which indicates that the laser energy is distributed evenly along the direction of plate thickness,which is extremely powerful for the welded joint.Figure 2 shows microstructure of welding material far from welding seam.As can be seen from the figure, the welding base material remains pure austenite phase at a distance from the weld, and no ε martensite is found.

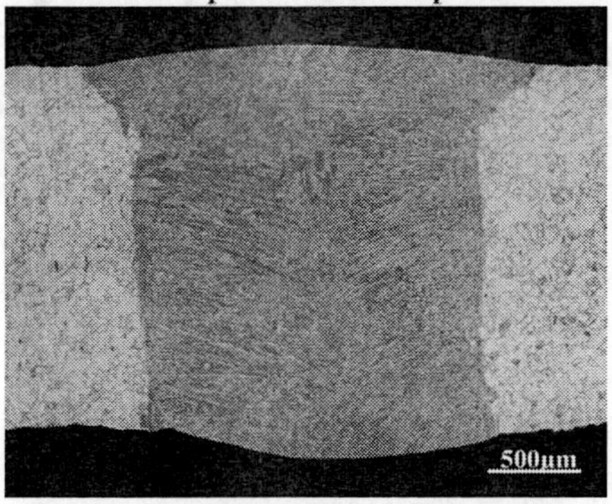

Figure 1. Microstructure of cross section of A alloy welding seam

Figure 2. Microstructure of welding material far from welding seam

Figure 3 shows microstructure of fusion zone of A alloy welding seam.As can be seen from the figure,there is a fusion zone with a width of about 80 microns between the weld and the base metal,and the microstructure on both sides of the fusion zone is very different.Further magnification shows that the structure of the base metal on the left side of the fusion zone is composed of the raustenite and a small amount of epsilon martensite,as shown in figure 4. This is because during laser welding,the weld metal undergoes the melting and solidification process,which leads to expansion and contraction of the weld metal,and such deformation is constrained by the surrounding cooler metal,resulting in the occurrence of residual stress in welding [7].The results show that $\gamma \rightarrow \varepsilon$ martensite transformation occurs in the welding seam of A alloy and its adjacent tissues due to laser welding.This is because of the residual stress generated in the welding process,and this kind of ε martensite transformation and its deformation relaxes the residual stress of the weld (i.e.,the stress self-adaptability),thus affecting the mechanical properties of the weld.

IOP Conf. Series: Materials Science and Engineering **711** (2020) 012027 doi:10.1088/1757-899X/711/1/012027

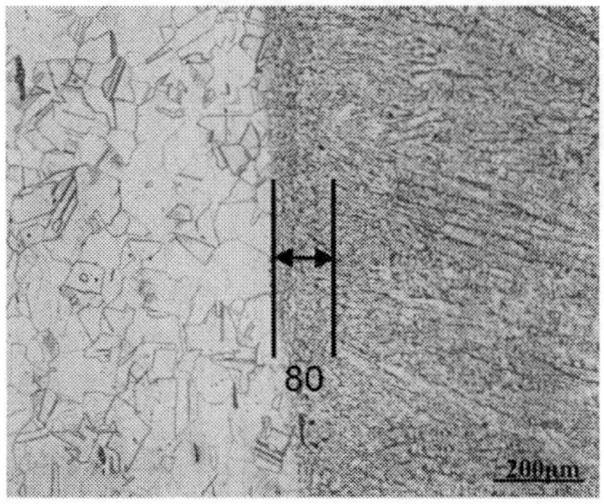

Figure 3. Microstructure of fusion zone of A alloy welding seam

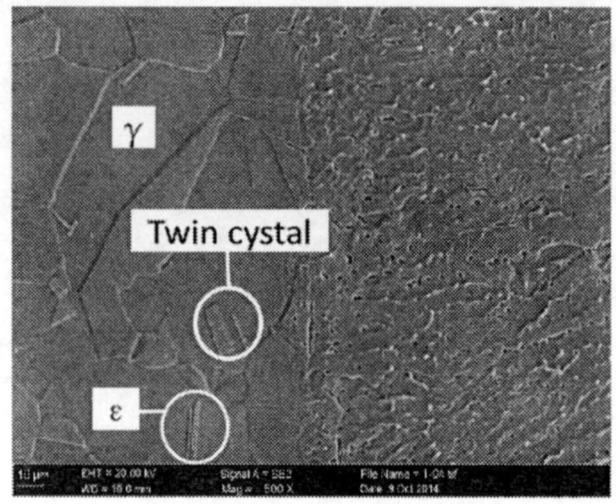

Figure 4. SEM micrographs of fusion zone

Figure 5 shows XRD patterns of A alloy welding seam and figure 6 shows XRD patterns of welding materials far from welding seam.After comparison,it is found that there is ε martensite in the welding seam microstructure after laser welding,as shown in figure 5,which indicating that the weld metal had ε martensite transformation during the welding process, which is caused by the $\gamma \rightarrow \varepsilon$ martensite transformation induces by residual stress of laser welding.The base material far away from the weld still retains a single γ austenite phase,as shown in figure 6.This is consistent with the results of the previous experiment.

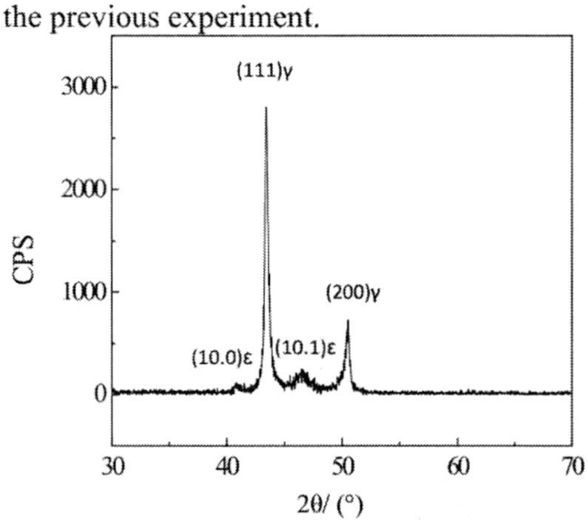

Figure 5. XRD patterns of A alloy welding seam

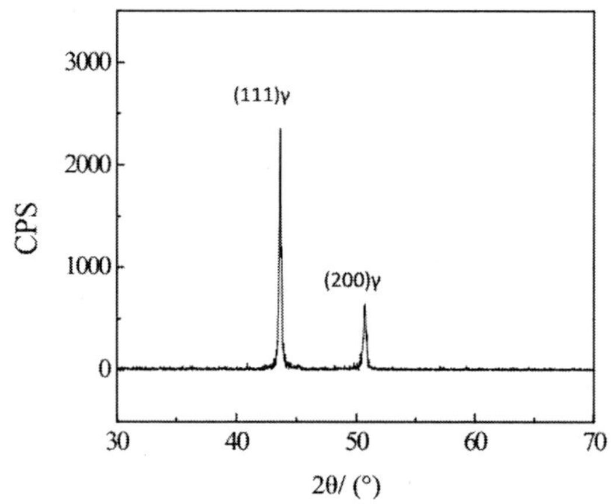

Figure 6. XRD patterns of welding materials far from welding seam

Figure 7 shows microstructure of A alloy welding seam center.As can be seen from the figure, the weld microstructure is composed of a large number of afterbrith-like twig crystals in the side of the weld near the fusion zone,and the crystals gradually change from afterbrith-like twig crystals to twig crystals from the fusion zone to the weld center. The afterbrith-like crystals structures in the weld area are equivalent to refining the bulky columnar crystals,which are very beneficial to the strength and toughness of the weld area.However,the neighboring afterbrith-like crystals belonging to the same twig crystals have the same phase.Therefore,in the future,the phase of ε martensite formed in the same twig crystals are the same,when the stress induced $\gamma \rightarrow \varepsilon$ martensite transformation occurs though denaturation,belonging to the same variant,which can maintain good crystallography reversibility when the heat is reversed and be beneficial to shape memory effect [8].

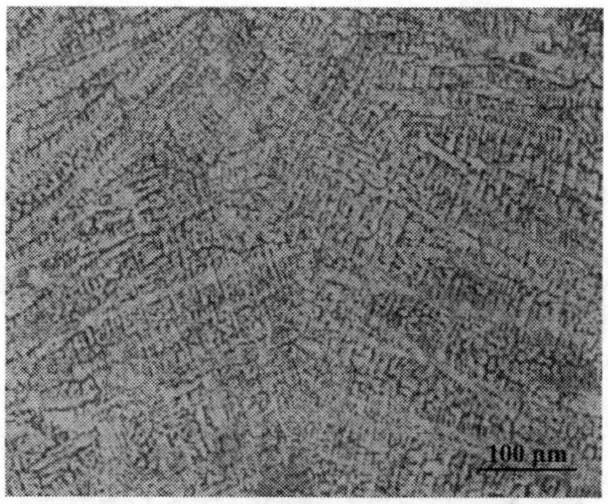

Fig. 7 Microstructure of A alloy welding seam center

4. Conclusion

(1) The welding seam formation of welding joint of Fe-Mn-Si memory alloy is well, and the cross section morphology of the weld is T-shaped. The welding base material remains pure austenite phase at a distance from the weld,and no ε martensite is found.

(2) The welding seam microstructure of Fe-Mn-Si memory alloy crystallizes symmetrically from the fusion zone on both sides to the weld center. The width of the fusion zone is about 80μm, and a large number of tiny afterbrith-like crystals are distributed in the fusion zone. From the fusion zone to the weld center, the crystals gradually change from afterbrith-like twig crystals to twig crystals.

(3) The $\gamma \rightarrow \varepsilon$ martensite transformation occurs in the welding seam and its adjacent tissues due to laser welding.This is because of the residual stress generated in the welding process, and this kind of ε martensite transformation and its deformation relaxes the residual stress of the weld (i.e.,the stress self-adaptability), thus affecting the mechanical properties of the weld.

References

[1] Lu W.W., Chen Y.H., Huang Y.D..(2014) Microstructure and mechanical property analysis about NiTiNb laser welding joint around heat treatment. Chinese journal of lasers, 41:1-5.

[2] Li H. M., Sun D. Q., Dong P..(2012) Analysis and prevention of cracks in laser-welded joint of TiNi shape memory alloy and stainless steel. Transations of the china welding institution, 33:41-45.

[3] Yang C. G., Shan J. G., Ren J. L..(2014) Effect of aging treatment on reverse phase transformation temperature of TiNi alloy laser welded metal. Materials science and technology, 22:19-23.

[4] Wang H.Y.. (2017) Development and prospect of laser welding technology. Engineering technology and application, 1:52-53.

[5] Xia W. G., Wang X., Wu X. Q..(2014) Phase transformation of NiTi shape memory alloys under the action of laser-induced shock wave. Acta armamentar, 35:198-202.

[6] Yang J., Chen P., GUO Z. B..(2015) Effect of laser welding on the microstructure and properties in weld zone of Fe-Mn-Si based shape memory alloys. Journal of functional materials, 46:14130-14133.

[7] Yang C.G.,Shan J.G.,Ren J.L..(2013) Phase transformation temperature control of weld metal of laser welded TiNi shape memory alloy joint.Acta metallurgica sinica, 49:199-206.

[8] Ueda Y, Kim Y C, Yuan M G. (1989) A prediction method of welding residual stress using source of residual stress. Jour. Trans. of JWRI., 18(1):135-138.

[9] Qiao Z.X., Li Z.M., Liu Y.C.. (2007) Study on microstructure and shape memory effect in weld joints of an Fe-Mn-Si-Cr-Ni alloy. Journal of functional materials, 38:1645-1647.

Protection of magnesium melt by hexafluoropropylene/air atmosphere

Hongli Jiang[1], Hukui Chen[1*], Jie Huang[1], Xiaxia Mu[1], Yu Jiang[1] and Yangyang Jie[2]

[1]College of Chemistry and Chemical Engineering, Baoji University of Arts and Sciences, Baoji, Shaanxi, 721013, China

[2]Xi'an Institute of Space Radio Technology, Xi'an, Shaanxi, 710000, China

[*]Corresponding author's e-mail: hk7115@yahoo.com

Abstract. The protection of hexafluoropropylene/air mixed gas for molten magnesium at temperatures from 670 °C to 800 °C has been investigated. The morphology and the composition of the protective film formed on the surface of the melt at 700 °C for holding time of 30 min and 60 min were investigated by scanning electron microscopy (SEM), energy dispersive spectrometer (EDS) and X-ray diffraction (XRD). The results showed that the protective effect of the mixed gas on molten magnesium increased with the increasing of hexafluoropropylene concentration, the prolongation of holding time and the decrease of temperature. The protective film on the surface of the melt was uniform and continuous, and mainly composed of MgF_2 and small amount of MgO.

1. Introduction

Due to the low density, good damping capacity and ease manufacturing, magnesium application scope has been expanding from aerospace, automotive, biomedicine and electronic communications industries to other fields [1,2]. However, magnesium has a high affinity for oxygen, a high vapor pressure in the molten state, and an oxide layer formed on the surface of the melt in air is a loose porous structure, which causes magnesium to violently oxidize and burn at a melting temperature in air. Therefore, in order to process molten magnesium securely and validly, the melt is currently protected by many methods, such as covering with salt flux, inert gas or changing the nature of the surface oxide with a cover gas, to slow the oxidation rate down to an acceptable level [3, 4].

Many studies [5-7] have shown that during the melt handling and casting process, the cover SF_6 gas protection is the most effective method to protect molten magnesium from catastrophic oxidation or even combustion and has been widely used in industrial production. The application of SF_6 in magnesium metal casting was first investigated in 1970 by Fruehling [5]. Later, a lot of researches have been done about this gas and found that its protection effect is very good. However, the use of SF_6 has come under raised superintend in virtue of its serious greenhouse effect (GWP=23900), which has led magnesium industry to seek alternatives for SF_6. Several reports [8-11] have shown that 1,1,1,2-tetrafluoroethane (CF_3CH_2F, HFC-134a) can be used as a substitute for SF_6. But its greenhouse effect (GWP=1300) is still high. Therefore, it is necessary to find a new alternative gas to SF_6.

A new gas, hexafluoropropylene (HFP), was found suitable for this situation [12]. It was reported that HFP gas shows a well protection on molten magnesium alloy from combustion. HFP, whose chemical formula described as C_3F_6, is a colorless and tasteless gas. It is nontoxic at room temperature, non-corrosive and security, and has a lower greenhouse effect (ODP=0, GWP=2) than SF_6. It does not burn in the air and is usually harmless to the clay [12, 13]. However, in the protection process of magnesium smelting by HFP gas, the factors affecting the protection effect is not well understood, and the composition and structure of the film formed on magnesium melts has not been characterized. In this paper, the protection of HFP/air mixed gas for molten magnesium at temperatures between 670 °C and 800 °C has been investigated. The morphology and the composition of the protective film formed on the surface of the melt at 700 °C were investigated. The purpose of this paper is to provide some reference for the re-search on gas protection for magnesium melt.

2. Experiment

A magnesium rod with a purity of 99.99% was used as the raw material. The main impurities were Si 4.10 ppm, Al 5.85 ppm, Mn 6.40 ppm, Fe 7.82 ppm, Zn 8.09 ppm. The gas used for the experiment was the mixture of HFP (HFP> 99.99%, oxygen $\leq$ 20 ppm, water $\leq$ 10 ppm) and air.

An electrical resistance furnace (the working chamber size of Φ100mm × 150mm) was employed to smelt the magnesium. A cover of low carbon steel (with the size of Φ270 mm×150 mm) with an inlet and exit hole and an observation port was placed above the electrical resistance to prevent air from entering. The gases of HFP and air mixed evenly through a 5000mL gas cylinder entered the electrical resistance furnace by the inlet channel. A low carbon steel crucible (the inside diameter of 60 mm and the depth of 75 mm) coated with boron nitride was put into the resistance furnace and preheated to 300 °C for 30 min. Then, about 120 g of pure magnesium specimens were added to the low steel crucible and a certain concentration of HFP/air mixture gases were introduced into the working chamber of the resistance furnace at a flow rate of 3 L/min. After purging the gas mixture inside the chamber for at least 60 min, the specimens were heated. When temperature reached the desired temperature, an original oxide film on the surface of melt magnesium was removed by use of a surface-film-pick-up device and the fresh protective film was formed in the atmosphere, which was recorded by camera after 10 min, 30 min, 60 min, 90 min and 120 min, respectively. After reaching the required holding time, the sample with a new surface protective film was taken out and cooled down for analysis of its composition and microstructure. The melt temperature was 670 °C, 700 °C, 730 °C, 760 °C and 800 °C, the concentration of HFP in mixed gases was 0.1%, 0.3%, 0.5%, 1%, 3% and 5%, which was adjusted successively in order to determine the minimum requirement of HFP for effective protection of magnesium melts without sparks.

X-ray diffraction (X Pert Pro MPD) with a Cu-Kα source operated at 40 kV and 40 mA was used to identify the phases in the surface film. The surface morphology and elemental composition of the surface films were studied by using a Quanta FEG 250 field emission scanning electron microscope equipped with an EDAX Genesis APEX energy dispersive spectroscopy system.

3. Results and Discussion

3.1. Protection effects

3.1.1. Effect of HFP concentration

Figure 1 shows the protective effects of the atmospheres containing different concentrations of HFP on magnesium melt at 700 °C for 10 min. It can be seen that in the atmospheres containing 0.1%, mars and cauliflowers appeared on the surface of molten magnesium (Figure 1a), and almost no protective film was formed on the surface of the melt. In 0.3% HFP/air, magnesium melt near the crucible wall was oxidized slightly, white particles were formed on the surface of the melt, and protective film was formed in part of the melt (Figure 1b). In 0.5% HFP/air, a surface with a partial metallic gloss appeared, which shows the formation of a protective film (Figure 1c). As the concentration of HFP rose to 1%, the whole surface presented a silvery white metallic luster (Figure 1d). Further increased the HFP concentration to 3% and 5%, the silvery white surfaces were similar (Figure 1d-f), which was relatively smooth and continuous. These results indicate the protective effects of HFP/air for magnesium melt increased with increasing of the concentration of HFP. In this experiment, 0.3% HFP was the minimum protective concentration.

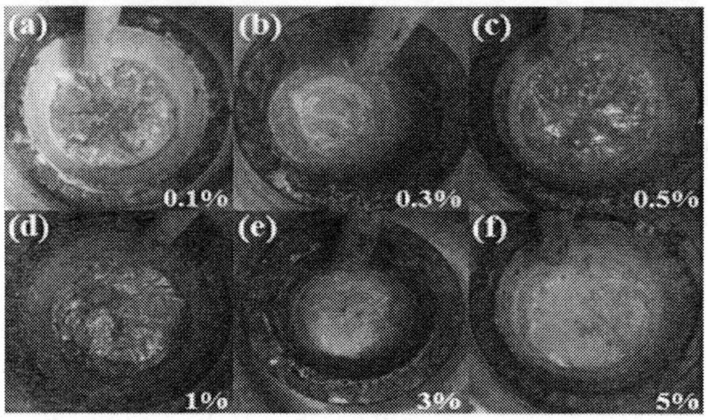

Figure 1. Protective effects of the gases containing different concentrations of HFP on magnesium melt at 700 °C for 10 min.

3.1.2. Effect of temperature

The protective effects of 1% HFP/air on magnesium melt at different temperatures for 10 minutes are presented in Figure 2. As can be seen, there was a smooth and uniform metallic luster surface on the melt at 670 °C and 700 °C. At 730 °C, the melt surface still has a certain metallic luster, but some areas produced white oxide. When temperature reached 760 °C, the metallic luster on the surface of the melt disappeared and a large amount of white oxide was produced in the central region. As temperature rose to 800 °C, the amount of white oxide in the central region increased. These results suggested that with the increase of temperature, the protective effect of HFP/air on magnesium melt became weak, and the oxidation of magnesium melt accelerated.

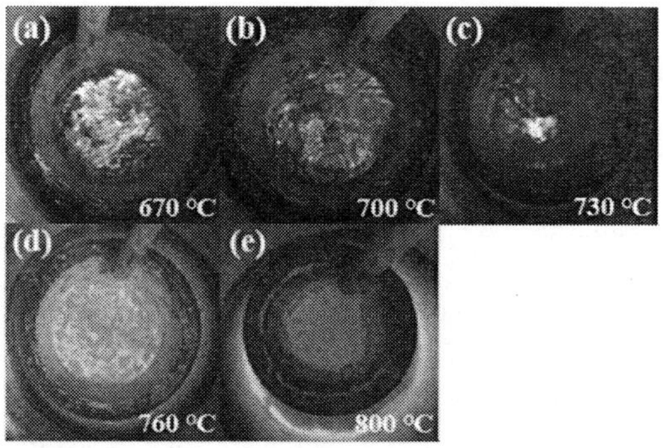

Figure 2. Protective effects of 1% HFP/air on magnesium melt at different temperatures for 10 min.

3.1.3. Effect of holding time

The protective effects of 1% HFP/air on magnesium melt for different times at 700 °C are shown in Figure 3. It can be found that when molten magnesium was exposed to the atmosphere within 60 min, its surface had a shiny metallic appearance and became flatter with increasing time (Figure 3a-d). As the holding time increased to 120 min (Figure 3e), the surface turned dark red and its metallic luster disappeared. This may be due to the thickening of the film over time.

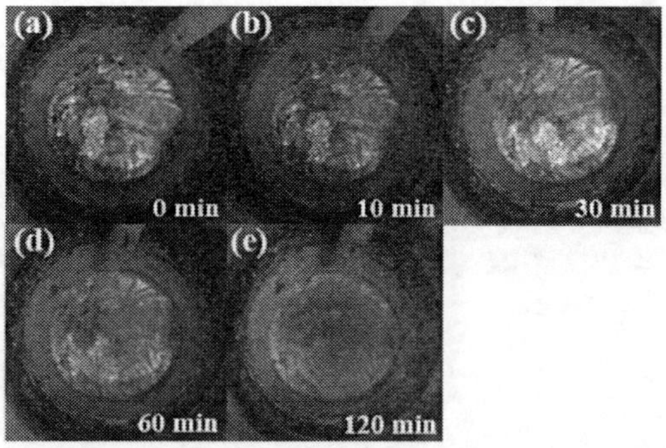

Figure 3. Protective effects of 1% HFP/air on magnesium melt at 700 °C for different times.

3.2. Morphologies and composition of the protective film

Figure 4 is the SEM image and EDS mapping results of the surface film produced in 1% HFP/air at 700 °C for 30 min. It shows that the surface was basically flat and uniform, but there were also some pores and cracks in the film. These pores and cracks may act as channels for transporting magnesium vapor outward to react with HFP and oxygen. The EDS analysis displayed that the surface chiefly included Mg, F, O and Si elements. Mg and F were the dominant elements and the contents were 57.392% and 36.233%, respectively. The content of O element was 6.028%, which was relatively small, and the content of C element was 0.347%. The results indicate that there was a relatively flat and uniform protective film on the surface of magnesium melt in HFP/air atmosphere, which was consisted of MgF_2 and a bit of MgO. It can be seen from the mapping diagram that the overall distribution of Mg, O, C and F elements on the surface film was relatively uniform, which also indicates that the surface film was homogeneous and a composite film. However, the local distribution of Mg and F elements was not very even.

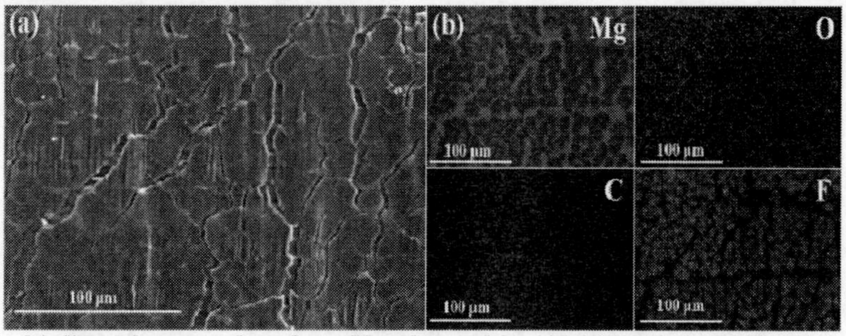

Figure 4. (a) SEM image of surface film formed in 1% HFP/air at 700 °C for 30 min, (b) mapping results of Mg, O, C and F on the surface film.

Figure 5 is the cross-sectional view of the sample after held in 1% HFP/air at 700 °C for 30 min and 60 min. The EDS line scanning results of the cross-section of the sample after held in 1% HFP/air at 700 °C for 60 min are shown in Fig. 5(c) and (d). It indicates that there was a protective film on the magnesium surface. The continuity of the film was not very good and there were some holes in the film for 30 min. For 60 min, there was a continuous film on the surface of magnesium and the film seems to be uniform in thickness. Further observation showed that the thickness of the film was between 2.011 and 2.448 μm. The reason why the film became uniform with time may be that with the increase of the holding time, the magnesium vapor from the pore continuously reacted with HFP, which made the surface film more continuous and denser. The fluorine element was not uniformly distributed in the film, and its content had a maximum near the matrix.

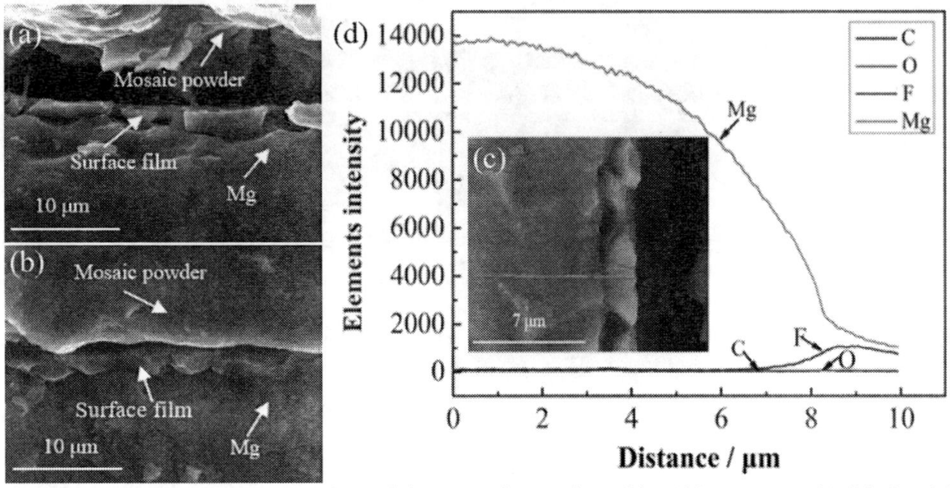

Figure 5. SEM image of the cross-section of the sample produced in 1% HFP at 700 °C for (a) 30 min and (b) 60 min, EDS line scanning results of the sample for 60 min showed at (c) and (d).

For the sake of identifying the reaction products on the surface of molten magnesium, X-ray diffraction analysis was carried out. Figure 6 indicates the XRD analysis results of magnesium after melted in air containing 1% HFP at 700 °C for 30 min. It can be seen that there were the diffraction peaks of MgF_2, MgO and Mg on the surface. MgO peak was not very obvious and the intensity of the peak was very weak. The intensity of MgF_2 peaks was high. The appearance of Mg peaks may be due to the X-ray breakdown of the film to the matrix. Therefore, it can be considered that the surface film consisted of MgF_2 and a bit of MgO.

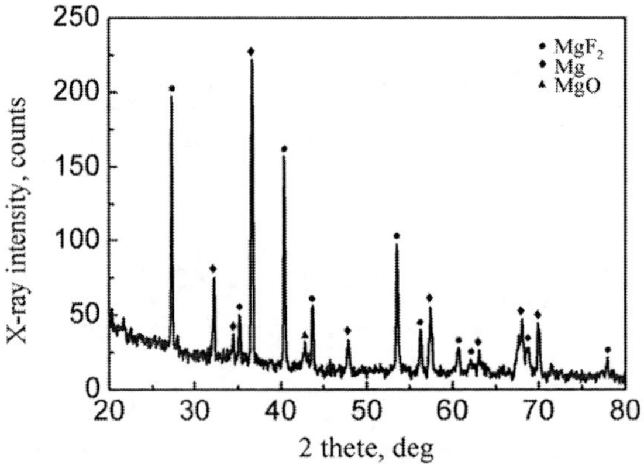

Figure 6. X-ray diffraction patterns of the film formed in the atmosphere of air containing 1% HFP at 700 °C for 30 min.

Based on the descriptions above, the protection mechanism of HFP/air atmosphere for molten magnesium can be described as follows: Due to the large amount of oxygen present in the protective atmosphere, and oxidation does not require complex decomposition processes, magnesium will first react with oxygen to form a thin MgO film on its melt surface, which is loose and porous, at the melt surface according to the reaction:

$$Mg\ (l) + 1/2O_2\ (g) = MgO\ (s) \qquad (1)$$

When the concentration of HFP in the atmosphere is low, there are no reactions between magnesium and HFP. The loose and porous MgO film cannot prevent further oxidation of magnesium, so in this case, the mixed gas has little protection on magnesium melt. When the concentration of HFP in the atmosphere is increased, magnesium will react with HFP or F_2, a decomposition product of HFP, to form MgF_2 in the pores of MgO film. The reactions are as follows:

$$3Mg(l) + C_3F_6(g) + 3O_2(g) = 3MgF_2(s) + 3CO_2(g) \quad (2)$$

$$Mg(l) + F_2(g) = MgF_2(s) \quad (3)$$

The formation of MgF_2 makes the film on the surface of the magnesium melt dense, which prevents further oxidation of magnesium. Therefore, the mixed gas of high HFP concentration can provide effective protection to magnesium melt, and the higher the HFP concentration is, the better the protection of the magnesium melt will be.

4. Conclusions

The protection of HFP/air mixed gases for molten magnesium at temperatures between 670 °C and 800 °C has been investigated. The results show that molten magnesium can be protected from excessive oxidation and burning by the mixed gases of HFP/air. With the increase of the concentration of HFP, holding time and the decrease of melt temperature, the protective effectiveness of HFP/air mixed gases increased. There was a uniform coherent film on the surface of molten magnesium, which contained MgF_2 and MgO. The thickness of the film was between 2.011 and 2.448 μm in 1% HFP/air at 700 °C for 30 min.

Acknowledgments

This work was supported by the National Natural Science Foundation of China (Grant No. 51471003) and Natural Science Basic Research Plan in Shaanxi Province of China (Program No. 2018JM5035).

References

[1] Mirak A. R., Davidson C. J., Taylor J. A. (2014) Characterisation of fresh surface films formed on molten Mg-Nd alloy protected by different atmospheres. Appl. Surf. Sci., 301: 91-98.
[2] Bian D., Deng J. X., Li N. et al. (2018) In vitro and in vivo studies on biomedical magnesium low-alloying with elements gadolinium and zinc for orthopaedic implant applications. ACS Appl. Mater. Inter., 10: 4394-4408.
[3] Emami S., Sohn H. Y., Kim H. G. (2014) Formation and evaluation of protective layer over magnesium melt under SF_6/air atmospheres. Metall. Mater. Trans. B, 45:1370-1379.
[4] Mirak A., Davidson C. J., Taylor J. A. (2010) Characterisation of fresh surface oxidation films formed on pure molten magnesium in different atmospheres. Corros. Sci., 52: 1992-2000.
[5] Fruehling J. F. Ph.D. Thesis. (1970) University of Michigan (USA).
[6] Shih T. S., Liu J. B., Wei P. S. (2007) Oxide films on magnesium and magnesium alloys. Mater. Chem. Phys., 104: 497-504.
[7] Jafari H., Idris M. H., Ourdjini A. (2013) A review of ceramic shell investment casting of magnesium alloys and mold-metal reaction suppression. Mater. Manuf. Process, 28: 843-856.
[8] Ricketts N. J., Cashion S. P. (2001) Hydrofluorocarbons as a replacement for sulphur hexafluoride in magnesium processing. In: Magnesium Technology. New Orleans, Louisiana, U.S.A, pp. 31-36.
[9] Ha W., Kim Y. J. (2006) Effects of cover gases on melt protection of Mg alloys. J. Alloy Compd., 422: 208-213.
[10] Chen H. K., Liu J. R., Huang W. D. (2008) Influence of diluent gas on the oxidation behavior of molten magnesium in atmospheres containing HFC-134a. Rare Metal Mat. Eng., 37: 913-1918. (in Chinese)
[11] Chen H. K. (2010) Effect of melt temperature on the oxidation behavior of AZ91D magnesium alloy in 1,1,1,2-tetrafluoroethane/air atmospheres. Mater. Charact., 61: 894-898.
[12] Xiong S. M., Chen X. (2008) China Patent, CN101302588A [P].
[13] Kang F., Chen H. K., Luo Z. X., Chang L., Liu J. R. (2017) High temperature thermal decomposition properties of hexafluoropropylene as cover gas of magnesium melt. Foundry Technology. 38: 128-137. (in Chinese)

Double-objective optimization simulation and orthogonal experiment of shredding roller for silage machine

Yingshuai Jiang[1], Guoping Li[1] and Xiangfang Zhang[1]

[1] School of Mechanical Engineering University of Jinan, Shandong Province, 250022, China

*Corresponding author's e-mail: 727515509@qq.com

Abstract. At present, the main factor affecting the quality of the silage of the silage machine is the shredding roller. Therefore, a theoretical model based on single-page hyperboloids and a shredding roller with a herringbone arrangement are designed. Then, the theoretical analysis of each key parameter was carried out. The consumption of energy and throwing height of the shredding roller can be targeted, and the orthogonal simulation was performed on each parameter combination. The optimal parameter combination was obtained as the rotational speed of 1426r/min, the rake angle is 53°, the sliding-cutting angle is 8°, and the influence of each factor and level on the consumption of energy and the throwing height is analyzed successively. The verification of experiment that the shredding loss rate was reduced by 5% compared with the pre-optimization, the optimized of broken stems is 25% higher than pre-optimization, the throwing height was 30mm higher than the pre-optimization, and the consumption of energy was reduced by 3.1% compared with the pre-optimization. The optimized of objective was within the prescribed range.

1.Introduction

The shredding roller is a key part of the design of the whole machine. According to research, shredding performance is affected by many factors such as the positional relationship between the moving knives and the fixed knife, the structural parameters of the moving knives (sliding-cutting angle, knife edge angle) and installation angle, the shape of the knife and the rotation speed of the shredding roller. The common shredding roller mainly include plane table-cutter, helical hob cutter, and slanting installing cutter with straight edge.

Xianglan Pu [1] et al, studied the energy of straight-edged oblique-mounted shredder to shred forage, selecting the lowest energy of consumption is the optimization goal, the energy of From the experiment we can see that the sliding-cutting is more labor-saving than cutting consumption was reduced and the performance was improved under the constraint condition; Yongqing Liu [2] et al, found out the optimal combination of sliding-cutting angle, fixed cutter configuration height and speed of roller by carried out an experimental study on the shredding energy of the helical hob cutter in cutting hay under different working conditions; Wu Qiaomei[3] et al, theoretically analyzed the parameters of the plane table-knife shredding conveyor, finite element analysis of the moving knives, and optimized size structure of moving knives; Zhao Xuelin[4] studied the stem shredding recovery device and designed a self-sharpening blade model.

According to the above analysis, only the power has been studied. The height of the throwing determines the parameters of the shredding roller. Therefore, a double-objective optimization analysis of

IOP Conf. Series: Materials Science and Engineering **711** (2020) 012029 doi:10.1088/1757-899X/711/1/012029

power and throwing height was proposed, thus, the most suitable parameter combination of the shredding roller was obtained.

2.Establishment of a single-leaf double-curved straight-knife oblique-mounted shredding roller model

The install of straight-edged oblique-loaded cutter is tilt, and the track is a single-leaf hyperboloid when the knife edge rotates around the rotary axis. As shown in Fig. 1, it can be seen that the edge line of AB is rotated around the rotation axis of OZ. The equation is

$$(x^2 + y^2)/a^2 - z^2/c^2 = 1 \tag{1}$$

In formula (1), a is the real half-axis of single-leaf double-curved; c is the virtual half axis of single-leaf double-curved.

In Fig. 1, taking the above-mentioned single-leaf hyperboloid with the section x=a, two straight lines can be obtained, and the equation is:

$$\begin{cases} x = a \\ z = cy/a \end{cases} \begin{cases} x = a \\ z = -cy/a \end{cases} \tag{2}$$

Fixed knife have two mounting positions to achieve equal gap cutting: One is a straight line that coincides with the edge of the moving knives at a certain rotation angle, the other is the conjugate straight line of the moving edge line at a rotation angle[5], but the installation of fixed knife with straight edge and the adjustment of cutting gap will be more difficult, so the fixed knife adopts straight edge and is installed horizontally.

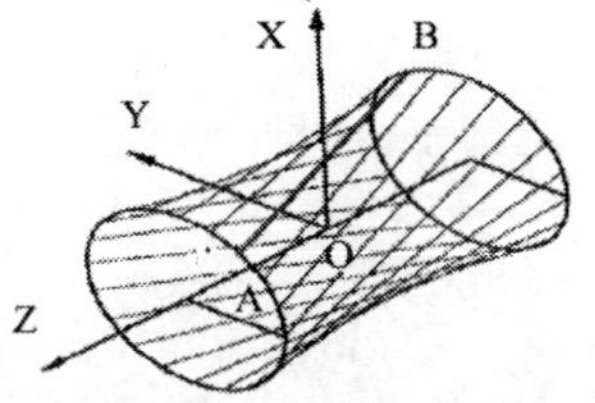

Figure. 1 The position of moving knives edge

Figure. 2 Diagram of herringbone moving knives

Cutting the forage by the way of equal gap will not be guaranteed when the fixed knife is installed with straight edge. For this reason, there are two ways to make the middle gap smaller with the premise of ensuring uniform cutting. One is to increase the number of knives; the second is to extend the length of the roller to achieve the same effect, but this way will increase the mechanical vibration and this way is limited by the model of silage machine.

Therefore, the moving knives adopt herringbone arrangement, as shown in Fig. 2, and appropriately increases the number of knifes to achieve uniform cutting.

2.1 Push angle and friction angle

As shown in Fig. 3, the moving knives edge is AB, fixed knife edge is CD, the angle between the cutting edge line and the fixed cutting edge line is the pushing angle, it is closely related to forage type and material of moving knives and fixed knife. Due to the horizontal installation of the fixed knife, the pushing angle is unchanged during the rotary cutting process[6]. The angle between the material and the knife is δ. If $\psi \geq 2\delta$, the resultant force F pushes the material outward along the knife when cutting. If $\psi \leq 2\delta$, the resultant force F is inward, the material must be stable and the force F needs to be inward.

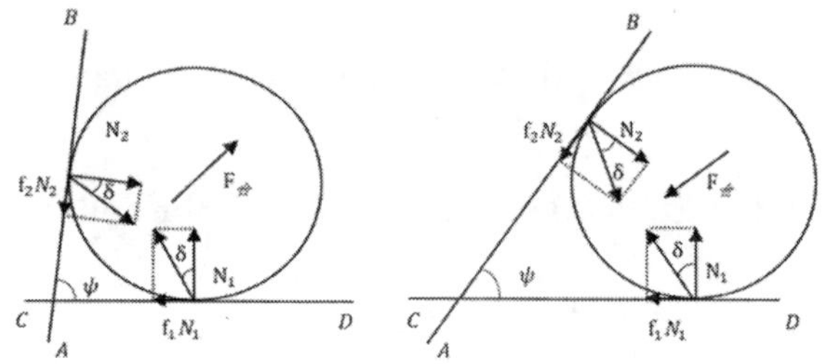

Figure. 3 Cutting force at different push angles

The pressure of the knife to forage is N_1, the friction is $f_1 N_1$. The pressure of moving knives to forage is N_2, the friction is $f_2 N_2$. The condition of stable cutting is:

$$f_1 N_1 \geq N_2 sinx - f_2 N_2 cosx \tag{3}$$

$$N_1 = f_2 N_2 sinx + N_2 cosx \tag{4}$$

$f_1 = tan\delta_1, f_2 = tan\delta_2$, the formula is simplified as follows:

$$tanx\psi \leq \tan(\delta_1 + \delta_2) \tag{5}$$

In the formula, ψ is the pushing angle; δ_1 is the friction angle of the material and the moving knives edge; δ_2 is the friction angle between the material and the fixed knife edge.

The dynamic friction coefficient of moving knives, fixed knife and silage corn with water content of 65% is 0.45, and the materials of moving and fixed knife are same. According to the above formula, so that the maximum pushing angle of cutting green corn is 48°.

2.2 Advantages and calculations of sliding-cutting angle

As shown in Fig. 4, AB is the moving edge line, v is the speed of moving knives, and v can decompose into the velocity v_n that perpendicular to the cutting edge and the velocity v_t that parallel to the cutting edge. v_n and v_t are called cutting speed and sliding-cutting speed, the ratio of the above two speeds is tangent, from Fig. 4 we can see that $tan\,\tau = \frac{v_t}{v_n}$, called the sliding-cutting coefficient, τ is called the sliding-cutting angle, when $\tau > 0$ and does not exceed the moving edge and the friction angle δ of the stem is called slip cutting. From the above analysis, it is known that for a single-leaf hyperbolic flat knife-type shredder, a suitable sliding-cutting angle needs to satisfy the condition: $0 < \tau < \delta$.

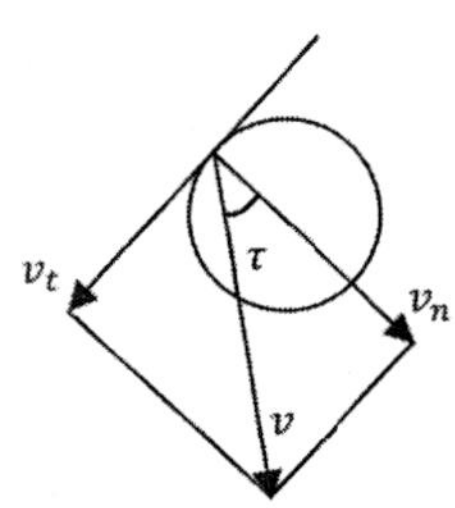

Figure. 4 Diagram of the sliding-cutting angle

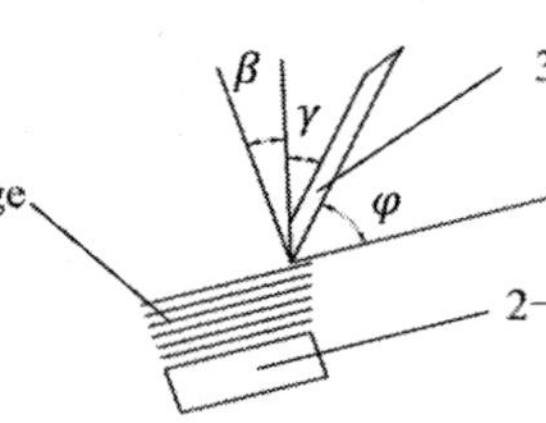

Figure. 5 Diagram of Angle Relations

2.3 Moving knives installation Rake angle

In order to reduce the friction between the front plane of the knife and the cutting surface, the installation of the knife should have a gap angle. The installation gap angle of the knife is the normal angle between the moving knives and the cutting plane. As can be seen from the Fig. 5, the rake angle φ, the knife angle β and the installation gap angle γ of the knife have the following relationship:

$$\varphi = 90^0 - \beta - \gamma \tag{6}$$

It can be seen that the pre-cutting angle determines the height of the forage, if the rake angle too large, the throwing height will not be reached, and the silage can't be throwing into the crushing device for crushing operation; on the contrary, if the knife installation gap angle γ is too large, the corresponding cutting rake angle φ becomes small, which increases the windward area of the material, thereby increasing the energy of consumption. Therefore, in the case of satisfying the throwing condition, a smaller installation gap angle should be selected as much as possible. In general, $\gamma \leq 48° \sim 58°$.

3. Orthogonal experiment to determine the parameters of the shredder roller

From the above analysis, the factors affecting consumption of energy mainly include sliding-cutting angle, rake angle and speed. From the above analysis, the factors affecting consumption of energy mainly include sliding-cutting angle, rake angle and speed. At the same time, these factors also affect the height of the centrifugal transfer of the shredding roller. According to the design requirements of the whole machine, the speed range is from 1200 to 1700 r/min[7], there are two objectives of throwing height and consumption of energy, a combination of low consumption of energy and higher throwing height is needed, the data is shown in the table 1.

Table1. Three Factors and Three Levels

Level \ Factor	Rotational Speed A (r/min)	Slip angle B (°)	Rake angle C (°)
1	1206	8	48
2	1426	16	53
3	1646	24	58

In order to save cost, and get the best parameter combination quickly and accurately, orthogonal experiment with the $L_9(3^4)$, the last column was deleted[8], The test data was obtained through simulation analysis, as shown in Table 2.

Table 2. Orthogonal Experimental

Times \ Factor	1	2	3	Results of each objective	
	A	B	C	Cutting force of peak(N)	Speed of peak (m/s)
1	1	1	1	8331	98
2	1	2	2	8796	86
3	1	3	3	9797	81
4	2	1	2	7196	134
5	2	2	3	7396	106
6	2	3	1	8872	136
7	3	1	3	8721	99
8	3	2	1	11196	137
9	3	3	2	10416	126

Table 3. Double-Objective Range

Range \ Objective	Cutting force of average			Speed of average		
K_1	26924	24248	28399	265	318	371
K_2	23464	27388	26408	376	329	346
K_3	30333	29085	25914	362	343	286
k_1	8975	8083	9466.	88	106	113
k_2	7821	9129	8803	125	110	115
k_3	10111	9695	8638	121	114	95
Range	2290	1612	828	37	8	20
Excellent solution	A_2	B_1	C_3	A_2	B_3	C_2

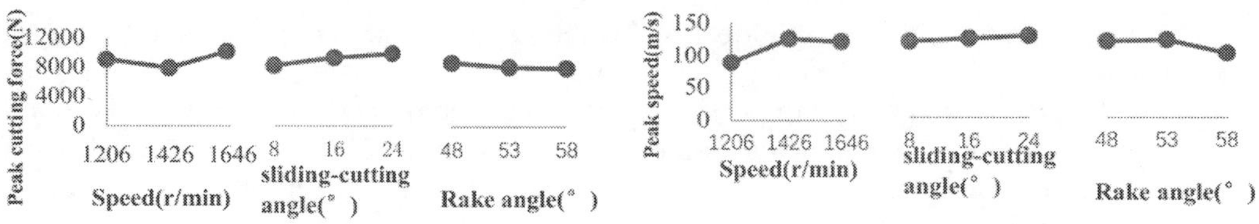

Figure. 6 Scatter plot of each indicator with factor level

From the above table, it is analyzed:

- The effect of speed A on each objective, as seen from Table 3, for energy of consumption, it is one of the most important factors about energy of consumption. It is best to take the second level. For the throwing height, although the range of the speed is the smallest, it is also taken the second level is the best, so for both indicators, 1426 is the best speed for double-objective .
- The influence of the sliding-cutting angle B on each objective, the forage of throwing height is reflected by the speed obtained after the forage is cut off. As seen from Table 3, for energy of consumption, the range is less than the speed, but much larger than the rake angle， it is best to take the first level. For the throwing height, the range is the smallest, it shows that the effect of this factor on the throwing height is the smallest. As seen from Fig. 6 that the first level is not too bad for the sliding-cutting angle. For the two objectives, the best sliding-cutting angle is 8.
- The influence of the rake angle on each objective, as seen from Table 3, for energy of consumption, the difference of the rake angle is the smallest, that is, the influence of consumption of energy is the smallest, and the level of 3 is the best. For the throwing height, the range of the front angle is between the speed and the sliding-cutting angle, the second level is best. For energy of consumption, the effect of the rake angle is minimal. It can be seen from Fig. 6, taking the second level of energy of consumption is not too high, so, it is better to take the second level.
- Through a comprehensive analysis of the impact of various factors on each objective, a better experiment plan is:
  A_2: speed, second level, 1426;
  B_1: sliding-cutting angle, first level, 8;
  C_2 : rake angle, first level, 53.

4. VERIFICATION OF EXPERIMENT

In order to verify whether the optimal combination of parameters meets the requirements, the analysis results were experimented. The indicators of experiment are:

4.1. Rate of Shredding loss

Taking 1Kg of shredded green corn stems. In the shredded stems that less than 20mm in length and larger than 35mm are considered to be losses. The formula is:

$$z_1 = x_1/x_2 \times 100\% \tag{7}$$

In the formula, z_1 is the rate of shredding loss; x_1 is the number of broken stems whose length is not between 20-30 mm; x_2 is total number of samples.

4.2. Average of cut length

Measuring the length and quantity of the sample and calculate the average .

4.3. Throwinging height

The detection of the throwing height needs to be measured separately before assembling the machine, place the gauze over the shredding roller, and adjust the height of gauze, then, observing the gauze whether have broken stems.

4.4.Energy of consumption ratio

The energy of the working device is derived from the engine, and is calculated by the distribution ratio of each working member. The formula is:

$$N = qv^2/(1-f) \times 10^{-3} \tag{8}$$

In the formula, the N is power, the unit is Kw; the q is feed quantity, the unit is $\mathrm{Kg/s}$; the v is circumferential speed, the unit is $\mathrm{m/s}$; the f is coefficient of comprehensive friction.

Three repeated experiment were carried out to compare the parameters before optimization, the pre-optimization speed is 1646 r/min, the sliding-cutting angle is 24°, and the rake angle is 48°, the experiment results are shown in Table 4.

Table 4. Optimal combination experiment result

Parameter	Shredding loss rate (%)	Broken stems length (mm)	Throwing height (mm)	Energy of consumption ratio (%)
Theoretical value	3	25	630	15
Value of experiment	3	24.43	630	15.9
Pre-optimization value of experiment	8	18.16	600	19

5. CONCLUSION

The experiment of double-objective orthogonal simulation is a very efficient experimental method. Analyzed the reasons of poor forage quality and improved the structure of the shredding roller, and on the basis of this structure, the energy of consumption and the throwing height are taken as the objectives, and the parameters are optimized by the experiment of double-objective orthogonal simulation. The results of experiment proved the simulation results fulfill the design requirements, and the following conclusions are drawn: the shredding loss is reduced by 5% compared with the pre-optimization, and the consumption of energy is reduced by 3.1% compared with the pre-optimization, the throwing height is 30mm higher than before optimization.

REFERENCES

[1] Pu Xianglan, Li Dezhu, Zhang Jinsong, et al. Ptimum design of straight edge slanting installing, rolling and cutting device[J]. *Journal of Agricultural Science Yanbian University*, 1999(3): 165-171.

[2] Liu Yongqing, Lin Gongzhen, Mao Jianhua, et al. The cutting energyof forage chopper on the cylinder - type cutterhead[J]. *Journal of Luoyang Institute of Technology*, 1997(4): 45- 49.

[3] Wu Qiaomei, Study and optimization of the parameter of slab cutter cylinder type shredding mechanism[D]. *Xian: Northwest A&F University*, 2010.

[4] ZHAO Xuelin. Design and research for the stalk chopping and recovery device of corn combine reaping both stalk and spike[D]. *Beijing: Chinese Academy of Agricultural Mechanization Sciences*, 2014.

[5] Liu Yang, design and finite element analysis of roller shredding mechanism for silage maiz harvester[D]. *Xian: Northwest A&F University*, 2008.

[6] Zhou Chunyan. finite element analysis and optimal design of important parts of cylinder-type corstalk shredder[D]. *Xian: Northwest A&F University*, 2012.

[7] Ren Dezhi, Bai Xuewei, Liu Dejun, et al. Design and Experiment of Separating Rolling and Uncovering Roller of Corn Stalk Skin[J]. *Transactions of the Chinese Society of Agricultural Machinery*, 2018, **49**(1): 93-99.

[8] Chen Kui. *Experimental design and analysis*[M]. 2rd ed. Beijing: Tsinghua University Press. 2005: 78-81.

Nonlinear Conductivity of ZnO Materails Induced by X Ray

Xiaoning Yang[1 2 *], Yuming Liu[1 2], Zhihao Wang[1], Yu Li[1], Jinghu Wang[1] and Yongtai Zhang[1]

[1] Beijing Institute of Spacecraft Environment Engineering, Beijing 100084, P. R. China

[2] Science and Technology on Reliability and Environment Engineering Laboratory, Beijing 100084, P. R. China

[*]Corresponding author's e-mail: yangxiaoning2014@126.com

Abstract. Due to high energy electron radiation in space, Spacecraft internal charging will occur, which will cause deep dielectric electrostatic discharge and cause the most serious anomalies, that is, those that have resulted in the loss of mission. Because spacecraft internal charging is accompanied with X-ray bremsstrahlung, it is possible to mitigate inner dielectric electrostatic discharge by using nonlinear conductivity material induced by X-ray radiation. It shows that the conductivity of ZnO materials increases with the increase of X-ray radiation dose rate. At low X-ray radiation dose rate, the conductivity of ZnO increases linearly with the increase of X-ray dose rate. At high X-ray radiation dose rate, the conductivity of ZnO is saturated gradually. The conductivity of ZnO with high original resistivity varies more widely with the dose rate of X-ray, which is more suitable for spacecraft internal charging protection.

1. Introduction

Spacecrafts will encounter various radiation environments in space, such as the earth radiation belt, solar high-energy particles, cosmic rays. Among them, space high-energy electrons will penetrate the surface structure of spacecrafts, chassis and other equipment shells, and deposit in dielectrics such as insulating cables and circuit boards. Because of the high resistivity of dielectrics, the electrons deposited would accumulate continuously and establish a high internal electric field in the dielectrics. When the electric field strength exceeds the dielectric breakdown field, discharge phenomenon would occur, which is called deep dielectric charging effect [1-3]. A large number of on-orbit experiments and theoretical analysis show that deep dielectric charging effect in the spacecrafts will cause serious anomalies [4,5]. So highly insulating materials should be avoided in spacecraft charging environments

An effective way to mitigate deep dielectric charging effect in spacecrafts is that when deep dielectric charging effect in spacecraft occur, the conductivity of the dielectrics is nonlinear and can increase to release the accumulated charges, prevent the development of large electric fields internal to the dielectrics and thereby prevent them from developing electrostatic discharge pulses [6]. Because the process of high-energy electrons injecting into spacecraft will produce X-ray bremsstrahlung, it is possible to develop a dielectric with nonlinear conductivity induced by X-ray radiation to reduce the risk of deep dielectric charging effect.

ZnO is a wide band gap semiconductor material, which has important applications in ultraviolet photoelectric detection. In recent years, it has been found that ZnO materials also have a great response to X-rays [7-10]. It is possible to use ZnO materials as nonlinear conductivity dielectrics to mitigate deep dielectric charging effect. Here, the conductivity of zinc oxide semiconductor materials has been studied under different X-ray radiation conditions, which is very important to develop deep dielectric charging effect mitigation.

2. experimental

Three kinds of zinc oxide powders with different particle sizes of 20 nm, 30 nm and 1 micron were used respectively. The ZnO powders were characterized by scanning electron microscopy (SEM) and X-ray diffraction (XRD). The ZnO powders were pressed into thin flakes and each side of the thin flake was clamped with aluminum foil. The thickness of ZnO flake was about 1 mm. Aluminum foil acted as test electrode. The conductivity of zinc oxide flakes induced by X-ray was tested at National Institute of Metrology of China and Beijing Normal University, respectively. The schematic of the test process is shown in Figure 1. The current–voltage measurement was performed using Keithley 6517B, which can provide a maximum voltage of 1000V and a current measurement range of 1fA to 20mA.

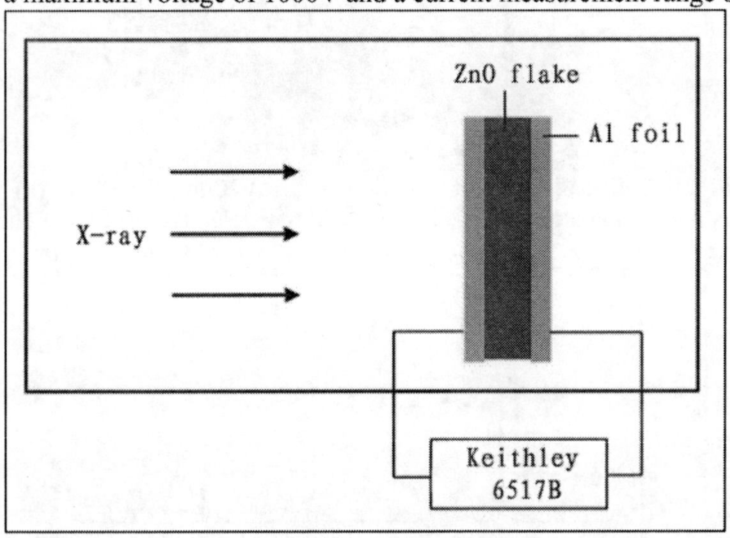

Figure 1. The schematic of the test process for conductivity measurement of ZnO flakes in X-ray radiation environment

3. result and discussion

The SEM images of ZnO powders with different particle sizes are shown in Figure 2. The diameters of the particles are 20 nm (Figure2. a), 30 nm (Figure 2. b), and 1 μm (Figure.2 c) respectively. The particle sizes of ZnO are relatively uniform, and each particle has edges and corners.

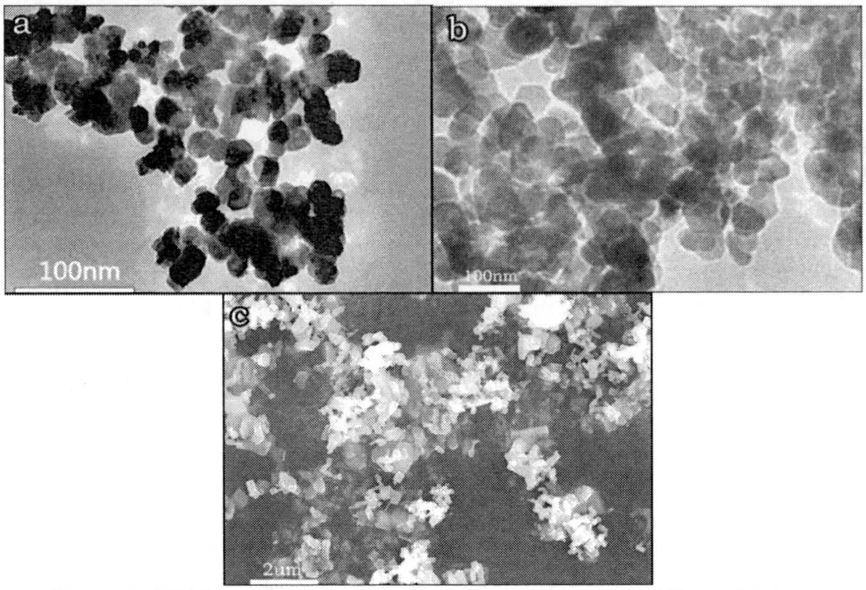

Figure 2. SEM images of ZnO powders. (a) 20nm, (b) 30nm, (c) 1μm

Figure 3 shows the XRD pattern of the sample with particle size of 20nm. The XRD spectra are in agreement with the standard spectra. It shows that the samples obtained are pure wurtzite zinc oxide.

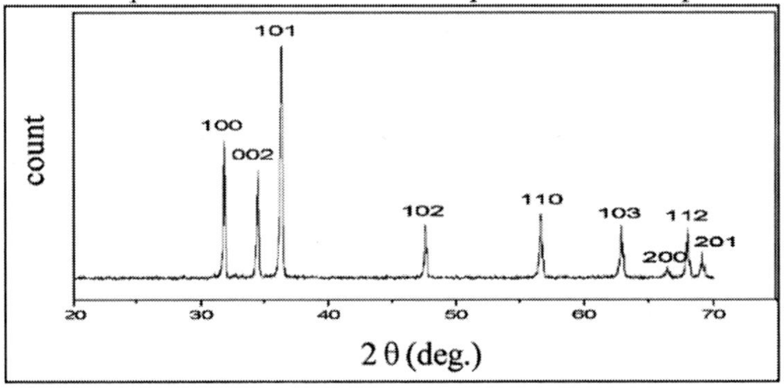

Figure 3. XRD pattern of 20nm ZnO powder

Table 1 shows the experimental results of the current of different particle size ZnO materials varying with X-ray radiation dose rate under the same voltage. Because the conductivity is proportional to the current under the same voltage, the change of conductivity can be reflected directly by measuring the current.

Table 1. X-ray radiation dose rate versus current of ZnO flakes

X-ray dose rate (mGy)	Current of 20nm ZnO under 200V (μA)	Current of 30nm ZnO under 1000V (μA)	Current of 1μm ZnO under 500V (μA)
0	208	41	115
0.81	207	38	126
1.2	208	40	131
10	209	43	170
100	205	42	214
500	209	41	243

It shows that the nonlinear conductivity phenomenon induced by X-ray radiation only occurs in 1μm ZnO particles. When the dose rate reaches 500 mGy, the current of ZnO increases from 115 μA to 243 μA, which means that the conductive of ZnO can be doubled by X-ray radiation.

Figure 4 shows the curve of the current versus X-ray dose rate for 1μm ZnO particles. It shows that the current of ZnO tends to be stable when the X-ray dose rate is high enough.

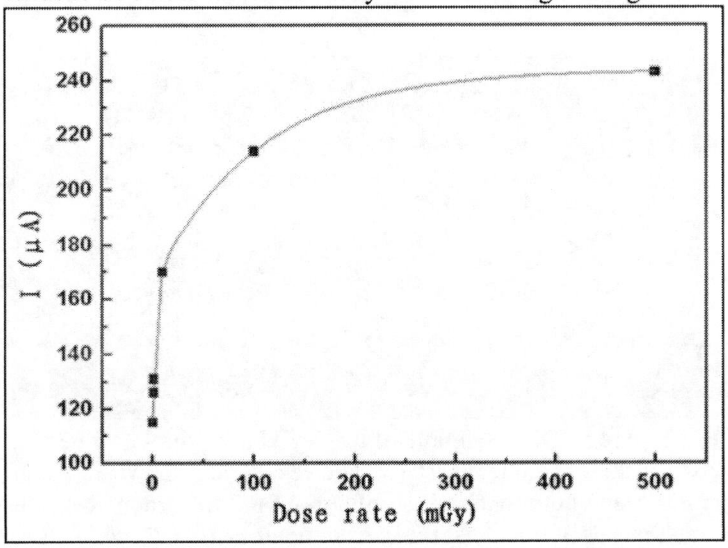

Figure 4. Current versus dose rate for 1 μm ZnO flake

The variation of the current of ZnO with the radiation dose rate can be well fit with double exponential function. The current changing equation with radiation dose rate is in accordance with the following formulas:

$$I_{1\mu m} = 243 - 78\exp^{-r/78} - 50\exp^{-r/3} \tag{1}$$

Where $I_{1\mu m}$ is the current of 1 μm ZnO flakes under applied voltage of 500 V, and r is X-ray radiation dose rate.

It is presented that the change of radiation-induced nonlinearly conductivity of ZnO is gradually stable with the radiation dose rate. The mean reason is the saturation of electron hole pairs in ZnO. In semiconductors, electrons are concentrated in a very small energy gap near the bottom of the conduction band. Because the energy gap is very small, it can be simply considered that the energy levels are occupied by these electrons on the bottom of the conduction band (Ec). The number of them can be expressed by the effective density of states (Nc) at the bottom of the conduction band. So the electron concentration in the conduction band (n) is given by:

$$n = Nc \cdot \exp\left(-\frac{E_C - E_F}{kT}\right) \tag{2}$$

Where $\exp\left(-\frac{E_C - E_F}{kT}\right)$ is the Boltzmann distribution representing the probability that electrons occupy the bottom of conduction band. E_F is Fermi level.

When ZnO semiconductor is in X-ray radiation environment, some electrons at the bottom of conduction band will jump to valence band, but the total amount of excitated electrons is limited by Nc. So the conductivity of ZnO will be saturated with the X-ray dose rate.

It seems that the change of conductivity induced by X-ray is related to the particle size of ZnO. More obvious change of radiation-induced conductivity can be obtained in ZnO with larger the particle size. So ZnO bulk material may present non-linear conductivity induced by X-ray much more clearly. Single crystal ZnO bulk material was fabricated by SunYat-sen University. Nonlinear conductivity induced of single crystal ZnO bulk material by X-ray was tested as shown in Figure 5.

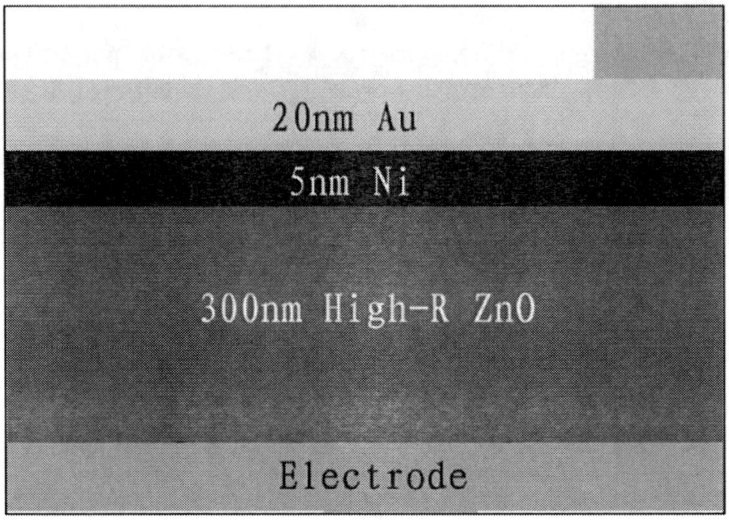

Figure 5. The schematic of the ZnO bulk sample for testing

The voltage between the electrodes is 10 V. The results are shown in Figure 6. It shows that the current of single crystal ZnO bulk material is only about 0.2 nA when X-ray dose rate is zero. When the dose rate of X-ray reaches 370 mGy/s, the current reaches 1.9 nA, which means the conductivity of single crystal ZnO bulk material increases by 8.5 times. Nonlinear conductivity of single crystal ZnO bulk material induced by X-ray is more obvious than ZnO particles, and single crystal ZnO bulk material is more suitable for mitigating deep dielectric charging effect.

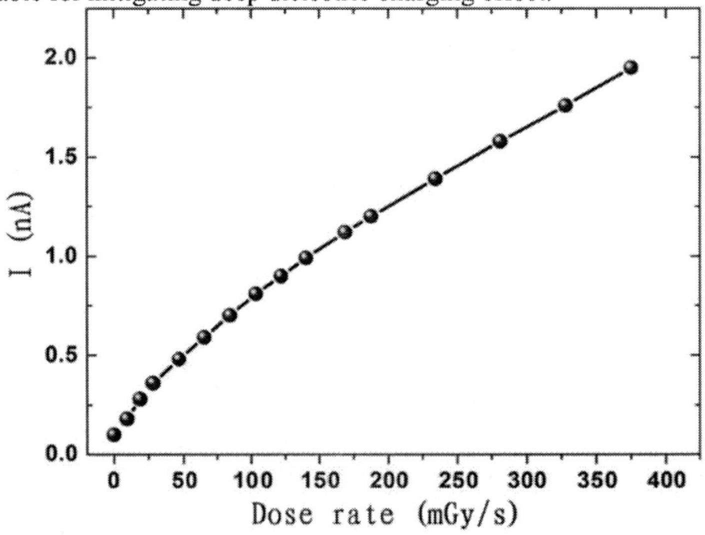

Figure 6. Current versus X-ray dose rate of single crystal ZnO bulk material

4. Conclusion

Non linear conductivity of ZnO semiconductor materials induced by X-ray will occurs changes under certain conditions. It seems that the phenomenon of radiation induced nonlinear conductivity of ZnO semiconductor materials is related to the particle size of ZnO. Radiation induced nonlinear conductivity of larger particle size ZnO or single crystal ZnO bulk material is more obvious. The variation of radiation-induced conductivity with X-ray dose rate can be described by exponential function, that is, there is a saturation value for the change of radiation-induced conductivity of ZnO. The conductivity of high-resistivity single crystal ZnO varies in a much larger range under the same X-ray radiation conditions and may be more suitable for deep dielectric charging effect mitigation.

References

[1] Frederickson, A. R., Benson, C. E., Cooke, E. M. (2000) Gaseous discharge plasmas produced by high-energy electron-irradiated insulators for spacecraft. Plasma Science IEEE Transactions on, 28(6): 2037–2047.

[2] Kim, W., Jun, I., Kokorowski, M. (2010) Internal electrostatic discharge monitor. IEEE Transactions on Nuclear Science, 57(6): 3143–3147.

[3] Jian-Guo, H. (2004) A study of characteristics for deep dielectric charging on spacecrafts. Acta Physica Sinica, 53(3): 961–966.

[4] Paulmier, T., Dirassen, B., Payan, D., Eesbeek, M.V. (2009) Material charging in space environment: Experimental test simulation and induced conductive mechanisms. IEEE Transactions on Dielectrics and Electrical Insulation, 16(3): 682–688.

[5] O'Brien, T. (2009) SEAES-GEO: A spacecraft environmental anomalies expert system for geosynchronous orbit. Space Weather, 7(9): 396–412.

[6] Frederickson, A. R., Dennison, J. R. (2003) Measurement of conductivity and charge storage in insulators related to spacecraft charging. IEEE Transactions on Nuclear Science, 50(6): 2284–2291.

[7] Gao, X., Kang, Q. S., Yeow, J. T. W., Barnett, R. (2010) Design and evaluation of quantum dot sensors for making superficial x-ray energy radiation measurements. Nanotechnology, 21:285502.

[8] Endo, H., Chiba, T., Meguro, K., Takahashi, K., Fujisawa, M., Sugimura, S., Narita, S., Kashiwaba, Y., Sato, E. (2011) Fabrication and characterization of a ZnO X-ray sensor using a high-resistivity ZnO single crystal grown by the hydrothermal method. Nucl. Instrum. Methods in Phys. Res. Section A, 665: 15–18.

[9] Zhao, X. L., Kang, X., Chen, L., Zhang, Z. B., Liu, J. L., Ouyang, X. P., Peng, W. B., He, Y. N. (2014) Study of ZnO photoconductive X-ray detector. Acta Phys. Sin., 63(9): 098502.

[10] Liu, K., Sakurai, M., Aono, M. (2010) ZnO-Based Ultraviolet Photodetectors. Sensors, 10(9): 8604–8634.

IOP Conf. Series: Materials Science and Engineering **711** (2020) 012031 doi:10.1088/1757-899X/711/1/012031

Analysis and Research of Molten Copper Sloshing in Mould of Anode Based on VOF Method

Jie Zheng[1], Guoping Li[1,2], Xinyi Teng[1] and Zefeng Qiu[1]

[1] School of Mechanical Engineering University of Jinan, China

[2] 727515509@qq.com

Abstract. In the production process of anode, the quality of anode is affected by the sloshing of molten copper caused by casting wheel during operation. Therefore, it is significant to study the sloshing of molten copper and the motion model of casting wheel for improving the quality of anode. Using VOF method and UDF to study the sloshing of molten copper and the motion characteristics of casting wheel. Establishing the $\mathbf{\Delta H - a}$ model through the method of FLUENT and MATLAB. The optimal motion model of casting wheel is S-model.

1. Introduction

The sloshing of copper liquid in the mould will be caused by the rotating process of casting wheel. Because the temperature of the mould is lower than that of the molten copper, when the molten copper sloshes, the molten copper will adhere to the mould and form copper, forming raised edge and burrs. When the raised edge is large or burr is much, the scrap anode is formed and recasted. The study of molten copper sloshing and motion characteristics of casting wheel is very important in the industry. It can improve the quality of anode, reduce the scrap rate and cost.

VOF refers to the volume function of the fluid, which is defined as the ratio of the volume of the target fluid to the volume of the grid in each grid of the flow field. As long as we know the value of this function on each grid, we can track the moving interface[1,2]. At present, many domestic scholars such as Yutai Huang[3] and Xicong Zou[4] have studied the fluid-solid coupling dynamics of aircraft fuel tank sloshing, Di Yu[5] and Zhuqing Cheng[6] have studied the liquid impact in automobile fuel tank, and analyzed the influence of fuel tank filling ratio and fuel tank baffle on liquid sloshing. The effect is well. However, there are few studies in the field of sloshing of molten metals. The software called ICEM CFD[7] has better mesh quality in fluid region, analysis and optimization of motion model using UDF[8], the sloshing of molten copper can be accurately traced by VOF.

2. Establishment of temperature-viscosity model of molten copper and its fluidity analysis

Molten copper is different from water. To analyze the sloshing of molten copper, it is necessary to study the viscosity characteristics of molten copper. The viscosity of molten copper is different at different temperatures. The type of mould and the average flow rate of molten copper will affect the fluidity of molten copper. The viscosity and fluidity affect the results of sloshing analysis.

2.1. Establishment of temperature-viscosity model of molten copper

The fluidity of molten copper is equal to the reciprocal of viscosity. The viscosity of molten copper affects the sloshing. In the industry, the main factor affecting the viscosity is temperature. The temperature of molten copper affects the sloshing. The relationship between viscosity and temperature:

$$\eta = B_0 e^{[E_n/(RT)]} \tag{1}$$

In formula (1), η is the viscosity, B_0 is the coefficient and RT is the temperature in a certain range. The melting point of copper is 1083°C. The fluidity of molten copper is better when the temperature is 1200°C. The temperature of molten copper studied in this paper is 1083°C - 1350°C, which is converted to Kelvin temperature of 1356 K - 1623K.

Referring to literature [9], two viscosities at different temperatures were obtained:$T_1 = 1373\text{K}, \eta_1 = 3.996 \times 10^{-3} \text{pa} \cdot \text{s}$; $T_2 = 1573\text{K}, \eta_2 = 2.638 \times 10^{-3} \text{pa} \cdot \text{s}$.

The relationship between temperature and viscosity of liquid copper is obtained by substituting temperature into formula (1):

$$\eta = 10^{1950/T - 3.8186} \tag{2}$$

The relationship between viscosity and temperature of molten copper is shown in Fig. 1.

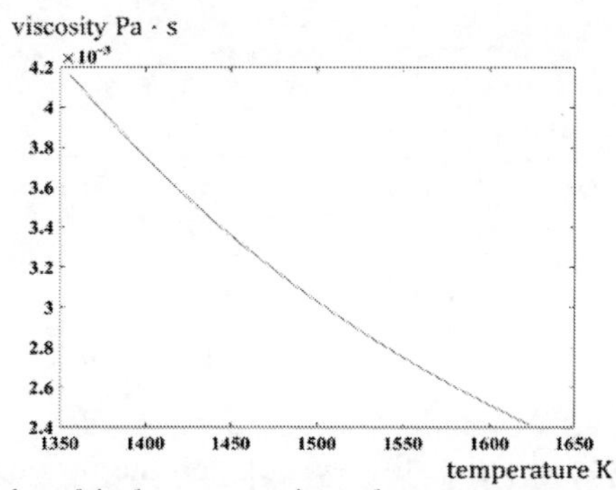

Figure 1. The relationship between viscosity and temperature of molten copper

2.2. Fluidity analysis of molten copper in mould

Reynolds is a dimensionless number that can be used to represent fluid flow. The formula for calculating Reynolds number is:

$$\text{Re} = \rho v d / \eta \tag{3}$$

Formula (3): ρ is density, v is velocity of fluid, d is length, η is viscosity.

The relationship between viscosity and temperature (2) is substituted into formula (3), and the Reynolds formula is obtained:

$$\text{Re} = \rho v d \, / \, (10^{1950/T - 3.8186}) \tag{4}$$

In the simulation of molten copper sloshing, the length of mould d = 1140 mm. The density of molten copper ρ=8.920 g/cm³.

The molten copper velocity V is similar to the linear velocity v_1 of the mould, $v = v_1$. According to the production requirements of copper smelter, the production task of 18-mode single casting wheel is x=30 tons/hour. Taking 275 kg anode as an example, the weight of each anode is G=275 kg, the modulus of casting wheel is S=18, the diameter of casting wheel is D=10 m, and the time of casting process is C=15 s. The results show that the average velocity v_1 of each position is:

$$v_1 = \frac{\pi D}{st} \tag{5}$$

In formula (5), t is the time for mould to rotate a position:

$$t = \frac{3600G}{1000x} - C \tag{6}$$

IOP Conf. Series: Materials Science and Engineering **711** (2020) 012031 doi:10.1088/1757-899X/711/1/012031

The average flow velocity $v = v_1 = 0.097m/s$ was obtained. The Reynolds formula was introduced to calculate the minimum value of the movement of molten copper in the mould, which was 309624.

3. The simulation of the molten copper sloshing

The diameter of the casting wheel is 10m. The velocity and acceleration of the outer ring is the highest, and the arc length of the outer ring is 1.74m. The rotational characteristics of casting wheel can be understood as the movement characteristics of the wheel.

3.1. The simulation model of molten copper level

It is necessary to understand the shape of molten copper in the mould to establish the simulation model. The specification of 275kg anode is shown in Fig. 2. L_1=1140mm, L_2=1110mm, L_3=960mm, D_1=770mm, D_2=200mm, D_3=620mm,R_1=75mm, R_2=105mm, H=45mm.

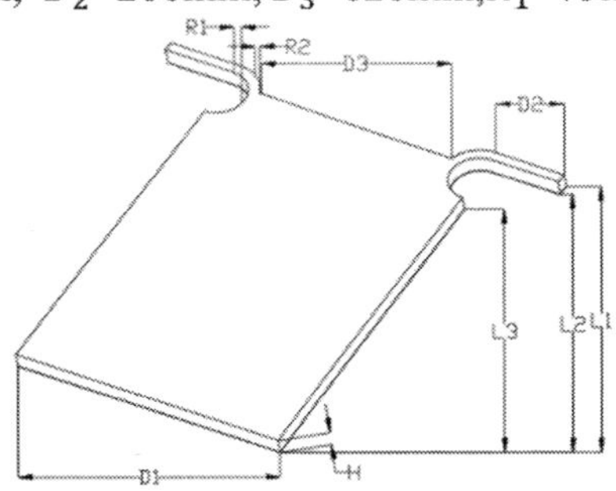

Figure 2. Specification of anode

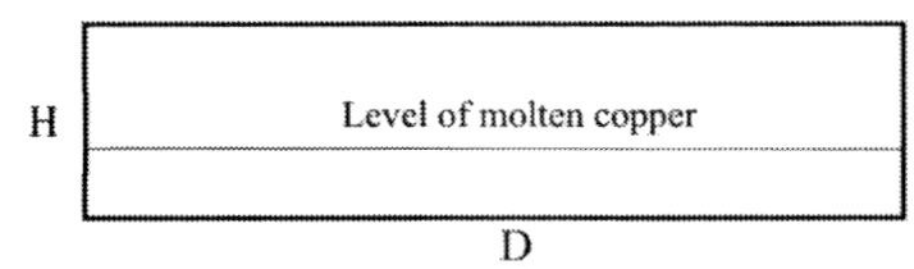

Figure 3. simulation model of molten

According to the specifications of anode, the shape of anode is about cuboid. Sloshing height is mainly caused by the tangential force of casting wheel, and the influence of centrifugal force is small. Therefore, the molten copper is simplified into rectangle, H = 100mm, D = 770mm, and the molten copper level is 45mm, as shown in Fig. 3.

3.2. Establishment of $\Delta H - a$ model and simulation

When the acceleration is 0 m/s^2, the level of molten copper is 0 mm. According to the experience, the acceleration a_1=0.1 m/s^2 and a_2=0.04 m/s^2 are selected, and the maximum liquid level height in the mould under the action of a_1 and a_2 is calculated by FLUENT simulation. The $\Delta H - a$ model was established by MATLAB curve fitting.

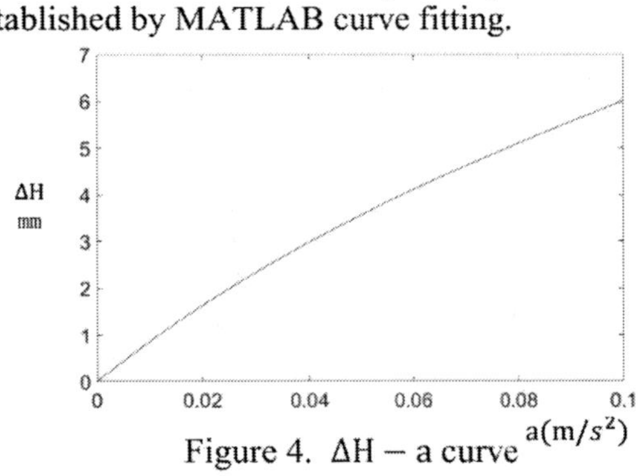

Figure 4. $\Delta H - a$ curve

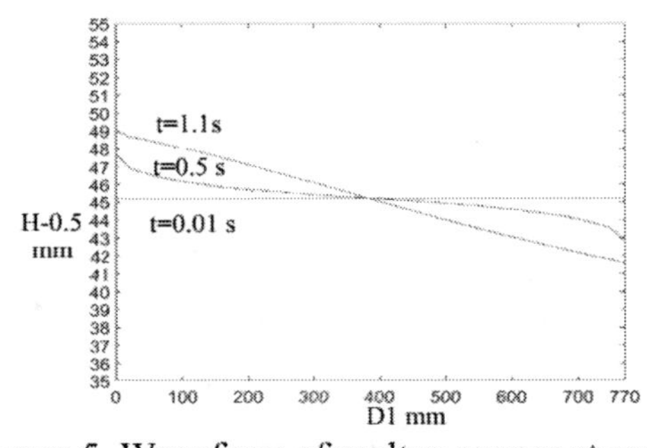

Figure 5. Waveform of molten copper at a=0.056

The curves are shown in Fig. 4. The model equation is:

$$\Delta H = 8832 \times (\sin(a - \pi)) - 446.1(a - 10)^2 + 4.461 \times 10^4 \quad (7)$$

Formula 7:$0m/s^2 < a < 0.1m/s^2$. When ΔH= 4mm, the a=0.056 m/s^2. Under the condition that the casting wheel rotates for the shortest time, the maximum sloshing height is 4mm.

In order to verify the correctness of the ΔH − a model, further simulation shows that when the acceleration is $0.056\ m/s^2$, the maximum molten copper level is 49mm, and the height difference ΔH=4mm, as shown in Fig. 5.

4. Establishment of moving model of casting wheel

4.1. Triangular motion model

The triangular motion model is shown in Fig. 6. The molten copper is accelerated in opposite direction, and the potential energy is transformed into kinetic energy. The level at 770mm will exceed 49mm. Therefore, the model does not meet the design requirements.

4.2. Trapezoidal motion model

Aiming at the problem that the sloshing high in the deceleration stage of triangular model, the uniform phase is added in the middle period. The sloshing tends to be stable through the uniform phase. At the same time, the temperature of molten copper decreases continuously, the viscosity increases further, and the resistance to sloshing increases. Under the same acceleration, the level will decrease. The trapezoidal motion model is designed as shown in Fig. 7. The relationship between speed and time is as follows:

$$v = \begin{cases} 0.056t \quad (0 \le t \le 3) \\ 0.168 \ (3 < t \le 10.4) \\ -0.056t \quad (10.4 < t \le 13.4) \end{cases} \tag{8}$$

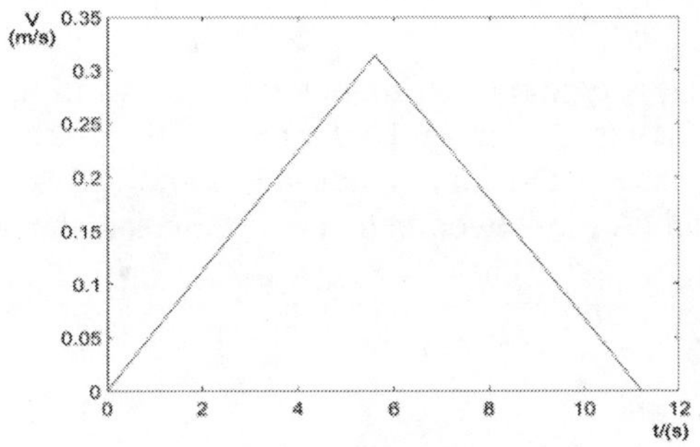

Figure 6. Triangular Motion Model

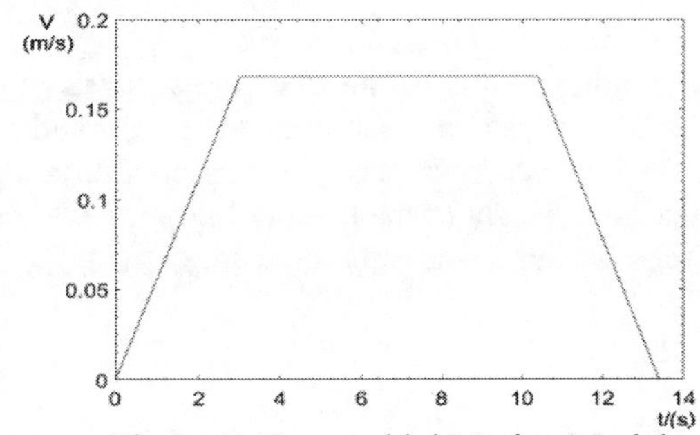

Figure 7. Trapezoidal Motion Model

4.3. The simulation of trapezoidal motion model

In order to study trapezoidal model, the time step is continuously lowered to increase the accuracy of simulation. The simulation results of acceleration section are shown in Fig. 8.

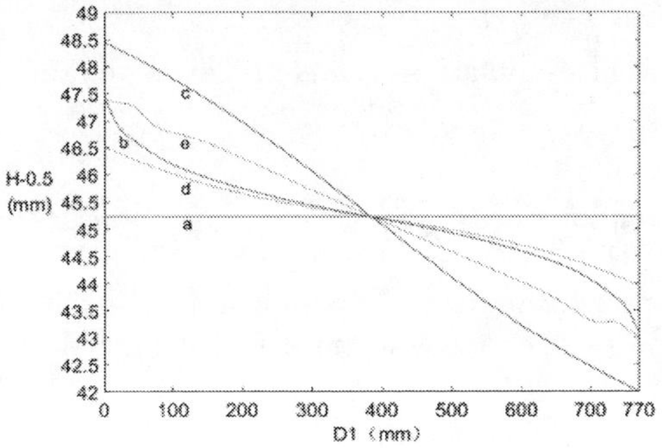

Figure 8. Waveform in acceleration section

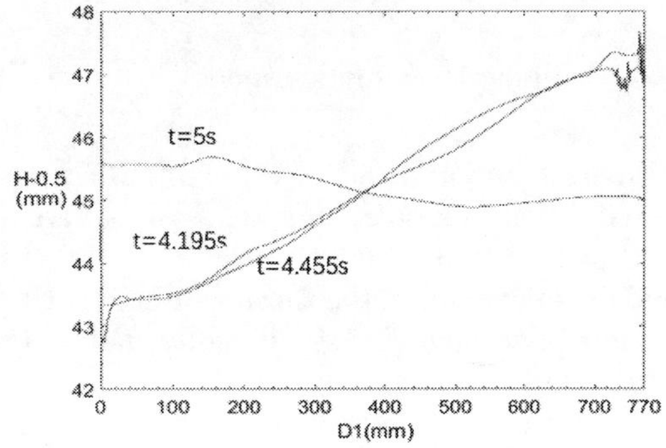

Figure 9. Waveform in uniform velocity section

Curve a represents the level of molten copper before acceleration. Curve b represents the level at t=0.5s. Curve c represents the level at t=1.165s. Curve d represents the level at t=2.445s. and curve e represents the level after acceleration for 3s. The level reaches 48.5 mm. Considering the grid error, the maximum range is 48-49 mm, the range of the sloshing height is 3-4 mm, which meets the design requirements and production requirements.

Simulation results of the level in uniform speed range are shown in Fig. 9. When the time t=4.195s, there are burrs on the boundary, and the maximum height of burrs is 47.5mm. When the time t =4.455s, the level reaches the 47.7mm.

4.4. Optimizing the moving model of casting wheel

The trapezoidal motion model is further optimized in order to reduce the height of the sloshing edge and the number of burrs. During the operation of casting wheel, the burr is caused by the sloshing of molten copper caused by the sudden change of acceleration. In order to reduce the frequency of the sloshing, the curve optimization method is used to improve the sudden change of acceleration. The optimized model is S model, as shown in Fig. 10. The motion model formula is as follows:

$$v(t) = \begin{cases} 0.056t, 0 < t \leq 2 \\ -0.014t^2 + 0.112t - 0.056, 2 < t \leq 4 \\ 0.168, 4 < t \leq 10 \\ -0.014t^2 + 0.28t - 1.232, 10 < t \leq 12 \\ -0.056t + 0.784, 12 < t \leq 13 \\ 0.014t^2 - 0.42t + 3.15, 13 < t \leq 15 \end{cases} \tag{9}$$

4.5. Verification of simulation results

The simulation results of the first five seconds of the S-type motion model are shown in Fig.11. The sloshing state of molten copper in curves a, b and c at t=2s, 3s and 4s. Under the action of deceleration, the sloshing slows down. The sloshing condition of curve d and curve d at t=4.48s and 5S respectively. The maximum height of the level is 46.4 mm, which is 1.3 mm lower than that of trapezoidal model, and the number of burrs is greatly reduced.

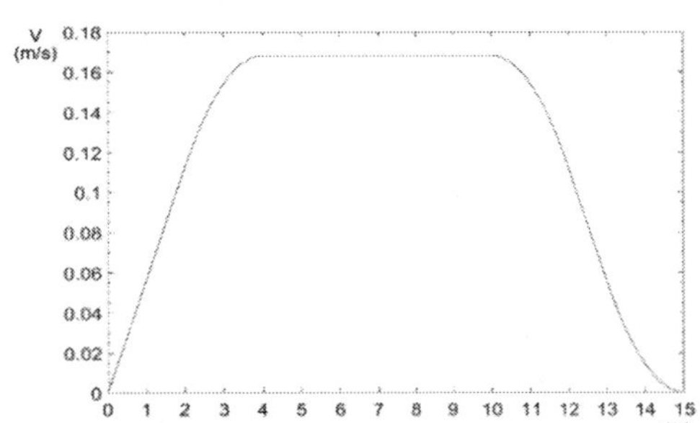

Figure 10. S-motion model

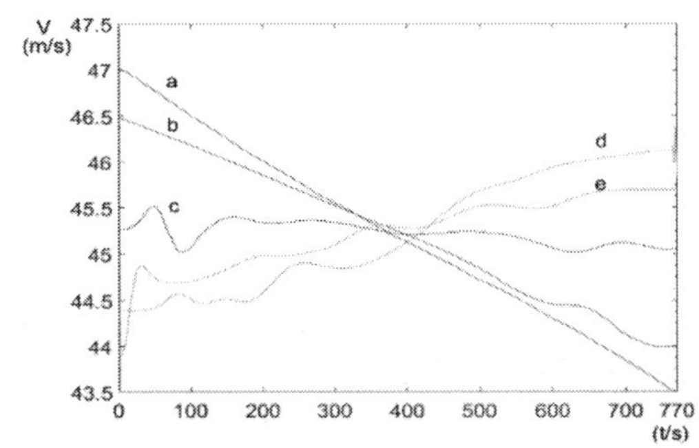

Figure 11. Simulation results of S-type motion

5. Verification and analysis of experimental results

T-type and S-type motion models are used to drive the rotation of casting wheel in a smelter. Fifty anode produced in each furnace were randomly sampled from five furnace to measure the height of sloshing edge and the number of burrs. As shown in Fig. 12, the proportion of anode with less than 40 burrs in S-type motion model increased significantly, and the proportion of anode with more than 40 burrs decreased from 70.2% to 15.6%.

Fig. 13, S refers to the serial number of the anode, and H refers to the height of the sloshing edge. The curve A above represents the height of sloshing edge of the anode produced under the trapezoidal motion model. The average height is 4.8mm. The maximum sloshing height of the anode is 5.8 mm. The curve

B represents the height of sloshing edge of the anode manufactured under the optimized S-type motion model. The average height is 3.6 mm. Maximum height of the sloshing edge is 4.2 mm. Compared with T-type model, the average height of sloshing edge decreases 1.2 mm, and the number of burrs decreases significantly.

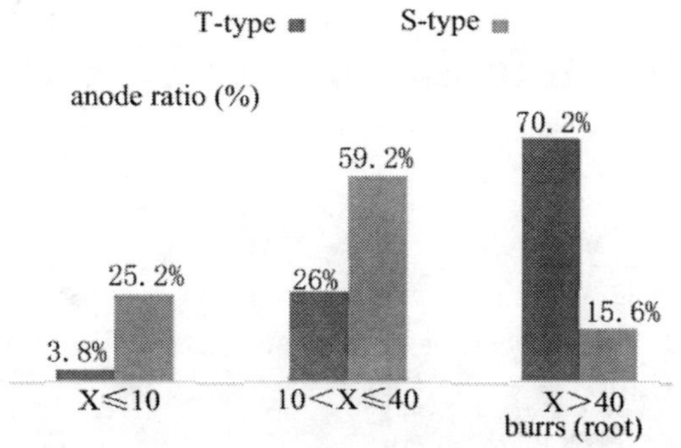

Figure 12. Proportion of anode in different burrs

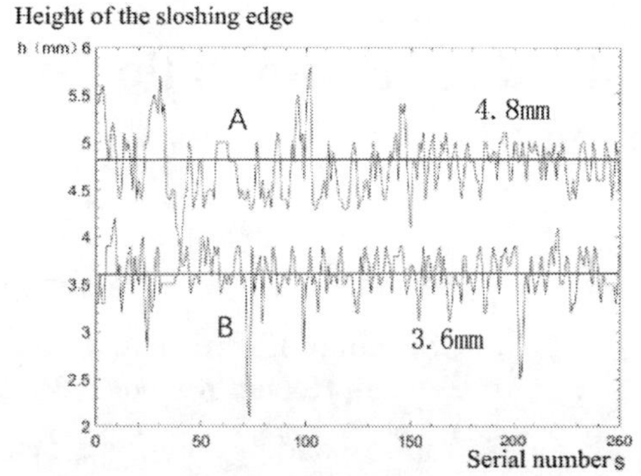

Figure 13. The height of sloshing edge under T-type and S-type motion models

6. Conclusion

When the acceleration is 0.056 m/s^2, the height of molten copper sloshing is 4 mm. The correctness is verified by FLUENT simulation. The motion model of casting wheel was analyzed and optimized. The sloshing caused by different motion models was compared by FLUENT simulation. The optimal motion model was S-type motion model. The experimental show that the average height of S-type model is 1.2 mm lower than that of T-type model, and the proportion of anode with more than 40 burrs decreases from 70.2% to 15.6%.

References

[1] Xin Lv, Qingping Zou, Yong Zhao, Dominic Reeve. A novel coupled level set and volume of fluid method for sharp interface capturing on 3D tetrahedral grids[J]. Journal of Computational Physics, 2009, 229(7): 2573-2604.

[2] Deizhi Dai. Numerical simulation of gas-liquid two-phase flow based on VOF method[D]. Nanjing: Nanjing University of Aeronautics and Astronautics, 2014.

[3] Yutai Huang. Dynamics research on the fluid-structure Interaction sloshing system in aircraft fuel tank[D]. Guangzhou: South China University of Technology, 2015.

[4] Xicong Zou. Fuel sloshing analysis in tanks of aircraft[D]. Harbin: Harbin University of Technology, 2012.

[5] Di Yu. Research on liquid sloshing non-linear model and driving stability of partially-filled tank trucks[D]. Changchun: Jilin University, 2016.

[6] Zhuqing Cheng. Simulation Research on Lateral Liquid Sloshing in Tank Vehicles [D]. Changchun: Jilin University, 2015.

[7] Peng Qiu, Gao Zheng. Research on Modeling of Marine Propeller Based on ICEM [J]. Mechanical Engineer, 2019 (06): 92-94.

[8] Ying Xin. Application of FLUENT UDF method in the study of numerical wave flume [D]. Dalian: Dalian University of Technology, 2013.

[9] Honggui Li. Metallurgical principle [M]. Beijing: Science Press, 2005.

Modeling and Simulation of Vibrating Online Viscometer System

Pengcheng Li[1], Liheng Wang[1]

[1] School of Electrical and Information Engineering, Wuhan Institute of Technology, Wuhan, Hubei Province, 430200, China

*Corresponding author's e-mail: 1753510997@qq.com

Abstract. Aiming at the present situation of the difficulties in the analysis and research of the vibration type on-line viscometer system, the structural characteristics and mathematical model of the vibrating online viscometer are analyzed and studied. Based on this, the closed-loop control principle is further elaborated. At the same time, the simulation model of the vibrating online viscometer system was built by closed-loop control technology. The MATLAB simulation results verify the accuracy of the closed-loop control and the effectiveness of the simulation model, which provides a theoretical basis and guiding ideas for the system design and debugging of the vibrating online viscometer.

1. Introduction

The vibrating viscometer studied is based on the viscous damping effect. When the vibrating structure is subjected to torsional vibration in the fluid to be tested, the amplitude will be attenuated. The greater the viscous damping, the greater the degree of attenuation. In order to maintain the amplitude, It is necessary to provide the energy consumed by the viscous damping from the outside, and the viscosity of the fluid at this time has the following correspondence with the energy provided by the outside[1][2]:

$$E \propto \sqrt{\eta\rho} \tag{1}$$

Where η is the viscosity of the fluid and ρ is the density of the fluid. The amplitude is kept constant through closed-loop control so that the fluid viscosity can be converted to the drive current. This closed-loop instrument improves response speed and reduces nonlinearity.

2. System Model

The simplified model of the vibrating online viscometer is shown in Figure 1. During the viscosity measurement process, the sensitive component is subjected to the viscous damping of the fluid, and the change in the magnitude of the damping reflects the change in viscosity, and the vibration amplitude of the torsion bar is changed by the spindle. The amplitude measuring circuit measures the change value of the amplitude, the single chip receives the amplitude change signal and changes the magnitude of the driving current, and controls the driving mechanism to maintain the amplitude of the torsion bar before the viscosity changes, and finds the relationship between the driving current and the viscosity of the fluid, and can calculate the viscosity of the fluid medium[3].

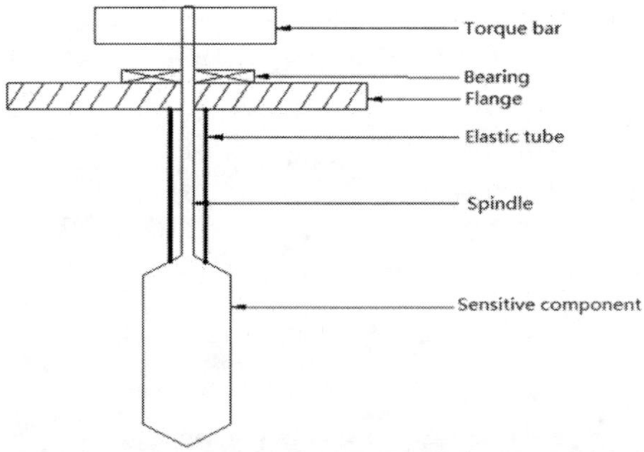

Figure 1. Vibration viscometer simplified model.

Separating the spindle, sensor and drive mechanism parts from the sensor.The equivalent model of the system is shown in Figure 2. It consists of a damper, an inertial disc and an elastic rod.

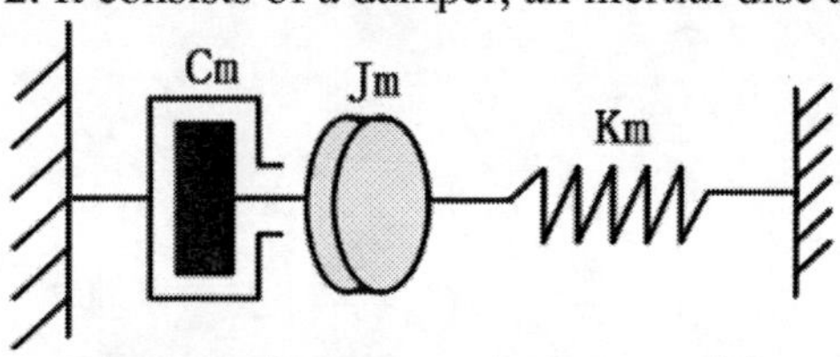

Figure 2. System equivalent model.

In order to analyze the motion characteristics of the vibration system, the system needs to be simplified into a dynamic model to determine the motion mode and degree of freedom of the system. The system is a torsional vibration system. The torsion angle can be used as the independent coordinate of the vibration system to determine the motion state of the system, and the torsional vibration system model with feedback control can simplify the system into a single-degree-of-freedom second-order closed-loop control system[4][5]. The dynamic equation is:

$$M_m = K_m \cdot \theta + C_m \cdot \frac{d\theta}{dt} + J_m \cdot \frac{d^2\theta}{dt^2} \tag{2}$$

In the formula, M_m is the driving torque, K_m is the equivalent stiffness, C_m is the equivalent damping, J_m is the equivalent moment of inertia, θ is the torsion angle. Among them, the natural frequency of the system is $\omega_0 = \sqrt{K_m / J_m}$.

3. Closed Loop Control Method

According to the fluid mechanics, when the sensitive component of the viscometer is subjected to torsional vibration in the viscous medium, a viscous moment is generated under the action of shear stress, and the size is (3), which hinders the relative movement of the sensitive component in the medium. Further, the amplitude of the drive crossbar is reduced.

$$M = \mu \int_\Omega K\omega ds \tag{3}$$

Where Ω is the contact area of the sensitive element with the medium, and K is the proportionality factor.

When the viscosity of the medium is constant, a constant driving current is maintained, so that the sensitive component is torsionally vibrated at a resonant frequency in the viscous medium under the driving of the driving crossbar, and maintain a certain vibration amplitude, and measuring amplitude by amplitude detection circuit. when changing the medium, the viscosity of the medium becomes larger or smaller, which will affect the amplitude of the torsional vibration to become smaller or larger.

IOP Conf. Series: Materials Science and Engineering **711** (2020) 012032 doi:10.1088/1757-899X/711/1/012032

If the magnitude of the drive current is controlled by closed-loop feedback so that the torsional vibration amplitude remains the same, the relationship between the drive current and the viscosity of the fluid can be found, The closed loop control process is shown in Figure 3.

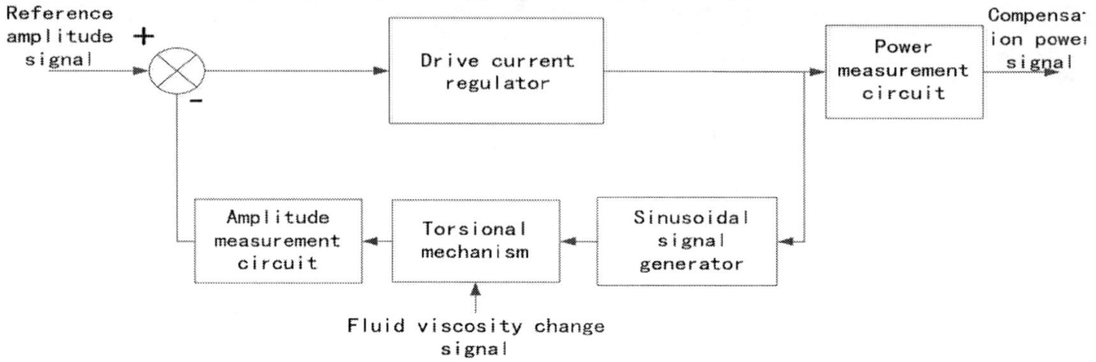

Figure 3. Closed loop control flow char.

4. Simulation analysis

4.1. Model construction

The system dynamics model can be seen from the previous analysis:

$$J_m \cdot \frac{d^2\theta}{dt^2} + C_m \cdot \frac{d\theta}{dt} + K_m \cdot \theta = M_m \tag{4}$$

If the input excitation current is $I = A\sin\omega t$, the system dynamics model can be written as:

$$J_m \cdot \theta''(t) + C_m \cdot \theta'(t) + K_m \cdot \theta(t) = H \cdot I(t) \tag{5}$$

If the input current is used as the excitation and the torsion angle is the response, by Laplace transformation of the above formula, we can get:

$$J_m \cdot s^2 \cdot \theta(s) + C_m \cdot s \cdot \theta(s) + K_m \cdot \theta(s) = H \cdot I(s) \tag{6}$$

So the transfer function of the system is:

$$G(s) = \frac{\theta(s)}{I(s)} = \frac{H}{K_m} \cdot \frac{K_m}{J_m \cdot s^2 + C_m \cdot s + K_m} = \frac{H}{K_m} \cdot \frac{K_m / J_m}{s^2 + C_m / J_m \cdot s + K_m / J_m} \tag{7}$$

The general expression of the second-order oscillation link transfer function is:

$$G(s) = \frac{K}{T^2 s^2 + 2\xi T s + 1} = \frac{K \cdot \omega_n^2}{s^2 + 2\xi\omega_n s + \omega_n^2} \tag{8}$$

So the system scale factor is $K = \frac{H}{K_m} = \frac{n\pi BDL}{2K_m}$, natural angular frequency is $\omega_n = \sqrt{K_m / J_m}$, time constant is $T = 1/\omega_n = \sqrt{J_m / K_m}$, damping ratio is $\xi = C_m / 2\sqrt{K_m J_m}$.

The system is accurately calculated, system stiffness is $K_m = 337\ N/m$, system moment of inertia is $J_m = 56 \times 10^{-6} kg \cdot m^2$, system damping is $C_m = 1.91 \times 10^{-5} \mu$, system drive torque is $M_m = 7.85I$, The time constant obtained by further calculation is $T = 1.3 \times 10^{-3}$, damping ratio is $\xi = 6.95 \times 10^{-5} \mu$, proportional coefficient is $K = 0.023$, so the system transfer function is as follows, the simulation diagram shown in Figure 4.

$$G(s) = \frac{23000}{1.66 s^2 + 5.56\mu \cdot s + 1000000} \tag{9}$$

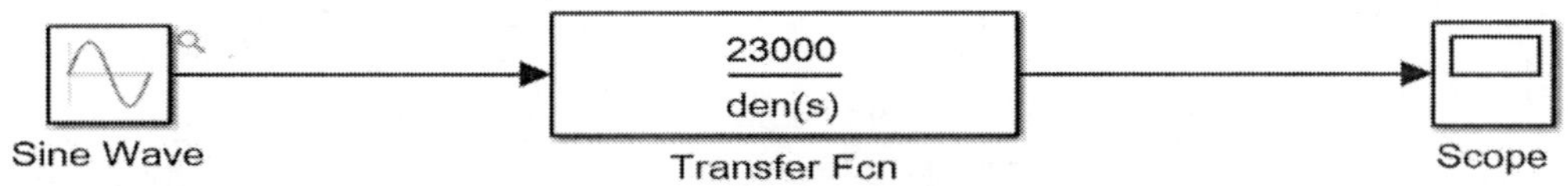

Figure 4. Simulation diagram.

4.2. Simulation analysis results

When the input is $I = \sin t$ and the medium is water, the dynamic viscosity of water is known to be 1 and the output amplitude is 0.04.

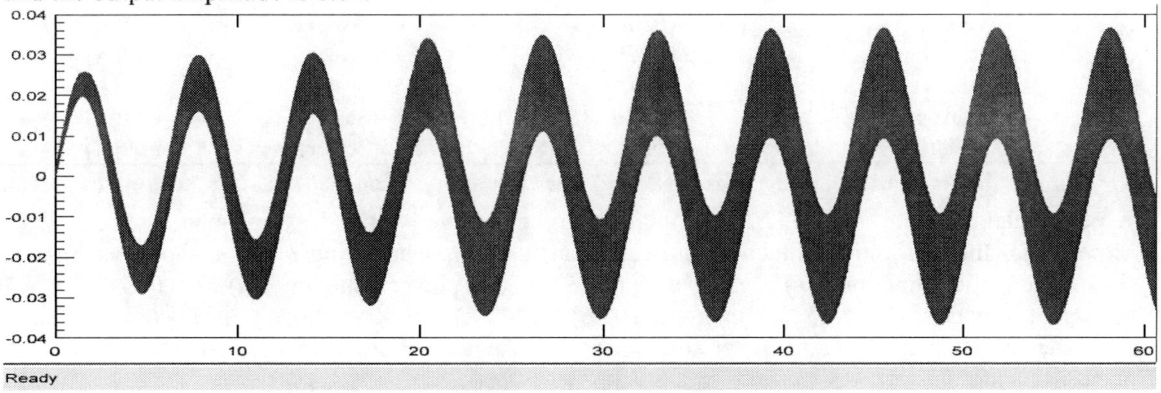

Figure 5. Media water current amplitude 1 output.

If the fixed medium is for honey μ=$3Pa\cdot s$, change the input current to a magnitude of 1.5 and the output amplitude is 0.035.

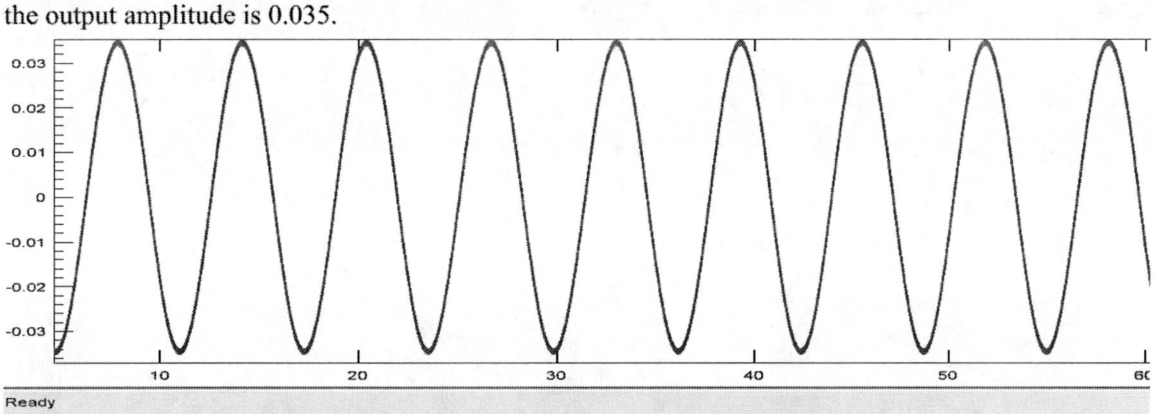

Figure 6. media honey current amplitude 1.5 output results.

According to this idea, when an output amplitude is fixed, the input current amplitude can be changed to reach the output amplitude for different viscosity media. The above simulation reflects the relationship between viscosity and excitation current amplitude. To find out the relationship between viscosity and current, the amplitude should be converted into the actual current size.The amplitude A in the excitation current reflects the magnitude of the sinusoidal variation. Because the value of periodic current is different at every moment, the effective value is often used to measure the role of periodic current in the circuit. The effective value of the periodic current is:

$$W = \int_0^T Ri^2(t)dt \quad (10)$$

$$I = \sqrt{\frac{1}{T}\int_0^T {I_m}^2 \sin^2(\omega t + \psi)dt} = \frac{I_m}{\sqrt{2}} = 0.707I_m \quad (11)$$

If the amplitude of the output is fixed at 0.04, the relationship between dynamic viscosity and current can be obtained by simulation in MATLAB using the viscosity of the known substance, as shown in Table 1.

Table 1. MATLAB simulation results for different media.

Viscous substance	Dynamic viscosity μ	Amplitude A	Electric current I
Head oil	5($Pa \cdot s$)	1.73(Scalar)	1.22(A)
Pear pulp	4($Pa \cdot s$)	1.72(Scalar)	1.21(A)
Paper glue	3($Pa \cdot s$)	1.70(Scalar)	1.20(A)
Semi-inverted syrup	2.4($Pa \cdot s$)	1.69(Scalar)	1.19(A)
Glycerol	1.5($Pa \cdot s$)	1.66(Scalar)	1.17(A)
Ketchup	1($Pa \cdot s$)	1.65(Scalar)	1.15(A)
Concentrated orange juice	0.63($Pa \cdot s$)	1.62(Scalar)	1.14(A)
Chocolate milk	0.28($Pa \cdot s$)	1.59(Scalar)	1.12(A)
Emulsion	0.2($Pa \cdot s$)	1.57(Scalar)	1.11(A)
Egg	0.15($Pa \cdot s$)	1.55(Scalar)	1.09(A)

Taking the current I as the abscissa and the dynamic viscosity μ as the ordinate, put the experimental data into the MATLAB simulation, and the curve is fitted by three mathematical models: exponential function, power function and Gaussian function. The fitting result is shown in Figure 7. The exponential function f(x) = 1.12*exp(1.128*x), the power function f(x) = 0.01347*x^29.78, Gaussian function f(x) = 528.2*exp(-((x-10.26)/4.134)^2). By comparison, it is known that the Gaussian function fits the curve best and can be used as an important theoretical basis for viscosity measurement.

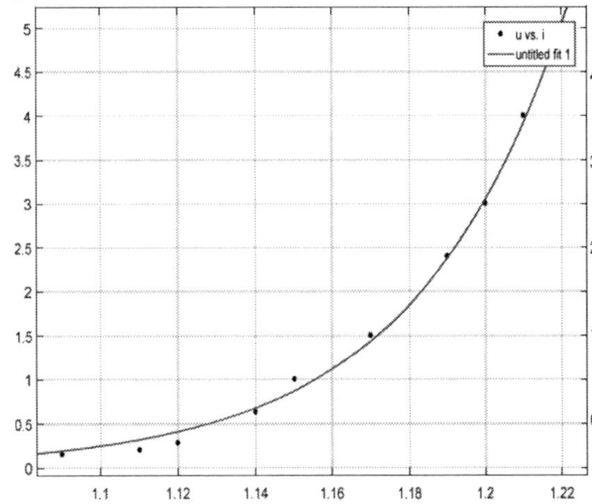

(a)Exponential function

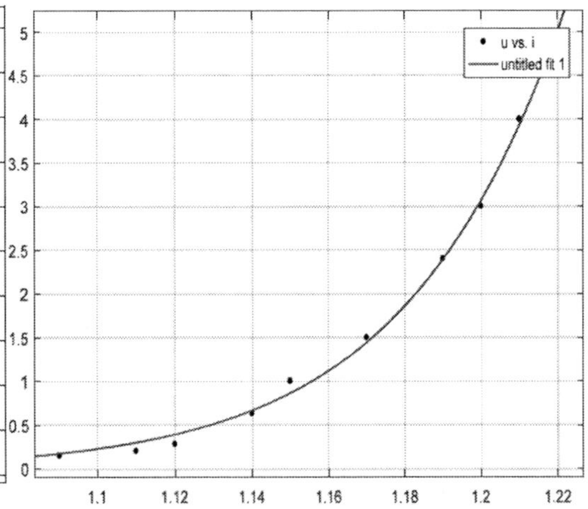

(b)Power function

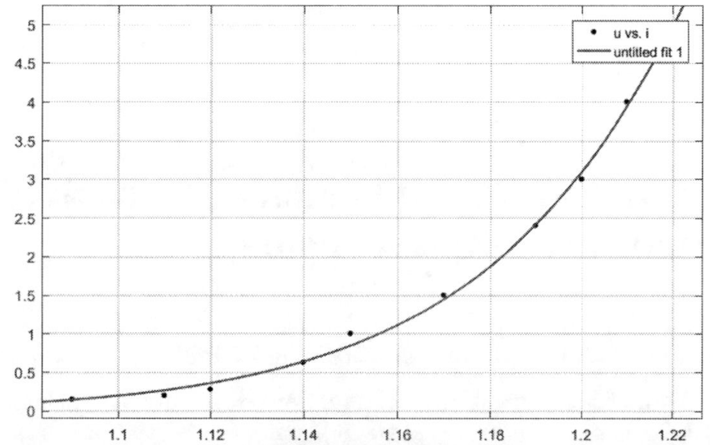

(c)Gaussian function

Figure 7. MATLAB fitting curve results.

5. Conclusion

In this paper, the structural characteristics and mathematical model of vibrating online viscometer are analyzed and researched, and the simulation model of vibrating online viscometer system is built by closed-loop control technology. The accuracy of closed-loop control and the effectiveness of the simulation model are verified by MATLAB simulation experiments. The relationship between viscosity and compensation current is established by Gaussian function model, which provides an important theoretical basis and guidance for system design and debugging of vibrating online viscometer.

References

[1] Chen Gang, Zhu Zhengang, Shui Jiapeng. Principle of measuring liquid viscosity coefficient by forced vibration torsion method[J].Acta Physica Sinica, 1999(03): 40-44.

[2] Margaret Stautberg Greenwood,Judith Ann Bamberger. Measurement of viscosity and shear wave velocity of a liquid or slurry for on-line process control[J]. Ultrasonics,2002,39(9).

[3] He Zheng, Li Changxi. Research on vibration viscometer [J]. Chemical automation and instrumentation, 2010,37 (12): 73-75.

[4] Hu Yanru, Chi Guoxuan.Calculation of Natural Frequency of Single Degree of Freedom Vibration[J].Journal of Jiamusi Institute of Technology,1991(04):230-236.

[5] Paul Van den Hof. Closed-loop issues in system identification[J]. Annual Reviews in Control,1998,22.

IOP Conf. Series: Materials Science and Engineering **711** (2020) 012033 doi:10.1088/1757-899X/711/1/012033

Implementation Of Medical Pendant Interactive System Based On Virtual Reality Technology

Hanyang li[1], Dini Duan[2] and Zhisheng Zhang[1*]

[1] School of Mechanical Engineering, Southeast University, Nanjing, Jiangsu Province, 210000, China

[2] School of Mechanical Engineering, Southeast University, Nanjing, Jiangsu Province, 210000, China

[*]Corresponding author's e-mail: oldbc@seu.edu.cn

Abstract. Nowadays, virtual reality is in the stage of rapid development. Rational application will improve the immersive feeling of human-computer interaction. This paper discusses the construction of interactive experience learning system from two perspectives of platform interaction experience and user demand. This system can provide the operation and practice platform of medical tower related equipment, and build a realistic medical environment based on Unity3D and HTC Vive. The base station of the equipment realizes the real-time positioning of users through the signals transmitted by the handle, in order to ensure the effectiveness and immersion of user interaction, and providing information such as equipment introduction, operating instructions, precautions and emergency operation tips of the medical tower. In order to improve the learning efficiency of users, the learning progress of users is recorded in real time, the tactile feedback device on the handle is used to prompt users when they make mistakes. This system provides an interactive platform for medical staff and professionals to learn and train medical hanging towers. It is also expected to serve as the basis for further research on virtual reality technology in the medical industry and improve people's interactive experience in the virtual environment.

1. Introduction

In recent years, the medical pendant has become more and more important in the application of various departments in hospitals. The medical pendant is widely used in the operating room, intensive care room, emergency room and other departments of the hospital, and it is one of the indispensable medical equipment for medical staff to treat patients. Borsci et al. [1] proposed that the working efficiency of medical staff decreases due to the unreasonable operation mode design of medical devices and the unfamiliarity of medical staff with the use process. Blandford[2] et al. analyzed the probability of safety accidents caused by medical staff unfamiliar with the operation of medical pendant through the use of data records of medical pendant equipment and cases of medical accidents. Lin et al [3] found that many of the reasons for medical events are the operating failures of medical devices.

Nowadays, virtual reality technology presents a booming trend. With the continuous development and upgrading of equipment and technology, virtual reality technology has been applied in many fields. Zhang Yuan [4] analyzed the application of virtual reality technology and proposed that it is precisely because users can experience the sense of actual use of products in virtual scenes and effectively improve product usability that the application field of virtual reality technology shows a trend of continuous expansion. Feng Yuliang [5] introduced in detail the successful application of virtual reality in different fields, such as improving visual effects in real estate sales exhibition, assisting high-end machine maintenance knowledge learning and new interactive forms in social software.

The above research can represent that the application of virtual reality technology in the field of medicine can exert great potential. From the above problems existing in the use of medical pendant, the combination of virtual reality technology can not only help people solve the problem of medical pendant learning and training, but also bring good interactive experience for medical staff.

2. Related Research

Sapkaroski et al [6] proposed an iterative development method and found that medical students using simulation tools had higher perceived scores for clinical and technical skills improvement than medical students who did not receive training.

Jimenez et al [7] studied radiation therapy-based medical device virtual environment (VERT) operational learning systems, for radiotherapy education aimed at providing learners with access to knowledge and skills in an interactive, risk-free environment opportunity.

Zhang Liyu et al [8] proposed that combining virtual reality for equipment use training not only effectively improves the nursing effect of medical nursing equipment on patients, but also helps improve the operation of medical staff. Liu Xiaohong et al [9] validated the effectiveness of virtual reality in medical teaching. In summary, the application in the medical field shows that virtual reality technology has played a huge role in improving medical standards.

3. Medical pendant interactive system based on virtual reality technology

3.1. User research of the system

This system is a usability platform for helping users to effectively learn and train. To enhance the user experience, the system design should be user-centered, and the products should be better served by in-depth research on target users. This paper conducts user interviews and combines the characteristics of different user model types, and finally obtains two user models of the system: professional user and inexperienced user. The characteristics of different user models are shown in Table 1:

Table 1. User characteristics of different users.

User model type	User characteristics
Professional user	The user's purpose is clear, and hope to enrich the knowledge of medical pendant through exercises; Want to standardize the level of the operating medical pendant; Require the detailed description and richness of the equipment description information in the system; Be able to record learning progress; Interested in virtual reality technology;
Inexperienced user	Want to learn the operation method of the device in the system; Want to contact different types of equipment; Hope to improve the operation level through practice; More enthusiasm and initiative;

3.2. User task flow

The system is divided into a server system and a user information interface. The server system is mainly divided into two categories: user operation input and userspace positioning. The user's real-time location in space is provided by the system-bound HTC Vive base station. The user operation flow chart in the system is shown in figure 1. It can be known from figure 1 that users can choose the operation module to enter according to their own requirements, and they can choose the learning or practice module. Therefore, the information navigation page should be at the first level of the system.

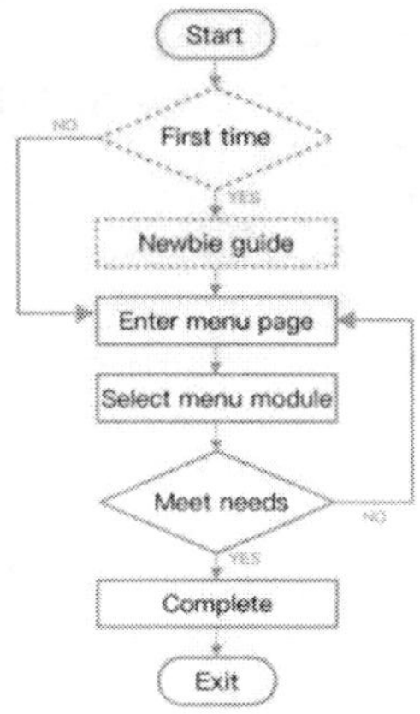

Fig. 1. The first time the user enters the system will first enter the novice guide page and then enter the information navigation page. When the user enters the function module, they can return to the navigation page at any time if this interface does not meet their requirements.

3.3. Information architecture design

According to the different needs of users, the system provides different function modules on the home page. For example, novice users are more likely to choose learning module and device explanation module, and provide users with learning progress query. Figure 2 shows the information structure of the system, which provides five modules, each containing different types of information forms.

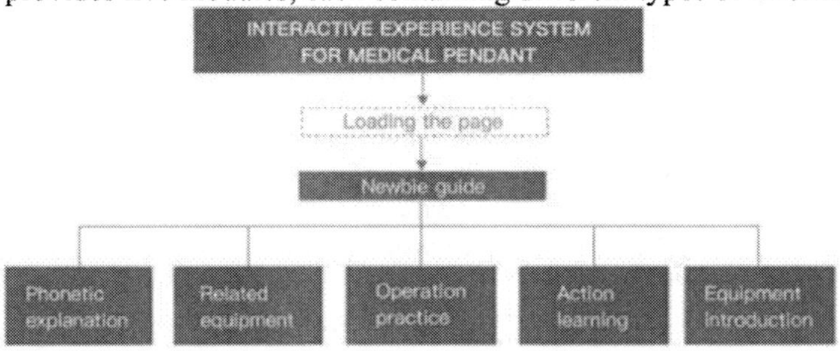

Fig. 2. Structure of system information.

4. System Implementation

The system is connected to the HTC Vive through a computer program generated by the Unity3d. Figure 3 shows the system implementation process. By using Unity 3D for system development, the implementation process can be roughly divided into two parts: virtual reality system construction part and virtual information interface design part. Virtual reality system construction is to import the 3d model of the medical pendant and other related products into Unity3D and conduct the matching of operation keys on the HTC VIVE handle with the target object. The part of virtual information interface design mainly refers to the interactive design of the information display page based on virtual reality technology in the system.

IOP Conf. Series: Materials Science and Engineering **711** (2020) 012033 doi:10.1088/1757-899X/711/1/012033

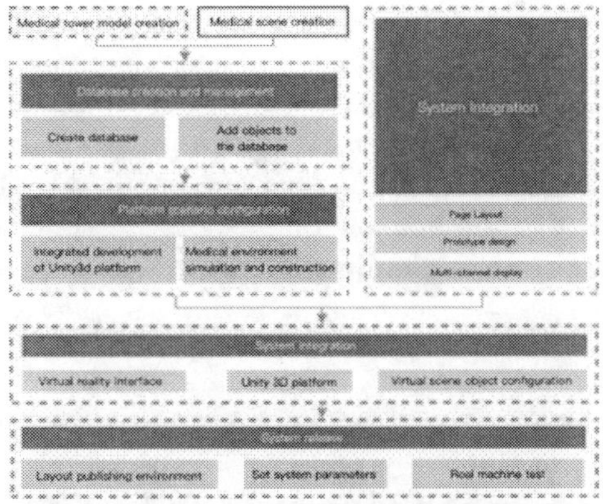

Fig. 3. System implementation process.

Figure 4 shows the system user interface. According to Mike Alger [10], in the virtual reality environment, the presentation of interface information is the most concentrated part of the user's visual perception. The focus area of the user's vision takes up one-ninth of the 360° virtual environment and is located in the center of the interface, with a size of 1200×600 pixels. The design style of navigation interface adopts card layout, which is placed in the center of the user's field of view.

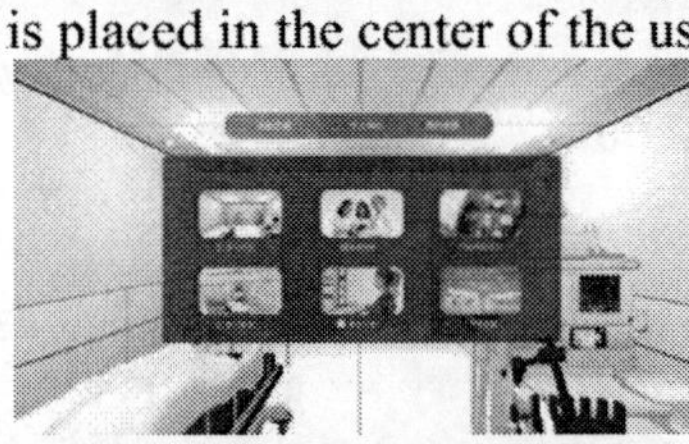

Fig. 4. Interface of menu navigation page.

Figure 5 is the system toolbar show up interface; The system toolbar is placed in the lower-left corner of the user, so as not to affect the user's main view. The introduction information of medical equipment is displayed on the right side of the object, in line with the visual browsing path of the user from the object to the information; The operation prompt message appears at the top of the interface center and does not affect the interactive view during the user's operation.

Fig. 5. When the user raises the right wrist, the toolbar page will pop up automatically.

The main interface functions of the system are as follows:

a. Interaction between users and medical pendant.The user selects the medical pendant part to operate in the scene, and when the user selects and presses the button, the part will automatically play the model animation bound by the part, and the user can understand the operational approach of the part object through the interactive animation played.

b. The user moves in the scene.Users can walk around in virtual space and browse according to their own preferences. In addition, when the user wants to quickly browse the spatial information in the scene, the position of the ray direction can be positioned by the handle button, and the fast movement can be realized by the instantaneous transmission.

5. Conclusion

This paper concerns a medical pendant interaction experience system based on virtual reality technology, which helps medical personnel to carry out the operation learning and practice of medical pendants, and provides new training methods for medical staff. In this paper, the interactive experience system provides an immersive and interactive teaching and training approach for medical staff, which can carry out human-computer interaction operations such as medical pendant training. In the system, advanced virtual reality interaction equipment can be used to realize real-time positioning and tactile interaction feedback for the operator. During the operation of each model of the medical pendant, the user can experience a more realistic three-dimensional environment through the handle operation. This system also provides an effective learning method for the students in the nursing major of medical college. The effective application of the medical interactive experience system to the teaching field has important practical significance for reducing the occurrence and prevention of medical accidents.

References

[1] Borsci, S., Buckle, P., Uchegbu, I., Ni, M., Walne, S., & Hanna, G. B.(2017) Integrating human factors and health economics to inform the design of medical device: a conceptual framework. In EMBEC & NBC 2017, 49-52.
[2] Blandford, A.(2013) Interactions "in the wild": Explorations in healthcare. Contemporary Ergonomics and Human Factors, 3-6 .
[3] Lin, Laura, Kim J. Vicente, and D. John Doyle.(2001) Patient safety, potential adverse drug events, and medical device design: a human factors engineering approach. Biomedical informatics 34(4), 274-284 .
[4] Zhang, Y.(2007) The study and application of usability in product design based on virtual reality. (Doctoral dissertation). .
[5] Feng Y.L. (2017)VR virtual reality design research. Chifeng College (Natural Science Edition) (05), 131-132.
[6] Sapkaroski, D., Baird, M., McInerney, J., & Dimmock, M. R. (2018) The implementation of a haptic feedback virtual reality simulation clinic with dynamic patient interaction and communication for medical imaging students. Medical radiation sciences 65(3),218-225.
[7] Jimenez, Y. A., Hansen, C. R., Juneja, P., & Thwaites, D. I. (2017) Successful implementation of Virtual Environment for Radiotherapy Training (VERT) in Medical Physics education: The University of Sydney's initial experience and recommendations. Australasian physical & engineering sciences in medicine, 40(4), 909-916.
[8] Zhang, L.Y.,Luo Y. M.(2017). The development of virtual reality in nursing education. General Surgery 15(16), 1931-1934
[9] Bi, H. J., Wang, H. L.(2016) Thoughts on Integrating Medical Virtual Reality Teaching into Diversified Instructional Design. Psychologist 22(24), 263-264 .
[10] Alger, M. (2015) VR interface design pre-visualisation methods. Online video clip. YouTube, 4.

Analysis of Weak Point of Fluid Material Tank Based on FLUENT

Yongqiang Li[1], Zhiyuan He[2]

[1]Naval Aeronautical University, No.188, Erma Road, Zhifu District, Yantai, Shandong, China

[2]Naval Aeronautical University, No.188, Erma Road, Zhifu District, Yantai, Shandong, China

[1] lyq281214@foxmail.com

Abstract. In view of the high reliability requirements of fluid material tanks and the difficulty of quality inspection, FLUENT is used to construct the tank model as a set of physics variable values on a finite number of discrete points. The original continuous physical field (pressure field and velocity field, etc.) of the fluid in the time system and the space system is replaced by certain laws and principles. At the same time, an algebraic equations are established for the relationship between these finite number of discrete points, and then the basic method of finite element method is used to solve the algebraic equations, and then the relative approximation of the stress acting on the tank in the time and space systems is obtained. After a continuous accelerated simulation, the weak link of the tank is finally determined.

1. Introduction

At present, the fluid material leakage monitoring device based on electrochemical sensor has various advantages and has been widely used. But its shortcomings are also quite obvious, such as short life (usually two to three years), longer response times, and the need for frequent calibration. Moreover, considering that the sloshing problem of the fluid level is unsteady flow, the law of conservation of the physics is followed in the non-inertial coordinate system, and the liquid inside the container is subjected to mass force and gravity. If the acceleration of the container changes with time, it is subject to the quality of the force will also be constantly changing. With FLUENT, the appropriate reference frame can be selected to convert the continuous physical quantity field into a quantitative algebraic equation, which is a powerful complement and optimization for the monitoring of electrochemical sensors. Among them, Shang Chunyu and Zhao Jincheng used the examples of linear motion of water with equal acceleration and basic self-vibration of rectangular water tank, combining with the research results of Tait et al, and the feasibility and superiority of solving the problem of page sloshing in rigid containers were verified.

2. Establishment of non-static process model of fluid materials

The non-static process of fluid material mainly refers to the flow of fluid. The basic conservation laws of fluid flow include: mass conservation law, momentum conservation law, energy conservation law. In this paper, due to the interaction between different components, it is also necessary to add component conservation law and the turbulent transport equation, the specific equations are as follows;

Mass conservation equation:

$$\frac{\partial_{\rho}}{\partial_{t}}+\frac{\partial}{\partial_{xi}}(\rho \mathrm{u}_{i})=S_{m} \tag{1}$$

Momentum conservation equation:

$$\frac{\partial}{\partial_{t}}(\rho u_{i})+\frac{\partial}{\partial_{x_j}}(\rho u_{i}u_{j})=-\frac{\partial_{p}}{\partial_{x_i}}+\frac{\partial \tau_{ij}}{\partial_{x_i}}+\rho_{g_i}+F_{i} \tag{2}$$

p is static pressure, τ_{ij} is the stress tensor, F_i and g_i are the external volume force and gravitational volume force in the direction.

Energy conservation equation:

$$\frac{\partial(\rho \mathrm{T})}{\partial_{t}}+div(\rho \mathrm{uT})=\mathrm{div}(\frac{k}{c_{p}}\mathrm{gradT})+\mathrm{S}_{T} \tag{3}$$

c_p is specific heat capacity, T is temperature, k is fluid heat transfer coefficient, S_T is fluid viscous dissipation term.

Turbulent transport equation:

$$\frac{\partial(\rho Y_{i})}{\partial_{t}}+\nabla(\rho \bar{v}\,\mathrm{Y}_{i})=-\nabla\bullet\overline{J_{i}}+R_{i}+S_{i} \tag{4}$$

$\overline{J_i}$ is $\overline{J}$'s diffusion flux, R_i is the net amount of chemical reaction for the i th substance ,and S_i is the rate of generation of the user-defined source and discrete items.

3. Numerical simulation of fluid material movement and tank stress

In the long-term storage process, due to transportation, technical preparation, maintenance and repairment, the tank will be subjected to different degrees of external impact, which in turn will drive the fluid material for various irregular movements. In this paper, the non-static process of liquid propellant is simplified into a combination of sinusoidal motion and sloshing, and the two irregular motions are integrated into a motion function to numerically simulate the fluid material tank.

Establish a geometric model for non-static processes of fluid material, as shown below:

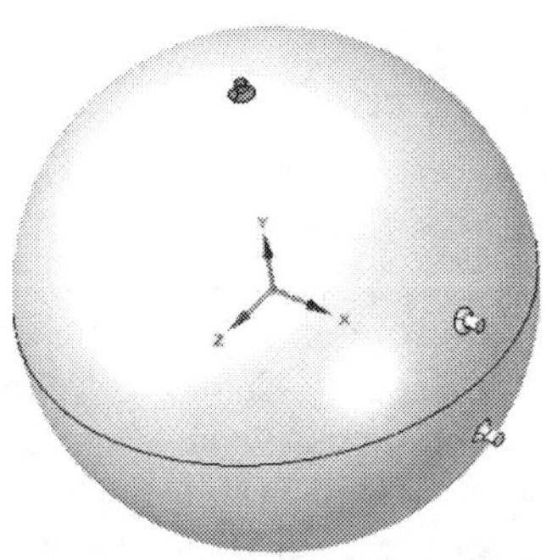

Figure 1. Fluid material tank geometry model

This paper mainly simulates the flow of fluid material during the irregular movement of the tank. It is necessary to pay attention to the force on the tank wall during the movement of the fluid material. At the same time, the speciality of the connection between the filler port and the tank cannot be ignored. Therefore, non-uniform meshing is required, and the grid is encrypted where the tank and the filler are connected, resulting in more accurate results. The specific mesh is as shown below.

IOP Conf. Series: Materials Science and Engineering **711** (2020) 012034 doi:10.1088/1757-899X/711/1/012034

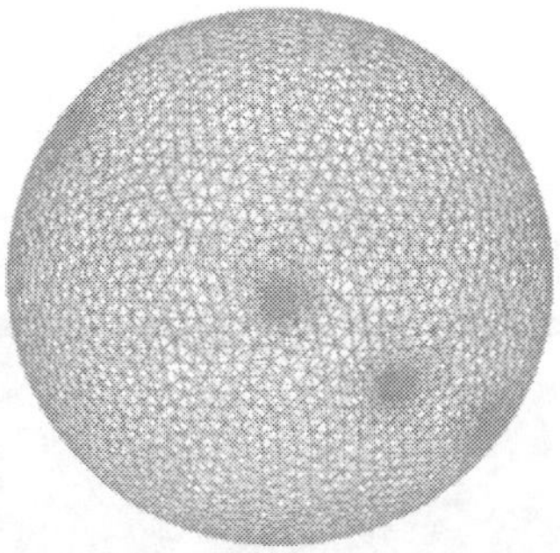

Figure 2. Meshing of geometric model of fluid material tank

The pressure-based solver does not directly solve the Navier-Stoke equation, but performs pressure correction on the momentum equation to achieve the purpose of the solution, and the compatibility of the multi-phase flow model is better. The program flow chart of the solver is as follows:

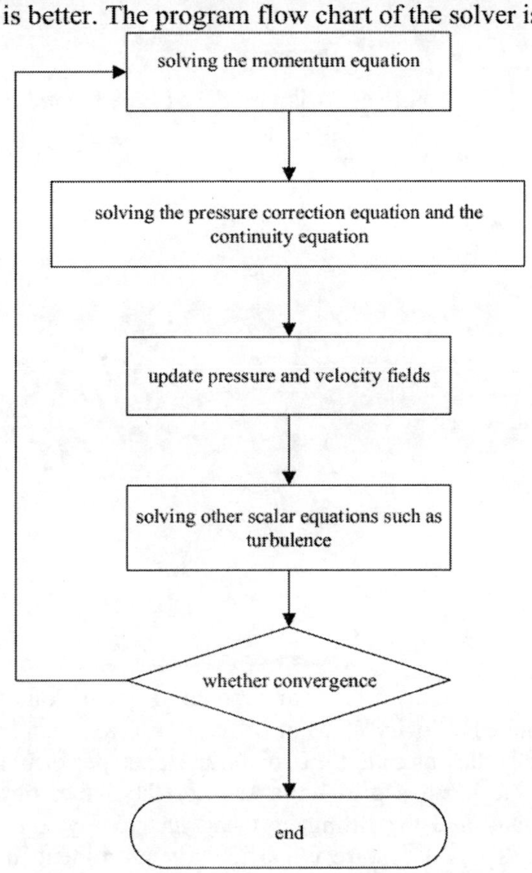

Figure 3. Pressure base solver flow chart

In this paper, for the motion simulation of the fluid material tank model, the simulation data of the 7-day and 300-day time nodes are selected respectively, and the CFD-post software in Fluent is used to convert the pressure and velocity cloud map. The simulation results of the tank are as follows:

4. Tank model simulation analysis

speed cloud map (XY plane)

IOP Conf. Series: Materials Science and Engineering **711** (2020) 012034 doi:10.1088/1757-899X/711/1/012034

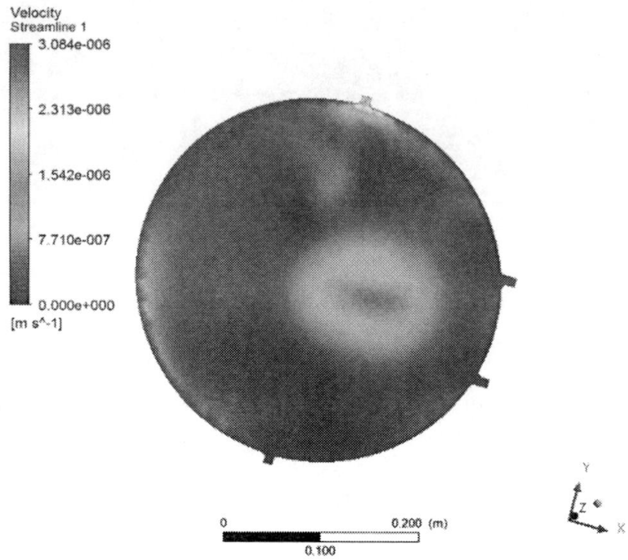

Figure 4. 7-day model of speed cloud

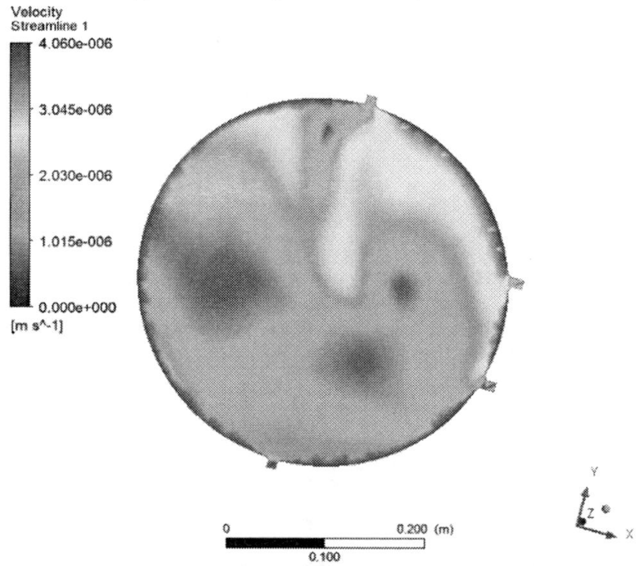

Figure 5. 300-day model of speed cloud

From the above-mentioned velocity maps at different times, it can be seen that during the random motion of the tank model, the internal fluid materials also perform irregular movements, and the velocity distributions in the various grid units are not the same, but a common feature is in the connection between the tank and the filling port and its vicinity, the speed and rate of change are greater than other locations'. This feature is particularly prominent in the 300-day simulated speed cloud.

Pressure Cloud Diagram

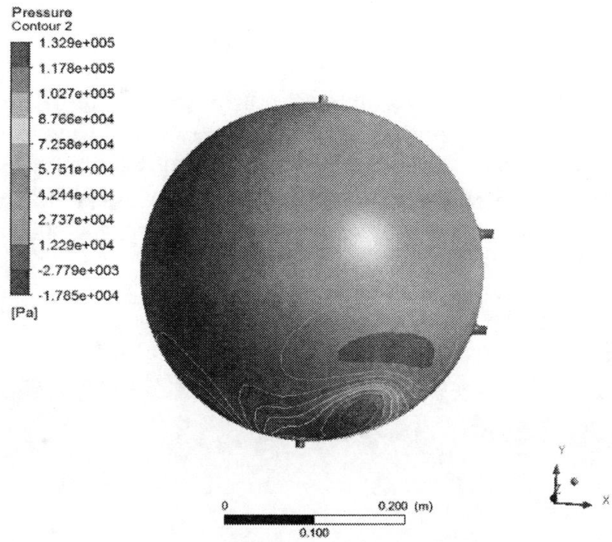

Figure 6. 7-day model of pressure cloud

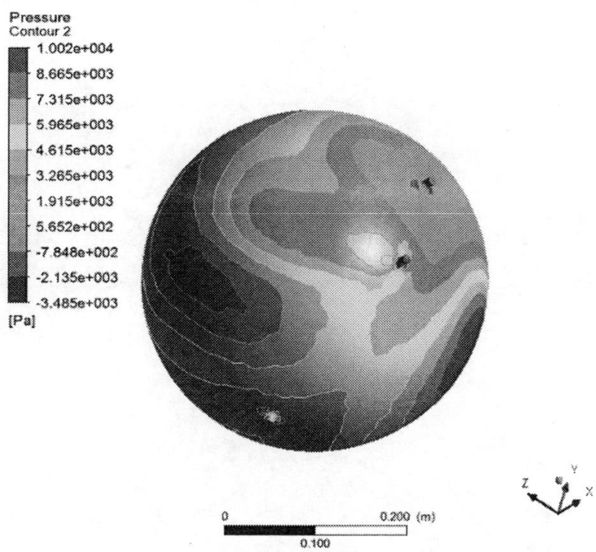

Figure 7. 300-day model of pressure cloud

In the pressure cloud diagram, a similar situation to that in the velocity cloud diagram can be found, that is, the pressure of the fluid on the connecting portion is slightly larger than that in the surrounding area. In the real fluid material storage tank, the tank and the filling port are welded and connected, and there are a large number of welding points in the joint portion. Unlike the overall structure, the strength of these solder joints is not large, and when it becomes a stress concentration point, it is easy to cause damage, which leads to leakage of fluid material.

5. Conclusion

In this paper, the random motion of the fluid material tank model is simulated, and the non-static state of the fluid material is simulated, which corresponds to the use of the fluid material storage tank in the long-term storage process. The simulation results show that under different motion modes, the fluid material has different forces on the tank, and the local force obviously makes the tank have the risk of being destroyed. By using FLUENT to establish a numerical simulation model, it can accurately find the weak link of the fluid material tank for more precise monitoring and management in the long-term storage station.

References

[1] Yue Chun-guo,Zhang Yu-xiang,Zhang You-Kong,Chang Xin-long,Zhang Wei.Some Problems Discussed of Liquid Propellant Long-term Storage and Management[C].2011 3rd World Congress in Applied Computing,Computer Science,and Computer Engineering Advances in Communication Technology,2011.

[2] K.W.Woodis,Leak detector for use in space environment,Marshall Space Flight Center,NASA TN D-5841,1970.

[3] Chutjian A.,Darrach M R,etc.A miniature Quadrupole Mass Spectrometer Array and GC For Space Flight:Astronaut EVA and Cabin-Air Monitoring. OOICES-202:1-10.

[4] Perez E.F.,Oliveira N G,Tanaka A A,et al.Electrochemical sensor for Hydrazine based on silica modified with nickel tetrasulfonated Plrthalocyanine[J].Electroanalysis,1998,10(2):111-115.

Research and Implementation of Indoor Location for Home Service Robot

Haifei Si[1,2*], Xingliu Hu[1*], Li Xu[1]

[1]College of Intelligent Science and Control Engineering, Jinling Institute of Technology, Nanjing, Jiangsu, 211169, China

[2]College of Automation, Harbin Engineering University, Harbin, Heilongjiang, 150001, China

[*]Corresponding author's e-mail: sihaif@jit.edu.cn, jimmy080@jit.edu.cn

Abstract. At present, the trend of population aging in the world is becoming more and more serious. Many countries have implemented many measures to solve this problem. Various high-tech products to help the elderly and disabled have also entered the lives of the elderly. In order to ensure the safety of the elderly or disabled at home, this paper designs a room using SLAM algorithm of lidar sensor under ROS (Robot Operating System) environment.Wheeled home service robots with internal positioning, mapping and multi-point continuous navigation can collect and analyze environmental data such as temperature, humidity, smoke and rainwater outside the windows and broadcast them by voice while navigation.After many debugging, this design has realized the function of indoor autonomous navigation, and timely alarm and control of abnormal environmental data in the home.

1. Introduction

According to the definition of the International Federation of Robots, service robots are robots that can help people realize semi-autonomous or fully autonomous operation of equipment other than production, processing and manufacturing[1]. Service robots can be classified as home service robots and special service robots [2].

The development of home service robot technology can alleviate the old-age problem caused by the aging population, reduce the pressure of young people's life, and promote social and economic development[3]. Therefore, the research of home service robots will be of great significance to many aspects of society. At present, most countries in the world are developing robots, of which about 25 countries are involved in the research and development of service-oriented robots. In the field of service robots, Japan, Korea, the United States and Germany are at the forefront of the world. Compared with Japan, the United States and other countries, the research and development of service robots in China started late, but developed rapidly. Home robot localization is a hot issue in the field of mobile robots research, and it is the premise and key of autonomous intelligence for robots to achieve various complex tasks. Sensor positioning is a common method of positioning[4]. Sensors for positioning include ranging sensors (such as sonar and laser sensors) and visual sensors. Vision is the most direct means for human to perceive the noisy environment. Therefore, the development of visual perception ability for robot positioning is conducive to improving its intelligence. Most of the existing localization algorithms rely on sensors carried by mobile robots. In the process of localization, the environment of the robot is unpredictable, such as the environmental changes caused by human movement and furniture movement. Therefore, it is easy to mismatch and fail to locate only according to the local perception information of the robot body. In order to achieve effective positioning, the robot needs to carry a large number of sensors, which not only increases the load of the robot, but also increases the computational burden of the robot processor. Through the study of the above problems, the SLAM algorithm based on lidar sensor is proposed to realize indoor positioning of service robot.

2. Signal Acquisition of Home Service Robot

2.1. Architecture Design of Home Service Robot

The home service robot is composed of two parts: the moving control system, the environment acquisition and broadcasting system [5]. In the moving control system, the Ubuntu operating system on the raspberry motherboard is used as the software carrier, and the combination of SLAM algorithm in ROS and lidar sensor is used to map and navigate indoors. The structure block diagram is shown in figure 1. The part of environment collection and broadcasting is also divided into two parts: host and slave. Host refers to the part installed in the body of the robot. Random robot moves while slave is fixed in a fixed position at home. The main part includes power supply, motor and motor drive, infrared sensor, ultrasonic, voice broadcasting, camera image transmission, controller and Zigbee wireless communication module. The slave consists of temperature and humidity, smoke and rain sensor module, control module, Zigbee communication module and relay output module.

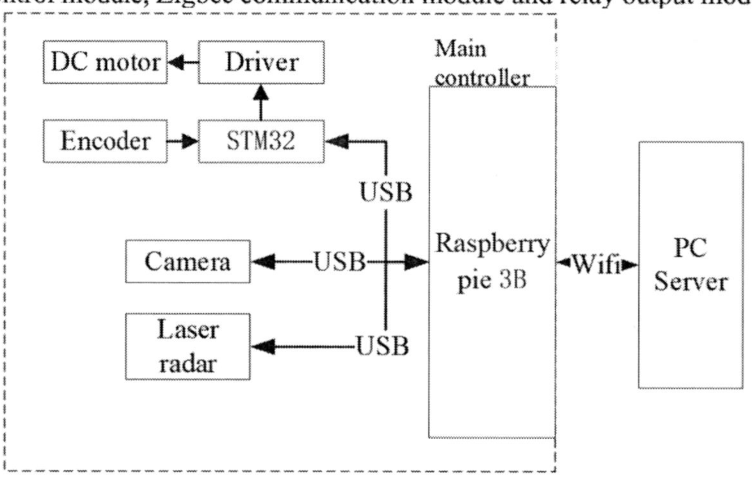

Figure 1. Hardware system block diagram of robot traveling.

2.2. Manual Indoor Mapping

Before the home service robot works properly, indoor construction map needs to be done manually. The meaning of the map is to scan the contour map of the room and let the robot have specific reference when it moves on its own. Then it can set coordinates for the robot to navigate by itself and take video and grab objects in real time. Users can also monitor the video at home remotely by mobile App. Home environment data collected by sensors are sent to the host through Zigbee network. The host analyses the data and broadcasts it. When a certain data value exceeds its normal threshold, the host sends the output instructions to the slave and the slave relay acts.The schematic diagram of the environment acquisition and broadcasting part is shown in figure 2.

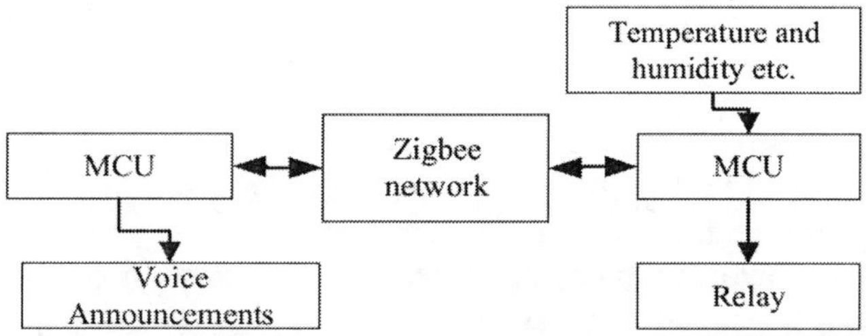

Figure 2. Block diagram of environmental acquisition system.

The basic RPLIDARA1 two-dimensional lidar developed by Shanghai SLAMTEC Company is selected as the lidar sensor.This object is shown in figure 3.Through high-speed rotation, scanning is carried out on a two-dimensional plane with a radius of 360 degrees and a radius of 6m at the center of rotation. Map information of plane point clouds in space is generated by software.This information is often used in map building, robot positioning and navigation applications.

Figure 3. RPLIDAR A1 physical map.

RPLIDARA1 uses laser triangulation ranging technology and SLAMTEC developed high-speed visual acquisition and processing mechanism, which can achieve more than 2000 ranging actions per second [6]. In each ranging process, the lidar transmits modulated infrared laser signal, and the lidar visual acquisition system receives after the laser signal irradiates the target object [7-9]. Through real-time calculation of the DSP processor embedded in the lidar, the distance between the reflected target and the lidar and the current angle information are output from the communication interface. The schematic diagram of its working principle is shown in figure 4.

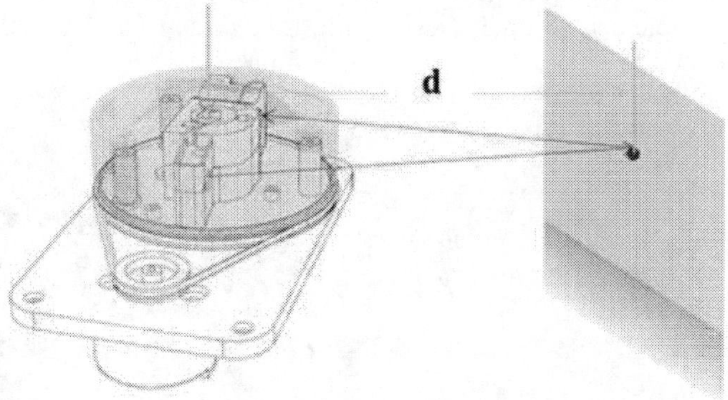

Figure 4. Schematic diagram of RPLIDAR A1 working principle.

RPLIDARA1 has its own speed detection and adaptive system. Radar scanning frequency will be automatically adjusted according to the actual speed of the motor. It does not need to provide a complex power supply system for RPLIDARA1. The core of RPLIDARA1 distance measurement is to rotate clockwise under the drive of motor mechanism, and to realize scanning and ranging in 360 degree omni-directional environment.

2.3. SLAM algorithm

SLAM (Simultaneous localization and mapping or simultaneous positioning and mapping) is a concept that the robot wants to determine its position and posture relative to the surrounding objects by continuously scanning and comparing the features of the objects around it.Starting from the unknown position in unknown environment, the feature contours of objects obtained by moving along with themselves are compared and combined to realize location and mapping [10-11]. At present, the two kinds of sensors that achieve better results of SLAM algorithm are lidar and camera. This topic chooses lidar. When the distance between the robot and the object is within the effective range of lidar, the mapping effect is the best. The map contour shows the phenomenon of a straight line without a dotted line. The effect is shown in figure 5.

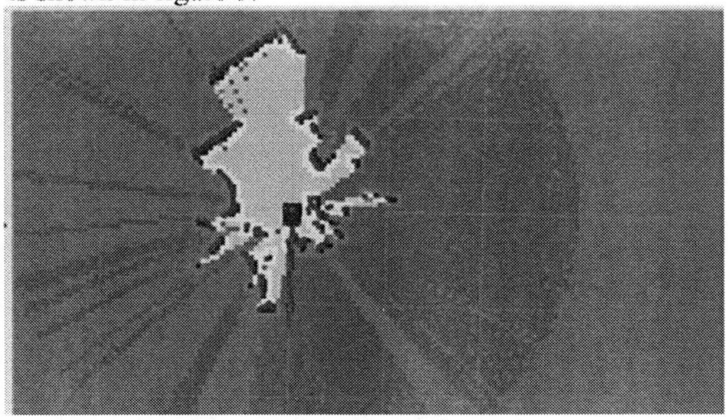

Figure 5. Drawing effect map.

3. Software Design of Home Service Robot

3.1. ROS System Framework

The main controller of the robot, remote connection tools and PC (Ubuntu system) are connected to the same WIFI. Node managers and keyboard debugging, SLAM and path planning nodes are established on the PC, and vision, lidar, mileage and basic controller nodes are created on the main controller of the mobile robot.After registering all nodes in the node manager and being managed by the manager, TCP/IP communication can be carried out through the end-to-end topology of the same network to achieve effective communication between nodes on different hosts. Finally, SLAM and path planning can be implemented on PC using the three-dimensional visualization tool RVIZ.The structure of the frame node of ROS system is shown in figure 6.

Figure 6. ROS System Framework Diagram.

3.2. Programming of Position Control

Before navigating, you need to put the map you built before into the maps folder and modify the map name in the map calling program.After the modification is completed, enter roslaunch pibot_navigation view_nav.launch in the terminal of the virtual machine to view the map. This command will open the rviz tool. By setting the 2D Nav Goal in rviz, the single navigation of the robot can be realized. The interface is shown in figure 7. The black border is the outline of the built room map, and the red border is the outline of the room that the robot scans in real time.

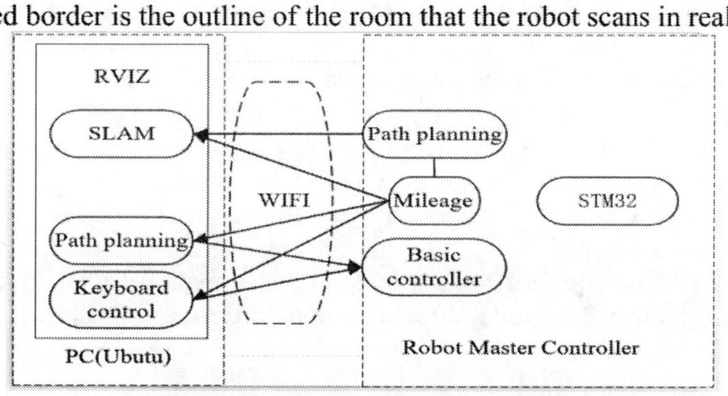

Figure 7. rviz navigation interface.

As ROS node, the position control module firstly analyses the linear velocity and angular velocity transmitted by the basic controller node, and obtains the linear velocity, angular velocity and displacement required by the left and right wheels of the robot.Then, by calculating the actual motion speed, angular velocity and the displacement of the left and right wheels of the robot measured by the encoder, the errors of the motion speed and direction of the robot are obtained.The PID algorithm is used to precisely control the speed of the left and right wheel motors, so that the robot can move according to the path planning trajectory.Finally, the attitude parameters of the robot, such as linear velocity, angular velocity and displacement, are released to the mileage node [12-14].The flow chart of the program is shown in figure 8.

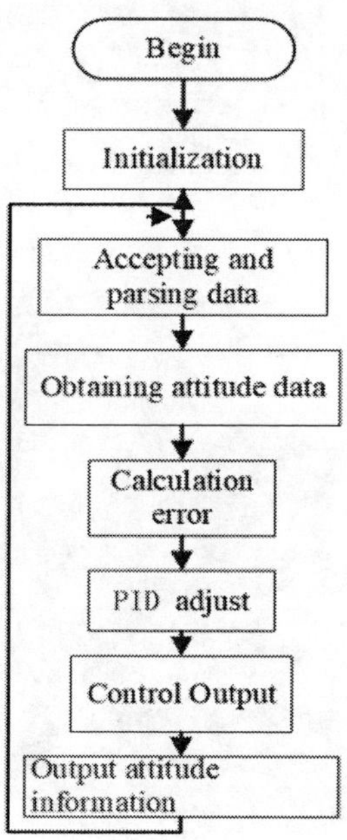

Figure 8. Flow chart of position control program.

This design uses SLAM algorithm of lidar sensor under ROS environment to produce a wheeled home service robot which can locate, map and navigate continuously in the indoor environment. While navigating, it can collect and analyze the environmental data such as temperature and humidity, smoke and rainwater outside the window, and broadcast them by voice.After many debugging, this design has realized the function of indoor autonomous navigation, and timely alarm and control of abnormal environmental data in the home.

4. Conclusion

This paper chooses the SLAM algorithm of lidar as the indoor positioning method.The positioning, mapping and navigation of the robot in the indoor environment are realized. The indoor positioning function of the robot is realized by lidar sensor and motor encoder, and the indoor map is constructed by RVIZ and mapping function package in ROS. Then the initial and target coordinates of the robot are set to realize autonomous navigation through lidar and coder.

In this paper, the robot has four layers. The chassis is made of iron sheets with a diameter of 35 cm and a thickness of 2 mm, so as to ensure that the whole robot can bear enough load. The larger weight objects such as batteries and motors are installed on the bottom to improve the stability of the vehicle. The second layer is equipped with voltage-reducing power supply module, motor drive module and sensor module, the third layer is equipped with control board, communication module and camera module, and the top layer is equipped with a three-axis manipulator. Each layer is connected by four 10 cm long connecting pillars. The robot effect is shown in figure 9. The physical objects produced are shown in figure 10 and figure 11.

Figure 9. Robot rendering.

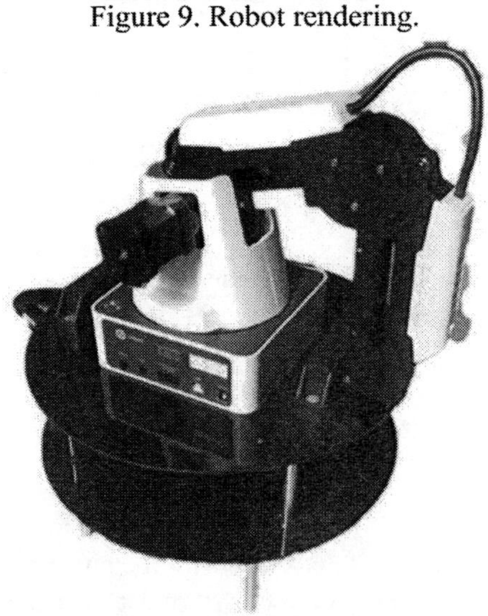

Figure 10. Robot Level 3 and Top Robot Arm.

Figure 11. Robot Bottom and Second Layer.

Nowadays, the satellite positioning technology for outdoor positioning is mature, but the convenient positioning method for indoor consumption level has not been truly popular.For example, people use the most frequently mobile phones, mobile phones can easily locate their location accurately outdoors, but when mobile phones enter a more closed indoor, their positioning function almost fails or deviates greatly.This problem is not only the difficulty in the design of this subject, but also the difficulty that science and technology need to break through. Due to the problems of indoor positioning, motion accuracy and product cost, it is still difficult for most of the service robots to really enter the family and integrate into the family.So I hope that in the future, the development of robots and science and technology will promote each other and realize the civilianization of home service robots as soon as possible.

Acknowledgments

This work was supported in part by the Jiangsu Province's Natural Science Foundation under Grant BK20171114, Youth Academic Leader in Qinglan Project of Jiangsu University in 2017, A joint prospective industry-University-Research Collaboration project of Jiangsu Province under Grant BY2016012-02, National Science Foundation of the Higher Education Institutions of Jiangsu Province under Grant 15KJB520010, Ministry of Education's Cooperative Education Project under Grant 201701056001,201801112066, Jinling Institute of Technology's Education Reform Project under Grant 2017JYJG03, Jinling Institute of Technology Talent Introduction Project under Grant Jit-rcyj-201604.

References

[1] Intellectual Research Consulting Group. (2018)In-depth Market Research and investment prospects analysis report of service robotics industry in China from 2017 to 2022. http://www.chyxx.com/research/201610/458911.html.

[2] Zeng, Y. (2013)Trends and Prospects of CPP Film Market in Europe in 2017. Plastic Packaging, 23 (2): 53-54.

[3] Wu, W. (2018)At the end of last year, 17.3% of the elderly over 60 years old in China accounted for nearly 500 million by 2050. http://news.sina.com.cn/c/2018-02-26/doc-ifyrvnsw9319158.shtml.

[4] Li, Q. (2010)Design and Implementation of Memory Library for Smart Home Service Robot. Harbin: Harbin University of Technology.

[5] Chikurtev, D., Yovchev, K., Chivarov, N., & Rangelov, I. (2019)Indoor Navigation Using Existing Infrastructure for Professional Service Robots. In International Conference on Robotics in Alpe-Adria Danube Region, pp. 231-239.

[6] Xue, Y., Lu, H., Xie, D. (2016)Performance of economical 2D lidar for mobile robot obstacle avoidance. Mechatronics, 22 (7): 7-9.

[7] Xu, Z., Zhang, Y., Dong, W. (2016)Design and implementation of positioning system for air-bearing platform based on lidar. Journal of Sensing Technology, 29 (12): 1924-1930.
[8] Zhao, J., Zhang, H., Wang, H., et al. (2017)Research on the construction of indoor environment map by lidar. Mechanical design and manufacture, 2017 (5): 135-137.
[9] Gu, W., Chen, B., Hu, Z., et al. (2015)Design and implementation of speech recognition-based home controller. Science and Technology Innovation Report, 2015 (32): 139-142.
[10] Maolanon, P., Sukvichai, K., Chayopitak, N., & Takahashi, A. (2019)Indoor Room Identify and Mapping with Virtual based SLAM using Furnitures and Household Objects Relationship based on CNNs. In 2019 10th International Conference of Information and Communication Technology for Embedded Systems (IC-ICTES), pp. 1-6.
[11] Yan, W., Pan, L., Tao, H., et al. (2011)SLAM scheme for mobile robot based on feature map. Computer measurement and control, 19 (4): 961-965.
[12] Li, J., Liu, X., Li, X., et al. (2017)Design of open source mobile robot system based on ROS. Mechatronics Engineering, 34 (2): 205-208.
[13] Zhang, J. (2012)Open Source Robot Operating System. Beijing: Science Publishing.
[14] Castro-Antonio, M. K., Carmona-Arroyo, G., Herrera-Luna, I., et al. (2019)An Approach based on a Robotics Operation System for the Implementation of Integrated Intelligent House Services System. In 2019 International Conference on Electronics, Communications and Computers (CONIELECOMP) , pp. 182-186.

Research on path planning of locally added path factor dijkstra algorithm for multiple AGV systems

Guorong Wu[1a], **Xuan Sun**[1b*]

[1]School of Mechanical Engineering, University of Jinan, Jinan, Shandong, 250022, China

*Corresponding author, e-mail: [a] 724855201@qq.com , [b] me_sunx@ujn.edu.cn.

Abstract. Aiming at the problem of relative encounter path conflict in multi-AGV systems under special working conditions, a path planning method is proposed to limit the running direction of AGV by adding path factor to Dijkstra algorithm. This method can realize simple traffic rule formulation. Effectively reduce the conflict of opposite encounter paths and achieve the purpose of orderly operation of multiple AGVs.

1. Introduction

The intelligent logistics system in industrial production workshops is an important part of intelligent manufacturing [1]. The AGV, which has the advantages of low cost, high efficiency, good flexibility and high reliability, is responsible for the main material handling tasks in intelligent logistics [2]. One of the key issues that multi-AGV has to deal with in intelligent logistics is the multi-AGV collision-free path planning problem [3]. If this problem is not handled properly, it will cause problems such as AGV blockage, deadlock and collision, which will seriously reduce the operating efficiency of multi-AGV systems [4]. In response to this problem, many scholars have proposed different solutions. For example, Guo Jinchao [5] et al. proposed a multi-AGV scheduling method based on traffic control; Dimitri Antakly [6] et al. proposed a delay by appropriate application. Time to avoid the method of emergency conflict; Toshiyuki Miyamoto [7] et al. proposed a local/random search method. These methods can better solve the path conflict problem of multiple AGVs in specific situations. However, not all factory conditions can use these methods to efficiently resolve conflicts. Different working conditions have their own unique characteristics, and it is necessary to take a unique method to solve the problem. A Dijkstra algorithm path planning method for locally adding path factors for this condition is proposed for the working conditions of the paper.

2. Working condition analysis

2.1. Workshop Overview

The multi-AGV operating conditions of the research project are as follows: there are 8 production lines in the workshop, the end of the production line is the product placement point, the work line is provided on both sides of the production line, and the stacking point of the stacking machine for loading and unloading goods is provided at the three-dimensional warehouse. When performing the task, the AGV needs to transport the products of the coded point to the warehouse for storage or to transport the materials from the warehouse to the station for use by the station. When the task

IOP Conf. Series: Materials Science and Engineering **711** (2020) 012036 doi:10.1088/1757-899X/711/1/012036

sequence has no tasks or the AGV needs to be charged, the AGV needs to return. Wait for the task or charge at the standby point. The working conditions of the workshop are shown in Figure 1.

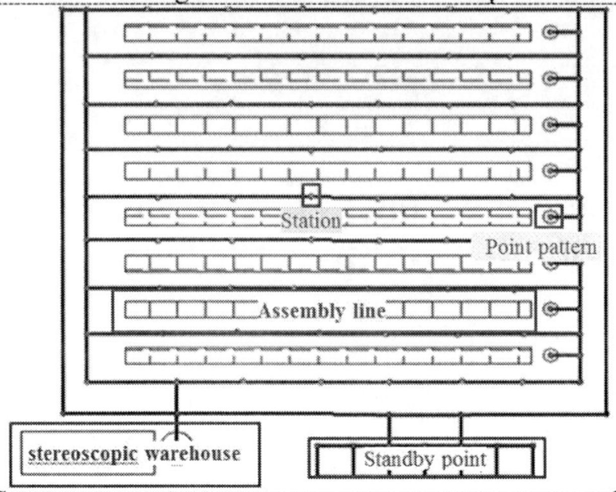

Figure 1. Schematic diagram of the working conditions of the AGV workshop

According to the logistics needs of the workshop and the needs of the AGV itself, the scheduling tasks of multiple AGVs are divided into five categories: feeding task, warehousing task, charging task, returning standby point task and fault task.

The AGV used in this model is an omnidirectional AGV for laser navigation. In order to better establish the motion model of the multi-AGV system, it is assumed that the AGV operates at the same speed and the acceleration and deceleration time is extremely short.

2.2. Conflict Analysis

In the current multi-AGV system, there are four types of path conflicts: catch-up collision, node occupancy conflict, vertical encounter conflict and opposite encounter conflict [8]. The most complicated and difficult to solve is the conflict of opposite encounters. These conflicts involve the problem of mutual occupation. One of the vehicles needs to be avoided to make the vehicle pass smoothly. That is to say, one of the vehicles needs to travel around the road. The operation efficiency of the multi-AGV system is greatly reduced, so the occurrence of the phase conflict is minimized in the operation of the multi-AGV system.

3. Solutions to reduce conflicts in the opposite direction

3.1. Formulate traffic rules

For the plant model under study, traffic rules can be used to reduce the possibility of conflicts in the opposite encounter path. The five types of tasks mentioned above are divided into two categories: uplink tasks and downlink tasks. The uplink task mainly refers to the task of the AGV transporting the raw materials from the three-dimensional warehouse to the upper station; the downward task mainly refers to the task of the AGV to transport the products from the palletizer to the three-dimensional warehouse, in addition to the task. Charging tasks and going back to standby tasks.

According to the characteristics of the factory model, the traffic rules are set: the AGV is allowed to go up (the up, down, left and right here refers to the up, down, left, and right in Figure 2, the same below), that is, when the uplink task is performed, the left lane is taken as far as possible; when the AGV is walking down That is, when the downlink task is executed, the right lane is taken as far as possible; at the same time, when the task is assigned, the AGV that has just completed the feeding task is preferentially assigned to the inbound task, and the AGV that has just executed the inbound task is preferentially assigned to the feeding task, so that the whole can be made. The running path of the system forms a one-way ring, which greatly reduces the conflict of opposite encounters.

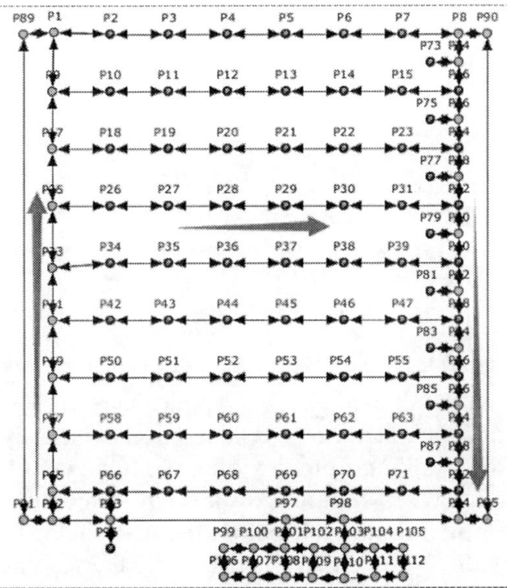

Figure 2. Schematic diagram of AGV running direction

3.2. Adding a path factor to the Dijkstra algorithm locally

The system uses Dijkstra algorithm as the basic algorithm of path planning. When using the Dijkstra algorithm for path planning, a path factor is added to some of the paths in the model to achieve the purpose of formulating traffic rules.

As shown in Figure 3, when the system plans the path of the uplink task, add a path factor of 2 to the unidirectional path of P66% P67, P93% P97, and P66% P65. In this case, you can improve the execution of the uplink task to make the AGV take the left path. Possibility, for example: When the AGV performs the feeding task from P96 to the station P39, whether it is the shortest path from the left path or the right path, there is a 50% probability that it is going from the right side, it is likely to be executed from the right side. The AGV of the downlink task has a conflict of opposite encounters. After adding a path factor to the partial path in one direction, the AGV performing the uplink task can be restricted from traveling on the right path.

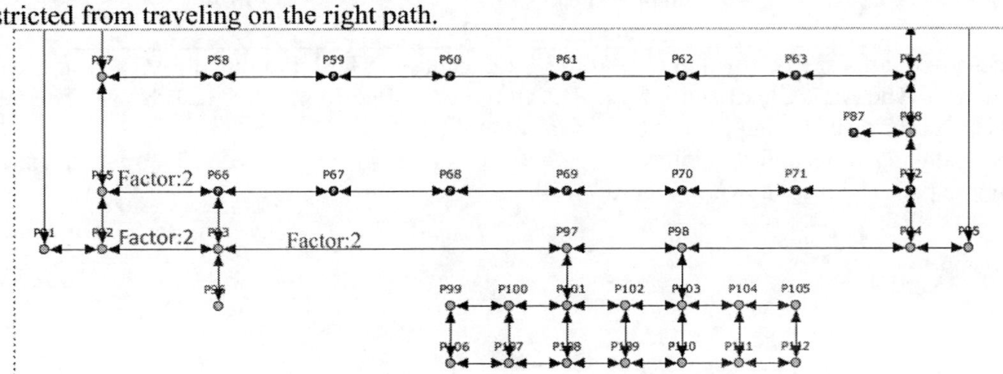

Figure 3. Adding path factors to the upstream task

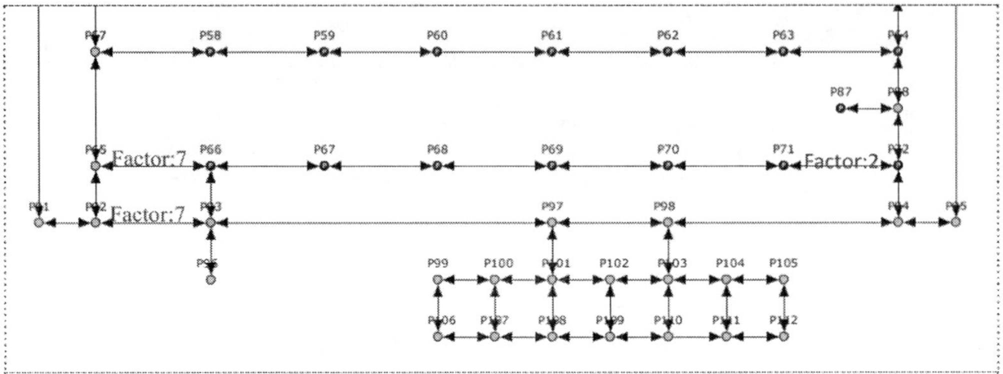

Figure 4. Adding path factors to the downlink task

When the system performs the downlink task, a path factor of 7 is added to the one-way side of P65% P66 and P92% P93, and a path factor of 2 is added to the one-way side of P72% P71. At this time, the AGV can be restricted from walking from the left path. For example, the AGV finishes the feeding task and reaches the P42 point. When returning to the P96 to continue the feeding task, the shortest path planned by the algorithm is from the left before the path factor is added. When the path is gone, the probability of collision with the AGV that is performing the feeding task is relatively high, and the addition of the path factor will restrict the AGV from descending from the left path, which meets the requirements of the traffic rules proposed above, and achieves the purpose of reducing the conflict of opposite encounters.

3.3. Case Analysis

Simulation experiments were carried out on the path direction of multiple AGV systems, and Open-TCS was used as the simulation platform. There are five tasks to be processed, three feeding tasks and two inbound tasks at the same time. The specific tasks are shown in Table 1.

Table 1. Task list

Serial number	Task type	Task start point	Task end point
1	Feeding task	P96	P55
2	Feeding task	P96	P39
3	Feeding task	P96	P71
4	Inbound task	P81	P96
5	Inbound task	P85	P96

These five tasks are assigned to five AGVs for execution. The 1st to the 5th AGVs are executed for tasks 1 to 5. The AGV No. 3 starts from P42 and goes to P96 to perform task No. 3. The remaining AGVs are at the task starting point. Start performing the task.

The path planning before adding the path factor is shown in Figure 6. The path planning after adding the path factor is shown in Figure 7.

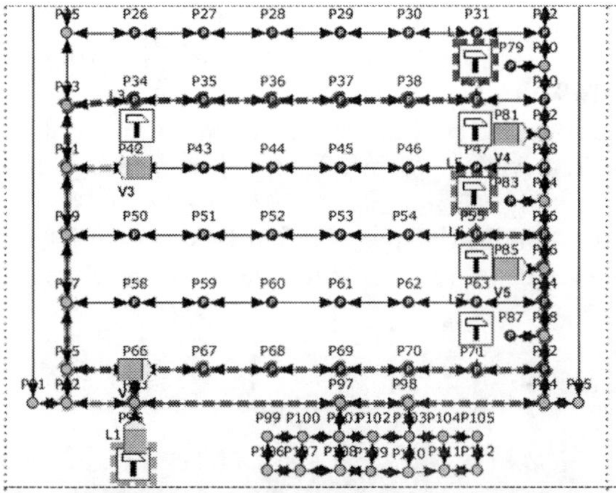

Figure 5. Schematic diagram of path planning before path factor addition

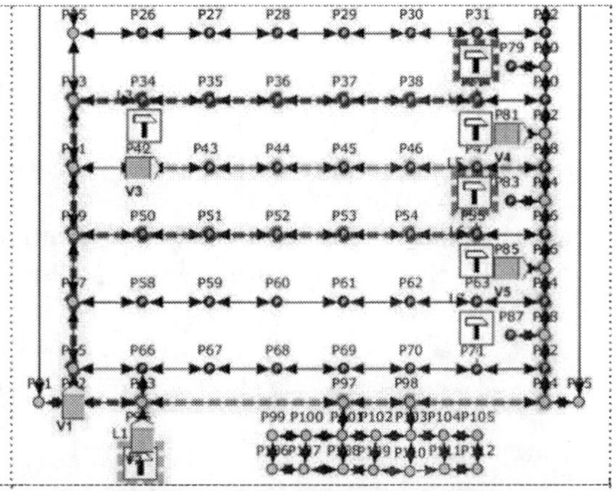

Figure 6. Schematic diagram of path planning after adding path factor

It can be seen from the above two figures that before the path factor limitation is added to the Dijkstra algorithm, the uplink task travels from the left side to the right side, for example, the task 1 goes to the P55 point V1AGV and executes the task in FIG. 2 Go to the V2AGV at point P39. At this time, the V1AGV planning path has a longer coincidence path with the route planned by V4 and V5AGV for performing the inbound task. If V1AGV meets the other two in this section, it will inevitably lead to an encounter conflict. At this time, V3AGV is going to P96 to perform task 3. The shortest path planned is to run from the left side, which will also have a high probability of conflict with the V2AGV that is performing the uplink task. Adding the Dijkstra algorithm After the path factor is limited, the AGV operates according to specific traffic rules. The path planning is shown in Figure 7, which can greatly improve the order of the multi-AGV system.

4. Conclusion

By adding the path factor to some of the paths during the path planning of Dijkstra algorithm, the running direction of the AGV is limited, and the system traffic rules are indirectly realized, so as to reduce the conflict of opposite encounters. At the same time, the model after adding the path factor does not lose the original bidirectionality of the lane. In some special cases, the AGV can also drive in the reverse direction. It is undeniable that the multi-AGV system with the added path factor will extend the certain running path for the driving path, but for the whole system, it can reduce the conflict of opposite encounters and improve the order of the system.

Acknowledgment

This work is supported by the national natural science foundation project (the project number is 51875250)

References

[1] Wang Y., Xie F.,Liu Y.,Zhu X.(2019)Research on Path Planning of Automatic Guided Vehicle Based on Improved Potential Field Ant Colony Algorithm.Manufacturing Automation,41(07):70-74.

[2] Cao X.,Zhu M.(2019)Multi-agv Collision Avoidance Decision Optimization Method Based on Conflict Prediction. http://KNS.cnki.net/kcms/detail/11.5946.tp.20190812.0851.002.html.

[3] Guo J.,Zhang F.,Lan D.,Cao H.,Wang P.(2019)gwm-based multi-agv path conflict resolution algorithm.http://KNS.cnki.net/kcms/detail/41.1437.ts.20190823.1019.010.html.

[4] Chen Z.,lu S.(2017)Research on agv Path Based on Probability Model under Time Window Constraint.Logistics Engineering and Management,39(10):65-67+32.

[5] Guo J.,Zhang F.,Lan D.,Cao H.,Wang P.(2019)Design of Multi-agv Scheduling System Based on Traffic Control.Electrical Technology,2019(16):147-149.
[6] Dimitri A.,Jean J.L.,Rosa A.(2017)A Temporised Conflict-Free Routing Policy for AGVs.IFAC-PapersOnLine,50:11169-11174.
[7] Toshiyuki M.,Kensuke I.(2016)Local and random searches for dispatch and conflict-free routing problem of capacitated AGV systems.Computers & Industrial Engineering, 91:1-9.
[8] Liu J.(2018)Research and Implementation of Multi-agv Path Planning in Automated Warehouse Scheduling System. Chinese Academy of Sciences University, Shenyang.

Analysis of circular steel tubular columns under the coupling of fire and impact

Tingwei Yuan [1*], Chunting Li[1], Qiuyue Zhang[1]

[1]College of Civil Engineering and Architecture, Shandong University of Science and Technology, Qingdao, Shandong, 266590, China

[*]Corresponding author's e-mail: skyuantingwei@163.com

Abstract: Aiming at the coupling of a pair of sliding circular steel tubular columns with one end fixed by fire and impact, this paper analyzes the model by finite element software ABAQUS. The impact load is applied to the mid-span of the test piece. Through the analysis of the displacement of the mid-section of the test piece, the following three points are clarified: 1) The circular steel pipe column absorbs energy through integral plastic deformation and local plastic deformation; 2) The displacement of the end of the circular steel tubular column and the displacement and displacement of the upper and lower surfaces of the span depend on the axial compression ratio, the impact velocity and the temperature; 3) Under the coupling of fire and impact, the safe area of the column member depends on the temperature. And axial compression ratio; the greater the temperature and axial compression ratio, the smaller the safety zone. In order to study the concrete-filled steel tube under the coupling of fire and impact.

1. Introduction

Concrete-filled steel tubular members may be subjected to impact loads due to natural or deliberate human factors. In recent years, domestic and foreign scholars have shown that the concrete-filled steel tubular members have excellent impact resistance[1-4].Bambach et al.[5] carried out impact experiments on three different square steel tubes with width-to-thickness ratios of 31, 22 and 13 respectively. It was found that under the impact load, the round steel tube column would have plastic hinges at the ends and spans, between adjacent plastic hinges. There is basically no deformation, and the local deformation of the mid-span is accompanied by the bending deformation of the whole round steel tubular column, and the tearing damage occurs on the upper surface of the end first when the event occurs. Wang et al.[6] found through experiments and simulations that the concrete-filled steel tubular specimens undergo ductile failure when the ferrule coefficient is large, absorb more energy, and brittle fracture occurs when the ferrule coefficient is small, and the energy absorbed is less. The energy required to break the specimen increases as the axial compression ratio increases. Wang et al.[7] and Aghdamy et al.[8] found that the hollow ratio can significantly change the peak value of the impact force time history curve through the experiment and simulation of double-walled steel tube concrete. The peak value of the peak stage is related to the hollow ratio and the transverse deformation of the concrete filled steel tubular specimen. The size is related to the hollow ratio, the yield strength of the steel, and the compressive strength of the concrete.

The static loading test and the impact test of the existing steel pipes are all studies on the material properties and impact resistance at normal temperature. The fire test is a study on the impact resistance of the test piece during static heating and the cooling after cooling. However, after the fire in the structure of the project, the bearing capacity of the structural members is reduced, the upper heavy objects fall, and the lower members need to bear the actual working conditions of the coupling of the fire load and the impact load.

In this paper, numerical analysis of circular steel tubular columns is carried out by experiments and ABAQUS finite element software. Through the parameter analysis, the effects of impact kinetic energy and temperature on the dynamic response mode of the circular steel tubular column under the impact and fire coupling environment were investigated. This paper provides a reference for the study of the response mode of CFST under the coupling of fire and impact.

The assumptions made in this paper are:

1) The deformation of the circular steel tubular column conforms to the flat section assumption, ignoring the influence of shear force on the yield of the section;

2) The steel material is an ideal elast-plastic material, ignoring the strain rate effect of the material;

3) The impactor is a rigid block, ignoring the deformation effect during the impact.

2. Finite element analysis under impact load

2.1. Calculation parameters

The ABAQUS/Explicit solver was used for the analysis of the impact load alone. The steel pipe adopts S4R unit, the inner diameter of the round steel pipe column is R=23mm, the wall thickness is t=2mm, and the length is L=1100mm. The round steel pipe column adopts the boundary condition of one end sliding and one end fixed. The load is applied by a certain mass and impact speed of a given rigid block in the mid-span of the test piece, as shown in Fig.1. The material constants of steel materials are shown in Table 1.

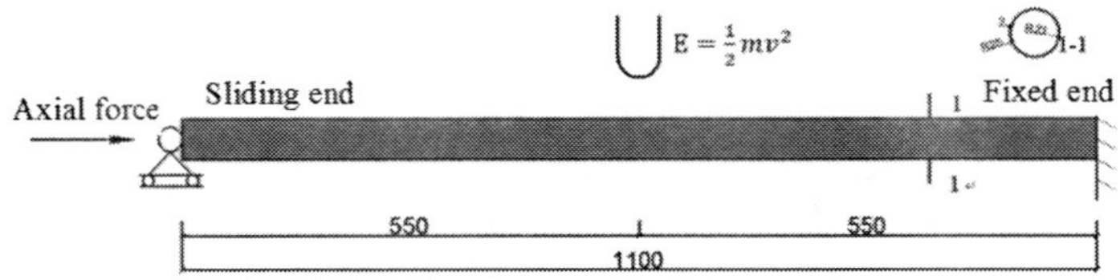

Fig.1 Schematic diagram of impact force of circular steel tubular column

Table 1 Steel material constant

$f_y/(MPa)$	$E_s/(GPa)$	v_s	$\rho_s/(kg/m^3)$
300	205	0.3	7850

2.2. Calculation results

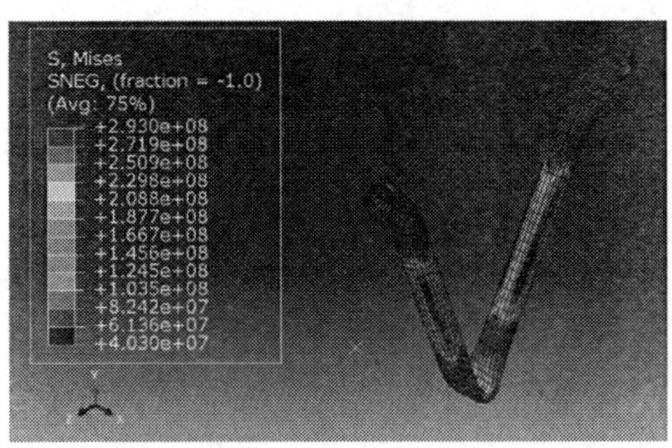

Fig.2 Collapsed cloud diagram of round steel tube column at normal temperature ($\mu_N = 0.1$, $v = 10m/s$)

Figure 2 is the final deformation cloud diagram of the round steel tubular column under $\mu_N = 0.1$, $v = 10m/s$ conditions. Under the impact load, the circular steel pipe column will produce overall bending deformation. Due to the existence of axial force, the round steel pipe column cannot maintain

IOP Conf. Series: Materials Science and Engineering **711** (2020) 012037 doi:10.1088/1757-899X/711/1/012037

its own stability on the basis of large overall deformation, and collapse occurs. It can be seen from the stress cloud diagram that most of the area of the round steel pipe column at the time of collapse has reached its own yield strength.

2.3. End displacement and mid-span displacement

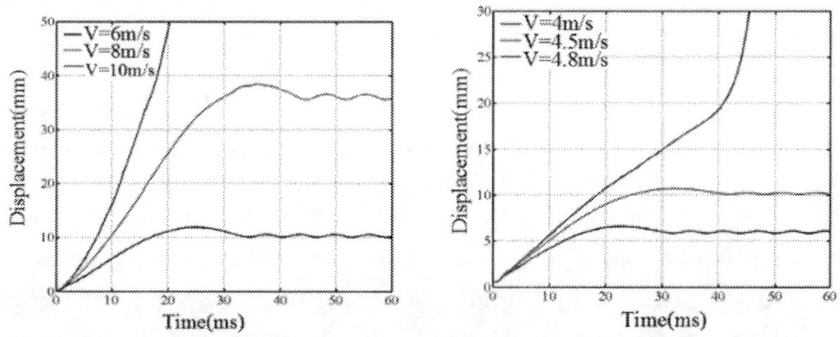

a b

Fig.3 End displacement of a circular steel pipe column with different axial compression ratios (a: $\mu_N = 0.1$; b: $\mu_N = 0.3$)

Figures 3a and b are displacement diagrams of the end of a circular steel pipe column with axial compression ratios $\mu_N = 0.1$ and $\mu_N = 0.3$, respectively. It can be seen from the figure that when the axial compression ratio is the same, the displacement of the end increases with the increase of the impact velocity; when the velocity is at a certain critical value, the displacement of the end gradually converges to a constant value after a sharp increase. The final displacement depends on the plastic bending deformation of the circular steel tubular column; however, when the velocity exceeds a certain critical value, the end displacement always increases with time and does not converge. Comparing Fig.3a, b, it can be seen that when $\mu_N = 0.1$ is used, the speed at which the end displacement does not converge is 10m/s, and when $\mu_N = 0.3$ is 4.8m/s, which is significantly smaller than the former. From this, we conclude that the end displacement of the round steel pipe column is highly dependent on the impact velocity and the end axial force.

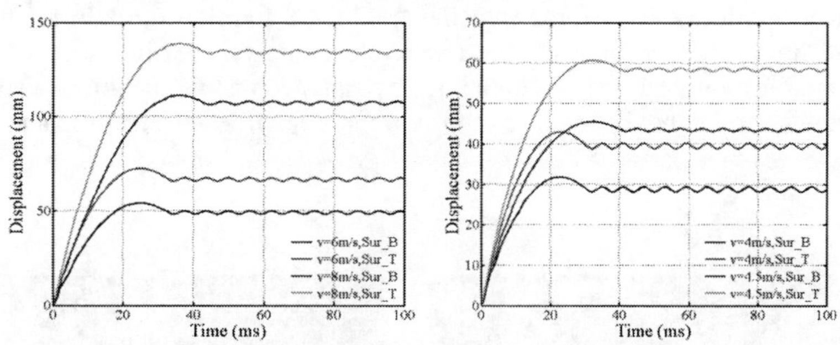

a b

Fig.4 Displacement of the upper and lower sections of the round steel pipe column at different axial compression ratios (a: $\mu_N = 0.1$; b: $\mu_N = 0.3$)

IOP Conf. Series: Materials Science and Engineering **711** (2020) 012037 doi:10.1088/1757-899X/711/1/012037

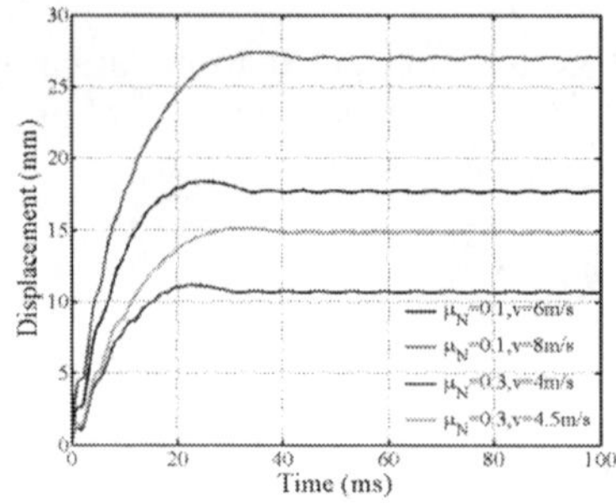

Fig. 5 Displacement difference between the upper and lower sections of a round steel tubular column at different axial compression ratios

Figures 4a and b are displacement diagrams of the upper and lower surfaces of a round steel tubular column in axial compression ratios $\mu_N = 0.1$ and $\mu_N = 0.3$, respectively. In the figure, Sur_B is the mid-span surface, and Sur_T is the mid-upper surface. It can be seen from Fig. 5a and b that at the same axial compression ratio, as the impact velocity increases, the displacement of the upper and lower surfaces of the round steel tubular column increases, indicating that the overall plastic deformation is greater, thereby absorbing more energy.

Figure 5 shows the difference in displacement between the upper and lower sections of a round steel tubular column at different axial compression ratios. It can be seen from the figure that with the same axial compression ratio, as the speed increases, the displacement difference increases, that is, the local plastic deformation increases, so that more energy can be absorbed. Referring to Figures 4 and 5, we conclude that the circular steel tubular column absorbs energy mainly by integral plastic deformation and local plastic deformation.

3. Finite element analysis of circular steel tubular columns under the coupling of fire and impact

3.1. *Calculation parameters*

Using the ABAQUS software to analyze the finite element analysis of the round steel pipe column after the first warming, firstly add the axial force analysis in the ABAQUS/Explicit module, and then introduce the model with the axial force into the ABAQUS/Explicit module for heating through the transfer results technique. After analysis, the temperature analysis results were again introduced into the ABAQUS/Explicit module for impact analysis by transfer results technique. The properties of the steel materials in this section are the same as those in the case of individual impact.

3.2. *Calculation results*

3.2.1. *Stress cloud map*

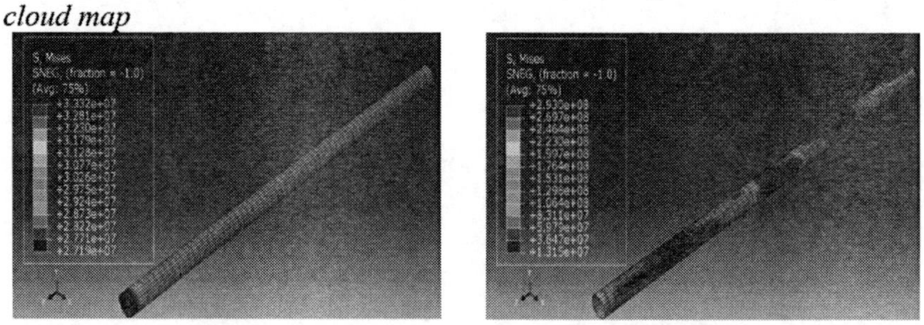

t=0ms t=1.25ms

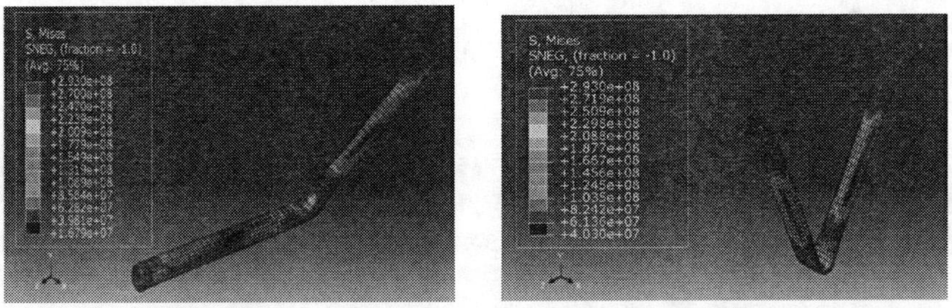

t=40ms t=100ms

Fig.6 Stress cloud diagram of impacted empty steel pipe column after first adding fire
($\mu_N = 0.1$, $v = 9m/s$, $T = 400°C$)

Fig.6 is a stress cloud diagram of the collapsed force when the axial force is applied and the impact after the fire. From Fig.6a: t = 0 ms, it can be seen that stress concentration occurs at the fixed end position due to the axial force and temperature. It can be seen from the figure that similar to the impact of the round steel pipe column at normal temperature, when the impact body is in contact with the circular steel pipe column, stress concentration occurs on the upper surface of the impact region, and then local deformation of the impact region occurs. When the internal force of the impact region is transmitted to the two fixed ends, a tensile force is generated at the upper end of the fixed end, and a tensile force is also generated at the bottom of the impact region, and the sliding end is displaced along with the overall deformation of the circular steel tubular column.

That is, the circular steel pipe column is accompanied by local deformation of the impact region, and overall bending deformation occurs, and finally the test piece collapses.As can be seen from Fig.6, the deformation of the round steel tubular column at the time of impact at a high temperature is similar to that at normal temperature.

3.2.2. End displacement

Fig.7a, b and c are the displacements at the ends of hollow steel tube columns with T=20℃,T=400℃ and T=600℃at $\mu_N = 0.1$, respectively. Comparing the three figures, it can be seen that although the temperature rise will change the critical speed at which the round steel column collapses, the change trend of the end displacement is similar.

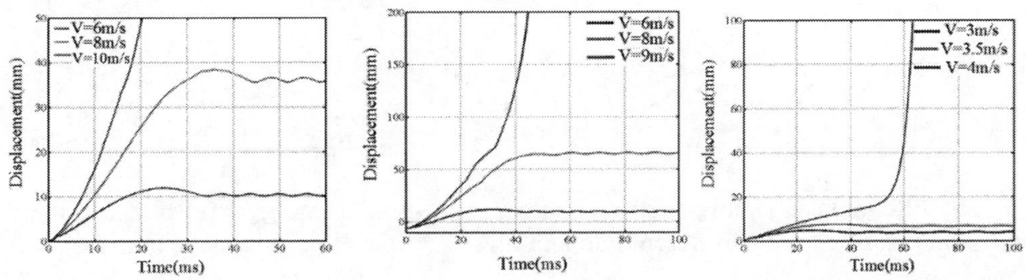

a b c

Fig.7 End displacement of empty steel pipe column at different temperatures when $\mu_N = 0.1$
(a: $T = 20°C$; b: $T = 400°C$; c: $T = 600°C$)

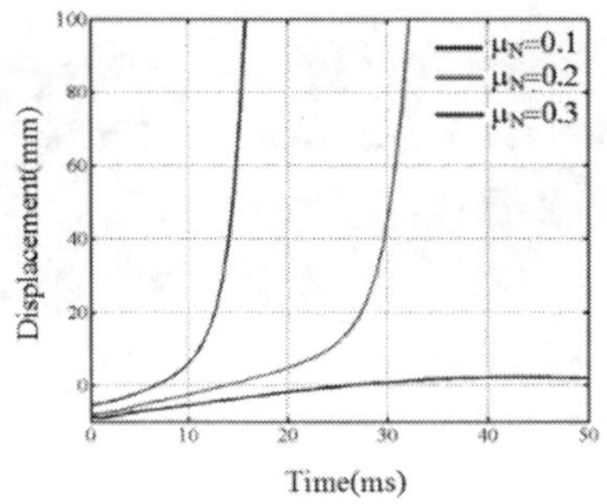

Fig. 8 End displacement of empty steel pipe column under different axial compression ratios ($T = 600°C$, $v = 5m/s$)

Figure 8 is the displacement of the end of the steel pipe column at different axial pressure ratios at $T = 600°C$, $v = 5m/s$ .It can be seen from Fig.8 that in the case where the temperature and the impact velocity are the same, an increase in the axial pressure ratio causes the displacement of the end portion of the circular steel pipe column to change from convergence to non-convergence. From Fig.7 and Fig.8, it is concluded that the axial compression ratio and temperature can significantly change the end displacement of the round steel tubular column.

3.3. *Critical energy ratio*

The critical speed is the speed at which the cylindrical steel pipe column just collapses when the axial pressure ratio and temperature are constant. Symonds and Fleming [8] et al. proposed energy ratios to describe the relevant analytical parameters.

$$R = {GV_0^2 EI}\Big/{M_0^2 L}$$

Where V_0 is the impact velocity of the impactor, M_0 is the plastic limit bending moment, and L is the length of the test piece. Through simulation analysis, we draw Figure 9 and Figure 10.

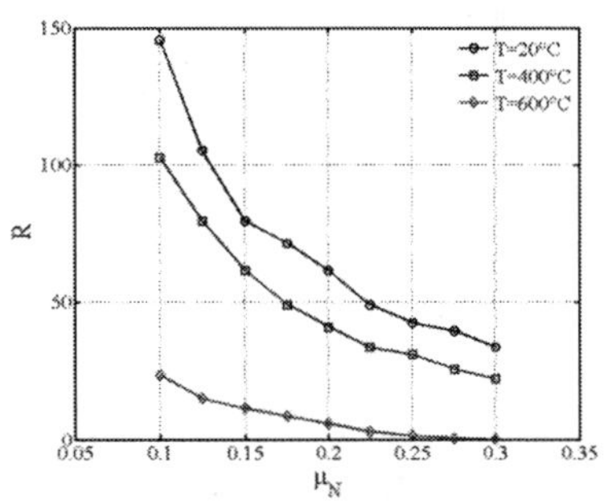

Fig.9 Relationship between energy ratio and axial pressure ratio at different temperatures

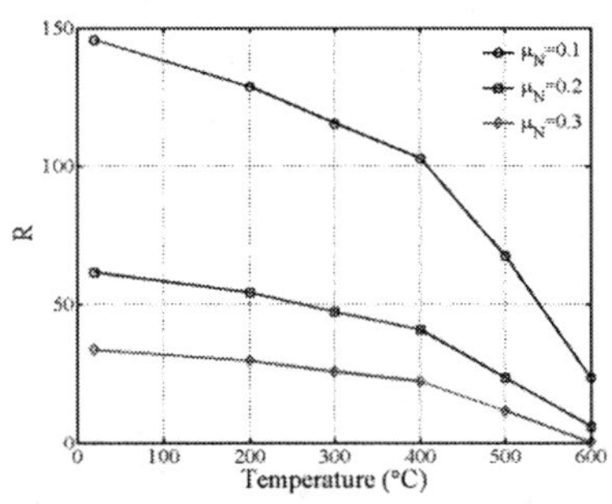

Fig.10 Energy ratio versus temperature for different axial compression ratios

Figure 9 is a graph showing the relationship between energy ratio and axial pressure ratio at different temperatures. It can be seen from the figure that when the temperature is constant, the R value decreases as the axial pressure ratio increases. Comparing the three curves in the figure, it shows that the increase in temperature leads to degradation of material properties, which reduces the R value. Figure 10 is a graph showing the energy ratio versus temperature for different axial pressure ratios. It can be seen from the figure that when the axial pressure ratio is constant, the R value decreases as the temperature increases. Comparing the three curves in the figure, it shows that the increase of the axial pressure ratio reduces the value.

In Fig.9 and Fig.10, each curve is bounded by a certain area with the coordinate axis, which is a safe area, that is, in the safe area, the column member does not collapse; outside the safety area, the

IOP Conf. Series: Materials Science and Engineering **711** (2020) 012037 doi:10.1088/1757-899X/711/1/012037

column collapse occurred. From Fig.9 and Fig.10, as the temperature and axial pressure ratio increase, the safety area becomes smaller and smaller. We can conclude that the temperature and axial pressure ratio significantly affect the safe area of the column member.

4. Conclusion

This paper analyzes the finite element software ABAQUS and draws the following conclusions:

1) The circular steel pipe column absorbs energy through integral plastic deformation and local plastic deformation;

2) Only under impact load, the end and mid-span displacement of the round steel tubular column, as well as the difference in displacement between the upper and lower surfaces, depends on the axial compression ratio and the impact velocity;

3) Under the coupling of fire and impact, the displacement of the end of the round steel tubular column and the displacement and displacement of the upper and lower surfaces of the span are dependent on the axial compression ratio, the impact velocity and the temperature;

4) Under the coupling of fire and impact, the safe area of the column member depends on the temperature and axial compression ratio; the larger the temperature and axial compression ratio, the smaller the safety area. The bearing capacity of the column members should be designed in a safe area.

Acknowledgments

Thanks to my classmates Chunting Li and Qiuyue Zhang, this article was done with their help. I also thank my tutor for helping me with this article.

References

[1] Deng Y H, Tuan C Y, Xiao Y.(2012) Flexural behavior of concrete-filled circular steel tube beams under high-strain rate impact loading.J.Journal of Structural Engineering.138: 449-456.

[2] Han L H, Hou C C, Zhao X L, Rasmussen K J R.(2014) Behaviour of high-strength concrete filled steel tube beams under transverse impact loading.J, Journal of Constructional Steel Research.92: 25-39

[3] Aghdamy S, Thambiratnam D P, Dhanasekar M, Saiedi S.(2015) Computer analysis of impact behavior of concrete filled steel tube beam columns.J.Advances in Engineering Software.89: 52-63

[4] M R Bambach.(2011)Design of hollow and concrete filled steel and stainless steel tubular columns for transverse impact loads.J.Thin-Walled Structures.49: 1251-1260

[5] Bambach M R, Jama H, Zhao X L, Grzebieta R H.(2008) Hollow and concrete filled steel hollow sections under transverse impact loads.J.Engineering Structures.30: 2859-2870

[6] Wang R, Han L H, Hou C C.(2013)Behavior of concrete filled steel tubular (CFST) members under lateral impact: Experiment and FEA model.J.Journal of Constructional Steel Research.80: 188-201

[7] Wang R, Han L H, Zhao X L, Rasmussen K J R.(2016)Analytical behavior of concrete filled double steel tubular (CFDST) members under lateral impact.J.Thin-Walled Structures.101: 129-140

[8] P. S. Symonds, W. T. Fleming.(1984)Jr, Parkes revisited: on rigid-plastic and elastic-plastic dynamic structural analysis. International Journal of Impact Engineering.2: 1-36

Bearing mechanism and bearing capacity analysis of bidirectional crimping uplift pile

Haichao Wang[1], Qiuyue Zhang[1*], Zhou Fang[1]and Zhuang Zhang [1]

[1]College of Civil Engineering and Architecture, Shandong University of Science and Technology, Qingdao, Shandong, 266590, China

[*]Corresponding author's e-mail: 13615426280@163.com

Abstract. Through the analysis of the design, mechanism of action and mechanics principle of bidirectional crimping uplift pile, the scientificalness and feasibility of method are further expounded. The basic concept, basic principle and method of the secondary structure function design are given, Emphasis is placed on the research and application of structural functions in the later stages of deformation to optimize structural design and the potential for structural play. The research shows that the compressive and pull-up bearing capacity of the bidirectional crimping uplift pile is not lower than that of the muti-plate-pile. Compared with the same volume of concrete square pile, the uplift capacity is increased by 170%; the construction process is more reasonable and extensive. Applicability, in the context of industrial building products, has significant economic benefits.

1. Introduction

The internal force effects in structural members are usually tensile, compression, bending, shearing and twisting.Under various loads, the design of tension, compression, bending, shearing and torsion of structural members can be carried out directly. For structural members with different mechanical properties, the process of reaching the limit state is generally different after the load is applied. Due to the mechanical properties of the structural members themselves, or due to human intervention in this process, the phenomenon of reducing or increasing their resistance may occur. The so-called concomitant effect is the effect of the non-direct action target in the process of load action. For example, under the action of load, the lateral deformation of concrete in steel tube is restrained, so that the internal concrete is in the state of three-dimensional compression, and the mechanical properties of concrete materials are changed[1]. The second-order effect is the transfer of dangerous section when the column is under eccentric load. The former shows the improvement of carrying capacity while the latter decreases the carrying capacity.

In both cases, the mechanical behavior opposite to the structural functional objectives of the design occurs during the increasing load effect. Confining the concrete to improve the bearing capacity is completely achieved by design. The idea of this design is to take advantage of the changes in the mechanical properties of the constrained concrete.

This paper studies "bidirectional crimping uplift pile design method" [ZL201310007835.5] [2], aiming at effectively improving the pile's uplift bearing capacity, which is characterized by a set of two piles, which rotate in opposite directions. The pile-forming method of the pile driving is realized and consolidated at the pile cap. When there is an external load, the opposing twists and turns are mutually constrained, and the generated torque is balanced as an internal force at the cap. Thereby, the pressure resistance and the pull-out bearing capacity are improved, and the pull-out bearing capacity is prominently improved. Its main technical advantages are reflected in the load, and the way to change the bearing mechanism is designed. The outstanding features of this method are reliable pull-out bearing capacity, extremely convenient construction and low cost.

2. Secondary structure function design method based on Concomitant effect

2.1. basic concepts

The so-called secondary structural function accompanying the effect of action means that under the action of the load (or external environment), the internal action of the structure will have an effect, and the original bearing mechanism is changed or partially changed due to the material of the structure and the internal structural structure. In accordance with the new bearing mechanism, the structural function is significantly improved within a certain range, and the effect of improving the structural function is improved. We are concerned with the part that enhances the structural function, that is, the advantageous secondary structure function. Has the following characteristics:

- Don't change the external mode of action.
- The effect is increasing.
- The secondary function of the designed structure is continuously strengthened in a certain range during the process of increasing the effect. Some of the enhanced structural functions are accompanied by effects.

The design method can be realized through two channels: one is the impact on the structural function reduction caused by the additional bending moment, we have enough safety by improving the safety reserve of the structure; the second is to change the mechanical environment of the material. Improve its mechanical properties or change its structural bearing mechanism.

2.2. Design of bidirectional crimping uplift pile

This section applies the design theory of secondary structure function to design the two-way compression-resistant uplift pile.

Bidirectional crimping uplift pile is a kind of pull-out pile designed by the author in the long-term research and practice according to the idea that the bearing capacity of the pile increases continuously with the effect of load action[2]. The shape of the pile is formed on the basis of a square pile (as shown in Fig. 1a), which is surrounded by a spiral surface formed by rotation. As shown in the figure: Since the friction angle of most soils is between 60°and 70°, the angle of the cutting angle is 60°, and when the rotation is 90°, the height is tan60°=1.732b, square. The length of the side. Rotate 360° to the height of a pitch: There are four spirals, and the equation is:

$$x = r\cos\theta,\ y = r\sin\theta,\ z = \frac{6.928b}{2\pi}\theta \tag{1}$$

Among them, θ is the helix radial rotation angle and r iotation radius. When $\theta < 0$, the spiral equation for counterclockwise rotation is obtained. The pile tip is a square pyramid that is rotated and is divided into two directions. The pile body is prefabricated with high-strength concrete. Since the modulus of concrete is much larger than the modulus of soft soil, piling into the soil is completely feasible. When the pile is piled, the pile load is continuously applied. When the effect is greater than the bearing capacity of the soil in contact with the curved surface, the spiral surface is rotated by the passive earth pressure. The piles that rotate in both directions are combined into one group, and are fixed by the cap after piling. The piling equipment adopts a static press machine, and a plane ball joint is installed on the top of the pile cap to facilitate the free rotation of the pile body when it is rotated downward.

In such a combined design of piles, when there is external load, the pile will still rotate under passive earth pressure, whether it bears vertical pressure or uplift force. However, due to the locking effect of the cap, the rotating pile body will be fixed, and the reverse torsion produced by the two piles will be balanced at the cap (as shown in Fig. 1b and 1c). At this time, the bearing capacity of the pile has the same effect as that of the screw pile, and its bearing capacity can be continuously improved with the load.

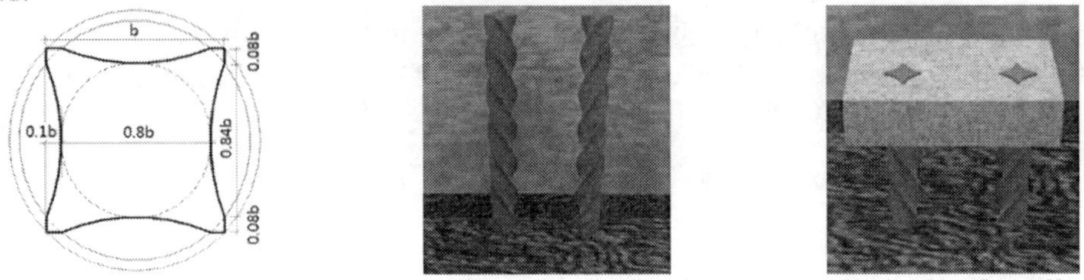

(a) Section design drawing (b) Bidirectional crimping uplift pile position (c)Pile body after pile caps are fixed

Fig.1. Cross section design and simulation effect diagram of bidirectional crimping uplift pile

The shape of the pile based on the parametric design is shown in Figure 2.

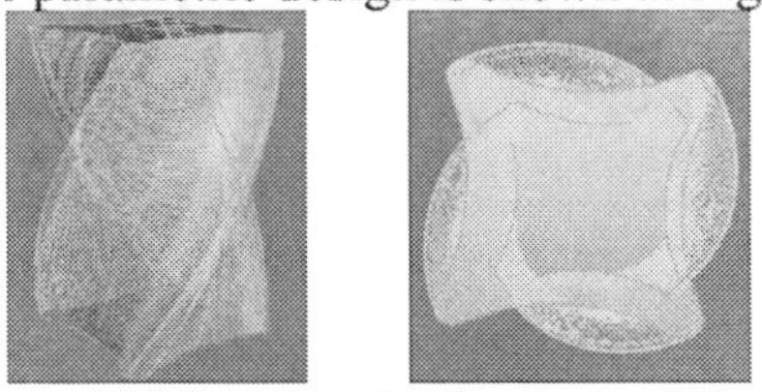

Fig.2. Model diagram of bidirectional crimping uplift pile wire frame

3. Study on bearing capacity of Bidirectional crimping uplift pile

3.1 Single-pile lateral resistance of Bidirectional crimping uplift pile

The lateral resistance of a conventional linear cast-in-situ pile is:

$$Q_{sk1} = u\sum q_{sik} l_i = 4b\sum q_{sik} l_i \tag{2}$$

Where: q_{sik} is the standard value of the ultimate side friction of the i-th layer of the pile side; u is the circumference of the pile; r is the diameter of the pile; l_i is the diameter of the pile; d is the thickness of the pile through the soil of the i-th layer.

For the two-way compression-resistant uplift pile with a real length of L, the pile side area increases by 0.134bL. the lateral resistance of the two-way compression-resistant pile is:

$$Q_{sk2} = u\sum q_{sik} L_i = 4b\sum q_{sik} L_i = 4.536b\sum q_{sik} l_i \tag{3}$$

Comparing formula (2) with formula (3), the lateral resistance of the two-way compression-resistant pile with a real length of L is 1.134 times that of the ordinary square pile.

3.2 Bearing capacity of single pile in Bidirectional crimping uplift pile

For the production of the two-way compression-resistant pile, the vertical height of a pile unit is $H = 1.732b$, and within a length of H, the spatial oblique surface of the pile unit body can be horizontally projected into a disk. Therefore, in the pile unit of the H length, it is considered that there is a disc, and the bearing capacity is estimated by the method of analogizing the muti-plate-pile.

According to the force mechanism of the multi-disc cast-in-place pile, the pile end, the pile bearing plate and the side resistance bear the load together. According to the Technical Specification for Building Pile Foundation, the formula [3], [4]:

$$Q_{uk} = Q_{sk} + Q_{pk} = u\sum q_{sik} l_i + \sum \eta_{pi} q_{pjk} A_{pj} + \psi_p \eta_p q_{pk} A_p \tag{4}$$

Among them: Q_{uk} is the standard value of vertical ultimate bearing capacity of single pile. Q_{sk} is the standard value of total limit side resistance of single pile. Q_{pk} is the standard value of total end resistance of single pile. q_{sik} is the limit side resistance standard of pile section in ith layer soil Value. q_{pjk} is the standard value of the limit end resistance of the jth branch or disk. q_{pk} is the standard value of the ultimate end resistance of the chassis. it can be calculated according to local experience. If there is no local experience, the "Technical Specifications for Building Pile Foundation" can be checked. u is the circumference of the pile body. l_i is the thickness of the pile body in the i-layer soil (this thickness is the value after removing the section or the thickness of the disc). A_{pj} and A_p respectively are the j-th branch or disc except the pile body area Area and chassis area. the manufacturing principle and upper sectional view of the two-way compression-resistant pile can be obtained: A_{pj}=0.333bL=0.577b^2, η_{pj}, η_p is the correction coefficient of the limit value of the limit end resistance of the disk. see the value Table 1, ψ_p is the chassis size effect coefficient.

Table 1. Modification coefficient of pile end resistance

Type of soil	Hard clay	Hard plastic clay	Plastic clay	Silty soil	Silty sand	Fine sand	Coarse sand
Correction factor	0.5~0.7	0.8~1.0	1.0~1.4	1.4~1.6	0.8~1.0	0.6~0.7	0.4~0.5

According to the formula (4), the bidirectional crimping uplift pile can be simplified into multi-disc pile perfusion for general analysis and calculation. The section obtained by the cross-section projection is a plane, and the thickness of the support disc is negligible. Zhu Qingdong [5] and others have carried out the calculation and comparison of actual engineering. The comparison results show that the bearing capacity of the circular piles with piles is 1.14 times higher than that of ordinary round piles.

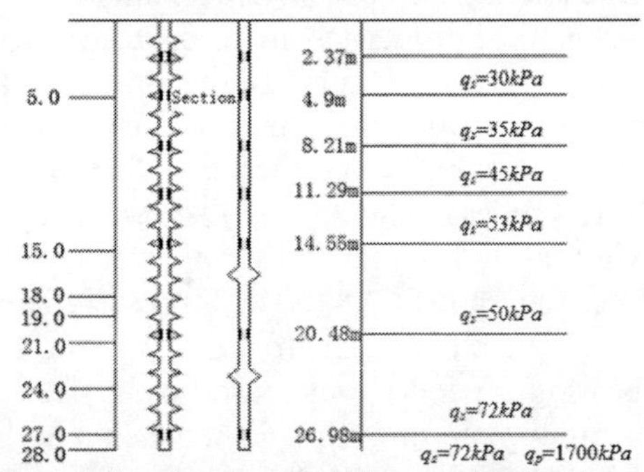

Fig.3. Pile size and Soil layer distribution

In order to facilitate the calculation, we quote the geological conditions in the actual engineering case studied by Ju Yuwen [6] and others. In the actual project, multi-disc piles are used for construction. The experiment shows that the bearing capacity is obviously improved compared with the ordinary pile foundation. However, the construction of the support pile is complicated and the economy is relatively poor. Therefore, we use the two-way compression-resistant pile as the pile foundation in the same geological conditions, and carry out the calculation of the compressive and pull-out bearing capacity, and the same volume. The square piles and the pile body are square branch piles for bearing capacity comparison.

The distribution of the soil layer and the test pile size are shown in Fig.3 below. In the figure, the distance between the support plates of bidirectional crimping uplift pile is the distance of one unit. The support disc distance of the multi-disc cast-in-place pile is the same as the original text.

In order to facilitate the calculation and comparison, we can take the common square pile length of 700mm and the pile length to be 24m. According to the "Technical Specifications for Building Pile Foundation", the formula can be used to calculate.

$$Q_{uk} = Q_{sk} + Q_{pk} = u\sum q_{sik} l_i + q_{pk} A_P \tag{5}$$

The length of the two-way compression-resistant pile is also 700mm, and the pile length is 24m. According to the principle of pile formation, the pile plate projected by each unit is a square circumcircle, so the diameter is 790mm, and the formula (4)is used. Calculation, calculation results are shown in Table 2.

Table 2. Comparison table between the uplift bearing capacity of bidirectional crimping uplift pile and the uplift bearing capacity of ordinary square piles

	Side lengt h/mm	length /m	Branch diameter /mm	Branch height/ mm	Frictional resistance /kN	End bearing force /kN	Bearing capacity of Branch/kN	Bearing capacity calculation/kN	percentage
Ordinary square pile	700	24	—	—	3684	—	—	3684	100
muti-plate-pile	700	24	1800	1650	2840	—	4285	7125	193%
Bidirectional crimping uplift pile	700	24	790	—	4178	—	5824	10002	270%

The comparison results show that the uplift capacity is 170% higher than that of the ordinary square pile. The piles are increased by 77%; the multiples of different geological conditions may fluctuate. In general, the resistance to pull-out of two-way compression-resistant piles is significantly improved. Compared with the muti-plate-pile, it can not only make up for the difficulties of construction of the muti-plate-pile, but also has better bearing capacity improvement than the muti-plate-pile, which has a very broad prospect.

4. Conclusion

(1) The secondary structure function design method of the bidirectional crimping uplift pile proposed in this paper is realized by changing the bearing mechanism of the structure. During the structural load-bearing process, the mechanical properties under the direct action effect are changed.

(2) Through the theoretical analysis and comparative study of the bidirectional crimping uplift pile, it is concluded that compared with the ordinary straight-type cast-in-place pile, the single-pile is due to the large contact surface area between the bidirectional crimping uplift pile and the foundation soil. The pile side resistance is 1.134 times of the ordinary square pile.

(3) The bidirectional crimping uplift pile are simplified into multi-disc piles for ultimate bearing capacity calculation. The results show that the compressive bearing capacity of the bidirectional crimping uplift pile is 130% higher than that of ordinary square piles. 54% higher than that of the muti-plate-pile. the uplift bearing capacity is increased by 170% compared with ordinary square piles, and 77% higher than that of muti-plate-pile. Studies have shown that bidirectional crimping uplift pile have superior mechanical properties and wide applicability.

Acknowledgments

Thanks to my tutor Professor Wang Haichao, this article was completed under the careful guidance of Teacher Wang. Thanks to my classmates for their valuable comments and thanks to all the teachers who have helped the paper.

References

[1] Li Y, Zhang L. (2005)The investigation and application of self-compacting expensive concrete filled steel tube. J. Journal of Beijing University of Technology. 31(5): 496-499.

[2] Wang H. (2013)A bidirectional crimping uplift pile design method: 201310007835.5.P.

[3] Wu Y, Zheng G, Yan P. (2000)A settlement prediction method for multi-under-reamed cast-in-situ pile foundation. J. Chinese Journal of Geotechnical Engineering. 22(5): 528—531.

[4] Xu Z, Zhang X. (2002)Application of Extruding-expanding multi-brace-disk pile in Engineering. J. Building Structure. 32(7): 13-16.

[5] Zhu Q. (2004)On design method of cast-in-place concrete pile with supporting-disc. J. Yangtze River. 35(10): 29-32.

[6] Ju Y, Liang R, Bai X, Zhang S. (2003)Experimental study of bearing and deformation behavior of squeezed branch pile and its bearing capacity calculation.J. Engineering mechanics. 6: 34-38.

The damping effect Analysis of viscous damper in eccentric concrete frame structure

Zhou Fang[1],Qiuyue Zhang[1] , Chongge Wang [1*]and Zhichuang Liu[1]

[1] School of Civil Engineering and Architecture, Shandong University of Science and Technology, Qingdao, Shandong 266590, China

*Corresponding author's e-mail: chgewang@163.com

Abstract: The application of viscous dampers in structures can effectively reduce the inter-layer displacement of structures, especially they can reduce the torsional effects of structures in earthquakes. We can simplify the eccentric structure into a series rigid plate model and divide it into three types: a model of the viscous dampers are provided at position 1 of the eccentric structure, and a model of the steel supports are set at position 1 of the eccentric structure and a original structural model . Then analyzing the earthquake resistance of a 13-layer eccentric concrete structure model using SAP2000,It is concluded that the model of setting the steel supports can effectively reduce the horizontal displacement of the structure compared with the original structural model, but the effect of reducing the torsional displacement is not obvious enough; the model of setting the viscous damper can not only effectively reduce the inter-layer displacement of the structure, but also effectively reduce the torsional displacement of the structure; Setting the viscous dampers can effectively reduce the inter-layer writhing displacement of the structure than setting the steel support model on the premise of reducing the same level of horizontal inter-layer displacement.

1. The research status at home and abroad and the content of this paper

The eccentric building structures are asymmetrical building structures relative to the axial building structures. They are different from the regular building structures, and have no simple body shapes, no regular planes and three-dimensional stereoscopic, and no relatively uniform mass and stiffness distribution. However, the eccentric building structure damages are more serious in the earthquake. Studies have shown that such structures as frame structures, frame shear wall structures and so on, they are subject to severe earthquake damages, mostly due to construction quality or severe structural irregularities [1]. Therefore, reducing the damage degree of eccentric buildings as much as possible has become a key concern of experts and scholars.

In recent years, the energy dissipation and vibration reduction technology has been increasingly valued and favored by scholars and engineers at home and abroad, and promoted to a large number of practical projects [2-4]. Abdelouahab Ras, Nadir Boumechra, Luca Landi Tong Guo et al [5-7] studied the seismic design of structures with additional viscous dampers, and analyzed and calculated the BouMeldS earthquake using SAP2000 software, compared the seismic effects between linear damping, nonlinear damping and unsupported structures by showing linearity in the form of tables and graphs. And analyzed the damper support models and damper placement strategies. Jia Bin, Luo Xiaoqun et al. [8-9] used a three-way seismic wave input to perform nonlinear time-history analysis on a gymnasium, and studied the mechanism and effect of viscous damper on seismic strengthening of large-span spatial structures. Tang Yuchuan, Zhang Yuliang, Guo Yuchen et al. [10-11] studied the nonlinear dynamic analysis of the viscous damper damping structure in the high-rise building structural dynamic analysis program HBTA. Through analysis, the damping effect of the viscous damper on the structure is investigated.

Based on the above literature research, we can simplify the eccentric structure into a series rigid plate model and divide it into three types: a model of the viscous dampers are provided at position 1 of the eccentric structure, and a model of the steel supports are set at position 1 of the eccentric structure and a original structural model . Through the seismic analysis of the thirteen-layer eccentric concrete structure, comparing the horizontal displacement and rotational displacement between layers between original structure with the supporting structure, and setting the horizontal displacement ,it is concluded that the provision of the viscous damper can effectively reduce the torsion of the structure. Furthermore, it provides a reference for the earthquake resistance of eccentric building structures.

2. Research method for setting viscous damper in eccentric concrete frame structure

Using the method assumed by the rigid floor, and representing each floor covered by a rigid piece. For the eccentric structure, the center of the upper and lower horizontal rigid sheets will deviate due to the existence of eccentricity, and the line formed by connecting the center of the rigid sheet must be a curve, and the formed rigid sheet system becomes a tandem rigid sheet system. The horizontal displacement and rotational displacement of the rigid piece can fully reflect the displacement generated by the eccentric structure under the action of earthquakes.

3. Seismic response analysis of an eccentric concrete frame structure

3.1. Engineering overview of eccentric structures

The model selects a 13-storey hotel with a flat structure shape of “L” . The north-south span is 48.00m and the east-west span is 36.00m. The north-south and east-west intersections have large cylinders diameter of 0.6m and small cylinders diameter of 0.4m. The column spacing is 6.00m and the building height is 3.3m. Figure 3.1 is a plan view of the building and dampers layout. Since the building is a typical type of planar irregularity, it is eccentric in both the vertical and horizontal directions, and the building will have severe torsional responses under the action of earthquakes.

3.2. Design parameters of eccentric structure

The structure is reinforced concrete structure, the concrete strength of the column and the plate is C30, the size of the side column is 800x1000mm, the size of the inner column is 1200x1200mm, the size of the corner column is 1300x1300mm, and the beam section is 450x750mm. The thickness of the curved hall is 120mm, and the thickness of the concrete slab is 100mm; the site is an 8 degree seismic zone, the site category is III, and the design earthquake group is the first group.

3.3. Arrangement of viscous dampers and steel supports in eccentric structures

3.3.1 Arrangement of viscous dampers

According to the SAP2000 operation result, the east end of the east-west span is set to the “a”end (live end), and the north end of the short span is set to the “b” end (solid end). For the eccentric structure, under the action of earthquake, the inter-layer displacement (horizontal displacement and rotational displacement) of the living end (“a”end) is larger than the displacement of the solid end (“b” end). Therefore, the steel supports and the viscous dampers should be placed at the live end of the structure. According to the analysis results of the original structure in SAP2000, the steel supports and the viscous dampers are placed at the sixth layer. The layout of the structural plane of the arrangement of the steel supports and the viscous dampers in the original structure is shown in Figure 1, and the vertical arrangement is shown in Figure 2.

IOP Conf. Series: Materials Science and Engineering **711** (2020) 012039 doi:10.1088/1757-899X/711/1/012039

3.3.2 Arrangement of steel supports in the structure

In order to compare with the case that the structure with viscous dampers, the arrangement position of the steel supports in the structure should be consistent with the arrangement position of the viscous dampers, while the other conditions remain unchanged. Steel support optional I-beam, the material is Q235 steel.

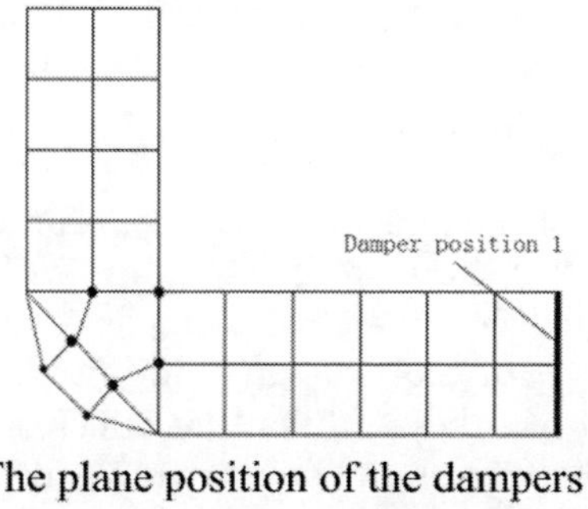

Fig.1 The plane position of the dampers layout

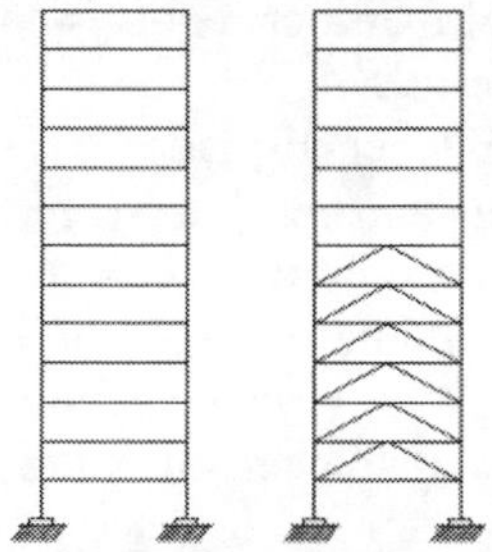

(a) (b)

Fig.2 Schematic diagram of original structure and the layout damper structure
(a)The original structure (b)The structure of the damper

3.4. The displacement ratio comparison

The structure with only the viscous dampers and the displacement ratio of the structure with only the supports and the original structure are compared, including the horizontal displacement ratio between layers as shown in Table 1, and the structure displacement ratio is shown in Figure 3.

Table.1 Structural displacement ratio

Number of layers	Original structure	Add Viscous dampers structure	Add supports structure
1	1.137	1.111	1.077
2	1.118	1.101	1.064
3	1.112	1.102	1.059
4	1.108	1.098	1.051
5	1.103	1.097	1.046
6	1.099	1.095	1.042
7	1.094	1.092	1.038
8	1.091	1.091	1.036
9	1.088	1.089	1.033
10	1.084	1.088	1.030
11	1.081	1.085	1.027
12	1.078	1.083	1.023
13	1.074	1.080	1.020

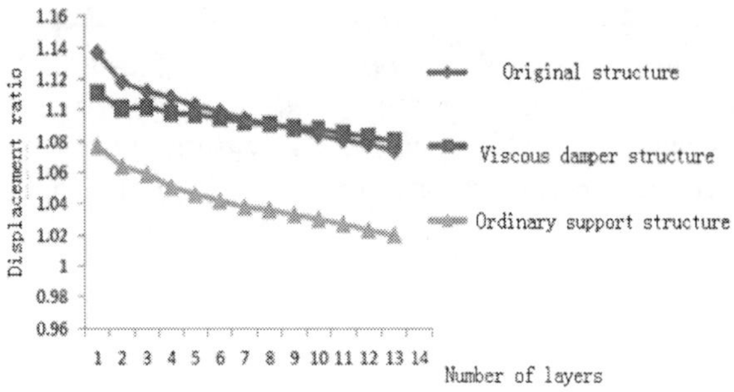

Fig.3 Structure displacement ratio curve diagram

It can be seen from Table 1 that the displacement ratio of the first layer of the structure after adding the viscous damper is reduced by 2.6% compared with the original structure, and the displacement ratio of the first layer of the structure with the steel support is lower than that of the original structure. About 5.3%. It can be seen that the effect of adding steel support to reduce the displacement between layers is better than adding a viscous damper. However, the addition of the ordinary support as part of the structural member requires consideration of the support deformation to the main component in engineering practice.

As can be seen in Figure 3, the inter-layer displacement ratio of the structure is increased faster in the 1-2th layer after adding the damper, and the inter-layer displacement ratio is slowed down from the 2nd-5th layer (the floor without the damper). The 5th to 6th layers descended faster, but the displacement ratio between the layers of the 6th-10th layer (the floor without the damper) slowed down, while the effect of the 10th-13th layer was accelerated, and the overall appearance was stepped; With the steel-supported structure, the displacement ratio between layers of 1-6 layers decreases rapidly, and the ratio of inter-layer displacement decreases from the sixth layer, and the step decreases. Analysis of the above data can be obtained. In the case of adding damping or support, the inter-layer displacement ratio of the structure can be reduced, especially in the floor where the viscous damper or the support is provided, and the displacement ratio reduction effect is remarkable.

Under the action of earthquakes, the torsional effect of eccentric structures must be taken seriously. The inter-layer rotational displacement ratio is shown in Table 2. The contrast curve between the viscous damper structure, the common support structure and the original structure is shown in Figure 4.

Table.2 Comparison of inter-layer rotational displacement in three cases

Number of layers	Original structure R1	Add supports structure R1	Add Viscous dampers structure R1
1	0.00037	0.00014	0.0006
2	0.00063	0.00019	0.0001
3	0.00079	0.0002	0.00012
4	0.00087	0.00018	0.00013
5	0.0009	0.00014	0.00013
6	0.00092	0.00012	0.00013
7	0.00093	0.00013	0.00014
8	0.00094	0.00018	0.00015
9	0.00095	0.00025	0.00017
10	0.00097	0.00031	0.00019
11	0.00098	0.00036	0.00022
12	0.00098	0.00038	0.00023
13	0.00098	0.00039	0.00024

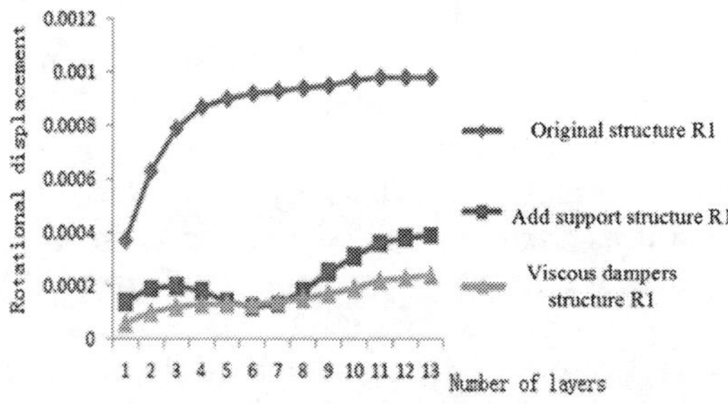

Fig.4 Comparison of the rotational displacement between the structure of viscous damper structure, steel support structure and the original structure.

As can be seen in Table 2, the viscous damper is reduced by about 83% compared with the original structure, and the rotational displacement of the bottom layer of the structure added with the ordinary steel support is lower than that of the original structure. 62%. It can be concluded that the addition of a damper is more significant.

It can be concluded from Figure 4 that in this example, although the building structure with the viscous damper has a lower effect on the reduction of the inter-layer displacement ratio than the building structure with the common support, the viscous damping is added in the anti-torsion effect. The building structure consumes more energy than the building structure with just ordinary support. It can be concluded that under the action of earthquake, the damper has excellent energy dissipation function for the eccentric structure which may generate large inter-layer displacement, and can reduce the structural torsion effect caused by eccentricity as much as possible.

4. Conclusions and recommendations

In this paper, the original structure is analyzed by analyzing specific engineering examples, and the viscous damper and steel support are respectively arranged at position 1, and the vibration mode decomposition reaction spectrum method is used to carry out the inter--layer displacement, inter-layer displacement ratio and torsional displacement of the original structure. Compared. The following main conclusions were obtained through research:

(1) By arranging the viscous damper at position 1, the viscous damper and the support can be effectively reduced when the viscous damper and the support are compared with the original structure and the inter-layer displacement of only the support structure. Inter-layer displacement. In the case of reducing the similar horizontal displacement, the viscous damper contributes more to reducing the torsional effect of the structure.

(2) Using the time-history analysis method in SAP2000, the structure of the viscous damper is compared with the torsional effect of the original structure, and the viscous damper can effectively reduce the torsional displacement of the structure.

Acknowledgement

I would like to thank Professor Wang for the ideas and directions that I have provided to me. I would like to thank my classmates for their help during the simulation period.

References

[1] Tsinghua University, Southwest Jiaotong University, Beijing Jiaotong University, (2008)civil engineering structure expert group. Journal of Building Structure .

[2] Ma L, Chen Y, Zhao G. (2009)Vibration reduction design of the main tower of Yintai Center using liquid viscous damper [J].Construction.39(5): 23-28.

[3] Wang D, Chen J, Bao L, et al. (2010)Research on energy dissipation support system and its application in Expo center engineering .J. Journal of Building Structures.31(5): 117-123.

[4] Deng W. (2012)Research progress on seismic strengthening methods for tall buildings[J]. China Water Transport.12(6): 256-257.
[5] Abdelouahab Ras, Nadir Boumechra. (2014)Study of Nonlinear Fluid Viscous Dampers Behaviour In Seismic Steel Structures Design .J.Arabian Journal for Science and Engineering.Vol.39 (12), pp.8635-8648.
[6] Luca Landi, Cristina Vorabbi, O. Fabbri, Pier Paolo Diotallevi. (2017)Simplified probabilistic Seismic assessment of RC frames with added viscous dampers .J. Soil Dynamics and Earthquake Engineering. Vol. 97 (7), pp. 277-288.
[7] Tong Guo, Jia Xu, Weijie Xu, Zhiqiang Di. (2005)Seismic Upgrade of Existing Buildings with Fluid Viscous Dampers: Design Methodologies and Case Study J. Journal of Performance Of Constructed Facilities.
[8] Jia B, Luo X, Zhang Q, Ding J. (2014)Vibration Control Effect of Viscous Damper on Spatial Structure[J]. Earthquake Engineering Newspaper. 36(01): 39-46.
[9] Jia B, Luo X, Ding J, Zhang Q. (2014)Study on damping effect of viscous damper on space truss structure.J.Action. 33 (6): 124-130.
[10] Tang Y, Zhang Y, Zhang T. (2004)Nonlinear Dynamic Analysis of Damping Structure of Viscous Damper.J.Engineering Mechanics.21(1): 67-71.
[11] Guo Y, Yang X. (2016)Research on shock absorption of viscous damper in concrete-filled steel tubular frame shear structure.J.Construction.48(6): 16-20.

Green Intelligent Logistics Sorting System in Big Data Environment

Xinyue Gao, Lei Liu*

School of Mechanical Engineering, Southeast University, Nanjing, Jiangsu, 211189, China

*Corresponding author's e-mail: liulei@seu.edu.cn

Abstract. By referring to the current sorting process of logistics sorting equipment, combined with emerging technologies, green concepts and big data statistics, it is possible to predict the trend of future sorting systems and integrate existing technologies. First, investigate and research existing logistics warehousing. Next, the intelligence and automation of existing intelligent sorting are investigated. After investigating, on the basis of the above, this study analyzes the trend of the future sorting system and combines with the concept of green logistics. It optimized classification under principles of green warehousing conceptual to make the classification more efficient, more space-saving and more convenient; it uses the big data statistics technology to collect logistics information, provides information to the logistics company to re-plan the sorting point, and feedback to the electronic merchants to obtain the location of different kinds of online shopping products in each region, and make it possible to purchase goods in a targeted manner.

1. Introduction

With the rapid development of e-commerce, the scale of network shopping has also shown a spurt type growth. The rapid development of Internet technology is accompanied by the popularization of smart mobile devices in various regions, which together promote the vigorous expansion of e-commerce worldwide. At the same time, the scale of network shopping continues to expand, and the contribution of logistics economy to China and even the world economy also increases year by year [1]. As can be seen from the above, with the rapid development of logistics industry, the industry competition is increasingly fierce. Logistics efficiency has become an important factor for consumers to choose logistics companies, and it is also the core competitiveness of logistics companies [2]. The sorting of goods is the link of the whole logistics with strong variability and easy integration with advanced science and technology and various disciplines. What is most worth thinking about is how to combine artificial intelligence with traditional sorting to create a logistics system that is more efficient, more economical and more environmentally friendly.

2. The Status Quo of Logistics Sorting

In the whole link of the supply chain, logistics plays a crucial role in it. The sorting system studied in this paper is the core part of the logistics warehousing center, which has attracted more and more attention from the public. Whether the allocation of cargo space is reasonable or not will affect the efficiency of the overall warehousing operation of the logistics storage center, and thus affect the rate of enterprises' profit acquisition [3]. Logistics intelligent sorting machine is a flexible transportation unit in the postal hub system, the working principle is to accept the master computer tasks, automatic handling mail sorting standards, and make the results and data back to the main control computer of the machine so that storage, query and printing are available in computer system, reducing errors and improving efficiency. The characteristics of intelligent postal equipment system are accurate, reliable and stable operation [4].

The above data analysis shows that the four elements of warehousing classification, intelligent distribution, green logistics and big data, which can be combined by the intelligent sorting system.

The commonly used automatic sorting equipment includes: belt sorting ma-chine, pallet sorting machine, surface sorting machine, horizontal moving roller table sorting machine, push block sorting machine and rocker sorting machine. However, the above sorting equipments can only sort goods of the same category or a few less varieties. They can only identify product varieties, but cannot identify customer orders. Therefore, they cannot quickly and intelligently sort goods of different varieties according to customer orders. Data cannot be extracted to quickly and intelligently respond to the sorting needs of different customers [5].

Intelligent sorting is to use the sorting system to sort the express according to certain rules and deliver them to a specific location, so as to achieve the effect of efficiency, speed and space saving. Some main results are: Martin Christopher embed RF technology (RFIII) sorting system, to build a mathematical model to find the optimal model, make the automatic sorting system more efficient, intelligent [6]. According to Daria Battini et al., sorting operations cost a lot of man-power and time in logistics activities to find out the optimal sorting strategy, reasonably allocate storage location and reduce the walking path of sorting process. Tricia Hyland, who has established ring sorting model first, designed the express distribution information control system. Ulrich Beck studied the tracking technology of express area, optimized and updated the feeder subsystem in the automatic sorting system with the mathematics of discrete parallel system, analyzed the upper process of static model and dynamic model in the stack-block sorting system respectively, optimized the feeder machine in the stack-block sorting system, and improved its feeder efficiency [7].

Now, the sorting devices formally applied in the factory is Caocao robot independently developed by Ali in Tmall supermarket. The robot can quickly locate the distribution of goods in the warehouse and plan the optimal picking path. After picking the goods, it will automatically send them to the packing desk such as Fig. 1. Another sorting robot, called minions, is being used by logistics companies such as EMS, JD and STO. It has a built-in intelligent identification system, which can recognize the information on the express side of the package, and then find the best way to stop delivery in the previously set route such as Fig.2.

Figure 1. Caocao robot.

Figure 2. Sorting robot called minions.

3. Green logistics and logistics big data

Different scholars and organizations have given different definitions of green logistics according to their different understandings. For example, H.J.Wu and S.Dunn (1995) believe that green logistics is a logistics system responsible for the environment. The United States reverse logistics executive committee (RLEC) considers green logistics as the logistics process with minimal impact on the eco-logical environment. Compared with traditional logistics, green logistics pays more attention to protecting the ecological environment. Green logistics refers to the scientific development concept of logistics based on the general principles of economics, sustainable development theory, ecological economics theory, eco-logical ethics theory, internalization theory of external cost and logistics performance evaluation. Green logistics USES advanced logistics technology to plan and implement logistics activities of transportation, warehousing, handling, circulation processing, packaging, distribution and other operational processes. The connotation of green logistics includes the following five aspects: intensive re-sources, green transportation, green storage, green packaging and waste material logistics. How to integrate information to realize green logistics, this time should use big data thinking to realize the optimization of logistics sorting.

The so-called big data of logistics refers to the data and information involved in the logistics links such as transportation, warehousing, handling, packaging, circulation and processing. At present, the scale of global online shopping has reached an unprecedented huge level, which also brings great pressure to logistics, and the demand for information on each node is also increasing. Through big data analysis, transportation and distribution efficiency can be improved, logistics costs can be reduced, and customer service requirements can be met more effectively. Nowadays, big data has penetrated into all links of logistics and be-come the cornerstone of innovation in the logistics industry. In the future, the logistics industry's demand for big data will have a broader prospect. The industry reform and cross-boundary integration of big data, including supply chain, is under way. The classic definition of "big data" can be summarized into four V's: volume of data, Rapid data flow and velocity, Variety, Huge data value.

Where can the information and data collected in the logistics process be used? Intelligent distribution of logistics outlets, optimization of transportation routes, improvement of loading rate, optimization of the last kilometer, decision-making at the company level, intelligent recommendation of deliverers, etc. From point to surface, gradually improve the level of intelligence and intelligent logistics will show its incomparable advantages in terms of efficiency, cost and user experience.

4. Green logistics is combined with big data in the sorting process

The first step is to judge the effective information under the guidance of green logistics rules, and then accurately and timely restore the business through big data technology, that is, to collect the data in the logistics sorting process in a timely and accurate manner and present it according to different levels of requirements. For the logistics system, it is very important to have a graphical dis-play, usually a picture is worth a thousand words. In the time dimension, the pro-duction capacity of each node and the difference between adjacent nodes can be displayed in real time, which can well control the business. In the process of sorting, the flow of goods per unit time, regional statistics, commodity statistics, volume classification of goods per unit time, as well as the delayed stock of goods exported by the sorting machine all need to be restored in a timely and accurate manner and displayed separately at different levels of demand. Logistics is the combination of commodity flow, material flow, capital flow and information flow. Therefore, it is very important to display information in different dimensions.

The second step is to evaluate the logistics information extracted from the periodic sorting of the unit through big data. In the era of big data, we can conduct business evaluation based on socialized data, and make use of the Internet gray scale test method to conduct process optimization evaluation. These methods can be fed back to logistics sorting to optimize its rules so as to conduct more reason-able sorting. For such laborintensive industries as logistics, the use of real-time data into the industry ranking can also play a good role in stimulating the scene.

The third step is to use big data to predict logistics after real-time monitoring and accurate evaluation of cargo information. Prediction has always been the core of big data applications, but also the most valuable place. Accuracy of pre-diction, also known as confidence in traditional statistics, is an important measure of prediction. Big data predictions, many of which make use of correlation, are easier to predict if the application is more tolerant of accuracy. For the logistics industry, it is very meaningful to predict the business volume in advance for resource scheduling, which can not only realize better timeliness, but also meet the requirements of green logistics to avoid waste. For example, in a certain peri-od of time, the logistics sorting points will be fed back to the forecast of big data to realize the manpower and material resources scheduling of warehousing sorting.

The fourth step is to make intelligent decisions based on big data. Largely de-pend on the business for the accuracy of the prediction accuracy and comprehensiveness, it can be concluded that the four quadrant, for business high prediction accuracy and comprehensiveness, the more likely it is to realize intelligent decision, the best way is still the man-machine combination, using big data and artificial intelligence technology to provide artificial auxiliary decision-making and letting human decision-making more reasonable. For example, according to the delivery location planning of logistics storage location and scope to intensive resources, according to receive location information to route planning which conforms to the green transportation, information planning different depending on the type of goods storage space so as to conform to the green storage, according to the volume of information and commodity type to scientifically design the packaging which conforms to the green packaging, according to the quantity of the goods and the size of the organization to focus on waste material logistics.

5. Conclusion

To sum up, the wisdom logistics sorting is based on large data processing technology. It combines equipment to better use software system which is guided by the principles of green logistics and improves system intelligent level to let people and equipment can play their respective advantages to reach the best state of the system and evolves constantly. Logistics intelligent sorting, as a new social infrastructure, optimizes the classification rules with the concept of green storage, making it more efficient, more space-saving and more convenient. It collect logistics information by using big data statistics technology and give feedback to logistics companies to facilitate the re-planning of sorting points, then give feed-back to electronic merchants to obtain the distribution of online shopping items in various regions so that they can purchase and sell goods in a targeted way, so as to optimize the supply chain further. Therefore, the development of intelligent logistics has broad prospects.

References

[1] Sunhui, Wang Xiangqian, Li Huizong& He Yerong. (2017) Analysis of the New Distribution Mode of the Last Kilometer——Taking the Intelligent Extraction Cabinet of Anhui University of Science and Technology as an Example. Journal of Liaoning University of Technology,19(04):49-52.

[2] Fu Xiankai. (2015) Design of Intelligent Express Transceiver System Based on Internet of Things. Shandong Industrial Technology, (01):207.

[3] Wang Yaping. Research on Logistics Location Optimization of Logistics Warehouse in Machinery Company. Changchun, Jilin,2016.

[4] Lu Jingchun, Gao Huaixiao, Xie Jianping, Chengzhuang. (2000) Postal transportation robot control and integrated information management system. Machinery and electronics, (01):31-32.

[5] Murong. Research on Sorting Strategy of Vertical Intelligent Logistics Picking Equipment. Tianjin, China, 2013.

[6] Martin Christopher. (1999) Logistics and Supply Chain Management: Strategies for Reducing Cost and Improving Service (Second Edition). International Journal of Logistics Research and Applications, 2(1).

[7] Antonella Meneghetti,Luca Monti. (2013) Sustainable storage assignment and dwell-point policies for automated storage and retrieval systems. Production Planning & Control, 24(6).

Research on Bearing Fault Diagnosis of Submersible Pump Motor Based on LMD and SVDD

Zhengtao Yuan[1], Fangzhen Song[1,2] and Renjie Dou[1]

[1] School of Mechanical Engineering, University of Jinan, Jinan, Shandong Province, 250022, China

[2] me_songfz@ujn.edu.cn

Abstract. The motor is a key component of the submersible pump. The health of the motor would greatly affect the safety and efficiency of the submersible pump. The bearing fault is one of the most common faults in motors. Therefore, detection and diagnosis of bearing faults are essential in the condition monitoring of pumps. In this paper, the local average decomposition (LMD) method is used to analyze the bearing vibration signals of submersible pump motor and extract feature vectors. A fault diagnostic model is established by the support vector data description (SVDD) to determine whether the submersible pump motor is faulty. The developed model exhibits practical significance in condition monitoring of submersible pump motor bearings.

1. Introduction

The bearings of the submersible pump motor are wearing parts, coupled with long-term overload and harsh working environment and other factors. Bearings are more prone to failure. Bearing faults will increase the load current of the submersible pump motor and abnormal working conditions will be induced. If the occurrence, type, and location of fault in submersible pump can be detected and diagnosed in real time, proper maintenance and repair can be carried out in the early stage of the fault, then the service life of the submersible pump can be extended, economic loss can be reduced, and accidents can be avoided.

In general, the reasons for the abnormal vibration of the submersible pump bearings are diversified, such as the misalignment of the rotor, the failure of the impeller, etc. However, these failure patterns can be reflected in the vibration signals. For example, periodic impulses with different frequency occur when the outer race, the inner race, the rolling elements and the cage fail. Therefore, according to the occurrence of the impulses and their frequency, the bearing fault types and locations can be judged. This paper developed a diagnostic model by combing the local average decomposition (LMD) and the support vector data description (SVDD) to detect and diagnose the failure patterns in bearing vibration signals, which aims for the online condition monitoring of submersible pumps.

2. Bearing failure analysis and vibration signal acquisition of submersible pump motor

2.1. Bearing failure

The unbalanced force in the motor, or the harsh operating environment, or the long-term running may induce different bearing failures. There are also various diagnostic methods for bearing faults: torque measurement method, rotational speed measurement method, temperature measurement method, oil analysis method and vibration method. Among them, the vibration method has the best applicability and effectiveness. The

IOP Conf. Series: Materials Science and Engineering **711** (2020) 012041 doi:10.1088/1757-899X/711/1/012041

vibration signal processing is simple and intuitive. The fault diagnostic method used in this paper is the vibration method. When a bearing fails, the corresponding characteristic frequency will appear in the vibration spectrum. By observing the waveform, it is easy to judge whether the bearing is faulty, but it is impossible to locate the faulty position. Therefore, the LMD method is used for signal processing and feature extraction, and a diagnostic model is established by the SVDD method. These two methods are combined to locate the faulty position of the bearing.

2.2. Bearing vibration signals acquisition of submersible pump motor
The experimental object of this paper is a BQS30-30-5.5 type mine explosion-proof vertical submersible pump. Its rated working voltage is 660V, power is 5.5KW, and its synchronous speed is 3200 r/min. The bearings are 6305 rolling element bearings. Their inner diameters are 25 mm, the outer diameters are 62 mm and the heights are 17 mm.

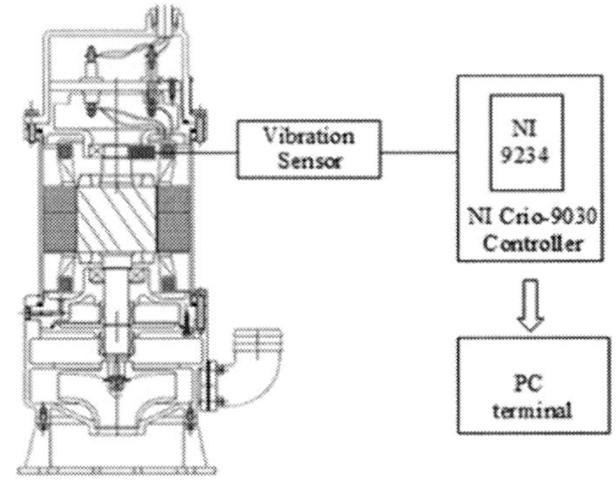

Figure 1. The vibration signal acquisition scheme

The vibration signals in this paper were collected by a NI cRIO-9030 controller. By adding a vibration sensor to the pump bearing, the vibration signals were collected under normal conditions and simulated fault conditions. The NI-9234 Quad C Series Dynamic Signal Acquisition Module in the NI cRIO-9030 controller can simultaneously acquire vibration signals at a frequency of 51.2 kHz per channel.

3. Research on fault diagnosis method based on LMD and SVDD

3.1 Local Mean Decomposition
The local mean decomposition (LMD) method is a new time-frequency analysis method for adaptive non-stationary signals analysis [1]. It has good adaptability and can decompose a complex nonstationary multi-component signal into several instantaneous frequency product functions with physical significance [2]. In terms of feature extraction, the bearing vibration signals were firstly decomposed by LMD, and then the feature vectors were constructed in the form of energy entropy. These feature vectors contain the vibration energy in different frequency bands, which can be used to distinguish various bearing faults [3].

3.2 Support Vector Data Description
Support Vector Data Description (SVDD) is a single-valued classification method. Suppose $\{\alpha_l$, l = 1,2... , n} is the training of input space sample set, n is the number of samples, The basic principle is to find a hypersphere defined with a spherical center μ and radius R in a high-dimensional feature space, and the hypersphere should contain as much target samples as possible [4]. Considering that, in the absence of faulty samples, only the normal samples are available, and a hypersphere can be trained be able to recognize the normal working condition while detect the abnormal conditions. It is suitable for processing high dimensional and small sample data and has been successfully applied in the field of condition monitoring [5].

In order to allow for outliers in the training sample, a variable ξ_1 is introduced to punish the anomaly point away from the hypersphere. Then the SVDD is described by the following equation:

$$\begin{cases} \min\limits_{R,\mu,\xi 1} R^2 + C\sum\limits_{l=1}^{n} \xi_1 \\ \left\|\varphi(x_l) - \mu\right\|^2 \le R^2 + \xi_1, \xi_1 \ge 0 \end{cases} \tag{1}$$

Which, C is used to achieve the balance between the number of misclassified samples and the volume of the hypersphere. In order to achieve nonlinear mapping, the sample of the input space is mapped to the high-dimensional feature space $\varphi(\alpha_l)$.

In order to find the optimal solution for equation (1), a dual form can be used:

$$\begin{cases} \max\limits_{\alpha_1} = \sum\limits_{l=1}^{n} \alpha_1 K(x_l, x_l) - \sum\limits_{l,k=1}^{n} \alpha_1\alpha_k K(x_l, x_k) \\ \sum\limits_{l=1}^{n} \alpha_l = 1, \quad 0 \le \alpha_l \le C, \quad l = 1, 2, \cdots, n \end{cases} \tag{2}$$

Solving equation (2), the sample α_l corresponding to $\alpha_l > 0$ is the support vector. The kernel function $K(x_l, x_k) = \langle \varphi(x_l), \varphi(x_k) \rangle$ represents the inner product of two vectors in the feature space. A basis Gaussian kernel function in the following form can be used:

$$K(x_l, x_k) = \exp\left(-\frac{\left\|x_l - x_k\right\|^2}{2\sigma^2} \right) \tag{3}$$

where the 'σ' is the width parameter of the radial basis Gaussian kernel function.

The radius R of the hypersphere can be expressed by the distance from the hypersphere sphere μ to any sample α_s located on the hypersphere:

$$R^2 = K(x_s, x_s) - 2\sum_{l=1}^{n} \alpha_l K(x_l, x_s) + \sum_{l,k=1}^{n} \alpha_l\alpha_k K(\alpha_l, \alpha_k) \tag{4}$$

For any sample z to be tested (which may be a known type or a position type), by comparing the distance from the center of the sphere to the $\alpha(z)$ and the radius R in the feature space, it can be determined whether the sample is a normal sample. The decision formula is:

$$f(z) = \left\|\alpha(z) - \mu\right\|^2 - R^2 \tag{5}$$

If $f(z) \le 0$, it means that the sample falls in the hypersphere and is a normal sample, otherwise it is a faulty sample.

IOP Conf. Series: Materials Science and Engineering **711** (2020) 012041 doi:10.1088/1757-899X/711/1/012041

3.3 Diagnostic process based on LMD and SVDD

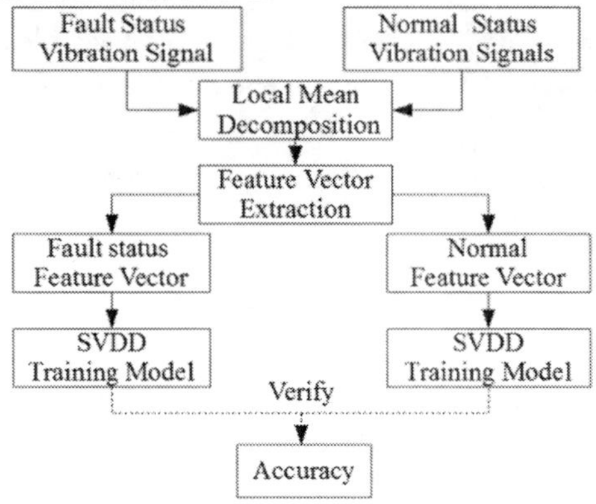

Figure 2. The flow chart of diagnostic principle

First, take a set of normal signals (no limit on sample number) for LMD decomposition and feature vector extraction, and then use the SVDD method to train the extracted feature vectors to build the diagnostic model. Then take the same number of fault signal samples, and build the diagnostic model following the same procedures. The developed diagnostic models are to be verified with the acquired vibration signals.

4. Analysis of experimental results

The acquired bearing vibration signals of the submersible pump were from normal and the outer race fault conditions, and a triaxial vibration sensor was used to acquire the bearing vibration in X, Y and Z. The signal sampling frequency is set to 10 KHz. Figures 3 and 4 show the motor bearing vibration signals under the normal and outer race fault condition, respectively.

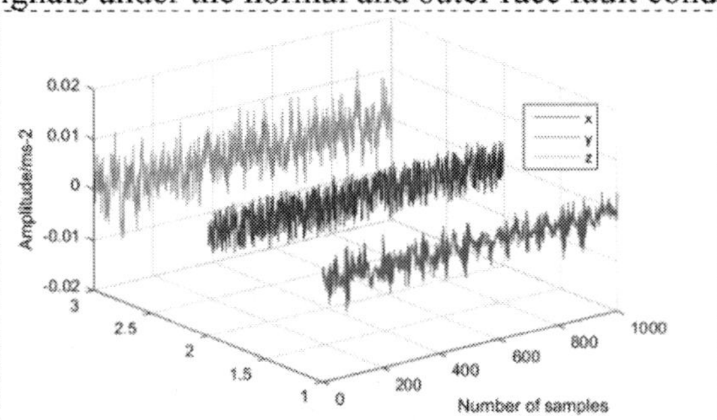

Figure 3. Motor bearing vibration signals under the normal condition

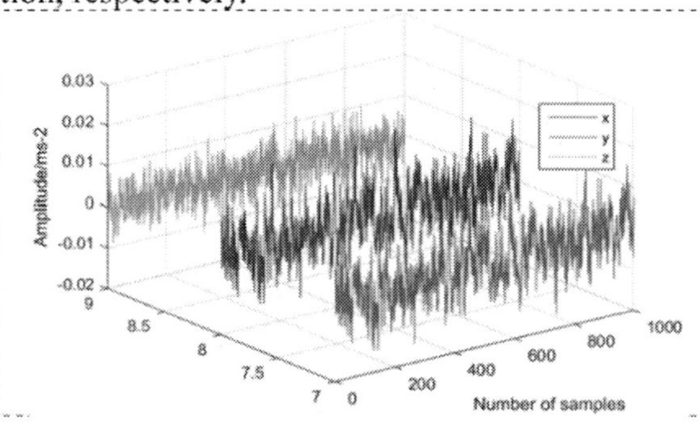

Figure 4. Motor bearing vibration signals under the outer race fault condition

In this paper, the acquired data were analyzed and verified by the developed diagnostic model. Firstly, a single classification model is established with the vibration signals under the normal conditions. The steps for establishing the model are as follows:

- 100 groups (1000 data points for each group) of vibration from the normal condition were selected as the training set, and the other 100 groups (1000 data for each group) of vibration signals under the normal condition were selected as the test set.
- Energy entropy feature vectors were extracted from the training set and test set by the LMD method.
- A minimum hypersphere was obtained for the training set by training with the SVDD method.
- The accuracy of the classification effect of the established diagnostic model was tested.

The LMD method is was used to process the data of the training set and the test set respectively. The processing result is shown in Fig. 5 and Fig. 6. The fifth-order PF component of the training set and the test set is used as the feature quantity of the original vibration signal.

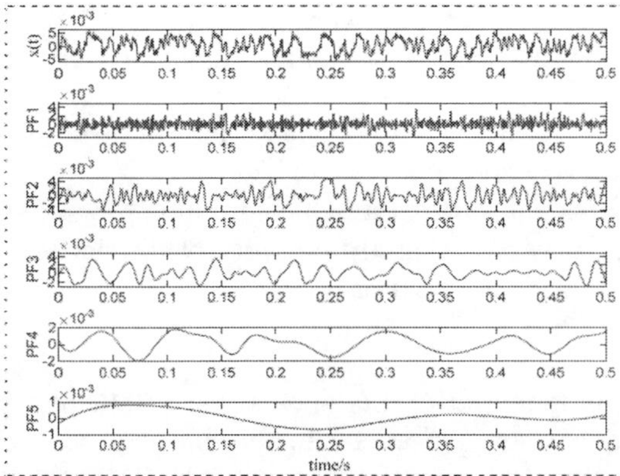

Figure 5. Training set LMD decomposition results

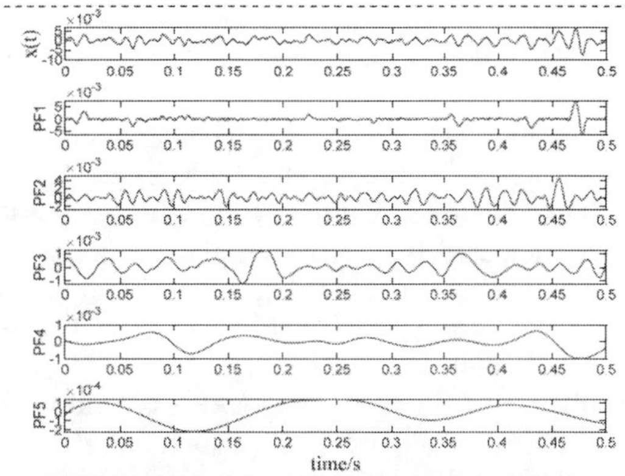

Figure 6. Training set and test set LMD decomposition results

The calculated energy of the 5th-order PF component of each vibration signal under normal condition and outer race fault are shown in the Fig.7 and Fig.8.

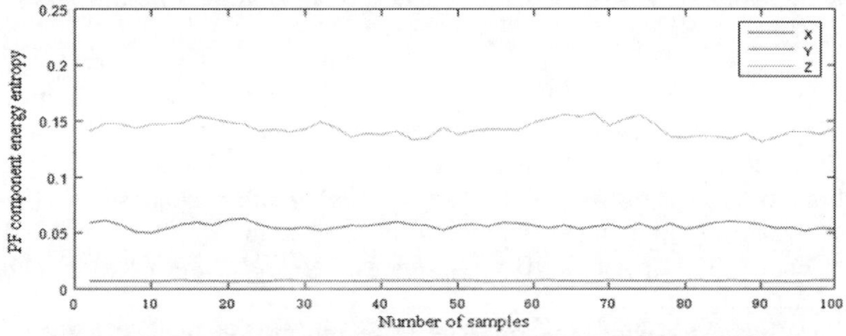

Figure 7. Feature extraction results from the normal condition.

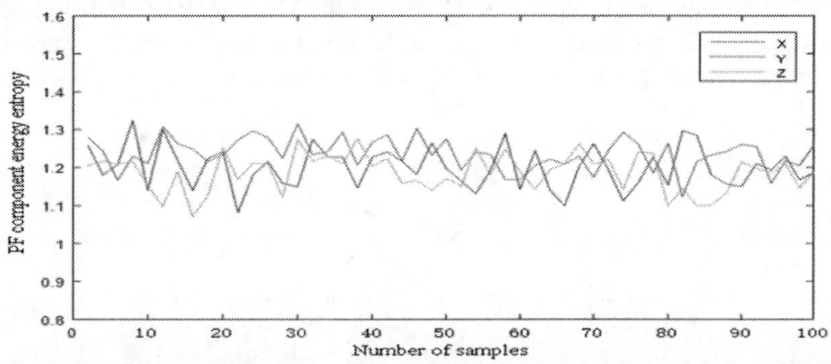

Figure 8. Feature extraction results in the bearing outer race fault condition

The kernel function used in this paper was the Gaussian kernel and the penalty factor is 1. A SVDD model of the bearing under normal condition was established. The model was applied on the test set, and the accuracy is 98%.

The 100 groups of outer race fault data were processed with the developed diagnostic model in the same way. The data under normal condition was predicted by the trained model, and the accuracy rate is 0. Then, use the 100 groups of data under outer race fault as the training set, and take the other 100 groups of data under outer race fault as the test set, and the test result is 98%.

Table 1. SVDD based bearing status classification results

Bearing status	Kernel function type (t)	Penalty factor (c)	Accuracy
normal status	Gaussian core	1	98%
Outer race fault status	Gaussian core	1	0

Finally, 100 groups of bearing data under inner race fault condition were selected as the unknown bearing fault condition. The LMD decomposition and feature extraction were performed in the same way. Then, the accuracy of the test set under the normal condition and the outer race fault condition was predicted.

The predicted results were 0% and 17%, respectively, so it can be concluded that the bearing is neither normal nor outer race fault, a new fault condition was detected.

It can be concluded that if the fault type of the bearing is unknown, after a new training process, the newly detected fault type can also be diagnosed. In this way, the diagnostic ability of the developed model can be evolved during the practical condition monitoring application.

5. Conclusion

In this paper, a diagnostic model is developed by combining the LMD method and the SVDD method for condition monitoring of the submersible pump motor bearings. Vibration signals were acquired from the submersible pump motor bearings, and the vibration signals were processed by the LMD method to obtain the 5th-order PF. The energy entropy of each component was used as the feature vector, and the SVDD method was used to train a hypersphere for fault condition recognition. The results show that the method based on LMD and SVDD can effectively diagnose the known fault condition of submersible pump motor bearing, and can evolve its diagnostic ability in condition monitoring.

References

[1] Yaqiang Jin. Vibration signal filtering method and its application in fault diagnosis of rolling bearings [D]. *University of Electronic Science and Technology*, 2018.

[2] Liuyang Peng. Research on fault diagnosis and state recognition method of rolling bearing based on vibration signal [D]. *China University of Mining and Technology*, 2019.

[3] Guangquan Zhao, Zedong Jiang, Cong Hu, et al. Bearing Fault Diagnosis Based on Wavelet Packet Energy Entropy and DBN [J]. *Journal of Electronic Measurement and Instrument*, 2019, **33**(02): 32-38.

[4] Lingjun Li, Zhousuo Zhang, Zhengjia He. Research on Mechanical Fault Diagnosis Based on Support Vector Data Description [J]. *Journal of Xi'an Jiaotong University*, 2003(09): 910-913.

[5] Nantian Huang, Lihua Fang, Yuqiang Wang, et al. Mechanical Condition Monitoring of High Voltage Circuit Breakers Based on Local Mean Decomposition and Support Vector Data [J]. *Advanced Technology of Electrical Engineering and Energy*, 2017, **36**(01):73-80.

Creep rupture strength of F12 steel welded joint after long-term service

Zhongwen Zhang, Xinmei LI*, Duqing ZHANG, Baoshuai Du

State grid shandong electric power research institute, Shandong jinan,250002,China;

*Corresponding author's e-mail: zzwh1013@163.com

Abstract: Creep rupture strength of F12 steel welded joint which was used as the main steam pipe and serviced 165000 hours was investigated by using high temperature creep rupture testing. Microstructure analysis results show that the microstructure of F12 steel welded joint became embrittlement significantly after long-term service. The precipitates mainly consist of M23C6 carbide and Laves phase, and the amount of M23C6 phase is more, which precipitated along the grain boundaries in bulk or chainlike with larger size. At the same time, there are larger inclusions with directional distribution, in which these inclusions have poor bonding with substrate and easily become crack source during the process of high temperature creep. The high temperature enduring strength of welded joint of F12 steel decreased significantly, it is necessary to enhance the examination of the microstructure and life assessment of pipeline in the next overhaul period.

1.Introduction

F12 standard martensitic steel which contains 12% Cr exhibits a typical high density martensite organization with some fine dispersed precipitates. The special structure of F12 steel contributes to its excellent properties such as high temperature creep strength, good obdurability, corrosion resistance and high temperature oxidation resistance and therefore expands its applications in power plant components such as the main steam pipes. [1-3].

After a long-term service on severe operating conditions such as high temperature and pressure, microstructure of F12 steel will be changed and in turn affect its performances. According to the relevant literatures [4-7], the microstructure and the properties of F12 steel will degrade significantly after its service. Specifically, the creep resistance will decrease significantly when the operating temperature exceeds 550 °C. As for the weakest link parts, the welded joints always exhibit coarse columnar grain structure due to the difficulty to achieve process conditions of the matrix when welding the steel. Moreover, the structure and properties of the HAZ will be changed obviously after one heating cycle. However, the details of the changing processes need to be investigated and whether the welded steel still meet its application requirements after these changes need to be considered. The investigated F12 steel welded joint was collected from the main steam pipe of #2 unit in a specific power plant after servicing 165000 hours at 540 °C. High temperature creep rupture testing together with SEM analysis technology was employed to investigate the high temperature creep strength of F12 welded joint after long-term high temperature operating. The detailed creeping structure damage and the performance degradation were analyzed and the service life predicition of F12 steel mian steam pipe was conducted in order to guarantee the safety of the unit.

2.Materials and testing methods

The size of the investigated F12 steel is φ355.6mm × 40mm. SPECTROLAB quantitative spectroscopy system was employed to analyze chemical components of F12 steel after servicing. The compositions (mass ratio, %) are C 0.19, Mn 0.59, Si 0.25, Cr 12.24, Mo 1.05, V 0.22, Ni 0.67, S 0.019 and P 0.014.

F12 steel and welded joint samples were prepared according to the standard procedures [8]. High-temperature endurance tests of the samples were conducted by using RC-1130 high-temperature creep rupture testing machine, under the testing temperature of 550 °C and the pressures of 220Mpa, 200 Mpa, 180 Mpa, 160 Mpa and 150 Mpa, and the testing temperature of 600 °C and the pressures of 160Mpa, 140 Mpa, 120 Mpa. Field emission scanning electron microscope SUPRA55 combined with an EDS system were both employed to analyze the microstructure and the main alloying elements of samples.

3.Results and discussion

3.1 Creep rupture strength testing

Figure 1 shows the creep rupture strengths of F12 steel and its welded joint. There is a good linear relationship between creep rupture time and pressure. Results in Figure 1 indicate that the creep rupture strengths of F12 steel and its welded joint decrease with increasing the temperature. Moreover, the creep rupture strengths of the welded joint are

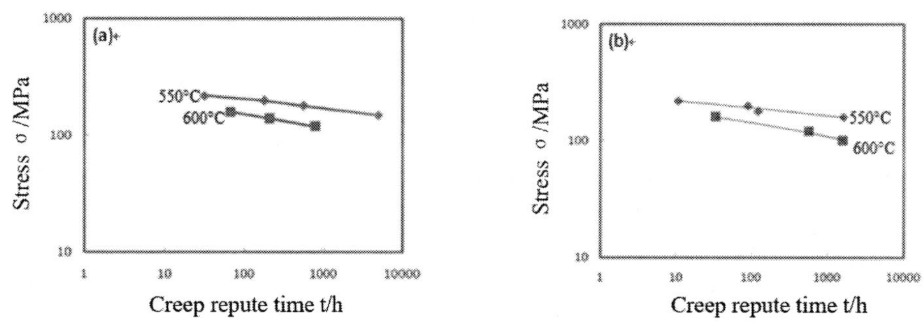

(a)F12 steel; (b) welded joint

Fig.1 Creep rupture strengths of F12 steel and its welded joint after servicing

lower than those of the matrix at both 550°C and 600°C testing temperature.

According to the Isothermal method, the creep rupture strengths of F12 steel and its welded joint were numerically analyzed as follows: The creep rupture strengths of F12 steel after servicing at 550°C and 600°C can be obtained according to equations (1) and (2), respectively. The R^2 are 0.989 and 0.999, respectively, indicating that excellent fittings can be obtained between the experimental and numerical data.

$$550℃\ \ \sigma=290.9\tau^{-0.07} \quad (1)$$

$$600℃\ \ \sigma=260.2\tau^{-0.11} \quad (2)$$

The creep rupture strengths of F12 steel welded joint after servicing at 550℃ and 600℃ can be obtained according to equations (3) and (4), respectively with R2 of 0.945
and 0.983, respectively, also indicating that excellent fittings can be obtained.

$$550°C\ \ \sigma=262.2\tau^{-0.07} \quad (3)$$

$$600°C\ \ \sigma=244.4\tau^{-0.11} \quad (4)$$

According to DIN17175 the creep rupture strengths of F12 servicing for 10^5 hours at 550°C and 600°C should be higher than 128Mpa and 59MPa, respectively. In this study, the longest creep rupture testing time of F12 steel at 550°C is approximately 5000 hours, and the testing pressure is 150 MPa, which is higher than the required value in the standard. According to equation (1), when τ = 50000h, *σ = 136MPa*, which is also higher than 128 MPa, indicating that F12 steel can keep on working for

50000 hours at 550 °C. When the working temperature is 600 °C, the longest creep rupture testing time is approximately 780 hours and the testing pressure is 120MPa. According to equation (2), when $\tau = 7800h$, $\sigma = 97MPa$, which is also higher than the required value of 59 MPa, indicating that F12 steel can keep on working for 7800 hours at 600 °C. Similarly, according to equation (3) and (4) we can observe that F12 steel welded joint can keep on working for 17000 hours and 16000 hours at 550 °C and 600 °C. However, when the working temperature is 550 °C, the strength of the welded joint is close to the limited value according to the standard.

3.2 Morphological observation

Different levels of necking occurred around the fracture. The fragments showed plastic deformation rather than entire creep deformation due to the relatively short testing time.Figure 2 shows the SEM microscopic photographs of the F12 welded joint creep fractures. The working temperature, the pressure and the working time of the analyzed samples were 550°C, 160MPa and 1621.3h, respectively. As can be seen in Figure 2, vast creep pores exist around the creep fracture. Moreover, extensive precipitates appeared in the tissure and distributed along the force direction.

3.3 Joint microstructure of F12 welded joint after servicing

As can be seen in Figure 3, F12 steel welded joint was still lath martensite structure after 165000hs' servicing, but these martensite already decomposed into fragments, leading to the increase of the amount and size of the precipitated phases. The XRD analysis results were shown in Figure 4, showing that both F12 steel and the welded joint consist of Fe-Cr solid solution and $M_{23}C_6$ phase. According to the chemical compositions of F12 steel, it is also easily to know that VC phase should be existed in the matrix, but no VC phase was detected in the XRD result probably due to the lack of the VC phase.

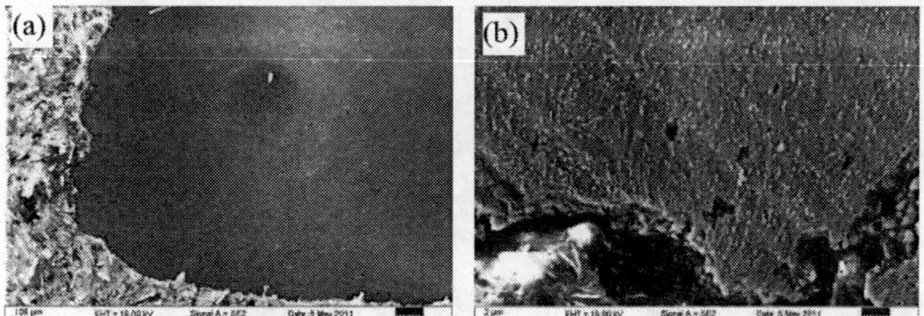

(a) Microtopography of the creep fissure; (b) Creep pores around the creep fracture
Fig.2 Microtopography of the creep fissure of F12 welded joint after servicing

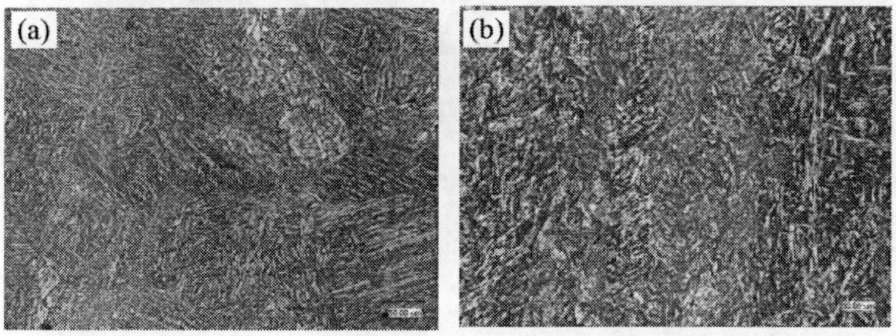

(a) F12 steel; (b) welded joint
Fig.3 Microtopography of of F12 steel and the welded joint after servicing

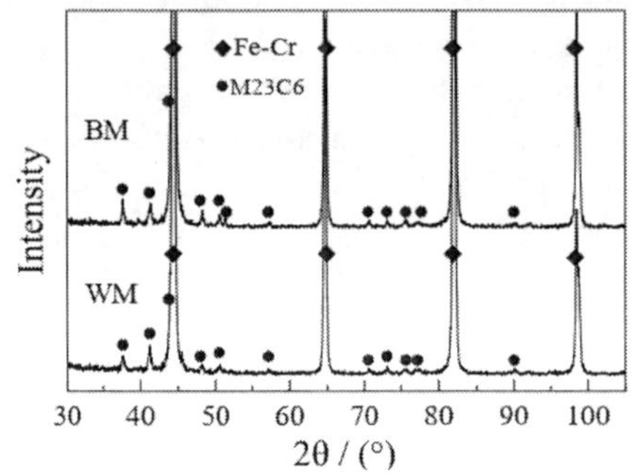

Fig.4 XRD result of F12 steel

3.4 High temperature creep rupture property analysis

Figure 5 shows the micromorphology of F12 welded joint, Table 1 shows the EDS analysis results of the precipitation phase. Figure 5 (a) and (b) are the secondary electron image and backscattered electron images of the same region, respectively, and the results show that the grain boundary precipitates grow up together and distributed as separate particles. The EDS results show that the main alloying elements are C , Cr, Fe, Mo, V. Figure 5 together with Figure 4 show that the main precipitation phase is $M_{23}C_6$. According to the backscattered electron images, trace level of Laves phase exists in the precipitation phases [12]. The contrast in the back-scattered electron image appears light due to the particular structure of Laves phase.

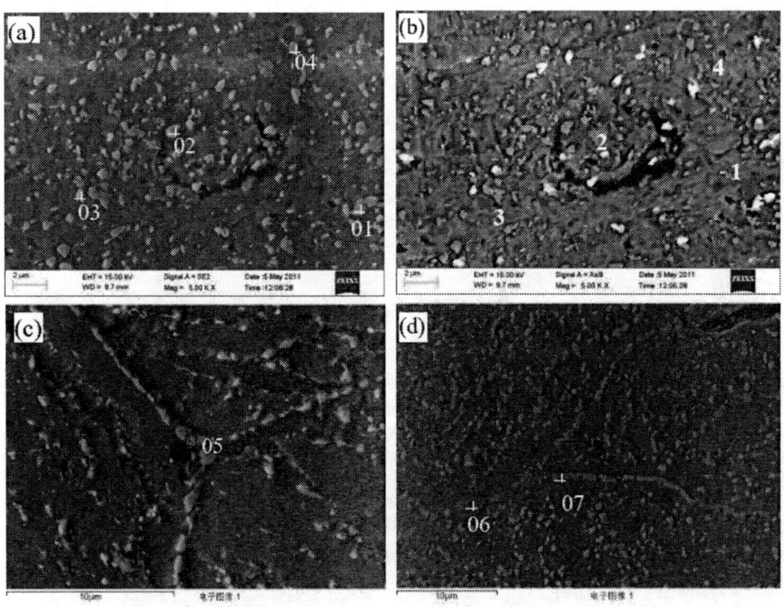

(a) and (c) secondary electron image; (b) and (d) backscattered electron image
Figure 5 Creep organizations and spectroscopy of F12 welded joints after running

Table 1 EDS analysis of microstructure of creep F12 welded joint

NO.	Contents of the main elements, wt%								
	C	Si	P/S	Cr	Mn	Fe	Ni	V	Mo
01	0.85	1.78	0.36/	13.20	0.76	63.95	1.36	/	17.73
02	1.03	/	/	26.67	/	68.33	0.69	0.58	2.71

03	0.78	0.34	/0.33	11.13	0.75	85.68	0.99	/	/
04	1.36	/	/	24.62	/	69.79	0.82	0.60	2.80
05	0.76	/	/	41.47	51.17	/	0.85	0.60	5.16
06	2.43	/	/	50.67	/	39.27	/	0.95	6.69
07	1.18	/	/	21.93	/	73.73	0.85	/	2.31

As can be seen in Figure 6, the amount and size of the precipitates of F12 welded joints both increased apparently, especially in the grain boundary and branch grain boundary, the $M_{23}C_6$ carbides nodulized, became rough and distributed along the grain boundary. The size of $M_{23}C_6$ carbides is large, leading to a low combination force with the matrix. When creep deformation occurred, pores were firstly formed in the boundary (as shown in Figure 6 (b)). Moreover, $M_{23}C_6$ with large grain size is also a main place to form pores (as shown in Figure 6 (a)).

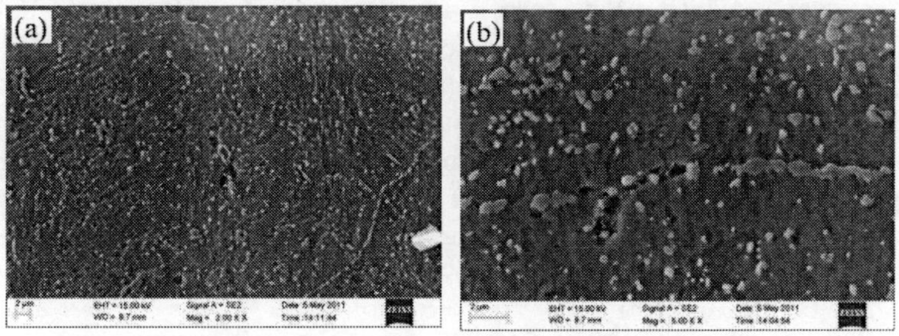

(a)×2000; (b)×5000

Figure 6 Creep deformation organization analysis of F12 welded joint

A certain number of strip-like phases with large grain size were observed in the matrix of F12 steel HAZ and showed directional distribution. The bar-like phases should be Al and Si inclusions. The binding force between bar-like inclusions and the matrix is relatively weak, so the inclusions are likely to detach from the substrate, leading to the formation of large pores and cracks. Together these pores and cracks with the precipitates become the weak point of the organization and exist as the precursor of creep deformation.

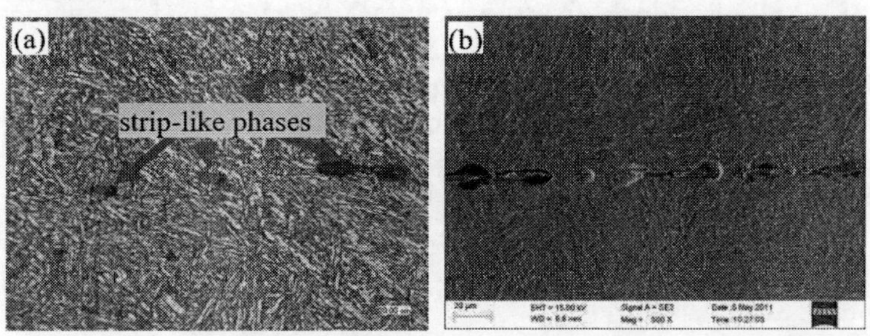

(a) metallographic structure; (b) SEM morphology

Figure 7 Morphology of the bar-like phase in F12 steel organization

4. Analysis and Discussion

When F12 steel welded joints which already served for 165000h were imposed a second high temperature creep rupture testing, their textures consist of martensite matrix and precipitate phase, and the lath structure of the martensite matrix remained clear, but the number of precipitates increased significantly and the precipitation strengthening effect increased. Due to the relatively low absolute values of molar formation of free energy of carbides (Cr, W, Mo) and the stability of carbides, $M_{23}C_6$ was easily to occur Ostwald ripening and prone to grow under the conditions of high temperature and creep rupture. Moreover, $M_{23}C_6$ precipitates out firstly near the grain boundary due to its low nucleation energy. With the continuous precipitation and growth of $M_{23}C_6$ during the operation, some precipitations with fine particle size were generated and distributed in the grains of the matrix. Due to the segregation of some elements such as Cr, W, Mo and there high defect concentration near the dendritic of primary dendrite, the quantity of $M_{23}C_6$ is large than the matrix and $M_{23}C_6$ prone to precipitate and grow rapidly. Together $M_{23}C_6$ phase, Laves phase with the bar-like inclusions (i.e. Al, Si phases) contribute to the generation of pores due to their poor plasticity and adhesion with the matrix at high temperature and with some applied stress. These formed pores serves as precursors of cracks. The above results indicate that some detections and supervisions on the texture changes are necessary and the formation of pores and cracks in the structure need to be monitored regularly.

5. Conclusions

(1) The dominate precipitations in the creep rupture structure of F12 welded joints are $M_{23}C_6$ and Laves phase. The particle sizes of the precipitations are large and the adhesions between the precipitations and the matrix are low, contributing to the formation of cracks. Some Al, Si inclusions were also observed and detached during the creep rupture process, contributing to the generation of large pores and cracks. These indicate that regular detections and supervisions on the texture changes are necessary and the formation of pores and cracks in the structure also need to be monitored.

(2) By inference, F12 steel welded joint can keep on working for 17000 hours. However, when the working temperature is high, the structures and properties of the materials changed apparently, especially considering that the creep rupture strength of the welded joint is lower than that of the matrix. It indicates that thorough life evaluations need to be conducted in the next maintenance.

References

[1]Bendick W., Harrmann K., Welnitz G., Zschau M. (1993)Properties of 9% to 12% chromium steels and their behaviour under creep conditions. VGB Kraft, 73(1):73-79.
[2] DIN17175, (1979) Heat-resisting seamless steel tube.
[3] Hu Z.F., Yang Z.G. (2003) Development and application of high chromiun heat-resiatant steel. Journal of iron andsteel research, 15(3): 60-64.
[4] Yang Z.G., Zhang J., Hu Z.F. (2006) Study on performance degradation of main steam pipe of X20CrMoV12.1 steel for long-term serive.Electric power, 38(2): 22-25.
[5] Hu Z.F., Yang Z.G. (2003) Embrittlement of X20CrMoV12.1 heat-resistant alloy steel for long term exposure at elevated temperature. Materials transactions, 12:14-17.
[6] Hu Z.F., Yang Z.G. (2007) Residual life assessment to X20CrMoV12.1 straight main steam pipe in service. Boilertechnology, 38(3): 14-18.
[7] DL/T 868, (2004)The code of welding procedure qualification..
[8] GB/T 2649, (1989) Methods of sampling for mechanical properties tests of welded joint.
[9] GB/T229, (2007) Metallic materials-Charpy notch impact test.
[10] GB/T2650, (2008) Methods of impact test for welded joint.
[11] GB 2039, (2012) Metallic materials-Creep and stress-repture test in tension.
[12] Yang Y.K., Wu T.L., Jiang X.M. (1986) High temperature strength of metal and test. Science & technology press, Shanghai.

Stress Rupture Properties of Welded Joints of Austenitic Heat-resistant Super304H Steel at 650 °C

XinmeiLI[1], ZhongwenZHANG[1], BaoshuaiDU[1], ShuaiSuo[2], WenLI[2]

[1]State Grid Shandong Electric Power Research Institute, Jinan 250002, China;

[2] Shandong Power Industry Boiler & Pressure Vessel Inspection Center Co. Lid., Jinan 25002, China

* Corresponding author, Email: zzwh1013@163.com

Abstract: With the help of metalloscope, scanning electron microscope and its energy dispersive spectroscopy, this paper investigates the stress rupture properties at high temperature and microstructure changes of the welded joints of austenitic heat-resistant Super304H steel at 650 °C as well as the effects of microstructure changes on stress rupture properties by high temperature creep rupture testing. The results show that at 650 °C, the welded joint of Super304H steel has high breaking strength at elevated temperature, and the fracture position is on the base material away from the fusion line, and the fracture type is mainly a transcrystalline rupture. The microscopic process of creep rupture of Super304H steel welded joints is the nucleation and growth of the void and crack, while the large bulk Nb(C,N) is the main nucleation site of void. As the careep time increases, M23C6 on the grain boundary is also the main nucleation site of the void.

1. Introduction

Super304H steel tube is mainly used in heating surface tubes of ultra supercritical steam generators. It is based on TP304H steel by adding about 3% Cu, 0.45% Nb and a certain amount of N. This allows the steel to pass through the precipitates during its service, including Nb (C, N), Cu-rich phase and NbCrN, endowing the steel with high temperature stress strength [1-6]. Its characteristics are grain refinement and precipitation hardening [7]. However, in welding conditions, these two characteristics in the welded joint will have some changes. Under high temperature, the difference in original tissue state will lead to different changes in tissue and performance [8-10], so it is particularly significant to study the creep rupture strength of welded joints of Super304H steel. In this paper, the microstructure changes of welded joints at 650 °C are studied by high temperature stress rupture testing of Super304H steel. Moreover, this paper explains the reasons for the change of stress rupture properties of Super304H steel welded joints, and provides reference for the technical supervision of Super304H steel parts in ultra supercritical steam generators.

2. Test materials and methods

The Super304H steel pipe for testing is supplied by Japan Sumitomo Corporation and its specification is Φ45 mm×9 mm. The welding wire is a matching argon arc welding wire #T-304H supplied by the Japanese company, and the specification is Φ2.4mm. The chemical composition of steel pipes and wires measured by SPECTROLAB quantitative spectrometer can be seen in Table 1.

Table 1 Chemical composition of Super304H steel pipe and welding wire (wt,%)

Material	C	Si	Mn	P	S	Cr	Ni	Mo	Cu	Nb
Super304H steel	0.10	0.22	0.85	0.033	0.006	18.4	8.56	0.26	2.41	0.48
# T-304H	0.09	0.22	3.44	0.013	0.002	18.1	15.8	0.76	3.06	0.65

Super304H steel pipe butt joints use V-shaped groove with an angle of 60° and the blunt edge of 1mm, and the root gap is 2mm-2.5mm. It uses a gas tungeten arc welding (GTAW), placed in a 45° fixed position for multi-layer block welding. The welding current is 70A-100A, the interlayer temperature is controlled below 150 °C, and the root is filled with argon protection.

The stress rupture specimens at elevated temperature of the Super304H steel joint are processed according to the standard [11], and the high temperature stress rupture testing is carried out on the RC-1130 stress rupture strength testing machine. The temperature of the test is set to 650 ° C, and the stresses are respectively 280 MPa, 260 MPa, and 240 MPa. The metallographic specimen preparation is etched with aqueous solution of ferric chloride hydrochloric acid solution, while the metallographic tissue is observed under an Olympus Model BX51M optical microscope. The creep rupture fracture and the microstructure are observed by a JSM-6380LA scanning electron microscope, and the main alloying elements of the precipitated phase in the tissue are tested by an energy dispersive spectroscopy.

3. Test results and analysis

3.1 Data analysis of stress rupture testing

The welded joint of Super304H steel is subjected to high temperature stress rupture test, and the specimens are all broken on the base metal away from the welded seam. At this point the temperature strength of the welded joint is still high. With the help of the Method of Isotherm, the creep rupture strength of the Super304H steel joint is extrapolated [12]. By creep rupture and fracture equation [13],

$$\tau=A\sigma^{-B} \tag{1}$$

Where σ is the stress level (MPa) of the specimen loading; τ is the fracture time of the specimen (h); A, B are the constants related to the test temperature and material. The logarithm function is used on both sides of equation (1), we get

$$lg\tau=lgA-Blg\sigma \tag{2}$$

It can be seen that $\lg\sigma$ has a linear relationship with $\lg\tau$, so linear regression can be performed by the least squares method. Test parameters A and B in the equation can be obtained by experiments with different stress levels at the same temperature.

Linear regression is performed on the experimental data and the results are as follows.

$$\sigma=426.49\tau^{-0.0714} \tag{3}$$

The correlation coefficient R^2 is 0.7506 at 650 ° C, indicating that the data has a good fitting effect. The stress rupture strength data of Super304H steel welded joints is plotted as Figure 1. It can be seen that the double logarithmic curve of creep rupture time and stress share a positive linear relationship.As a rule of thumb, it is reliable to extrapolate an order of magnitude for stress rupture strength data to derive data. When $\tau = 10000$, the result is Super304H steel $\sigma = 220$ MPa.

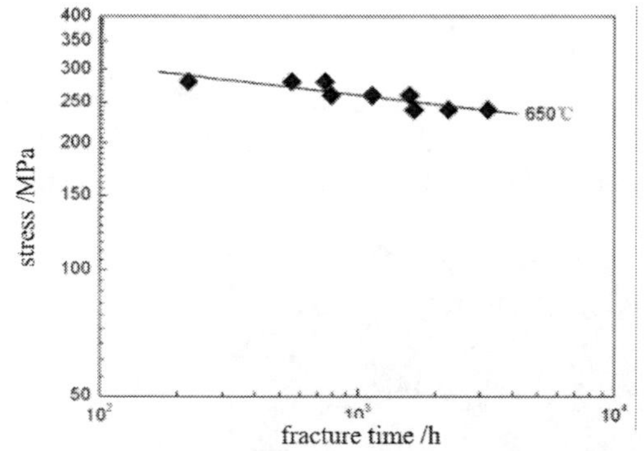

Figure 1 Stress rupture strength curve of Super304H steel joint

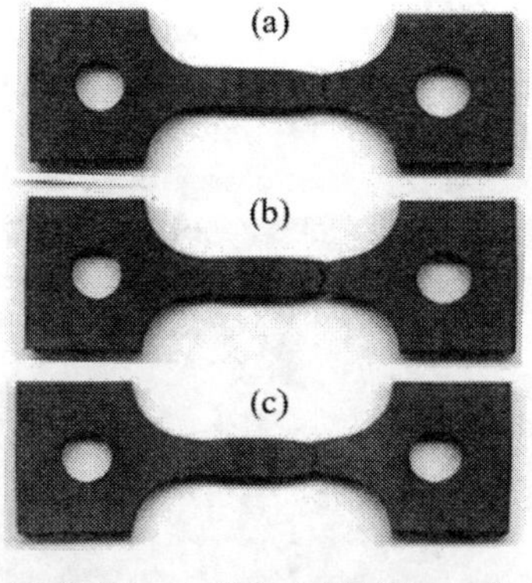

（a）240MPa；（b）260MPa；（c）280MPa
Figure 2 Photo of stress specimens of Super304H steel joint

3.2 Creep fractography analysis

The stress specimens of the Super304H steel welded joints are all fractured on the base metal away from the fusion line. A macro photograph of parts of the specimens after fracture is shown in Fig. 2. It can be seen that different degrees of necking occurs at the specimen fracture. Since the test time is relatively short, the specimens are not completely creeped before the fracture, but a certain plastic deformation is generated.

Figure 3 shows the characteristics of the creep fracture of the welded joint of Super304H steel. Under high stress, the fracture of the specimen is mainly plastic deformation. As the stress is different, and the degree of plastic deformation will also be different. Under the creep condition of 650 °C / 240MPa (Figure 3(a)), the creep fracture is mainly in the form of transcrystalline rupture. There is a circular cleavage plane in the grains, and there are a large number of tear ridges between the grains. This indicates that the material is a ductile transcrystalline rupture, but a small number of short secondary cracks are observed. Under the creep condition of 650 °C / 280 MPa (Figure 3 (c)), the fracture morphology is also mainly the transcrystalline rupture. In the meanwhile, the tear ridges are significantly increased, and a small amount of secondary crack also occurs.

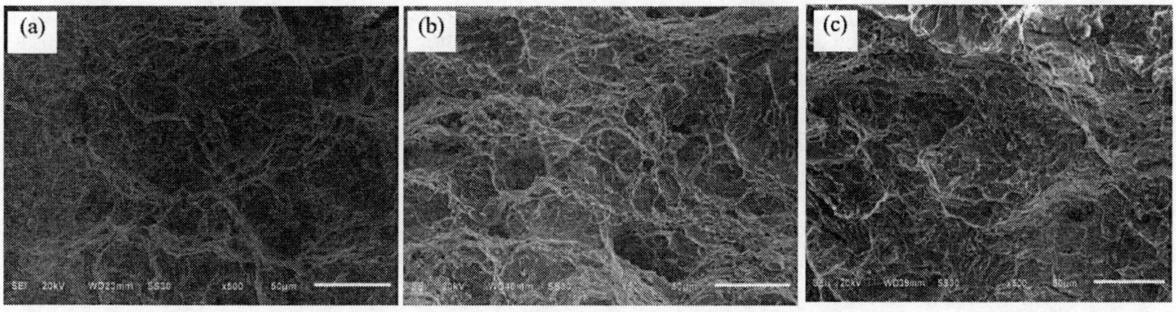

（a）650°C/240MPa/2253.4h；（b）650°C/260MPa/1590.6h；（c）650°C/280MPa/555.6h；
Figure 3 SEM morphology of the stress fracture

3.3 Tissue analysis of creep specimens

At 650 °C, the morphology and microstructure of the Super304H steel joint creep fracture are shown in Figure 4. The grain of the fracture specimen undergoes significant plastic deformation, and it is severely elongated (Figure 4(a) and (b)). It is identified as a transcrystalline, and precipitates along the grain can be observed at the boundary. More importantly, however, cracks will preferentially occur on large bulk precipitates. As described in the literature [5][10], the presence of the bulk Nb(C,N) in the base metal will adversely affect the creep rupture strength of the material. Moreover, the more continuous and larger the precipitated phase on the grain boundary, the more unfavorable the creep rupture strength. The void in the photo reflects the decrease in the binding force of the precipitated phase to the matrix during the elongation of the grain boundary. During the specimen preparation, the precipitated phase is peeled off, and the deformation becomes larger. Lower bonding force is further developed to connect the voids to form cracks.

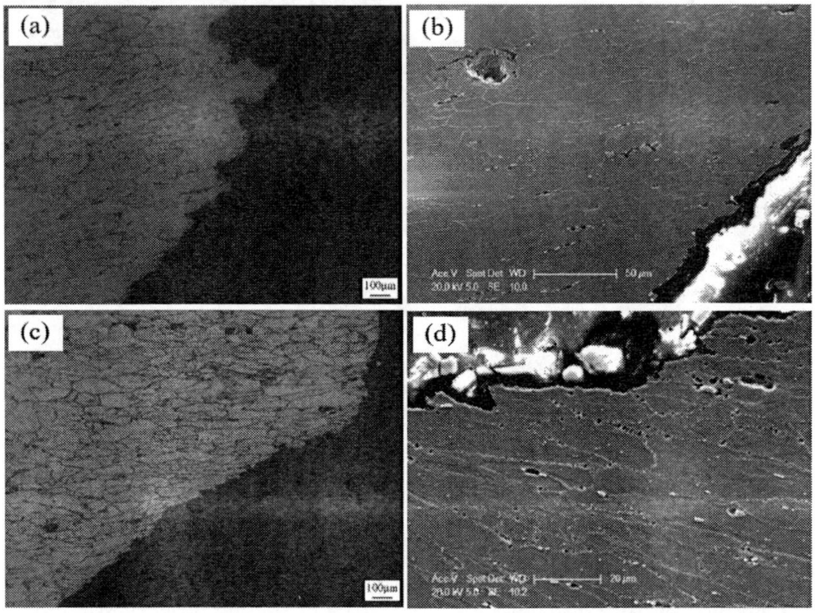

(a) and (b) 650°C/240MPa/2253.4h; (c) and (d) 650°C/260MPa/1138.7h;
Figure 4 Scanned photo of the creep fracture specimens at 650 ° C

The precipitated phase in the creep specimen mainly has large-sized bulks and fine particles. The bulk precipitates are mainly distributed near the grain boundaries, and the results of the spectroscopy analysis (Figure 5 and Table 2) indicate that the main alloying elements are Nb. The number of fine particle precipitates is significantly increased compared to that in the supply state. Especially at the grain boundary, the longer the creep time, the more the number of precipitates on the grain boundary. The precipitation phase on the grain boundary is almost continuously distributed after creeping at 650 ° C/240 MPa for 2253.4 h. Spectroscopy analysis shows that the precipitates on the grain boundaries are mainly Cr and Fe compounds, and some of the grain boundary precipitates contain a certain amount of Nb.

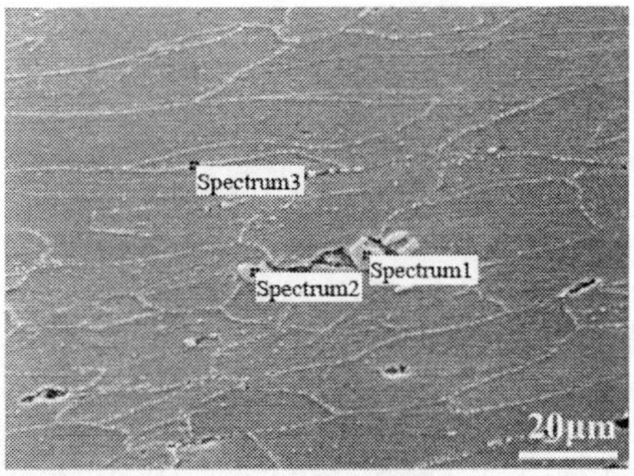

Figure 5 Spectroscopy analysis of the precipitates at about 50μm from the fracture of the creep specimen

Table 2 Spectroscopy analysis results (by weight, %)

Measuring point	Main alloying element content					
	Cr	Mn	Fe	Ni	Cu	Nb
Spectrum 1	2.26	-	2.59	-	-	95.15
Spectrum 2	3.21	-	5.00	-	-	91.79
Spectrum 3	22.43	1.34	67.04	7.11	2.09	-

4. Analysis and discussion

Under high stress, the material will generate a large amount of dislocation during the deformation. At high temperatures, the dislocations are affected by the interaction of stress and thermal activation. The creep behavior of Super304H steel belongs to the second type of creep behavior. The creep rate is controlled by the climb, and the self-diffusion energy of the matrix is similar to the creep activation energy [14][15]. Therefore, as the temperature goes up, the self-diffusion coefficient increases, causing the creep rate to increase significantly. Under high stress, the grain boundary sliding accounts for a small proportion in the creep deformation. The grain boundary is elongated along with the grain and exhibits the feature of transcrystalline upon fracture. As the stress decreases, the proportion of grain boundary sliding deformation increases in creep deformation. To a certain extent, a void is formed at the grain boundary to develop a crack, which gradually transforms into an intercrystalline feature. As is known, the slidable interface of the precipitated phase and the matrix transmits only normal stress and cannot transmit tangential stress, and the plasticity of the precipitated phase is relatively poor. When there is a precipitated phase large in size in the grain boundary and the grain, cracking occurs to then become a crack. Therefore, it can be observed in the tissue that the larger-sized precipitate phase first produces crack.

Under the test circumstance, the creep fracture of the welded joint of Super304H steel is located on the base metal away from the fusion line, and the fracture morphology is mainly transcrystalline rupture. Besides, a certain degree of necking of the specimen occurs at the fracture site. Analysis of the microstructure of the specimen reveals that after high temperature creep, the matrix structure is still austenite and precipitate. The results of spectroscopy shows that the bulk precipitates are mainly Nb compounds. The precipitates on the grain boundary are transformed from particles or strips to net, chain and isolated particles. In addition, the precipitates on the grain boundary are mainly Cr and Fe compounds, and some precipitates are analyzed to contain element Nb. The second phase precipitate plays an important role in the void nucleation. Observing the characteristics of the creep fracture, it can be found that the creep damage of Super304H steel mainly appears as the formation of voids at the second phase precipitate, especially at the large Nb (C, N) . In addition, the coarse precipitates on the grain boundary are also the main position of the void nucleation.

5. Conclusion

(1) Under high stress, the creep fracture of the welded joint of Super304H steel is located on the base

IOP Conf. Series: Materials Science and Engineering **711** (2020) 012043 doi:10.1088/1757-899X/711/1/012043

metal away from the fusion line, and the fracture morphology is mainly transcrystalline rupture.

(2)The microscopic process of creep rupture of Super304H steel welded joints is the nucleation and growth of the void and crack, while the large bulk Nb(C,N) is the main nucleation site of void._As the creep time increases, $M_{23}C_6$ on the grain boundary is also the main nucleation site of the void.

Reference

[1] Zhou Q.W., Ping S.B., Meng X.B., Wang R.K., Gao Y. (2017) Precipitation kinetics of M23C6 carbides in the Super304H austenitic heat-resistant steel. Journal of materials engineering and performance, 26(12): 6130-6139.

[2] Qi Y.H., Wu Z.Q., Zhang X.M., Ma C.Y. (2018) Microstructure and phases of deposited metal of SUPER304H steel under high temperature Persistent stress. Scientific reports, 8(1):2618.

[3] Kan T, Sawaragi Y, Yamadera Y. (1998) Properties and Experiences of a New Austenitic Stainless Steel Super 304H (0.1C-18Cr-9Ni-3Cu-Nb-N) Tubes for Boiler Tube Application. Materials for advanced power engineering, 441-450.

[4] Ou P., Xing H., Sun J. (2015) Precipitation of Nanosized MX at Coherent Cu-Rich Phases in Super304H Austenitic Steel. Metallurgical and Materials Transactions, 46(1):1-5.

[5] Du B.S., Zhang Z.W., Liu Q.L., Li X.M., Deng H.L. (2017) Microstructure and Properties of Super304H Steel Used in Ultra Supercritical Unit after Long-term Service. Hot Working Technology,46(24):147-150.

[6] Igarashi M. (2004) Creep Properties of Heat Resistant Steels and Superalloys. Springer Berlin Heidelberg, Springer:

[7] Li X.M.，Zou Y.，Zhang Z.W.，Zou Z.D. (2010) Microstructure and property of Super304H austenite stainless steel. Materials science＆technology, 18(2): 281-286.

[8] Zhang Z.W., Li X.M., Du B.S., Zou Y.，Wei Y.Z. (2011) Effect of aging temperature on microstructure and toughness of welded metal of Super 304H steel. Transactions of the China welding institution, 32(7):43-46.

[9] Zhang Z.W., Li X.M., Du B.S., Li Z.M., Li W. (2018) Creep Ruputre Properties and Aging Microstructure of Super304H Welding Joint, Materials for mechanical engineering,42(11): 62-66.

[10] Peng X.N., Ma C.Y., Yin Q., Peng Y., Tian Z.L. (2015) Microstructural evolution of Super304H weld steel during stress rupture test at elevated temperature. Transactions of materials and heat treatment, 36(7): 133-138.

[11] GB/T 2039. (2012) Metallic materials—Creep and stress-rupture test in tension.

[12] Zhao J. (2011) Statistical analysis and reliability prediction of creep rupture property of heat resistant steel. Science press，Beijing.

[13] Mao W.M., Zhu J.C., Li J. (2008) The Structure and Properties of Metallic Materials. Tsinghua University Press , Beijing.

[14] Zhang J.S. (2007) Material deformation and fracture in high temperature. Science Press, Beijing.

[15] Li X.M., Zou Y., Zhang Z.W., Zou Z.D. (2009) Effect of aging temperature on precipitation of Super304H steel. Transactions of materials and heat treatment, 30(6): 51-56.

IOP Conf. Series: Materials Science and Engineering **711** (2020) 012044 doi:10.1088/1757-899X/711/1/012044

Design and Fabrication of a High Precision Constant Current Source Current Generator

XiulianLi

Technician department, ShanDong Labor Vocational and Technical College, Jinan, ShanDong Province, 250022, China

*Corresponding author's e-mail: lixiulian2005@163.com

Abstract: A high precision constant current generator with current sources is designed. It adopts precise operational amplifier circuits, high precision constant current sources and FPGA chips. It can generate sine waves, triangle waves and square waves (the duty cycle is adjustable) with adjustable frequency and high precision. The output current ranges from 0 to 4000mA, which can be set as required. It has digital display and strong anti-interference ability, which can provide the current with 0.05% precision. In this paper, the working principle and design idea of the circuit are analyzed. The whole program is compiled under the environment of Lab VIEW, and the assembly, welding and debugging of the circuit are completed.

1.Introduction

At present, considering economic benefits, most generators sold in the market are designed to use overload current simulation to test electrical products because most of these products are of high automation, large volume, heavy weight and high cost; or they directly use visual inspection method by increasing load, but the load is not controllable and adjustable during the inspection, which makes it hard to get the required current value quickly and accurately. In short, it is difficult to widely use this kind of current generator in the field. The current generator with constant current source introduced in this paper mainly has the advantages of high precision, low cost, small volume, light weight, easy to operate, and the equipment has a hanging hole, which can be hung at the nearby site through the hanging rope, etc. The operation of setting the required current is simple, and the output current value is displayed in real time by using the digital tube.

This current generator with constant current source can be widely used in a variety of occasions: the test and verification of standard power frequency current transformer, the calibration and setting of electrical equipment in civil building site, the field debugging and setting of electrical fire monitoring system, and the adjustment of development, production, testing and maintenance of various electrical equipment in workshops and laboratories. It is easy to carry and low in price, but it can accurately, safely and quickly acquirecurrent ranging in 0-4000mA, which has high practical value.

2.Design scheme

Through the analysis of its working principle, further analysis of the main parameter design, determine the design scheme.

2.1 Analysis of design principle

It adopts the design of precise operational amplifier circuit, high-precision constant current source, charging voltage of 7.5V and keyboard input of 4 × 4 matrix. The main control chip uses

FPGA-XC3S1200E to generate high-precision sine wave. The frequency and phase are adjusted by software programming and sent to the digital to analog converter (DAC). The data is converted by DA and filtered by first-order high-pass filter and second-order active low-pass filter. The sine wave amplitude is adjusted by voltage to obtain the required current valuewhich is displayed in real time by digital tube. The RS232 serial port of the computer completes the communication functions such as waveform signal type selection, output current range selection, etc. The functional block diagram is shown in Figure 1.

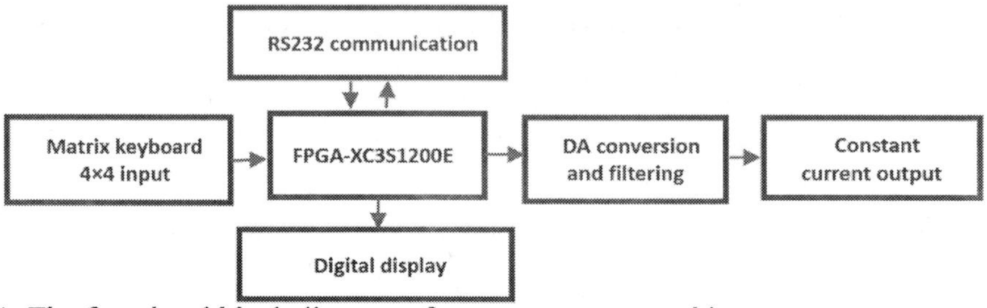

Figure 1. The functional block diagram of current generator with constant current source

2.2 Design of main parameters

The generator uses lithium battery pack as power supply, with charging voltage of 7.5V; it uses FPGA chip to generate frequency-adjustable and high-precision sine wave, triangle wave and square wave (duty cycle adjustable), with output current of 0-4000mA and step sizeof 0.01mA for setting;it can provide current of 0.05% accuracy, with single output time of 5-300 seconds and step sizeof1 second for setting; its output circuit impedance is less than 0.5Ω. See Table 1 for main indexes of design technology.

Table 1. main indexes of design

Name	paramete
Output current	0-4000mA, step size 0.01 mA can be set
Output accuracy	0.05%
Output loop impedance	<0.5Ω
Single output time	5-300 seconds, step size 1 second can be set
Charging voltage	DC ±7.5V

3.Circuit design

From the power supply circuit, keyboard input circuit, current shaping amplification, data DA conversion to digital display to describe the design ideas.

3.1Design of voltage stabilizing circuit for power supply

The product is portable, with large capacity lithium battery pack as power supply, which has long working life. The power supply is connected in parallel with Im7805 and Im7905 to obtain stable voltage, which can provide current ≤ 4A. The circuit diagram is shown in Figure 2.

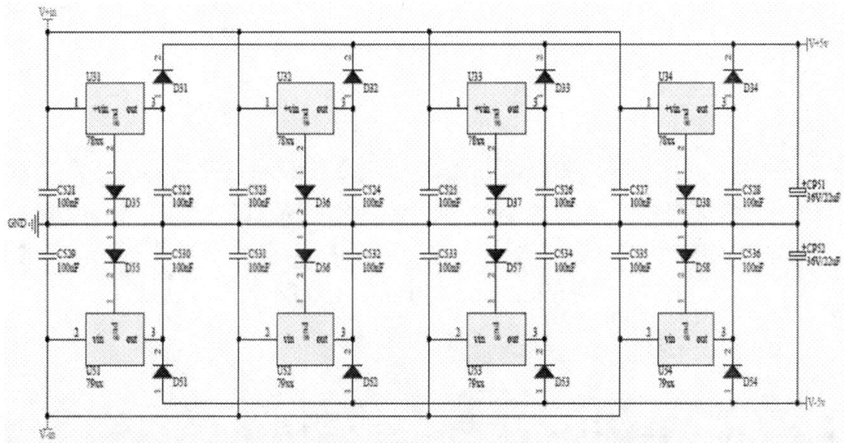

Figure 2 .The circuit diagram of voltage stabilizing circuit for power supply

3.2 Design of keyboard input circuit

The keyboard is input in 4 × 4 matrix mode, with simple operation. Press the key to input the current value and output duration required, which can be output after the confirmation by one key. The circuit diagram of the keyboard input is shown in Figure 3.

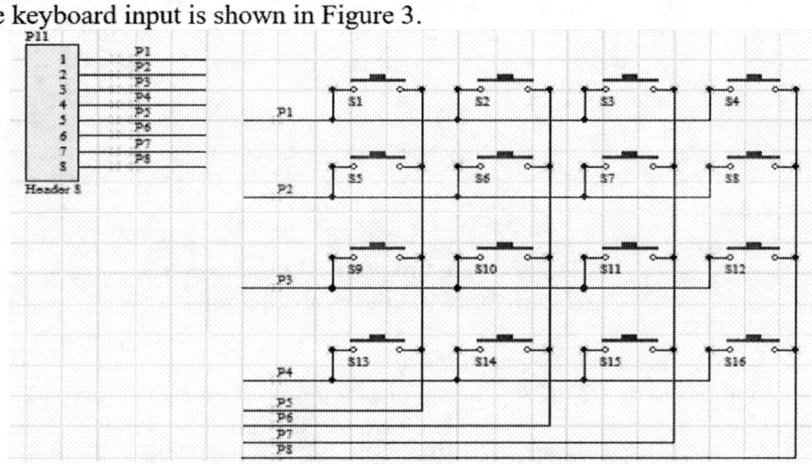

Figure 3. The circuit diagram of the keyboard input

3.3 Design of DA conversion circuit

There is serial communication between chip FPGA SPI and the communication interfaces of DA converter and chip LTC2624. The gain of each amplifier is set to 8-bit command word, which is composed of two 4-bit parts, and the highest B3 is sent first. The amplifieroutput AMP_DOUT responds to the original gain setting. When the DAC_CS of FPGA is set to low, the SPI bus starts to transmit data. The amplifier collects the output serial data AMP_DOUT at the falling edge of the clock signal SPI_SCK. The voltage output formula $V_{OUT} \times V_{REFERENCE}$is configured as 5V here, as shown in Figure 4.

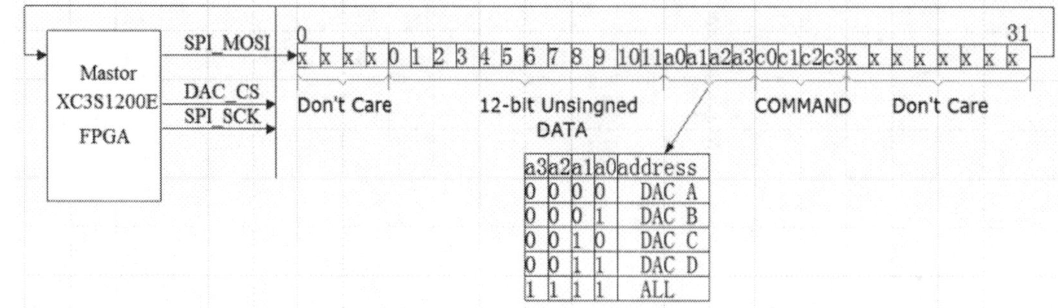

(a)SPI Communication Protocol to LTC2642 DAC

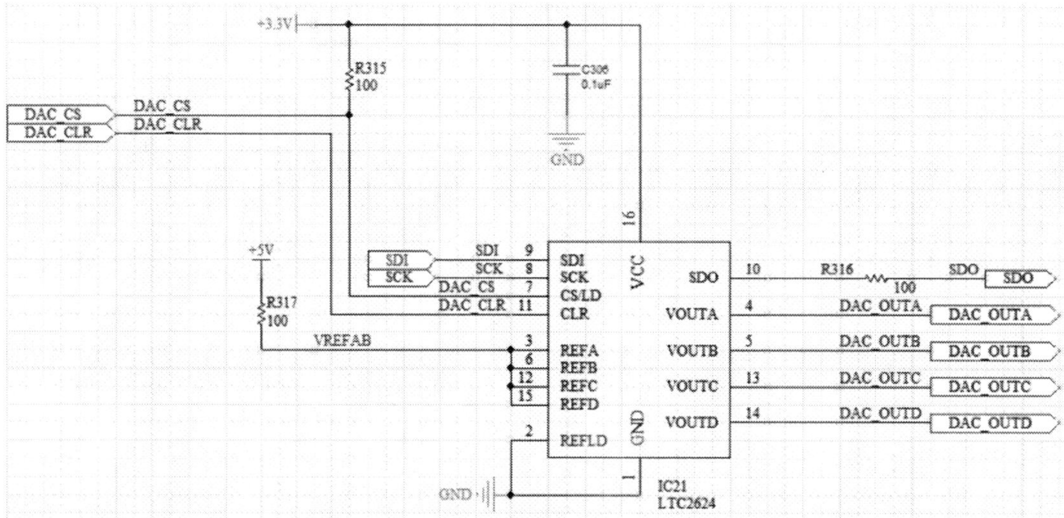

(b) Schematic diagram of data DA conversion

Figure 4. DA conversion circuit

3.4 The circuit of current amplification and filtering

The main control chip FPGA-XC3S1200E generates high-precision sine wave. The frequency and phase are adjusted by software programming, and sent to the digital to analog converter (DAC) of SPI interface. The data is converted by DA and filtered by first-order high-pass filter and second-order active low-pass filter. The signal generation mode is selected and the output current is adjusted to obtain the required current value. Combined with precise operational amplifier circuit and high-precision constant current source, the current required for output is guaranteed.

The large current power operational amplifier OPA548 is powered by 30V dual-power supply, and the gain is determined by negative feedback resistance. Taking R1 = 51K and R4 = 10K, the system can reach the accuracy index of 0.01, and can change the value of load voltage to make the change range of output current meet the requirements of $\leq$ Q1% + 1mA.

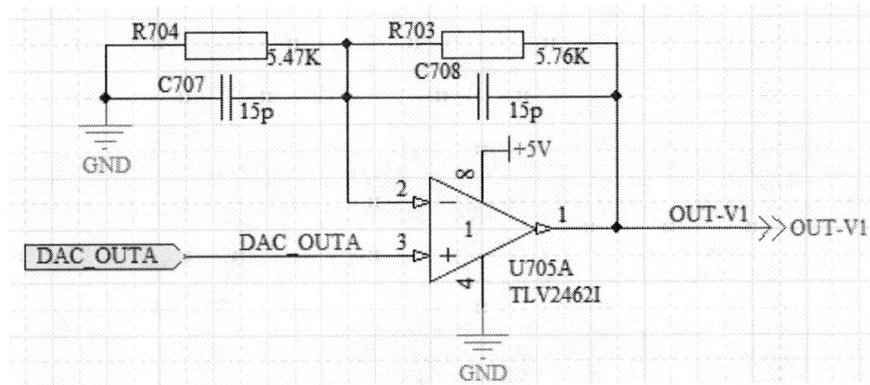

(a) Amplification and shaping

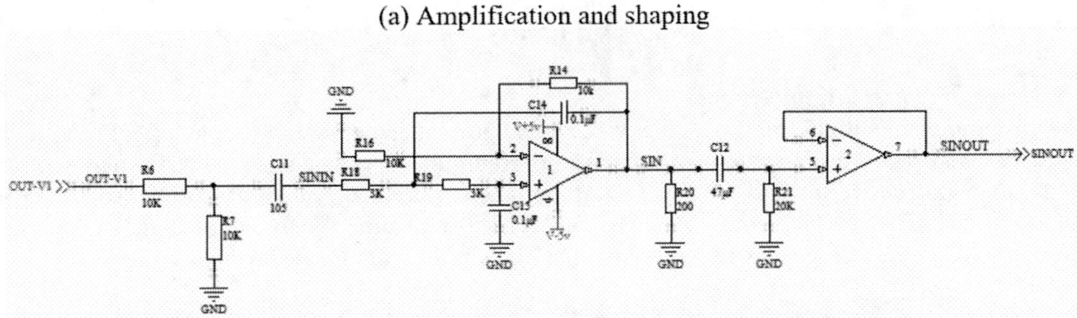

(b) Zero-crossing shaping and dual-power conversion circuit

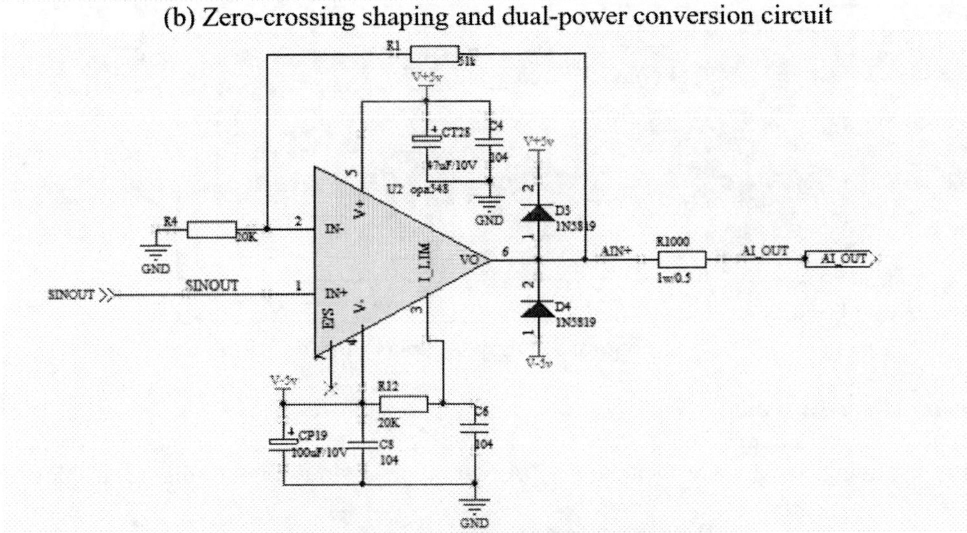

(c) Current amplification output circuit

Figure 5. The circuit of current amplification and filtering

3.5 Serial communication RS232

There is a serial port communication function of computer RS232, the circuit design of waveform signal form selection, output current range control, etc. of which is completed, as shown in Figure 6.

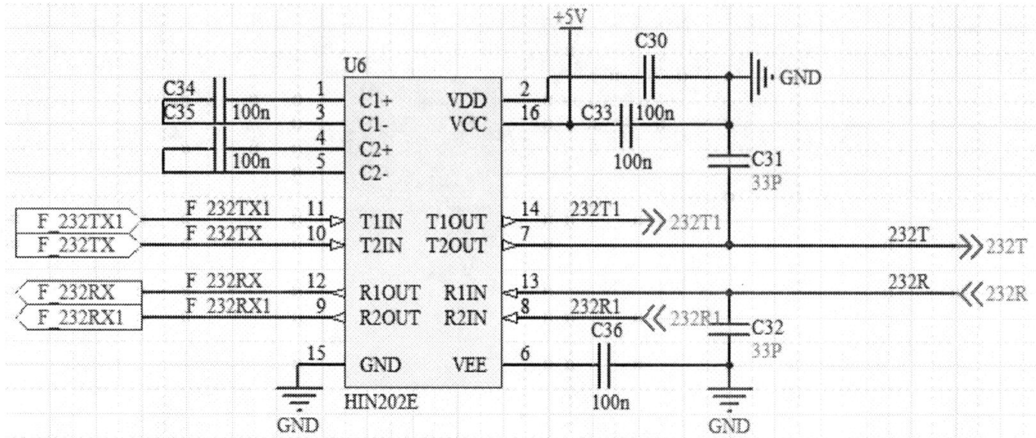

Figure 6. RS232 communication circuit

3.6 Design of digital display circuit
The programmable high-precision sine wave generator is generated by the chip FPGA, and the frequency and phase are adjusted by software programming. Here, it is set as 50Hz and sent to the digital to analog converter (DAC) of SPI interface. The required current value is obtained by adjusting the amplitude. The output current value is displayed in real time by five-digit digital tube, and the minimum display accuracy can be set to 0.01mA. The circuit diagram is shown in Figure 7.

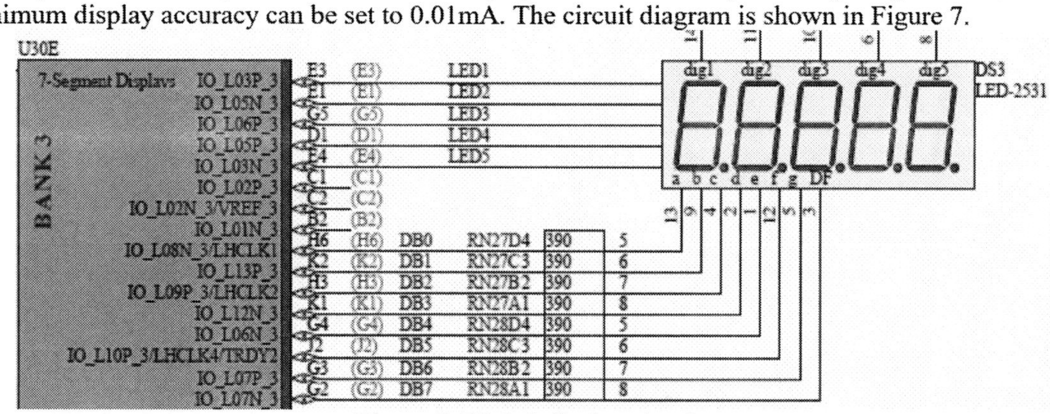

Figure 7. Digital display circuit

4.Program block diagram

The whole program is completed in the Lab VIEW environment. Figure 8 is part of the program block diagram, including sine wave generator, DA data conversion output, digital display and other programs.

IOP Conf. Series: Materials Science and Engineering **711** (2020) 012044 doi:10.1088/1757-899X/711/1/012044

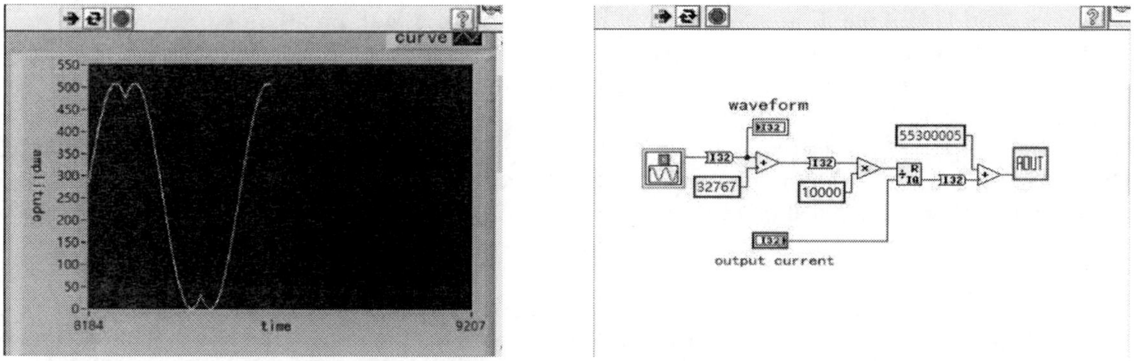

Figure 8. Partial program block diagram of FPGA of current generator with constant current source

5. Manufacture and debugging

After the completion of the circuit design, the simulation debugging is carried out. The circuit board is made according to the schematic diagram designed, and then the corresponding electronic components and chips are purchased according to the circuit diagram to assemble and weld the circuit board. The main control chip FPGA-XC3S1200E is used to complete the welding of circuit board with the peripheral circuits of voltage stabilizing circuit for power supply, keyboard input circuit, digital display circuit, filtering and current amplification circuit. After checking, download the program. After the joint debugging of software and hardware, carry out data test. The measurement diagram of 50Hz and 2000mA voltage signal output is shown in Figure 9. It can connect with computer through RS232 communication serial port, select signal generation mode and adjust output current, and download to flash for operation, or operate independently through keyboard, as shown in Figure 10.

Figure 9. The measurement diagram of 50Hz and 2000mA voltage signal output

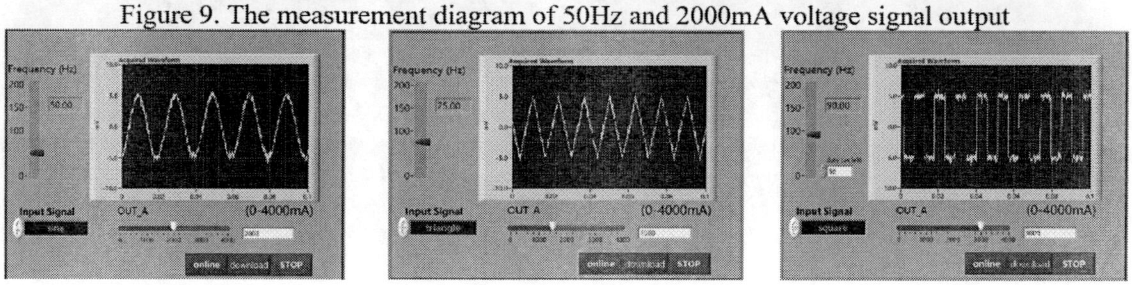

(a) Sine wave (b) Triangular wave (c) Square wave

Figure 10. Waveform signal selection and output diagram

IOP Conf. Series: Materials Science and Engineering **711** (2020) 012044 doi:10.1088/1757-899X/711/1/012044

6.Conclusion

This paper introduces the design process and manufacture of a high-precision current generator with constant current source. The whole program is completed in the Lab VIEW environment. The output current can be set according to the needs. With digital display, convenient operation, high stability, it can meet the requirements of a variety of constant current source circuits. Through the test, the feasibility of it has been verified. Now it has been mass produced and used in many occasions. The product design is a breakthrough and improvement of conventional products, with high cost performance and good application prospects.

Reference

[1] Li, Z.P., Wang, S.K., Zhang, N., Zhao, Q. (2017)Design and experiment of current source based on single chip microcomputer. Research and exploration in laboratory, 36 (1): 49-53.
[2] Shao, W.H., Gu, A., Sui, J.J. (2015)Design and implementation of STM32-based portable constant current source. Automation & Instrumentation, 30 (12): 29-32.
[3] Xie, Z.Y., Gong, Z.G., Yang, X., Wu, X.Y. (2013)The design of a high stable and digitally controlled alternating constant-current source. Electrical measurement & instrumentation, 50 (567): 102-106.
[4] Mi, W.W., Yang, F., Xu, L.L. (2012)Design and Manufacture of High Precision Constant Current Source. Electronic test, 2012 (12): 65-71.
[5] Miao, X.F. (2011)Digital Constant Current Source Based on the Adjustable Voltage Regulator. Electronic measurement technology, 34 (7): 13-15.
[6] Yang, Y.H., Yan, X.C., Guo H. (2009)The design of AC high precision constant current source. Electrical measurement & Instrumentation, 46 (10): 72-75.
[7] Jiang, H.Y.(2008) Research of a new type of constant current source with high-precision and high temperature stability. Integrated circuit and materials, 277 (14): 3-6.
[8] Tao, L.W., Wang, Y.M., Lei, K.Z. (2007)Precision numerical control constant current source based on microprocessor. Application of electronic technique, (7): 138-140.
[9] Zhai, Y.W., Ai X.Z., Yang, X. (2002)Design of practical constant current source circuit. Electronic measurement technology, 26 (5): 25-26.
[10] Xu, Y.X., Hou, G.L., Zhang, J.H. (1999)Design of high loaded constant current circuit. Modern electric power, 16 (8): 71-72.
[11] Zheng, Q.X. (2010)Xilinx Fpga Digital Circuit Design. Science Press, Beijing.
[12] Lin, J., Lin, Z.Y., Zheng, F.R.(2010)Lab VIEW Virtual Instrument Programming. People post and Telecommunications Publishing House, Beijing.
[13] Tian, G., Xu, W.B.(2008) Xilinx FPGA Development Practical Guid. Tsinghua University Press, Beijing.

IOP Conf. Series: Materials Science and Engineering **711** (2020) 012045 doi:10.1088/1757-899X/711/1/012045

Optimization design of gear reducer based on multi-attribute decision making and reliability sensitivity

CaoTong[1*], Zhuang Tian[2]

[1] School of Mechatronics Engineering, Shenyang Aerospace University, Shenyang, Liaoning, 110136, China

[2] School of Mechanical Engineering and Automation, Shenyang Institute of Technology, Fushun, Liaoning, 113122, China

[*]Corresponding author's e-mail:tongcao19@163.com

Abstract. In order to solve the complexity and implicitness of lightweight optimization of the gear reducer, this paper constructs a multi-objective reliability optimization mathematical model of the gear reducer. and to obtain multi-objective Pareto solution, artificial intelligence algorithm is used to optimize structural parameters. Then, through repeatedly changing the structural parameters of the reducer, the experimental simulation analysis is carried out to obtain the maximum working stress. Based on the idea of the weight reduction and miniaturization, the stress state and reliability index values under different structural parameters are compared and analyzed. In order to determine the optimal solution of structural parameters, multi-attribute decision theory is used to determine the solution. Finally, in order to effectively reveal the inherent variation of attribute values and decision results and quantitatively reflect the impact of various structural parameters on system reliability, reliability sensitivity analysis of the gear reducer is carried out by the response surface-Monte Carlo and DPS methods. Thereby, an optimized structure that meets the safety reliability and the appropriate economy is obtained, and this study produces a balanced and improved optimized design that is coordinated with cost, quality, volume, safety and stability.

1. Introduction

In order to reduce the volume of the gear reducer and reduce the weight of the product, the modern method is to calculate and check the empirical formula of its strength and stiffness, so the cabinet design of the reduction gearbox tends to be conservative. With the development of virtual test technology, the virtual simulation technology is used to establish the full-body model of the gear reducer, and the stress distribution and deformation of the whole part are analyzed to optimize the overall structure of the product. The overall strength and rigidity of the reducer are guaranteed, and the weight of the reducer is optimized.

The optimization of gear reducer has been discussed in many literatures [1-4]. Tao [1] established the mathematical model of the optimization design of the gearbox of the tillage machine, which guarantees the given center distance range, transmission power, structure and installation. In the case of the case, the gear transmission parameters are reasonably selected to minimize the use of the gearbox. Zhu [2] taked the number of pinion gears, the modulus of the normal surface and the transmission ratio of the first two stages in the transmission of the spur gear reducer as the design variables, and establishes the third level with the minimum sum of the center distances of the gears of the reducer as the objective function. The mathematical model of the optimization design of the spur gear reducer has achieved good results. Zhu [3] called Matlab to optimize the empirical mathematical formula in Visual Basic, and optimize the design of single-stage spur gear transmission. Although the accuracy is not high, the method is simple and easy. Row. Gai [4] comprehensively considered the factors such as displacement coefficient, tooth profile coefficient, equal strength and lubrication conditions, and establishes the optimization model with the minimum center distance as the target, and optimizes the transmission ratio of JS40-25 three-stage conical cylindrical gear reducer for coal mine. Distribution.

However, due to the complexity and implicit nature of multi-stage gear reducers, most studies have difficulty in considering their system reliability during the optimization process. In the past literature, although the previous literature has initially determined the optimal structural design parameters and basically met the specified reliability and light weight requirements, the previous methods are not the optimal solution and has some limitations because it does not consider the system reliability sensitivity factor. In order to get the optimal design solution more accurately, this paper combines each attribute with reliability sensitivity analysis. Based on the Pareto solution set, the multi-attribute decision theory is used to optimize the design of typical structural products.

2. Multi-attribute decision analysis theory

2.1. Decision matrix normalization method

In the multi-attribute decision problem, $P=\{P_1, P_2, \ldots, P_m\}$ represents a set of m schemes, each scheme has n attributes, and $I=\{I_1, I_2, \ldots, I_n\}$ represents a set of attributes. The jth attribute value of the ith scheme is represented by $a_{ij}(i=1, 2, \ldots, m; j=1, 2, \ldots, n)$, and then $A=(a_{ij})_{m\times n}$ is the original decision matrix. The line number represents the scheme and the column number represents the attribute.

At present, there are many methods normalizing decision matrices. Linear proportional transformation method, range transformation method and gravity transformation method are commonly used [5]. Considering that an extreme value in an attribute is regarded as a constant, other attribute values are normalized. Therefore, the weighting method is used to normalize the original decision matrix:

When I_j is a benefit attribute,

$$x_{ij}=\frac{a_{ij}}{\sum_{i=1}^{m}a_{ij}} \tag{1}$$

When I_j is a cost-type attribute,

$$x_{ij}=\frac{1/a_{ij}}{(\sum_{i=1}^{m}1/a_{ij})} \tag{2}$$

And then Normalized decision matrix $X=(x_{ij})_{m\times n}$ is obtained from (1) and (2)。

2.2. Method for determining attribute weights

In multi-attribute decision making, we need to consider the mutual importance of the attributes. Usually, we use weights or weight coefficients to indicate the mutual importance of attributes, that is, weights are measures of the importance of attributes or indicators. Different weights will bring different evaluation results. Scientific and reasonable attribute weights may produce reliable and correct multi-attribute decision-making results. Therefore, weight determination or weight coefficient allocation is an important issue in the evaluation scheme selection process.

2.3. Multi-attribute decision-making evaluation method

Simple linear weighting is one of the most concise and multi-attribute evaluation methods that people often use. The attribute weight vector corresponding to the attribute set $I=\{I_1, I_2, \ldots, I_n\}$ is $W=\{w_1, w_2, \ldots, w_n\}$, where $\sum_{i=1}^{n} w_{ij} = 1$, $0 \le w_{ij} \le 1 (j = 1,2,\ldots,n)$. Then the comprehensive evaluation value of the scheme i is

$$y_i = \sum_{j=1}^{n} w_j x_{ij}, i = 1,2,\ldots,m \tag{3}$$

According to the size of y_i, you can get the result of the scheme sorting. If $y_1' > y_2' > \cdots > y_m'$, the order of the scheme is $P_1' > P_2' > \cdots > P_m'$.

3. Reliability sensitivity analysis method

3.1. Response surface method

The response surface method is an effective method for performing implicit functions of mechanical reliability analysis. The idea is to simulate a real limit state surface by fitting a response surface through a series of deterministic experiments. The response surface can be simulated with a mathematical equation, assuming the system response r and the random parameter vector affecting the structure $\boldsymbol{X} = [X_1, X_2, \ldots, X_{NR}]$ can be described by a quadratic function with cross terms.

$$r = a_0 + \sum_{i=1}^{NR} a_i X_i + \sum_{i=1}^{NR}\sum_{j=i}^{NR} a_{ij} X_i X_j \tag{4}$$

Where a_0、a_i、a_{ij}($i = 1\ldots NR$; $j = i\ldots NR$)are undetermined coefficients.

The NS sample points of the random parameters are numerically calculated to obtain NS output points ($z_1,z_2,\ldots,z_{NS}$), and the data are subjected to regression analysis by least squares method [5],

$$s = \sum_{i=1}^{NS} \varepsilon^2 = \sum_{i=1}^{NS}\left[z_i - \left(a_0 + \sum_{i=1}^{NR} a_i x_i + \sum_{i=1}^{NR}\sum_{j=i}^{NR} a_{ij} x_i x_j \right)\right]^2 \tag{5}$$

Where ε is the error term. The response surface function is used to replace the real response for th e reliability analysis.

3.2. Sensitivity analysis based on response surface-Monte Carlo and PDS

To obtain the influence of random variables on structural reliability, the sensitivity of the limit state function to structural parameters (such as size, diameter, width, etc.) is characterized and explained, thus providing guidance and reference value for improving the structure.

Through the reliability sensitivity analysis module of ANSYS Probabilistic Design System (PDS), the influence of each structural variable on the limit state equation is obtained. The analysis file of the response surface function of the component is written by APDL, and the PDS module of ANSYS can be used effectively. Analyze the influence of various factors on the limit state equation. In the analysis process, each random parameter is described as an uncertainty variable obeying a certain probability density distribution, and the distribution characteristics, influence relationship and influence degree of each response parameter are analyzed through a large number of sampling points. Parameter sensitivity. Main steps is following as: select the experimental design scheme, virtual simulation test, establish response surface function, select analysis tool, specify reliability analysis file, cycle sampling, and post-processing result.

4. Reliability sensitivity analysis and optimization design of gear reducer

4.1. Multi-attribute decision making optimization based on Pareto solution set

After the multi-objective genetic algorithm is optimized, the Pareto solution set of the typical structural product gear reducer is obtained, which is a non-inferior solution that satisfies the reliability index and lightweight. It is used as the initial decision matrix of multi-attribute decision making, and the example optimization design and sensitivity analysis, table 1 shows the initial multiple sollutions.

Table 1 Pareto optimal solution set

x_1 (mm)	x_2 (mm)	x_3 (mm)	x_4 (mm)	x_5 (mm)	x_6 (mm)	x_7 (mm)	x_8 (mm)	x_9 (mm)	f_1 (10^7mm^3)	f_2
34.0815	33.0442	31.5264	2.4626	28.2383	24.0321	27.0750	37.1806	8.3164	0.80759	0.99988
34.0835	32.0734	33.8462	1.7626	27.2383	24.0321	27.0750	37.1806	7.3486	0.80808	0.99926
32.0815	31.3642	31.7466	2.0128	25.0663	22.3033	25.9131	37.0002	7.6874	0.79808	0.99929
30.0112	31.7546	31.0086	1.8035	23.0638	23.3687	27.0642	34.9423	6.9468	0.80199	0.99943
32.0582	32.1346	33.8726	2. 0121	24.7856	23.6472	27.0021	38.1648	7.3648	0.80057	0.99937
31.0086	32.9436	32.4367	2. 0227	22.6984	23.3857	27.7841	36.8639	6.9431	0.79867	0.99955
33.0385	34.9647	32.4697	2.1689	23.5770	22.3756	28.067	37.0003	6.9738	0.79270	0.99935
32.4364	32.4673	32.2486	2.0735	22.8731	22.0479	27.9468	38.2487	7.4367	0.78892	0.99929

The reducer system volume f_1 is taken as the cost type attribute, and the system reliability f_2 is taken as the benefit attribute. The original decision matrix of the quality evaluation problem is

$$A=\begin{bmatrix}0.8076\times10^7,0.8081\times10^7,0.7981\times10^7,0.8020\times10^7,0.8006\times10^7,0.7987\times10^7,0.7927\times10^7,0.7889\times10^7;\\0.99988,0.99926,0.99929,0.99943,0.99937,0.99955,0.99935,0.99929\end{bmatrix}^T$$

The normalized matrix obtained by normalizing the original decision-making annihilation using equations (1) and (2) is

$$\boldsymbol{X}=\begin{bmatrix}1.0154,1.0275,1.0000,1.0243,1.0267,1.0344,1.0394;\\0.12506,0.12498,0.12500,0.12499,0.12502,0.12499,0.12498\end{bmatrix}^T$$

The attribute vector given by experts according to experience is W=[0.6, 0.4]. The comprehensive evaluation value of each scheme calculated by equation (3) is y_1=0.12556, y_2= 0.12558, y_3= 0.12464, y_4=0.1267, y_5= 0.12488, y_6= 0.12471, y_7= 0.12414, y_8= 0.12379, so $y_4 > y_2 > y_1 > y_5 > y_6 > y_3 > y_7 > y_8$, the scheme is sorted to $P_4 > P_2 > P_1 > P_5 > P_6 > P_3 > P_7 > P_8$, the optimal scheme based on multi-attribute decision making is shown in Table 2.

Table 2 the optimized structural parameters

Type	x_1 (mm)	x_2 (mm)	x_3 (mm)	x_4 (mm)	x_5 (mm)	x_6 (mm)	x_7 (mm)	x_8 (mm)	x_9 (mm)	f_1 (10^7mm^3)	f_2
Multiple attribute decision	30.0 112	31.7 546	31.0 086	1.8 035	23.0 638	23.3 687	27.0 642	34.9 423	6.9 468	0.80199	0.99943

Due to the Pareto solution set based on multi-objective genetic optimization algorithm, the structural parameters are all decimals. In order to facilitate the convenience of the processing technology, the improved optimized structural parameters are shown in Table 3.

Table 3 the rounding optimized structural parameters

Type	x_1 (mm)	x_2 (mm)	x_3 (mm)	x_4 (mm)	x_5 (mm)	x_6 (mm)	x_7 (mm)	x_8 (mm)	x_9 (mm)	f_1 (10^7mm^3)	f_2
Multiple attribute decision	30	32	31	1.8	23	23	27	35	7	0.81066	0.99939

Compared with the virtual simulation results, it is found that except for the comparison between the x_3 parameters and the virtual simulation results, the x_2, x_4 and x_5 parameters have slight differences, and the remaining parameters are within a reasonable range.

4.2. Establishment of response surface function

The reducer is the most important power and motion transmission equipment, and the gear transmission is the main part of the reducer. It is known from the FMECA failure analysis, the gear failure is the main failure mode of the system, and the system response surface function is established by the gear force characteristics. From the analysis results of the virtual reliability test, the reliability of the secondary transmission gear is the lowest one, which is the main failure mode of the typical structural product.

The finite element calculation of the pinion gear is carried out by ANSYS/LS-DYNA. The calculation result is shown in Fig. 1. The maximum contact stress of the gear tooth surface is 396.743 MPa. The values of the first 10 sample points (55 in total) calculated. According to the Latin hypercube sampling method and calculating its corresponding response value, the values of the first 10 sample points (55 in total) are listed in Table 4.

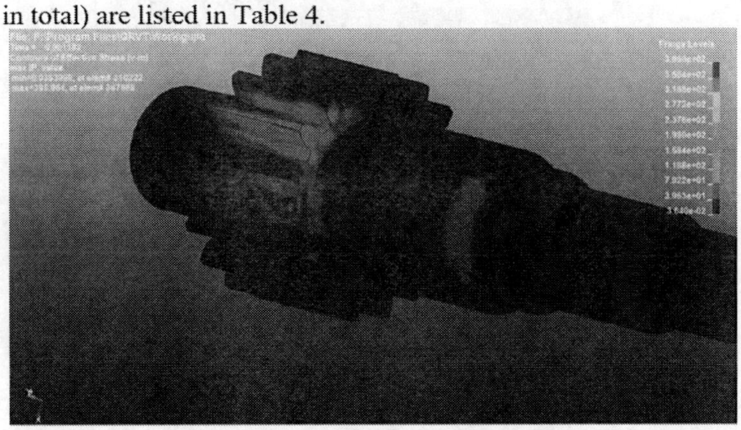

Fig.1. Gear dynamic contact stress

Table 4 Latin hypercube sampling sample points and response values

Sample point	x_1 (mm)	x_2 (mm)	x_3 (mm)	x_4 (mm)	x_5 (mm)	x_6 (mm)	x_7 (mm)	x_8 (mm)	x_9 (mm)	r_1 (Mpa)
1	32.2798	33.4707	32.5310	2.1386	27.3180	24.9273	27.2937	36.9625	8.3741	399.1719
2	34.7682	31.4077	33.7937	1.7073	26.7761	24.6343	27.9916	37.1524	8.0908	399.1608
3	33.1683	32.6973	31.6474	2.0719	24.8727	22.9054	27.4393	37.4772	8.6963	392.9833
4	31.0792	31.2857	33.6953	2.4009	22.4355	24.2992	25.7169	36.3087	6.3664	406.5987
5	31.4895	32.6653	32.9997	1.8356	28.5827	23.7147	26.1053	36.8197	7.3591	392.2891
6	33.6839	31.2073	31.3155	2.3650	28.8820	24.0189	27.1204	38.5628	8.8849	395.5237
7	32.7056	31.5278	32.4862	2.2462	27.6757	23.2670	28.8590	37.2840	7.4801	393.1715
8	33.3324	33.8571	33.1030	2.1530	22.8733	22.6144	28.7572	36.7840	7.0974	391.4969
9	33.7812	32.4385	33.5762	2.3455	24.9844	23.8279	26.5286	38.2749	8.9942	398.4493
10	31.5623	31.3419	32.2004	1.7349	25.2649	24.2481	28.9484	36.5426	6.4174	387.8587

4.3. System Reliability Sensitivity Analysis

The input coefficient and the correlation coefficient between the input variables is selected and defined according to the response surface function. The maximum contact stress of the gear tooth surface is regarded as a random output variable, and the distribution type, distribution function and its parameters is specified for each input variable. According to equation (4), the limit state function is established as G. If $G<0$, it is invalid. In the probabilistic design process, the 2-axis pinion analysis module is sampled for 50,000 cycles to obtain the reliability and the sensitivity value of each random parameter. Fig. 2 shows the limit state variable G sample history curve, which reproduces the value of the limit state variable G in 50,000 samples.

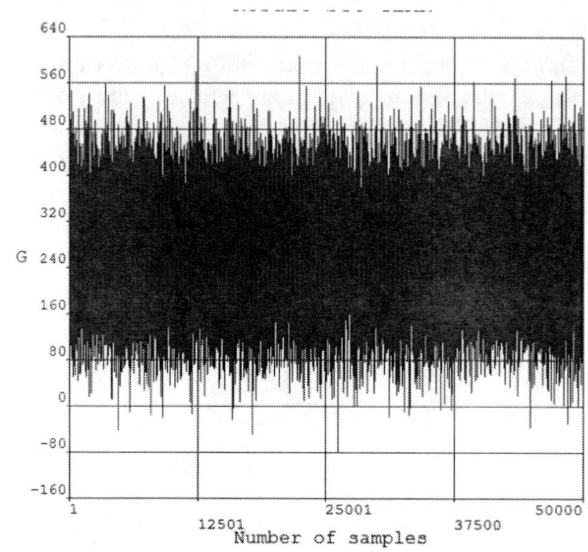

Fig.2. Limit state variable *g* history curve

For those sensitive factors, do not change the size of the structure design as much as possible. In processing and production, it is necessary to strictly control the precision. For those parameters that are less sensitive, in order to improve the economy, the requirements can be appropriately widened. No strict control is required. The above numerical calculation results can provide a basis for the engineering designer to accurately design the reducer gear. The above calculation results are completely consistent with the usual qualitative analysis results, which further demonstrates the comprehensiveness and correctness of the sensitivity matrix to the analysis of various factors of the reducer gear.

Fig. 3(a) shows the sensitivity by the bar graph. Fig. 3 (b) is the sensitivity by the pie chart. As can be seen from the figure 3, the sensitivity of the gear modulus x_4 is 1.5216, and the direct x_5、x_6、x_7 and x_8 of each axis are 0.22204, 0.37733, 0.34559, and 0.19341, respectively. The sensitivity values of the other parameters are very small. It indicates that these parameters have less impact on system reliability. The sensitivity of the tooth width is negative, indicating that the system reliability will decrease as the tooth width of the secondary gear pair increases. This is probably due to the mutual influence of the bending and torsion deformation of each shafting system, resulting in the tooth surface deviation. Load, and thus local stress is too large, resulting in a reduction in system reliability, so this is consistent with the gear analysis we know. The sensitivity of all other parameters is positive, indicating that the reliability of the gear reducer system is increased by increasing the parameters. For the parameters of the gear modulus and the shaft diameter, the larger the value, the higher the bearing capacity, so the reliability is also The higher the accuracy is, therefore, the theoretical analysis is performed to verify the correctness of the results of the reliability sensitivity analysis.

From the previous reliability sensitivity analysis, the change of gear modulus has the greatest influence on the reliability of the system. Therefore, in the optimization design and evaluation of the gear, the change of the gear modulus parameter should be strictly controlled and obtained by the multi-objective genetic algorithm. In the optimal Pareto solution set, only the modulus of scheme 2 and the current optimal scheme 4 are small. Therefore, if the factors of the structural parameters of the gear reducer are changed due to human factors or factors of expert weighting in multi-attribute decision analysis, scheme 2 is a better choice.

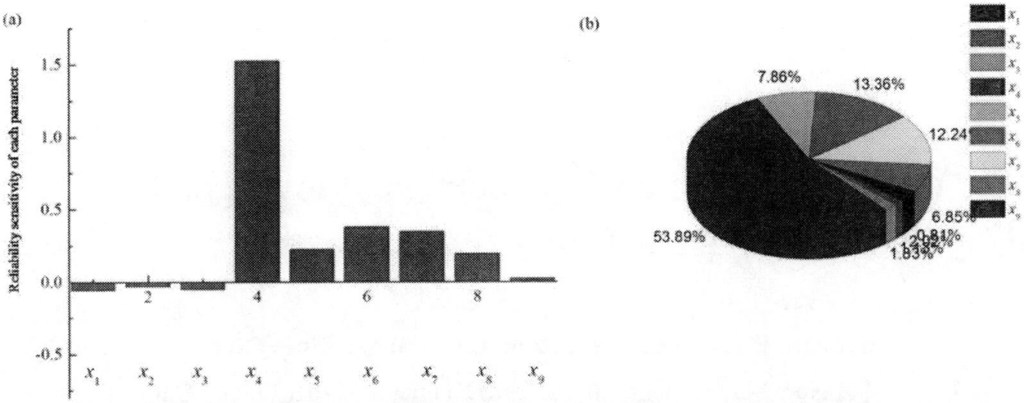

Fig.3. Sensitivity map of contact fatigue failure of random parameters

5. Conclusion

Aiming at solving the complexity and implicit problems in the optimization design of gear reducer, this paper proposes an optimization design method based on the multi-attribute decision-making and reliability sensitivity. Through analysis of the reducer's example, the optimal structure with the lightweight and high reliability are obtained. The design scheme provides an important theoretical basis for the application of the industrial gear reducers.

Acknowledgments

This research was financially supported by the Young Doctor Scientific Research Foundation of College (Grant No. 19YB27) and Liaoning Provincial Natural Science Foundation (Grant No. 2018010334-301). Their financial supports are gratefully acknowledged.

References

[1] Tao, D.C., Sun S.L., Lu Y.E. (2001) Optimization design of gearbox of cultivator. Transactions of the Chinese Society for Agricultural Machinery, 32(2): 118-20.

[2] Zhu, L.L., Wang G.X. (2005) Optimization design of parameters of three-stage cylindrical gear reducer. Journal of Dalian Jiaotong University, 26(2): 20-24.

[3] Zhu Z.B., Pan D.L. (2005) Optimization Design of Cylindrical Gear for Mine ZGB Car Based on Matlab. Machinery Design & Manufacture, (12): 72-73.

[4] Gai M.M., Ren J., Wei B.Y., Guo Q. (2015) Optimization Design of JS40 Three-stage Conical Cylindrical Gear Reducer Drive System. Mining & Processing Equipment, 43(02):105-109.

[5] Hwang C.L., Yoon K. (1981) Multiple attribute decision making: methods and applications . Berlin: Springer-Verlag.

[6] Yan M., Sun Z.L., Yang Q. (2007) Reliability sensitivity method based on response surface method. Journal of Mechanical Engineering, 43(10)： 67-71.

Study on Preparation and properties of low sensitive polycarboxylate superplasticizer

Mengqin Guan, Xianghe Li, Zhuojun Jiang*, Xinqi Guo

KZJ New Materials Group Co., Ltd. Xiamen 361101, Fujian, China

*Corresponding author, E-mail: jzj0407@126.com

Abstract. Esterified macromonomers (P1) were obtained from methacrylic acid(MAA), maleic anhydride(MAn) and methoxy polyethylene glycol monomethyl ether (MPEG). A low sensitive polycarboxylate superplasticizer (P-DMG) was prepared by copolymerization of Esterified macromonomers (P1) with acrylic acid (AA), hydroxyethyl acrylate (HEA), acryloyloxyethyl trimethyl ammonium chloride (DAC) and 4-hydroxybutyl vinyl polyoxyethylene ether (VPEG). Compared with the conventional polycarboxylate superplasticizer, P-DMG can reduce the sensitivity to concrete dosage, water consumption, temperature and raw materials, and has good workability. It can be widely used in various concrete projects.

1. Introduction

The domestic research on polycarboxylate superplasticizer began in the late 1990s and early 2000s. As the third-generation concrete admixture, compared with the second-generation admixture, the polycarboxylate superplasticizer has the advantages of high water reduction, low dosage and high slump retention[1-3]. It is widely used in highways, railways, bridges, nuclear power, dams, tunnels, marine engineering and high-rise buildings and other engineering fields.

With the application of polycarboxylate superplasticizer becoming more and more extensive, the differences of raw materials, regional environment, and the limitations of technical personnel's understanding, habits and theoretical knowledge level in the process of ready-mixed concrete, some problems have arisen in the process of using polycarboxylate superplasticizer, which directly affects the use effect of polycarboxylate superplasticizer. At present, the mechanism of action and self-sensitivity of polycarboxylate superplasticizer in China can not be completely controlled. Especially in practical projects, due to various kinds of cement and improper construction operation of workers, when polycarboxylate superplasticizer is directly applied to concrete, the problems such as poor adaptability to cement[4-8], unsatisfactory performance of concrete and possible large collapse often occur, which lead to the failure to meet the requirements of construction site. Therefore, it is of great significance to develop a low sensitive polycarboxylic acid product. Esterified macromonomer (P1) is obtained by esterification of methacrylic acid, maleic anhydride and methoxypolyethylene glycol monomethyl ether (MPEG). And copolymerization of P1, acrylic acid, hydroxyethyl acrylate, acryloxyethyl trimethyl ammonium chloride (DAC) and 4-hydroxybutyl vinyl polyoxyvinyl ether (VPEG) to prepare a low sensitive polycarboxylate superplasticizer (P-DMG).

2. Experimental

2.1.Materials

2.1.1.The main synthetic experimental raw materials. 4-hydroxybutyl vinyl polyoxyvinyl ether (VPEG), industrial grade; Methoxypolyethylene glycol monomethyl ether (MPEG) , industrial grade; From methacrylic acid(MAA), industrial grade; maleic anhydride(MAn), industrial grade; Acrylic acid (AA), industrial grade; p-toluene sulfonic acid (PTSA), industrial grade; hydroquinone (HQ), industrial grade; hydroxyethyl acrylate (HEA), industrial grade; Hydrogen peroxide (H_2O_2), industrial grade; Sodium hypophosphite(SHP), industrial grade; Acryloxyethyl trimethyl ammonium chloride(DAC), industrial grade; 3-mercaptopropionic acid (MPA), industrial grade; Sodium hydroxide (NaOH), 30% aqueous solution, industrial grade.

2.1.2.Main performance test raw materials for experiment. Cement(C1): Hua Run P.O 42.5 Cement; Cement(C2): Min Fu P.O 42.5 Cement; Cement(C3): Hong Shi P.O 42.5 Cement; Cement(C4): Hai Luo P.O 42.5 Cement; Cement(C5): Ji Dong P.O 42.5 Cement. Sand: river sand with a fineness modulus of 2.7, containing 2% mud; Small stones: crushed stones with a diameter of 5-20 mm; Large stones: crushed stones with a diameter of 16-31.5 mm. Flyash: class ii; P-JS:Standard polycarboxylate superplasticizer, products of KZJ New Materials Group Co., Ltd.

2.2. Esterification
The esterification product P1 was obtained by adding metered MPEG, MAH, MAA and HQ into the reactor, opening the nitrogen-opening device, heating up to 60°C, adding metered PTSA, slowly heating up to 120°C, constant temperature for a certain time, and cooling down.

2.3. Copolymerization
Metered water, AA and VPEG were added to the reactor, heated and stirred until VPEG was dissolved. After the system temperature reached 10°C, SHP, AA, H_2O_2, HEA, DAC and MPA were added to the reactor respectively. The dropping was controlled within 2 h, then the constant temperature was 0.5 h, and the pH value was adjusted to 5.0-6.0 by adding 30% sodium hydroxide, and then P-DMG was obtained.

2.4. Performance test method

2.4.1. Concrete performance test. Concrete according to GB/T50080-2016" Standard for test method of performance on ordinary fresh concrete". And according to "Technical Specification for Sensitivity Evaluation of polycarboxylate superplasticizer ", that compiled by Kezhijie New Materials Group Co., Ltd. The concrete mix ratio (kg/m3) is m (cement): m (fly ash): m (small stones): m (large stones): m (sand): m (water) = 260:80:525:525:790:170.

2.4.2. Gel chromatography test. Waters 1515 was used. The column was connected in series by UltrahydragelTM 250 and UltrahydragelTM 500. The mobile phase was 0.1 mol/L sodium nitrate aqueous solution at a flow rate of 0.8 mL/min. The standard calibration curve was sodium polyacrylate and the running time was 60 min.

3. Experimental results and discussion

3.1. Gel chromatography
Gel chromatography was used to test P-DMG. The content of polymer in the spectrogram reflects the conversion rate of P-DMG monomer. The spectrum of P-DMG is shown in Figure 1.

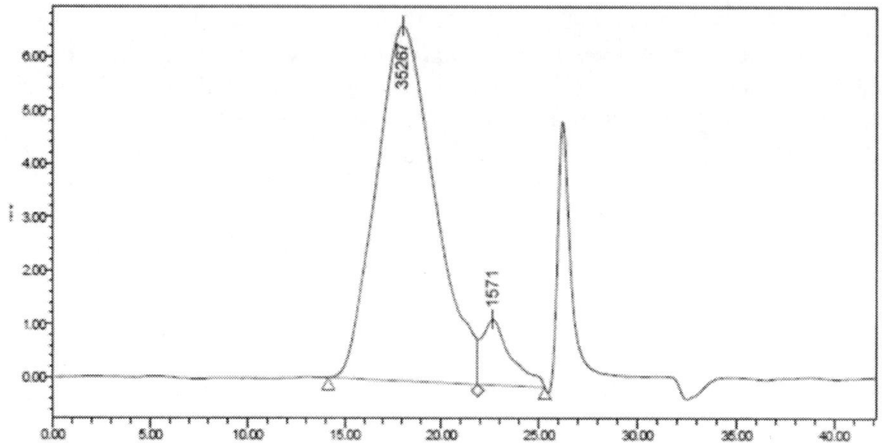

Figure 1. Gel chromatogram of P-DMG

From the experimental results in Figure 1, it can be seen that the content of the polymer is as high as 91.62%, indicating that the monomer conversion rate of P-DMG is high, which can meet the requirement of application.

3.2. Sensitivity test of concrete

3.2.1. Sensitivity of dosage. The laboratory temperature is 20±2°C, keeping the mix proportion of concrete unchanged. By adjusting the dosage of admixture, the initial slump of concrete at (400±10) mm is taken as the starting point A0, and the initial slump of concrete at (550±10) mm is taken as the end point A1. The ratio of the end point of the admixture to the amount of the starting point is used to evaluate the sensitivity of the polycarboxylate superplasticizer (Admixture dosage width Rc= A1/ A0). The larger the Rc is, the less sensitive the content of the polycarboxylate superplasticizer. The results of dosage sensitivity test of P-JS and P-DMG are shown in Fig. 2.

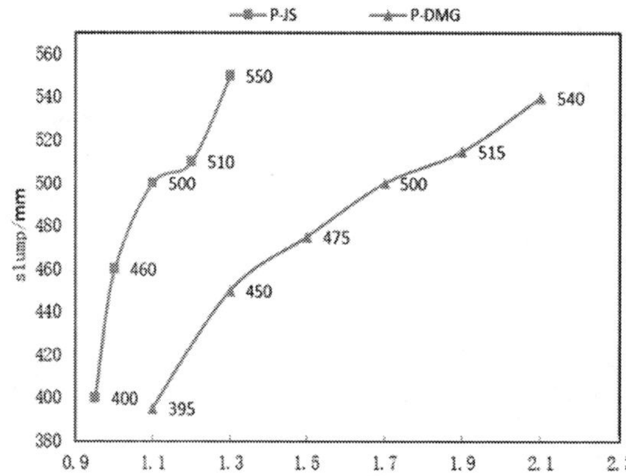

Figure 2. Evaluation results of dosage sensitivity of P-JS and P-DMG

As shown in Figure 1, the content of P-JS is 0.95%-1.3%, RcP-JS is 1.37; the content of P-DMG is 1.0%-2.1%, Rc(P-DMG) is 2.1, RcP-DMG > Rc(P-JS). It can be seen that P-DMG is less insensitive to the dosage.

3.2.2. Sensitivity of water consumption. Keep the concrete mix ratio unchanged and adjust the admixture content. Taking the initial slump of concrete is (450±10)mm as the reference amount, and then adjusting the water consumption of concrete mix proportion, the change of concrete slump with polycarboxylate superplasticizer under different water consumption was investigated. The test results are shown in Fig. 3.

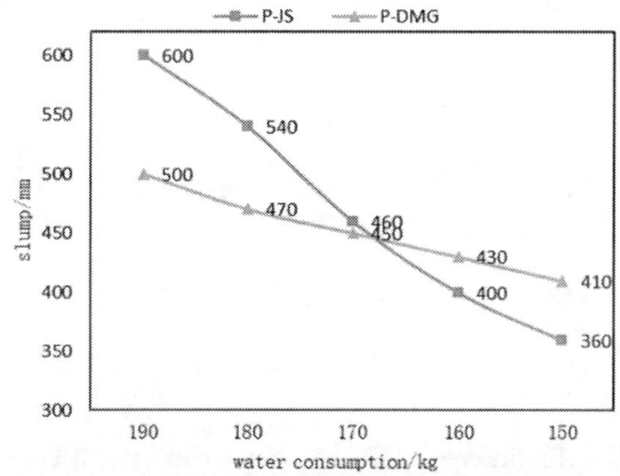

Figure 3. Evaluation results of water consumption sensitivity of P-JS and P-DMG

It can be seen from Figure 3 that for every 10 kg of water added, the concrete slump with P-JS changes by 40-80 mm. For every 10 kg of water added, the slump degree of concrete mixed with P-DMG varies by 20-30 mm. It can be seen that P-DMG is less sensitive to water consumption.

3.2.3. Sensitivity of temperature. Keep the mix ratio of concrete unchanged and adjust the dosage of admixture. Taking the initial slump of concrete is 550±10mm as the reference value, adjusting the laboratory temperature, the slump change of concrete mixed with polycarboxylic superplasticizer at different temperatures was studied. The test results are shown in Figure 4.

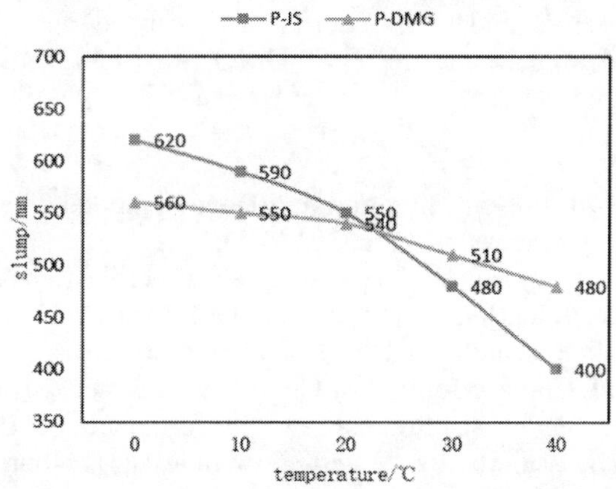

Figure 4. Evaluation results of temperature sensitivity of P-JS and P-DMG

It can be seen from Fig. 4 that at a temperature of 0°C to 40°C, the concrete slump of mixed P-JS is changed by 30 mm to 80 mm for every 10°C increase in temperature. And the concrete slump degree of mixed P-DMG is changed by 10mm~30mm.

3.2.4. Sensitivity of Raw material. Min Fu cement is used to keep the mix ratio of concrete unchanged and adjust the admixture dosage. Taking the initial slump of concrete is 550±10mm as the reference value. Then replacement of different kinds of cement for testing. And the slump change of concrete mixed with polycarboxylic superplasticizer at different cement varietywas studied. The test results are shown in Figure 5.

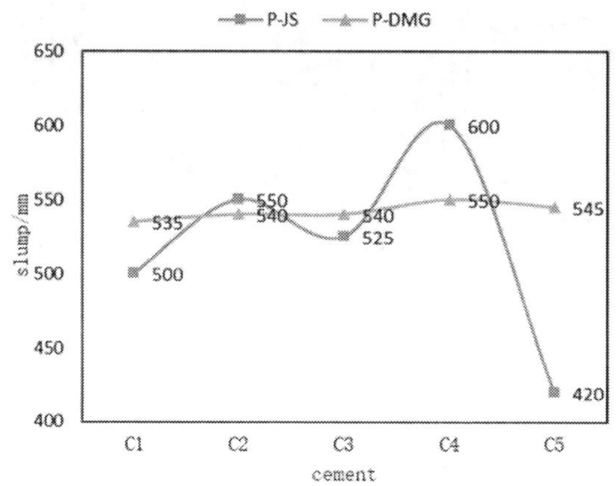

Figure 5. Evaluation results of cement sensitivity of P-JS and P-DMG

It can be seen from Fig. 5 that with 5 different cements, the concrete slump with P-JS is 420-600mm and that with P-DMG is basically unchanged.

4. Conclusions

(1) The low sensitive polycarboxylate superplasticizer P-DMG was synthesized in this experiment. The content of the product polymer obtained by this synthesis method is as high as 91.62%. It shows that the monomer conversion of P-DMG is high and it can meet the requirements of production and application.

(2) The sensitivity test of P-DMG was carried out, and the sensitivity evaluation of the dosage, water consumption, temperature and raw materials was carried out. The results showed that the low-sensitive polycarboxylate superplasticizer P-DMG was less sensitive to the dosage, water consumption, temperature and raw materials, which was superior to P-JS.

References:

[1] Miao C W, Ran Q P, et al. Present Situation and Developmental Trends of Polycarboxylate-Type Superplasticizers[J]. Materials China, 2019(11).

[2] Fang Y H, Ma X X, et al. Impact of polycarboxylate superplasticizer's molcular structure on the performance of concrete[J]. Concret, 2010(12):80-82+151.

[3] Gao Y J, Li S K, et al. Research on Highly Adaptability Crosslinking Polycarboxylate Slump Retention Agent[J]. Chinese Journal of Colloid & Polymer, 2017(01).

[4] Wang D M, Zhang L R, et al. Process and Performance of Ester Based Polycarboxylate Superplasticizer by Microwave-Assisted Synthesis[J]. Journal of Building Materials, 2017(03).

[5] Jin Z Q, Qiu C H, et al. Study on Synthesis and Compound Properties of Functional Polycarboxylic Water Reducer[J]. Ready-Mixed Concrete, 2016(09).

[6] Guo X Q. Study on Preparation of Low Hydration Thermal Polycarboxylate Superplasticizer by Low Temperature Esterification[J]. IOP Conference Series: Materials Science and Engineering, 2018, 392:1-6.

[7] Jiang J Y, Gai G Q,Kong L W, et al. Research progress of compatibility between raw materials composed of concrete and polycarboxylate superplasticizers [J]. Concrete, 2015, (11):69-74.

[8] Huang X Q, Li X R, Zhang D W, et al. Application of Polycarboxylate Superplasticizer in the Concrete[J]. Materials Science Forum, 2017, 898:2076-2080.

Quantitative analysis of Satellite counterweight and Influencing Factors

Zheng Jian dong, Zou Chen, Pi Xiao li, Xue Bi jie, Zhong Hong xian, Zhang Lei, Zou Shuang, Yang Wen

(Institute of Telecommunication Satellite， China Academy of Space Technology， Beijing 100094， China)

Abstract: Counterweight will increase with the use of more and more load equipment. More counterweights in satellite will reduce weight percentage of load equipment, consume fuel and short the lifetime of satellite on orbit. Firstly, this paper analyzes the relative factors with counterweight, and the relationships between them. That is, the relationships between counterweight and designed centroid from satellite layout, centroid error from product and computational error, installation scheme and error of 490N engine, as well as the disturbance torque value. Then the way to reduce counterweight is approached.

1. Introduction

Counterweights for some platform satellite stay high. From the date analyses of counterweight will increase with the use of more and more load equipment, such as deployable antennas. More counterweights in satellite will reduce weight percentage of load equipment, consume fuel and short the lifetime of satellite on orbit.

2. Mechanism analyses

If the thruster vector does not point instantaneous centroid during 490N engine igniting, disturbance torque is produced (thrust is a physical quantity that makes satellite rotate around its centroid). If disturbance torque exceeds the required value of control subsystem, more counterweights and adjusting engine pointing will be taken to meet that requirement. So, counterweight depends on satellite centroid, 490N engine thrust (attachment point, value and direction) and disturbance torque value. That is, counterweight is related to designed centroid from satellite layout, centroid error from product and computational error, installation scheme and error of 490N engine, as well as disturbance torque value.

3. Relationships between counterweight and various factors

3.1 Influence of satellite layout, product and computational error on eccentricity

Satellite layout is affected by many factors, so the satellite centroid is difficult to ensure no eccentricity. More serious eccentricity cost more counterweights (invalid weight) to align satellite centroid. Measurement value of satellite centroid is divided into designed centroid value, product from satellite layout and computational error. Computational error of quality characteristic is related with product error, and product error is main source of computational error. Secondary source of computational error is numerous fasteners, auxiliary materials of thermal and assembly which difficult to count.

3.1.1 Influence of satellite layout on eccentricity

(1) For example, maximum value of designed centroid from satellite layout is 24.0mm, product and computational error is 5.3mm. Maximum value of designed centroid from satellite layout is apparently higher than centroid error from product and computational error. Because satellite layout is limited by many factors, centroid value maybe large, while product and computational error is relatively limited.

(2) From these two aspects, maximum values of theoretical centroid of x coordinate are 24.0mm respectively. Maximum values of theoretical centroid of y coordinate are 13.6mm. This indicates that maximum value of x coordinate is apparently higher than the value of y coordinate, while the centroid value of y coordinate is relatively limited. To find out its cause, equipment and devices are mainly located at north and south panel, and the amount of which is nearly up to 500 sets. The amount and weight are basically balanced, so the total error is relatively limited (this will be explained in detail in the next section). Main reasons for serious eccentric of x coordinate are those heavy and remote antennas and cameras from satellite centroid. Because counterweight can change centroid value of x and y coordinates, so simply take the maximum value of two directions. Maximum and average values of theoretical centroid are equal to 48.0kg and 19.0kg counterweights, respectively.

(3) Maximum and average value of centroid changing caused by product and computational error are 5.3mm and 3.2mm respectively, equaling to 10.6kg and 6.4kg counterweights.

3.1.2 Influence of product and computational error on centroid error

Influence of product and computational error is analyzed using model data in the last section. theoretical analysis and comprehensive consider is conducted then. In order to facilitate analysis, satellite is divided into three parts: generic product (subdividing into equipment and device, cable and waveguide, structure and thermal, solar array), big part outside the satellite such as antenna, camera etc., and propellant.

For general product, both weights of equipment and device in south and north panel are 550kg according to construction standard. Taking equipment of some satellite and weight error from construction standard for example, assume that the error of single sample is random distribution, and the total error (weight difference of north and south panel) is more than 3kg. Because per time the total error is different, so 100 times random sampling is in progress in this paper. The maximum error is 2.88kg. Assume that cable and waveguide affects counterweight of 1kg. For east and west antenna and camera, the counterweight needed is 10.6kg. For those propellant eccentricity caused by processing and installation technology of central column shell and tank, the maximum eccentricity of oxidant and fuel of some satellite equaling to 2.4kg and 2.9kg counterweights. The above-mentioned is based on that both propellant eccentricities are the maximum and at same direction.

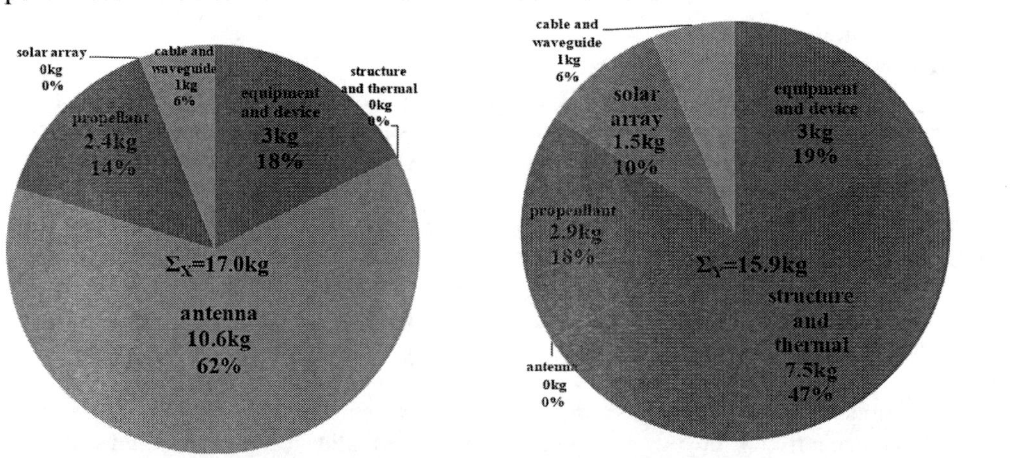

(a) Error source and influence in the x-direction (b)Error source and influence in the y-direction

Fig.1 source and influence of product and computational error

According to above analysis, influence factor of product is shown amply in figure 1. Theoretically, the influence factors of centroid error are east and west antenna, equipment and device, structure and thermal, propellant. In the x direction, main factors are east and west antenna (62%, equaling centroid changing caused by 10.6kg counterweight), equipment and device (18%, 3kg), propellant (14%, 2.4kg), cable and waveguide (6%, 1kg) in sequence, of which the first three accounted for 96%. In the y direction, main factors are structure and thermal (47%, 7.5kg), equipment and device (19%, 3kg), propellant (18%, 2.9kg), solar array (10%, 1.5kg), cable and waveguide (6%, 1kg) in sequence, of which the first three accounted for 84%.

Take the maximum one, because counterweight can change centroid in the both x and y direction at the same time. So, east and west antenna, propellant errors are main factors that cause product eccentricity. Centroid error will reach a maximum when it is limit error. From preliminary theoretical hypothesis and analysis, the maximum product error equals counterweight of 17kg, which causes eccentricity of 8.5mm.

The influence of product error is analyzed above; the influence of computational error of quality characteristic will be analyzed then. According to some model rules, the maximum centroid errors in the x and y direction are both ±2mm. It will cost counterweight of more than 4kg to offset those errors.

3.2 Influence of installation scheme and error of 490N engine

3.2.1 Influence of installation scheme

Because the time of engine installation align is before quality characteristic test, so there is only computation value of quality characteristic at present.

The installation scheme of 490N engine (pointing and location) is divided into present method and optimism method. The difference between these two is thrust pointing. Present method will make the inclined angle of thruster vector and z axis approach to 0; while optimism method is based on the assuming that centroid computation value is dependable, and make the thrust vector to point centroid around during orbit transfer.

Next, the influence of installation scheme is analyzed through a calculation example. Assume that the eccentricities in the x and y direction are both 5.0mm.

Tab 1 disturbance torque value（unit：N.m）

Disturbance torque Tx ,Ty	Old method		Optimism method
	(counterweight 0Kg)	(counterweight 30Kg)	
Maximum value	-4.7	0.8	0.8
Minimum value	**-7.8**	-0.8	-0.8

It can be find out from disturbance torque value in tab 1 that more counterweight, more eccentricity, if present method is taken. When counterweight is 0, the disturbance torque reaches 7.8N.m. That will need nearly 30kg counterweight to minimize the disturbance torque (if the eccentricity of 5.0mm is about a single direction, that will need counterweight of 5000kg×5mm/1000mm=25kg). While the optimism method makes thruster vector to point to computational centroid nearby, so wherever the computational centroid is, there will be no more counterweight for the computational centroid, which is a key approach to achieve the target of non-counterweight. The actual centroid distributes a spherical region which center is computational centroid. So it is apparently rationality that thruster vector points to computational centroid.

3.2.2 Influence of installation error

In assembly error, because the position error of antennas etc. is small, so antennas etc. have less influence on the centroid and counterweight, on which main factor of influence is the installation error of 490N engine; In order to offset maximum installation error which technology requirement allows, it will cost counterweight of 4kg.

IOP Conf. Series: Materials Science and Engineering **711** (2020) 012047 doi:10.1088/1757-899X/711/1/012047

3.3 Summary

To avoid undervalue, for large counterweight, worse theoretical upper limit is chosen. To avoid overestimate, the average value of model is chosen. According to relationships between counterweight and design method, eccentricity of satellite layout, product and computational error, installation error of 490N engine, and total counterweight is shown in fig.2.

1) The maximum and average centroid values from satellite layout are 24.0mm and 9.5mm respectively, equaling counterweights of 48.0kg and 19.0kg respectively.

2) The maximum and average centroid values from product error are 8.5mm and 2.2mm respectively (average computational error is deducted), equaling counterweights of 17.0kg and 4.4kg respectively.

3) The maximum and average centroid values from computational error (in the x and y direction) are ±2.0mm and ±1.0mm respectively, equaling counterweights of 4.0kg and 2.0kg respectively.

4) The maximum and average counterweights using to offset 490N engine are 4.0mm and 2.0mm respectively.

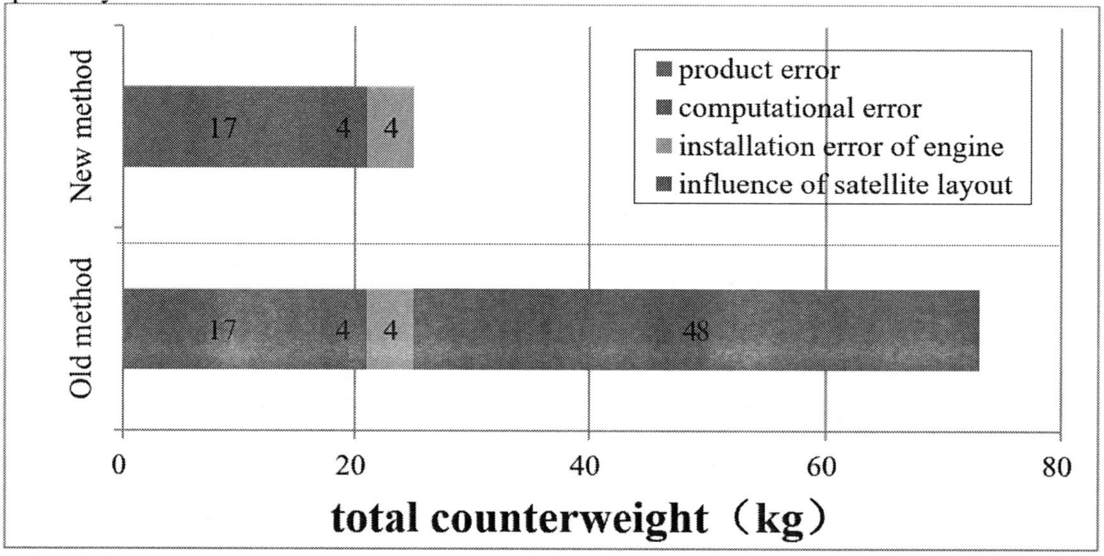

Fig.2 Relationships between counterweight and design method, eccentricity of satellite layout, product and computational error, installation error of 490N engine

In summary, following conclusions are observed:

(1) In the worst condition: if present method is taken, total counterweight will be 51.6kg. This value is twice as much as the maximum counterweight of some model. If optimism method is taken, total counterweight will be 3.6kg.

For actual models, if weight allowance is large, a certain amount of counterweight is usually added to satellite to reduce disturbance torque value. If the disturbance torque is controlled among 2.2Nm (average value of models), it will need counterweight of 13.9kg, and the average counterweight value of 10 satellites is 12.7kg, both are in very good agreement, which also proved that the analysis method is reasonable and accurate.

(2) The influence from different design methods (installation scheme). If present method is taken, more eccentricity, more counterweight. If optimism method of minimizing the disturbance torque (making the thrust vector to point centroid around) is taken, it will overcome the shortcomings of present method. Designed eccentricity itself will not add counterweight, which is the main way to reduce counterweight.

Model development for many years makes us accumulating certain data. The purposes of this paper are to mine relations and laws, and to enhance design level.

4. How to reduce counterweight

In summary, counterweight is decided by designed centroid of satellite layout, product and computational error, installation scheme and error of 490N engine, disturbance torque index. According to influence extent, installation scheme of 490N engine, designed centroid of satellite layout, centroid error from product error are three main factors to influence counterweight. Computational error of quality characteristic and installation error of 490N engine are two secondary factors, as well as the disturbance torque.

For main factors, satellite layout is limited by many factors, so it is hard to reduce designed centroid error. Control level of product error will not be remarkably improved. Installation scheme of 490N engine is the main factor, and also the breakthrough of improvement. so, optimism method is the key way to reduce counterweight.

For secondary factors, computational error and Installation error of 490N engine have little influence on counterweight, so reduce those errors only play a quite limited addition role in reducing counterweight. Otherwise, the disturbance torque is decided by configuration and performance of control subsystem, which this paper will not discuss, only suggested range for designing is present.

It can see that, if the optimization algorithm is taken, the ranges, The maximum component of disturbance torque is symmetry distribution, and tends to minimize, which realizes the target of optimization design. This method will reduce counterweight of satellite, save fuel and improve life on orbit.

5. Summary

In the case, counterweight stays high without action. Firstly, this paper analyzes the relative factors with counterweight, and the relation between them. Then, the way to reduce counterweight is approached. Finally, the optimization method is present.

Reference:

[1] MATLAB optimization algorithm case analysis and application, Tsinghua university press, Yu shengwei, 2014.
[2] Hu Wei，Jin Huanghuang，Wang Qinghua．Research on Optimizing Mounting Position of Satellite 490N Thruster[J]．Spacecraft Engineering，2013，22(6):44-47 (in Chinese)
[3] Xu Chunsheng，Cui Jiatao，Zhang Lei，et al．Application Research of Genetic Algorithm in Installation Adjusting Optimization of Satellite Engine[J]．Spacecraft Engineering，2014，23(2):58-63 (in Chinese)
[4] Peng Wei, Lu Xi-cheng, ． A hybrid genetic algorithm for function optimization [J]． Journal of Software． 1999，10(8)：819-823
[5] Zhang Liang，Wang Ling． Genetic ordinal optimization for Stochastic traveling salesman problem[C]//Proceedings of the World Congress on Intelligent Control and Automation(WCICA)．Beijing：Dept．of Autom．，Tsinghua University，2004(3)：2086-2090
[6] Abe M，Matsumoto H，Kuroda C． An artificial neural network optimized by a genetic algorithm for real time flow-shop rescheduling[J]． International Journal of Knowledge-Based Intelligent Engineering Systems，2002，6(2)：96—103
[7] Zhao Guang，Fan Luhai，Feng Shuo． Redundancy switching solution of spaceborne TWTA using Genetic Algorithm[J]．Spacecraft Engineering，2011，20(5)：93—98(in Chinese)
[8] Wang Yuying，Li Yunze，Liu Dongxiao． Optimized design of heat insulation layers"thickness and radiator surfaces for nano—satellite[J]．Spacecraft Engineering，2010，19(2)：46—51(in Chinese)
[9] KIRKPATRICK S, GELATT C D, VECCHI M P. Optimization by simulated annealing science[J]. Science,1983, 220(4598): 671-680.
[10] Costa A L, Cunha M C C C, Coelho P A, et al. Application of the Simulated Annealing Algorithm

for Transport Infrastructure Planning[M]//Metaheuristics and Optimization in Civil Engineering. Springer, 2016:235-253.

[11] Kulturel-Konak S, Konak A. A large-scale hybrid simulated annealing algorithm for cyclic facility layout problems [J]. Engineering Optimization, 2015, 47(7):963-978.

[12] Geng J, Huang M, Li M, et al. Hybridization of seasonal chaotic cloud simulated annealing algorithm in a SVR-based load forecasting model [J]. Neuro computing, 2015,151:1362-1373.

Analysis of Vehicle Dynamic Performance under MCAT Excitation

Wenbiao Huo[1,2], Sheng Yang[1*], Pingbo Wu[1]

[1] State Key Laboratory of Traction Power, Southwest Jiaotong University, Chengdu, 610031, China

[2] CRRC Tangshan Co., Ltd., Hebei, Tangshan, China

[*]Corresponding author's e-mail: yangcius@qq.com

Abstract. In order to validate the running safety on track of railway vehicle, a numerical analysis is performed with a multi-body vehicle dynamics model built. Follow the vehicle-track interaction (VTI) evaluation declared in north America standard declared by Federal Railway Association (FRA), a Minimally Compliant Analytical Track (MCAT) worthiness analysis is employed. In this investigation, the analytical track geometry irregularities which consist of surface, vertical and short warp are defined. Then the running performances concerning the derailment safety of the vehicle is discussed by evaluating the wheel unloading and derailment coefficients. Vibration acceleration of the carbody is also monitored to evaluate the ride comfort. By changing the track irregularities MCAT, the impact of track wavelength, gauge and cant deficiency on the dynamic performance of the vehicle can be assed.

1. Introduction

As the running speed of high-speed trains increases, we make higher requirements for trains' dynamic performance. Many countries have put forward their own vehicle dynamics standards which differ in ride comfort, running safety and stability of the vehicle. Trans for transnational intermodal must meet different dynamic standards. Shi[1] carried out a study of the American dynamics standard system. Hou[2] [3]analyzed dynamic performance of High-speed train on MCAT. In addition, the existing research mainly focuses on the study of track irregularities limit[4][5], the research of vehicle dynamic performance evaluation under American standards VTI and MCAT excitation have not been investigated yet.

For a certain type of high-speed train whose bogie is widely used in the United States, a multi-body dynamic model is established by SIMPACK. The model is made up of several rigid bodies such as carbody, frames and wheelsets. Wheel-rail contact geometry, wheel-rail force and nonlinear suspension characteristics contribute to the nonlinear elements of the model. In this simulation, the excitation of the track is set as ideal geometric irregularitits (MCAT). Calculating method and working conditions are introduced. Running safety and ride comfort of the vehicle under different track irregularities are evaluated according to the American dynamic standard VTI (Vehicle-Track Interaction).

2. Multi-body Vehicle Modeling

Railway vehicle is a complex multi-body system. It has not only the interaction force and relative motion between components, but also the interaction between wheel and rail. The following assumptions are made when establishing the mathematical model of vehicle system:

(1) The elastic vibration of wheelset, bogie and carbody is much smaller than that of suspension system. The elastic deformation of each component is neglected.

(2) Regardless of the interaction between adjacent vehicles, single vehicle is adopted in this model.

The model of the vehicle is shown in Figure 1. The degree of freedom contained in each part of the model is shown in the Table 1. The specific modeling method of vehicle model refers to the literature[6][7].

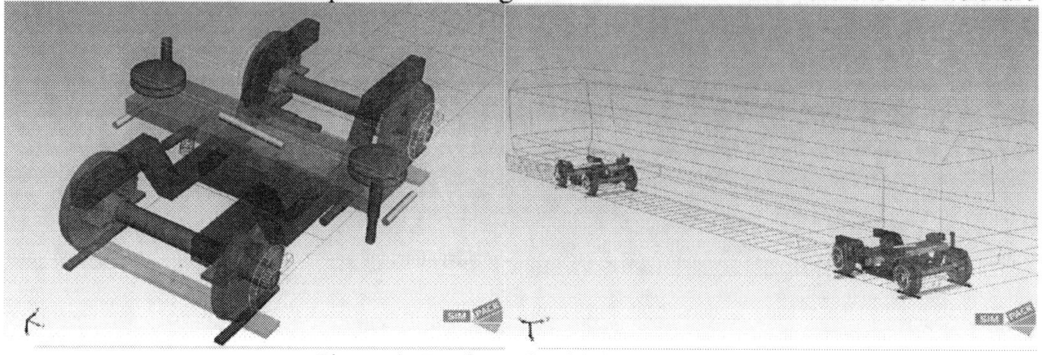

Figure 1. Bogie and vehicle model

Table 1. Degree of freedom

rigid name	longitudinal	lateral	vertical	roll	pitch	yaw
carbody	X_c	Y_c	Z_c	Φ_c	Φ_c	ψ_c
frame	$X_{b1\sim2}$	$Y_{b1\sim2}$	$Z_{b1\sim2}$	$\Phi_{b1\sim2}$	$\theta_{b1\sim2}$	$\psi_{b1\sim2}$
wheelset	$X_{w1\sim4}$	$Y_{w1\sim4}$	$Z_{w1\sim4}$	$\Phi_{w1\sim4}$	$\theta_{w1\sim4}$	$\psi_{w1\sim4}$
axlebox					$\theta_{a1\sim2}$	
traction beam						$\psi_{t1\sim2}$

2.1 Nonlinear wheel-rail interaction and force elements

Geometry of wheel-rail contact, can be expressed as a table of wheelset displacement. The median value is approached by linear interpolation, which can be derived by the wheel-rail contact program of SIMPACK. The creep force between wheel and rail is calculated by Kalker nonlinear simplified creep theory program in SIMPACK. Vertical dampers in primary and secondary suspension, lateral stop as well as yaw dampers in this model are considered as nonlinear force elements, and their damping characteristic curves are piecewise linearized[7].

2.2 Rubber spring elements

The static and dynamic characteristics of rubber spring are considered in the model by using N0.42 force element (Dynamic Bushing/Hydromount). The static stiffness of spring in longitudinal, lateral and yawing are simulated by spring force element. The parameters of the force element come from the static stiffness test of rubber spring. Dynamic characteristics are implemented by spring-damping series model. The spring stiffness in series model is 20% of the static stiffness. The damping coefficients are determined by limiting the transmission frequency to 2 Hz.

2.3 Parameters of wheel and rail

Wheel-rail contact plays an important role in vehicle dynamic performance. The parameters of wheel and rail in this model are shown in Table 2.

Table 2. Parameters of wheel and rail

parameter	value
wheel back distance	53.39 inch

distance between wheel back and nominal rolling circle	2.84 inch（ATPA 220） 2.76 inch（ATPA 340）
rolling circle diameter	New wheel：36 inch
wheel tread profile	ATPA 220
rail profile	AREMA 136RE
rail cant	1:40
gauge	56.5 inch
sliding friction coefficient	0.35
saturation scaling factor of Kalker creep coefficient	1.0

3. Numerical Simulation

Minimally compliant analytical track (MCAT) is a common used excitation in American dynamic standard. As shown in Figure 2, irregularities are set as 9 different types such as hunting perturbation, gauge narrowing, gage widening, repeated surface, repeated alinement, single surface, short warp and combination perturbation in MCAT.

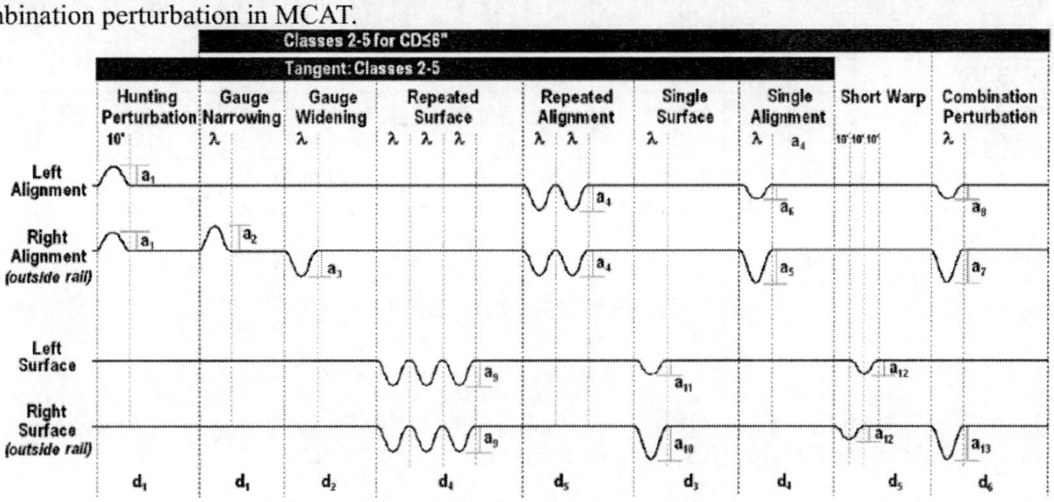

Figure 2. Irregularities in MCAT[8]

In Figure 2, a stands for excitation amplitude and d is the length of each section. The length of d_3 section is 1500ft, and that of other sections is 1000ft. λ is wavelength of each excitation, among which wavelength of hunting perturbation and short warp are set to 10ft and 20ft respectively. According to FRA 49 CFR 213.333[8], the VTI limit is used to evaluate the results of simulation under MCAT excitation.

3.1 Evaluation indice

VTI indice for derailment coefficient and acceleration limit is shown in Table 3.

Table 3. Wheel-rail force and acceleration limit

parameter	limit	filter/window
single wheel vertical load ratio V/V_0	≥0.15	25Hz low-pass/ 5ft
single wheel L/V ratio	$\leq \frac{\tan\delta - 0.5}{1+0.5\tan\delta}$	25Hz low-pass/ 5ft
net axle lateral L_2/V_a ratio	$\leq 0.4+5/V_a$	25Hz low-pass/ 5ft
bogie side L_3/V_3 ratio	≤0.6	25Hz low-pass/ 5ft t

carbody lateral	≤0.65g peak-peak	10Hz low-pass/ 1s
carbody vertical	≤1.0g peak-peak	10Hz low-pass/ 1s

In Table 3, V and V_0 stand for actual and vertical normal vertical load of single wheel, respectively. L is the lateral force on the same wheel. L_2 is the net axle lateral force, which is exerted by the axle on the rail. Accordingly, V_a is the static vertical load of the axle. L_3 and V_3 are the lateral and vertical forces that the wheels on one side of the bogie exert on the rail. When processing the acceleration of the carbody, any peak lasting less than 50ms should be excluded.

3.2 numerical cases

As shown in Table 4, the vehicle speed increases from 30mph to 105mph at intervals of 5mph, and the track class increase from 2 to 5 correspondingly[8].

Table 4. Vehicle speed condition

Track	speed(mph)															
Class	30	35	40	45	50	55	60	65	70	75	80	85	90	95	100	105
Class2	1	2														
Class3		3	4	5	6	7	8	9								
Class4								10	11	12	13	14				
Class5												15	16	17	18	19

When the working conditions contain curves, the speed and the cant deficiency need to be matched. Therefore, it is necessary to calculate the radius of the curve at different speeds. The calculation formula is as follows:

$$R = \frac{C}{2\sin\left(D_c / 2\right)} \quad (3.1)$$

$$D_c = \frac{h_0 + E_u}{0.0007V^2} \quad (3.2)$$

In the Eq. (3.1) and Eq. (3.2), C is chord length, 100ft; D_c is curvature, deg; R is curve radius; E_u is cant deficiency, inch; V is speed, mph; h_0 is curve radius, inch.

4. Results and discussions

Given space limitations, this simulation is carried out under the condition of air spring inflation, new wheel and AW0 loading. When the wavelength of the track excitation turns to 31ft, the results of class 4 and class 5 are summarized in Table 5.

Table 5. VTI indice of class 4 and class 5

parameter	limit	condition	Class 4					Class 5				
		inch	65mph	70mph	75mph	80mph	85mph	85mph	90mph	95mph	100mph	105mph
single wheel V/V_0 ratio	15.0%	Tan.	0.271	0.056	0.389	0.333	0.222	0.228	0.198	0.115	0.036	0.067
		3" CD-56.5"	0.255	0.261	0.272	0.282	0.345	0.278	0.257	0.269	0.278	0.288
		3" CD-57.0"	0.329	0.336	0.362	0.351	0.478	0.261	0.269	0.277	0.281	0.269
single wheel L/V ratio	1.128	Tan.	0.043	0.043	0.044	0.044	0.044	0.044	0.045	0.045	0.045	0.046
		3" CD-56.5"	0.351	0.363	0.377	0.397	0.424	0.425	0.446	0.456	0.484	0.501
		3" CD-57.0"	0.683	0.681	0.793	0.797	0.897	0.469	0.477	0.454	0.447	0.404
net axle L_2/V_a ratio	0.546	Tan.	0.033	0.033	0.033	0.034	0.033	0.034	0.034	0.034	0.035	0.036
		3" CD-56.5"	0.130	0.130	0.143	0.156	0.175	0.145	0.149	0.162	0.179	0.194
		3" CD-57.0"	0.230	0.233	0.299	0.300	0.372	0.153	0.166	0.164	0.168	0.155

bogie side L_3/V_3 ratio	0.600	Tan.	0.028	0.028	0.029	0.029	0.030	0.030	0.030	0.031	0.031	0.033
		3" CD-56.5"	0.227	0.243	0.255	0.267	0.291	0.277	0.288	0.292	0.300	0.308
		3" CD-57.0"	0.408	0.407	0.457	0.459	0.515	0.300	0.297	0.284	0.278	0.252
carbody P-P acc. Y	0.650g	Tan.	0.043	0.044	0.046	0.049	0.050	0.050	0.052	0.054	0.056	0.058
		3" CD-56.5"	0.074	0.076	0.070	0.076	0.084	0.080	0.067	0.067	0.078	0.090
		3" CD-57.0"	0.122	0.124	0.127	0.120	0.137	0.073	0.075	0.079	0.077	0.075
carbody P-P acc. Z	1.000g	Tan.	0.011	0.012	0.013	0.013	0.013	0.013	0.013	0.014	0.014	0.014
		3" CD-56.5"	0.018	0.018	0.016	0.017	0.020	0.018	0.016	0.016	0.018	0.021
		3" CD-57.0"	0.030	0.031	0.032	0.032	0.035	0.019	0.021	0.021	0.020	0.020

Normalizing data in Table 5, that is to say, data in Figure 3 is the ratio of data in Table 5 to their respective limit. In this way, multiple indice can be drawn together to be comparable and we can analyze multiple dynamic indice in one picture, as show in Figure 3. It shows that when the speed increases from 65 to 85mph in class 4, the VTI indice is getting closer to the limit. When the speed exceeds 85mph in class 5, the VTI indice decreases rapidly. Dynamic performance of the vehicle tends to be stable at the speed of 90mph. In general, the vehicle has a good dynamic performance in this condition and the VTI indice is far below the limit in most cases. 3 inch cant deficiency and widening gauge will worsen the dynamic performance of the vehicle.

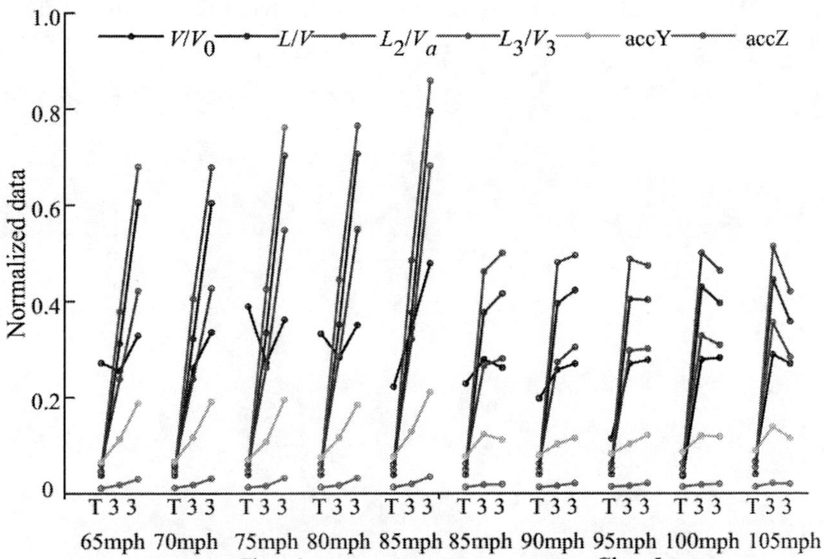

T: straight line; 3: 3 inch cant deficiency; 3: 3 inch cant deficiency with widened gauge

Figure 3. Normalized data for VTI indice

5. Conclusions

In this paper, a multi-body dynamic model is established to evaluate the dynamic performance of the high-speed train whose bogie is widely used in the United States. The modeling of the multi-rigid vehicle model with various nonlinear elements is introduced. The VTI standard, which is applied to American trains, is the standard of vehicle dynamic performance evaluation in this paper. As the excitation of the simulation, MCAT provides us a variety of geometric irregularities of the track. We can make a complete assessment of the vehicle dynamic performance with the variation of curve, gauge and cant deficiency. In this simulation, the vehicle has a good dynamic performance and the VTI indice is far below the limit. Cant deficiency and widening gauge have an adverse influence on the dynamic performance of the vehicle in most cases.

Acknowledgement
This work was partially supported by National Key R&D Program of China (#2018YFB1201702).

Reference

[1] Shi, H.L. (2019) Research on American dynamics standard system. Report: State Key Laboratory of Traction Power, Chengdu, 1-41.
[2] Hou, M.L. (2016) Research on characteristic spectrum of railway vehicle dynamics simulation. Master thesis: Southwest Jiaotong University, Chengdu, 1-78.
[3] Hou, M.L., Chi, M.R. (2016) Analysis of dynamic performance of high-speed passenger train on track irregularity. Mechanical Engineering & Automation, 5:34-36.
[4] Cai, C.B., Zhai, W.M., et al. (1995) Study on allowable safety criterion of track geometric irregularities. Journal of the China Railway Society, 4:82-87.
[5] Chen, G., Zhai, W.M., et al. (2001) Safety management of track irregularities of 250 km/h high-speed railway. Journal of Southwest Jiaotong University, 36(5):495-499.
[6] Luo, R., Li, R., et al. (2015) Dynamic analysis of high-speed train with stochastic parameters. Journal of Mechanical Engineering, 51(24):90-96.
[7] Teng, W.X., Luo, R., et al. (2019) Dynamics behavior of high-speed train at the low temperature of -40°C. Journal of Mechanical Engineering, 55(04):165-170.
[8] FRA 49 CFR213.333. (2013) Vehicle/Track Interaction Safety Standard; High-Speed and High Cant Deficiency Operation. Standard: Federal Register, Washington.
[9] Kalker, J. (1991) Simulation of the development of railway wheel profile through wear. Wear, 150(1-2):355-365.

IOP Conf. Series: Materials Science and Engineering **711** (2020) 012049 doi:10.1088/1757-899X/711/1/012049

Numerical research on sealing performance of high speed and high pressure centrifugal compressor seal

Jun Wu*, Zhenxing Zhao, Lu Dai, Xingsheng Lao, Weijian Lv

Science and Technology on Thermal Energy and Power Laboratory, Wuhan Second Ship Design and Research Institute. Wuhan, Hubei Province, 430205, China

*Corresponding author's e-mail: wu_jun1986@126.com

Abstract. In this paper, two seal schemes of stepped labyrinth seal and straight through-hole seal are selected for the shaft end seal of S-CO_2 centrifugal compressor with the output of 100kW. The full three-dimensional CFD numerical simulation method is adopted to analyze and compare the sealing characteristics of two S-CO_2 seals in S-CO_2 centrifugal compressor under different outlet back pressures and sealing gaps. The results show that the leakage flow rate of S-CO_2 stepped labyrinth seal and through-hole seal increases with the decrease of back pressure. The leakage flow rate of the stepped labyrinth seal and the through-hole seal reaches the corresponding critical value when the back pressure is 1.0 MPa and 3.0 MPa respectively. The critical leakage flow rate of the stepped labyrinth seal and the through-hole seal under the condition of 0.1 mm sealing gap is 0.158 kg/s and 0.128 kg/s respectively. Compared to the through-hole seal, the stepped labyrinth seal can withstand greater pressure differentials at the same sealing gap and has superior S-CO_2 sealing performance.

1. Introduction

Due to the high density, low viscosity and low critical point parameters of the S-CO_2 working fluid, the S-CO_2 Brayton cycle power system can achieve the goal of simultaneous optimization of efficiency, weight and volume. The S-CO_2 compressor is the core equipment for the system thermal power conversion process. The harsh environment of high pressure difference, high density, high speed and complex physical properties will cause problems such as leakage, blast loss, flow excitation and wear failure of the S-CO_2 compressor shaft end seal. As a key component of the S-CO_2 compressor, the shaft end seal controls the leakage flow of the working fluid at the gap passage between the rotor and stator of the compressor. Moreover, the unsteady airflow excitation force at the seal has a significant impact on compressor operating efficiency and operational stability[1-3].

Limited to complex experimental conditions and severely varying physical parameters, the traditional experimental measurement methods and numerical prediction methods have been greatly challenged in the field of S-CO_2 compressor sealing technology. At present, there are relatively few experimental and numerical studies on the leakage characteristics of compressor shaft end seals in the S-CO_2 operating environment. SANDIA National Laboratory evaluates the sealing performance of at least 16 labyrinth seals in a 125kW S-CO_2 turbomachinery component performance test rig. It is finally determined that the stepped labyrinth sealed with a floating wearable static ring can meet the sealing requirement[4].

In this paper, the scheme design of stepped labyrinth seal and straight through-hole seal is performed for the S-CO_2 centrifugal compressor with the output of 100kW. The full three-dimensional

CFD numerical method is adopted to analyze and compare the sealing performance of two S-CO_2 shaft end seals under different outlet back pressures and sealing gaps.

2. Scheme design of the S-CO_2 compressor shaft end seal

Based on the main installation dimensions of the compressor shaft end seal shown in Figure 1 and Table 1, two seal schemes of stepped labyrinth seal and straight through-hole seal are designed separately.

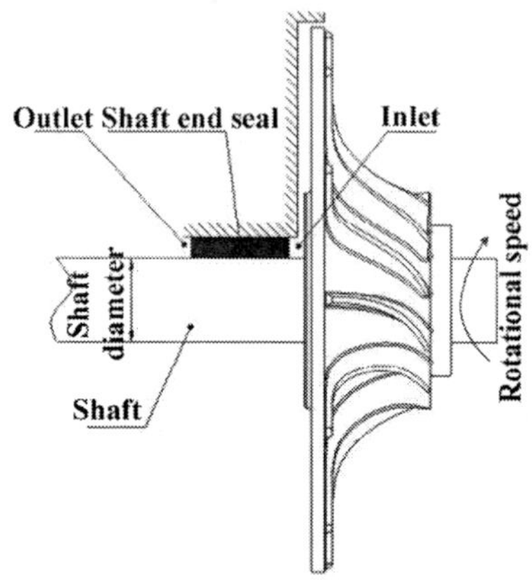

Figure 1. Schematic diagram of the S-CO_2 compressor rotor.

Table 1. Structural parameters of the S-CO_2 compressor rotor.

Number	Variable	Unit	Value
1	Working fluid type	—	S-CO_2
2	Inlet pressure of the shaft end seal	MPa	12.0
3	Inlet temperature of the shaft end seal	°C	100
4	Shaft diameter	mm	30
5	Rotational speed	r/min	30000

2.1. Stepped labyrinth seal

The labyrinth seal has the advantages of simple structure, low cost, convenient installation and replacement, and large tolerance to changes in heat and pressure difference. It has been widely used in turbomachinery. Figure 2 shows the schematic diagram of the S-CO_2 stepped labyrinth seal. Table 2 shows the labyrinth seal geometry parameters. Considering the installation size requirements, the stepped labyrinth seal is designed with the axial length of 52 mm and the diameter of 30 mm. The stepped labyrinth seal has 12 sealing teeth. The height of the upstream step of each tooth is 0.5 mm. Two sealing gaps of 0.1 mm and 0.2 mm are designed to study the influence of the sealing gap on the sealing performance.

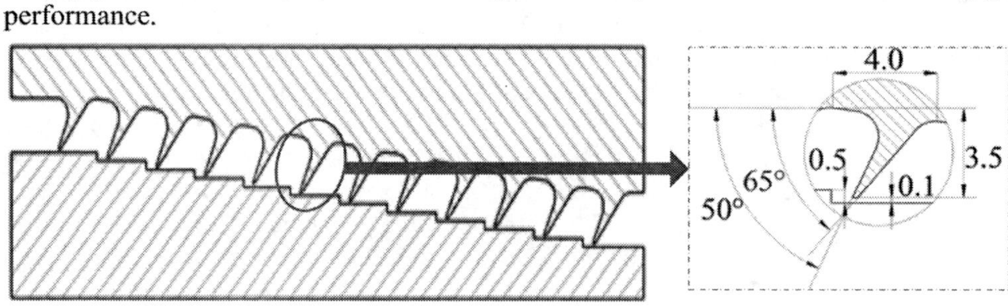

Figure 2. Schematic diagram of the stepped labyrinth seal structure.

Table 2. Geometry parameters of the stepped labyrinth seal.

Number	Variable	Unit	Value
1	Length of the seal	mm	52.0
2	Sealing gap	mm	0.1/0.2
3	Sealing tooth number	—	3.5
4	Tip thickness of the sealing tooth	mm	0.2
5	Slope angle of the sealing tooth	°	50/65
6	Chamber length of the seal	mm	4.0
7	Step height	mm	0.5

2.2. Through-hole seal

In order to achieve the purpose of improving the dynamic characteristics of flow excitation, the through-hole seal has a large roughness by changing the surface of the smooth stator. The round holes in the through-hole seal are machined on the metal material by a milling cutter or an electric spark. Therefore, it has the advantages of simple manufacturing process, low cost, good structure controllability, and wear resistance. Figure 3 shows the schematic diagram of the S-CO_2 through-hole seal. Table 3 shows the through-hole seal geometry parameters. Considering the installation size requirements, the through-hole seal is designed with the axial length of 60 mm and the diameter of 30 mm. The number of axial holes is 23 and the number of holes is 1150. Two seal gaps of 0.1 mm and 0.2 mm are designed to study the influence of the seal gap on the sealing performance.

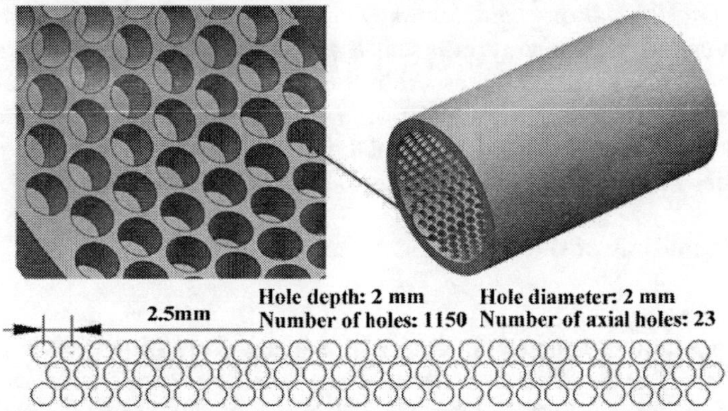

Figure 3. Schematic diagram of the through-hole seal structure.

Table 3. Geometry parameters of the through-hole seal.

Number	Variable	Unit	Value
1	Length of the seal	mm	60
2	Radial gap	mm	0.1/0.2
3	Hole diameter	mm	2.0
4	Hole depth	mm	2.0
5	Distance between the centre of adjacent holes	mm	2.5
6	Number of holes	—	1150
7	Number of axial holes	—	23

3. Numerical calculation model and method for shaft end seal

3.1. Numerical calculation model

Figure 4 and figure 5 show three-dimensional computational grids of the labyrinth seal and the through-hole seal. The software ICEM CFD is applied to generate the multi-block structured grid for the shaft end seal. A total of 25-35 nodes are arranged at the sealing gap. Based on the leakage flow rate, the numerical calculation model is tested for grid independence. Finally, the number of mesh nodes for the labyrinth seal and the through-hole is 640,000 and 1.64 million respectively.

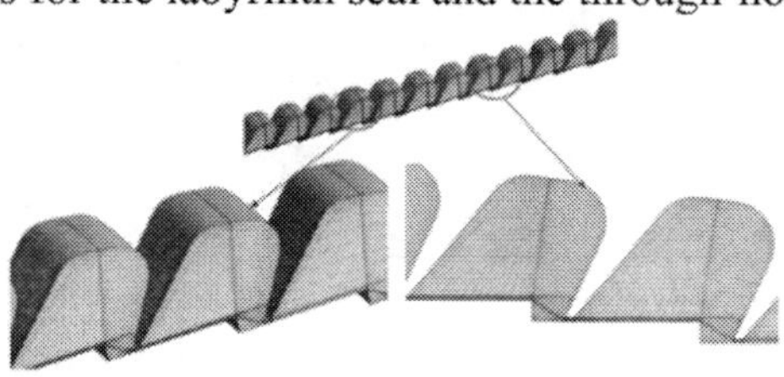

Figure 4. Labyrinth seal meshing.

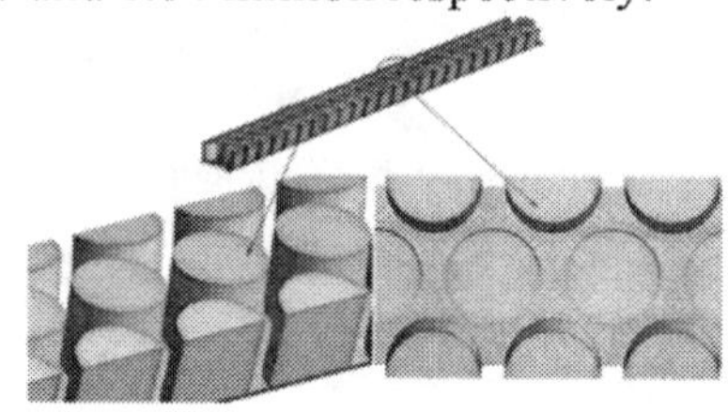

Figure 5. Through-hole seal meshing.

3.2. Numerical calculation method

The RANS equation is solved numerically using the commercial software ANSYS-CFX 16.0. The total pressure and total temperature are set at the inlet and the average static pressure is set at the outlet. The SST turbulence model is selected for the labyrinth seal and the through-hole seal calculation. The solid walls are all set to a smooth adiabatic boundary, and the near wall is treated by the improved wall function method. The labyrinth seal and the through-hole seal are solved in the static domain, and the rotor surface is set to the rotational speed of 30,000 rpm.

In order to evaluate the validity and accuracy of the numerical method for the prediction of the leakage characteristics, the S-CO_2 labyrinth seal test piece with two seal teeth in the literature [5] is analyzed. The leakage flow rate at 13 pressure ratios (0.3 - 0.9) is calculated and compared with the experimental results for an inlet pressure of 10 MPa, a temperature of 55 °C, and a rotational speed of 0 rpm. The results show that the numerical method used in this paper can accurately predict the leakage characteristics of the S-CO_2 labyrinth seal. The leakage prediction error is less than 6%.

4. Numerical result analysis of the sealing performance for shaft end seal

4.1. Stepped labyrinth seal

Figure 6 shows the variation of the stepped labyrinth seal leakage flow rate with back pressure under different sealing gaps. The leakage flow rate of the labyrinth seal increases as the back pressure decreases. When the seal gap increases from S = 0.1 mm to S = 0.2 mm, the critical leakage flow rate of the labyrinth seal increases from 0.158 kg/s to 0.265 kg/s.

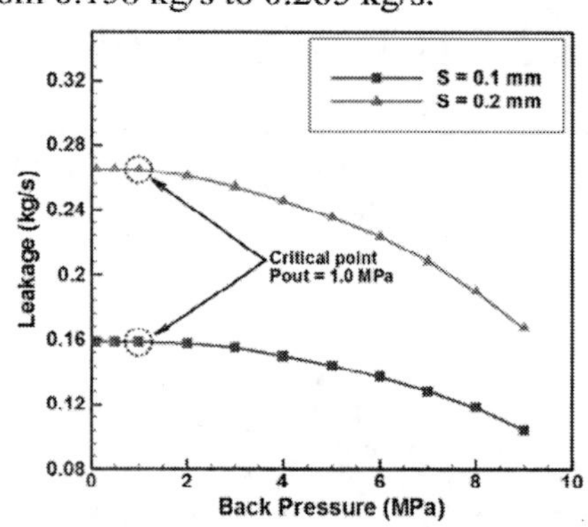

Figure 6. Variation of labyrinth seal leakage flow rate with back pressure.

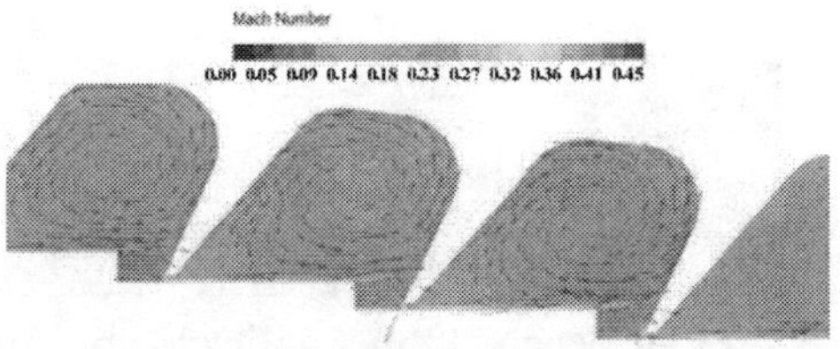

Figure 7. Mach number and velocity map of downstream chamber. (Pout=9 MPa)

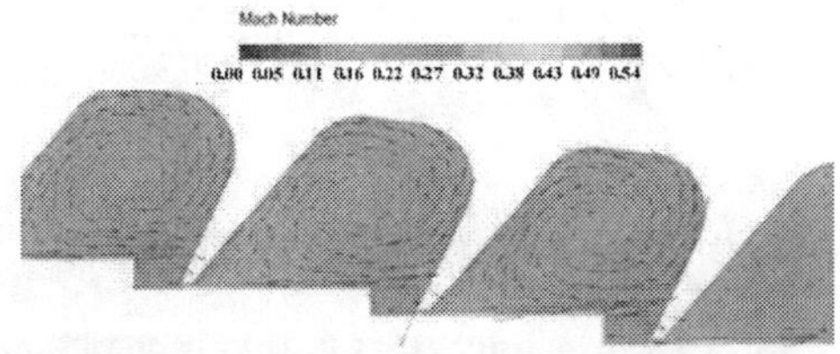

Figure 8. Mach number and velocity map of downstream chamber. (Pout=7 MPa)

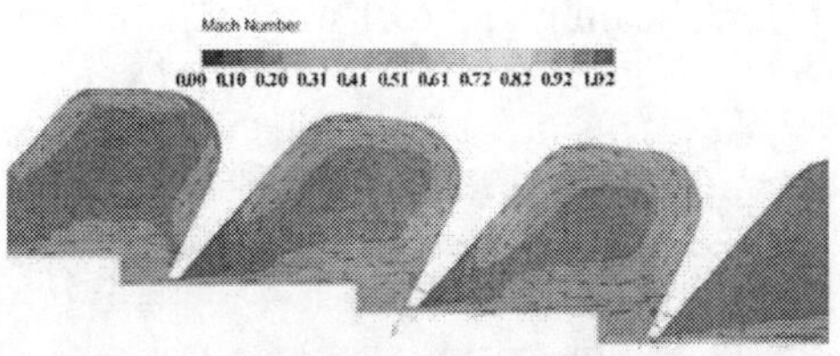

Figure 9. Mach number and velocity map of downstream chamber. (Pout=3 MPa)

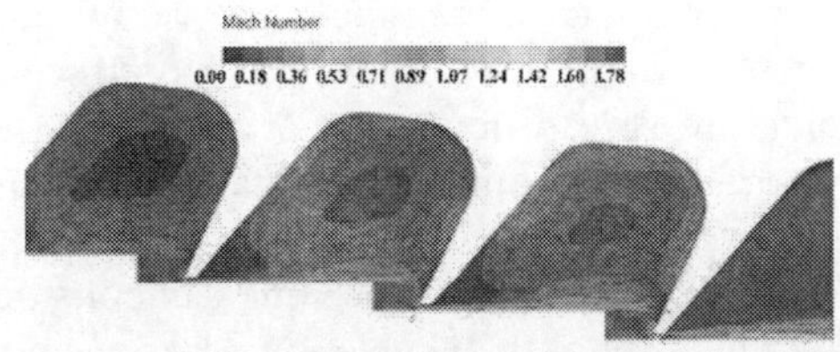

Figure 10. Mach number and velocity map of downstream chamber. (Pout=1 MPa)

When the labyrinth seal back pressure is reduced to 3.0 MPa, the local area near the rotor surface at the last seal tooth gap reaches the critical flow state (Ma = 1.0). When the labyrinth seal back pressure is further reduced to 1.0 MPa, the last seal tooth gap is completely occupied by the supersonic flow. The flow reaches the critical state and the leakage flow rate reaches its maximum.

4.2. Through-hole seal

Figure 11 shows the variation of the through-hole seal leakage flow rate with back pressure under different sealing gaps. The leakage flow rate of the through-hole seal increases as the back pressure decreases. When the seal gap increases from S = 0.1 mm to S = 0.2 mm, the critical leakage flow rate of the through-hole seal increases from 0.128 kg/s to 0.330 kg/s. When the through-hole seal back pressure is further reduced to 3.0 MPa, the last seal tooth gap is completely occupied by the supersonic flow. The flow reaches the critical state and the leakage flow rate reaches its maximum.

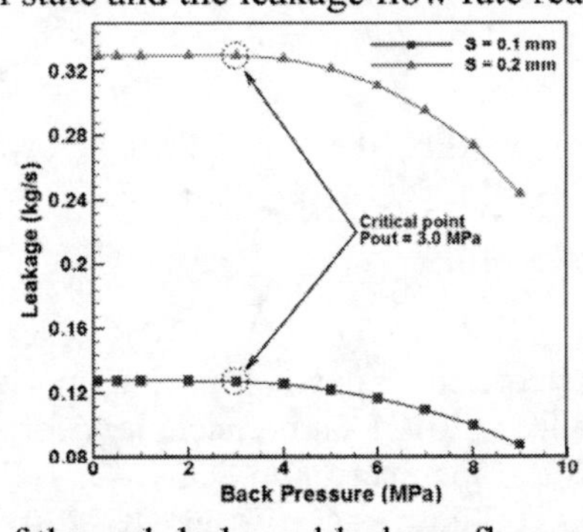

Figure 11. Variation of through-hole seal leakage flow rate with back pressure.

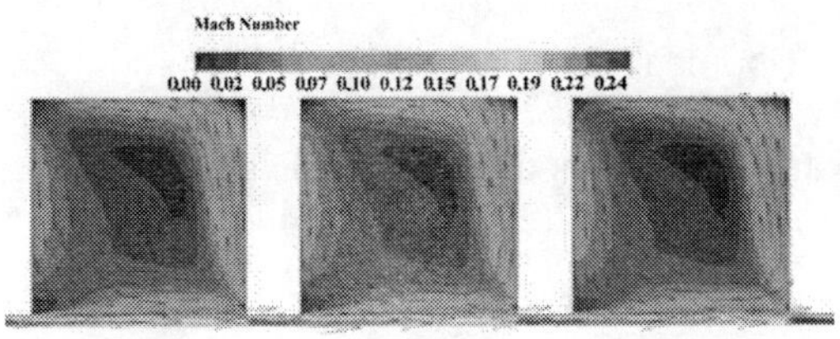

Figure 12. Mach number and velocity map of downstream chamber. (Pout=9 MPa)

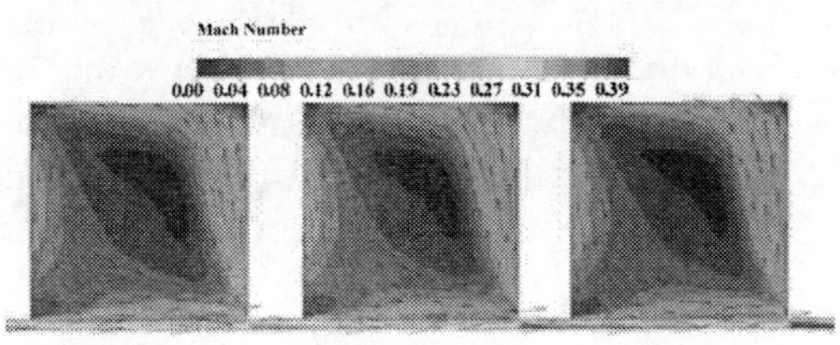

Figure 13. Mach number and velocity map of downstream chamber. (Pout=7 MPa)

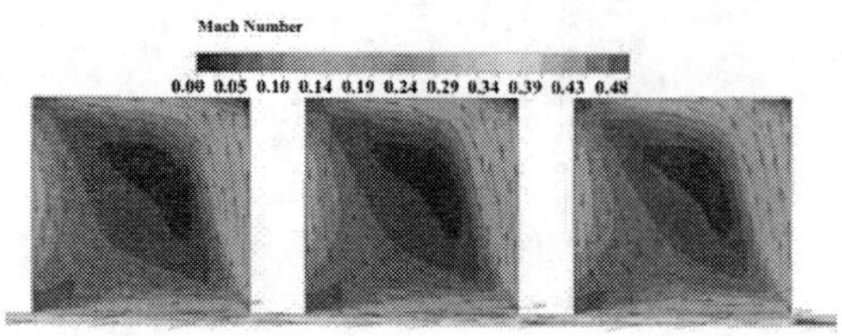

Figure 14. Mach number and velocity map of downstream chamber. (Pout=6 MPa)

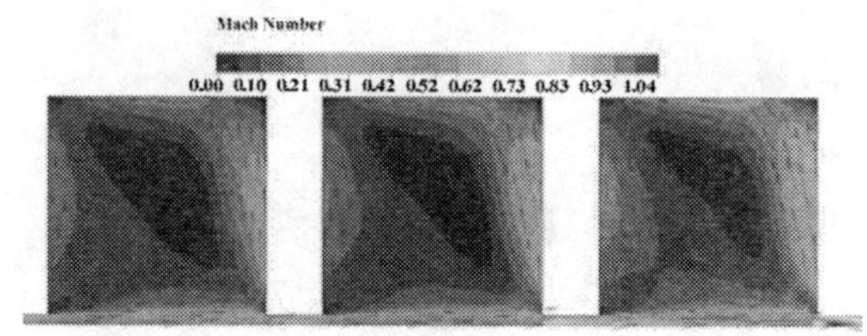

Figure 15. Mach number and velocity map of downstream chamber. (Pout=3 MPa)

According to the above analysis, it can be concluded that the S-CO_2 through-hole seal has better sealing performance than the S-CO_2 labyrinth seal at a small sealing gap (0.1 mm). The labyrinth seal has better sealing performance at larger sealing gap (S = 0.2 mm). The labyrinth seal can withstand greater pressure differentials at the same sealing gap. As a result, the stepped labyrinth seal has superior S-CO_2 sealing performance compared to the through-hole seal. Considering the dynamic characteristics of the seal flow excitation under high inlet pre-rotation conditions, the S-CO_2 compressor shaft end seal scheme can consider the combination of the stepped seal tooth rotor with the sealing hole stator to achieve low leakage and high damping sealing performance.

5. Conclusions

In this paper, the scheme design of stepped labyrinth seal and straight through-hole seal is performed for the S-CO_2 centrifugal compressor. Moreover, the sealing performance of two S-CO_2 shaft end seals under different outlet back pressures and sealing gaps are analyzed. The main conclusions are as follows:

1) The critical leakage flow rate of the labyrinth seal increases from 0.158 kg/s to 0.265 kg/s as the seal gap increases from 0.1 mm to 0.2 mm. The critical leakage flow rate of the through-hole seal increases from 0.128 kg/s to 0.330 kg/s as the seal gap increases from 0.1 mm to 0.2 mm.

2) The stepped labyrinth seal has superior S-CO_2 sealing performance compared to the through-hole seal.

3) In order to achieve the requirements of low leakage and high damping sealing performance, the S-CO_2 compressor shaft end seal scheme can consider the combination of the stepped seal tooth rotor with the sealing hole stator.

Acknowledgments

This research was financially supported by the Hubei Province Natural Science Foundation of China (NO. 2016CFA019, NO. 2017CFB325).

References

[1] Feng, Z.P., Zhao, H., Zhang, H.Z., Deng, Q.H. (2016) Research progress on supercritical carbon dioxide power cycle system and its power unit. Thermal Turbine, 45(2): 85-94.

[2] Fuller, R., Preuss, J., Noall, J. (2012) Turbomachinery for supercritical CO_2 power cycles. In: ASME Turbo Expo 2012. Copenhagen. pp. 735-742.

[3] Allision, T.C., Moor, J.J., Wikes, J.C., Brun, K. (2017) Turbomachinery overview for supercritical CO_2 power cycles. In: The 46th Turbomachinery & 33rd Pump Symposia. Houston. pp. 355-363.

[4] Conboy, T. (2010) Gas bearings and seals development for supercritical CO_2 turbomachinery. https://www.sandia.gov.

[5] Fairuz, Z., Ingo, J. (2016) The influence of real gas effects on the performance of supercritical CO_2 dry gas seals. International Journal of Tribology, 102(2016), 333-347.

IOP Conf. Series: Materials Science and Engineering **711** (2020) 012050 doi:10.1088/1757-899X/711/1/012050

Experimental Research on Fracture Toughness and Tensile Strength of Red Sandstone under Different Moisture Content

Chunting Li[1*], Tingwei Yuan[1]

[1]College of Civil Engineering and Architecture, Shandong University of Science and Technology, Qingdao, Shandong, 266590, China

*Corresponding author's e-mail: 996910097@qq.com

Abstract: In order to study the variation and correlation of mode I fracture toughness and tensile strength of red sandstone under different moisture content conditions, Brazilian splitting and NSCB tests were carried out on the red sandstone specimens with moisture content of 0%, 1.63% and 3.33% respectively by using the CMT53 test loading system, and the test data obtained were fitted and analyzed. The test results show that: 1. The tensile strength and fracture toughness of red sandstone gradually decrease with the increase of moisture content, and the softening effect of water on red sandstone is obvious, with the reduction coefficients of 0.72 and 0.66, respectively. 2. With the increase of moisture content, the dispersion of tensile strength and fracture toughness of red sandstone increases. 3. The fitting formulas of water content and tensile strength and fracture toughness are obtained by data fitting, and the fitting correlation is relatively high. The piecewise function is used to describe the relationship between tensile strength and fracture toughness. It is convenient to estimate the fracture toughness of red sandstone through its tensile strength and it has a certain reference value for studying the relationship between tensile strength and fracture toughness of red sandstone.

1. Introduction

The destruction and fracture of rocks are inseparable. A large number of practices and studies [1-3] show that the failure of rocks is often related to the microcracks inside the rock and the instability of macroscopic cracks. The ability to resist crack instability and resist brittle fracture is expressed by the fracture toughness K_{IC}, which is an inherent property of the material and one of the important mechanical properties of the rock. It plays an important role in evaluating the stability and safety of geotechnical engineering.

M. D. Kuruppu et al [4-5] considered that NSCB samples have more advantages in testing the mode I fracture toughness of rock materials under quasi-static loading conditions. R. J. Fowell et al [6-8] compared the three samples of SR, CB and CCNBD, and considered that the CCNBD sample test method is simple and can be used for the fracture toughness test of various fracture modes. Zhang Sheng et al [9] determined the fracture toughness of marble by using a central circular hole platform disc specimen with a single length of cracks of different lengths, and based on Bazant's scale law theory, the results were fitted according to the results of finite element analysis. A functional expression of the dimensionless energy release rate. Yang Jianfeng et al [10] on the mudstone experimental study, showing that the fracture toughness of mudstone has obvious changes under different soaking effects. Zuo Jianping [11-12] conducted real-time thermal cracking and tensile fracture tests on Beishan granite. It was found that in the range of 25°C ~ 100°C, Beishan granite is dominated by brittle failure, and the temperature is in the range of 100°C. The fracture behavior has little effect. Zhao Yixin et al [13] studied the tensile and fracture properties of coal under quasi-static loading conditions, and obtained that the J-integrated fracture toughness is more suitable for evaluating the fracture properties of coal than the plane strain fracture toughness K_{IC}.

At present, some researches have been made on the relationship between fracture toughness and strength parameters of rock at home and abroad, but few studies have been conducted on rock mode I fracture toughness and tensile strength under different water content conditions. In view of this, this paper selects red sandstone as the research object for Brazilian splitting and NSCB test, aiming to investigate the influence of water on tensile strength and mode I fracture toughness of red sandstone and the relationship between tensile strength and fracture toughness. Through the research, the fitting formulas of tensile strength, mode I fracture toughness and moisture content of red sandstone and tensile strength and mode I fracture toughness are obtained. It is easy to estimate the mode I fracture toughness by tensile strength of red sandstone. It provides some reference value and experimental data support for the correlation between rock mode I fracture toughness and tensile strength.

1.1. Sample preparation and test plan

According to the Brazilian disc preparation method recommended by ISRM [14], 36φ50mm×25mm disc specimens were processed, and 27 Brazilian discs were cut to form 27 SCB specimens, which were divided into A, B and C. Then three groups of NSCB specimens with groove width of 1 mm and depth of 4, 7 and 10 mm were prepared by slitting at the bottom center of group A, B and C, three samples in each group. Group A specimens were dried (water content: 0%), which was placed in an oven at a temperature of 105° for 24h. Group C was subjected to saturation treatment (water content: 3.33%), which was immersed in water for 48 hours. Group B was first dried, and then immersed in water until the water content was about 1/2 of C group (water content: 1.63%). The test piece parameters are detailed in Table 1. In the same way, the remaining 9 Brazilian discs were finally divided into three groups of D, E and F, which were respectively dried, semi-saturated and saturated for the Brazilian splitting test. The parameters of the test pieces are shown in Table 2.

Table 1. Parameters of NSCB specimens.

	Number	D (mm)	t (mm)	m (g)	Cutting depth a (mm)	a/R	Moisture content w (%)
A	A4-1	50.12	25.36	60.82	3.9	0.16	0
	A4-2	50.05	25.14	60.02	3.94	0.16	0
	A4-3	49.94	25.02	58.21	3.92	0.16	0
	A7-1	49.92	26.07	60.71	6.87	0.28	0
	A7-2	50.05	25.81	60.44	6.98	0.28	0
	A7-3	49.96	26.15	61.18	7.04	0.28	0
	A10-1	50.08	25.56	60.04	9.88	0.40	0
	A10-2	50.15	25.59	60.32	9.9	0.40	0
	A10-3	50.07	25.49	59.91	9.95	0.40	0
B	B4-1	49.95	25.24	59.17	3.94	0.16	1.80
	B4-2	50.08	25.57	60.24	3.97	0.16	1.63
	B4-3	50.06	24.94	58.98	3.93	0.16	1.67
	B7-1	50.03	25.14	59.94	7.05	0.28	1.65
	B7-2	50.12	24.83	60.25	7.03	0.28	1.60
	B7-3	50.17	25.52	59.93	7.01	0.28	1.41
	B10-1	50.06	25.59	59.91	9.99	0.40	1.54
	B10-2	50.07	25.73	61.42	10.00	0.40	1.73
	B10-3	50.10	25.42	60.07	9.92	0.40	1.66
C	C4-1	50.03	24.86	58.26	3.94	0.16	3.50
	C4-2	50.08	24.96	59.69	3.98	0.16	3.39
	C4-3	49.95	25.25	59.32	4.00	0.16	3.29
	C7-1	49.98	25.12	58.04	6.96	0.28	3.37
	C7-2	49.90	24.68	56.72	7.02	0.28	3.27
	C7-3	50.12	25.61	60.39	7.06	0.28	3.30
	C10-1	50.04	25.44	59.64	10.04	0.40	3.51
	C10-2	50.14	25.41	60.76	10.00	0.40	3.10
	C10-3	50.03	25.12	58.27	10.02	0.40	3.23

Table 2. Parameters of Brazilian disc specimens.

	Number	*D* (mm)	*H* (mm)	Moisture content *w* (%)
D	D1	50.02	25.02	0
	D2	50	25.7	0
	D3	49.98	25.68	0
E	E1	50.08	25.8	1.71
	E2	50.02	25.2	1.68
	E3	50.04	26.12	1.73
F	F1	50.06	25.26	3.38
	F2	50.06	26.08	3.40
	F3	50.08	26.02	3.47

ISRM recommends that when using the NSCB method to test the mode I quasi-static fracture toughness of rock materials, the distance S between the support points and the diameter D of the sample should satisfy $0.5 \leq S/D \leq 0.8$ [4]. In this paper, S/D=0.8 is selected for testing. The geometry of the sample required for the test is shown in figure 1.

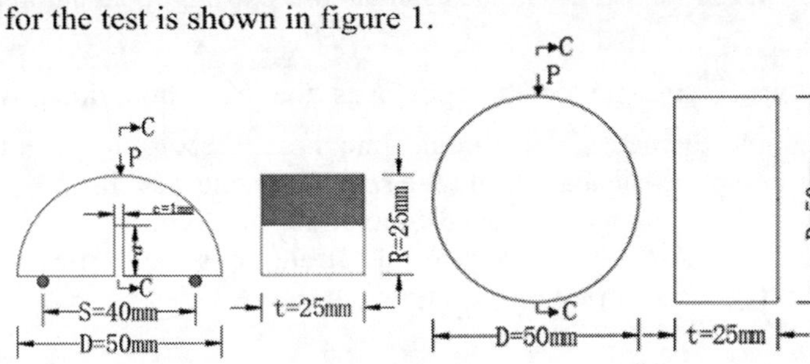

Figure 1. Geometric figure of NSCB and Brazilian disc specimens.

1.2. Experimental scheme

The Brazilian splitting and NSCB tests were carried out with the CMT53 electronic universal test loading device in Shandong key laboratory of Shandong university of science and technology. All tests were displacement control loading with a loading rate of 0.1mm/min, which was in line with the requirements of static crack growth (the loading rate was no more than 0.2mm/min) [4]. The system automatically recorded the load and displacement data when the testing machine was loaded. The red sandstone NSCB sample and the Brazilian disc sample are placed as shown in figure 2.

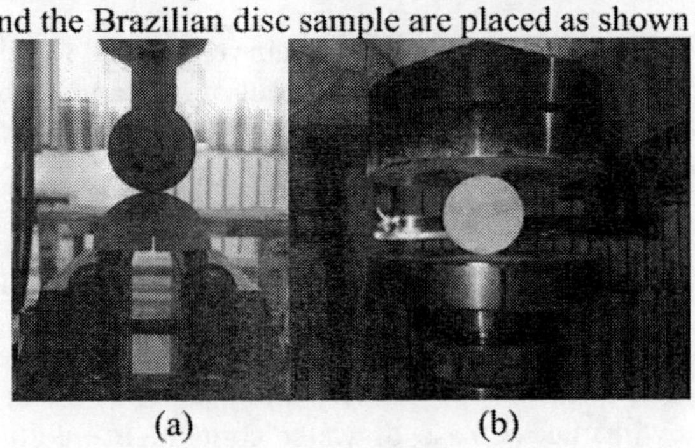

(a) (b)

Figure 2. Placement of red sandstone NSCB and Brazilian disc specimens.

2. Red sandstone-like BD splitting test

Table 3 gives the parameters such as the diameter, height and failure load of the red sandstone Brazilian disc specimen.

Table 3. BD splitting test results.

Number		D (mm)	H (mm)	P (N)	σ_t (MPa)	σ_{ta} (MPa)
	1	50.02	25.02	11029.01	5.61	
D	2	50	25.7	10625.55	5.26	5.5
	3	49.98	25.68	11366.74	5.64	
	1	50.08	25.8	9214.24	4.54	
E	2	50.02	25.2	9305.98	4.7	4.48
	3	50.04	26.12	8623.02	4.2	
	1	50.06	25.26	8400.11	4.23	
F	2	50.06	26.08	8195	4	3.95
	3	50.08	26.02	7410.87	3.62	

The tensile strength of red sandstone can be calculated according to equation [15]:

$$\sigma_t = 2P / \pi DH \tag{1}$$

Where, σ_t is the tensile strength of the rock; P is the maximum load when the specimen is destroyed; D is the sample diameter; H is sample thickness. According to equation (1), the tensile strength of red sandstone can be calculated, and the specific results are shown in table 3.

According to the tensile strength of red sandstone under different water content, a scatter diagram with water content as the horizontal axis and tensile strength as the vertical axis was drawn, and a quadratic polynomial fitting was carried out, as shown in figure 3.

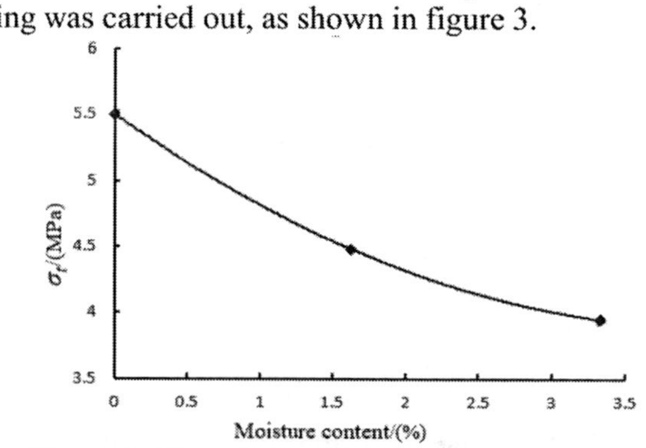

Figure 3. The relationship curve between tensile strength and moisture content of red sandstone.

Through data fitting, the composite formula of tensile strength and water content of red sandstone is obtained. Correlation coefficient R^2=1, and the fitting formula is:

$$\sigma_t = 0.0842w^2 - 0.7412x + 5.5 \tag{2}$$

By observing figure 3 and table 3, it can be found that the softening effect of water on the tensile strength of red sandstone is relatively obvious, and the reduction coefficient of the tensile strength is 0.72. With the increase of water content, the tensile strength gradually decreases. Secondly, the slope of the fitting curve decreases with the increase of water content, indicating that the change rate of tensile strength decreases with the increase of water content. In addition, the dispersion of tensile strength is also significantly correlated with water content. Dry red sandstone has the highest tensile strength, with the average tensile strength of 5.5MPa and the dispersion coefficient of 0.031. The tensile strength was 4.48MPa and the dispersion coefficient was 0.047. The saturated tensile strength is the lowest, 3.95MPa, and the discrete coefficient is 0.064. It shows that the higher the water content is, the greater the discreteness of the tensile strength of red sandstone is, which indicates that the measured data deviate from the stable value.

3. Fracture toughness test of red sandstone NSCB

The test results of 25 specimens (A4-3 and A7-2 specimens were not obtained) were measured. According to the test results, it was found that: the prefabricated slit depth and water weakened the peak strength of the red sandstone sample significantly, and the average peak load of the specimen gradually decreased with the increase of slit depth and water content.

The fracture toughness of specimens depends on peak load, incision length and specimen thickness. At present, the commonly used calculation formula of K_{IC} of plane strain fracture toughness of NSCB samples of rock materials is as follows [16]:

$$K_{IC} = Y'P_{\max}\sqrt{\pi a}/(2Rt) \tag{3}$$

$$Y' = -1.297+9.516[s/(2R)]-\{0.47+16.457\cdot[s/(2R)]\}\beta+\{1.071+34.401[s/(2R)]\}\beta^2 \tag{4}$$

Where, β is the slit depth of dimension one, $\beta = a/R$; R is the specimen radius (mm); a is slit depth (mm); t is the thickness of the sample (mm); Y' is a dimensionless stress intensity factor. According to the formula (3) and (4) available red sandstone mode I fracture toughness values, the specific measured parameters and the calculation results are shown in table 4.

Table 4. Calculations of NSCB Fracture Toughness Test for Red Sandstone.

	Number	a/R	Peak load (N)	Y'	Plane strain fracture toughness value (MPa·mm$^{1/2}$)	Average value of plane strain fracture toughness (MPa·mm$^{1/2}$)
	A4-1~4-3	0.16	1526±11	4.88	20.65±0.03	
A	A7-1~7-3	0.28	1277±83	4.74	21.7±1.3	21.46±0.59
	A10-1~10-3	0.40	952±24	5.31	22.03±0.29	
	B4-1~4-3	0.16	1491±72	4.87	20.40±0.97	
B	B7-1~7-3	0.28	999±98	4.73	17.62±1.92	19.29±1.2
	B10-1~10-3	0.40	840±62	5.41	19.85±1.49	
	C4-1~4-3	0.16	964±105	4.87	13.25±1.42	
C	C7-1~7-3	0.28	651±16	4.74	11.52±0.22	13.35±1.55
	C10-1~10-3	0.40	637±37	5.43	15.30±0.99	

As shown in figure 4, the mode I fracture toughness K_{IC} along with the change of moisture content of quadratic polynomial fitting, the correlation coefficient R^2=1, the fitting formula is:

$$K_{IC} = -0.6523w^2 - 0.2647w + 21.46 \tag{5}$$

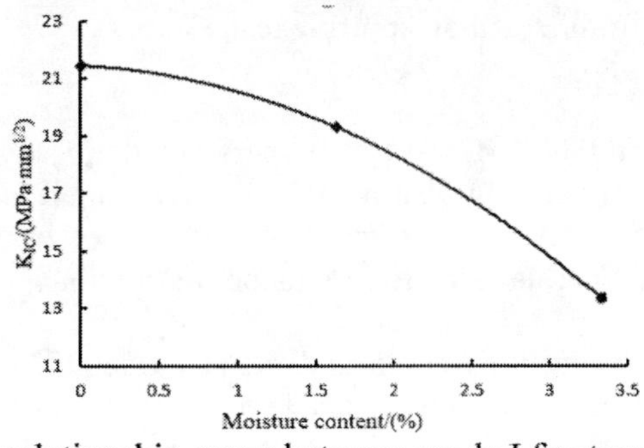

Figure 4. The relationship curve between mode I fracture toughness K_{IC} and moisture content of red sandstone.

Type can be found: the mode I fracture toughness K_{IC} decreases with increasing moisture content and slope increases gradually. Water content has a certain influence on the discreteness of plane strain fracture toughness of red sandstone. The average of mode I fracture toughness under dry condition is 21.46MPa·mm1/2 and the dispersion coefficient is 0.027; the second under semi-saturated condition is 19.29MPa·mm1/2 and the dispersion coefficient is 0.062; while the lowest under saturated condition is 13.35MPa·mm1/2 and the dispersion coefficient is 0.116. Second, the water is obtained by calculation of red sandstone fracture toughness of softening coefficient is 0.66, showing that water in I red sandstone mode fracture toughness K_{IC} softening effect is obvious.

4. Correlation analysis between mode I fracture toughness and tensile strength of red sandstone

The linear fitting and polynomial fitting of the obtained experimental data show that the linear fitting correlation coefficient is less than 0.82, and the polynomial fitting does not satisfy the general rule that the fracture toughness increases with the increase of moisture content, and cannot accurately describe the relationship. Therefore, a piecewise straight line is used to describe the relationship between mode I fracture toughness K_{IC} and tensile strength (as shown in figure 5.). The broken line diagram is divided into two straight lines. The formulas for each straight line are as follows:

$$K_{IC} = 0.6345w + 1.1035(3.95 \le w \le 4.48), K_{IC} = 0.9279w - 0.6237(4.48 < w \le 5.5) \qquad (6)$$

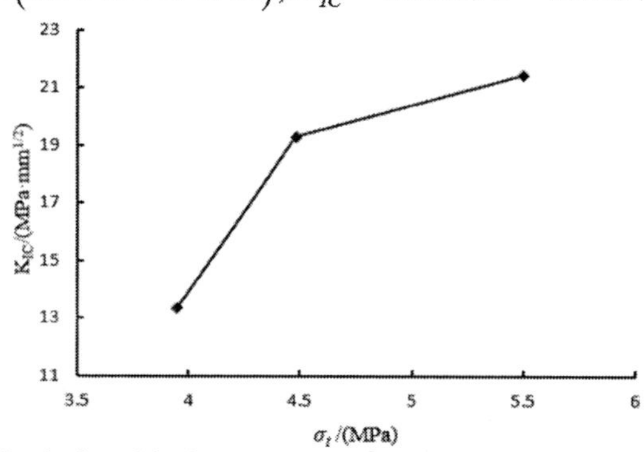

Figure 5. The line chart of relationship between mode I fracture toughness K_{IC} and tensile strength.

It can be found from the line chart that the K_{IC} of red sandstone increases with the increase of tensile strength, that is, the tensile strength and fracture toughness of red sandstone reach the highest under dry conditions, and reach the lowest under saturated conditions. Secondly, the slope of the line in the interval [3.95,4.48] is significantly larger than the slope of the line in the interval (4.48,5.5), indicating that the rate of change of the mode I fracture toughness K_{IC} increases with the increase of the tensile strength. The segmental straight line is used to describe the correlation between the K_{IC} of the red sandstone mode I fracture toughness and the tensile strength under different moisture content conditions, which is more suitable for the data obtained from the test. When the moisture content of red sandstone is known, it will be more convenient to calculate the mode I fracture toughness K_{IC} by the tensile strength obtained from Brazilian splitting test.

5. Conclusion

(1) Brazilian disc splitting and NSCB tests were performed on dry, semi-saturated and saturated sandstone samples. The test results show that the tensile strength and mode I fracture toughness of red sandstone decrease with the increase of water content, and the softening effect of water on the tensile strength and mode I fracture toughness of red sandstone is obvious, and the reduction factor is 0.72 and 0.66.

(2) The fitting formulas of tensile strength and mode I fracture toughness of red sandstone with water content are obtained. The correlation coefficients are R^2=1. According to the fitting formulas, the tensile strength and mode I fracture toughness of red sandstone under different moisture content can be calculated.

(3) With the increase of moisture content, the tensile strength and fracture toughness of red sandstone are more and more discrete.

(4) Based on the experimental data, the break-line diagrams of mode I fracture toughness K_{IC} and tensile strength of red sandstone under different moisture content are obtained, and the formula of piecewise straight line is obtained. Through the formula, it can be found that the mode I fracture toughness of red sandstone increases with the increase of tensile strength, and the linear slope in the interval [3.95,4.48] is obviously larger than that in the interval (4.48,5.5). Therefore, the piecewise function is more suitable for describing K_{IC} and tensile strength of mode I fracture toughness of red sandstone.

References

[1] Zhu Weishen, Chen Weizhong, Shenjin. (1998) Model test of goose-shaped crack growth and fracture mechanics mechanism [J]. Journal of Solid Mechanics, 19 (4): 355-360.
[2] RAMSAY J G. (1967) Folding and fracturing of rocks[M]. New York: Mc Graw-Hill.
[3] Ma Yongshan, Chen Weizhong, Yang Diansen, et al. (2017) Brittle rock failure test based on three-dimensional digital image correlation technology [J]. Geotechnical Mechanics, 38 (1): 117-123.
[4] KURUPPU M D, OBARA Y, AYATOLLAHI M R, et al. (2014) ISRM-suggested method for determining the mode I static fracture toughness using semi-Circular bend specimen[J]. Rock Mechanics and Rock Engineering, 47(1): 267–274.
[5] KURUPPU M D, CHONG K P. (2012) Fracture toughness testing of brittle materials using semi-circular bend (SCB) specimen[J]. Engineering Fracture Mechanics, 91: 133–150.
[6] FOWELL R J. (1995) Suggested method for determining mode I fracture toughness using cracked chevron notched Brazilian disc (CCNBD) specimens[J]. International Journal of Rock Mechanics and Mining Sciences, 32(1): 57–64.
[7] OUCHTERLONY F. (1988) Suggested methods for determining the fracture toughness of rock[J]. International Journal of Rock Mechanics and Mining Sciences, 25(2): 71–96.
[8] IQBAL M J. MOHANTY B. (2007) Experimental calibration of ISRM suggested fracture toughness measurement techniques in selected brittle rocks[J]. Rock Mechanics and Rock Engineering. 40(5): 453–475.
[9] Zhang Sheng, Wang Qizhi. (2007) Determining the fracture toughness of marble by single size specimen with variable cracks [J]. Engineering Mechanics, 24 (6): 31-35.
[10] Yang Jianfeng, Liang Weiguo, Chen Yuedu, et al. (2017) Experimental study on fracture mechanical properties of mudstone under different water damage degrees [J]. Journal of Rock Mechanics and Engineering, 36 (10): 2431-2440.
[11] Zuo Jianping, Zhou Hongwei, Fangyuan, et al. (2012) Experimental study on thermo-mechanical coupled fracture characteristics of Beishan granite with double notches [J]. Journal of Rock Mechanics and Engineering, 31(4): 738-745.
[12] Zuo Jianping, Zhou Hongwei, Fangyuan, et al. (2011) Experimental study on thermal cracking of deep granite in Beishan area, Gansu Province [J]. Journal of Rock Mechanics and Engineering, 30 (6): 1107-1115.
[13] Zhao Yixin, Gong Shuang, Jiang Yaodong, Han Chao. (2016) Study on tensile and fracture properties of coal samples based on semi-circular bending test [J]. Journal of Rock Mechanics and Engineering, 35 (06): 1255-1264.
[14] ZHOU Y X. XIA K. LI X B. et al. (2012) Suggested methods for determining the dynamic strength parameters and mode–I fracture toughness of rock materials[J]. International Journal of Rock Mechanics and Mining Sciences. 49(1): 105–112.
[15] Cai Meifeng, He Manchao, Liu Dongyan. (2013) Rock Mechanics and Engineering [M].: Science Press, Beijing.
[16] KURUPPU M D, CHONG K P. (2012) Fracture toughness testing of brittle materials using semi-circular bend (SCB) specimen. Engineering Fracture Mechanics[J], 91: 133–150.

IOP Conf. Series: Materials Science and Engineering **711** (2020) 012051 doi:10.1088/1757-899X/711/1/012051

Effect of mechanical juice extraction method on the quality of fresh-squeezed apple juice

Zhengchen Wang[1], Huanyong Cui[1*] and Shichao Fan[1]

[1] School of Mechanical Engineering University of Jinan, China

[*]Corresponding author's e-mail: 747750112@qq.com

Abstract. In order to study the effect of different mechanical juice extraction methods on the quality of fresh-squeezed apple juice, the optimal fresh-squeezed apple juice processing method was selected. The fresh-squeezed apple juice is processed by three kinds of mechanical juice extraction methods: squeezed juice, spiral juicer and broken juice. Five main evaluation indexes of soluble solid content, titratable acid content, value of b*, value of a* and 625nm transmittance of fresh-squeezed apple juice obtained by three mechanical juice extraction methods were determined. Through the analysis and research of the data obtained from the five indicators, the influence of different mechanical juice extraction methods on the quality of fresh-squeezed apple juice was determined, and the mechanical juice extraction method suitable for processing fresh-squeezed apple juice was determined.

1. Introduction

Apple is widely distributed all over the world and is one of the world's major fruits [1]. Apples are rich in minerals and vitamins and have high nutritional value. According to statistics from the Food and Agriculture Organization of the United Nations, China's apple production in 2017 was 41.129 million tons, accounting for 49.79% of the world's apple production. China is the world's largest apple producer [2]. At the same time, China is also the world's largest producer and exporter of concentrated apple juice. With the progress and development of society, consumers' attention to health is increasing. Fresh-squeezed apple juice, which retains apple nutrition to a high degree, and has a good natural flavor, is increasingly popular among consumers. In the United States and Europe, fresh-squeezed apple juice sales account for more than 40% of the apple juice market share [3].

The core process of fresh-squeezed apple juice is juicing, and different methods of juicing directly affect and determine the quality of fresh-squeezed apple juice. At present, the research on the processing methods of fruit and vegetable juices at home and abroad mainly focuses on large industrial juicers such as hydraulic tube juicers and belt juicers, and there are few studies on small household juicers [4]. The research on the influence of processing technology on the quality of fruit and vegetable juice is mainly concentrated on the finished fruit and vegetable juice, and the research on the quality of fresh-squeezed fruit and vegetable juice is less [5]. In recent years, the domestic fresh-squeezed fruit and vegetable juice market has continued to expand, and the sales of household fruit juicers have been increasing. At the same time, various vending fresh-squeezing juice machines have appeared in cities. Studying the effects of different processing techniques on the quality of fresh-squeezed apple juice is of great significance for maintaining the natural nutrient composition and functional properties of fresh-squeezed apple juice.

At present, there is no domestic evaluation standard for fresh-squeezed juice. National Standard GB/T 31121-2014 "Fruit and Vegetable Juices and Beverages" stipulates that raw juices (non-rejuvenated juices), which are made from fruits, are fermentable but unfermented and unconcentrated juice product made directly by mechanical means. Raw juice (non-reconstituted juice) made by non-heat treatment or pasteurized sterilization can be called fresh juice [6]. Jiyun Nie [7] preliminarily constructed an fresh-squeezed apple juice evaluation system through factor analysis, probability distribution and analytic hierarchy process. Jiankang Deng [8] established a comprehensive evaluation model of apple (substandard fruit) juice quality based on analytic hierarchy process and gray correlation method. Among them, the five main evaluation indexes for fresh-squeezed apple juice quality are: soluble solid content, titratable acid content, value of b*, value of a*, and 625nm transmittance.

This study is mainly aimed at the three kinds of mechanical juice extraction methods of common broken juice, squeezed juice and spiral juice on the market. According to the five main evaluation indexes affecting the quality of fresh-squeezed apple juice, the effects of the above three mechanical juice extraction methods on the quality of apple fresh juice were studied. By comparing the experimental results and data, the advantages and disadvantages of different mechanical juice extraction methods are analyzed, which provides a scientific basis for the selection of apple mechanical juice extraction methods.

2. Materials and Methods

2.1. Materials and reagents

Fuji apple, the origin of Yantai, Shandong, has a complete appearance, no mechanical damage, and is stored in a 4 ° C cold storage. Sodium hydroxide, Damao Chemical Reagent Factory. ascorbic acid, phenolphthalein, Sinopharm Chemical Reagent Co., Ltd.

2.2. Equipment and instruments

Squeezing juicer, SZ-18 Baijie stainless steel manual juicer, Deqing Baijie Electric Co., Ltd.. spiral juicer, Shuangmeijian Juicer, Linyi Shuangjian Food Machinery Co., Ltd.. Broken juicer, TL-780B green baby broken wall nutrition conditioning machine, Foshan Shunde District Good Life Electronic Technology Co., Ltd.. HT119ATC refractometer, WGZ-200BS desktop turbidity meter, Shenzhen Yuanhengtong Technology Co., Ltd.. WSC-2B portable precision color difference Instrument, Shanghai Yidian Physical Optical Instrument Co., Ltd.. V-5000 spectrophotometer, Shanghai Yuanxi Instrument Co., Ltd.. YP20002 electronic balance, FA2004B electronic balance Shanghai Youke Instrument Co., Ltd.. LXJ-IIB low speed large capacity multi-tube centrifugation Machine, Shanghai Anting Scientific Instrument Factory.

2.3. Experimental method

Take the experimental apple, wash it with ultrapure water, peeling and denucleating, and divide the pulp into rectangular pieces of no more than $1cm^3$. The apples were juiced after the splitting using a squeeze juicer, a spiral juicer and a broken juicer. The weight of the pulp and the weight of the obtained juice were recorded, and the juice yields of the three juice extraction methods were obtained (Table 1). Add ascorbic acid to the fresh-squeezed apple juice (addition amount 0.2g/kg). Put the juice into the centrifuge, centrifuge at 4500 rpm for 10 minutes. Filter 8 layers of gauze. Seal the filtered juice into a refrigerator at 5 °C stand-by.

The fresh-squeezed apple juice was sampled by three different juice extraction methods, and the soluble solid content was determined by HT119ATC refractometer. The titratable acid content was determined by the indicator titration method [9]. The value of b* and value of a* was measured by WSC-2B portable precision colorimeter. The 625nm transmittance (T625) was measured using a V-5000 spectrophotometer. Three times for each experiment, the results were averaged (Table 2).

Table 1. Juice rate

	Broken juice	Squeeze juice	Spiral juice
Average weight of pulp	385.5g	423.26g	269.39g
Average juice weight	113.38g	162.59g	126.73g
Juice rate	29.4%	38.4%	47.0%

Table 2. Five main evaluation indicators

	Broken juice	Squeeze juice	Spiral juice
Soluble solids	13.0	12.5	16.0
Titrate acid	1.7152	2.0904	1.5812
value of b*	2.32	2.28	4.49
value of a*	0.82	1.86	1.22
625nm transmittance	34.1	81.1	23.9

3. Result analysis

The soluble solids in fresh-squeezed apple juice are mainly soluble sugars. The higher the soluble solids value, the higher the sugar content in fresh-squeezed apple juice. The titratable acid content in fresh-squeezed apple juice (calculated as malic acid), the higher the value indicates the higher malic acid content. The sweetness in the juice is related to the content of soluble solids and titratable acid. When the value of a* is positive, the value of a* is larger, the color is redder. When the value of a* is negative, the absolute value of the value of a* is larger, the color is greener. When the value of b* is positive, the value of b* is larger, the color is yellower. When the value of b* is negative, the absolute value of the value of b* is larger, the color is bluer. The color of the apple juice affects it's value of a* and value of b*. The greater the transmittance at 625nm, the better the light transmittance, which is an important indicator reflecting the quality of apple turbid juice.

3.1. Broken juice

According to the data in Table 1 and Table 2, it can be seen that the broken juice method has the lowest juice yield. The soluble solid content of the fresh-squeezed apple juice obtained is lower than that of the spiral juicer, which is higher than the squeezed juicer. The titratable acid content (calculated as malic acid) and the light transmittance at 625nm are lower than the squeezed juice extraction method, which is higher than the spiral juice extraction method. The sweetness of the obtained juice is between the other two juice extraction methods. The value of b* is slightly larger than the squeezed juice method, much smaller than the spiral juice extraction method. The value of a* is the smallest. The resulting juice color is the worst. The main juicer of the broken juice is the rotating blade (Fig. 1). During the work, the apple pulp is cut by the high-speed rotation of the rotating blade, and the larger fruit is cut into finely divided apple pulp, which increases the contact area between the apple pulp and the air and oxidative browning speed [10]. The apple juice is obtained by mixing the apple pulp with the apple juice (Fig. 2), which needs further separation by high speed centrifugation, and is not suitable for processing fresh-squeezed apple juice.

Figure 1. Rotating blade.

Figure 2. Apple pulp.

3.2. Squeeze juice

According to the data in Table 1 and Table 2, it can be seen that the juice extraction rate of the squeezed juice method is higher than that of the broken juice, which is lower than that of the spiral juice. The separation of the pulp and the juice during the work does not require high-speed centrifugation. The obtained fresh-squeezed apple juice has the lowest soluble solid content. The titratable acid content (calculated as malic acid) and transmittance at 625nm are the highest. The juice is relatively acidic. The value of b* is the smallest and the value of a* is the largest, so the resulting juice has the best color. The main juice extraction mechanism of the squeezed juice is a conical upper cover and a juice net (Fig. 3). The flesh is placed between the conical top cover and the juice net. When juicing, the conical top cover and the juice net are close to each other to squeeze the apple pulp in the middle. Only one action is required to complete, and the juice speed and efficiency are high. Due to the quickest juice extraction process, the juice has the least oxidative browning and the juice color is the best.

Figure 3. Conical upper cover, juice net.

3.3. Spiral juice

According to the data in Tables 1 and 2, the spiral juice extraction method has the highest juice yield. During the working process, the pulp and juice can be automatically separated without high-speed centrifugation. The obtained fresh-squeezed apple juice has the highest soluble solid content. The titratable acid content (calculated as malic acid) and transmittance at 625nm are the lowest. The juice is relatively sweet. The value of b* is the largest and the value of a* is between the broken juice and the squeezed juice, so the color of the juice obtained is between the other two juice extraction methods. Since the spiral juicer structure is the screw rod (Fig. 4), the pitch on the screw is decreasing, the pressing force is increasing when juicing, and the pulp is squeezed layer by layer, and the juice yield is the highest. However, the juice extraction process increases the contact time and area of the juice with the air, resulting in a higher degree of oxidative browning of the freshly squeezed apple juice than the squeeze juicer, but lower than the crushed juicer.

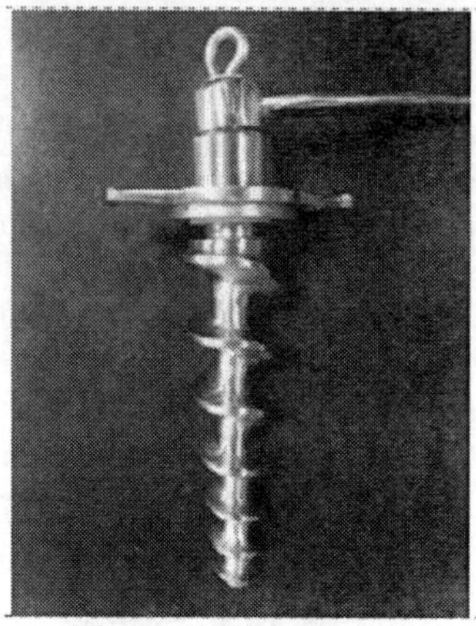

Figure 4. Screw rod.

4. Conclusion

It can be seen from the above experimental results that different mechanical juice extraction methods will have a greater impact on the quality of fresh-squeezed apple juice. In terms of the quality of the juice and the juice yield, the broken juice has the worst effect, and the pulp obtained after the juice extraction needs further separation, so it is not suitable for the fresh juice processing of apples. The quality of the fresh-squeezed apple juice obtained by the squeeze juicer is optimal without considering the juice rate. the juice yield of the fresh-squeezed apple juice using the spiral juicer is the best regardless of the quality of the juice. If the juice rate and juice quality are combined, the best overall performance is the squeeze juicer, which takes into account both the quality and the juice yield. It is the most suitable mechanical method for fresh-squeezed apple juice production.

References

[1] Jiang, A.M. (2007) Food Materials Science. Southeast University Press, Nanjing. 10-43.

[2] FAO Statistical Database, (2019). http://www.fao.org/faostat/en/?#data/QC.

[3] Li, J., Gao, H.S., Wang, Z.F., et al. (2006) Prediction model of survival / death probability of three pathogens in fresh apple juice. Chinese Journal of food science, (04): 11-18.

[4] Ma, S.B., Zhang, S.Y., Wang, J. (2016) Design and testing machine verification of a new continuous stacking juicer. Food and Machinery, 32(05): 77-80+224.

[5] Chen, X.H., Qin, W.D., Ma, L.H., et al. (2014) Study on the influence of processing conditions on the quality of fruit and vegetable juices. Science and Technology of Food Industry, 35(01):355-362.

[6] GB/T 31121-2014, fruit and vegetable juices and their drinks.

[7] Nie, J.Y., Qi, Y.L., Li, H.F., et al. (2013) Construction of apple fresh juice quality evaluation system. Chinese Agricultural Science, 46(08): 1657-1667.

[8] Deng, J.K., Liu, X., Wu, X.H., et al. Evaluation of apple (equivalent fruit) juice quality based on analytic hierarchy process and grey correlation degree 0 method. Journal of Chinese Institute of Food Science, 17(04): 197-208.

[9] GBT12456-2008, Determination of total acid in food.

[10] Liang, Y.N., Ye, F.Y., Lei, L., et al. (2018) Research progress on brown juice control technology of apple juice. Food and Fermentation Industries, 44(03): 280-286.

Study on the preparation high performance microfiber glass blanket with high water resistance for buildings

Xiaotian Hao[1,2], Qingjiang Wang[1], Hailong Wang[3,*]

[1] College of Urban Construction Engineering, Chongqing Technology and Business Institute, Chongqing 401520;

[2] Key Laboratory of Roads and Railway Engineering Safety Control (Shijiazhuang Tiedao University), Ministry of Education, Shijiazhuang 050043;

[3] School of Civil Engineering, Hebei University of Architecture, Zhangjiakou 075000;

Abstract: The high performance microfiber glass blanket with high water resistence is mostly used as insulation and sound absorption materials in the construction field, such as roof insulation materials, wall insulation materials. The high performance microfiber glass blanket was prepared by centrifugal method with spraying binder. The influence of fiber diameter and aging time on the water resistance, strength, filtration efficiency and resistence was investigated. The results show that the optimum kiln pressure is 24±2Pa, kiln temperature is 490±5°C and leakage current is 787±5mA); the optimum dosage of phenolic resin compound and water-resistant agent is 7% and 0.4% respectively; the fracture strength of the prepared samples in the direction of length and width is better than that of the same products abroad by using the above optimum preparation conditions. The microfiber glass mat was prepared by the flame method with spraying binder at the same time. The preparation condition of microfiber glass mat was mainly investigated and optimized. Shing currency was 787±5mA. The optimal added content of phenolic resin adhesive and water resistant agent was 7% and 0.4%, respectively. Through utilizing the optimal conditions, the prepared microfiber glass mat obtained more excellent mechanical properties than the overseas products.

1. Introduction

Microfiber glass blanket with high water resistence is mostly used as insulation and sound absorption materials in the construction field, such as roof insulation materials, wall insulation materials, auditorium and other building materials. Oversea 85% of microfiber glass blanket is used for insulation and sound absorption in building materials. Microfiber glass has the characteristics of fine diameter (<4μm), high porosity, small bulk density and non-flammability so that microfiber glass blanket has excellent sound insulation and thermal insulation properties. Microfiber glass blanket is a loose and porous mat material which is made of fiberglass cotton by bonding the fibers together and then cured by heating. It has excellent insulation, heat preservation and sound absorption performance, and is the basic raw material for sound insulation in aviation, high-speed rail and vacuum insulation panel (VIP) and other fields [1-3].

At present, the filtration microfiber glass blanket is mainly prepared by centrifugal sizing method and the diameter of the fiber is distributed in the range of 0.5 to 4μm. Compared with the ultra-fine fiberglass cotton fabricated by flame method, the diameter of fiberglass cotton fabricated by centrifugal method is thicker, which results in the poor sound insulation performance of the prepared felt. At the same time, because the glass fiber cotton prepared by centrifugal method is longer and the sizing is not uniform, the glass fiber cotton felt produced has high hardness, high bulk density and poor comprehensive mechanical properties. Therefore, the manufacture of ultra-fine glass fiber cotton felt with finer fiber diameter and uniform sizing has become a research hotspot in the field of thermal insulation [4-6].

In this paper, The high performance microfiber glass blanket was prepared by centrifugal method with spraying binder. The influence of fiber diameter and aging time on the water resistance, strength, filtration efficiency and resistence was investigated.

2. Experimental method

2.1 Sample preparation

23.5 quartz sand, 11 soda, 4.5 potassium feldspar, 5.5 albite, 2.5 calcite, 8.5 borax, 6 dolomite, 1.5 barium carbonate and 1 zinc oxide were selected according to the weight ratio and mixed evenly in the kiln to melt the glass liquid with uniform composition and transparent impurities. Then the glass liquid was thrown out by high-speed rotating centrifugal disc. Ultrafine glass fibers, 95% of the diameter of the prepared Ultrafine glass fibers are normally distributed at 2.5-3μm, and 90% of the length of the fibers are normally distributed at 20-30mm. Acrylic acid mixed with 1% KH560 silane coupling agent is sprayed on the surface of Ultrafine glass fibers evenly, in which the injection pressure is controlled at 0.4Mpa and the injection pressure is controlled at 0.4Mpa. The flow rate is controlled at 200Kg/h, and then sucked into the cotton collector by negative pressure air induction to form dry composite ultra-fine glass fiber cotton. Finally, the dry composite microfiber glass blanket is cured at 600°C for 8mins.

The surface and cross-section microstructure of the prepared microfiber glass sample was observed and characterized by KYKY 2800B scanning electron microscope (SEM). Since the glass sample is not electrically conductive, it is necessary to process the gold spray treatment before the observation.

For characterization of filtration performance of the prepared microfiber glass blanket, according to Washburn's formula, the through hole is regarded as a capillary with different apertures and lengths. Due to the effect of surface tension, if a completely wetted liquid is used, it will naturally fill the pore. At this time, another medium is used to overcome the surface tension by increasing pressure. To replace the liquid, when the pressure rises to enough to remove the infiltrating liquid with the largest pore diameter, that is to say, the bubble point pressure is reached and the corresponding pore diameter is the largest pore diameter. If the pressure continues to increase, there will be smaller pores gradually opened. Through the monitoring of the flow variation of the two curves of thousand sample and wet sample, the pore diameter fraction can be calculated. Gas permeability can also be calculated by dry curve with cloth, minimum aperture and average flow aperture.

3. Experiment and discussion

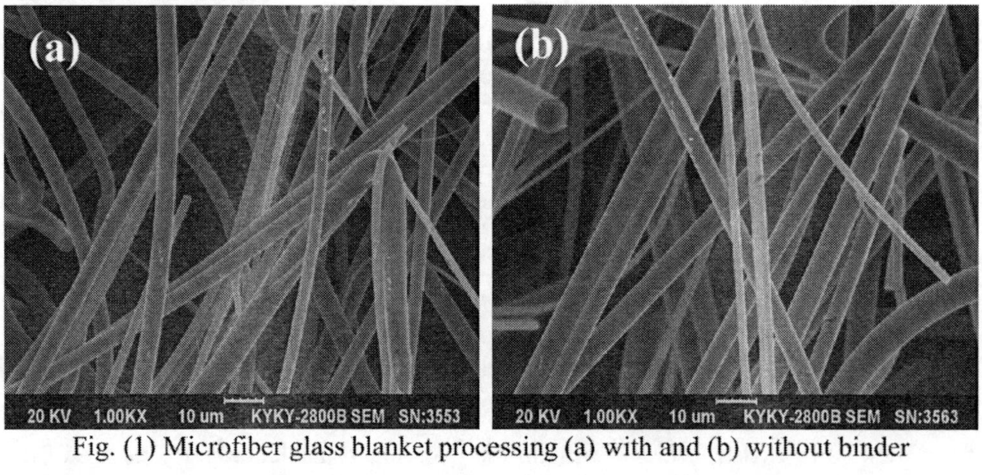

Fig. (1) Microfiber glass blanket processing (a) with and (b) without binder

The microstructures of the samples were observed by SEM. The measurement was carried out by cutting three samples with nominal size of 200*70mm from the prepared microfiber glass blanket. Make sure there is no caking, folding, folding or other protrusion on the surface of the material. When sampling, it makes sure to take one edge from the middle of the prepared blanket and the third one from the center. The length direction of 200mm is parallel to the length direction of the glass fiber coil. The other two pieces are taken from

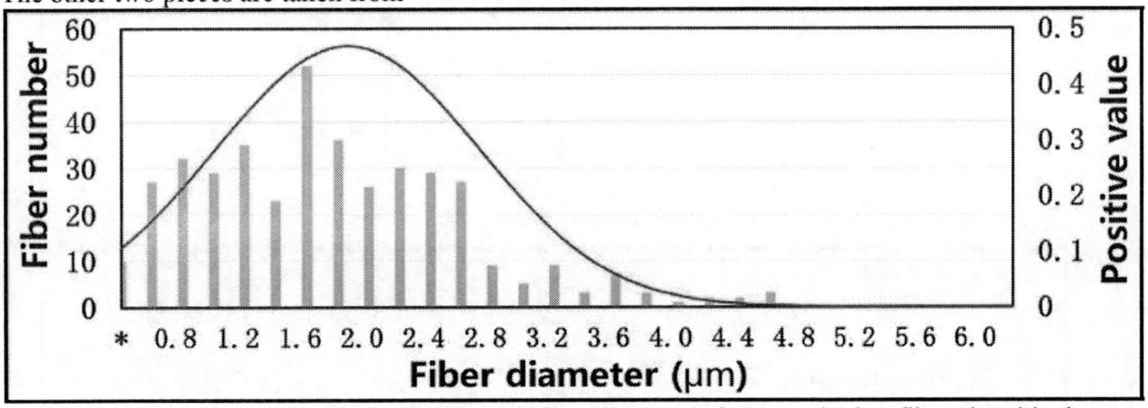

Fig. (2) The normal distribution of average fiber diameter of prepared microfiber glass blanket

positions equal in width to both sides. Place the sample flat, flat but not tight. The average diameter of each specimen was recorded to 0.01mm. The arithmetic mean values of all five specimens were recorded as the diameter of the material. By comparing the with Fig. (1a) and without Fig. (1b) of the fiberglass precursor in the ultra-fine microfiber glass blanket, it can be observed that the glue has been uniformly sprayed on the surface of the fiberglass, and the sticky nodules are formed at the nodes of the fiberglass network structure, which makes the prepared microfiber glass blanket have better mechanical properties. As shown in Fig. (2), the normal distribution of average fiber diameter of prepared microfiber glass blanket were sampled and analyzed by SEM. The average value of the prepared samples was 1.98μm, the standard deviation was 0.24, and the range of values was 1.83-2.45μm.

The effect of microfiber diameter on the water resistence, strength, resistence and filtration efficiency properties of prepared microfiber glass blanket is shown as Fig. (3), as can be seen, the filtration efficiency of prepared microfiber glass blanket increases firstly with increasing the diameter of microfiber glass. Firstly, from the aspect of fiber fineness, the single fibers of the filter mat are arranged irregularly, and the smaller the linear density of the fibers is, the smaller the pore formed by them interlacing in disorder. At this time, the easier it is to intercept the dust passing through, the greater the filtration efficiency. So when the mixing ratio is small, the ultra-fine microfiber glass plays a leading role. The filtering efficiency of prepared microfiber glass blanket decreases slightly as the diameter microfiber glass gradually increasing; Secondly, from the aspect of cross-section of fibers, the specific surface area of fibers with special cross-section is larger, so the probability of collision between fibers and particles is larger, and the particles are easier to be trapped; when the filtering conditions are the same, the collision opportunity between fibers and dust particles is proportional to that of fibers. The cross-section effect makes the filtering efficiency of prepared microfiber glass blanket gradually improve.

Fig. (3) shows that the thicker the fiber, the easier the fiber structure unfolds, resulting in more effective fiber space structure. Therefore, with the decrease of the content of superfine fibers and the incorporation of microfiber glass with higher linear density, the surface of prepared microfiber glass blanket is smoother, and the resistance of strength decreases.

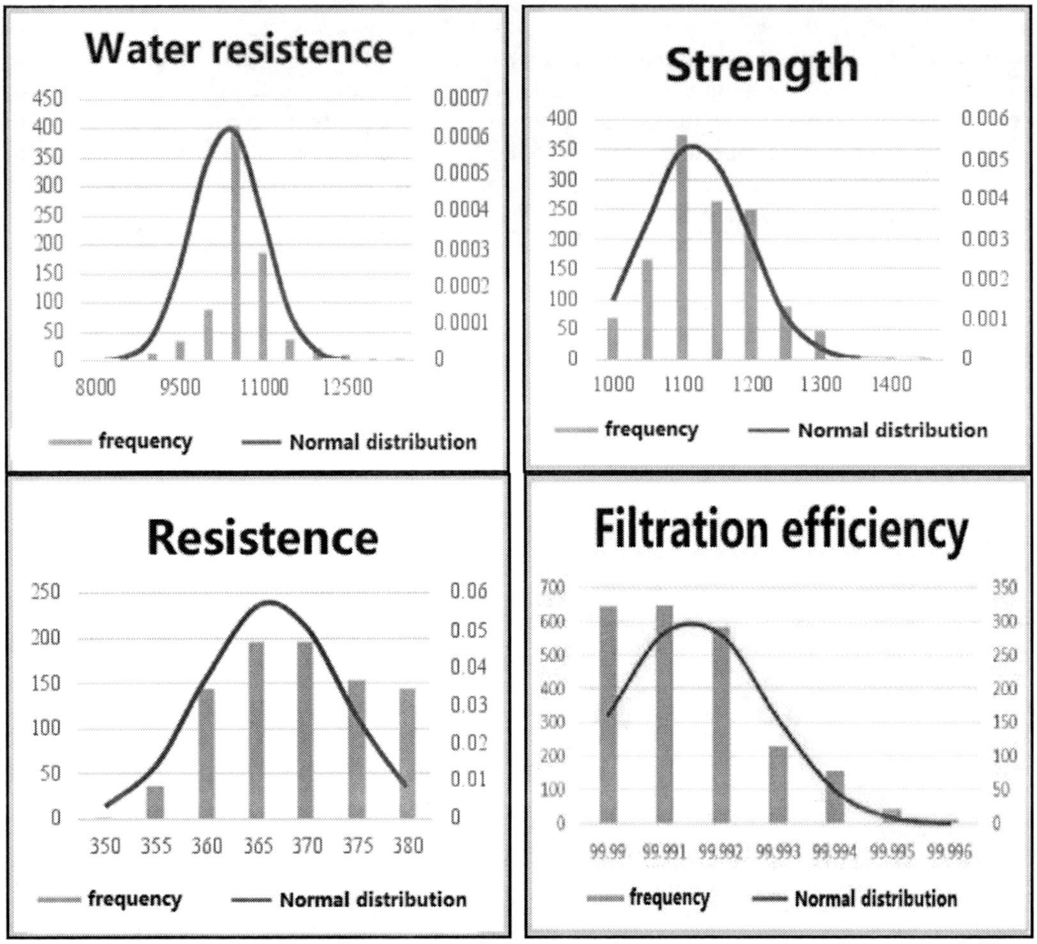

Fig. (3) The effect of microfiber diameter on the water resistence, strength, resistence and filtration efficiency properties of prepared microfiber glass blanket

The effect of aging time on the strength, resistence and filtration efficiency properties of prepared microfiber glass blanket is shown in Fig. (4), as can be seen, for the adhesive prepared microfiber glass blanket, the aging time increases greatly after the temperature rises to 75°C followed by the particle size of 0.17μm, the filtration efficiency of prepared microfiber glass blanket was almost stable. But the efficiency decreases greatly, followed by the particle size of 0.39μm and the particle size of 1.22μm cannot see the change of m because the efficiency is close to 100%. For non-adhesive prepared microfiber glass blanket, after the temperature rises to 75°C, it still follows the previous trend of change and there is no sudden change. The MPPS point of the efficiency curve moves right to about 0.17μm. The MPPS point moves to the right and the efficiency curve of the rubber filter paper changes abruptly, the strength and resistence value increase with the aging time rising. Therefore, when the test is carried out according to the standard conditions, the efficiency value obtained by MPPS method is on the safe side. The primary glass filament strand is formed by glass liquid passing through the leaky plate. The primary glass filament strand is melted and pulled into ultra-fine glass fibers under the action of high temperature and high speed flame airflow. The solution containing adhesives and processing aids was evenly sprayed on the surface of superfine glass fibers and evenly dispersed on the forming net.

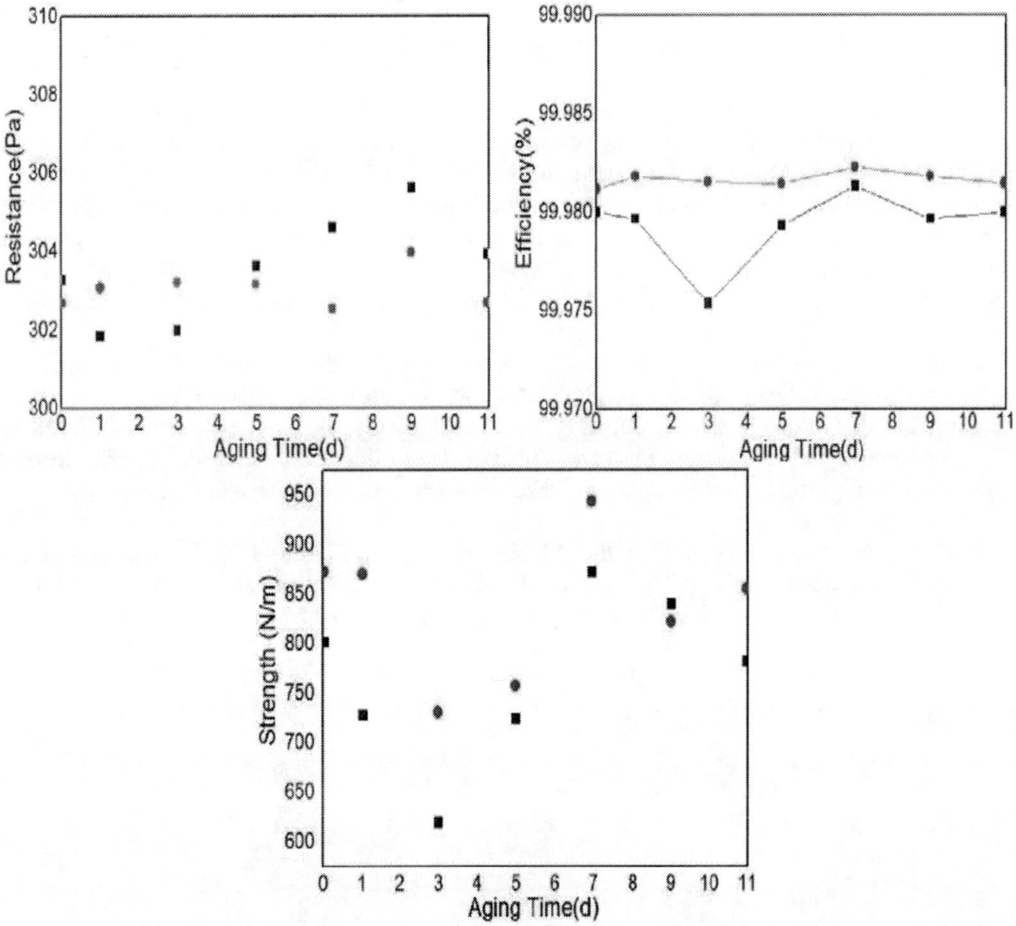

Fig. (4) The effect of aging time on the strength, resistence and filtration efficiency properties of prepared microfiber glass blanket

4. Conclusion

The glue has been uniformly sprayed on the surface of the fiberglass, and the sticky nodules are formed at the nodes of the fiberglass network structure, which makes the prepared microfiber glass blanket have better mechanical properties. The average value of the prepared samples was 1.98μm, the standard deviation was 0.24, and the range of values was 1.83-2.45μm. The thicker the fiber, the easier the fiber structure unfolds, resulting in more effective fiber space structure. With decreasing of the content of superfine fibers and the incorporation of microfiber glass with higher linear density, the surface of prepared microfiber glass blanket is smoother, and the resistance of strength decreases. The solution containing adhesives and processing aids was evenly sprayed on the surface of superfine glass fibers and evenly dispersed on the forming net.

Acknowledgement

This work is supported by Key Laboratory of Roads and Railway Engineering Safety Control (Shijiazhuang Tiedao University), Ministry of Education. No. STKF201718.

Reference

[1] Les S. Chapter 3-Hydraulic unit equipment features. Hydraulic Rig Technology and Operations, 2019, 59-187.
[2] Xu B B, Chen D Y, Li H H, Zhuang K Y, Hu X, Li J W, Hans I, Kong J H, Edoardo P. Priority

analysis for risk factors of equipment in a hydraulic turbine generator unit. Journal of Loss Prevention in the Process Industries, 58 (2019) 1-7.
[3] Babak K, Gholamreza J, Kamran S, Mohammadreza N. Scaling and thermal-hydraulic design of a test loop for the VVER-1000 reactor. Progress in Nuclear Energy, 113 (2019) 18-27.
[4] Ghiban B, Safta C A, Ion M, Crina E C, Grecu M C. Structural aspects of silt erosion resistant materials used in hydraulic machines manufacturing. Energy Procedia, 112 (2017) 75-82.
[5] Yu, H, Sneha S. Fibrillated fibers for liquid filtration media. US Patent 9511330. Dec. 19, 2013.
[6] Cambo W H, Huang M T. Pleatable composite filter media. US Patent 8545607. Oct. 1, 2013.
[7] Dorrer C, RÜhe J. Wetting of silicon nanograss: From superhydrophilic to superhydrophobic surfaces. Advanced Materials, 20 (2008) 159-163.
[8] Si Y, Guo Z. Superwetting materials of oil–water emulsion separation. Chemistry Letters, 44 (2015) 874-883.
[9] Zhu Y Z, Wang D, Jiang L, Jin J. Recent progress in developing advanced membranes for emulsified oil/water separation. NPG Asia Materials, 6 (2014) 101-111.
[10] Huang M, Si Y, Tang X, Zhu Z, Ding B, Liu L, Zheng G, Luo W, Yu J. Gravity driven separation of emulsified oil-water mixtures utilizing in situ polymerized superhydrophobic and superoleophilic nanofibrous membranes. Journal of Materials Chemistry A, 45 (2013) 14071-14074.
[11] Lin X D, Heo J W, Jeong H J, Choi M Y, Chang M W, Hong J K. Robust superhydrophobic carbon nanofiber network inlay-gated mesh for water-in-oil emulsion separation with high flux. Journal of Materials Chemistry A, 4 (2016) 17970-17980.

IOP Conf. Series: Materials Science and Engineering **711** (2020) 012053 doi:10.1088/1757-899X/711/1/012053

Performance Study on Acrylic Pressure-Sensitive Adhesive Modified with Metakaolin

Junhao Zhang[1], Xuntian Zheng[1]*, Quanzhi Ma[2]

[1]Departments of materials science and engineering, Wuhan University of technology, 430070, China

[2]Departments of materials science and engineering, Wuhan University of technology, 430070, China

[3]Departments of materials science and engineering, Wuhan University of technology, 430070, China

Email of all the authors: junhaozhang@whut.edu.cn, xuntian@whut.edu.cn,1570915381@qq.com

Corresponding Author: Zhongkui Wu; email:zhongkuiwu@163.com; phone:027-87653405;

Abstract: This paper takes butyl acrylate(BA) and 2-ethylhexyl acrylate(2-EHA) as the soft monomers, vinyl acetate(VAC) as the hard monomer of pressure-sensitive adhesive and acrylamide(AM) and acrylic acid(AA) as the raw materials of modified monomer to synthesize pressure-sensitive adhesive to carry out Fourier Transform Infrared Spectroscopy(FTIR)characterization. In order to satisfy some special requirements and improve its adhesive property, this paper uses metakaolin and metakaolin modified with silane couling agent KH570 to carry out mechanical blending with pressure-sensitive adhesive respectively and prepare composite acrylic pressure-sensitive adhesive. This paper tests the viscosity of composite pressure-sensitive adhesive with metakaolin and modified metakaolin. It is found that with the increase of doping content, the viscosity will be increased accordingly and the viscosity of composite pressure-sensitive adhesive with modified metakaolin has a larger growth. The adhesive property indicates that composite pressure-sensitive adhesive with KH570 modified metakaolin has an obviously better adhesive property than that with unmodified metakaolin, and when 0.8wt% of modified metakaolin is added, inorganic/organic composite acrylic pressure-sensitive adhesive has the best comprehensive adhesive property.

1.Introduction

With the high-speed economic development, all walks of life can not be separated from packaging while packaging can not leave the adhesive[1]. Science and technology have been developing with each passing day, which provides the possibility for the development of pressure-sensitive adhesive. Pressure-sensitive adhesive [2] (PSA) is a kind of special viscous-elastic material which can rapidly wet the adherend without heating, adding solvent or curing agent and other methods, and form a solid adhesive strength at the interface between adhesive and adherend while not destroying the surface structure of adherend once it has been laminated or applied to the bonding plane. There is a serious shortage of functional pressure-sensitive adhesives in the fields of electronics, aviation, military affairs and so on. It has become an urgent need of the current society to develop new functional monomers, improve polymerization technology and select new materials to produce new pressure-sensitive adhesives to expand their application fields[3,4]. Acrylic pressure-sensitive adhesive is a kind of artificially synthesized resin polymer adhesive formed by free radical polymerization of various acrylic acid monomers. Polyacrylic acid pressure-sensitive adhesive has numerous advantages such as excellent low temperature resistance and high temperature resistance, no harmful gas volatilization in the application process, etc[5-7] , which has made acrylic acid pressure-sensitive adhesive become the most widely used application product in the pressure-sensitive tape industry. Jin X[8] is used as the acrylic acid pressure-sensitive tape for liquid crystal display under the synthesis condition of taking 2-ethylhexyl acrylate as the viscous monomer, methyl methacrylate as the cohesive monomer, acrylic acid as the crosslinking monomer and benzoyl peroxide as the initiator.

Due to the poor properties of solvent acrylic acid pressure-sensitive adhesive such as weather fastness, thermal endurance, durability and so on, it needs to be carried out modification. Inorganic nanoparticles have many excellent properties, such as unique high temperature resistance, high hardness, high strength, solvent resistance and so on. In addition, there are microscopic quantum size effect, surface effect and photo-electromagnetic effect that make them have a complete development. A homogeneous and heterogeneous material which has not only the rigidity and dielectricity of inorganic matter, but also the transparency, workability, plasticity and other advantages of organic matter is obtained through organic/inorganic nanocomposite, so that composite material has become a new material which has a great development prospect in the field of material science. The inorganic nanoparticles are mixed into the acrylic acid pressure-sensitive adhesive, and the composite acrylic acid pressure-sensitive adhesive with inorganic nanoparticles is obtained by combining their advantages.

2. Material and Methods

2.1 Experiment

BA, 2-EHA and VAC are blended in ethyl acetate with the ratio of 1:6:1 and the reaction temperature is 82°C. 2.5% of initiator BPO, functional monomer AA and AM are added into the reaction system in twice. The reaction lasts 8 hours and the material is obtained after cooling.

Modification of Metakaolin

This experimental design uses silane coupling agent γ-methyl propyl trimethoxy silane to carry out surface modification for nano-metakaolin(MK) and the modified nano-metakaolin(mMK) is obtained.

Preparation of Inorganic/Organic Composite Pressure-Sensitive Tape

Metakaolin with the ratio of 0.2wt%, 0.4wt%, 0.6wt%, 0.8wt% and 1.0wt% and modified metakaolin are carried out blending reaction with acrylic acid pressure-sensitive adhesive for 2h. The reaction temperature is 40°C and the pressure-sensitive tape is produced by 35 um coating.

2.2 Test

Fourier Transform Infrared Spectroscopy(FTIR) is used to carry out characterization analysis of basic pressure-sensitive adhesive and metakaolin. Surface tension meter is used to test the contact angle of modified and unmodified metakaolin and DJ-1 rotational rheometer is used to test the viscosity of basic pressure-sensitive adhesive.

The initial adhesion, permanent adhesion and 180° peel strength of the composite pressure-sensitive adhesive are tested according to the international standard.

3. Result and Conclusion

3.1 FTIR Characterization of Basic Pressure-Sensitive Adhesive

Figure 1 is the FTIR characterization diagram of basic pressure-sensitive adhesive. From the test results and characteristic peak values of each vibration group in characteristic peaks of 3200-3600cm-1, 2960cm-1, 2870cm-1and so on, we can draw the conclusion that five monomers have copolymerization and produce the copolymer acrylic acid pressure-sensitive adhesive.

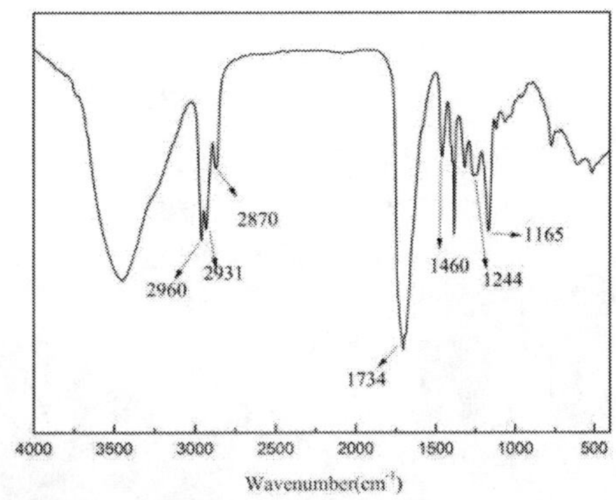

Figure 1. Infrared spectrogram of basic pressure-sensitive adhesive

3.2 FTIR Analysis of Modified Metakaolin

Figure 2 is the infrared spectroscopic analysis chart of unmodified and modified metakaolin. Through observing the modified metakaolin spectrogram, we find that there are two small peak values in large peak 3000-2800cm-1 compared with unmodified metakaolin. It is speculated that it should be the peak value of methylene-CH2. The KH570 molecules contain methylene, which indicates that the two peaks of the modified metakaolin comes from KH570, thus it can be explained that KH579 has successful modification to metakaolin. It is found that there is a larger peak value pf modified and unmodified metakaolin in the wave number 3442cm-1, which shows that it should be the stretching vibration peak of hydroxyl groups on the surface of particles. Through comparison, it is found that the peak value of modified metakaolin is decreased and it is inferred that silane coupling agent has a chemical reaction with some hydroxyl groups of metakaolin, but most reactions should be physical absorption.

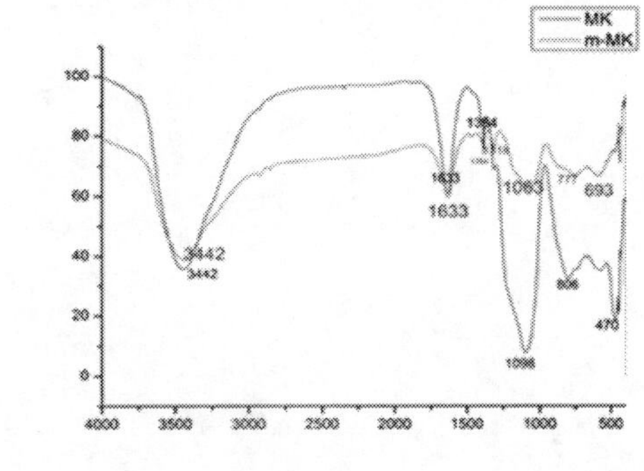

Figure 2. Infrared spectrogram between metakaolin modified by KH570 and unmodified metakaolin

3.3 Water Contact Angle Analysis of Modified Metakaolin

Figure 3 and Figure 4 are water contact angle diagrams of unmodified and modified metakaolin. The contact angle changes from 19° to 81.12°. The larger the angle, the stronger the hydrophobicity, which indicates that the modification is successful. This is because the unmodified metakaolin has a large surface, abundant hydroxyl groups and excellent hydrophilia, while KH570 has a soft and long alkyl chain with hydrophobicity and has low interface energy. Therefore, through the modification of me-

IOP Conf. Series: Materials Science and Engineering **711** (2020) 012053 doi:10.1088/1757-899X/711/1/012053

takaolin by KH570, metakaolin has long-chain alkyl groups with hygrophobicity by surface grafting, which makes modified nano-metakaolin have better hydrophobicity, so the contact angle becomes larger.

Figure 3. Water contact angle of unmodified metakaolin

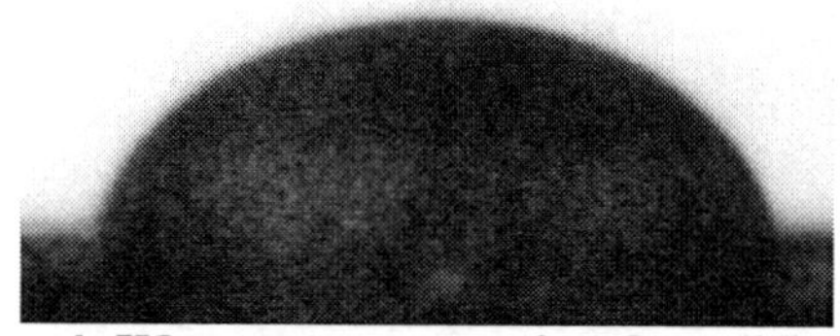

Figure 4. Water contact angle of modified metakaolin

3.4 Viscosity Analysis of Composite Pressure-Sensitive Adhesive

As shown in Figure 5, as the addition volume of inorganic particles is increased, the viscosity is enlarged and the viscosity of modified metakaolin is larger. Rheology believes that metakaolin with inorganic particles is the undeformable solid granule with poor fluidity. Its addition encroaches on a part of free volume of the macromolecular chain segment, and the reduction of free volume will make the segment motion of macromolecular chain become worse, while the modified metakaolin will disperse more uniformly in the system, so that the acting force between the modified metakaolin and the hydrogen bond of molecular chain of acrylic acid pressure-sensitive adhesive will be strengthened. The increase of viscosity can indirectly reflect the storage modulus and loss factor of the polymer will become larger, which will also reflect the viscoelasticity of polymer. Viscoelasticity has a close relation with cohesiveness of polymer.

IOP Conf. Series: Materials Science and Engineering **711** (2020) 012053 doi:10.1088/1757-899X/711/1/012053

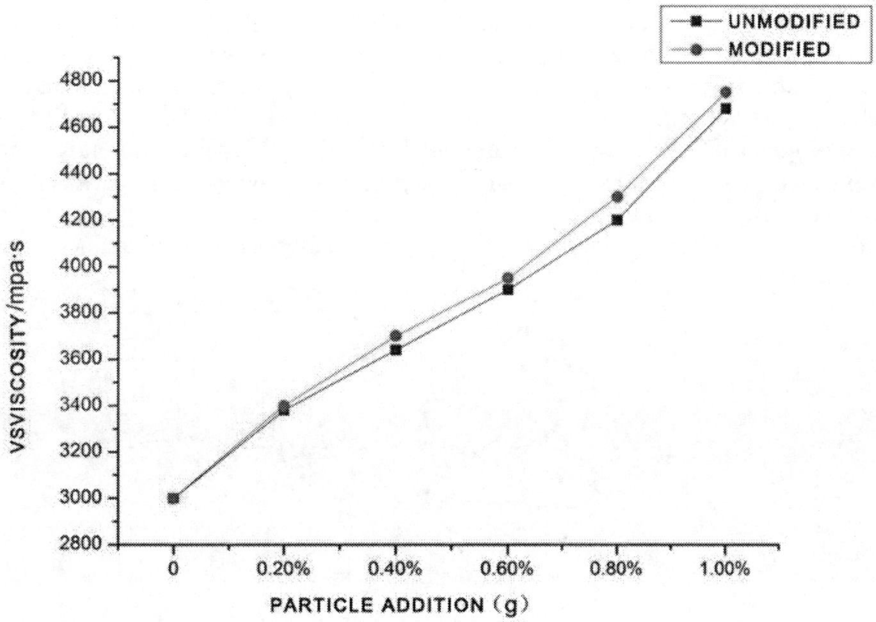

Figure 5. The impact of adding modified and unmodified metakaolin on viscosity

3.5 Adhesive Property Analysis

3.5.1 Initial Adhesion Analysis

As shown in Figure 6, whether the modified metakaolin by KH570 or the unmodified metakaolin, the initial adhesion shows a variation tendency of rising at first and declining later with the increase of addition volume. However, pressure-sensitive adhesive with modified metakaolin has better cohesiveness. It is analyzed that the initial cohesion of metakaolin modified by KH570 is significantly better than unmodified metakaolin, and has the best property when the addition volume is at 0.8wt%. This is because the modified metakaolin makes the surface introduce the oily groups, thus the oil phase composite improves the affinity between organic matters and it, which makes it have better dispersibility in the acrylic acid pressure-sensitive adhesive. In addition, because the metakaolin with surface modification will introduce more polar groups, the modified metakaolin surface will introduce double bonds after modification, so that the molecular polarity of pressure-sensitive adhesive will be enhanced, which will form more hydrogen bonds with adherend and have a better adhesive property.

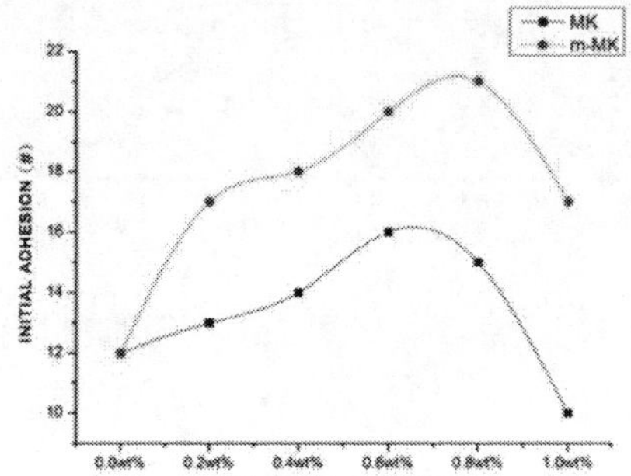

Figure 6. The impact of metakaolin on the initial cohesion of solvent pressure-sensitive adhesive

IOP Conf. Series: Materials Science and Engineering **711** (2020) 012053 doi:10.1088/1757-899X/711/1/012053

3.5.2Permanent Cohesion Test

As shown in Figure 7, with the increase of content, the permanent cohesion maintains the tendency with continuous growth, and the metakaolin modified by KH570 has obvious increase effect of permanent cohesion. The increase of permanent cohesion should be taken into account the cohesion factor, which should be caused by the increase of cohesion. Considering that metakaolin is a kind of rigid particle, the addition of metakaolin will inevitably lead to the increase of its modulus. In addition, the increase of metakaolin is bound to form more crosslinking of hydrogen bonds, which will improve the cohesion of acrylic acid pressure-sensitive adhesive and increase the permanent cohesion. The metakaolin modified by KH570 has better dispersibility and stronger polarity. The hydrogen bonds formed are more uniformly distributed, which obviously makes the permanent cohesion have a better increase effect.

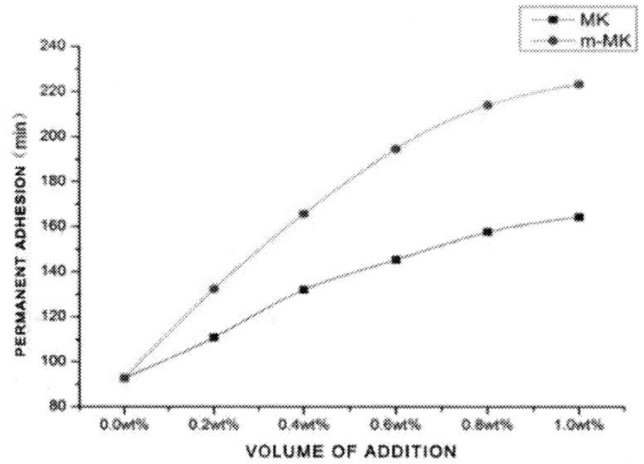

Figure 7 The impact of metakaolin on the permanent cohesion of solvent pressure-sensitive adhesive

3.5.3 180° Peel Strength Test

Figure 8 is the impact effect diagram of metakaolin on the 180° peel strength of solvent pressure-sensitive adhesive. It is found that its change law is similar to the initial cohesion, which shows the tendency of increasing at first and decreasing later, and the modified metakaolin by KH570 has a better effect than that without modification. The reason should be similar to that of initial cohesion, which can be explained by free volume, hydrogen bond, better dispersibility after modification and introduction of polar groups.

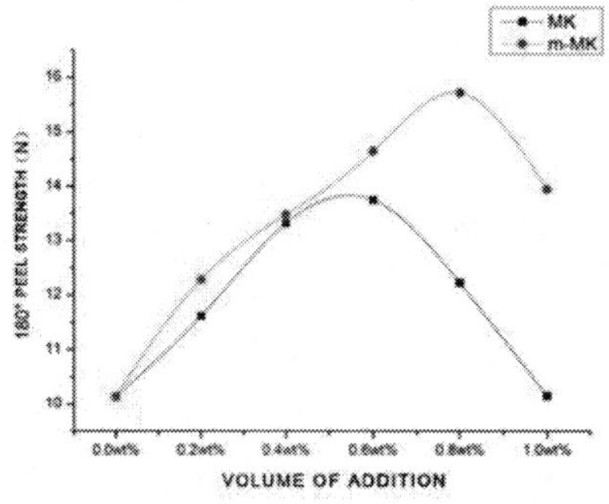

Figure 8. The impact of metakaolin on the 180° peel strength of solvent pressure-sensitive adhesive

4.Conclusion

This paper analyzes the adhesive property between composite pressure-sensitive adhesive with unmodified metakaolin and modified metakaolin by KH570. The results show that the adhesive property of composite pressure-sensitive adhesive with modified metakaolin by KH570 is significantly better than that with unmodified metakaolin. When the addition volume of modified metakaolin is less than 0.8wt%, the curves of peel strength and initial cohesion of composite pressure-sensitive adhesive rise with the addition volume increase of modified metakaolin. While its admixture amount is more than 0.8wt%, the peeling strength and initial cohesion show a decreasing tendency with the increase of its admixture amount, but the permanent cohesion is enlarged with the increase of its admixture amount. To sum up, when the content of modified metakaolin reaches 0.8wt%, inorganic/organic composite acrylic acid pressure-sensitive adhesive has the best comprehensive adhesive property.

Acknowledgements
This research was financially supported by National college students innovation and entrepreneurship training program fund.

References
[1] Baek S S, Jang S J, Hwang S H. Preparation and adhesion performance of transparent acrylic pressure sensitive adhesives: effects of substituent structure of acrylate monomer[J]. International Journal of Adhesion & Adhesives, 2016, 64:72-77
[2] Yi Zhang, Huashu Qi, Guochen Duan,etc. Study on the process of pressure-sensitive adhesive of solvent-based polyacrylate[J]. Chinese adhesive, 2011, 20(4):44.
[3] Lakrout H, Sergot P, Creton C. Direct Observation of Cavitation and Fibrillation in a Probe Tack Experiment on Model Acrylic Pressure-Sensitive-Adhesives[J]. Journal of Adhesion, 1999, 69(3-4):307-359.
[4] Baek S S, Jang S J, Hwang S H. Preparation and adhesion performance of transparent acrylic pressure sensitive adhesives: effects of substituent structure of acrylate monomer[J]. International Journal of Adhesion & Adhesives, 2016, 64:72-77.
[5] Awajaa F, Gilbert M,Kellya G, et al. Adhesion of polymers[J]. Progress in Polymer Science, 2009, 34:948-968.
[6] Xu C, Jiang L, Guan A, et al. Tailoring the Viscoelasticity of HEMA-Functionalized StyreneAcrylate Copolymers Using Coagulants[J]. Ind.eng.chem.res, 2014, 53(49):19175-19183.
[7] Lim J C, Lee M C, Lim T K, et al. Synthesis of sorbital based nonionic surfactants and characterization of interfacial and adhesive properties for waterborne pressure sensitive adhesives[J]. Colloids & Surfaces A Physicochemical & Engineering Aspects, 2014, 446(5):80-89.
[8] Chen J H, Zhao Q, Wang Y, et al. Improvement in mechanical properties and thermal stability of solvent - based pressure - sensitive adhesives based on triazine heterocyclic monomer[J]. Journal of Applied Polymer Science, 2016, 133(15):n/a.

IOP Conf. Series: Materials Science and Engineering **711** (2020) 012054 doi:10.1088/1757-899X/711/1/012054

System Design of Small-scale Seawater Desalination Device

Lingzhi Feng[1], Yuanzhou Zheng[1*], Aobo Zhang[2], Tianfu Lu[3], Tongyao Miao[4], Yunhui Peng[5]

[1]Departments of Materials science and engineering, Wuhan University of technology, 430070, China

[1*]Departments of School of navigation, Wuhan University of technology, 430070, China

[2]Departments of School of navigation, Wuhan University of technology, 430070, China

[3]Departments of School of navigation, Wuhan University of technology, 430070, China

[4]Departments of School of navigation, Wuhan University of technology, 430070, China

[5]Departments of Materials science and engineering, Wuhan University of technology, 430070, China

* Corresponding Author's e-mail: flz1115674048@whut.edu.cn, zhengyuanzhou0909@163.com, 414420285@qq.com, 1035104274@qq.com, 312263702@qq.com, 852440944@qq.com

Abstract. Water is of great significance to the subsistence and development of human beings. As we all know, fresh water issues have always been a major problem to the ocean navigation, as the existing desalination devices generally have the disadvantages of large volume, high-energy consumption, which would bring pollution to the environment. Basing on the study results in existence, this paper presents a new system design of small-scale seawater desalination device, which founds on the enhanced heat transfer processes of fluid phase change, including loop heat pipe technology, new type of heat-sink type spoiler evaporator and integrated soaking plate fin condenser to implement the low-power and low-temperature system. This system covers an area of only 1.8m2, as the energy consumption is 62.5% of the reverse osmosis desalination method and 20% of the multistage distillation desalination method under the same conditions. What's more, the system consists of the loop heat pipe based on the phase change thermal principle, integrated soaking plate fin condenser and the new type of heat-sink type spoiler evaporator, improves the heat transfer efficiency, besides it has low pollution.

1. Introduction

Plenteous fresh water is the basis and guarantee of the subsistence of human beings. Nowadays, as the explosive growth of the population and improvement of the industrialization level, the demand for water increases day by day. Meanwhile, with the development of the shipping industry, the navigation distance is getting further and further and the operation time is becoming longer and longer, resulted in the

IOP Conf. Series: Materials Science and Engineering **711** (2020) 012054 doi:10.1088/1757-899X/711/1/012054

correspondingly increasing demands to fresh water. Generally, it is suggested by statistical analysis that the large ships consume about 200 tons of fresh water per day.

Due to the low economic benefits and low safety of water transportation and storage of ocean-going ships. Seawater desalination becomes an important supplement of the fresh water to these ships. Conventional commercial marine desalination mainly uses reverse osmosis method and distillation method. The reverse osmosis method doesn't need to provide heat resource, but the cost of operation and maintenance is high. Moreover, it requires high quality of seawater. Besides, the brine contains a large number of chemical agents, and the direct emissions of which can lead to water pollution. The distillation method has low requirements for seawater quality and the produced water is in high quality. Although the initial investment cost is high, the post-maintenance cost is low. For these reasons mentioned above, distillation has become the mainstream marine desalination technology[6]. At present, the research on seawater desalination mainly uses system calculation and small-scale experimental methods to determine the performance of seawater desalination system. And there are few studies on the structure characteristics of the desalination device itself and the designed devices are large in size and insufficient in compactness. In view of the above problems, a new system design of small-scale seawater desalination device is presented, which founds on the enhanced heat transfer processes of fluid phase change, including loop heat pipe technology, new type of heat-sink type spoiler evaporator and integrated soaking plate fin condenser. This system has lots of advantages, such as small in size and high conversion efficiency and so on.

2. System design

2.1. Design ideas

As shown in figure 1, aiming at improving the heat transfer efficiency, this paper innovatively puts forward a small-scale seawater desalination device, including a loop heat pipe that can transfer a large amount of heat over a long distance with a small temperature difference, which adopts a plate design. In order to make the device more simplified and efficient, this paper designs a desalination device including an evaporation system, a condensation system, and a fresh water collection system. The seawater evaporates in the evaporation system and produces vapor, which is then condenses in the condensation system to produce fresh water. And after that the fresh water is collected through the fresh water collection system, and the outflow of the waste liquid is discharged in the feedback system.

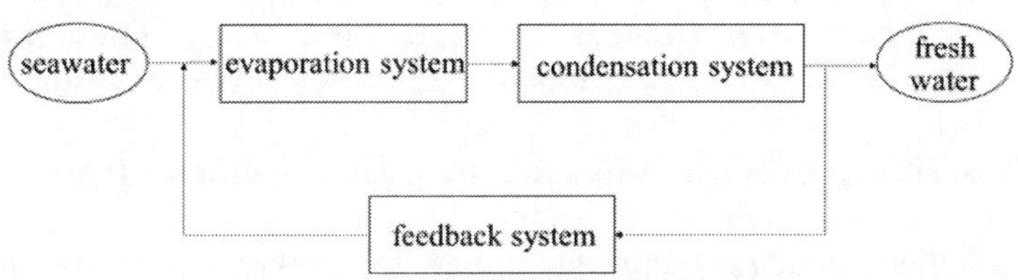

Figure 1. Flow chart of the design.

2.2. Integral design

This work mainly consists of evaporation system, condensation system, vacuum maintenance system, circulating sprinkler system, fresh water collection system, including power-supply and control system and a base and limit device are also designed, as shown in Figure 2. This desalination device has the advantages of low energy consumption, low cost, small footprint, energy and environmental protection, and it can be used for vessels and water-deficient islands as well as for certain waterways. Our domestic islands are mainly distributed in the South China Sea, where there exists long illumination time and strong radiation, which is suitable for the development and promotion of solar desalination devices and the application of solar energy is non-polluting and in line with national environmental policy guidelines[7]. The chassis and the limit device can improve the stability of the platform by improving the buoy type desalination device.

Figure 2. Sketches of the integral design.

3. System introduction

3.1. Evaporation system

The evaporation system is composed of a plate type loop heat pipe[5], a solar heat collecting plate, a hot plate and a heat-sink type spoiler evaporator. The heat collecting systems is formed by the solar collector plate and the electric heating plate, which are distributed on both sides of the plate type loop heat pipe, and the electric heating plate is disposed on the backlight side. The heat source is mainly based on solar heat collection. The electric heating plate provides electric auxiliary heating when necessary. The heat transfer medium in the loop heat pipe is transformed to a heat-sink type spoiler evaporator placed inside the tank to evaporate seawater[3][4].

- Design of plate type loop heat pipe. Utilizing the phase change heat transfer principle, the evaporator absorbs heat during operation, so that the liquid medium inside the capillary core evaporates, and the evaporated gas diffuses through the fine pores on the capillary core and passes through the evaporation loop to the condensation end. The vapor is liquefied by at the condensation end, and the liquefied medium is returned to the capillary core by the capillary force to realize heat transfer. Compared with traditional single-phase heat transfer, two-phase heat transfer can improve the heat transfer efficiency and reduce the heat loss during heat exchange[8][4].
- Design of heat sink type spoiler evaporator. In the evaporation system, the evaporator adopts 15 single-piece heat sinks placed in parallel at intervals of 2cm gap, and the total size is 30cm×30cm×20cm, which is connected in parallel by the main tubes, and finally fixed by the heat sink frame into groups, and the single-chip heat sink is disturbed. The monolithic fins spoiler evaporator is produced by laser welding and bulging technology. The fluid flow path in the evaporator has a U-shaped structure, and small particles are distributed on the U-shaped path to spoil the flow path, so that the inner wall of the evaporator is in full contact with the fluid heat conduction medium, thereby the heat transfer efficiency can be proved. The structure is shown in Figure 3[1][2].

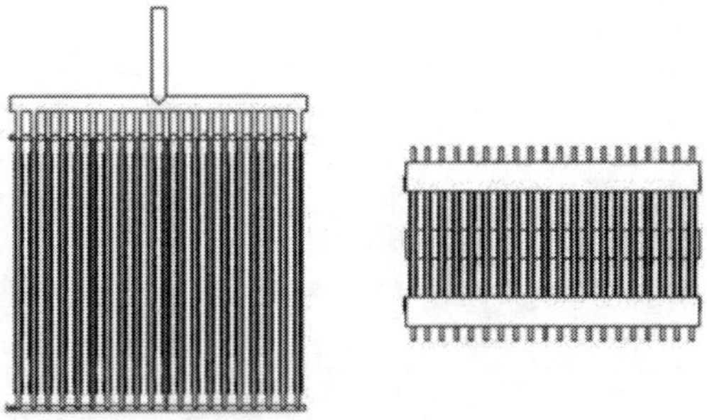

Figure 3. Sketches of the heat sink spoiler evaporator.

3.2. Design of condensation system

The main part of the condensation system is composed of cold-water inlet, cold water outlet and an integrated uniform temperature plate finned condenser. The flow rate of the cold-water inlet and outlet is controlled by the throttle valve. The integrated uniform temperature plate finned condenser is autonomously processed by aluminium, and is composed of ribs, a uniform temperature plate and a self-made spoiler pipe, as shown in Figure 4.

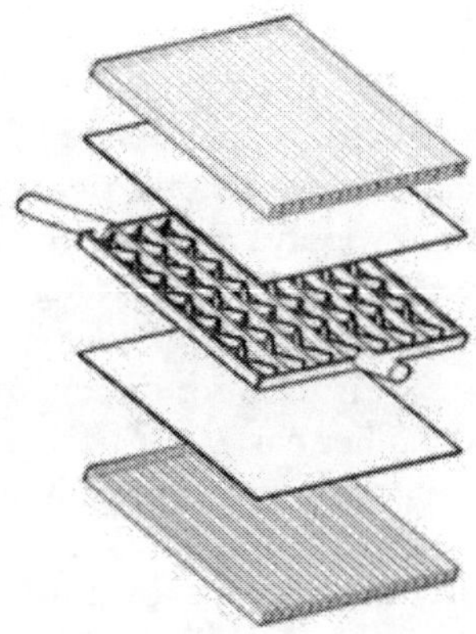

Figure 4. Sketches of the integrated uniform temperature plate fin condenser.

Vapor chamber is shown in Figure 5. The sintered mesh structure is attached to the inner side of the cavity, and each of the upper and lower layers has a layer spacing of 200μm. The liquid medium at the bottom of the vacuum chamber absorbs the heat from the water vapor to the fin at the bottom of the temperature equalizing plate during operation. The vapor evaporates and diffuses into the vacuum chamber, and transfers the heat to the condensation pipe. The condensed liquid is then refluxed through a sintered web attached to the inner wall of the temperature equalizing plate by capillary force.

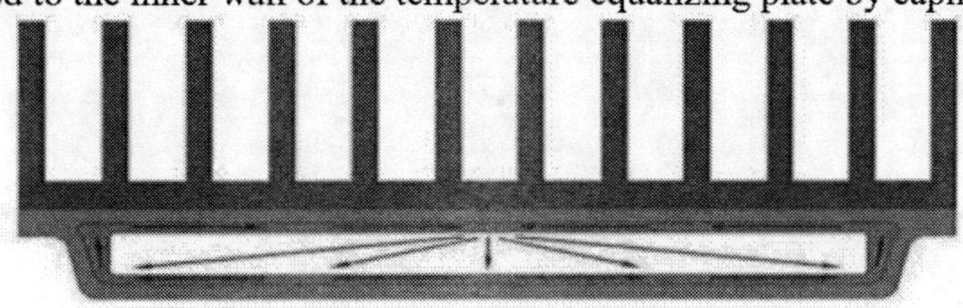

Figure 5. Sketches of the vapor chamber.

*The red arrow in the figure is the vapor medium after evaporation and the blue arrow is the liquid medium after condensation.

The vapor chamber is similar to the heat pipe in principle, but the conduction mode is different. The heat pipe is a bunch of linear heat conduction, and the vapor chamber is conducted on a two-dimensional surface, so it is more power-efficient and energy-efficient.

The spoiler pipeline is composed of the outer wall pipeline, five isolation cross plates and six spoilers. The specific structure is shown in Figure 6. The six isolated horizontal plates are evenly distributed in parallel, and the seven spoilers are interspersed between the isolated horizontal plates to flow around. And the cold seawater inside the spoiler pipe is in full contact with the upper and lower vapor chamber to increase the heat exchange efficiency.

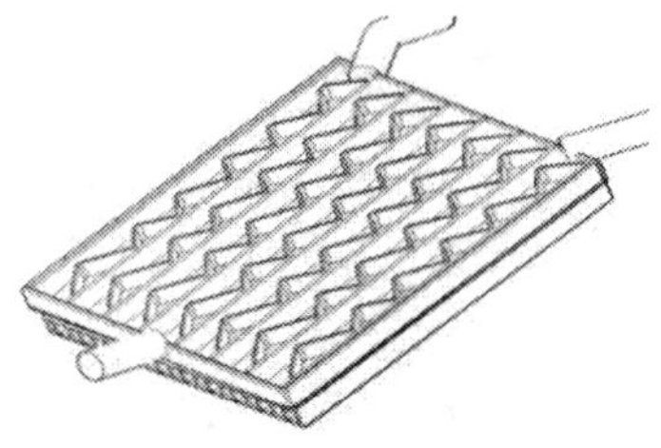

Figure 6. Sketches of the spoiler pipeline structure.

The design of the fin structure greatly increases the contact area between the condenser and the vapor, which also increases the heat transfer rate. The condensed liquefied vapor adheres to the surface of the fin and then drops to enter the fresh water collecting tank by gravity.

4. Circuit design

This system chiefly uses environmental energy to generate electricity, and then forms a stable charging current through MPPT charger controller to charge the powered lithium ion battery through this current. Aimed to create a 'reservoir' of electrical energy, the electrical energy is stored when external sources of energy are sufficient, so that the normal supply of electricity can be realized when external sources of energy are insufficient. After voltage stabilization, the storage battery and the built-in inverter of the conduct current boost and inversion to obtain 220V AC power supply; After the inversion is completed, the circuit is divided into two parts: one part is used for direct power supply of AC electrical apparatus such as high-power centrifugal pump; the other part is for power supply of the DC logic control circuit and DC sensor after rectifying by the switching power supply and step-by-step isolation and voltage stabilization.

The design sketches of the overall circuit model is shown in Figure 7.

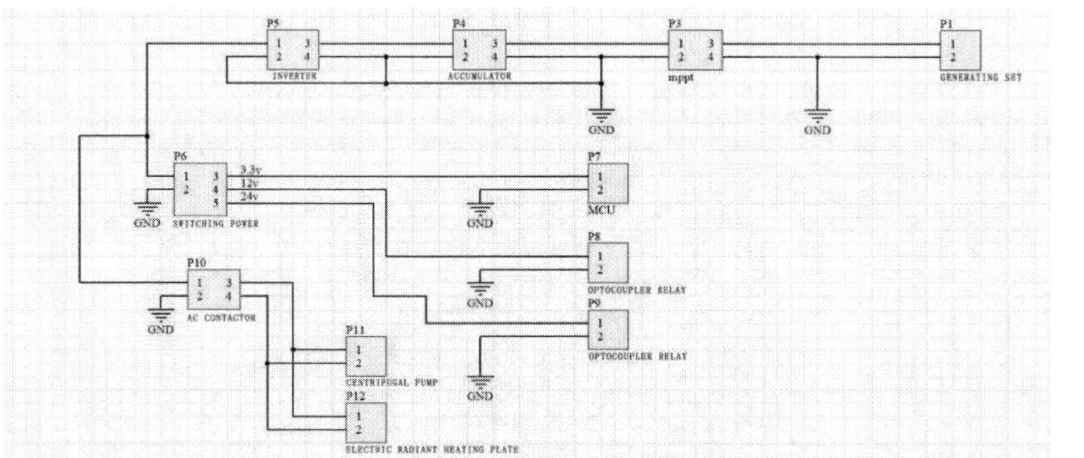

Figure 7. Overall circuit model connection diagram.

Some specific modules are demonstrated as follows:

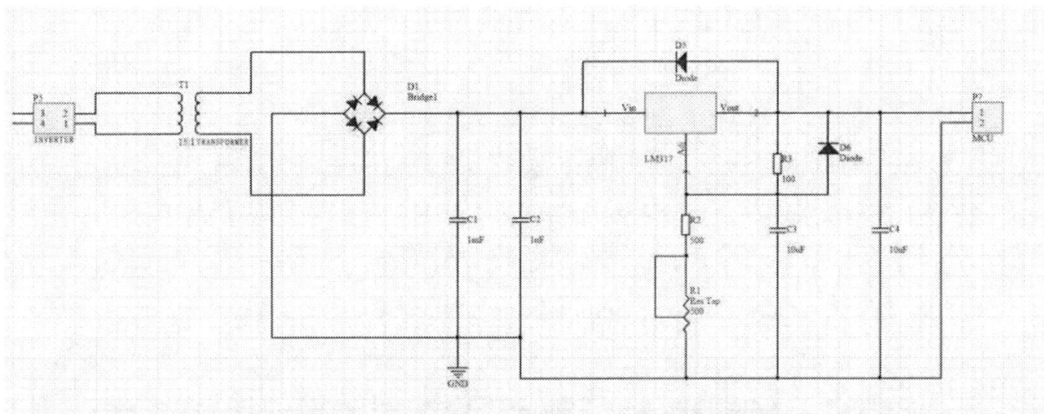

Figure 8. Sketches of switching power supply

As shown in Figure 8, this is a switching power supply design principle diagram. 220V alternating current provided by the inverter, after reduction voltage by passing transformer and rectification by passing rectifier bridge, will be changed into direct current, then through a capacitor filter, using the LM317 voltage chip, by adjusting the resistance of variable resistor R1, this module can realize the stable and adjustable direct current output, from 0V to 24V.

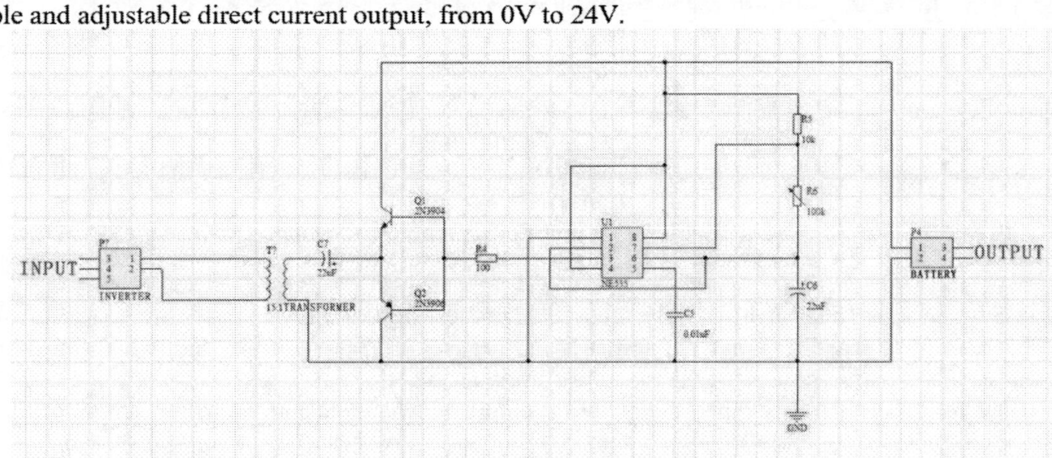

Figure 9. Sketches of inverter.

As shown in Figure 9, this is the design schematic diagram of the inverter. The 12V direct current provided by the storage battery is changed into square wave through the NE555 chip, and through two complementary triodes, it is changed again into alternating current, under the effect of the booster transformer, 220V alternating current will be output.

5. Theoretical analysis of seawater desalination

Table 5-1. The meanings and values of the constants.

symbols	meanings	values
m	quality of a single water molecule	3×10-26kg
kB	Boltzmann constant	1.38×10-23J/K
NA	Avogadro constant	6.02×1023
R	ideal gas constant	8.314J·mol-1·K-1

S*1	relative heat exchange area of the evaporation section	0.327
S*2	relative heat exchange area of the condensation section	0.340
T1	operating temperature of the evaporation section	90°C
T2	operating temperature of the condensation section	15°C

Table 5-2. The meanings and values of the symbols in theoretical analysis.

variable symbol	variable	meaning
[a]	sea water	evaporates involved in the reaction in a seawater desalination unit
[b]	water vapor	water vapor that involves in the reaction of seawater evaporation in a seawater desalination unit
[c]	fresh water	fresh water produced after condensation
[b']	other gases	gases that do not involve in the reaction in a seawater desalination unit
ya	water vapor ratio	proportion of water vapor in all gases in seawater desalination plant
yb	other gases ratio	ratio of other gases in seawater desalination plant in all gases
x	solute ratio	ratio of other solutes in evaporation

The steady process of seawater input, the theoretical process of heat transfer-based kinetics is as follows:

The amount of seawater is maintained at $[A]_0$, that is, $[A]\sim N([A]_0, \sigma)$ is stable, and the analytical process of the dynamic process of the reaction system is solved.

The seawater is continuously input, the quantity remains unchanged, and the series system of the first-order reaction is:

$$\frac{d[A]}{dt} = e$$

$$\frac{d[B]}{dt} = k_1[A] - k_2[B]$$

$$\frac{d[C]}{dt} = k_2[B] + l(t)$$

Suppose the amount of seawater remains unchanged at the level of $[A]_0$, that is, $[A]\sim N([A]_0, \sigma)$, then solve the above equation, here is the result

$$[B] = -\frac{k_1[A]_0}{k_2} e^{-k_2 t} + \frac{k_1[A]_0}{k_2}$$

The final differential equation of desalinated water is:

$$\frac{d[C]}{dt} = -k_1[A]_0 e^{-k_2 t} + k_1[A]_0$$

The time-varying function of desalinated water after solving the equation above is:

$$[C] = \frac{k_1[A]_0}{k_2} e^{-k_2 t} + k_1[A]_0 t - \frac{k_1[A]_0}{k_2}$$

Given the nature of the analytic solution function, it is observed that the total amount of seawater in the device has been stable, the water vapor stabilizes at $k_1[A]_0/k_2$ over time, and stable yield of fresh water will be realized at $k_1[A]_0$.

According to (Table 5-2), substitute the parameters into

$$k = \frac{y_A P N_A}{2R} \sqrt{\frac{2k_B T}{\pi m}} S^*$$

the result is $y_A = 710$ on the basis of (Table 5-2). The reaction rates of the evaporation segment and the condensation segment are as follows:

$$k_1 = 0.274, \quad k_2 = 0.116$$

Given the corresponding reaction rate, substitute them into the kinetic model, the theoretical result is 1.096l /10min in terms of stable fresh water yield. Next, the system will be analysed as a whole with the help of the parameters just calculated. Through simulation analysis of the obtained calculation, it is showed that, as demonstrated in Figure 10, the final seawater desalination tends to be stable, indicating that the designed device is stable and reliable.

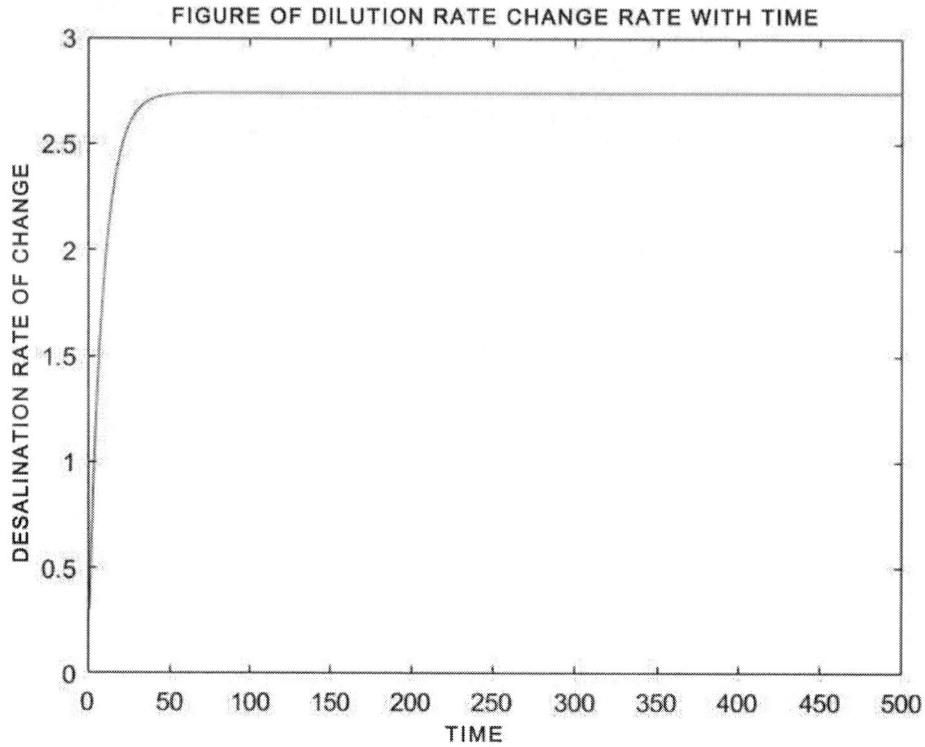

Figure 10. Time-varying rate of seawater desalination efficiency

6. Conclusion

This device could be used on ships or islands with less space occupancy by reducing the volume of seawater desalination device. On this basis, it could be modified into a buoy-like seawater designated device, enabling it to be used on the channel. At the same time, a plate condensing heat exchanger is provided, and the spoiler is inserted between the isolation plates to realize the full contact between the cold seawater inside the spoiler pipeline and the upper and uniform plates, so as to increase the heat exchange efficiency. Based on mathematical analysis of Kinetics model, theoretical analysis and calculation and empirical numerical simulation were carried out by using the ideal gas-wall collision probability formula and Raoult's theorem in physics, and it could be concluded that, the fresh water yield would be stable at 1.0961 /10min.

Acknowledgements

This research was financially supported by New CPL capillary pump type low temperature table top seawater desalination system.

References

[1] Kaya Tarik, Goldak John.Numerical analysis of heat and mass transfer in the capillary structure of a loop heat pipe. Int.J .Heat Mass Transfer, 2006 , 49 (18):3211-3220
[2] Ren Chu an , Wu Qin gsong , Hu Maobin. Heat transf er with flow and evaporation in loop heat pipe' s wick at l ow or moderat e heat flu xes.Int .J .Hea t Mass Transf er , 2007 , 50 (12):2296-2308.
[3] Yan Y H , Ochterbeck J M .Numerical investigation of the steady-stat e operation of a cylindrical capillary pumped loop evaporat or .Journal of Electronic Packaging , 2003 , 125 (2):251-260
[4] Han Yanmin , Liu Wei , Huang Xiaoming .The numerical simulation for the unsteady heat and mass transf er process in capillary pumped loops evaporator . Journal of Astronautics , 2003 , 24

(4):397-403
[5] Zhang Jiaxun. The state-of-the-art of loop heat pipe[J]. Journal of Engineering Thermophysics, 2004(4): 682～684.
[6] Pan Xianhui et al. Development and application of marine seawater desalination plant [B]. State Oceanic Administration, Tianjin Institute of Seawater Desalination and Comprehensive Application, 2015.
[7] Zhang Yaoming, Zou Ningyu. Seawater desalination technology and solar energy utilization [J]. Chinese Engineering Science, 2015, (11).
[8] Maydanik Y F. Loop heat pipes[J]. Applied Thermal Engineering, 2005(25): 5～6, 635～657.

IOP Conf. Series: Materials Science and Engineering **711** (2020) 012055 doi:10.1088/1757-899X/711/1/012055

Effect of calcium chloride on hydration kinetics and pore structure of hydrated tricalcium silicate

Qinfei Li[1, 2, 3, a], Wang Yang[1, b], Heng Chen[1, 3, c*], Pengkun Hou[1, 3, d], Xin Cheng[1, 3, e]

[1] Shandong Provincial Key Laboratory of Preparation and Measurement of Building Materials, University of Jinan, Jinan, P. R. China

[2] State Key Laboratory of Green Building Materials, Beijing, P. R. China

[3] Department of Materials Science and Engineering, University of Jinan, Jinan, P. R. China

[a]hitqfli@gmail.com, [b]wangyangyang1226@163.com, [c]mse_chenh@ujn.edu.cn, [d]pkhou@163.com, [e]chengxin@ujn.edu.cn

Abstract. Chemical admixtures are frequently used to regulate the setting and strength development of concrete materials. In this study, tricalcium silicate (C_3S) was used as a model of the cement system, and the influence of calcium chloride, an extremely useful accelerator, on C_3S hydration and the pore structure of hardened C_3S paste were investigated by the combination of the techniques of differential scanning calorimetry (DSC) and the N_2 adsorption (BET). The results indicated that the addition of calcium chloride would significantly shorten the pre-induction and induction periods and enhance the specific surface area and porosity of hardened C_3S paste. However, the presence of $CaCl_2$ has little effect on the pores, with a width ranging from 2.5 nm to 5 nm. DSC technique has an advantage of measuring continuously the process of C_3S hydration by changes of free water in hydrated C_3S.

1. Introduction

Calcium chloride ($CaCl_2$) is an extremely useful accelerator of tricalcium silicate (C_3S) and cement hydration[1]. This accelerator is still frequently used in the unreinforced concrete. The higher the dosage of $CaCl_2$, the more significant the accelerating effect. $CaCl_2$ could promote the dissolution of cement, accelerate the formation of hydration products, and increase the heat flow rate of C_3S hydration [2-4].

According to the heat flow curve, the hydration of C_3S is generally divided into four stages:(1) pre-induction period; (2) induction period; (3) acceleration period; (4) deceleration period[5]. The hydration of C_3S would release abundant heat, and produce amorphous calcium silicate hydrates and crystalline calcium hydroxide[6, 7]. The evolution of hydration heat flow is frequently used to represent the hydration kinetics of C_3S [8].

Calorimetry is a commonly used and effective method to measure the exothermic properties of cement hydration reaction, including heat of solution calorimetry, semi-adiabatic/adiabatic calorimetry, isothermal conduction calorimetry and differential scanning calorimetry (DSC)[9]. The approach for DSC is to measure the changes of free water in C_3S and cement hydration, which can continuously measure the hydration regardless of the amount of heat released. In the study, with the same method, the influence of $CaCl_2$ on C_3S hydration was investigated.

With the progress of C_3S hydration, the number of gel pores from C-S-H gels[1, 10, 11] would increase, whereas that of capillary pores would decrease. Since the capillary pores are the main channel of water transportation, it closely related to the deterioration of cement-based materials caused by ingress of foreign ions [12]. Therefore, the study of the pore structure of hydrated C_3S is helpful to understand the C_3S hydration products, microstructure and performance.

2. Materials and Methods

Sample Preparation. Triclinic C_3S was purchased from CTL Group, Skokie, IL, U.S.A and C_3S particles were passed through the sieve of 45μm. Anhydrous C_3S powder was mixed with deionized water with a liquid-to-solid ratio of 0.4 and 0.5, and $CaCl_2$ was added with 1% wt of anhydrous C_3S. Hydrated C_3S pastes were placed into CryoTube vials (1.8mL) under N_2-gas protection at room temperature and then were sealed in a vacuum bag. The samples of hydrated C_3S without and with $CaCl_2$ were respectively dried in a vacuum oven at 23°C after 1, 3, 7 and 28-day hydration.

X-ray Diffraction. The X-ray diffraction tests of samples were conducted on a PNAalytical X'Pert Pro diffractometer equipped with a Co-target X-ray tube (λ=1.79 Å) and rapid X'celerator detector. The 2-theta ranges from 5° to 55°with a step of 0.02°. X-pert High Score Plus software was used to analyze the XRD patterns of the samples.

Differential Scanning calorimetry (DSC). Differential scanning calorimetry measurements were performed using a DSC Q2000 from TA Instruments. Around 50 mg of sample was mounted into an aluminum pan (ϕ5.4*2.0 mm), and then was sealed with the appropriate cover to avoid the leaking of water. Then, to the samples were equilibrated to -30°C and kept at this temperature for 1 min, after which the samples were heated from -30°C to -12°C at 20°C/min and -12°C to +25°C at 4°C/min. The collected data were analyzed by a Q series software (version 5.5.22).

N_2 adsorption (BET). The pore distribution of samples was measured by an accelerated surface area and porosimetry system (ASAP 2020) from Micromeritics (U.S.). The test temperature is around room temperature.

3. Results and discussion

XRD Patterns. Hydration reaction of C_3S hydration would form amorphous C-S-H and crystalline calcium hydroxide (Portlandite), which generally follows the following formula:

$$C_3S + (3\text{-}x\text{-}y)\ H_2O \rightarrow C_xSH_y + (3-x)\ Ca\ (OH)_2 \quad (1)$$

Allen *et al.* [13] determined x and y values in the above formula by using small-angle X-ray scattering and small-angle neutron scattering, and obtained the following equation:

$$C_3S + 3.1H_2O \rightarrow C_{1.7}SH_{1.8} + 1.3Ca\ (OH)_2 \quad (2)$$

It can be seen from Fig. 1 that CH and C-S-H (a hump at the d-spacing of 3.08 Å) are the only hydration products, and that CH was formed after 1-day hydration and the C_3S were almost consumed after 28 days. This means that the addition of $CaCl_2$ did not change the types of hydration products, and no new substances were produced.

Hydration kinetics of C_3S. In this paper, the evolution of free water content (Free Water Index, FWI) was chosen to represent the hydration kinetics of C_3S, by the technique of DSC which allows the quantitative analysis of free water content by integrating the melting peak of free water[9, 14]. Fig. 2 presents the heat flow curves of hydrated C_3S pates as a function of hydration time. With these heat flow curves, FWI can be calculated according to the following formula:

$$FWI=\Delta H_{exp}/\phi_w \Delta H_{theor} \quad (3)$$

where ΔH_{exp}——the area of the water melting peak, J/g;

ϕ_w——the weight fraction of water in the paste;

ΔH_{theor}——the theoretical value of melting enthalpy of water in the C_3S/water paste, 333.4J/g;

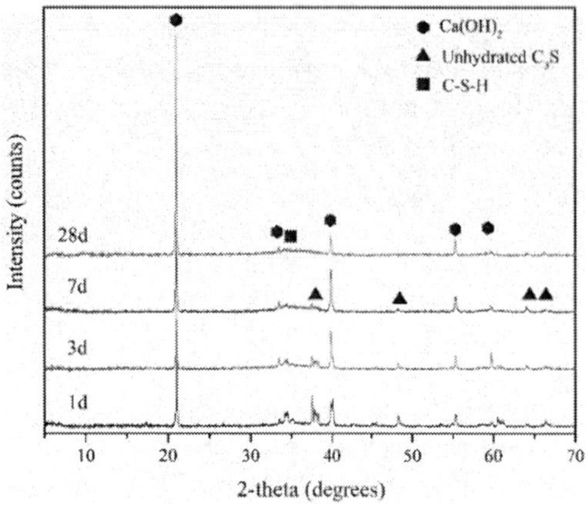

Fig. 1 XRD patterns of phase change from hydrated C_3S (*w/c*=0.5) in the presence of $CaCl_2$ during different hydration time[4].

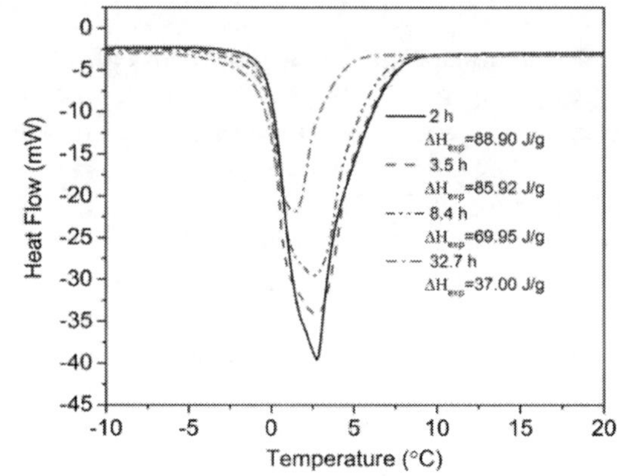

Fig. 2 Heat flow for sample C_3S (L/S=0.5) hydrating at 20°C

The changes of free water in C_3S pastes of the melting peak of C_3S pastes with w/c of 0.4 and 0.5 over time is shown in Fig. 3. It can be seen that the addition of $CaCl_2$ would shorten the pre-induction and induction period, which is consistent with the results in the literature. In the absence of $CaCl_2$, where would be an induction period of around 3 hours, this period would almost disappear in the presence of $CaCl_2$. For the system with a w/c of 0.4 as shown in Fig. 3b, similar results can be found that the addition of $CaCl_2$ would accelerate the assumption of free water. Interestingly, it can be found that the FWI in the systems with a w/c of 0.4 are lower than those with a w/c of 0.5 regardless of the presence of $CaCl_2$.

$CaCl_2$ can significantly increase the nucleation rate of hydration products on the surface of C_3S particles and has little influence on the growth rate[15]. The addition of would shorten the pre-induction and the induction period as shown in Fig. 3. This means that $CaCl_2$ would promote the dissolution of C_3S particles and the release of Ca^{2+} and SiO_4 into the pore solution, and finally accelerate the saturation or supersaturation of hydration products. The critical effect of $CaCl_2$ is to trigger the initiation of acceleration, to accelerate the hydration products of nucleation[16].

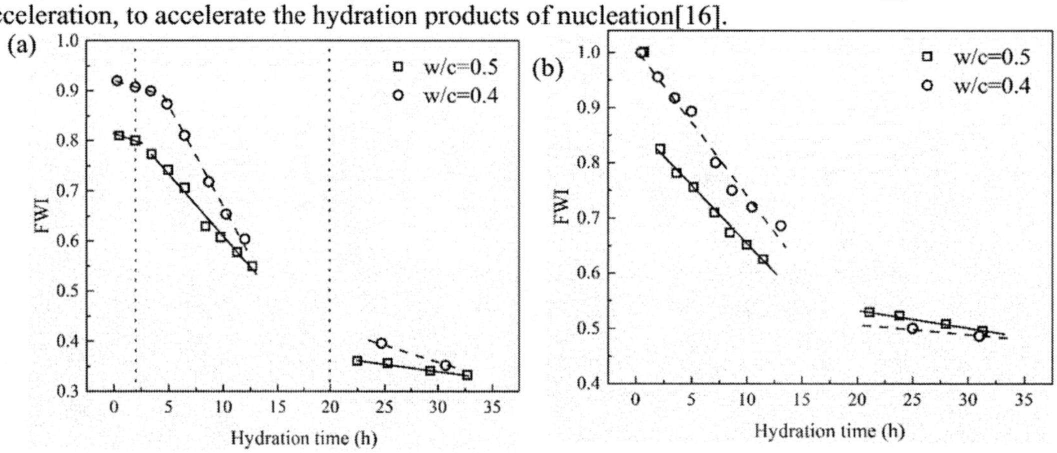

Fig. 3 Hydration kinetics of C_3S with 0.5 and 0.4 of w/c at 23°C in the absence (a) and presence (b) of $CaCl_2$

Analysis of Pore structures. N_2 adsorption is a commonly used method to measure the specific surface area and pore size distribution of cement-based materials, especially for the determination of relatively small gel pores[17]. Hydrated C_3S paste is porous, in which there are plenty of pores with the diameters ranging from nanometers to microns, such as gel pores and capillary pores.

Fig. 4 presents the adsorption isotherm of C_3S pastes in the absence and presence of $CaCl_2$. From figure 4, it can be seen that the maximum adsorption of $CaCl_2$-free samples reaches up to 58.8 cm3/g STP at 28 days, while the 7-day maximum adsorption amount of samples is ranging from 20 cm3/g STP to 30 cm3/g STP. C-S-H gel has a high specific surface area, and these amorphous gels stack and is rolled to form gel pores. As the hydration time increases, more foil-like and layer-structure C-S-H gel are generated, increasing N_2 adsorption. According to the International Union of Pure and Applied Chemistry (IUPAC) classification of N_2 adsorption isotherms, the shapes of hysteresis loops of the samples with different hydration time are shown in Fig. 4. It can be seen that hysteresis loop type of the pores in the hydrated C_3S sample transform from H4 type to H3 type with hydration. Hysteresis type H4 is associated with non-rigid aggregates of plate-like particles (e.g., calcium hydroxide), while hysteresis type H3 is associated with being filling with micropores, aggregated hydration product of C-S-H gels. The pores in C_3S samples hydrated for 28 days show H3 hysteresis ring, and the adsorption amount was enormous when the relative pressure was high. The slit pores formed by the accumulation of flake particles are mostly laminated Ca $(OH)_2$ crystals and C-S-H layered structure.

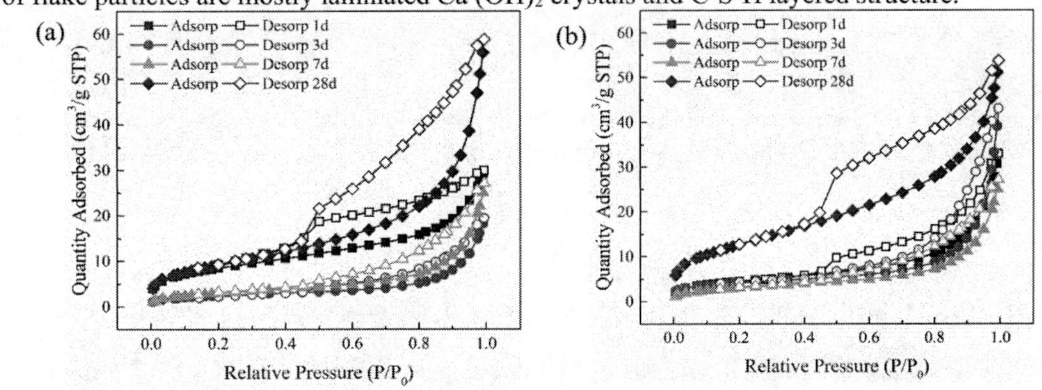

Fig. 4 The adsorption isotherm of C_3S in the absence (a) and presence (b) of $CaCl_2$

The specific surface areas of Langmuir, BET, t-plot and BJH adsorption/desorption are shown in Table 1. The specific surface area of C_3S pastes gradually increases with hydration time, which reflects the increase of C-S-H hydration products in the sample. Noticeably, the specific surface area of C_3S pastes with $CaCl_2$ is all larger than that without addition after the same hydration time. Also, Langmuir specific surface area is higher than the BET specific surface area. Theoretically, Langmuir adsorption is based on single-layer adsorption and focuses on measuring smaller pores, while BET specific surface area is based on multi-layer adsorption, which is suitable for testing larger pores.

In Fig. 5, the pore diameter of the hydrated C_3S is mainly distributed between 2.5 nm and 5nm at different hydration times. Furthermore, regardless of the presence of $CaCl_2$ or not, the average pore diameter of the gel pores in the hydrated C_3S samples are about 3.5nm. Although $CaCl_2$ can accelerate the hydration reaction of C_3S and produce more hydration products in the early stage, it has little effect on the size of the gel pores.

Table 1 Specific surface area of the pores in hydrated C_3S with different hydration time (m^2/g)

Samples	BET	Langmuir	t-plot	BJH adsorption	BJH desorption
C_3S_1d	10.46	14.97	11.36	10.33	13.98
C_3S_3d	9.16	12.92	8.88	7.72	11.61
C_3S_7d	11.29	16.26	12.96	11.93	16.58
C_3S_28d	33.52	47.19	34.17	35.72	56.78
C_3S_$CaCl_2$_1d	16.33	22.96	16.26	16.673	24.31
C_3S_$CaCl_2$_3d	10.63	14.94	10.11	9.97	15.52
C_3S_$CaCl_2$_7d	14.85	20.65	12.64	12.59	19.99
C_3S_$CaCl_2$_28d	46.58	65.72	48.72	48.30	65.00

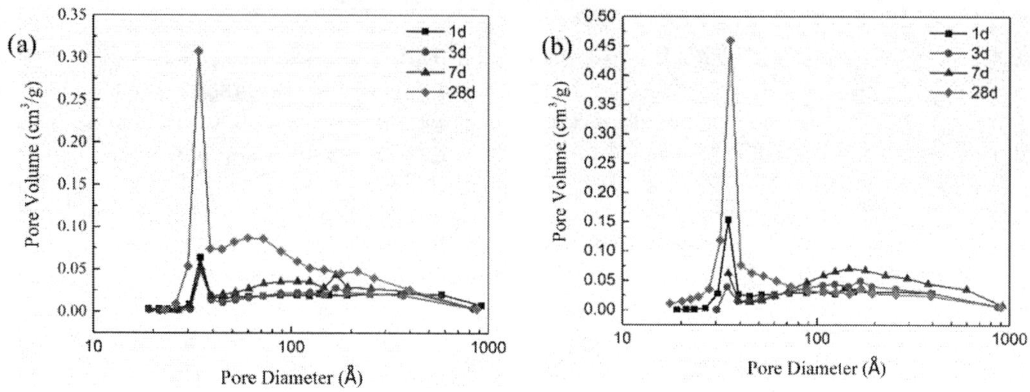

Fig. 5 Pore size distribution of hydrated C_3S with different hydration time in the absence (a) and presence (b) of $CaCl_2$

4. Summary

Differential scanning calorimetry offers the possibility of studying the effect of $CaCl_2$ on the hydration kinetics of C_3S hydration. Changes in free water in hydrated C_3S pastes can easily be determined continuously by DSC.

The presence of $CaCl_2$ does not change the types of hydration products, which is mainly C-S-H gel and calcium hydroxide. $CaCl_2$, as an inorganic chloride salt, can shorten distinctly the pre-induction period and induction period during the hydration reaction.

As the progress of hydration, the specific surface area of hydrated C_3S pastes gradually increase due to the formation of hydration products, such as C-S-H gel and calcium hydroxide, and the addition of $CaCl_2$ would accelerate this increasing rate. Based on the shapes of H4 hysteresis loops of hydrated C_3S pastes, the pores in hydrated C_3S are of slit pores ranging from macropores to mesopores due to the generation of C-S-H gel with a layer structure and plate-like crystal calcium hydroxide. However, the presence of $CaCl_2$ does not affect the pore width of mesopores of hydrated C_3S, which is ranging from 2.5 nm to 5 nm.

Acknowledgments

Funding for this project was provided by the National Natural Science Foundation of China (No. 51802112 and No. 51761145023). Supports from the 111 Project of International Corporation on Advanced Cement-based Materials (No. D17001) is much appreciated. The funding is also supported by the Opening Project of State Key Laboratory of Green Building Materials and by the Taishan Scholars Program.

References

[1] P.K. Mehta, P.J.M. Monteiro, Concrete: Microstructure, Properties, and Materials, 4th ed., McGraw-Hill, New York, 2014.
[2] E. Gartner, J. Young, D. Damidot, I. Jawed, Structure and performance of cements, Vol. 13 (2002) p. 978-970.
[3] S. Bae, M. Kanematsu, D. Hernández-Cruz, J. Moon, D. Kilcoyne, P. Monteiro, Materials, Vol. 9 (2016) p. 976.
[4] Q. Li, Y. Ge, G. Geng, S. Bae, P.J.M. Monteiro, Journal of Nanomaterials, Vol. 2015 (2015) p. 10.
[5] J.W. Bullard, H.M. Jennings, R.A. Livingston, A. Nonat, G.W. Scherer, J.S. Schweitzer, K.L. Scrivener, J.J. Thomas, Cem. Concr. Res., Vol. 41 (2011) p. 1208-1223.
[6] J.J. Thomas, H.M. Jennings, J.J. Chen, J. Phys. Chem. C, Vol. 113 (2009) p. 4327-4334.
[7] S. Bae, R. Taylor, D. Kilcoyne, J. Moon, P. Monteiro, Materials, Vol. 10 (2017) p. 131.
[8] J.J. Thomas, J. Am. Ceram. Soc., Vol. 90 (2007) p. 3282-3288.

[9] F. Ridi, E. Fratini, P. Luciani, F. Winnefeld, P. Baglioni, J. Colloid Interface Sci., Vol. 364 (2011) p. 118-124.
[10] S. Mantellato, M. Palacios, R.J. Flatt, Mater. Struct., Vol. 52 (2019) p. 5.
[11] G. Goracci, M. Monasterio, H. Jansson, S. Cerveny, Scientific Reports, Vol. 7 (2017) p. 8258.
[12] M.C. Garci Juenger, H.M. Jennings, Cem. Concr. Res., Vol. 31 (2001) p. 883-892.
[13] A.J. Allen, J.J. Thomas, H.M. Jennings, Nat. Mater., Vol. 6 (2007) p. 311.
[14] A. Damasceni, L. Dei, E. Fratini, F. Ridi, S.-H. Chen, P. Baglioni, J. Phys. Chem. B, Vol. 106 (2002) p. 11572-11578.
[15] J.J. Thomas, A.J. Allen, H.M. Jennings, J. Phys. Chem. C, Vol. 113 (2009) p. 19836-19844.
[16] V. Robin, B. Wild, D. Daval, M. Pollet-Villard, A. Nonat, L. Nicoleau, Chem. Geol., Vol. 497 (2018) p. 64-73.
[17] F.F. Ataie, K.A. Riding, Construction and Building Materials, Vol. 106 (2016) p. 274-281.

Study of pulping technology on disposable tableware using withered leaves as raw material

Liu bo-fan[1], Liu Ming-dan[1]*, Jin Peng[2], Zhong Yi-xin[3]

College of Mechanical and Electrical Sichuan Agricultural University 625014, Yaan Sichuan, China

e-mail: 276591355j@qq.com, 2296482900@qq.com,

1048721651@qq.com, 844929149@qq.com

TEL: 13086317002

Abstract. Disposable tableware is widely used around world due to its advantages such as sanitation, thermal preservation and low price. However, more and more countries are restricting the use of disposable tableware due to its difficulties in recycling and degradation in the natural environment. If disposable tableware is made of photobiodegradable PP plastic or high-impact polystyrene (HIPS), it is difficult to degrade and has a great impact on the earth environment. In this study, four kinds of withered leaves were selected as raw materials and mechanical chemical pulping method and caustic soda pulping method were used for experiments. The main processes include pulp preparation, adsorption and drying. The experiment indicated that withered leaves of magnolia grandiflora was the best raw material for making recyclable tableware.

1. Introduction

At present, there are four types of disposable degradable tableware on the market: degradable polypropylene tableware, pulp moulded tableware, all-starch tableware and vegetable fiber tableware [1]. The raw material of degradable polypropylene tableware is still plastic. By adding degradable material such as and photosensitizer or degradable starch material, plastic can be degradable [2]. However, Its degradability is constrain by many aspects: long degradation phase, non-completing degradation; pulp-molded tableware faces limited sources and pricey material with severe pollution during manufacture, which make it unpopular in China; Although all-starch tableware possess advantages such as completing degradation, high food security, its low thermal stability, high flexibility, difficulty in preservation and high price make it unfavored in market[3]. Thus, plant fiber tableware is the direction of this study because of its recyclable, degradable, and reusable characteristics [4, 5]. However, if the raw materials come from crops, other substances like pesticide residues and animal waste are inevitable. In recent years, Tan Wei and others have also carried out an analysis and Research on papermaking process using corn straw, rice husk, Pineapple leaves and banana fiber as raw materials [6, 7, 8], which proves the universality of plant fiber materials. Recently, a German team degraded a plate made of leaves in 28 days [9], which inspired us to use dead leaves as a disposable dish material. In this study, fallen leaves around the school were used as the raw material and mechanochemical pulping method was used to carry out the single-factor experiment to compare disposable plates made from different falling leaves under different reaction conditions [10, 11], were compared to find out ideal material and parameter for pulp preparation.

2. Material and Methods

2.1 Material

The experimental materials collected from dry leaf of yulan, sycamore, palmetto and bamboo in Sichuan agricultural university.

2.2 Instrument

Steaming and boiling container, induction cooker, multi-functional breaker, dryer (model: DHG-9070A, Shanghai yiheng scientific instrument co. LTD), electronic scales, 0.45mm stainless steel mesh, gauze, compressor (hydraulic multi-functional testing machine with microcomputer display, model WEW-600D, Jinan hengsi shanda instrument co. LTD)

2.3 Reagent

Concentrations were 6%, 16%, 26% and 36% sodium hydroxide solution, 30% hydrogen peroxide solution, 15% sodium sulphide solution, Anthraquinone reagent.

2.4 Method

Both mechanical and chemical pulping method were used in this experiment to ensure enough quantity of pulping while removing more lignin to ensure pulping quality [12].

Chemical pulping adopts caustic soda pulping in paper manufacturing [13]. Leaves are streamed after drying (dry conditions of 75 °C, 30 min), then using sodium hydroxide to decompose mesophyll cell obtaining vein nest in leave. In this process, lignin is extracted from fibrous cell walls and intercellular layers as result of disconnection of chemical bond in cellular lignin molecule by adding defused OH- SH- ions in the cooking liquid to lignin in leaves lignin and expansion of cellular wall caused by cooking liquid. Leaving vein nest become fibrous after delignification process. HS- and S2- from dissolved sodium sulphide can accelerate delignification while protecting cellulose from strong alkali. During steaming and boiling, Small amount of cellulose and hem-cellulose are dissolved and lost due to Alkaline hydrolysis and peeling reaction caused by alkali. Therefore, the addition of anthraquinone [14], in cooking liquid, which equivalent to 0.1% fibrous material weight can prevent harmful peeling reaction and facilitate delignification.

Mechanical pulping is to obtain single strain fiber from raw material that are chemical treated withered leaves. The withered leaves are broken by breaking machine to separate leave veins and remove mesophyll.breaking machine also create heat and friction to soften intracellular. By mechanical treatment, lignin in leave veins become single strain fiber. After dring (80 °C drying conditions, 80 min), the weight of fibers is quantity of pulp. The final quantity of pulp is showed by using mass fraction W (%)

$$w = \frac{m_1}{m_2} \times 100\%$$

m1- quantity of dried leaf (75 °C, 30 min)
m2- quantity of dried pulp (80 °C, 210 min) [15]

2.5 Experimental Design

The experiment was carried out in three different trials. Trial one is the comparison between pulp yield rate of different leaves. Selected dry leaf of yulan, sycamore, palmetto and bamboo were put into forced hot air oven for 30 min. After drying, 100g of each dry leave were cleaned and put into boiling container with 2.5 L distilled water, 500ml 16% sodium hydroxide solution, 250ml 15% sodium sulphide solution and 0.1% Anthraquinone reagent for 2 hours.

Trial two is to compare influences of different sodium hydroxide solution concentration on the leaf pulp yield rate.

The selected leaves of yulan were put into forced hot air oven for 30 minutes. The dried leaves were divided into four groups, each group had 100g cleaned dry leaves. Each group were put into individual boiling containers with 2.5L distilled water, 500ml sodium hydroxide solution, 250ml 15% sodium sulphide solution and 0.1% anthraquinone reagent for 2 hours. In each boiling container, the concentrations of sodium hydroxide solution were 6%, 16%, 26% and 36%.

Trial three is to compare influences of boiling time on pulp yield rate.

Selected leaves of yulan were put into force hot air oven for 30min. Dried leaves were divided into four groups, each group was 100g cleaned dry leaves. Every group were put into individual boiling containers with 2.5 L distilled water, 500ml 16% sodium hydroxide solution, 250ml 15% sodium sulphide solution and 0.1% anthraquinone reagent. The boiling time of each group were 1h,2h,3h and 4h.

After boiling trial, pulp was cleaned by using distilled water, broken by breaker and screen by mesh to remove leave vein residue. After that, pulp was boiled and bleached with 1L water and 67ml 30% hydrogen peroxide solution for 30min, then pass through 0.45mm sieve and molded by mold in 80°C forced hot air oven for 4 hours to produce paper fiber. The final products produced by compressing paper fiber using multifunctional compressing machine.

2.6 Degradation experiments

The common disposable dishes on the market at present: plastic dishes, Degradable Polypropylene dishes, paper dishes, and withered leaf dishes made in this experiment were used as sample materials. Cut 40*40 mm square pieces on the flat part of the bottom of the plate as experimental materials, and put them into 80 mm soil to simulate natural degradation for degradation test. Degradation was recorded every two days.

3. Results

3.1 Influences of different leaves (withered leaves) on pulp yield rate

Table 1. comparison of pulp yield rate among different leaves

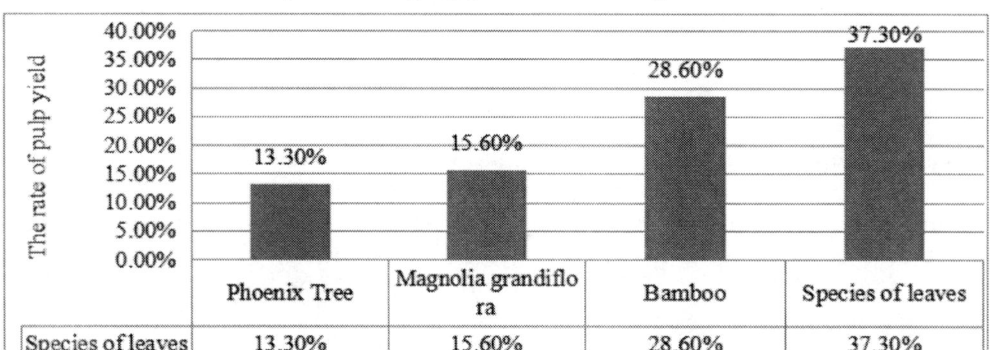

From experiment, pulp was successfully extracted from withered leaves of four different plants. Based on this figure, the pulp yield rate from high to low are palmetto, bamboo, yulan, sycamore. Although palmetto had highest pulp yield rate, its fiber was thick. Meanwhile, bamboo was hard to collect.

3.2 Influences of different sodium hydroxide concentration on pulp yield rate

Table 2. influences of different sodium hydroxide concentration on pulp yield rate

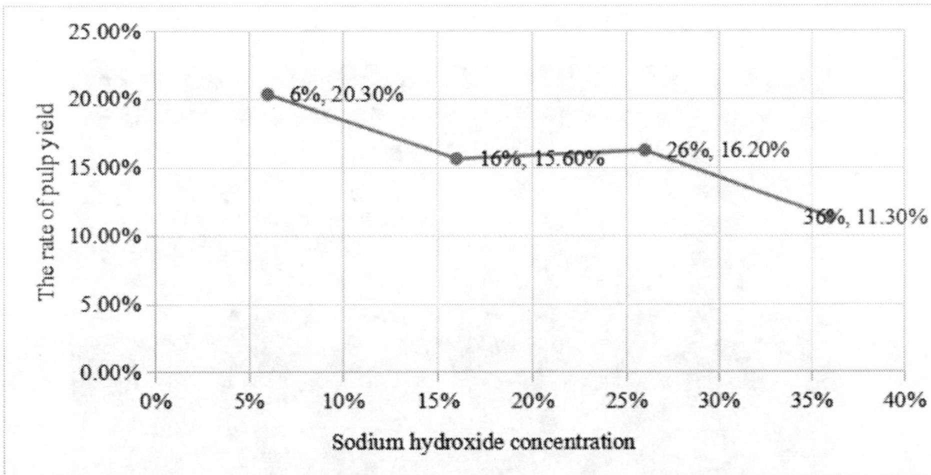

Based on figure 3, when sodium hydroxide concentration rised, the pulp yield rates declined. Thus, sodium hydroxide concentration highly impacts pulp yield rate. When sodium hydroxide concentration fell into range from 6% to 16%, the yield rate had obvious decline and fiber in pulp was too thick. When sodium hydroxide concentration fell into range from 16%-26%, the change of yield rate was moderate while diameter of fiber was favorable. When sodium hydroxide concentration fell into range from 26% to 36%, the declination of yield rate increased while diameter of fiber was thinner. To save reagent and reduce pollution, the concentration of sodium hydroxide should be 16% to 36%. Therefore 16 % sodium hydroxide was used in next trial.

3.3 Influences of different boiling time on pulp yield rate

Table 3. influence of different boiling time on pulp yield rate

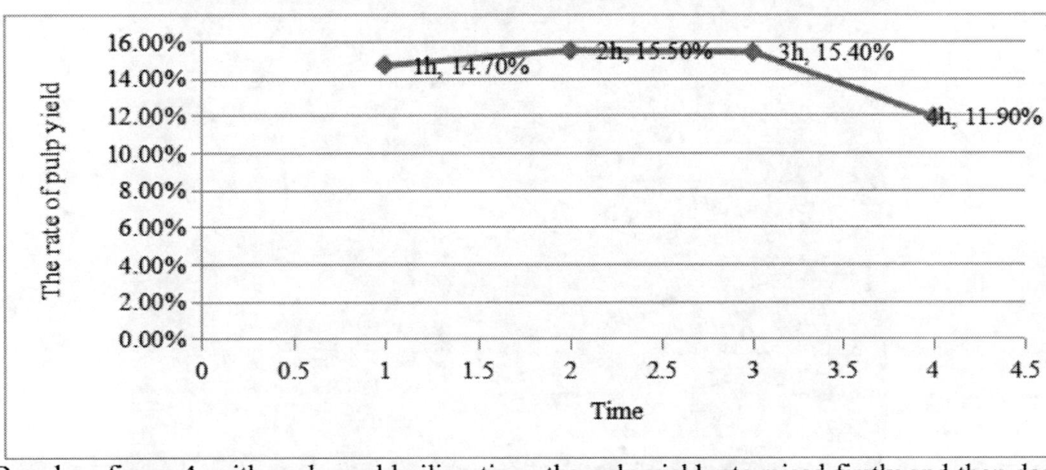

Based on figure 4, with prolonged boiling time, the pulp yield rate raised firstly and then decline. When boiling time was 1-2 hours, the pulp yield rate had a minor improvement. In this period, the diameter of fiber was thick initially, but with gradually lignin precipitation, the fiber became thinner. When boiling time was 2-3 hours, the diameter of fiber was ideal. When boiling time was 3-4 hours, due to long time reaction, lignin was largely precipitated, thus fiber was too thin and pulp yield rate dropped quickly. Therefore, the most ideal boiling time is 2 hours: both saving time and ensuing yield rate.

IOP Conf. Series: Materials Science and Engineering **711** (2020) 012056 doi:10.1088/1757-899X/711/1/012056

3.4 Analysis of Degradation Experiments
Degradation experiments showed that the material of the withered leaf plate degraded significantly on the 5 day, and almost completely (fusion with soil) on the 14 day. Because the whole plate is not taken for testing, the actual degradation time should be increased, but the degradation rate of the leaves made into disposable dishes is much faster than other materials on the market at present, and the withered leaves will not cause ecological damage.

Figure 1. Analysis of Degradation Experiments
(From left to right are paper plate, dead leaf plate, Degradable Polypropylene plate and plastic plate)

Figure 2. Degradation after 14 days (fusion with soil)

4. Discussion
Although, livistona chinesis leaves has highest pulp yield rate, the thickest fiber leads to fragile disposable tableware, thus it is impractical. Sterculiaceae and Magnolia leaves are easy to collect and

has ideal pulp yield rate, thus is best option for producing pulp for disposable tableware.

Sodium hydroxide concentration has obvious influences on pulp yield of Magnolia leaves. 36% Sodium hydroxide concentration has low yield rate with thick fiber diameter, while 6% Sodium hydroxide concentration has higher yield rate with thin fiber diameter. Therefore, 16% Sodium hydroxide concentration has both high yield rate and suitable fiber diameter.

To some degree, boiling time has influences on pulp yield of Magnolia leaves. When boiling time is 1 or 4 hours, the fiber quality is poor with low yield rate. When boiling time is 2 or 3 hours, the fiber quality is much better with higher pulp yield rate. To save time, 2 hour boiling time is recommended.

5. Conclusions

In this paper, dry leaf of yulan, sycamore, palmetto and bamboo were selected, and the crude fiber was obtained by mechanical chemical pulping and caustic soda pulping. After processes of breaking, compressing, drying etc., the disposable food container that is environment-friendly and easy to degrade was produced. Also, optimistic processing condition is studied in experiment. It was found that, in four kind of leaves, that yulan dry leaf was the best material for the type of disposable dish. Boiling under 100 ℃ with 16%NaOH, 15% sulfidation parameter, 0.1% anthraquinone for 2 h has the best pulp yield rate and fiber quality, which process enough tenacity to meet strength requirement. This study provides a fundamental basis for developing disposable tableware using plant-sourced fiber.

Acknowledgements

This research was financially supported by my instructor, Liu Ming-dan (Corresponding author), a respectable, responsible and resourceful scholar, who has provided me with valuable guidance in every stage of the writing of this paper. Thanks to my team, we find and solve all kinds of problems together, meet setbacks and never give up, and work unremittingly for common goals.

Corresponding author: Liu Ming-dan [1],male, (1965—), professor.Major in interests include theory teaching, application and control of motor , College of Mechanical and Electrical Sichuan Agricultural University Ya an Sichuan，625014
E-mail: 276591355j@qq.com, TEL:13086317002
The first author: Liu bo-fan[1] Major in agricultural mechanization, sichuan agricultural university
Layout fees are paid by the correspondent author Liu Ming-dan

Reference

[1] Cai Huijuan, Yan Nahan, Su Qinpo, Analysis of disposable tableware use in college students [J]. Journal of Henan Science and technology,2013(17):170.

[2] Zhou Logan, Discussion on the development of disposable tableware [J]. Technology development of interprise,2001(03):4-5.

[3] Chen Yanling, Not easy popular for environment-friendly tableware [J]. Rural agriculture and farmers,2018(05):41-43

[4] Liu Tianshu, Li Shujun, Zhou Shuhui, Disposable ware technology in China [J]. Farm Machinery,2011(12):9-11.

[5] Shi Chang. Disposable tableware with plant fiber is sold at home and abroad [J]. National business,2001(Z2):8

[6] Tan Wei, Changjiang, Bian Zhiqi, Zheng Qi, Liu Chang, Guo Yue. Analysis and Research of Corn Straw Papermaking Process [J]. Journal of Harbin Commercial University (Natural Science Edition), 2015, 31 (04): 458-459+464.

[7] Amit Ramdhonee, Pratima Jeetah. Production of wrapping paper from banana fibres[J] Journal of Environmental Chemical Engineering5 (2017) 4298–4306

[8] Stephen Sibaly, Pratima Jeet. Production of paper from pineapple leaves [J]Journal of Environmental Chemical Engineering, 5 (2017) 5978–59

[9] From 730000 to 28 days: Leaf trays shorten the degradation time of tableware [J]. China Social Organization, 2016 (20): 52-55.
[10] Jia Jingyu, Wu Ting, Han Xiaohui, Miao Li. Leaf reuse pattern of urban garden in new period [J]. Agriculture of Jilin[J].,2018(05):57.
[11] Chen Kefu, Zhang Hui. Green progress and engineering technology of China's paper industry [M] ,China light industry press 2016（01）.136-173.
[12] Huang Junyan, Tian Yuanyuan, Zhu Tingting, Guo Runlan, Packaging paper processing and inspection technology[M], Printing industry press, 2009（01）:24
[13] Li Xushengs. Studies and progress on anthraquinone pulping mechanism [J]. East China Pulp & Paper Industry 2016（04）:30-33
[14] ZhanHuaiyuany, LiuQiujuan, Jinfuming, Pulping technology[M]. China light industry press,2012(01:37-45)
[15] Wunan,Wulei, Pulp and Paper Mill Laboratory Test Methods[M]. China light industry press,2009(01:253-256)

IOP Conf. Series: Materials Science and Engineering **711** (2020) 012057 doi:10.1088/1757-899X/711/1/012057

Optimum Design of Transmission Coil for Giant Magnetostrictive Actuator

Zeng Fan [1,a] ,Shi Feng [1,b] and Hui Ying[1,c]

[1]1 Navy Submarine Academy, Qingdao,China, 266042

[a]zhilinwenchu777@sina.com, [b]47012518@qq.com, [c]yhcathy@163.com

Abstract. The driving coil is optimized by replacing the magnetic field with ampere turns. For a variety of wire diameter, select the appropriate trade-off between the ampere-number maximization and the power minimization when the volume is closest to its allowable value and does not exceed the maximum output current of the power supply. It is calculated that 1mm diameter coil has the minimum power consumption. Therefore, choose 1mm diameter enameled wire to wound coil.

1. Introduction

The giant magnetostrictive actuator has great advantages in terms of response speed, load carrying capacity and conditions of use. In particular, it has been studied in detail for the structure, model and control theory of giant magnetostrictive actuator. The application of vibration with Submarine machinery and equipment is of great significance for our military development.

The magnetostatic actuator is driven by a magnetic field. The design of magnetic field includes driving the magnetic field and the bias magnetic field. The driving magnetic field is usually generated by the input current of the hollow cylindrical electromagnetic coil. Therefore, the driving coil is a carrier of an electric-magnetic conversion. The magnetic induction strength with current excitation determines the magnitude of the output of the giant magnetostrictive actuator, so the design of the drive coil becomes critical [4-6]. At the same time, the geometric size of the coil is an important factor affecting the magnetic field strength and the efficiency of a magnetic conversion. It is also the main influence factor of the GMA volume. In practical application, the wound wire on the coil inevitably has the resistance loss, so the temperature of the GMA (giant magnetostrictive actuator) is also an important impact. In this paper, the drive coil is optimized from three aspects: coil volume, driving capacity and power consumption.

2. Optimization of Drive Coil

Optimization of drive coil magnetic circuit is considered from the following points: the magnetic field is uniform on telescopic rod; high magnetic field strength produced by coil; coil drive current does not exceed the maximum output drive current; coil heat is small, that is to say, low coil power dissipation. Geometric parameters are known: skeleton length, diameter, wire diameter. Electrical parameters are laid out: coil maximum allowable current, coil power supply voltage. The number of layers of the coil

$$n = \frac{D - d_a}{4d} \tag{1}$$

IOP Conf. Series: Materials Science and Engineering 711 (2020) 012057 doi:10.1088/1757-899X/711/1/012057

D is the outer diameter (m) of the coil, da is the inner diameter (m) of the coil, and d is the diameter of the coil wire (m). Actuator frequency response range around the coil, the thread and the tail are at the same end of the skeleton, so the number of layers should be even

$$M = 2n \tag{2}$$

The geometrical size determines the maximum volume of the coil, since the diameter of the wire can only be a limited amount of dispersion, which determines that the number of layers can only be discrete. When is optimizing, the coil volume is not necessarily the largest volume, and the maximum values are used as candidate volumes. Therefore, the number of layers is used as the loop argument to calculate the parameters. The drive coil is as shown in Figure 1.

$$N = \frac{MH}{d} \tag{3}$$

$$D = d_a + 2Md_a \tag{4}$$

$$L = \frac{4\pi N^2 R_a (\pi Rk - t(0.693 + c))}{H} \tag{5}$$

$$t = \frac{D - d_a}{2} \tag{6}$$

$$R_a = \frac{D + d_a}{4} \tag{7}$$

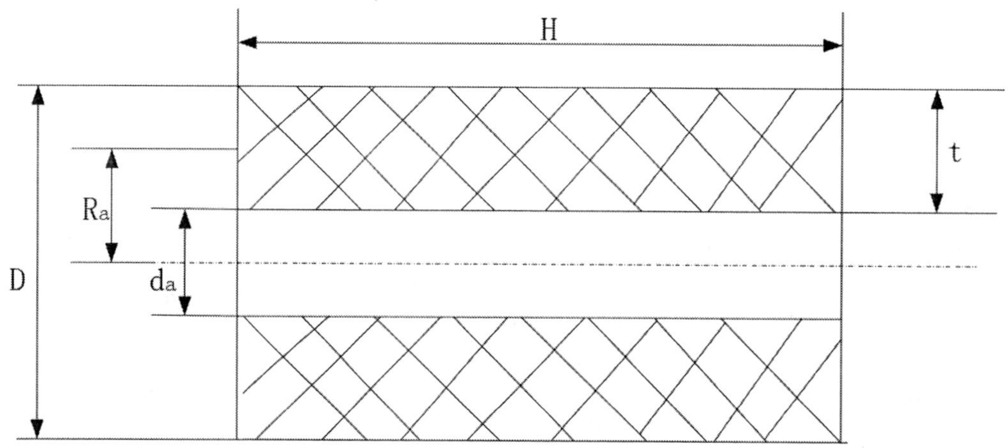

Fig. 1 Schematic diagram of the drive coil

L is the inductance (H) of the coil, R is the average radius of the coil (m), H is the height of the coil (m), N is the total number of turns of the coil, t is the thickness of the coil (m), k is the length (Determined by 2R / l), c is the self-inductance (coefficient determined by l / t).

$$l = \frac{M(2d_a + 2Md_a)H}{2\pi d} \tag{8}$$

$$R = lR_0 \tag{9}$$

$$Z = \sqrt{(2\pi fL)^2 + R^2} \tag{10}$$

$$I = \frac{E}{Z} \tag{11}$$

$$P = I^2 R \tag{12}$$

$$NI = \frac{MHI}{d_a} \tag{13}$$

Where l is the total length of the coil, Z is the coil impedance, f is the excitation frequency, I is the coil current, E is the supply voltage, P is the coil power, and NI is the total current of the coil.

The performance of the actuator depends on the driving magnetic field, and the intensity of the driving magnetic field is determined by the number of turns of the coil. Since the relationship between the driving magnetic field and the ampere-turns is complicated, and the larger the ampere-turns is, the greater the magnetic field strength. With ampere turns instead of magnetic field strength, the drive coil is designed optimistically. For a variety of wire diameter, select the appropriate trade-off between the ampere-number maximization and the power minimization when the volume is closest to its allowable value and does not exceed the maximum output current of the power supply. The objective function is:

$$F = aNI + (1-a)P \tag{14}$$

3. Optimization calculated of Drive Coil Diameter

Using MATLAB software to write the drive coil optimization program, the program flow chart is as shown in Figure 2.

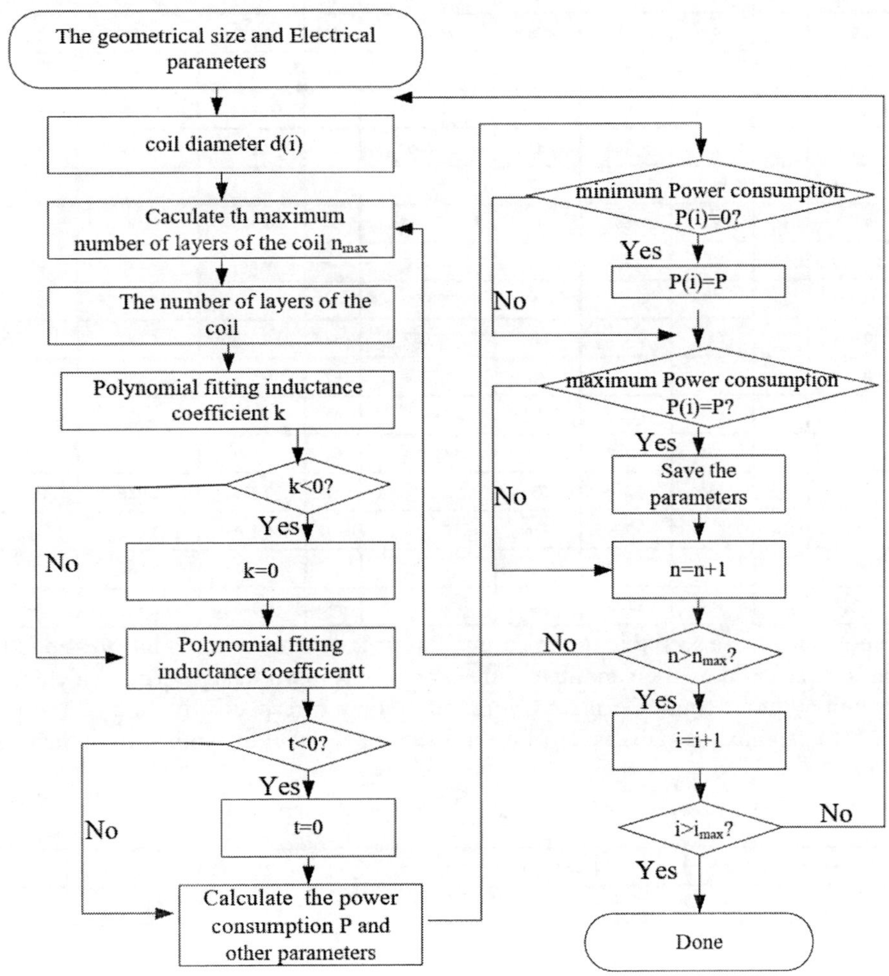

Fig. 2 Program flow chart

When the outer diameter of the coil is 63mm, the optimized parameters are listed in Table 1. Where D is the coil diameter (mm), R0 is the conductor resistance (Ω / m), M is the number of coils, N is the number of coils, I is the coil current (A), NI is the coil ampere, l is the length of coil(m), D is the coil diameter (mm), R is the total resistance of the coil (Ω), L is the coil inductance (mH), Z is the coil impedance (Ω), and P is the coil power consumption (W).

Table 1 Optimize the results of coil maximum diameter of 63mm

d	R_0	M	N	I	NI	l	D	R	L	Z	P
0.50	0.087	44	7920	0.026	208	1020.138	63	88.752	642.800	1372	0.061
0.53	0.078	42	7132	0.104	739	924.475	64	71.638	526.483	347	0.769
0.60	0.061	36	5400	0.037	201	688.763	62	41.643	294.246	966	0.058
0.67	0.049	32	4299	0.041	175	546.108	62	26.481	185.294	886	0.044
0.71	0.043	30	3803	0.438	1664	481.460	62	20.789	144.235	82	3.979
0.75	0.039	30	3600	0.323	1164	469.354	64	18.159	135.373	111	1.899
0.85	0.030	26	2753	0.684	1884	355.458	63	10.710	77.963	53	5.016
0.90	0.027	24	2400	0.115	276	306.117	62	8.225	58.123	313	0.109
0.95	0.024	24	2274	0.057	129	298.577	65	7.202	54.617	634	0.023
1.00	0.022	22	1980	2.065	4089	255.035	63	5.552	40.175	17	23.685
1.12	0.017	20	1607	0.367	590	209.028	64	3.627	26.877	98	0.490
1.18	0.016	18	1373	3.971	5451	173.557	61	2.713	18.755	9	42.767
1.25	0.014	18	1296	0.700	907	168.967	64	2.354	17.544	51	1.153
1.32	0.013	16	1091	0.154	168	137.499	61	1.717	11.787	233	0.041
1.40	0.011	16	1029	0.090	93	133.778	64	1.485	11.009	398	0.012
1.50	0.010	12	720	0.137	99	83.692	55	0.831	4.526	262	0.016
1.62	0.009	14	778	0.682	530	101.844	64	0.866	6.362	53	0.402
1.68	0.008	12	643	0.190	122	79.087	59	0.626	3.941	189	0.023
1.74	0.007	12	621	3.580	2222	77.764	61	0.573	3.780	10	7.344
1.88	0.006	12	574	0.726	417	75.005	64	0.473	3.455	50	0.250
1.95	0.006	12	554	0.167	93	73.774	66	0.433	3.315	215	0.012
2.02	0.006	8	356	0.216	77	39.371	51	0.215	1.026	167	0.010
2.10	0.005	10	429	0.761	326	53.856	61	0.273	1.811	47	0.158
2.26	0.004	8	319	0.287	91	37.112	55	0.162	0.889	125	0.013
2.44	0.004	10	369	1.048	386	50.291	68	0.189	1.526	34	0.207

From Table 1, it can be seen that the maximum number of ampere turns is 4089 and the diameter is 1mm, which can be used as the candidate wire diameter for coil design. Then calculate the relevant parameters with 60mm, 61mm, 62mm, 64mm, 65mm, 66mm respectively, and obtain the parameters corresponding to the maximum diameter of the ampere turns under different coil diameters, which is as shown in Table 2.

Table 2 Optimization results of different coil outer diameter

	60	61	62	63	64	65	66
d	1.120	1.120	1.180	1.000	1.250	1.250	1.250
R0	0.017	0.017	0.016	0.022	0.014	0.014	0.014
M	18	18	18	22	18	18	18
N	1446	1446	1373	1980	1296	1296	1296

I	4.178	2.772	4.796	2.065	5.618	5.639	5.639
NI	6044	4009	6584	4089	7280	7309	7309
l	177.947	177.947	173.557	255.035	168.967	168.967	168.967
D	59	59	61	63	64	64	64
R	3.087	3.087	2.713	5.552	2.354	2.354	2.354
L	19.949	19.949	18.755	40.175	17.544	17.544	17.544
Z	9	13	8	17	6	6	6
P	53.901	23.718	62.384	23.685	74.274	74.853	74.853

As can be seen from Table 2, within the allowable volume range, the diameter is 1.25mm and the ampere-turns maximum is 7309, but at the same time its power consumption is big; Determined the from the coil power, minimum is 23.68W and line Diameter is 1mm. Respectively, take three different sets of ampere-turns, and re-check the above four kinds of diameter coil power consumption, the results are as shown in Table 3.

Table 3 Check of the coil power consumption with different ampere turns

NI	d(mm)	N	I(A)	R(Ω)	P(W)
7500	1	1980	3.79	5.55	79.66
	1.12	1446	5.19	3.09	83.06
	1.18	1373	5.46	2.71	80.94
	1.25	1296	5.79	2.35	78.82
5750	1	1980	2.9	5.55	46.8
	1.12	1446	3.98	3.09	48.86
	1.18	1373	4.19	2.71	47.53
	1.25	1296	4.44	2.35	46.26
4000	1	1980	2.02	5.55	22.66
	1.12	1446	2.77	3.09	23.64
	1.18	1373	2.91	2.71	23.02
	1.25	1296	3.09	2.35	22.42

It can be seen from Table 3, with the same ampere turns different diameter power consumption is not very different, which provide a certain degree of flexibility for the purchase of wire diameter, but 1mm diameter is the common diameter. Finally, we get the conclusion: the coil power consumption is minimal with 1mm diameter, so choose 1mm diameter of the enameled wire to wound drive coil.

4. Summary
The drive coil is optimized from three aspects: coil volume, driving capacity and power consumption. Considering ampere turns instead of magnetic field strength and power consumption, we get the best optimization choice of coil diameter, which establishes the base for the generation of the driven magnetic field, the giant magnetostrictive actuator of dynamic response and the cooling of the drive coil.

References
[1] Zhang Xuhui. Study on Crucial Technology of Active Vibration Isolation System Based on Structure-Optimized Giant Magnetostrictive Actuator [D]. Beihang University, Dissertation Submitted for the Degree of Doctor, 2008:20-25
[2] Wu Yijie, Liu Chuhui. Study on Design Approach of Giant Magnetostrictive Actuator [J]. Journal of Zhejiang University (Engineering Science), 2004, 38(6):747-751

[3] Zhifeng Tang. Fundamental Theory and Experimens Study of Giant Magnetostrictive Actuator, Zhejiang University [D], Dissertation Submitted for the Degree of Doctor, 2005, 34-39
[4]]Lu Quanguo. Research and Application of Micro-Actuation Based on GMM [D]. Wuhan University of Technology, Dissertation Submitted for the Degree of Doctor, 2007:23-28
[5] Yu Peiqiong, Mei Deqing and Chen Zichen. Design of a Micro-displacement Actuator Based on Giant Magnetostrictive Material [J]. Transaction of the Chinese Society for Agricultural Machinery, 2003, 34(3):105-108
[6] Yang Wang. The Design and Optimization of Straight-GMM Transducer Used in Vibration Welding [D]. Zhejiang University, A Thesis Submitted for Master Degree, 2007:32-34

Corrosion behavior of 316L stainless Steel under Cl^- corrosion medium

xiejunjun[1, a], ningyuheng[2, b], sunxu[3, c], yangzhanjun[4, d]

[1]Datang Northeast Electric Power Test and Research Institute Co., Ltd., No. 3195, Ulsan Road, High-tech Zone, Changchun City, Jilin Province, China

[2]Datang Northeast Electric Power Test and Research Institute Co., Ltd., No. 3195, Ulsan Road, High-tech Zone, Changchun City, Jilin Province, China

[3]Datang Northeast Electric Power Test and Research Institute Co., Ltd., No. 3195, Ulsan Road, High-tech Zone, Changchun City, Jilin Province, China

[4]Datang Northwest Electric Power Test and Research Institute Co., Ltd., No.155, Fengcheng 7th Road, Economic Development Zone, Xi'an, Shaanxi, China

[a]1271165624@qq.com, [b]nyh3570@sina.com

[c]380541075@qq.com, [d]617578204@qq.com

Abstract. 316L stainless steel is widely used in the fields of thermonuclear industry and petrochemical industry due to its excellent comprehensive mechanical properties and corrosion resistance. Especially in the petroleum industry, 316L stainless steel is commonly used as oil pipelines and storage tanks because of its high corrosion resistance to media such as CO_2. However, 316L stainless steel will pitting in a high Cl^- concentration environment, posing a potential risk to these devices. At present, there is not enough data to show that whether 316L stainless steel will bring about pitting or stress corrosion cracking under the synergistic effect of high stress and high Cl^- concentration, this brings higher risks to the application and promotion of this material. In response to this phenomenon, the paper uses constant stress corrosion cracking method, and analyzes the fracture by scanning electron microscope (SEM). Researching 316L Stainless Steel corrosion behaviors and corrosion creeps in different Cl^- concentration solutions.

1. Introduction

Austenitic stainless steel has been widely used in many traditional industries such as petrochemical industry hydrometallurgical industry nuclear industry and so on because of its excellent comprehensive mechanical properties good weldability and corrosion resistance. As a typical austenitic stainless steel, 316L stainless steel is widely used in thermal nuclear industry, petrochemical industry and other fields, due to long-term work in harsh environments, the combined effects of temperature, corrosive media, and mechanics will cause corrosive damage to these devices [1-3]. At present, there is not enough data to show that whether 316L stainless steel will bring about pitting or stress corrosion cracking under the synergistic effect of high stress and high Cl^- concentration, this brings higher risks to the application and promotion of this material, which requires us to conduct further research on it.

2. Method for preparation and test of sample

The test material is 316L stainless steel, it is an austenitic stainless steel from the point of view of the metallurgical organization, the standard grade of our country is 022Cr17Ni12Mo2. The specific chemical composition is shown in Table 1, and the mechanical properties of 316L stainless steel are shown in Table 2.

Table1. 316L stainless steel chemical composition (wt%)

Element	C	Cr	Mn	Ni	Si	P	S	Ti	Fe
Element percent	0.097	17.75	1.19	9.31	0.53	0.03	0.0064	0.56	allowance

Table2. 316L stainless steel mechanical properties

Mechanical property	Tensile strength (MPa)	Yield Strength (MPa)	Elongation rate (%)
Index	660	360	30

The sample preparation method is mainly based on GB/T 15970[1].4-1995 stress corrosion uniaxial tensile sample preparation, its shape and size shown in Fig1. Before the test, the sample was polished to 400# by using a metallographic sandpaper, and then washed with an ethanol-acetone mixture, rinsed with deionized water, and dried.

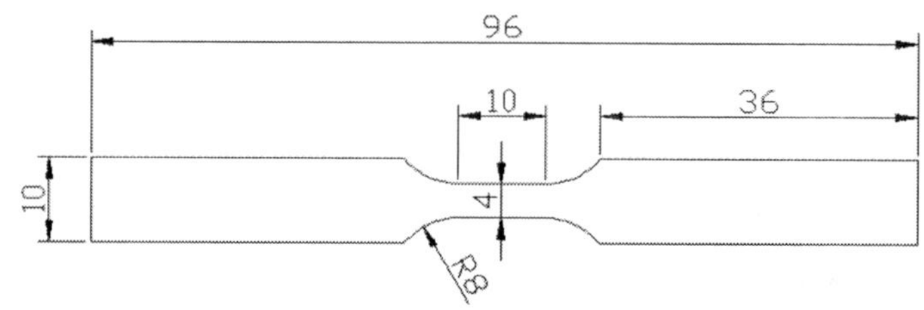

Fig1. Configuration and dimension of the specimen

The test method adopts a constant - load stretching pattern method, and the test medium is NaCl solution at different concentrations under room temperature. In order to speed up the progress of the experiment and shorten the experimental period, a certain amount of $FeCl_3$ was added to the corrosion solution. Table 3 shows the corrosive media parameters for corrosion and mechanical behavior of 316L stainless steel under constant stress conditions.

Table3. 316L stainless steel Corrosive media parameters under constant stress corrosion and Conditions of the mechanical behavior

group indication	Cl^- potency (ppm)	Fe^{3+} potency	pH value
1	141480	1%	5
2	70740	1%	5
3	14148	1%	5

3. Test results and discussion

3.1 Stress corrosion cracking of 316l stainless steel under constant stress

The indicators of stress corrosion cracking susceptibility are diverse. the first indicator used to evaluate is the fracture time of the sample. There are some deficiencies in this evaluation index, such as the fracture toughness of the material, the number of crack initiation, and the width and thickness of the sample, affect the fracture time. This section mainly evaluates the stress corrosion cracking susceptibility of 316L stainless steel by three aspects of fracture time and elongation, and the morphology of corrosion fractures. Fig2 shows the deformation-time curve of 316L stainless steel in different Cl^- concentration NaCl solutions, Fig3 shows the fracture time and elongation curves of 316L stainless steel in different Cl^- concentration NaCl solutions.

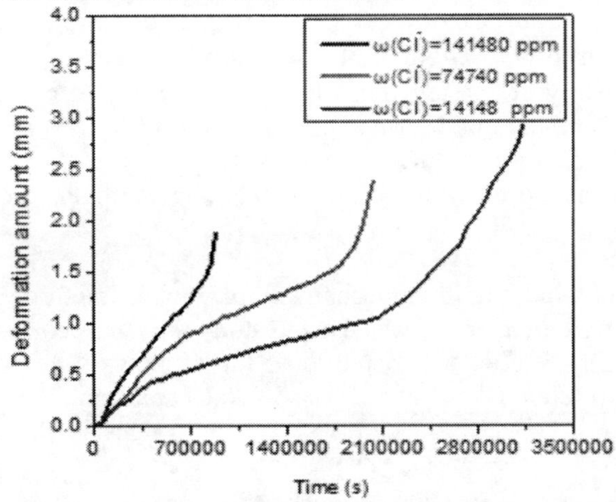

Fig2. 316L stainless steel deformation-time curve in different Cl^- concentrations

Fig2. shows that the deformation-time curves of 316L stainless steel in different Cl^- concentration NaCl solutions are roughly divided into three stages: the first stage is the rapid deformation stage of 316L stainless steel at the beginning of the tensile stress, which is a short period of time; the second stage is the slow deformation stage of 316L stainless steel under the action of tensile stress and corrosive media; the third stage is the crack propagation process of 316L stainless steel under the synergistic action of tensile stress and corrosion medium, which accelerates the deformation of the specimen until the fracture stage.

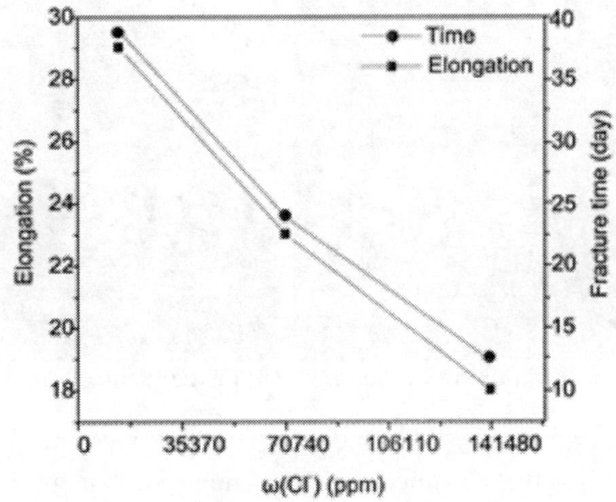

Fig3. 316L stainless steel fracture time and elongation curves in different Cl^- concentrations of NaCl solutions

It can be seen from Fig3. that when the concentration of Cl^- in the corrosion solution is 14148 ppm, the fracture time of the 316L stainless steel sample is approximately 37 days, the variation is 29%, the plastic loss is 4%, and the Cl^- concentration is 70740 ppm, the sample fractures. the time was about 23 days; the line variation was 23%, the shaping loss was 24%; when the concentration of Cl^- was 141,480ppm, the fracture time of the sample was about 12 days, the line variation was 18%, the plastic loss was 40%; It can be seen that with the increase of the concentration of the corrosion solution Cl^-, the fracture time of the sample decreases and the plastic loss increases, indicating that the stress corrosion sensitivity of the 316L stainless steel is enhanced.

This is mainly due to the anodic dissolution of 316L stainless steel under the combined action of tensile stress and corrosive media. As the concentration of Cl^- in the solution increases, Cl^- can be preferentially adsorbed on the surface of the 316L stainless steel sample at some active sites, and binds with the metal ions to form soluble chlorides. These chlorides develop into small particles， the pitting corrosion will largely destroy the passivation film on the surface of the metal sample. And with the increase of Cl^- concentration, the more serious the passivation film cracks, pitting pits [4-5]. The pitting pits form a large stress concentration under the action of tensile stress, which leads to a decrease in the plasticity of the material, the more pitting pits, the greater the depth and the greater the plastic loss.

3.2 Fracture morphology

An important means for studying the mechanism of metal fracture is micro fracture analysis. JMS-5600 scanning electron microscope was used to study the stress corrosion fracture morphology of 316L stainless steel under different Cl^- conditions. Fig.4 shows the fracture morphology of 316L stainless steel in air and different Cl^- concentrations of NaCl solution.

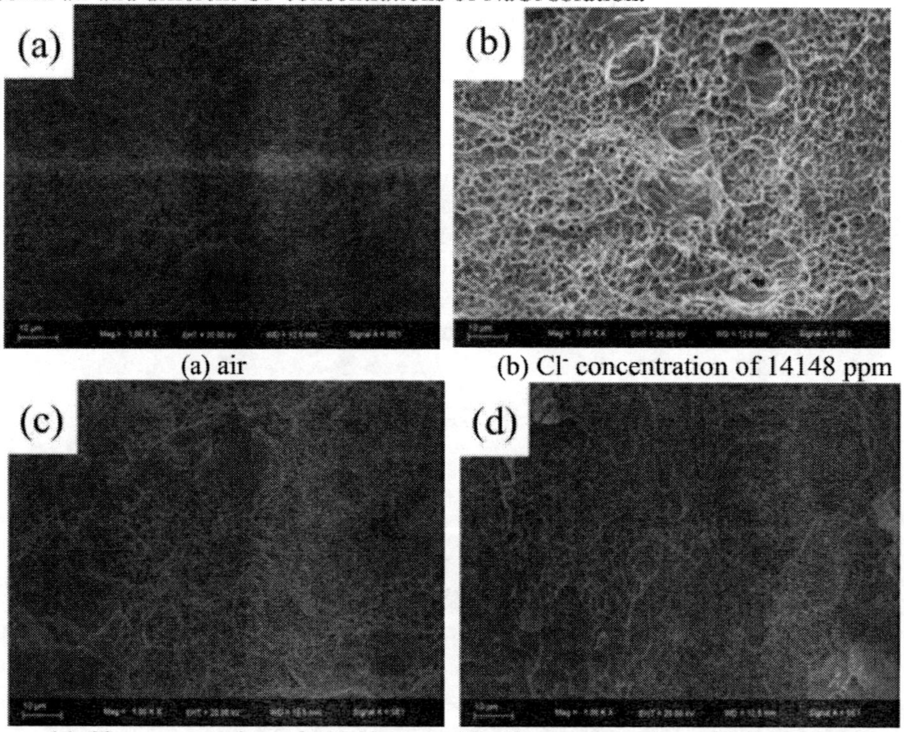

(a) air (b) Cl^- concentration of 14148 ppm

(c) Cl^- concentration of 70740ppm (d) Cl^- concentration of 141480ppm

Fig4. 316L stainless steel fracture morphology in air and different Cl^-concentration of NaCl solutions

It can be seen from Fig.4 that the microscopic fracture morphology of (a) is composed of many small dimples, that is, 316L stainless steel is stretched in the air to be a pronounced ductile fracture, and (b) the microscopic fracture morphology is also more The dimples and fractures are mainly ductile fractures, but compared to (a), the number of dimples is decreasing. (c) Microscopic fracture morphology The dimples are larger, and the number of dimples is also decreasing. The dimple

IOP Conf. Series: Materials Science and Engineering **711** (2020) 012058 doi:10.1088/1757-899X/711/1/012058

fractures are still dominant, but the number of dimples in the (d) diagram is further reduced, the dimples are further enlarged, and they resemble the patterns of river patterns, the fracture mode is dimple fracture, but it also showed some quasi-clearance fractures. The above results show that with the increase of Cl^- concentration of corrosion solution, the fracture mode of 316L stainless steel changes from ductile fracture to brittle fracture, and the corrosion sensitivity increases. When the concentration of Cl^- is 141480ppm, the fracture mode of the sample presents a certain quasi-solution and show a certain degree of brittleness.

3.3 Corrosion creep of 316L stainless steel under constant stress

Material in the more corrosive medium and constant load under long-term plastic deformation will occur slowly, that is, corrosion creep phenomenon. In this paper, the variation of corrosion creep rate in NaCl solution of 316L stainless steel with different Cl^- concentration was studied. Fig.5 shows the corrosion creep rate curve of 316L stainless steel in different Cl^- concentration NaCl solution.

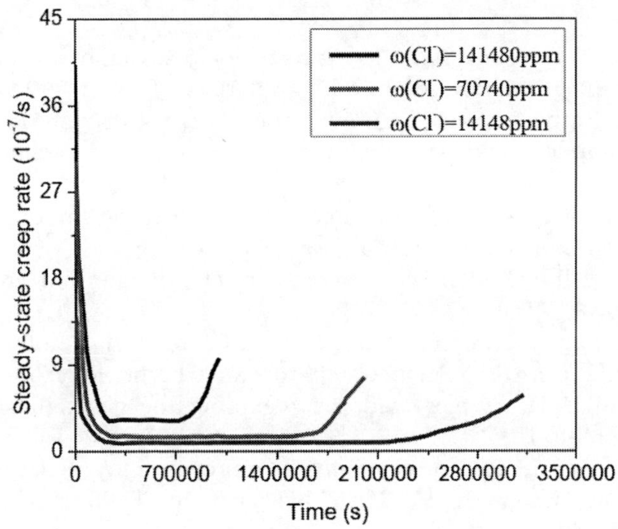

Fig5. 316L stainless steel corrosion creep rate curve

It can be seen from Fig.5 that the corrosion creep rate curves of 316L stainless steel in different Cl^- concentration NaCl solutions undergo three stages: deceleration creep stage—steady state creep stage—accelerated creep stage. When the Cl^- concentration of the solution is 141480ppm, the steady-state creep rate is about $4\times10\text{-}7 m.s^{-1}$; when the Cl^- concentration is 70740ppm, the steady-state creep rate is about $2\times10\text{-}7\ m.s^{-1}$, and the Cl^- concentration At 14148 ppm, the steady state creep rate is approximately 1.8 x 10-7 $m.s^{-1}$. It shows that the steady-state creep rate of 316L stainless steel decreases with the decrease of Cl^- concentration. It can also be observed that the steady-state creep phase increases when the Cl^- concentration decreases.

The main reason for the creep phenomenon is that 316L stainless steel undergoes anodic dissolution and cathodic hydrogen evolution reaction under the action of stress and corrosive media. And with the increase of Cl^- concentration, the more severe the passivation film cracks, pitting pits are formed. The pitting pits form a large stress concentration under the action of tensile stress, which also reduces the strength of the material and promotes corrosion creep. It is also theorized that the vacancies produced by anodic dissolution promote dislocation climbs and promote corrosion creep. Anodic dissolution will form a corrosion product film on the metal surface, and these corrosion product films will produce an additional tensile stress, which is equivalent to increase the creep stress, and further promote the corrosion creep process [6-7].

3.4 Conclusions

(1) The deformation-time curves of 316L stainless steel in NaCl solution with different Cl^- concentration show that the fracture time decreases from 37 days to 12 days and the plastic loss increases from 4% to 40% with the increase of Cl^- concentration in the solution. The sensitivity of 316L stainless steel to stress corrosion is enhanced.

(2) The observation of fracture surface morphology shows that with the increase of Cl^- concentration in the corrosion solution, the dimple size of the micro fracture morphology increases, the number decreases, and the appearance resembles that of a river pattern, showing a brittle fracture trend. That is, the fracture mode of 316L stainless steel changed from ductile fracture to brittle fracture with the increase of Cl^- concentration.

(3) 316L stainless steel under the action of stress and corrosive medium occurs Corrosion creep phenomenon. Corrosion creep rate experiences three stages: deceleration creep stage, steady state creep stage and accelerate the creep phase. With the decrease of Cl^- concentration in solution, the steady-state creep rate decreased from 4×10-7 m.s-1 to 1.8×10-7 m.s-1, and the steady-state creep stage increased.

References

[1] R. K. Singh Roman, Wai Hoong Siew. Stress corrosion cracking of an Austenitic Stainless Steel in Nitrite-Containing Chloride Solutions[J]. Materials, 2014, 7:7799-7808.

[2] Rokuro Nishimura, Yasuaki Maeda. SCC evaluation of type 304 and 316 austenitic stainless steels in acidic chloride solutions using the slow strain rate technique[J]. Corr Sci, 2004, 46(3):769-785.

[3] Scully J C, Powell D T. Stress corrosion cracking mechanism of titanium alloys at room temperature[J]. Corr sci.1970,10(10):719~733.

[4] Parkins R N, Greenwell B S. Interface between corrosion fatigue and stress corrosion cracking[J]. Material science,1977,11(8):405~413.

[5] Wu Ying,Jiang Yiming,Liao Jiaxing,et al. Effect of Cl^- on critical pitting temperature of 304,316 stainless steel[J].Corrosion Science and Protection Technology,2007,27(1):16-19.

[6] Revie R W, Uhlig H H. Further evidence regarding the dezincification mechanism[J]. Corr Sci,1972,12:669-671.

[7] Leinonen H, Hanninen H. SCC susceptibility of nitrogen alloyed stainless steels in 50% $CaCl_2$ solution[J]. In: Hanninen H. High Nitrogen Steel 98. Switzerland:Trans Tech Pub Ltd,1999:545-549.

Thermal Analysis of Double Mass Flywheel Secondary Flywheel

Zhengfeng Yan[1, a], Dale Yin[1, b], Kunpeng Zhou[1, c], Lei Chen[2, d, *]

[1]School of Automotive and Transportation Engineering, Hefei University of Technology, Hefei, 230009, China

[2]School of Mechanical and Electronic Engineering, Wuhan University of Technology, Wuhan, 430070, China

*Corresponding author, [a]zf.yan@hfut.edu.cn, [b]2017170682@mail.hfut.edu.cn, [c]zhoukunpeng2011@163.com, [d]chenlei811001@163.com

Abstract–To solve the problem of thermal overload failure in dual-mass flywheel (DMF), we performed temperature field calculation and thermal analysis of the secondary flywheel in DMF. Based on the results of this analysis, we performed thermal stress analysis on the secondary flywheel. The heat transform model was built for the secondary flywheel using the thermodynamics theory. Finite element model (FEM) was built for the secondary flywheel using ABAQUS 6.12-3 software. This model was solved to obtain the distributions of the temperature field and thermal stress in the secondary flywheel after it completed five engagement-disengagement cycles. Finally, thermal shock test was conducted for the secondary flywheel on the thermal shock test bench. We performed a targeted test to verify the accuracy of thermal analysis of secondary flywheel. The analysis method can be used to optimize product parameters such as the product thickness.

1. Introduction

Dynamic transmission system is an important component located between the engine and drive wheels of an automobile [1]. Dual mass flywheel (DMF) is one of the best torsional vibration dampers in the transmission system, and it is widely used in automobiles [2]. Secondary flywheel is a key component of DMF, which engages with the clutch friction disc and transmits power to the speed transmission system *via* friction [3]. The transmission system may get shocked or overloaded during the process of power transfer, especially in bad working conditions. In such situations, tremendous heat is generated due to the relative sliding friction between the clutch pressure plate and secondary flywheel of DMF. This heat cannot be dissipated in time because of the closed shell, leading to a higher thermal stress. The pressure plate and the secondary flywheel ablate and break under the effect of this high thermal stress. When the secondary flywheel breaks or ablates, the damping system of DMF fails.

In this paper, the secondary flywheel in DMF was used as the research object. Referring to the thermo-mechanical coupling model of the clutch, the temperature field model and the thermal conduction and convection formulae were built according to the engaging characteristics of the dual mass flywheel and the clutch. Based on the heat conduction model of the dual mass flywheel and the heat-structure coupling model of the clutch pressure plate, we developed the thermal-stress finite element model of the secondary flywheel to study the distributions of the temperature field and thermal stress in the secondary flywheel of DMF during five engagement-disengagement cycles. Finally the accuracy of thermal analysis of secondary flywheel was verified by a test.

2. Temperature field calculation of secondary flywheel

The heat transfer process is a very complex phenomenon, it is usually divided into the following basic heat transfer methods: heat conduction, heat convection and thermal radiation. Heat is transmitted from the friction pair to the secondary flywheel by thermal conduction. The heat of secondary flywheel is dissipated into air via thermal convection and thermal radiation. In this paper, heat radiation and the heat transferred by the secondary flywheel to other components are not considered. Therefore, we developed a separate model for the heat transfer of the secondary flywheel. We can use formula (1) to calculate heat conduction [4]:

$$Q = \lambda A' \frac{\Delta t}{\delta} \, . \tag{1}$$

Here, A': cross-sectional area perpendicular to the direction of heat transfer; δ: Thickness of the conductor; Δt: Temperature difference between the two sides of the conductor; λ: Coefficient of thermal conductivity.

2.1. *Determination of heat flux density of the secondary flywheel*

The friction pair generated heat, which was then transmitted in the form of heat flux density into the secondary flywheel.

Figure 1 is a schematic flowchart for calculating the heat flux density on the working surface of secondary flywheel.

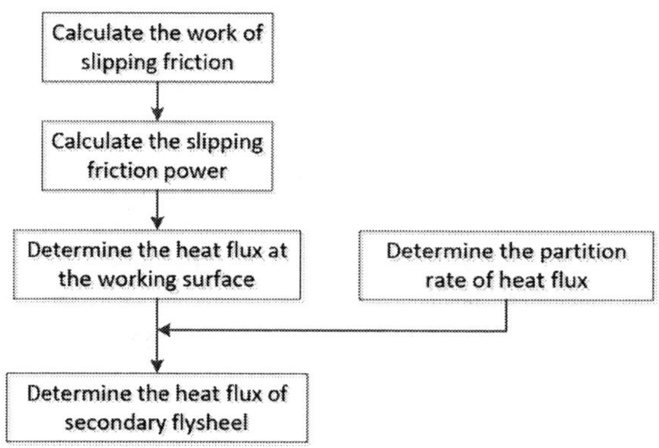

Fig. 1 Flowchart for heat flux density calculation.

The secondary flywheel tightly engages with the friction disc during clutch slipping. The work of slipping friction is calculated using formula (2)[5]:

$$L = T_{emax} \cdot \frac{J_a \omega_0^2}{\frac{2}{3}T_{emax} - T_\psi} \cdot \left[\frac{\frac{7}{3}h+1}{6} - \frac{T_{emax} - \frac{15}{8}T_\psi}{3T_{emax} - \frac{9}{2}T_\psi} \right] \, . \tag{2}$$

Where, T_{emax} is the maximum output torque $(\mathrm{N \cdot m})$; J_a is the equivalent moment of inertia converted from the gross mass of vehicle $(\mathrm{kg \cdot m^2})$; h is the coefficient constant, which are 2.2 for gasoline engine; ω_0 is the angular velocity of engine at the beginning of clutch slipping $(\mathrm{r/min})$; T_ψ is the moment of road resistance $(\mathrm{N \cdot m})$.

The temperatures of friction disc and secondary flywheel will increase after absorption of heat. The larger the sliding friction radius, the higher the sliding friction speed. The greater the sliding friction torque, the more heat is generated [6]. As shown in Figure 2, the friction surface can be simplified into a concentric circle with an inner diameter r_1 and outer diameter r_2.

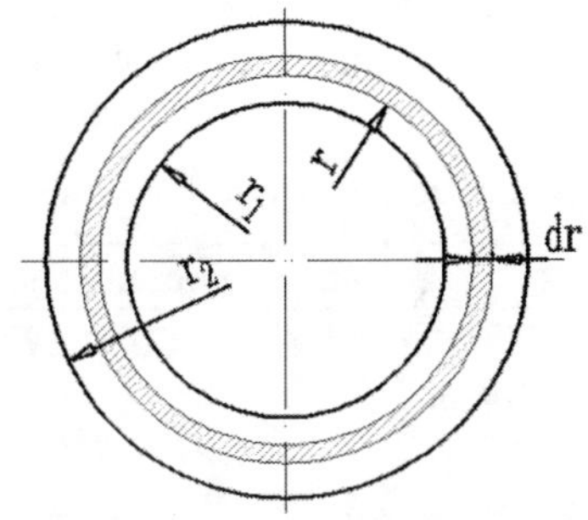

Fig. 2 Concentric circle representing the friction surface.

The heat flux density of the friction surface with radius r is given by the following equation:

$$q = \frac{d_N}{d_s} = \frac{3N_0 r}{2\pi\left(r_2^2 - r_1^2\right)}\left(1 - \frac{t}{t_0}\right). \tag{3}$$

The thermal conductivity of secondary flywheel and friction disc varies according to the material used to manufacture them [7]. Slipping friction generates heat flux density, which is then partitioned between the secondary flywheel and the friction disc; the partition coefficient is calculated by the formula (4)[8]:

$$k = \frac{q_{SF}}{q_{FD}} = \sqrt{\frac{\lambda_{SF} \cdot c_{SF} \cdot \rho_{SF}}{\lambda_{FD} \cdot c_{FD} \cdot \rho_{FD}}}. \tag{4}$$

Here, λ_{SF} is the coefficient of thermal conductivity (W/(m · K)); C_{SF} is specific heat capacity of the secondary flywheel (J/ (kg·K)); ρ_{SF} is the density of secondary flywheel(kg/m^3); λ_{FD} is the coefficient of thermal conductivity of friction disc (W/(m · K)); C_{FD} is the specific heat capacity of friction disc (J/ (kg·K)); ρ_{FD} is density of friction disc (kg/m^3).

The sum of heat flux density transmitted to the secondary flywheel and friction disc is the total heat flux density generated from slipping friction. The heat flux density of secondary flywheel is given by formula (5):

$$q_{SF} = q\frac{k}{1+k}. \tag{5}$$

The simultaneous equation is built by combining formulae (3) and (5). At a given time t, the heat flux density of the secondary flywheel with radius r is as follows:

$$q_{SFr} = \frac{3N_0 r}{2\pi\left(r_2^2 - r_1^2\right)}\left(1 - \frac{t}{t_0}\right) \cdot \frac{k}{1+k}. \tag{6}$$

2.2. Determination of heat transfer coefficient of convection

Surfaces other than the friction surface of the secondary flywheel will dissipate heat when they come into contact with air during engagement. When the secondary flywheel separates from the friction disc, all surfaces will dissipate heat as they come into contact with air. Convection is the main mode of heat transfer in this system. Therefore, the heat transfer coefficient of convection is estimated by the following formula:

$$h = 0.023\frac{K_a}{d}R_e^{0.8}P_r^{0.4}. \tag{7}$$

Here, K_a is the coefficient of thermal conductivity of air (W/(m · K)); R_e is Reynolds number; d is the diameter of secondary flywheel (m); P_r is Prandtl number.

There is a very small difference between the convection coefficients at consecutive temperatures. In order to reduce the amount of data input during simulation, this paper takes 25°C as the starting temperature and takes 80°C, 130°C, and 180°C as the intermediate interpolation temperature. According to the relevant parameters of "The Table of Physical Properties of Dry Air", the convective heat transfer coefficient of each temperature node is calculated and used as the input data of the FEM simulation calculation formulae.

Table 1. Heat transfer coefficients of convection at different temperatures.

Temperature (°C)	Heat transfer coefficients ($W/(m^2·°C)$)
25	72.07
80	65.72
130	60.63
180	573.3

3. Finite modeling of secondary flywheel

In this experiment, we performed thermal analysis of secondary flywheel after it completed five engagement-disengagement cycles; the analysis was conducted by referring to the engineering experiences encountered while performing relevant experiments on clutches. The flow chart of thermal analysis of secondary flywheel is shown in Figure 3.

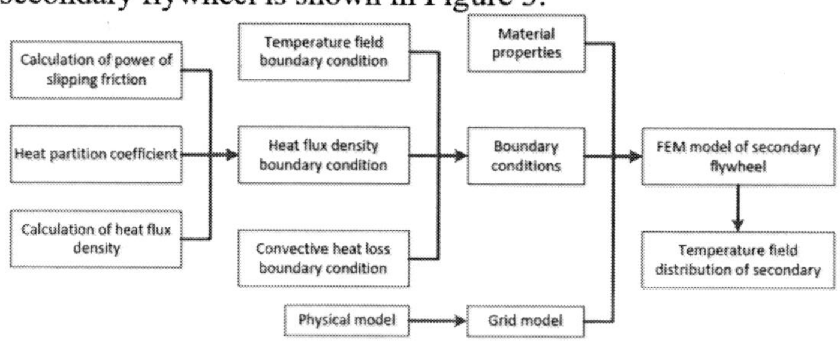

Fig. 3 Flow chart for the thermal analysis of secondary flywheel.

3.1. Establishment of 3D model and grid generation

The 3D physical model of the secondary flywheel was built using Pro/E software and meshed using ABAQUS software. Figure 4 illustrates the grid model of secondary flywheel.

Fig. 4 Grid model of secondary flywheel.

3.2. Temperature field boundary condition

Before starting, we placed the DMF in a room and ensured that the temperature of DMF was equivalent to the room temperature. The temperature-field boundary condition of the secondary flywheel was set to the room temperature (25°C). When the clutch is engaged and slipped, the surface temperature of secondary flywheel surface rises continually. On the one hand, the secondary flywheel absorbs heat flux and transfer to the other parts, on the other hand, the secondary flywheel also dissipates heat into the surrounding air by convection.

3.3. Analysis step configuration of FEM model

We determined the distributions of the temperature field and thermal stress in the secondary flywheel after it had undergone five continuous cycles of engagement and disengagement. We defined 10 analysis steps corresponding with the five disengagement/engagement cycles. The analysis steps were used for carrying out transient heat transfer analysis.

The engagement and disengagement between secondary flywheel and the friction pair were considered to be equivalent to the engagement and disengagement between secondary flywheel and clutch friction disc. The engagement time was 2.25s, whereas the disengagement time was 1s.

3.4. Heat density flux boundary condition

During engagement, slipping friction occurs between the friction disc and the working surface of secondary flywheel. The slipping friction generates heat, which is then absorbed by the friction disc and secondary flywheel in the form of heat flux density.

The heat flux density boundary condition was constrained for the working surface of secondary flywheel in the load module of ABAQUS software; the expression of heat flux density has been presented in formula (6).

3.5. Displacement constraints condition

On the condition of the displacement constraints of the auxiliary flywheel, this paper is determined through the installation of dual mass flywheel. In general, secondary flywheel is installed on the first flywheel through bearing, connected with damping flange through rivets. The clutch assembly is fixed on the secondary flywheel. Only the peripheral orifice of the auxiliary flywheel is a fixed connection. Therefore, the displacement constraint is restricted and the displacement freedom is restricted.

3.6. Convective heat-loss boundary condition

Except for the working surface of the secondary flywheel, surfaces dissipate heat when they come into contact with air during engagement. When the secondary flywheel separates from the friction disc, all the surfaces dissipate heat as they come into contact with air.

Convective heat-loss boundary condition was configured using the parameters presented in Table 1. The input ambient temperature was the room temperature of 25°C. Five interactions were configured in ABAQUS software and convective heat transfer was configured for each step.

3.7. Configuration of material properties

With reference to the standard ISO1083-2004, we presented the material parameters of QT450 in Table 2. The material parameters were input in the property module of ABAQUS software to establish the material mode.

Table 2. Parameters of QT450.

Material parameter	Value
Density	$7.06 \times 10^3 \mathrm{kg/m^3}$
Coefficient of thermal conductivity	$4.7 \times 10^{-3}\ \mathrm{W/(m \cdot K)}$
Young's modulus	$1.69 \times 10^{11} \mathrm{N/m^2}$
Specific heat capacity	$5.1 \times 10^3 \mathrm{J/(kg \cdot K)}$
Coefficient of thermal expansion	$1.01 \times 10^{-5}\ \mathrm{K}$
Poisson's ratio	0.257

4. Distributions of temperature field and thermal stress in secondary flywheel

4.1. Temperature field distribution of secondary flywheel

To obtain the temperature field distribution of secondary flywheel, we solved the FEM model using ABAQUS software. The above finite element model is presented to solve the temperature field distribution of the vice flywheel, and the results are presented in the form of temperature field distribution nephogram. Figure 5 is the secondary flywheel temperature fields after the first to the fifth engagement-disengagement cycles, respectively.

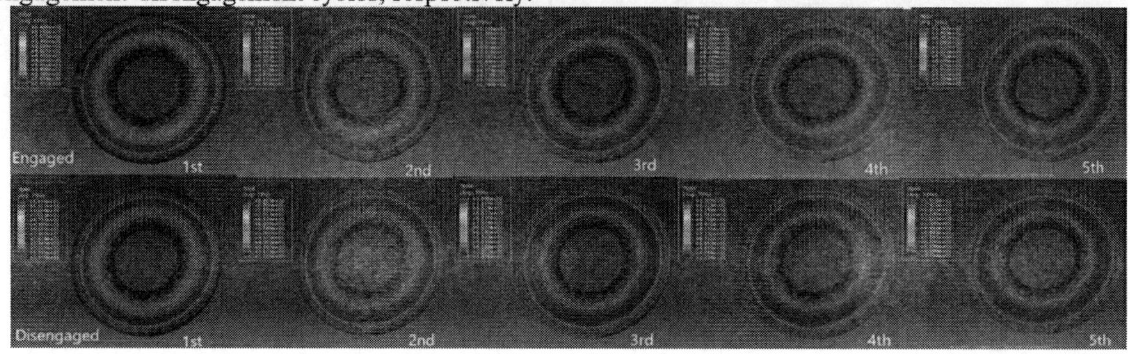

Fig.5 Temperature field distribution of the secondary flywheel in the five engagement/disengagement.

The temperature field of secondary flywheel had a radial distribution pattern. The temperature of the secondary flywheel increased significantly when it was engaged with the friction disc. Moreover, the temperature of the secondary flywheel did not drop apparently when the secondary flywheel separated from the friction disc. After the fifth engagement, we observed that the temperature of the secondary flywheel was the highest at 158.7 °C. Figure 6 shows the variation trend of temperature from the first to the fifth cycles of engagement and disengagement. For the secondary flywheel, the average temperature rise was 32.4°C.

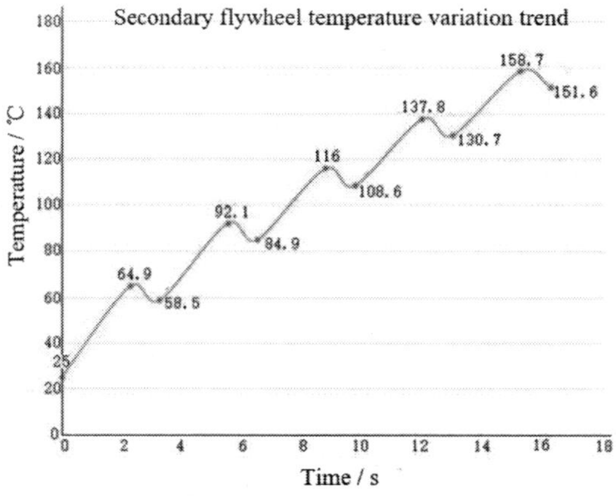

Fig. 6 Secondary flywheel temperature curve after consecutive 5- engagement and disengagement.

4.2. Thermal stress distribution of secondary flywheel

Due to the action of temperature field and external displacement constraints, thermal expansion occurs in the secondary flywheel. This thermal expansion induces loads and causes thermal stress in secondary flywheel. This belongs to a heat-mechanics coupling issue [9].

Thermal analysis model was built for the secondary flywheel using ABAQUS software. The results of temperature field analysis were the load inputs of the model. Sequentially heat-mechanics analytical method was used for performing thermal stress analysis of secondary flywheel [10]. Figure 7 shows the results of thermal stress analysis of secondary flywheel.

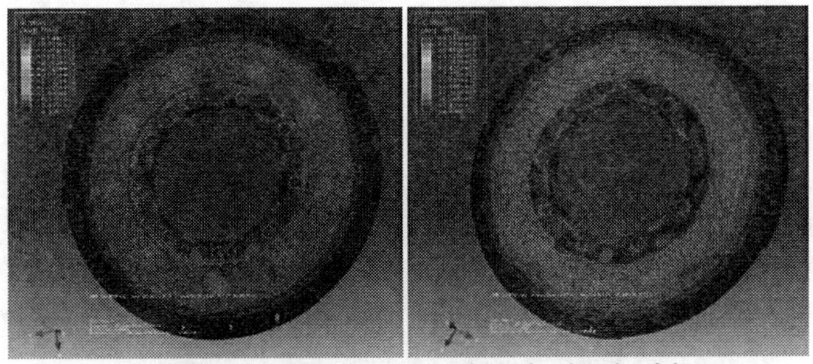

(a) The front side (b) The back side

Fig. 7 Thermal stress distribution of secondary flywheel.

The results indicate that the thermal stress distribution of secondary flywheel agreed well with the temperature field distribution of secondary flywheel. There was apparent stress concentration at the inner ring of the secondary flywheel. The maximum thermal stress was 414Mpa, so it was less than the tensile strength (450Mpa) of the material. Therefore, the secondary flywheel did not show fracture failure.

5. Thermal shock test for secondary flywheel

Thermal shock test was performed on secondary flywheel to verify the results of thermal stress analysis. The equipment, procedures, and results of thermal shock test are as follows:

5.1. Thermal shock test bench

The bench primarily consisted of the following components: power system, hydraulic servo system, and loading system. Figure 8 illustrates a schematic diagram of the thermal shock test bench.

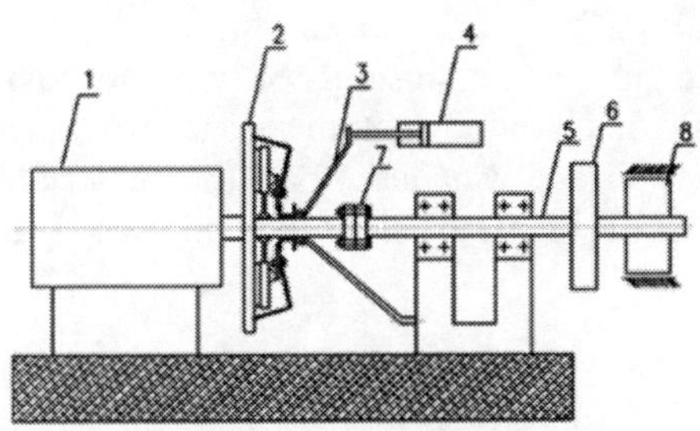

1. Inverter motor;
2. Specimen;
3. Disengagement level;
4. Hydraulic servo system;
5. Loading shaft;
6. Loading disc;
7. Coupler;
8. Brake

Fig. 8 Schematic diagram of the thermal shock test bench.

The bench was used to perform thermal shock test on DMF. The working conditions of the test were as follows:

(1) Ambient temperature: Room temperature (25°C).

(2) Rotational speed: 3000 ± 100rpm; torque: 380Nm.

(3) Ventilation was maintained during the test.

(4) Loading energy was determined using the above-mentioned formulae. Its calculated value was 117KJ in this experiment.

5.2. Thermal shock test

According to the requirements of the thermal shock test for DMF, we performed the thermal shock test for DMF.

(1) First, turn on the variable frequency motor and accelerate to 3000 RPM. Then the hydraulic servo system was shut down, and the clutch was engaged with the secondary flywheel for 2.25 s.

(2) Hydraulic servo system was turned on to separate the clutch from the secondary flywheel for 1s. The brake was applied to keep the specimen motionless.

(3) The above steps were repeated five times.

5.3. Results and analysis

Thermal shock test was performed on DMF using the above-mentioned procedure. As shown in Figure 9, the DMF was removed from the bench and detected.

Fig. 9 The secondary flywheel after thermal shock test

According to the internal standard requirements of the relevant enterprise, secondary flywheel should neither break nor develop cracks after completion of the thermal shock test. Figure 9 clearly illustrates that the surface of the secondary flywheel is smooth without any cracks or breaks. The results agree well with the simulation analysis results of section 5.2, which is presented earlier in this paper.

According to the test standard provided by the manufacturer, an intuitive method is adopted to judge the material strength limit. When the thermal stress is greater than the material limit, the specimen will crack and damage. When the thermal stress is less than the material limit, the specimen will not crack. In the second chapter, the maximum thermal stress of the secondary flywheel under the above temperature field is 414Mpa less than its tensile strength of 450 Mpa. After the test, the test piece was disassembled, and the second flywheel found that the surface integrity of the secondary flywheel was not produced, and its thermal stress was less than the material strength limit. The feasibility of the reliability analysis of the flywheel is verified by the above comparison.

6. Conclusion

We used ABAQUS software to analyze the distributions of temperature field and thermal stress in the secondary flywheel of DMF. We developed the FEM model of secondary flywheel using the thermodynamic theory. We analyzed the distributions of temperature field and thermal stress after five cycles of engagement and disengagements between the secondary flywheel and friction disc were solved. Thermal shock test was performed to verify the thermal analysis. We arrived at the following conclusions:

(1) The temperature of the secondary flywheel increased gradually from the first to the fifth cycle of engagement. The temperature of the secondary flywheel apparently decreased following each disengagement. After the fifth engagement, the temperature of the secondary flywheel was the highest at 157.8 °C. For the five cycles of engagement and disengagement, the average temperature rise was 32.4 °C.

(2) Thermal stress distribution of secondary flywheel basically corresponded with the temperature field distribution of secondary flywheel. Thermal stress concentration was observed at the inner ring of the secondary flywheel during engagement. The maximum thermal stress was 414Mpa, so it was less than the material's tensile strength (450Mpa). Therefore, the secondary flywheel did not show any signs of fracture failure, which was verified by thermal analysis. Our findings can serve as theoretical and experimental basis for the design of DMF.

References

[1] Y.F. Hu, L. Tian, B.Z. Gao and H. Chen, "Nonlinear gearshifts control of dual-clutch transmissions during inertia phase," Isa Transactions, vol. 53(4), pp. 1320-1331, 2014.

[2] L. Chen, M.R. Deng and Z.F. Jiang, "Research on the method of performance parameter optimization for dual-mass flywheel," Chinese Society for Internal Combustion Engines, vol. 30(3), pp. 277-283, 2012.

[3] D.F. Wang, J.Q. Song and B. Li, "Simulation analysis and matching optimization of automotive dual mass flywheel performance," Chinese Journal of Mechanical Engineering, vol. 21(9), pp. 1128-1133, 2010.

[4] V. Depau, V. Rossi and S. Ponticelli, "Torsional vibration protection system," Proceedings of the ASME Turbo Expo, vol. 6, pp. 1077-1084, 2011.

[5] Y.Y. Wang, J.Y. Fang and X.L. Deng, "A research on the engagement characteristics of worn clutch in the condition of start," Automotive Engineering, vol. 36(5), pp. 613-615, 2014.

[6] Y.Y. Wang and N. Li, "Slipping friction load control of dry clutch on semi linkage manipulation," Journal of Tongji University (Natural Science), vol. 43(10), pp. 1570-1574, 2015.

[7] J.L. Zhang and B. Ma, "Study on the factors affecting temperature fields and stress fields of the wet shift clutc," Journal of Beijing Institute of Technology, vol. 30(6), pp. 660-663, 2010.

[8] Y.F. Chen and L. Li, "FEM simulation for thermal analysis of brakes," Automotive Engineering, vol. 3(34), pp.236-240, 2012.

[9] Y.C. Liu, W.Z. Wang, H. Wang and Y.Z. Hu, "Determination of heat flux partition coefficient in frictional heat transfer," Tribology, vol. 4(22), pp. 58-62, 2002.

[10] X.Y. Zhang, W.Y. Liu, P. Wang and T.Q. Yang, "Analysis of temperature rise of clutch friction plate," Lubrication Engineering, vol. 10(2), pp.16-19, 2000.

IOP Conf. Series: Materials Science and Engineering **711** (2020) 012060 doi:10.1088/1757-899X/711/1/012060

Analysis of Factors Influencing the Resistance of a Type of Air Filter Paper Based on SPSS

Qi Peiyun [1, a], Yan Chunxiao [1*, b], Zang Chuanqin [1, c], Xu Jiayu [1, d], Huang Xingqi [1, e], Dai Xuezhi [1, f], Jin Yanren [2, g] and Zhao Ting [2, h]

[1]Institute of NBC Defese, Beijing, China, 102205, [2]SHANXI XINHUA CHEMICAL CO., LTD, Taiyuan, China, 030008

[a]1533357869@qq.com, [b]yanchunxiao1964@sina.com, [c]zangchuanqin2004@163.com, [d]983816788@qq.com, [e]daixuezhi@126.com, [f]972205784@qq.com, [g]jin.yr@163.com, [h]zt870319@163.com

Abstract: The thicknesses, basis weights and resistances of a type of air filter paper randomly sampled were tested by experiment. The statistical analysis of the test data was carried out by multiple linear regression analysis of the statistical software SPSS19.0. The conclusion were as follows: when the paper had a basis weight variation from 55.2 g/m^2 to 59.8g/m^2 and a thickness variation from 0.385mm to 0.406mm, α=0.05, basis weight variation on resistance existed significant influence, the regression equation was Δp=2.551w, correlation coefficient R^2=0.83; thickness variation on resistance did not exist significant influence.

1. Introduction

The air filter paper can remove the toxic and harmful aerosol particles by using its porous structure formed by the fine, soft and interlacing fibers, so as to purify the air. When air filter paper is used as personal protective materials, it is usually used for anti-mite masks, gas masks, etc [1]. When used as collective protective materials, it is usually used in air filters for air conditioning systems, clean workshops, and civil air defense fortifications [2,3]. The filter resistance of air filter paper is inevitably an important indicator due to the respiratory resistance of the person wearing a mask and the intake resistance of the air filter.

For air filter paper, its resistance is related to many of its structural parameters. Papermaking process limitations often result in structural heterogeneities among different points on the same type of air filter paper, then resulting in uneven resistance. Therefore, it is necessary to carry out research on air filter paper, find out the cause of resistance variation, and try to provide reference for the control of resistance uniformity in papermaking process, and further support the development of paper industry technology [4~6].

2. Data Acquisition

Variable Determination. The factors influencing the resistance of air filter paper are extremely complicated. The resistance of air filter paper is calculated as follows:

$$\Delta p = A_1 L. \tag{1}$$

In the formula (1), Δp represents resistance, Pa; L represents thickness, cm; A_1 represents a constant related to the fiber structure of air filter paper, air viscosity and gas flow rate. When A_1 is constant, the resistance increases linearly with the increase of the thickness; when the thickness, the viscosity of the air, the specific velocity of the air flow are constant, the resistance is related to the fiber structure of the air filter paper, and the basis weight *(ie, the mass per unit area of the paper, also called the surface density)* is a parameter that expresses the fiber structure of air filter paper. When the fiber density *(paper basis weight)* increases, the amount of fibers contained in per unit volume of the air filter paper increases. When the airflow passes through the air filter paper, there are more obstacles in per unit volume, the pressure drop loss increases, and the resistance increases [7,8].

Therefore, this paper selects thickness and basis weight of air filter paper as the independent variables, and studys their correlation with the dependent variable resistance.

Data Test. On a certain type of air filter paper, 25 circular paper sheets of d=6 cm were randomly cut with tweezers, and the thickness, the basis weight and the resistance were measured for each circular paper. The thicknesses of the air filter papers were tested by a thickness meter, and each round paper took three points for testing and then averaged; the round papers were weighed by an electronic balance, and the weight was divided by the area to obtain a basis weight value; the resistances of the air filter papers were measured by an inclined-tube micro-manmeter, the velocity of the airflow was 0.32L/min/cm^2, the effective test area was 28 cm^2. The resulting 25 sets of test data was listed in Table 1.

Table 1 Test results of thickness, basis weight and resistance of air filter paper

Serial numer	Thickness[mm]	Basisweight[g/m^2]	Resistance[Pa]
1	0.385	57.9	157
2	0.386	55.2	151
3	0.387	56.5	153
4	0.388	55.2	147
5	0.389	58.4	155
6	0.390	59.0	157
7	0.391	56.7	154
8	0.391	58.7	156
9	0.392	57.7	158
10	0.393	55.7	148
11	0.394	59.7	161
12	0.395	55.6	147
13	0.395	56.2	153
14	0.395	59.1	160
15	0.396	56.8	155
16	0.396	57.5	155
17	0.396	58.1	159
18	0.397	57.4	158
19	0.397	56.9	156
20	0.397	55.4	151
21	0.398	57.3	154
22	0.398	59.8	161
23	0.406	56.8	154
24	0.406	58.0	157
25	0.406	59.6	162

It can be seen from Table 1 that the air filter papers had a thickness of 0.385 ~ 0.406 mm, a basis weight of 55.2 ~ 59.8 g/m^2, and a resistance distribution of 147 ~ 162 Pa.

3. Analysis Methods

Multivariate statistical analysis studys how to effectively organize and analyze data that is randomly affected and to infer or predict the problem under investigation, then help us grasp the characteristics and changing rules of the event as a whole. In many real problems, the relationship among variables is complicated, it is difficult to give an accurate quantitative, but it can be seen from the data that there is a correlative relationship. Linear regression analysis is a mathematical statistics method for dealing with variable correlation. It assumes a linear relationship between the dependent variable and the independent variables, and uses a certain linear regression model to fit the data of the dependent variable and the independent variables. The regression equation is obtained by determining the model parameters, and then the correlation between the variables can be analyzed by this regression equation [9,10]. Now we need to study the relationship between the variations of thickness and basis weight and resistance of a certain type of air filter paper, so we use multiple linear regression to solve.

The regression analysis based on the optimal combination of multiple independent variables to predict the dependent variable is called multiple regression. Assume the number of independent variables influencing the dependent variable Y is p, and record independent variables as x_1, x_2, ...x_p, record the partial regression coefficients of the variables as β_1, β_2, ...β_p, then the model for multiple linear regression analysis is:

$$Y=\beta_0+\beta_1x_1+\beta_2x_2+...+\beta_px_p+\varepsilon. \quad (2)$$

Among them, $\varepsilon \sim N(0, \sigma^2)$, it must meet the four hypotheses including normality, unbiasedness, homoskedness, and independence, and independent variables x_1, x_2, ...x_p must satisfy the assumption that there is no multicollinearity. When building a multiple linear regression model, if there is a linear correlation between two or more independent variables, multiple collinearity will occur. If the tolerance is less than 0.1, the variance expansion factor VIF *(reciprocal of the tolerance)* is greater than 10, indicating that the data has multiple collinearity [11].

At the same time, it is necessary to test the significance of the linear relationship between the dependent variable and the independent variables, in other words, to test the validity of the multiple linear regression equation. Regression analysis method is used to determine the degree of correlation of various factors. The correlation coefficient R^2 is generally used to evaluate the correlation degree or the effect of fitting. $0 \le R^2 \le 1$, the closer R^2 is to 1, the higher the correlation and stronger the significance is; the closer to 0, the lower the correlation and weaker the significance is, which means that the variables outside the model are more effective [12]. Using the statistic F *(F = Regression Mean Square / Residual Mean Square)*, the significance of the entire regression equation can be tested, and the statistic t *(t = partial regression coefficient / standard error of regression coefficient)* can be used to test the regression coefficient [13]. According to the significance level of α=0.05, when the probability P value of the F statistic or t statistic is less than 0.05, the significance test can be passed, the equation or the regression coefficient has statistical significance; otherwise, it cannot pass significance test, equation or regression coefficient was not statistically significant.

SPSS *(Statistical Package for the Social Science)*, the sociological statistical software package is one of the most famous statistical softwares in the world and is widely used in various fields of social sciences and natural sciences. The software package is rigorous in theory, and its statistical analysis functions are complete which cover almost all statistical analysis functions from descriptive statistics, exploratory data analysis to multivariate analysis. It has been gradually popularized in China.

4. Analysis Process

The test data was subjected to multiple linear regression analysis of SPSS. The analysis results are shown in Table 2 to Table 6 and Figure 1 to Figure 3.

Table 2 shows the way in which the independent variables enter the model. Two independent variables including thickness and basis weight are forced into the regression model.

Table 2 Input/Remove variables[a]

Model	Input variable	Variable removed	Method
1	Thickness, basis weight[b]	.	Enter

a. dependent variable: resistance;
b. all requested variables have been entered.

Table 3 shows that the correlation coefficient R^2=0.828, indicating that there is a close linear correlation between the dependent variable resistance and the selected two independent variables. At the same time, it can be seen that the Durbin-Watson test value is 1.838. Generally speaking, the Durbin-Watson test value is from 0 to 4, the closer to 2, the more likely the observations are independent of each other. That is, the observations of multiple linear regression in this study are independent of each other.

Table 3 Model summary[b]

Model	R	R square	Adjusted r square	Standard estimated error	Durbin-Watson
1	.910[a]	.828	.813	1.797	1.838

a. predictor variables: (constant), thickness, basisweight;
b. dependent variable: resistance.

As can be seen from Table 4, the test statistic F value of the model is 53.035, and the significance level P value is 0. 000. According to the significance level of α=0. 05, the model is significant overall. It can also be seen that the regression coefficient of at least one independent variable is not zero, indicating that the established regression model is statistically significant(α=0. 05).

Table 4 Anova[a]

Model		Sum of square	Degree of freedom	Mean square	F	Significant
1	return	342.353	2	171.176	53.035	.000[b]
	residual	71.007	22	3.228		
	total	413.360	24			

a. dependent variable: resistance;
b. predictor: (constant), thickness, basis weight.

Table 5 gives the regression coefficients, t-test results, and collinearity test results in the regression model. The tolerance value is 0.941, which is greater than 0.1, indicating that there is no multicollinearity in the data. The t values are -0.874, 9.622, and 1.235 respectively, and the P values are 0.392, 0.000, and 0.230 respectively. According to the significance level of α=0.05, the analysis shows that there is a significant linear relationship between the basis weight and the resistance, no significant linear relationship between thickness and resistance, and the constant term cannot pass the significance test and should not be included in the equation. Therefore, the established multiple regression equation is:

$$\Delta p = 2.551w. \tag{3}$$

In the formula (2), Δp represents resistance, Pa; w represents basis weight, g / m^2.

IOP Conf. Series: Materials Science and Engineering **711** (2020) 012060 doi:10.1088/1757-899X/711/1/012060

Table 5 Coefficient[a]

Model		Non-standardized coefficient		Standard coefficient			Collinear statistics	
		B	Standard error	Beta	t	Significant	Tolerance	VIF
1	(Constant)	-23.360	26.727		-.874	.392		
	Basis weight	2.551	.265	.876	9.622	.000	.941	1.062
	Thickness	81.225	65.789	.112	1.235	.230	.941	1.062

a. dependent variable: resistance.

It can be seen from Table 6 that the absolute value of the normalized residual has a maximum value of 1.482, which does not exceed the default value of 3, indicating the result has no abnormality, that is, the residual satisfies the unbiasedness.

Table 6 Residual statistics[a]

	Minimum value	Maximum value (x)	Average value	Standard deviation	Digital
Predictive value	148.83	161.68	155.16	3.777	25
Standard predicted value	-1.676	1.727	.000	1.000	25
Residual	-3.583	2.662	.000	1.720	25
Standard residual	-1.994	1.482	.000	.957	25

a. dependent variable: resistance

Figure 1 is a cumulative probability plot of model normalized residuals. It can be seen from Figure 1 that the distribution of each point is close to the diagonal, indicating that the cumulative residual probability of observation fits normal distribution. Figure 2 is a scatter plot of the regression residuals, where the points are evenly distributed, indicating that the data conforms to the equal variance.

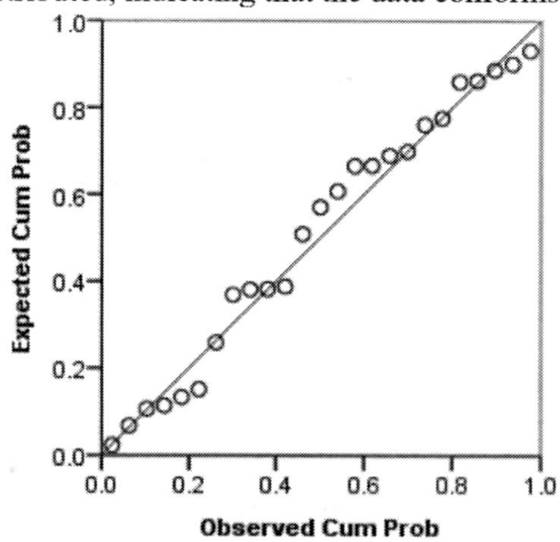

Figure 1 Cumulative probability plot of normalized residuals

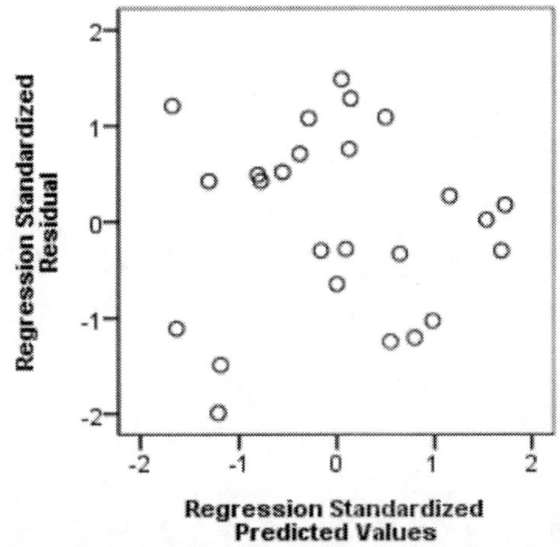

Figure 2 Scatter plot of regression residuals

5. Summary

Based on the relevant research results, this paper combed out the factors influencing the resistance of air filter paper (including thickness and basis weight), and extracted 25 samples of a certain type of air filter paper by randomly sampling to conduct experimental tests, applyed multivariate regression analysis method in SPSS19.0 software to analyze the test data. The conclusion were as follows: when the basis weights varied from 55.2g/m^2 to 59.8g/m^2, the thicknesses varied from 0.385 mm to 0.406mm, α=0.05, basis weight variation had a significant effect on resistance, and its regression equation was Δp =2.551w, correlation coefficient R^2=0.83; thickness variation had no significant effect on resistance.

References

[1] ZHANG Chen, HU Zhi-jun, WANG Song-ling, etc. Preparation and Properties of Glass Fiber Air-Filtration Paper[J]. Paper and Paper Making, 2014, 33[10]: 49~52.
[2] Kang Ding-xue. ULPA Filter Paper[J]. China Pulp & Paper, 1997, 2: 15~18.
[3] CHENG Yi-min, WANG Song. Comparative Analysis of Glass Fiber Air Filter Paper Standards[J]. Paper and Paper Making, 2016, 35[9]: 28~33.
[4] LU Zhao-qing, YANG Bin, ZHANG Mei-yun, etc. Uniformity Testing of Para-Aramid Paper by MATLAB Software[J]. Paper and Paper Making, 2012, 31[12]: 31~34.
[5] HAN Hong-sheng. Quality Control of Base Papers for Coated Grades[J]. China Pulp & Paper, 2008, 27 (8): 48~51.
[6] YANG Bo-jun. Evenness of paper sheet[J]. China Pulp & Paper Industry, 2004, 25[3]: 32~35.
[7] ZHONG Zhao-lin, LIANG Yun, WANG Gui-zhou, etc. Effect of Trilobal Viscose Fibers on the Performance of Filter Paper[J]. China Pulp & Paper, 2011, 30[12]: 20~23.
[8] Chen Yang, Wang Ying, Hu Xiaoxia, etc. Influence of Thickness on Filtration Performance of Fiberglass Filter Paper[J]. Fiber Glass, 2012, 5: 24~26.
[9] WEI Zhi-jing, LIUXi-yu, ZHAO Qing-zhen,etc. The Analysis Based on Statistic Software SPSS and Multiple Linear Regression Aanalysis[J]. Information Technology and Informatization, 2006, 2: 107~109.
[10] FU De-yin. Discussion on the Establishment of Statistical Test System of Multivariate Statistical Analysis[J]. JOURNAL OF BUSINESS ECONOMICS, 2009, 208[2]: 76~80.
[11] WANG Hong-qiang. Analysis of Impact of Raw Materials on Concrete Cracks Based on SPSS Multi-elemental Regression Analysis[J]. YUNNAN WATER POWER, 2010, 26[4]: 3~5.

[12] Xiao Tiqiong, He Chunxia, Chen Yongsheng, etc. Research on Influencing Factors on Agricultural Mechanization in Jiangsu Province Based on SPSS Multiple Regression Analysis [J]. Journal of Chinese Agricultural Mechanization, 2014, 35(3): 263~267.
[13] Wu Changyou. The Research and Application on Neural Network [D]. Northeast Agricultural University, 2007.

IOP Conf. Series: Materials Science and Engineering **711** (2020) 012061 doi:10.1088/1757-899X/711/1/012061

Theoretical Analysis of Residual Head Velocity after the Jet Penetrating the Multi-layer Metal-Liquid Composite Structure

ZHAO Changfang[a], REN Jie[b], KUAI Tengfei, ZHOU Zhitan, ZHANG Kebin, ZU Xudong, LE Guigao

School of Mechanical Engineering, Nanjing University of Science and Technology, Nanjing, 210094, China

[a]E-mail: lackychang@njust.edu.cn, [b]Correspondence author e-mail: renjie@njust.edu.cn

Abstract. The residual head velocity after the jet penetrating a target plate is an indication of the residual penetration capability of the jet, which changes obviously with the penetration time. In this study, the residual head velocity after the jet penetrating through the n-layer metal and the n-layer liquid was deduced by using the quasi-steady penetration model theory. The influence of shock wave to the residual head velocity of jet was analyzed. The calculation results show that the residual head velocity of the jet decreases with the increase of the number of layers of the composite structure, and the shock wave obviously affects the velocity of the residual head of the jet but not very strong. According to the difference between the experimental value and the theoretical value of the jet penetrating the single-layer composite structure, a method of modifying the theoretical mathematical model by using constant C as the calibration value is proposed. The mathematical model can be extended to calculate the residual head velocity of the jet penetrating the multi-layer metal-liquid composite structure with different materials and thickness.

1. Introduction

Penetration and protection have always been the relationship between spears and shields, but also mutual promotion. It is because of this relationship that there has been continuous scientific and technological progress and innovation on penetration and protection. Brikhoff et al put forward the theory of jet forming for the first time, and established a one-dimensional hydrodynamic penetration model [1]. But this theory is only suitable for continuous jet penetration. For this reason, combined with Bernoulli equation, Pack and Evans give the penetration equation with jet fracture [2]. Considering the effect of target plate strength on jet penetration, Eichelberger introduced the plastic deformation impedance term [3], which more accurately describes the penetration process of ideal in-compressible viscous steady state jet. In practice, however, the velocity of the jet's head, tail and diameter is constantly changing. Based on this, Allison and Vitali put forward a quasi-steady penetration model with virtual origin as the origin point [4]. Furthermore, on the basis of virtual distance, Dipersio and Simon give the penetration equation of jet in three different cases [5], which is closer to the actual description of jet penetration. Recent years, Wang Jing et al get the virtual origin by using the method of the least square fitting [6]. Chen Chuang et al established a double virtual origin calculation model by referring to the improved PER theory, which more accurately described the penetration process of the jet [7].

At present, there are many researches on jet penetration into multi-layer spacer targets and composite targets, such as aluminum foam, ceramics, rubber, explosives and other composite armor [8-12]. White, Andersson, Shi Jin-wei, Zhang Xian, Gao Zhen-yu, Shan Feng and others have carried out studies on the anti-jet penetration performance of mono-layer metal-liquid composite structures [13-22], and obtained some results that can be used for reference. However, the multi-layer metal-liquid composite structure has not been analyzed, which is worth studying about the unknown effects. In order to solve the above problems, the equation of jet penetrating multi-layer metal-liquid composite structure of is established, and the residual head velocity after jet penetrating into metal-liquid composite structure under the influence of shock wave is analyzed, and the mathematical model of residual head velocity modified by calibration method is proposed.

2. Theoretical Analysis

2.1 Quasi-steady Penetration Model

According to the virtual origin theory redefined by Dipersio and Simon, the jet is formed through the virtual origin and penetrates the 1st layer of metal plate after passing through the air. The penetration model is shown in Fig. 1. After the shaped charge is initiated, the detonation wave axial collapses the charge cover, and the residual head velocity of the jet after forming is very high. The quasi-steady penetration model can be used to describe both jet and target as fluid before the residual head velocity of the jet is not reduced to the v_{jmin}, also known as hydrodynamic model. The quasi-steady penetration model is the most widely used penetration model and the most basic penetration model. Therefore, we can reference this theory to analysis our study.

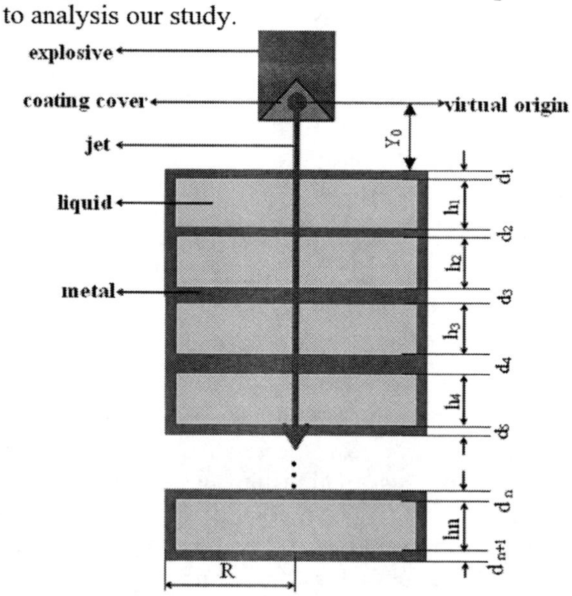

Fig. 1 Geometric model of jet penetration into metal-liquid composite structure

Assuming the jet flow is an ideal steady and in-compressible non-viscous fluid and the interaction process with metal is a steady process, and the diameter and velocity of the micro-element of the jet do not change. We assume the velocity (v_{j0}) of the head after jet forming as the residual head velocity of the jet before penetration, that is, the influence of air and other external factors is not considered. The residual head velocity (v_{j1}) after jet penetration into the 1st metal plate is

$$v_{j1}=\frac{v_{j0}\cdot Y_0^{\sqrt{\frac{\rho_{m1}}{\rho_j}}}}{(Y_0+d_1)^{\sqrt{\frac{\rho_{m1}}{\rho_j}}}} \tag{1}$$

Where, Y_0 is the distance between the virtual origin and the fist layer of metal, ρ_{m1} is density of the fist layer of metal, ρ_j is density of the jet, d_1 is thickness of the fist layer of metal. So, v_{j1} is the initial head velocity of the jet penetrating the 1st layer of liquid.

According to the equation (1), the residual head velocity (v'_{j1}) after the jet penetrating the 1st layer of liquid is

$$v'_{j1}=v_{j1}\left(\frac{Y_0+d_1}{Y_0+d_1+h_1}\right)^{\sqrt{\frac{\rho_{f1}}{\rho_j}}}=\frac{v_{j0}\cdot Y_0^{\sqrt{\frac{\rho_{m1}}{\rho_j}}}\,(Y_0+d_1)^{\frac{\sqrt{\rho_{f1}}-\sqrt{\rho_{m1}}}{\sqrt{\rho_j}}}}{(Y_0+d_1+h_1)^{\sqrt{\frac{\rho_{f1}}{\rho_j}}}} \tag{2}$$

Where, v'_{j1} is the initial head velocity of jet penetration into the second layer of metal plate, ρ_{f1} is density of the fist layer of liquid, h_1 is thickness of the fist layer of liquid.

Similarly, the residual head velocity (v_{j2}/v'_{j2}) after the jet penetrates the second layer of metal plate and the second layer of liquid are respectively as follows

$$v_{j2}=v'_{j1}\left(\frac{Y_0+d_1+h_1}{Y_0+d_1+h_1+d_2}\right)^{\sqrt{\frac{\rho_{m2}}{\rho_j}}}=\frac{v_{j0}\cdot Y_0^{\sqrt{\frac{\rho_{m1}}{\rho_j}}}\,(Y_0+d_1)^{\frac{\sqrt{\rho_{f1}}-\sqrt{\rho_{m1}}}{\sqrt{\rho_j}}}}{(Y_0+d_1+h_1+d_2)^{\sqrt{\frac{\rho_{m2}}{\rho_j}}}}\bullet(Y_0+d_1+h_1)^{\frac{\sqrt{\rho_{m2}}-\sqrt{\rho_{f1}}}{\sqrt{\rho_j}}} \tag{3}$$

$$v'_{j2}=v_{j2}\left(\frac{Y_0+d_1+h_1+d_2}{Y_0+d_1+h_1+d_2+h_2}\right)^{\sqrt{\frac{\rho_{f2}}{\rho_j}}}=\frac{v_{j0}\cdot Y_0^{\sqrt{\frac{\rho_{m1}}{\rho_j}}}\,(Y_0+d_1)^{\frac{\sqrt{\rho_{f1}}-\sqrt{\rho_{m1}}}{\sqrt{\rho_j}}}}{(Y_0+d_1+h_1+d_2+h_2)^{\sqrt{\frac{\rho_{f2}}{\rho_j}}}}\bullet(Y_0+d_1+h_1)^{\frac{\sqrt{\rho_{m2}}-\sqrt{\rho_{f1}}}{\sqrt{\rho_j}}}\,(Y_0+d_1+h_1+d_2)^{\frac{\sqrt{\rho_{f2}}-\sqrt{\rho_{m2}}}{\sqrt{\rho_j}}} \tag{4}$$

Where, d_2 is thickness of the second layer of metal, ρ_{m2} is density of the second layer of metal, ρ_{f2} is density of the second layer of liquid, h_2 is thickness of the second layer of liquid.

And so on, according to the above iterative law, the residual head velocity (v_{jn}/v'_{jn}) after the jet penetration into the n-layer metal plate and the n-layer liquid are obtained

$$v_{jn}=\frac{v_{j0}\prod_{n=1}^{n}\left[(Y_0+\sum_{i=1}^{n}d_{i-1}+\sum_{i=1}^{n}h_{i-1})^{\frac{\sqrt{\rho_{mn}}-\sqrt{\rho_{f(n-1)}}}{\sqrt{\rho_j}}}\right]}{(Y_0+\sum_{n=1}^{n}d_n+\sum_{n=1}^{n}h_{n-1})^{\sqrt{\frac{\rho_{mn}}{\rho_j}}}}\prod_{n=1}^{n}\left[(Y_0+\sum_{i=1}^{n}d_{i-1}+\sum_{i=1}^{n}h_{i-2})^{\frac{\sqrt{\rho_{f(n-1)}}-\sqrt{\rho_{m(n-1)}}}{\sqrt{\rho_j}}}\right] \tag{5}$$

$$v'_{jn}=\frac{v_{j0}\prod_{n=1}^{n}\left[(Y_0+\sum_{i=1}^{n}d_{i-1}+\sum_{i=1}^{n}h_{i-1})^{\frac{\sqrt{\rho_{mn}}-\sqrt{\rho_{f(n-1)}}}{\sqrt{\rho_j}}}\right]}{(Y_0+\sum_{n=1}^{n}d_n+\sum_{n=1}^{n}h_n)^{\sqrt{\frac{\rho_{fn}}{\rho_j}}}}\prod_{n=1}^{n}\left[(Y_0+\sum_{i=1}^{n}d_i+\sum_{i=1}^{n}h_{i-1})^{\frac{\sqrt{\rho_{fn}}-\sqrt{\rho_{mn}}}{\sqrt{\rho_j}}}\right] \tag{6}$$

Where, d_n is thickness of the n-layer of metal, ρ_{mn} is density of the n-layer of metal, ρ_{fn} is density of the n-layer of liquid, h_n is thickness of the n-layer of liquid.

2.2 The Influence of Shock Wave

When the jet formed by explosion runs through the 1st layer of metal plate, the acoustic velocity c_{f1} of the undisturbed liquid is far lower than v_{j1}, so the motion of the jet in the liquid is supersonic. According to Gao Zhen-yu's study [23], the cone-like shock wave produced by the jet will move the liquid through the side face of the closed metal structure and cause the liquid to converge in the radial direction while the jet penetrates the liquid radial reaming hole. This will interfere with the jet, and lead the residual head velocity of the jet to drop so that reduces the ability to continue to penetrate, so it is necessary to modify the above equation.

The micro-elements at the point where the jet head interacts with the liquid in the hole at a certain time are taken for analysis. According to the momentum conservation principle, the jet head is transmitted to the liquid because of the interference of the liquid, which causes the liquid to move, and the velocity is u_n. Define the direction of the jet head motion is positive, the momentum conservation differential equation is

$$v_{jn} dm_{jn} = u_n dm_{fn} \cos\theta + v_{jn\text{-shock wave}} dm'_{jn} \tag{7}$$

Where, dm_{jn} is the mass differential element of jet, dm_{fn} is the mass differential element of liquid, dm'_{jn} is the mass differential element of jet after the jet penetrating liquid, θ is the angle between u_n and v'_{jn}, $v_{jn\text{-shock wave}}$ is the residual head velocity of the jet under the influence of shock wave.
According to the relationship between mass and volume, equation (7) can be rewritten to

$$v_{jn} \rho_j dV_{jn} = u_n \rho_{fn} dV_{fn} \cdot \cos\theta + v_{jn\text{-shock wave}} \rho_j dV'_{jn} \tag{8}$$

Where, dV_{jn} is the volume differential element of jet, dV_{fn} is the volume differential element of liquid, dV'_{jn} is the volume differential element of jet after the jet penetrating liquid.

Due to the liquid is in-compressible, the shape of the micro-element changes and the volume remains unchanged, taking the unit volume micro-element, there is

$$v_{jn} = \frac{u_n \rho_{fn} \cos\theta + v_{jn\text{-shock wave}} \rho_j}{\rho_j} = \frac{\rho_{fn} \cos\theta}{\rho_j} u_n + v_{jn\text{-shock wave}} \tag{9}$$

Thus, the residual head velocity of the jet under the influence of shock wave is obtained

$$v_{jn\text{-shock wave}} = v_{jn} - \frac{\rho_{fn} \cos\theta}{\rho_j} u_n \tag{10}$$

It can be seen that the residual head velocity will decrease after the jet is disturbed by liquid, and the decrease is related to ρ_{fn}, ρ_j and u_n.

For obtaining the penetration velocity u, we can only consider the effect of the first shock wave because the time of interaction between jet and liquid is very short. And the velocity of the shock wave is assumed to be constant and the attenuation of the wave is not taken into account. According to Xiao Qiang-qing's study [24], the penetration velocity (u_n) is

$$u_n = \frac{1}{2[\lambda_n \rho_j - (1+\lambda_n)\rho_{fn}]} \left\{ (2\lambda_n \rho_j v_{jn} - \rho_{fn} c_{fn}) - \sqrt{4\lambda_n \rho_j \rho_{fn} v_{jn} [(1+\lambda_n) v_{jn} - c_{fn}] + 8\lambda_n R_{tn} [\lambda_n \rho_j - (1+\lambda_n)\rho_{fn}] + \rho_{fn}^2 c_{fn}^2} \right\} \tag{11}$$

Where, λ_n is the Hugoniot parameter of the liquid and R_{tn} is the dynamic yield strength of the liquid, c_{fn} is the acoustic velocity of the *n*-layer undisturbed liquid.

When the penetration condition is $v_{jn} > c_{fn}$, the effect of shock wave should be taken into account. The initial velocity of jet penetration into the head of the n-layer metal plate should be calculated by using the $v'_{jn\text{-shock}}$ wave which associated with u_n. The effect of $v_{jn} < c_{fn}$, shock wave on penetration condition is negligible, and the initial head velocity of jet penetration into the n-layer metal plate should by using v'_{jn}.

Note that the condition of the above analysis is that $v_{jn} > v_{jmin}$, and $v_{jn} > c_{fn}$. When $v_{jn} < v_{jmin}$, the strength of the metal plate should be considered. And when $v_{jn} < c_{fn}$, the motion is no longer supersonic, and the penetration will be basically stopped because the velocity of the head is too low.

3. Calculation Example and Model Correction

In order to simplify the study, we take one to five layers metal-liquid composite structure to study. Define that the thickness of each layer of metal and liquid is the same, and all metals are LC4 aluminum alloy, all liquids are diesel oil.

Where, d=5 mm and h=12.5 mm, Y_0=28.85 mm and v_{j0}=6510 m/s are known. According to the material parameters in Table 1, ρ_{mn}=2770 kg/m^3, ρ_{fn}=837 kg/m^3, c_{fn}=1775 m/s, R_t=150 Pa, ρ_j=8960 kg/m^3, λ_n=1.725. In combination with the formula (1)-(11), the theoretical results about one to five layers metal-liquid composite structure are calculated by programming.

According to the experimental data in reference [11], the residual head velocity is 6038 m/s after the jet penetrating the mono-layer metal-liquid composite structure, and the theoretical result of this paper is 5110.77 m/s, the relative error is 15.4%. In order to improve the accuracy of the mathematical model in the theoretical analysis, the experimental data can be used as the calibration value to modify the results obtained from the equations (5) and (6). Define the calibration value as

$$C = v_e - v_{j2} \tag{12}$$

Where, v_e is the experimental value of residual head velocity of the jet penetrating the mono-layer metal-liquid composite structure.

The function of the calibration value C is to make the original theoretical calculation result (the relation curve between n and v) to translate C units along the positive direction of the velocity axis. That is, C is an intercept correction value. Through this method, the theoretical results modified by the calibrated value C=927.25 are shown in Fig. 2.

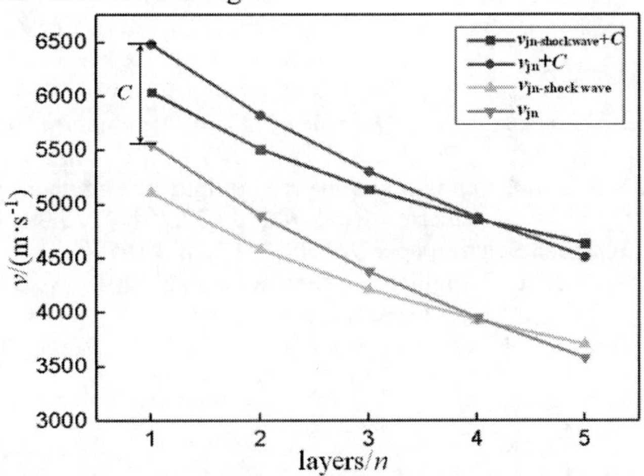

Fig. 2 Jet residual head velocity change trend under quasi-steady penetration model and under the influence of shock wave

From the theoretical calculation results, v decreases with the increase of n under the influence of quasi-steady penetration model and shock wave. However, the v under the influence of shock wave is lower than the quasi-steady value before a certain critical point, and then it higher than the quasi-steady value. The critical position is the 4th layer composite structure.

4. Summary

The liquid-filled unit cell was studied by Gao Zhen-yu and Zhang Xian [11,16-23], which was characterized by only one layer or many mono-layer with transverse arrangement. The disadvantage is that only some of the composite structures can play a protective role when the jet penetrates vertically, so the protective ability is limited. In contrast, the innovation of this paper lies in the multi-layer axial arrangement of the composite structure, which can effectively and continuously interfere with the jet.

Acknowledgments
This work was financially supported by the Postgraduate Research & Practice Innovation Program of Jiangsu Province (KYCX19_0327).

References

[1] G. Birkhoff, D.P. Mcdougall, E.M. Pugh, et al. Journal of Applied Physics Vol. 19(1948), p. 563
[2] W.M. Evans, D.C. Pack. Proceedings of the Physical Society Vol. 64(1951), p. 298
[3] R.J. Eichelberger. Journal of Applied Physics Vol. 27(1956), p. 63
[4] F.E. Allison, R. Vitali. Ballistic Research Laboratory Report No. 1184(1963)
[5] J. Simon, R. Dipersio, R.J. Eichelberger. Armyballistic Research Lab Aberdeen Proving Ground MD (1959)
[6] J. Wang, C. Wang, J.G. Ning. Acta Armamentarii Vol. 29(2008), p. 1409
[7] C. Chen, X.M. Wang, W.B. Li, et al. Acta Armamentarii Vol. 35(2014), p. 604
[8] K.B. Yan, X.D. Zu, Z.X. Huang, et al. Journal of Ballistics Vol. 25(2013), p.59
[9] B.B. Liu, Z.X. Huang, X.D.Zu, et al. China Sciencepaper Vol. 10(2015), p. 488
[10] L.Y. Yue, W. Li, X.D. Zu, et al. MATEC Web of Conferences Vol. 39(2016), No. 01012
[11] Z.Y. Gao, Z.X. Huang, M. Guo, et al. Journal of Vibration and Shock Vol. 35(2016), p. 176
[12] B. Ma, Z.X. Huang, Q.Q. Xiao, et al. IEEE Transactions on Plasma Science Vol. 45(2017), p. 875
[13] J.J. White, J.M. Wahll. Orlando, FL: International Ballistics Society (1981), p. 305
[14] G. Andersson, S. Karlsson, A. Watterstam. Midrand, South Africa: The South Africa Ballistics Organisation (1998), p. 183
[15] J.W. Shi, X.B. Luo, J.W. Jiang, et al. Chinese Journal of Energetic Materials Vol. 24(2016), p. 213
[16] X. Zhang, X.D. Zu, Z.X. Huang, et al. Explosion and shock wave Vol. 37(2017), p. 1101
[17] X. Zhang, X.D. Zu, Z.X. Huang, et al. Ordnance Industry Automation Vol. 36(2017), p. 70
[18] X. Zhang. Nanjing University of Science and Technology (2017), p. 15
[19] F. Shan, C.F. Zhao, X.D. Zu. Journal of Ordnance Equipment EngineeringVol. 39(2018), p. 68
[20] Information on http://kns.cnki.net/kcms/detail/61.1234.TJ.20171113.0924.024.html.
[21] C.F. Zhao, X.D. Zu. China Sciencepaper Vol. 12(2017), p. 2565
[22] C.F Zhao, F. Shan, J. Yu, et al. Engineering Blasting Vol. 23(2017), p. 22
[23] Z.Y. Gao. Nanjing University of Science and Technology (2016), p. 15
[24] Q.Q. Xiao, Z.X. Huang, X.H. Gu. Chinese Journal of High Pressure Physics Vol. 25(2011), p. 333

IOP Conf. Series: Materials Science and Engineering **711** (2020) 012062 doi:10.1088/1757-899X/711/1/012062

A SEM Study on Space Environment Effects of Carbon Nanotube Arrays

Dongbo TIAN[1,2,a], Yuming LIU[1,2,b] and Man LI[1]

[1]Beijing Institute of Spacecraft Environment Engineering, Beijing 100084, P. R. China

[2] Science and Technology on Reliability and Environment Engineering Laboratory, Beijing 100084, P. R. China

[a] Tiandongbo@126.com, [b] Lyming2005@126.com

Abstract. Due to their unique properties, carbon nanotubes will have great potential applications in many fields, particularly in space applications. But in development of carbon nano-materials intended for operation in aerospace, one of the most important problems is assurance of their resistance against the impact of the AO which is one of the predominant environment in space. Here we present the AO erosion effects on pure carbon nanotube arrays by SEM analysis. It is shown that the morphologies of CNT arrays are quite different from those before AO experiments. Different morphologies of CNT arrays after AO exposure are shown by SEM images. It is presented that shrinkage and crack formation of CNT arrays are caused by bombardment force.

1.Introduction

Carbon nanotubes (CNTs) with extremely high strength, high stiffness, low density, good chemical stability, and high thermal and electrical conductivities will have great potential applications in many fields, particularly in space applications [1]. Atomic oxygen (AO) is considered the most erosive particle to spacecraft materials in low earth orbit (LEO), which can etch most materials of spacecraft. In development of carbon nanotubes intended for operation in aerospace, one of the most important problems is assurance of their resistance against the impact of the AO [2]. A few studies have been carried out to present the AO effects on CNTs[3-5], such as CNT composites, CNT wires and CNT films. We also have studied the AO erosion effects on CNT arrays [6, 7]. All these works focus on the erosion yields or the mechanical properties of CNTs in AO environment.

In our previous works, it was found that some "magic" patterns on CNT arrays had formed after AO exposure [6, 7]. As a stable nanotube foams, CNT arrays can be elastically deformed. Here we will give some more SEM analysis results, which may be helpful to understand the mechanics of AO erosion effects on CNT arrays.

2.Experimental method

As was reported before [7], well aligned CNT arrays were synthesized by a common method of chemical vapour deposition of ethylene gases in the presence of iron catalysts on the substrate of silicon. Three AO exposure experiments were carried out at Beijing Institute of Satellite Environment Engineering of China using an AO simulation facility. The facility produces a 5 eV neutral AO beam by placing a metal plate in contact with magnetically confined AO plasma. The AO plasma was produced by a microwave electron cyclotron resonance ion source driven by a 0.6 kW radio frequency at 2.45 GHz. The CNT array was placed on a holder in the AO simulation facility, and a directed AO beam impacted CNT arrays. The incident direction of AO beams was parallel to the radial direction of CNTs in the arrays. The AO flux was about 5.5 $\times$ 10^{15} atom/(cm^2 • s). The AO fluence was 1.5×10^{20} atom/cm^2. The morphologies of CNT arrays before and after AO exposure were characterized by scanning electron microscopy (SEM, S-4800, Japan).

3.Result and discussion

Fig.1 shows typical SEM images of CNT arrays before AO exposure, which have been reported before [7]. The surface of the as-grown CNT arrays is quite flat in macro-scale. After AO exposure, the surface of CNT array isn't flat and continuous any more, and CNTs are completely etched away in some regions.

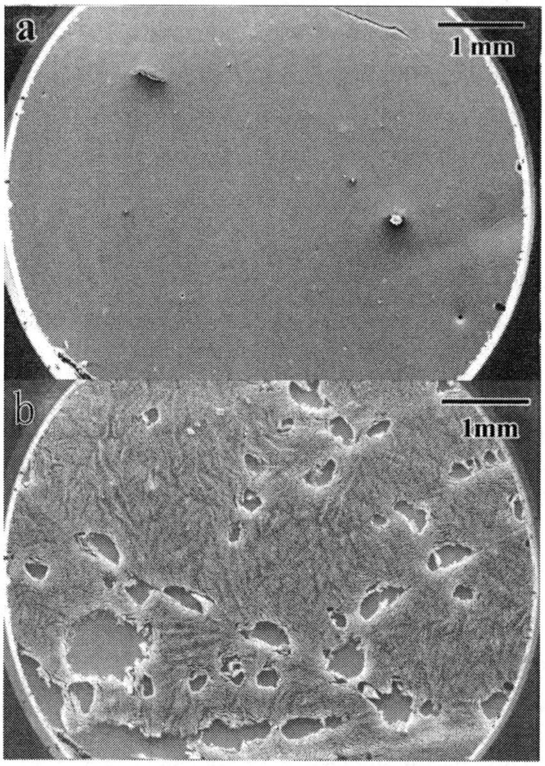

Fig. 1 Overview images of CNT arrays before (a) and after (b) AO exposure.

Some "magic" lines can be clearly seen as pointed out in Fig.2. It seems that these lines are higher than other zones. And the blank zones are located on the side of the lines. Few blank zones crossing the lines can be observed.

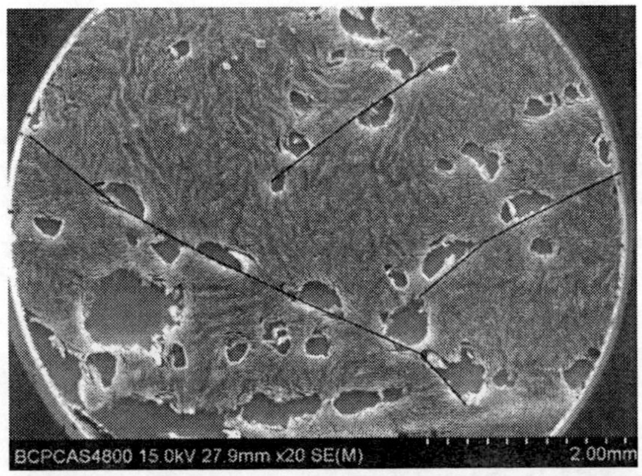

Fig. 2 "Magic" lines formed after AO exposure

Enlarged images show that the deformed surface of CNT arrays is like a mountain range (Fig. 3 a). There are many ridges and ravines. The "magic" line is relatively straight and continues, while other ridges are broken and continued. It shows that all the ridges are composed of compressed CNTs, on the top of which some amorphous carbon can be observed (Figure 3 b c). Along the "magic" line, there are some holes which may be the beginning of the blank zone formed during the AO exposure.

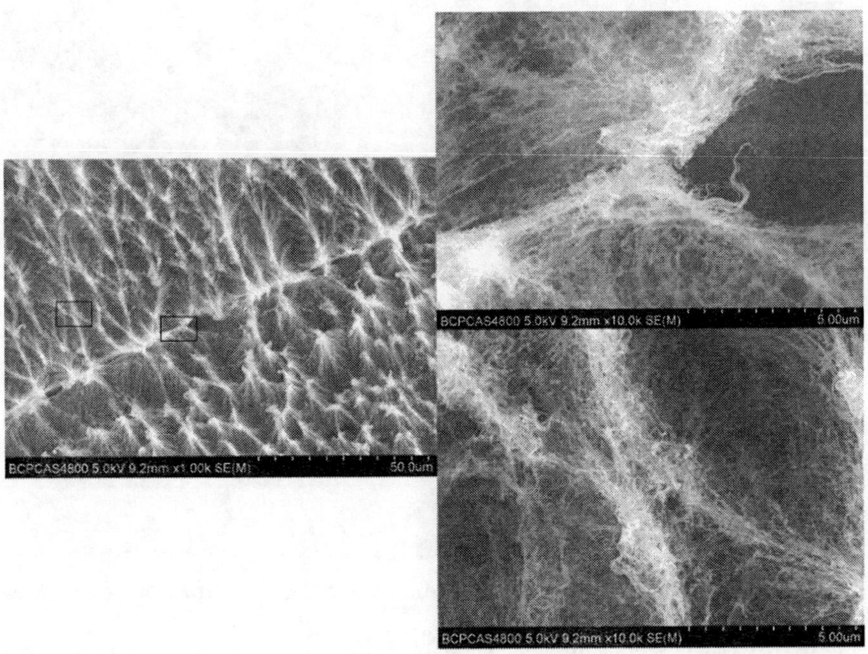

Fig. 3 Enlarge images of CNT arrays near "magic" line zone after AO exposure

We analyzed the morphologies of the "magic" line and other ridges carefully by SEM, but no difference had been observed. They all had shrinkage and crack formation formed by carbon nanotube aggregates, on the top of which some amorphous carbon could be observed. It is confused that how the "magic" line are formed and what's the difference of the carbon nanotubes along the lines. It seems that the "magic" line is like a wall to resist the AO beams, and the blank zone in which all the carbon nanotubes are etched away can only extend to one side.

IOP Conf. Series: Materials Science and Engineering **711** (2020) 012062 doi:10.1088/1757-899X/711/1/012062

The edges of the blank zone are presented in Fig. 4. There are three typical morphologies as shown in Fig. 4 a b c. One is carbon nanotube slices standing straightly as indicated in Fig.4 a. This phenomenon is non common. The other morphology is smooth edge as shown in Fig.4 b. The edge is vertical to the substrate. The SEM image taken from the side of the edge shows that the CNTs are oriented and not be bended and condensed, which means that they are not damaged by AO beams yet (Fig.4 c). Another morphology is carbon nanotube slices falling down toward the blank zone (Fig.4 d). The sides of the slices would be impacted by AO beams. The SEM image confirms that the side of the slices is deformed by AO beams as shown in Fig.4 e. Some big fibers composed of compressed carbon nanotubes play an important role in supporting the slices.

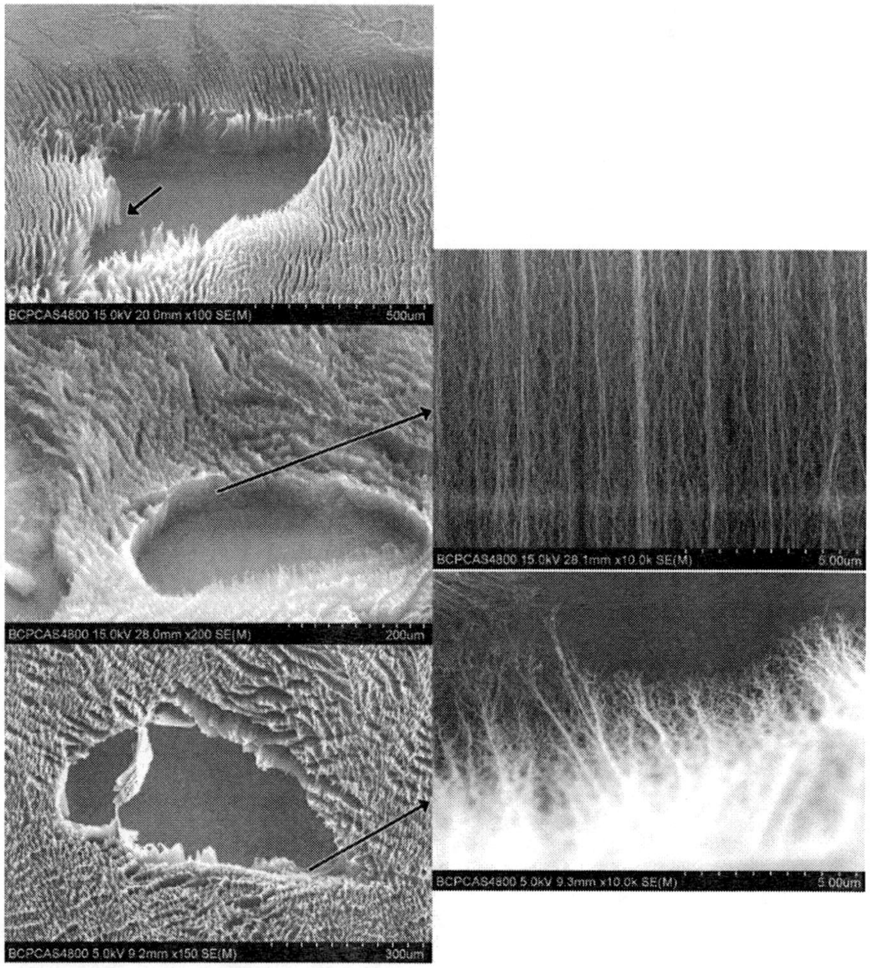

Fig. 4 SEM images of blank zones in CNT arrays after AO exposure

The formation of CNT arrays after AO exposure is a result of their unique properties and the bombardment effect of AO beams. It is known that CNT arrays are aligned, narrow spacing, high aspect ratios and flexibility. This indicates that AO bombard not only the surface but also the interior of the CNT arrays. When AO impacts on the wall of CNTs in the interior of the CNT arrays, the impact force would squeeze the CNTs on both sides. The surface of CNT arrays seems to have been cut apart. So a mountain range–like morphology has been obtained. Subsequently the AO would gather in the ravines, so the AO beam is like a beam knife to cut the CNT arrays. As a result, some slices are formed. Van der Waals interactions between condensed nanotubes are not the same in the whole CNT arrays, which may be the reason for different morphologies. And the low-density regions or vacancies of carbon nanotube arrays may also play an important role in the formation of the pattern. But there are still some questions: why are there some “magic” lines appeared? What’s the different between the blank zone and other zone in CNT arrays?

4.Summary

The effects of AO on CNT arrays have been studied by SEM analysis. The morphologies of CNT arrays are also quite different from those before AO exposure. The deformed surface of CNT arrays is like a mountain range with many ridges and ravines. Some ridges (so called “magic” line) are straight and continue with all the blank zones beside them. The blank zones have different morphologies, which may be due to the different Van der Waals interactions between condensed nanotubes. Though many questions still exist, it is interesting to find out how an anisotropic material would change during AO beam exposure.

References

[1] O. Gohardani, M. C. Elola, C. Elizetxea, Potential and prospective implementation of carbon nanotubes on next generation aircraft and space vehicles: A review of current and expected applications in aerospace sciences, Progress in Aerospace Sciences, 70(2014) 42-51.

[2] S. L. Koontz, K. Allyn, L. J. Leger, Atomic oxygen testing with thermal atom systems: a critical evaluation, J. Spacecraft Rockets, 33(1996) 8-14.

[3] S. B. Jin, G. S. Son, Y. H. Kim, C. G. Kim, Enhanced durability of silanized multi-walled carbon nanotube/epoxy nanocomposites under simulated low earth orbit space environment, Composites Science and Technology, 87(2013) 224-231

[4] H. E. Misak, V. Sabelkin, S. Mall, P. E. Kladitis, Thermal fatigue and hypothermal atomic oxygen exposure behavior of carbon nanotube wire, Carbon, 57(2013) 42-49.

[5] L. Jiao, Y. Gu, S. Wang, Atomic oxygen exposure behaviors of CVD-grown arbon nanotube film and its polymer composite film, Composites: Part A, 71(2015) 116-125.

[6] Y. M. Liu, M. Li, X. P. Liu, Erosion of carbon nanotube arrays induced by atomic oxygen irradiation, Chinese Journal of Materials Research, 30(2016) 795-800.

[7] Y. M. Liu, X. N. Yang, L. Man, Y. T. Zhang, C.Q. Zhao, Q. Wei, Erosion effects of atomic oxygen on carbon nanotube arrays in different incidence direction, Surf. Interface Anal. 50(2018) 592–598.

IOP Conf. Series: Materials Science and Engineering **711** (2020) 012063 doi:10.1088/1757-899X/711/1/012063

Diamond Forming Tool Innovative Design Based on Segments Thickness Dimension Compensation

Baoliang Xing [1, a], Chunying Zheng [2, b], Jing Wang [1, c], Shanshan Xue [1, d], Huiying Cao [1, e] and Zhengmei Zhang [1, f*]

[1] Shandong Labor Vocational and Technology College, Jinan 250022, China

[2] Shandong Institute for Product Quality Inspection, Jinan 250100, China

[a]xbl-19690504@163.com, [b]918654210@qq.com, [c]wangjing@sdlvtc.cn, [d]xueshanshan@sdlvtc.cn, [e]caohuiying2012@126.com, [f]zmzhang907@163.com

Abstract. Based on machining experiments the diamond forming tool wear degrees of irregular area were studied, the tool wear rate mathematical model was deduced by regression analysis and the improvement project of tool was innovatively designed. Through the diamond forming tool segments thickness dimension compensation, the tool wear of irregular area can be uniformly achieved, the performance of diamond forming tool has been increased highly and the diamond forming tool life is greatly extended.

1. Introduction

The granite special-shaped products have exquisite artistic shapes, excellent physical properties and other special characteristics, they are widely used in architectural decoration, living appliances, precision machinery, scientific instruments, craft products and so on. The manufacturing processes of granite special-shaped products are mainly the special surface processing. To simplify shaping movement, reduce equipment costs, increase productivity, the forming processing methods are generally used. Currently, the diamond Segments forming cutting tools are widely used to produce granite special-shaped products as a cost-effective processing methods in stone enterprises.

It is the diamond particles that play the cutting role in manufacturing granite products, it is the grinding process by the diamond particles controlled by diamond segments [1]. Because the granite is very hard stone, so the diamond tool is seriously grinded, a large number of diamond particles are consumed, which is due to the high processing cost for manufacturing granite products. The current research focus are abrasive processes and forms of diamond particles in diamond tool. Some researchers have established a diamond grinding single particle model [2]. The state of diamond particle was divided into first appearance, fully exposed, removed, local collapse, micro-crushing, macro-crushing and particle detachment [3]. Other scholars have studied the wear processes of diamond circular saw [4-6].

The key technology of manufacturing granite special-shaped products is the forming processing, namely, how to use the shortest possible time to remove the machining granite allowance, rough machining the required irregular surface. Take the processing of quarter arc granite surface as an example shown in Figure 1, the cross-section *EDF* is the removed region. Because the diameter of each section in arc region of diamond tool is different, so he diamond particles in each arc section are different to machining allowance, processing parameters and cutting loads, the wear degree of each part of the tool is also different. If the local region of diamond tool was excessive abraded, even if other parts of the tool are good, the cutting tool may be scrapped due to decrease in machining accuracy of granite special-shaped products. Therefore, the uniform wear for diamond forming tool is a key technology to ensure accuracy, extend tool life and reduce material consumption.

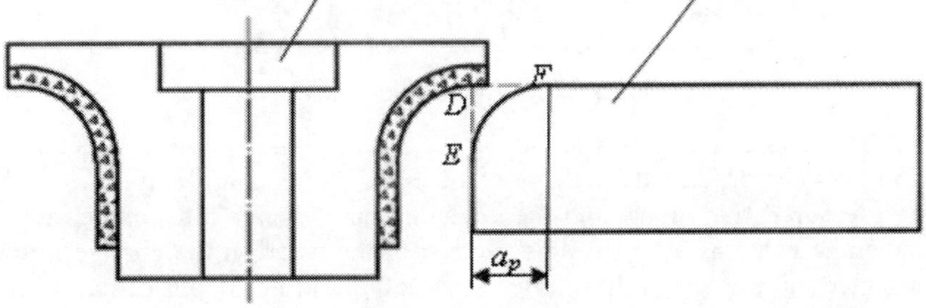

Fig. 1 Quarter arc granite surface schematic diagram

Therefore, the diamond forming tool to meet the conditions of processing performance, as far as possible to achieve uniform wear to ensure accuracy, extended tool life, reduce material consumption.

The research program for diamond forming tool was proposed based on segments thickness dimension compensation. Through experimental data analysis the wear trend of different diamond area was calculated, and then compensated for the thickness of the segments in order to achieve the uniform wear for diamond forming tool.

2. Experimental Procedure

Equipment and Material. The test machine is SPEED Y03. Cutting tool is diamond forming tool (grinding wheel ASS10105, produced in Italy. The detecting device is made up of microscope system (Leica DMLM, 2500 times), digital height gauge (0.01mm), digital vernier calliper (0.01mm) and so on. Qilu red granite (G3754) was selected as experimental materials, the main physical properties of Qilu red granite was shown in Table 1. The test device was shown in Figure 2.

Table 1 Qilu red granite of the main physical properties

Physical Properties	Unit	Value
Bulk Density	[kg/m^3]	2661
Water absorption	[%]	0.11
Dry compressive strength	[MPa]	224.0
Compressive strength of water saturated	[MPa]	218.6
Dry bending strength	[MPa]	16.9
Flexural strength of water	[MPa]	16.2
Dry Modulus of Rupture	[MPa]	19.6
Modulus of Rupture of water saturation	[MPa]	16.8
Abrasion resistance	[1/cm^3]	54

Fig. 2 Experimental device

Five groups tool diameters of five cross-section 1-1, 2-2, 3-3, 4-4, 5-5 were selected to measure in Figure 3, the section between separated by distance of 6mm, 4mm, 3mm and 2mm. For each set of measurements were measured five times, and then calculate the change after processing the value of tool diameter, the average value as a tool in this section of the wear on the change in diameter. Test parameters were selected cutting speed 30m/s, feed rate 500mm/min, depth of cut 15mm, length of processing 60m.

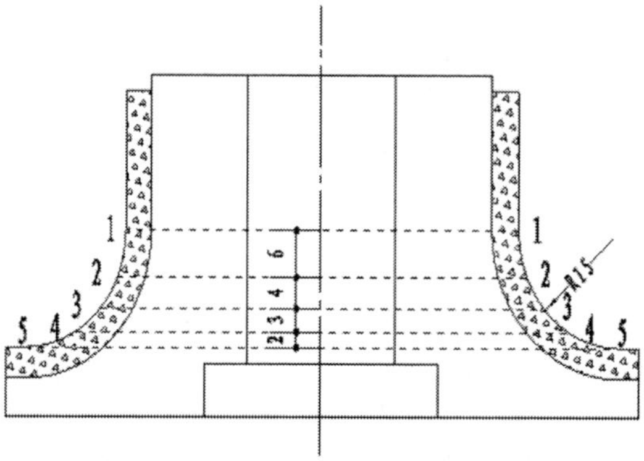

Fig. 3 Schematic diagram of experimental data selected

Analysis of test results. As the processing of the arc radius is 15mm and depth of cut is 15mm, so the 1-1 cross-section of diamond particles on the base did not participate in cutting, there is no changes in diameter. With 1-1 cross-section as measuring datum, the test measurement results were shown in Table 2. The trend chart of measured sections diameters of diamond tool was shown in Figure 4.

Table 2 Experimental measurements

Measurement section	1-1	2-2	3-3	4-4	5-5
Section corresponds to length dimension (mm)	0	6	10	13	15
Variation of section diameter (mm)	0	0.038	0.112	0.22	0.57

IOP Conf. Series: Materials Science and Engineering **711** (2020) 012063 doi:10.1088/1757-899X/711/1/012063

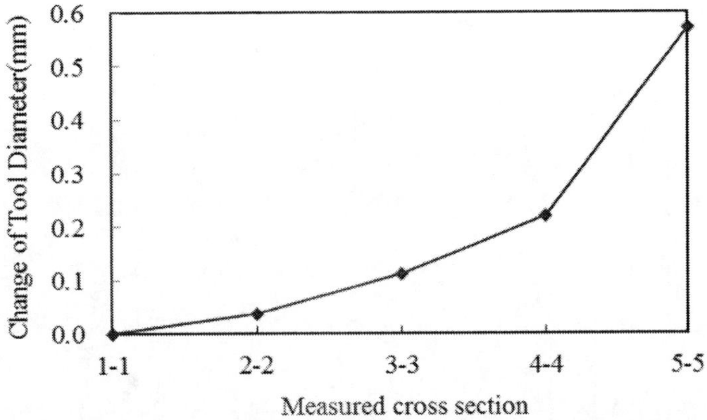

Fig. 4 The measured cross-section diameter of trend chart

The trend in Figure 4 can be seen, the tool wear rate arc region can be fitted parabola. The mathematical model can be establishment

$$y = a_0 + a_1 x + a_2 x^2 \tag{1}$$

In the formula: x is 1-1 as the base of cross-section to correspond to the length dimension, y is the change in cross-section diameter.

By the regression statistic in Table 2, the equation of diamond forming tool wear rate obtained is

$$y = 0.0184132 - 0.032443x + 0.0042932x^2 \tag{2}$$

3. Improved design tools

To achieve uniformity wear of diamond forming tool, it is necessary to compensate the diamond thickness dimension based on Figure 4. Making the basic segment thickness dimension 3mm, then the compensated thickness dimension of each section of segment was shown in Table 3, the schematic diagram of the thickness of segment was shown in Figure 5.

Table 3 Section thickness of diamond segment

Measurement section	1-1	2-2	3-3	4-4	5-5
Section corresponds to the length of size(mm)	0	6	10	13	15
Agglomeration section thickness(mm)	3	3.63	4.87	6.67	12.5

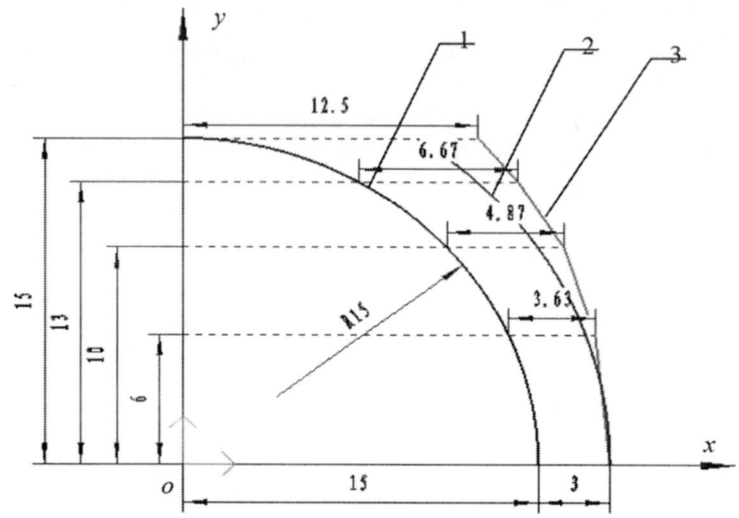

1—15mm radius arc segment 2—3mm thickness of arc segment
3—The connection section dimension compensation

Fig. 5 Schematic diagram of thickness of segment

The ligature 3 in Figure 5 was fitted by three-point arc method, the segment thickness dimension compensation cross-section was shown in Figure 6. In Figure 6, *AB* is the segment of the outer circular arc of radius 15mm, the thickness of area *ABCD* is no compensation for the segment dimension, *ABED* is the segment compensated area, *CED* is the dimension of compensation area.

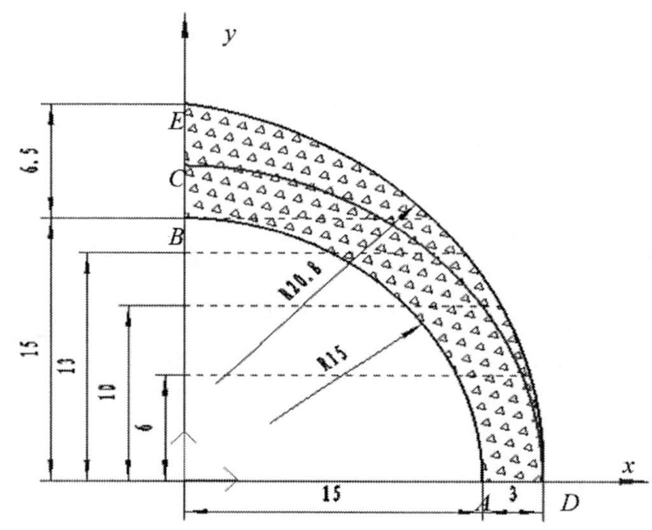

Fig. 6 Schematic diagram of diamond segment thickness compensation

Based on diamond segment compensated, a new diamond forming tool was innovatively designed as shown in Figure 7.

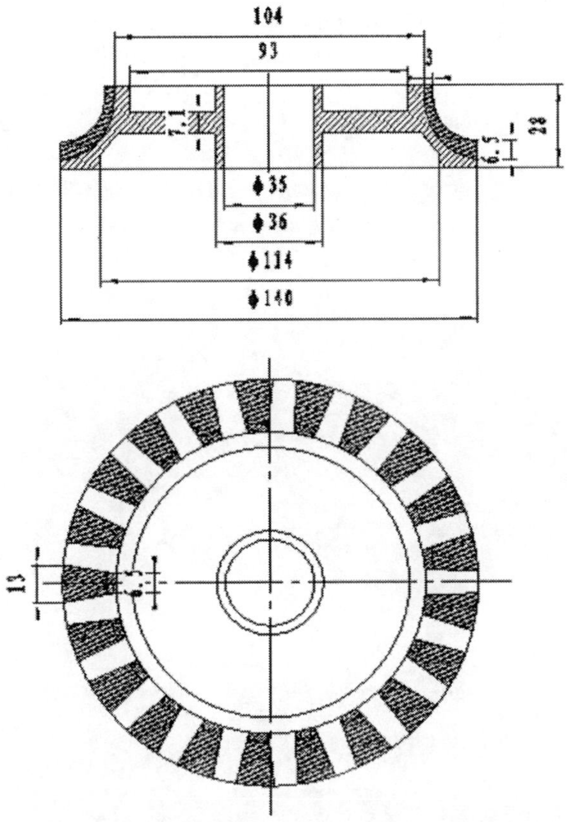

Fig. 7 The new designed diamond forming tool

To test the new diamond forming cutting tools, the machining parameters were selected as cutting speed 30m/s, feed rate 500mm/min, depth of cut 14.2mm, processing length 30mm. In these parameters combination, the new tool cutting efficiency was 19.4385cm^3/min, tool wear rate was 0.0812mm/dm^3. The tool wear rate was decreased significantly, the arc area uniform wear was obtained, the new designed diamond forming tool has been achieved the desired results.

4. Conclusion

(1) Established by experiments of diamond forming tool quarter circular cross-section area of the mathematical model of wear rate.

$$y = 0.0184132 - 0.032443x + 0.0042932x^2$$

(2) Through the diamond forming tool segments thickness dimension compensation, the tool wear of irregular area can be uniformly achieved, the performance of diamond forming tool has been increased highly and the diamond forming tool life is greatly extended.

References

[1] Xu Xi-peng 2003 Diamond sawing of rock material progress *Mechanical Engineering Press* vol 39 P 17-22.

[2] Antoniomaria Di Ilio, Antonio Togna 2003 A theoretical wear model for diamond tools in stone cutting *International Journal of Machine Tools and Manufacture* vol 43 p 1171-1177.

[3] Arkadeep Kumar and Shreyes N. Melkote 2018 Diamond Wire Sawing of Solar Silicon Wafers A Sustainable Manufacturing Alternative to Loose Abrasive Slurry Sawing *Procedia Manufacturing* vol 21 p 549-566.

[4] Deniz Tumac and Aydin Shaterpour-Mamaghani 2018 Estimating the sawability of large diameter circular saws based on classification of natural stone types according to the geological origin *International Journal of Rock Mechanics and Mining Sciences* vol 101, p 18-32.
[5] Benoit B. Mandelbrot, The fractal geometry of nature. Macmillan. New York, 2012.
[6] A.Ersoy, S.Buyuksagic and U.Atici 2005 Wear characteristics of circular diamond saws in the cutting of different bard abrasive rocks *Wear* vol 258, p 1422-1436.

Research on Influencing Factors on Rheological Properties of Chitosan/Starch Film Slurry

Kaiqiang Sun[1, 2, a], Jianyong Li[1, 2, b, *], Fangyi Li[1, 2, c], Xin Zhou[3, d], Zunyi Wang[3, e], Lei Wang[1, 2, f], Xu Sun[1, 2, g], Jinfeng Cui[1, 2, h], Chuanwei Zhang[1, 2, i]

[1] Key Laboratory of High Efficiency and Clean Mechanical Manufacture (Ministry of Education), School of Mechanical Engineering, Shandong University, Jinan 250061, China

[2] National Demonstration Center for Experimental Mechanical Engineering Education, Shandong University, Jinan 250061, China

[3] China heavy duty truck group co. LTD

[a]17865196539@163.com, [b]ljy@sdu.edu.cn, [c]lifangyi@sdu.edu.cn, [d]houxin2018@163.com, [e]angzunyi2018@163.com, [f]anglei2018@163.com, [g]unxu_1995@qq.com, [h]jf445101427@163.com, [i]hangchuanwei3722@163.com

Abstract: In order to improve the industrial production efficiency of chitosan/starch film, reducing slurry rheology is an important measure. This paper explored that the temperature, chitosan starch ratio, plasticizing different components had a great impact on the rheology of the slurry. As the starch content and temperature increased, the rheology of the slurry first decreased and then increased. Finally, it was determined that when the slurry was transported at 60°C with chitosan: starch =1:1.5 and only the chitosan was plasticized, the viscosity was the minimum. The viscosity of the slurry was related to the hydrogen bond formed by the slurry. The more hydrogen bonds, the greater the viscosity. Plasticizing chitosan destroyed the crystal structure and facilitated the formation of the network structure, which was conducive to the shear flow of the slurry.

1. Introduction

Packaging materials, prepared by petroleum derivatives, has caused huge environmental pollution in various industries, such as the plastic rubber and so on. It becomes solid waste and eventually causes great trouble to people's production and life with the aging of use, [1]. Therefore, the current research is focused on the development of degradable green packaging materials. In packaging materials, antibacterial packaging due to its inhibition of microorganisms was known as a promising development direction for packaging materials [2]. In the previous study, the chitosan has been widely used in antimicrobial packaging materials because its proven biocompatibility [3]and antimicrobial activity [4]. And the starch has been regarded as the frequently-used degradable materials because of its convenient preparation and abundant reserves [5]. We usually mix these two materials to overcome the shortcomings of the bad mechanical properties of the chitosan film and the starch film. [6]. Therefore, it has a good application effect and a broad market prospect by regarding the chitosan-starch mixed film as antibacterial packaging material.

The industrial production of chitosan-starch film often adopts the technology of jostling, blowing and molding. The slurry was often attached to the surface of the pipe, which made it difficult to control in the process of transportation. In addition, the rheological property is an important parameter of slurry in the industrial production of composite products, and also one of the main physical quantities to calculate the friction resistance of pipeline in the design of automatic transportation equipment of slurry production line [7]. In order to better improve the processability of composite films, it was of great significance to study the rheological property of mixed slurry. The research results of many scholars at home and abroad have confirmed that temperature and self-concentration have an effect on the rheological property of slurry [8].

In order to improve the performance of the composite film as well as make a better compatible of starch and chitosan, the researchers are also exploring different modification methods such as glycerol [9], glutaraldehyde [10], mustard oil [11]and polydopamine [12] to optimize the film properties.

Many scholars have already made a lot of explorations on the slurry of biomass material. Willett et al. found that the melting behavior of starch in the extruder obeyed the power law formula, and the melt viscosity decreases with the increase of temperature and moisture content [13]. A Silva-Weiss et al. studied the effect of murta leaf on the rheology property of chitosan-starch mixed slurry [14]. Aichholzer et al. used pseudoplasticity and viscosity to describe the flow behavior of thermoplastic starch during extrusion. It was concluded that no matter what kind of thermoplastic starch, it showed the flow behavior of adhesive wall in the presence of a large amount of glycerol or other small molecular additives [15]. Zhang et al. studied the effect of plasticized oxidation composite modification on the rheological property of biomass slurry, and the effect of composite modification on the rheological property of starch slurry was introduced in detail from the microscopic point of view [16].

In summary, scholars mainly explore the rheological behavior of composite slurry from the point of view of quantitative model of rheological behavior, external factors and rheological optimization. There was a shortage of studies analyzing the effects before and after modification on what the molecular structure change has made to the rheological property and people don't make an in-depth inquiry on the root causes for the change of rheological property of plasticizing modification. Therefore, the main purpose of this article is to explore the effect of single-whole plastic modification on the rheological properties of chitosan-starch film by means of plasticization and separation. And the effects of modification of single/integral plasticization on the rheological properties of the slurry were characterized by scanning electron microscopy, infrared spectroscopy, X-ray diffraction(The following three test torches were referred to as SEM/XRD/FTIR)and other experimental methods to explore the mechanism of the influence of plasticizing modification on the rheological properties of the mixed slurry.

2. Experiment

2.1. Experimental instruments and materials

Glycerin and ethylene glycol (99% purity) were purchased as plasticizers from Tianjin Fuyu Fine Chemical Co., Ltd.; Corn starch, reagent grade, average particle diameter 70nm, Hebei Huachen Starch Sugar Co., Ltd.; Chitosan (95% deacetylation degree), Shandong Dongrun Biotechnology Co., Ltd.; Urea, AR, purity ≥99.0%, Tianjin Zhiyuan Chemical Reagent Co., Ltd.; NaOH powder, purity ≥96.0%, Cangzhou Manyue Chemical Sales Co., Ltd.

HHS-2 electronic constant temperature water bath, Shanghai Kanglu Instrument Equipment Co., Ltd.;JJ-1 precision booster, Changzhou Boyuan Experimental Analytical Instrument Factory; Electronic balance, 2000g/0.01g, Shanghai Huachao Electric Co., Ltd.; Electric thermostat blast drying oven, Shanghai Jinghong Experimental Equipment Co., Ltd.; Infrared spectrometer, BRUKER VERTEX-70, Bruker, Germany; i-Strentek 1510 electronic universal testing machine, Jinan Blu-ray Electromechanical Technology Co., Ltd.; Scanning electron microscope, model FEG250, Beijing Oubo Tong Optical Technology Co., Ltd.;YS20B nickel-copper filter radiator, Taizhou Yishun Valve Co., Ltd.%, Tianjin Zhiyuan Chemical Reagent Co., Ltd..

2.2. Slurry preparation
To prepare the chitosan solution, we mixed and stirred 8 g of NaOH, 4 g of urea and 88 g of water to obtain the chitosan dissolution system with 8% NaOH and 4% urea [17]. Then, the chitosan powder was dispersed into the mixture, which was stirred for 20 min to obtain a chitosan suspension. Then we chilled this suspension to -30°Cand stirred it three times during this time. Finally, it was thawed at room temperature to obtain the transparent chitosan solution.

To prepare the gelatinized starch, we poured 100 g of distilled water and 24 g of starch into beakers. The mixture was then stirred. A uniform suspension of the starch was obtained, and the beakers were placed in a water bath at a constant temperature of 80 °C. The contents were stirred using a blender for 30 min to obtain gelatinised starch.

2.3 Test concerning the effect of Slurry viscosity
(1) Experimental study on the effect of formulation concentration on slurry viscosity

Take five 500ml round bottom beakers, the prepared chitosan and starch slurry were mixed and disposed in 5 beakers in a ratio of 1:0, 1:0.5, 1:1, 1:1.5, and 1:2. Then set the precision magnetic stirrer rotation speed to 80r/min and stirred for 30 minutes, stood for 24 hours at room temperature. Then, the viscosity was measured by using a viscometer, and the shear rate of the rotor was set to 100 s - 1, with the slurry temperature was 20 °C at room temperature. The experimental data was collected every 0.2 seconds to record the stable viscosity data. Repeated the experiment five times and recorded the experimental data, then removed the collected data from the collected data points and drew the viscosity characteristic curve.

(2) Experimental study on the effect of Temperature on slurry viscosity

Take five 500ml round bottom beakers, the prepared chitosan and starch slurry were mixed and disposed in 5 beakers in a ratio of 1:1.5. Then set the precision magnetic stirrer rotation speed to 80r/min and stirred for 30 minutes, stood for 24 hours at room temperature. The five beakers were placed in a constant temperature water bath at a temperature of 20°C, 40°C, 60°C, 80°C, and 100°C. After one hour, the viscosity of the mixed slurry was measured at the corresponding temperature. The experimental data was collected every 0.2 seconds, to record the stable viscosity data. Repeated the experiment five times and recorded the experimental data, then removed the collected data from the collected data points and drew the viscosity characteristic curve.

(3) Experimental study on the effect of Single/integral plastic modification on slurry viscosity

100 g of starch slurry and 100 g of chitosan slurry were treated with glycerol 1.875 g, respectively. Took three 500ml round bottom beakers and mixed plasticized chitosan and common starch (PC-S), plasticized starch and common chitosan (PS-C), plasticized chitosan and plasticized starch (PCS)according to the ratio of starch: chitosan=1:1.5 (The above three kinds of slurries were simply referred to as "PC-S/PS-C/PCS" hereinafter), and then the above process was repeated for viscosity measurement.

2.4. FTIR test
The different plasticized mixed slurry was dried in a 100°C incubator for 24 h.5g sample and 300mg KBr powder were mixed and fully ground to a particle diameter of less than 2.5. The ground mixture was transferred to a tableting mold and applied a pressure of about 12 to 14 MPa to a transparent or translucent sheet. Place the sheets on the holder of the infrared spectrometer separately and set the sampling frequency to 400-4000cm-1. Then run the program to scan the sample, printed and saved the spectrum. Finally, we analyzed peak shape attribution.

2.5.XRD test
The remaining powder of the infrared spectrum test was taken 5 g, which was ground with an agate grinding dish until the particle size was 195 mesh. The test was carried out at room temperature by using a graphite curved crystal monochromator and a nickel copper filter radiator. During the sample test, the sample and the glass frame must be flush and the sample should be flat. The index of the seam

system was DS/RS/SS =1°/0.16 mm/1°. The test used Cu-Kα ray, Ni filter, the tube pressure was set to 35 KV, the tube flow was set to 25mA sweep, the drawing speed was 5 (°)/min, the analysis angle (2q) and the interval was 10-90°.

2.6.SEM test

All the samples were mounted on the aluminum short post with a ribbon, and a gold water coating was sputtered on the surface of the slurry to make it thermally conductive. Then the inside of the PC-S/PS-C/PCS mixed slurry was observed by using a scanning electron microscope. The electron microscope needed to have an accelerating voltage of 10kv.

3. Experimental analysis

3.1 Effect of temperature on slurry viscosity

The rheology of a material was usually characterized by viscosity. The lower the viscosity of the material, the better its rheological properties. The viscosity test results of the mixed slurry at different temperatures were shown in Fig.1. The viscosity of the slurry decreased firstly and then increased with the increase of temperature. The viscosity of the slurry reached a minimum at 60°C and was 10,425 cp. At the beginning, due to the rise of temperature, the thermal motion of the molecules inside the slurry was accelerated, the energy of the microscopic particles or groups in the slurry was increased, the binding force between the molecules was reduced, the mixed slurry was easily stirred, and the viscosity was reduced. When the temperature continued to rise to a certain critical value, the molecular thermal motion was too intense that the microscopic particles collided with each other, the energy was reduced, and the viscosity was reduced.

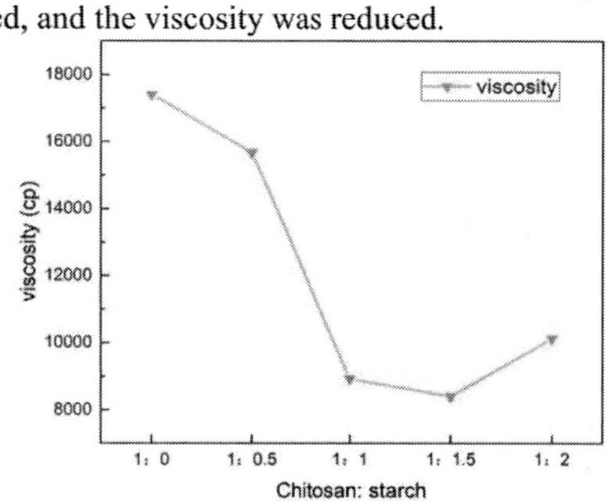

Fig.1 Viscosity curve with temperature

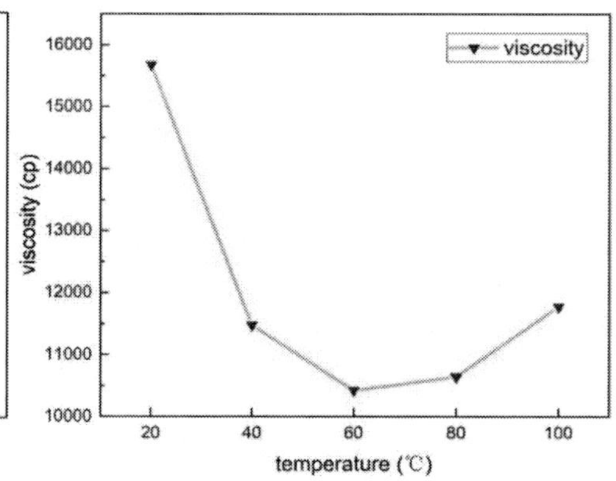

Fig2.Viscosity curve with starch concentration

3.2 Effect of concentration on slurry viscosity

The results of slurry viscosity test when chitosan and starch were mixed according to different proportions were shown in Fig.2. The results showed that the viscosity of the mixed slurry decreased firstly and then increased with the increase of starch content. The viscosity of the slurry after mixing was reduced compared to the viscosity of the individual slurry. The chitosan molecule and the starch molecule were connected by hydrogen bonding as shown in Figure 4, starch and chitosan contained a large amount of hydroxyl groups. During the mixed process, the starch molecules and the chitosan molecules were connected by hydrogen bonds, resulted in a decrease in the binding force between the molecules [18]. The viscosity was reduced. Moreover, as the starch content was increased, agglomeration occurred between the starch molecules and the viscosity increased correspondingly.

3.3 Effect of Single/integral plastic modification on slurry viscosity

The results of viscosity test when plasticizing the mixed slurry of different modification methods showed that PS-C had the highest viscosity, PCS viscosity was the second, and PS-C had the lowest viscosity. The results showed that the viscosity of the slurry would be greatly reduced when the slurry was plasticized. The viscosity of the co-plasticized slurry was reduced by 51.4% compared to the viscosity of the unplasticized slurry. It was concluded that the plasticizing treatment was beneficial to reduce the rheology of the slurry.

3.4. FTIR results

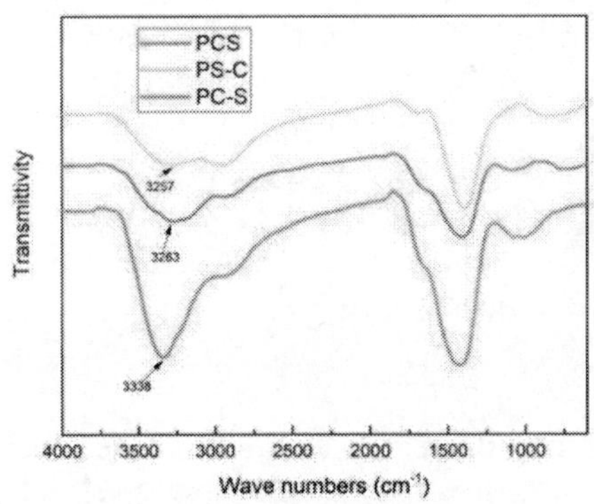

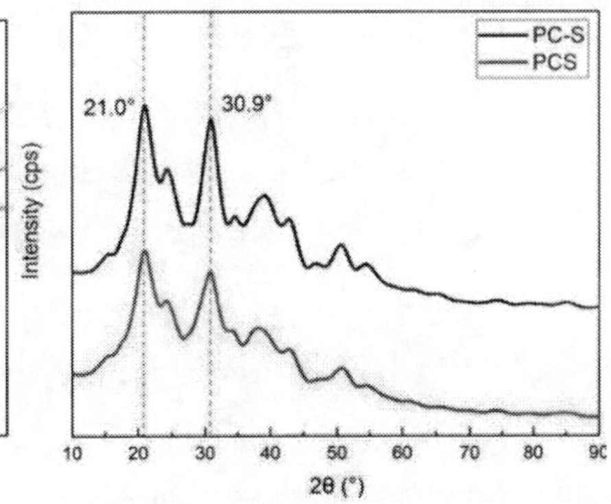

Fig.3 The infrared spectra of PCS/PS-C/PC-S Fig.4 X-ray diffraction curves of PC-S and PCS

In order to explain the change of the viscosity of the slurry from a microscopic point, the three kinds of slurries were tested by infrared spectroscopy. In this study, the O-H stretching vibration absorption peak in the FTIR spectrum was located at 3000-3600 cm-1, and the lower the hydroxyl peak frequency, indicating the formation of new hydrogen bonds [19]. An intermolecular hydrogen bond shown in Fig.3 was formed between chitosan and starch. Figure 5 showed the FTIR spectrum when plasticizing different components. The results showed that the hydroxyl absorption peak frequency of PS-C was the lowest, and the number of hydrogen bonds formed was the largest, which also led to the highest viscosity among the three treatment methods. Followed by PC-S, and finally PC-S. The FTIR test results showed that the order of the number of hydrogen bonds formed was the same as the order of the viscosity size of the slurry, which indicates that the formation of hydrogen bond would affect the rheology of the slurry, and the more the number of hydrogen bond formation was, the greater the viscosity of the slurry.

3.5 XRD results

In order to further verify the results of the FTIR test, XRD was tested on both PC-S and PCS slurries. The test results were shown in Fig.4. The results showed that PC-S had a peak at $2\theta=20.96°$, $30.98°$, whereas PCS had a peak at $2\theta=21.04°$, $31.02°$. The peak intervals of the two were similar, which indicated that PC-S and PCS had similar crystal microscopic morphology. The crystallinity of a material was usually characterized by a crystallinity index, and the greater the crystallinity index of the material, the higher its crystallinity. The processing and analysis by Jade software showed that the crystallinity index of PCS was significantly smaller than that of PC-S, which indicated that the crystallinity of PCS was reduced when it was modified. Polymer materials such as starch and chitosan were composed of a crystalline region and an amorphous region. The degree of crystallinity was used to some extent to indicate the size of the crystalline region. The lower the crystallinity, the smaller the crystalline region of the material, the larger the amorphous region, and the more the free branch in the amorphous region, which was more conducive to the combination between starch and chitosan. So the combination between the two was closer, which resulted in a higher viscosity.

3.6 SEM result

To further explain the correctness of FTIR and XRD, SEM analysis was performed on the three mixed slurries. The results were shown in Fig.5, in which Fig. (a) was a PS-C structure diagram, Fig. (b) was a P slurry structure diagram, and Fig. (e) was a slurry structure diagram of PCS. The internal structure of PS-C was a block structure with tight bonding, few pores, and the highest viscosity of the slurry; The internal structure of PC-S was a network structure with many pores and low viscosity; When chitosan and starch were co-plasticized, the structure of the slurry was a reticulated joint block structure. The void was less than PC-S, so its viscosity was greater than the viscosity of PCS.

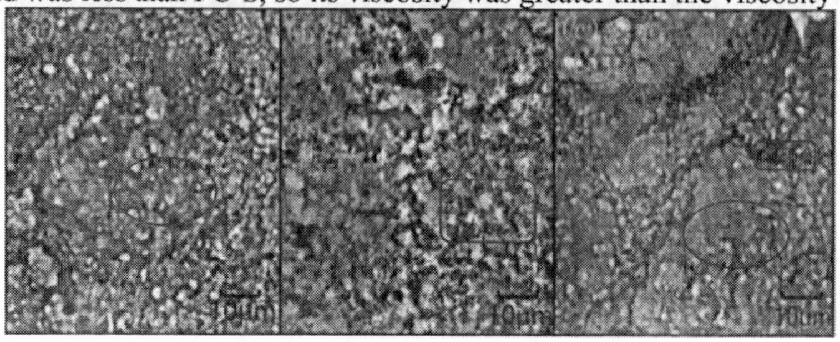

Fig.5 SEM of (a)PS-C(b)PC-S(c)PCS

4. Conclusions

The viscosity of the chitosan/starch film mixed slurry decreases at first and then increases with the increase of temperature and starch content, the viscosity reaches the lowest at 60 °C and at the chitosan: starch=1:1.5. Effect of single/integral plasticization modification on slurry viscosity: PS-C>PCS>PC-S. The FTIR test of the single/integrated plasticized slurry shows that the hydroxyl absorption peak of PS-C is closer to the low frequency, then the PCS, and finally the PC-S. The amount of hydrogen bonds formed is related to the rheology of the slurry. The XRD test results shows that the crystallinity of PS-C is significantly smaller than that of PCS, which makes it less rheological. SEM test results shows that the internal structure of PS-C is a block structure with tight bonding and few pores; The internal structure of PC-S is a network structure with many pores; When chitosan and starch were co-plasticized, a sheet-like structure is formed on a large area, and a network structure is generated between the sheets.

References

[1] Jambeck J R, Geyer R, Wilcox C, et al. Plastic waste inputs from land into the ocean[J]. Science, 2015, 347(6223):768-771.

[2] M. Aider, Chitosan application for active bio-based films production and potential in the food industry: review, LWT-Food Sci. Technol. 43 (2010) 837-842.

[3] Park, S. I., Daeschel, M. A., & Zhao, Y. (2004). Functional properties of antimicrobial lysozyme–chitosan composite films. Journal of Food Science, 69(8), M215–M221

[4] Kong M, Chen X G, Xing K, et al. Antimicrobial properties of chitosan and mode of action: A state of the art review[J]. International Journal of Food Microbiology, 2010, 144(1):51-63.

[5] Avérous, L., Fringant, C., & Moro, L. (2001). Starch-based biodegradable materials suitable for thermoforming packaging. Starch/Stärke, 53(8), 368–371.

[6] Pelissari, F. M., Grossmann, M. V., Yamashita, F., & Pineda, E. A. (2009). Antimicrobial, mechanical, and barrier properties of cassava starch–chitosan films incorpo- rated with oregano essential oil. Journal of Agricultural and Food Chemistry, 57(16), 7499–7504.

[7] Li gang, li fangyi, guan kaikai, et al. Experimental study on preparation and performance of biomass buffer packaging materials [J]. Functional materials, 2013, 44(13):0-0.

[8] Li Xingke, Jiang Qixing, Xia Wenshui. Rheological properties and application of chitosan solution[J]. Science and Technology of Food Industry, 2011, 32(2): 65-68.

[9]Y.M. Yusof, M.F. Shukur, H.A. Illias, M.F.Z. Kadir, Conductivity and electrical properties of corn starch-chitosan blend biopolymer electrolyte incorporated with ammonium iodide, Phys. Scr. 89 (2014) 035701.

[10] Li H, Gao X, Wang Y, et al. Comparison of chitosan/starch composite film properties before and after cross-linking[J]. International Journal of Biological Macromolecules, 2013, 52(none):275---279.
[11] Tuhin M O, Rahman N, Haque M E, et al. Modification of mechanical and thermal property of chitosan–starch blend films[J]. Radiation Physics & Chemistry, 2012, 81(10):1659-1668.
[12] Min Z, Qin L, Wu S, et al. Starch/chitosan films reinforced with polydopamine modified MMT: Effects of dopamine concentration[J]. Food Hydrocolloids, 2016, 61:678-684.
[13] Willett J L, Jasberg B K, Swanson C L. Rheology of thermoplastic starch: Effects of temperature, moisture content, and additives on melt viscosity[J]. Polymer Engineering and Science, 1995, 35(2):202-210.
[14] Silvaweiss A, Bifani V, Ihl M, et al. Structural properties of films and rheology of film-forming solutions based on chitosan and chitosan-starch blend enriched with murta leaf extract[J]. Food Hydrocolloids, 2013, 31(2):458-466.
[15] Aichholzer W, Fritz H G. Rheological Characterization of Thermoplastic Starch Materials[J]. Starch - St?rke, 1998, 50(2-3):77-83.
[16] Zhang C W, Li F Y, Li J F, et al. Research on Rheological Behavior of Biobased Composite Slurry Composed of Sisal Fiber and Thermoplastic Oxidized Starch[J]. Journal of Biobased Materials & Bioenergy, 2017, 11(2):119-124.
[17] Hu X, Du Y, Tang Y, et al. Solubility and property of chitin in NaOH/urea aqueous solution[J]. Carbohydrate Polymers, 2007, 70(4):451-458.
[18] Chen S, Li F Y, Li J F, et al. Effects of single-modification/cross-modification of starch on the mechanical properties of new biodegradable composites[J]. RSC Advances, 2018, 8(22):12400-12408.
[19] M. A. Garcia, A. Pinotti, N. E. Zaritzky. Physicochemical, Water Vapor Barrier and Mechanical Properties of Corn Starch and Chitosan Composite Films[J]. Starch-Starke, 2006, 58(9):453-463.

IOP Conf. Series: Materials Science and Engineering **711** (2020) 012065 doi:10.1088/1757-899X/711/1/012065

Simulation and Calculation of Interior Ballistic of A Large Caliber Artillery

Dan-yu WANG[1, a], Feng-qiang NAN[1,b,*], Ping DU[1,c], Xin-min ZHAO[2,d], Zheng-lun WANG[1,e]

[1]School of Chemical Engineering, Nanjing University of Science and Technology, Nanjing, Jiangsu, China

[2]Xi'an Shengying Co., Ltd., Xi'an, Shanxi, China

[a]wdyxy1109@163.com, [b]nanfq@163.com, [c]dp1314@163.com, [d]feaplus@163.com, [e]wangzl@mail.njust.edu.cn

Abstract. Based on the classical interior ballistic model, the model is improved according to the actual situation. Relying on the newly established model, the related software is compiled. The maximum bore pressure, muzzle velocity and detonation temperature of a 155mm caliber gun are simulated by using relevant software. The maximum bore pressure is 338.2748 MPa, muzzle velocity is 855.1096 m/s and detonation temperature is 3468.402 K. After comparing the calculated results with the experimental data, it is found that the software is more practical and can be applied in engineering practice.

1. Introduction

The calculation of interior ballistic model is the core of interior ballistic theory, which plays an important role in gun design, improvement and problem solving. The traditional interior trajectory solution methods include empirical method, analytical method, tabular method and graphic method, which are simple but limited. With the application of computer in ballistics, numerical methods have appeared. Although they are approximate solutions, the accuracy of calculation can meet the actual needs, especially some softwires such as MATLAB, FLUENT, CFD (Computational Fluid Dynamics), which provide more possibilities for numerical calculation. For example, LUO Qiao and Zhang Xiao-bing applied FLUENT software to Interior Ballistic Analysis of Ultrahigh Firing Rate Guns artillery.[1] Liu Lin and Fan Cheng-jun used Simulink to simulate the interior trajectory of mixed charge, and the results met the actual engineering requirements.[2]

In this paper, the interior ballistic simulation model is built and calculated based on relevant software. Comparing the calculated values with the experimental results, the feasibility of this method in interior ballistic calculation is verified.

2. Establishment of Interior Ballistic Model

2.1 Basic Assumptions

Internal ballistic calculation is developed on the basis of classical interior ballistic theory, so it has certain similarity and inheritance in the construction of interior ballistic model. Based on the classical theory of interior ballistics, this paper establishes relevant models. Based on the classical theory of interior ballistics, the following basic assumptions are put forward:

a) The combustion of propellant obeys the law of geometric combustion.

b) The combustion of propellant and the movement of projectile are carried out under the condition of average pressure, and the flow of gas in bore follows Lagrange hypothesis.

c) The combustion of propellant obeys the law of exponential combustion.

d) During and after combustion, the combustion products of propellant remain unchanged, i.e. the propellant power f and the residual volume of propellant gas α are treated as constants.

e) When the projectile moves in the bore, it is well sealed and there is no air leakage.

f) When the projectile shoots out of the muzzle, the gunpowder gas is continuously flowing out and the gas pressure in the bore is constantly changing. The process of flowing out is carried out under the adiabatic condition.[3]

2.2 *Fundamental Equations*

a) The commonly used interior ballistic model is the case of porous propellant, and its shape functions of gun powder is as follows:

$$\psi_i = \begin{cases} \chi Z_i\left(1+\lambda Z_i+\mu Z_i^2\right) & (Z_i<1) \\ \chi_s Z_i\left(1+\lambda_s Z_i\right) & (1\le Z_i<Z_k) \\ 1 & (Z_i \ge Z_k) \end{cases} \tag{1}$$

Among that, ψ is the burned percentage of fast burning propellant; Z is the relative thickness of burning propellant; χ , λ , μ , χ_s and λ_s are the shape characteristic quantity of the main charge propellant; Z_k is the relative thickness of burned propellant after burning.

b) Burning Rate Equation

$$\frac{dZ_i}{dt} = \begin{cases} \dfrac{u_1}{e_{1,i}} p_i^n & (Z_i<Z_k) \\ 0 & (Z_i \ge Z_k) \end{cases} \tag{2}$$

Among that, t is time, u_1 is burning rate coefficient, e_1 is arc thickness of the main charge propellant, p is pressure, and n is burning rate-pressure exponent.

c) Travel Equation

$$v_i = \frac{dl_i}{dt} . \tag{3}$$

Among that, v is the velocity of the projectile and l is the travel of the projectile in the controlling tube.

d) Motion Equations of The Projectile

$$\varphi m \frac{dv}{dt} = Sp . \tag{4}$$

Among that, S is the maximum cross-section area of the projectile, φ is the secondary work coefficient, m is the mass of the projectile, p is the pressure in the chamber.

e) Energy Balance Equation

$$Sp(l_\psi + l) = f\omega\psi - \frac{\theta}{2}\varphi m v^2 . \tag{5}$$

Among that, l_ψ is equivalentlength of the free volume of the first stage, f is the gunpowder force, By combining the above equations, we can get:

$$\begin{cases} \psi_i = \begin{cases} \chi Z_i\left(1+\lambda Z_i+\mu Z_i^2\right) & (Z_i<1) \\ \chi_s Z_i\left(1+\lambda_s Z_i\right) & (1\le Z_i<Z_k) \\ 1 & (Z_i \ge Z_k) \end{cases} \\ \dfrac{dZ_i}{dt} = \begin{cases} \dfrac{u_1}{e_{1,i}} p_i^b & (Z_i<Z_k) \\ 0 & (Z_i\ge Z_k) \end{cases} \\ v_i = \dfrac{dl_i}{dt} \\ \varphi m \dfrac{dv}{dt} = Sp \\ Sp(l_\psi + l) = f\omega\psi - \dfrac{\theta}{2}\varphi m v^2 \end{cases} \quad (6)$$

Eq. 6 is the classical interior ballistic equations used in this paper.[4]

3. Calculation of Simulation Parameters of Interior Ballistics

In this paper, a 155mm large caliber gun is taken as the calculation object, and the propellant of 23/19 propellant is used to simulate. Next, the relevant parameters are calculated.

3.1 *Secondary Work Coefficient*

$$\varphi = K + \frac{1}{3}\frac{\omega}{m}. \quad (7)$$

Among that, K is a constant related to weapons. The experience shows that the secondary work coefficient of the large caliber gun is 1.02.

3.2 *The Gunpowder Force*

For propellant composed of C, H, O and N, and the unit mass element is $C_aH_bO_cN_d$. The combustion reaction formula of propellant can be expressed as follows:

$$C_aH_bO_cN_d \rightarrow xCO_2 + y\mathrm{CO} + z\mathrm{H_2}O + uH_2 + vN_2$$

The formulas for calculating specific heat, detonation heat, detonation temperature and the gunpowder force are as follows:

$$W_1 = 22.41n = 22.41\left(a + \frac{b}{2} + \frac{d}{2}\right) = 22.41(x+y+z+u+v). \quad (8)$$

$$Q_V = xq_{CO_2} + yq_{CO} + zq_{H_2O} - q + (x+y+z+u+v)RT_0. \quad (9)$$

$$T_V = \frac{Q_V}{\sum n\bar{C}_V} + T_0. \quad (10)$$

$$f = nRT_V = \frac{P_1W_1}{273}T_V = \frac{103.3W_1}{273}T_V. \quad (11)$$

Based on the above equation, the calculation results are as Table 1 shows.

IOP Conf. Series: Materials Science and Engineering **711** (2020) 012065 doi:10.1088/1757-899X/711/1/012065

Table 1. Parameters of Artillery Charge

Name	Symbol	Value	Unit
force capacity	f	1040000	J/kg
Specific volume	W_1	0.9484	m^3/kg
detonation heat	Q_V	4407000	J/kg
explosion temperatue	T_V	3071	K

3.3 Covolume

The residual volume is the incompressible volume (m^3/kg) of the gas molecule of 1 kg propellant. The residual capacity of gunpowder is:

$$\alpha = \sum_i y_i \alpha_i . \tag{12}$$

According to experience, we take α= 1 in the calculation.[6]

3.4 Other Relevant Parameters

Referring to the relevant information, other parameters of interior ballistics are obtained as shown in Table 2 below.

Table 2. Other Relevant Parameters of Internal Ballistic

Name	Symbol	Value	Unit
Projectile Quality	m	45.0	kg
caliber	d	0.155	m
density	δ	1.65	g/cm^3
charge	ω	15.6	kg
burning rate coefficient	u	2.3e-10	$m^3/(s\cdot kg)$
Burning Rate Index	n	0.921	
Secondary work coefficient	φ	1.02	
chamber volume	V_0	0.02398	m^3
projectile trip	l_g	7.6	m
Characteristic Quantity of Powder Shape	χ	0.75	
	λ	0.12	
	μ	0	
	χ_s	1,696	
	λ_s	-0.4104	
propellant web size	e_1	2.3	mm

4. Result

By substituting the above values into the program, the data simulation results are shown in Table 3, and the output interior ballistic performance curves are shown in Fig. 1 to Fig. 3.

Table 3. Simulation Results of Interior Ballistic Performance of Artillery

T [s]	p [MPa]	v [$m\cdot s^{-1}$]	T [K]
0.00005	31.07375	0.6244	3468.402

0.00025	35.61718	3.3476	3467.982
0.0053	327.5129	355.8887	3164.382
0.0605	338.2748	459.3889	3075.322
0.0097	165.3699	855.1019	2573.933

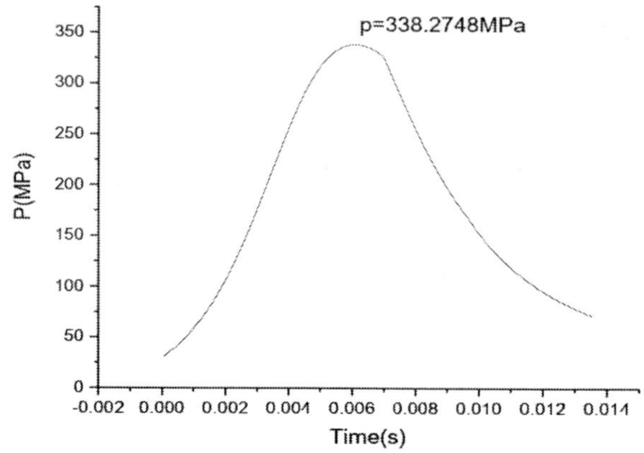

Fig. 1. Simulation Curve for Calculating the Maximum Bore Pressure of Artillery

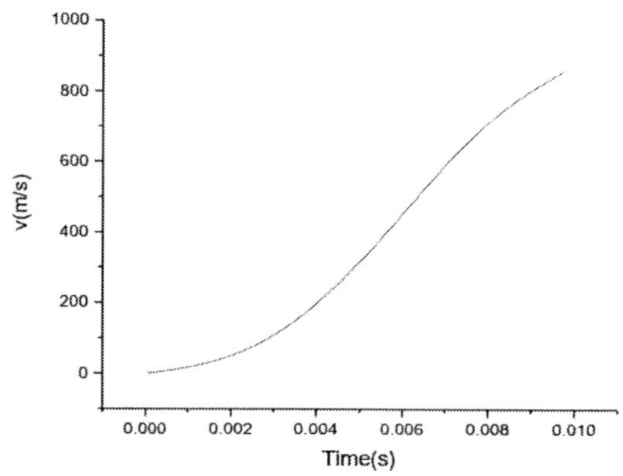

Fig. 2. Muzzle Velocity Calculating Simulation Curve

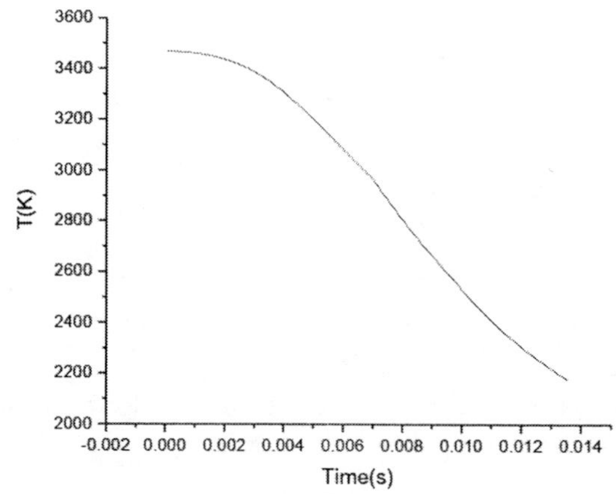

Fig. 3. Simulation Curve of Gun Gas Temperature Calculation

5. Comparisons between calculation results and experiments

In order to verify the correctness of the simulation results, the results are compared with the experimental data. The results are shown in Table 4 below.

Table 4. Comparison of simulation results with experimental results

the Maximum Bore Pressure p_m [Mpa]			Muzzle Velocity v_0 [m·s^{-1}]			detonation tempe Tv [K]		
Calculated	Experiment	Error/%	Calculated	Experiment	Error/%	Calculated	Experiment	Error/%
338.2748	340	0.507	855.1096	458	86.705	3468.402	3071	12.940

6. Conclusions

Comparing the simulation results with the experimental data, it can be found that the fitting results of the maximum bore pressure are better and the error is less than 1%. However, the fitting results of muzzle velocity are much larger than the actual measurement results, which may be due to the friction resistance between projectile launching and internal ballistic in the actual measurement process or the decrease of muzzle velocity due to the increasing pressure and air resistance.

References

[1] LUO Qiao, ZHANG Xiao-bing. Simulation for Launch Process of Ultrahigh Firing Rate Guns Based on Two-way Coupling of FLUENT and Interior Ballistic Model [J]. ACTAAR MAMENTARII, 2016, 37(10). (in Chinese)
[2] LIU Lin, FAN Cheng-jun, HU Jing-kun, et al. Simulation and Research on Multi-parameter According Calculation of Mixed-charge Interior Ballistic [J]. JOURNAL OF GUN LAUNCH & CONTROL, 2014(1):48-51. (in Chinese)
[3] CHENG Lin, BA Shu-hong, JIANG Da-qian, et al. Interior Ballistic Parameters Calculation and Simulation of Large Caliber Gun [J]. Journal of Ordnance Equipment Engineering, 2017(11):75-78. (in Chinese)
[4] JIANG Shu-yuan. Research and Simulation of the Discrete Model of the Large Caliber and High Speed Davis Gun [D]. Nan Jing University of Science & Technology, 2017. (in Chinese)
[5] ZHANG Xiao-bing. Interior Ballistics of Guns [M]. Beijing: Beijing Institute of Technology Press, 2014: 11-80. (in Chinese)
[6] WANG Ze-shan, HE Wei-dong, Xu Fu-ming. Principle and Technique for Gun Propellant Charge Design [M]. Beijing: Beijing Institute of Technology Press, 2014:52-72. (in Chinese)

Study on environment-friendly passivation for hot-dip galvanized steel with high corrosion resistance

Dong Xueqiang[1, a], **Guo Taixiong**[1, b], **Zhouyilin**[2, c]

[1] Pangang Group Research Institute Co., Ltd., State Key Laboratory of Vanadium and Titanium Resources Comprehensive Utilization, Panzhihua 617000, Sichuan, China

[2] PanGang Group Panzhihua Steel & Vanadium Co. Ltd., Panzhihua 617000, Sichuan, China

[a]dongxueqiang123@126.com, [b]guotaixiong@126.com, [c]pzhzyl@163.com

Abstract. In order to improve the corrosion resistance of hot-dip galvanized steel, experimental study developed a new trivalent chromium passivation film with high corrosion resistant. The preparation method of passivation coating with high corrosion resistant was that the CrO_3 was reduced to trivalent chromium by sufficient organic compound with reduction, then added colloidal SiO_2, the matter containing radical phosphate and other substances after reaction completely. The passivation coating did not contain hexavalent chromium, and can meet the requirement of RoHS directive, so the film on hot-dip galvanized steel after passivation was environment-friendly. After coated on the hot-dip galvanized steel, the coating could be fast drying at 100 °C about 10s to 20s. The film was uniform, and did not appear floating, falling and other abnormal phenomenon, so that the passivation coating had good coating performance. The passivation reaction of the new trivalent chromium coating on the hot-dip galvanized steel was compound processes of electrochemical reaction, precipitation reaction and physical deposition. The passivation film produced no corrosion basically after 120h in the neutral salt spray test according to the Chinese national standards GB/T10125 and corroded area is less than 5% after 216 h, thus the corrosion resistance of the passivation film was excellent. In addition, the film has excellent heat resistance and fingerprint resistance, so it can meet the requirements of the user.

1. Introduction

The hot-dip galvanized steel has been widely used in household appliances, automobiles, Building and other industrial fields owing to its excellent mechanical properties. The GI appear corrosion easily in the moist environment, so the hot-dip galvanized steel need to be treated with passivation [1-2]. At present, the passivation coating for the hot-dip galvanized steel contain the trivalent chromium passivation, the hexavalent chromate passivation and the chromate-free passivation. The hexavalent chromate passivation has some advantages, such as simple process, good adhesion, high corrosion resistance, self-repaired ability. But because the hexavalent chromate that is a carcinogen has serious harm to the human body and the environment, it is mainly used for the construction. The toxicity of trivalent chromium is only one percent of hexavalent chromate, the passivation mechanism [3-4] is clear, and the corrosion resistance of the trivalent chromium passivation film after proper closed treatment [5], has been more than six chromium passivation, so the trivalent chromium passivation is mainly used in some field of household appliances. The chromium-free passivation technology [6] has been researched at home and abroad. But no matter how the development of chromate-free passivation, it was unable to reach the standard of chromate passivation. With the corrosion resistance of the hot-dip galvanized steel the user required is more and more high as the expansion of the market, the distance between users and factories increased and the deterioration of the global environment. One user required the corrosion area of the hot-dip galvanized steel in NSST/200h is less than or equal to 10% that the hexavalent chromate and chromium-free passivation were far short of this requirement. Therefore, this paper study a new trivalent chromium passivation coating with high corrosion resistance for the hot-dip galvanized steel, then discuss the surface quality, performance of the coating, and the film-forming mechanism.

2. Experimental

Preparation of passivation coating. First of all, make up the solution contained trivalent chromium. The CrO_3 as oxidant and the organic acid reaction react in acidic solution. In order to make hexavalent chromate completely translate into trivalent chromium, the organic acid must be excess and the reaction time was at least five hours. Secondly, add an appropriate amount of deionized water into the trivalent chromium solution in order to make the content of chromium appropriate, then add SiO_2 colloid, phosphoric substances, surface active agent and so on. At last, stir evenly.

The passivation treatment. The substrate is the oiled the hot-dip galvanized steel without passivation treatment producted by Panzhihua iron and Steel Group Company, that is DX51D+Z and the Specifications for that is 600mm*100mm*0.8mm. The technological process of former processing is: polish (to the edge burr) skim washing dry.

The hot-dip galvanized steel after cleaning was coated by the trivalent chromium passivation coating used the US RSD4# scraping stick, then dryed and cured in the BYK gradient oven about 10s to 20s at the temperature of 100°C.

The new passivation coating did not contain hexavalent chromium, and can meet the requirement of RoHS directive, so the film on hot-dip galvanized steel after passivation was environment-friendly. Do not print page numbers: each sheet toward the middle near the bottom (outside the typing area) should be numbered with a soft pencil.

Performance test.

1) The corrosion resistance of the passivation film. Usually the neutral salt spray tests could be used. Test conditions and methods accorded to the GB/T10125, and the test instrument is the YWX/Q-020 salt fog corrosion test box. In the test, made the NaCl solution with a concentration of 50 g/L and pH value of 6.5 as corrosion material, the temperature was Set in 33~37 °C, specimen was placed with the vertical direction about 15° to 25°. Then spray Continuedly for 120h and 216h, and observated the white rust percentage of the specimen surface.

2) The microstructure and composition of the passivation film. The microstructure of was observed by JSM-5600LV scanning electron microscope (SEM), and the composition of the coating was analyzed by the energy dispersive spectroscopy (EDS).

3) The heat resistance of the passivation film. Bake the sample in the hot blast stove at 300 °Cfor 20 minutes. And measure the color value of the sample by BYK color difference meter before and after baking, then determine the heat resistance according to the value of ΔE derived from the formula $\Delta E = (\Delta L^2 + \Delta a^2 + \Delta b^2)^{1/2}$.

4)The fingerprint resistance of the passivation film. Coat some Vaseline on the specimen surface with a clean soft cloth. And measure the color value of the sample by BYK color difference meter before and after coating, then determine the heat resistance according to the value of ΔE derived from the formula $\Delta E = (\Delta L^2 + \Delta a^2 + \Delta b^2)^{1/2}$.

3. Results and Discussion

3.1 The film-forming mechanism.

The microstructure of the passivation film on the hot-dip galvanized steel was shown in Fig. 1, and the composition is shown in Fig. 2. Thus, the passivation film on the hot-dip galvanized steel is a transparent film, no undesirable phenomena such as flooding and falling, and the film is uniform and compact. The main components of the passivation film is O, Si, P, Cr and Zn that consist with the passivation coating.

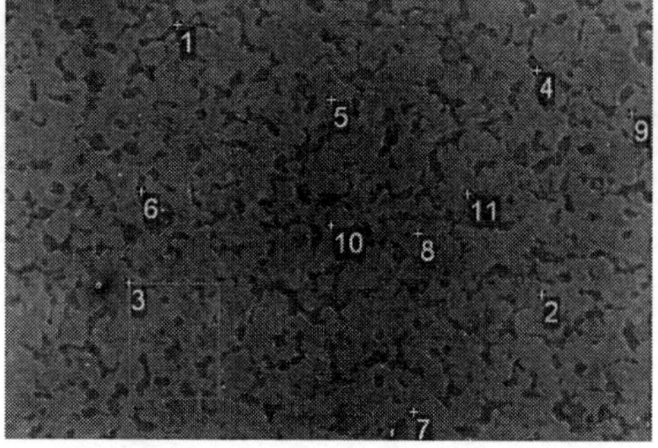

Fig. 1 The microstructure of the passivation film

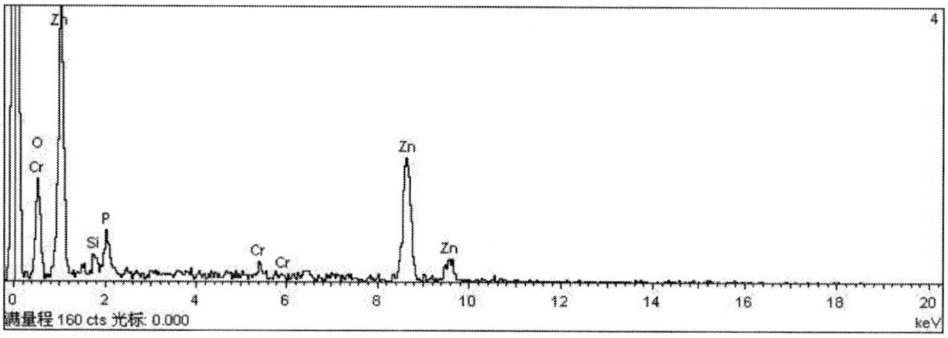

Fig.2 The composition of the passivation film

The surface morphology of the passivation film half of which was soaked in 5%NaOH for 10 min is shown in Fig.3, and the composition is shown in Table 1. The film has been completely dissolved after being soaked by 5% NaOH from Fig.3 and table 1. And the passivating film soaked in water for 1h did not change significantly. So the main component of the film can be dissolved by NaOH solution, but not by water. Color of the passivation film was light green, and the main component of trivalent chromium coating contain Cr^{3+}, H_2PO^{4-}, the organic acid and SiO_2, and the coating is highly acidic, so can infer that the main component of the passivation film is $CrPO_4 \cdot 3H_2O$ and the film may contain a small amount of $CrPO_4$ and $CrPO_4 \cdot 2H_2O$. In addition, the film also contain $Zn_3\ (PO_4)_2$ and $ZnHPO_4$, because the Zn on the hot-dip galvanized steel sheet would react with the trivalent chromium coating.

Fig. 3 The microstructure of the passivation film after being soaked by 5%NaOH

Table1 The composition after being soaked by 5%NaOH%

Area	O	Si	P	Cr	Zn
1	28.34	2.08	6.93	2.29	60.36
2	3.41	-	-	-	96.59
3	4.38	-	-	-	95.62
4	25.31	2.04	6.74	1.91	64.00

It follows that the passivation reaction of trivalent chromium coating on the hot-dip galvanized steel were compound processes of electrochemical reaction, precipitation reaction and physical deposition.

In the passivation coating, there is the ionization equilibrium shown in equation (1).

$$H_2PO^{4-} \Leftrightarrow HPO_4^{2-} + H^+ \Leftrightarrow PO_4^{3-} + 2H^+ \quad (1)$$

In the film-forming process, the first of all, the electrochemical reactions were started, the anodic reaction was shown in equation (2) and cathodic reaction was shown in equation (3). Then the pH value of the interface reaction increased with large consumption of H^+ of the cathode region , and the ionization in equation (1) turn to the right continuously, then PO_4^{3-} increase constantly, so many precipitation reaction take place. The reaction was shown in equation (4), (5), (6).

$$Zn - 2e \rightarrow Zn^{2+} \quad (2)$$

$$2H^+ + 2e \rightarrow H_2 \quad (3)$$

$$Cr^{3+} + PO_4^{3-} \rightarrow CrPO_4 \quad (4)$$

$$3Zn^{2+} + 2PO_4^{3-} \rightarrow Zn_3(PO_4)_2 \quad (5)$$

$$Zn^{2+} + HPO_4^{2-} \rightarrow ZnHPO_4 \quad (6)$$

In the drying and curing process, $CrPO_4$ will continue to lose crystal water, thus $CrPO_4 \cdot 3H_2O$, $Zn_3(PO_4)_2$, $ZnHPO_4$ together constitute the skeleton of the passivating film. Also in the curing process, the SiO_2 occur cross-linking reaction by - OH and - O - bond, and form network structure though the covalent bonds about Si-O-Si and O-Si-O, then cover the fill in the skeleton of the passivating film. Due to sealing effect of SiO_2, the passivation film is very dense.

3.2 The Properties of Passivation film

The corrosion resistance. The passivation film with high corrosion resistance can be combined tightly with the hot-dip galvanized steel due to formation of chemical reactions, so that the passivation film is dense and uniform, which can effectively block the hot-dip galvanized steel react with water and oxygen in the air, thus prevents the hot-dip galvanized steel corrosion occur. The corrosion resistance of passivation film the was shown in Fig. 4. In the Fig. 4, the passivation film had no corrosion basically affer NSST/120h, and the area of corrosion only 2% after NSST/216h. At present, the corrosion resistance of all passivation films is required that the corrosion area in NSST/72h are less than 5%. so this film has excellent corrosion resistance.

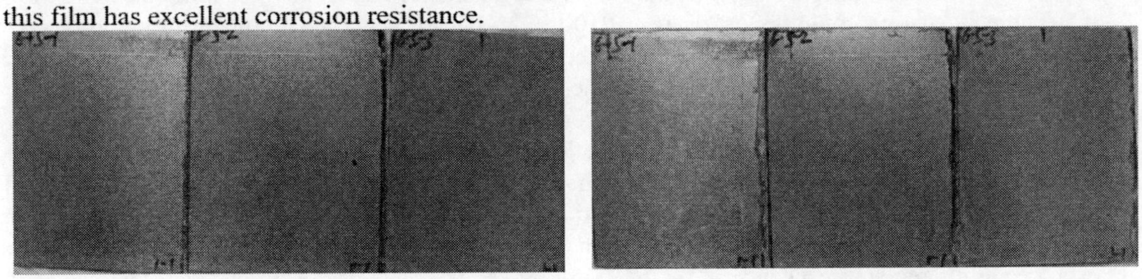

a) NSST/120h b) NSST/216h

Fig. 4 The corrosion resistance

The heat resistance. In order to ensure the film on the hot-dip galvanized steel not come into being color change when single-sided coating and baking, The passivation film must has good heat resistance. Experimental study show that the film has excellent heat resistance when the $\Delta E < 4$. The results are shown in Fig. 5 and Table 2. Thus, the passivating film with high corrosion resistance after baking did not appear obvious discoloration and the ΔE were less than 3. Because this passivating film is mainly inorganic, containing only a small amount of organic components. Therefore, it has very excellent heat resistance.

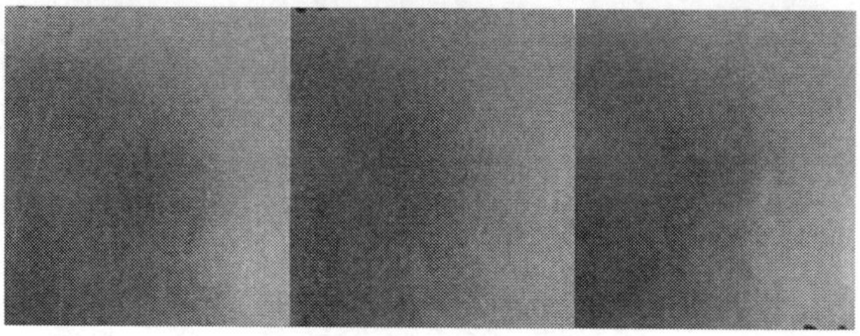

Fig.5 The heat resistance

Table 2 Change of the color value

Samples	ΔL	Δa	Δb	ΔE
1	-1.53	0.29	2.54	2.98
2	-0.78	0.22	2.37	2.50
3	-1.22	0.25	2.23	2.56

The fingerprint resistance. In order to avoid the operator stay fingerprints on the surface of part that can affect coating adhesion or appearance quality in the parts production process, user require the film on the hot-dip galvanized steel has the fingerprint resistance. Experimental research shows the film has excellent fingerprint resistant when $\Delta E < 1$. The result are shown in Table 3. Thus, the film with high corrosion resistance has excellent fingerprint resistance.

Table 3 The fingerprint resistance

Samples	ΔL	Δa	Δb	ΔE
1	0.07	-0.04	-0.06	0.10
2	-0.34	-0.10	-0.01	0.35
3	-0.38	0.04	-0.01	0.38

4. Conclusions

(1) The preparation method of the new trivalent chromium passivation coating was that the CrO_3 was reduced to trivalent chromium by sufficient organic compound with reduction, then added colloidal SiO_2, the matter containing radical phosphate and other substances after reaction completely.

(2) The new coating could be fast drying at 100 °C in 10s to 20s. The film is uniform and compact, and did not appear floating and falling or other abnormal phenomenon. The film can meet the requirement of RoHS directive, so was environment-friendly.

(3) The passivation reaction of the new trivalent chromium coating on the hot-dip galvanized steel was compound processes of electrochemical reaction, precipitation reaction and physical deposition.

(4) The new passivation film on the hot-dip galvanized steel produced no corrosion basically after 120h in the neutral salt spray test and corroded area is 2% after 216 h. So this film has excellent corrosion resistance. In addition, the new film with high corrosion resistance has excellent heat resistance and fingerprint resistance.

References

[1] Cowiesn D R, Scholefield A R. Passivation of Tin-Zinc Alloy Coated Steel[J]. TransIMF, 1985, 63(2): 98.

[2] Robert Berger, A comparative study of the corrosion protective properties of chromium and chromium free passivation methods[J]. Surface & Coatings Technology, 2007, 202.

[3] BELLEZZE T, ROVENTI G, FRATESI R. Electrochemical study on the corrosion resistance of Cr III-based conversion layers on zinc coatings [J]. Surface and Coatings Technology, 2002, 155 (2/3): 221-230.

[4] Tang P T, Bech-Nielsen G, Moller P M. Molybdate Based Alternatives to Chromating as a Passivation Treatment for Zinc[J]. Plat & Surf Fin, 1994, 81(11):20.

[5] A.Baral, R.D. Engelken. Chromium-based Regulations and Greening in Metal Finishing Industries in the USA[J]. Environ. Sci.Policy.2002, 5(2): 121-133.

[6] UPTON P. The effect of the sealers on increase of corrosion resistance of chromate-free passivates on zinc and zinc alloys [J]. Plating and Surface Finishing, 2001, 88 (2): 68-72.

Effect of basalt fiber on chloride ion penetration of Reactive Powder Concrete

ZHAO HaiLong[1,2,a], WANG LuLiang[1,2,b], WANG Dehong[3,c], BAI Junfeng[3,d] and FAN Yunting[3,e]

[1]Electric Power Research Institute, Hainan Power Grid Co.,Ltd., Haikou 570100, China;

[2]Key Laboratory Of Physical And Chemical Analysis Of Hainan Power Grid, Haikou 570100, China;

[3]School of Civil Engineering and Architecture, Northeast Electric Power University, Jilin 132012, China

[a]zhaohl3@hn.csg.cn, [b]93732309@qq.com, [c]hitwdh@126.com, [d]13844647177@163.com, [e]270263869@qq.com

Abstract. In order to investigate the influence of the basalt fiber on the durability of reactive powder concrete (RPC), chloride ion penetration test of nine groups of RPC containing basalt fiber and three groups of RPC without fiber was carried out, carbonization test of one group of basalt fiber RPC was performed. The result showed that, the electrical flux of RPC without basalt fiber varied from104-120C, the chloride ion permeability was very low. The electrical flux of basalt fiber RPC was less than 100C, so the chloride ion permeability can be ignored. When water-to-binder ratio (W/B) was 0.22 and volume fraction of basalt fiber was 0.10%, the chloride ion permeation resistance of specimens was the best. The basalt fiber RPC had good carbonization resistance, and the carbonation depth was zero in 28 days.

1. Introduction

Concrete materials are widely used in civil engineering construction because of their good economics. In the coastal and salty lake areas, the chloride ion content in the environment is high, and chloride ions can invade into the concrete through penetration, diffusion, and electrochemical migration, causing corrosion of the steel bars, volume expansion, and increase of pores and cracks, affecting the structure. Durability, which is one of the main reasons for the failure of reinforced concrete structures in coastal areas. In addition, ordinary concrete has low tensile strength and it is prone to cracking, which is another important reason for its lack of durability. The addition of short fibers to concrete can effectively improve its crack resistance. Reactive Powder Concrete (RPC) is an ultra-high-performance fiber reinforced cementitious composite. For long-term corrosion-resistant engineering structures in salt lakes, saline soils, and marine engineering, chemical fibers with good corrosion resistance may be a better choice for improving concrete durability. Moreover, instead of metal fibers using of chemical fibers can significantly reduce the cost of concrete materials. Yew carried out experimental research on polypropylene fiber-concrete with different contents. Polypropylene fiber can improve the tensile strength, flexural strength and elastic modulus of concrete,

IOP Conf. Series: Materials Science and Engineering **711** (2020) 012067 doi:10.1088/1757-899X/711/1/012067

but it will lead to the drop of a slump. Basalt fiber is a new type of environmentally friendly high-performance fiber with excellent mechanical properties. The tensile strength of basalt fiber is 3000-4800 MPa, which is higher than that of metal fiber such as steel fiber; the elastic modulus is about 80-120 GPa, which is between other chemical fibers such as steel fiber and polypropylene fiber. The addition of basalt fiber can improve the bending strength, toughness and fracture energy of cement-based materials. At present, the research on the properties of basalt fiber concrete is still limited. This paper intended to study the effect of basalt fiber on the chloride ion penetration resistance of RPC and determine the suitable fiber volume fraction.

2. Test design

2.1 Materials and mix proportion.

The basalt fiber reactive powder concrete of this test consisted of the following materials: cement, silica fume, quartz sand, basalt fiber, high-efficiency water reducer and water. The cement was made of P.O 42.5 ordinary Portland cement. The silica content in the silica fume was 82.2%. The quartz sand had a particle size range of 0-1.25 mm. The basalt fiber (BF) was a short-cut basalt fiber with a fiber, which 12 mm length and 13 μm diameter and 80-110GPa elastic modulus. The high-efficiency water reducing agent was a new non-naphthalene-based, high-performance water reducer AN3000 with the water reduction rate of 29%. This experiment mainly studied the effect of W/B and basalt fiber volume fraction on the durability of RPC. considering three different W/B and three basalt fiber volume fraction. At the same time, to compare the improvement effect of basalt fiber on the durability of RPC, the RPC contrast test piece without any fiber was designed. The test mix ratio is shown in Table 1. In the mixed number, A, B, and C respectively represent the two W/B of 0.20, 0.22, and 0.24, and the latter numbers indicate the percentage of basalt volume fraction. For example, "A0.10" indicates that the water-to-binder ratio of the mixture is 0.2. Basalt fiber volume fraction is 0.10%.

Table 1 Mix proportion of basalt fiber RPC anti-chloride ion permeation test specimen

Number	W/B	Fiber length	Cement	Silica fume	Coarse sand	Medium sand	Fine sand	Fiber quality (volume fraction)	Water reducer	Water
A0	0.20	—	762.89	228.87	743.97	374.12	170.89	0	68.66	150.60
A0.05	0.20	12	762.53	228.76	743.62	373.94	170.81	1.2 (0.05%)	68.63	150.51
A0.10	0.20	12	762.16	228.65	743.26	373.76	170.72	2.4 (0.10%)	68.59	150.45
A0.15	0.20	12	761.79	228.54	742.90	373.58	170.64	3.6 (0.15%)	68.56	150.39
B0	0.22	—	756.89	227.07	738.12	371.18	169.54	0	68.12	169.08
B0.05	0.22	12	756.52	226.96	737.76	371.00	169.46	1.2 (0.05%)	68.09	169.02
B0.10	0.22	12	756.16	226.85	737.41	370.82	169.38	2.4 (0.10%)	68.05	168.93
B0.15	0.22	12	755.80	226.74	737.06	370.64	169.30	3.6 (0.15%)	68.02	168.84
C0	0.24	—	750.98	225.29	732.36	368.28	168.22	0	67.59	187.28
C0.05	0.24	12	750.62	225.19	732.00	368.10	168.14	1.2 (0.05%)	67.56	187.19
C0.10	0.24	12	750.26	225.08	731.65	367.93	168.06	2.4 (0.10%)	67.53	187.10
C0.15	0.24	12	749.89	224.97	731.29	367.75	167.98	3.6 (0.15%)	67.49	187.04

2.2 Test methods.

The chloride ion permeation test of the basalt fiber RPC of this test used the electric flux method. The chloride ion penetration test used a cylindrical test piece which had a diameter of 100 mm and a height of 50 mm. The test piece was made by cutting and drilling a core piece of 150 mm × 150 mm × 150 mm to ensure uniformity of the test piece, as shown in Fig. 1.

Fig. 1 Anti-chloride ion permeability test specimen

Before the test, the sides of the cut test piece were sealed with paraffin, and the holes in the coating were filled, and then vacuumed. The vacuum water retention test firstly reduces the absolute pressure in the vessel to 1-5kPa within five minutes, keeps the pressure for 3 hours, and then adds enough distilled water while keeping the vacuum pump running continuously until the test piece is completely submerged by distilled water. The parts were returned to normal pressure after being immersed for 1h, and the test piece was continuously immersed in a water immersion machine for 18h. The prepared test piece was installed in the test tank and clamped, and then the gap between the test piece and the test tank was sealed with paraffin, and the sealing performance between the test piece and the test tank was examined with distilled water.

In the chloride ion permeation test, a constant voltage of 60V was applied to both ends of the test piece, and a test solution on one side of the test piece was a NaCl solution having a mass concentration of 3.0%, and the other side was a NaOH solution having a molar concentration of 0.3mol/L. The test data is automatically collected by the concrete electric flux intelligent measuring instrument, as shown in Figure 1. The initial current reading I_0 was recorded every 30 minutes, and tested for 6h.

3. Results and discussion

Fig. 2 shows the relationship between current and time of different water-binder ratio basalt fiber RPC specimens. It can be seen from the test results in the figure that the current value of basalt fiber RPC increases with time, and the growth rate of current value at the initial stage of the test is higher than that of the later current value. The current value of the active powder concrete is significantly higher than that of the basalt fiber RPC, and the current of RPC grows faster with time, while the current growth rate of the basalt fiber RPC is relatively stable.

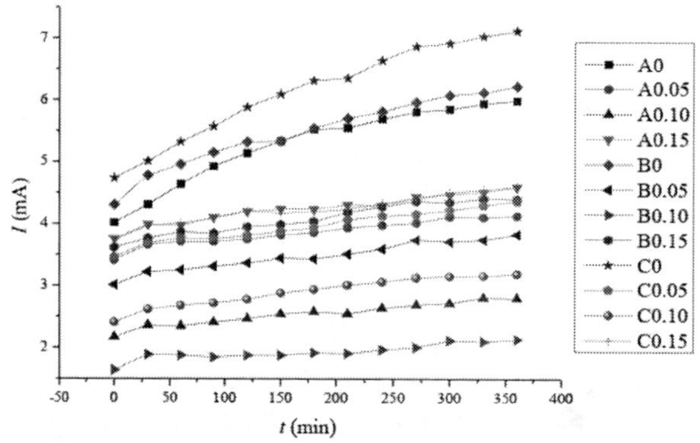

Fig. 2 Current value-time relationship of different W/B basalt fiber RPC

According to the current data, the electric flux value of the basalt fiber RPC specimen can be calculated from the simplified calculation Eq. 1 (as shown in Table 2).

$$Q = 900(I_0 + 2I_{30} + 2I_{60} + \cdots + 2I_t + \cdots + 2I_{300} + 2I_{330} + 2I_{360}) \quad (1)$$

Where Q is the total electrical flux (C) through the test piece, I_0 is the initial current amount (A) of the test piece, I_t is the current (A) at time t (min).

The test piece for the chloride ion penetration resistance of concrete specified in the *Standard for test methods of long-term performance and durability for ordinary concrete* is a test piece with a diameter of 95 mm. Since the test piece used in this paper is a cylindrical test piece with a diameter of 100 mm, it is necessary to test piece. The measured data is converted according to Eq. 2 (as shown in Table 2):

$$Q_s = Q_x \times (95/x)^2 \quad (2)$$

Where Q_s is the electric flux through the 95mm diameter test piece (C), Q_x is the electric flux (C) of the test piece with a diameter of x (mm), x is the actual diameter of the test piece (mm).

Table 2 Test results of electric flux of different basalt fiber content RPC specimens

Number	W/B	Fiber volume fraction (%)	Electric flux (C)	Converted electric flux (C)
A0	0.20	0	114.687	103.505
A0.05	0.20	0.05	83.61	75.458
A0.10	0.20	0.10	55.089	49.718
A0.15	0.20	0.15	91.746	82.801
B0	0.22	0	118.827	107.241
B0.05	0.22	0.05	75.096	67.774
B0.10	0.22	0.10	41.94	37.851
B0.15	0.22	0.15	88.335	79.722
C0	0.24	0	133.056	120.083
C0.05	0.24	0.05	85.761	77.399
C0.10	0.24	0.10	62.874	56.744
C0.15	0.24	0.15	91.539	82.614

It can be seen from the test results that the electric flux of the prime RPC increases with the increase of the water-to-binder ratio, which is because the increase of the water-to-binder ratio leads to the decrease of the RPC compactness, thereby increasing the electric flux. For the basalt fiber RPC of the same volume, when the water-to-binder ratio is 0.22, the electrical flux of the test piece is the lowest. For the RPC with a water-to-binder ratio of 0.2 and 0.22, the electric flux is close to that, and the addition of basalt fiber reduces the workability of RPC, which results in its compactness being affected. High performance concrete also exhibits similar properties. Thus, the basalt fiber RPC having a water-to-binder ratio of 0.2 has a chloride ion permeability lower than that of a basalt fiber RPC having a water-to-binder ratio of 0.22. The water-to-binder ratio of 0.24 fiber RPC has the highest electric flux, which is caused by the decrease in the compactness of RPC.

When the water-to-binder ratio is constant, the electric flux of basalt fiber RPC decreases first and then increases with the increase of fiber volume fraction. When the basalt fiber content changes between 0-0.10%, the anti-chloride ion permeability of basalt fiber RPC gradually increases with the enhance of fiber volume fraction. When the dosage exceeds 0.10%, the anti-chloride ion of basalt fiber RPC the permeability decreases as the fiber volume fraction is increased. The anti-chloride ion permeability of basalt fiber RPC is better than that of RPC. The basalt fiber volume is 0.10%, and the water-to-binder ratio is 0.22. The RPC specimen has the highest chloride ion permeability. The flux value is only 35.3% of the prime RPC.

The chloride ion penetration level of concrete can be evaluated by the electric flux value, and its evaluation index is shown in Table 3.

Table 3 Chloride ion permeability grade of concrete [13]

Conductivity / Coulomb	Chloride ion permeability
> 4000	high
2000-4000	medium
1000-2000	low
100-1000	very low
< 100	ignore

It can be seen from the test results that the basalt fiber RPC has a certain degree of improvement in chloride ion penetration resistance compared with the RPC, and the permeability of the basalt fiber RPC varies with the different fiber volume fraction. The 6h electric flux measured by the anti-chloride ion permeation test of ordinary concrete is usually between 1000C and 1500C, in contrast, the anti-penetration performance of RPC is much better, and its electric flux value not exceeding 150C indicates that its very low chloride ion permeability, because RPC composition contains a variety of cementitious materials, and does not contain coarse aggregates such as stones, so its compactness and crack resistance has been greatly improved, cracks and The pore phase is much less than that of ordinary concrete, so water and chloride ions are less likely to penetrate into the concrete, which in turn greatly improves the chloride ion penetration resistance of RPC. The electric flux of basalt fiber RPC is less than 100C, which can be regarded as negligible in engineering application, and basalt fiber can improve the chloride ion penetration resistance of RPC, which is due to the disordered phase distribution of basalt fiber in RPC. By bridging the internal cracks, preventing the initial crack development, reducing the number of connected cracks, resisting the infiltration of chloride ions, thus further improving the compactness of RPC, improving the internal structure of RPC, and thus improving its resistance to chloride ion penetration ability. This also shows that basalt fiber plays a certain role in improving the durability of RPC.

4. Summary

In this paper, the anti-chloride ion penetration test of basalt fiber RPC was carried out, and the effects of different water-to-binder ratio and basalt fiber content on the anti-penetration performance of RPC were determined. The results show that the electrical flux of the nine groups of basalt fiber RPC specimens designed in this paper are all lower than 100C, and the chloride ion permeability is negligible. It can be used in the corrosive environment such as saline soil, coastal and deicing salt. When the water-to-binder ratio is 0.22 and the basalt fiber volume fraction is 0.10%, the basalt fiber RPC has the best chloride ion penetration resistance.

References

[1] Afroughsabet V, Biolzi L, Ozbakkaloglu T. High-performance fiber-reinforced concrete: a review[J]. Journal of Materials Science, 2016, 51(14): 6517-6551.
[2] Yuan Y, Ji Y, Shah S P. Comparison of two accelerated corrosion techniques for concrete structures[J]. Aci Structural Journal (2007), 104(3): 344-347.
[3] Wang D, Ju Y, Shen H. Crack Resistance Properties of HPFRC Beam-Column Joints under Cyclic Load[J]. Advances in Materials Science & Engineering (2019): 1-11.
[4] Shaikh F U A. Review of mechanical properties of short fibre reinforced geopolymer composites[J]. Construction and Building Materials (2013): 37-49.
[5] Alberti M G, Enfedaque A, Gálvez J C. Fibre reinforced concrete with a combination of polyolefin and steel-hooked fibres[J]. Composite Structures (2017), 171: 317-325.
[6] Habel K, Viviani M, Denarié E, Brühwiler E. Development of the mechanical properties of an Ultra-High Performance Fiber Reinforced Concrete (UHPFRC)[J]. Cement and Concrete

Research (2006), 36(7): 1362-1370.

[7] Roux N. Experimental Study of Durability of Reactive Powder Concretes[J]. Journal of Materials in Civil Engineering (1996), 8(1): 1-6.

[8] Yew M K, Mahmud H B, Shafigh P, Ang B C, Yew M C. Effects of polypropylene twisted bundle fibers on the mechanical properties of high-strength oil palm shell lightweight concrete[J]. Materials and Structures (2015), 49(4): 1221-1233.

[9] Yew M K, Mahmud H B, Ang B C, Yew M C. Influence of different types of polypropylene fibre on the mechanical properties of high-strength oil palm shell lightweight concrete[J]. Construction and Building Materials (2015), 90: 36-43.

[10] Branston J, Das S, Kenno S Y, Taylor C. Mechanical behaviour of basalt fibre reinforced concrete[J]. Construction and Building Materials (2016), 124: 878-886.

[11] Wang D, Ju Y, Shen H, Xu L. Mechanical properties of high performance concrete reinforced with basalt fiber and polypropylene fiber[J]. Construction and Building Materials (2019), 197: 464-473.

[12] Ralegaonkar R, Gavali H, Aswath P, Abolmaali S. Application of chopped basalt fibers in reinforced mortar: A review[J]. Construction and Building Materials (2018), 164: 589-602.

[13] Yao Y, Wang L, Tian P. High performance concrete[M]. Beijing: Chemical Industry Press, (2006).

[14] Cong X H, Cong X Q, Zhang Y D, Bai J Q. Experimental study on the resistance of chloride concrete to chloride ion penetration by electric flux method[J]. Journal of Hebei Institute of Civil Engineering and Architecture (2017), 35(01): 48-51.

Study on the Film of Ti-Si-C-O Oxides and Its Application in Memristor

Xin Zhao[a], Yuhan Yuan[b], Cheng Liu[c], Dongyang Li[d], Wei Li[e*]

School of Optoelectronic Science and Engineering, University of Electronic Science and Technology of China, Chengdu 610054, China

[a]xinzhao0314@qq.com, [b]yhyuan94@qq.com, [c]18200151542@163.com, [d]open_please@163.com, [e*]wli@uestc.edu.cn ([*]corresponding author)

Abstract. The thin films of Ti-Si-C-O oxides are fabricated via direct-current magnetron co-sputtering using a Ti_3SiC_2 target with oxygen (O_2). The X-ray photoelectron spectroscopy is used to verified the composition of the Ti-Si-C-O thin films. A memristor is designed as the structure of Pt/TSCO/ITO/glass, and the *I-V* curves as well as the switching behaviors are studied for the first time. The results indicated that there appear typical pinched hysteresis *I-V* loops and continuous resistance changes under positive and negative sweeping bias, and that the device conductance can be modulated by consecutive potentiating and depressing programming. The carrier transport mechanism of the device is also investigated and it is found that the Ohmic connection is the leading factor in the linear low electric field region while Schottky emission in the non-liner high electric field region.

1. Introduction

As early in 1971, Prof. Chua at the University of California pointed out from the perspective of symmetry that besides resistors, capacitors and inductors, there is a missing circuit element, namely memristor [1]. In 2008, Strukov etc. at HP Labs published a paper named *the missing memristor found* in Nature and have realized a memristor predicted by Chua 30 years ago based on the $Pt/TiO_2/Pt$ structure, and observed the resistance switching behavior in this device, which signalled a new stage in the progress of memristors [2]. With the development of memristors, a variety of materials have been used in these *missing* devices which show obvious differences in the memristive behaviors like ON/OFF ratio, retention time and endurance characteristics, etc. Depending on the type and characteristics, the materials can be divided into the following categories: binary oxides [3-5], perovskite-type complex oxides [6], solid electrolyte [7], organic chemicals [8], silicon-based semiconductors [9], etc., in which resistance materials have become increasingly popular because of its promise for use in many emerging applications such as neuromorphic devices, photovoltaic device, bioelectronic devices, and memristors[10, 11].

As a ternary layered compound with excellent metallic and ceramic properties, polycrystalline Ti_3SiC_2 alloy has better thermal, electrical and mechanical properties [12]. However, many researches on Ti_3SiC_2 are mainly focused on the preparation of bulk or powder materials, and the mechanical or tribological properties as well as comprehensive performance improvements[13, 14]. Up to now, there are few reports that use Ti_3SiC_2 as memristive materials.

In this paper, we report the fabrication of Ti-Si-C-O (TSCO) oxides films by direct-current (DC) magnetron co-sputtering using a Ti_3SiC_2 target and verify its composition by using of the X-ray photoelectron spectroscopy. A memristor is designed as Pt/TSCO/ITO/glass structure shown as in Fig. 2(a), and the *I-V* curves, voltage sweeps and response currents of the memristor are studied to characterize the memristive behavior. The effect of the compliance current I_{cc} on the set voltage V_{set} and the carrier transport mechanism of the device are also studied.

2. Material and methods

As shown in Fig. 2(a), the memristor was designed as Pt/TSCO/ITO/glass structure. The glass substrates (15×15 mm^2) with a 110-nm-thin ITO film were cleaned in a sonication bath with acetone and ethylalcohol for 15 minutes respectively, and the ITO film with a resistivity of about 6.5 Ω/□ was taken as the bottom electrode. A TSCO thin film was fabricated on the above substrates via direct-current (DC) magnetron co-sputtering using a Ti_3SiC_2 target in an atmosphere of argon (Ar) and oxygen (O_2). The chamber was pumped to a base pressure of 7×10^{-4} Pa, and the current was set to 0.3 A. Before sputtering, a flux of O_2 and Ar at the ratio of 1:20 was mixed. Argon flow served as working gas was kept at 15 sccm and the mixed gas was kept at 20 sccm to keep the chamber pressure at 0.5 Pa. The TE patterns were made through a photolithography and lift-off technique where the 100-nm thin metal layer was deposited using DC magnetron sputtering under vacuum environment of 8×10^{-4} Pa. The contact area of the individual device was about 200 μm.

All electric measurements were performed based on a Keithley 2636B apparatus and a probe system. During measuring, the voltage was applied on top Pt electrode and the bottom ITO electrode was grounded.

3. Results and discussion

The XPS measurement is carried out to verify the composition of the TSCO thin films. In order to eliminate the effect of surface pollutants, all spectra are calibrated with C1s peak at the binding energy of 284.6 eV. Fig. 1(a) shows the full survey spectrum of one TSCO thin film, in which Ti, Si, C and O elements can be observed as we expected, indicating that oxygen has already oxidated with Ti, Si or C in the film during magnetron sputtering. Fig. 1(b), Fig. 1(c), Fig.1(d) and Fig. 1(e) shows the spectrum of Ti2p, Si2p, C1s and O1s, differently. The Ti2p peaks are perceived at the binding energies of 458.31 eV and 464.08 eV (Fig. 1(b)). These two peaks are $Ti2p_{1/2}$ and $Ti2p_{3/2}$, and the aperture between them is 5.77 eV. Compared with the standard XPS spectra, Ti mainly exists in the as-deposited TSCO thin film in the form of +4 valence. The Si2p peaks locate at 101.3 eV and 101.8eV (Fig. 1(c)). The C1s peaks are positioned at 284.1 eV and 288.2eV (Fig. 1(d)), and the O1s peaks are positioned at 529.8eV and 531.3eV (Fig. 1(e)). O mainly exists in the form of Ti-O bonds and Si-O bonds. Si always combines with C or O and generates SiO, SiC and SiO_2. The possible chemical components in TSCO thin film are C, TiC, SiC, SiO_2 and TiO_2.

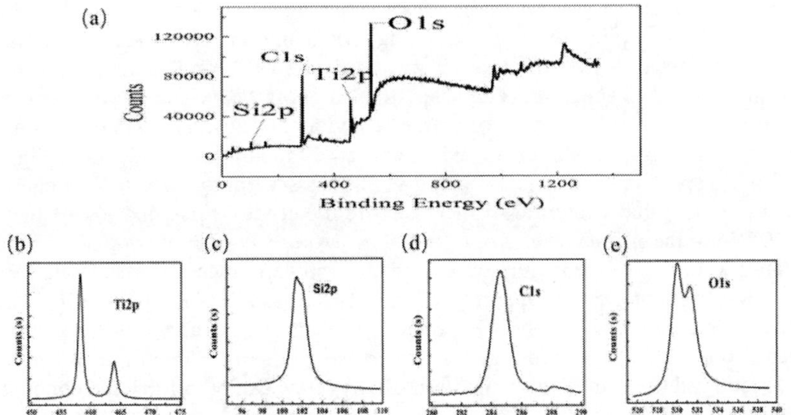

Fig. 1 Full XPS survey (a) and Ti2p (b), Si2p (c), C1s (d) and O1s (e) individual XPS spectrum in a TSCO thin film.

As shown in Fig. 2(b), the first current-voltage (*I-V*) characteristic of the Pt/TSCO/ITO/glass memristor is a pinched hysteresis loop, which is a typical feature of memristor device, exhibiting a characteristic pinched hysteresis loop at the 1st and 3rd quadrant of the *I-V* plane. The device exhibits standard nonlinear pinched hysteresis, which is consistent with the reported scanning curves of memristors [15]. The sweeping voltage from 0 V to 6 V and 0 V to – 6 V is used for the first voltage scan. In addition, consecutive positive and negative DC-sweeps are also used to study the *I-V* characteristic, as shown in Fig. 2(c) and Fig. 2(d). When the device is loaded first with 5 consecutive positive voltages (0 V to +8 V) sweeps, the amplitude of the current increases gradually (Fig. 2(c)). Also, when 5 consecutive negative voltages (0 V to -8.0 V) sweeps are loaded to the device next, the current value gradually decreases, and depression occurs (Fig. 4(d)). We take the measured current as an indirect expression of the device conductivity with the scanning voltage. It is shown that the conductivity of the device decreases gradually with the decrease of forward scanning voltage under DC voltage scanning. It is a typical feature for memristor device and also is crucial for emulating biological synapse functions in the electronic synapse.

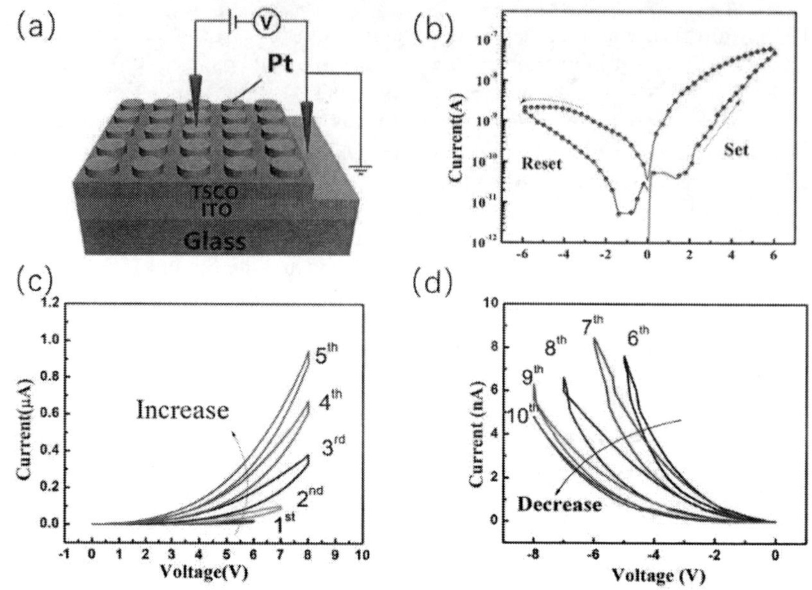

Fig. 2 The structure of the Pt/TSCO/ITO device (a), typical *I–V* curves (b), 5 positive DC-sweeps (0 V to 8 V) (c) and 5 negative DC-sweeps (0 V to -8 V) (d).

The resistive behavior of a memristor is mainly reflected in its *I-V* curve, and the effect of the compliance current (I_{cc}) on the set voltage V_{set} can be studied by means of *I-V* curve as Fig. 3. There are two different modes of resistive switching: bipolar resistive switching (BPS) and unipolar resistive switching (UPS). As in earlier studies, the asymmetric MIM structure always performs as BPS. But in fact, a lot of devices with symmetric MIM structure also show the characteristics of the BPS. Generally, a memristor requires an forming process to establish a subsequent switching behavior. During the process a compliance current is needed to protect the device from being brokendown caused by the high current. As in Fig. 3, the I_{cc} is set as 1 μA, 2 μA and 4 μA respectively and the sweeping voltage is set from 0 V to 6 V. It can be seen from Fig. 3(a), Fig. 3(b) and Fig. 3(c) that when the I_{cc} increases from 1 μA to 2 μA, the V_{set} of the device decreases from 2.6 V to 2 V. While the I_{cc} increases from 2 μA to 4 μA, the V_{set} increases from 2 V to 4 V. Under the same scanning voltage, the V_{set} of the memristor decreases firstly and then increases with the growth of I_{cc} . It can be interpreted that the free electron quantity of the TSCO-based memristor is affected by I_{cc} under the forward bias. With the increase of I_{cc}, there are enough free electrons which would be generated in the TSCO film to fill the traps, so the V_{set} of the memristor will decrease gradually at this time. When the I_{cc} is increased, a higher voltage is required to promote the migration of oxygen vacancy to form the conductive filaments, so the V_{set} of the memristor will increase.

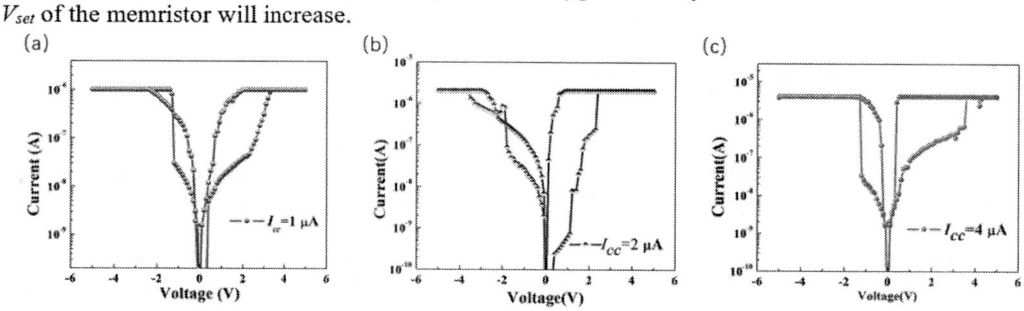

Fig. 3 *I–V* curves under different compliance current: (a) I_{cc}=1 μA; (b) I_{cc} =2 μA and (c) I_{cc} =4 μA.

In order to identify the carrier transport mechanisms of the TSCO-based memristor, the nonlinear (high electric field) region of the *I-V* curve is plotted in log-log scale (Fig. 4(a)), while the low electric field section is fitted linearly which exhibits an Ohmic conduction mechanism (Ohmic slope = 1). The nonlinear region of the *I-V* curves is generally categorized by three different carrier transport mechanisms: Schottky emission (SE), Polle-Frenke (P-F) and Space Charge Limited Current (SCLC) [16]. The SE mechanism occurs when electrons obtain sufficient thermal activation energy to pass over the energy barrier at the metal oxide interface [17]. SE is the most common carrier transport mechanism in metal oxides [18]. The Polle-Frenkel (P-F) emission is very similar to SE, the thermally activated electrons may jump over from trap level into metal-oxide conduction bands [18- 20]. In P-F emission, the barrier height varies with the change of trap levels. The SCLC mechanism can be explained by the trapping and de-trapping process of the charge carriers. The non-linear part for I–$V^{½}$, (I/V)–$V^{½}$ and I–V^{2} are re-plotted respectively, as the reported before [21] (Fig. 4(b), (c) and (d)). The Linear fitting lines plotted and the fitting equations of each graph are calculated. For SE, P-F and SCLC, the slope of the fitting equation is 16.6, -5 and 0.19, respectively. The highest linearity is SE, followed by P-F and SCLC. Obviously, SE is the dominant carrier transport mechanism in the non-linear high electric field region. The weaker linear dependency of the P-F and the SCLC mechanisms are poorly represented. As a result, the linear region of the *I-V* curve points to Ohmic conduction, while the non-linear region of the *I-V* curve of the TSCO-based memristor device demonstrates an SE-type conduction mechanism.

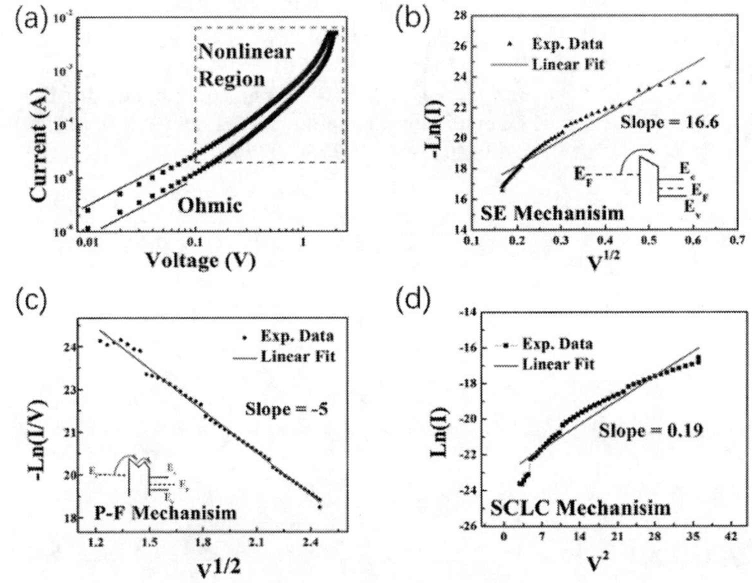

Fig. 4 *I-V* plots of Pt/TSCO/ITO memristor: (a) log *I*-log *V*, (b) ln I-$V^{½}$ for SE, (c) ln (I/V)-$V^{½}$ for P-F and (d) I-V^{2} for SCLC.

4. Summary

In this paper, the TSCO thin films are fabricated by direct-current (DC) magnetron co-sputtering using a Ti_3SiC_2 target in the atmosphere of argon (Ar) and oxygen (O_2). The composition of the TSCO thin film is verified by means of X-ray photoelectron spectroscopy (XPS). The TSCO thin film is used as memristive layer and the TSCO-based memristor is fabricated as Pt/TSCO/ITO/glass structure for the first time. The *I-V* curves, voltage sweeps and response currents of the memristors are studied to characterize the switching behavior of the device. The results show that there appear pinched hysteresis *I-V* loops in the TSCO-based memristor, and continuous resistance changes could be obtained under positive and negative sweeping bias, indicating that the TSCO-based memristor could be used as an electronic synapse. With the increase of compliance current I_{cc}, the set voltage V_{set} decreases first and then increases. It is found that Ohmic mechanism is the dominant factor in the linear low electric field region while Schottky emission (SE) is the dominant factor in the non-liner high electric field region.

Acknowledgment
This work was partially sponsored by the National Natural Science Foundation of China (Grant No. 61421002).

References

[1] Chua, L: IEEE. T. Circuits. SYST Vol. 18 (1971), p. 507-519
[2] Strukov, Dmitri B., Snider, Gregory S., Stewart, Duncan R.and Williams, R. Stanley: Nature Vol. 459 (2009), p. 1154-1154

[3] B. Gao, B. Sun, H. Zhang, Lifeng Liu et al: IEEE Electr. Device L. Vol. 30 (2009), p. 1326-1328
[4] Y. Wang, Q. Liu, S. Long, et al: Nanotechnology Vol. 21 (2010), p. 045202
[5] Y. Yang, P. Gao, S. Gaba, et al: Nature Commun. Vol. 3 (2012), p. 732
[6] X. G. Chen, X. B. Ma, Y. B. Yang, et al: Appl. Phys. Lett. Vol. 98 (2011), p. 122102-122102-3
[7] Z. Xu, Y. Bando, W. Wang, et al: ACS Nano Vol. 4 (2010) 2010, p. 2515-2522
[8] Y. S. Lai, C. H. Tu, D. L. Kwong, et al: Appl. Phys. Lett. Vol. 87 (2005), p. 5655-5660.
[9] S. H. Jo, W. Lu.: Nano Lett. Vol. 8 (2008), p. 392-397.
[10] Amouzadeh Tabrizi M, Shamsipur M, Saber R, et al: Biosen. Bioelectron. Vol. 98 (2017), p.113-118.
[11] Kawamura K, Tsuchiya T, Takayanagi M, et al: Jpn. J. Appl. Phys. Vol. 56 (2017), p. 06GH01
[12] N. Tzenov, M.W. Barsoum, T. El-Raghy.: J. Eur. Ceram. Soc. Vol. 20 (2000), p. 801-806
[13] T. El-Raghy, M.W. Barsoum: J. Am. Ceram. Soc. Vol. 82 (1999), p. 2855-2860
[14] T. El-Raghy, A. Zavaliangos, M. Barsoum, et, al: J. Am. Ceram. Soc. Vol. 80 (1997), p. 513-516
[15] Jo S H, Chang, Ting: Nano Lett. Vol. 10 (2010), p. 1297-1301
[16] Lampert, Murray A.: Phys. Rev. Vol. 103 (1956), p. 1648-1656
[17] Dementjev A P, Ivanova O P, Vasilyev L A, et al: J. Vac. Sci. Technol. A. Vol. 12 (1994), p. 423-427
[18] Lim E W, Ismail R.: Electronics, Vol. 4 (2015), p. 586-613
[19] SZE, Simon M.: *Physics Of Semiconductor Devices* (Physics of Semiconductor Devices, 2014)
[20] Joshua Yang J, Miao F, Pickett M D, et al: Nanotechnology, Vol. 20 (2009), p. 215201
[21] Gul F, Efeoglu H.: Ceram. Int. Vol. 43 (2017), p. 10770-10775

Experimental Study of Transmission Tower Damage Detection Using Time Series Model

Huiyong Guo[1, a], Zhihua Wang [1]

[1] School of Civil Engineering, Chongqing University, Chongqing 400045, P. R. China

[a] ghy267@hotmail.com

Abstract. In order to solve damage detection problem of transmission tower, a damage index based on inter-storey stiffness and cepstral distance is presented. First, transmission tower model is established and the tower is divided into six sub-structures. Thus, we only need to identify the sub-structure location where the damaged rods are included. Then, autoregressive model is introduced, the cepstral distance between autoregressive models under damaged and undamaged states is given, and a damage index based on inter-storey stiffness and cepstral distance is proposed. Finally, the experiment of transmission tower model is conducted to evaluate the effectiveness of detecting the damaged layer of transmission tower. The analysis results of the experimental signals show that the damage index based on inter-storey stiffness and cepstral distance is an available method for damage detection of transmission tower, and the damage feature can be well extracted with the damage index.

1. Introduction

In the process of developing ultra high voltage (UHV) transmission line, safety monitoring of transmission tower has become more and more important for electricity transmission system. Structural damage identification testing based on dynamic response is an available method for safety monitoring of transmission tower. The dynamic response of acquisition structure after driving is the focus of the damage identification method with wide use [1]. One of the reasons why this method is widely used is that it is intertwined with many technologies, such as sensing technology, signal acquisition, data processing, information management, data communication and so on. Many researchers used dynamic response to detect structural damage [2].

Structural dynamic response can be analyzed by using time series model. Time series model possesses these advantages, such as simplicity of modeling and sensitivity to conditional variation, which can effectively reflect some intrinsic characteristics of structure. Time series model has been widely used in feature exaction to realize structural damage detection. AR, ARX and ARMA models have been utilized to extract damage features of simple structures like shear structure [3-5]. However, transmission tower is a three-dimensional complex structure, which will bring challenges to the damage detection. Therefore, in this paper a novel damage index based on AR model and cepstral distance is proposed to identify damage location. An experiment study is also conducted to realize damage detection via practical vibration signal such as acceleration dynamic response.

2. Transmission Tower Model

The transmission tower model under investigation is a steel tower, which is shown in Fig. 1. The main frame of the transmission tower model is composed of four circular steel tubes with a certain angle to the horizontal platform. The interface of each layer of the structure consists of four solid circular pipe crossbars, which are welded with circular steel tubes through steel plane. The crossover braces are composed of solid square aluminum bars, which are bolted to circular steel tubes. The experimental model has a height of 2200 mm and a base width of 600 mm. The transmission tower model can be divided into 6 sub-structures, which mean that we can just identify the location of damaged sub-structure. Thus, time series models become available method for damage detection of three-dimensional structure. Structural response should be acquired to establish the time series models. The basic dynamical equation of transmission tower can be written as

$$M\ddot{x}+C\dot{x}+Kx=F \tag{1}$$

where, M, C and K are structural mass matrix, damping matrix and stiffness matrix, respectively. F is driving force. $\ddot{x}$, $\dot{x}$ and x are acceleration, velocity and displacement, respectively. Thus, we can obtain the acceleration response and then utilize the response to detect damaged sub-structure location via numerical simulated response signal or practical measured signal.

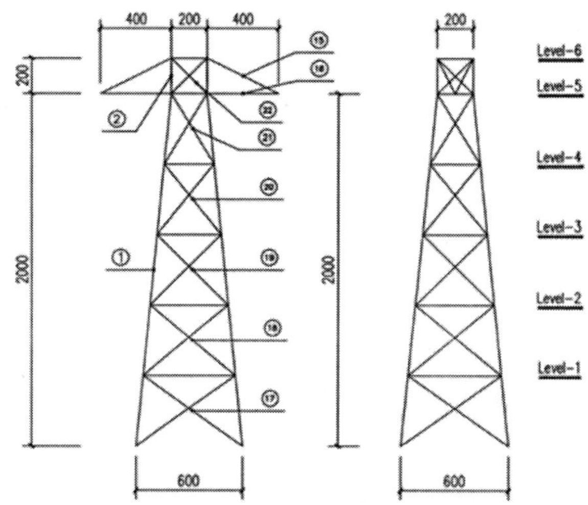

Fig. 1 Transmission tower model size

3. Time Series Model and Its Damage Index

Autoregressive Model. Autoregressive(AR) model is a simple model to simulate acceleration time domain response. A typical AR(p) model can be written as

$$y_t=c+\sum_{i=1}^{p}\varphi_i y_{t-i}+\varepsilon_t \tag{2}$$

in which y_t is the time series data, c is a constant, p is the order of autoregressive model, φ_i is autoregressive coefficient, ε_t is the error. In general, the error terms should meet the homoscedastic assumption.

Cepstral Distance. Cepstral distance[6] of AR model can be obtained through autoregressive moving-average (ARMA) model. it can be simply described as follows:

An autoregressive moving-average model is given by

$$y_t=c+\sum_{i=1}^{p}\varphi_i y_{t-i}+\sum_{i=1}^{q}\Phi_i\varepsilon_{t-i} \tag{3}$$

in which φ_i and Φ_i are the model coefficients, p and q are the model orders, and ε_t is the error. The system function of ARMA model in the Z domain is given by

$$H(z)=\frac{\sum_{i=0}^{q}\Phi_j z^{-i}}{\sum_{i=0}^{p}\varphi_i z^{-i}}=\frac{\prod_{i=1}^{q}\left(1-\beta_i z^{-1}\right)}{\prod_{i=1}^{p}\left(1-\alpha_i z^{-1}\right)} \tag{4}$$

in which α_i is the pole of ARMA model and β_i is the zero of ARMA model. The cepstrum of transfer function $H(z)$ is the inverse Fourier transform of the logarithm of its power spectrum $P(z)$ [7]:

$$\log P(z)=\log\left[H(z)\bar{H}(1/z)\right]=\sum_{n\in z}C_n z^{-n} \tag{5}$$

in which C_n is cepstrum parameter. Substituting equation (4) into (5), equation (5) becomes [7]

$$\log P(z)=-\sum_{i=1}^{p}\log\left|z-\alpha_i\right|^2+\sum_{i=1}^{q}\log\left|z-\beta_i\right|^2+\log\sigma^2 \tag{6}$$

Then, the power cepstrum can be written according to the poles and zeroes as follows:

$$C_n=\begin{cases}\dfrac{1}{|n|}\left[\sum_{i=1}^{p}\alpha_i^{|n|}-\sum_{i=1}^{q}\beta_i^{|n|}\right] & n\neq 0\\ \log\sigma^2 & n=0\end{cases} \tag{7}$$

We assume that the cepstrum coefficient of model M is C_n and the cepstrum coefficient of model M' is C_n'. Thus, the cepstral distance can be defined as follows:

$$d(M,M')=\left[\sum_{n=1}^{\infty}n\left|C_n-C_n'\right|^2\right]^{1/2} \tag{8}$$

Due to the fact that the cepstral distance d is a Euclidean distance, the ARMA models have the following property

$$d(MM'',M'M'')=d(M,M') \tag{9}$$

The system functions of ARMA models M and M' can be expressed as $M=B/A$ and $M'=B'/A'$, respectively. In addition, the third ARMA model M'' is the transfer function $M''=1/BB'$. Thus, the corresponding AR models N and N' with system functions can be formed as follows:

$$N=MM''=\frac{1}{AB'} \tag{10}$$

$$N'=M'M''=\frac{1}{A'B} \tag{11}$$

Depending on the property (9), we have

$$d(M,M')=d(N,N') \tag{12}$$

Thus, we can find that it is sufficient to just consider AR models N and N'. For AR models N and N' with the order p and p', along with the poles α and α', the cepstral distance of AR models can be given by

$$d(M,M')=\left[\log\frac{\prod_{i=1}^{p}\prod_{j=1}^{p}\left(1-\alpha_i\bar{\alpha}_j'\right)\prod_{i=1}^{p}\prod_{j=1}^{p}\left(1-\alpha_i'\bar{\alpha}_j\right)}{\prod_{i=1}^{p}\prod_{j=1}^{p}\left(1-\alpha_i\bar{\alpha}_j\right)\prod_{i=1}^{p}\prod_{j=1}^{p}\left(1-\alpha_i'\bar{\alpha}_j'\right)}\right]^{1/2} \tag{13}$$

Damage Index. For an n storey structure the normalized cepstral distance index of the j storey can be written as

IOP Conf. Series: Materials Science and Engineering **711** (2020) 012069 doi:10.1088/1757-899X/711/1/012069

$$V_j = \frac{d_j(M,M')}{\sum_{j=1}^{n} d_j(M,M')} \tag{14}$$

Generally, the cepstral distance index can just identify the change of degrees of freedom, which cannot directly find the damage site. So, a damage index based on inter-storey stiffness and cepstral distance is proposed. The index is given by

$$C_j = \frac{V_{j-1} + V_j}{2 - V_n} \tag{15}$$

If j=1, then $V_{j-1}=V_0=0$. The damage index based on inter-storey stiffness can directly detect damage site. (1)

4. Experimental Study

Experiment is utilized to validate the proposed damage detection method. The experimental model is shown in Fig. 2. Load is applied by means of a shaking table to the base floor. The system is instrumented with six accelerometers mounted at the node of each floor along the vibration direction. The acceleration response data at all floors, including the first floor have been are measured and recorded. Damage can be simulated through removing the crossover bars. There are six damaged cases, which are listed in Table 1.

Fig. 2 Transmission tower model

Table 1 damaged cases of transmission tower model

Cases	Damaged	The number of removed
Case 1	The 2nd layer	2
Case 2	The 2nd layer	4
Case 3	The 3rd layer	2
Case 4	The 3rd layer	4
Case 5	The 5th layer	2
Case 6	The 5th layer	4

In general, we can utilize Akaike information criterion to estimate the order of AR model. Here, the order p of AR model is 35. The AR models under undamaged and damaged states are utilized to establish cepstral distance. Here, the procedure of damage identification was introduced in Section 2. The measured acceleration responses under undamaged and damaged states are applied to damage detection of transmission tower. For the six damaged cases, the damage index based on time series model and cepstral distance are used to find damaged layer or sub-structure.

For Cases 1 and 2, two and four crossover bars of the second layer are removed, respectively. The damage index based on inter-storey stiffness and cepstral distance is applied to damage detection. The detection results of the experiment are depicted in Figs. 3 and 4 with the layer number plotted against the damage index value. The higher the index value is, the greater the damage probability becomes. From Figs. 3 and 4, we can see that the damage value of the second layer is obviously higher than those of the other five layers, which means the damage index based on inter-storey stiffness and cepstral distance can well find damaged site.

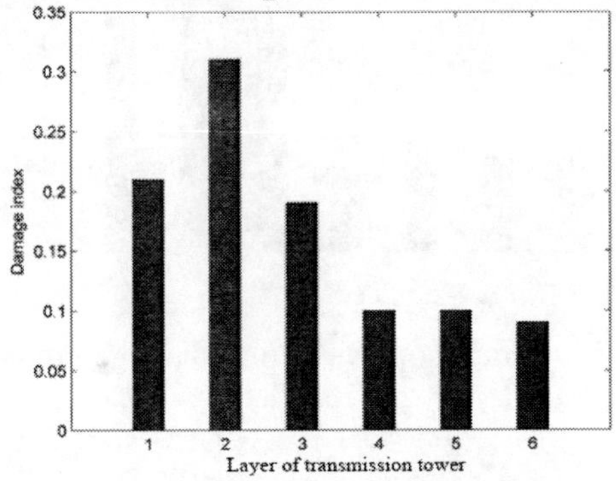

Fig. 3 Detection results of two damaged rods in the second layer

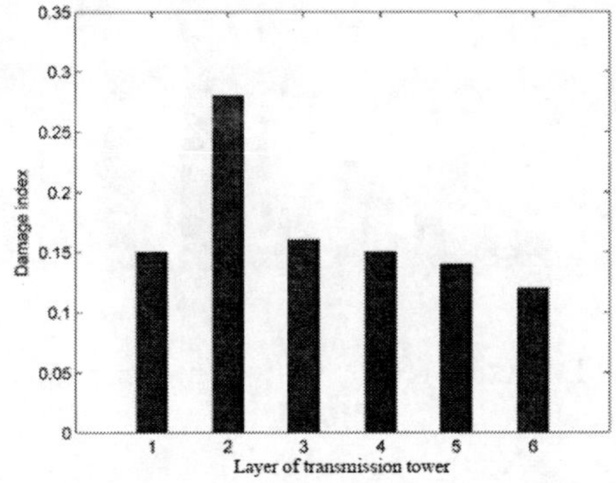

Fig. 4 Detection results of four damaged rods in the second layer

For Cases 3 and 4, two and four crossover bars of the third layer are removed, respectively. The damage index based on inter-storey stiffness and cepstral distance is applied to damage detection. The detection results of the experiment are depicted in Figs. 5 and 6 with the layer number plotted against the damage index value. From Figs. 5 and 6, we can observe that the damage value of the third layer is greater than those of the other five layers. Therefore, the proposed index has the potential to detect the damaged site of transmission tower.

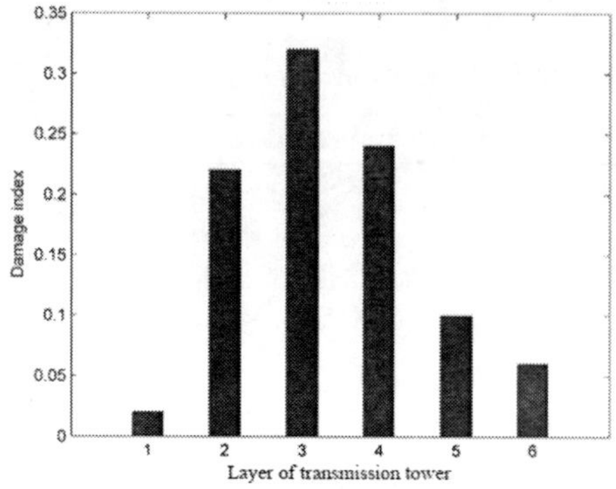

Fig. 5 Detection results of two damaged rods in the third layer

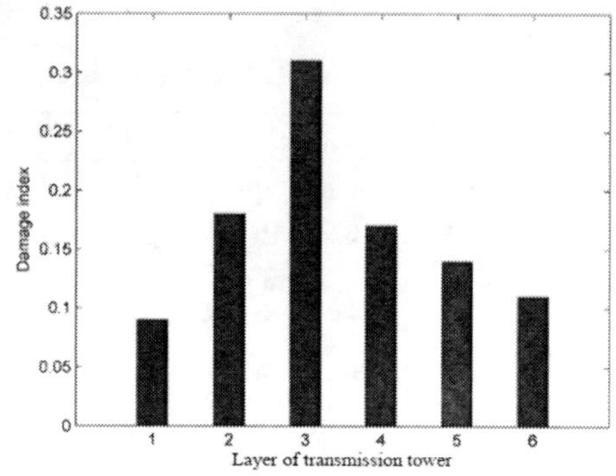

Fig. 6 Detection results of four damaged rods in the third layer

For Cases 5 and 6, two and four crossover bars of the fifth layer are removed, respectively. The proposed damage index is also utilized to realize the damage diagnosis. The identification results of the index are shown in Figs. 7 and 8 with the layer number plotted against the damage index value. From Figs. 7 and 8, it can be observed that the values of the fifth layer are obviously higher than those of the other five layers. So, the proposed index can well identify the damaged layer of transmission tower.

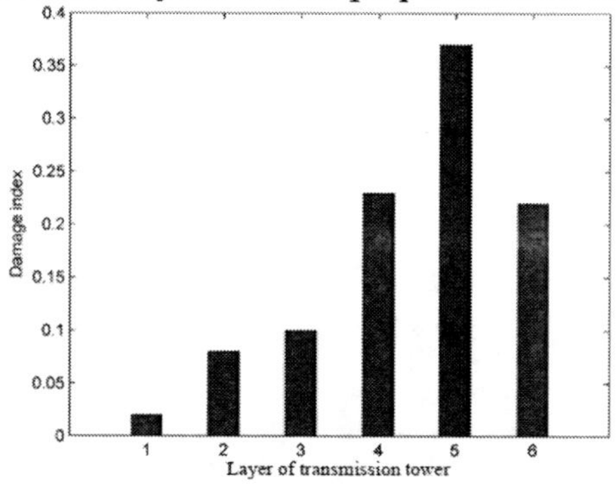

Fig. 7 Detection results of two damaged rods in the fifth floor

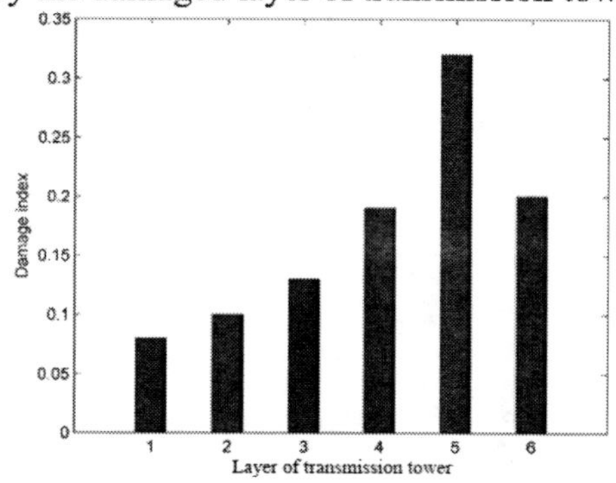

Fig. 8 Detection results of four damaged rods in the fifth floor

5. Conclusions

In this paper, a damage index based on inter-storey stiffness and cepstral distance is presented. First, transmission tower model is established and the tower is divided into six sub-structures. Thus, we only need to identify the sub-structure location where the damaged rods are included. Then, the cepstral distance of AR models is analyzed and a damage index based on inter-storey stiffness and cepstral distance is proposed. Finally, the experiment of transmission tower model is conducted to evaluate the effectiveness of detecting the damaged layer or sub-structure location. The experimental results demonstrate that the proposed damage index is available to detect the damage of transmission tower, and the calculated values of the damaged layer are higher than those of the other undamaged layers.

Acknowledgment

This work was supported by the National Natural Science Foundation of China (Grant No. 51578094).

References

[1] D. Feng, M.Q. Feng: Computer vision for SHM of civil infrastructure: From dynamic response measurement to damage detection–A review. Engineering Structures Vol. 156 (2018), p. 105-117

[2] X.Y. Li, L.X. Wang, S.S. Law et al: Covariance of dynamic strain responses for structural damage detection. Mechanical Systems and Signal Processing Vol. 95, Issue 10 (2017), p. 90-105

[3] E.P.Carden, J.M.W. Brownjohn: ARMA modeled time-series classification for structural health monitoring of civil infrastructure. Mechanical Systems and Signal Processing Vol.22, Issue 2 (2008), p. 295-314

[4] D.Bernal, D. Zonta, M.Pozzi: ARX residuals in damage detection. Structural Control & Health Monitoring Vol. 19, Issue 19 (2012), p. 535–547

[5] F. Eloi, F. Joaquim, P. Gyuhae et al: Influence of the autoregressive model order on damage detection. Computer–Aided Civil and Infrastructure Engineering Vol. 26, Issue 3 (2011), p. 225-238

[6] R.J. Martin: A metric for ARMA processes. IEEE Transactions on Signal Processing Vol. 48, Issue 4 (2011), p. 1164-1170

[7] Z. Xing, A. Mita: Locating the damaged storey of a building using distance measures of low-order AR models. Smart Structures & Systems Vol. 6 (2011), p. 991-1005

Preliminary Study on the Innovative Design of Original Bamboo Furniture Based on the Coordination Evolution Rules of Subsystems of TRIZ Theory

Wenying Dong[1,2,a], Xiangdong Dai[1,b], JinJin Yao[2,c], Yi Xiong[2,d]

[1]Departments of furniture and art design, Central South University of Forestry and Technology, 410000, China

[2]College of forestry, Sichuan Agricultural University, 611000, China

Email: 10775660@qq.com,13875882611@163.com,
1147127322@qq.com,1945730676@qq.com
TEL: 13875882611

Abstract: The original bamboo furniture is green and environmentally friendly and is a good substitute for wood. But at present, the design of original bamboo furniture develops slowly and its added value is not very high. By applying the coordinated evolution principle of subsystems in TRIZ technology evolution theory, this research makes a preliminary exploration on the innovative design methods of original bamboo furniture in form and structure in the purpose of bringing inspiration and new thinking to the innovative design of original bamboo furniture. In this way, promoting the development of original bamboo furniture design faster and better.

1. Introduction

Today, when we strongly advocate sustainable development, bamboo plays an important role in the sustainable development of the environment. Green furniture is a trend in furniture design and development. Green furniture design attaches importance on the environmental attributes of products (removability, recyclability, maintainability, reusability, etc.) and takes them as design objectives. While meeting the requirements of environmental objectives, it also ensures the proper functions, service life, quality and other requirements of furniture. Raw bamboo furniture is mainly made of bamboo because bamboo has the advantages of fast growth speed, high yield and good wood properties. The use of bamboo can effectively relieve the environmental pressure caused by wood shortage. In addition, the original bamboo furniture is recyclable, and the materials can be reused after recycling. However, at present, the original bamboo furniture in China is simple in shape, single in product category and lack of innovation in design. By using the TRIZ theory coordinated evolution rule, it provides innovative ideas for the design of original bamboo furniture, which can promote the development and utilization of it, increase the added value of original bamboo furniture design, enrich the variety of bamboo furniture products, and at the same time alleviate the shortage of wood resources pressure.

2. The Overview of the TRIZ System Coordination Evolution Rules

The evolution of technology systems is moving in a more coordinated direction with each other. That is, the various components of the system fully utilize their respective functions while maintaining coordination, which is also a necessary condition for the entire technical system to perform its functions.

The manifestation of coordinated evolution is: shape coordination; structural coordination; coordination of performance parameters; coordination of work rhythm, frequency; coordination of location, etc. The coordination between each subsystem of the technical system, the parameters, and the parameters of the system and the parameters of the super system are the basic conditions for the system to realize its functions.

In the field of furniture design, product improvement, appearance design, structure and function all follow the same biological systems process of production, growth, maturity and aging, which has its intrinsic rule to follow. The coordinated evolutionary rule of TRIZ system is one of the core contents of TRIZ theory, and it is also the main theory to reveal the laws of various components in the technical system. The technological system has always been in evolving situation. Most of the new furniture products in the world are developed on the basis of the old products. So, it is practical to analyze the existing furniture market and make a prediction of the future design direction for the purpose of promoting the development and design of a new generation of furniture products. This paper not only makes a correlation analysis of the original bamboo furniture design cases, but also summarizes the innovative methods in shape and structure coordination based on the the coordinated evolutionary rule of TRIZ system.

3. Application Analysis of Original Bamboo Furniture Innovative Design

With the improvement of quality of life, people's aesthetic requirements for furniture are getting higher and higher. More and more designers have endowed original bamboo furniture with different connotations through their understanding of the design. They have brought new vitality to the innovative design of the original bamboo furniture. In the design of existing original bamboo furniture, there are some innovative design cases at home and abroad in many aspects. Most of these cases have carried out innovative design from the combination and application of the structure, form, culture and materials of the original bamboo furniture. For example, the bamboo stool named "Ching" (Figure 1, 2 and 3) designed by Taiwan designers Lin Ta-Chih and Hsieh Yi-Fan integrates the two most common materials of bamboo and cement. Taking the toughness and structure of bamboo as the breakthrough point, a new configuration was designed. The knot inside the bamboo is also applied to the small structure of strengthening the stool, which not only innovate the shape of the stool, but also guarantees the stool's function of carrying weight. The curve on the bamboo stool is natural and graceful, and the bamboo parts are also very suitable which gives people a very comfortable visual feeling. In addition, regarding the original bamboo furniture design form, Taiwan designer Chengzong Fan and master craftsman Gaoming Chen designed a bamboo chair named "Flow" (Figure.4). The designer made use of the bending characteristics of bamboo to create a natural flowing feeling of the furniture, making it look more vital and dynamic.

Figure 1. Ching (1)

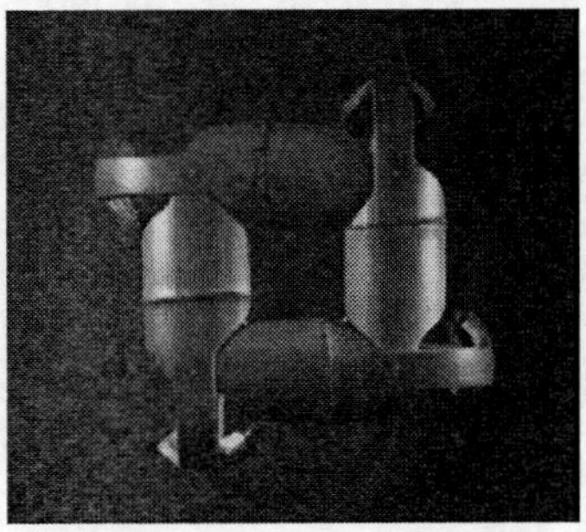

Figure 2. Ching (2)

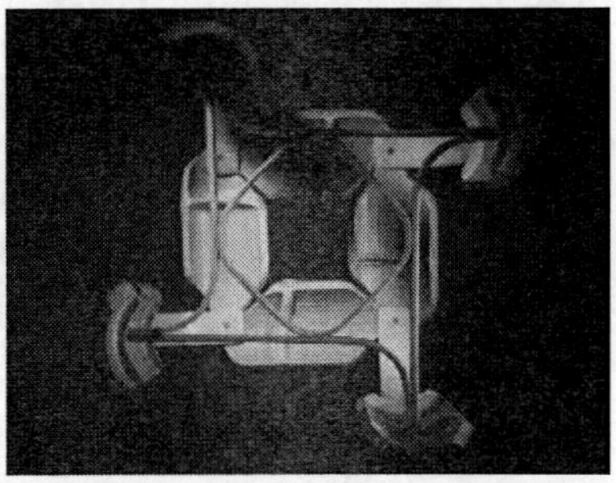

Figure 3. Ching (3)

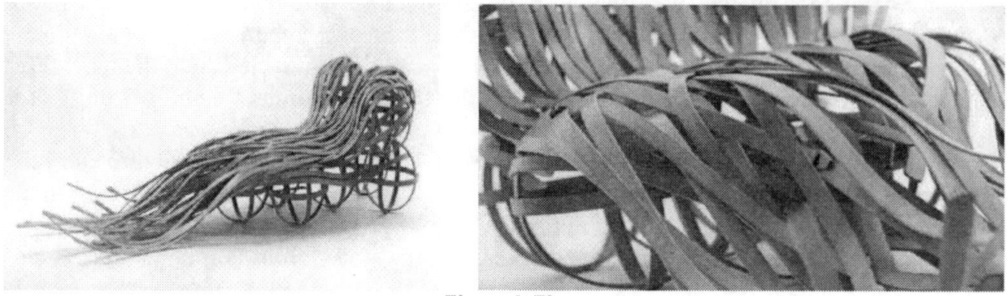

Figure 4. Flow

Through the analysis of the above original bamboo furniture cases, it is found that the original bamboo furniture design attaches importance to the use of the cultural meaning of bamboo and the original form of bamboo, while considering the characteristics of the bamboo itself, such as exploring the internal composition and structural form of the original bamboo for the purpose of seeking the possibility of developing and utilizing the different parts of bamboo under their different physical states. Therefore, we should recognize the basic characteristics of bamboo as a furniture material, and at the same time make use of the aesthetic characteristics of bamboo's elegance and grace to design furniture products conforming to modern life.

4. The Application of Coordination Evolution Rule in Original Bamboo Furniture Innovation

In the above-mentioned original bamboo furniture cases, the shape design of furniture cases not only takes into consideration the characteristics of the bamboo itself, but also combines points, lines and surfaces to form a harmonious relationship between the shape, material, function and structure of the furniture shape design, which is in line with the coordinated evolution rule of the TRIZ system. In the analysis of furniture design cases and the coordination rules, we should pay attention to how to achieve such a systematic coordination relationship, so that we can design the better original wood furniture. TRIZ system coordinated evolution method is a universal tool for leading innovation. In the design process of the original bamboo furniture, according to the essential requirements of the original bamboo furniture design, it can be classified into two aspects as showed below.

4.1 The Coordination of original bamboo furniture design

The elements of the original bamboo furniture form include points, lines and surfaces. In geometry, the position of a point can be determined as long as coordinates are given, so the point can exist in any shape and size. Line is the basic element of furniture outline in the original bamboo furniture design. Different expressions of line will compose different shapes; lines make up the surface. In the design of original bamboo furniture, there are many forms of surface, such as bamboo woven chair back, chair seat made of strips of bamboo board ...By using certain design methods, the design of original bamboo furniture applies these design elements to achieve coordination. After the summary of innovative design methods, we can carry out innovative design thinking from the aspects of line-surface mutual transformation, curved surface transformation, mix and match of materials, asymmetry, change of texture, etc. so as to achieve the coordination of original bamboo furniture design. The following is a detailed analysis of line-surface mutual transformation.

Line-surface mutual transformation: line-surface mutual transformation is to use the material of bamboo to transform the shape of the furniture to achieve the aim of enhancing the beauty of the furniture. By using regular or irregular methods, Lines and surfaces follow a certain rule to create new beauty form, presenting a novel, peculiar visual perceptions and regular fluctuating, rhythmic rhythms and organized dynamics feeling for people. For example, bamboo is processed into bamboo chips and strips after following the characteristics of bamboo, and the strips are regularly arranged and combined to form different patterns (Figure 5). at the same time, the innovative design work can be achieved by giving the bamboo a sense of volume (Figure 6).

Curve surface: Furniture form may consist of various facets. In the design of original bamboo furniture, curve parts are used instead of straight parts. The curve parts are connected, staggered and overlapped to form a surface, so that the position, combination relationship and size ratio of the surface meet the design requirements. Furniture made of curve surface is simple and light. (Figure 7).

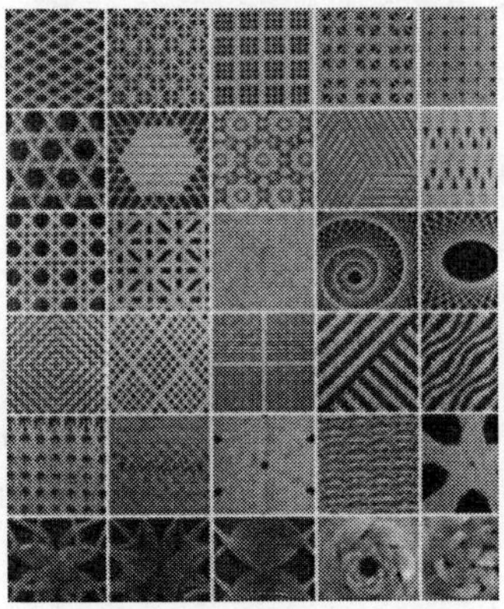

Figure 5, bamboo woven in different patterns

Figure 6, A casual chair with a bamboo

Figure 7, Curved bamboo

Mix and match materials: On the basis of compliance with wood properties, bamboo and other materials are mixed to show materials and techniques to enhance the artistic quality of products. In the appearance design of original bamboo furniture, furniture materials can be mixed and matched. By using bamboo as the main material and combining wood, fabric, metal and other materials, a contrast or relatively coordinated visual effect can be produced. Bamboo is relatively easy to coordinate with other materials. Local ornament can be made of metal, cloth, etc. (Figure. 8).

Change texture: Different treatments were used to change the surface texture of bamboo to achieve the desired performance. For example, natural rhythm can be expressed by using natural bamboo knots (Figure. 9). Through the smooth treatment of the bamboo epidermis, it can have a refreshing and clean texture. Processing bamboo into different strips and adopting different knitting techniques or combination rules can make different texture effects of bamboo pattern.

Asymmetry: By combining with its functions, we can design products of different shapes. As people pursue the differentiation and individualization of furniture products, the inspiration and space tension of furniture asymmetric design are loved by more and more people. In the design of original bamboo furniture, simple and flexible lines can be used to stagger all sides of the furniture.

Figure 8, Bamboo frame cloth screen

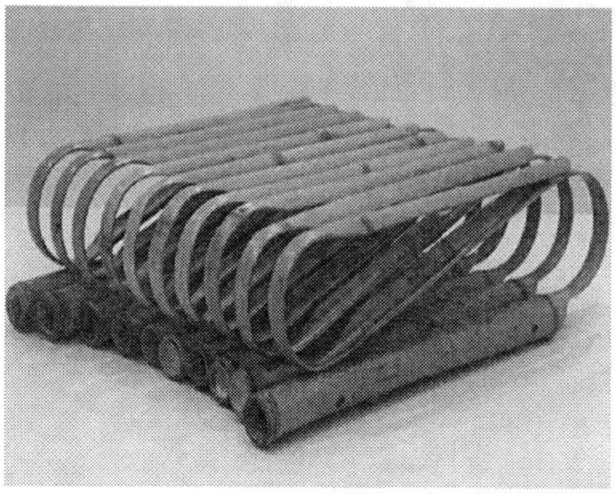

Figure 9, "The Question of Bamboo"

Figure 10, Interlaced bending of table top

4.2 The Coordination in Design Structure of Original Bamboo Furniture

The original bamboo furniture has different designs, but its basic structure is similar and can be divided into two types: one is frame structure; The second is the plate-like component structure. The coordination of the original bamboo furniture design structure is mainly presented in the rationality, aesthetics and technical feasibility of the structure. In the design of the original bamboo furniture structure, in order to ensure the coordination of the structural design, the following methods can also be used to auxiliary design.

Bamboo-wood socket joint: original bamboo and wood are used as an intermediary to realize the effective connection between them. In the structural design of the original bamboo furniture, the frame of the bamboo furniture should not only consider the appearance of the furniture, but also consider the main force-bearing components of the frame. The frame structure has two major types: bending combination and straight material combination. The bending combination methods are divided into heating bending and slotting bending. After bending, the bamboo is combined with other components to form the frame structure. Generally, the commonly used connection methods are rod-shaped docking, T-joint, cross-joint, L-joint, parallel joint, embedded joint, seam joint, etc. In order to ensure the firmness of the connection, auxiliary materials such as glue, iron nails, bamboo nails and the like are also used. This connection method has good load-bearing performance but poor stability and is easy to fall off at the joint. However, the method of bamboo-wood socket joint not only maintains the advantages of good load-bearing performance, but also effectively solves the shortcomings of poor stability, greatly improving the coordination of the structural design of the original bamboo furniture.

Bending curved: Bamboo has a simple structure, vascular bundles and bamboo rafts are arranged in parallel, the fibers are straight, and have a good property of bending. Under the action of external force, the bamboo can be permanently deformed without destroying its integrity, and it is easy to manufacture bending and peculiar works. By using this property of bamboo, it is possible to design unique and special works of various combinations of bending shapes. (As showed in Figure 7).

When designing a piece of original bamboo furniture, by combining the above several design methods, it can provide different ideas for product design and add creativity. The following table is an analysis and summary of the original bamboo furniture design method.

Table 1. Analysis and summary of original bamboo furniture design method

Coordinated evolutionary rules	Corresponding Elements of Original Bamboo Furniture Design	Innovative Design Method of Original Bamboo Furniture	Explanation

Morphological **coordination**	Original bamboo diameter; length; diameter; different forms; color; volume; texture; bamboo wall thickness; flexibility	line-surface mutual transformation	The mutual transformation and flexible application between lines and surfaces. Different lines form different surfaces. When using line-surface mutual transformation, it is necessary to consider the different forms of bamboo strips, flexibility and other elements.
		Curved surface	According to the bending characteristics of bamboo, the curved surface can be processed into different arcs, which can be used in the turning point of the original bamboo furniture. The application of curved surface design method in the design of the original bamboo furniture should consider the bamboo joint length (and the position of bamboo joint), bamboo flexibility, bamboo wall thickness and other factors.
		Material mix and match	The combination of bamboo with another different materials will make the best use of each advantages
		Asymmetry	When the original bamboo furniture is designed asymmetrically, the left and right sides present different shapes, which can be changed differently through bamboo volume or texture.
		Texture Change	The texture of bamboo can be shown as delicate and smooth skin, vertical and smooth cross-section, delicate and fine bamboo weaving, etc. The texture of bamboo, thickness of bamboo wall, flexibility and other factors need to be considered in the design of changing texture.
Structural coordination	Compressive strength; Tensile strength; Torsional strength; Bamboo hardness; Bamboo density; Shear strength; Glue curing; bending	bamboo and wood socket joint	According to the structural characteristics of the product, the bamboo and the wood are joined together. The design needs to take into account the compressive strength of the bamboo, the hardness of the bamboo, etc., and sometimes the bamboo will be bent or glued and cured in reality.
		Bending curved	By conforming to the bamboo lines and bending it into different shapes, it can enhance the innovation of the product. The most important thing to pay attention to when designing the curved design is the torsional strength of bamboo.

5. Conclusion

In the process of furniture design, designers need to face many kinds of problems and different objects. Through the analysis of existing furniture cases and the coordination evolution rule of TRIZ system, designers can be helped to fully understand the existing technology, improve the design efficiency and increase the added value of product design. The TRIZ theory system coordination evolution rule is applied to guide furniture design, systematize and smooth furniture design process, which help designers shorten design cycle and save cost.

Original bamboo furniture is endowed by nature with unique aesthetic feeling and superior material characteristics, which plays an important role in improving the quality of human living environment. In the study of original bamboo furniture, many innovative methods of original bamboo furniture are summarized, which provides help for the original bamboo furniture design based on TRIZ system coordination evolution rule. It is hoped that more designers can pay attention to the innovative manufacture of original bamboo furniture and carry out value added innovative design in line with the material characteristics, regional characteristics, national resources, development trends and other factors, so as to promote the development of original bamboo furniture design in China.

Acknowledgements

This research was financially supported by the project of Undergraduate Scientific Research Interest of

Sichuan Agricultural University (2019234), and the Art Research Project of 211 Support Plan of Sichuan Agricultural University (2019).

References

[1] Xia Yu. Study on Manufacturing Technology of Original Bamboo Furniture [J], Journal of Bamboo, 2017, 36(5): 64-67,43.
[2] Zhang Buyong, Liu Zuhua. Study on the Connection Mode of Bamboo and Wood in the Original Bamboo Building[J], Journal of Jiamusi University (Natural Science Edition) [J], 2013.31 (41).
[3] Lin Yihuang. Analysis of Traditional Round Bamboo Furniture and New Material of Bamboo Furniture [J], Journal of longyan university, 2009.29(5).
[4] Xiao Jun. Application of Resource Conception in TRIZ Theory to Furniture Design[J], journal of Northeast Forestry University, 2010,38(4): 117-118.
[5] LI Junwei. Features and Production Technology of Bamboo Glued Furniture[J]. Wood Processing Machinery, 2011(4): 47-49.
[6] Chen Mengyao, Zhang Zhongfeng. Application of material characteristics in furniture design[J]. Packaging Engineering, 2017, 38(2): 141-145.
[7] Lei Da, Chen Doudou. Analysis of the form of bamboo furniture [J], Journal of bamboo research, 2004, 23(3): 31-33
[8] Xu Minxiao. Design and manufacture of bamboo ecological furniture[J]. Furniture of bamboo, 2008, 6(6): 37-41.
[9] Wu Zaixing, Tu Yongning. Relying on bamboo resources to develop bamboo-based furniture[J]. World bamboo and rattan, 2005, 3(4): 43-44
[10] Wang Lianjun. The critical analysis of bamboo product design[J]. Journal of bamboo research, 1996, 15(4): 43-51.
[11] Liu Xiaoxu. Material mix and match in furniture design[J]. Furniture, 2016, 37(1): 57-60.

IOP Conf. Series: Materials Science and Engineering **711** (2020) 012071 doi:10.1088/1757-899X/711/1/012071

Research on dynamic inverse method of surrounding rock mechanical parameters of large underground caverns in excavation

Shengjie Di [1,2,a], Ying Zhang [1,2,b], Heng Zhou [1,2], Jing Liu [1,2]

[1]Northwest Engineering Corporation Limited, Power China, Xi'an, Shanxi, 710065, China

[2]High Slope and Geological Hazard Research & Management branch, National Energy and Hydropower Engineering Technology R&D Center, Xi'an, Shanxi, 710065, China

[a]14276804@qq.com, [b]zhangying919@126.com

Abstract: Rock mechanical parameters are important factors to affect the stability of surrounding rock in underground engineering. With the development of underground cavern excavation progress, the mechanical parameters of rock mass will change significantly. In order to reflect the change of parameters during excavation in real time based on uniform design and optimization algorithm technology, dynamic inverse method of surrounding rock parameters during excavation of underground caverns is carried out. Considering the location, elevation and period of excavation, the parameters inversion are divided into excavation sections and staged. Verified by actual engineering, the calculated displacement values of the inversion agree well with the measured.

1.Introduction

The determination of rock mechanics parameters is the precondition for geotechnical engineering calculation. Due to the complexity of engineering topography, geology and measurement, rock mechanical parameters are difficult to measure accurately. In order to overcome the inaccuracy of the parameters, the dynamic surrounding rock mechanics parameters inversion was taken up. A large number of studies [1-7] have shown that it is a good way to invert the mechanical parameters by using on-site monitoring displacement data. The monitoring data of the underground cavern is used to invert the parameters, and the stability of the surrounding rock is analyzed and evaluated accordingly. It plays an important role in ensuring the safety of construction and avoiding the risk of major accidents in underground engineering.

Based on the research of the underground powerhouse engineering of a hydropower station, this paper come up with the method to inverse the rock parameters. The uniform design and artificial neural network technology was used to inverse the parameters in the excavation of the underground powerhouse, and the inversion parameters are used to calculate and analyze of the current excavation layer. The dynamic inversion method may optimize the design and construction of underground engineering effectively.

2.Rock mechanical parameters inversion method

As the surrounding rock parameters are the control factors for analyzing rock stability and supporting design. Dynamic surrounding rock parameter inversion of large-scale underground cavern group excavation process plays an important role to get accurate parameters. The calculation flowchart of displacement inversion analysis is shown in Fig. 1.

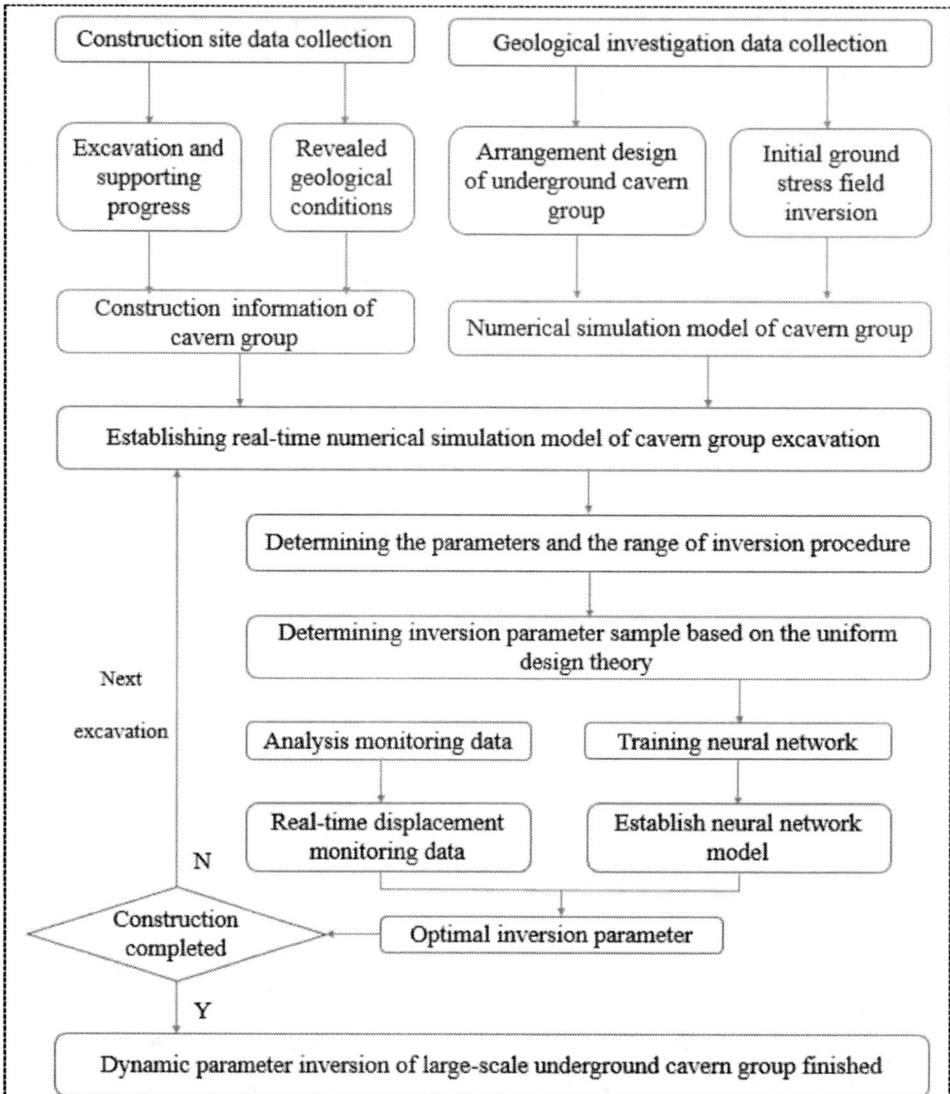

Figure 1. Calculation and analysis flowchart of dynamic inversion

The key technologies to be considered in inversion method are as follows:

1. According to the measured value of the multi-point displacement meter and the geological investigation data of the cavern group, the factors which affecting the deformation are analyzed. The sensitivity analysis is used to select the strong sensitivity parameters to inverse. In addition, based on the test results and the monitoring data, the range of the parameters are determined.

2. The uniform design theory is used to carry out multiple sets of feature analysis samples, then the genetic algorithm is used to establish the initial weight of neural network, then substitute the sample for neural network training, jointly obtain the approximation parametric solution of the error function minimum by genetic and gradient descent algorithm.

3. Constructing a closed-loop inversion system for the simulation of the construction process of underground caverns, the optimal parameters are obtained. Combined with positive analysis to obtain

IOP Conf. Series: Materials Science and Engineering **711** (2020) 012071 doi:10.1088/1757-899X/711/1/012071

a new set of samples closer to the measured values, which will be put into the original sample group and substituted into the system. The cyclic method can automatically increase the effective samples gradually when the training samples are insufficient, continuously improve the simulation precision, reduce the error, and ensure the high-efficiency and high precision output the optimal parameters.

4. The obtained optimal mechanical parameters are used for the forward analysis, and predict the response of the displacement for the next excavation.

3.Analytical procedure for engineering application

Calculation and analysis model. Taking the large underground power station caverns engineering as an example, its three-dimensional finite element numerical model is shown in Fig. 2. The underground powerhouse cavern includes the terrain, faults, strata and underground plant excavation of the hydropower station right bank. And the model consists of six stratums, from left to right are ⑩ P2a-Ss+Sb, ⑨ P2a-Sb, ⑧ P2a-Ss, ⑦P2a-Sb+Ss, ⑥ P2a-Ss, ⑤ P2a-Sb+Ss separately, the stratified excavation and computing mesh of FEM model are shown in and Fig. 3.

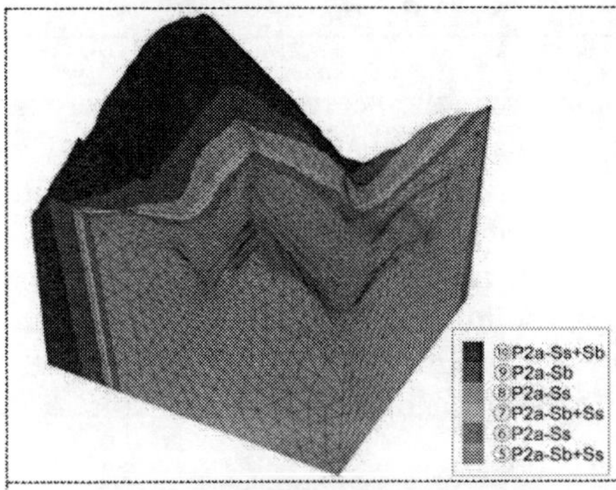

Figure 2. Three-dimensional numerical mesh model

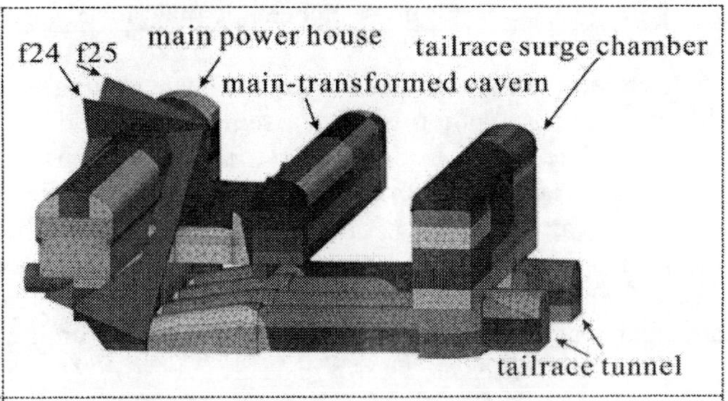

Figure 3. Cavern group numerical mesh model

Parameters range of dynamic inversion. The 11 parameters with high sensitivity have been identified, based on the sensitivity analysis of lithologic parameters various types, actual test data and numerical analysis results of previous scientific research, which are (1) the young's modulus E_1, cohesion c_1 and friction angle φ_1 of sand and slate interbedded: ⑤ P2a-Sb+Ss, ⑦ P2a-Sb+Ss and ⑩ P2a-Ss+Sb; (2) the young's modulus E_2, cohesion c_2 and friction angle φ_2 of sandstone: ⑥ P2a-Ss and ⑧ P2a-Ss; (3) the young's modulus E_3, cohesion c_3 and friction angle φ_3 of slate: ⑨ P2a-Sb; (4) the equivalent modulus E_f and equivalent cohesion c_f of fault. The inversion parameters data range are shown as in Table 1.

Table 1. Data ranges of inversion parameters

Inversion parameters	E_1 / GPa	E_2 / GPa	E_3 / GPa	E_f / GPa	c_1 / MPa	c_2 / MPa	c_3 / MPa	c_f / MPa	φ_1 / °	φ_2 / °	φ_3 / °
Lower limit	8.00	10.00	7.00	0.50	0.40	0.60	0.20	0.05	35.00	40.00	30.00
Upper limit	20.00	25.00	18.00	2.00	1.50	1.80	1.20	0.30	50.00	55.00	45.00

Considering the unloading relaxation and mechanical properties of surrounding rock caused by excavation blasting, according to the depth of surrounding rock loose circle obtained by geophysical testing, the surrounding rock is divided into strongly disturbed zone, weakly disturbed zone and undisturbed zone, which is shown in Fig. 4. Based on the results of parameter sensitivity analysis, the mechanical parameters of the surrounding rock in the unloading zone are divided.

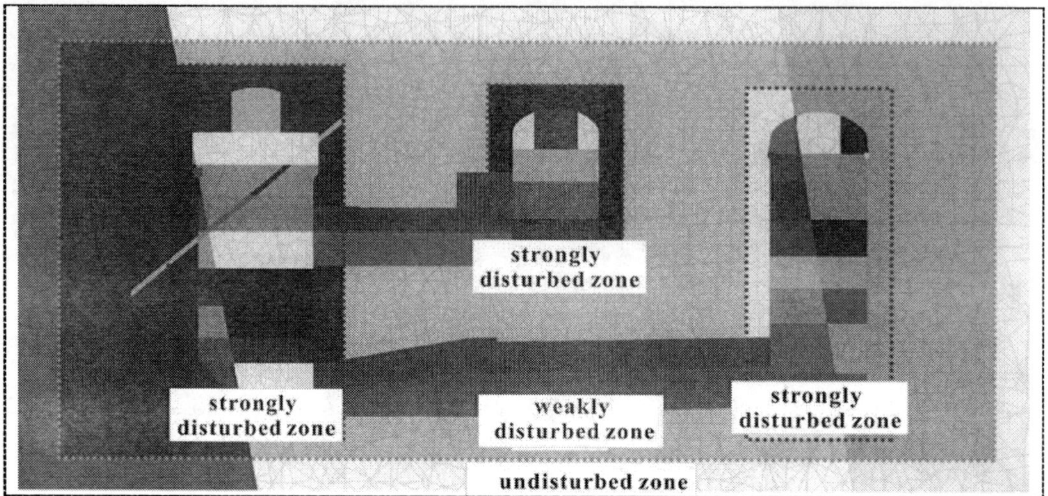

Figure 4. Schematic diagram of surrounding rock parameters

Inversion process and results. The training sample and the corresponding incremental displacement are combined into a sample pair, and the rock mass mechanical parameter inversion analysis system is introduced for training, and the nonlinear mapping relationship between the parameters and the corresponding incremental displacement is obtained. Then the actual corresponding incremental displacement is input to the trained neural network to obtain the rock mechanics parameter value. The parameter value obtained by the inversion is taken as a new sample, and the rock mass mechanical parameter inversion analysis system is retrained. Loop like this, artificial neural network error gradually decreases, until the error does not change, the calculation ends. The inversion results are shown in Table 2.

Table 2. The inversion results of parameter

Young's modulus of strongly disturbed zone / GPa			Young's modulus of weakly disturbed zone / GPa		
$E_{1\text{-}1}$	$E_{2\text{-}1}$	$E_{3\text{-}1}$	$E_{1\text{-}2}$	$E_{2\text{-}2}$	$E_{3\text{-}2}$
7.27	8.13	6.89	7.44	8.51	7.16

The $E_{1\text{-}1}$, $E_{2\text{-}1}$, $E_{3\text{-}1}$ and $E_{1\text{-}2}$, $E_{2\text{-}2}$, $E_{3\text{-}2}$ are the young's modulus of different lithology in strongly and weakly disturbed zone respectively. The inversion parameters are input into the positive analysis model for calculation, and the obtained calculated displacement values are compared with the actual monitored values. Taking construction stage VI to VII as an example, the comparison results are shown in Table 3. The results show that the calculated displacement values are not much different from the actual monitoring, indicating that the parameters of the group are reasonable and can be used as the input values of the surrounding rock mechanical parameters of the next excavation layer.

Table 3. Calculated displacement and measured comparison of excavation stage VI to stage VII

Typical monitoring point	Surveying Depth / m	Calculated values / mm			Measured values / mm			Difference value / mm
		VI	VII	VII-VI	VI	VII	VII-VI	
1	0	15.51	16.32	0.82	14.57	15.33	0.76	-0.06
	2	12.56	13.25	0.69	13.13	13.70	0.57	-0.12
2	7	11.24	12.24	1.01	3.44	4.87	1.43	0.42
	15	6.64	7.43	0.79	0.79	1.32	0.53	-0.26
3	7	11.25	13.16	1.92	8.37	9.95	1.57	-0.34
4	15	6.06	9.60	3.55	5.48	8.78	3.30	-0.25
5	15	19.72	26.82	7.10	8.52	15.54	7.02	-0.08
6	0	19.74	23.01	3.27	11.76	14.42	2.65	-0.62

	7	13.87	16.80	2.94	10.71	13.11	2.39	-0.54
7	15	8.73	9.95	1.23	0.11	1.19	1.09	-0.14

4.Conclusions

1. An algorithm based on inheritance, gradient descent and neural network is proposed, which can improve and supply the effective sample automatically, and solves the problem of inaccurate value of rock mechanics parameters;

2. According to displacement monitoring and geophysical prospecting results, inversion dynamic method achieve the purpose of surrounding rock parameters divided by sections and staged, which reflect the dynamic changes and develop process of surrounding rock mechanical behaviour.

3. The engineering application example shows that the mechanical parameters obtained by the inversion are acceptable, which verified the rationality and operability of this method.

References

[1] S. Sakurai, K. Takeuchi: Rock Mechanics and Rock Engineering. Vol 16 (1983), p. 173-180.
[2] S. Sakurai, I. Kawashima, T. Otani: Journal of Geotechnical Engineerning. (1994), p. 189-195.
[3] H.B. Zhao, X.T. Feng: Chinese Journal of Rock Mechanics and Engineering. Vol 22 (2003), p. 1618-1622.
[4] Q. Jiang, X.T. Feng, G.S. Su, et al.: Chinese Journal of Rock Mechanics and Engineering. Vol.26 (2007), p. 2654-2662.
[5] S. Won, J. W. Wilson: Rock Mechanics and Rock Engineering. Vol 32 (1999), p. 101-122.
[6] X. G. He, F. Dai, N. W. Xu, et al.: Advanced Materials Research. (2014), p. 878-882.
[7] R. Bhasin, T. Pabst: Ksce Journal of Civil Engineering. Vol 19 (2015), p.725-732.

IOP Conf. Series: Materials Science and Engineering **711** (2020) 012072 doi:10.1088/1757-899X/711/1/012072

Measurement of in-plane thermal conductivity of self-suspended thin films

Liqiang XU[1,a]

School of Optics and Photonics, Beijing Institute of Technology, beijing, 100081, China

xlq940105@163.com

Abstract A method for in-plane thermal conductivity measurement of self-suspended thin films by steady-state infrared thermal thermography is introduced. Based on the one-dimensional heat conduction equation, a theoretical model of temperature distribution on the surface of thin film at steady state is established. The simulation results show that the error between in-plane thermal conductivity and theoretical value is less than 3% when the temperature rise of the film is less than 5K. By measuring the boundary temperature and thickness of the film, the in-plane thermal conductivity and emissivity can be obtained simultaneously without measuring the absorption of visible light by the film. Polyimide film with the thickness of 550 nm is experimentally investigated by this method. We find that the in-plane thermal conductivity is 2.15W/mK and emissivity is 0.93 by fitting temperature distribution curve. These results validate the method developed.

1. INTRODUCTION

Thin film materials have been widely used in microelectronics, information, aerospace technology and other high-tech fields. With the development of technology, more and more attention has been paid to the study of thermophysical properties of thin film materials. Because the thermophysical properties directly determine the research and application of thin films, it is of great significance to accurately measure the thermophysical parameters of thin films.

Transient 3ω method[1] is less restricted in terms of the type of film investigated, but sample preparation is very demanding. Time-resolved optical pump-and-probe measurements[2] depend on calibration and smooth film surfaces without diffuse reflection. Micro-bridge method[3] is simple, but the influence of thermal radiation losses and contact thermal resistance is ignored. These method cannot easily provide information of the thermophysical properties of thin film materials.

In this paper a method is introduced that in-plane thermal conductivity measurement of self-suspended thin films by steady-state infrared thermal thermography. The measurement is a non - contact methods, we cannot consider the contact thermal resistance and the sample preparation is simple. We will introduce the measurement principle and simulation and present the necessary experimental setup. We apply the method to the polyimide film.

2. MEASUREMENT PRINCIPLE AND SIMULATION

2.1 Theoretical model

When visible light illuminates the surface of the film, the film absorbs visible light energy and exchanges heat with the surrounding environment through conduction, convection, and thermal radiation.

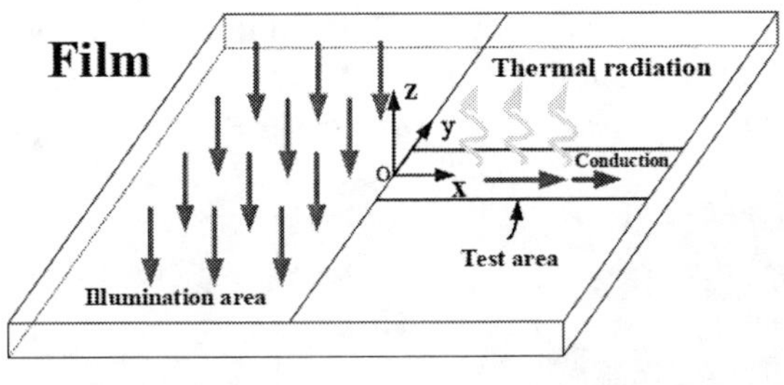

Fig. 1 schematic diagram of the measurement

Fig. 1 shows the schematic diagram of the method. Z direction is the thickness direction of the film. The film surface is located in the X-Y plane. The model is divided into two parts. The left side is the illumination area, and the right middle position is the test area. Sample is placed over a Vacuum cavity, heat exchange caused by convection can be neglected, Owing to the thickness of the film is small enough, it can be considered that there is no temperature gradient in Z direction [4]. When visible light irradiates the film, Y direction is heated evenly. Thus the heat conduction equation of the film can be simplified into a one-dimensional heat conduction equation along the x-axis [5]:

$$kd\frac{\partial^2 T(x)}{\partial x^2} - \sigma\varepsilon(T(x)^4 - T_\infty{}^4) = C_p d\rho\frac{\partial T(x)}{\partial t} \tag{1}$$

With the in-plane thermal conductivity k, the ambient temperature T_∞, the thickness d, the density ρ, the heat capacity c_ρ, the emissivity ε and the Stephen Boltzmann constant σ and the temperature distribution T(x) at a distance x from the x=0 along the x-axis.

At steady-state conditions($\frac{\partial T\ (x)}{\partial t} = 0$), This leads to

$$kd\frac{\partial^2 T(x)}{\partial x^2} - \sigma\varepsilon(T(x)^4 - T_\infty{}^4) = 0 \tag{2}$$

When the temperature difference between the thin film and the ambient temperature is small, approximate conditions $\varepsilon\sigma(T(x)^4 - T_\infty{}^4) \approx 4\varepsilon\sigma T_\infty{}^3(T(x) - T_\infty)$ can be adopted, and Eq. 2 can be simplified to:

$$kd\frac{\partial^2 T(x)}{\partial x^2} - 4\sigma\varepsilon T_\infty{}^3 T(x) = 0 \tag{3}$$

The energy of visible light is a fixed value, and the heat flux is given. It belongs to the Neumann boundary condition

$$-k\left.\frac{\partial T}{\partial x}\right|_{x=0} = q \tag{4}$$

Merging Eq. 1 - Eq. 4, and applying the boundary condition $T\big|_{x=\infty} = T_\infty = T_{substrate}$, leads to the temperature distribution

$$T(x) = \frac{q}{k\sqrt{\alpha}}e^{-\sqrt{\alpha}x} + T_\infty \tag{5}$$

Expected for the Theoretical model investigated here. With $\alpha = \frac{4\sigma\varepsilon T_\infty^3}{kd}$.

In Eq. 5, the thermophysical information of the thin film is included. the in-plane thermal conductivity and emissivity can be obtained simultaneously by fiting thermal diffusion curve of film.

2.2 Simulation analysis

In the theoretical model, the linear approximation of the thermal radiation term is made by assuming that the temperature difference between the thin film and the ambient temperature is small. In order to study the error induced by the assumed conditions, the finite element method is used to establish the simulation model of the thin film.

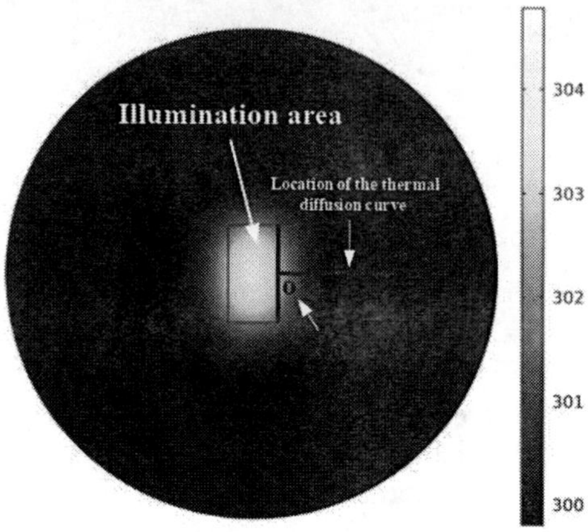

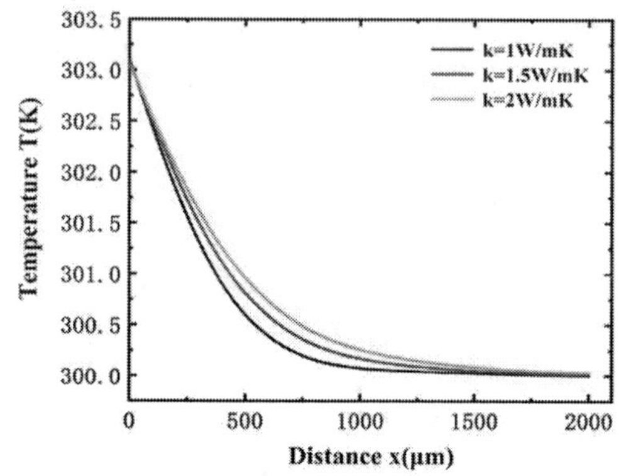

Fig. 2 Temperature distribution of thin film Fig. 3 Thermal diffusion curves of simulation

In the simulation model, radius is set to 10mm and thickness is set to 1 micron. The illumination area is a rectangle with a size of 4×2mm located in the center. The ambient temperature, initial temperature and boundary temperature are set to 300K. Fig. 2 shows the temperature distribution of thin film at steady-state. In-plane thermal conductivitys are set to 2 W/mK, 1.5W/mK, 1W/mK and emissivity is set to 0.9, Thermal diffusion curves of simulation are shown in Fig. 3.

With the thickness and emissivity of the thin film as known conditions, the in-plane thermal conductivity k can be obtained by fitting the temperature diffusion curve of the simulation model. By comparing with the thermal conductivity values set in the simulation model, we can obtain the the error induced by the assumed conditions.

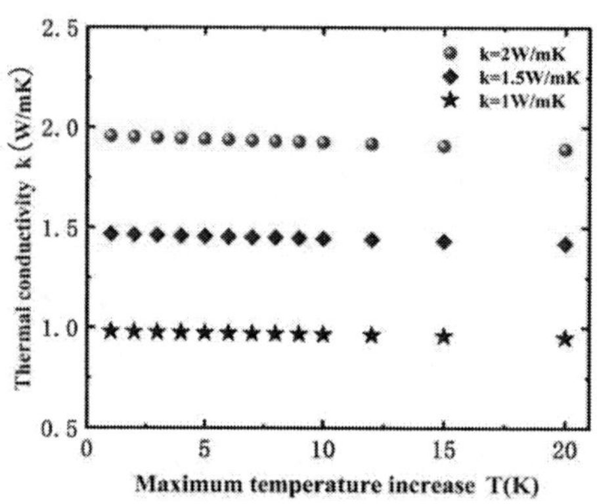

Fig. 4 Values of k obtained by fitting temperature diffusion curve

Fig.4 shows the values of k obtained by fitting temperature diffusion curve. The range of maximum temperature increase is from 1K to 20K. With the rise of the maximum temperature increase of thin film, error of in-plane thermal conductivity obtained by fitting temperature diffusion curve increases. When maximum temperature increase is 5K, we obtain k_2=1.94W/mK, $k_{1.5}$=1.46W/mK and k_1=0.97W/mK. The errors induced by the thermal radiation approximation are 3%, 2.7% and 3%.

Simulation results show that when the maximum temperature increase of the film is less than 5 K, the error induced by the thermal radiation approximation is less than 3%.

3. EXPERMENTAL STUDY

3.1 SETUP DETAILS

Integral parts of the setup are the IR camera for the spatially resolved temperature measurement and the vacuum cavity, shown in Fig. 5, for the reduction of the losses by air.

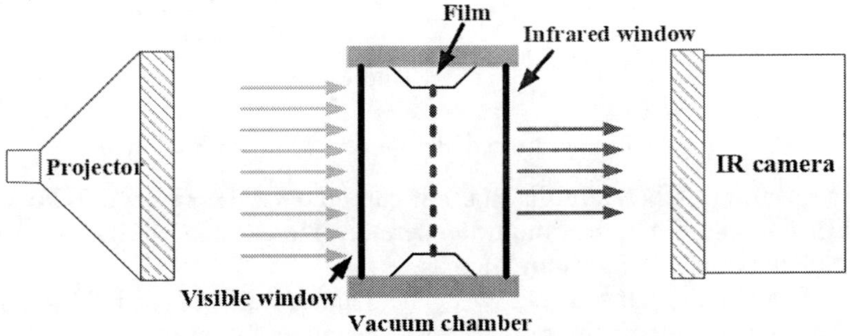

Fig. 5 Schematic diagram of Setup

In this paper, The polyimide self-suspension film samples with diameter of 3 inches and thickness of 550 nm were measured by this measuring system. Fig. 6 is the picture of the polyimide film sample.

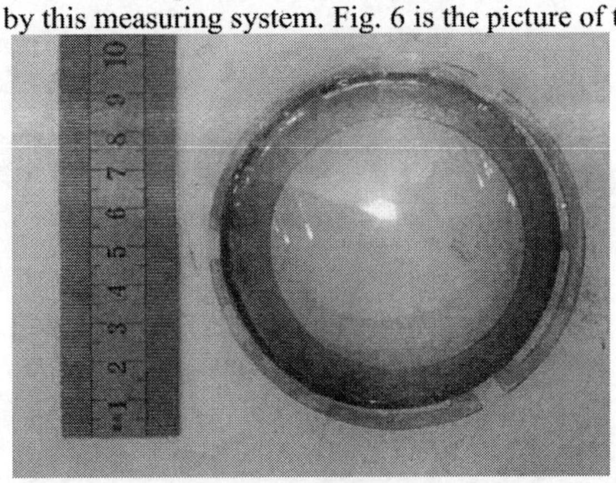

Fig. 6 Samples of polyimide films

The sample is illuminated via an Projector, the illumination area is a rectangle with a size of 4×2mm located on the film. Temperature rise of thin films after absorption of visible light. At the steady-state conditions, temperature distribution of the thin films is captured via IR camera and the boundary temperature of thin films is measured by the thermocouple.

3.2 RESULTS AND DISCUSSION

In the experiment, the vacuum degree of the vacuum chamber was kept at 4×10^{-3} Pa, and the steady-state boundary temperature of the sample measured by the thermocouple was 307.89K.

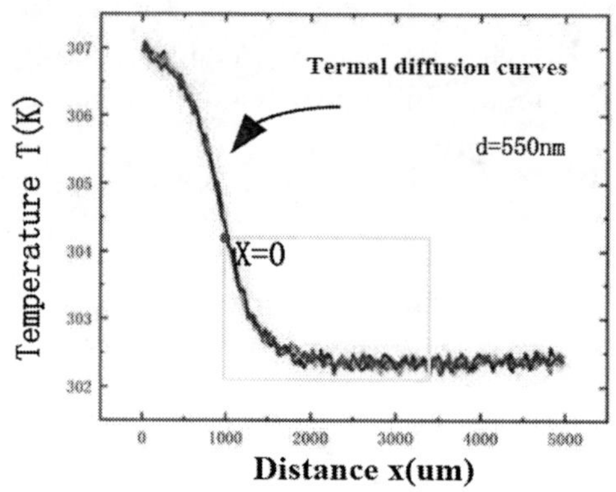

Fig. 7 Temperature distribution of thin film surface

Fig. 7 shows the film temperature distribution captured via IR camera. Take x=0 as the starting point of film diffusion curve, the maximum temperature increase is 1.86K, less than 5K, Within the range of simulation maximum temperature increase.

According to Stephen Boltzmann's Law[6], The relationship between the equivalent blackbody temperature and the actual temperature of the film surface is as follows.

$$E_{radition} = \sigma T_r^4 = \sigma \varepsilon T_{film}^4 \tag{6}$$

Applying the Eq. 6, we get the radiation intensity distribution of thin film are shown in Fig. 8. Merging Eq.5 and Eq.6 ,we get

$$E_r = \sigma\varepsilon\left(\frac{q}{k\sqrt{\alpha}}e^{-\sqrt{\alpha}x} + T_\infty\right)^4 \tag{7}$$

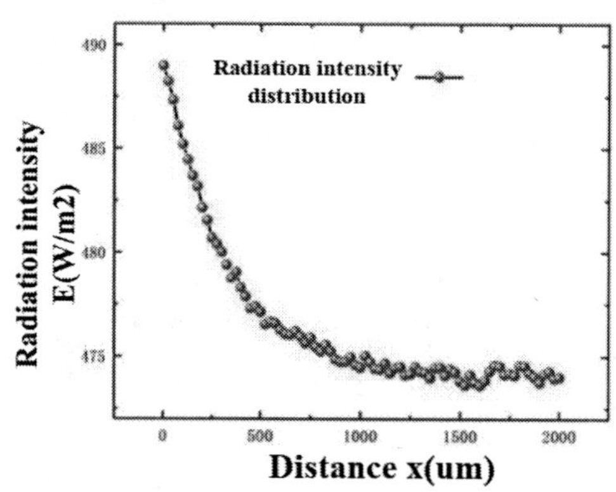

Fig. 8 Radiation intensity distribution

By measuring film thickness by Step profiler and the boundary temperature by thermocouple, we get d=550nm and $T_{substrate}$=307.89K. Applying the Eq. 7, by fitting the curve of radiation intensity distribution of thin film, we obtain k=2.15W/mK, ε=0.93 and heat flux $q=1.75\times10^4 W/m^2$.

4. CONCLUSION

In this paper, we demonstrate a simple and quick method for the in-plane thermal conductivity measurement of the self-suspended thin films. The measurement method was validated by using the finite-element simulation results as the measurement results. The simulation results indicated that theoretical value of k is less than 3% when the temperature rise of the film is less than 5K. By

IOP Conf. Series: Materials Science and Engineering **711** (2020) 012072 doi:10.1088/1757-899X/711/1/012072

measuring the polyimide film with thickness of 550 nm, we obtain the in-plane thermal conductivity is 2.15W/mK and emissivity is 0.93 at room temperature.

REFERENCES

[1] Kommandur S, Yee S, A suspended 3-omega technique to measure the anisotropic thermal conductivity of semiconducting polymers, J. Sci. Review of Scientific Instruments. 89(2018).
[2] Greppmair A, Stoib B, Saxena N, Measurement of the in-plane thermal conductivity by steady-state infrared thermography, J. Sci. Review of Scientific Instruments. 88(2017).
[3] YANG Junyou, Research Advances in the Measurement for the Thermal Conductivity of Thin Solid Films, J. 24(2010)103-107.
[4] LI Zhuo, Study on Thermal Transfer Properties of Polyimide Thin Films, J. 33(2013) 650-652.
[5] Yu X Y, A membrane for visible-to-infrared transducer, J. Sci. 51(2002)1307-1311.
[6] WANG Houhua, Heat Transfer Science, China.ChongQing University Press, (2006).

IOP Conf. Series: Materials Science and Engineering **711** (2020) 012073 doi:10.1088/1757-899X/711/1/012073

Study on the Outburst Scheme of a Shallow Buried Bias Tunnel in Yunnan

Jiaxian Wang[1,a], Ke Li[2,b], Hongyan Guo[2,c], Xiangyang Cui[3,d]

[1]Shidian Baoshi Expressway Investment and Development Co.,Ltd, Baoshan, 678000, China;

[2]China Merchants Chongqing Communications Technology Research & Design Institute Co., Ltd, Chongqing,400000, China;

[3]Chongqing Jiaotong University, Chongqing, 400000, China;

[a]e-mail: 1224976566 @qq.com, [b]e-mail: 364842342@qq.com, [c]be-mail: 717692502@qq.com, [d]be-mail: 1559039300@qq.com

Abstract. Improper construction of the tunnel entrance is likely to cause collapse and affect construction safety and damage the ecological environment. Aiming at the poor stability of surrounding rock at the exit position of Xiaoguanzhuang tunnel, on the basis of investigating the geology and driving scheme near the tunnel, the asymmetric distribution of axial force of cross-section structure under shallow eccentric pressure is obtained through numerical simulation, and the maximum axial force is applied at the left arch waist and right arch foot. According to the analysis results, the surrounding rock should be reinforced in time, and the import section should be dealt with in the design stage to ensure construction safety and efficiency, and provide reference for future construction.

1. Introduction

The explosive growth of expressways has led to more and more tunnels passing through the mountains and mountains. The entrances or exits of tunnels are often located in the anticlines or skews of the mountains, making the construction of the tunnels likely to cause a risk of collapse. Therefore, it is important to analyze the excavation method at the hole to stabilize the surrounding rock and smoothly enter the hole to ensure construction safety [1-3].

Domestic and foreign scholars have used the theory and numerical simulation to do the following research on tunnel tunneling: Yan Tao et al. used the principle of limit equilibrium for the left and right tunnel tunnel gates to obtain the bias rate and displacement of the surrounding rock. And the stress is larger than in the case of no slope [4]. Song Qing will use numerical simulation method to analyze a shallow buried tunnel, and obtain the stress and deformation characteristics of the supporting structure at the hole, and propose corresponding treatment measures [5].

Based on the Xiaoguanshi Tunnel as an engineering example, this paper analyzes the geological environment of the site and analyzes the stress characteristics of the tunnel structure during the construction process by numerical simulation method, so as to propose targeted treatment measures for construction safety and normal highway [6-9].

2. General situation of Engineering

Xiaoguanshi tunnel starts and ends mileage ZK1+830~ZK4+005, length is 2175m, K1+805~K3+990, length is 2185m, long tunnel. The maximum buried depth of the tunnel is 136.52m and the right hole is 134.87m. The tunnel clearance is 15~17m, and the section form is a separate tunnel. As shown in Figure 1, the elevation of the tunnel area is 1935~2093m, and the relative height difference is 158m. It is a low-mountain landform with a structurally denuded terrain. The terrain is steep and the tunnel has a rural entrance. The dirt road leads to the inconvenient transportation; the entrance section is lush vegetation, and many pines and other forests, while the mouth section is mountainous, and the vegetation is less developed.

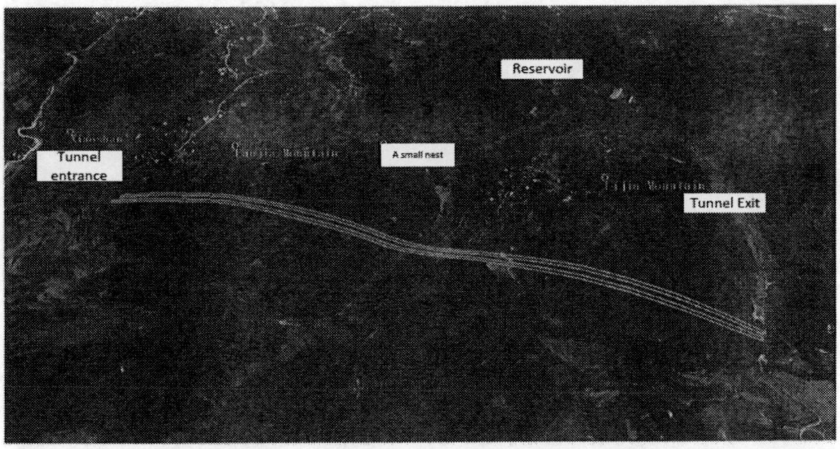

Figure 1 Topography of Xiaoguanshi Tunnel NO.3

2.1 Engineering geological conditions.

According to the geological data, the area where the line is located in the composite part of the frontal arc of the Yunnan-type mountain structure and the middle section of the eastern branch of the Qinghai-Tibet-Burma-type structure system. The geological structure in the area is complex, and the tunnel crossing area is dominated by Devonian limestone. The upper part of the Upper Cambrian Formation of the Cambrian is dominated by limestone, which is a low-middle mountainous landform of dissolution and erosion. The tunnel mountain is a dome mountain. The top of the mountain is smooth and round, the slope of the entrance section is gentle, and the natural slope is 15~30°. Steep, natural slope 25~45°, large terrain fluctuations .

The tunnel area belongs to the low-latitude mountain monsoon climate dominated by the mid-subtropical zone. The rainfall is abundant, the dry and wet seasons are distinct, and the climate is mild and humid. The average annual rainfall is 944.5mm. The abundant rainfall in the survey area is the main source of groundwater recharge, and due to seasonal changes, the groundwater also has obvious dynamic changes with the seasons. The annual average temperature is 17.6 °C. The annual average maximum temperature is 32 °C, and the annual average minimum temperature is -3.2 °C. The vertical zoning of the climate is obvious, and the areas at different altitudes are very different. The valley area is quite hot, with no frost all year round; alpine areas are colder, and there is often a brief snow in late winter and early spring. In general, regional climatic conditions have less adverse effects on tunnel construction and operation.

2.2 Hydrogeology

The project area is located in the Nujiang River system, in which the K3+200 is located in the Shidian River Basin. The Donghe River is the upper reaches of the Baltic River, a tributary of the Nujiang River. It flows from the north to the south through the Baoshan Basin. The Shidian River originates from the Nangao Mountain Area of the basin. The north flows through the basin, and flows from the sharp corner of Wang Street to the southwest into the Nujiang River. The basin is manually refurbished and straightened, the longitudinal slope is small, the flow in the rainy season is large, and the dry season is cut off.

2.3 Analysis of the design of the hole section

The exit section of the exit section of the tunnel is steep, with a natural slope of 25-45°. The exposed rock is limestone, the rock is weathered, the bedrock is bare, the joints are relatively developed, and the rock mass is relatively intact. The slope of the slope of the exit section of the tunnel is steep, and the natural slope of the slope is stable. The projection analysis of the natural slope of the slope on the slope of the exit section of the tunnel is shown in Figure 3. The existing natural slope is stable, and the slope is steeped by the slope of the natural slope (1: 0.50), where the occurrence C and the joint J1 are combined. The J2 composite structure is unfavorable to the stability of the slope, and may form a mold slip. It is necessary to pay attention to the fact that the exit section is the JI top slope, which needs to be emphasized. Therefore, when using s: 0.50 for grading, it is recommended to use the anchor cable frame.

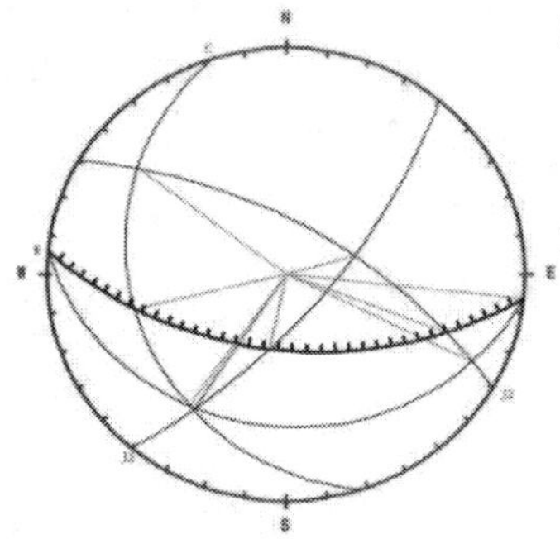

Figure 2. The red slope projection of the natural slope of the slope on the exit section of the tunnel

3. Numerical simulation analysis

3.1 Parameter selection

In this thesis, the model is simulated by flac3d. In the simulation process, the surrounding rock adopts the Mohr-Coulomb constitutive model, and the support adopts the elastic constitutive model. The initial branch adopts the shell unit, and the second lining adopts the solid element as shown in Figure 3. As shown, the top of the numerical simulation model is a free boundary, a fixed constraint is applied to the bottom, and a displacement constraint is applied around.

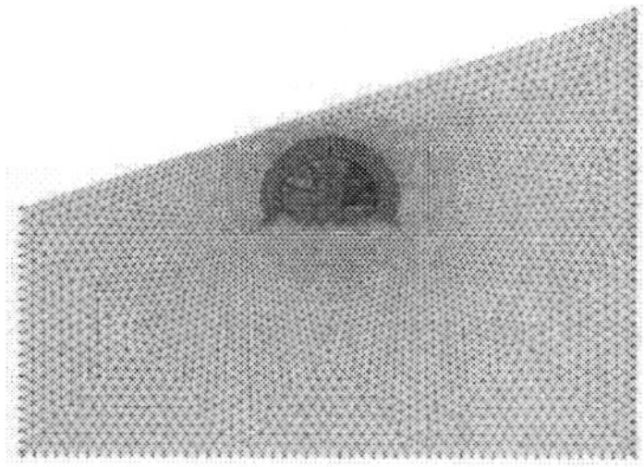

Figure 3. Calculation model diagram

The surrounding rock and support parameters are shown in Table 1.

Table 1. Mechanical parameters of surrounding rock and support

Surrounding rock level	Gravit γ (kN/m³)	Modulus of deformation E(GPa)	Poisson's ration μ	Friction φ (°)	Cohesion c(MPa)
V	20	1.3	0.35	22	0.12

3.2 Analysis of results

According to the actual engineering geological conditions, the construction simulation is carried out at the entrance of the hole. For the construction method of the CRD method, the axial force variation of the surrounding rock during the construction process is shown in Figure 4. During the construction process by CRD, with the excavation of the left and right guide pits, there is obvious bias phenomenon at the hole, and the axial force of the left arch waist and the right arch of the tunnel structure is larger. After the tunnel is completely excavated, the axial force of the tunnel biasing portion is greatly affected.

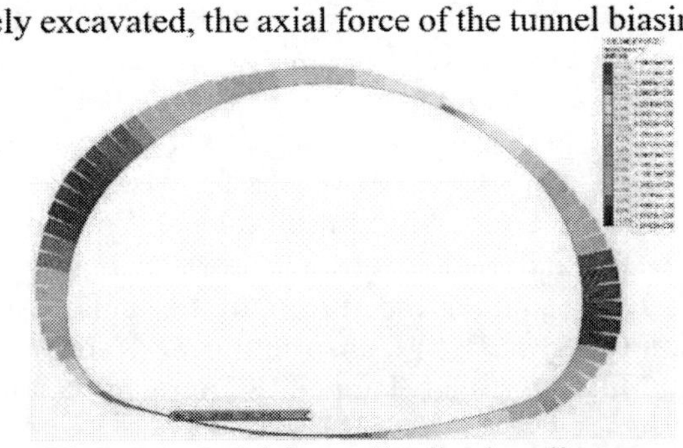

Figure 4. Axial force distribution diagram

In summary, it can be concluded that the biasing part of the tunnel is affected by the thickness of the soil during tunnel excavation. The tunnel structure is mainly affected by the tension and the force is obvious. In the construction stage, the surrounding rock reinforcement treatment of the tunnel entrance section should be strengthened early and designed. Ensure the structural bearing capacity of the biasing section to ensure the safety of tunnel construction.

4. Conclusion

Through the comprehensive analysis of the Xiaoguanshi tunnel, the optimization of the construction scheme of the double-lane shallow-buried bias tunnel is studied. Through the numerical simulation of the excavation section, the following conclusions are drawn:

(1) The construction of the tunnel section of Xiaoguanshi tunnel uses the CRD method to reduce the slope of the tunnel entrance, avoiding the damage to the ecological environment at the entrance, speeding up the construction schedule, reducing the disturbance to the surrounding rock and ensuring the construction safety.

(2) The construction of the hole should control the deformation of the surrounding rock, and the rock mass of the tunnel should be broken. The tunnel should be short-stroke, multi-cycle, weak blasting, strengthen the initial support, and timely lining to prevent the collapse of the landslide and affect the stability of the slope. The design of the top slope and the cave door foundation should be strengthened in advance to avoid landslides due to poor stability of the surrounding rock at the entrance.

Acknowledgments

The authors would like to express their appreciation to the National Natural Science Foundation of China (41601574) , the Chongqing Basic and Frontier Research Project (cstc2015jcyjBX0118) , the Chongqing Science and Technology Innovation Leading Talent Support Program (CSTCCXLJRC201715) , and the Chongqing Social Undertakings and Livelihood Security Science and Technology Innovation and Special Program (cstc2017shmsA30010) for providing funding for this research.

Reference

[1] Huang W.X., Wu H., Wang J., Xia M. (2019) Comparative study on the excavation sequence of weakly biased surrounding rock tunnels[J].Journal of Engineering Geology,27(02):277-285.
[2] Li S., Lu F., Shi R.R. (2019) Study on the Influence of the Offset Angle of Shallow Tunnel on

the Safety of the Back Cavity Lining[J]. Highway, 64(03): 310-314.
[3] Liu R.T. (2019) Construction technology and safety management of large-span shallow buried collapsible loess tunnel[J]. Chinese and Foreign Architecture, (03): 172-174.
[4] Yan T., Li K.J., Qi Z.H., Zhang J.X, Yu L., Li Q., Li G.Q., Peng T. (2019) Study on the Analytical Method of Surrounding Rock Pressure of Shallow Buried Bias Tunnel under Variable Slope Conditions[J/OL]. Journal of Southwest Jiaotong University: 1-7.
[5] Song Q.H. (2019) Analysis and Treatment of Construction Geological Hazards in Shallow Buried Bias Tunnel Openings[J]. Journal of Highway and Transportation Technology (Apps & Technology Edition), 15(03): 93-96.
[6] Li K., Liu K. (2018) Analysis and remediation of track bed uplift in a passenger tunnel[J]. Tunnel Construction (Chinese and English), 38(11): 1910-1818.
[7] Yu Q.Y., Liu W., Yan L., Dai S.L. (2018) Characteristics of Bias Stress Ratio of Biased Tunnel[J].Journal of Jilin University: Earth Science Edition,48(06):1797-1803.
[8] Gao X., Wang H.N. (2019) Analytical solution for formation response of shallow buried tunnel under slope[J]. Water Resources and Hydropower Engineering, 50(01): 1-9.
[9] Deng B.Z, Ma W., Song Y.M. (2018) Construction Technology of Bias and Long Distance Shallow Buried in Shimalongshan Tunnel[J]. Highway, 63(10): 195-199.

IOP Conf. Series: Materials Science and Engineering **711** (2020) 012074 doi:10.1088/1757-899X/711/1/012074

Study on Key Construction Procedures of CRD Method for Re Shuitang No.3 Tunnel

Zhongwei Wang[1,a], Ruihuai Yang[1,b], Hongyan Guo[2,c], Ke Li[2,d]

[1]Shidian Baoshi Expressway Investment and Development Co.,Ltd,Baoshan, 678000, China;

[2]China Merchants Chongqing Communications Technology Research & Design Institute Co., Ltd,Chongqing,400000, China;

[a]e-mail: 3250706322 @qq.com, [b]e-mail: 1679755524 @qq.com,[c]e-mail: 364842342@qq.com, [d]be-mail: 717692502@qq.com,

Abstract. Controlling the critical construction steps of the CRD method plays an important role in ensuring construction and ensuring construction safety. When using the CRD method for the construction of the Re Shuitang No. 3 tunnel, through the numerical simulation of the initial support structure and the temporary support structure during the tunnel construction process, it is concluded that the left upper step, the upper right step and the temporary support structure are obviously stressed. Construction steps. According to the analysis results, the surrounding rock of the vault should be strengthened in advance, and the second lining structure should be applied after the surrounding rock is stabilized.

1. Introduction

When using CRD method in construction, due to the complexity of the construction process, more attention should be paid to each construction step, control the key construction steps, and ensure the construction safety in the parts with poor stability of surrounding rock. Therefore, the analysis of the key construction procedures using the CRD method can guide the key monitoring of key processes during the construction process, while ensuring construction safety has important significance [1-3].

The domestic and foreign scholars used the theory and numerical simulation to do the following research on the construction of the CRD method: Zhou Baosheng analyzed the traditional CRD construction technology and analyzed the surrounding rock based on the analysis of the supply and drainage pipelines of Shenzhen Railway Line 9. Optimization of grouting stable working face, optimization of temporary inverting arch and installation steps of profile steel [4]. Wang Zhijian analyzed the mechanized construction of the large section of Zhengwangao Iron and Steel Co., Ltd., using on-site investigation and theoretical analysis methods to optimize the control of the face surface, optimization of support parameters and construction method during mechanized construction [5].

This paper relies on the Re Shuitang No. 3 tunnel as an engineering example. The numerical simulation method is used to analyze the construction process when using the CRD method. Through the analysis and comparison of the construction steps of the excavation process, the key construction procedures are obtained, which is the construction safety and control. Rock deformation provides research support [6-10].

2. General situation of Engineering

The starting point of the Re Shuitang No. 3 tunnel is K9+780~K10+335, the length is 555m, the buried depth is 66.69m, the left hole is ZK9+765~ZK10+335, the length is 570m, the short tunnel is buried, the depth is 67.65m. The tunnel has a clear distance of 22~25m and the cross section is a separate tunnel. The elevation of the tunnel area is 1759~1830m, and the relative height difference is 73m. It belongs to the dissolution and erosion low mountain landscape. The terrain is steep, the tunnel entrance and exit has no road access, the traffic is inconvenient, the import section is now forest land, and the vegetation is relatively developed.

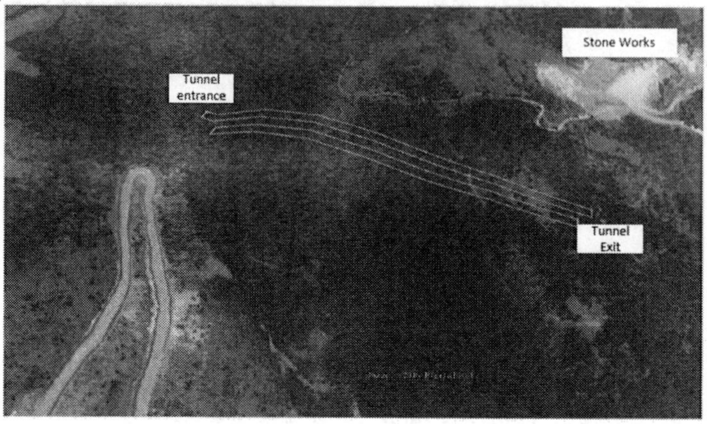

Figure 1 Topography of Re Shuitang Tunnel NO.3

2.1 Engineering geological conditions.

According to the regional geological data of Yunnan Province, the area where the line is located is located in the composite part of the Yunnan-type mountain front structure and the middle branch of the eastern branch of the Qinghai-Tibet-Burma-type structure system. The geological structure is complex in the area, and the tectonic geological structure is the backbone structure. The long-term activity of the zonal structure, the meridional structure is a prominent structural system in the area, which belongs to the meridional tectonic system of the Sichuan-Yunnan system, mainly consisting of the compressive, torsional fracture and small amount of the same direction of the meridional tectonic system.

The upper part of the tunnel crossing area is dominated by limestone in the upper part of the Baoshan Formation of the Cambrian system. It belongs to the dissolution and erosion low-mountain landform. The tunnel mountain is a dome mountain. The top of the mountain is smooth and round, and the slope of the entrance section is steep. The natural slope is 20~30. °; the exit section is slow, the natural slope is 15~20°, and the terrain is undulating.

2.2 Hydrogeology

The project area is located in the Shidian River Basin. The Shidian River originates from the Nangao Mountain Area of the basin. It flows through the basin from the south to the north, and flows into the Nujiang River from the sharp corner bend of Wang Street to the southwest. The basin channel is manually repaired. Straight, the longitudinal slope is small, the flow in the rainy season is large, and the dry season is cut off.

3 Numerical simulation analysis

3.1 Parameter selection

In this paper, the Midas GTS is used to numerically simulate it. As shown in Figure 2, in the simulation process, the surrounding rock adopts the Mohr-Coulomb constitutive model, the support adopts the elastic constitutive model, and the second lining adopts the plane strain model.

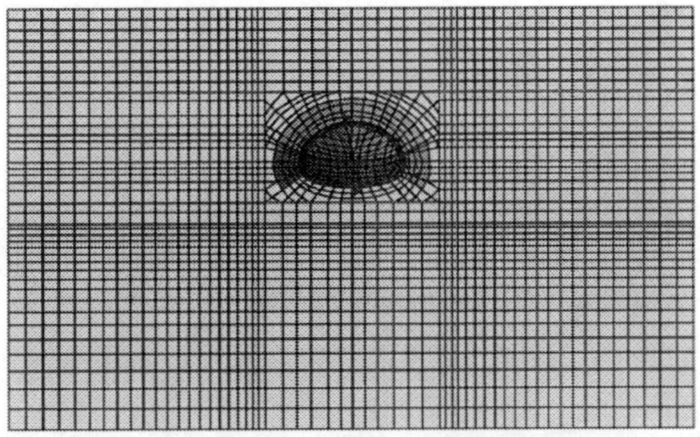

Figure 2. Calculation model diagram

The surrounding rock and support parameters are shown in Table 1.

Table 1. Mechanical parameters of surrounding rock and support

Rock and soil	Density (g/cm3)	Bearing capacity of foundation (kPa)	friction φ (。)	Cohesion (MPa)
Silty clay	18	230	13.4	0.04

3.2 Parameter selection

The whole process of construction is carried out under the condition that the surrounding rock depth is 20m. The excavation and support sequence are as follows: 1 step excavation on the left side wall and full ring support → 2 left wall lower step excavation and full ring closure Support → 3 steps on the right side wall excavation and full ring closure support → 4 right side wall step excavation and full ring closure support → 5 full section initial support closure.

3.3 Analysis of results

During tunnel excavation, with the excavation of each step, the stress change of the shotcrete is shown in Figure 3. The initial support structure is gradually increased. The temporary support of the left guide pit is the main load-bearing structure, and the horizontal pressure. The largest, lateral support is the main load bearing unit. The excavation of the right guide pit makes the bearing capacity of the surrounding rock completely supported by the lateral and vertical supports. Therefore, the supporting structure should be applied in time. When the surrounding rock is stabilized, the support should be strengthened to monitor the deformation of the surrounding rock.

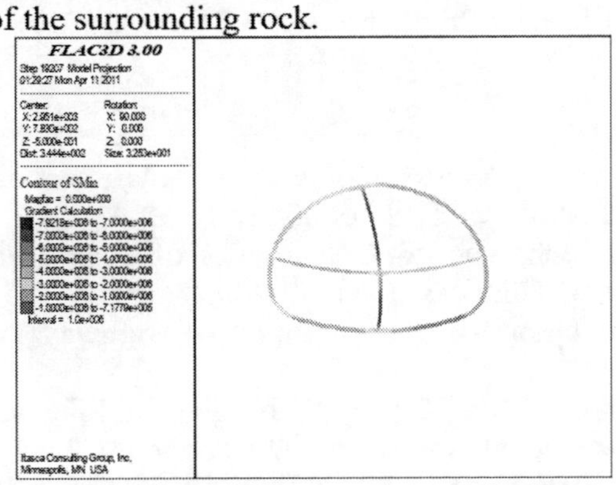

Figure 3. Initial branch structure stress distribution map

When the CRD method is applied, the supporting structure bears the deformation of the surrounding rock before the second lining is applied, and the temporary support is the main force-bearing unit, and the joints of the supporting structure are concentrated, as shown in Fig. 3.4. The stress at the node is at most 0.5 MPa, meeting the design requirements.

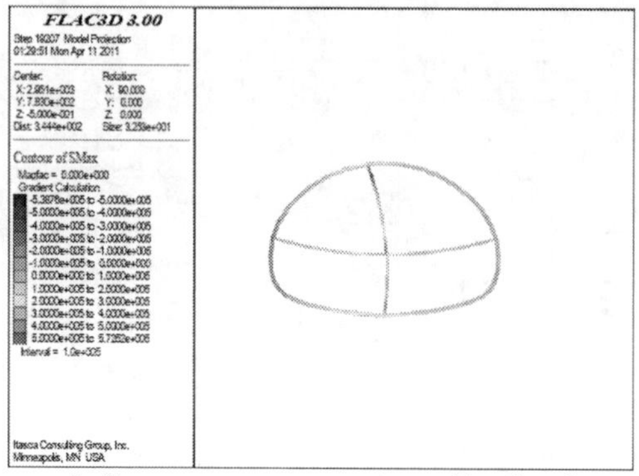

Figure 4. Support structure stress distribution map

In summary, when using the CRD method for construction, attention should be paid to the upper step excavation on the left side, the excavation on the right side, and the temporary support demolition, to avoid the disturbance of the surrounding rock due to the removal of the temporary structure, and the construction takes place.

4. Conclusion

Through the comprehensive analysis of the Re Shuitang No. 3 tunnel, the influence of the construction of the two-lane tunnel on the soil above the tunnel is studied. Through the numerical simulation of the excavation section, the following conclusions are drawn:

(1) The amount of explosives should be strictly controlled during the construction of tunnel blasting excavation, and the possibility of damage to the top of the building caused by blasting vibration should be considered.

(2) The tunnel is a short tunnel. Due to the fracture of the rock mass and the requirements of the construction period, the distance between the two working faces should be strictly controlled to avoid the superposition effect of the two-hole stress release ring.

Acknowledgments

The authors would like to express their appreciation to the National Natural Science Foundation of China (41601574) , the Chongqing Basic and Frontier Research Project (cstc2015jcyjBX0118) , the Chongqing Science and Technology Innovation Leading Talent Support Program (CSTCCXLJRC201715) , and the Chongqing Social Undertakings and Livelihood Security Science and Technology Innovation and Special Program (cstc2017shmsA30010) for providing funding for this research.

Reference

[1] Zhai Z.J. (2019) Analysis of key technologies for small-section tunnel construction [J/OL]. Journal of Hubei Institute of Technology, (02): 39-42.

[2] Chen A.G. (2019) Key Techniques for Construction of Huayingshan High-pressure Water-rich Tunnel[J]. Southwest Highway, (01): 39-44.

[3] Guo Z.F. (2019) Research on Key Technologies of Highway Tunnel Safety Construction[J]. Traffic world,(08):87-88.

[4] Wang Z.J. (2018) Research on Key Technologies of Mechanized Construction of Large Section of Zhengwan High-speed Railway Tunnel[J]. Tunnel Construction (Chinese and English), 38(08): 1257-1270.

[5] Zhou B.S. (2018) Optimization of Construction Technology of Super Shallow Buried Tunneling by CRD Method[J].Modern Tunnel Technology,55(04):186-190.

[6] Fan R.M. (2018) Key Techniques for Construction Quality Control of Long Railway Tunnels[J]. Journal of Shijiazhuang Railway Vocational and Technical College,17(04): 9-13.

[7] Li F.G., Zhao Y.H. (2018) Key Techniques for Subsurface Excavation Construction of Urban Shallow Tunnel with Ultra-shallow Buried Extra Large Section and Straight Wall Tunnel[J]. Tunnel Construction (Chinese and English),38(11): 1868-1877.
[8] Xian Guo, Yin J.W. (2018) Key Technologies and Practices of Chenglan Railway Tunnel Construction[J]. Tunnel Construction (Chinese and English), 38(10): 1741-1752.
[9] Li Y.L., Zhang L., Tao Z.Z., Liu H.P. (2018) Key Techniques for Construction of Small Clear Distance Tunnels in Weak Surrounding Rocks in Southwest Mountainous Areas[J]. Highway, 63(10): 190-194.
[10] Zheng J. (2018) Key Construction Technology of Shield Tunneling in Railway Station Area[J]. Journal of Shijiazhuang Railway University (Natural Science Edition),31(03): 29-35.

Experiment and finite element analysis study on the deflection of aluminum extruded 6063-T5 hollow structural beam

Minghsiung Ho[1,a], Pinghui Lee[1,b], Chao Ye[2,c], Chipang Lin[1,d] and Chenfu Huang[1,d]

[1] Department of Mechanical Engineering, Nanya Institute of Technology, Jhongli, Taiwan, R.O.C.

[2]Engineering & Development, KingStone Machinery Co., Ltd., Changshu City, Jiangsu Province, China.

[a]morrisho@nanya.edu.tw, [b]phui@nanya.edu.tw, cyechao@ksmsh.com, [d]jefflin@kimo.com, [e]hf9456@nanya.edu.tw

Abstract. In this study, the tensile tests were performed on samples extruded from aluminum beam to confirm the material properties of the various parts. Then, the beam was cut into lengths of 1160 mm and 680 mm to perform the bending test, and the beam loading point deformation was recorded. At the same time, the amount of beam deformation caused by three type beam fixing methods was compared. The Solidwork 3D CAD software used to build the beam model and perform finite element analysis. Among them, the analysis results of different boundary conditions were compared with the experimental results. The experimental load conditions were used to calculate the theoretical deformation formula amount through the beam theory deformation formula of material mechanics. It is to compare beam deformation data of experiments, finite element analysis, and theoretical formulas, and explore possible causes of differences.

1. Introduction

The aluminum alloy material is a kind of green metal, which can be recycled and remanufactured. Although the aluminum alloy strength is smaller than that of the steel material, the same volume is lighter than the iron metal and corrosion-resistant, so it is widely used in the aerospace and vehicle industries. Since 1950, aluminum extrusion molding technology has made significant progress in mold design and process precision due to high productivity and low-cost requirements. The development trend of aluminum alloy extrusion technology can be divided into two directions: "high strength" and "high precision." How to improve the precision of products, and thus improve their efficiency, or achieve high-quality and high-strength, has been regarded as one of the foremost research topics. High-precision aluminum alloy extrusion products, the main appeal is that the performance of the extruded product can be improved. From the beginning to the end of the extrusion, the yield of extruded products can reach more than 90%, and the dimensional accuracy and material quality of the head, middle and tail materials need to be uniform. The hollow extruded aluminum beam product has a standardized appearance, interchangeable end face dimensions, and standard cross-sectional dimensions of various lengths. The inner section of the hollow aluminum forming beam is designed with reinforcing ribs to meet the high rigidity requirement. The shape of the rib is directly related to the structural strength and the high efficiency of material use, and the purpose is to achieve the highest rigidity with a minimum material cost.

About beam deformation and experimental research, Wang and his coworker[1] studied large deformation cantilever beam under point load at the free tip. They investigated a homotopy analytic method (HAM) to calculate the vertical and horizontal displacements of a cantilever beam with large deformation. Cruz [2] used large geometric deformations and assuming a Galerkin approach to model gas turbine fans, wind turbine blades, springboards dynamic behavior. Lee [3] had researched the elastic and buckling loads of the nonlinear elastic tapered cantilever columns subjected to an axial load at the free end. Lee [4] obtained a numerical solution by using Butcher's fifth-order Runge–Kutta method and presented in a tabulated form. Ho [5] had analyzed of the aluminum alloy structure beam of 5086-H32, 6061-T6, 7005-T6 under cantilever loading conditions.

The purpose of the study was to understand the mechanical properties of aluminum extruded beams. Through structural experiments and finite element analysis, the effects of structural beam length and restraint fixation on deformation are studied. Moreover, find out the relationship between the deformation formula of the cantilever beam and the experimental method.

2. Methods

Material. The 6063-T5 aluminum extrusion material used in the experiment was provided by KingStone Machinery (Changshu) Co., Ltd. The length of the extruded beam was 4 m, and the beam section size was 80x80 mm. The dimensions of the extruded beam section profile are shown in Fig. 1. The 4 m beam was cut and machining in two types of beams, as shown in Fig. 2 and 3. The beam of length 1160 as model A. The beam of length 680 as model B. In Table 1, the geometric properties of the test and analysis of the two beams were listed.

Experiment. There are two experiment methods in this study. The first is the tensile test to understand the material properties of aluminum extruded beams. Secondly, the bending tests are performed on the beam structure to test the structural rigidity.

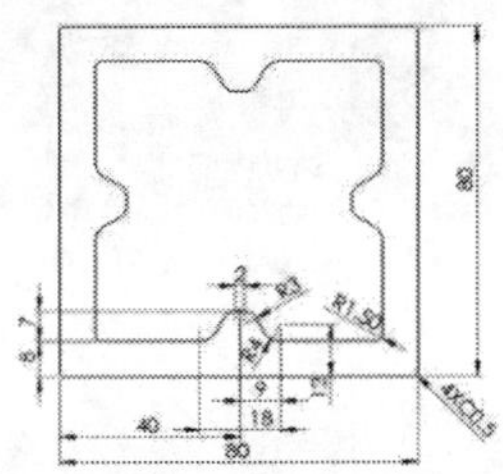

Figure 1. The section dimension of analysis models.

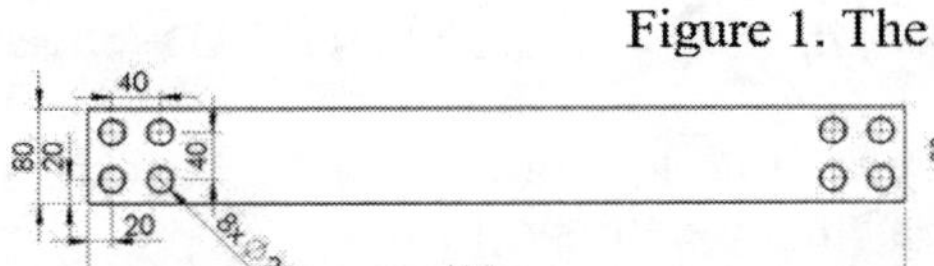

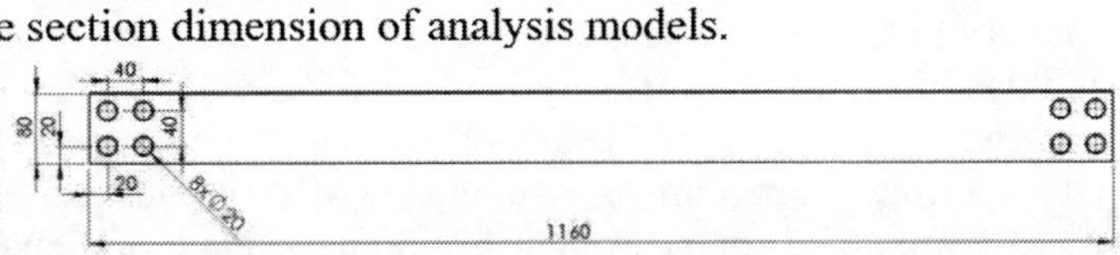

Figure 2. The dimension of beam length 680 mm Figure 3. The dimension of beam length 1160 mm

Table 1. The geometric properties of the test and analysis beams provided.

Model	Length	Model weight	Actual weight	Section modulus	Cantilevered loading length	3 points bending length
Unit	mm	kg	kg	m^4	mm	mm
A	1160	7.776	7.640	2.132E-06	1010	940
B	680	4.442	4.365	2.132E-06	530	

Tensile test. The mechanical properties of the extruded beam at the axial and lateral position was studied. The tensile test specimens were sampled in the front, middle, and end positions of the 4m extrusion beam. The samples were machined, and the dimensions of the test specimens are shown in

Fig. 2. The tensile tests would follow ASTM E8M specifications. The experiments used displacement control with a speed of 6 mm/min. The data of the force and strain are recorded during the tensile process.

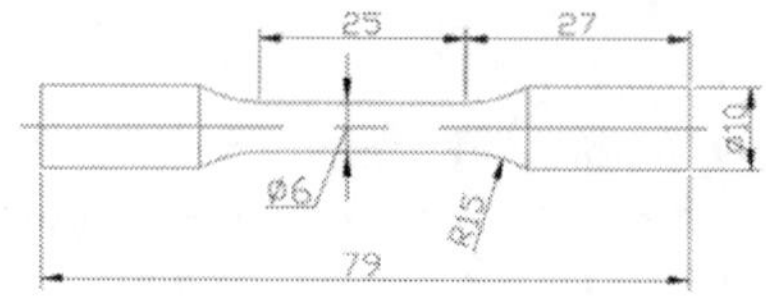

Figure 2. The specimen dimensions.

Beam bending test. The test is to let the designer understand the deformation state of the beam under the load. The experiment uses the end load bending and three-point bending experimental method. The test bench experimental architecture of the two experimental methods are shown in Fig. 3. The beam experiment would use three kinds of endpoint restraint fixation methods as shown in Fig. 4 to compare the amount of deformation at the same loading condition. The test lengths of the beam are list in Table 1. The dial gauge is used to measure the deformation of the beam at the loading position.

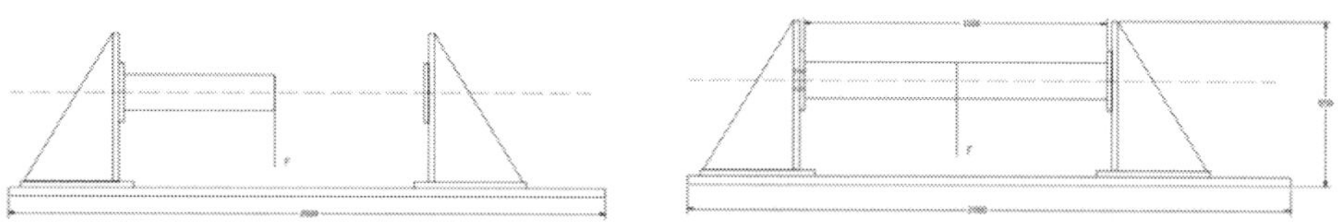

(a) Cantilevered beam test (b) Three points bending test

Figure 3. Two types of tests bench layout.

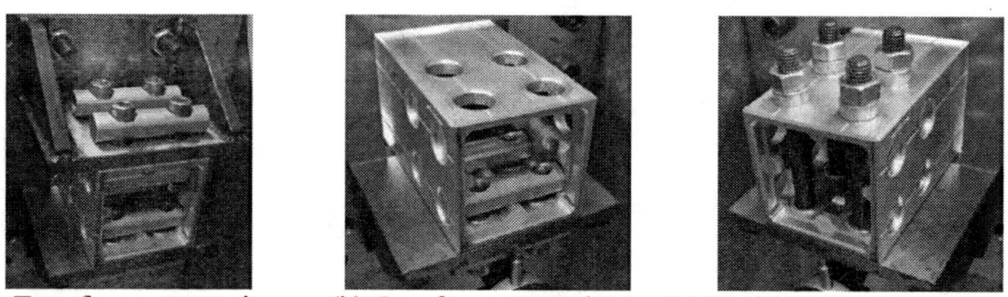

(a) Two faces constrain. (b) One face constrain. (c) Position pin constrain.

Figure 4. Constrain types.

Finite element analysis. The purpose of study by CAE is to use the finite element method to compare with the experimental results. The 3D models were built using Solidwork 2017 CAD software, and then the deformation analysis was performed.

Model. The analyzed 3D models were based on the section of Fig. 1. In order to match the length of the experimental specimens, two lengths of 1160 mm and 680 mm were built.

Loading. The counterweight of the experiment and analysis was 201 N. There are two conditions for loading as shown in Fig. 4. One is fixed at one end and loaded on the cantilever end of the beam as shown in Fig 4(a). The other is that the ends of the beam are fixed and loaded at two intermediate positions as shown in Fig 4(b).

Constrains. According to the fixed method of the experiment, there are three kinds of constraints on the analysis model. The end face fixing methods of the beam were two faces restricted by the pin with the screw fixing as shown in Fig. 5(a), the lower faces restricted by the pin with the screw fixing as shown in Fig. 5(b), and the positioning pin and the screw lock the upper and lower face as shown in Fig. 5(c). The beam length of constraining face was 110 mm.

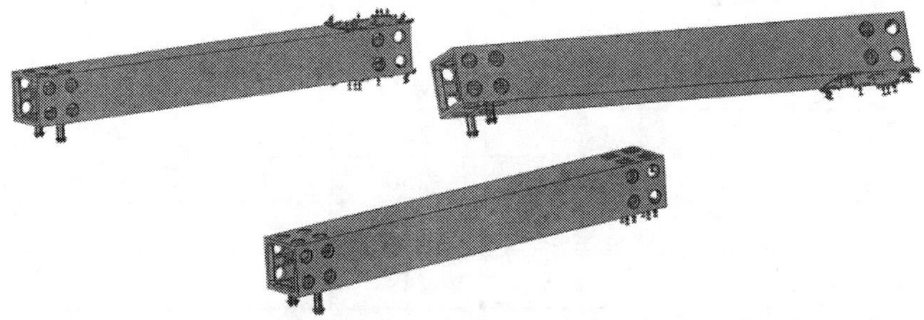

(a) Two faces constrain (b) One face constrain (c) Pin holes constrain
Figure 5. Models of constrain types.

3. Results and discussion

Material properties. The material properties of Young's modulus, yielding stress, and the ultimate stress were obtained through the stress and strain data of the tensile test. The yielding stress was obtained by the 0.2% strain offset method. The 6061-T5 aluminum extruded beam tensile stress-strain curve are shown in Fig. 6. In the axial direction, the head, middle and tail sampling positions, the tensile stress and strain curves are well superimposed in Fig. 6(a). However, in Fig. 6 (b), in the lateral direction, head, middle, and tail sampling positions, the tensile stress-strain curves are different. The sample sampled at the head position has high elongation. In Table 2, it was shown that three material properties in the axial direction, in which the yield stress and the ultimate stress are higher than the lateral direction except for the Young's modulus.

It is observed from the tensile properties that the material in the axial direction of the beam during the aluminum extrusion process has enough soothing space and can be cooled slowly during the process. However, the beam has a lateral width of only 80 mm. However, the beam has a lateral width of only 80 mm. Because of the effect of the restricted mold and rapid cooling, the Young's modulus is higher, but the lodging and ultimate stress properties are slightly worse. As a whole from Table 2, although the material properties of the axial direction and the lateral direction are directional due to extrusion. However, the difference in material properties is not too significant, so aluminum extruded beams could be considered as isotropic materials. The finite element analysis will use average material properties in both directions: E = 68.5 GPA, **σ**y = 222 MPa, **σ**u = 245.5 MPa.

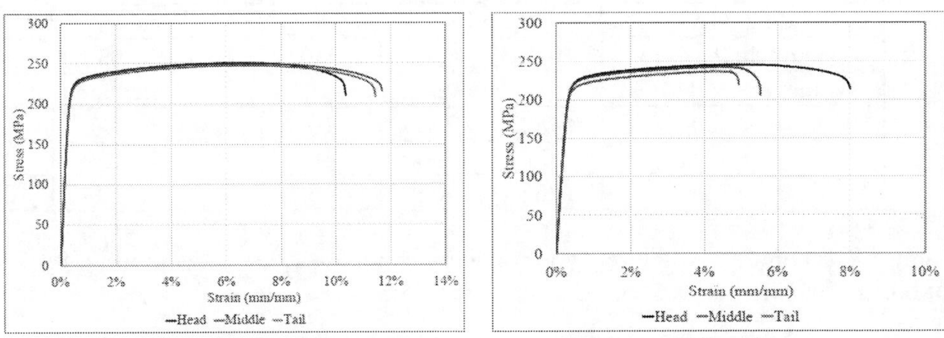

(a) Axial direction (b) Lateral direction
Figure 6. The tensile stress-strain curve of 6063-T5 aluminum extrusion beam.

Table 2. The material properties of 6063-T5 aluminum extrusion beam.

Sampling	Axial			Lateral		
Properties	E	σy	σu	E	σy	σu

	Young's Modulus	Yielding stress	Ultimate stress	Young's Modulus	Yielding stress	Ultimate stress
Position	GPa	MPa	MPa	GPa	MPa	MPa
Head	69	226	251	72	223	245
Middle	68	224	250	69	223	243
Tail	67	220	246	67	216	237
Average	**68**	**223**	**249**	**69**	**221**	**242**

Deflection analysis. The deflection analysis is achieved in two ways, one using experimental methods and the other using CAE computer simulation analysis.

The experiments were done on the test bench. According to the load position, it was divided into two experimental methods. One is that the cantilever beam was loaded 201 N at the end of the beam shown in Fig. 7(a), and the other end was fixed. Another experiment was loaded 201 N at the intermediate of the beam and fixed at both ends as shown in Fig. 7(b). The experimental results are shown in Table 3. The results were shown that the fixing method of the boundary condition (b) would maximize the bending deflection, followed by the boundary condition (c) and the boundary condition (a) is the smallest. This trend does not change due to the beam length and load position.

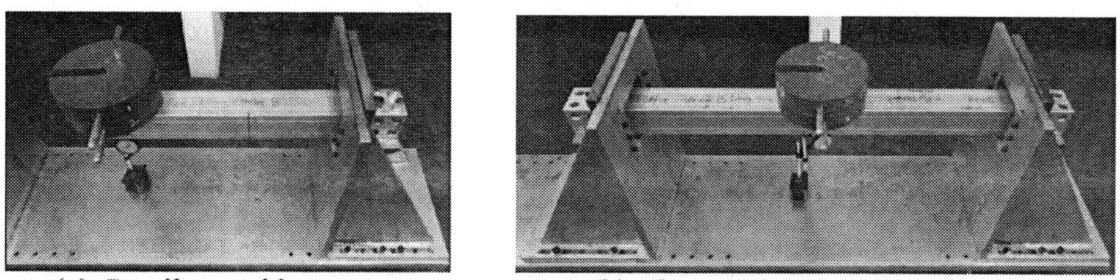

(a) Cantilevered beam test (b) Three points bending test

Figure 7. Two types of tests bench layout.

Table 3. Beam deflection analysis.

Method	Experiment test			CAE nonlinear analysis		
Loading position	End		Middle	End		Middle
Beam length (mm)	1010	530	940	1010	530	940
Boundary conditions	Deflection (μm)			Deflection (μm)		
a	1482	234	10	565	84	7.43
b	2335	466	18	770	139	13.74
c	1965	365	16	639	104	12.54

The finite element model with three type boundary conditions were shown in Fig. 8. The deflection is in Y axis direction.

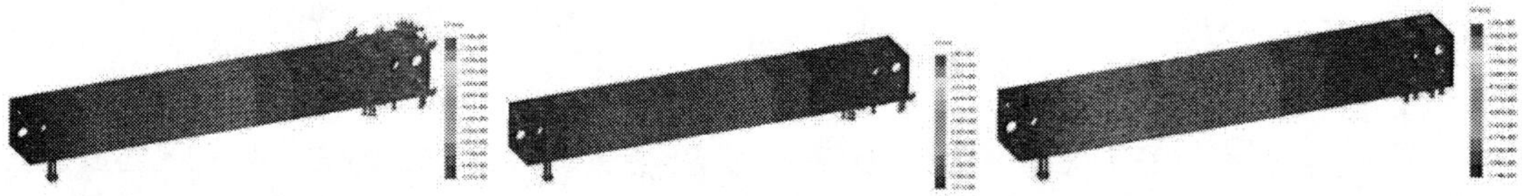

(a) Two faces constrain (b) One face constrain (c) Pin holes constrain

Figure 8. The three finite element analysis models with boundary conditions.

Fixed the beam both ends and with an intermediate load, the experimental method is similar to the condition of 3-point bending. In the meantime, the experimental and CAE analysis data Compared to the experimental results of the cantilever method, the deformation data is relatively small, and the experiment is similar to the CAE analysis data. From the experimental and analytical results data, it is close to the elastic analysis, not the phenomenon of large deformation.

Under the experiment of cantilever beam loading, the deflection of the beam is the largest. This experimental model is already in a state of large deformation. The three-point bending experimental method with fixed ends has the smallest deflection. Experimental data is near linear elastic deformation state. This state is because the structural rigidity of the beam will be better under the constraints of both sides.

From the experimental results of the cantilever beam load, the beam deflection formula Equ. 1 is applied. Among them, the index x of the beam length is used to compare the indices of different restraint conditions. The formula index from the known beam endpoint when fully fixed is x = 3.

$$v_{max} = \frac{PL^x}{3EI} \quad (1)$$

In Equ. 1, The values are known: P = 201 N, E = 68.5 GPa, I = 2.132×10^{-6} m^4, L = 1010 mm, and 530 mm. Since in equation (1), the independent variable is the beam length L, and the dependent variable is v_{max}, other parameters are known. By introducing the deformation values obtained for the two experimental lengths into equation (1), the index x value can be obtained. The experimental data of Table 3 are calculated to obtain the index x of Table 4. The index x represents the degree of restraint in the boundary conditions. The closer the index x is to 3, the closer it is to the ideal full restraint condition. The beam boundary condition (a) is close to complete restraint. The coefficient of CAE analysis in condition (a) is close to 3; this indicated that the model's boundary condition is nearly wholly constrained. Boundary condition (b) in test and analysis is fixed in one surface; the beam is slightly constrained. The index x at boundary condition (b) is the minimum value. Boundary condition (c) is fixed by the positioning pin at the upper and lower holes of the beam surface. The degree of restraint is between the boundary conditions (a) and (b).

Table 4. The index coefficients x of deflection formula.

Boundary conditions	Test	CAE
a	2.863	2.956
b	2.499	2.655
c	2.611	2.815

4. Conclusion

1. The material properties of the aluminum extruded beam are not much different from the tensile test results in the axial direction, and the lateral direction and can be regarded as an isotropic material.

2. The experiment of 3 point bending of the beam and the CAE analysis results show the elastic deformation phenomenon of small displacement, and the two deformation results are very close.

3. The large deformation analysis of the beam is analyzed by CAE nonlinear method, and the result is quite different from the experiment. Some factors are not included in the analytical model, which may be caused by assembly gaps, deformation of the fixture.

4. The index of the test length in the cantilever beam deformation formula can be used to check the degree of restraint of the experimental and analytical models. The model of the CAE boundary condition (a) is closest to the ideal constraint condition.

References

[1] J. Wang, J.K. Chen and S.J. Liao: J. Comp. App. Math. Vol. 212 (2008), p. 320
[2] G.Cruz, and H.R. Gilberto: Mech. Eng. Vol. 62 (2016), p. 187
[3] J . Leeand B. Lee: Eng. Solid Mech. Vol. 6 (2018), p. 39

[4] K.G. Lee: Mechanics Vol. 37 (2002), p. 439
[5] M.H. Ho, P.N. Wang and J.P. Yeh: AIP Conf. Pro., Adv. Eng Sci. Env. Eng Vol. 1829 (2017)

Study on the preparation and influence of alkali metal oxides R_2O on properties of microfiber glass with high water resistance

Xiaotian Hao[1,2], Qingjiang Wang[1], Hailong Wang[3,*]

[1]College of Urban Construction Engineering, Chongqing Technology and Business Institute, Chongqing 401520, China; haoxiaotian521@163.com

[2]Key Laboratory of Roads and Railway Engineering Safety Control (Shijiazhuang Tiedao University), Ministry of Education, Shijiazhuang 050043, China;

[3]School of Civil Engineering, Hebei University of Architecture, Zhangjiakou 075000, China; wanghailong-65@163.com

Abstract: The microfiber glass with high water resistant was prepared by the centrifugation with synchronous water-repellent atomization spray. The influence of the alkali metal oxide $Na_2O/(Na_2O+K_2O)$ ratio on the fiber forming ability, chemical stability, thermal and acoustical insulation properties of the Si-Ca-Mg-Al-Fe-B multi-glass fiber system was studied by combing with the "mixed alkali effect". It was found that when the ratio $Na_2O/(Na_2O+K_2O)=0.97$, the fiber forming ability and chemical stability of the prepared microfiber glass were optimized. The transmission loss and the sound absorption coefficient are distributed at 2000Hz at a maximum of 9.1dB and 0.88. Therefore, under this paper preparation conditions the optimal Na_2O and K_2O distribution ratio of Si-Ca-Mg-Al-Fe-B multi-glass fiber system is 16wt% and 0.5wt%, respectively. Each prepared fiber is tightly connected together by the surface treatment of the water repellent to form a stable fiber network structure, demonstrating the feasibility of producing high water resistant microfiber glass by the centrifugation with synchronous water-repellent atomization spray in practical production applications.

1. Introduction

Energy is the material basis for human survival and an important support and key strategic material for the country's economic development. At present, China is in a rapid stage of urbanization development so that the huge energy is demanded. Furthermore, this demand will become more urgent in the near future. Therefore, the development of green intelligent buildings towards smart ecological cities is the inevitable direction for solving the energy crisis, resulting in that the development of high-performance green building materials is becoming the core of development [1-3].

Microfiber glass is a new type of green building material, which possesses the excellent thermal insulation performance, non-combustible, light weight, electrical insulation and corrosion resistance to provide people a comfortable and safe living environment and reduce the overall energy consumption of building construction and operation by reducing CO_2 emissions [4-5]. In the composition system of microfiber glass, alkali metal oxide R_2O corresponding to Na_2O and K_2O has a great influence on the physical and chemical properties of microfiber glass [6], but the study on the effect of alkali metal oxide R_2O on the properties of microfiber glass by combing with the "mixed alkali effect" are rare reported [7].

In this paper, the microfiber glass was prepared by the centrifugation with synchronous water-repellent atomization spray. The effect of the alkali metal oxide $Na_2O/(Na_2O+K_2O)$ ratio on the fiber forming ability, chemical stability, thermal insulation performance and sound insulation performance of Si-Ca-Mg-Al-Fe-B microfiber glass system was researched, which has a guiding help for further improving the performance of microfiber glass used in the green building materials.

2. Experimental method

2.1 Sample preparation

The microfiber glass with high water resistant mentioned in this paper is prepared by the simultaneous process of centrifugation and atomization spray. First, according to the chemical designed composition of the microfiber glass shown in Table 1, selecting the appropriate amount of quartz sand, soda ash, potassium feldspar, albite, calcite, borax, dolomite, barium carbonate and zinc oxide raw materials, and then the raw material mixture was poured into the furnace with the temperature of 1300-1450°C for the high-temperature calcination in order to remove bubbles and impurities and melt the mixture into a uniform transparent glass liquid. Then the glass liquid flowed through the nickel-cobalt alloy bushing into the centrifuge, and the inflowing glass liquid (850-1000°C) was centrifuged by high-speed rotation spinner with the small holes (ø=0.8mm), resulting in taking out of microfiber glass with an average fiber diameter of ø=2.5μm. Meanwhile, the 2.5wt% hydroxy silicone oil water-repellent agent was sprayed uniformly on the surface of the prepared microfiber glass during the falling collection process. The surface-treated fiber was evenly adsorbed and laid on the mesh belt of the fiber collecting machine with a frequency of 32 to 35 Hz negative induced pressure according to a certain fiber arrangement. Finally, the prepared microfiber glass was dried at 150-175°C for 3-4mins to form a high water-resistant microfiber glass wool with an areal density of $2000g/m^2$。

Table 1 the chemical composition of microfiber glass

SiO_2	CaO	MgO	Al_2O_3	Fe_2O_3	B_2O_3	R_2O $(Na_2O + K_2O)$
64.5	6.2	3.0	3.2	0.13	6.5	X= $Na_2O/(Na_2O + K_2O)$ (X=0.85, 0.88, 0.91, 0.94, 0.97, 1)

2.2 Test characterization

First, a sample of microfiber glass wool with a size of 50×15mm was taken to weigh and record the value as m_1. Then the sample was soaked into a plastic container with 500g of pure water (pH=7±0.1) for 24 hours, and then its weight recording as m_2. So the hydrolysis rate ρ of sample can calculated by the following the formula (1),

$$\rho=(m_1-m_2)/m_2\times10^6 \quad (1)$$

Where: m_1: weight before soaking (g); m_2: weight after soaking (g); ρ: rate of hydrolysis (ppm).

The German Netzsch HFM 436 heat flow analyzer was used to test the thermal conductivity λ of the prepared microfiber glass sample at room temperature. Before testing, the sample was cut into a 300×300mm square and placed horizontally on the cold plates of analyzer, and the opposite hot plate was pressed down to compress the sample to a certain thickness before starting the test.

The coefficient of thermal expansion α of the prepared microfiber glass sample was measured using a German NETZSCH DIL 402C dilatometer. A strip sample with a size of 10×10×50mm was first prepared, and the test temperature was set to 25-300 °C.

The surface and cross-section microstructure of the prepared microfiber glass sample was observed and characterized by Japanese electronic JEOL JSM-6360 scanning electron microscope (SEM). Since the glass sample is not electrically conductive, it is necessary to process the gold spray treatment before the observation.

The acoustic-insulation performance test of prepared microfiber glass sample was carried out by the standing wave tube method. The test system is the sound absorption coefficient of AWA6290T transfer function.

3. Results and analysis

3.1 Analyzing fiber forming ability and chemical stability of prepared microfiber glass sample

Table 2 shows the effect of the ratio of alkali metal oxide $Na_2O/(Na_2O+K_2O)$ on the melting temperature T_m, forming temperature T_w, liquidus temperature T_l, forming temperature interval ΔT and hydrolysis performance ρ of Si-Ca-Mg-Al-Fe-B multi-glass fiber system. It can be seen from Table 2 that as the ratio of the alkali metal oxide $Na_2O/(Na_2O+K_2O)$ increases, the T_m, T_w, T_l and ρ of the prepared microfiber glass sample gradually decrease and ΔT gradually increases. When $Na_2O/(Na_2O+K_2O)$=0.97, the samples T_m, T_w, T_l and ρobtain the minimum value, respectively. ΔT takes the maximum value. When the ratio of alkali metal oxide $Na_2O/(Na_2O+K_2O)$ further enhances to be 1, corresponding to the alkali metal oxide in the ultrafine glass fiber cotton is all Na_2O, the T_m, T_w, T_l and ρ of sample rise again, and ΔT appears to fall.

Table 2 the effect of the ratio of alkali metal oxide $Na_2O/(Na_2O+K_2O)$ on the melting temperature T_m, forming temperature T_w, liquidus temperature T_l, forming temperature interval ΔT and the hydrolysis performance ρ of Si-Ca-Mg-Al-Fe-B multi-glass fiber system

Sample	X= $Na_2O/(Na_2O+K_2O)$	T_m/°C	T_w/°C	T_l/°C	ΔT/°C	ρ/ppm
1	0.85	1452	1141	1066	75	1198
2	0.88	1432	1128	1031	97	1122
3	0.91	1415	1098	986	112	1058
4	0.94	1403	1065	940	125	995
5	0.97	1382	1011	869	142	978
6	1.00	1398	995	873	122	1047

The variation curve of the forming temperature interval ΔT and the hydrolysis performance ρ of the Si-Ca-Mg-Al-Fe-B multi-component glass fiber system with the ratio of the alkali metal oxide $Na_2O/(Na_2O+K_2O)$ is shown as Fig.1. When $Na_2O/(Na_2O+K_2O)$=0.97, corresponding to the mass fractions of Na_2O and K_2O as be 16wt% and 0.5wt%, respectively, the forming temperature range reaches the maximum value ΔT_{max}=142°C, and the hydrolysis performance reaches the minimum value ρ_{min}=978ppm. The forming temperature interval ΔT reflects the difficulty in forming the glass fiber, resulting from the wider the forming temperature range, the easier the formation of the microfiber glass. In the process of preparing microfiber glass by the centrifugation and simultaneous atomization spray, since the large contact area and the long contact time between the centrifugal spinner and the glass liquid, when the melting temperature, the forming temperature and the liquid phase temperature are not properly controlled, leading to the increase of crystallization possibility of the glass liquid. Once the glass liquid appears submicron-crystal or full crystal, it is easy to cause clogging for microporous of the centrifugal spinner, which makes it difficult to prepare the microfiber glass by the centrifugation method. At the same time, since the centrifugal spinner is limited by the temperature resistance of the material, centrifugal fiberglass is required to be prepared at the liquidus temperature as low as possible. If the molten temperature of glass system is too high, the centrifugal spinner and the bushing will be eroded by the high temperature molten glass, which will greatly shorten their service life, thereby increasing the forming difficulty and production cost of the preparation of microfiber glass. The smaller ρ value of the prepared microfiber glass sample, the more stable the chemical properties of the sample. Both Na_2O and K_2O are the most common alkali metal oxides in high-silica glass system. The Na_2O provides the oxygen ion with high polarization for the glass fiber system, which mitigates the competition of Si^{4+} with metal ions such as Ca^{2+}, Mg^{2+} and Al^{3+} for oxygen ions, polarizes bridge oxygen and weakens the silicon-oxygen bond, resulting in loosening glass fiber structure. Therefore, adding Na_2O can reduce the high temperature viscosity of the glass system, and the chemical stability of the prepared microfiber glass is lowered. Due to the mixed alkali effect, when the Na_2O is partially replaced by the K_2O corresponding to the ratio of $Na_2O/(Na_2O+K_2O)$

decreasing, the forming temperature interval ΔT and the hydrolysis performance ρ don't change in a unidirectional linear relationship, but there is an inverse linear transformation and appears the extreme value. Due to that the K^+ radius is larger than Na^+ and its field strength is smaller than that of Na^+, so the agglomeration capacity for the surrounding non-bridge oxygen ions of K^+ is weaker than that of Na^+, resulting in that the excessive free oxygen makes the internal network structure destruction ability of the glass fiber system enhanced, the chemical bond is aggravated and the network structure becomes loose when K_2O partial replacement of Na_2O in glass fiber system compared with only pure Na_2O. Therefore, when the ratio of $Na_2O/(Na_2O+K_2O)$ is equal to 0.97, the optimum ratio of Na_2O and K_2O in the Si-Ca-Mg-Al-Fe-B multi-glass fiber system is 16 and 0.5wt%, respectively. The fiber forming ability and chemical stability of the prepared microfiber glass sample under the reported test conditions are optimized.

3.2 Analyzing of expansion coefficient and thermal conductivity of ultra-fine glass fiber cotton

Fig.2 is a graph showing the expansion coefficient α and the thermal conductivity λ of the Si-Ca-Mg-Al-Fe-B multi-glass fiber system as a function of the ratio of $Na_2O/(Na_2O+K_2O)$. As shown in Fig.2, as the ratio of $Na_2O/(Na_2O+K_2O)$ increases, the α and λ of the prepared microfiber glass sample gradually decrease. When $Na_2O/(Na_2O+K_2O)=0.97$ corresponding to the mass fraction of Na_2O and K_2O as 16 and 0.5wt%, the expansion coefficient and thermal conductivity obtain the minimum value $\alpha_{min}=86.283°C^{-1}$ and $\lambda_{min}=0.028W/(m\cdot K)$, respectively. But when the ratio of $Na_2O/(Na_2O+K_2O)$ further enhance to 1 representing the all alkali metal oxide in multi-glass fiber system as Na_2O, the α and λ value appear to be increasing.

The main reason for the above phenomenon is that when the K_2O partially replacing part of Na_2O in the glass fiber system, the repulsive force between the pure Na^+ is greater than that between Na^+ and K^+, so that the binding force of Na^+ and K^+ in the mixed alkali system with the glass fiber network is greater than that of the single alkali metal ions. When the free path of the gas molecule is larger than the pore diameter of microfiber glass wool, the gas molecules are generally in a static state, and most of the gas molecules are adsorbed on the pore walls inside the material and a large number of semi-connected or closed pore networks inside the microfiber glass wool. Meanwhile, the structural factors of microfiber glass wool make it difficult for gas molecules to undergo thermal convection. The layered of microfiber glass facilitates dense overlap and uniform dispersion between fibers and reduces heat convection. In addition, since heat conduction can only be conducted between solids, the thermal conduction can only pass through a small number fiber of each fiber layer due to the layered structure characteristics of microfiber glass wool.

With the decrease of the ratio of $Na_2O/(Na_2O+K_2O)$ corresponding to the alkali K_2O more substituting for Na_2O in fiber glass system, the migration of alkali metal ions is restricted the viscosity of the fiber glass system is limited, resulting in the

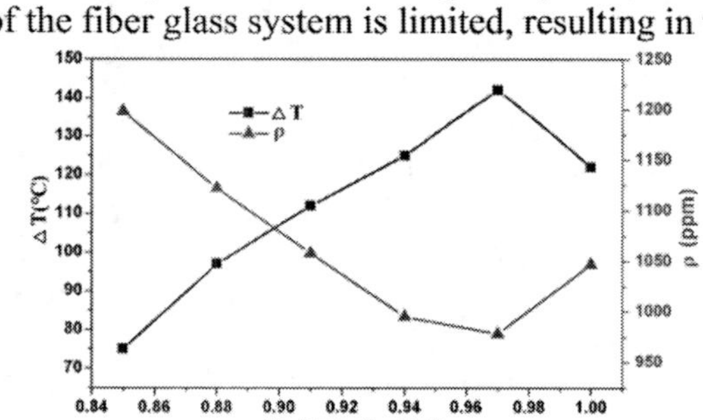

Fig.1 The effect of the ratio of $Na_2O/(Na_2O + K_2O)$ on the forming temperature interval ΔT and hydrolysis performance ρ

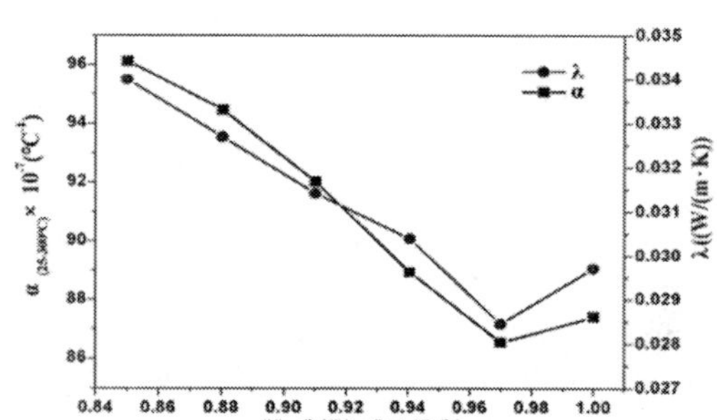

Fig.2 The effect of the ratio of $Na_2O/(Na_2O + K_2O)$ on the expansion coefficient α and thermal conductivity λ

continuous crystallization and an increase in thermal conductivity of preparedmicrofiber glass due to the increased thermal conductivity of the glass fiber skeleton. Therefore, the optimum ratio of Na_2O and K_2O in the Si-Ca-Mg-Al-Fe-B multi-glass fiber system is 16wt% and 0.5wt%.

3.3 Analyzing of sound insulation performance of prepare microfiber glass

Fig. 3 is shown a sound insulation characteristic curve of Si-Ca-Mg-Al-Fe-B multi-component fiber glass system with the ratio of $Na_2O/(Na_2O+K_2O)$. As shown as Fig. 3 (a, b), at the same noise frequency, the comparison of sound transmission loss and the sound absorption coefficient of prepare microfiber glass increase continuously with increasing the $Na_2O/(Na_2O+K_2O)$ ratio. When $Na_2O/(Na_2O+K_2O)=0.97$ corresponding to the mass fraction of Na_2O and K_2O as be 16wt% and 0.5wt%, respectively, the sound transmission loss and sound absorption coefficient of the sample are at 2000Hz frequency reach to the maximum values of 9.1 dB and 0.88, respectively. When the ratio of $Na_2O/(Na_2O+K_2O)$ is further increased to 1, that is, the alkali metal oxide in the prepared microfiber glass is all Na_2O, the sound transmission loss and the sound absorption coefficient of the sample appear to be declining.

The sound insulation performance of prepared microfiber glass sample mainly depends on the material's own characteristics (surface density, stiffness, damping and other parameters), the internal fiber network structure and the frequency of incident sound waves. Since this paper mainly studies the effect of alkali metal oxide $Na_2O/(Na_2O+K_2O)$ ratio on the sound insulation performance of microfiber glass sample, the sound insulation characteristic curve (stiffness damping control area) of the sample in the low frequency range (≤3000Hz) is selected as the research target. The sound insulation is proportional to the stiffness and increases with the stiffness of the prepared sample.

The accumulation ability of K^+ is weak for its surrounding non-bridged oxygen ions, resulting in more free oxygen in the mixed alkali fiber glass system and increasing the internal network structure damage and loose structure, finally causing chemical bond breakage and rigidity reduction for the prepared microfiber glass sample.

When sound waves in a certain frequency are incident on the surface of the prepared sample, the sound waves are divided into two parts to propagate in different directions: one part is reflected on the surface of the prepared sample and the other part is propagated into the inside of the sample. Fig. 3(a) is a partial SEM picture of a sample of prepared microfiber glass for the Si-Ca-Mg-Al-Fe-B multi-glass fiber system when the ratio of $Na_2O/(Na_2O+K_2O)$ is 0.97. It can be seen that the uniform

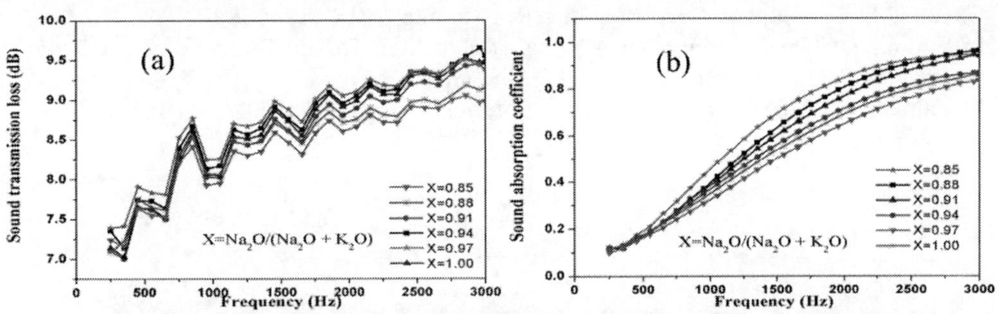

Fig. 3 The comparison of (a) sound transmission loss and (b) sound absorption coefficient of microfiber glass with different alkali metal oxides $Na_2O/(Na_2O + K_2O)$ ratio

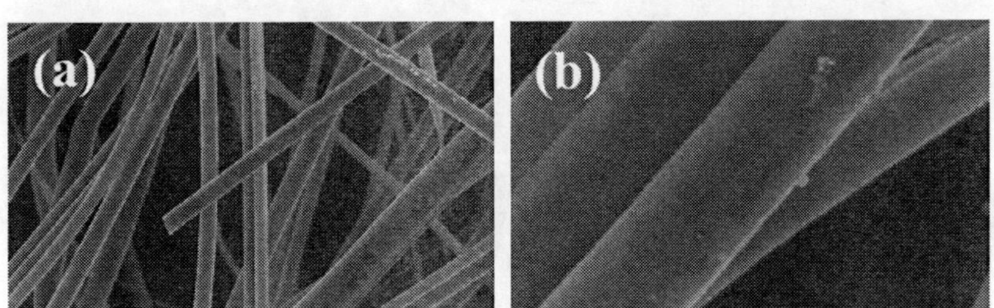

Fig. 4 The SEM pictures of sample with $Na_2O/(Na_2O + K_2O)$ ratio equal to 0.97

atomization sprayed 2.5wt% hydroxy silicone oil water repellent on the sample surface, and each microfiber glass is closely connected by the surface treatment of the water repellent to form a stable fiber network. The incident of lower frequency sound wave doesn't cause the damage of internal fiber structure of the prepared sample. Fig. 3(b) is a partially enlarged SEM image of the sample shown in Fig. 3(a), the surface-modified microfiber glass are bonded together by the action of the water repellent to improve the stability of the fiber structure in order to ensure a three-dimensional network structure. Therefore, the prepared microfiber glass sample sound transmission loss exceeds 9 dB at a frequency of 2000 Hz.

4. Conclusion

When $Na_2O/(Na_2O+K_2O)=0.97$, the forming temperature interval ΔT of the prepared microfiber glass sample reaches to the maximum value $\Delta T_{max}=142°C$; The hydrolysis performance reaches the minimum value $\rho_{min}=978ppm$; The expansion coefficient α and The thermal conductivity λ obtain the minimum values $\alpha_{min}=86.283°C^{-1}$ and $\lambda_{min}=0.028W/(m\cdot K)$, respectively; The acoustic loss and the sound absorption coefficient achieved maximum values of 9.1 dB and 0.88 at a frequency of 2000 Hz, respectively. Under the conditions of this paper, the optimal Na_2O and K_2O distribution ratio of Si-Ca-Mg-Al-Fe-B multi-component fiber glass system cotton was 16wt% and 0.5wt%, further demonstrating the feasibility of producing high water resistant microfiber glass by the centrifugation with synchronous water-repellent atomization spray in practical production applications.

Acknowledgment

This work is supported by the open project of Key Laboratory of Roads and Railway Engineering Safety Control (Shijiazhuang Tiedao University), Ministry of Education, (STKF201718), China.

Reference

[1] Wu Zezhou, Li Heng, Feng Yong, et a1. Building and Environment, 2019, 152: 50-58.
[2] Khoshnava S M, Rostami R, Valipour A. Journal of Cleaner Production, 2018 ,173： 82-99.
[3] Figen B. Journal of Material Cycles and Waste Management, 2015, 17 (2)： 336-349.
[4] Gayan C B, Christopher A H, Vincent T R. Talanta, 2018 , 176: 589-594.
[5] Li Yanming, Chen Zhaofeng, Yang Yong, et al. Cotton Textile Technology, 2017, 45(3): 5-8.
[6] Rodtsevich S P, Tavgen V V, Minkevich T S. Glass and Ceramics, 2007, 64 (7-8): 244-246.
[7] Megahed A A, Gohar I A, Atta A, et al. II Nuovo Cimento. D, 1992 ,14 (8): 789-793.

IOP Conf. Series: Materials Science and Engineering **711** (2020) 012077 doi:10.1088/1757-899X/711/1/012077

Influence of annealing temperature on photoelectronic properties and wettability of TiO_2 nanotube arrays

Lei Yang [1a], Yu Bao[1], Chaoqian Li[1], Yajuan Zhang[1], luyao Zhao, Zixiu Wang, Yiwen Sun, Minghuang Lin[1]

[1]Department of Chemistry and Materials Engineering, Hefei University, Hefei 230601, PR China.

[a]email: ylei531@163.com

Abstract. TiO_2 nanotube arrays (TNs) were prepared by anodization of titanium foil and then annealed at different temperatures (300°C, 400°C, 500°C, 600°C, and 800°C). The samples were characterized by X-ray diffraction (XRD), scanning electron microscopy (SEM), and ultraviolet-visible spectroscopy (UV-Vis). Effect of annealing temperature on crystalline structures, morphology, photoelectronic properties, and wettability has been investigated. The results indicate that low temperature (below 500°C) has no obvious influence on surface morphology, while amorphous phase changes into anatase phase and rutile phase. The ultraviolet absorbance decreases while the visible light absorbance slightly increases with the increase of annealing temperature. At 500°C, the TNs show the highest photocurrent and maximum light-induced contact angle change, which are ascribed to their good crystallization, mixed crystal structure, and nanotubular structure.

1. Introduction

Heterogeneous photocatalysis has received increasing attention since Fujishima' work in 1972 [1]. Of the many semiconductor metal oxides, TiO_2 is regarded as the important photocatalytic candidate because of its high chemical and optical stability, nontoxicity, low cost, and corrosion resistance [2-9]. Therefore, in recent years, particularly TiO2 nanotube arrays (TNs) have received considerable attention. [10].

It is well known that the physical and chemical properties of TiO_2 thin films strongly depend on the preparation methods and annealing conditions [11-13]. Therefore, it is necessary to investigate the effect of the annealing temperatures on the physical properties and photoelectronic properties of the films. So far, most of the works mainly focus on the effect of annealing conditions on the morphology and photocatalytic activity. In this work, TNs were prepared by anodization, and then annealed at various temperatures. The relation of crystalline structures, morphology, photoelectronic properties, and wettability were investigated systematically.

2. Methods

2.1 Preparation of TiO_2 nanotube arrays

TNs were prepared as the method described by Schmuki and co-workers [14]. Ti foil anode and a platinum foil cathode was connected to direct current power supply. The electrolyte contained 0.175 M NH_4F. All anodizations were performed at 30 V for 3 h. After fabrication, the samples were annealed at 300°C, 400°C, 500°C, 600°C and 800°C for 2h, respectively.

2.2 Characterization

The crystalline structures of the TNs were checked by XRD. The morphology observation was performed on a FE-SEM. Absorbance spectra of the samples were obtained by an ultraviolet-visible spectrophotometer.

2.3 Photoelectronic performance measurements

Photocurrent transient was measured in 0.1mol/L Na_2SO_4 solution with electrochemical workstation The samples were irradiated with a high pressure mercury lamp (λ=365nm).

2.4 Contact angle measurement

The wettability of TNs was evaluated by contact angle measurements using a contact angle detect system (SCI3000, hengda), 5 μL deionized water droplets were dropped on three different positions and the digital images of the droplet were captured with a CCD camera, the average value was taken for accuracy. A 15 W UV (254 nm) lamp was used to study the contact angles of the TNs.

3. Results and discussion

3.1 Microstructure and morphology

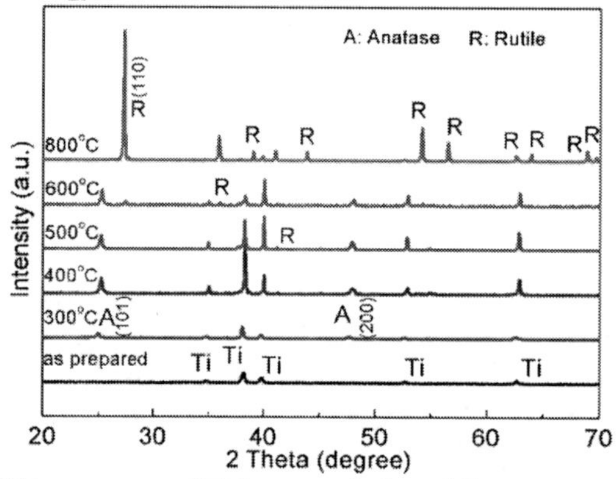

Fig. 1 XRD patterns of TNs annealed at different temperatures

Fig. 1 shows the XRD spectra of the TNs annealed at various temperatures. It is can be seen that only the diffraction peaks of the titanium substrate for the unannealed sample (JCPDS No. 44-1294), which means it is amorphous. For the sample annealed at 300°C, two weak peaks at 2θ = 25.3° and 48.1°, corresponding to (101) and (200) plane diffraction of anatase TiO_2 (JCPDS No. 21-1272) can be observed. With increasing annealing temperature from 300°C to 500°C, the magnitude of the anatase peaks increases. Annealed at 500°C, a weak peak of rutile appears (JCPDS No. 21-1276) (too weak to calculate average crystalline size below). At 600°C, the intensity of the titanium peaks decreases, whereas the intensity of the rutile peaks increases. This indicates that atmospheric O_2 was diffusing through the oxide layer and oxidizing the remaining Ti substrate [15]. For the sample annealed at 800°C, the diffraction peaks of the titanium substrate almost vanish while anatase phase changes into rutile phase completely, which is well accordance with Grimes' results [16].

3.2 Surface morphology analysis

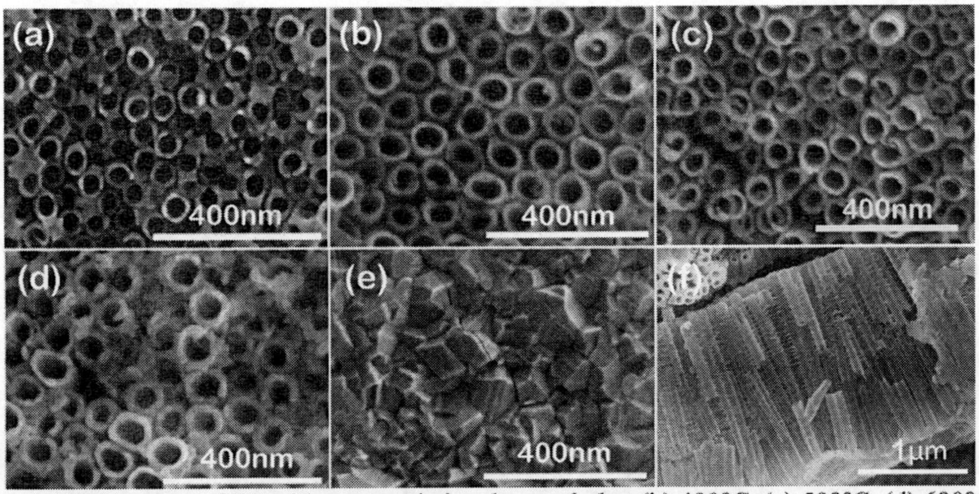

Fig. 2 SEM images of TNs (a) unannealed and annealed at (b) 400°C, (c) 500°C, (d) 600°C, and (e) 800°C, (f) cross-sectional image of the unannealed TNs

Fig. 2 presents SEM images of TNs annealed at various temperatures. All samples exhibit highly ordered TNs except the sample annealed at 800°C. Annealing below 600°C, it has no obvious effect on morphology of TNs. At 800°C, the TNs are destroyed and dense rutile film is observed. This is accordance with the mechanism of TNs crystallization reported by Grimes et al. [16] The transformation of anatase to rutile is constrainted by the tube walls whereas the transformation continues in the interface region. This condition persists until there is enough energy to form larger rutile crystallites and cause the tube walls to rupture.

3.3 Optical properties

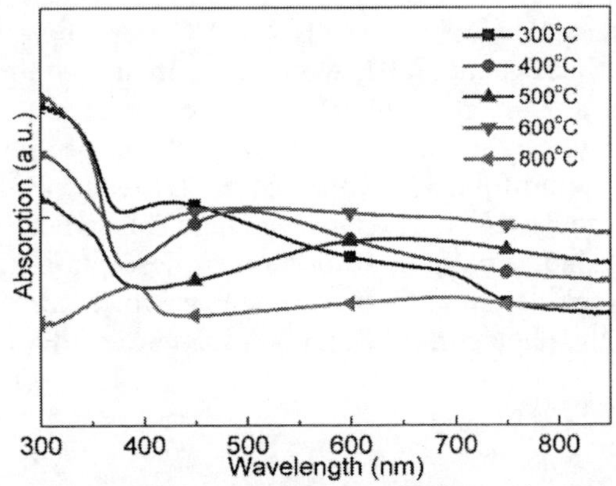

Fig. 3 Absorption spectra of TNs annealed at various temperatures

Fig. 3 shows the absorption spectra of TNs. As shown, with the increase of annealing temperature, the UV absorbance decreases while the visible light absorbance slightly increases, which is ascribed to the increase of rutile content at higher annealing temperature [17]. The absorption spectra of the 800°C sample is the lowest, which may be attributed to the collapsing of the nanotubular structure.

3.4 Transient photocurrent response

Photocurrent transients of the TNs were measured in a three-electrode system, shown in Fig. 4(a) (The photocurrent values of the TNs unannealed and annealed at 800°C are too small to superimpose). It can be seen that the photocurrent values rapidly decrease as soon as the light turns off, and then increase to a value when the light turns on, which have a good reproducibility. The photocurrent fluctuations can be ascribe to surface species absorption/desorption and appearance of defects [18]. It

IOP Conf. Series: Materials Science and Engineering **711** (2020) 012077 doi:10.1088/1757-899X/711/1/012077

is generally accepted that oxygen molecules are adsorbed onto the TNs surfaces by capturing free electrons [$O_2(g)+e^- \rightarrow O_2^-(ad)$], leading to a low-conductivity depletion region near the surface [19-20]. Upon UV illumination, e^--h^+ pairs are generated. The h^+ migrate to the surface and combine with oxygen ions, inducing desorption of oxygen from the TNs surface [$h^+ + O_2^-(ad) \rightarrow O_2(g)$]. This leads to an increase of the conductivity [20-21]. As shown in Fig. 4(a), the photocurrent increases with increasing annealing temperature, nevertheless, it decreases when the temperature exceed 500°C. At 500°C, the sample shows the biggest photocurrent due to its good crystallization, and nanotubular structure. At 600°C, with anatase turn into rutile, the photocurrent rapidly decreases. According to Seto [22], there are a large number of defects at the grain boundary due to incomplete atomic bonding. These defects will result in the formation of trapping states, which have the ability to trap carriers and immobilize them, hence reduce the number of free carriers for electrical conductivity. Therefore, the decrease of photocurrent may be attributed to the increase of to the grain boundary between anatase and rutile and the decrease of anatase content. At 800°C, there is a very low photocurrent due to the formation of dense rutile film.

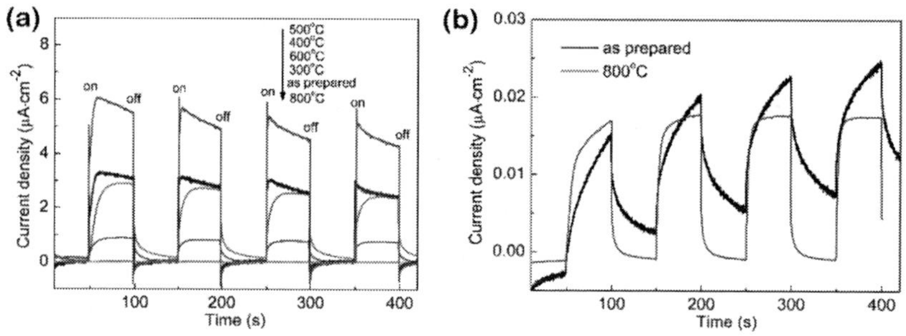

Fig. 4 (a) Photocurrent transients of TNs unannealed and annealed at 300°C, 400°C, 500°C, 600°C, and 800°C. (b) Photocurrent response curves of TNs unannealed and annealed at 800°C

In order to further investigate the effect of annealing on electrical conductivity, the photocurrent response curves of TNs unannealed and annealed at 800°C were replotted, as shown in Fig. 4(b), the photocurrent of unannealed TNs is about 0.02μA, which is about 300 times lower than that annealed at 500°C. It indicates that there are a large number of localized states in the amorphous TNs which act as traps and recombination centers. This reduces the TNs' photo-electronic performance [23-24]. It is noteworthy that the photocurrent of the TNs annealed at 800°C has a fast growth and decay while the response duration of the unannealed TNs is relatively long. This indicates that apart from the exchange process for carrier trapping due to oxygen absorption or desorption on the surface of TNs, there is another slower electron recombination at the native defects which are vanish after heating [25]. Thus, slow response and decay of the photocurrent occurs for unannealed TNs [26].

3.5 wettability of TNs

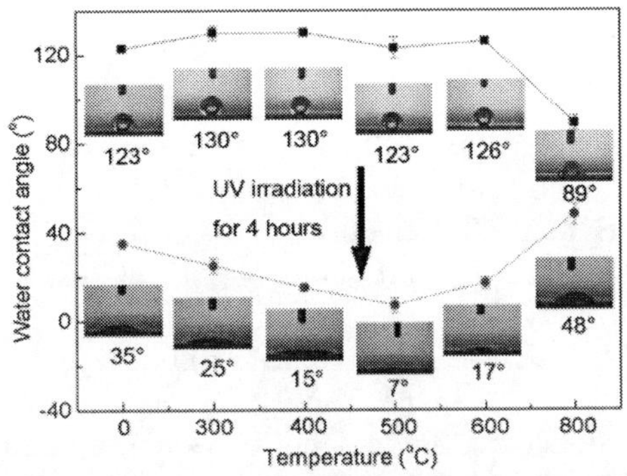

Fig. 5 Photographs of water droplets on the TNs annealed at different temperatures before and after UV irradiation for 4 hours

The water droplets showing the completely spread for each as-prepared TNs, this means the intrinsic state of TiO_2 film is hydrophilic [27]. After the samples stored in the dark for several weeks, they gradually transformed into hydrophobic. This wetting behavior may be ascribed to the adsorption of organic contaminants and the replacement of the chemisorbed OH group with oxygen in air [28-29]. As shown in Fig. 5, before UV irradiation, the water contact angle (WCA) of samples annealed below 600°C has no significant difference, within the range of 123°~130°. At 800°C, the WCA decreases to 89°. On illumination, the WCA value for TNs annealed at 800°C gave the least change whereas the WCA decreases significantly from 123° to 35° for TNs unannealed, from 130° to 25° for TNs annealed at 300°C, from 130° to 15° for TNs annealed at 400°C and from 126° to 17° for TNs annealed at 600°C. It is noticeable that maximal change was observed when TNs were annealed at 500°C, the change rate of WCA decreased 94.3%. It is known that wettability is related to the surface morphology and crystalline structure of films. At 800°C, the nanotubes are completely destroyed, a large number of compact rutile particles prevent the liquid permeates the surface result in the least change of WCA. It can be observed in Fig. 1 and 2, annealing below 600°C, it has no obvious effect on surface morphology, however, the crystal structure changed obviously. A widely accepted mechanism for the wettability is that, under UV illumination, photoexcited e^- is excited from the valance band to conduction band. These e^- will reduce Ti^{4+} to Ti^{3+}, in this process, oxygen atoms are desorbed, and oxygen vacancies are created. These oxygen vacancies are favorable for hydroxyl adsorption than oxygen adsorption that make the surface hydrophilic [30]. At 500°C, due to its good crystallization and nanotubular structure, the sample shows the most obvious hydrophobic-to-hydrophilic transition.

4. Conclusions

Highly ordered TNs were prepared by anodization of titanium foil in a mixed electrolyte solution of glycerol and NH_4F. The annealing temperature has significant influence on morphology, crystal structure, and photoelectronic properties. Photocurrent and photoinduced hydrophilicity increase with increasing annealing temperature, nevertheless, they decrease when the temperature exceeds 500°C. At 500°C, the sample shows the biggest photocurrent and the most remarkable hydrophobic-to-hydrophilic transition due to its good crystallization and nanotubular structure. At 800°C, the lowest photocurrent was observed, which can be attributed to the formation of full rutile phase and the collapsing of ordered nanotube. The photocurrent of the unannealed TNs has a slow growth and decay. This indicates that apart from the exchange process for carrier trapping due to oxygen absorption/desorption on the surface of TNs, there is another slower electron recombination at the native defects.

Acknowledgements

This work is supported by the National Natural Science Foundation of China (Nos. 61804039), and the Natural Science Foundation of Anhui Province (No. 1808085QE126), Major Natural Science Project of the Research and Development Fund of Hefei University (No. 18ZR04ZDA), University excellent talent program (gxyq2019115).

References and Notes

[1] A. Fujishima and K. Honda, *Nature*, **238,** 37 (1972)
[2] Y. Ma, K. Kobayashi, Y. Cao, and T. Ohno, *Appl. Catal. B*, **245,** 681 (2019).
[3] F. Navarro-Pardo, D. Benetti, J. Benavides, H. G. Zhao, S. G. Cloutier, V. M. Castano, A. Vomiero, and F. Rosei, *ECS J. Solid. State. Sc.*, **6**, N32 (2017).
[4] L. Yang, D. Chu, Y. Chen, W. Wang, Q. Zhang, J. Yang, M. Zhang, Y. Cheng, K. Zhu, J. Lv, G. He, and Z. Sun, *J. Electrochem. Soc.*, **163**, H1805 (2016).
[5] L. Xiao, T. Liu, M. Zhang, Q. Li, and J. Yang, *ACS Sustain. Chem. Eng.*, **7**, 2483 (2019).
[6] S. Prabhu, L. Cindrella, O. Joong Kwon, and K. Mohanraju, *Sol. Energy Mater. Sol. Cells*, **169,** 304 (2017).
[7] J. Low, B. Dai, T. Tong, C. Jiang, and J. Yu, *Adv. Mater.*, **31**, 1802981 (2019).
[8] M. Rosales, T. Zoltan, C. Yadarola, E. Mosquera, F. Gracia, and A. García, *J. Mol. Liq.,* **281,** 59 (2019).
[9] Q. Wang, J. Huang, H. Sun, Y. H. Ng, K.-Q. Zhang, and Y. Lai, *ChemSusChem,* **11**, 1708 (2018).
[10] K. Nakata and A. Fujishima, *J. Photoch. Photobio. C,* **13,** 169 (2012)
[11] A. V. Emeline, A. V. Rudakova, M. Sakai, T. Murakami and A. Fujishima, *J. Phys. Chem. C,* **117,** 12086 (2013)
[12] R. P. Antony, T. Mathews, S. Dash and A. K. Tyagi, *J. Phys. Chem. C,* **117,** 6851 (2013)
[13] O. K. Varghese, D. Gong, M. Paulose, C. A. Grimes and E. C. Dickey, *J. Mater. Res.*, **18,** 156 (2003)
[14] J. M. Macak and P. Schmuki, *Electrochim. Acta,* **52,** 1258 (2006)
[15] P. Roy, S. Berger and P. Schmuki, *Angew. Chem. Int. Ed.*, **50,** 2904 (2011)
[16] O. K. Varghese, D. Gong, M. Paulose, C. A. Grimes and E. C. Dickey, *J. Mater. Res.*, **18,** 156 (2003)
[17] M. M. Rahman, K. M. Krishna, T. Soga, T. Jimbo and M. Umeno, *J. Phys. Chem. Solids*, **60,** 201 (1999)
[18] X. Fang, Y. Bando, M. Liao, U. K. Gautam, C. Zhi, B. Dierre, B. Liu, T. Zhai, T. Sekiguchi, Y. Koide and D. Golberg, *Adv. Mater.*, **21,** 2034 (2009)
[19] M. Chen, L. Hu, J. Xu, M. Liao, L. Wu and X. Fang, *Small,* **7,** 2449 (2011)
[20] G. Liu, N. Hoivik, X. Wang, S. Lu, K. Wang and H. Jakobsen, *Electrochim. Acta*, **93,** 80 (2013)
[21] J. Zou, Q. Zhang, K. Huang and N. Marzari, *J. Phys. Chem. C*, **114,** 10725 (2010)
[22] J. Y. W. Seto, *J. Appl. Phys.*, **46,** 5247 (1975)
[23] G. Liu, K. Wang, N. Hoivik and H. Jakobsen, *Sol. Energy Mater. Sol. Cells,* **98,** 24 (2012)
[24] H. Tsuchiya, J. M. Macak, A. Ghicov, A. S. Räder, L. Taveira and P. Schmuki, *Corros. Sci.*, **49,** 203 (2007)
[25] A. Bera and D. Basak, *Appl. Phys. Lett.*, **94,** (2009)
[26] Y. He, W. Zhang, S. Zhang, X. Kang, W. Peng and Y. Xu, *Sensor Actuat. A*, **181,** 6 (2012)
[27] R.-D. Sun, A. Nakajima, A. Fujishima, T. Watanabe, and K. Hashimoto, *J. Phys. Chem. B,* **105**, 1984 (2001).
[28] Y. Haoyong, J. Zhensheng, Z. Shunli, W. Shoubin, and Z. Zhijun, *Sci. China Ser. B,* **45**, 625 (2002).
[29] R. Wang, K. Hashimoto, A. Fujishima, M. Chikuni, E. Kojima, A. Kitamura, M. Shimohigoshi, and T. Watanabe, *Adv. Mater.,* **10** , 135 (1998).
[30] M. Laurenti, V. Cauda, R. Gazia, M. Fontana, V. F. Rivera, S. Bianco, and G. Canavese, *Eur. J. Inorg. Chem.*, **2013**, 2520 (2013).

Fast wave front correction method of reception diversity laser communication based on the analysis of distorted wave front of different receivers

Chenlu Xu[1,a], Shiqi Hao[2,b], Yong Wang[3,c], Yujie Liu[4,d], Qingsong Zhao[5,e]

[1,2,3,5]State Key Laboratory of Pluse Laser Technology, National University of Defense Technology, Hefei, Anhui 230037, China

[4]Chinese people's liberation army unit 94654, Nanjing，Jiangsu 210046，China

[a]xcl951230@126.com, [b]Liu_hsq@126.com, [c]wyeei@126.com, [d]1007519965@qq.com, [e]Pine_zhao@yeah.net

Abstract. Reception diversity laser communication can improve the channel capacity as well as weaken light intensity scintillation caused by atmospheric turbulence. However, it can't reduce the wave front distortion. Wave front distortion is usually corrected by adaptive optics system, because that reception diversity laser communication has more than one receiver, it would take much more time as well as resources to correct the distortion of different receivers. Based on the analysis of wave front distortion of different receivers, we set the best correcting voltage of one receiver as the initial correcting voltage of the other receiver, it can be concluded from the simulation that using the method we propose need less iteration times than the traditional methods when achieving same correcting effect and it can also achieve higher correcting effect when setting same iteration times.

1. Introduction

Compared with traditional communication way, free space optical (FSO) communication has many advantages including wide bandwidth, high confidentiality, cost effectiveness, high data rates[1-3]. As the optical signal transmits in the atmospheric, the quality of the communication such as channel capacity and received signal-to-noise ratio(SNR) may be degraded because of intensity scintillation as well as phase distortion[4] caused by atmospheric turbulence.

Reception diversity is an important way to improve the quality of laser communication, it can not only improve the channel capacity without widening spectrum range or increasing transmitted power, but also weaken light intensity scintillation [5-6]. However, reception diversity can't reduce the wave front phase distortion, and there is no research about the wave front phase distortion of reception diversity laser communication.

Wave front distortion can be corrected through wave front senseless adaptive optics (AO) system, but when it's used to correct the distortion of reception diversity laser communication, correcting speed would be slow because it has more than one receiver, and it would also consume more resource. So the paper proposed a new method to correct the wave front distortion of reception diversity laser communication quickly based on the analysis of wave front characteristics.

The paper is organized as follows: In section 2, system model is shown. The characteristics of distorted wave front are analyzed in section 3. In section 4, a new method is proposed to correct the wave front distortion according to the characteristic of the wave front. And the simulating results are shown in section 5. The conclusions are concluded in section 6.

2. System model

Reception diversity is an important way to improve the quality of FSO communication, the more the receivers are, the better the communication quality. But when the number of receivers is up to a certain value, with the increase of receivers' number, the improvement of communication quality would be inconspicuous. At the same time, the more receivers would lead to higher cost. In this paper, reception diversity laser communication system with two receivers is taken as the example to analyze.

The schematic diagram of reception diversity laser communication system with two receivers is shown in the Fig.1, within the scope of the dashed line is the simulation area of the phase screen, d is the distance of the adjacent receiver, L_x, L_y is the simulation size of the phase screen, and we define the radius of the receiver is a.

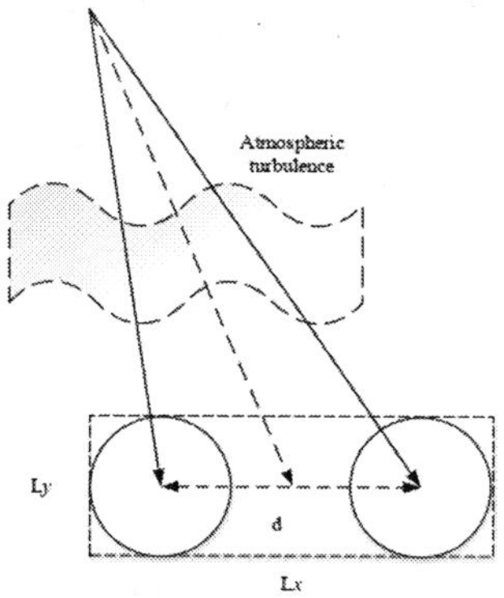

Fig. 1 Schematic diagram of reception diversity laser communication system with two receivers.

Wave front distortion is corrected through deformable mirror (DM)[7], in this letter, we use 32-element to correct the phase distortion. The layout of 32-element DM actuators is shown in Fig.2.

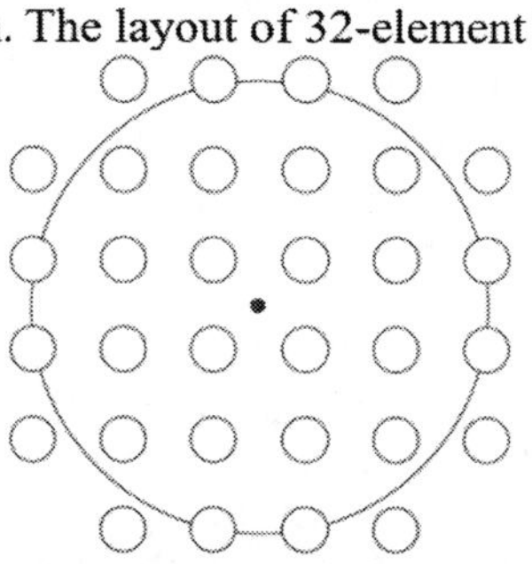

Fig. 2 The layout of 32-element DM actuators.

The influence function of DM can be described as[7]:

$$S_j(x,y) = \exp\left\{ \ln\omega \left[\frac{1}{d}\sqrt{(x-x_j)^2+(y-y_j)^2} \right]^{\alpha} \right\} \tag{1}$$

where $(x_j,\ y_j)$ is the position coordinates of jth actuator, ω is the coupling coefficient which defined as the ratio of the surrounding actuators center deformation value to the maximum deformation value of the working actuator. d is the interval of the actuators. The phase compensated by DM[7] is

$$u(x,y) = \sum_{j=1}^{32} v_j S_j(x,y) \tag{2}$$

where v_j is the controlling voltage of the jth actuator.

3. The analysis of distorted wave front corresponding to different receivers

The atmospheric turbulence is formed of many vortexes of a certain scale[8], which we call outer scale. When the distance of two points is shorter than the outer scale, the atmospheric turbulence has correlation, so the wave front distortion caused by atmospheric turbulence of the two receiving areas also has certain relationship. In order to Figure out the relationship of wave front distortion of different receivers, the phase screen used to simulate the phase distortion caused by atmospheric turbulence is generated through power spectrum inversion[9] and low frequency compensation[10]. We take modified von-Karman power spectrum[11] as atmospheric turbulence power spectrum, in this paper, we mainly consider the influence of atmospheric turbulence outer scale to the correlation of two receiving areas, so we set $l_0 = 0$.

Traditionally speaking, the distance between the receivers is shorter than outer scale, the atmospheric turbulence of different receivers has correlation with others, so the phase screens of different receivers shouldn't be generated independently. We generate the the phase screen of the area surrounded by the receivers, then we choose part of the generated phase screen to analyze the characteristic of the phase screens of the receivers.

Assuming that there exists a beam in the middle of the two beams, and its' tilt angle is φ_0, We firstly generate a phase screen corresponding to reception diversity laser communication system, and the phase screens of the receiving areas are the two corners of the system's phase screen. Considering the the tilt angle of the optical signals arriving at different receivers are different from each other, the phase screens need to be modified. From the equations of generating phase screen[8-9], we can draw a conclusion that the phase distortion has positive correlation with the square root of $\sec(\frac{\pi}{2}-\varphi)$, so the phase screen can be modified through multiplying $\sqrt{\sec(\frac{\pi}{2}-\varphi_0)/\sec(\frac{\pi}{2}-\varphi')}$.

We define the correlation coefficient between two phase vectors x_1 and x_2—the mth columns of two phase screens as follows:

$$r_m = \left\langle \frac{\sum_{i=1}^{n}(x_{1i}-\overline{x_1})(x_{2i}-\overline{x_2})}{\sqrt{\sum_{i=1}^{n}(x_{1i}-\overline{x_1})^2}\sqrt{\sum_{i=1}^{n}(x_{2i}-\overline{x_2})^2}} \right\rangle \tag{3}$$

Where, $\langle\cdot\rangle$ means ensemble average, n is the row number, and then we define the correlation coefficient of two phase screens is the average of correlation coefficient between phase vectors in corresponding position : $r = \overline{r_m}$

In this paper, we generate a phase screen of reception diversity laser communication system with 512x64 sample points, and we set d/a=14, which means that the size of the phase screen of the receiver is 64x64. Fig.3 shows the correlation coefficient of the phase screens of different receivers, and we can find that the correlation coefficient gets larger with the increase of the outer scale of atmospheric turbulence.

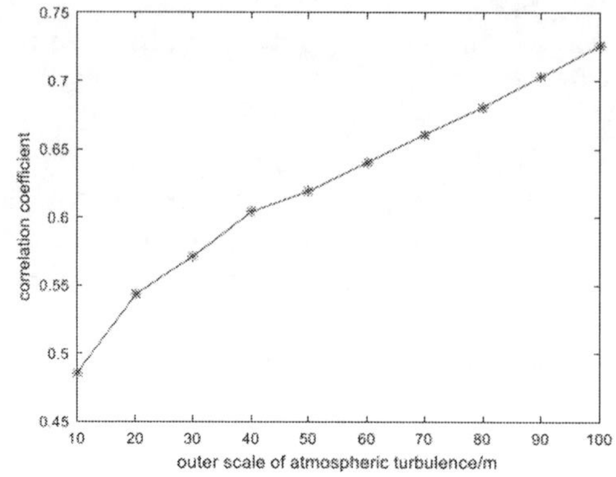

Fig. 3. The correlation coefficient of the phase screens of different receivers in the average of 1000 times calculation

We also compare the correlation coefficient of the phase screens of different receivers in different atmospheric turbulence coherence length r0. From the Figure.4, we can get the conclusion that the coherence length of atmospheric turbulence has little influence on the correlation coefficient.

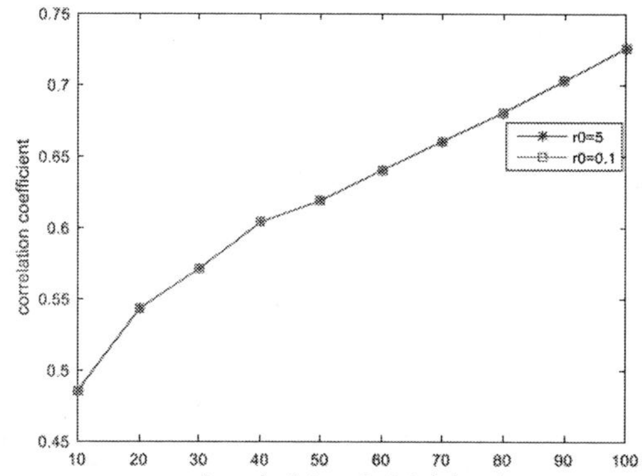

Fig. 4 The correlation coefficient of the phase screens of different receivers in different atmospheric coherence length

We simulate the correlation coefficient of the phase screens in different distances of the receivers, the result is shown in Fig.5, it can be find that the correlation coefficient increases with the distance of the receivers shortening.

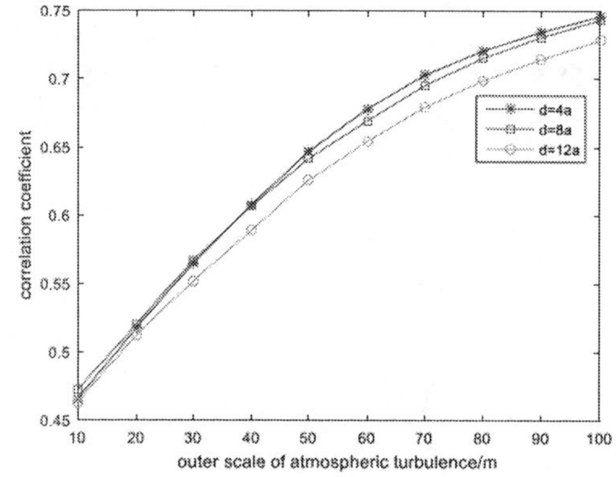

Fig. 5. The correlation coefficient of the phase screens of different receivers in different receivers' distance.

From Fig.3-Fig.5, we can draw a conclusion that the phase distortion in different receivers have correlation, so if the phase distortion in one receiver is corrected, the phase distortion of other receivers can be corrected quickly.

4. Wave front Correction

For reception diversity system has more than one receiver, correct wave front directly would influence correcting speed, and it's also a waste of resources. From the analysis above, we know that the phase distortion of different receivers has correlation, so we can correct phase distortion of other receivers more quickly if the phase distortion of one receiver has been corrected.

Particle swarm optimization(PSO) algorithm, proposed by Kennedy and Eberhart, has already been used in many fields[12]. PSO algorithm can meet the requirements of AO system because of its' advantages such as simple, quick convergence.

The principle of PSO algorithm can be described as follows[13]:

In moment t, $X_i = (x_{i1}, x_{i2}, ...x_{in})$ is the position of ith particle, n is the dimension of the particle, $V_i = (v_{i1}, v_{i2}, ...v_{in})$ is the speed, $P_i = (p_{i1}, p_{i2}, ...p_{in})$ is the best position, we set the number of the particle is M, $P_g(t)$ is the best position of all particle.

The evolutionary process of basic PSO algorithm is:

$$V_{ij}(t+1) = V_{ij}(t) + c_1 r_{1j}(t)[p_{ij}(t) - x_{ij}(t)] + c_2 r_{2j}(t)[p_{gj}(t) - x_{ij}(t)] \tag{4}$$

$$x_{ij}(t+1) = x_{ij}(t) + v_{ij}(t+1) \tag{5}$$

c_1 , c_2 are the accelerated factors , r_{1j}, r_{2j} are uniformly distributed random number in interval (0,1), $r_1 \in U(0,1), r_2 \in U(0,1)$.

The fitness value of is calculated to measure whether the position of ith particle at moment t is the best state. For wave front correction, we choose strehl ratio(SR)[14] as fitness function, SR is defined as $SR = \frac{I(0,0)_{aberrated}}{I(0,0)_{unaberrated}}$,where I is the intensity of the laser beam.

The relationship between wave front correcting parameters and PSO algorithm parameters is shown in table.1

Table.1 Corresponding relationship between wave front correction and PSO algorithm

PSO algorithm parameters	Wave front correction parameters
position	Controlling voltages of DM
dimension	The number of DM actuator
speed	The voltage increment
Fitness function	SR

After the wave front distortion of one receiver is corrected, we can set the best correcting voltage as the initial controlling voltage of the other receiver's DM because that the wave front distortion of different receivers have correlation.

The best correcting voltage is set as the initial controlling voltage, which means DM only need to correct the difference of wave front distortion of the two receivers.The flow chart of PSO algorithm in wave front correction is shown as Fig.6.

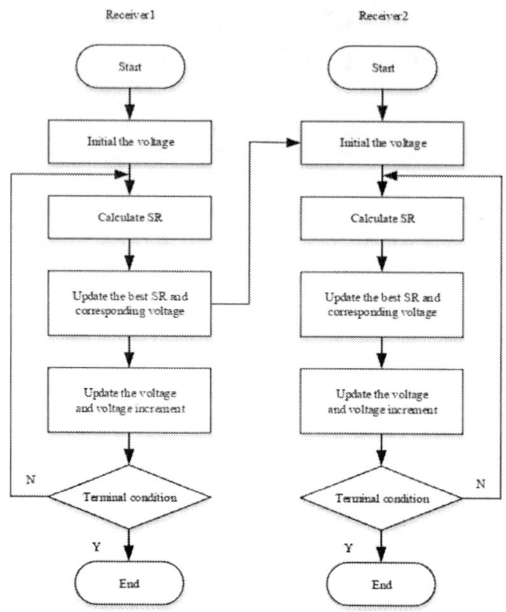

Fig. 6 The flow chart of PSO algorithm in wave front correction of reception diversity laser communication.

5. Simulations

The phase screens of two receivers generated by power spectrum inversion within low-frequency compensation are shown in Fig7, in order to correct the wave front distortion, we firstly use PSO algorithm to control DM voltages correcting Fig.7(a), the correcting result is shown in Fig.8(a). And the best correcting voltage is set as the initial voltage to correct Fig.7(b), the correcting comparison of random initial voltage and set initial voltage is shown in Fig.8(b).

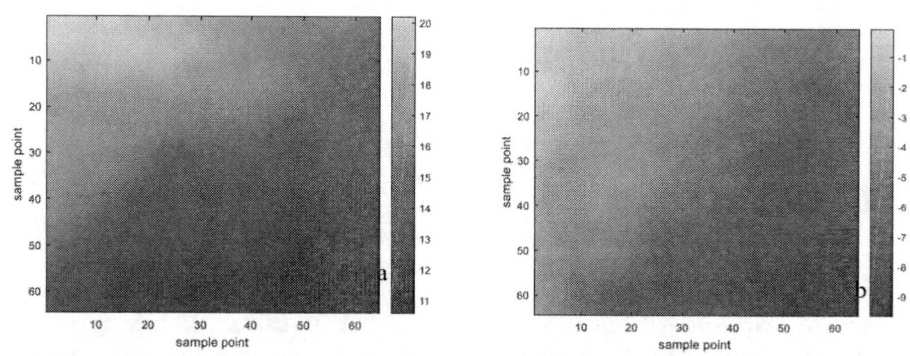

Fig. 7 The phase screen of (a) receiver1, (b) receiver2 generated by power spectrum inversion within low-frequency compensation when the laser wavelength is 1550nm, L_0=100m, r_0=0.2m .

When comparing the correcting effect of random initial voltage and set initial voltage, in order to eliminate the influence of randomness voltage increment of PSO algorithm, we simulate 1000 times to get the mean value of SR. We can conclude from the Fig.8(b) that after 1000 iterations, SR of random initial voltage is approximately equal to 0.89, while SR of set initial voltage achieves 0.935. So when the initial voltage of receiver2 is set as the best correcting voltage of receiver1, it would have a considerable decrease of correcting time if we set a certain SR, and it would also have a considerable improvement of correcting effect if we set a certain iteration number.

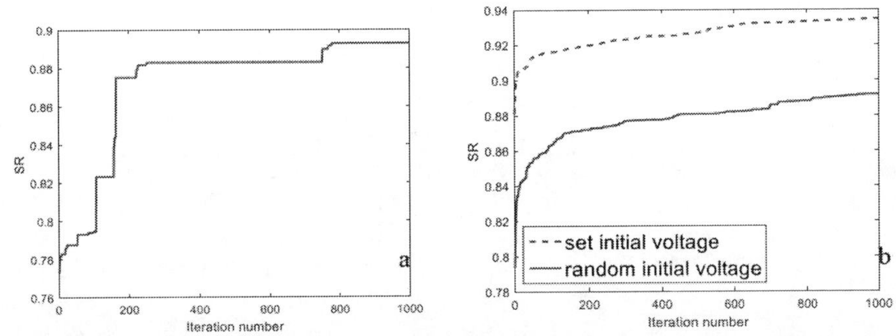

Fig. 8 (a) Correcting effect of receiver1; (b)comparison of correcting effect of set initial voltage and random initial voltage.

We compare the correcting effect of set initial voltage and random initial voltage in different atmospheric coherence length, the results are shown in Fig.9, it can be concluded that compared with large atmospheric coherence length, when atmospheric coherence length is small, the improvement of correcting effect is more obvious. Though the correlation coefficient is same in different atmospheric coherence length r_0, but when r_0 is small, the wave front distortion is more serious, random initial voltage is more difficult to correct the distortion, so the correcting effect of set initial voltage is more obvious than the condition in large r_0.

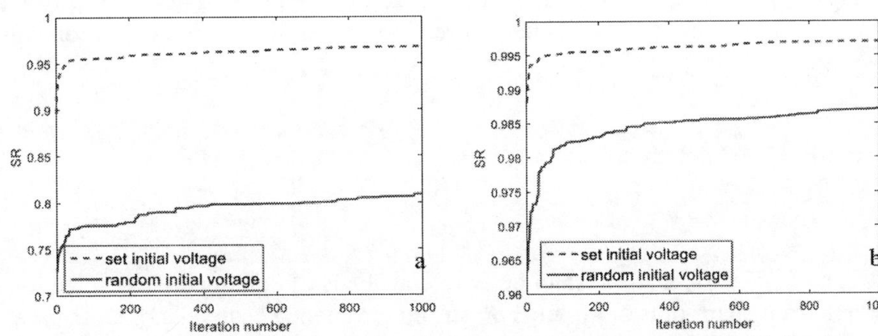

Fig. 9 Comparison of correcting effect of set initial voltage and random initial voltage in different atmospheric coherence length (a)r_0=0.2m,(b)r_0=1m.

The correcting effects of set initial voltage and random initial voltage in different atmospheric turbulence outer scale are compared, the results are shown in Fig.10, it can be concluded that compared with small atmospheric turbulence outer scale, when atmospheric turbulence outer scale is large, the improvement of correcting effect is more obvious. According to the conclusion: the correlation coefficient of different receivers' wave front distortion gets larger with the increase of the outer scale of atmospheric turbulence, so the larger outer scale is, the more wave front distortion of receiver2 would be compensated when set the initial voltage as the best correcting voltage of receiver1, which leads to better improvement in larger atmospheric turbulence outer scale. From the Fig.10, we can also find that when random initial voltage is used to correct wave front distortion, the larger outer scale is, the worse correcting effect is, which proves that the wave front distortion would be more serious in larger atmospheric turbulence outer scale.

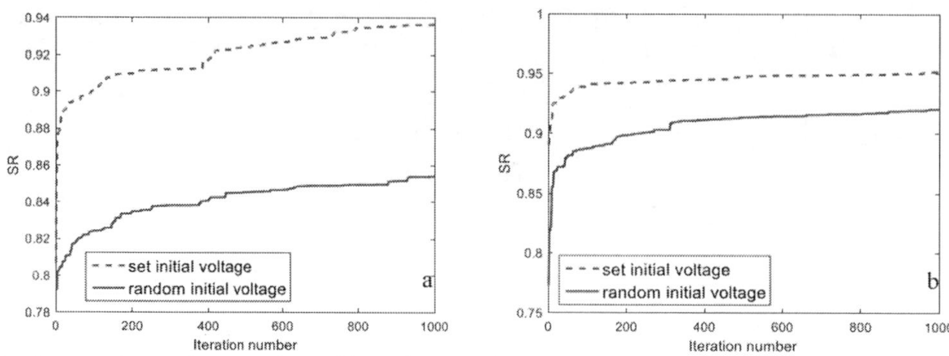

Fig. 10 Comparison of correcting effect of set initial voltage and random initial voltage in different atmospheric turbulence outer scale, (a)L_0=100m,(b)L_0=50m.

It can be concluded from the simulations that set initial voltage of receiver2 as the best correcting voltage of receiver1 would have advantages in correcting speed as well as correcting effect. And it can be found that the improvement is more obvious in large atmospheric turbulence and small atmospheric coherence length.

6. Conclusion

In this paper, the correlation coefficient of the wave front distortion of different receivers is analyzed, and based on the analysis, we propose a method to quickly correct the wave front distortion of different receivers, which would help increasing the correcting efficiency as well as reducing the cost of resources .

Summary

Phone number: +86 17398388480
Email: xcl951230@126.com

References

[1] Niu M , Cheng J , Holzman J F . Exact error rate analysis of equal gain and selection diversity for coherent free-space optical systems on strong turbulence channels[J]. Optics Express, 2010,

[2] Khalighi M A, Uysal M. Survey on Free Space Optical Communication: A Communication Theory Perspective[J]. Communications Surveys & Tutorials IEEE, 2014, 16(4):2231-2258.

[3] Zhang D , Hao S , Zhao Q , et al. Wavefront reconstruction method based on wavelet fractal interpolation for coherent free space optical communication[J]. Optics Communications, 2018, 410:723-729.

[4] Li Z , Cao J , Zhao X , et al. Atmospheric compensation in free space optical communication with simulated annealing algorithm[J]. Optics Communications, 2015, 338:11-21.

[5] Liang B , Chen W . Diversity reception and equalization techniques for laser communication in space[J]. Chinese Optics Letters, 2007, 5(4):197-200.

[6] Koca M , Tuğba Özbilgin. Polarization Diversity in MIMO Communication Over Atmospheric Turbulence Channels[J]. Journal of Lightwave Technology, 2016, 34(17):3981-3992.

[7] Dan Y , Haitao N , Ye L , et al. Fast correction approach for wavefront sensorless adaptive optics based on a linear phase diversity technique[J]. Applied Optics, 2018, 57(7):1650-.

[8] Panofsky H A, Dutton J A. Atmospheric turbulence. Models and methods for engineering applications[M]// Matrix Methods for Engineering /. 1984.

[9] Mcglamery B L . Restoration of Turbulence-Degraded Images*[J]. Journal of the Optical Society of America, 1967, 57(3):293-296.

[10] Lane R G , Glindemann A , Dainty J C . Simulation of a Kolmogorov phase screen[J]. Waves in Random Media, 1992, 2(3):209-224.

[11] Mohamed F, Chatterjee M R. Spectral and Performance Analysis for the Propagation and Retrieval of Signals from Modulated Chaos Waves Transmitted through Modified von Karman Turbulence[C]// Optical Society of America Fio. 2015.
[12] Peng Y , Chen J . Vehicle Routing Problem With Fuzzy Demands And The Particle Swarm Optimization Solution[C]// International Conference on Management & Service Science. IEEE, 2010.
[13] Tehrani K F , Kner P , Shen P , et al. Adaptive optics stochastic optical reconstruction microscopy (AO-STORM) by particle swarm optimization[J]. Biomedical Optics Express, 2017, 8(11):5087.
[14] Mahajan V N . Strehl ratio for primary aberrations: some analytical results[J]. Journal of the Optical Society of America, 1982, 71(9):1258-1266.Niu M , Cheng J , Holzman J F . Optics Express, 2010, 18(13):13915-13926.

IOP Conf. Series: Materials Science and Engineering **711** (2020) 012079 doi:10.1088/1757-899X/711/1/012079

Research on laser water vapor concentration detection technology in the air-sea flux observation

Ke-ke ZHANG[1, a], Xue-yong ZHENG[2, b], Shi-zhe CHEN[1, c], Xiao-zheng WAN[1, d], Ji-ming ZHANG[1, e], Qiang ZHAO[1, f], Bo WANG[1, g], Liang QI[1, h]

[1]Institute of Oceanographic Instrumentation, Qilu University of Technology (Shandong Academy of Sciences), Qingdao, Shandong 266061, China

[2]Qingdao Kernel Spectrum Technology Co., Ltd, Qingdao, Shandong 266043, China

[a]18661475909@163.com, [b]flykernel@126.com, [c]hcszom@163.com, [d]18661817997@163.com,[e]jimingzhangwld@163.com, [f]zqhero9494@163.com, [g]bob80.wang@hotmail.com, [h]563498491@qq.com

Abstract. In the observation of water vapor flux at air-sea interface, the collection frequency of water vapor concentration is more than 10 Hz. High-speed and high-precision measurement of water vapor concentration is an urgent problem to be solved in the observation of water vapor flux at air-sea interface. The ratio of the peak-peak value of second harmonic to the average value of first harmonic is used to measure water vapor concentration based on tunable diode laser absorption spectroscopy. The sensitivity of laser water vapor concentration detection equipment caused by marine salt fog and dust is reduced through precise optical design, and the water vapor concentration at the air-sea interface is calculated using a binary polynomial fitting algorithm. The response frequency of the laser water vapor concentration analyzer is 40 Hz. The water vapor concentration values from the laser water vapor concentration analyzer and LI-7500 CO_2/H_2O gas analyzer are collected synchronously in real time using LabVIEW software. The comparison results show that they have good consistency in the trend and response speed of water vapor concentration, and the measurement deviation of water vapor concentration is within ±1%. The experimental results show that it can meet the needs of high-speed, high-precision measurement of water vapor concentration in the air-sea interface water vapor flux observation.

1. Introduction

Air-sea interface flux is an important parameter for describing the interaction between the atmospheric boundary layer and the surface layer of the ocean. Water vapor flux at air-sea interface is the exchange of water vapor between the ocean and the atmosphere. Water vapor flux is closely related to ocean salinity change, atmospheric water vapor content, ocean and land precipitation and so on. High-speed and high-precision measurement of water vapor concentration is very important for the observation of water vapor flux at air-sea interface. Tunable diode laser absorption spectroscopy (TDLAS) is a gas measurement technology based on molecular absorption spectroscopy. It has the advantages of high sensitivity, high precision and fast response[1-3]. Domestic research institutes represented by Anhui Institute of Optics and Precision Machinery, Chinese Academy of Sciences have made some achievements in measuring water vapor concentration by tunable diode laser absorption spectroscopy. Yanan Cao studied water vapor concentration in a closed glass container using tunable diode laser absorption spectroscopy[4]. Ying He measured water vapor concentration in large-scale area using tunable diode laser absorption spectroscopy, and compared it with the measurement data of LI-7500 gas analyzer in the same site vorticity correlation observation system[5]. Wei Nie carried out the measurement of water vapor absorption spectrum parameters at low temperature based on TDLAS technology[6].

Although many scholars have made a lot of fruitful research on the measurement of water vapor concentration using TDALS technology, this technique has not been applied to air-sea flux observation. At present, non-dispersive infrared spectroscopy (NDIR) is used in the field of water vapor flux measurement. In this paper, TDLAS technology is applied to the laser water vapor concentration measurement of air-sea flux observation, and a high-speed and high-precision laser water vapor concentration detection equipment is developed to replace expensive imported products. It is of great significance to improve the level of marine monitoring and marine scientific research in China.

2. Detection Principle

Tunable diode laser absorption spectroscopy is a gas measurement technology based on the theory of molecular absorption spectroscopy. When a laser beam passes through the measured gas, if the laser emission spectrum is overlapped with the gas absorption spectrum, the output light intensity will be weakened by the absorption of the gas, and the Bill-Lambert law is satisfied between the input light intensity and the output light intensity.

When narrow-band light passes a gas cell, the transmitted intensity $I(v)$ at a specific optical frequency v is given by Beer-Lambert law[7-8]:

$$I(v) = I \cdot \exp\left[-\alpha(v)CL\right]. \tag{1}$$

where I is the incident intensity; $I(v)$ is the output intensity; v is the laser frequency; $\alpha(v)$ is the absorption coefficient; C is the gas concentration; L is the length of the gas cell.

The laser driving circuit is composed of a DC drive signal, a low-frequency triangular wave signal and a high-frequency sine wave signal. Because of the influence of the intensity modulation in the process of wavelength modulation[9], the laser power can be represented as:

$$I = I_0\left(1+p_\Omega \gamma x - p_\omega m\gamma \cos\omega t\right). \tag{2}$$

where I_0 is the incident intensity; γ is the half width at half maximum (HWHM); m is the width-normalized frequency modulation amplitude; p_Ω and p_ω describe the laser power variations as a function of the optical frequency originated from the low frequency triangle-wave modulation and the high frequency cosine modulation. According to Eq.2 and Eq.1, it is obtained that:

$$I(v) = I_0\left(1+p_\Omega \gamma x - p_\omega m\gamma \cos\omega t\right)\exp\left[-\alpha(v)CL\right]. \tag{3}$$

Considering the nonlinearity of the intensity modulation, the odd harmonic component and the high-order harmonic component are introduced into the second harmonic, which can distort the second harmonic[10]. The second harmonic signal is asymmetric[11], and the peak- peak value of the second harmonic can be expressed as:

$$V_{2,PP} = \frac{2S_{2,\max} - S_{2,\min1} - S_{2,\min2}}{2} \propto I_0 CL. \tag{4}$$

where $S_{2,\max}$ is the peak value of the second harmonics; $S_{2,\min1}$ and $S_{2,\min2}$ are the two valley values of the second harmonics respectively.

According to Eq.4, the peak-peak value of the second harmonic is directly proportional to the water vapor concentration and it can be used to measure the water vapor concentration. However, such factors as dust, window pollution and laser power attenuation change will change the transmittance, and then affect the peak-peak value of the second harmonics. When the transmittance of the system changes, in order to eliminate the influence of the change of the transmittance on the instrument linearity, an influencing factor that synchronously changes with the peak-peak of the second harmonic needs to be found.

According to reference[12], the first harmonic signal is singularly symmetric. According to Eq.3, the average value of the first harmonic can be expressed as:

$$V_{1,\mathrm{AVER}} = I_0 p_\omega m \gamma. \tag{5}$$

The average value of the first harmonic is proportional to the light intensity I_0, and it has nothing to do with the water vapor concentration to be measured. The ratio of the peak-peak value of the second harmonic to the average value of the first harmonic can be expressed as:

$$S_{\mathrm{out}} = \frac{V_{2,\mathrm{PP}}}{V_{1,\mathrm{AVER}}} \propto CL. \tag{6}$$

According to Eq.6, the ratio of the peak-peak value of second harmonic to the average value of first harmonic is proportional to the measured water vapor concentration and the length of absorption path, which can be used to measure water vapor concentration.

The second harmonic peak-peak value and the first harmonic average value come from the high frequency modulation sine wave, which are completely synchronized with the light intensity change, the dark current, the circuit and so on. The ratio of the two is not related to the light intensity and the transmittance attenuation, which can be used to eliminate the influence of laser output intensity fluctuation and window contamination on water vapor concentration measurement.

3. System Design

The principle block diagram of TDLAS water vapor concentration detection is shown in Fig. 1. According to HITRAN 2016 molecular spectroscopy database[13], there is a strong absorption line of water molecules near 1392 nm, and it can be used to detect water vapor concentration. The vertical cavity surface emitting laser (VCSEL) operating at 1392 nm is used as the light source.

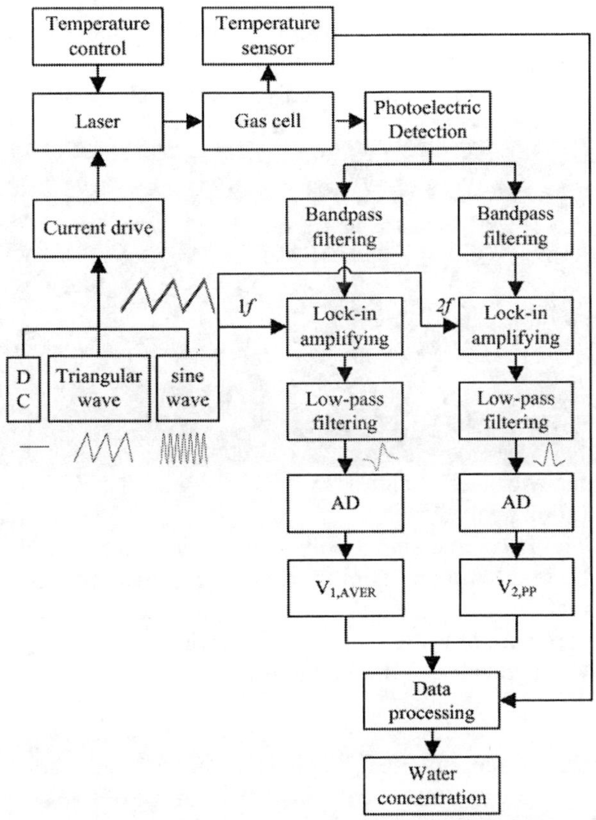

Fig.1 Principle of water vapor concentration detection

Laser is the core device of the laser water vapor concentration analyzer. The stability of the working current and temperature of the laser is very important for the core components[14]. The current drive of the laser consists of DC, low-frequency triangular wave and high-frequency sinusoidal wave. The temperature control circuit and current drive circuit control the diode laser to emit the laser. The tunable laser generates spectral absorption in the open gas cell. Then the photoelectric detector is used to convert the absorbed signal. After photoelectric conversion, the signal is band-pass filtered. The second harmonic peak-peak signal and the first harmonic average signal are obtained by using phase-locked amplification technology. A temperature probe is set in the absorption gas cell to correct the temperature of the detected harmonic signal. The temperature data and harmonic signals are fed into the data processing unit, and the water vapor concentration is calculated according to the fitting equation.

The laser and the photodetector are placed on the same side of the open gas cell, and a concave mirror is set on the other side. The laser emitted by the laser is reflected by the concave mirror and received by the photodetector. Reflective cell can reduce noise overlap and line-to-line interference caused by transmission cables, which can effectively reduce the number of connections without using transmitting lens, receiving lens and diaphragm. The wedge-shaped window with a certain angle is set outside the laser and photodetector, and the reflected light of the wedge-shaped window will deviate from the direction of the original tunable diode laser, eliminating the influence of reflected light on the laser. The inclined angle design of the wedge window can ensure that the equipment has better water avoidance effect. The sensitivity of the laser water vapor concentration analyzer to light source fluctuation and window pollution caused by sea water splash, salt mist and dust can be effectively reduced by precise optical design combined with the first harmonic normalization method.

In the measurement of water vapor flux, the collection frequency of water vapor concentration should be above 10 Hz[15-16], which requires a higher response frequency of laser water vapor concentration analyzer. The frequency of low frequency triangular wave scanning signal and high frequency sinusoidal wave modulation signal are 40 Hz and 30 kHz respectively. Fig.2 shows triangular wave signal and second harmonic wave signal. Data acquisition is carried out along the descending edge of triangular wave. Data processing is completed at the rising edge of triangular wave. The response frequency of laser water vapor concentration analyzer is 40 Hz, which can meet the needs of high dynamic measurement of water vapor flux at air-sea interface.

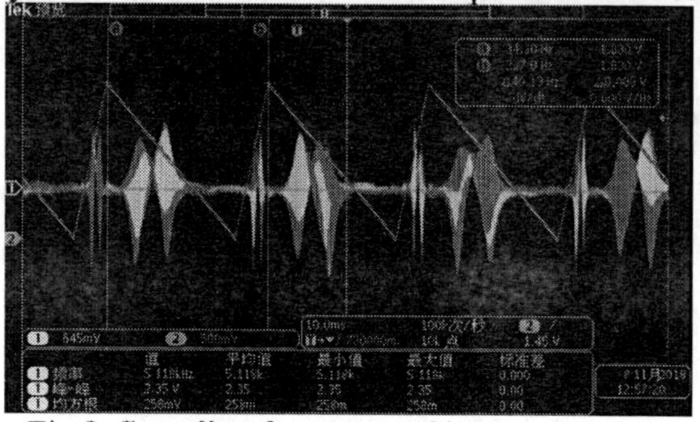

Fig.2 Sampling frequency of laser water vapor concentration analyzer

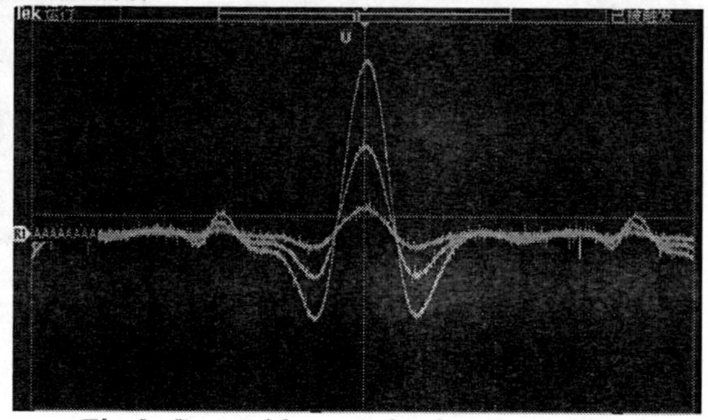

Fig.3 Second harmonic signals at different water vapor concentrations

Fig. 3 shows the second harmonic signal collected under different water vapor concentration. The peak-peak value of the second harmonic is proportional to the water vapor concentration. The system can effectively suppress the influence of light intensity fluctuation, optical lens window pollution and other factors on the measurement of water vapor concentration by the ratio of the peak-peak value of the second harmonic to the average value of the first harmonic.

4. Data Fitting

For water vapor concentration measurement using TDLAS harmonic detection method, the change of temperature and water vapor concentration will cause the change of line strength and line broadening of water vapor absorption spectrum. Therefore, the influence of temperature and water vapor concentration on the measured value of water vapor concentration should be considered in the process of measuring water vapor concentration at air-sea interface[17].

Water vapor concentration calibration experiment was carried out by using temperature and humidity standard box and high precision cold mirror dew point meter. By controlling temperature and relative humidity in temperature and humidity standard box, temperature and water vapor concentration were changed. When the temperature and relative humidity in the temperature and humidity standard box are stable, the temperature and water vapor concentration measured by the cold mirror dew point analyzer are recorded, and the temperature and AD values of the laser water vapor concentration analyzer are also recorded. Using matlab cftool function to fit binary polynomial data, the fitting curve is shown in Fig. 4. The fitting equation is as follows:

$$C = p_1 + p_2\cdot x + p_3\cdot y + p_4\cdot x^2 + p_5\cdot x\cdot y + p_6\cdot y^2 + p_7\cdot x^2\cdot y + p_8\cdot x\cdot y^2 + p_9\cdot y^3 + p_{10}\cdot x^2\cdot y^2 + p_{11}\cdot x\cdot y^3 + p_{12}\cdot y^4 + p_{13}\cdot x^2\cdot y^3 + p_{14}\cdot x\cdot y^4 + p_{15}\cdot y^5 \quad (7)$$

where p_1=-1202; p_2=-36.73; p_3=6.82; p_4=-1.677; p_5=0.1413; p_6=-0.003138; p_7=0.004002; p_8=-0.0001112; p_9=1.732e-06; p_{10}=-6.676e-07; p_{11}=2.623e-08; p_{12}=-3.588e-10; p_{13}=1.673e-11; p_{14}=-1.675e-12; p_{15}=2.601e-14; x is temperature; y is AD value; C is water vapor concentration.

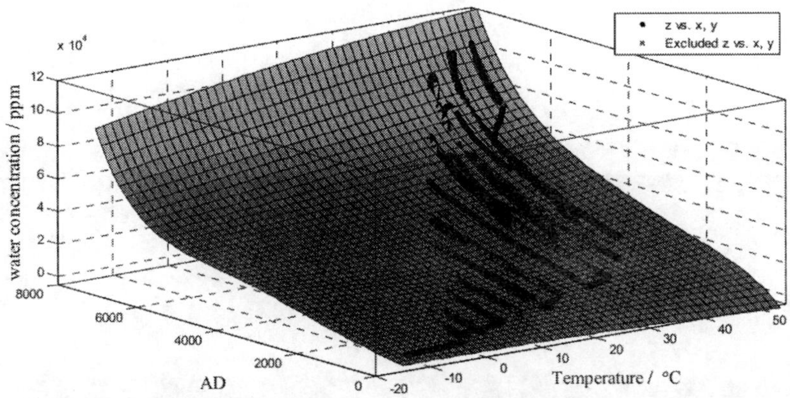

Fig.4 Fitting curves of temperature, ad value and water vapor concentration

The coefficient R-square of the fitting function is 1, Adjusted R-square is 1, RMSE is 54.75, and the polynomial fitting effect is better. Laser water vapor concentration analyzer collects temperature and AD values in real time. The model can be used to retrieve water vapor concentration at air-sea interface.

5. Comparative Experiment

The LI-7500 series of LI-COR open-circuit CO_2/H_2O gas analyzer has been widely used in the field of global vorticity covariance carbon flux monitoring. LI-7500 uses non-dispersive infrared spectroscopy to measure the density of CO_2 and H_2O in the air. The comparative experiment was carried out between laser water vapor concentration analyzer and LI-7500, as shown in Fig. 5.

Fig.5 Comparative experiment between laser water vapor concentration analyzer and LI-7500

After fitting the data of laser water vapor analyzer, the equipment was compared with LI-7500 of LI-COR Company. Laser water vapor concentration analyzer and LI-7500 are placed in the temperature and humidity standard box at the same time. By changing the temperature and water vapor concentration, the laser water vapor concentration analyzer and LI-7500 water vapor concentration are collected synchronously in real time using LabVIEW software.

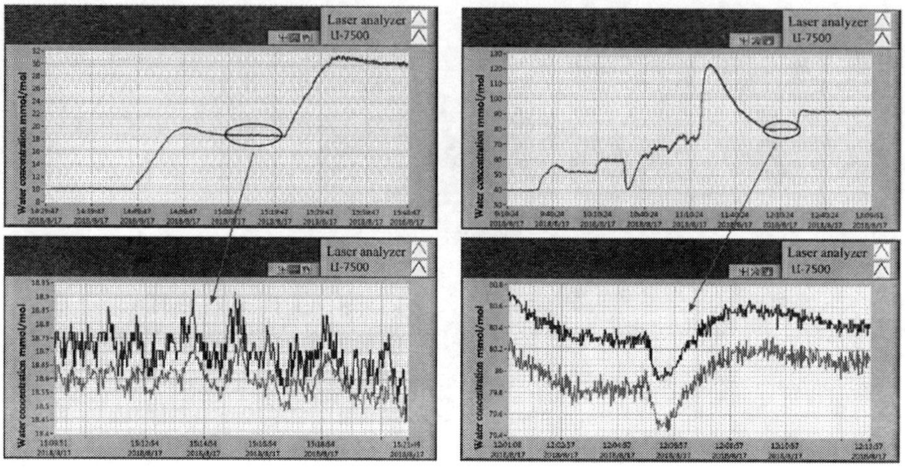

(a)0~30mmol/mol (b)40~130mmol/mol

Fig.6 Measurement data of laser water vapor concentration analyzer and LI-7500

Fig.6 shows the output values of laser water vapor analyzer and LI-7500 in the range of 0~130 mmol/mol. From the figure, it can be seen that the laser water vapor analyzer and LI-7500 have good consistency in the trend of water vapor concentration change and response speed and the measurement error of the two analyzers is less than ±1%.

6. Summary

In this paper, a prototype of laser water vapor concentration analyzer is developed based on tunable semiconductor laser absorption spectroscopy. The peak-peak value of the second harmonic and the average value of the first harmonic are used to measure the water vapor concentration at the air-sea interface. The influence of ocean salt fog and dust on the window is reduced through precise optical design. The water vapor concentration values of laser vapor analyzer and LI-7500 CO_2/H_2O gas analyzer are collected synchronously. The measurement results show that the deviation of water vapor concentration is less than ±1% in the whole range, and the response frequency of the device can reach 40 Hz. Laser water vapor concentration analyzer based on tunable diode laser absorption spectroscopy technology provides a new way for high-speed and high-precision measurement of water vapor concentration in air-sea interface flux.

Acknowledgment

This project is supported by the national key research and development program of china (Grant No.2017YFC1403303), and the natural science foundation of shandong province (Grant Nos.ZR2017QD015, ZR2019MF022), and the shandong provincial key research and development program (Grant No. GG201703130110, 2019GHY112056).

References

[1] YANG Ya-han, LI Guo-lin, LI Xiao-peng, et al. Partial least squares algorithm application in TDLAS based trace H2S analyses in natural gas. Acta Photonica Sinica, Vol.46(2017), p.0230002

[2] ZHANG Ke-ke, YAN Xing-kui, ZHENG Xue-yong, et al. Research on laser humidity detection technology based on TDLAS. Optoelectronic technology, Vol.36(2016),p.23-27

[3] CHEN Hao, JU Yu, HAN Li, et al. Effects of temperature of laser shell on background signals for trace gas detection in TDLAS. Spectroscopy and spectral analysis, Vol.38(2018), p.1670-1674

[4] CAO Ya-nan, WANG Gui-shi, TAN Tu, et al. Concentration and pressure measurement of water vapor in sealed glass containers based on tunable diode laser absorption spectroscopy. Acta Photonica Sinica, Vol.65(2016), p.084202

[5] HE Ying, ZHANG Yu-jun, WANG Li-ming, et al. Study on large-scale regional laser detection methods for water vaper concentration. Spectroscopy and spectral analysis, Vol.33(2013), p.608-612
[6] NIE Wei, KAN Rui-feng, XU Zhen-yu, et al. Measuring spectral parameters of water vapor at low temperature based on tunable diode laser absorption spectroscopy. Acta Photonica Sinica, Vol.66(2017), p.204204
[7] WANG Di, NI Zi-yan, WANG Ming-ji, et al. Data processing of direct absorption spectroscopy for laser gas detection under modulated noise. Acta Photonica Sinica, Vol.48(2019), p.0307001
[8] MO Xiao-bao, XIANG Liu, HUANG Zhi, et al. Study on Evaporation from a Free Water Surface Using Laser Absorption Spectroscopy. Chinese Journal of Lasers, Vol.42(2015), p.1015003
[9] Schilt Stéphane, Thévenaz Luc and Robert Philippe. Wavelength modulation spectroscopy: combined frequency and intensity laser modulation. Applied Optics, Vol.42(2003), p.6728-6738
[10] LI Han, LIU Jian-guo, HE Ya-bai, et al. Simulation and analysis of second-harmonic signal based on tunable diode laser absorption spectroscopy. Spectroscopy and spectral analysis, Vol.33(2013), p.881-885
[11] Seyedali Hosseinzadeh Salati and Alireza Khorsandi. Apodized 2f/1f wavelength modulation spectroscopy method for calibration-free trace detection of carbon monoxide in the near-infrared region: theory and experiment. Applied Physics B, Vol.116(2014), p.521-531
[12] CAO Jia-nian, ZHANG Ke-ke and WANG Zhuo. New scheme of methane concentration detection with tunable diode laser absorption spectroscopy. Chinese Journal of Scientific Instrument, Vol.31(2010), p.2597-2602
[13] I.E. Gordon, L. S. Rothman, C. Hill, et al. The HITRAN2016 molecular spectroscopic database[J]. J. Quant. Spectrosc. Radiat. Transfer, Vol.203(2017), p.3-69
[14] YAN Wan-hong, ZHOU Yan-wen, YU Di, et al. Temperature control system of semiconductor device and application for infrared gas detection. Acta Photonica Sinica, Vol.48(2019), p.0312002
[15] WANG You-heng, JING Yuan-shu, GUO Jian-xia, et al. Contrast research on flux correction methods for eddy-covariance measurement. Meteorological Science and Technology, 2011, Vol.39(2011), p.363-368
[16] ZHAO Jia-yu， ZHANG Mi， XIAO Wei, et al. Greenhouse Gas Fluxes at Water-Air Interface in Small Pond Using Flux Gradient Method Based on Spectrum Analyzer. Environmental Science, Vol.38(2017), p.41-51
[17] LI Zheng-jui, YAO Shun-chun, LU Wei-ye, et al. Study on temperature correction method of CO_2 measurement by TDLAS. Spectroscopy and spectral analysis, Vol.38(2018), p.2048-2053

IOP Conf. Series: Materials Science and Engineering **711** (2020) 012080 doi:10.1088/1757-899X/711/1/012080

Performance of Free Space Optical Communication Base on Orbital Angular Momentum with Pointing Errors

Hubang Wang[1,a]

[1] National Defense University of Science and Technolog, Hefei, China

[a]13965101659@163.com

Abstract. Due to the misalignment of the receiver, the spiral spectrum appears to be extended when the Laguerre-Gaussian beam is applied to atmospheric laser communication. This paper investigates the spiral spectrum spread of the Laguerre-Gaussian beam in the case of receiver misalignment. The analytical expression of Laguerre-Gaussian spiral spectrum under the influence of alignment error is studied by mathematical derivation and simulation. The results show that the detection probability of the original orbital angular momentum (OAM) state varies with the lateral displacement distance and the inclination angle. The bit error rate and channel capacity of each channel were studied to verify the effects of misalignment in multiple OAM communication systems.

1.Introduction

Since Allen [1] discovered the beam carrying OAM in 1992, more and more researches have been made on this type of beam. This type of beam is called a vortex beam and is mainly used in quantum information coding, optical traps, optical manipulation, biomedical and optical communication. The Laguerre-Gaussian (LG) beam is one of the most common vortex beams that can carry angular momentum in different modes. Angular momentum can provide a good carrier for mode multiplexing in atmospheric laser communication.

The OAM mode is an arbitrary integer, which can be regarded as a component of the infinite Hilbert space, which constitutes the physical basis of OAM in the field of information transmission [2]. The communication transmission scheme that first proposed with OAM was derived in [3]. This scheme proves the large capacity and security of OAM information transmission. However, when an OAM beam is applied to a free-space optical communication system, atmospheric turbulence causes amplitude fluctuations of the optical signal, and the phase of the signal will vary randomly on the one hand, resulting in degradation of communication quality. system. On the other hand, random jitter occurs due to the swing of the platform during transmission and reception [4]. Alignment errors caused by these factors can also have a large impact on the communication link. Due to the effects of turbulence and alignment errors, the OAM mode of the vortex beam changes in the atmosphere, which affects the performance of the system.

To clearly describe the changes in the OAM mode, Torner et al. used a linear superposition of spiral harmonic functions to describe the spiral spectrum of a vortex beam [5]. At present, there are some studies on the influence of atmospheric turbulence on the spiral spectrum of vortex beams [6, 7]. Reference [8] studied the transmission characteristics of LG beam in turbulent atmosphere, and derived the analytical formula of beam spiral spectrum. When the LG beam is transmitted in a weak turbulent flow, the reference [9] mainly analyzes the spiral spectrum spread of the LG beam, as well as the probability of detection, the probability of crosstalk, etc. in communication. In [10], the effect of OAM beams on the helical spectrum spread in non-Kolmogorov turbulence is studied. When there is an alignment error between the transmitting end and the receiving end of the vortex beam, the spiral spectrum of the beam is studied in [11]. In [12], the influence of distance is analyzed based on the research of [11], and the analytical formula of spiral spectrum expansion is given.

This paper mainly studies the performance of OAM transmission under atmospheric turbulence and alignment error. Based on the Rytov approximation, the LG beam spiral spectrum model of turbulence effect and alignment error is established. Considering the four factors of turbulence intensity, propagation distance, lateral displacement and angular dip angle, the channel capacity and bit error rate of free-space optical communication system based on vortex are studied.

2. Spiral spectrum of LG beam with pointing error

2.1. Spiral spectrum

As the most common light carrying OAM, the LG beam's electric field function in cylindrical polar coordinates can be described as

$$\begin{aligned} u_0(r,\varphi,z) = A\frac{w_0}{w(z)}\exp\left[-\frac{r^2}{w^2(z)}\right]\left[\frac{\sqrt{2}r}{w(z)}\right]^s \times L_p^s\left[\frac{2r^2}{w^2(z)}\right]\exp(is\varphi) \\ \times\exp\left\{i(2p+s+1)\tan^{-1}\left(\frac{z}{z_0}\right)-i\left[\frac{r^2}{w^2(z)}\right]\left(\frac{z}{z_0}\right)+ikz\right\} \end{aligned} . \quad (1)$$

where $w(z)=w_0\sqrt{1+(z/z_0)^2}$. z is transmission distance and w_0 is beam waist, φ is the azimuthal angle, r is the radius of the cylindrical coordinate, s and p are the azimuthal and radial mode indices respectively, z_0 is Rayleigh distance, and $L_p^s[\cdot]$ is associative Laguerre polynomials. In this paper, s=0. Using the spiral spectrum harmonic function, the equation (1) can be decomposed into following,

$$u_0(r,\varphi,z)=\frac{1}{\sqrt{2\pi}}\sum_{m=-\infty}^{\infty}a_m(r,z)\exp(im\varphi) . \quad (2)$$

Where

$$a(r,z)=\frac{1}{\sqrt{2\pi}}\int_0^{2\pi}u_0(r,\varphi,z)\exp(-im\varphi)d\varphi . \quad (3)$$

The energy of the beam can be expressed as $U=2\varepsilon_0\sum_{-\infty}^{\infty}C_m$, where $C_m=\int_0^{\infty}|a_m(r,z)|^2 rdr$. The normalized intensity of the spiral spectrum of the m mode can be obtained as

$$A_m=\frac{C_m}{\sum_{-\infty}^{\infty}C_q} . \quad (4)$$

2.2. Lateral displacement error

Keeping the LG beam axis and the receiving antenna completely coaxially aligned is difficult for an actual communication system. Especially for long distance transmissions, this requirement is extremely difficult to satisfy. The error in alignment first causes loss of received optical power, and more importantly, it causes expansion of the spiral spectrum. For the l -order LG beam, the lateral offset of the beam is (x_0,y_0) . Equation (1) can be expressed as,

$$
\begin{aligned}
u_l(r,\varphi,z) &= A_l\left[1+\left(\frac{z}{z_0}\right)^2\right]^{-0.5} \times\exp\left[-\frac{r^2+d^2-2rd\cos(\varphi-\rho)}{w^2(z)}\right] \\
&\times\left[\frac{\sqrt{2}\left(re^{i\varphi}-de^{i\rho}\right)}{w(z)}\right]\times\left\{L_p^s\left[2\frac{r^2+d^2-2rd\cos(\varphi-\rho)}{w^2(z)}\right]\right\}^2 \\
&\times\exp\left[i(s+1)\arctan\left(\frac{z}{z_0}\right)-i\frac{r^2+d^2-2rd\cos(\varphi-\rho)}{w^2(z)}\frac{z}{z_0}\right]
\end{aligned}
\quad . \qquad (5)
$$

where d is the lateral offset distance, and ρ denotes the beam axis offset direction. Since $\exp(ix\sin\theta)=\sum_{m=-\infty} J_m(x)e^{im\theta}$, and $I_m(x)=\exp(-im\frac{\pi}{2})J_m(ix)$, where $J_m(\cdot)$ is the first kind Bessel function with m order, and $I_m(\cdot)$ is the modified first kind Bessel function with m order, the above equation can be decomposed into

$$
\begin{aligned}
u_l(r,\varphi,z) &= A_l\left[1+\left(\frac{z}{z_0}\right)^2\right]^{-0.5} \times\exp\left[-\frac{r^2+d^2-2rd\cos(\varphi-\rho)}{w^2(z)}\right] \\
&\times\left\{L_p^s\left[2\frac{r^2+d^2-2rd\cos(\varphi-\rho)}{w^2(z)}\right]\right\}^2\times\left[\frac{\sqrt{2}\left(re^{i\varphi}-de^{i\rho}\right)}{w(z)}\right] \\
&\times\exp\left[i(s+1)\arctan\left(\frac{z}{z_0}\right)\right]\times\sum_{m=-\infty}^{\infty} I_m\left(\frac{2rd}{w^2(z)}\right)\times\exp\left[im(\varphi-\rho)\right] \\
&\times\sum_{n=-\infty}^{\infty} J_n\left(\frac{2rdz/z_0}{w^2(z)}\right)\exp\left[in(\varphi-\rho+\pi/2)\right]
\end{aligned}
\quad . \qquad (6)
$$

2.3. Angular deflection error

Lateral offset and angular tilt are usually generated at the same time. In order to describe the angular deviation effect, it is equivalent to the beam passing through a phase wedge with a projection function $exp\left(i\beta r\cdot cos(\varphi-\eta)\right)$, where $\beta=k\cdot sin\gamma$, η denotes the beam azimuth angle and γ denotes the tilt angle[12]. The expression of the light beam at the propagation distance z is

$$
u_l'(r,\varphi,z)=\sum_{m=-\infty}^{\infty} D_{l,m}(r,d,z)\exp\left[im\varphi-i(m-s)\rho\right]. \qquad (7)
$$

$$
\begin{aligned}
D_{l,m}(r,d,z) &= A_l\left[1+\left(\frac{z}{z_0}\right)^2\right]^{-0.5}\times\left[\frac{\sqrt{2}\left(re^{i\varphi}-de^{i\rho}\right)}{w(z)}\right]\times\exp\left[i(s+1)\arctan\left(\frac{z}{z_0}\right)\right]^2 \\
&\times\exp\left[-\left(1+i\frac{z}{z_0}\right)\frac{r^2+d^2}{w^2(z)}\right]\times\left\{L_p^s\left[2\frac{r^2+d^2-2rd\cos(\varphi-\rho)}{w^2(z)}\right]\right\} \\
&\times\sum_{q=0}^{s}\left\{C_s^q r^q(-d)^{s-q}\sum_{n=-\infty}^{\infty}\left\{J_n\left(\frac{2rdz}{z_0w^2(z)}\right)\exp\left[\frac{in\pi}{2}\right]\right.\right. \\
&\left.\left.\times\sum_{p=-\infty}^{\infty} I_{m-q-n-p}\left(\frac{2rd}{w^2(z)}\right)J_p(\beta r)\exp[ip(\rho-\eta+\pi/2)\right\}\right\}
\end{aligned}
\quad . \qquad (8)
$$

3. Numerical simulation

3.1. Bit-Error Rate analysis

The pointing errors will cause the OAM state dispersion, which can take crosstalk into the communication. In this section, the detection probability of the received state is called p_{mn}, where m is the transmitting state and n is the receiving state.

The signal-noise ratio(SNR) is defined as the ratio between the transmit power times the squared detection probability p_{mm} and the total noise power. The noise power contains two parts. The crosstalk signal modelled as independent Gaussian source is the one part, and the other part is the received noise power N. The SNR can be expressed as

$$SNR = \frac{p_{mm}^2}{\sum_{n\neq m} p_{mn}^2 + N/P} . \tag{9}$$

For OOK or binary PPM modulation format, the bit-error rate(BER) can be written as

$$b_m = \frac{1}{2} erfc\left(\sqrt{SNR/2}\right) . \tag{10}$$

In Fig.1, the BER performances under different lateral distance errors and tilt angle errors are shown. The Δ is the distance between the original center and lateral offset center. $\Delta^2 = \left(x_0^2 - x^2\right) + \left(y_0^2 - y^2\right)$.

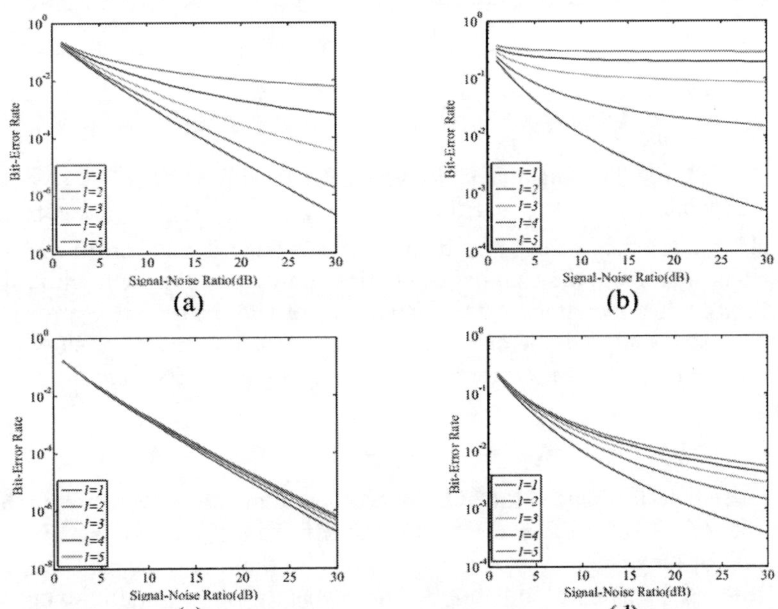

(a) (b)

(c) (d)

Fig. 1 BER versus SNR with $l = 1,2,3,4,5$

(a) Δ=$0.2\omega_0$, $z = 1000m$, (b) Δ=$0.4\omega_0$, $z = 1000m$, (c) γ=$2mrad$, $z = 1000m$, (d) γ=$4mrad$, $z = 1000m$

It is evident that when the lateral offset distance error is small, the BER of each channel has a good performance, which is shown in Fig.1 (a) and Fig.1 (b). As the lateral distance increases, the BER of each OAM state would be worse. For the same lateral offset distance, the smaller states show better performance. Similar conclusions can be drawn for tilt errors from Fig.1 (c) and Fig.1 (d).

3.2. Channel capacity analysis

Channel capacity it the measure of the whole system performance. In this section, the channel capacity of the mode multiplex communication system is analyzed. The capacity of each channel can be presented as

$$C(b_m) = 1 + b_m \log_2 b_m + (1 - b_m)\log_2\left(1 - b_m\right) . \tag{11}$$

IOP Conf. Series: Materials Science and Engineering **711** (2020) 012080 doi:10.1088/1757-899X/711/1/012080

The channel capacity can be presented as

$$C_M = \arg\max \sum_{m \in S} C(b_m) \ . \tag{12}$$

where M is the number of the desired states, and S is the number of all the channels. In Fig.2, the BER performances under different lateral offset distance errors and tilt angle errors are shown.

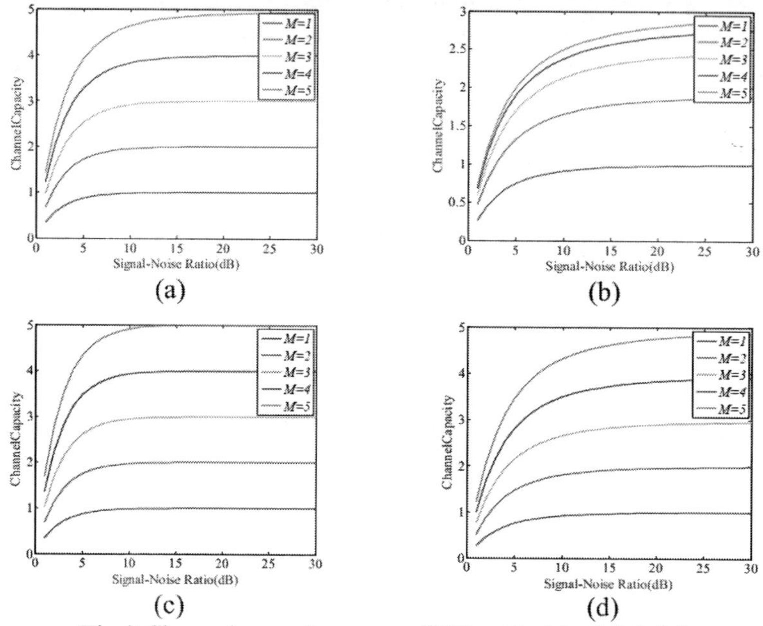

Fig.2 Channel capacity versus SNR with $M = 1,2,3,4,5$

(a) $\Delta=0.2\omega_0$, $z = 1000m$,(b) $\Delta=0.4\omega_0$, $z = 1000m$,(c) $\gamma=2mrad$, $z = 1000m$,(d) $\gamma=4mrad$, $z = 1000m$

In Fig.2 (a) and Fig.2 (c), the lateral offset distance errors and tilt angle errors are small, the crosstalk among OAM states is small, and the channel capacities show a good growth with SNR. In contrast, in Fig.2 (b) and Fig.2 (d), when the lateral offset distance error and the tilt angle error increase, the OAM state crosstalk increases, and the increase in channel capacity becomes slower. Therefore, in a multiplexed system, the larger the misalignment error, the larger the SNR required to achieve the same channel capacity.

4. Conclusion

In this paper, the system performance of the OAMFSO communication system with the alignment error is analyzed. The lateral offset error and the angular tilt error are considered respectively, and the analytical expression of the detection probability is given in these two cases. Then, through numerical simulation, the BER of the system and the channel capacity of the multiplexing system under the influence of two kinds of errors are analyzed.

References

[1] Allen L, Beijersbergen M W, Spreeuw R J C, and Woerdman, J.P. Orbital angular momentum of light and the transformation of Laguerre-Gaussian laser modes[J]. Physical Review A Atomic Molecular & Optical Physics, 1992, 45(11):8185.

[2] Nenad Bozinovic Y Y, Yongxiong Ren M T, PoulKristensen, et al. Terabit-scale orbital angular momentum mode division multiplexing in fibers [J]. Science, 2013, 340(1545): 1545-1548

[3] Gibson, G., Courtial, J., Vasnetsov, M., Padgett, M. J., Frankearnold, S., and Barnett, S. M. Free-space information transfer using light beams carrying orbital angular momentum.[J]. Optics Express, 2004, 12(22):5448.

[4] Djordjevic I B. Deep-space and near-Earth optical communications by coded orbital angular momentum (OAM) modulation[J]. Optics Express, 2011, 19(15)：14277-89.
[5] Torner L, Torres J P, Carrasco S. Digital spiral imaging[J]. Optics Express, 2005, 13(3):873-81.
[6] Courtial J, Padgett M J. Performance of a cylindrical lens mode converter for producing Laguerre–Gaussian laser modes[J]. Optics Communications, 1999, 159(1–3)：13-18.
[7] Qian Xian-Mei, Wen-Yue Zhu, Rao Rui-Zhong. Intensity distribution properties of Gaussian vortex beam propagation in atmospheric turbulence[J]. Chinese Physics B, 2015, 24(4):202-206.
[8] Zhang L, Zhang Y, Zhu Y, and Hu, Z. Effects of non-Kolmogorov turbulence on the spiral spectrum of Hypergeometric-Gaussian laser beams[J]. Optics Express, 2015, 23(7):9137-9146.
[9] Padgett M, Courtial J, Allen L. Light's orbital angular momentum[C]// Quantum Electronics Conference, 1998. IQEC 98. Technical Digest. Summaries of papers presented at the International. IEEE, 2004：139-140.
[10] Zhu Y, Chen M, Zhang Y, and Li Y. Propagation of the OAM mode carried by partially coherent modified Bessel-Gaussian beams in an anisotropic non-Kolmogorov marine atmosphere[J]. Journal of the Optical Society of America A Optics Image Science & Vision, 2016, 33(12):2277.
[11] Vasnetsov M V, Pas'Ko V A, Soskin M S. Analysis of orbital angular momentum of a misaligned optical beam[J]. New Journal of Physics, 2005, 7(1):46.
[12] Zhao, Qingsong , et al. "Orbital angular momentum detection based on diffractive deep neural network." Optics Communications 443(2019):245-249.
[13] Mair A, Vaziri A, Weihs G, and Zeilinger A.Entanglement of the orbital angular momentum states of photons[J]. Nature, 2002, 412(6844):313-6.

IOP Conf. Series: Materials Science and Engineering **711** (2020) 012081 doi:10.1088/1757-899X/711/1/012081

Study on damage of polyimide target irradiated by Pulsed CO_2 Laser

Jiacheng Wu[1,3,a], Yue Hu[2,b]and Xi Wang[*3,c]

[1]Anhui University of Chinese Medicine, Hefei 230012, China;

[2]Hefei Second Sanatorium for Retired Cadres of Anhui Provincial Military Command, Hefei 230061, China;

[3]State Key Laboratory of Pulsed Power Laser Technology, National University of Defense Technology, Hefei 230037, China

[a]wjc_@163.com, [b]641008068@qq.com, [c]eastangus@126.com

Abstract. In this paper, the PI damage caused by pulsed CO_2 laser was simulated. The theoretical model of PI damaged by pulsed CO_2 laser was established. The temperature and stress distribution in PI samples were numerically analyzed by finite element method. The simulation results show that under the condition of low incident laser energy, serious melting damage has occurred on the PI surface, and there exists thermal stress damage. The damage of PI by pulsed CO_2 laser is mainly thermal damage, and the melting phenomenon will be obvious.

1. Introduction

The use of unmanned aerial vehicles(UAVs)is becoming more and more popular. A large number of "black flights" of UAVs have caused serious security risks to air traffic safety and social security, endangering the safety of people's lives and property. Laser irradiation will become effective in controlling "black flight". By laser irradiation, the UAV is effectively physically damaged and shot down.UAV susually consist of some composite materials, which contain organic polymer material, polyimide(PI).PI is a special engineering plastics with excellent properties, which can be widely used in aerospace, machinery, electronics and other high-tech fields.[1]Therefore, it is of theoretical and practical significance to study the damage law of PI under laser irradiation. Researches on PI by domestic and foreign scholars mostly focus on its own properties and laser etching of PI-based composite materials[1-7].So far, there is no published literature on the damage of PI by pulsed CO_2 laser.

In this paper, the simulation results of single pulse laser damage PI with fixed pulse width output by pulsed CO_2 laser were studied. The temperature and stress distribution on the damaged sample surface were simulated by using ANSYS finite element analysis software. The damage mechanism of PI under pulsed CO_2 laser irradiation was discussed from the point of view of thermal damage and stress damage.

2. Theoretical model

In cylindrical coordinates, the irradiation effect of pulsed CO_2 laser on PI was studied, as shown in Fig.1. We selected a disc PI sample with a radius of r_s=20mm and a thickness of h=2mm.Pulsed laser is incident perpendicularly on the PI surface (z=2mm).The Z axis is opposite to the direction of laser irradiation. The radius of the spot incident on the surface of the sample is R, and the beam center coincides with the sample center. Usually, the output laser beam of pulsed CO_2 laser is multi-mode flat-topped beam, which is approximately uniform distribution. So we set the uniform distribution of laser intensity to I_0 in the range of spot. Because the distribution of sample and incident light intensity is axisymmetrical, the two-dimensional plane physical model was selected for the system.

Figure 1.Physical Model of PI Sample.

Considering that the pulse width of a pulsed CO_2 laser is generally in the order of microsecond, the interaction time with PI is very short. Air is a bad conductor of heat, so adiabatic approximation is considered, i.e. no convection heat transfer. The loss of thermal radiation energy is very small compared with the absorption of laser energy. The influence on the temperature field can be neglected, and all the laser energy absorbed by PI is converted into heat energy. So the temperature distribution on the sample irradiated by laser is also axisymmetric.

$$\rho c\frac{\partial T}{\partial t}=k\left(\frac{\partial^2 T}{\partial r^2}+\frac{1}{r}\frac{\partial T}{\partial r}+\frac{\partial^2 T}{\partial z^2}\right)+Q \quad (1)$$

In the formula,ρ,$c$and$k$represent the density, specific heat capacity and thermal conductivity of the material respectively, and Q represents the bulk heat source absorbed by the laser beam in depth. For the heat conduction equation (1), the initial and boundary conditions must be determined before the problem of definite solution of the equation can be discussed.Initial and general boundary conditions can be written as

$$T(r,z,0)=T_0 \quad (2)$$

Underside: $$-k\frac{\partial T}{\partial n}\bigg|_{z=0}=0 \quad (3)$$

Side: $$-k\frac{\partial T}{\partial n}\bigg|_{r=r_s}=0 \quad (4)$$

In the aboveformulas , T_0 is the initial temperature, h is the thickness of the sample, and r_s is the radius of the sample.

3. Key Problems in ANSYS Modeling

3.1 Establishment of Finite Element Model and Mesh Generation

We selected a two-dimensional thermo-mechanical coupled axisymmetric rectangular element PLANE13, and the mesh size of the element was divided into 0.2mm.For a finite element simulation analysis, meshing is a very important problem. Reasonable meshing is the key to the success of simulation. By comparison, the mesh size of the selected element is 0.2mm, and the finite element model of PI sample was established as shown in Fig. 2.

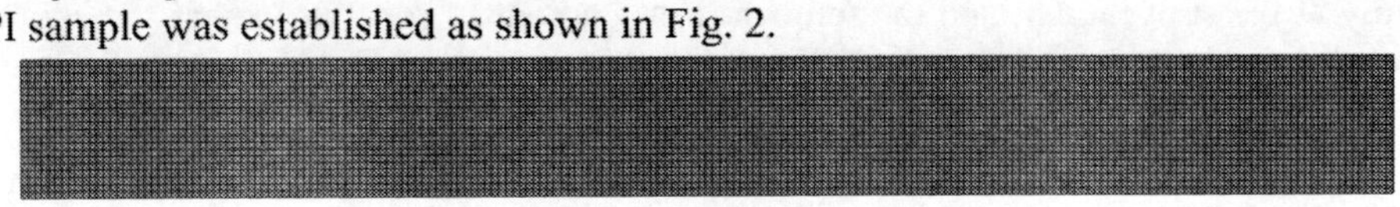

Figure 2.Sample finite element model .

3.2 Pulsed Laser Loading

The loading of pulsed laser is the key problem of numerical simulation.Generally, there are two types of loading for thermal effects of laser irradiation.One is to load the energy or intensity absorbed by laser irradiation on the material surface as a surface heat source, that is, there is an external heat source on the surface of the material.There is no heat source in the material,so $Q=0$ in formula (1).The other initial and boundary conditions remain unchanged.The boundary conditions of the upper surface are added as follows:

Upper surface: $-k\left.\frac{\partial T}{\partial n}\right|_{z=2} = I_0$ (5)

Another type of loading is to treat the absorbed laser energy or intensity as a bulk heat source, which exists on a very thin surface of the material.At present, most of the literature reports use surface heat source, which is loaded on the upper surface of the material in the form of heat flux. PI material is opaque and laser energy is almost absorbed except reflection. We chose the form of surface heat source to load pulsed laser.The working time is pulse widthτ.We used the parametric design language APDL in ANSYS to compile the variation of laser heat source with the loading position.

3.3Thermophysical parameters of materials

It is considered that the initial temperature distribution of PI sample is uniform, $T_0=22^0$C.The mechanical boundary condition is that the horizontal displacement of all particles on the axis is 0.The thermophysical and mechanical parameters required for calculation are shown in Table 1.

Table 1.Parameters of PI for analysis[1,7]

property	**symbol**	**value**
Density (kg/m^3)	ρ	1530
Specific heat(J/kg.K)	c	1090
Thermal conductivity[W/(m.K)]	k	0.12
Tensile strength (MPa)	σ_b	100
Compressive strength (MPa)	σ_{bc}	276
Melting point (^{0}C)	T_m	170
Young modulus(N/m^2)	E_X	4×10^9
Poisson ratio	μ	0.3
Linear expansibility(K^{-1})	β	2.0×10^{-5}

4. Calculation results and analysis

4.1Temperature and stress field distribution

Choosing the incident energy of the laser as E=1J, the spot radius as R=5mm, and the pulse width as τ=10us.Temperature and stress distributions on the surface are shown in Figs. 3 and 4.σ_r,σ_z and $\sigma_\theta$are radial stress, axial stress and circumferential stress,respectively.Compressive stress is negative and tensile stress is positive.

As shown in Figs. 3 and 4,the surface temperature in the spot area is the same, about175.4^{0}C. The melting point temperatureof PI(170^{0}C) has been reached and melting damage will occur.Temperature drops rapidly at the spot radius, and the temperature gradient is very large near the spot radius.The Z component of thermal stress is basically 0.From the center to the edge of the sample, the radial stress σ_r gradually tends to 0 from the maximum compressive stress (about 17.2 MPa), which is much less than the PI compressive fracture strength(276MPa).It reaches its maximum (about 0.1395 MPa) at the outer edge of the spot radius (about r=5.8246 mm), and then decreases gradually to 0.The maximum

IOP Conf. Series: Materials Science and Engineering **711** (2020) 012081 doi:10.1088/1757-899X/711/1/012081

circumferential stress (0.13596 MPa) does not exceed the PI tensile fracture strength (100 MPa), so there will be no stress damage on the surface.

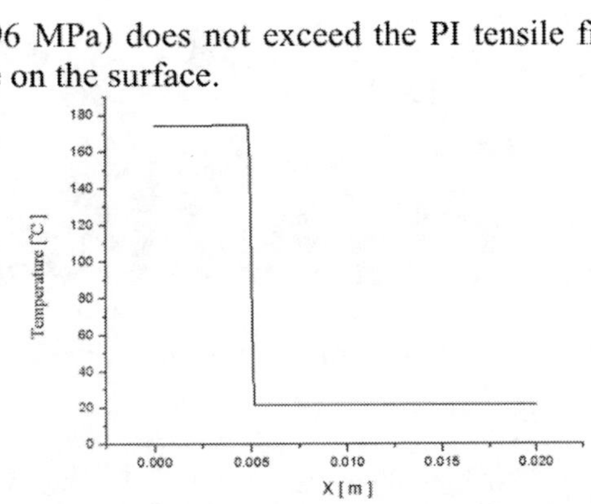

Fig. 3Temperature distribution on sample surface(t=10μs， E=1J)

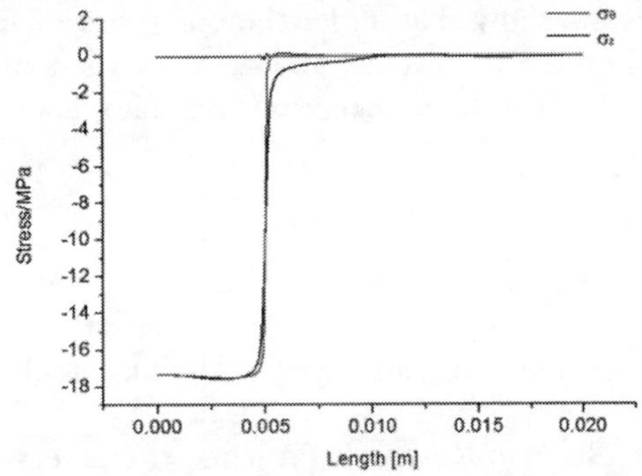

Fig. 4 Stress distribution on sample surface(t=10μs, E=1J)

4.2 Overall morphology of laser damage

Setting the end of the pulse as a time reference point,the temperature and circumferential stress distribution of the sample at this time were calculated.Then we obtained the overall morphology of laser damage in the sample.Taking single pulse energy E=1J, spot radius R=5mm and pulse width τ=10μs, we obtained the equivalent nephogram of temperature and circumferential stress distribution of the sample, as shown in Figs. 5 and 6.

Fig. 5 shows that PI is a material with poor thermal conductivity, and there is only a temperature change in the laser irradiated area.The maximum surface temperature here is 175.4^{0}C, reaching the melting temperature of PI 170^{0}C, which will cause melting damage.

Fig. 6 shows that the circumferential stress shows tensile stressin a part of the sample parallel to the spot position. The maximum tensile stress is 1.181 MPa, which is much less than the tensile strength. In addition, there are circumferential stresses in a part of the surface area adjacent to the spot area. However, once the incident laser energy increases further, the stress value in this region will also increase further, which will exceed the tensile strength and cause surface stress damage.

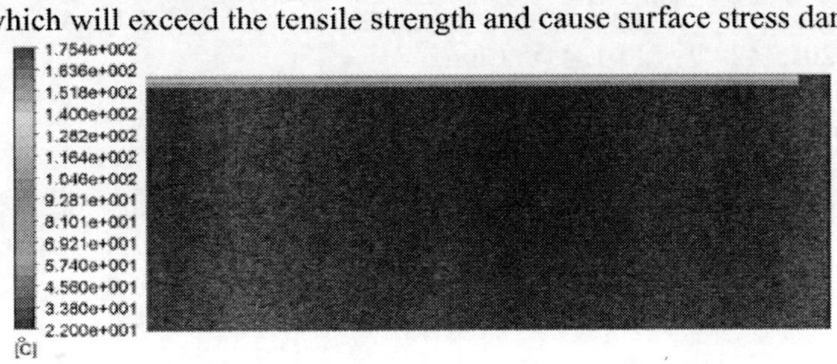

Fig. 5 Equivalent nephogram of sample temperature distribution（E=1J，t=10μs）

Fig. 6 Equivalent nephogram of circumferential stress distribution（E=1J， t=10μs）

5. Conclusions

Based on the heat transfer theory, a theoretical model of PI irradiated by pulsed CO_2 laser wasestablished.The establishment of model, meshing, loading of pulsed laser, setting of initial and boundary conditions, and setting of thermophysical and mechanical parameters of materials were discussed. The simulation results show that under the condition of low incident laser energy, serious melting damage has occurred on the PI surface, and there exists thermal stress damage.Therefore, the damage of PI by pulsed CO_2 laser is mainly thermal damage, and the melting phenomenon will be obvious.

Acknowledgements

This research was supported by the National Natural Science Foundation of China(Projecet No.61871 389), Anhui Province Natural Science Foundation (Projecet No.1908085MF222), the Research Plan Project of the National University of Defense Technology(Projecet No.ZK18-01-02), the Foundation of the State Key Laboratory of Pulsed Power Laser Technology (Projecet No. SKL2019ZR04), Key Natural Science Research Projects in Universities of Anhui Province (Projecet No.KJ2019A0438), and Key Natural Science Research Projects of Anhui University ofChinese Medicine (Projecet No.2019zrzd12). The authors would like to express their gratitude to Prof. Jinsong Nie from the State Key Laboratory of Pulsed Power Laser Technology for his assistance in the discussion.

References

[1] Cui Yong-li, Zhang Zhong-hua,Jiang Li,et al.Applications and Characteristics of Plastics Pipes for Hot Water Installations[J].Plastics Science and Technology,2005(03):50-53+64.(in Chinese)
[2] Zhou Ji,Jin Hai-lan.Research Progress of Polyimide Fiber Materials[J].Paper and papermaking,2017,36(02):14-17.(in Chinese)
[3] Zhang Xiu-ju, Chen Ming-cai, HuangYu-hui, et al. Properties, Usages and Development of the Polyimide[J].Guangzhou Chemical Industry,1998(03):58-66. (in Chinese)
[4] Wang Wei,Li Cheng-xiang,Zhang Guo-bin,Sheng Liu-si.Synthesis and Studies ofPolyimide[J].Materiasl Reports,2008(02):119-120+128.(in Chinese)
[5] Liu Xiao-li,Xiong Yu-qing,Yang Jian-ping,et al.Simulation ofTemperature Field for Laser Etching of Aluminum Thin Films on Polyimide Substrate[J].Chinese Journal ofLasers,2015,42(07):120-124.(in Chinese)
[6] Liu Xiao-li,Xiong Yu-qing,Zhou Li-cheng,et al.Technological Study on Laser Etching of 2μm Aluminum Film/PI Material System[J].Surface Technology,2018,47(10):321-327.(in Chinese)
[7] Zhang Hong-wei,Ren Ni,Xue Hong-tao,et al.Temperature Distribution for Laser Etching of Metal Thin Films on Polyimide Substrate[J].Chinese Journal of Lasers,2016,43(05):100-107.(in Chinese)

Study on a New Measurement Method of Small Polarization Rotation Using the Common-Path Heterodyne Interferometry and a Half-Wave Plate

Ming-Hung Chiu [1] and Chih-Ting Tsai [1*]

[1] Department of Electro-Optical Engineering, National Formosa University, No.64, Wunhua Road, Huwei, Yunlin 632, Taiwan

[*] d0677102@gm.nfu.edu.tw

Abstract. This thesis proposes a novel method for small polarization rotation measurement using a half-wave plate and a high-stability common-path heterodyne interferometry. The phase with azimuth angle 22.5° of a half-wave plate has a distinct change resulting from the small polarization rotation of the incident light. The optical rotation properties of a material, such as glucose solution, can be determined in this phase from the concentration of the solution. Finally, the phase can be measured using a lock-in amplifier, and the relevant formula between the glucose concentration and the polarization rotation can also be identified with a recording table.

1. Introduction

In the optical measurement domain, common-path heterodyne interferometry is a critical technology that features numerous advantages such as stability, precision, high linearity, and minimum environmental interferences. Many models that measure phase delay of birefringence materials are based on common-path heterodyne interferometry.

The objective of this study was to identify a new simple method that is easy to operate to measure slight phase delay in a sample and further derive the birefringence or concentration properties of the material. Therefore, this study combined common-path heterodyne interferometry and half-wave plates to create a measurement system.

Many methods for measuring wave plate phase delay have been proposed [1]–[8]. Feng et al. [9] developed an optical heterodyne interferometer by adding two acousto-optic modulators (AOMs) to a Mach–Zender interferometer to measure the concentration of glucose. The advantage of this method is that its phase measurement is related only to solution concentrations and irrelevant to the index of refraction of the sample. Moreover, the left-handed and right-handed polarized lights share one polarization path and are unlikely to be affected by the environment. Cameron and Coté [10] designed a closed loop control system that uses common-path heterodyne interferometry and two Faraday rotator polarization controllers to modulate polarization vectors. They also used a lock-in amplifier to measure phase shifts. Chiu [11] employed an electro-optic modulator and lasers with two wavelengths to measure the wave plates' phase delay and the direction their principal axis. The advantage of such a system is the ability to measure wave plates as well as phase delay and axial position of ordinary birefringent materials. Márquez et al. [12] proposed a technique for measuring phase difference between positive and negative eigenvectors of liquid crystal spatial light modulator and added sawtooth wave voltage into an electro-optic modulator to provide time-varying phase shift and measure phase delay of two eigenvectors. The technique was defective for being unable to accurately measure actual phase delay of

IOP Conf. Series: Materials Science and Engineering **711** (2020) 012082 doi:10.1088/1757-899X/711/1/012082

any eigenvector alone. Lo and Hsu [13] proposed a method to measure birefringence by driving electro-optic modulators through a sinusoidal signal in the common-path heterodyne interferometer. This technique uses wider modulation frequencies than those of conventional techniques. Wang and Oakberg [14] developed a device to measure low-level linear birefringence of optical materials. This device uses the polarization of a modulated laser in a photoelastic modulator, and has been proved to be able to measure different low-level linear birefringence with high sensitivity and favorable repeatability. Ohkubo and Umeda [15] developed a near-field scanning optical microscope with birefringence contrast imaging. This system uses right and left circularly polarized lights and can simultaneously measure the angle of the principal axis and the phase delay of linear birefringence using a Zeeman laser.

The present study combined half-wave plates and common-path heterodyne interferometry to develop an experimental model. Because a half-wave plate at an azimuth of 22.5° exhibited considerable phase shifts, it was used to measure the phase delay of birefringent materials to determine a simple and easy-to-operate method.

2. Principle

2.1 Common-path heterodyne interferometry

The principles of a common-path heterodyne interferometer involve using a heterodyne source with difference frequency. When the source passes through a sample, s- and p-polarized lights generate different phase delay variations, which are related to the physical quantity of the sample. Fig. 1. illustrates the principles. An assumption is made to have the heterodyne source wave proceed along the z-axis. When the source passes through the sample, its incident ray enters the analyzer ANt (45°). Then, electric field strength along the x and y-axes can be obtained to determine interference. The interference signal can be converted into digital signals of periodic cosines I_{out}. Reference signals that have not passed through the sample and the test signals that have passed through the sample are simultaneously input into a lock-in amplifier or a phase meter to determine the phase difference values.

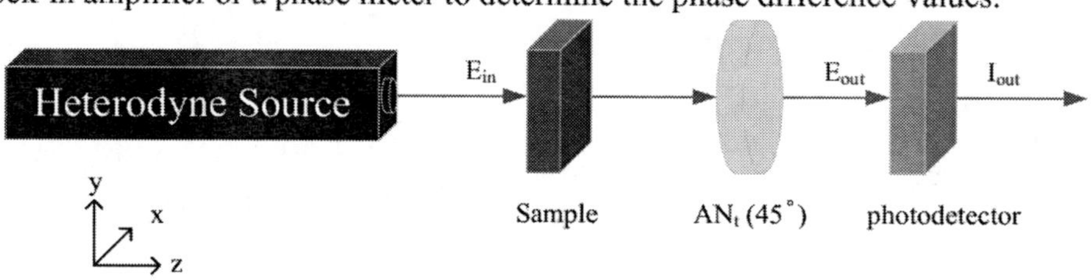

Fig. 1 Diagram of common-path heterodyne interferometer.

As depicted in Fig. 1, if the source proceeds along the z-axis, the electric field of difference frequency light generated through heterodyne sources along the x-axis (parallel to the optical table) can be represented as $\boldsymbol{E_x e^{i(\omega_1 t-\phi x)}}$; the electric field along the y-axis (vertical to the optical table) can be represented as $\boldsymbol{E_y e^{i(\omega_2 t-\phi y)}}$. The Jone's matrix of the incident light into the sample can be written as

$$E_{in}=\begin{bmatrix} E_x e^{i(\omega_1 t-\phi x)} \\ E_y e^{i(\omega_2 t-\phi y)} \end{bmatrix} \tag{1}$$

The x-axis and y-axis polarized light from the sample induces phase delays as $\boldsymbol{\phi_1}$ and $\boldsymbol{\phi_2}$. After the light passes through the analyzer at 45° ($A\boldsymbol{N_t}$), the photodetector ($\boldsymbol{D_t}$) detects an electric field ($\boldsymbol{E_{out}}$) as follows:

$$E_{out}=\frac{1}{2}\begin{bmatrix} 1 & 1 \\ 1 & 1 \end{bmatrix}\begin{bmatrix} A_x e^{-i\phi_1} & 0 \\ 0 & A_y e^{-i\phi_2} \end{bmatrix} E_{in} \tag{2}$$

where $\boldsymbol{A_x}$ and $\boldsymbol{A_y}$ are, respectively, reflected coefficients and transmission coefficients of amplitudes of p-polarized light and s-polarized light passing through the sample. The final interference signal strength $\boldsymbol{I_{out}}$ is as follows:

$$I_{out}=(A_xE_x)^2+(A_yE_y)^2+2A_xA_yE_xE_y cos\left[2\pi(f_1-f_2)t-(\phi_x-\phi_y)-(\phi_1-\phi_2)\right] \tag{3}$$

As revealed in Eq. (3), the interference signal is a cosine signal where $(f_1 - f_2)$ represents the difference frequency of two orthogonal linear polarization lights, $(\phi_1 - \phi_2)$ represents the phase difference of the sample, and $(\phi_x - \phi_y)$ represents the initial phase. Because the physical quantity of the sample is related to the phase variation of the sample, this study compared the reference signal that had not passed through the sample with the interference signal at the same frequency in terms of phase. The phase difference was $(\phi_1 - \phi_2)$, which was further used to determine the physical quantity of the sample. Compared with conventional interferometry, the proposed method was more convenient and accurate, making it suitable for applications related to precision measurements.

2.2 Acousto-optic modulator [16]–[18]

An AOM uses acousto-optical effects. When a piezo-electric transducer produces periodically varying vibrations under a radio frequency, grating is formed inside a crystal because of different indices of refraction; following the movement of grating, the incident ray is deflected. As shown in Fig. 2, when the incident ray passes through the AOM, it generates rays of +1, 0, and −1 orders according to the Doppler effect, yielding frequencies $f - f_s$, f, and $f + f_s$. The frequencies range between dozens of MHz and hundreds of MHz. In this study, the AOM was used as the heterodyne source for the measurement experiment.

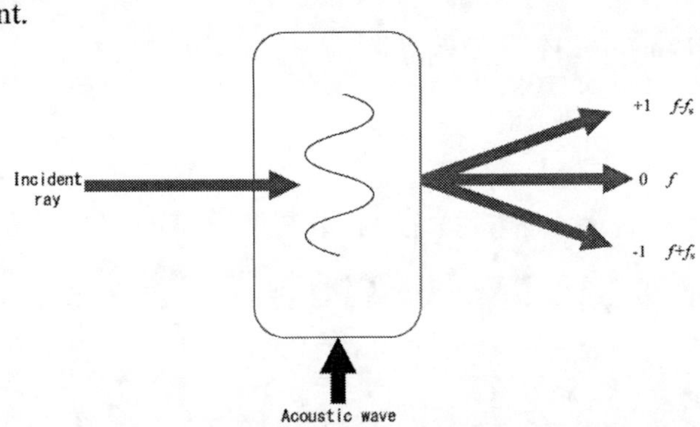

Fig. 2 Diagram of the common-path heterodyne interferometer.

In this experiment, the AOM was used to generate the necessary beat frequency for the experiment. The principle is to affect light transmission by forming grating, which can be achieved by using different indices of refraction of a crystal during acoustic wave transmission.

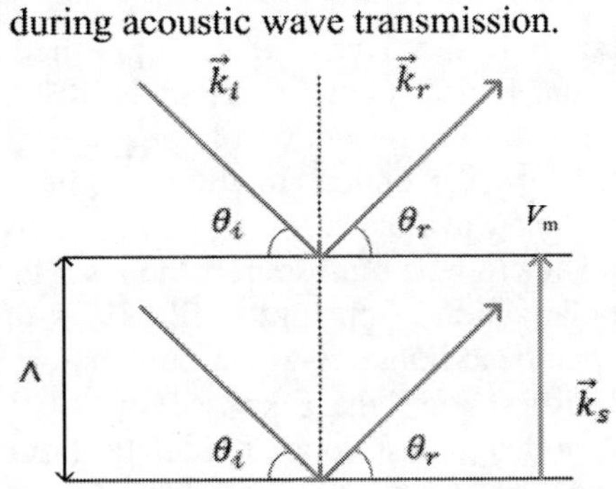

Fig. 3 Internal structure of an AOM.

Fig. 3. shows the movement of an incident ray in an AOM, where $\vec{k}_i$ is the wave phasor of the incident ray, $\vec{k}_r$ is the wave phasor of the emerging light, $\vec{k}_s$ is the wave phasor of the acoustic wave, and Λ is the wavelength of the acoustic wave.

According to the vector space, the relationship between $\vec{k}_r$, $\vec{k}_i$, and $\vec{k}_s$ is written as follows:

$$\vec{k}_s + \vec{k}_i = \vec{k}_r \tag{4}$$

The vector strength of $\vec{k}_s$ can be expressed as Eq. (5):

$$k_s = k_i sin\theta_i + k_r sin\theta_r \tag{5}$$

Because $k_i \cong k_r = \frac{2\pi}{\lambda}$, $\theta_i \cong \theta_r = \theta$, and $k_s = \frac{2\pi}{\Lambda}$, Eq. (5) is revised as

$$\frac{1}{\Lambda} = \frac{2}{\lambda} sin\theta \tag{6}$$

When the incident ray is of a +1 order, then $\Lambda = \frac{V_m}{f_m}$ and Eq. (6) is revised as

$$sin\theta = \frac{\lambda f_m}{2V_m} \tag{7}$$

where V_m is the velocity of the acoustic wave and f_m is the modulating frequency of the acoustic wave.

To determine the frequency of the ray of +1 order, the relationship between V_i, V_r, and V_m must be identified first.

$$\textit{Relative velocity of } V_i \text{ against } V_m = V_i + V_m sin\theta \tag{8}$$

$$\textit{Relative velocity of } V_r \text{ against } V_m = V_r - V_m sin\theta \tag{9}$$

$$\textit{Relative velocity of } V_r \text{ against } V_i = -2V_m sin\theta \tag{10}$$

Eq. (7) is substituted into Eq. (10)

$$\textit{Relative velocity of } V_r \text{ against } V_i = -\lambda f_m \tag{11}$$

Because $V_r = \lambda f_r$, it is substituted into Eq. (11) as

$$V_r = \lambda f_i - \lambda f_m = \lambda(f_i - f_m) \tag{12}$$

$$f_r = f_i - f_m \tag{13}$$

2.3 Experiment framework

The experimental framework is depicted in Fig. 4. Common-path heterodyne interferometry was combined with a half-wave plate to measure a sample with an internal diameter of 1 cm that contained glucose solution.

The motivation of this thesis is to discover a new simple method to measure small polarization rotation. Due to a large phase change near the azimuth angle of 22.5° of a half-wave plate, the small polarization rotation from the test medium generates a large phase change between the two orthogonal polarizations of the test light. Thus, the optical rotation can be measured using the common-path heterodyne interferometry and a half-wave plate.

First, in order to generate the heterodyne light source, the laser light is divided into *p*-polarized light and *s*-polarized light using a polarization beam splitter (PBS1), with the optical setup as presented in Fig. 4. The frequency of the *p*-polarized light passing through a acousto-optic modulator (AOM1) is of f_0−80.00MHz for the +1 order diffraction light. Likewise, the *s*-polarized light after reflection by the mirror (M2), and incident into another acousto-optic modulator (AOM2), the +1 order diffraction light has frequency f_0−80.01MHz, where f_0 represents the optical frequency of the laser. Finally, the two lights are combined through another polarization beam splitter (PBS2) to obtain a heterodyne source with a frequency difference of 10kHz.

A beam splitter (BS) can split a heterodyne light source into transmitted and reflected light. The reflected light passes through an analyzer (ANr) with an azimuth angle of 45°, and is received by a photodetector (Dr). The detected signal is the reference signal I_r. The transmitted light passing through a sample, a half-wave plate ($W_{\lambda/2}$) and another analyzer (ANt) at an azimuth angle of 45°, and detected

IOP Conf. Series: Materials Science and Engineering **711** (2020) 012082 doi:10.1088/1757-899X/711/1/012082

by a photodetector (Dt), is called the. Finally, the phase difference between the reference signal I_r and the test signal I_t can be captured using a lock-in amplifier (SR830).

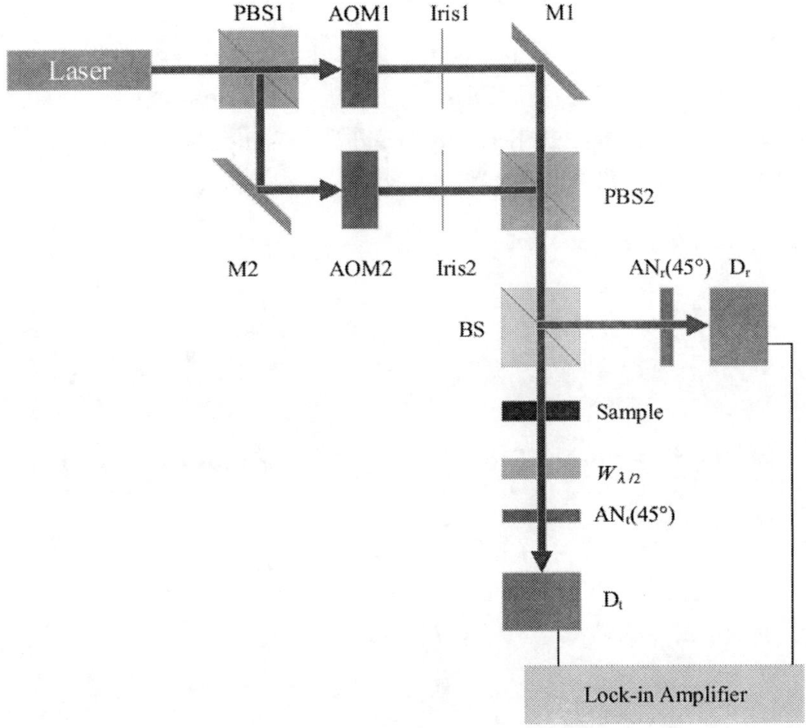

Fig. 4 Experimental system architecture

2.4 Formula development

This section mainly involves formula development as depicted in Fig. 5. During the formula development process, the optimal measurement conditions were identified to obtain an innovative measurement method.

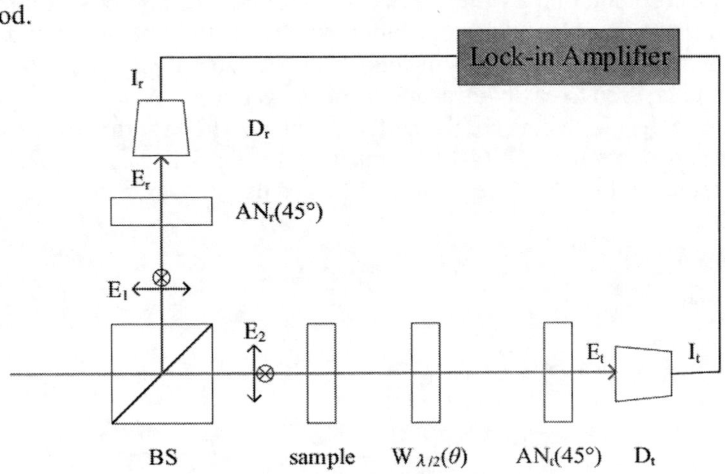

Fig. 5 Experimental framework.

In Fig. 5, the Jone's matrix of the reference signal can be written as

$$E_r = AN_r(45^o)E_1 = \frac{1}{2}\begin{bmatrix} A_x e^{j(\omega_1 t)} + A_y e^{j(\omega_2 t)} \\ A_x e^{j(\omega_1 t)} + A_y e^{j(\omega_2 t)} \end{bmatrix} \tag{14}$$

IOP Conf. Series: Materials Science and Engineering **711** (2020) 012082 doi:10.1088/1757-899X/711/1/012082

$$I_r = |E_r|^2 = \frac{1}{2}A_x^2 + \frac{1}{2}A_y^2 + A_x A_y \cos(\omega t) \tag{15}$$

where A_x and A_y are electric field amplitudes along the x and y axes, respectively; $\omega = \omega_2 - \omega_1$.

If $\Gamma \neq \pi$, the Jone's matrix of the test signal can be expressed as

$$E_t = AN_t(45^o)\ W_{\frac{\lambda}{2}}(\theta)SE_2 = \frac{1}{2}\begin{bmatrix} MA_x e^{j(\omega_1 t+\delta_1)} + NA_y e^{j(\omega_2 t+\phi_0+\delta_2)} \\ MA_x e^{j(\omega_1 t+\delta_1)} + NA_y e^{j(\omega_2 t+\phi_0+\delta_2)} \end{bmatrix} \tag{16}$$

where ϕ_0 is the phase delay difference between E_2 and E_1; δ is the phase delay difference of the sample ($\delta = \delta_2 - \delta_1$), and θ is the azimuth of the wave plate.

$$\mathrm{M} = cos\frac{\Gamma}{2} - j\,(cos\,2\theta + sin\,2\theta)\,sin\frac{\Gamma}{2} \tag{17}$$

$$\mathrm{N} = cos\frac{\Gamma}{2} + j\,(cos\,2\theta - sin\,2\theta)\,sin\frac{\Gamma}{2} \tag{18}$$

$$I_t = |E_t|^2$$
$$= \frac{1}{2}\Big\{\cos(\tfrac{\Gamma}{2})^2\,(A_x^2 + A_y^2) + \sin(\tfrac{\Gamma}{2})^2\,[(\cos 2\theta + \sin 2\theta)^2 A_x^2 + (\cos 2\theta + \sin 2\theta)^2 A_y^2] + 2\gamma\cos(\alpha - \beta)\Big\} \tag{19}$$

where

$$\gamma = \sqrt{\left(\cos(\tfrac{\Gamma}{2})^2 - \sin(\tfrac{\Gamma}{2})^2\cos 4\theta\right)^2 + (\sin\Gamma\cos 2\theta)^2 A_x A_y} \tag{20}$$

$$\alpha = \omega t - \phi_0 + \delta \tag{21}$$

$$\beta = tan^{-1}\left(\frac{sin\,\Gamma\,cos\,2\theta}{cos(\frac{\Gamma}{2})^2 - sin(\frac{\Gamma}{2})^2\,cos\,4\theta}\right) \tag{22}$$

3. Results and Discussions

This study used glucose solution as the analyte sample. Because glucose solution is birefringent, a change in the concentration of the glucose solution results in slight polarization rotation. These results were observed in p- and s-polarized lights in common-path heterodyne interferometry. Consequently, a lock-in amplifier can be used to easily measure the phase variation. A half-wave plate was used without the sample to measure the phase delay of the half-wave plate. The experimental results are shown in Fig. 6. The solid line represents the MATLAB simulation results when Γ = +177°, and the dotted line represents the measurement results. The results confirmed that the phase delay (Γ) of the half-wave plate was +177°.

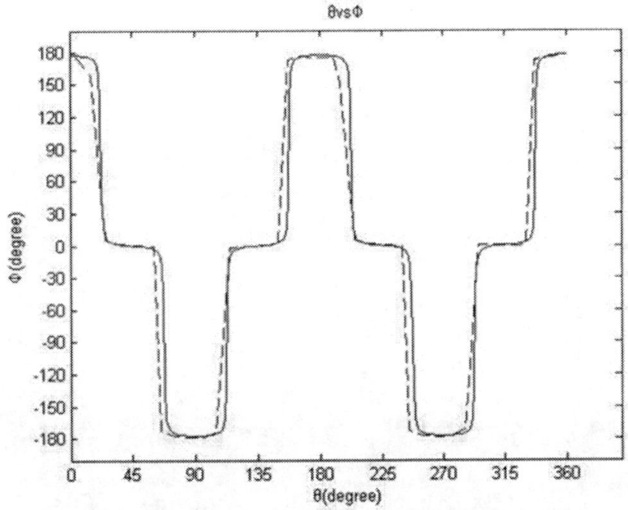

Fig. 6 Experimental result of the half-wave plate.

When the azimuth angle of the half wave plate is 22.5°, the average slope of phase in the azimuth angle segment in the range [22.3°, 22.7°] is −103.29 (°/°) for a phase retardation of 177° of the wave plate. The phase change because of the optical rotation is greater than the retardation of the specimen. The degree of polarization rotation is defined as the azimuth rotation. This method measures the polarization change accurately, because it has a very high phase slope property, and a normal material has very small polarization rotation.

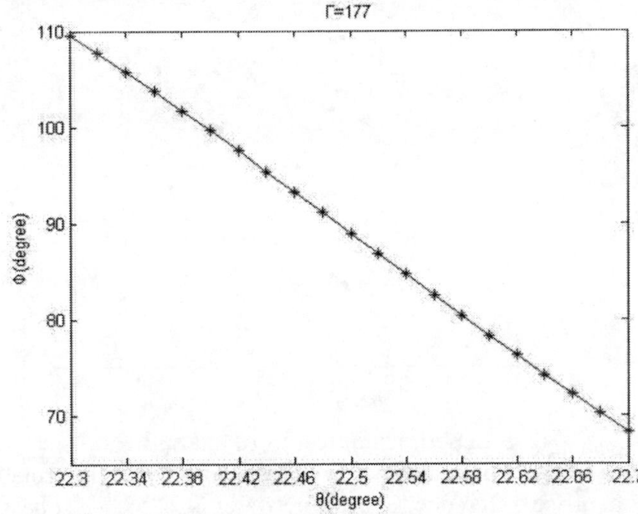

Fig. 7 Γ=177° degrees in the graph of 22.3° to 22.7°

The measured result when the half-wave plate is at $\theta = 0°$ is the phase retardation. Fig. 8. depicts the measurement results. The phase difference fluctuates, indicating that the change occurs only in the phase retardation, not in the polarization. The phase slope due to the polarization rotation at $\theta = 0°$ is zero. The phase distribution fluctuates with no correlation between the concentration of the sugar and polarization rotation, indicating that the phase change is only related to the axis of the birefringence of the sugar solution. The phase appears to fluctuate wildly when the speed axis changes irregularly and randomly.

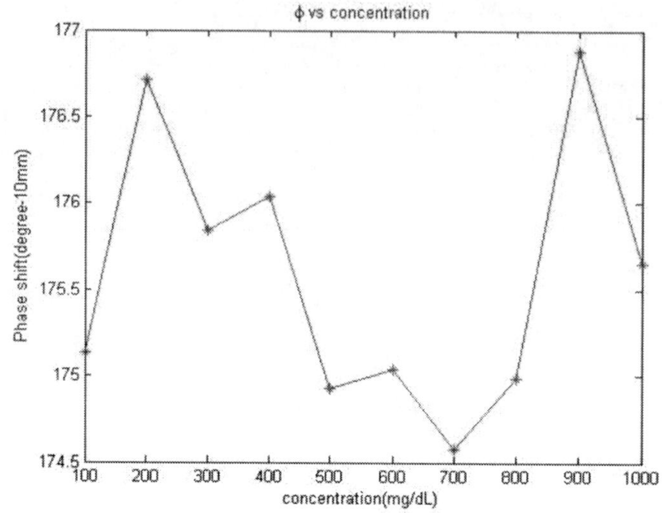

Fig. 8 Experimental results with azimuth of 0°

The azimuth angle of half wave plate for the small polarization rotation measurement is θ=22.5°. Fig. 9. shows the measurement results, showing an increasing phase change as the glucose concentration increases. The polarization rotation of the glucose solution causes this phase change is caused by, and is proportional to the phase difference.

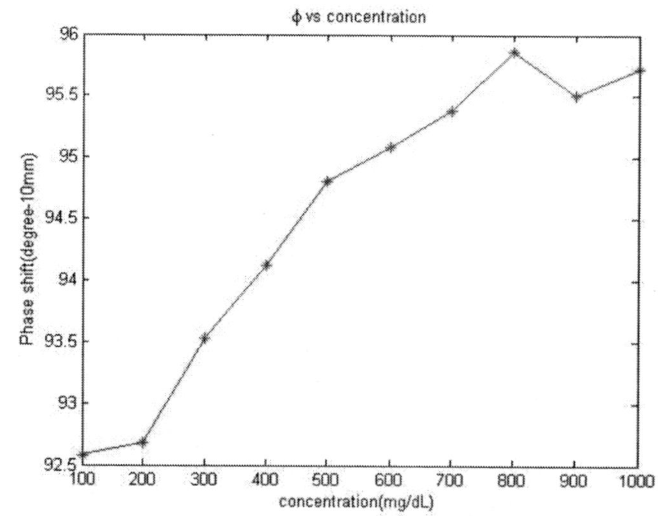

Fig. 9 Experimental results of azimuth 22.5°

Fig. 10. displays the measurement results of the degrees of polarization rotation of different glucose concentrations. The initial phase difference for the pure water is 92.5°. The phase change from the initial phase is divided by the average phase slope of −103.29 (°/°) is the level of polarization rotation. Thus, the relationship between the glucose concentration and the polarization rotation is obtained. If the phase resolution is 0.01°, then the optimal azimuth resolution (polarity rotation) $=\frac{0.01}{103.29} = 9.68 \times 10^{-5\circ} = 0.35''$.

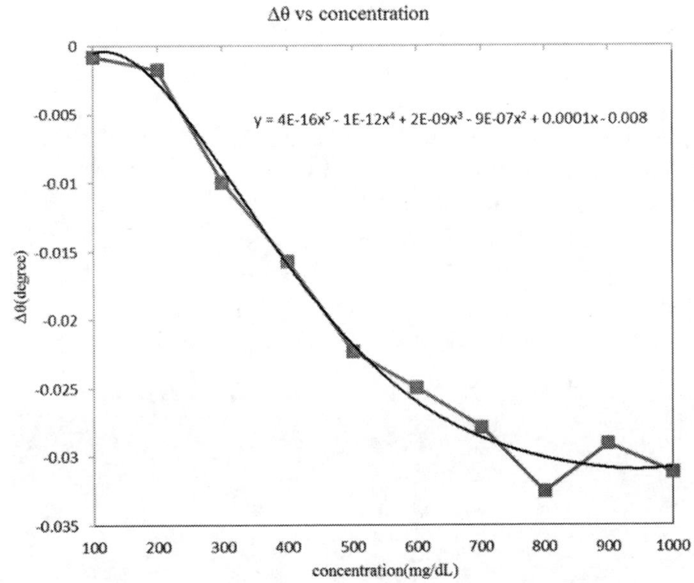

Fig. 10 Relationship between glucose concentration and polarization angle

The polarization rotation can be regarded as the azimuth rotation. From Fig. 10, the negative value of the azimuth rotation indicates that the wave plate is equivalent to clockwise rotation, so the direction of the polarization rotation is inverse. In other words, the polarization has right-hand rotation.

4. Conclusions

If an optical rotation system with improved sensitivity can be used in future studies, glucose solutions with lower concentrations will be able to be measured, thereby elevating the analysis resolution to identify minute degrees of optical rotation. The application of such a technology in medical science can help measure the blood glucose concentration in patients with diabetes. In addition, the system can be applied in high-precision measurement to determine the optical rotation of materials.

This study demonstrates the feasibility of the proposed small polarization rotation measurement method. This method successfully measures small polarization variations of 0~0.035° for the phase resolution of 0.01°. The resolution of polarization rotation can reach to 0.35″, so can be analyzed for some research purposes that require the micro-rotation of polarization.

Acknowledgment

The authors would like to thank the Ministry of Science and Technology of the Republic of China, Taiwan, for supporting this research under Contract No. MOST107-2637-E-105-006-.

References

[1] C.M. McIntyre and S.E. Harris: J. Opt. Soc. Ame. Vol. 58 (1968), pp. 1575-1580.
[2] H.B. Serreze and R.B. Goldner: Rev. Sci. Instrum. Vol. 45 (1974), pp. 1613-1614.
[3] L.H. Shyu, C.L. Chen and D.C. Su: Appl. Opt. Vol. 32 (1993), pp. 4228-4230.
[4] M. Sypek: Opt. Laser. Technol. Vol. 23 (1991), pp. 42-44.
[5] Y. Lin, Z. Zhou and R. Wang: Opt. Let. Vol. 13 (1988), pp. 553-555.
[6] S. Nakadate: Appl. Opt. Vol. 29 (1990), pp. 242-246.
[7] W. Liu, M. Liu and S. Zhang: Appl. Opt. Vol. 47 (2008), pp. 5562-5569.
[8] P. Zhang, Y.D. Tan, W.X. Liu and W.X. Chen: Sci. China Technol. Sci. Vol. 56 (2013), pp. 1155-1164.
[9] C.M. Feng, Y.C. Huang, J.G. Chang, M. Chang and C. Chou: Opt. Commun. Vol. 141 (1997),

pp. 314-321.
[10] B.D. Cameron and G.L. Cote: IEEE. Trans. Biomed Eng. Vol. 44 (1997), pp. 1221-1227.
[11] M.H. Chiu, J.Y. Lee, and D.C. Su : Appl. Opt. Vol. 36 (1997), pp. 2936-2939
[12] A. Márquez, M. Yamauchi, J.A. Davis and D.J. Franich: Opt. Commun. Vol. 190 (2001), pp. 129-133.
[13] Y.L. Lo and P.F. Hsu, 2002:Opt. Eng. Vol. 41 (2002), pp. 2764-2767.
[14] B. Wang and T.C. Oakberg, 1999: Rev. Sci. Instrum. Vol. 70 (1999), pp. 3847-3854.
[15] S. Ohkubo and N. Umeda: Sens. Mater. Vol. 13 (2001), pp. 433-443.
[16] M.J. Ehrlich, L.C. Phillips and J.W. Wagner: Rev. Sci. Instrum. Vol. 59 (1988), pp. 2390-2392.
[17] M.G. Gazalet, M. Ravez, F. Haine, C. Bruneel and E. Bridoux: Appl. Opt. Vol. 33 (1994), pp. 1293-1298.
[18] P. Dirksen, J.V.D. Werf and W. Bardoel: Precis. Eng. Vol. 17 (1995), pp. 114-116.

Measurement of raised pavement marker's photometric performance

He Huayang[1], Leng Zhengwei[2], Xue Yingqi[2], Zhou Yishu[1*]

[1]Research Institute of Highway, Ministry of Transport, Beijing, China

[2]National Center of Metrization for Equipments of Roads and Bridges, Beijing, China

*Corresponding Author: ys.zhou@rioh.cn

Abstract. Large error exists in the photometric performance test of raised pavement marker. This paper reports a set of retroreflective measurement standard devices built to measure raised pavement markers based on the principle of direct measurement, and the photometric performance of the raised pavement markers was tested. The study also involved calibration of a portable meter using the standard device. According to the experimental comparison, the method proposed in this paper provides a new direction for the development of retroreflection measurement technology. The results measured by the proposed method showed high accuracy. This method can solve the large measurement error problems with indication errors of 2% at most.

1. Introduction

Raised pavement markers on roadways are mainly used to provide guidance to drivers as supplements to regular markings [1]. The basic performance of raised pavement markers is critical to road safety and mainly refers to photometric performance [2]. Photometric performance is usually quantified using the coefficient of luminous intensity [3]. The coefficient of luminous intensity is defined as the ratio of the reflected radiant or luminous flux to the incident flux within the narrow confines of incident and reflected geometrical conditions [4].

This paper proposed a photometric performance measurement method of raised pavement markers and calibration method for the portable retroreflectometer. General procedures for measuring the basic performance of raised pavement markers and retroreflective devices were described. Our own retroreflective device along with our measurements are presented, and we take advantage of these new measurements to extend the study of raised pavement markers and test methods.

2. Materials and Methods

Raised Pavement Marker. Raised pavement markers consist of small retroreflective structures, with many versions existing. A common type of retroreflective structure based solely on light reflection is the so-called cube corner reflector [5]. The cube corner reflector normally consists of three mutually perpendicular planar surfaces. This type of retroreflector can be mass fabricated by, for instance, embossing microprisms into polymers.

The photometric performance of raised pavement markers in the field of traffic safety refers to the coefficient of luminous intensity. The coefficient of luminous intensity is used to specify the performance of retroreflective devices. It considers the retroreflected luminous intensity as a function of the perpendicular illuminance incident on the device. The coefficient of luminous intensity can be

IOP Conf. Series: Materials Science and Engineering **711** (2020) 012083 doi:10.1088/1757-899X/711/1/012083

used to describe the performance of roadway delineators. White and yellow raised pavement markers were selected. The coefficients of luminous intensities were in the range of the requirements of GB/T 24725-2009 raised pavement markers. According to the different coefficients of luminous intensity, the raised pavement markers were divided into two levels. They are shown in Figure 1.

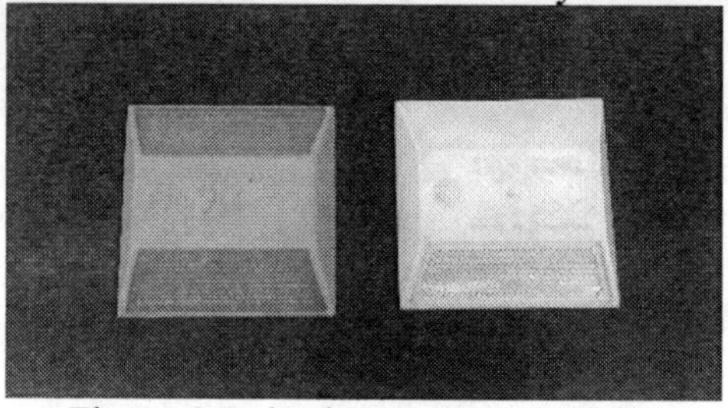

Figure 1 Raised pavement markers

Experimental Devices. The experimental devices consist of a standard device and a portable raised pavement marker retroreflectometer.

The standard device involves a light projector source, a photometric scale, an observation angle positioner and a test specimen holder in a suitably darkened laboratory (as shown in Figure 2). The specimen center is 15 m away from the photometric scale entrance aperture. Black curtains and other means are used to eliminate stray light [6].

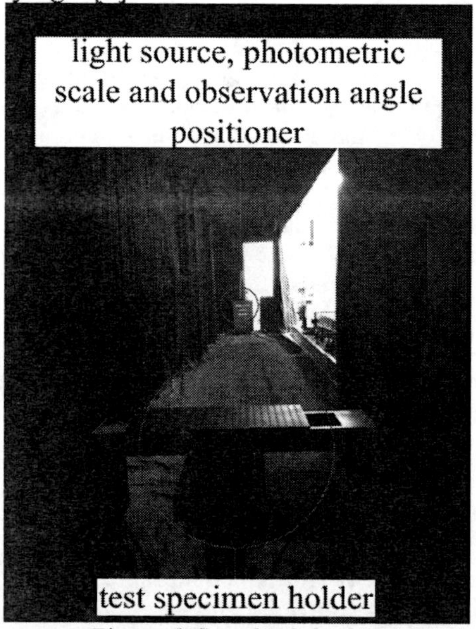

Figure 2 Standard device

The raised pavement marker retroreflectometer (as shown in Figure 3) contains an internal light source and photoreceptors. It relies on the method of substitution calibration. The testing principle is the ratio method described in JT/T 690 test method for photometric characteristics of retroreflectors, which belongs to the class of indirect measurement methods.

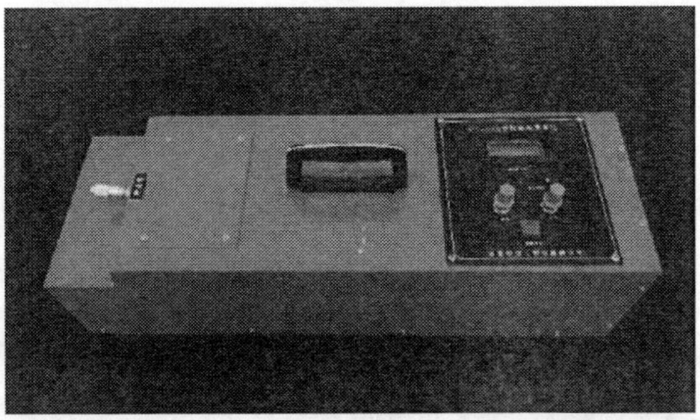

Figure 3 Raised pavement marker retroreflectometer

Experiments. First, standard plates were calibrated with the standard system. The standard system simulates the use of the raised pavement marker. When a vehicle is driven at night, the car headlights illuminate the raised pavement markers on the road surface, and the light is reflected by the raised pavement markers and observed by the driver sitting in the vehicle (as shown in Figure 4). Second, a black plate was used to adjust the zero point of the raised pavement marker retroreflectometer. Then, the retroreflectometer was calibrated using the standard plates. Finally, under the geometrical condition specified in the tables below, the reference plate's coefficient of luminous intensity was measured along the reference direction with the calibrated raised pavement marker retroreflectometer.

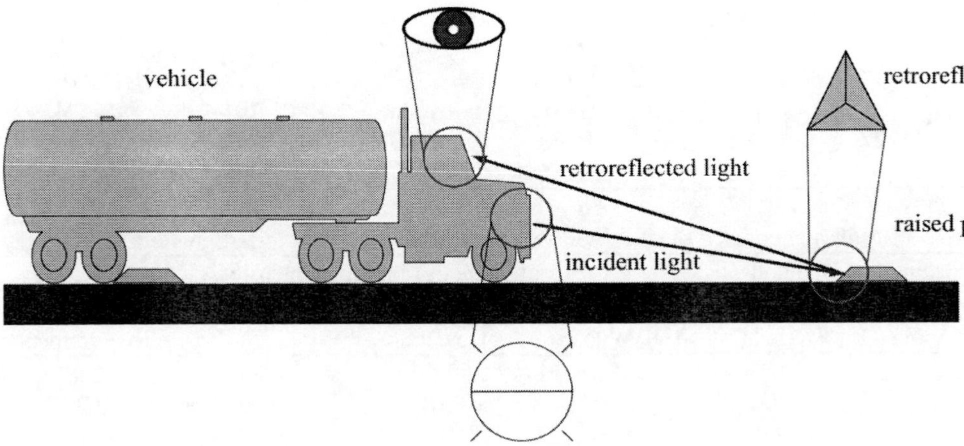

Figure 4 Service process of the raised pavement marker

3. Results and Discussion

Table 1 presents the measurement results of standard plate A obtained with the standard system.
Table 2 and Table 3 presents the average value, standard deviation and variance of standard plate A obtained with the standard system.

Table 1 Measurement results of standard plate A obtained with the standard system

Color	Geometric Conditions		Coefficient of Luminous Intensity [mcd/lx]									
	Observation Angle	Entrance Angle	1	2	3	4	5	6	7	8	9	10
White	0.2°	0°	834.5	835.3	835.6	835.8	836.2	836	836.3	835.9	836.3	836.4
		20°	530.1	530.6	530.6	530.3	530.5	530.5	530.7	530.8	531.1	530.8
		-20°	530.1	530.6	530.6	530.3	530.5	530.5	530.7	530.8	531.1	530.8
	0.33°	5°	754.2	754.3	754.5	754.2	754.5	754.9	754.7	755.4	755	755.1
		-5°	772.4	772.5	773	773.2	773.1	773.2	773.5	773.2	773.6	773.2

IOP Conf. Series: Materials Science and Engineering **711** (2020) 012083 doi:10.1088/1757-899X/711/1/012083

	1°	10°	271.2	271.4	271.5	271.7	271.4	271.4	271.9	271.3	271.6	271.6
		-10°	278	277.9	278.4	277.9	278.2	278.2	277.9	278.4	277.9	278
	2°	15°	16.8	16.8	16.79	16.79	16.79	16.76	16.77	16.77	16.77	16.78
		-15°	18.89	18.86	18.87	18.85	18.87	18.85	18.85	18.85	18.85	18.86
Yellow	0.2°	0°	320.6	321.7	322	322	322.2	322.3	322.2	322.2	322.2	322.3
		20°	208.3	208.8	209	209.1	209.1	209.2	209.3	209.2	209.1	209.3
		-20°	208.3	208.8	209	209.1	209.1	209.2	209.3	209.2	209.1	209.3
	0.33°	5°	277.7	278.2	278.4	278.4	278.6	278.7	278.5	278.7	278.6	278.7
		-5°	271.9	272.3	272.6	272.5	272.6	272.8	272.7	272.7	272.6	272.8
	1°	10°	86.68	86.87	86.89	87.03	86.97	87.05	87	87.04	87.08	86.95
		-10°	85.54	85.68	85.8	85.74	85.73	85.9	85.78	85.76	85.77	85.83
	2°	15°	15.59	15.6	15.6	15.57	15.57	15.58	15.58	15.57	15.6	15.58
		-15°	15.47	15.5	15.48	15.47	15.46	15.47	15.49	15.47	15.49	15.46

Table 2 Results of standard plate A(white) obtained with the standard system

Color	Geometric Conditions		Coefficient of Luminous Intensity [mcd/lx]			
	Observation Angle	Entrance Angle	Average Value	Standard Deviation	Variance	Minimum Requirement
White	0.2°	0°	835.83	0.58	0.34	580
		20°	530.60	0.28	0.08	272
		-20°	530.60	0.28	0.08	272
	0.33°	5°	754.68	0.41	0.17	472
		-5°	773.09	0.38	0.14	472
	1°	10°	271.50	0.21	0.04	74
		-10°	278.08	0.2	0.04	74
	2°	15°	16.78	0.01	0	11.8
		-15°	18.86	0.01	0	11.8

Table 3 Results of standard plate A(yellow) obtained with the standard system

Color	Geometric Conditions		Coefficient of Luminous Intensity [mcd/lx]			
	Observation Angle	Entrance Angle	Average Value	Standard Deviation	Variance	Minimum Requirement
Yellow	0.2°	0°	321.97	0.51	0.26	348
		20°	209.04	0.3	0.09	163.2
		-20°	209.04	0.3	0.09	163.2
	0.33°	5°	278.45	0.31	0.1	283.2
		-5°	272.55	0.27	0.07	283.2
	1°	10°	86.96	0.12	0.01	44.4
		-10°	85.75	0.1	0.01	44.4
	2°	15°	15.58	0.01	0	7.08
		-15°	15.48	0.01	0	7.08

Table 4 presents the measurement results of standard plate B obtained with the standard system and retroreflectometer.

Table 4 Results of standard plate B obtained with the standard system and retroreflectometer

Color	Geometric Conditions		Coefficient of Luminous Intensity [mcd/lx]				Indication Error	
	Observation Angle	Entrance Angle	Standard System		Retroreflectometer			
			Level 2	Level 1	Level 2	Level 1	Level 2	Level 1
White	0.2°	0°	657.8	1767.3	661.1	1768	0.50%	0.04%
		20°	444.5	1364.5	445.2	1347	0.16%	-1.28%

		-20°	444.5	1364.5	449.9	1388	1.21%	1.72%
	0.33°	5°	673.6	744.7	670.7	731	-0.43%	-1.84%
		-5°	628.1	833.3	630.4	839	0.37%	0.68%
	1°	10°	226.9		224.8		-0.93%	
		-10°	247		245.5		-0.61%	
	2°	15°	13.8		13.8		0.00%	
		-15°	13.7		13.7		0.00%	
Yellow	0.2°	0°	283.2	747.2	283.5	750	0.11%	0.37%
		20°	185	641.1	188.1	640	1.68%	-0.17%
		-20°	185	641.1	185.6	644	0.32%	0.45%
	0.33°	5°	246.5	445.4	249.8	447	1.34%	0.36%
		-5°	240.5	459.9	239.4	464	-0.46%	0.89%
	1°	10°	74.2		74.6		0.54%	
		-10°	74.3		74.3		0.00%	
	2°	15°	12.4		12.4		0.00%	
		-15°	12.2		12.2		0.00%	

The data from Table 1,

Table 2 and Table 3 indicate that the standard deviation of the standard system when measuring the coefficient of luminous intensity is less than 0.6 mcd/lx. Obviously, the standard system has good measurement repeatability.

According to the requirements of JJG (traffic) 059-2004 retroreflectometer, the indication error of the retroreflectometer should be at most ±8%. The experimental results show that with the calibration method, the instrument has an indication error of less than ±2%. The calibration method can meet the requirements of the regulation.

4. Conclusion

This paper introduced a basic performance measurement method for raised pavement markers, for which a standard system was built. According to the experimental comparison, the method proposed in this paper provides a new direction for the development of retroreflection measurement technology. The results measured with the improved method have high accuracy. The method can solve the problem of large measurement error. The indication errors are at most 2%.

References

[1] Lu Q, Guo L, Yu B. Design Improvements of Retroreflective Raised Pavement Markers Based on Quantification of Their Physical Properties[R]. 2016.

[2] Nguyen T T, Spehr J, Lin T H H, et al. Fused Raised Pavement Marker Detection Using 2D-Lidar and Mono Camera[C]// IEEE International Conference on Intelligent Transportation Systems. 2015.

[3] Eckerle, K. L., and J. J. Hsia. "Retroreflectance MAP (Measurement Assurance Program) service for coefficient of luminous intensity." Final Report National Bureau of Standards, Washington, DC. National Measurements Lab. (1984).

[4] Zhao F H , Qiao W D , Zhao M , et al. Measurement for Coefficient of Retroreflected Luminance by Direct Luminance Method[J]. Jiliang Xuebao/Acta Metrologica Sinica, 2017, 38(5):580-583.

[5] Zhou L , Kahn J M , Pister K S J . Corner-cube retroreflectors based on structure-assisted assembly for free-space optical communication[J]. Journal of Microelectromechanical Systems, 2003, 12(3):233-242.

[6] He H, Zhou Y, Zhang J, et al. Measuring method and standard system for retroreflective traffic marking's photometric characteristic[C]//Optical Measurement Systems for Industrial Inspection XI. International Society for Optics and Photonics, 2019, 11056: 110563L.

IOP Conf. Series: Materials Science and Engineering **711** (2020) 012084 doi:10.1088/1757-899X/711/1/012084

Effective Backbone Network for 3D Object Detection in Point Cloud

Jun Xu[1, a], Yanxin Ma[2, b]*and Songhua He[1, c]

[1] College of Information Science and Engineering, Hunan University, Changsha, China

[2] College of Meteorology and Oceanography, National University of Defense Technology, Changsha, China

[a]15738335219@139.com, [b]mayanxin@nudt.edu.cn, [c]13973132618@139.com

Abstract. Three-dimensional (3D) object detection is composed of object classification and object localization, and has been used in many applications, such autonomous driving and mobile robot. However, the accuracy of classification and localization is greatly affected by the depth of the network. Shallow networks tend to cause poor classification, but as the depth of the network increases, the network degradation will become more obvious, which is not conducive to the training of network. To solve this problem, a novel Backbone Network is proposed in this paper for 3D object detection in point cloud, which consists of multiple residual modules. Experimental results on the KITTI 3D object detection benchmarks show that Backbone Network proposed can effectively improve the accuracy of 3D object detection.

1. Introduction

Three-dimensional (3D) object detection is a challenging subject in the field of computer vision, and is widely used in various practical applications, such as autonomous driving [1, 2], video surveillance [3], mobile robot [4] and so on. It can be divided into two sub-tasks: object classification and object localization. Object classification is to accurately determine the category of the object, object localization is to give an accurate 3D bounding box of the object in a given scene. In recent years, deep learning has become an important research direction of computer vision with its powerful feature learning ability, and has made substantial progress in many fields, including 3D object detection.

Sufficient depth plays a key role in the success of deep learning model in various tasks. The main module of deep learning network structure is a standard non-linear transformation module, which consists of convolution layer, pooling layer and activation layer. The deeper the model is, the stronger its non-linear expression ability is, allowing it to learn more complex transformations and fit more complex feature inputs [5]. Although deepening the network can improve the performance of the model to a certain extent, it does not mean that the performance of the network is directly proportional to the depth, which is mainly reflected in the improvement of performance and optimization. First of all, when the depth of the network reaches a certain level, the performance of network tends to be saturated and will not increase with the increase of depth. In this case, the deepening of the network can only bring expensive time cost. Secondly, the deepening of the model may also lead to the decline in the learning ability of some shallow levels, which limits the learning of some deep levels. Last but not least, the problem of network degradation caused by deep network always exists. Although it can be alleviated, it cannot be eliminated [6]. Therefore, it is possible that with the deepening of the network, performance will begin to decline.

Network degradation means that in the process of increasing network depth, the training accuracy tends to be saturated gradually, and the training accuracy will decline if we continue to increase the depth of network. Moreover, this reduction in accuracy is not caused by overfitting. To solve this issue in 3D object detection network, we use the residual learning model proposed by He *et al.* [6], and its basic structure is shown in Fig. 1. The experimental results show that the use of residual module ensures good network performance when training deeper networks.

2. Related Work

3D Object Detection. The existing 3D object detection methods can be divided into three categories by the data, *i.e.* monocular image based, 3D point cloud based and multi-model fusion based methods.

Monocular Image Based Detection. Nowadays, monocular image based two-dimensional (2D) object detection has achieved superior performance in both speed and accuracy. However, due to the lack of spatial structure information, 3D object detection still faces great challenges. Even so, some studies [7, 8] are still devoted to the research of 3D object detection in monocular image, since it is a basic problem and the equipment for obtaining image is much cheaper. DS3D [8] first uses 2D object detection technology to obtain 2D candidate boxes from the image. Then each candidate box is inferred into 3D space by combining the prior information of the object. Next, three view features of the 3D candidate box are extracted from the Bird's-Eye View (BEV), the Front View (FV) and the side view. Finally, these features are merged and the merged feature is used to refine the 3D bounding box to improve the performance of detection. Unfortunately, the accuracy of 3D object detection in monocular image is far from meeting the needs of applications such as autonomous driving.

3D Point Cloud Based Detection. The 3D point cloud is irregular data. To apply Convolutional Neural Network (CNN) whose input must be regular, there are mainly two ways to process point cloud: projection [1, 9] and voxelization [2, 10]. Although projecting point cloud onto one or more 2D planes can improve computational efficiency, it is easy to lose spatial structure information. Therefore, many technologies [2, 10] convert point cloud into regular data by rasterizing it into 3D voxel grids, and then the regular data is input to the CNN.

In addition, some studies have proposed direct processing of point cloud [11, 12]. PointRCNN [12] is a two-stage object detector. In the first stage, each point is classified using PointNet [11] and a large number of candidate boxes are generated. In the second stage, redundant candidate boxes are discarded according to the category information and the remaining candidate boxes are refined. Since all input points of PointRCNN [12] to be classified and produce corresponding candidate boxes, the detection is slow, although PointRCNN [12] achieves good performance on the KITTI dataset [13]. Considering the requirements of accuracy and speed, voxelization is used to convert point cloud into regular data in this paper.

Multi-model Fusion Based Detection. Some techniques [14 - 17] have achieved good detection performance by fusing RGB image with point cloud. These algorithms use CNN to extract the appearance information of RGB image and structure information of point cloud respectively, and then use the fused feature to determine the final detection results. There are also some studies to improve the accuracy of 3D object detection by fusing RGB image and depth image [18, 19]. Lukas *et al.* [18] designed an object detection framework, which consists of Network-in-Network (NIN) [20] and GoogLeNet [21]. In detail, NIN [20] uses Multi-Layer Perceptron (MLP) instead of traditional convolution layer, the nonlinearity of MLP makes NIN [20] more generalization. Combining multiple models tends to improve detection accuracy, however, the speed is relatively slow, since the large number of features to be processed.

Residual Network. The depth of the network is critical for the performance of the network. However, simply stacking networks will lead to inevitable network degradation. To solve this issue, He *et al.* proposed the residual network [6] which consisting of several residual blocks. The architecture of the residual block is shown in Fig. 1. The network is designed as $H_{(X)} = F_{(X)} + X$ through a short connection, so a residual function $F_{(X)} = H_{(X)} - X$ to be learned by the network. This method will not cause additional parameters and calculations. The most important is that when the network is deepened, the residual block can solve the problem of network degradation very well. Consequently, it has been applied to various tasks, such as object classification [6], object detection [18] and so on.

3. Our Network

The architecture proposed will be introduced in this section. As shown in Fig. 2, it consists of four parts: (1) Data preprocessing; (2) Feature Learning Network; (3) Backbone Network; (4) Header Network.

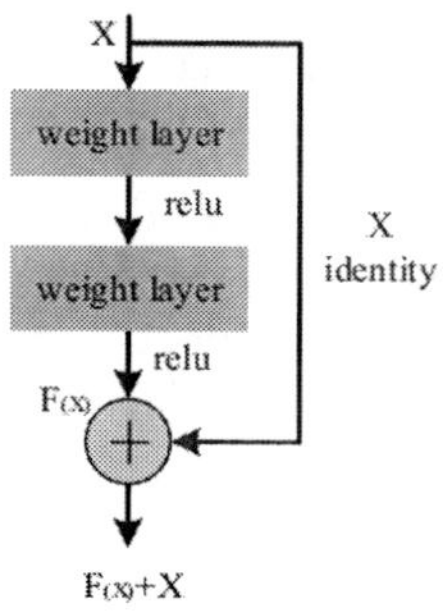

Figure 1. Residual learning: a building block.

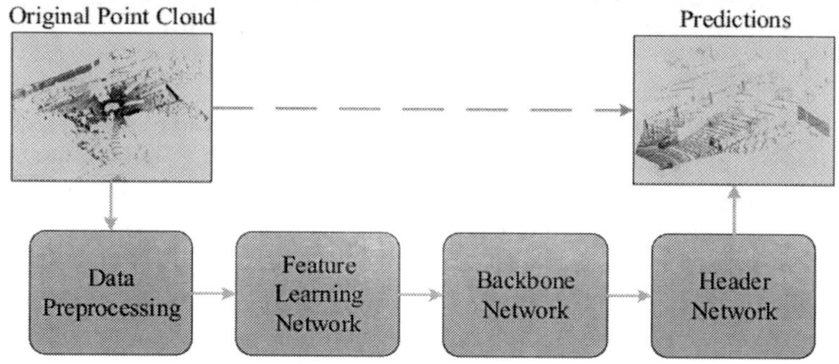

Figure 2. Network overview. The main components of the network are a Feature Learning Network, Backbone and Header Network.

Data Preprocessing. To apply CNN, it is necessary to convert the point cloud into regular data. Firstly, we crop point cloud using a fixed dimension $L \times W \times H$ m^3 in the directions of X, Y and Z axes. Then the cropped point cloud is discretized into an evenly spaced grid in the x-y plane with size D_x and D_y. Consequently, a total $L_1 \times W_1$ pillars are generated, where $L_1 = L / D_x$, $W_1 = W / D_y$. In addition, to reduce the sampling deviation between the pillars, we use random down-sampling to fix the number of points in each non-empty pillar to a number N.

Feature Learning Network. To obtain a more expressive feature, any point in a non-empty pillar is represented as a vector $p_i=\{x, y, z, r, x\text{-}\Delta x, y\text{-}\Delta y, z\text{-}\Delta z, x\text{-}x_c, y\text{-}y_c\}$ $(i \leq N)$, where (x, y, z, r) represents the coordinates and reflectance of the point itself, $(\Delta x, \Delta y, \Delta z)$ represents the mean coordinates of all points in each pillar, and (x_c, y_c) represents the x, y center of the pillar. Then, a simplified version PointNet is used in each non-empty pillar to learn pillar-wise feature. In detail, the simplified PointNet consists of a linear layer, a Batch Normalization (BN) layer and Rectified Linear Unit (ReLU) layer. After that, each non-empty pillar is represented as a 128D vector. And the empty pillars are filled with a 128D zero vector. As a result, the point cloud can be represented by a pseudo-image of size $L_1 \times W_1 \times 128$.

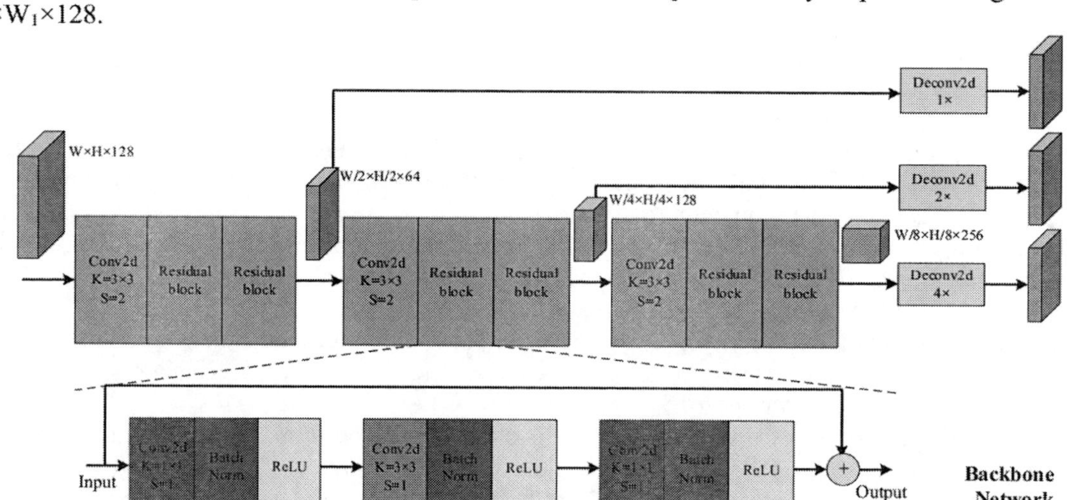

Figure 3. Backbone Network: Conv2d is the 2D convolutional neural network, and K and S are the convolutional kernel and stride respectively. Deconv2d is the 2D deconvolutional neural network.

Backbone Network. Some studies have used multiple convolutional layers in the Backbone Network to obtain better classification results [2, 10, 22]. However, the deepening of the network is likely to cause network degradation. In order to end the problem, we propose a novel Backbone Network for 3D object detection in the point cloud, and its structure is shown in Fig. 3. It is the unidirectional feed-forward computation of 2D CNN, which computes the pyramidal 2D feature hierarchy. Every network layer consists of a convolution layer and a residual block. In particular, the structure and parameters of residual block in each network layer are the same, as shown in Fig. 3. Each residual block is composed of three convolutional layers, and each layer is followed by a BN layer and a ReLU layer. In addition, to aggregate high-level features and high-level features, three deconvolution layers are applied to the output of three convolution blocks. After that, three feature maps with different levels but the same size are obtained, and then these feature maps are inputted to the Header Network for classification and regression.

Header Network. The aim of Header Network is to detection objects from feature maps generated by the Backbone Network. All the 2D feature map will be fused to concatenated together to form a stronger semantic feature map for classification and regression tasks.

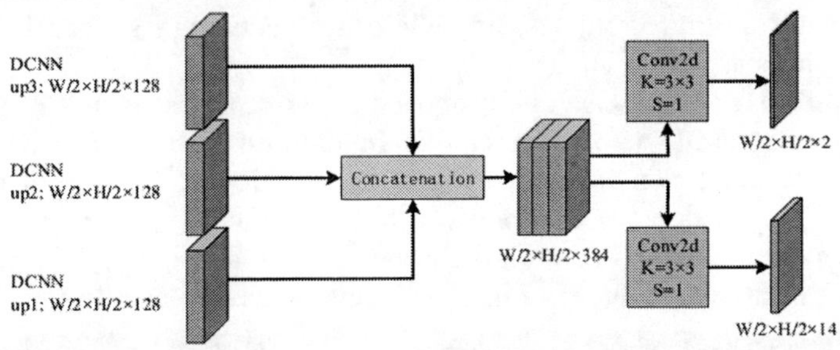

Figure 4. Header Network: DCNN is the 2D deconvolutional neural network, Conv2d is the 2D convolutional neural network, and K and S are the convolutional kernel and stride respectively.

4. Experiment and Results

All experiments are performed on the challenging KITTI benchmarks [13], which contains 7481 training point clouds and corresponding RGB images, covering three categories: Car, Cyclist and Pedestrian. Because the label of the testing dataset is not available, the training dataset is divided into training set (3712) and validation (3769). Moreover, the loss function and the parameters of network will be described in this section.

Loss function. The loss of network consists of two parts: classes classification and 3D bounding box regression. As shown in Eq. 1, different weights are used to balance relative importance, where $\rho=1$ and $\mu=2$.

$$L_{all} = \rho L_{cls} + \mu L_{reg}. \tag{1}$$

Classes Classification L_{cls}. For the object classification loss, focal loss proposed by Lin *et al.* [23] is used, which is defined as follows:

$$L_{cls} = -\alpha(1-p)^{\gamma}\log p. \tag{2}$$

3D Bounding Box Regression L_{reg}. For the 3D bounding box regression task, we use the same loss functions defined in PointPillars [22]. Anchors and ground truths are represented by 7 parameters, *i.e.* central coordinates, dimensions and rotation angles around the Z axis. Consequently, the regression targets are defined as a vector, (Δx, Δy, Δz, Δl, Δw, Δh, $\Delta\theta$), and details are as follows:

$$\Delta x = (x^g - x^a)/d^a. \quad \Delta y = (y^g - y^a)/d^a. \quad \Delta z = (z^g - z^a)/h^a.$$
$$\Delta l = \log(l^g/l^a). \quad \Delta w = \log(w^g/w^a). \quad \Delta h = \log(h^g/h^a). \tag{3}$$
$$\Delta\theta = \sin(\theta^g - \theta^a).$$

$$L_{reg} = \sum_{r\in(x,y,z,l,w,h,\theta)} \mathrm{SmoothL}_1(\Delta r). \tag{4}$$

where x^g and x^a are the ground truth and anchor respectively, and $(d^a)^2 = (l^a)^2 + (w^a)^2$.

Setting. Considering that the dimensions of Car, Cyclist and Pedestrian are quite different, different sizes are used for cropping and rasterization of point clouds in data preprocessing.

Car Detection. The x, y, z range of point cloud is L × W × H = [0, 70.4] × [-40, 40] × [-3, 1]m^3, and the pillar size is set to be $D_x = D_y = 0.16$m. Following the PointPillars [22], the dimension of the anchor is set to (3.9m, 1.6m, 1.5m) with a z center of z = -1m.

Cyclist and Pedestrian Detection. The x, y, z range of point cloud is L × W × H = [0, 47.36] × [-19.84, 19.84] × [-2.5, 0.5]m^3, and the pillar size is set to be $D_x = D_y = 0.16$m. The dimension of the cyclist anchor is set to (1.76m, 0.6m, 1.73m) while the pedestrian anchor is (0.8m, 0.6m, 1.73m). Moreover, all the anchors are centered at z = -0.6m.

Results and Qualitative Analysis. The proposed network is tested on the KITTI verification set, and we compare it with other 3D detection methods. In addition, to better prove the effectiveness of our Backbone Network, some ablation experiments are conducted with the basic network, *i.e.* PointPillars [22]. The 3D and BEV detection results are given in Tab. 1 and Tab. 2, respectively. The number in parentheses indicates the number of convolutional layers with a stride of 1 in each convolution block. For better visualization, (a), (b) and (c) in Fig. 5 give several examples of three categories respectively.

(a) Car (b) Cyclist (c) Pedestrian

Figure 5. Visualization of 3D object detection on KITTI validation dataset. All the 2D and 3D bounding boxes in RGB images are obtained from the result of detection. In the point clouds, red 3D boxes represent ground trues while teal 3D boxes indicated that have been successfully detected.

Table 1. 3D detection performance: Average precision (AP) (in %) for 3D boxes in the KITTI validation set.

Method	Modality	Car			Cyclist			Pedestrian		
		Easy	Mod.	Hard	Easy	Mod.	Hard	Easy	Mod.	Hard
MV3D	Img &Lidar	71.09	62.35	55.12	N/A	N/A	N/A	N/A	N/A	N/A
AVOD	Img &Lidar	81.94	71.88	66.38	64.00	52.18	46.61	50.80	42.81	40.88
F-PointNet	Img &Lidar	81.20	70.39	62.19	71.96	56.77	50.39	51.21	44.89	51.21

VoxelNet	Lidar	77.47	65.11	57.73	61.22	48.36	44.37	39.48	33.69	31.50
SECOND V1	Lidar	85.99	75.51	68.25	70.51	53.85	46.90	51.07	42.56	37.29
SECOND V1.5	Lidar	86.79	75.92	74.32	77.11	60.54	53.74	58.08	49.66	43.27
PointPillars (3)	Lidar	85.30	76.20	69.58	77.94	58.12	55.03	60.57	57.46	52.30
Our (1×3)	Lidar	87.08	76.62	69.64	82.35	62.97	58.78	**67.69**	**61.03**	**55.57**
PointPillars (6)	Lidar	86.44	**77.23**	73.49	82.24	60.37	55.52	57.53	55.05	49.96
Ours (2×3)	Lidar	**87.66**	77.17	**74.33**	**83.90**	**63.05**	**59.84**	63.68	58.60	52.70

Table 2. BEV detection performance: Average precision (AP) (in %) for BEV boxes in the KITTI validation set.

Method	Modality	Car			Cyclist			Pedestrian		
		Easy	Mod.	Hard	Easy	Mod.	Hard	Easy	Mod.	Hard
MV3D	Img.&Lidar	86.02	76.90	68.48	N/A	N/A	N/A	N/A	N/A	N/A
AVOD	Img.&Lidar	88.53	83.79	77.90	68.09	57.48	50.77	58.75	51.05	47.54
F-PointNet	Img.&Lidar	88.70	84.00	75.33	75.38	61.96	54.68	58.09	50.22	47.02
VoxelNet	Lidar	89.35	79.26	77.39	66.07	54.76	50.55	46.13	40.74	38.11
SECOND V1	Lidar	89.23	86.25	78.95	73.67	56.04	48.78	55.10	46.27	44.76
SECOND V1.5	Lidar	89.61	85.73	78.51	73.61	63.58	56.16	57.62	49.35	46.57
PointPillars (3)	Lidar	89.87	87.36	84.98	81.76	61.07	58.20	68.94	64.72	60.18
Our (1×3)	Lidar	90.10	87.40	85.04	83.54	64.57	60.98	**73.85**	**68.87**	**63.47**
PointPillars (6)	Lidar	90.11	87.16	84.98	83.30	62.84	59.55	66.45	63.35	58.71
Ours (2×3)	Lidar	**90.27**	**88.03**	**86.49**	**84.96**	**64.83**	**61.76**	72.19	67.47	61.79

Car Detection. For car detection task, as shown in Tab. 1 and Tab. 2, the detection performance of the proposed Backbone Network on KITTI validation set is significantly better than other algorithms listed, both of 3D and BEV. Compared with the basic network PointPillars [22], our network achieves better results with respect to average precision (AP).

Cyclist and Pedestrian Detection. Compared with the task of car detection, the accuracy of cyclist and pedestrian detection is slightly poor. There are two main reasons: firstly, the network training is insufficient due to the lack of cyclists and pedestrians. In the training set, there were 10520 cars, while only 594 cyclists and 2104 pedestrians. Another reason is that the sizes of cyclists and pedestrians are small and their distribution is dense, which makes detection difficult. This is more evident in pedestrian detection task. However, the accuracy of our method is much higher than other algorithms, especially for the pedestrian, the 3D detection performance of our method is 3.74% higher than that of PointPillars [22] for the *Hard* level. In addition, we can see from the results of the ablation experiment that shallow network is more suitable for pedestrian detection.

5. Summary

To solve the problem of degradation caused by network deepening, a novel Backbone Network is proposed in this paper, which consists of several residual modules. The results of the ablation experiments on the challenging KITTI validation set show that the performance our network is better than that of PointPillars [22] under the same network depth. In addition, the results of comparison with other methods also fully demonstrate that the proposed Backbone Network is more suitable for 3D object detection.

References

[1] L. Bo, Z. Tianlei, X. Tian. Vehicle detection from 3D lidar using fully convolutional network. arXiv 2016, arXiv:1608.07916, (2016).

[2] Z. Yin, O. Tuzel. VoxelNet: End-to-End Learning for Point Cloud Based 3D Object Detection. In Proceedings of the IEEE Conference on Computer Vision and Pattern Recognition, pp. 4490–4499, (2018).

[3] J. Chengbin, L. Shengzhe, T. D. Do, H. Kim. Real-Time Human Action Recognition Using CNN Over Temporal Images for Static Video Surveillance Cameras. Advances in Multimedia Information Processing – PCM 2015, pp. 330-339, (2015).
[4] F. Pomerleau, F. Colas, R. Siegwart. A Review of Point Cloud Registration Algorithms for Mobile Robotics. Foundations and Trends® in Robotics, (2015).
[5] M. Bianchini, F. Scarselli. On the complexity of neural network classifiers: A comparison between shallow and deep architectures[J]. IEEE transactions on neural networks and learning systems, pp. 1553-1565, (2014).
[6] H. Kaiming, Z. Xiangyu, R. Shaoqing, S. Jian. Deep Residual Learning for Image Recognition. The IEEE Conference on Computer Vision and Pattern Recognition, pp. 770-778, (2016).
[7] C. Xiaozhi, K. Kundu, Z. Ziyu, M. Huimin, S. Fidler, R. Urtasun. Monocular 3d object detection for autonomous driving. In the Proceedings of the IEEE Conference on Computer Vision and Pattern Recognition, pp. 2147–2156, (2016).
[8] L. Buyu, O. Y. Wanli, L. Sheng, Z. Xingyu, W. Xiaogang. GS3D: An Efficient 3D Object Detection Framework for Autonomous Driving. The IEEE Conference on Computer Vision and Pattern Recognition, pp. 1019-1028, (2019).
[9] Y. Bin, L. Wenjie, R. Urtasun. PIXOR: Real-time 3D Object Detection from Point Clouds. The IEEE Conference on Computer Vision and Pattern Recognition, pp. 7652–7660, (2018).
[10] Y. Yan, M. Yuxing, L. Bo. SECOND: Sparsely Embedded Convolutional Detection. Sensors (2018).
[11] C. R. Qi, S. Hao, K. Mo, Guibas, L. J. Guibas. PointNet: Deep Learning on Point Sets for 3D Classification and Segmentation. In Proceedings of the IEEE Conference on Computer Vision and Pattern Recognition, pp. 77–85, (2017).
[12] S. Shaoshuai, W. Xiaogang, L. Hongsheng. PointRCNN: 3D Object Proposal Generation and Detection from Point Cloud. The IEEE Conference on Computer Vision and Pattern Recognition, pp. 770-779, (2019).
[13] Kitti 3D Object Detection Benchmark Leader Board. Available online: http://www.cvlibs.net/datasets/kitti/eval_object.php?obj_benchmark=3d (2018).
[14] C. Xiaozhi, M. Huimin, W. Ji, L. Bo, X. Tian. Multi-view 3D object detection network for autonomous driving. In Proceedings of the IEEE Computer Vision and Pattern Recognition, Honolulu, Volume 1, pp. 1907-1915, (2017).
[15] C. R. Qi, L. Wei, W. Chenxia, S. Hao, L. J. Guibas. Frustum PointNets for 3D Object Detection from RGB-D Data. In Proceedings of the IEEE Conference on Computer Vision and Pattern Recognition, Salt Lake City, pp. 918–927, (2018).
[16] R. Huitl, G. Schroth, S. Hilsenbeck, F. Schweiger, E. Steinbach. TUMindoor: An extensive image and point cloud dataset for visual indoor localization and mapping. IEEE International Conference on Image Processing, (2012).
[17] J. Ku, M. Mozifian, J. Lee, A. Harakeh, S. L. Waslander. Joint 3D Proposal Generation and Object Detection from View Aggregation. In Proceedings of the IEEE/RSJ International Conference on Intelligent Robots and Systems, pp. 1–8, (2018).
[18] L. Schneider, M. Jasch, B. Frohlich, T. Weber, U. Franke, M. Pollefeys, M. Rätsch. Multimodal Neural Networks: RGB-D for Semantic Segmentation and Object Detection, pp 98-109, (2017).
[19] D. Zhuo, L. J. Latecki. Amodal Detection of 3D Objects: Inferring 3D Bounding Boxes from 2D Ones in RGB-Depth Images. In Proceedings of the IEEE Conference on Computer Vision and Pattern Recognition, pp. 5762-5770, (2017).
[20] L. Min, C. Qiang, Y. Shuicheng. Network in Network. arXiv 2013, arXiv:1312.4400. (2013)
[21] C. Szegedy, L. Wei, J. Yangqing, P. Sermanet, S. Reed, A. Dragomir, D. Erhan, V. Vanhoucke, A. Rabinovich. Going Deeper with Convolutions. The IEEE Conference on Computer Vision and Pattern Recognition, pp. 1-9, (2015).

[22] A.H. Lang, S. Vora, H. Caesar, Z. Lubing, Y. Jiong, O. Beijbom. PointPillars: Fast Encoders for Object Detection from Point Clouds. The IEEE Conference on Computer Vision and Pattern Recognition, pp. 12697-12705, (2019).

[23] T.-Y. Lin, P. Dollar, R. Girshick, H. Kaiming, B. Hariharan, S. Belongie. Feature Pyramid Networks for Object Detection. CVPR, pp. 936–944, (2017).

Design of large field of view curved optical system based on ZEMAX

Yuanlin Zhao[1,a], Xiangyang Peng[1,b], Shengli Chang[2,c], Wuming Wu[2,d]

[1]School of Physics and Optoelectronic Engineering, Xiangtan University, Xiangtan 411105, China;

[2]School of Physics and Microelectronics, Hunan University, Changsha 410012, China

[a]email: 201721521326@smail.xtu.edu.cn, [b]email: xiangyang_peng@xtu.edu.cn, [c]email: changshli@163.com, [d]email: 286682722@qq.com

Abstract. Structural design solved the problem in the paper that the large field of view and high resolution of the compound eye optical system can not be realized simultaneously. A curved compound eye system including a 3-ring lens group is designed. The compound eye system is divided into two parts: the central sub-eye system and the edge sub-eye system. The edge sub-eye system reduced the complexity of processing because of the same lens group. In the design process, we systematically studied the field of view between the single sub-eye system and the compound eye system, and obtained the arrangement of the sub-eyes. Then, the geometrical model of a single sub-eye system stitching is presented which is based on the conjugate relations between the incident and exit window. The whole field of the compound eye system designed by this method could reach up 79.5°, the angular resolution of the center system is 0.01°, and the angular resolution of the edge system is 0.017°. Compared with traditional single aperture optical system, the image quality of edge field of view is greatly improved, and the ability to acquire the target orientation is improved. Finally, the correctness of the optical system is validated with the optical design software ZEMAX. And this research could further promote the application of compound eye imaging system.

1. Introduction

With the development of high-tech into the direction of integration and miniaturization, the existing traditional monocular optical system can not meet the practical needs of people due to many factors, especially in the field of target recognition, positioning, tracking and other issues. For example, the contradiction between large field of view and high resolution [1]. The comparison of the resolution between compound eye and single eye is shown in Fig.1, where the yellow portion is the shared value portion.

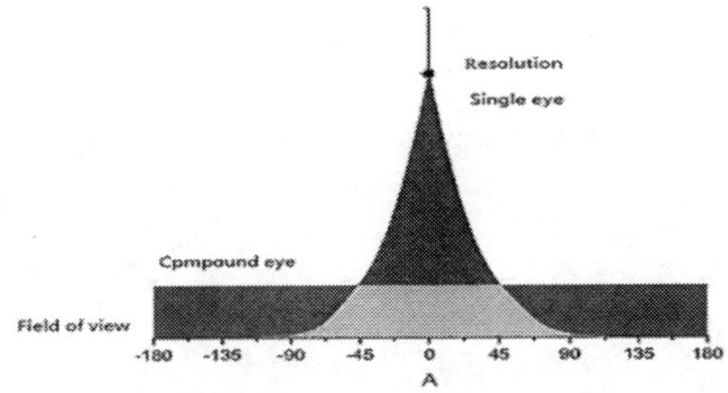

Fig.1 Comparison of the resolution between compound eye and single eye

In recent years, biologists have found that insect compound eyes had the characteristics of high sensitivity, small size, large field of view and sensitivity to moving targets. After that, the researchers made a deep research on the bionic compound eye theory. At present, many foreign research groups have proposed a number of bionic compound eye systems which was based on micro-lens array. The Japanese Tanida team has prepared the classic compound eye system TOMBO [2-4]. The German research team proposed an APCO similar to the TOMBO layer, which was based on the Bionic compound eye structure [5-11]. However, the image field of these planar compound eye structures is poor in image quality and the field of view is small, so it is a feasible choice to simulate the compound eye system on the curved surface.

To solve the above problems, an array spherical bionic compound eye system is designed in this paper. The relationship between the subsystems on each ring and the field of view angle of the compound eye system is analyzed. The mechanical arrangement among the subsystems is calculated systematically, which makes the system more compact on the mechanical mechanism and ensures that there is no blind area in the detection area. Compared with the traditional large field of view single hole optical system, the system greatly reduces the chromatic aberration and distortion of the system, resolves the contradiction between large field of view and high resolution, and increases the aperture value of the sub-eye. In addition, the detector array with curved surface distribution improves the acquisition ability of target azimuth information.

2. Mathematical model of field of view stitching

According to the design, each sub-eye is considered as an independent optical system. Firstly, the number of loops in the system is determined by the angle of view in the meridian (X) direction, and then the number of subsystem cycles in each ring is determined by the arc-sagittal non-blind area splicing. Secondly, the overall structure of compound eye system is established. Finally, the sub-eye system is designed and optimized.

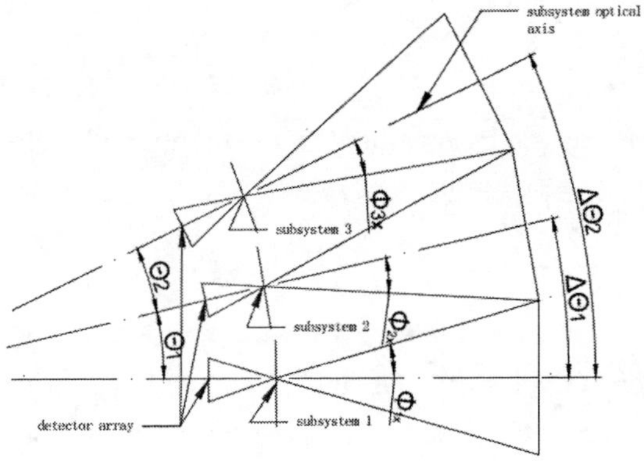

Fig.2 X-Direction field of view stitching diagram

The relationship of the field of view of the system in the meridional direction is shown in Fig.2. The meridional field of view of the three-ring lens group is taken as an example. Subsystem 1 represents the central optical system, subsystems 2 and 3 represent the primary subsystem and the secondary subsystem. The total field of view of the system is 2ω, the angles of view of subsystems 1, 2 and 3 in the meridional direction are Φ_1, Φ_2 and Φ_3, θ_1 and θ_2 are the angles of the optical axes of adjacent subsystems. In the target positioning, in order to meet the field of view without blind spots. Considering the mechanical spacing, the above parameters should meet the following conditions [12,13];

$$\begin{cases} \Phi_1 + \Phi_2 \geq \theta_1 \\ \Phi_2 + \Phi_3 \geq \theta_2 \\ \omega = \theta_1 + \theta_2 + \Phi_3 \end{cases} \quad (1)$$

According to the determined elements in meridian direction, the arc-sagittal field of view splicing is realized by the circular array of subsystem lenses. The field of view splicing can be guaranteed without blind area under the critical condition. Assuming that the size of the selected detector is $x \times y$, and the space size of the field of view of the image square with the working distance L is $\mathrm{X} \times \mathrm{Y}$, according to the conjugate relation between the object and the image:

$$\begin{cases} \mathrm{X} = 2L \tan \Phi_{ix} \\ \mathrm{Y} = 2L \tan \Phi_{iy} \end{cases} \quad (2)$$

The X -direction edge field of view angle of compound eye i-ring array is θ_{ix}, then

$$W_{ix} = \Delta\theta_i + \Phi_{ix} \quad (3)$$

When the distance is L, the field of view of the compound eye system is a circular. The spherical radius is R.The spherical circumference is C.

$$R = L \sin W_{ix}$$

$$C = 2\pi R = 2\pi L \sin W_{ix} \quad (4)$$

The edge field of view of the subsystems on each ring is connected to each other, which as a critical splicing condition of the Y field of view;

$$n_i = \frac{2\pi L \sin W_{ix}}{Y} = \frac{2\pi L \sin W_{ix}}{2L \tan W_{iy}} \quad (5)$$

According to the basic principles of the geometric optics,

$$\begin{cases} \tan W_{ix} = \dfrac{x_i}{2f'} \\ \tan W_{iy} = \dfrac{y_i}{2f'} \end{cases} \quad (6)$$

According to the W_{ix}, W_{iy} given by formula (6), the number of cycles is

$$n_i = \frac{\pi \sin(\Delta\theta_i + \arctan(\dfrac{x_i}{2f'}))}{\dfrac{y_i}{2f'}} \quad (7)$$

When the actual number of cycles of each ring $N_i \geq n_i$ [12], the entire compound eye system is successively combined in the sagittal direction, and there is no blind zone in the field of view.

3. System design

According to the above relationship, the structural design is carried out, and an industrial camera with a parameter value of $1/2''$ is selected. The image size of the detector is 6.4×4.8 mm, and the size of the selected camera casing is $50 \times 50 \times 42$ mm. The optical system design lens has a total length of approximately 54 mm, and the camera projection has a width of 12mm on the torus of the front view, then the total length is 66mm. As shown in Fig.3, where B is a side view of the compound eye system, A is a torus front view of the subsystem 2. We treat the lens group as a straight line and the camera as a rectangular. If the lens system does not coincide, the area of the rectangular projected by the camera on the ring, which does not coincide. According to the geometric relationship, the torus radius (the radius of the A circle) on the subsystem 2 is approximately 160 mm. From the radius of the A circle and the angle between adjacent systems, it can be determined that the outer spherical radius is approximately 662mm and the inner radius is 653mm. In the actual operation, a structure larger than the above critical parameters can be adopted.

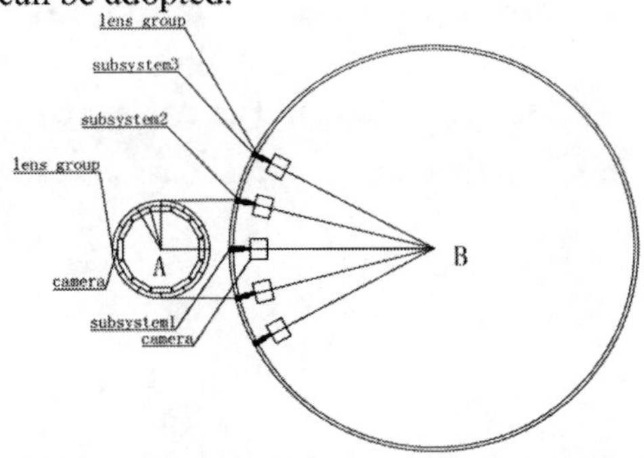

Fig.3 Side view of the compound eye system and front view of the torus

Based on the above mathematical model and system structure, a set of bionic compound eye optical system for visible target observation is designed. When the actual number of cycles is greater than the theoretically calculated value $(n_1 = 9.8)$, it can be calculated that there is no blind zone in the field of view from the above mathematical model. The accuracy of the mathematical model is proved by geometric methods. We take the total length of the central optical system as the radius, According to the formula (2), $Y = 12.25$ can be obtained. Then we take the full length of the edge optical system as the edge object image spacing, $R = 18.38$ can be obtained by the formula (4), and the image field space $X \times Y$ is sequentially connected as the condition of the blind field splicing in the field of view, the rotation half angle is $t = YR/2 = 18.43^\circ$, the number of cycles is $n = 2\pi/2t = 9.76$. Since the number of cycles can only take integers, the results of geometric analysis are consistent with the mathematical model, which proves the accuracy of the mathematical model.

In the design process, we refer to the functional partitioning characteristics of the insect compound eye system, this design reduces the system aberration by improving the traditional single lens compound eye system. The sub-eye of the system is divided into two parts, which include the central sub-eye system and the edge sub-eye system. From the resolution formula $\tan \Delta\varphi \approx s/f'$ (s is the detector pixel size), it can be seen that the central field of view is small, the resolution is high, and the target is highly accurately recognized, the edge system has a large field of view and low resolution, targets can be searched and captured over a wide range. The parameters of each eye are shown in Table 1.

Table 1. The main parameters of sub-eye

	Central optical system	Edge optical system
Detector size	$6.4 \times 4.8mm$	$6.4 \times 4.8mm$
Focal length f'	$36\,mm$	$20\,mm$
Field of view	$5.08^\circ \times 3.81^\circ$	$9.09^\circ \times 6.84^\circ$
F number	3	2.6

The pixel size of the detector is $6\,\mu m$. According to the resolution formula, the limit angle resolution of the central optical system is $\Delta\varphi_c \approx 0.00955^\circ$, the edge optical system is $\Delta\varphi_e = 0.017^\circ$. The structural diagrams of the optical system are shown in Fig.4 , Fig.5 and Fig.6 are used to evaluate the image quality of the optical system, which include spot diagram, modulation transfer function(MTF). In the spot diagram, the RMS radius of each field of view is less than $6\mu m$ pixel size of detector. The maximum geometric(GEO) radius is less 3 times than the root mean square(RMS) value. The detector cut-off frequency is $\upsilon_d = 1mm/(2\times s) = 83lp/mm$. In the modulation transfer function (MTF), the total sub field MTF of the central sub-eye is greater than 0.5 at 75 lp/mm, the full field of view MTF is greater than 0.3 at 144 lp/mm. The full field of view of the edge sub-eye is greater than 0.5 at 54 lp/mm, the full field of view MTF is greater than 0.3 at 78 lp/mm, so the designed system meets the requirements.

Using the relationship between the system's field of view and the sub-eye system, the structure of the compound eye system is designed which is combined with the formulas (1) to (7). To achieve a field of view of 79.5° for the full field of view, the central field of view is chosen to be $\Phi_1 = 4^\circ$ The edge field of view is Φ_2、$\Phi_3 = 10^\circ$ for design. The angle of the optical axis in the system is $\theta_1 = 12.9^\circ$ and $\theta_2 = 17.76^\circ$ respectively, and the total angle of view is $2W = 2(\theta_1 + \theta_2 + \Phi_{3x}) = 79.5^\circ$.

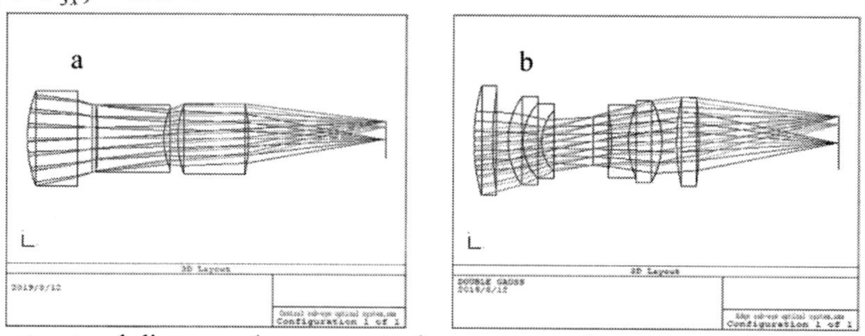

Fig.4 The structural diagrams.(a represents the center eye,b represents the edge of the eye)

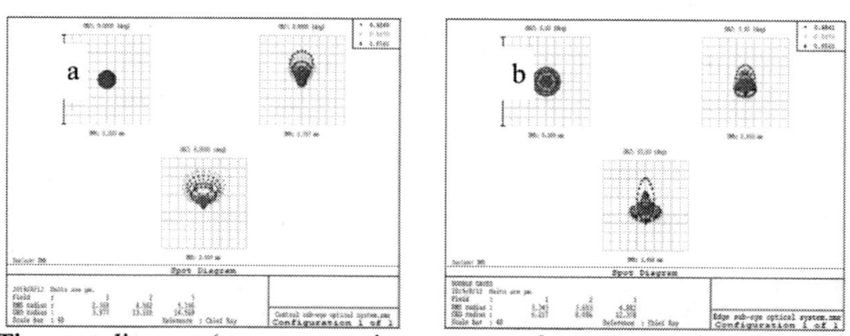

Fig.5 The spot diagram.(a represents the center eye,b represents the edge of the eye)

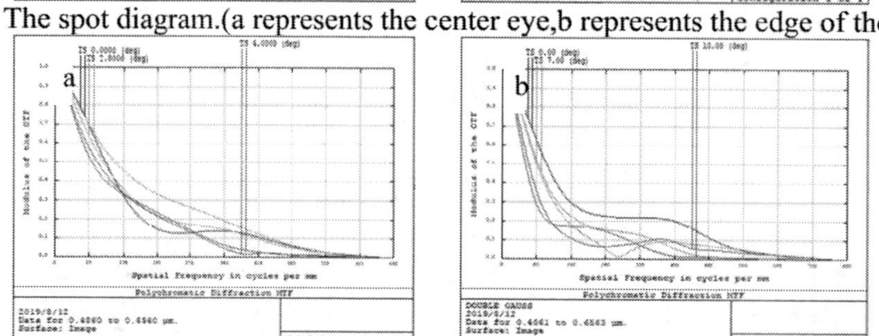

Fig.6 The modulation transfer function(MTF).(a represents the center eye,b represents the edge of the eye; the x axis is Spatial Frequency in cycles per mm, y axis is Modulus of the OTF)

4. Conclusion

According to the mathematical model between the compound eye and the sub-eye system, a set of compound eye system including a 3-ring lens group is designed. The problem that the traditional single aperture lens has a large field of view and high resolution is difficult to achieve at the same time is solved, and the accuracy of the mathematical model is proved by the geometric structure design. The designed compound eye system can achieve a full field of view of 79.5° degrees, we can increase or decrease the number of system lens rings according to different field of view requirements. The simulation results show that there is no blind zone in the field of view within a certain working distance range. Compared with the traditional large field of view lens, the image quality of the edge field of the subsystem is more excellent. The mathematical model and geometric structure proposed in the paper provide a theoretical reference for the design of large aperture bionic compound eye system.

References

[1] Yang J J, Flores A, Wang M R. Achromatic hybrid refractive-diffractive lens with extended depth of focus[J]. Applied Optics, 2004, 43(30):5618.

[2] Tanida J, Kumagai T, Yamada K, et al. Thin Observation Module by Bound Optics (TOMBO): Concept and Experimental Verification[J]. Applied Optics, 2001, 40(11):1806-1813.

[3] Jacques Duparré, Dannberg P, Schreiber P, et al. Artificial apposition compound eye fabricated by micro-optics technology[J]. Applied Optics, 2004, 43(22):4303-10.

[4] Jacques Duparré, Schreiber P, André Matthes, et al. Microoptical telescope compound eye[J]. Optics Express, 2005, 13(3):889-903.

[5] Jeong K H, Kim J, Lee L P. Biologically Inspired Artificial Compound Eyes[J]. Science, 312.

[6] Jacques Duparré, Radtke D, Andreas Tünnermann. Spherical artificial compound eye captures real images[J]. Proceedings of SPIE - The International Society for Optical Engineering, 2007, 6466.

[7] Floreano D, Pericetcamara R, Viollet S, et al. Miniature curved artificial compound eyes.[J]. Proc Natl Acad Sci U S A, 2013, 110(23):9267-9272.

[8] Camara R P, Vila G B, Lecoeur J, et al. Miniature artificial compound eyes for optic-flow-based robotic navigation[C]// Workshop on Information Optics. IEEE, 2014.

[9] Di S, Du R X. Optimal design of single-layer spherical compound eye imaging system[J].Opto-Electronic Engineering, 2010, 37(2):27-31

[10] Li F, Chen S,Luo H,et al. Curved micro lens array for bionic compound eye[J].Optik-International Journal for Light and Electron Optics, 2013, 124(12):1346-1349

[11] Song Y M, Xie Y Z, Viktor.M, Xiao J L,et al. Digital cameras with designs inspired by the arthropod eye.[J]. Nature, 2013, 497(7447).

[12] Fu Y G, Zhao Y, Liu Z Y, et al. Design of the Bionic Compound Eye Optical System Based on Field Splicing Method [J].Chinese Journal of Scientific Instrument, 2015, 36(2):422-429. (in chinese)

[13] Yan F, Guo Y Z, Shi L F, et al. Research of Image Mosaic Algorithm Based on Bionic Compound Eye System [J]. Aero Weaponry, 2017(6):49-53. (in Chinese)

New Method for Eliminating Background Noise in Characteristic Spectral Imaging

Xiaoli Xie [1, a], Zheng Wang[1, b], Wuming Wu[1, c] Chujun Zhao[1, d], Shengli Chang[1, e]

[1]School of Physics and Microelectronics, Hunan University, Hunan Province, China

[a]1251866295@qq.com, [b]624604976@qq.com, [c]286682722@qq.com
[d]cjzhao@hnu.edu.cn, [e]changshli@126.com

Abstract. We proposed a method to eliminate the background noise, which is based on subtraction of two different central wavelength filters according to the characteristic spectral imaging. In the experiment, the sodium yellow light with 589 nm wavelength is selected as the characteristic wavelength of spectral imaging experiment, and the visible light industrial camera was used to collect diffuse reflective wall images. Firstly, two narrow-band filters with a center wavelength of 589 nm and 579 nm are placed in front of the camera separately, and two images of the signal light containing the narrowband characteristic wavelength are acquired. Then, the surf algorithm is used to find the feature points of the two images, and the two images are matched, geometrically transformed and cropped to obtain images of the same size and the same angle. After the histogram matching, the background gray values in the two images are close, and the two images are subtracted, which can obtain an image with removed background noise. The experimental results show that this method can provide a simple, high-speed method to effectively eliminate the background noise of the characteristic spectrum.

1. Introduction

In recent years, research on spectroscopy has attracted more and more extensive and widespread attention. A characteristic spectrum is a light emitted by a certain element or light source passing through an element to display a specific bright band or a dark band on the spectrum [1]. According to the range of the spectrum, it is divided into ultraviolet light, visible light and infrared light [2]. The spectrum of the material is unique, and its properties can be identified [3]. The background light contains direct sunlight, scattering and reflection of atmospheric clouds, and transmission and reflection of various particles in the air [4,5,6]. The full spectrum of the target under ambient light contains a variety of noises that the naked eye cannot distinguish from the background. Moreover, the full spectrum image information data is huge and the algorithm is complex with poor real-time performance. When the background radiation intensity is similar or stronger than the target spot, it is indistinguishable to the naked eye.

The method of feature spectrum elimination background can be applied in various applications, such as criminal investigation. The visible light spectrum imager is used to collect multi-spectral images of sweat fingerprints in two cases with severe background interference. The spectrometer is in the range of 400 to 720 nm. By comparing the fingerprint and background spectra, the two images are divided and then processed. The image is stretched linearly to obtain enhanced fingerprints and weakened interference background [7]. In the non-invasive biochemical detection of near-infrared spectroscopy, some people used the blood flow volume spectral subtraction method to eliminate the interference of tissue background and the effectiveness of the method.[8] The hyperspectral imager performs spectral

IOP Conf. Series: Materials Science and Engineering **711** (2020) 012086 doi:10.1088/1757-899X/711/1/012086

data acquisition on blood fingerprints in different complex backgrounds and selects regions of interest. People use the spectral properties of blood to spectrally decompose the image to achieve segmentation and extraction of the target blood fingerprint image. The method can effectively eliminate complex background interference and has widely applications. [9] The characteristic spectrum is also very important for the national defense. With different empirical models, such as atmospheric transmission law, unique spectral line type, and target grayscale defocusing principle, the characteristic spectral images containing specific targets have been investigated [10,11]. However, the difference between the background and the target spot is obvious and easy to distinguish on the spectrum. According to the characteristics of target and background, simple and high-speed spectral imaging method needs to be explored to eliminate background noise, which can not only improve the signal-to-noise ratio, but also ensure the real-time performance.

Here, it is proposed a method to eliminate the background noise based on subtraction of two different central wavelength filters according to the characteristic spectral imaging. With the sodium yellow light with 589 nm wavelength and the visible light industrial camera, the method has been validated experimentally. The experimental results show that this method can provide a simple, high-speed method to effectively eliminate the background noise of the characteristic spectrum.

2. Signal collection

A clean whitewashed wall, the white wall resembling Lambert, and a room with good shading performance constitute the basic laboratory environment. The camera is an industrial camera produced by Daheng Image Company (model MER-131-210U3M) with a resolution of 1280*1024, a spectral range of 350-950 nm, and a lens with focal length of 25 nm. The spectrometer is produced by Ocean Optics, model USB4000, with a wavelength range of 400-1100 nm and a resolution of 2 nm. The power-adjustable tungsten lamp simulates natural light. The sodium lamp is an ND20 with low voltage power and its wavelength is 589 nm. A set of Convex lens and beam splitter system are coupled to the spectrometer for aligning the target spot. The center wavelengths of the narrow band filters are 579 nm and 589 nm. The sodium lamp power is not adjustable, so the light is illuminated onto the concave mirror and then reflected onto the wall. The size and brightness of the sodium spot are adjusted by controlling the distance between the concave mirror and the wall. The focal length of the spectroscope system is 100mm.The spectroscope system transmission end is connected to the input fiber of the spectrometer. The reflection end is connected to an industrial camera. The system is fixed on a tripod 5 meters from the wall measuring the relative radiation intensity of the tungsten and sodium lamp launching walls. Another tripod with industrial camera can be placed 5 m away from the wall.

IOP Conf. Series: Materials Science and Engineering **711** (2020) 012086 doi:10.1088/1757-899X/711/1/012086

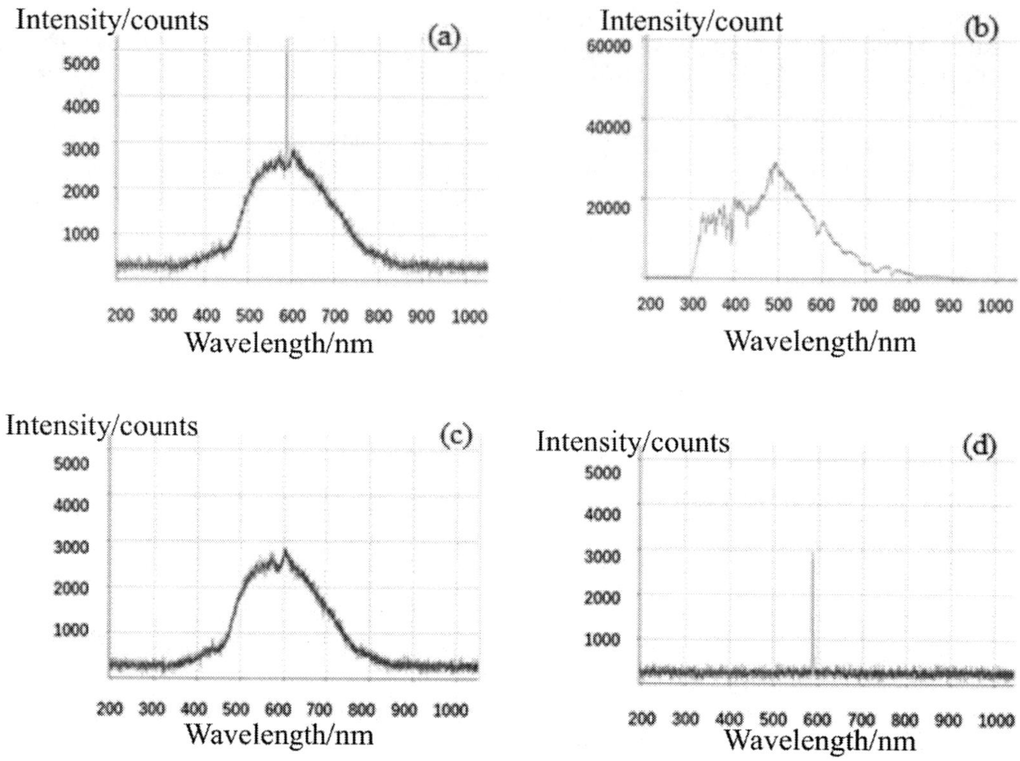

Fig.1. Spectra of different kinds of light. (a)Sodium lamps and tungsten lamps; (b) Solar background; (c)Tungsten lamp; (d)Sodium lamp.

The wall is smooth and the light will cause specular reflection. The actual background environment is similar to the Lambertian body, and the incident light is uniformly reflected in all directions, also known as diffuse reflection. A slightly rough A4 paper is attached to the wall to prevent specular reflection, which is similar to the actual environment and the results are more reliable. Connecting the spectrometer, computer and convex lens and beam splitter system, we turn on the tungsten lamp and the sodium lamp. The position of the tungsten lamp and the concave mirror is adjusted to control the spot illuminance on the white wall. It is mentioned that the two spots are approximately the same at 589 nm, as shown in Fig. 1(a). The measured solar background spectrum is shown in Fig. 1(b). The spectrum of the tungsten lamp of the simulated background is still different from the background spectrum in the details, but the envelope is roughly the same, as shown in Fig.1(c). Figure 1(d) is a spectrum of a sodium lamp, which is a narrow band spectrum. According to the optical signal-to-noise ratio formula

$$\mathrm{OSNR}=10\lg(P/N)+10\lg(B/B_r) \quad (1)$$

where P is the signal power, N is the noise power in the range of equivalent noise bandwidth B, B_r represents optical bandwidth, and B is the noise equivalent bandwidth. The ambient light should be minimized into the narrow-band filter, and a sodium lamp with a narrow wavelength range should be selected to simulate the target, which is beneficial to improve the signal-to-noise ratio.

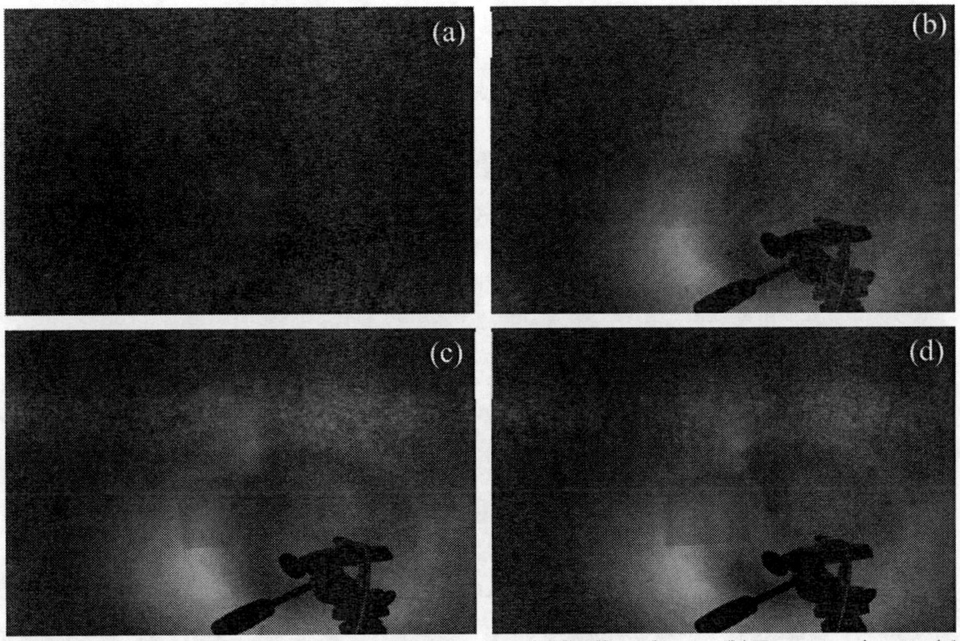

Fig.2. Grayscale image acquired by industrial cameras. (a)Sodium lamp; (b)Tungsten lamp; (c)Sodium and tungsten lamps with 579 nm filter; (d)Sodium and tungsten lamps with 589 nm filter.

The feature points are the points where the gray value changes drastically or the curvature on the edge of the image changes greatly. This experiment is carried out in front of a single-color wall without edges and corners. It is impossible to collect enough feature points. Therefore, when the image is collected, a camera with a sharp angle is placed in front of the lens to facilitate the extraction of feature points for later image processing. The spectrometer is turned off and the computer is connected to an industrial camera. Then the industrial camera to capture the picture, the steps are as follows:

(1) When the entire laboratory only the sodium lamp is turned on, the industrial camera without the filter is used to take a picture of the spot on the white wall, which is shown in Fig. 2(a);

(2) After turning off the sodium lamp and turning on the tungsten lamp, the camera without the filter is used to take a picture of the spot on the white wall, as shown in Fig. 2(b);

(3) By simultaneously turning on the sodium and tungsten lamps, we install the filter with center wavelength of the 579 nm, and take a picture of the spot on the white wall, as shown in Fig.2(c);

(4) The camera is installed with a center wavelength of 589 nm filter, and repeat step (3) to obtain Fig. 2(d).

3. Background noise elimination method

The purpose of eliminating background noise is to improve the signal-to-noise ratio of the characteristic spectral imaging. In the characteristic spectral imaging experiment, the difference between the signal light and the background noise intensity in the characteristic spectrum segment is not large, resulting in poor imaging contrast, which greatly affects the characteristic spectral imaging results. In order to eliminate the influence of background noise, a method based on two different central filters for subtraction to eliminate background noise is proposed. In the smaller wavelength range, it is considered that the background intensity is independent of wavelength.

We proposed the method to eliminate the background as follows. The images taken by the filter industrial cameras are stored in the image set format, and the two gray images are used to find the feature points of the two images by the surf algorithm. The surf algorithm extracts feature points by comparing the grayscale differences of surrounding pixels. Each pixel in the image can find a Hessian matrix whose matrix is,

$$H(f(x,y)) = \begin{bmatrix} \frac{\partial^2 f}{\partial x^2} & \frac{\partial^2 f}{\partial x \partial y} \\ \frac{\partial^2 f}{\partial x \partial y} & \frac{\partial^2 f}{\partial y^2} \end{bmatrix} \quad (2)$$

The matrix describes the local curvature of the function.

The scale space of the surf algorithm consists of several groups. Each group contains several layers in the scale space. Each group selects adjacent three-layer Hessian determinant images. For each Hessian determinant value of the middle layer, it can be used as the point to be compared. In the space, 26 points around the point are selected for comparison. If the point is larger than the other 26 points, the point is a feature point. The two images are matched according to the obtained feature points. The image is geometrically transformed by the matching result. The image is cropped to a given size of the overlap. In order to reduce the grayscale difference, the two images are used histogram matching, and thus the signal-to-noise ratio is improved. Figure 3(a) is the image of experimental results without histogram matching, and its background is a little gray, indicating that there are grayscale differences between the two images. There is a large area of white area around the sodium lamp, and the illumination range of the sodium lamp is smaller than the actual one, and the noise has a large influence. Finally, the two images are subtracted and the contrast is not noticeable. This formula (3) is used to enhance the contrast, and the result is shown in Fig.3(b), where y is the value obtained after the difference between the two figures, and z is the value after the result is enhanced.

$$z = \frac{255}{[\max(y)-\min(y)]*[y-\min(y)]} \quad (3)$$

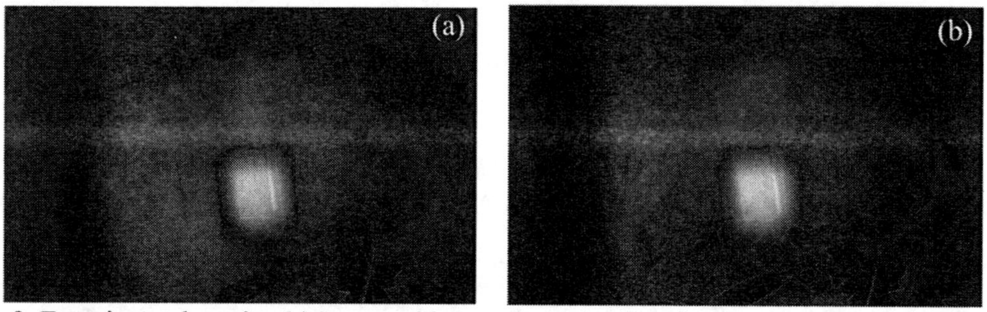

Fig.3. Experimental results. (a) Image without noise reduction; (b) Histogram matching processed image.

4. Analysis of results

The sum of gray value of Fig. 3(a) is 895556, and there are 854 gray values of pixels exceeding 110. The sum of gray value of Fig. 3(b) is 557736, and there are 861 gray values of pixels exceeding 110. The gray value range of the sodium lamp is approximately between 90 and 150. This means that the background gray value of Fig. 3(b) is closer to 0, that is, eliminating background noise is more superb. The histograms of Fig. 3(a) and Fig. 3(b) are shown in Fig. 4(a) and (b). There are a large range of gray scales and prominent details in Fig. 4(a). The grayscale range of image Fig. 3(b) is mainly concentrated above 150, and the contrast is strong, and irrelevant details and noise are removed. It is shown that the histogram matching effectively adjusts the difference in background gray values. There are many experimental steps and the filter is changed frequently, which cause some noise. The algorithm also introduces some noise, but the final imaging results are reasonable and the signal-to-noise ratio is significantly improved. Comparing Fig. 2(a) with Fig. 3(b), the result is close to the ideal case, but the background has a small range of distortion.

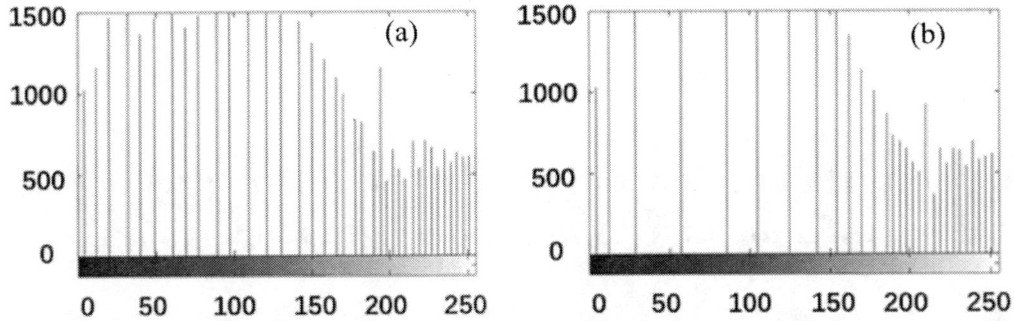

Fig.4. (a) Histogram of Fig. 3(a); (b) Histogram of Fig. 3(b). (The abscissa represents the gray level and the ordinate represents the frequency.)

5. Conclusion

A method is proposed to eliminate the background noise, which is based on the subtraction of two different central wavelength filters according to the characteristic spectral imaging. Feature spectrum imaging eliminates background noise making full use of the characteristics of light. The difference of light radiation cannot be distinguished between background and target. But it is distinguished by its spectral. Two filters are used to extract the background and the target respectively. The match and noise reduction processing are performed to suppress the noise. The background image is subtracted from the target image, which can enhance the background and target contrast greatly. On the whole, the influence of background noise of the image is eliminated. Therefore, the signal-to-noise ratio of the image is improved. The experimental results show that the method can provide a simple and high-speed method for feature spectral imaging to eliminate background noise.

References

[1] Jiang Li. Detection of mature pomegranate under natural light conditions based on human visual characteristics [D]. Nanjing Agricultural University, 2009.

[2] Huang Yaosong. Research on flame combustion stability detection based on radiation spectroscopy [D]. North China Electric Power University, 2013.

[3] Jiang Ping. Experimental study on multi-spectral detection technology of underwater targets [D]. Huazhong University of Science and Technology, 2013

[4] Shi Lei. Research on channel characteristics of near spacecraft [D]. Xi'an University of Electronic Science and Technology, 2012.

[5] Yang Chunping. Modeling and simulation of spectral characteristics of sky background [D]. University of Electronic Science and Technology, 2008.

[6] Hong Weiwei. Study on transmission characteristics and polarization distribution of near-infrared light in clouds[D]. Hefei University of Technology, 2013.

[7] Qi Minzhen, Chen Yihua, Wang Xinquan. Research on the Method of Removing Fingerprint Interference Background by Spectral Imaging Technology[J]. Criminal Technology, 2016, 41(04): 299-301.

[8] Ding Haiquan, Lu Qipeng,Gao Hongzhi. Study on background subtraction in non-invasive biochemical detection by spectral subtraction method[J]. Spectroscopy and Spectral Analysis,2012,32(09):2347-2350.

[9] Gao Guoming, Li Xue. A New Method for Eliminating Raman Spectral Fluorescence Background and Its Application[J]. Acta Optica Sinica, 2013, 33(02): 265-273.

[10] Huang Da, Huang Shucai, Wu Jianfeng. Summary of space-based early warning detection for missile boosting segment [J]. Laser and Infrared, 2018, 48(08): 945-950.

[11] Huang Da, Huang Shucai, Zhao Wei, Lu Wei. Fuzzy Recognition of Missile Tail Flame in Characteristic Spectral Images[J]. Acta Optica Sinica, 2018, 38(02): 385-391.

IOP Conf. Series: Materials Science and Engineering **711** (2020) 012087 doi:10.1088/1757-899X/711/1/012087

Small target detection and window adaptive tracking based on continuous frame images in visible light background

Zheng Wang[1, a], Jianping Zeng[1, b], Xiaoli Xie [1,c], Wuming Wu[1,d] and Shengli Chang[1,e]

[1]School of Physics and Microelectronics, Hunan University, Hunan Province, China

[a]624604976@qq.com, [b]1251866295@qq.com, [c]1239156751@qq.com, [d]286682722@qq.com, [e]changshli@126.com

Abstract. Due to the difficulty in distinguishing the target and the noise in the visible light background and the loss of the target for a short period of time, a scheme of track association and target tracking for the small target in the sky is carried out. Multi-feature similarity is taken into account, such as the maximum intensity, the maximum brightness, the size and the moving speed. Firstly, the trajectory prediction is carried out by the least square prediction method. By updating the real-time data to deduce the trajectory parameters, the target position in the next frame image is predicted. Based on the trajectory confidence test, all current recorded trajectories are traversed. The over-confidence trajectory is detected and the trajectories of continuous lost points are deleted, as well as the trajectories below the threshold of confidence. For the trajectory of the over-confidence, the target trajectory without the significant motion feature is deleted by judging the magnitude of the moving speed, so that the effective trajectory can be displayed while the false target is eliminated. Then, the target tracking is carried out, which is based on the LOG filter and the constant false alarm rate segmentation. Considering the temporal and spatial continuity of the target trajectory and the randomness of the amplitude and position of the noise, the false alarm rate can be reduced by eliminating the high amplitude noise by using the difference between the two in terms of temporal and spatial correlation. Owing to that the setting of the division threshold is not less than 5 times the background variance of the window image, the generation of the noise by the constant false alarm rate segmentation can be avoid when there is no target. In addition, the target coordinates may be abrupt due to sudden changes in target speed and the jitter of the image acquisition equipment. Therefore, we propose a dynamic adjustment scheme for the tracking window size to enable window adaptive tracking, which can set the length and width of the rectangular window to k times of the original size according to actual needs. At the same time, the number of points of the relatively stationary target in each frame of the image can be counted in real time. Finally, the reliability of the method is verified by simulation experiments, and the running speed is relatively fast and the recognition accuracy is relatively high.

1. Introduction

As an important technology, weak target detection technology has been widely used in search and tracking, early warning systems, precision guidance and large field of view target detection [1, 2]. In the visible light background, the image sometimes has a low contrast, the target is weak and small, and there is inevitably a large amount of noise interference. Therefore, weak target detection in the visible light background has always been a difficult and hot topic in the field of computer vision.

A multi-feature similarity is integrated, such as maximum intensity, maximum brightness, size and motion speed. And a scheme for trajectory correlation [3] and target tracking [4] for small targets in the sky is proposed. In terms of trajectory correlation, the algorithm mainly includes least squares prediction method [5, 6, 7] and trajectory confidence test [8]. In terms of target tracking [9], it mainly relies on LOG filtering [10, 11] and constant false alarm rate segmentation [12] to realize window adaptive tracking [13] while real-time counting the number of points of relatively stationary targets in each frame of image. Finally, the method is proved to be reliable, the running speed is fast, and the recognition accuracy is relatively high.

2. Trajectory association

Trajectory correlation is an effective means to perform target detection and noise filtering by using the continuity of state changes in the space-time domain. It can establish correlation between multiple short tracking segments of the tracking target to obtain a continuous and smooth tracking trajectory of the tracking target. The process is as shown in Figure. 1.

In the calculation of the predicted points, we use the least squares prediction method, which can predict the position of the target in the next frame by recursive trajectory parameters. The trajectory of the confidence is detected, which is based on the trajectory confidence test. And the trajectory of consecutive lost points and the trajectory lower than the lower confidence threshold are deleted. For the trajectory of the confidence level, the target trajectory without significant motion features is deleted by judging the magnitude of its motion speed.

Considering the continuity of the target motion, we use the least squares method to fit

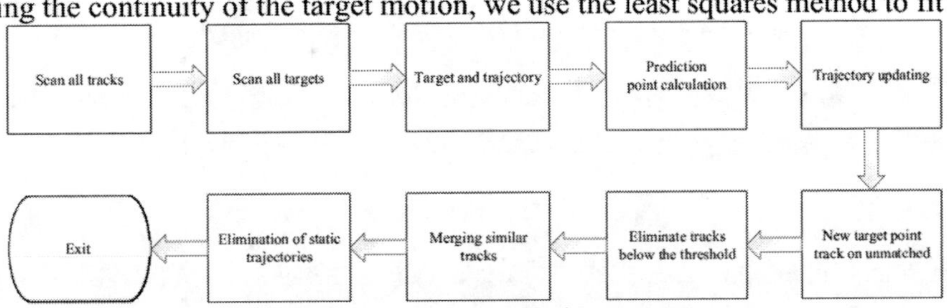

Fig. 1 Trajectory association process.

the polynomial to approximate the trajectory, and the trajectory equation can track the target. In the process of tracking the target, the moving speed and the direction of the target will change with time according to a certain law. In order to extract the useful signal, the target motion law should be assumed first. The least squares method is a widely used optimal estimation method. The basic idea is to choose a polynomial $P(x)$ to approximate $f(x)$ with a minimum mean square error.

The trajectory confidence detection traverses the trajectories of all the current records and finds the trajectory of the confidence. First, a confidence test is performed only for tracks whose length can be greater than the set length. The trajectory with confidence higher than the high threshold is counted and recorded. For the trajectory with confidence lower than the low threshold, it is set as the trajectory to be deleted, and will be deleted in subsequent processing. Similarly, if there is a phenomenon of consecutive lost points in the trajectory, it is also set as the trajectory to be deleted. Next, the merging of the overlapping trajectories is performed. The criterion for judging the overlap of two trajectories is to calculate the trajectory between the current position of the two trajectories and the position of the trajectory points before the three frames. If the distance is smaller than the threshold value we set, it is determined that the two trajectories overlap. For the way of overlapping trajectories, we choose to keep the trajectory with high confidence and delete the other trajectory. Finally, we judge the motion speed of the confidence track based on the total offset of the X and Y directions and delete the target track with obvious motionless features. Whether the target moves by the total number of relatively stationary targets, if there are more than five frames of relatively stationary targets in the same frame picture, it is determined that the target trajectory is a clutter.

3. Target tracking

The main task of target tracking is to estimate the position of the target object in the first frame of the video image, and to estimate the state information of the target in the next multi-frame image through the appearance model and the motion model. The target tracking process is as shown in Figure. 2.

LOG filtering is a combination of Gaussian filtering and Laplacian filtering. The LOG operator obtained by finding the second-order derivative of the two-dimensional Gaussian function is denoted as $\nabla^2 G(x, y, \sigma)$, where σ is the mean square error of the Gaussian function.

$A = \left(x^2 + y^2\right)/\sigma^2$, and K is a scale factor.

The Gaussian part G can smooth the image and effectively eliminate all image intensity changes that are much smaller than the Gaussian distribution factor(σ). To select a suitable and orientation-independent operator, we choose the lowest order isotropic differential operator, Laplacian ∇^2, which can reduce the computational complexity to a large extent.

$$G(x, y, \sigma) = \left(1/\left(2\pi\sigma^2\right)\right)\exp(-(1/2)A) \tag{1}$$

$$\nabla^2 G(x, y, \sigma) = K(2 - A)\exp(-(1/2)A) \tag{2}$$

In the signal detection, the Newman-Pearson criterion is often used in the detection because the probability that the target appears under certain conditions is not known in advance and the loss caused by the missed detection is not known. This criterion can effectively deal with hypothesis testing problems when both prior probabilities and costs are difficult to determine. Set the false alarm rate $P(D_1 \mid H_0) = P_{fa}$ to a constant value to maximize the detection probability. For a detection system, the input signal and the discrimination result can be represented.

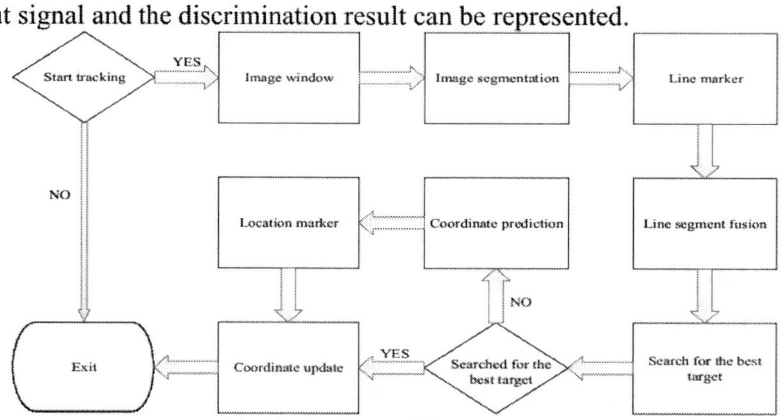

Fig. 2 Target tracking process.

When $x(t) = n(t)$, the situation is judged as H_0, indicating that there is only noise and no signal. When $x(t) = n(t) + s(t)$, it means that there are both signals and noise.

Let the probability density function be $p(x \mid H_0)$ only under the condition of noise, and the probability that the noise level exceeds the discriminating threshold V_T to cause false alarm is

$$P(D_1 \mid H_0) = \int_{V_T}^{+\infty} p(x \mid H_0)dx \tag{3}$$

V_T is decided by $P(D_1 \mid H_0) = P_{fa}$ .Similarly, under the condition that the signal and noise are simultaneously input, the probability density function of the signal plus noise is $p(x \mid H_0)$, and the probability that the discrimination result is H_1 is

$$P(D_1 \mid H_1) = \int_{V_T}^{+\infty} p(x \mid H_1)dx \tag{4}$$

The following criteria are given.

When $\frac{p(x \mid H_1)}{p(x \mid H_0)} > a_0$, $(a_0 \propto V_T)$, the discrimination result is H_1.

When $\frac{p(x \mid H_1)}{p(x \mid H_0)} < a_0$, $(a_0 \propto V_T)$, the discrimination result is H_0.

4. Results and analysis

Firstly, the least squares prediction method is used to predict the trajectory. The corresponding optimal approximation mode is determined according to different trajectory lengths. The trajectory parameters are recursively updated by updating the real-time data to predict the target position in the next frame image. Based on the trajectory confidence test, all current recorded trajectories are traversed, from which the trajectory of the confidence is detected and the trajectory of consecutive lost points and the trajectory lower than the lower confidence threshold are deleted. For the case where two tracks appear, if the distance between the current frame and the track point before the 3 frames is lower than the threshold, it is determined that the two tracks coincide, and only one with a high degree of confidence is reserved. For the trajectory of the confidence level, the target trajectory without the significant motion feature is deleted by judging the magnitude of the motion velocity, thereby displaying the effective trajectory while rejecting the false target.

The trajectory association matching is to match the trajectories of the candidate targets in each stage, and try to eliminate the false target trajectory and obtain the real target trajectory. A potential target trajectory within a phase can be obtained through a dynamic programming approach. Due to the low signal-to-noise ratio of the target, the target trajectory may be lost and the false alarm trajectory may exist. Therefore, we match the potential target trajectory according to the trajectory association condition to obtain important information such as the direction, velocity, position and gradation accumulation of the target.

During the target tracking process, there may be a sudden change in the spatial position of the target due to a sudden change in speed or a sudden change in the target coordinates in the adjacent acquisition frame due to burst jitter of an image acquisition device such as a camera. If the tracking window is too small, it may cause the tracking target to be lost within the shooting range. If the tracking window is too large, it will slow down the target lock. To this end, we propose a dynamic adjustment scheme for the tracking window size. The scheme can take the target centroid position in the image of the previous frame lost as the target as the center of the rectangular tracking window, and set the length and the width of the rectangular window to the original k times according to actual needs until the target is re-locked into the tracking window. If two frames of data are continuously lost, the tracking target is lost, and the target tracking process is terminated, the target detection recognition process is waited for the tracking to be turned on again. At the same time, it is also necessary to judge whether the target identified by the detection and the target position of the current tracking are consistent. If they are inconsistent, the tracking process has an error and accumulates the number of errors. When it exceeds the upper limit of the set number of errors, the tracking is turned off. The details are shown in Figure. 3 and Figure. 4.

The image we captured with the camera has a resolution of 768*517 (width 768, height 517). A total of four frames of images from 6361 to 6364 were collected, and the two-dimensional coordinates of the target centroid pixels were obtained by calling the image coordinate information function. Qualitative analysis and modeling of coordinates help us to perform target motion state analysis and image acquisition device jitter research. The obtained target pairs are shown in Table 1.

We randomly selected four frames of images, and obtained the number of points of relatively stationary targets in each frame by algorithm. Comparing it with the actual relative static point, the accuracy of the algorithm is obtained, as shown in Table 2 and

Figure 5. The results show that our algorithm obtains 100% accuracy of relative static points. Therefore, the algorithm can count the number of targets and suspected targets in each frame in real time and effectively, which has certain practical value.

Figure 6(a) is a picture taken by the camera in which the target is mixed with noise and the difference between the two is not large. The three-dimensional effect of the brightness intensity is shown in Figure 6(b).

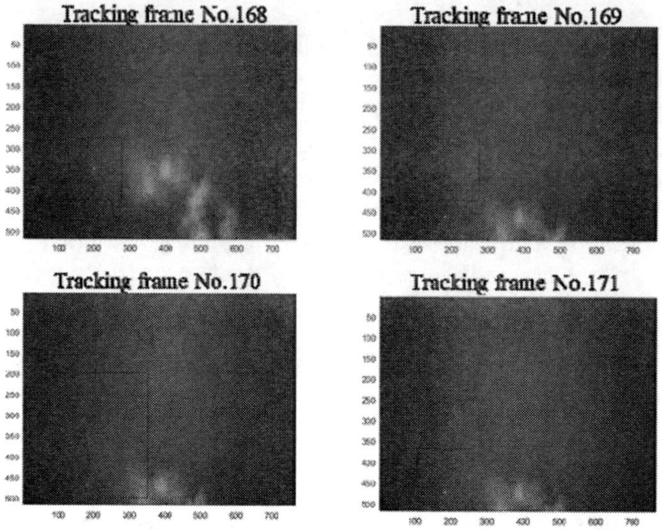

Fig. 3 Tracking effect when the dynamic adjustment multiple of tracking window is 2 times.

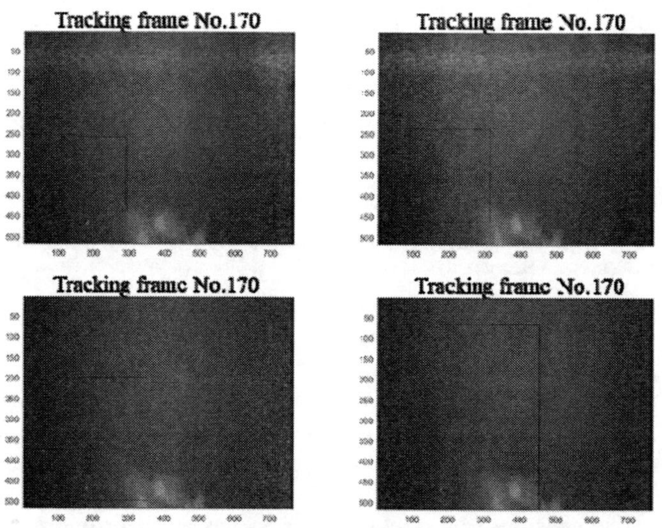

Fig. 4 Tracking effect when the dynamic adjustment multiple of tracking window is 1.2 times, 1.5 times, 2 times and 3 times respectively.

Table. 1. Image coordinate value of 4 consecutive frames.

Frame number	6361	6362	6363	6364
Pixel info:(X,Y)	(199,346)	(182,460)	(182,467)	(182,470)

Table. 2. The calculation of the accuracy of relative stationary point number.

Frame number	6188	6222	6288	6388
Relative static points (obtained by algorithm)	3	6	5	5
Relative static points (actual)	3	6	5	5
The algorithm obtains the accuracy of relative static points (percentage)	100%	100%	100%	100%

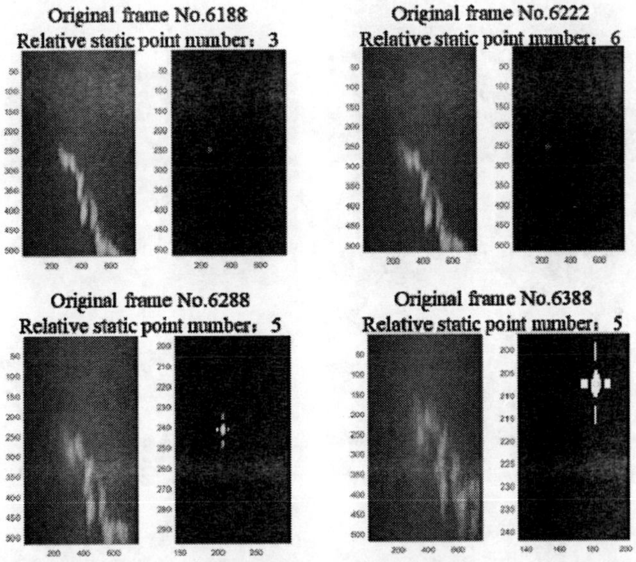

Fig. 5 Statistics of relative static points.

It can be seen that the intensity of the noise is up and down and fluctuating in the two-dimensional coordinate distribution. Thus, to a certain extent, we can distinguish between the target and the noise to be identified by having a transition phase. However, the results obtained in this way are not intuitive enough. In order to better improve the recognition speed and recognition accuracy, we use LOG filtering and image multiplication. Through the combination of the two, we can concentrate the value of the target and its surrounding high-intensity pixel points on the same pixel while filtering out the noise, and then distinguish the target from many noises. As shown in Figure 6(c), the obtained three-dimensional intensity effect is shown in Figure 6(d). Although Figure 6(d) can visually represent the two-dimensional coordinate position of the target, it is difficult to find the target position only by visually observing the Figure 6(c). To this end, we use the image coordinate information acquisition function to find the coordinate value of the pixel point 3 of the target centroid through the Figure 6(c) and then enlarge the target in the Figure 6(c) several times to obtain the Figure 6(e), and then draw the three-dimensional The strength effect is shown in Figure 6(f). It has been found that the coordinates of the target centroid pixels in the Figure 6(d) and the Figure 6(f) obtained by our algorithm are the same. At the same time, it also verifies the feasibility of the method.

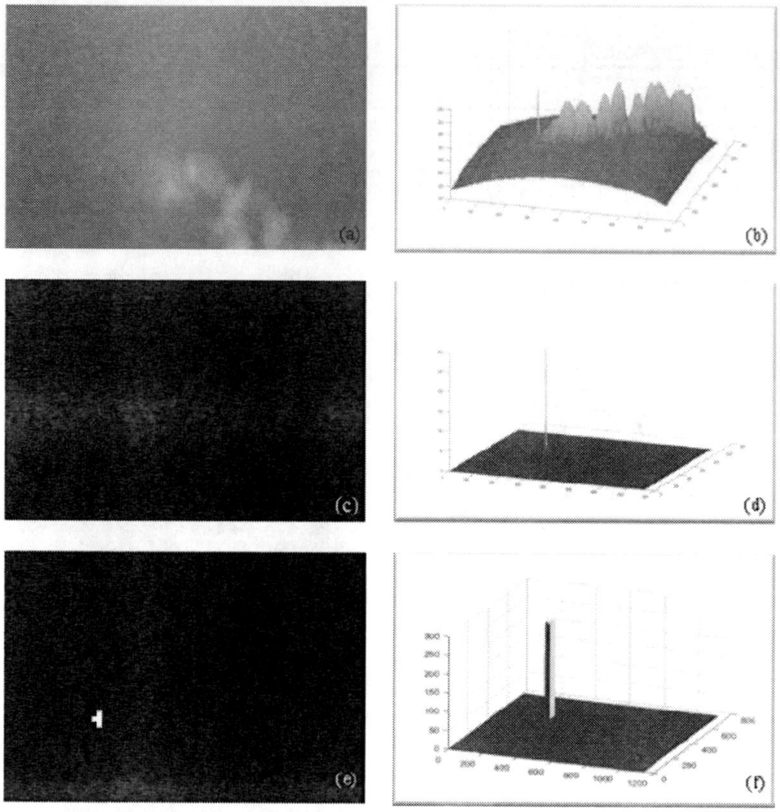

Fig. 6 Pictures before and after filtering, the corresponding 3D images.

5. Summary

A method of trajectory correlation and target tracking of small targets in the sky under visible light background is proposed. Firstly, the least square prediction method is used to predict the trajectory, and all the tracks are traversed based on the trajectory confidence test, from which the trajectory of overconfidence is detected and the trajectory of continuous lost points and the trajectory below the threshold of confidence are deleted. Then, the target tracking is carried out based on LOG filtering and constant false alarm rate segmentation. The feasibility of the filtering effect is verified by comparing the filtered images and their corresponding 3D images. In addition, due to the sudden change of target speed and image acquisition equipment jitter, we also proposed a tracking window size dynamic adjustment scheme to reduce the impact of target coordinate mutation. The two-dimensional coordinate values in each frame of the image and the number of points of the relatively stationary target are also counted in real time.

References

[1] Viola P, Jones M. Rapid object detection using a boosted cascade of simple features[C]// IEEE Computer Society Conference on Computer Vision & Pattern Recognition. 2003.

[2] Ren S, Girshick R, et al. Faster R-CNN: Towards Real-Time Object Detection with Region Proposal Networks[J]. IEEE Transactions on Pattern Analysis & Machine Intelligence, 2017, 39(6):1137-1149.

[3] Mori S, Barker W, Chong C Y, et al. Track association and track fusion with nondeterministic target dynamics[J]. IEEE Transactions on Aerospace and Electronic Systems, 2002, 38(2):659-668.

[4] Leibe B, Schindler K, Cornelis N, et al. Coupled Object Detection and Tracking from Static Cameras and Moving Vehicles[J]. IEEE Transactions on Pattern Analysis & Machine Intelligence, 2008, 30(10):1683-1698.

[5] Janik L J, Skjemstad J, Shepherd K, et al. The prediction of soil carbon fractions using mid-infrared-partial least square analysis[J]. Australian Journal of Soil Research, 2007, 45(2):73-81.

[6] Baylar A, Hanbay D, Batan M. Application of least square support vector machines in the prediction of aeration performance of plunging overfall jets from weirs[J]. Expert Systems with Applications, 2009, 36(4):8368-8374.

[7] Pemberton J. Exact least square multi-step prediction from nonlinear autoregressive models[J]. Journal of Time Series Analysis, 2008, 8(4):443-448.

[8] Breitenstein M D, Reichlin F, Leibe B, et al. Robust Tracking-by-Detection using a Detector Confidence Particle Filter[C]// IEEE 12th International Conference on Computer Vision, ICCV 2009, Kyoto, Japan, September 27 - October 4, 2009. IEEE, 2009.

[9] Caron F, Emmanuel Duflos, Denis Pomorski, et al. GPS/IMU data fusion using multisensor Kalman filtering: introduction of contextual aspects[J]. Information Fusion, 2017, 7(2):221-230.

[10] Chen H C, Fang L T, Lee L, et al. LOG-filter-based inspection of cluster Mura and vertical-band Mura on liquid crystal displays[C]// Machine Vision Applications in Industrial Inspection XIII. International Society for Optics and Photonics, 2005.

[11] Liqin S L S, Dinggang S D S, Feihu Q F Q. Edge detection on real time using LOG filter[C]// International Symposium on Speech. IEEE, 1994.

[12] Baldygo W, Brown R, Wicks M, et al. Artificial intelligence applications to constant false alarm rate (CFAR) processing[C]// Radar Conference. IEEE, 1993.

[13] Arulampalam M S, Maskell S, Gordon N, et al. A Tutorial on Particle Filters for Online Nonlinear/Non-Gaussian Bayesian Tracking[J]. IEEE Transactions on Signal Processing, 2002, 50(2):174-188.

IOP Conf. Series: Materials Science and Engineering **711** (2020) 012088 doi:10.1088/1757-899X/711/1/012088

Infrared contrast characteristic of satellites

LV Xiang-yin[1,2,3], Jin Wei[1,2,3]

[1]Key Laboratory of Infrared and Low Temperature Plasma of Anhui Province, Hefei 230037, China

[2]Advanced Laser Technology Laboratory of Anhui Province, Hefei 230037, China

[3]State Key Laboratory of Pulsed Power Laser Technology, Hefei 230037, China

Abstract. Using infrared characteristic of space target for detecting and identifying the space target is a kind of effective method. Infrared radiation contrast study of satellite has great significance for both detecting and identifying the satellite, and monitoring satellite condition. In this paper, the method to calculate the radiant heat flux of satellite is given firstly. Then, the mathematical model of the surface temperature of satellite is obtained and the numerical calculation is done. The calculated numerical data is similar to the telemetry data, which proves that the orbit flux and temperature models are correct. Finally, the infrared radiant contrast of every surface on the satellite's body and solar panels were calculated and analyzed. The results show that, on the one hand, satellite infrared radiation in long wave infrared band is larger than that in medium wave infrared band. From the point of view of the range coverage, use the long wave infrared is more advantageous to detect satellite. And on the other hand, satellite infrared contrast in medium wave infrared band is much larger than that in long wave infrared band. From the point of view of the resolution of image detail, use the medium wave infrared is more advantageous to identify satellite.

1. Introduction

As a main space target, satellite detection and recognition has become an important research topic. The research on infrared characteristics of satellites is the basic technology for identifying and acquiring information of satellites. The calculation and analysis of satellite infrared contrast is an important part of studying satellite infrared characteristics. It has certain reference value for satellite detection to judge whether the satellite is invalid or not, and infrared recognition of true and false satellites.

For example, satellite solar panels act as generators. Its normal work is the premise of payload operation such as electronic equipment on satellite. It is of great significance to detect and judge whether the satellite solar panels are in normal working state. The change of infrared radiation characteristics of solar panels can provide an important basis for detecting and judging whether they are in normal working state. The infrared characteristics of solar panels can also provide an important reference for judging whether the satellite is invalid or not.For different surfaces of the satellite body, the infrared radiation of the payload is different when the payload is in different working conditions.However, the detection of the absolute value of infrared radiation on any surface is easy to be affected by many complex factors, resulting in errors in judgment. However, the contrast between different surfaces can provide a good basis for judging.

Up to now, there are many studies on temperature field related to satellite thermal control, and few studies on satellite infrared characteristics. In China, Shurui[1] and Li Ming[2] have established

IOP Conf. Series: Materials Science and Engineering **711** (2020) 012088 doi:10.1088/1757-899X/711/1/012088

theoretical models of infrared radiation of satellites, but no numerical calculation has been carried out. Han Yuge and others have studied the infrared characteristics of satellites, especially the radiation surface[3]. Sun Chengming and others have studied the modeling methods of infrared characteristics of satellites[4-5]. Guo Ming and others have carried out infrared modeling and analysis of satellites under the background of stars simulation[6]. None of the above studies involves the infrared dynamic characteristics related to satellite orbit operation, nor the band characteristics and contrast of satellite infrared.

From the calculation of orbital heat flow, this paper presents a calculation method for calculating the transient temperature field on the satellite surface changing with the satellite orbit operation and carries out numerical calculation. By comparing the temperature calculation data with the telemetry data, the correctness of the orbital heat flow and temperature model is verified. On this basis, contrast of infrared radiation in mid-infrared band 3-5um and far-infrared band 8-14um between different satellite surfaces are calculated and analyzed.

2. Calculations of orbital heat flow

The infrared characteristics of satellites mainly depend on the temperature distribution of the target itself and the background radiation characteristics. And the temperature distribution depends on the energy exchange relationship between the satellite and the surrounding background besides the structure of the satellite. That is to say, it depends on the orbital heat flow, specifically solar radiation, earth reflection and earth radiation.

2.1 Solar Radiation Received on Satellite Surface

The solar radiation received per unit area of the satellite surface is

$$E_{\mathrm{sundir}} = \zeta E_{sun} F_{\mathrm{sun}} \tag{1}$$

In the formula, $\zeta = 1 + 0.33\cos(360n/370)$, the corrected value caused by the distance between the sun and the earth (n is the number of days in a year), $E_{sun} = 1353\,\mathrm{W/m^2}$, called the solar constant, F_{sun} the angle factor of the direct solar radiation from the target surface.

2.2 Earth Reflections Received on Satellite Surface

The solar radiation is incident on the earth and its atmosphere, and the reflected part reaches the surface of the satellite, which constitutes the reflection of the earth.In calculating the earth reflections, the earth surface should be divided into several faces, and the solar radiation reflected from each face to the satellite surface should be calculated separately. Then the solar radiation reflected from each face to the satellite surface can be superimposed to obtain the earth reflections to the satellite surface.[7]

The radiation energy density of the earth reflection radiation on a satellite surface can be deduced theoretically.

$$E_{sunref} = \frac{\rho_e E_{sun}}{\pi} \int \frac{\cos\eta \cos\alpha_1 \cos\alpha_2}{l^2} dA_E \tag{2}$$

In the formula, l is the distance between the earth surface dA_E and the satellite surface, η is the incident angle of the sun to the earth surface dA_E, α_1 is the angle between l and the earth surface dA_E normal, α_2 is the angle between l and the satellite surface normal. The dA_E integral region is the visible range of the satellite surface to the earth surface irradiated by the sun.

2.3 Earth Radiation Received on Satellite Surface

In fact, the earth is an irregular sphere. In order to simplify the calculation model, we regard it as a regular sphere with a radius of 6379 km, and take the annual average of the radiant exitance of the earth $M_{earth} = 237\,\mathrm{W/m^2}$. The radiation energy density of the earth received from a certain surface of the satellite is as follows

$$E_{earth} = \frac{M_{earth}}{\pi}\int \frac{\cos\alpha_1 \cos\alpha_2}{l^2}dA \tag{3}$$

In the formula, the integral range of the earth surface is the visible range of the satellite surface to the earth surface.

The above radiation heat flux received by satellite will vary with the change of relative position of satellite with the sun and the earth as time goes on in orbit. It should be calculated in combination with six elements of satellite orbit and coordinate transformation which describe satellite orbit equation[8]. The process is rather complicated, and the process is not described in detail here.

3. Calculation of Satellite Surface Temperature

The infrared radiation of the target is determined by the surface temperature and emissivity. Emissivity is an inherent property of materials. Usually, for a specific target, once the surface material is fixed, the emissivity is determined. The surface temperature of an object is affected by various complex factors. The determination of the target surface temperature is the key to study the infrared radiation characteristics of the target.

3.1 Differential equation of heat conduction

The surface temperature of a satellite depends on its heat exchange with its surroundings and with itself. Thermal convection can be neglected in the space outside the atmosphere, so only heat conduction and radiation are considered.

The equation describing the law of temperature distribution in a heat conducting body is called a heat conduction differential equation.For satellites, the thermal conductivity differential equation is used to calculate the temperature distribution of satellites with space and time under the radiation and heat transfer boundary conditions on the outer surface, and the temperature distribution and its change with time on the satellite surface are obtained.

The heat conduction differential equation is based on the law of conservation of energy and Fourier law. Its general form in Cartesian coordinate system is as follows

$$\rho c\frac{\partial T}{\partial \tau} = \frac{\partial}{\partial x}\left(k\frac{\partial T}{\partial x}\right) + \frac{\partial}{\partial y}\left(k\frac{\partial T}{\partial y}\right) + \frac{\partial}{\partial z}\left(k\frac{\partial T}{\partial z}\right) + \Phi_v \tag{4}$$

In the formula, ρ is density; c is specific heat capacity; τ is time; k is thermal conductivity; Φ_v is heating power per unit volume of microelement.

3.2 boundary condition

The boundary condition guides the connection or interaction of heat exchange between the thermal object on its boundary surface and the external environment. For unsteady heat conduction, it is often the external driving force for the process to occur and develop. For any surface of a satellite, its boundary conditions can be written as

$$k\frac{\partial T}{\partial n}\Big|_{\text{boundary surface}} = \alpha_{sundir}E_{sundir} + \alpha_{earth}E_{earth} + \alpha_{sunref}E_{sunref} - M_{out} \tag{5}$$

In the formula, α_{sundir}, α_{earth}, α_{sunref} are the absorptivity of solar radiation, earth reflection and earth radiation on the satellite surface, M_{out} is the radiation energy per unit area of the satellite surface into space, $M_{out} = \varepsilon\sigma T^4$, and in the formula, ε is the emissivity of the outer surface of the satellite, σ is Stefan-Boltzmann constant.

3.3 Solution of Satellite Surface Temperature

The temperature field on the satellite surface at any time can be determined by the initial value given by the thermal conductivity differential equation and boundary conditions.In the calculation, a three-axis stabilized satellite is taken as an example, which is mainly composed of solar panels and satellite body. The shape of the satellite body is cuboid, and there is no heat conduction between the surfaces. Because of the complexity of the problem, the computer is usually used to solve it, and the finite difference method can be used in the process of solving it. For this reason, the satellite should be spatially discretized and divided into several discrete elements, each of which is called a node. For each node, the energy conservation equation for volume is established. The key is to establish the energy conservation equation for boundary nodes according to boundary radiation conditions. By combining the energy conservation equations of each node, the distribution of temperature field can be obtained by solving the equations.[9]

For interior nodes, the difference form of thermal conductivity differential equation can be used directly. For boundary nodes, different boundary nodes have different expressions. This is because different nodes have different heat exchanges with the outside world. Specifically, according to the energy conservation method, the equations should be established one by one. By solving the equations, the temperature distribution on each surface of the satellite can be obtained. Fig. 1 shows the temperature variation of satellite solar panels with time. Fig. 1 (a) is the calculated data of two periods of satellite operation, and Fig. 1 (b) is the telemetry data of one period of satellite stable operation from literature.[10]

As can be seen from Fig. 1, the temperature of satellite solar panels is related to the initial temperature during a short period of initial operation, and then the influence of the initial temperature is quickly eliminated. Thereafter, the change of temperature mainly depends on the change of orbital heat flow. The calculated values in a period after stable operation are basically consistent with the temperature variation law of telemetry data, which shows the correctness of the calculation model of track heat flow and temperature. At the same time, there is a certain deviation between the temperature calculation value and the telemetry data. This is mainly due to the limitation of the related parameters, such as material physical parameters and orbit parameters, given by the literature which provides the telemetry data, which results in the deviation between the temperature calculation value and the telemetry data.

(a)calculated temperature (b)telemetry temperature

Fig.1 The temperature of the solar panels surface to the sun

4. Computation of Satellite Infrared Contrast

4.1 Calculation of Satellite Infrared Radiation

The infrared radiation of the target is mainly composed of two parts: the emitted infrared radiation and the reflected infrared radiation. In order to simplify the calculation, the target is regarded as diffuse grey body. Since any amount of radiation can be calculated by the calculation of radiance, this paper takes the calculated value of radiance as the result.

1) self-radiation

After calculating the temperature of a certain surface of the target, the radiation brightness of the target in the band $\Delta\lambda$ ($\lambda_1 \sim \lambda_2$) can be calculated as follows

$$L_s = \frac{1}{\pi}\int_{\Delta\lambda} \varepsilon \frac{c_1}{\lambda^5} \frac{1}{e^{\frac{c_2}{\lambda T}} - 1} d\lambda \tag{6}$$

2) Reflective radiation

The reflection radiation of space objects to the environment is mainly composed of the direct radiation of the sun, the radiation of the earth and the reflection of the earth.

(a) Reflection of direct solar radiation

The reflective brightness of a satellite surface to direct solar radiation in the band $\Delta\lambda\,(\lambda_1 \sim \lambda_2)$ is as follows

$$L_{sundir} = \frac{\rho_{\Delta\lambda} k_{\Delta\lambda-sun} E_{sun} F_{sun}}{\pi} \tag{7}$$

In the formula, $\rho_{\Delta\lambda}$ is the reflectance of a satellite surface in the band and $k_{\Delta\lambda-sun}$ is the ratio of solar radiation energy in the band to solar radiation energy in the whole band.

(b) Reflection of Earth radiation

The reflective brightness of a satellite surface to the earth radiation is

$$L_{earth} = \rho_{\Delta\lambda} \frac{k_{\Delta\lambda-earth} M_{earth}}{\pi} \int \frac{\cos\theta_1 \cos\theta_2}{r^2} dA \tag{8}$$

In the formula, $k_{\Delta\lambda-earth}$ is the ratio of the earth radiation energy in the band $\Delta\lambda$ to the total radiation energy in the whole band, and dA integral range of the earth surface is the visible range of the satellite surface to the earth surface.

(c) Reflection of the Earth reflection

The reflectance of indirect solar radiation from a satellite surface is as follows

$$L_{sunref} = \rho_{\Delta\lambda} \frac{\rho_e k_{\Delta\lambda-sun} E_{sun}}{\pi} \int \frac{\cos i_e \cos\theta_1 \cos\theta_2}{r^2} dA \tag{9}$$

In the formula, dA integral range is the visible range of the satellite surface to the earth surface irradiated by the sun.

The radiance of the target reflection is

$$L_r = L_{sundir} + L_{earth} + L_{sunref} \tag{10}$$

In summary, the total radiation brightness of the target is

$$L = L_s + L_r \tag{11}$$

4.2 Analysis of Satellite Infrared Contrast

Radiation contrast is defined as the ratio of the difference between target and background radiance to background radiance. This paper mainly analyses and calculates the infrared radiation contrast between the six planes of the satellite body and the solar panels facing the sun. According to the definition, the radiation contrast between the six planes of the satellite body and the solar panels is as follows

$$C = \frac{L_i - L_s}{L_s} \tag{12}$$

In the formula, L_i is the radiant brightness of the satellite surface, and L_s is the radiant brightness of the solar panel towards the sun surface.

Figure 2 shows the calculation results of contrast in 3-5um bands, and Figure 3 shows the calculation results of contrast in 8-14um bands. From figs. 2 and 3, it can be seen that the contrast of solar panels is less than zero in most of the time, whether in 3-5um or 8-14um bands. This shows that the radiation of solar panels is higher than that of the satellite body in most of the time. However, in a part of the time, the contrast of radiation between individual surfaces of the body and the surface of solar panels will be greater than zero, indicating that the radiation of individual surfaces of the body is sometimes higher than that of the sun energy battery board. This usually happens when the surface is close to the solar direction. Because the backside scattering of the solar panel is higher than the internal scattering of the body surface, the temperature of the body surface is higher than that of the solar panel. It can also be seen from the figure that although the radiation of each surface of the satellite in 8-14um band is much higher than that in 3-5um band, the radiation contrast between each surface of the satellite and solar panels in 3-5um band is more significant than that in 8-14um band with time. This shows that medium wave infrared imaging is more suitable for satellite imaging with higher contrast.

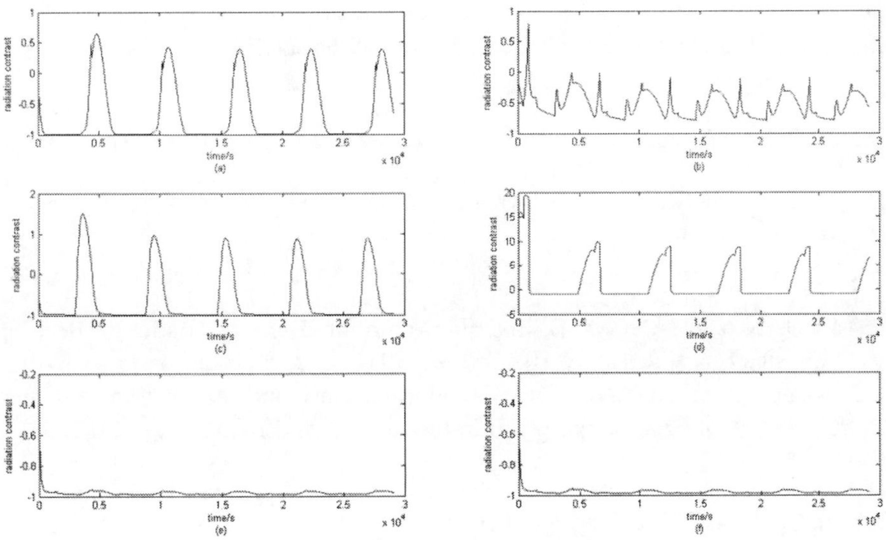

Fig.2 The radiation contrast of every surface on satellite's body to the surface to the sun of solar panels (3-5um)

(a. surface to space, b. surface to earth, c. forward surface, d. backward surface, e. left surface, f. right surface)

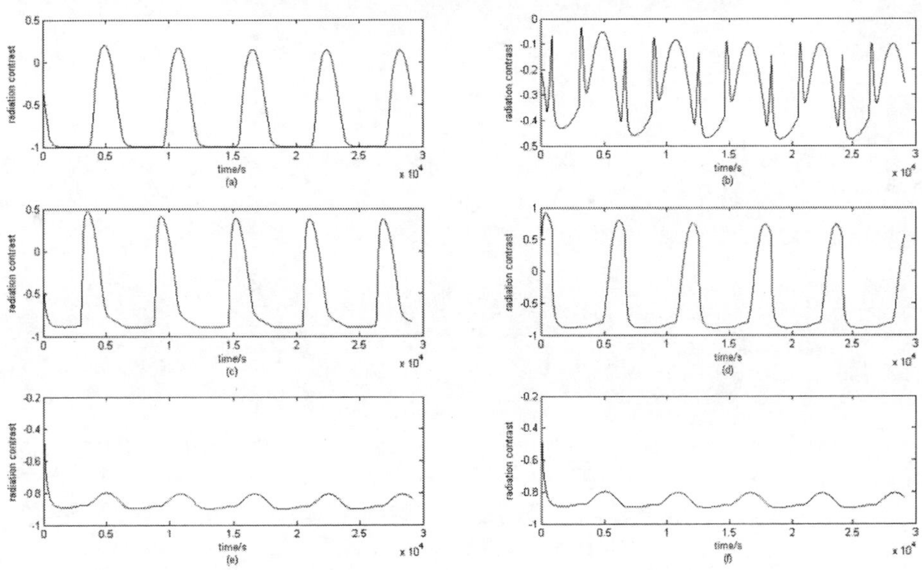

Fig.3 The radiation contrast of every surface on satellite's body to the surface to the sun of solar panels (8-14um)

(a. surface to space, b. surface to earth, c. forward surface, d. backward surface, e. left surface, f. right surface)

5. Concluding remarks

The infrared radiation of different surfaces of satellites in a certain orbit varies significantly with time, and the infrared contrast of different surfaces is also very obvious. When the infrared system can imagine the satellite, it is more advantageous to use the infrared imaging in the 8-14 um band of long wave to increase the operating distance, but in the aspect of contrast, using the infrared imaging in the 3-5 um band of medium wave will be easier to distinguish the shape and contour of the satellite and other details, which is conducive to further recognition of the satellite.

References

[1] SHU R, ZHOU Y P,TAO K Y. The study of infrared spectrum of space target [J]. Optical Technique, 2006, 32(2): 196-199.(in Chinese)

[2] LI M,DU X P. The analysis of infrared spectrum of space target [J]. Aircraft Design, 2011, 31(1): 67-70.(in Chinese)

[3] HAN Y G,XUAN Y M. Infrared feature of the satellite [J]. Infrared and Laser Engineering, 2005, 25(1): 119-127.(in Chinese)

[4] SUN C M,YUAN Y,ZHANG X B.Modeling of infrared characteristics of deep space target[J]. Acta Physica Sinca, 2010, 59(10): 7523-7530.(in Chinese)

[5] SUN C M,YUAN Y,HUANG Z F， etal..Modeling and simulation on infrared imaging characteristics of space target[J]. Infrared and Laser Engineering, 2012, 41(3): 563-568.(in Chinese)

[6] GUO M,WANG X W. IR modeling and simulation of space target/star and space environment [J]. Infrared and Laser Engineering, 2010, 39(3): 399-404.(in Chinese)

[7] Dan D. V. Bhanderi, Thomas Bak. Modeling earth albedo for satellite in earth orbit [C]. AIAA 2005-6465, 2005

[8] XIAO Y L. Aviation spacecraft motion model building[M]. Beijing: Beihang University Press.2003. (in Chinese)

[9] YANG SH M,TAO W Q. Heat transfer[M].Beijing:Higher Education Press, 2010. (in Chinese)

[10] QIN W B,CHENG H E,LI P. Thermal analysis of solar Array with rigid substrate on spacecraft in orbit [J]. Journal of Harbin Institute of Technology, 2008, 40(5): 827-831.(in Chinese)

IOP Conf. Series: Materials Science and Engineering **711** (2020) 012089 doi:10.1088/1757-899X/711/1/012089

Automatic Classification for Ground Targets under Complex Background Based on Bag of Words Model

Wei Jin[1,2,a], **Yunsong Feng**[1,2,b]

[1]State Key Laboratory of Pulsed Power Laser Technology, College of Electronic Countermeasures, National University of Defense Technology, Hefei 230037, China

[2]Key Laboratory of Infrared and Low Temperature Plasma of Anhui Province,

College of Electronic Countermeasures, National University of Defense Technology, Hefei 230037, China

[a]kingvee@163.com, [b]fyseei@163.com

Abstract. In order to improve the performance of the automatic classifier for ground targets under complex background, a new classifier which was based on bag of words model and support vector machine (SVM) was build. Firstly, the keypoints of a target image were extracted into vectors by using Scale-Invariant Feature Transform algorithm, and vectors which were extracted from all trained images were merged into large vectors. Secondly, the large vectors were clustered into small vectors which included K vectors by using K-means clustering algorithm. To analyze how different values of K affected the classifying results, K was set to several different numbers. Thirdly, vectors which were extracted from all trained images were classified, and K-dimensional numerical vectors were obtained. Finally, the classifier was build with SVM algorithm, and the K-dimensional vectors were as input features of the classifier. The training images were randomly chosen from 90% of images dataset which included missile launch vehicles and tanks, and the rest 10% were as the testing images. The trained classifier is tested, and the f1-score value is 0.79. The results show that the new classifier can perform well.

1. Introduction

The complex background does not have a standard definition, mainly based on a single, simple background. The complex background can generally be understood as: firstly, there are some interference including changeable background, ground objects blocking and shadow; secondly, the quality of images are affected by these factors including changeable sun light, cloudy day, fog and haze; thirdly, the difficulty of detecting will be increased, because the shape and scale of the targets image are changed with the many types of targets, different shooting angles and positions. Automatic detection and classification for ground targets are important for military field. It can be played a key role in knowing dynamic deployment, identification and tracking, precisely striking for vehicle targets including missile launch vehicle, tank and infantry fighting vehicle, etc[1].

Generally, there are two ways in classifying targets images. Firstly, targets images are matched to the categories templates which can identify the category of target. When the targets are complex and the amount of templates is large, it will cost very large amount of time in calculating[2,3]. Secondly, the images are preprocessed to separating and choosing targets boxes, and extracting target features; the classifier is built by using the machine learning or deep learning methods[4-8]. The second way is adopted in this paper. The bag of word model is used to automatically classify missile launch vehicle and anti-aircraft gun under complex background.

2. Bag of the word model

The bag of word model was first applied to automatic document classification[9], and then was used to image classification[10-12]. Similar to document classification, large amount of local invariant features are clustered into small amount of visual words. A image is expressed as a histogram of these visual words. There are four steps to implement the bag of word model: extracting images features, generating a visual dictionary, building a histogram,and creating a classifier.

2.1 Extracting images features

The image features include visual features, global invariant features and local invariant features. Using local invariant features to identify and classify targets in a complex environment can better solve background noise and local occlusion[13]. Local invariant features include corner points, SIFT (scale-invariant feature transform) features, and contour descriptions.

SIFT which first proposed by Lowe.D in 2004 is the most commonly used image local feature descriptor[14]. SIFT has rotational invariance, scale invariance and illumination invariance. Therefore, it may eliminate the effects of noise, affine transformation, viewing angle changes. SIFT is adopted to extract images features in this paper.

The SIFT feature extraction algorithm is mainly divided into four steps: scale space extreme value detection, key point positioning, specifying direction parameters and key point descriptors. Through the above four steps, an image is converted into a numerical matrix

$$m_i = \begin{pmatrix} t_{1,1} & t_{1,2} & \cdots & t_{1,127} & t_{1,128} \\ t_{2,1} & t_{2,2} & \cdots & t_{2,127} & t_{2,128} \\ \vdots & \vdots & \vdots & \vdots & \vdots \\ t_{n-1,1} & t_{n-1,2} & \cdots & t_{n-1,127} & t_{n-1,128} \\ t_{n,1} & t_{n,2} & \cdots & t_{n,127} & t_{n,128} \end{pmatrix} \quad (1)$$

m_i donates the ith image, and n donates the amount of key point descriptors. A key point descriptor is described by a 128-dimensional numerical vector.

SIFT features of a missile launch vehicle and an anti-aircraft gun image are extracted and shown in Fig. 1. One circle donates a SIFT feature.

Fig.1 SIFT features of a missile launch vehicle and an anti-aircraft gun image

When the SIFT features of all images have been extracted, all features numerical vectors are concatenated into a bigger matrix

$$T = \begin{pmatrix} t_{1,1} & t_{1,2} & \cdots & t_{1,127} & t_{1,128} \\ t_{2,1} & t_{2,2} & \cdots & t_{2,127} & t_{2,128} \\ \vdots & \vdots & \vdots & \vdots & \vdots \\ t_{N-1,1} & t_{N-1,2} & \cdots & t_{N-1,127} & t_{N-1,128} \\ t_{N,1} & t_{N,2} & \cdots & t_{N,127} & t_{N,128} \end{pmatrix} \quad (2)$$

N donates the amount of the SIFT features of all images.

2.2 Generating a visual dictionary

The process of generating a visual dictionary is the process of clustering feature vectors. The center of each cluster is regarded as a visual word, and all visual words constitute a visual dictionary. K-means clustering algorithm is highly efficient and is applied frequently for clustering large-scale data.

K-means clustering uses K as a parameter, and divides N samples into K clusters. It makes samples in the same cluster have higher similarity, and samples in the different clusters have lower similarity. The process is in the following.

(1) Setting the number of clusters K. K is a natural number. To compare the effects of different K values on results of classification, K is taken a value of 5 every interval from 10 to 50, and K =(10,15,20,……,45,50).

(2) Calculating the center of each cluster. The center of a cluster is updated by calculating the means of samples in the cluster:

$$\mu_i = \frac{1}{N_i} \sum_{p_i \subset C_i} p_i \quad (3)$$

C_i donates a cluster, N_i donates the number of SIFT features in the cluster C_i, p_i donates a SIFT feature vector.

(3) Calculating clustering error. The Euclidean distances between samples of each cluster and the corresponding cluster center are all added, which is used as an error criterion function:

$$J = \sum_{i=1}^{K} \sum_{p_j \subset C_i} \left| p_j - \mu_i \right|^2 \quad (4)$$

(4) Determine converge or not. An expected error δ is set firstly. If $J < \delta$, the clustering is over; else repeating steps (1) ~ (4) until $J < \delta$.

K 128-dimensional numerical vectors which represents the center of each cluster constitute a visual dictionary, such as:

$$D = \begin{pmatrix} \mu_{1,1} & \mu_{1,2} & \cdots & \mu_{1,127} & \mu_{1,128} \\ \mu_{2,1} & \mu_{2,2} & \cdots & \mu_{2,127} & \mu_{2,128} \\ \vdots & \vdots & \vdots & \vdots & \vdots \\ \mu_{K-1,1} & \mu_{K-1,2} & \cdots & \mu_{K-1,127} & \mu_{K-1,128} \\ \mu_{K,1} & \mu_{K,2} & \cdots & \mu_{K,127} & \mu_{K,128} \end{pmatrix} \quad (5)$$

2.3 Building a histogram

All SIFT features of a image are mapped to the closest visual words, and how many times all features of an image appear on each visual word is counted. k-Nearest Neighbor (kNN) classification algorithm is used to solve the problem.

The words of the visual dictionary ***D*** are regarded as samples of which categories are 1 to K respectively. Suppose an image has n SIFT features. If a feature belongs to a category by using *k*NN the times corresponding to the category will be increased 1. It can be expressed as a vector:

$$h=(h_1,h_2,\cdots,h_{K-1},h_K) \tag{6}$$

The sum of all elements inside the vector h equals to n. When all SIFT features of N images are classified by *k*NN, it can be expressed:

$$H=\begin{pmatrix} h_{1,1} & h_{1,2} & \cdots & h_{1,K-1} & h_{1,K} \\ h_{2,1} & h_{2,2} & \cdots & h_{2,K-1} & h_{2,K} \\ \vdots & \vdots & \vdots & \vdots & \vdots \\ h_{N-1,1} & h_{N-1,2} & \cdots & h_{N-1,K-1} & h_{K-1,K} \\ h_{N,1} & h_{N,2} & \cdots & h_{N,K-1} & h_{N,K} \end{pmatrix} \tag{7}$$

2.4 Creating a classifier

The support vector machine (SVM) classification algorithm is the most commonly used classification algorithm in supervised learning. Its advantage is that it can obtain better classification results without many samples, and can deal with the problem that the samples are linearly inseparable. The basic idea of the algorithm is to establish an optimal decision hyperplane, so that the distance between the two types of samples on the two sides of the plane from the plane is maximized, thus providing a good generalization ability for the classification problem.

For the two-category problem, after some preliminary derivation, the key to the support vector machine is to solve a constrained optimization problem, which is expressed as follows:

$$\begin{aligned} &\min_{\omega,b,\xi} \frac{1}{2}\omega^T\omega + C\sum_{i=1}^{l}\xi_i \\ &s.t. \quad y_i(\omega^T\phi(x_i)+b) \geq 1-\xi_i \\ &\qquad \xi_i \geq 0, i=1,\dots,l \end{aligned} \tag{8}$$

The meaning of each variable in the formula can be referred to [15].

3. Experiments

3.1 Extablishing targets images database

Two ground targets, missile launchers and anti-aircraft guns, were collected from the open Internet, numbering 447 and 501 respectively, and some images are shown in Fig.2. It is easy to see from the images that the background environment of these targets are more complicated, including: sky, buildings, trees and other military targets. In addition, the targets in the images are also different, including: inconsistent models, different shooting angles, and different lighting. These factors will affect the results of the target classification.

(a) Missile launchers

(b) Anti-aircraft guns

Fig.2 Images of two ground targets

3.2 Experimental conditions

The program is written in Python. *K*-means clustering, *k*-Nearest Neighbor and SVM are implemented by calling the *sklearn* package respectively, and SIFT features are extracted by using *OpenCv* package. All experiments were conducted in a Windows7 PC with an Intel Cerelon G1630 2.80 GHz processor and 4.0G RAM.

3.3 Results

When the experiments begin, the validation dataset is randomly chosen from 10% of the entire dataset. When *K* equals to 20, the results are shown in Table 1. the average value of precision, recall and f1-score equal to 0.79、0.78 and 0.79 respectively.

Table 1 Results when *K*=20

	precision	recall	f1-score	support
0	0.68	0.72	0.70	35
1	0.84	0.82	0.83	66
avg / total	0.79	0.78	0.79	101

To compare the effects of different *K* values on results of classification, when *K* get value from (10,15,20,……,45,50) in turn, the change of f1-score value is shown in Fig.3. The maximum value of f1-score is 0.79 when *K*=20, the minimum value is 0.66 when *K*=35.

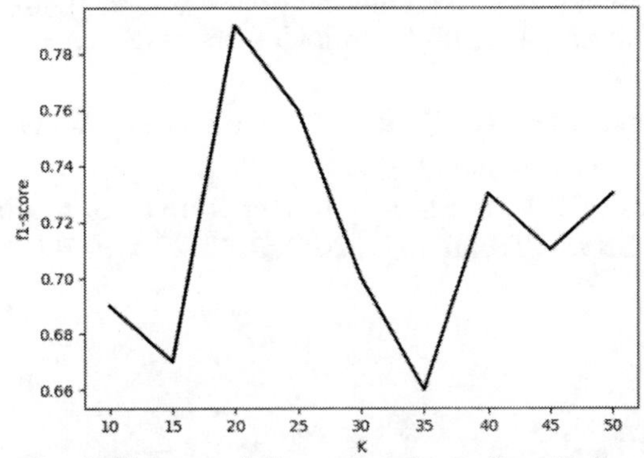

Fig. 3 Change of f1-score value when *K* changes

4. Conclusion

In this paper, the bag of word model was used to automatically classify missile launch vehicle and anti-aircraft gun under complex background. Firstly, SIFT features of target images were extracted. Secondly, SIFT features were clustered supervised with *K*-means clustering algorithm. Thirdly, a histogram was obtained with *k*-Nearest Neighbor. Finally, histograms of all targets images were classified with SVM. When *K*=20, the classification results were the best, and f1-score = 0.79. The results show that the bag of word model can perform well when it is used to automatically classify ground targets under complex background.

References

[1] XIE Xiaozhu, HE Cheng. Review of vehicle recognition in complex environment background [J]. Journal of Ordnance Equipment Engineering, 2017, 38(6): 90-94.
[2] Zhu Xufeng, Ma Caiwen, Liu Bo. Aerial target automatic classification based on improving bag of words model [J]. Journal of Infrared and Laser Engineering, 2012, 41(5): 1384-1388.
[3] LIN Yuchi, CUI Yanping, HUANG Yinguo. Study on edge detection and target recognition in complex background [J]. Journal of Optics and Precision Engineering, 2006, 14(3): 509-514.
[4] LIU Tianyu, FENG Quan. Detecting grape leaves based on convolutional neural network [J]. Journal of Northwest University (Natural Science Edition), 2017, 41(4): 505-512.
[5] WANG Hong, MO Yumeng, YIN Lijun, et al. An image recognition algorithm on small arms shooting target in complex field background [J]. Journal of Fire Control& Command Control, 2009, 34(7): 60-62.
[6] DENG Liu, WANG Zijie. Deep convolution neural networks for vehicle classification [J]. Application Research of Computers, 2016，33(3):930-932.
[7] ZHANG Feiyun. Car detection and vehicle type classification based on deep learning [D]. Zhengjiang: Jiangsu University, 2016.
[8] CAI Yingfeng, WANG Hai, CHEN Long, et al. Robust vehicle recognition algorithm using visual saliency and deep convolutional neural networks [J]. Journal of Jiangsu University (Natural Science Edition), 2015, 36(5): 331-336.
[9] YU Minjie. Bayesian network classifier and applications [D]. Kunming: Yunnan University of Finance and Economics, 2012.
[10] Fei-Fei L, Perona P. A Bayesian hierarchical model for learning natural scene categories [C] // Proceedings of IEEE Computer Society Conference on Computer Vision and Pattern Recognition, 2005: 524 – 531.
[11] Roberto Toldo, Umberto Castellani, Andrea Fusiello. The bag of words approach for retrieval and categorization of 3D objects [J]. The Visual Computer: International Journal of Computer Graphics-Special Issue on 3D Object Retrieval, 2010, 26(10): 1257-1268.
[12] XIAO Zhe, QIN Zhiguang, DING Yi, et al. Efficient method for image classification based on low-scale bag of word model [J]. 2016, 45(6): 997-1001.
[13] CAO Jian, CHEN Hongqian, MAO Dianhui, et al. Survey of image object recognition based on local features [J]. Journal of Central South University (Science and Technology), 2013, 44(2): 258-262. (in Chinese)
[14] LOWE D. Distinctive image features from scale-invariant keypoints [J]. International Journal of Computer Vision，2004，60(2): 91-110.
[15] C. -C. Chang, C. -J. Lin. LIBSVM: a library for support vector machines [J]. ACM Transactions on Intelligent Systems and Technology, 2011, 2(3): 27:1-27:27.

IOP Conf. Series: Materials Science and Engineering **711** (2020) 012090 doi:10.1088/1757-899X/711/1/012090

Security study of laser measuring methods on high explosives

Zheng-Ming GAO[1, a], Juan ZHAO[2,b] and Su-Ruo LI [1,c]

[1]School of Computer Engineering, Jingchu University of Technology, Jingmen 448000, China

[2]School of Electronics and Information Engineering, Jingchu University of Technology, Jingmen 448000 China

[a]gaozming@jcut.edu.cn, [b]ajuan323@163.com, [c]2410500736@qq.com

Abstract: Interaction activities between the laser photons and the molecular of high explosives were described; the threshold of the laser beams' power on the surface of the high explosives is achieved limiting the interaction probability between molecules of the high explosive and the photons, their changes of properties with the mathematical norm commonly used in engineering. Research confirm the security of common active optical measuring methods on high explosives, and prove the capability of their wide applications in measuring the properties and state characteristics of high explosives.

1. Introduction

Most of the propelling energy of the modern weapons is provided by the high explosives; therefore, the physical and chemical properties including the state of the high explosives have great influences on the managements and storage of such weapons. With demands to ensure the uniform state of the loaded high explosive, maintenance of its mechanical capability and the stability of the chemical components, etc. called a variety of measuring methods based on the acoustics, electrics and other theories keeping up with the development of science and technology. Being an important active method of the optical measuring methods, the laser measuring method provides a highly precise and automated ways to verify the state and properties of the high explosives. Lasers are a kind of pure light with high coherence, orientation and luminance; therefore, it can be used in measuring the existence of the explosives, the size of the loading surface [1], loading displacements, characteristics of aging and damage near the surface [2], etc.. However, the lasers are composed by energy photons and the high explosives are sensitive to thermal loads; this leads the lasers to be treated as ground weapons detonating the high explosives in coming weapons, or as spatial orientated weapons to attack the satellites and block the missiles [3] in military applications, it was seldom to be used in measuring the state and properties of high explosives [4], as well as the radiation detection methods [5].

According to the interaction activities between the molecule of the high explosives and the photons, the paper calculated the interaction probability of a single molecular of high explosives when the laser beams' power varied and thermal field of the radiated areas was simulated. Aiming to prove the security of its usage for measuring the state and properties of the high explosives, the paper obtained the limit magnitude of the laser beams' power based on the mathematical norm commonly used in engineering and the time-temperature superposition principle.

2. Theoretical Background

Causing that some of the energy of the photons is absorbed and the photons vanish or change, some photons are scattered by the molecules of matters, the intensity of the photon's flows will be decreased when it is penetrating matters. The energy of the laser beams is absorbed by the molecules and changed to another form: kinetic energy. Being scattered by the molecules, the motion directions of the photons in the laser beams are changed and some of their energy is transferred to the molecules collided.

Interaction activities between the photons and matters. The main interaction activities of the photons with matters are the Photon-electric Effect, the Compton Effect, the Pair Production Effect and the Photoionization Effect. The Rayleigh scattering effect should be considered when the energy of the photons is quite low, furthermore, other activities such as photonuclear reactions and nuclear resonance reactions happens with little probability.

When the photons interact with the constrained electrons around the nucleus, the total energy of the photons is transferred to some constrained electrons, the electrons emit while the photons vanished, this phenomenon is called the Photon-electric Effect, the emitted electrons are called photoelectrons. The atoms absorb the energy; part of them acts as the ionization energy the electrons needed to escape and part of them acts as kinetic energy of the photoelectrons. When the energy of the photons is large, the photons collide non-elastically with the electrons around the nucleus. Part of the energy of the photons is transferred to the electrons and they recoil. Homogenously, the energy of the photons as well as their motion directions change; this phenomenon is called the Compton Effect. As the energy being increased, the photons will change to a pair of electrons carrying positive and negative charge respectively in the coulomb fields around the nucleus; this is called the Pair Production Effect. The neutral atoms are ionized to electrons and positive ions when the energy of the photons is larger than the ionization energy threshold of the gaseous molecules, such a phenomenon is called the Photoionization Effect.

The interaction probability between the photons and the matters is quantified by cross sections, which is defined as the probability of a single target nucleus in the unit area reacting with a particle in given energy and vertical directions. The unit of the cross sections is barns, equaling to $10^{-24} cm^2$. Each reaction channel has its own cross sections, e.g. the photon-electrical cross sections σ_{ph}, the coherent scattering cross sections σ_{cs}, the incoherent cross sections σ_{is}, and the pair production cross sections σ_{pp}, etc.

High Explosives. The high explosives are a kind of material composed of the Carbon, hydrogen, nitrogen and oxygen elements with high-energy density. The modern typical high explosives are mainly composited by HMX (or called Octogen; the molecular formula is （CH_2）$_4$（NO_2）$_4$), They are composed of insensitive or plasticized materials such as TATB, BDNPA, etc. (The general high explosives, and their constitutions are listed in Table. 1).

Table. 1 molecular formula and constitutions of some high explosives

High explosives	Molecular formula	constitutions			
		C	H	N	O
HMX	296.2	4	8	8	8
TATB	258.2	6	6	6	6
TNT	227.1	7	5	3	6
HNS	450.3	14	6	6	12

3. Security Analysis

Application security of the high explosives has been paid great attention. In addition to the effect on human bodies, the security that the high explosives is blocked by alloys in weapons engineering is researched in detail and the limit of the laser beams' power was also made clear. Due to some unknown problems, little efforts were down to monitor the change of the high explosives when they were exposed to laser beams with moderate power.

The carbon, hydrogen, nitrogen and oxygen elements that high explosives compose of interact with the photons and the Photon-electric Effect, scattering or photoionization effect takes place, absorbing photons' energy and producing positive holes or electrons (or pairs of electrons), such effects can break the chemical bonds of high explosives and speed up the thermal decomposition. Therefore, the measuring security can be evaluated by the microscopic interaction probability of a single molecular and the increased temperature macroscopically when the high explosives are radiated by lasers.

The Interaction Probability of a Single Molecular of high explosives with Laser Beams. The electrons around the nucleus in the molecule of the high explosives will escape or be excited after they interacted with the photons and thus the molecule change from the original state, the physical and chemical characteristics, to some degree, will change with it. Their interaction activities with photons can break the chemical bonds and speed up the thermal decomposition process. Therefore, in microscopic view, the interaction probability of the molecular of the high explosives with the photons should be small enough to remain their macroscopic properties.

Illustrating with HMX, the total cross-section of a single HMX molecular is the linear sum of their component elements:

$$\sigma_t = \sum_i m_i \sigma_{t,i}, \ \sum_i m_i = 1, \ \sigma_{t,i} = \sigma_{i,ph} + \sigma_{i,cs} + \sigma_{i,is} + \sigma_{i,pp} \tag{1}$$

Where σ_t is the cross-section of the HMX molecular to photons, $\sigma_{t,i}$ is the total cross sections of carbon, hydrogen, nitrogen and oxygen, m_i is the mass percentage. According to the EPDL dataand equation(1), the cross sections of hydrogen, nitrogen and oxygen elements based on the energy numbers in the cross section data of carbon element could be interpolated linearly and then the total cross section data of a single molecular of HMX would be obtained(see in Fig. 1)

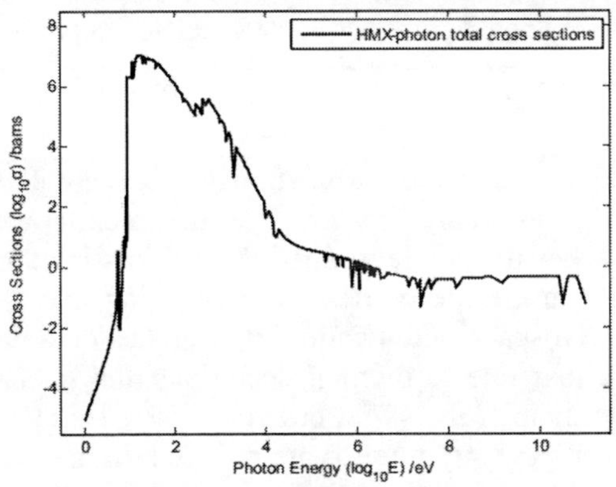

Fig. 1 total cross-section of a single HMX molecular with photons

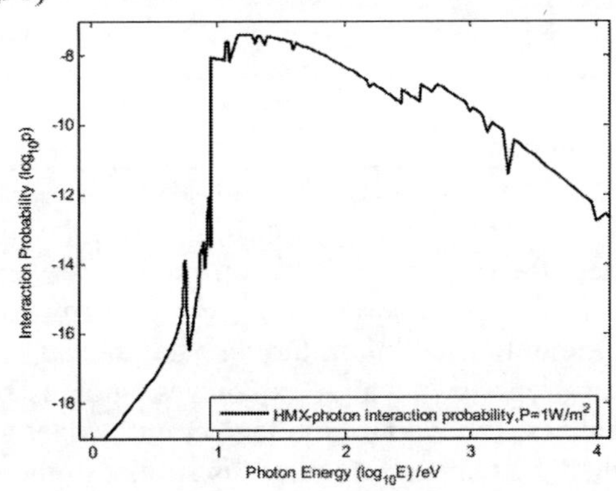

Fig. 2 interaction probability of a single HMX molecular with laser beams

Supposing the power of the laser beams: P= 1W/m^2, the interaction probability of a single molecular of HMX with photons: $\eta = N\sigma_t = P\sigma_t/E$ (E: energy of a photon of the laser beams) (see in Fig. 2).

Nowadays, the range of the wavelength of the lasers developed is 0.12nm$\leq\lambda\leq$3mm, and the photon energy $E = h\upsilon = hc/\lambda$ (h is the Prandtl number, 6.6260693×10^{-34} J·s), that is: 4.136×10^{-4}eV$\leq E \leq$10.339keV. And the common lasers used in industry are visible, the photon energy: $1.6325 \leq E \leq 3.265$ (eV). Conclusions could be drawn that:

(1) The interaction probability of a single HMX molecular with photons of industrial lasers is quite small, and it rises quickly when the lasers' frequency grown up, the maximum cross section is up to 10^7 in magnitude. However, the probability goes down when it passes over the maximum with the frequency being increased. The cross section of a single HMX molecular interacting with laser photons decreased steadily and stabilized itself in about 10^0 barns in magnitude.

(2) The interaction probability of HMX molecule is high when it is exposed to industrial lasers with power of P=1W/m^2, the magnitude varies in $10^{-10}\sim10^{-8}$. According to the mathematical norms in engineering, the interaction probability of a single molecular of HMX with photons is supposed to be less than 1‰, then the exposed power of lasers on the high explosives should be less than 10^5W/m^2, 10^7W/m^2 in some cases. The interaction probability decreases when the laser frequency was increased, which means the interaction probability of HMX could be deceased when the laser frequency had been raised while the laser beams power kept unchanged.

Energy Absorption by High Explosives of Lasers. A laser beam is composed of many photons with given energy. Considering the photons are wave-corpuscle duality, the energy transferring procedure can be explained in two ways: the photon energy is transferred to the high explosives while the molecules of high explosives interacting with the photons, or molecule of the high explosives are excited under the electro-magnetic fields produced by the laser photons and then they move faster. The high explosives absorb energy explained from either side of the wave-corpuscle duality of photons.

Supposing the energy absorption ratio of high explosives to lasers is $\eta=0.3$. According to the thermodynamic principle, the temperature field of the high explosives exposed to laser beams satisfies the flowing equation

$$\rho c_p \frac{\partial T}{\partial t} = \lambda\left(\frac{\partial^2 T}{\partial x^2} + \frac{\partial^2 T}{\partial y^2} + \frac{\partial^2 T}{\partial z^2}\right) + \eta P \quad (2)$$

Where ρ—— the density of the high explosive, c_p—— specific heat, T—— temperature, t——time, x,y,z ——distance between a random point to the original planes in global coordinates systems, P—— power of the laser beams.

Thermal simulation analysis could be carried on with the finite element method. (Thermal properties are listed in Tab. 2), the initial temperature T_0=20°C and the conduction coefficient h=15W/（m^2·K）(See Fig. 3 and Fig. 4).

Tab. 2 thermal properties of HMX

	Density (ρ)	Thermal-conductivity coefficient (λ)	Specific heat (c_p)
unit	Kg/m^3	W/（m·K）	J/（Kg·K）
numbers	1790	0.346	1050

Conclusion can be drawn that: the increased temperature of the high explosives is ignorable when the exposed power of laser beams is quite low, e.g. far less than 10^2W/m^2.The increased temperature rises when the power grows up and it is very apparent when the power reached to 10^5 W/m^2 in magnitude.

It is very important for the high explosives to remain its properties steady causing that they are commonly used in military or vital industry such as aerospace and aviation. Researches demonstrated that properties of high explosives, such as the mass loss due to thermal decomposition, mechanical changes, follow the time-temperature superposition principle, the slight effect of viscoelastic material when the temperature raised is similar to the effect when the strain ratios are reduced. In mathematics, for a random mechanical property E, there exists:

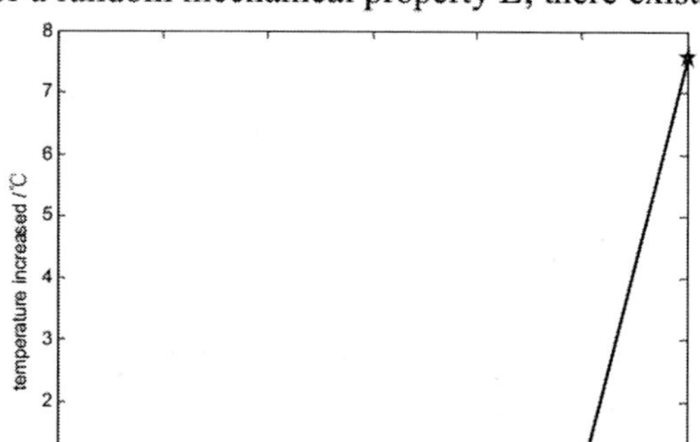

Fig. 3 Powers of laser beams versus the maximum increased temperature of HMX

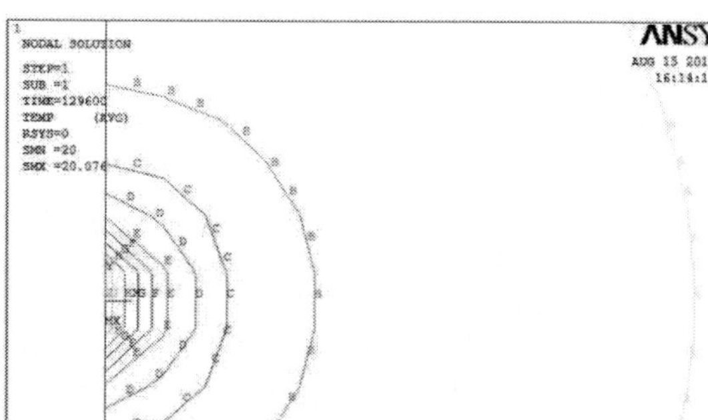

Fig. 4 Temperature fields of the high explosives exposed to laser beams with power P= 10^5W/m^2

$$E(T,t) = E[T_0, t/\alpha_t(T)] \quad (3)$$

Where α_t（T） —— the shift factor in given temperature, t/α_t（T） —— the shift time. Shift factors can be solved by Williams-Landel-Ferry (WLF) equations[6]

$$log\ \alpha_t\ (T) = -\frac{C_1(T-T_0)}{C_2+(T-T_0)} \quad (4)$$

Where C_1、C_2 are constant numbers of given materials, T_0 is chosen as the reference temperature. To some high explosive, researches showed that: (1) the mass loss follows the time-temperature superposition principle and shift factors are C_1=-11.9232、C_2=-342.7419; (2) the extended modulus and extended power follows the time-temperature superposition principle and shift factors are C_1=-73.8942、C_2=-426.8054[7].

Supposing the properties of high explosives could not change over 5% in given temperature when they were measured with laser beams, that is $1/\alpha_t(T) \leqq 1.05$, then the maximum temperature increased of the high explosives calculated with the shift factors is 0.6080K and 0.1224K respectively, which means the maximum magnitude of lasers power is 10^5W/m^2 (see Fig. 3).

4. Conclusions and Discussions

Optical measuring methods provide an accurate way to measure the properties and state characteristics of high explosives and it could be upgraded automatically and intelligently. Being an active optical measuring method, the laser measuring method might cause some security problems because the energy of lasers could be absorbed and the high explosives are thermal sensitive. The security problems could be evaluated by means of limiting the interaction probability between the high explosives and the laser photons, or the changes of their properties. Results show that:

(1) The interaction probability of a single molecular of HMX with the laser photons is quite small when the wavelength is in the range of visible lights, and the frequency increases, the probability grows up rapidly to the maximum and then decreases slowly. If the interaction probability is limited to 1‰, the power of the laser beams exposed to the high explosives should not exceed 10^5W/m^2, in some cases, 10^7W/m^2 in magnitude.

(2) The energy absorbed by the high explosives could be transferred quickly when the power of the lasers was very small, e.g. when the power was less than 10^2W/m^2, the increased temperature is not evident. The increased temperature grows up while the power was raised. The temperature change is assignable when the power of laser beams exceeds 10^6W/m^2 in magnitude. Furthermore, the properties of high explosives change with the temperature, if the property value is not supposed to change exceeded over 5%, the power of the laser beams exposed to the high explosives should be less than 10^5W/m^2 in magnitude according to the time-temperature superposition principle.

Conclusions were drawn that: (1) it is secure of the high explosives when they are measured by lasers commonly used in industry, the interaction probability with laser photons is quite low and the change of the properties is not remarkable. (2) The power of the laser beams is expected to be less than 10^5W/m^2 in magnitude to remain the state of the high explosives stabilized when they are measured with active optical measuring methods.

Acknowledgements

This work was supported by Hubei Provincial Natural Science Foundation of China (No. 2019CFB661), the cultivatable science foundations of Jingchu university of technology (No. PY201903), and it was also supported in part by the key research and development project of Jingmen (No. 2019YFZD009), and the research project of Hubei Provincial Department of Education (No. B2019213).

References

[1] Yang, S.-W., Lin, C.-S. and Lin, S.-K. 3D surface profile measurement of unsymmetrical microstructure using Fizeau interferometric microscope. *Optics and Lasers in Engineering*, 51, 4 (4// 2013), 348-357.
[2] Fitch, M. J. *12 - Terahertz (THz) spectroscopy*. Woodhead Publishing, City, 2014.
[3] Agoston, T. high energy laser weapons. *Academic and Applied Research in Military Science*, 9, 2

(2010.2), 327-341.

[4] GAO, Z.-m. and ZHAO, J. Safety Evaluation Method of Lasers Measuring and Radiographic Technologies Applied in Specific Components. .*Journal of Ordnance Equipment Engineering*, 40, 1 (2019.01), 188-191.

[5] GAO, Z.-m., HE, S.-p. and LIU, J. Security Study of X-Ray and γ-Rays Detection of High Explosives. *Transactions of Beijing Institute of Technology*, 34, 8 (2014.08), 843-847.

[6] Kim, J., Byron, T., Sholar, G. and Kim, S. *development of dynamic modulus testing and data interpretation capability*. state materials office, State of Florida, 2008.3.

[7] YANG, G.-m. *the time-temperature superposition principle and the properties of high explosives*. China Academy of Engineering Physics, Mianyang, 2004.6.

IOP Conf. Series: Materials Science and Engineering **711** (2020) 012091 doi:10.1088/1757-899X/711/1/012091

Study on the Interaction between Channels of Image Intensifier

Yuntao Xie[1, a],Yang liu[2,b], Xiaoquan Sun[3,c]

[1,3] State Key Laboratory of Pulsed Power Laser Technology, National University of Defense Technology, Hefei 230037, China

[2] Educational Insurance Departmen, National University of Defense Technology, Hefei 230037, China

[a]wsqijian@126.com, [b]343040226@qq.com, [c]sun_xq@vip.163.com

Abstract: The experiment of mutual interference between channels of the UV imager was carried out. The plateform consists of jamming light with large area and target light with small area was established. The curve of the gray value of target light in image intensifier with the intensity of interference light was measured. Experimental results show that there is mutual interference between channels when the UV imager is irradiated by continuous laser and the mutual effects will benefit to suppress the target signal by reducing the gain of the adjacent channels. Theoretical mechanism of this phenomenon was studied. The electrical coupling model between channels was established, and the influence of transverse resistance between adjacent channels on the gain characteristics of adjacent channels was studied by nodal analysis. Numerical simulation results show that the transverse resistance between channels is the main cause of mutual interference between channels, which greatly influence the gain of the adjacent channels.

1. Introduction

Ultraviolet imager has the characteristics of high gain, high temporal and spatial resolution. It is not only widely used in civil fields such as lunar base plasma detection[1], high-voltage corona discharge detection[2], but also has important military applications[3-4], as a core component of ultraviolet imaging warning device on aircraft and satellite platforms.

Laser has the characteristics of high peak and good directivity. It is a high brightness light source and very suitable for long-distance transmission. Laser interference and damage to optoelectronic imaging system has always been a research hotspot at home and abroad[5]. In 2007, RIC (H.) M. A. Schleijpen studied the relationship between different laser parameters and saturation area of infrared imaging system[6]. In 2010, Titterton D H and others analyzed the performance of four different interference evaluation indicators[7]. In 2011, Anne Dureuc studied the interference effect of laser with different parameters on HgCdTe focal plane array detector[8]. Domestically, in 2011, Zhang Zhen et al. interfered the TDI CCD with repetitive laser, and explained the phenomenon of equal interval black lines in images theoretically[9]. In 2012, Zhang Zhenyun et al. used the continuous laser to irradiate the linear array CCD, and studied the mechanism of three interference spots in the image[10]. In 2014, Kang Wenyun et al. used repetitive laser to interfere with push-broom camera, and compared it with staring camera[11]. In 2016, Shengliang et al. studied the phenomenon of pixel inversion when CMOS camera was irradiated by continuous laser, and explained the jamming mechanism[12]. It can be seen that the research on laser interference and damage optoelectronic system mainly focuses on the interference and damage mechanism of CCD or infrared thermal imager, and the research on ultraviolet imager has not been reported publicly.

In this paper, we observed the gain saturation of MCP and the mutual interference between channels of ultraviolet imager while it was irradiated by continuous laser. The mechanism of channel interference was successfully explained by constructing a mutual interference model between MCP channels.

2. Experimental System

As shown in Figure 1, if the projection area of the ultraviolet tail flame of an incoming missile is small, the target can be seen as a strong point light source on the photocathode of a ultraviolet imager while it approaches the ultraviolet warning device from a long distance. At this time, the region of the image intensifier irradiated by the target is in a gain saturation state.

When the intensity of the interference light is weak and the projection area on the photocathode is much larger than that of the target light, the interference light wraps the target in the center of the field of view of the ultraviolet imager. Since the intensity of the interfering light is weaker than that of the target light, can the interfering light affect the imaging of the target light? To answer this question, the following experiments are designed.

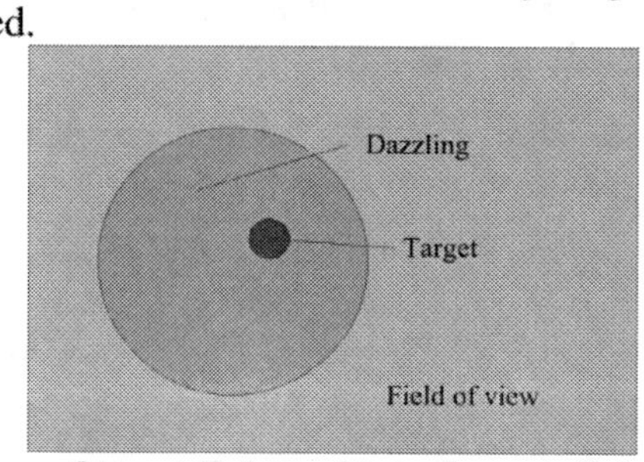

Figure 1 Projection distribution of target light and interference light on photocathode surface

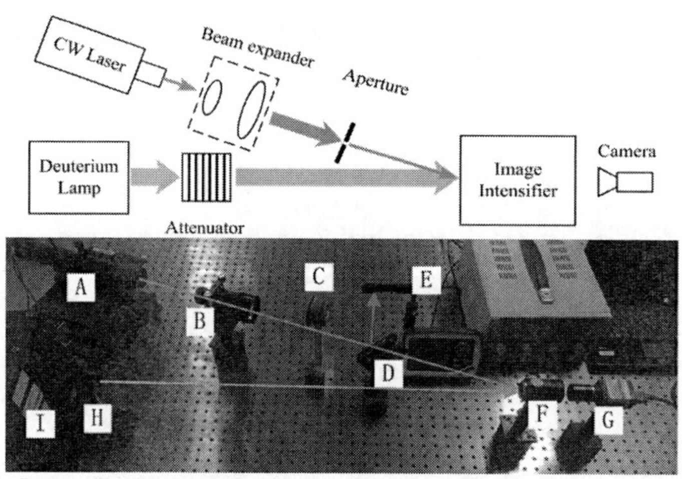

A: He-Ne Laser; B: Beam expander; C: Aperature; D: Spliter; E: Power meter;
F: Image intensifier; G: Industrial camera; H: Attenuator; I: Deuterium lamp
Figure 2 Schematic and physical diagrams of the experimental device for inter-channel interference

The experimental device is shown in Fig. 2. It consists of two beams of light, one is the target light and the other is the interference light. The target light comes from a continuous He-Ne laser with a wavelength of 632.8 nm. A beam expander with magnification of 10 times and a small hole with diameter of 0.1 mm are placed in the target light path. The small hole is located behind the beam expander. The interference light source is a deuterium lamp, and its spot is large enough to fill all the field of view of the image intensifier. The neutral density filter is placed in front of the deuterium lamp, and the interference intensity is changed by adjusting the OD number of the filter. The lenses of the image intensifier are removed, and the image intensifier window is illuminated directly by the two light beams. The industrial camera is used to take the image intensifier fluorescent screen directly, and the image gray value represents the light intensit.

3. Experimental Results

Setting the integral time of CCD to 3.342 ms .When the power of the target light is 1.73×10^{-10} W， the average power density of the target light at the photocathode plane of the image intensifier is $2.2\times10^{-6}\,\mathrm{W/cm^2}$ calculated from the spot diameter of 0.1 mm. At this time, the target spot is shown in Fig.3(a). It can be seen that the spot is circular distribution, and the maximum gray value of the spot center is 205. Fig.3(b) and Fig.3(c) are the imaging conditions of the image intensifier when the interference illumination is $1.1\times10^{-7}\,\mathrm{W/cm^2}$ and $4.4\times10^{-6}\,\mathrm{W/cm^2}$, respectively. When the interference illumination is $2.2\times10^{-7}\,\mathrm{W/cm^2}$, the corresponding gray value of the interference light is 45, and the corresponding gray value of the target light is 195. When the interference illumination is $8.8\times10^{-6}\,\mathrm{W/cm^2}$, the corresponding gray value of the interference light is 65, and the corresponding gray value of the target light is 137. It can be seen that the gray value of the target light decreases with the increase of the intensity of the interference light.

(a) The interference light intensity is 0 and the target light intensity is $2.2\times10^{-6}\,\mathrm{W/cm^2}$

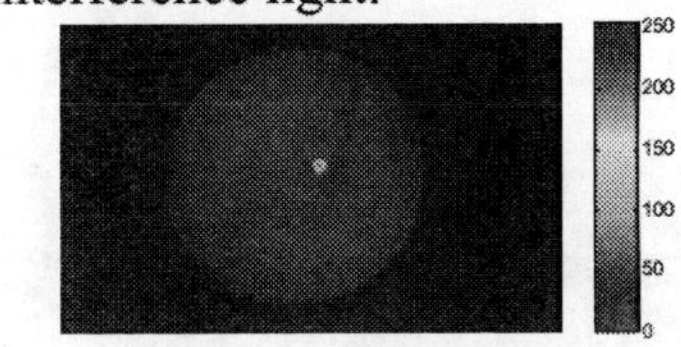

(b) The interference light intensity is $2.2\times10^{-6}\,\mathrm{W/cm^2}$ and the target light intensity is $1.1\times10^{-7}\,\mathrm{W/cm^2}$

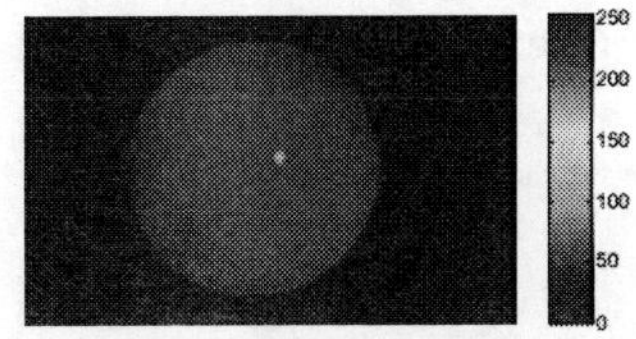

(c) The interference light intensity is $2.2\times10^{-6}\,\mathrm{W/cm^2}$ and the target light intensity is $8.8\times10^{-7}\,\mathrm{W/cm^2}$

Figure 3 Gray distribution of image intensifier under intense light irradiation

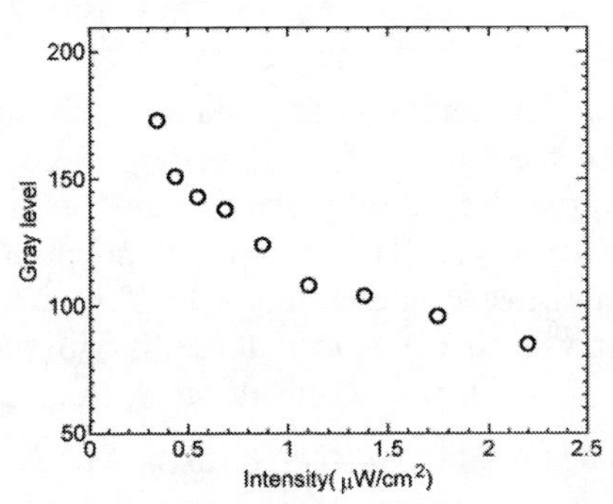

Figure 4 Changes of gray value of target spot with interference illumination

On the premise that the intensity of the interfering light is always weaker than that of the target light, the gray value of the target spot is obtained as shown in Fig.4 as the intensity of the interfering light is continuously improved. It can be seen that the gray value of the target light decreases continuously with the increase of the intensity of the interference light. It is assumed that the gain of the MCP channel corresponding to the target spot (hereinafter referred to as the activation channel) is affected by the surrounding channel (hereinafter referred to as the proximity channel). As the radiation intensity of the adjacent channel increases, the adjacent channel gradually enters gain saturation. The electrons originally supplied by the adjacent channel can not reach the activation channel in time,

IOP Conf. Series: Materials Science and Engineering **711** (2020) 012091 doi:10.1088/1757-899X/711/1/012091

resulting in the decrease of the gain of the activation channel. To verify this conjecture, the conduction mechanism of MCP is analyzed, and the current supplement model between MCP channels is established to study the effect of signal intensity of adjacent channels on the gain of activated channels.

4. Theoretical Analysis Model

The coupling between channels originates from the resistance connection between channels. In order to explain the coupling effect between channels, it is necessary to model the mutual coupling of current and potential between channels. The potential distribution in the channel is analyzed when different intensity signals are incident, and then the channel gain is determined by the potential distribution.

The two-channel coupling model is shown in Figure 5. Each channel is composed of n current sources and several resistance elements. Each element corresponds to the channel length of $dz=L/n$, where L is the total length of the channel, and the current is the average value of the signal multiplier current in the channel in dz.The signal current I_{in} enters the channel at the channel entrance $z=0$, and I_{n} is the output current at the channel exit $z=L$. The potential difference between the entrance and the exit is U_0.

The boundary of the resistance unit in the equivalent circuit is the insulation layer around the channel: the transverse resistance R_1 is the resistance between the emitter layer resistance and the conductive layer, the longitudinal resistance R_2 is the channel conductive layer resistance, and the transverse resistance R_3 is the coupling resistance between the channels. In this figure, the magnitude of the element resistance is: $r_1=R_1(n\text{-}1)$, $r_2=R_2/n$, $r_3=R_3(n\text{-}1)$.

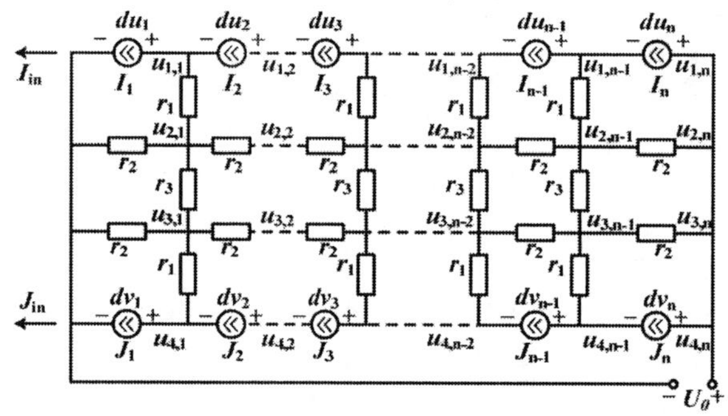

Figure 5 Equivalent circuit diagram of resistance coupling between two channels

Literature [13] has elaborated the method for calculating the current gain of single channel under saturated and unsaturated conditions. Based on the above theory, the channel electric field distribution under uncoupled conditions can be solved. The current and potential distribution in the channel can be solved by the iterative method in the case of coupling. The steps are as follows:

(1) Based on the circuit analysis theory, the voltage distribution of each node of the equivalent circuit is calculated according to the current distribution I_{i} and J_{i} in the channel. $u_{\text{i,j}}$ is taken as the solution object, which contains a total of $4(n\text{-}1)$ unknowns (where $u_{\text{i,n}}=V_{\text{s}}$ is known). By using node analysis method[14], a total of $4(n\text{-}1)$ independent equations can be obtained for each node in the circuit, which can be used to solve the potential value of each node in the circuit.

(2) Based on the voltage distribution calculating method, the current distribution in the channel can be obtained. The current gain of each cell in the channel is related to the potential difference at both ends of the cell. The multiplied current value of each cell in the channel can be solved by the following formula.

$$I_i = I_{in}\exp\left(\frac{i}{n}\ln M_0\right)\exp\left(\sum_0^i \ln(1+\frac{1}{v_s}\frac{d\Phi(i)}{dz})\right) \quad (1)$$

In the formula, M_0 is the gain under unsaturated condition and $\Phi(z)$ is the channel voltage and offset $\Phi(i)=u_i - V_s i/n$.

In the first iteration, the current distribution in the channel under unsaturated conditions can be simulated. When unsaturated, the channel potential distributes uniformly along the axis. From the formula（1）, the channel current increases exponentially along the axis.

The coupling model between channels and the calculation method of the input and output currents of each channel under the coupling condition are given above. An example of the effect of the coupling between channels on the imaging process of image intensifier is given below.

5. Result Analysis

The parameters of the microchannel board are set as Table 1.

Table 1 Performance parameters of microchannel plates

Parameter	Value	Unit
Channel length(L)	320	μm
Pore diameter(d_c)	5	μm
Center to center distance(h)	6	μm
R_1	8×10^{10}	Ω
R_2	8×10^{15}	Ω
Emission layer thickness	0.01	μm
Resistive layer thickness	0.2	μm

The transverse resistance R_3 between channels depends on the conductivity of the channel material and can be calculated by the following formula[15]:

$$R_3 = \frac{\rho}{L}\left(\frac{2h}{\sqrt{h^2-d_c^2}}\arctan\sqrt{\frac{h+d_c}{h-d_c}} - \frac{\pi}{2}\right)^{-1} \quad (2)$$

The longitudinal resistance R_4 between channels can be calculated by the following formula:

$$R_4 = \rho L\left(d_c h - 0.25\pi d_c^2\right) \quad (3)$$

In the formula, ρ is the resistivity, which is substituted with the parameters of microchannel board. $R_3 \approx 8.25\rho\Omega$ and $R_4 \approx 3.2\times10^5\rho\Omega$ can be obtained. The unit of ρ is $\Omega\cdot\text{cm}$. From the analysis of the structure of microchannel plate, each channel is resistive shunt with resistance value $2R_4$. Assuming that the shunt current does not exceed 3%[15]of the channel current, $2R_4 \geq 20R_2$ is required. Substituting (3), you can get: $\rho \geq 2.5\times10^{11}\Omega\cdot\text{cm}, R_3 \geq 2\times10^{12}\Omega$.

Figure 6 illustrates the relationship between channel gain and resistance R_3 in the case of two channels. The two groups of curves in the figure are weak saturation state $I_{in}=2J_{in}=2\times10^{-16}$A and strong saturation state $I_{in}=2J_{in}=2\times10^{-16}$A respectively. It can be seen that when $R_3 \geq 2\times10^{17}\Omega$, the gain of channels is independent of R_3, and there is no coupling between channels. As the coupling resistance decreases, the coupling between channels becomes stronger and stronger. The gain of large

IOP Conf. Series: Materials Science and Engineering **711** (2020) 012091 doi:10.1088/1757-899X/711/1/012091

signal channel is increased while that of small signal channel is decreased, and the gain of two channels is closer.

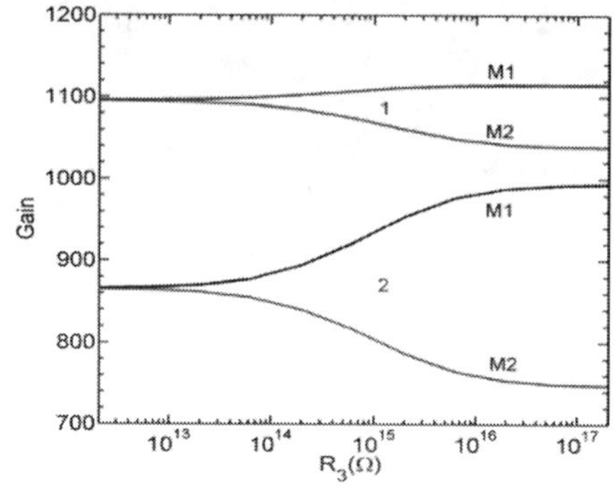

Figure 6 The relationship between channel gain and coupling resistance, $1—I_{in}=2J_{in}=2\times10^{-17}A$, $2—I_{in}=2J_{in}=10^{-16}A$

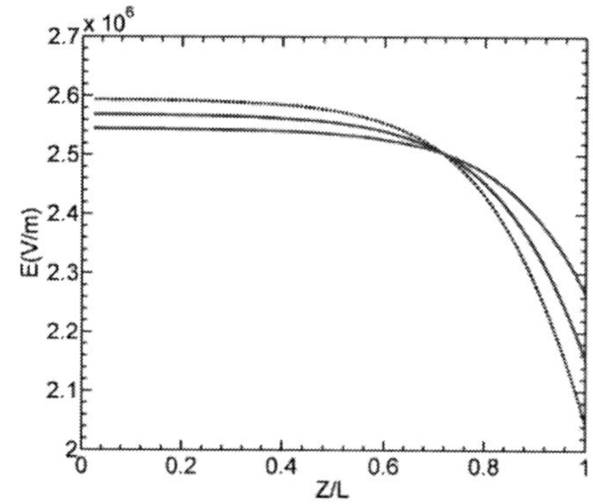

Figure 7 Electric field distribution in the channel: solid line- $R_3=2\times10^{17}\Omega$, dotted line- $R_3=2\times10^{12}\Omega$

Fig. 7 shows the electric field distribution in uncoupled ($R_3=2\times10^{17}\Omega$) and strongly coupled ($R_3=2\times10^{12}\Omega$) channels when $I_{in}=2J_{in}=10^{-16}A$. In the case of strong coupling, the maximum difference between the electric fields of the two channels is less than 3% because the longitudinal resistance r_2 in the channel is connected by the virtual resistance r_3, which indicates that the potential and electric field distribution in the two channels tend to be the same in the case of coupling.

The above results show that the transverse resistance between channels has a great influence on the channel gain.

6. Conclusion

Starting from the experimental phenomena, the ultraviolet imager is irradiated by continuous laser. The imaging characteristics of ultraviolet imager before and after irradiation are compared and analyzed. The gain saturation phenomenon of ultraviolet imager and the mechanism of channel interference are studied.

Channel interference results from the electrical coupling between channels of microchannel plate while ultraviolet imager irradiated by CW laser. The electrical coupling between channels helps to reduce the gain difference between channels and the effect of gain saturation on image quality. The coupling resistance improves the image quality obviously. Within the permissible range, the distributing resistance between MCP channels can be minimized, which is beneficial to improve the contrast transfer ability of microchannel plates.

References

[1]Yoshikawa I, Yamazaki A, Murakami G, et al. Telescope of extreme ultraviolet (TEX) onboard SELENE: science from the Moon[J]. Earth, planets and space, 2008, 60(4): 407-416.

[2]Fang Chen-Yan. Research on the Ultraviolet Features and Multi-Spectral Imaging Technology of High Voltage Air Discharge[D].University of Chinese Academy of Sciences

[3]Song Shan-Shan. Design of Solar Blind Ultraviolet warning optical system[D]. ChangCun University of Science and Technology, 2014. (in Chinese)

[4]Wang Hong. Optical System Study and Design Used for the Ultraviolet Warning Camera With Ultra-wide Field[J]. ACTA PHOTONICA SINICA,2014.(in Chinese)

[5]Liao Z M, Nostrand M, Carr W, et al. Modeling of laser-induced damage and optic usage at the National Ignition Facility[C]//Pacific Rim Laser Damage 2016: Optical Materials for High Power Lasers. International Society for Optics and Photonics, 2016: 998304-998304-6.

[6]Dimmeler A, Eberle B, van den Heuvel J C, et al. Laser dazzling of focal plane array cameras[C]//Optics/Photonics in Security and Defence. International Society for Optics and Photonics, 2007: 67380O-67380O-10.

[7]Titterton D H. Application of laser technology to optical countermeasures[J]. The Imaging Science Journal, 2010, 58(5): 286-294.

[8]Durécu A, Bourdon P, Fleury D, et al. Infrared laser irradiation breadboard: dazzling sensitivity analysis of a HgCdTe focal plane array[C]//SPIE Security+ Defence. International Society for Optics and Photonics, 2011: 81870K-81870K-10.

[9]Zhang Z, Cheng X, Wang R, et al. Dazzling effect of repetitive short pulse laser on TDI CCD camera[J]. Optics and Lasers in Engineering, 2011, 49(2): 292-296.

[10]Zhang Z, Cheng X, Jiang T, et al. A dazzling phenomenon of CW laser on linear CCD camera[J]. Optik-International Journal for Light and Electron Optics, 2012, 123(3): 223-227.

[11]KANG Wen-Yun,WEI Jun, SONG Xiao-Quan,WANG Yu. Jamming of laser to push-broom camera[J]. J. Infrared Millim. Waves, 2014, 33(6): 608-612.(in Chinese)

[12]Sheng Liang, Zhang Zhen, Zhang Jian-Min, Zuo Haoyi. Pixel upset effect and mechanism of CW laser irradiated CMOS camera[J]. Infrared and Laser Engineering, 2016, 45(6): 606004-0606004 (4). (in Chinese)

[13]Giudicotti L. Time dependent model of gain saturation in microchannel plates and channel electron multipliers[J]. Nuclear Instruments & Methods in Physics Research, 2011, 659(1):336-347.

[14]Charles K. Alexander and Matthew N. O. Sadiku.Fundamentals of Electric Circuits, 4th edition[M]. McGraw Hill, 2009

[15]Berkin A B, Vasil'Ev V V. How the interaction of the channels of a microchannel plate affects the image, contrast[J]. Journal of Optical Technology C/c of Opticheskii Zhurnal, 2008, 75(5):338-340.

System Design Of An Infrared Dual-band Seeker

Yan-Kai Mao, Zhen-Yu Zhao

Luoyang Institute of Electro-optical Equipment, AVIC, Luoyang 471009, China

myk2012cmy@163.com, chenmengyang1987@163.com

Abstract: Infrared imagine seeker is an important development direction of ground-to-groung ballistic missle terminal guidance technology,but it's derection probability is vulnerable to target/background characteristic、detector performance、field of view and other factors. In order to improve the ability of infrared seeker to adapt to complex battlefield environment,a new design principle of infrared dual-band seeker is introduced. The whole system adopts dual band design,and uses image fusion technology to enrich the information difference between target and background features. In addition,the design of large and small dual field of view can further improve the target interception and accurate recognition and tracking ability of seeker. The working principle of infrared seeker is analyzed,and the optical design idea and image fusion method are introduced.Through index analysis and calculation,the infrared seeker meets the system requirements,which can provide reference for future ballistic missle seeker design

1. Introduction

The key targets in the war are high-level command headquarters,airports,power plants,bridges and other strategic objectives. It is difficult for general subsonic weapons to attack and destroy them effectively. Ground-to-ground ballistic missiles with penetration capability have the function of attacking these strategic targets. It has the characteristics of high-speed, long range, maneuverable orbit change and difficult interception of existing weapons. It carries enough warheads to destroy aircraft carriers and deep underground shelters. According to the need of precise attack, the development trend of ground-to-ground ballistic missile is to install precise terminal guidance seeker on the basis of orginal inertial guidance,and to achieve precise hit by map matching or terrain matching system. Infrared imaging terminal guidance has become an important development direction of terminal guidance technology because of its high sensitivity, high spatial resolution,strong anti-jamming ability and quasi-all-weather characteristics.

With the development of modern war, jamming and anti-jamming technology is progressing in mutual confrontation. The battlefield environment faced by various tactical missiles is becoming more and more complex, and the confrontation is becoming more and more fierce. The operational effectiveness of single-band infrared imaging guidance weapon will be weakened day by day due to the effective information obtained. It will not be able to strike accurately in future wars, which also forces people to seek further development of infrared imaging guidance technology.

At present, there are many kinds of tactical missile equipments using infrared imaging guidance at home and abroad. Some of them adopt dual-band or dual-field of view design. They can distinguish targets and interference by using the characteristics of energy, shape, trajectory and spectrum, which greatly improves the detection ability and has strong anti-infrared decoy ability. For example, France's MICA-IR and South Africa's A-Darter use two-color line detectors to form a scanning imaging system. The US Navy's standard-3 interceptor uses a dual-band staring infrared detector, The Chaparral missile developed by Lall Company in the United States uses a long-wave detector with two wide and narrow field of view. Norway's new generation anti-ship missiles NSM uses two detectors, medium wave and long wave, and has two wide and narrow field of view. These seekers are more advanced in technology. They can adopt dual-band or dual-field of view for different application requirements, further enhance their ability to resist artificial and complex background interference, and improve target detection, interception and tracking performance. They are one of the development directions of infrared imaging guidance.

The infrared dual-band/dual-field-of-view seeker designed in this paper can detect and acquire scene information distribution in long-distance using large field of view after dual-band image fusion, and accurately recognize and track targets in short-distance using small field of view. The information between the two seekers works asynchronously in series. The results of long-range and large-field-of-view infrared imaging can support the follow-up recognition and tracking of small-field-of-view infrared targets, and improve the anti-jamming ability and confidence of infrared targets.

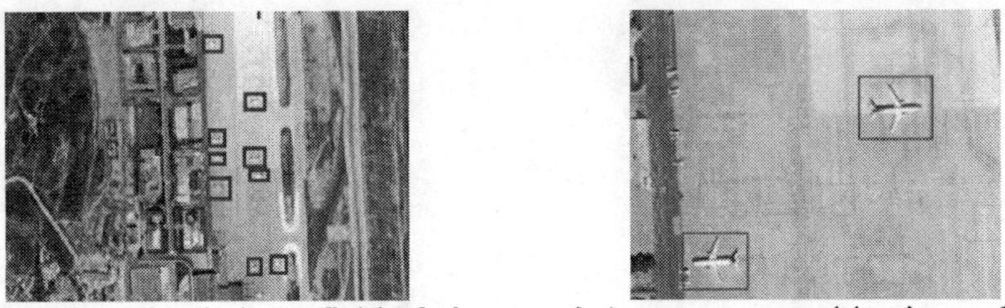

a Target detection in large field of view b Accurate recognition in small field of view

Fig.1 Aircraft target imaging diagram

2. System design

2.1 Summary of principles

Both medium-wave detector and long-wave detector adopt 640× 512 focal plane refrigeration detector. After photoelectric conversion of target infrared energy, the detector outputs an electrical signal containing target information. The detector processing unit amplifies, filters, processes the frame and converts the signal to analog and digital. Finally, two infrared digital image signals are output. Target information processing module fuses two images to complete image signal selection decision, feature extraction, intelligent recognition, judgment and tracking processing. The central processor can drive the field of view switching mechanism to complete the magnification switching between large and small field of view. Finally, the central processor acts as the main control device of the seeker, receives commands and data from the missile-borne computer, and reports the target deviation in real time.

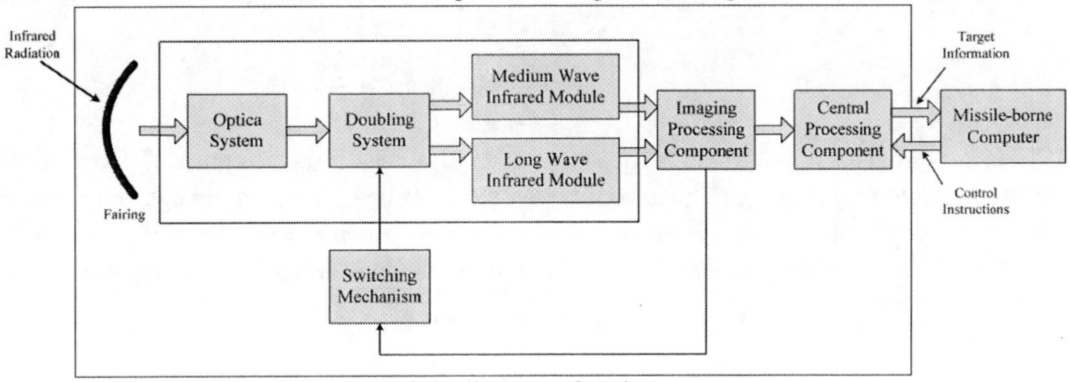

Fig.2 Principle diagram of seeker system

2.2 Optical design ideas

Infrared dual-band seeker is strapdown structure, using common-light path design, front-end sharing objective lens, back-end using prism splitting, long-wave transmission, medium-wave reflection, the two bands are separated, back-end for medium-wave, long-wave respectively optimized sub-light path, each imaging in the corresponding band detector. In order to achieve fast switching between large and small field of view, optical element switching and doubling method is adopted, which cuts optical elements into and out of the optical system to change the focal length field of the optical system. This method has fast speed and requires little axial space.

In order to ensure high imaging quality, the refrigeration optical system hopes to achieve 100% Cold-shield matching. In the optical design process, it is necessary to set the system diaphragm at the detector cold-shield. If the aperture diaphragm is far away from the front lens, the radial size of the front lens of the optical system will be enlarged more; if the cold aperture matching is satisfied, the optical aperture can be effectively reduced by using the second imaging method, which is conducive to the miniaturization of the system. The optical structure of the second imaging is shown below.

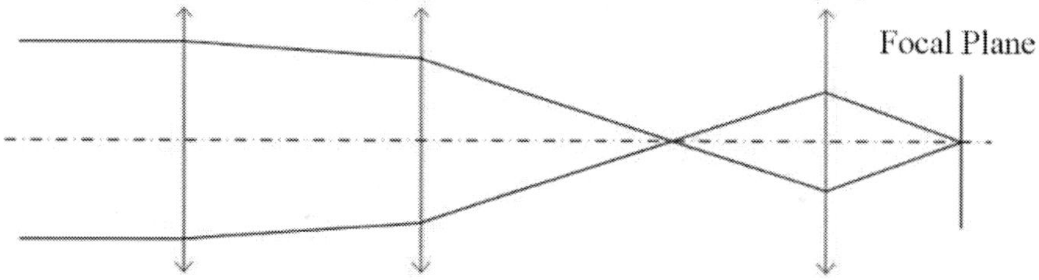

Fig.3 Secondary imaging optical structure

Infrared optical system has fewer kinds of materials available, and the performance of different bands varies greatly, which makes the system chromatic aberration correction difficult to design. Because the temperature coefficient of refractive index of infrared optical material is large, the change of curvature and thickness of infrared system caused by temperature change will cause the image plane of optical system to drift, and the focal plane of detector and optical system will no longer coincide, which will seriously affect the performance of the system.

In order to meet the requirements of high and low temperature environment, the system adopts optical passive non-thermal design. By choosing lens material reasonably, distributing light intensity, and making use of mutual compensation between lens material and lens mechanical structure material, the dual-band infrared optical system achieves the purpose of simultaneous thermal and achromatic. The material parameters of the lens should satisfy the requirements of both light focus, achromatism and calorific aberration.

$$\frac{1}{h_1\phi}\left(\sum_{i=1}^{n} h_i\phi_i\right)=1 \tag{1}$$

$$\left(\frac{1}{h_1\phi}\right)^2\sum_{i=1}^{n} h_i^2(\phi_i C_i)=0 \tag{2}$$

$$-\frac{1}{h_1^2\phi}\sum_{i=1}^{n} h_i^2\phi_i T_i=\alpha \tag{3}$$

In the formula, ϕ is the total optical focus of the system; ϕ_i is the optical focus of the ith lens; h_1 is the incident height of the first paraxial ray on the ith lens; C_i is the chromatic aberration coefficient of the ith lens, which is the reciprocal of Abbe number; T_i is the thermal difference coefficient of the ith lens; α is the thermal expansion coefficient of the barrel material. T_i is defined as:

$$T_i=\frac{1}{\phi_i}\frac{d\phi_i}{dt}=-\frac{dn_i/dt}{n_i-1}-\alpha_i \tag{4}$$

In the formula, n_i is the refractive index of the ith lens material, dn_i/dt is the index of change of refractive index with temperature, and α_i is the thermal expansion coefficient of the lens material.

The machinability, chemical stability and coating properties of optical materials were comprehensively analyzed. Finally, germanium, zinc selenide and zinc sulfide were selected as dual-band optical materials. The working wavelength of the long-wave detector is 7.7-10.3 um, the area array is 640×512, the pixel size is 15 um, the working wavelength of the medium-wave detector is 3.7-4.8 um, the area array is 640×512, and the pixel size is 15 um. Using CODE V software to optimize the design, the system F number is 2, in which the large field of view is 25° ×20°, the focal length is 22 mm, and the aperture is 11 mm; the small field of view is 10 °×8°, the focal length is 55 mm, and the aperture is 27.5 mm. The design results have high optical transmittance and transfer function, which meet the requirements of the system.

2.3 Image fusion method

Image fusion should be the sum of the common features and the unique features of two bands. According to the different fusion levels, image fusion can usually be divided into three levels: pixel level fusion, feature level fusion and decision level fusion . This system chooses to use pixel level fusion. The image fusion method proposed by LIS [1] and others is to decompose the image into detail layer and basic layer, respectively, to guide the filtering of detail layer and basic layer. The fusion method retains the complementary information of multiple source images [2]. The flow chart of image fusion algorithm based on guided filtering is shown in the following figure.

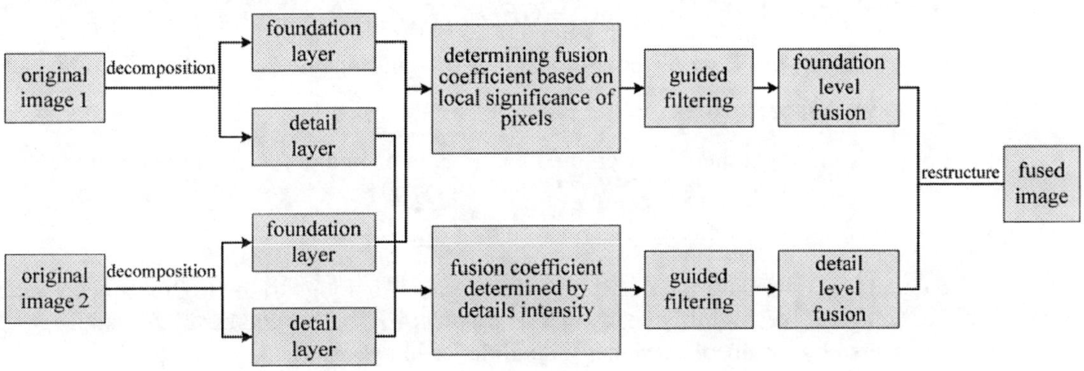

Fig.4 Flow chart of image fusion algorithm

2.3.1 Image decomposition

Suppose there are some images to be fused,such as $I_n(n=1,2,...,N)$, N is the number of images to be fused. The basic layer image is obtained by means of mean filtering for each image.

$$B_n = I_n * A \tag{5}$$

In the formula, A is a mean filter and $*$ is a convolution operation.

Further, the detail layer image is obtained by the difference between the original image and the base layer image.

$$D_n = I_n - B_n \tag{6}$$

In this way, through image decomposition, it is easy to obtain the basic layer and detail layer of each image. The basic layer expresses the general picture of the image and the gray level change on a larger scale, while the detail layer contains the details on a smaller scale.

2.3.2 Fusion coefficient

The fusion coefficients of the basic layer and the detail layer are constructed respectively. Firstly, the construction of fusion coefficient of foundation layer is introduced. The high-pass image is obtained by filtering the source image by Laplace operator[3].

$$H_n = I_n * L \tag{7}$$

L is 3 ×3 Laplacian operator:

$$L = \begin{bmatrix} -1 & -1 & -1 \\ -1 & 8 & -1 \\ -1 & -1 & -1 \end{bmatrix} \tag{8}$$

The local saliency of the pixels is represented by the absolute value of the high-pass image:

$$S_n^B = |H_n| \tag{9}$$

By comparing the saliency maps, the weighted mapping of the underlying image is obtained.

$$P_n^B = \begin{cases} 1 & if\ S_n^B = \max\limits_{n=1,\ldots,N} S_n^B \\ 0 & otherwise \end{cases} \tag{10}$$

Obviously, the fusion image formed by the weighted mapping will produce sawtooth and gray jump at the junction, which makes it difficult to maintain the spatial structure of the image and ensure the smooth gray transition of the image. Therefore, using the characteristics of guided filtering, the source image I_n is used as the guided image, and the guided image P_n^B is filtered:

$$w_n^B = GF(P_n^B, I_n, r_B, \varepsilon_B) \tag{11}$$

GF is for guiding the filtering operation, r is for guiding the size of the filtering window and ε is the regularization parameter.

w_n^B is normalized to get the final weighted parameters of image fusion at the basic level:

$$W_n^B = w_n^B / \sum_{n=1}^{N} w_n^B \tag{12}$$

For the detail layer, because the detail layer image itself belongs to the high frequency information of the source image, it has high frequency characteristics[4]. The local saliency of detail layer pixels is directly represented by the absolute value of detail layer image itself.

$$S_n^D = |D_n| \tag{13}$$

Similarly, weighted mappings are obtained and guided filtering and normalization are performed:

$$P_n^D = \begin{cases} 1 & if\ S_n^D = \max\limits_{n=1,\ldots,N} S_n^D \\ 0 & otherwise \end{cases} \tag{14}$$

$$w_n^D = GF(P_n^D, I_n, r_D, \varepsilon_D) \tag{15}$$

$$W_n^D = w_n^D / \sum_{n=1}^{N} w_n^D \tag{16}$$

2.3.3 Image fusion

According to the fusion coefficients obtained above, image fusion is performed at the basic level and the detail level respectively.

$$F_B = \sum_{n=1}^{N} W_n^B * B_n \tag{17}$$

$$F_D = \sum_{n=1}^{N} W_n^D * D_n \tag{18}$$

In the formula, * is a Point multiplication.

Then, the image of the base layer and the detail layer are added together to get the final fusion result: $F = F_B + F_D$。

The simulation results of dual-band image fusion are shown in the following figure. It can be seen that the fused image has important details of two original images, so it is more advantageous for target interception.

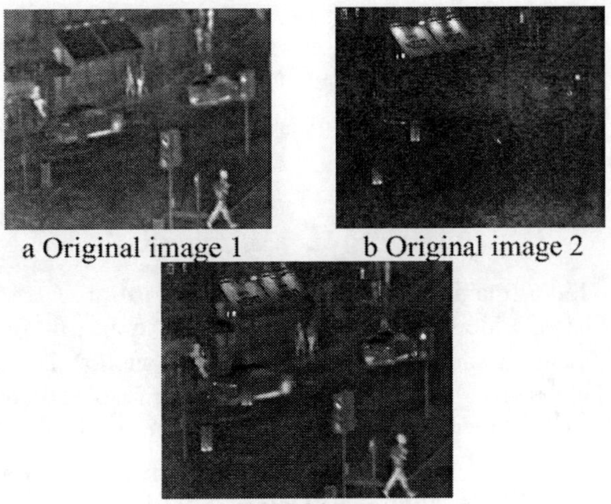

a Original image 1 b Original image 2

c Fused image

Fig.5 Simulation effect of image fusion

3. Analysis of design indicators

3.1 Spatial Resolution Analysis

The imaging schematic diagram of Strapdown infrared seeker is shown in the following figure. Its imaging model is perspective imaging model, so the resolution of pixels corresponding to different positions in the image plane is also different. For large surface ships and other targets, they can be approximated as plane targets at higher altitudes, and the size of the target in the image plane from different viewing angles is also different. The imaging field of view and resolution are analyzed separately below.

The sea surface area corresponding to the field of view is calculated as follows:

$$L = 2H \cdot [\tan(\omega_h / 2)] / \sin\varphi \tag{19}$$

$$W = H \cdot [\tan(\varphi_1) - \tan(\varphi_0)] \tag{20}$$

φ is the angle between the main optical axis of the imaging sensor and the horizontal direction, ω_h and ω_v represent the horizontal and vertical field of view, respectively. $\varphi_0 = \pi / 2 - \varphi - \omega_v / 2$, $\varphi_1 = \varphi_0 + \omega_v$。

Average pixel resolution in horizontal and vertical directions is: $\langle res_h \rangle = L / S_h, \langle res_v \rangle = W / S_v$。 $S_h \times S_v$ is image size.。

Vertical pixel resolution is:

$$res_v(m) = H\Delta\omega_v / \cos^2(\varphi_0 + m\Delta\omega_v) \tag{21}$$

Horizontal pixel resolution is:

$$res_h(n) = H\Delta\omega_h / \cos[(S_h / 2 - n)\Delta\omega_h] \tag{22}$$

IOP Conf. Series: Materials Science and Engineering **711** (2020) 012092 doi:10.1088/1757-899X/711/1/012092

Among them, $m = 1,2...S_v, n = 1,2...S_h, \Delta\omega_v = \omega_v / S_v, \Delta\omega_h = \omega_h / S_h$。

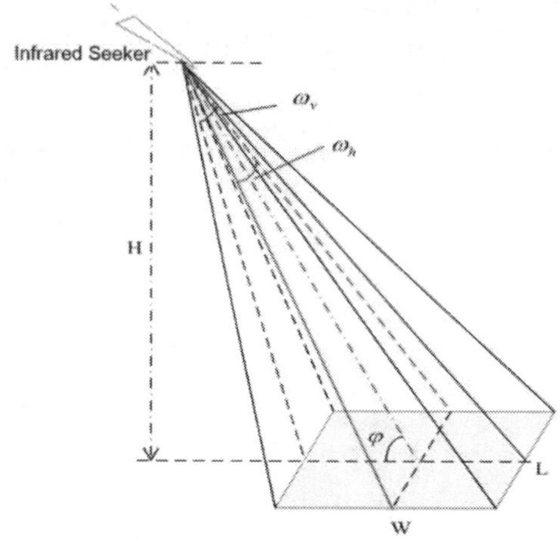

Fig.6 Imaging schematic diagram of infrared seeker

The infrared detector has a wide field of view of 25°×20°, a narrow field of view of 10°×8°and an image size of 640×512. For the typical time-sensitive targets such as large ships with 320m×70m and parked aircraft with 60m×15m size, the field of view, imaging resolution and image targets at different altitudes from 8km to 1km are calculated according to the above formulas. The number of pixels is shown in the table below.

Tab.1 Target pixel number in large field of view（$\varphi = -90°$）

Imaging Height H/m	Imaging Area (L, W)/m	Resolution Power (res)(res)	Large Ships (320m×70m) /pixel	Parking aircraft (60m×15m) /pixel
8000	(3547, 2821)	(5.5, 5.5)	(58, 12)	(10, 2)
7000	(3103, 2468)	(4.8, 4.8)	(66, 14)	(12, 3)
6000	(2660, 2115)	(4.1, 4.1)	(77, 16)	(14, 3.6)
5000	(2216, 1763)	(3.5, 3.5)	(92, 20)	(17, 4)
4000	(1773, 1410)	(2.7, 2.7)	(116, 25)	(21, 5)
3000	(1330, 1057)	(2.0, 2.0)	(154, 33)	(29, 7)
2000	(886, 705)	(1.4, 1.4)	(228, 50)	(42, 10)
1000	(443, 352)	(0.7, 0.7)	(457, 100)	(85, 21)

Tab.2 Target pixel number in small field of view（$\varphi = -90°$）

Imaging Height H/m	Imaging Area (L, W)/m	Resolution Power (res)(res)	Large Ships (320m×70m) /pixel	Parking aircraft (60m×15m) /pixel
8000	(1399, 1118)	(2.2, 2.2)	(146, 32)	(27, 6)
7000	(1224, 978)	(1.9, 1.9)	(167, 36)	(31, 7)
6000	(1049, 839)	(1.6, 1.6)	(195, 42)	(36, 9)
5000	(874, 699)	(1.3, 1.3)	(246, 53)	(46, 11)
4000	(699, 559)	(1.1, 1.1)	(290, 63)	(54, 13)
3000	(524, 419)	(0.8, 0.8)	(400, 87)	(75, 18)
2000	(349, 279)	(0.5, 0.5)	(640, 140)	(120, 30)
1000	(174, 139)	(0.27, 0.27)	(1185, 259)	(222, 55)

From the above results, it can be seen that the imaging field of view is proportional to the size of the image, the imaging resolution of the line direction is basically the same, and the resolution of each pixel does not change much. However, in the case of non-top-view attack, the imaging resolution of each pixel changes, especially along the view. The resolution of the line changes the most in the imaging direction. For a large ship, it can be considered as a planar target, and its outline in the image is an affine change. Therefore, for the follow-up detection and recognition processing, on the one hand, affine invariant statistical features need to be used. On the other hand, in the process of accurate target recognition and location, the calculation of target scale parameters also needs to be revised with the help of imaging line-of-sight angle, height and other parameters.

According to the number of pixels corresponding to the target at different heights in the table above, large ships can meet the recognition requirements in wide and narrow field of view, and parked aircraft can also meet the requirements in narrow field of view. When the target is too small or is disturbed by cloud cover, large landmarks can be selected near the target to add templates to ensure that it reaches the target. To meet the recognition requirements, 5 km operating distance requirements.

3.2 Temperature Sensitivity Analysis

For a target whose spatial frequency is f, the actual equivalent temperature difference between the target and the background is greater than or equal to the minimum resolvable temperature difference MRTD (f) of the system when it reaches the thermal imaging system through atmospheric attenuation. At the same time, the angle of the target to the system should be greater than or equal to the minimum angle of view required by the observation level. The verification of temperature sensitivity of infrared system can be calculated according to the following formula[5]:

$$\Delta T = \Delta T_e \bullet \tau_a(R) \geq MRTD \tag{23}$$

Among them, ΔT is the apparent temperature difference between target and background after atmospheric attenuation; ΔT_e is the actual equivalent temperature difference between target and background, taking 3K; $\tau_a(R)$ is the average atmospheric transmittance at R distance.

We chose mid-latitude summer with more stringent conditions, when the target distance is 5 km and the visibility is not more than 5 km and the relative humidity is not higher than 70%. Through the Modtran software, we can calculated the average atmospheric transmittance in mid-wave band is 0.473,ΔT =3K×0.473=1.419K； and the long-wave band is 0.638, ΔT =3K×0.638 =1.914K。

Considering the relevant performance parameters of optical system, target temperature and emissivity, background temperature and emissivity, and then calculating the equivalent temperature difference, the temperature resolution can be determined[6].

$$MRTD(f_T) = \frac{3NETD}{(\Delta f)^{1/2}} \frac{f_T}{MTF_S} \left(\frac{\alpha\beta}{\tau_d T_e F_n} \right)^{1/2} \tag{24}$$

In the formula: $NETD$ is equivalent noise temperature difference; Δf is bandwidth; f_T is target space (angular) frequency, its unit is cyc/mrad.

MTF_S is the total modulation transfer function of the system, with the medium wave 0.65, the long-wave short focal length 0.61 and the long-wavelength focal length 0.55; α and β are spatial resolution (mrad) in azimuth and pitch direction, the medium wave is 0.682 mrad, the long-wave short focal length is 0.622 mrad, the long-wavelength focal length is 0.273 mrad; τ_d is the detector integration time (s); T_e is the human eye integration time (s); F_n is Frame frequency.

After calculation, the MRTD of the system is 62 mk in the medium-wave infrared wide field of view, less than ΔT = 1.419K, 66 mk in the long-wave infrared wide field of view, 73 mk in the long-wave infrared narrow field of view, all less than ΔT = 1.914K. Considering a certain design margin, the system still meets the temperature sensitivity requirements.

4. Conclusion

Aiming at the limitation of single-band seeker for ballistic missile, a new Infrared Dual-band dual-field-of-view seeker is proposed in this paper. The difference between target and background information can be enhanced by means of dual-band image fusion and dual field of view switching, and the target interception and accurate recognition and tracking ability of seeker can be effectively improved. According to the working principle of the seeker, the optical design idea and image fusion method are introduced, and the Infrared Dual-band seeker imaging system which meets the requirements of the target is designed. The feasibility of the principle is proved, which has a certain reference value for the future development of new ballistic missile seeker.

Acknowledgements

This research was financially supported by"135"pre-research project of General equipment department (Grant No.41415020105).

References

[1] LI S,KANG X,HU J. Image fusion with guided filtering[J]. IEEE Transactions on Image Processing: a Publication of the IEEE Signal Processing Society, 2013,22(7):2864-2875.
[2] Haitao Yin,Shutao Li. Multimodal image fusion with joint sparsity mode[J].Optical Engineering, 2011,50(6):1301-1305.
[3] Ping Zhang,Yuchen Yuan,Chun Fei,Tian Pu,Shuhang Wang. Infrared and visible image fusion using co-occurrence filter[J]. Infrared Physics and Technology . 2018.
[4] Infrared and visible image fusion via gradient transfer and total variation minimization[J] . Jiayi Ma,Chen Chen,Chang Li,Jun Huang. Information Fusion . 2016
[5] Yang Zhao,,Minyao Mao. Optomechanical Uncooled Infrared Imaging System: Design, Microfabrication and Performance. Journal of Microbiology . 2002
[6] Eric J. Jumper,Edward J. Fitzgerald. Recent advances in aero-optics[J] . Progress in Aerospace Sciences . 2001 (3).

IOP Conf. Series: Materials Science and Engineering **711** (2020) 012093 doi:10.1088/1757-899X/711/1/012093

Low Noise VCSEL Driver for SERF Atomic Magnetometer

Feihu Wang[1,2], Zhaohui Hu[1,2,3,4,a] and Xin Liu[1,2]

[1] School of Instrumentation and Optoelectronic Engineering, Beihang University, Beijing 100191, China

[2] Science and Technology on Inertial Laboratory, Beihang University, Beijing 100191, China

[3] Research Institute of Frontier Science, Beihang University, Beijing 100191, China

[4] Beijing Advanced Innovation Center for Big Data-Based Precision Medicine, Beihang University, Beijing 100191, China

[a] huzh@buaa.edu.cn

Abstract. A new type of VCSEL (Vertical-Cavity Surface-Emitting Laser) which uses Pt1000 as a temperature sensitive component and non-magnetic heating membrane as a temperature actuator is designed for SERF (Spin-Exchange Relation-Free) atomic magnetometer, which can decrease the magnetic interference produced while heating or cooling VCSEL. Therefore a low noise driver for this kind of VCSEL is necessary. A driver for the novel VCSEL is designed in this paper. The output current is adjustable from 0 to 4.5 mA, and RMS (Root Mean Square) noise of current is less than 80 nA. The operating temperature of the VCSEL is adjustable between 30-110 °C and RMS noise of temperature is less than 0.6 mK. So that it can be ensured that the VCSEL outputs a laser with stable optical power and wavelength.

1. Introduction

High sensitivity magnetometry is demanded in many fields, such as biology, geology, and medical science[1]. Traditionally those applications is enabled by superconducting quantum interference device (SQUID) magnetometer[2, 3]. But in the recent years SERF (Spin-Exchange Relation-Free) magnetometer has been demonstrated and achieved a magnetic field sensitivity of $0.16\mathrm{fT}/\sqrt{\mathrm{Hz}}$ [1]. SERF magnetometer also opens the possibility of detecting magnetoencephalogram (MEG) and magnetocardiogram (MCG) with array of ultrahigh sensitivity magnetometers. In the operating of SERF magnetometer, laser is required to optically pump the gaseous alkali-metal atom in the cell and detect its polarization. VCSEL (Vertical-Cavity Surface-Emitting Laser) is often used to play the role. But the noise of wavelength and optical power can highly reduce the sensitivity of SERF magnetometer, which requires a low noise VCSEL driver.

VCSEL driver often contains two parts, stable current driver and temperature control system. For the current driver, there have been a lot of commercial products, such as LD1255R of Thorlabs company which has a ripple current smaller than 1 μA, Newport company's LDX-3210 with 2 μA, and Wavelength company's MPL-250 with 1 μA. For the temperature control system, commercial product can also be found. For example, Thorlabs company's TED200C temperature control system can achieve the temperature stability of 2 mK, so does Wavelength company's LFI-3751, Newport company's LDT-5910C-120V can even achieve 1 mK. Those advanced commercial products is stable enough for

SERF magnetometer, however it is hard to integrate these commercial products into compact SERF magnetometer control system because of their big sizes and incompatibility. And TEC (Thermoelectric Cooler) is often used to cool and heat in the temperature control system which can generate extra magnetic interference because of its large current. The extra magnetic interference will make it hard to keep the high sensitivity of SERF magnetometer. So it is more proper to heat VCSEL with the special non-magnetic heating membrane for SERF magnetometer.

In this paper we present a low noise VCSEL driver, which also contains current driver and temperature control system. For the current driver, the current can be tuned between 0 mA and 4.5 mA which can satisfy most VCSEL, and RMS noise of current is no more than 80 nA. For temperature control system, the non-magnetic heating membrane is used to heat the new VCSEL, temperature can change between 30°C and 110°C, and RMS noise of temperature is no more than 0.6 mK. Besides its interface for FPGA can easily be transplant into the control system of SERF magnetometer and magnetic interference produced by VCSEL can be highly reduced. The theory of VCSEL including the relation between current, operating temperature, wavelength and optical power of VCSEL will be discussed. Then we will describe the design of VCSEL driver, key technology and how it works. In the end experiments will be set up to check the performance of VCSEL driver and measure the stability of current and operating temperature.

2. Theory

The wavelength and optical power of VCSEL beam is highly affected by current flowing through VCSEL. Operating temperature of VCSEL can change the concentration of electron in the optical resonant cavity, the feature of optical resonant cavity and the threshold current, which in the end will also affect the wavelength and optical power of VCSEL. So it is feasible to achieve low-noise optical beam by making the current and operating temperature stable enough. In the following parts of this section, we will discuss the relation between wavelength, optical power, current and operating temperature.

The relation between current and optical power features like Fig. 1. When the current is less than threshold current, the optical power of VCSEL slowly increases with the current, which is mainly caused by spontaneous radiation. When the current is larger than threshold current, the optical power P of VCSEL has a linear correlation with current I, which is shown in Eq.1,

$$P = \eta_{\mathrm{d}} \frac{h\upsilon}{q}(I - I_{\mathrm{th}}),\ (I > I_{\mathrm{th}}). \tag{1}$$

Where η_{d} is external differential quantum efficiency of VCSEL; h is Planck constant; υ is the frequency of laser and q is unit charge.

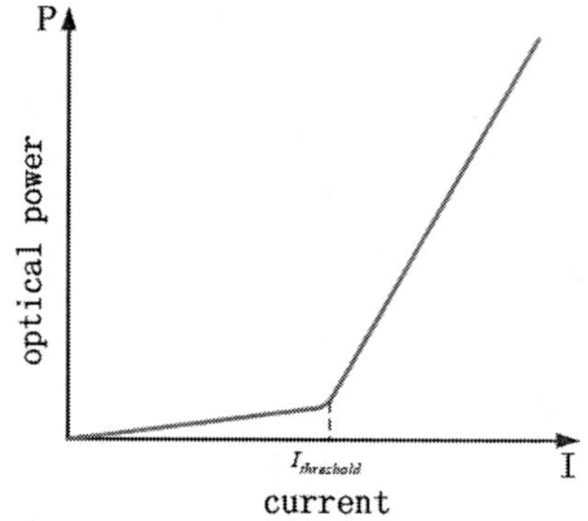

Fig. 1 The correlation between current and optical power

With the increasing of operating temperature, external differential quantum efficiency η_d decreases and threshold current get lager, which makes optical power of VCSEL smaller. So the optical power has a negative correlation with operating temperature.

Operating temperature can also affect the wavelength of VCSEL, higher temperature will make the wavelength larger. During the operation of VCSEL, more Joule heat will be produced with larger current, which equals to the increasing of operating temperature. Therefore larger current will move wavelength to the 'red side' (larger wavelength).

In most datasheets of VCSEL, the relation between current, operating temperature and wavelength is simplified to wavelength temperature coefficient and wavelength current coefficient. The relation between current, operating temperature and optical power is also approximately linear and it can be simplified to optical power temperature coefficient and optical power current coefficient which can be obtained through experiments. All the coefficients of VIXAR's 795 nm VCSEL is shown in Table 1.

Table 1 All coefficient of VIXAR's 795 nm VCSEL

Parameter	Typical	Units
Wavelength Temperature Coefficient	0.055(from datasheet)	nm/K
Wavelength Current Coefficient	0.3(from datasheet)	nm/mA
Optical Power Temperature Coefficient	-4.1×10^{-3} (from experiment)	mW/K
Optical Power Current Coefficient	0.324(from experiment)	mW/mA

3. Design of VCSEL Driver

VCSEL driver contains host computer system, FPGA software and hardware circuit. Figure 2 shows the block diagram of VCSEL driver. Host computer system is used to control the operating state of VCSEL such as current size, operating temperature and PID parameters. The setting command from host computer will be sent to FPGA, in which commands is transformed into electrical control signal. After receiving the electrical control signal from FPGA, hardware circuit generates the current for VCSEL and keeps the operating temperature which is set from host computer system.

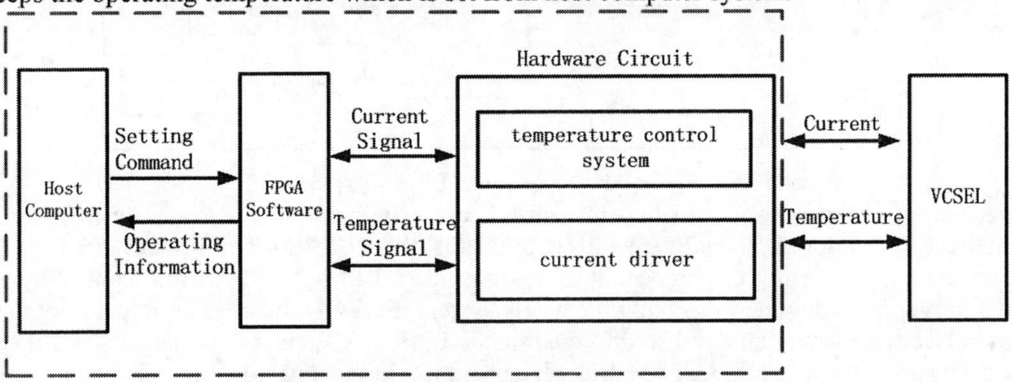

Fig. 2 Block diagram of VCSEL driver

3.1 Host Computer System

Host computer system is designed for user to control the operating state of VCSEL easily, in which current, operating temperature, sample frequency and PID parameters are set. And it is easy to monitor the actual operating state of VCSEL. Figure 3 shows the interface of host computer system.

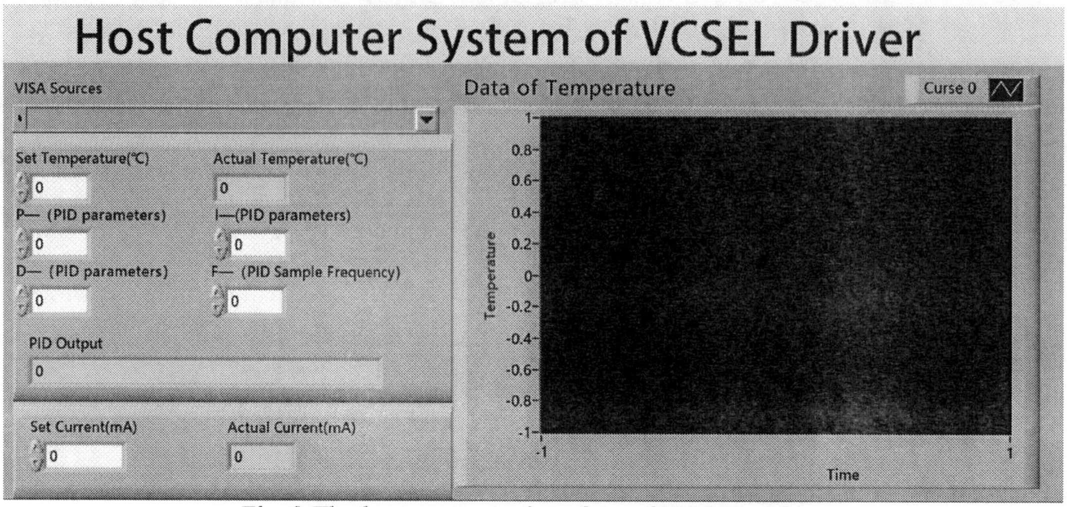

Fig. 3 The host computer interface of VCSEL driver

3.2 FPGA

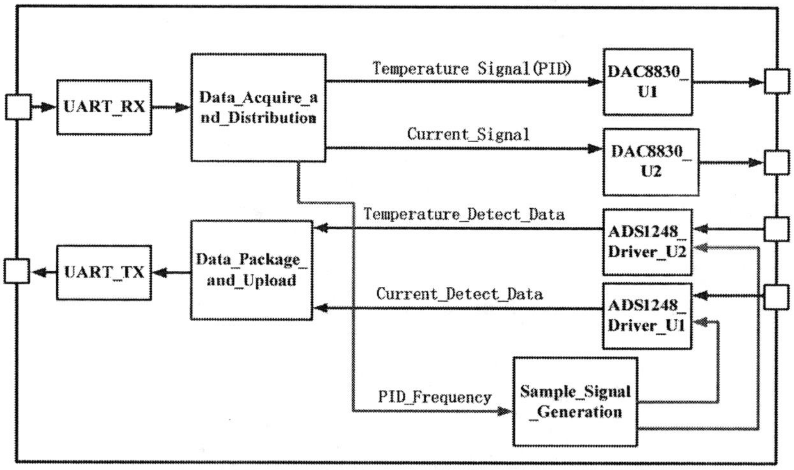

Fig. 4 The diagram of FPGA software

In order to be easily transplanted into the whole control system of SERF atomic magnetometer, a commerical FPGA development board is used to control the VCSEL driver instead of a FPGA core in the hardware circuit. Only the proper interfaces are needed from SERF atomic control system, can VCSEL driver be easily intergrated into the whole system. FPGA software is used to accept the control command from host computer system through module UART_RX, and the packaged control command is checked and distributed to DAC8830 driver to generate control signal of current and temperature. And the operating state of current and temperature is collected by ADS1248 driver, which is packaged and uploaded to host computer system through module UART_TX.

3.3 Hardware Circuit

The hardware circuit comprises power source, current driving circuit and temperature control circuit which are shown in Fig. 5. The power source circuit is used to generate the voltage of ±18V,±12V, ±5V, ±24V and ±15V, which will be used in other parts of circuit. The current driving part can be divided into a digital-to-analog circuit, voltage-controlled current source circuit, protective circuit, signal processing before ADC and analog to digital circuit. The temperature control circuit includes analog to digital circuit which is used to obtain the temperature, a digital to analog circuit and two-stage amplifying circuit which is used to heating the membrane. Then, the principle and function of some key parts will be described in detail.

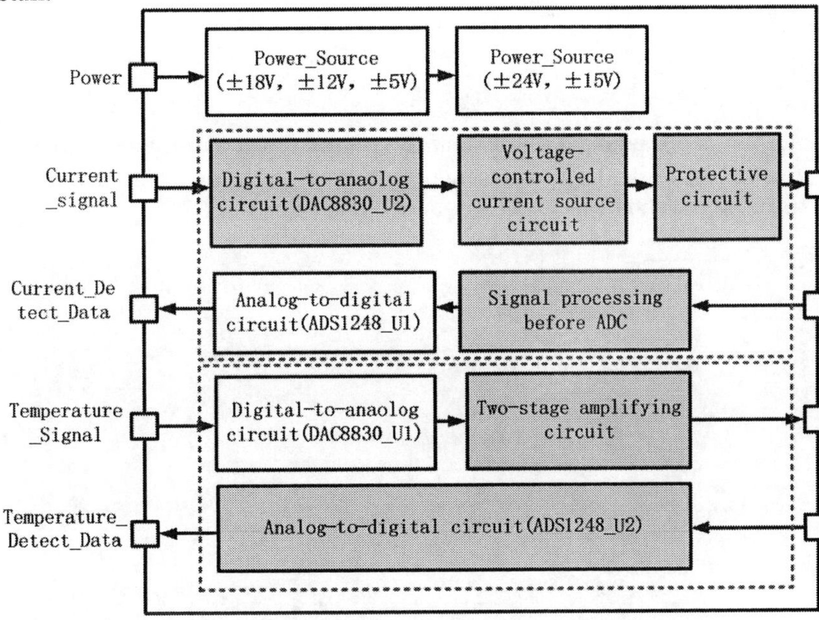

Fig. 5 Block diagram of hardware circuit

The digital-to-analog circuit uses DAC8830 to convert the digital control signal into an analog voltage. The DAC8830 is a 16-bit digital-to-analog converter with low noise and an integral nonlinearity error of 1LSB. In order to further reduce the noise of the DAC8830 output voltage, a low-pass filter is added in front of the power supply port, and a low-pass filter is added to the back end of the analog voltage output port, which enables the DAC8830 to output a more stable control voltage. The circuit diagram is shown in Fig. 6.

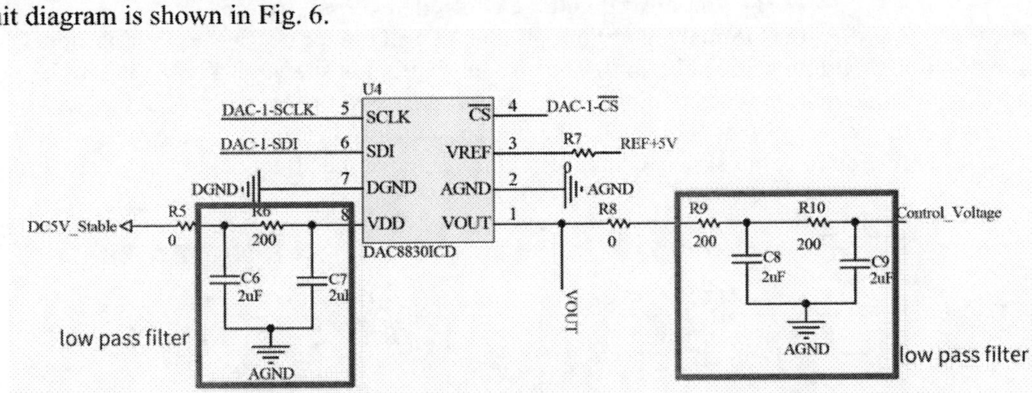

Fig. 6 Digital-to-analog circuit of current driver

The voltage-controlled current source circuit is implemented by a bipolar junction transistor. Based on the voltage-controlled current source in Ref. [4], some improvements have been made because VCSEL requires less current but higher stability[5]. The circuit diagram is shown in Fig. 7. The output current I_{DC} is proportional to the control voltage $V_{control}$.

$$I_{DC} = \frac{\beta_1}{1+\beta_1} \times \frac{\beta_2}{1+\beta_2} \times \frac{R_2}{R_3 R_4} \times V_{control}, \quad (2)$$

where β_1 is the current gain of the transistor Q1, β_2 is the current gain of the transistor Q2. According to the characteristics of the transistor, it can be known that β_1 and β_2 are much larger than 1, and R_2 equals to R_3 in the design. The above formula can be simplified to

$$I_{DC} = \frac{V_{control}}{R_4}. \quad (3)$$

The output current is only related to the control voltage. The control voltage is up to 5V (full-scale output of the DAC8830), while the maximum current required by the VCSEL does not exceed 4.5mA, usually about 2mA. So the resistance can be obtained from Eq. 4.

$$R_4 = \frac{V_{control}}{I_{dc\_max}} = \frac{5\text{V}}{0.0045\text{A}} = 1111.11\Omega. \quad (4)$$

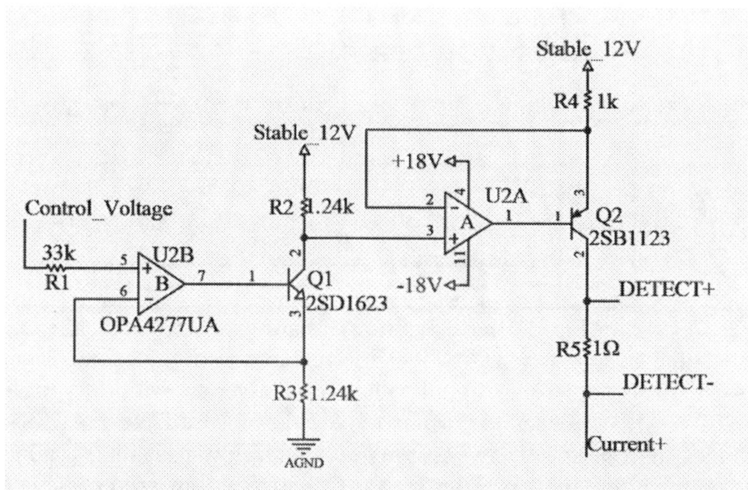

Fig. 7 Improved voltage-controlled current source

In order to ensure the normal operation of the VCSEL, a protective circuit is added behind voltage-controlled current source circuit[6]. The circuit diagram is shown in Fig. 8. When the power is turned on, the output current does not instantaneously increase to the set current, but slowly increases, which achieve the effect of soft start; when the circuit suddenly loses power, the current on the laser does not directly drop to 0, but slowly decreases, which achieve the effect of shutdown protection.

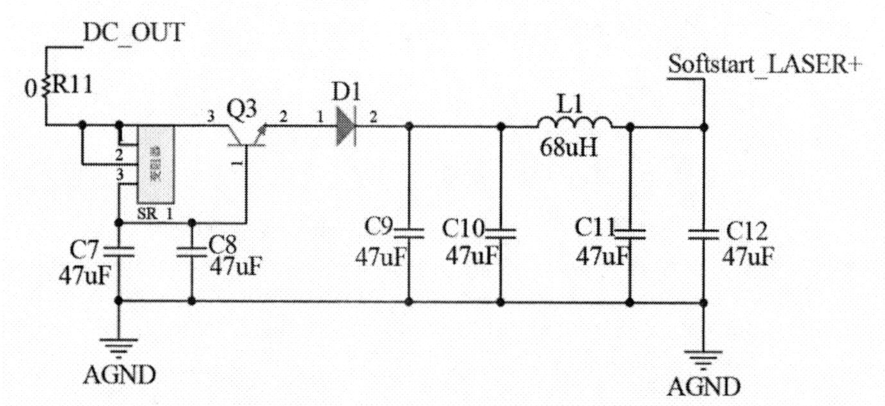

Fig. 8 Protective circuit for current driver

In order to detect the current in real time, the resistance R5=1Ω is added to the current path in Fig. 7, and the current can be known by observing the voltage on both sides. In order to ensure that the current monitoring process does not affect the output current, a voltage follower is added to isolate two parts. To improve the detection accuracy, a differential amplifying circuit is added to improve the signal-to-noise ratio of the detection signal. And a second-order Butterworth filter is added to avoid aliasing error during DAC conversion. Then ADS1248 converts analog signal into a digital signal, which is uploaded to the host computer system through the FPGA to display the current in real time. These parts are shown in Fig. 9.

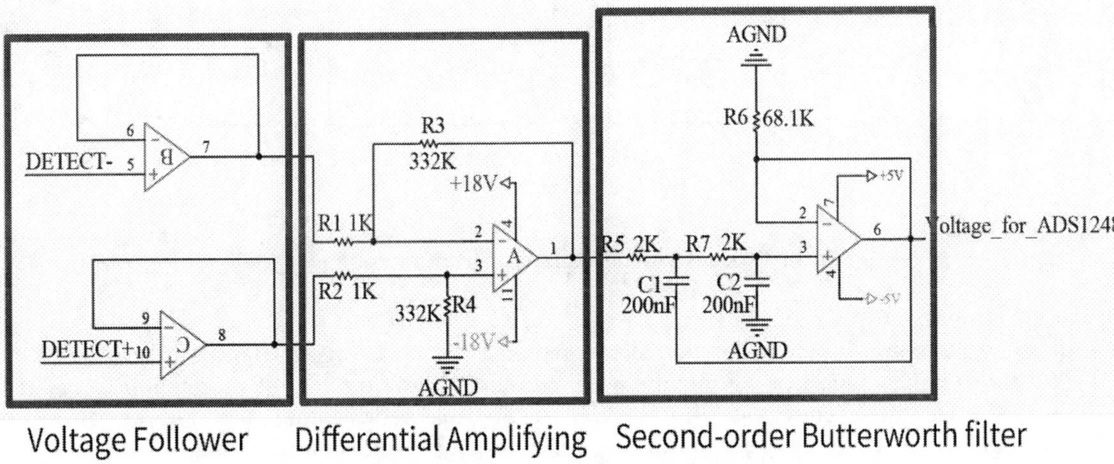

Fig. 9 Signal processing before ADC

The temperature control system includes analog-to-digital circuit, digital-to-analog circuit and two stage amplifying circuit. The analog-to-digital circuit is shown in Fig. 10. ADS1248 is used to collect the resistance value of PT1000 which can be used to calculate the temperature[7]. In order to reduce the influence of current drift on the acquisition accuracy, the reference voltage of ADS1248 is generated by constant current and high-precision resistance. PT1000 is also connected to the high-precision resistance[8].

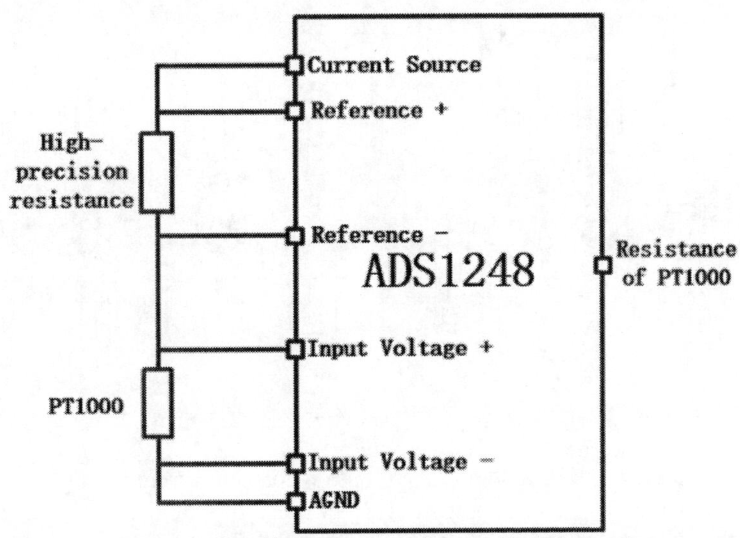

Fig. 10 The diagram of analog to digital circuit

The required heating voltage signal is generated after the calculation of host computer system with PID algorithm[9]. Then it is transmitted to the circuit board through the FPGA and converted into an analog voltage after DAC8830. Since the heating power required for the non-magnetic heating membrane is relatively large, it is necessary to pass through a two-stage amplifying circuit to amplify the signal and increase the carrying capacity. Process block diagram of heating voltage signal is shown in

Fig. 11.The amplified heating signal is then applied to both ends of the non-magnetic heating membrane for heating VCSEL.

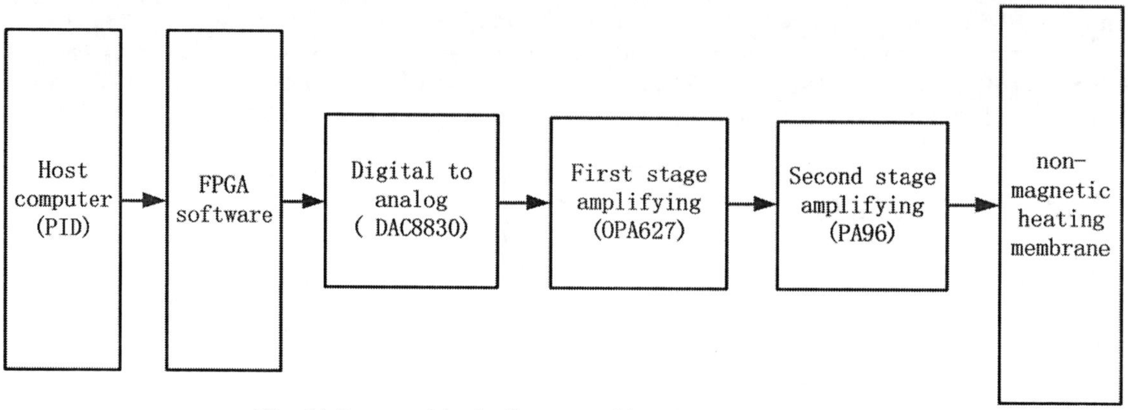

Fig. 11 Process block diagram of heating voltage signal

4. Experiment Setup and Results

An experiment setup is built to test the performance of VCSEL driver. Because the VCSEL with PT1000 and non-magnetic heating membrane is currently in the process of production, a traditional VCSEL is used to test the current stability. In addition, non-magnetic heating membrane and PT1000 are adhered together through silica gel to test the stability of temperature control system.

4.1 Test the stability of current

Different current values have been set from the host computer. A high-precision multimeter is connected to the current output to collect the current value within 30 minutes. The actual current data when set current is 3mA is displayed in Fig. 12.

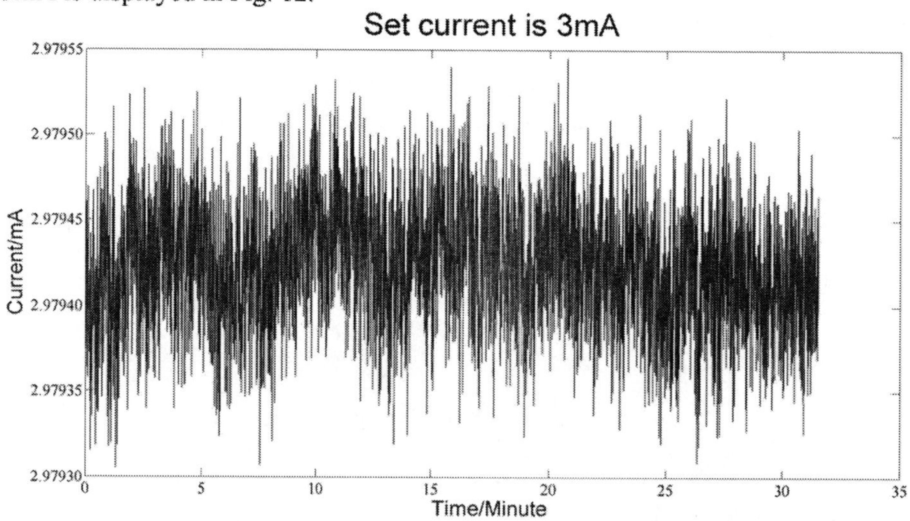

Fig. 12 The actual current data when set current is 3mA

After analyzing current data, mean value and RMS noise of actual current are shown in Table 2. It can be seen that the output current is adjustable within 0-4.5 mA, and the mean value of the output current differs little by the set current, which are less than 0.032 mA. The RMS noise is less than 0.08 μA.

Table 2 The result of current experiments

Set current[mA]	Mean value of actual current[mA]	Deviation between set current and mean value[mA]	RMS noise of current[μA]
0	0.00030	-0.0003	0.02
0.5	0.49679	0.00321	0.003
1	0.99338	0.00662	0.006
1.5	1.48982	0.01018	0.04
2	1.98639	0.01361	0.06
2.5	2.48298	0.01702	0.07
3	2.97947	0.02053	0.04
3.5	3.47584	0.02416	0.05
4	3.97209	0.02791	0.06
4.5	4.46835	0.03165	0.08

In order to determine the long-term stability of the current driver, the output current is set to 1.5 mA for 10 consecutive hours. The actual current in the 10 hours is displayed in Fig. 13. It can be seen that the RMS noise of the output current is less than 0.04 μA, and the drift of current is very small.

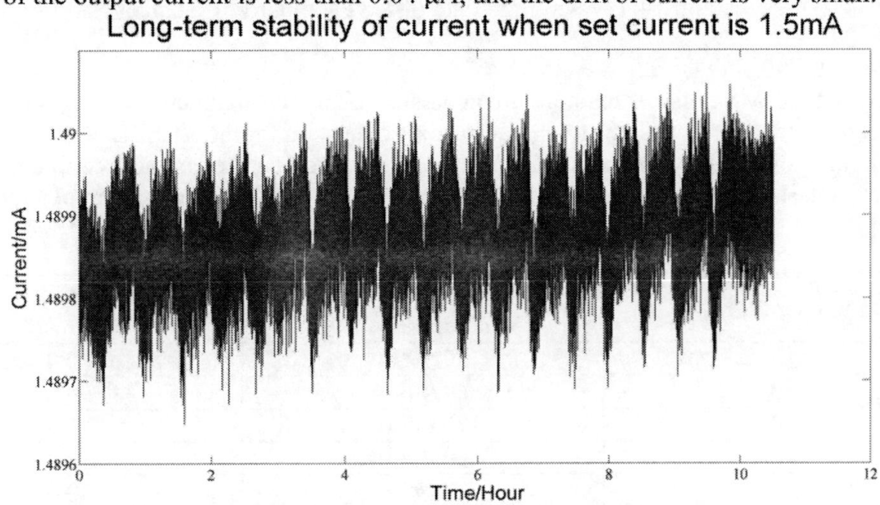

Fig. 13 Long-term stability of current

4.2 Test the accuracy of temperature detection

The stability of the temperature measurement is the premise for stable temperature control. And temperature is measured by detecting the resistance of PT1000. Then the resistance is reflected into temperature value. Therefore, the stability of the resistance measurement is directly related to the stability of the temperature control system. Here we use a 1kΩ resistor with an accuracy of 0.01% and a temperature drift of 1ppm as a standard component to test the stability of the temperature acquisition. The acquired data of standard resistance within 30 minutes is displayed in Fig. 14. It can be seen that for 1kΩ resistances, the resistance acquisition RMS noise is less than 1.3 mΩ. Because the coefficient between resistance and temperature of PT1000 is 3.9 Ω/K, the noise of resistance acquisition can be reflected to the temperature acquisition RMS noise of 0.33 mK, which is stable enough for VCSEL driver.

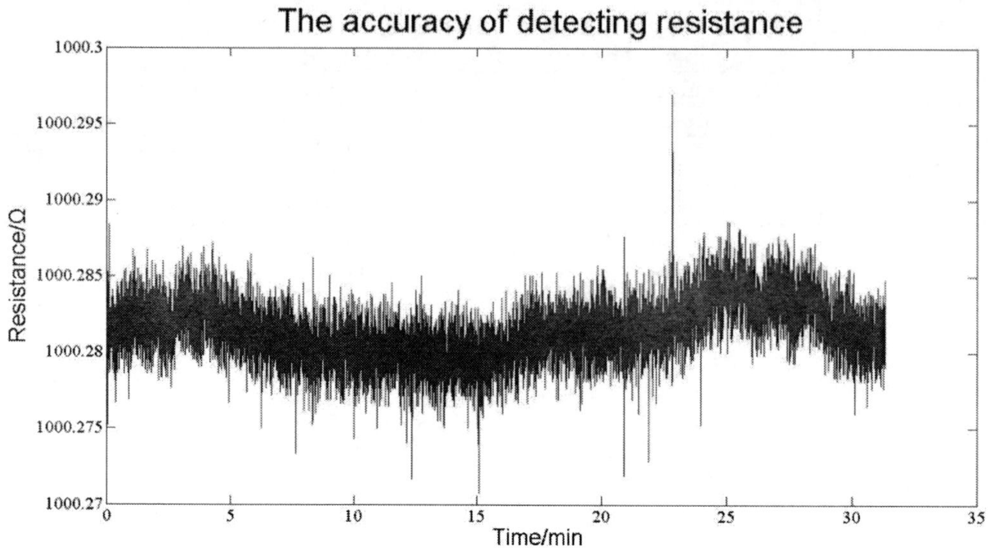

Fig. 14 The acquired data of 1kΩ standard resistance

4.3 Test the stability of temperature control

After checking the RMS noise of temperature acquisition, the non-magnetic heating membrane and the PT1000 are glued together to test the temperature control system for the VCSEL. Different temperatures are set from host computer system and temperature data is also acquired by host computer system. Each set temperature lasts for 30 minutes. The mean value and RMS noise of actual temperature are listed in Table 3.

Table 3 The stability of temperature control

Setting temperatutre（°C)	Mean value of actual temperature（°C)	RMS noise of temperature (mK)
30	29.9999	0.4
50	49.9999	0.5
70	70.0000	0.5
90	90.0000	0.6
110	109.9999	0.5

It can be seen that the operating temperature can change between 30-110°Cand RMS noise of the temperature control system is within 0.6 mK.

4.4 Predict the RMS noise of optical power and wavelength

According to the coefficients in Table 1, the RMS noise of optical power and wavelength can be predicted assuming the novel VCSEL has the same coefficients with VIXAR's 795 nm VCSEL. Because the noise from current and operating temperature is independent of each other, the RMS noise of optical power N_{power} and wavelength $N_{wavelenght}$ can be obtained from Eq. 5 and Eq. 6,

$$N_{power} = \sqrt{(N_{current} \times C_{current\_power})^2 + (N_{temperature} \times C_{temperature\_power})^2} \, . \quad (5)$$

$$N_{wavelenght} = \sqrt{(N_{current} \times C_{current\_wavelength})^2 + (N_{temperature} \times C_{temperature\_wavlenght})^2} \, . \quad (6)$$

where $N_{current}$ is the RMS noise of current and $N_{temperature}$ is the RMS noise of operating temperating. $C_{current\_power}$ means the coefficient between current and optical power, $C_{temperature\_power}$ means the coefficient between operating temperature and optical power. $C_{current\_wavelength}$ is the coefficient between current and wavelength. $C_{temperature\_wavlenght}$ is the coefficient between temperature and wavelength. The predicted RMS noise of optical power is 0.026 mW and RMS noise of wavelength is 4.08×10^{-5} nm according to Eq. 5 and Eq. 6.

5. Conclusion

For higher sensitivity of SERF atomic magnetometer we demonstrated a low noise driver for a new type of VCSEL that uses PT1000 as temperature sensitive component and non-magnetic heating membrane as the temperature actuator. The driver contains two parts: current driver and temperature control system. The current driver can generate a current of 0-4.5mA and the current RMS noise is less than 80nA. The effect of soft start and shutdown protection can be realized, so that the VCSEL can work safely. The temperature control part realizes that the operating temperature is adjustable between 30-110 °C, and temperature RMS noise is within 0.6 mK, which can meet the requirements of VCSEL for SERF atomic magnetometers.

Acknowledgements

The work is supported by National Key R&D Program of China (Grant No. 2017YFB0503100).

References

[1] H. B. Dang, A. C. Maloof and M. V. Romalis: Applied Physics Letters Vol. 97 (2010), p. 227
[2] V. K. Shah and M. V. Romalis: Physical Review A Vol. 80 (2009), p. 013416
[3] S. J. Seltzera and M. V. Romalis: Applied Physics Letters Vol. 85 (2004), p. 4804-4806
[4] Q. Lu, X. Zhang, X. Zhao, D. Yang, B. Zhou and Z. Hu: Optical Information & Network Vol. (2017), p. 2285035
[5] K. Libbrecht and J. L. Hall: Review of scientific instruments Vol. 64 (1993), p. 2133-2135
[6] C. Bradley, J. Chen and R. G. Hulet: Review of scientific instruments Vol. 61 (1990), p. 2097-2101
[7] J. Chauhan and U. Neelakantan: 2016 International Conference on Electrical, Electronics, and Optimization Techniques (ICEEOT) Vol. (2016), p. 3213-3215
[8] Y. Fang, K. Lei, Q. Zhang, Q. Huang and Z. Fan: 2013 IEEE International Conference of IEEE Region 10 (TENCON 2013) Vol. (2013), p. 1-4
[9] J. Zhang, H. Li, K. Ma, L. Xue, B. Han, Y. Dong, Y. Tan and C. Gu: IOP Conference Series: Materials Science and Engineering Vol. 322 (2018), p. 072020

Simultaneous Analysis on Phenolic Substance in Cloudy Apple Juice by HPLC

Danshi Zhu, Hao Fu, Yusi Shen, Xuehui Cao, He Liu*

College of Food Science and Technology, Bohai University; National & Local Joint Engineering Research Center of Storage, Processing and Safety Control Technology for Fresh Agricultural and Aquatic Products; Jinzhou, Liaoning, 121013, China

Abstract. Phenolic substances are important bioactive substances in apple juice. The phenolic substance extraction methods in 'Ralls' cloudy apple juice and the simultaneously detected conditions using high-performance liquid chromatography (HPLC) were investigated. The results showed that the extraction method for mixed phenolic substance by ethyl acetate was better than that extraction for neutral phenols and acidic phenols respectively. Most phenolic substances in apple cloudy juice could be detected at 280 nm, and which was the primary detection wavelength for simultaneous analysis on phenolic substance by HPLC. Chlorogenic acid, epicatechin and gallic acid were the main phenolic substance in 'Ralls' apple juice. Quantitative analysis showed that the content of chlorogenic acid was the highest, and it was the characteristic phenolic substance in 'Ralls' apple juice.

1. Introduction

Apple is one of the most widely cultivated fruits in the world. Nowadays, more than 40% of apples in the world are produced in China, which plays an indispensable role in the world apple industry. In recent years, the world's apple production had exceeded 80 million tons, showing an ever-growing trend [1]. Except for some apples used for fresh-eating, approximately 35% of them are used for processed food (juice, cider, vinegar and jam), of which apple juice accounts for 65%, ranking second for global consumption and production after orange juice[2, 3]. It is commonly served as clarify juice for better stability. However, the market for cloudy apple juice is growing at an unprecedented rate for the abundant nutrition. Cloudy apple juice is well known to be rich in polyphenols, which has potential effects on anti-cancer, inflammation, hyperglycaemia, lipids metabolism, anti-oxidation, anticardiac and vascular function[4, 5].

Some studies had shown that the number of polyphenols exceeds 5000, which was a secondary plant metabolite with diverse structures and extensive phylogenetic distribution[6]. At present, the detections of polyphenols mainly include spectrophotometry, chemiluminescence and high performance liquid chromatography (HPLC). Among these methods, HPLC has become the most common and accurate method for the separation and quantification of polyphenols in plants[6, 7]. However, the proper extraction method and accurate detection parameters depend on the properties of raw material. In this study, the effects of two common methods on the extraction of phenolic substance from cloudy apple juice were evaluated, and the effects of detection wavelengths on simultaneous analysis of phenolic substance using HPLC were also studied. Then, three characteristic phenolic compounds in apple juice were quantified after the analysis on limit of detection (LOD) and recovery rate of HPLC.

2. Materials and Methods

2.1 Materials and reagents

Materials. Mature apples (Malus pumila) cultivar 'Ralls' were hand-harvested from commercial maturity trees in local farm of Jinzhou city, Liaoning Province, China. Cored apples with ascorbic acid (0.1%, g/kg) as antioxidant were crushed with a crusher (DS-1, Wanhua Experimental Instrument Factory, Jiangsu, China). Then juice was hydrolyzed by enzyme, and then filtered by a filter cloth of 300 mesh to remove the fruit residue. After sterilization, the cloudy apple juice was stored at 4 °C for subsequent analysis.

standards and reagents. All standards, including rutin, anthocyanin, gallic acid, tannic acid, phlorin, chlorogenic acid and epicatechin were chromatographic pure, Sigma-aldrich.

2.2 Comparison of extraction methods of phenolic substance

Extraction method 1: Extraction for mixed phenolic substance. Cloudy apple juice (50 mL) were added in 25 mL ethyl acetate, and the mixtures were oscillated at room temperature for 10 min before separated in a separatory funnel. Then, repeated extraction twice with ethyl acetate. The supernatant extracted for the three times were combine collected, and frozen at -18°C for 2 hours to remove moisture. After that, the sample was filtered, and then concentrated using a rotary evaporator (RE-2000B, Chengdu Kangyu Instrument Co., Ltd. Sichuan, China) at 55 °C for 10 minutes. The concentrate was dissolved in methanol and diluted to 10 mL. The dilution was stored at 4°C, and filtered through a filter of 0.45 μm before HPLC analysis.

Extraction method 2: Extraction for neutral phenol and acid phenol. Cloudy apple juice was diluted with ultra-pure water until 11.5 °Brix. After adjusting the pH of the dilution (25 mL) to 7.0 with NaOH (1 mol/L), the dilution was mixed with 25 mL ethyl acetate. The mixtures were oscillate extracted at room temperature for 10 min, and then, the ester phase and the aqueous phase in the mixture were separated in a separatory funnel. The aqueous phase was re-extracted once with ethyl acetate, and the twice extracted ester phases were combined in a dark-colored flask. The ester phase was concentrated in a rotary evaporator, and the residue was dissolved in 25 mL of ultra-pure water to obtain neutral phenol. The pH of the aqueous phase was adjusted to 2.0 with HCl (6 mol/L), and repeat extracted twice with ethyl acetate, then, combined ester phases. The ester phase was concentrated, and the residue was dissolved in 25 mL of ultra-pure water to obtain acid phenol. The neutral phenol and the acid phenol were stored at 4 °C, and filtered through 0.45 μm filters respectively before HPLC analysis.

Preparation of standard solutions. A mixed standard reserve solution was prepared in methanol and stored in the dark at 4°C for further analysis. The standard solutions were filtered through a 0.45 μm filter prior to injection.

HPLC analysis method. The HPLC analysis was refer to the research of Gaurav Rajauria [6] and Adina Frum [8] with slight modifications using an HPLC system, 1206 series provided by Agilent Technologies. The column employed was the Zorbax Eclipse Plus C18 with the following dimensions: Analytical 250 mm×4.6 mm, 5-Micron. The mobile phase finally adopted was (A) ultra-pure water, (B) acetonitrile, (C) methanol and (D) purified water / formic acid (98/2, v/v) with flow rate at 0.6 mL/min and column temperature of 25°C. The gradient schedule was: 0-30 min, 5% B and 95% D; 30-45 min, 25% B and 75% D; 45-50 min, 40% B and 60% D; 50-51 min, 40% B and 60% D; 51-60 min, 5% B and 95% D; 60-65 min, 5% B and 95% D. The injection volume was 10 μL. UV absorbance at 280nm, 320nm and 360nm were recorded respectively.

Qualitative and quantitative analysis of phenolic substance in cloudy apple juice. The phenolic substances extracted in cloudy apple juices were detected by HPLC, and the main phenolic substances were quantified. Standard solutions (1mg/mL) of chlorogenic acid, epicatechin and gallic acid were prepared by methanol solution, respectively. For quantitative analysis, standard solutions are diluted to different concentrations with methanol. The calibration curve was drawn by plotting the peak area against the concentration of each standard solution, and then, the contents of three main phenolic compounds in cloudy apple juice were calculated.

IOP Conf. Series: Materials Science and Engineering **711** (2020) 012094 doi:10.1088/1757-899X/711/1/012094

3. Results and discussions

HPLC analysis method optimization. In HPLC analysis, elution conditions were established by testing several gradients of two different phases containing ultra-pure water/formic acid and acetonitrile. Phenolic substances, especially phenolic acids, are prone to decomposition, resulting in tailing, broadening and asymmetry of the chromatographic peaks. The mobile phase is acidified with an appropriate amount of acid inhibitor (formic acid and acetic acid, etc.) to avoid tailing of the chromatographic peak and improve peak shape[6]. Inappropriate pH value will damage the chromatographic column and reduce the efficiency, resulting in severe baseline drift [9]. Initially, various combinations of mobile phases were applied using total gradient system, which showed that mobile phase of (B) acetonitrile and (D) ultra-pure water/formic acid (98/2, v/v) showed better eluting profile than (A) ultra-pure water and (C) methanol for polyphenol extract in cloudy apple juice. In addition, the mobile phase of (B) and (D) also showed a clear separation of more peaks in gradient system. Therefore, acetonitrile and ultra-pure water/formic acid (98/2, v/v) were used as mobile phases. The qualitative analysis of phenolic substance was applied by comparing the RT with the standard of phenolic substance (Fig.1) detected by HPLC.

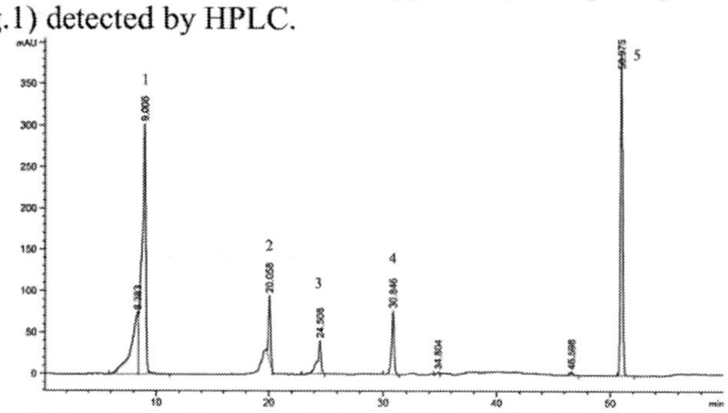

Fig.1. Mixed standards of phenolic substance detected by HPLC. Peaks:1. Gallic acid; 2. Chlorogenic acid; 3. Epicatechin; 4. Rutin; 5. Phlorin.

Comparison of extract methods of phenolic compounds. Phenolic compounds by different extract methods in cloudy apple juice were showed in Fig.2. From "Extraction method 1" (Fig.2-A), the detected peaks were better separated, the peaks shape were more symmetrical, and no tailing occurred. By comparing their retention times (RT) with phenolic substance standard (Fig.1) in HPLC analysis, the higher content phenolic compounds in 'Ralls' apple juice were chlorogenic acid and epicatechin. By "Extraction method 2", acidic phenol and neutral phenol were extracted. As shown in Fig.2-B, the extracted acid phenol has only one peak with good peak shape and strength, which was identified as chlorogenic acid. From Fig.2-C, there were three peaks as neutral phenols, however, their signal intensity were weaker. Therefore, "Extraction method 1" was used to extract mix phenolic substance from cloudy apple juice.

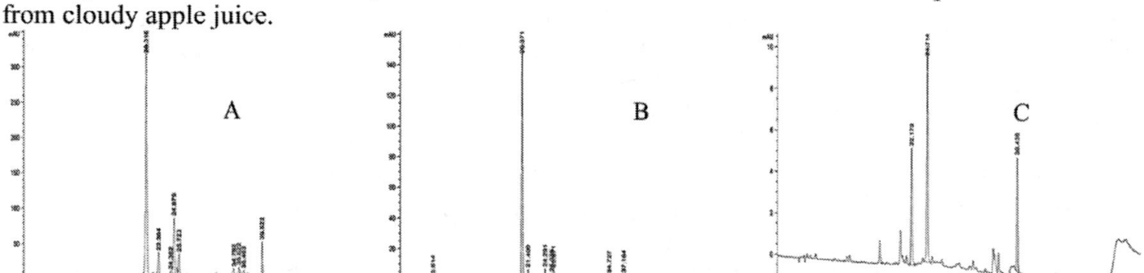

Fig.2. Phenolic compounds by different extract methods in cloudy apple juice detected using HPLC: extraction for mixed phenolic substance (A) by "Extraction method 1", extraction for acid phenol (B) and neutral phenol (C) by "Extraction method 2".

The effects of detection wavelength of HPLC. Different phenolic substances have different maximum absorption wavelengths. Research showed that the maximum absorption wavelength of phenolic substance is as follow: gallic acid (280 nm), catechin (280 nm), chlorogenic acid (320 nm), epicatechin (280 nm), rutin (360 nm), phlorizin (280 nm) and phloretin (280 nm) [10]. So, in this study, the three different wavelengths of 280 nm, 320 nm and 360 nm were chosen for sample analysis using HPLC (Fig.3) for evaluate effects of simultaneously analysis on phenolic substance. From Fig.3-A, the detected peaks were separated better, and the signal intensity was stronger. The phenolic substance in cloudy juice were mainly identified as chlorogenic acid (RT=19.599 min), epicatechin (RT=24.083 min), tannic acid (RT=35.340 min) and possibly isomer of chlorogenic acid (RT=21.505 min). Fewer phenolic compounds were detected at 320 nm (Fig.3-B), which were mainly chlorogenic acid (RT=19.863 min), rutin (RT=31.741 min) and tannic acid (RT=35.849 min). The signal intensity of chlorogenic acid was relative stronger. At 360nm (Fig.3-C), the signal intensity was the weakest, and the peaks were more interfered by baseline. Therefore, the detection wavelength of 280 nm was better adopted as the final detection wavelength of HPLC for simultaneously analysis on phenolic substance in cloudy apple juice.

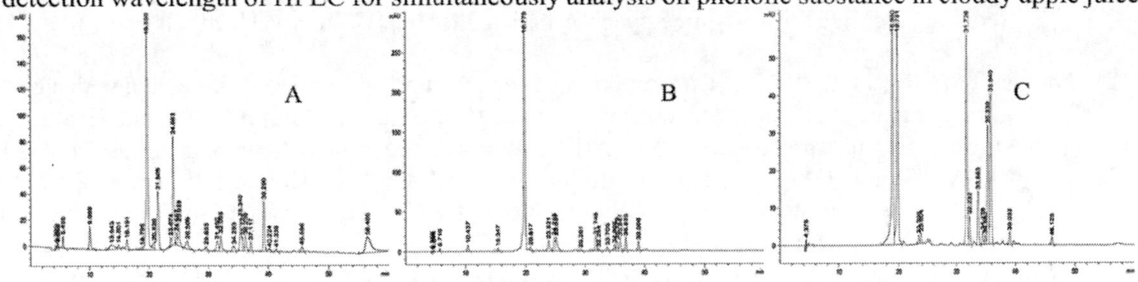

Fig.3. Phenolic compouds in cloudy apple juice analized by HPLC with different detection wavelength: A. 280nm, B. 320nm, C.360nm.

Quantification of characteristic phenolic substance in cloudy apple juice. It was shown in Fig.3-A, phenolic substance in 'Ralls' apple juice were chlorogenic acid, epicatechin and gallic acid, therefore, the three compounds were quantified. Calibration curves (Table 1) were plot using standard phenolic compounds, and the limit of detection (LOD) was also determined. In addition, the extraction of juice both added and didn't add standard substance were determined by HPLC, and the recovery rates of phenolic compounds were detected (Table 1). It was shown that, the recovery rates were above 90%, indicating that this method was reliable and relative accurate.

Table 1. Standard curve, LOD and recovery rate of main phenolic substances

Phenolic substance	Calibration curves	R^2	LOD [µg/mL]	Recovery rate [%]
Chlorogenic acid	Y=4471.5x-1419.9	0.9934	0.0638	102.6
Epicatechin	Y=11.305X-34.539	0.9995	0.0688	94.3
Gallic acid	y=38.403x-18.251	0.9997	0.0235	95.4

Table 2. Contents of main phenolic substance in cloudy apple juice

	Chlorogenic acid	Epicatechin	Gallic acid
Content [mg/L]	174.21±1.04	92.76±1.84	4.74±0.20

The values were calculated as means ± standard deviation of three replicates ($p<0.05$).

The contents of main phenolic substance in cloudy apple juice were determined under the optimized chromatographic conditions, and the results were shown in Table 2. The content of chlorogenic acid was the highest phenolic compound in 'Ralls' cloudy apple juice, which reached 174.21 mg/L, and was much higher than epicatechin and gallic acid.

4. Conclusion

Phenolic substance extraction method and HPLC detect conditions were important for simultaneous analysis on the phenolic substance in cloudy apple juice. In general, extraction for mixed phenolic substance of 'Extraction method 1' was more fit for cloudy apple juice. Most phenolic substances in apple cloudy juice could be detected at 280 nm, and which was the primary detection wavelength for simultaneous analysis on phenolic substances. Quantitative analysis showed that the content of chlorogenic acid was the highest, and it was the characteristic phenolic compound in 'Ralls' apple juice.

Acknowledgements
This work was financed by a grant from National Natural Science Foundation of China (31701618), Natural Science Foundation of Liaoning province (20180551103) in China, and LiaoNing Revitalization Talents Program in China. We thank Dr. He Liu (Bohai University, CN) for his critical review of this manuscript.

Reference
[1] Huang Z, Hu H, Shen F, Wu B, Wang X, Zhang B, et al. Relatively high acidity is an important breeding objective for fresh juice-specific apple cultivars[J]. Scientia Horticulturae, 2018, 233: 29-37.
[2] Nicklas TA, O'Neil CE, III VLF. Consumption of various forms of apples is associated with a better nutrient intake and improved nutrient adequacy in diets of children: National Health and Nutrition Examination Survey 2003–2010[J]. Food & Nutrition Research, 2015, 59(3): 25948.
[3] Zielinski AA, Zardo DM, Alberti A, Bortolini DG, Benvenutti L, Demiate IM, et al. Effect of cryoconcentration process on phenolic compounds and antioxidant activity in apple juice[J]. Journal of the Science of Food and Agriculture, 2019.
[4] Tian Y, Lijun S, Yali Y, Xiaoju G, Pengfei N, Yurong G. Changes in the physicochemical properties, aromas and polyphenols of not from concentrate (NFC) apple juice during production[J]. CyTA - Journal of Food, 16(1): 755-764.
[5] Rong T, Yang R, Young JC, Zhu H. Polyphenolic Profiles in Eight Apple Cultivars Using High-Performance Liquid Chromatography (HPLC)[J]. J Agric Food Chem, 2003, 51(21): 6347-6353.
[6] Rajauria G. Optimization and validation of reverse phase HPLC method for qualitative and quantitative assessment of polyphenols in seaweed[J]. J Pharm Biomed Anal, 2018, 148(4): 230-237.
[7] Bai X, Zhang H, Ren S. Antioxidant activity and HPLC analysis of polyphenol-enriched extracts from industrial apple pomace[J]. Journal of the Science of Food & Agriculture, 2013, 93(10): 2502-2506.
[8] Frum A. HPLC Determination of Polyphenols from Calendula officinalis L. Flowers[J]. Acta Universitatis Cibiniensis Series E: Food Technology, 2017, 21(2): 97-101.
[9] Bai L, Guo S, Liu Q, Cui X, Zhang X, Zhang L, et al. Characterization of nine polyphenols in fruits of Malus pumila Mill by high-performance liquid chromatography[J]. Journal of Food and Drug Analysis, 2016, 24(2): 293-298.
[10] Suarez B, Picinelli A, Mangas JJ. Solid-phase extraction and high-performance liquid chromatographic determination of polyphenols in apple musts and ciders[J]. Journal of Chromatography A, 1996, 727(2): 203-209.

IOP Conf. Series: Materials Science and Engineering **711** (2020) 012095 doi:10.1088/1757-899X/711/1/012095

Effects of Gradient Optimizer on Model Pruning

Yang Zhang[1, a], Bin Zhu[1, b], Qi Ma[1, c], Huanan Wang[2, d]

[1]State Key Laboratory of Pulsed Power Laser Technology, National University of Defense Technology, Hefei, Anhui, China

[2]92020 Troops, Qingdao, China

[a]ahjpzy@163.com, [b]zhubineei@163.com, [c]905303927@qq.com, [d]wecholer@126.com

Abstract. Deep convolutional neural networks deploy in applications are hindered by their large parameters and high computational cost. In this paper, we test different gradient optimizer's effect on YOLOv3 model pruning. This is achieved by enforcing channel-level sparsity in YOLOv3 network. The model trained with Adam optimizer get 5× reduction in model size only after training 60 epochs.

1. Introduction

Object detection is developing faster and faster with the improvement of computer performance and the algorithm optimization of convolutional neural networks. This computer vision task is applied in many fields and become more and more closely with life. For example, medical diagnosis [1], object tracking [2], autonomous driving [3], etc. From a practical point of view, high speed and high precision is the most important aspect.

In deep learning field, object detection is divided into two genres: one-stage detection and two-stage detection. Where the former frames the detection as a "complete in one step" process while the later frames it as to "coarse-to-fine". In generally, the one-stage detection algorithm is more advanced in speed, but accuracy is lower than the two-stage detection algorithm. In recent years, there appears several excellent one-stage object detection algorithm, such as YOLO [4, 5, 6], SSD [7], Retina-Net [8]. YOLOv1 is the originator of the end-to-end model in the area of object detection, its core idea is to use the whole picture as the input of the network, directly regress the position of the bounding box at the output layer and its associated class. Because two boxes are predicted in a grid, and belong to one class. Therefore, for objects that are closed to each other and small group, YOLOv1 can't detect well. YOLOv2 proposed a method to joint train on object detection and classification [5], the basic idea is that training the object detectors on both detection data set and the classification data set at the same time. The model learns object's exact position from detection data, and increases the classification quantity of the classification and improve the robustness from classification data. Besides, YOLOv2 applied some improvement measures, such as batch normalization, high resolution classifier, convolution with anchor boxes, dimension clusters. Compared with YOLOv2, YOLOv3 make some further improvements: (1) employ logistic regression to predict bounding box; (2) improve single-label prediction to multi-label prediction; (3) combine 3 different sizes of bounding boxes for prediction; (4) adopt Darknet-53 as the backbone compared to YOLOv2's Darknet-19.

YOLOv3 is currently the most popular one-stage object detection algorithm, many researchers do lots of works to improve the performance of YOLOv3. Mohammad Mahdi Derakhshani et al. [9] inspired from curriculum learning and proposed a simple and effective learning technique that feed in

IOP Conf. Series: Materials Science and Engineering **711** (2020) 012095 doi:10.1088/1757-899X/711/1/012095

localization information by exciting certain activation during training, and improve the mAP of YOLOv3 by 2.2% on MSCOCO dataset.

When large CNNs deployed in products applications, there are three limits[10]: 1) Model size: millions of trainable parameters which explain the strong representation power should be stored on disk and loaded into memory during inference time. Larger CNNs mean more parameters and larger model size, which is a big resource burden to embedded devices. 2) Run-time memory: The intermediate activations/responses of CNNs could even take more memory space than storing the model parameters, and many applications with low computational power are unaffordable. 3) Number of computing operations: The convolution operations are computationally intensive on high resolution images. It is normal for large CNN to take minutes to test one single image on mobile device which has low computing power.

In order to retain large CNN models' high efficiency, large works have been proposed to solve these problem. These works include two directions: 1) net-work quantization[11,12] and binarization[13,14], weight pruning[11], dynamic inference[15], etc. 2) sparsity the network[16, 17]. However these methods above require special software/hardware accelerators. Zhuang Liu et al. [10] proposed a simple yet effective network training scheme called *network slimming* achieving CNN models with up to 20× model-size compression and 5× reduction in computing operations of the original ones.

2. Related Work

Regularization. In the supervised learning, regularization has the vital effect on the model training. Regularization coefficientλbalanced the empirical error and the regularization. It is likely to be overfitting when the value ofλis too small, on the contrary, it may be underfitting ifλis too large. Therefore, parameterλhighly decide the model's performance in the training.

The form of common regularization is L1 norm and L2 norm. The L1 norm is the sum of the absolute values of the elements in the matrix and it is usually used for achieving parameter matrix sparse by Eq. 1.

$$\min \sum_{(x,y)} L(y, f(x; w)) + \lambda \|w\|. \tag{1}$$

In comparison with L1 norm, L2 norm is equal to the root after the sum of squares of each element in a matrix, refer to Eq. 2.

$$\min \sum_{(x,y)} L(y, f(x; w)) + \frac{\lambda}{2} \|w\|^2. \tag{2}$$

It minimize each element in parameter matrix and bring it infinitely close to 0 which in L1 norm is equal to 0. With the exception of L1 and L2, another famous method is named dropout which randomly discard some parameters to lighten or prevent overfitting. By setting an inactivation probability for each layer's neuron of the network, dropout will randomly discard some neuron when train a network. When apply dropout to convolution neural network, we can set the layers which have more neurons or easy to become overfitting with higher inactivation probability.

Model compression. Among the existing model compression methods, model pruning is one of the most popular methods. The method judge the importance of parameters by adding a sparsity-induced penalty on the scaling factors which is multiplied to the out of channel, and prune those channels with small factors at last. [18] proposed the pruning method based on weights, judge the importance by the value of weight. As for filter, calculating the sum of its absolute kernel weights and pruning the filter with low value. [19] defined APoZ(Average Percentage of Zeros) to measure the number of activation with 0 in each filter, and then judge the filter is important or not. [20] proposed an entropy-based pruning method which used entropy to judge the importance of filters. An energy-aware pruning was put forward by Yang T J in [21], the author calculate the energy consumption in each layer, and prune the layer with high energy consumption. In spite of model pruning, the sparsity of kernel is another direction to compress model. This method induce the updating of weights to become more sparser in the training, and it can be divide into regular method and irregular method. [22] presented a learning method named Structured Sparsity Learning which can learn a sparsity structure to reduce calculation of consumption, the learnt structure with sparsity would effectively accelerate in hardware. Guo, Yiwen proposed a dynamic model cutting method [23], this method add splicing process to restore the pruning weights which are important. Liu et al. [10] put forward a simple and effective model pruning approach called network slimming. The approach directly applied the scaling factors in batch normalization (BN) layers as channel-wise scaling factors and trained networks with L1 regularization on these scaling factors to obtain channel-wise sparsity. Channel pruning is a coarse-grained but effective approach that it is convenient to apply the pruned models without the requirements of dedicated hardware or software. In this paper, we follow in Liu's footstep, try several different hyperparameters setting and compare the performance.

3. Network slimming

Network slimming is a simple scheme to achieve channel-level sparse. The challenges of achieving channel-level sparse require pruning all the incoming and outgoing connections associated with the same channel [10]. A scaling factorγfor each channel is implemented when train the network weights, then prune those channels with small factors, and fine-tune the pruned network in the last. The theory is given by Eq. 3.

$$\mathrm{L} = \sum_{(x,y)} l(f(x,W),y) + \lambda \sum_{\gamma\epsilon\Gamma} g(\gamma). \tag{3}$$

Which is known as L1-norm and widely used to achieve channel sparse.

Sparse training with BN. Batch normalization is widely adopted in modern CNNs to achieve fast convergence and better generalization performance. BN layer normalize convolutional features using mini-batch statics, which is formulated as follow Eq. 4.

$$\mathrm{y} = \gamma \times \frac{x-\bar{x}}{\sqrt{\sigma^2+\varepsilon}} + \beta. \tag{4}$$

Where $\bar{x}$ and σ^2 are mean and variance of input features in a mini-batch, γ and β denotes trainable scale factor and bias.

Channel pruning. After training with sparsity, we obtain a model with many scaling factors near zero. Then the channels with near zero scaling factors are prune by removing all incoming and outgoing connections and corresponding weights. The threshold we prune channels is defined as a certain percentile of all the scaling factor values.

Fine-tuning. After channel pruning, the accuracy of model may descend a little, but perform fine-tuning on the pruned model could compensate the temporary degradation.

4. Experiments

We have tried different gradient optimizer and learning rate decay when train YOLOv3 model on oxford hand datasets. Our experiments were operated in the environment with python3.6 and pytorch1.1, the Linux server with Intel Xeon(R) CPU E5-2650 v4 @ 2.20GHz × 48, and TITAN V/PCIe/SSE2.

Sparse training. We trained with sparsity on Oxford hand datasets, and chose scale sparse rate with 0.01. The number of epochs were both 60, and learning rate attenuation mode was that reduce learning rate when loss has stopped descending. The training loss and validation mAP is shown in Fig. 1. The average mAP and model size with two different optimizer is shown in Table 1.

Pruning and Fine-tuning. We pruned the channels of models trained with sparsity and different gradient optimizers. The pruning percentage determine the pruning threshold, the maximum pruning percentage of model trained with SGD is 45%, and the model trained with Adam is 80%. After pruning, a fine-tuning process was applied in the pruned model with 80 epochs. Fig. 2 shows the fine-tuning performance.

5. Results

We compare the detection performance of two models which both trained with sparsity on validation set of Oxford hand in Fig. 1, and the models trained with SGD optimizer and Adam optimizer have the same model size in Table 1.

Table 1. MAP and Model size trained with different gradient optimizer

Optimizer	mAP	Model size
SGD(warm up + lr decay)	0.70	246.4MB
Adam(warm up + lr decay)	0.69	246.4MB

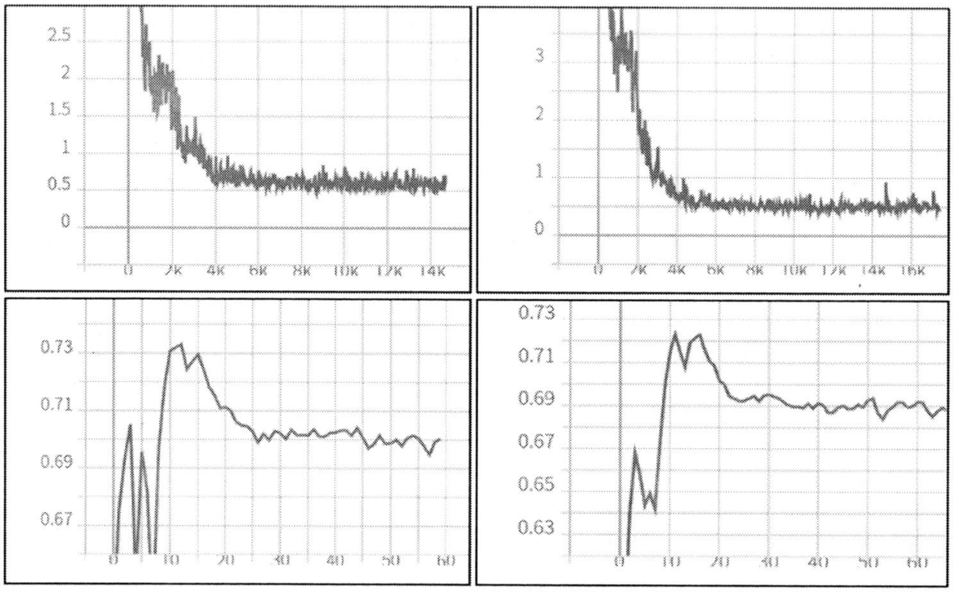

Fig. 1: The training loss and mAP of validation sets with SGD optimizer (left) and Adam optimizer (right)

Table 2. Model size pruned percentile with different gradient optimizer

Model	Sparse train epochs	Model size pruned	Pruned percents
Yolov3(SGD)	60	99.6MB	45%
Yolov3(Adam)	60	46.7MB	80%

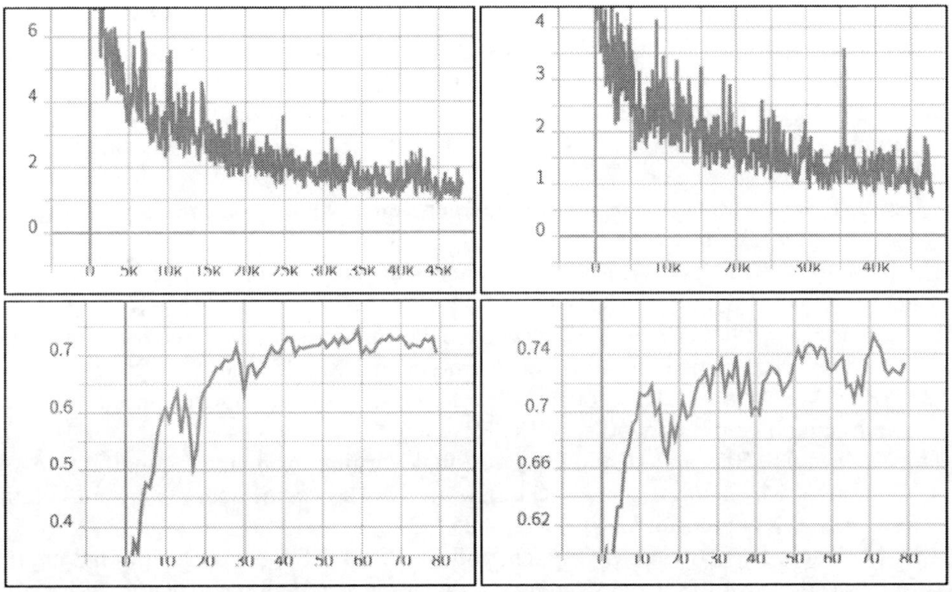

Fig. 2: Fine-tuning the pruned models which trained with different gradient optimizer, the train loss and validation mAP with SGD optimizer are the left and the right are Adam optimizer.

As shown in Table 2, we can get a more compact model trained with Adam optimizer in the same training epochs, and mAP on validation set is all around 0.72 after fine-tuning.

6. Conclusion

In this paper, we try different optimizer in training YOLOv3, and find that the model trained with Adam gradient optimizer is easier to get a compact model. Network slimming technique [10] directly impose sparsity-induced regularization on the scaling factors in batch normalization layers, and automatically identify unimportant channels during training and then pruned. In comparison with SGD, Adam optimizer is a better choice for sparse data and complex convolution neural network.

References

[1] Gurovich Yaron, Hanani Yair et al. Identifying facial phenotypes of genetic disorders using deep learning

[2] K. Kang, H. Li, J. Yan, X. Zeng, B. Yang, T. Xiao, C. Zhang,Z. Wang, R. Wang, X. Wanget al., "T-cnn: Tubelets withconvolutional neural networks for object detection fromvideos,"IEEE Transactions on Circuits and Systems for VideoTechnology, vol. 28, no. 10, pp. 2896–2907, 2018.

[3] Chen C, Seff A, Kornhauser A, et al. [IEEE 2015 IEEE International Conference on Computer Vision (ICCV) - Santiago, Chile (2015.12.7-2015.12.13)] 2015 IEEE International Conference on Computer Vision (ICCV) - DeepDriving: Learning Affordance for Direct Perception in Autonomous Driving [J]. 2015:2722-2730.

[4] Redmon J, Divvala S, Girshick R, et al. You Only Look Once: Unified, Real-Time Object Detection [J]. 2015.

[5] Redmon J, Farhadi A. [IEEE 2017 IEEE Conference on Computer Vision and Pattern Recognition (CVPR) - Honolulu, HI (2017.7.21-2017.7.26)] 2017 IEEE Conference on Computer Vision and Pattern Recognition (CVPR) - YOLO9000: Better, Faster, Stronger[J]. 2017:6517-6525.

[6] Redmon J, Farhadi A. YOLOv3: An Incremental Improvement [J]. 2018.

[7] Liu W, Anguelov D, Erhan D, et al. SSD: Single Shot MultiBox Detector[C]// European Conference on Computer Vision. 2016.

[8] Lin T Y, Goyal P, Girshick R, et al. Focal Loss for Dense Object Detection [J]. IEEE Transactions

on Pattern Analysis & Machine Intelligence, 2017, PP (99):2999-3007.
[9] Derakhshani M M, Masoudnia S, Shaker A H, et al. Assisted Excitation of Activations: A Learning Technique to Improve Object Detectors[C]//Proceedings of the IEEE Conference on Computer Vision and Pattern Recognition. 2019: 9201-9210.
[10] Liu Z, Li J, Shen Z, et al. Learning efficient convolutional networks through network slimming[C]//Proceedings of the IEEE International Conference on Computer Vision. 2017: 2736-2744.
[11] W. Chen, J. T. Wilson, S. Tyree, K. Q. Weinberger, and Y. Chen. Compressing neural networks with the hashing trick. In ICML, 2015.
[12] S. Han, J. Pool, J. Tran, and W. Dally. Learning both weights and connections for efficient neural network. In NIPS, pages 1135–1143, 2015.
[13] M. Courbariaux and Y. Bengio. Binarynet: Training deep neural networks with weights and activations constrained to+ 1 or-1. arXiv preprint arXiv:1602.02830, 2016.
[14] M. Rastegari, V. Ordonez, J. Redmon, and A. Farhadi. Xnornet: Imagenet classification using binary convolutional neural networks. In ECCV, 2016.
[15] K. He, X. Zhang, S. Ren, and J. Sun. Deep residual learning for image recognition. In CVPR, 2016.
[16] S. Changpinyo, M. Sandler, and A. Zhmoginov. The power of sparsity in convolutional neural networks. arXiv preprint arXiv:1702.06257, 2017.
[17] B. Liu, M. Wang, H. Foroosh, M. Tappen, and M. Pensky. Sparse convolutional neural networks. In Proceedings of the IEEE Conference on Computer Vision and Pattern Recognition, pages 806–814, 2015.
[18] Hao L, Kadav A, et al. Pruning Filters for Efficient ConvNets [J]. 2016.
[19] Hu H, Peng R, Tai Y W, et al. Network Trimming: A Data-Driven Neuron Pruning Approach towards Efficient Deep Architectures [J]. 2016.
[20] Luo J H, Wu J. An Entropy-based Pruning Method for CNN Compression [J]. 2017.
[21] Yang T J, Chen Y H, Sze V. Designing Energy-Efficient Convolutional Neural Networks Using Energy-Aware Pruning[C]// 2017 IEEE Conference on Computer Vision and Pattern Recognition (CVPR). IEEE, 2017.
[22] Wen W, Wu C, Wang Y, et al. Learning Structured Sparsity in Deep Neural Networks [J]. 2016.
[23] Guo Y, Yao A, Chen Y. Dynamic Network Surgery for Efficient DNNs [J]. 2016.

Effect of Al Content on the Stability of MgF_2 Powder in Air at High Temperatures

Hukui Chen[1, a], Jie Huang[1, b] and Hongli Jiang[1, c]

[1]Boji University of Arts and Science, Baoji 721013 PR China

[a]hk7115@yahoo.com, [b]402409628@qq.com, [c]361523354@qq.com

Abstract. The effect of Al on the stability of MgF_2 powder at 600–1000 °C in air has been investigated by XRD, EDS and SEM. It is found that the thermal stability of MgF_2 is strongly influenced by both Al content and temperature. With the increase of Al content and temperature, MgF_2 was gradually oxidized to MgO, that is, the thermal stability of MgF_2 was gradually lowered. The reason that the addition of Al causes the thermal stability of MgF_2 to decrease may be related to the oxidation of Al and the formation of $MgAl_2O_4$ at high temperatures.

1. Introduction

As an excellent optical coating material, MgF_2 is the preferred choice for preparing the film with high reflectivity and high damage threshold due to its wide transparent band, low refractive index, wide energy gap, high mechanical strength and laser damage threshold [1]. MgF_2 is also an important electronic ceramic material, which has attracted widespread attention in the development of metal-ceramic materials because of its good insulating performance. Besides, during processing of Mg and its alloys, MgF_2, as a main component of the composite protective film formed on the surface of Mg and its alloy melt, plays a crucial role in the protection of the melt against burning by SF_6 and other fluorine-containing gases, and its content determines the protective effect of SF_6 and other fluorine-containing gases for the melt [2~5].

MgF_2 is considered to be a stable compound, which generally does not decompose or change to other compounds in air. However, some studies have found that MgF_2 becomes less stable at high temperatures and it has a tendency to convert to MgO. Liu et al [6] examined the stability of MgF_2 in air by X-ray diffractometer and found that MgF_2 would be changed to MgO with the temperature above 800 °C, and the conversion rate increased with increasing temperature. Chen et al [7] researched the oxidation of MgF_2 in air and found that the reaction of MgF_2 converted to MgO could occur at high temperatures, and the reaction was accelerated by the increase of temperature and the presence of impurities. In the study of preparing MgF_2/Al composite films by electron beam evaporation deposition, Weimer et al [8] found that about 1% MgO was present in the MgF_2/Al composite film. They believed that the presence of MgO may be due to the 2% Na impurity contained in the raw material, which leads to the oxidation of MgF_2 at high temperature. Xu and co-workers [9] investigated the preparation process of a 30% Ag-MgF_2 cermet material. It was found that MgF_2 component in the 30%Ag-MgF_2 cermet material would turn into MgO in high oxygen partial pressure atmosphere at the temperature of over 600 °C. But pure MgF_2 was rarely oxidized in the atmosphere even though at 1000 °C, while Ag participated in, MgF_2 would be severely oxidized into MgO under this condition. Sun et al [10] investigated the oxidation behavior of MgF_2 in Ag-MgF_2 cermets. They also found that when Ag participated in, MgF_2 could be oxidized to MgO at high temperature and high oxygen partial pressure atmosphere or air; otherwise, MgF_2 was rarely oxidized or not oxidized at all under the condition. Obviously, these results are different from those of Liu et al.

Thus it can be seen that whether the participation of impurities has an effect on the oxidation of MgF_2 remains a debate issue. Since the conversion of MgF_2 to MgO reduces the quality of MgF_2 and seriously damages its practical performance, it is necessary to further study the effect of impurities on the stability of MgF_2 in the air. In this paper, the oxidation properties of MgF_2 powder with different Al contents in air at high temperatures were investigated. The purpose of the paper is to clarify the effect impurities on the stability of MgF_2 in air, which will provide a theoretical basis for the reasonable processing and application of MgF_2 as optical coating materials and electronic ceramic materials and for the optimization of protection conditions of fluorine-containing gases for Mg and its alloy melt.

2. Experimental procedure

The raw materials used for the present study were high purity MgF_2 and high purity Al powders. The chemical composition of MgF_2 powder (wt.%) is 99.99 MgF_2, 0.001 Na, 0.001 Fe, 0.002 Si, 0.002 Ca, 0.001 Pb, 0.002 SO_4^{2+} and 0.001 H_2O. The chemical composition of Al powder (wt.%) is 99.8 Al, 0.08 Fe, 0.06 Si, 0.01Cu, 0.05 H_2O.

The MgF_2 powders with a particle size of 72 μm and Al powders with a particle size of 45 μm were mixed in the R-1002 rotary evaporator with the weight ratios of Al to MgF_2 of 0%, 1%, 5%, 10%, 20%, 30% and 40%, respectively. The mixed MgF_2 and Al powder was placed in a stainless steel mold with a diameter of 20 mm and pressed at a pressure of 100 MPa for 15 min, then a disk sample with the size of 20mm×2mm was obtained.

The experiment was conducted in an SK-G05123K tube furnace and the experimental procedure was as follows: The disk sample was first placed in an alumina crucible and then moved into a quartz glass tube in the tube furnace. Next, the sample was heated to the desired temperature at the rate of 8 °C/min and held for 2 h. After the time was reached, the oxidized sample was cooled and taken out for X-ray diffractometer (XRD), scanning electron microscope (SEM) and energy dispersive spectroscopy (EDS) analyses. XRD analysis of the oxidized sample was carried out on a Rigaku Ultima IV X-ray diffractometer with a Cu-Kα source operated at 40 kV and 40 mA. The scanning speed was 4°/min, the step width was 0.02°, and the diffraction angle was 20° to 80°. The elemental composition and morphology of the oxidized sample were investigated by an EDAX Genesis APEX EDS and Quanta FEG 250 FESEM.

3. Results and discussion

Figure 1 shows the XRD patterns of 1%Al+MgF_2 samples after exposure to air at 600 °C–1000 °C for 2 h. It can be seen that at 600 °C and 700 °C, only the peaks of MgF_2 appeared, which indicates that MgF_2 was stable at the temperature and no reaction occurred. However, as the temperature reached 800 °C, in addition to the peaks of MgF_2, the peak of MgO appeared, which indicates that MgF_2 became unstable and the reaction of conversion of MgF_2 to MgO occurred at this temperature. With the further increase of temperature, the content of MgF_2 decreased rapidly, and the content of MgO increased sharply.

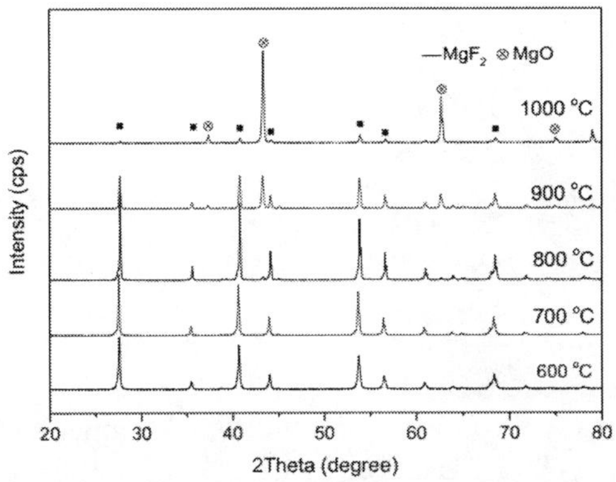

Figure 1. XRD patterns of MgF_2 sample containing 1%Al exposed to air at 600 °C–1000 °C.

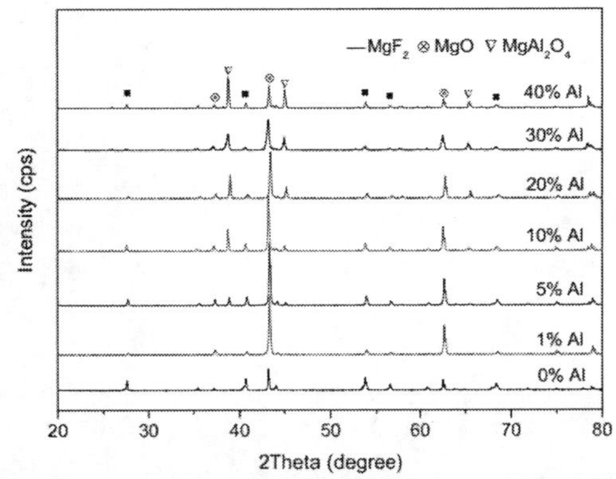

Figure 2. XRD patterns of MgF_2 samples with different contents of Al exposed to air at 1000 °C.

Figure 2 shows the XRD patterns of MgF_2 samples with different contents of Al after exposure to air at 1000 °C for 2 h. As can be seen, after exposure to air at 1000 °C for 2 h, the sample containing 1% Al was mainly composed of MgO and a small amount of MgF_2. For the sample containing 5% Al, a small amount of $MgAl_2O_4$ appeared, in addition to MgO and MgF_2. With the further increase of Al content in the sample, the content of MgO increased first and then decreased, the content of MgF_2 decreased slowly, and the content of $MgAl_2O_4$ increased gradually.

Figure 3 shows the SEM images of MgF_2 sample containing 1% Al after exposure to air at 600 °C–1000 °C for 2 h. As shown in Figure 3, the effect of temperature on the surface morphology of the samples is not very significant. As the temperature increased, the surface became rough and white particles increased.

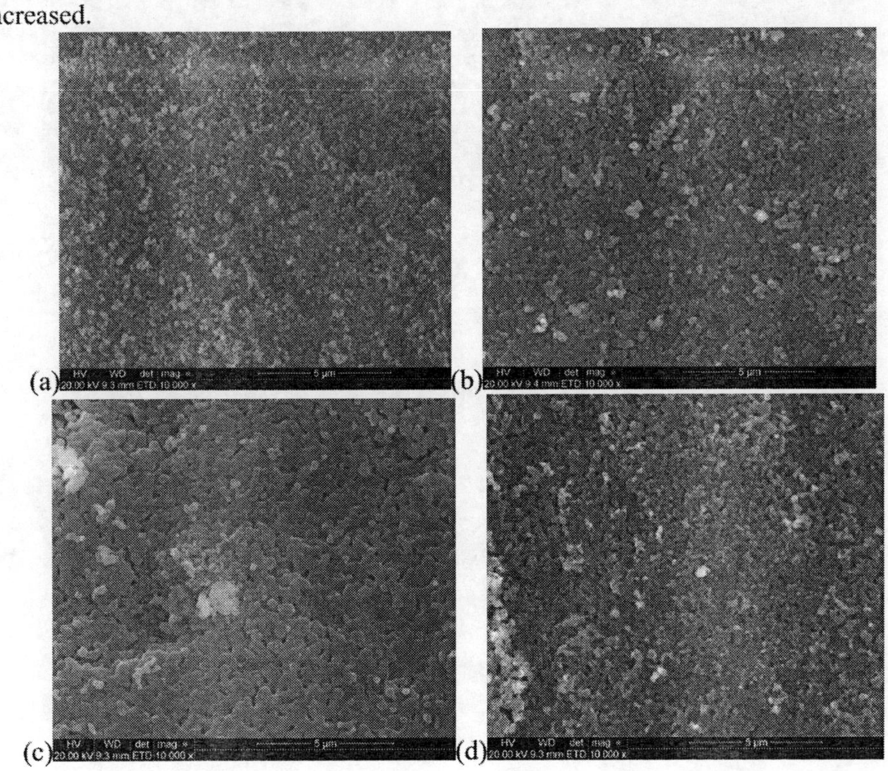

(e)

Figure 3. SEM images of MgF_2 sample containing 1% Al exposed to air at different temperatures: (a) 600 °C, (b) 700 °C, (c) 800 °C, (d) 900 °C and (e) 1000 °C.

Figure 4 shows the SEM images of MgF_2 sample with different contents of Al after exposure to air at 1000 °C for 2 h. It can be observed that as the Al content increased, the surface of the sample became more and more rough, white particles increased and tended to become larger.

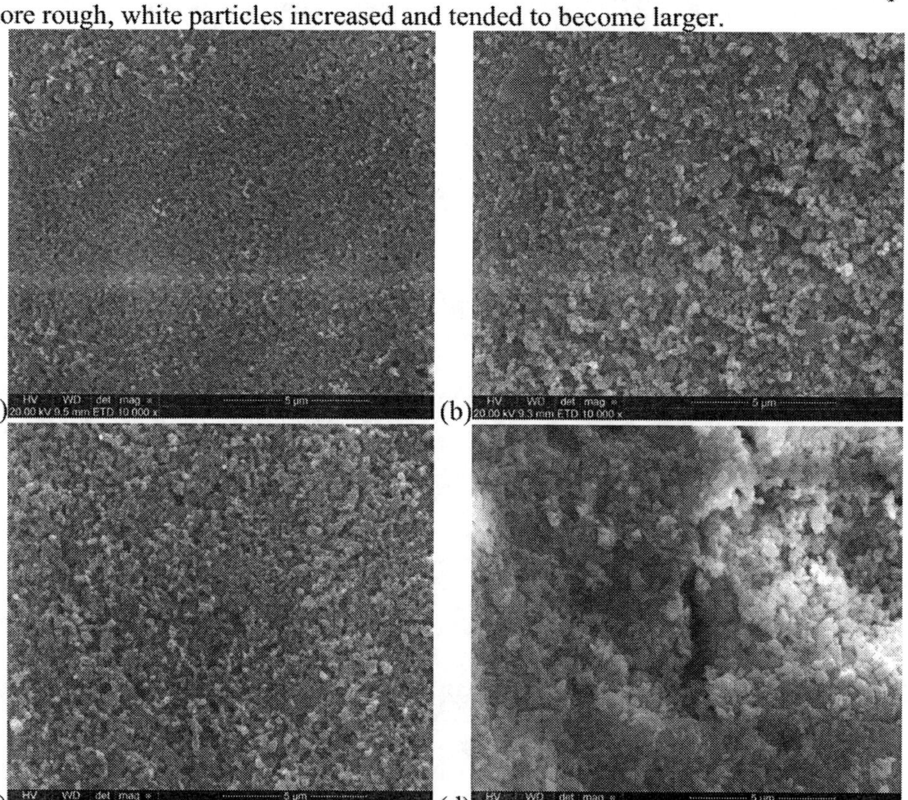

(a) (b) (c) (d)

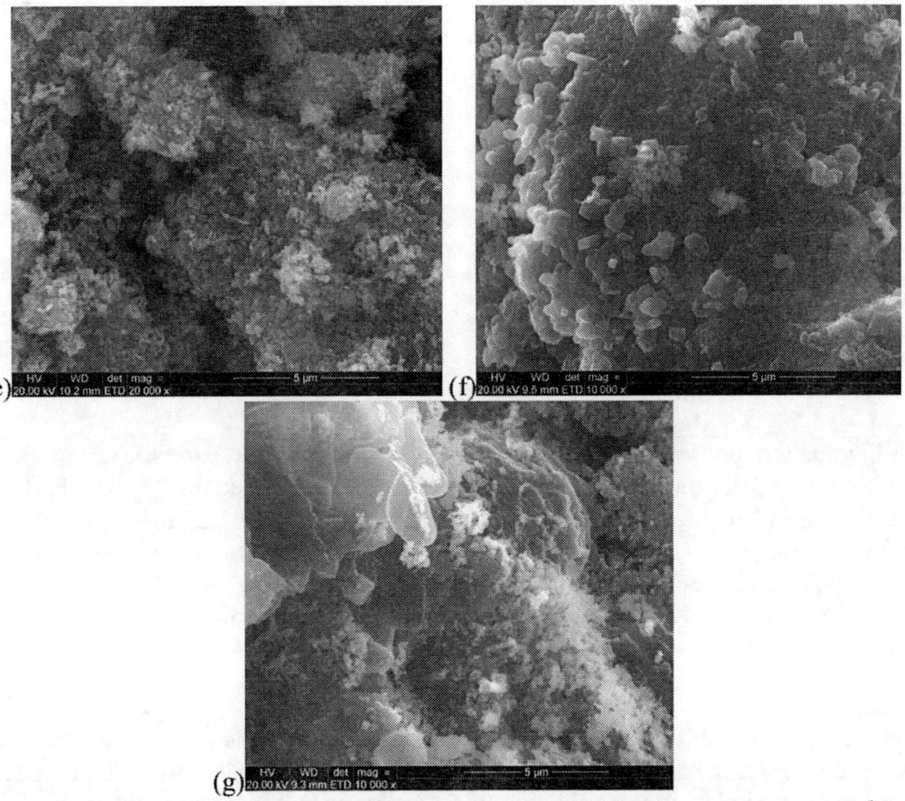

Figure 4. SEM images of MgF_2 sample with different contents of Al exposed to air at 1000 °C: (a) 0% Al, (b) 1% Al, (c) 5% Al, (d) 10% Al, (e) 20% Al, (f) 30% Al and (g) 40% Al.

The EDS analysis results of MgF_2 samples containing 1% Al after exposure to air at 600 °C–1000 °C for 2 h are listed in Table 1. According to Table 1, at 600 and 700 °C, the content of fluorine and oxygen did not change very much in the sample, and the oxygen content was small. As temperature increased from 800 to 1000 °C, the oxygen content decreased and the fluorine content increased. At 1000 °C, the oxygen content reached a maximum and the fluorine content reached a minimum. This result indicates that MgF_2 sample containing 1% Al undergoes significant oxidation at 800 °C. As the temperature increased further, the oxidation process intensified. At 1000 °C, MgF_2 was almost completely oxidized.

Table 1. EDS results of MgF_2 samples containing 1% Al exposed to air at 600 °C–1000 °C [wt.%]

Temperature [°C]	O	F	Mg	Al
600	1.92	53.97	41.90	2.21
700	1.18	54.91	43.34	0.57
800	9.34	43.46	46.44	2.75
900	28.89	9.23	60.88	1.00
1000	30.99	0.87	67.56	0.58

The EDS analysis results of MgF_2 samples with different contents of Al after exposure to air at 1000 °C for 2 h are shown in Table 2. It can be found that the content of fluorine was very small. As the aluminum content increased, the contents of fluorine and oxygen were generally reduced and the content of aluminum was increased.

Table 2. EDS results of MgF_2 samples with different contents of Al exposed to air at 1000 °C [wt.%]

Al [wt.%]	O	F	Mg	Al
0	43.12	2.59	54.29	-
1	30.99	0.87	67.56	0.58
5	32.15	1.68	65.56	0.61
10	34.58	1.60	55.82	8.00
20	31.24	0.53	51.46	16.76
30	30.65	0.95	47.17	21.23
40	30.35	1.43	36.52	31.70

The potential reactions that might occur in the MgF_2 + Al system in air are:

$$MgF_2(s) + 1/2O_2(g) = MgO(s) + F_2(g) \quad (1)$$

$$MgF_2(s) + H_2O\,(g) = MgO(s) + 2HF(g) \quad (2)$$

$$2Al\,(l) + 3/2O_2(g) = Al_2O_3(s) \quad (3)$$

$$MgO(s) + Al_2O_3(s) = MgAl_2O_4(s) \quad (4)$$

According to literature 11, thermodynamically, reaction (1) is not favorable (ΔG>0), other reactions are favorable (ΔG<0) in the temperature range of interest, and reaction (3) has the lowest ΔG value. Therefore, the mechanism of Al promoting the oxidation of MgF_2 at high temperatures may be as follows: Al can react with O_2 to produce Al_2O_3, the resulting Al_2O_3 then can react with MgO to form $MgAl_2O_4$, which consumes MgO and breaks the balance of reaction (2), thereby accelerating the reaction of MgF_2. It is clear that the formation of Al_2O_3 is the main reason for Al to accelerate the oxidation of MgF_2 in this study.

4. Conclusions

The effect of Al on the stability of MgF_2 was related to its content and temperature. With the increase of Al content and temperature, the thermal stability of MgF_2 was gradually lowered. The effect of Al on the thermal stability was linked to the oxidation of Al in the system, which produced Al_2O_3 and promoted the formation of $MgAl_2O_4$, resulting in the acceleration of the oxidation reaction of MgF_2.

Acknowledgements

This work was supported by National Natural Science Foundation of China (Grant No. 51471003) and Natural Science Basic Research Plan in Shaanxi Province of China (Program No. 2018JM5035).

References

[1] S.Z. Li, H. Liao, J.C. An, J.Q. Wang, R. Fu, X.J. Xiao, H.Y. Bai and M.D. Xie: J. Yunnan Normal Univ. (Nat. Sci. Ed.) (in Chinese) Vol. 33 (2013), p. 34.
[2] G. Pettersen, E. Øvrelid, G. Tranell, J. Fenstad and H. Gjestland: Mater. Sci. Eng. A Vol. 332 (2002), p. 285.
[3] S.P. Cashion, N.J. Ricketts and P.C. Hayes: J. Light Met. Vol. 2 (2002), p. 37.
[4] H.K. Chen, J.R. Liu and W.D. Huang: Corros. Sci. Vol. 52 (2010), p. 3639.
[5] J.R. Liu, H.K. Chen, L. Zhao and W.D. Huang: Corros. Sci. Vol. 51 (2009), p. 129.
[6] Y.M. Liu, Z.Y. Zhao, A.X. Li, K.T. Yang, J.H. Han and Z.Q. Sun: J. Anhui Univ. (Nat. Sci. Ed.) (in Chinese) Vol. 26 (2002), p. 46.
[7] H.K. Chen, Y.Y. Jie and L. Chang: IOP Conf. Series: Mater. Sci. Eng. Vol. 170 (2017) 012035.
[8] J.J. Weimer, J. Kim, M. Zkic and D.G. Torr: J. Vac. Sci.Technol. A Vol. 13 (1995), p. 1008.
[9] Z.Y. Xu, D.M. Sun, A.X. Li and Z.Q. Sun: J. Chin. Ceram. Soc. Vol. 30 (2002), p. 505.
[10] D.M. Sun, Z.Q. Sun, A.X. Li and Z.Y. Xu: Vacuum, Vol. 55 (1999), p. 39.
[11] H.K Chen, Y.Y.Jie and L.Chang: Foundry Technology, Vol. 39 (2018), p. 1325.

Research of photon correlation technology

Lianying Zou[1,a], Ning Gao[2,b], Qian Xiao[3,c]

[1]Wuhan Institute of Technology, Wuhan 430205, China

[2]Institute of Biophysics, Chinese Academy of Sciences, Beijing 100101, China

[a]zou_ly@126.com, [b]392047692@qq.com, [c]qianxiao@ibp.ac.cn

Abstract: Photon correlation technology is an important technology in the field of nanoparticle measurement at present. This paper expounds the research significance of photon correlation technology, introduces the research background and basic concepts of this technology, and studies and analyzes the research status of the correlation algorithm of the new photon correlator. Finally, prospect and application development trend of this technology are analyzed, from the aspects of practicability, application environment and hardware conditions, the difficulties faced by the photon correlation technology are put forward. There are some difficulties when the photon correlator needs to meet the requirements of accuracy, small sampling interval and large dynamic range, this article presents some possible solutions.

1. Introduction

Fluorescence Correlation Spectroscopy (FCS), counts the photon pulse data of detected photoelectric signals and performs correlation operations to obtain the light intensity function, which is used to detect the concentration, size and interaction between molecules under specific experimental conditions [1, 2]. By calculating the time correlation of light fluctuation of scattered by particles, the size of nanoparticles in solution can be obtained by photon correlation spectroscopy, and various motions of polymer chain can be deduced. The concept of Fluorescence Correlation Spectroscopy was put forward in the 1970s, as an interdisciplinary discipline consisting of physics, material science, biochemistry and electronic informatics, it didn't get the actual application and research progress. In the 1990s, with the rapid development of computer technology, laser focusing technology and optical detection technology, fluorescence correlation technology was gradually applied in various fields [3]. With the increasing demand of fluorescence correlation measurement devices in various research and application departments, the processing speed, accuracy and dynamic range of photon correlator are required more and more. As an interdisciplinary subject, the technical progress in this field can promote the development of multiple fields, but it is also restricted by other fields, such as the accuracy of photoelectric data acquisition.

Photon correlators require specialized hardware to obtain and calculate the intensity correlation of rapidly changing input signals, so photon correlators are usually relatively expensive, and most of the correlator instruments are used at the laboratory level. It has great practical significance that the fluorescence correlation spectrometers with independent core technologies are used at the industrial or commercial level. This paper analyzes and summarizes the recent researches in the field of photon correlator, and puts forward the direction of improving the performance of photon correlator from the perspective of the implementation method of photon correlator.

2. Basic concepts of photon correlation technology

2.1 Measurement of nanoparticles size by photon correlation technique

After the subjects are labeled with fluorescence, the fluorescence particles in the sample solution are subjected to Brownian motion by collisions of surrounding particles. Concentration of particle can be determined by measuring Brownian motion-related diffusion coefficients. When the laser is irradiated into the solution containing nano-fluorescence particles, scattering light is generated. Scattered light is converted into TTL level signal by photoelectric detection system, and the signal is processed by amplification, reshaping, etc, and the number of photons in each cycle is obtained by the photon counter. The number of photons in the sampling time is used as the input of the photon correlator to obtain the correlation function, which is used to obtain the concentration of fluorescent labeled particles result to the upper computer.

2.2 Photon correlator experimental device

The laser excited labeled fluorescence particles, and the optical path system transmitted the fluorescence signal reflected by the fluorescence particles to the photoelectric detection device, Photon correlator finally carries out photon counting and correlation operation [4, 5]. After the front-end processing of the signal, the photoelectric signal becomes a pulse string in the form of level, and the pulse number within the sampling time is the function of light intensity. Photon correlator deals with the real-time correlation algorithm of light intensity function and saves the result in the host computer. The composition of the experimental device of photon correlator is shown in Fig.1.

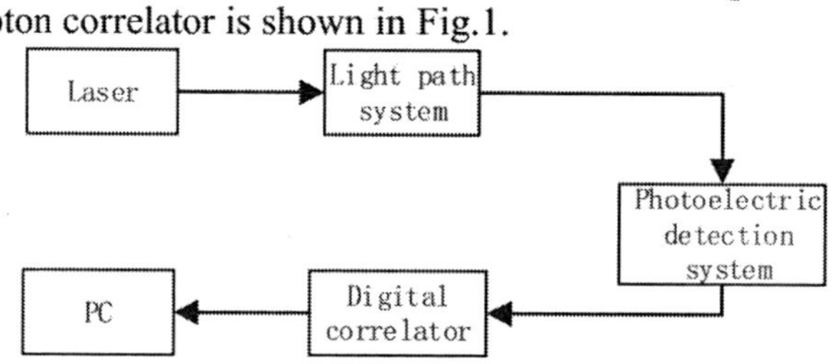

Fig.1 Diagram of experimental device of photon correlator

2.3 Light intensity autocorrelation function

The scattered light intensity signal varying with time can be obtained through the time autocorrelation of the light intensity function, refer with: Eq.1,

$$G(\tau) = \langle I(t)I(t+\tau)\rangle = \lim_{T\to\infty} \frac{1}{T}\int_0^T I(t)I(t+\tau)dt. \quad (1)$$

In Eq.1, $I(t)$ and $I(t+\tau)$ represents the scattering light intensity signal at t and $t+\tau$ moment, brackets < > represents the average operation, T represents the total experimental time. In the delay time is very small, $I(t)$ and $I(t+\tau)$ is almost equal, two signal correlation is very strong, the larger delay time τ is, the weaker the correlation is.

Eq.1 is the autocorrelation function of light intensity in the continuous case. In fact, it is in discrete form, refer with: Eq.2,

$$\hat{G}(i\Delta t) = \frac{1}{M}\sum_{j=1}^{M} n(j)n(j+i). \quad (2)$$

M is the total sampling times, the total experimental time is $M\Delta t$, Δt is the single sampling time, $n(j)$ is the number of photon pulses obtained by the photon correlating device during the j.th sampling time, and $n(j+i)$ is the number of photon pulses during the $j+i$.th sampling time. Eq.2 actually refers to the estimation of the light intensity autocorrelation function with $i\Delta t$ in a finite time, and the experimental time and the sampling time will affect the estimation error of light intensity autocorrelation function.

3. Improved photon correlator

3.1 FPGA based hardware correlator

The photon correlator processes the counting value obtained from the photon counter, and seperate the value according to the sampling period. The current sampling period is multiplied by the value of different delay sampling periods, and the results are added to finally complete the correlation operation [6]. The delay, multiplication, accumulation, and cache output of a particular delay signal consists of delay modules, multiplication modules, and accumulation modules, which are called a channel in the hardware implementation, and the photon correlator consists of these channels. The dynamic range of correlation calculation can be increased by increasing the number of correlator channels. As a high-speed integrated chip, the actual number of channels is limited by internal resources of FPGA, cannot blindly increase channels [7]. The delay unit is realized by the shift register. When the number of channels is fixed, increasing the sampling time can also increase the dynamic range of correlation calculation, but reduce the resolution of correlation function curve.

In 1985, Klaus Schatzel of Germany first proposed the multi-sampling time related technology [8]. The structure of photon correlator proposed by him can guarantee the resolution of correlation function curve and enlarge the dynamic range of correlator within a limited number of channels. Multi-delay photon correlation structure groups correlator channels according to sampling time. Each group of channels has the same sampling time, and different groups of channels have different sampling time. The whole correlator can be regarded as an organic connection of multiple linear correlation modules. In the structure of muti-sampleing-time correlator, the time for sampling number of photons remains the same in the group, and the time for sampling number of photons doubles between groups. There are 16 linear channels in group 0, and 8 linear channels in the other group, as shown in Fig.2.

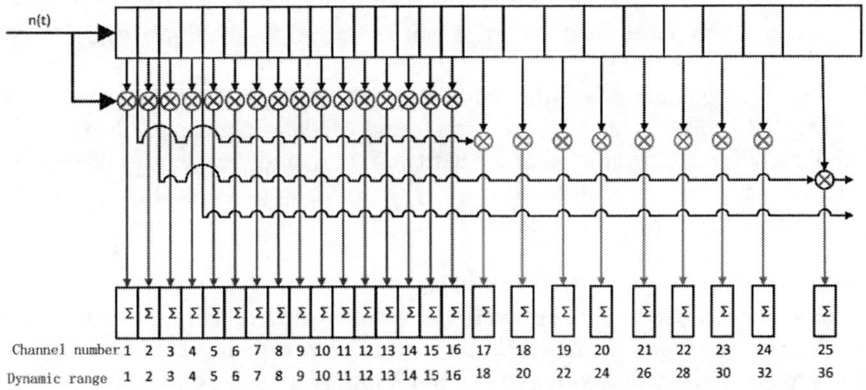

Fig.2 Multi-sampling time correlator structure diagram

At present, many scholars have made many improvements in the design of large dynamic range high-speed correlator based on FPGA. For example, Lu Wenling, Chen Yenting and others from Shandong University of Technology completed the design of high-speed correlator with large dynamic range based on DSP+FPGA [9, 10, 11]. They used DSP to be designed as the ring shift register and combined it with FPGA chip to achieve large dynamic range. The dynamic range is several orders of magnitude larger than using FPGA chips alone. Wang Lipeng from Qilu University of Technology completed the structural design of multi-delay adjustable correlator, solving the problem of adjustable dynamic range of photon correlator [12]. The ideal photon correlator should be able to dynamically adjust its dynamic range, adjust the delay time of each channel, and make the resolution of correlation function curve reach the best under different sampling time.

3.2 Software correlator based on Labview

In 2003, Davide Magatti and Fabio Ferri developed a multi-sampling time correlator using software [13, 14]. The correlator consists of a standard photon counting unit, a fast timer (6602-pci) and a PC unit.

IOP Conf. Series: Materials Science and Engineering **711** (2020) 012097 doi:10.1088/1757-899X/711/1/012097

When the delay time is small, the input signal has a great impact on the result of correlation function, records accurate time data. When the delay time is large, the accurate information of data is abandoned to improve the operation speed. The measurement system combining the two methods can calculate the correlation function online. Software has natural advantages in data processing, and software correlator is more flexible [15]. Software can be used to reduce the cost effectively. Limited by the performance of computer and electronic equipment, software correlator lacks competitiveness in running speed. However, with the continuous development of electronic equipment field, software correlator also has great development potential.

4. Development difficulties and solutions of photon correlator

4.1 Combination of software and hardware to achieve a large dynamic range

In radio astronomy, materials science and other application fields, the dynamic range of correlator is required to reach at least 10^{12}, and the experimental time can last more than 100 hours. In the implementation of photon correlator with large dynamic range, hardware correlator is expensive, and the number of channels and other parameters are fixed. The hardware resources cannot be extended indefinitely, and the delay time need to be quite large. The software correlator has slow operation speed, low precision and low performance. So far, the implementation of photon correlator is in a state of "hard or soft". Although some scholars have put forward similar Suggestions, there is no new low-cost photon correlator combining hardware and software.

Presented here based on muti-sampling-time correlation technology, combined with the advantage of software and hardware parts. When the sampling time is small, using the speed and performance advantage of the hardware implementation method to process the correlation function in real-time. When the sampling time is large enough to be minutes or seconds, some correlator channel calculation can be caculated by software, and can save cost of correlator, and improve the correlator flexibility. The hardware part of the photon correlator carries out real-time correlation operation and collection of photon counting signals. When the collection and calculation reach a certain time, the data will be transferred to the upper computer, and the hardware part itself will be cleared to start the next round of work again. The PC processes and stores each round of data it receives. In this way, high-density processing and operation are completed in the hardware part, and low-density processing and operation are completed in the upper computer software. However, there is a difficult control problem in the process of software and hardware interaction.

4.2 The development difficulty of software correlator

The photoelectric signal extraction of software correlator is different from that of hardware correlator. It needs to collect data to upper computer through a high-performance data acquisition card. There are errors in the optoelectronic pulse width from photomultiplier tube and the frequency of weak signal detection circuit, so there is an error in the photoelectric signal collected by the data acquisition card. Existing photomultiplier tubes have recovery time (or dead time), which affects the minimum sampling time of correlator. The minimum sampling time of existing correlators is generally 25ns, which requires a more sensitive photoelectric detection system to reduce the minimum sampling time. If the signal collected by the photon correlation system is the photon arrival signal, all the information collected by the photon signal can be retained, and part of the information will not be lost due to multiple pulses within two adjacent sampling times. How to improve the performance of the data acquisition system is a difficult problem.

5. Summary

Based on the development of photonic correlation technology in recent 20 years, this paper reviews the basic concept, implementation method and research status of this technology. Taking the new photon correlator as the research focus, this paper summarizes the key problems existing in the field of photon correlator technology from two aspects of hardware and software correlator, and puts forward possible

IOP Conf. Series: Materials Science and Engineering **711** (2020) 012097 doi:10.1088/1757-899X/711/1/012097

improvement of directions. The current development of photon correlation technologies requires more accurate and economical measurement methods to be commercialized or industrialized. It is hoped that this paper can provide some references and inspirations for researchers in the field of measuring instruments and photons.

References

[1] Liu tiegen, Zhang fan, Meng zhuo. Research status and development of methods for detecting the size and distribution of nanoparticles [J]. Optical technology, 2005 (01):96-100.

[2] Saffarian S, Elson E L. Statistical analysis of fluorescence correlation spectroscopy: The standard deviation and bias [J]. Biophysical Journal, 2003, 84(3):2030-2042.

[3] Elson E L. Fluorescence Correlation Spectroscopy: Past, Present, Future [J]. Biophysical Journal, 2011, 101(12):2855-2870.

[4] Liu W, Shen J, Sun X. Design of Multiple-Tau Photon Correlation System Implemented by FPGA[C]// International Conference on Embedded Software & Systems. IEEE, 2008.

[5] Lee H Y , Lin H Y , White J D , et al. High-speed low-cost correlator for single molecule fluorescence correlation spectroscopy[J]. Proceedings of SPIE - The International Society for Optical Engineering, 2009, 7185:71850R-71850R-8.

[6] Mocsár, Gábor, Kreith, Balázs, Buchholz J, et al. Multiplexed multiple-{\tau auto- and cross-correlators on a single FPGA [J]. Eprint Arxiv, 2011.

[7] Chen haibin. Design of hardware system of fluorescence related spectrum analyzer [D]. Shanghai jiaotong university, 2013.

[8] Klaus Schäzel, Martin Drewel, Sven Stimac. Photon correlation measurement at large lag times: improving statistical accuracy [J], Journal of Modern Optics, 1998, 35(4): 711-718.

[9] Cheng y t. research on multi-channel time-delay photon correlator [D]. Shandong university of technology, 2009.

[10] Cheng yanting, shen jin, liu wei, sun xianming, Yang yan. Normalization method of autocorrelation function in multi-tau photon correlator [J]. China laser, 2009,36(02):444-448.

[11] Cheng yanting, shen jin, liu wei, et al. A method and implementation of dynamic range adaptive adjustment of photon correlator [J]. China laser, 2008, 35(s2).

[12] Wang lipeng. Research on dynamic light scattering nanometer particle size measurement method based on FPGA [D]. Qilu university of technology, 2014.

[13] Davide Magatti and Fabio Ferri, Fast multi-tau real-time software correlator for dynamic light scattering[J], Optical Society of America, 2011

[14] Magatti D, Ferri F. 25 ns software correlator for photon and fluorescence correlation spectroscopy [J]. Review of Scientific Instruments, 2003, 74(2):1135.

[15] Schaub, Emmanuel. High countrate real-time FCS using F2Cor [J]. Optics Express, 2013, 21(20):23543.

Design and Research of Ground Control System of the UAV-borne Photoelectric Platform

Xinglong Zhao[1,a], Jingshan Zhao[2,b], XingZheng[3,c], JianCheng[4,d], Chuanmei Bao[5,e], Zhengwu Zhang[6,f], Xiaojin Fu[7,g] and YunLiu[8,h]

[1-8]Ordnance NCO Academy, Army Engineering University, Hubei,430075, China

[a]zhaotrueman666@126.com, [b]467726114@qq.com, [c]zhengxcn@foxmail.com, [d]10954822@qq.com, [e]279017167@qq.com,[f]picante@126.com, [g]6785562385@qq.com, [h]1253684078@qq.com

Abstract: According to the working principle of a UAV-borne photoelectric platform, as well as the main functions and characteristics of the photoelectric platform ground control system, the overall design scheme and target of this system are proposed with an introduction of its composition and working principle. On this basis, the design and research of the system are carried out from hardware and software. Specific schemes of the two aspects are given. Finally, the future of the photoelectric platform ground control system is predicted on the trend of UAV-borne photoelectric platform.

1. Introduction

Unmanned aerial vehicle (UAV) photoelectric platform is a high technology system including the fields as optics, mechanism and electricity. It is an important part of UAV airborne reconnaissance system. It mainly provides real-time battlefield television images, infrared images and platform specific parameters as the intelligence and reference for UAV battlefield surveillance and positioning [1].

2. Overall Design of Control System

As the photoelectric platform needs to be equipped with visible TV camera, forward-looking infrared instrument, laser rangefinder and other loads, therefore, in order to fully meet the requirements and purpose of stability control and function detection, the system needs to achieve these in design:

(1) Provide stable DC power supply, power addition and power failure control for UAV-borne photoelectric platform;

(2) Communicate directly with the control components of the UAV-borne photoelectric platform, completely control the system, receive and display the feedback information of each component;

(3) Control functions such as doubling and focusing of CCD sensors and infrared sensors in the UAV-borne photoelectric platform, switch and configure working states such as locking, tracking and scanning of the platform;

(4) Control the target tracking box to search the target, and achieve the target acquisition and tracking;

(5) Functions of sending and receiving right or wrong judgment instruction, fault location information display;

(6) Functions of real-time display, collection and playback of video images output from the photoelectric platform, video input and synchronous output capability;

(7) Cooperate with other test equipment to complete the performance test of the photoelectric platform, such as search angular velocity, search angular acceleration, angle measurement accuracy and stability accuracy;

(8) Control the vibration reduction lifting mechanism of the photoelectric platform to lift and down, and display the position in place;

(9) Display the power consumption of the photoelectric platform.

3. Hardware Design of Control System

The ground control system of UAV-borne photoelectric platform is mainly realized by a set of ground control cabinet.

3.1 Overall Design of Control Cabinet Hardware

The control cabinet adopts the structure of "main control computer +VXI(PXI) + desktop instrument + connector + adapter", supplemented by special adapters and control accessories[2]. The control cabinet is mainly composed of the cabinet body, main control computer, display, image acquisition module, power conversion module, communication conversion module, voltmeter, ammeter and device switch, as shown in figure 1.

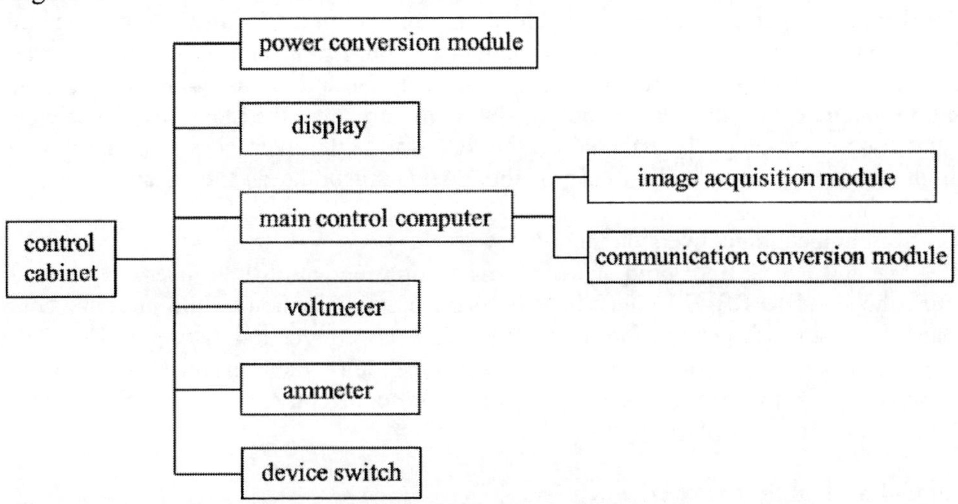

Fig.1. Control cabinet composition

Adopting the frame structure, internal control cabinet is divided into independent spaces with diaphragm plate. Main control computer, display, image acquisition module, power conversion module, communication conversion module, voltmeter, ammeter, and device switch are installed in different spaces. The diaphragm plate can be moved up and down to the right place before fixation. To make the front panel operating easily and nicely presented, voltmeter, ammeter, and device switch are in a horizontal line with logos above the switch.

3.2 Selection Analysis of Main Components of Control Cabinet

Main components of the control cabinet include main control computer, power conversion module, image acquisition module and communication conversion module.

(1) Main control computer

The main control computer of the control cabinet sends instructions issued by the control software to the photoelectric platform through the communication conversion module to realize the state control, system parameter transfer and scene search of the photoelectric platform.

IOP Conf. Series: Materials Science and Engineering **711** (2020) 012098 doi:10.1088/1757-899X/711/1/012098

The main control computer is the running platform of the control software with functions of data storage and image display. It adopts industrial control computer with stable performance, which can not only meet the requirements of software operation, but also avoid the characteristics of unstable performance of desktop, so as to realize the integration and reliability of the ground control system design [3]. The operator can control the photoelectric platform by operating the photoelectric platform control software of the main control computer in the cabinet.

The main control computer collects the images output from the photoelectric platform through the image acquisition module and outputs them to the display. When recording the images, it can record them through the image acquisition software for later playback.

(2) Power conversion module

The normal working voltage of the visible camera and forward-looking infrared instrument in the photoelectric platform is 28V DC (working voltage range is 24V ~ 29V), with a rated power of 250W and the maximum power of 350W. The power supply voltage is required to be stable to avoid the influence on the reliability and life span of some precision parts. In particular, the reliability of the power supply is highly required, otherwise the entire photoelectric platform will be damaged [4]. Therefore, it is necessary to design a power conversion module in the control cabinet to convert 220V/50Hz alternating current into 28V DC for the use of the visible camera and forward-looking infrared instrument. The power supply conversion module of the control cabinet can choose off-line AC 220V/28V DC switching power supply.

(3) Image acquisition module

To display output images from visible light cameras and the forward-looking infrared detector within a photoelectric platform, a general monitor can do. But its control system integration is not high. Since images are needed in some function test sessions for datum positioning and data acquisition, it is unable to handle data either. Image acquisition card can solve the above two problems. It is small in size and can be conveniently installed on the PCI slot of the main control computer. We can choose the high digital video acquisition card of the YuanGang brand as the image acquisition card of the control cabinet.

(4) Communication conversion module

RS-232 and RS-422 are both standard serial communication interfaces. In general photoelectric systems, the standard RS-422 interface is used for communication, while ordinary computers are the standard serial communication interface of RS-232, so conversion is required [5]. RS-422 / RS-232 communication interface converter has got mature technology and products, so a conversion interface card is enough. A bi-directional communication protocol converter of the ShengWei brand can be selected as the converter of the control cabinet.

4. Control System Software Design

4.1 Control Software Design

The control software is integrated into the control cabinet running in the main control computer. It is mainly composed of initialization subroutine module, human-computer interaction subroutine module, timing interrupt subroutine module and feedback processing subroutine module, as shown in figure 2.

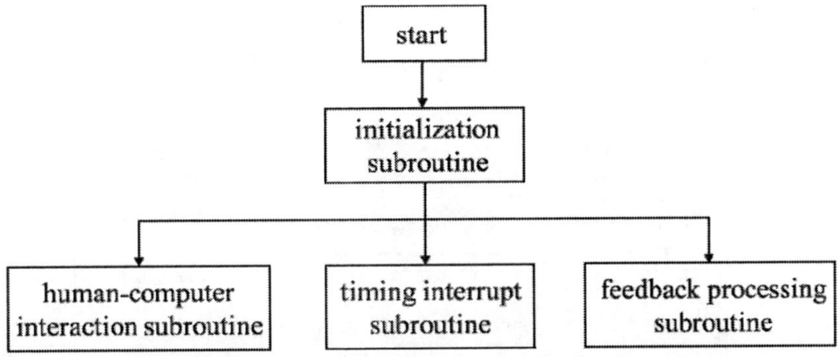

Fig.2. Process chart of control software program

The initialization subroutine mainly completes the initial settings of serial port control, human-computer interaction control, timer control and data variables. Specifically, it mainly includes the definition of serial port baud rate, port number, working mode, check mode, human-computer interaction mode and timer interrupt cycle, as well as the assignment of data variables.

Human-computer interaction subroutine mainly completes the human-computer interaction between ground control system and the task operator. When the task operator operates the software control interface through the mouse or keyboard, the control software program responds to the operation control through buttons, check boxes, text boxes and other controls. The control software will judge the correctness of the command, and identify the type of the command, as well as make corresponding processing according to different types of operation commands.

The timing interrupt subroutine is mainly used to complete the instruction transmission to the photoelectric platform. The timing interrupt subroutine interrupts every 80ms and sends the corresponding control instruction to the photoelectric platform by calling the serial port control. What needs to be explained here is that sending an instruction to the photoelectric platform and receiving a set of feedback messages from the photoelectric platform is called an instruction cycle, and the time of an instruction cycle is 80ms. In theory, the program runs until the user turns off the software control interface. The timing interrupt subroutine is used to determine whether human-computer interaction existed during the last 80ms. If there is human-computer interaction, the control instruction of task operator is sent to the photoelectric platform[6]. On the contrary, maintenance communication instructions will be sent to the photoelectric platform and displayed on the software control interface.

The feedback processing subroutine is mainly used to receive the feedback data from the photoelectric platform, store the feedback data into an array, and display the received feedback data on the software control interface. Meanwhile, status information corresponding to the photoelectric platform is displayed in text form on the software control interface.

4.2 Control Software Interface Design

The main interface of UAV photoelectric platform control software is shown in figure 3. The control interface is developed based on Visual C++ software, which mainly sends instructions to the photoelectric platform through the serial port of the main controller and collects the feedback information of the photoelectric platform.

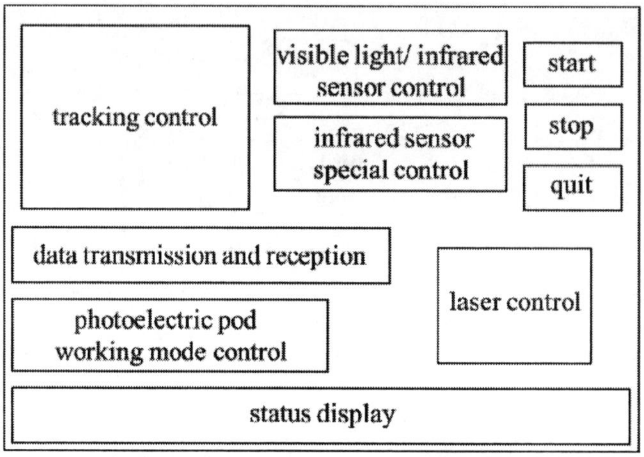

Fig.3. Main interface of control software

The main interface includes visible light/infrared sensor control, infrared sensor special control, photoelectric pod working mode control, data transmission and reception, tracking control, laser control and status display modules.

(1) The visible light/infrared sensor control part includes the adjustment of the focal length and field of view of the sensor;

(2) The infrared sensor special control part mainly aims at image gain and brightness adjustment, control mode setting, positive and negative image switching, image correction, etc., and can also control power supply and power off.

(3) The photoelectric pod working mode control part is used to switch operating modes of it: start, inertia, recovery, hold, lock, automatic scan, and setting azimuth and pitch angle under the lock and scan working state.

(4) The data transmission and reception part is used to display the data sent and received, which will increase after the system is powered up;

(5) The tracking control part is used to confirm the target and implement tracking or change the tracking target;

(6) The laser control part is used for laser ranging and laser emission period determination, etc.;

(7) The status display contents include video source type, focal length value, azimuth and pitch angle value of rotary table, timing, etc.

5. Conclusion

With the progress of science and technology and the development of modern war, UAV-borne photoelectric platform develops rapidly. It mainly advances towards the three-dimensional space, the diversification and miniaturization of structural forms, the integration of sensors, and the integration of observation and strike. Therefore, the ground control system of UAV-borne photoelectric platform needs to be developed in the following aspects: optimize the structure form, reduce the volume of control cabinet, and reduce the weight; It also needs to be easily compatible with a variety of photoelectric platforms; The stability accuracy of the control system gets gradually improved, and the integration and modularization become more prominent.

References

[1] Shi lixia. Research on target tracking jamming technology of airborne photoelectric countermeasure stabilized platform [D]. Changchun: changchun university of science and technology, 2009.

[2] Liu wei. Research on stable platform control system of airship photodetector [M]. Changsha: national university of defense technology, 2009.

[3] Xie haoyu. Research on optimization technology of stable platform structure of airship detector [M]. Changsha: national university of defense technology, 2008.
[4] Guan wei. Research on low speed and positioning of airborne photoelectric platform [M]. Changchun: changchun university of technology, 2007.
[5] Wu shuping. Development of vehicle triaxial stabilized platform control system [M]. Nanjing: nanjing university of science and technology, 2007.
[6] Zhang zhiyong. Research on key measurement and control problems of photoelectric stabilized servo mechanism [D]. Changsha: national university of defense technology, 2006.

IOP Conf. Series: Materials Science and Engineering **711** (2020) 012099 doi:10.1088/1757-899X/711/1/012099

Damage characteristics of VO_2 films under nanosecond/picosecond laser

LU Yuan[1,2,3,4, a], FENG Yunsong[1,2,3,4,b] , CHEN LeiLei[1,2,3,4,c] and WU Chang[1,2,3,4,d]

[1] Infrared and Low Temperature Plasma Key Laboratory of Anhui Province, Hefei 230037, China

[2] State Key Laboratory of Pulsed Power Laser Technology, Hefei 230037, China

[3] National Synchronous Radiation Laboratory of University of Science and Technology of China, *Hefei 230037,*

[4]Po box 904, 460 Huangshan Road, Hefei city, Anhui Province, P.R. China

[a]luyuanmail@163.com, [b]fyseei@163.com, [c]dianychen@sina.com, [d]boomswu@mail.ustc.edu.cn

Abstract. Vanadium dioxide films were prepared by magnetron sputtering method. An experimental apparatus for laser damaging VO_2 thin film was set up. Nanosecond laser and picosecond lasers were used to irradiate the thin film. The damage morphology of the thin film under these two kinds of laser was observed by metallographic microscope. Under the action of nanosecond laser, the damage of thin film is mainly thermal damage, which is mainly determined by the energy contained in the laser pulse itself, and the required damage threshold is relatively higher. The zero probability damage threshold of monopulse irradiate the film is about 0.13J /cm^2.Under femtosecond laser, the damage of thin film is caused by multi-photon ionization and avalanche ionization, and the damage threshold is relatively lower. The zero probability damage threshold of monopulse irradiate the film is about 10.2m J/cm^2.

1.Introduction

The vanadium oxide has many valence states. It will undergo a phase transition at a certain temperature and show different optical and electrical properties. In 1959, Miron, a scientist at bell LABS, first discovered that vanadium oxides have the characteristics of semiconductor-metal phase transition[1]. Among many valence states of vanadium oxide, the phase transition temperature of VO_2 is close to room temperature, which is about 68 °C. Near the phase transition point, the gap width of VO_2 will change, and the phase transition from insulator to metal will occur. The physical parameters of the material (such as refractive index, infrared transmittance, resistivity, etc.) will change, and the macroscopic manifestation is the abrupt change of optical and electrical properties before and after phase transition [2][3].

In recent years, new features of vanadium dioxide have been discovered continuously, making it a research hotspot[4][5]. Such properties make vanadium dioxide have many potential applications in intelligent Windows[6], laser protection[7][8], solar energy, camouflage[9] and other aspects.

In infrared band, vanadium dioxide thin film has quite a high transmittance change before and after phase transition, which makes it have good application prospect in laser protection field. In this paper, vanadium oxide thin films were prepared with sapphire as the substrate, and the damage characteristics of vanadium oxide thin films under nanosecond laser and picosecond laser were studied.

2. damage experiment of Laser IrradiatingVO_2 thin film

2.1 experimental equipment

Experimental device of laser irradiating damage vanadium oxide thin film is shown in FIG. 1. The laser emits a certain amount of energy, which is attenuated by the rotating wave plate and adjusted to the required energy value. Then the laser is focused by lens and irradiated on the vanadium dioxide thin film sample. The vanadium dioxide thin film samples are placed on the adjustable sample rack. The height and left and right positions of the sample rack are adjusted to make the laser shine on different points. The damage characteristics of these irradiated points under laser irradiation were observed. At the same time, a beam splitting mirror is used to output a part of the laser energy to an energy meter.

The thin film used in the experiment was prepared by magnetron sputtering method, and the base material was single crystal alpha-Al_2O_3, which's diameter was 2 inch, thickness of Al_2O_3 was 200nm and VO_2 film thickness was 400nm.

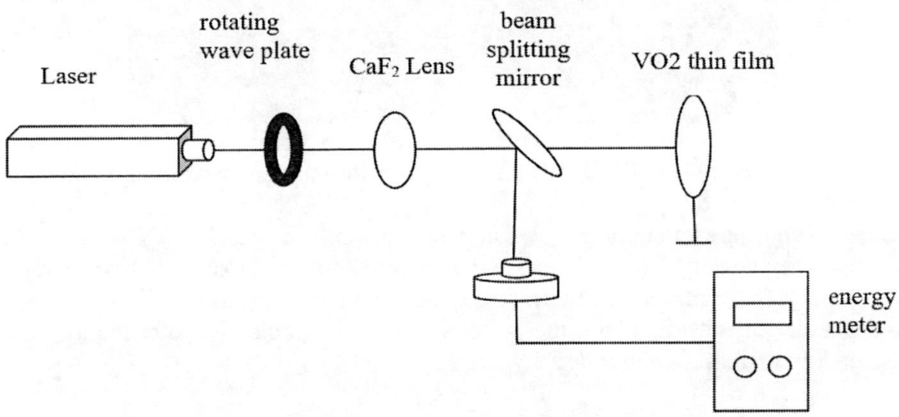

FIG. 1 schematic diagram of laser damaging VO_2 thin film test

Table 1 laser parameters in damage test

Nanosecond laser parameters		Femtosecond laser parameters	
Wave length	1064 *nm*	Wave length	850*nm*
Pulse repetition rate	1Hz	Pulse repetition rate	10Hz
Pulse width	10ns	Pulse width	50fs

3. Experimental results

3.1 damage threshold analysis

The laser irradiation point was directly observed by metallographic microscope. By observing the damage of each grade of energy density to the irradiation point on the thin film, the damage probability of the thin film under different pulse energy density was obtained.

After multi-point test, it was found that the 1-on-1 zero-probability damage threshold of VO_2 thin film under nanosecond laser was about 0.13J /cm^2. The zero probability damage threshold under femtosecond laser 1-on-1 irradiation is about 10.2m J/cm^2. The damage threshold under femtosecond laser is much lower than that under nanosecond laser. The damage threshold is related to the pulse width of the irradiation laser, and the narrower the pulse width is, the lower the damage threshold of the thin film is [10].

3.2 damage morphology and evolution law analysis

3.2.1 nanosecond laser damage morphology and evolution law

The nanosecond laser is used to irradiate vanadium oxide thin film, adjust the power of the incident laser, irradiate on different points of the film, observe the damage rule of each point.

It was magnified 200 times by a metallographic microscope for observation. The damage morphology is shown in figure 2.

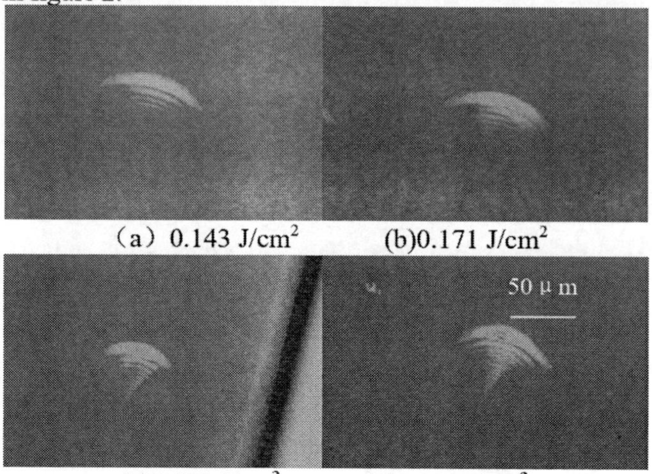

FIG. 2 micromorphologies of VO_2 film damage at different nanosecond laser power densities

It can be seen from figure 2 that the damage area of the thin film gradually expands with the increase of incident laser energy density. This is because the pulse width of nanosecond laser is relatively wide, and the main damage of thin film is thermal damage, which is mainly determined by the energy contained in the laser pulse itself. When the pulse width of laser pulse is nanosecond, the relationship between pulse energy and laser energy density is:

$$W_p = p_p \times \tau_p \times S_p \tag{1}$$

In the formula, Wp is the laser pulse energy film received, pp is incident laser energy density, τ_p and Sp is the pulse width and irradiated area respectively. When the spot radius and pulse width of the laser are the same, the higher the laser energy density is, the higher the pulse energy received by the film will be, and the larger the damage area will be under the effect of thermal diffusion.

By observing the damage morphologies at different energies, it can be found that: the damage morphologies of the thin films are all arc-shaped and wavy. With the increase of laser energy density, the damage morphologies gradually expand from inside to outside, and the outer arc length is smaller than the inner one.

This kind of arc ripple damage is caused by the melting of VO_2 film. When the incident laser energy reaches the damage threshold of the film, the film melts to form a accumulation layer. With the enhancement of laser energy, heat transfer to the outside causes the melting of the outer VO_2 film, leaving a wavy burn and melt trace in its heat transfer direction.

Since the nanosecond pulses used are gaussian lasers, the central energy is greater than the outside, resulting in the melting shape of VO_2 thin film showing a narrow arc with a wide middle and narrow sides. With the transfer of heat from inside to outside, the damage arc length becomes shorter gradually. It can also be found that for the same film, each wave bends in the same direction at different laser energies.This is because the pulse of the nanosecond laser is wider than the carrier-lattice energy transfer relaxation time of ps magnitude in VO_2. At this time, the damage is mainly caused by the lattice heat effect, and the ripple direction is related to the lattice orientation of the thin film itself. For films which have same lattice orientation, the wavy bending direction is same.

When the nanosecond laser energy density continues to increase to 0.322 J/cm^2, the damage morphology is shown in figure 3.

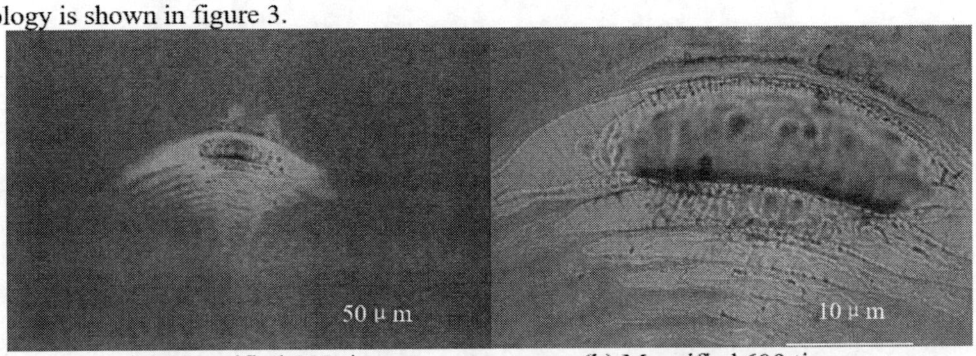

(a) Magnified 200 times (b) Magnified 600 times

FIG.3 damage morphology of VO^2 film under nanosecond laser of 0.322 J/cm^2

With the energy of the nanosecond pulses continued to increase, a dark pit about 20 microns deep appeared in the center of the spot. This is due to the excessive energy density of nanosecond laser pulse at this time, which causes the VO_2 film in the central region to be completely stripped and the laser continues to ablate Al_2O_3 substrate [11]. In FIG. 3b, the Al_2O_3 substrate exposed at the ablation center and the surrounding accumulation can be clearly seen. Many irregular polygonal squares in the accumulation also indicate that it has obvious melting and cooling traces.

3.2.2 damage characteristics of VO_2 films under picosecond laser

The femtosecond laser was used to irradiate vanadium oxide films, and the energy density of incident laser was adjusted to observe the effect of femtosecond laser energy density on the damage morphology of VO_2 films. With different power densities, damage morphologies were observed, as shown in FIG. 4.

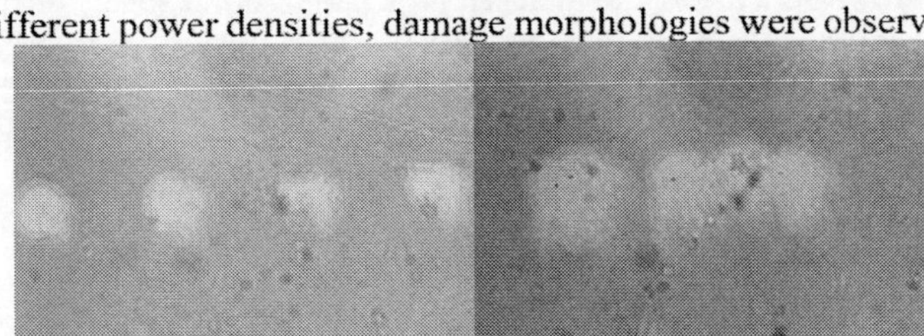

(a) 20 mJ/cm^2 (b) 40 mJ/cm^2

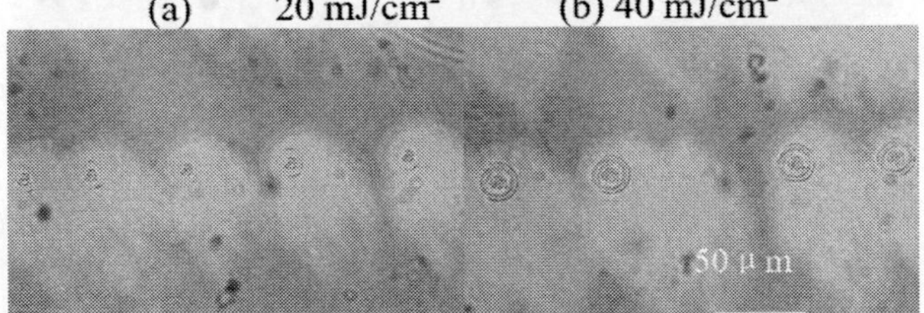

(c) 80 mJ/cm^2 (d) 140 mJ/cm^2

FIG. 4 damage morphology of VO2 films exposed by femtosecond laser

To observe subtle morphology, The image of the damaged area is enlarged more, as shown in FIG. 5.

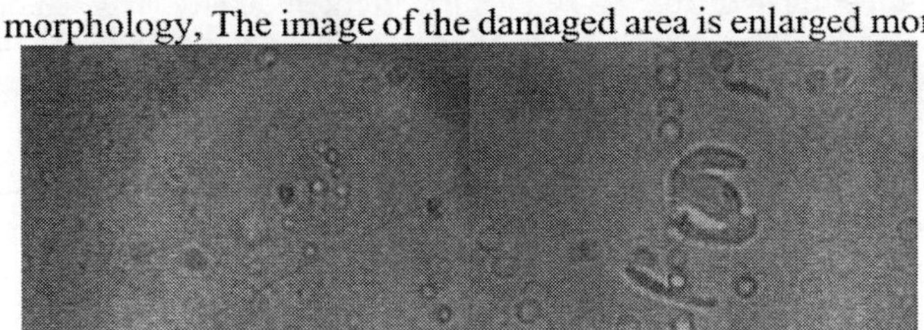

(a)20 mJ/cm^2 (b)40 mJ/cm^2

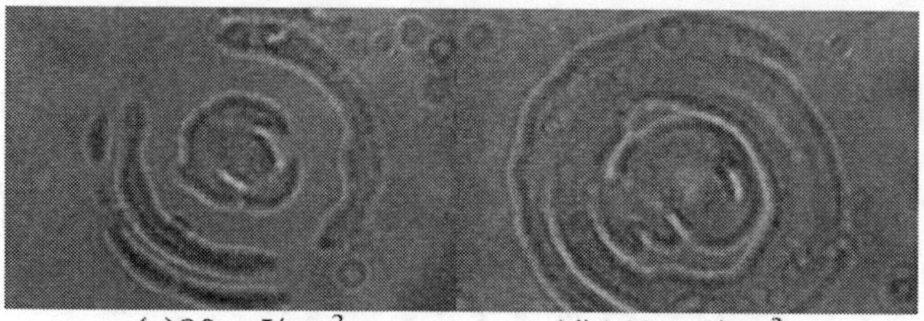

(c)80 mJ/cm^2 (d)140 mJ/cm^2

FIG. 5 details of damage morphology evolution of VO2 films under femtosecond laser irradiation

It can be seen from the experiments that the damage morphologies of VO2 films under nanosecond laser and femtosecond laser are quite different.Compared with the two, there are the following characteristics :(1) Damage sizes, the maximum diameter of femtosecond laser damage is only 20 microns, which is because the overall energy of femtosecond laser used is much smaller than that of nanosecond laser, and only the laser irradiation area is damaged. (2) Damage morphology: the damage morphology of nanosecond laser is an outward-inward semi-ripple, while that of femtosecond laser is an outward-extending concentric ring. When femtosecond laser energy is low, the circle of damage morphology does not close, and the damage only occurs in the direction parallel to the laser polarization. With the increase of femtosecond laser energy, the concentric ring gradually closes, which basically conforms to the geometric morphology of gaussian beam.This indicates that the damage morphology induced by femtosecond laser is closely related to its light intensity or energy density.

3.2.3 comparative analysis of femtosecond laser and nanosecond laser damage characteristics of VO_2 films

Femtosecond laser and nanosecond laser have different processes in the interaction between laser and vanadium dioxide film. When the incident laser is a nanosecond laser, its pulse is wider than the carrier-lattice energy transfer relaxation time (ps order) in VO_2, and most of the absorbed energy is used to increase the material temperature. At this time, the VO_2 film is a pure thermodynamic process, and the damage caused by nanosecond laser to the film is thermal damage, which requires a high damage threshold.

When femtosecond pulse laser is irradiated on VO_2 thin film, multi-photon ionization and avalanche ionization will occur, and the damage process is related to the increase of free electron density in VO_2 thin film.Under the irradiation of femtosecond laser, the electrons in VO_2 valence band are excited to the conduction band, and the free electron density is increased.When the free electron density in the system is greater than the critical value ($10^{21}cm^{-3}$), a large number of covalent bonds in VO_2 are destroyed, resulting in weakened binding force in the lattice of vanadium dioxide.VO_2 lattice tends to be disordered, causing damage.

The process time is very short (<100fs), which is far less than the lattice thermal effect relaxation time, so the damage will occur before the lattice heating.This process is called cold ablation.

The change rate equation of the free electron density in the material under the action of femtosecond laser can be expressed as follows [12] :

$$\frac{dN}{dt} = \alpha IN + \beta I^m - \frac{N}{\tau} - \Gamma\nabla^2 N \tag{2}$$

Where, N is the electron density, I is the pulse light intensity, $\alpha$is a constant representing the avalanche process, β is a constant representing the dissociation process of multiple photons, τ is the free electron density relaxation time, Γ is the free electron diffusion coefficient, m is the absorption coefficient of photons.

It can be seen from formula 2 that the change rate of free electron density dN/dt IN the material is mainly related to avalanche ionization process represented by αIN and multi-photon absorption process represented by beta βIm, both of which are related to laser intensity I.

Therefore, the mechanism of femtosecond laser thin film ablation will be different with the difference of laser intensity or laser energy density. It is pointed out that the avalanche ionization mechanism plays the leading role under the action of lower laser energy density, and the damage morphology is elliptic. At a higher laser energy density, multi-photon absorption and free electron diffusion dominate, and the ablation morphology is nearly circular [12].

4.Summary

Vanadium dioxide films were prepared by magnetron sputtering method. The damage morphology of the thin film was observed by nanosecond and picosecond laser irradiation. Under the action of nanosecond laser, the damage of thin film is mainly thermal damage, which is determined by the energy contained in the laser pulse itself. Under the action of femtosecond laser, the damage of thin film is caused by multi-photon ionization and avalanche ionization, and the required damage threshold is low.

References

[1] MORIN F J. Oxide which show a metal-insulator transition at the neel temperature [J]. *Phys Review Letter*, 1959, 13(1): 34-36.

[2] CHEN Zhang, CAO Chuan-xiang, LUO Hong-jie, et al. VO_2-based thermo chromic smart window: From energy From energy savings to generation [J],Chinese Science Bulletin，2016，61(15):1661-1678.

[3] Xu Kai, Lu Yuan, Ling Yongshun, Qiao Ya, Effects of oxidational annealing on properties of VO_2 thin films, Infrared and Laser Engineering, 2015,44(12),3724-3727

[4] Alexander Pergament;Andrey Velichko;Maksim Belyaev;Vadim Putrolaynen.Electrical switching and oscillations in vanadium dioxide[J].PHYSICA B-CONDENSED MATTER,2018,Vol.536: 239-248

[5] Information on https://www.nature.com/articles/s41427-018-0061-2

[6] Huang Zhangli, Chen Sihai, Lv Chaohong, et al. Infrared characteristics of VO_2 thin film for smart window and laser protection applications[J]. *Applied Physics Letters*, 2012, 101(19): 191905-(1-4).

[7] Lu Yuan, Ling Yongshun, Feng Yunsong, Zhang Peng, Analysis of VO2 thin film intelligent protection against pulsedpower infrared-laser [J], Infrared and Laser Engineering, 2012, 41(11), 2887-2891

[8] XU Kai,LU Yuan, LING Yong shun, et al. Finite element analysis of temperature rising in VO_2 thin films under irradiation of CO_2 pulse laser [J]. ACTA PHOTONICA SINICA，2015，44(5):0531003.

[9] MAO Z D,WANG W,LIU Y,et al. Infrared stealth property based on semiconductor (M) -to-metallic (R) phase transition characteristics of W-doped VO2 thin films coated on cotton fabrics[J]. Thin SolidR Films,2014，558(19):208-214.

[10] Dai Fu, Xiong Shengming，Experimental study on damage of optical coating irradiatedby high repetition rate DPL laser, Infrared and Laser Engineering,2014,43(7):2055-2060.

[11] ZHOU G.Y. Research on Damage Characteristics in Single-Crystal Silicon Irradiated by Long Pulse Laser[D], xihua university 2015,4

[12] Li Xiaoguang, Shen Jun, Research progress in laser induced damage on optical films, HIGH POWER LASER AND PARTICLE BEAMS, 2010 ,22 (10): 2237-2243.

IOP Conf. Series: Materials Science and Engineering **711** (2020) 012100 doi:10.1088/1757-899X/711/1/012100

A High-Efficiency and Real-Time Method for Quality Evaluation of PPG Signals

Hongxing Li[1, a], Shizhao Huang[2, b*]

[1]Electronic communication engineering college, Anhui Xinhua University, Hefei, Anhui, 230088, China;

[2]Electronic communication engineering college, Anhui Xinhua University, Hefei, Anhui, 230088, China;

[a]capricorn_l@qq.com, [b]735426836@qq.com

Abstract. The pulse oximeter (POM) is a device for real-time detection of HR and SpO2 based on PPG (Photo Plethyamo Graphy) signals, and its detection accuracy varies with diverse interferences. Therefore, it is presented in the paper a method for quality evaluation of PPG signals to detect wavebands of interference signals. Firstly, features of PPG signals are acquired by slope weighted sum function (SSF). Then, the normalized cross-correlation coefficient of each signal and its reference signal is calculated. Finally, signal quality index (SQI) is calculated based on nonlinear scaling function. In the research, 20 PPG signals with 1469 beats are chosen from the CB database to make the experiments of signal quality evaluation. The experimental results show that the method included can effectively identify useful signals from the mixed signals. The sensitivity and positive prediction value are 94.66% and 96.78%, respectively. Consequently, the method proposed has high recognition accuracy with key significance in clinical applications.

1.Introduction

SpO2 refers to the percentage of the capacity of HbO2 in the total capacity of Hb, that is, the concentration of blood oxygen in the blood, which is a key physiological parameter of respiratory and vascular system. The amount of light absorbed during arterial pulses is related to the amount of oxygen in the blood. The ratio of the two spectra (visible red spectrum (660 nm) and infrared spectrum (940 nm) absorbed is calculated to acquire the value of SpO2, which is the principles of POM[1]. Nowadays, POM has been a standard monitoring tool in many clinical applications.

However, the accuracy of POM is easily attacked by such interferences in practical applications as ambient light, physiological movements of test subjects, and low perfusion which reduces signal quality and parameter accuracy, thus resulting in false alarm[2]. Therefore, real-time calculation of PPG signal quality with high accuracy has a key significance in clinical blood oxygen monitoring.

At present, there are few algorithms for direct PPG signal quality (SQI) calculation. Most algorithms aim at removing bands of interference signals or minimizing the influence of interference on signals and they can be classified into the following categories.

(1) Statistical Method

ICA [3] or PCA [4] is adopted to separate the useful signals from the interfering signals. Due to high similarity between some interfering signals and the original pulse signals, the detection results often

IOP Conf. Series: Materials Science and Engineering **711** (2020) 012100 doi:10.1088/1757-899X/711/1/012100

do not meet the expectation.

(2) Frequency Analysis Method

The method includes FFT [5], DST [6], and so on. It has strong robustness that the useful signals can be accurately extracted with strong interference. But it requires too large amount of calculation.

(3) Evaluation Method Based on Wave Forms [7] [8]

Signal quality evaluation is obtained by feature recognition of wave forms and template matching. The method can receive good detection effect with a large amount of threshold detections.

The methods above have certain anti-interference performance, but they all require large computation. Real-time detection may be unavailable in dual-wavelength POM. Therefore, to improve calculation efficiency of detection with movement interferences, it is presented a new method for quality evaluation of PPG signals. Firstly, features of PPG signals are acquired by slope weighted sum function (SSF) to obtain the band of each beat. Then, the normalized cross-correlation coefficient of the signal band of each beat and its reference signal is calculated. Signal quality index (SQI) is calculated by nonlinear scaling function based on the cross-correlation coefficient obtained. Finally, the sensitivity (Se) and positive prediction value (+P) are applied to verify accuracy of the method. SQI ranges from 0 to 100, where 0 indicates poorest signal quality and 100 indicates best signal quality.

2.Method Design

Fig.1 shows the algorithm flow of the signal quality evaluation system. Firstly, the acquisition and preprocessing of original signals are implemented. Then, the SSF method is adopted to acquire the features of PPG signal in each wave band. Finally, signal quality evaluation algorithm is applied to calculate SQI value of each band.

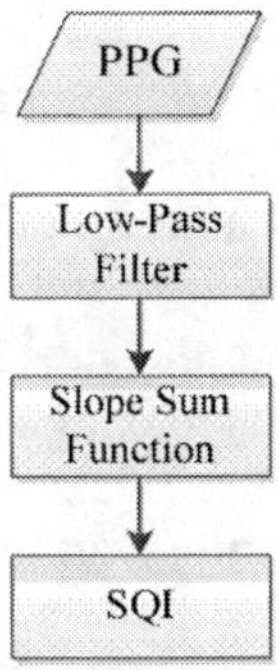

Fig.1 Flow Chat of the Method

2.1 Low-Pass Filter

Low-pass filter aims at filtering out the high frequency noises mixed in signals so as to improve the accuracy of signal features extracted. In the method, moving average filter is adopted to implement signal preprocessing. The moving average filter uses the mean value of the input signal values of a fixed time window as the output signal. A special case of FIR filter, moving average filter has the advantages of small calculation and easy implementation [9]. It can be expressed as equation (1), where, $y(n)$ is output signal, $x(n+L)$ stands for input signal, and L represents the width of time windows.

$$y(\mathrm{n}) = \frac{1}{L}\sum_{n=0}^{L-1} x(\mathrm{n}+\mathrm{L}) \qquad (1)$$

2.2 SSF Algorithm

The principle of SSF algorithm [10] is to sum up the rising slope values of signal within a fixed time window, while the rest of the signal return to zero, so as to enhance the information of the rising slope parts of the PPG signal and to suppress the rest parts of the signal. It has been expressed as equation

(2), where w is the width of analysis window, $\Delta y_k = y_k - y_{k-1}$, and y_k refers to the signal obtained by low-pass filtering processing of PPG signal. To maximize the enhancement effect of the parts of PPG signal in rising slopes by SSF, w should be as close as possible to the duration of contraction period of PPG signal. In the method, w is 128ms.

$$z_i = \sum_{k=i-w}^{i} \Delta u_k, \Delta u_k = \begin{cases} \Delta y_k, \Delta y_k < 0 \\ 0, \Delta y_k \geq 0 \end{cases} \quad (2)$$

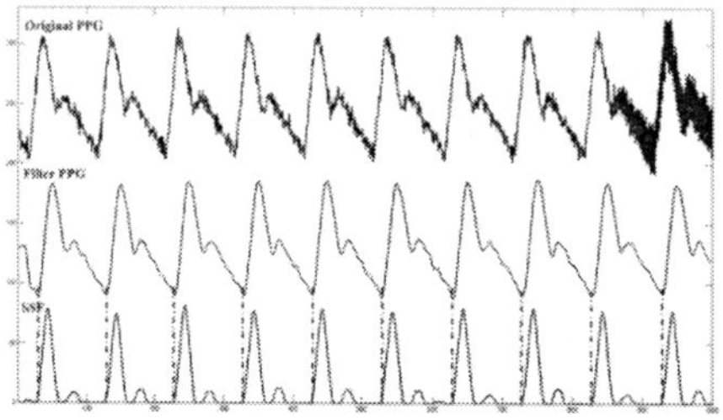

Fig.2 corresponding relation of PPG signal and SSF signal

Fig.2 shows the corresponding relation of PPG signal and SSF signal. Firstly, seek the points with the largest value in enhanced SSF signal which correspond to those points with largest rising slope in PPG signal. Then, seek the nearest zeros around the points with the largest value, and they are the origins and peaks of the pulse wave. The origins of original PPG signal are shown as dotted lines in Fig.2.

2.3 Calculation of SQI Value

PPG signal is collected by photoelectric sensors. Besides interferences from electronic devices, blood perfusion indexes, breathing and movement interferences, and spatial offset of sensors can all affect the wave form of PPG signal, reduce signal quality, and finally influence evaluation results. Therefore, separating the high-quality PPG signal from mixed signals has great clinical significance.

High-quality PPG signal refers to the signal stable within a period of time and containing no external interference factors that may change the subsequent processing results. Generally, there are two methods for evaluating signal quality: (1) making comparison with features of the former wave; (2) applying extra static evaluation algorithm [11]. Because the former method has the characteristics of self-adaptation and can completely show the variation trend of signal quality, it is adopted to calculated signal quality index (SQI) in the method of the paper.

Firstly, evaluation of PPG signal within a certain time is implemented to acquire the reference signal. The evaluation method is calculating normalized correlation number C between each beat signal and its former beat signal, where $C \in [-1,1]$. The calculation of C is shown in equation (3), where $x(n)$ and $y(n)$ respectively stand for two sets of PPG signals with the length of N to be analyzed [12]. It can be derived by Schwartz Inequality that $|C| \leq 1$. When $x(n)$ equals $y(n)$, C gets its maximum value. Conversely, when $x(n)$ is completely independent of $y(n)$, $C=0$. When there is a certain similarity between $x(n)$ and $y(n)$, C has its value between 0 and 1. Therefore, when C tends to be stable, the current signal is considered to be stable in this recording and chosen as a reference signal for subsequent analysis.

$$C = \sum_{n=1}^{N} x(n)y(n) / \sqrt{\sum_{n=1}^{N} x(n)^2 \sum_{n=1}^{N} y(n)^2} \quad (3)$$

Then, the normalized cross -correlation number between each heart beat signal subsequently recorded and the reference signal is calculated (Notice that, the reference signal is also updated in real time in the process of cross-correlation calculation of each band).

Finally, is applied to calculate the SQI value. Equation (4) shows the nonlinear scaling function, where e has the coefficient of 8 to ensure SQI value in the range of 0-100.

$$SQI = \left(\frac{e^{\frac{50\times(C+1)}{99}\times 8}}{e^{8}} \right) \times 100\% \qquad (4)$$

When SQI value approaches 0, it indicates that the current signal is greatly interfered and the signal quality is poor. When SQI value approaches 100, it indicates that the current signal quality is good and can be used for analysis and calculation. In addition, when there is an irregular transformation in heart rate, or when there is no signal for a long time or even cardiac arrest, the system automatically returns 0 for SQI.

3. Performance Evaluation Method

A set of standard biological signal databases recorded by the complex systems laboratory (CSL), Capnobase (CB) database is maintained by the school of electrical and computer engineering, Columbia University[13]. The subjects are aged 1-74 years old and data are acquired with all general anesthesia. Although the signal sampling rate was 100Hz, all of them were derived with 300Hz oversampling frequency. Moreover, CB database contained 124 data records of 120s and 42 data records of longer than 480s with all the dc components removed. Therefore, PPG signals of 1469 beats are chosen from CB data to verify the performance of the method in the paper. The performance of the signal quality evaluation method was qualitatively implemented by calculating the sensitivity Se and the positive prediction $+P$. The expressions are shown in equation (5) and Equation (6), where TP refers to the amount of true positive results, FN stands for the amount of false negative results, and FP represents the amount of false positive results.

$$Se = \frac{TP}{TP + FN} \qquad (5)$$

$$+P = \frac{TP}{TP + FP} \qquad (6)$$

4. Experimental Results

The signal quality evaluation uses the PPG signal data of 20 sets with a total of 1469 beats in CB database including normal PPG signals, signals with movement interferences, and signals with asystole. The accuracy of SQI calculation was verified by manual identification. The criterion is that when SQI > 50, it indicates that the current signal band is acceptable; otherwise, it indicates that the current signal band has no analytical significance. As is shown in table.1, the quality evaluation method has a good detection performance with the sensitivity and positive prediction values respectively to be 94.66% and 96.78%.

Table.1 Detection Results of the SQI Evaluation Method

Count	TP	FP	FN	Se (%)	+P (%)
1469	1348	45	76	94.66	96.78

Fig.3, Fig.4 and Fig.5 respectively show the effects of SQI evaluation method on three types of PPG signals (normal signal, signal with motion interferences and signal with asystole). The left coordinate represents the amplitude level of PPG signal, and the right one stands for the SQI value. ambient light

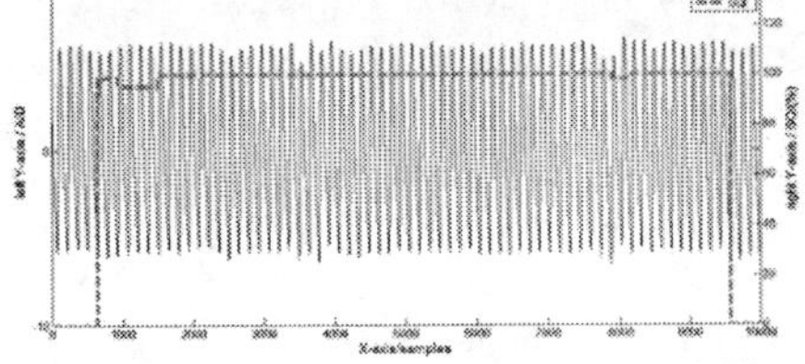

Fig.3 SQI Level of the Normal PPG Signal

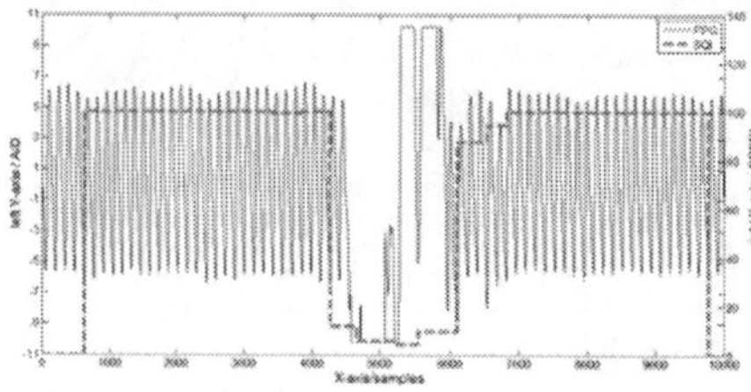

Fig.4 SQI Level of the PPG Signal with Movement Interferences

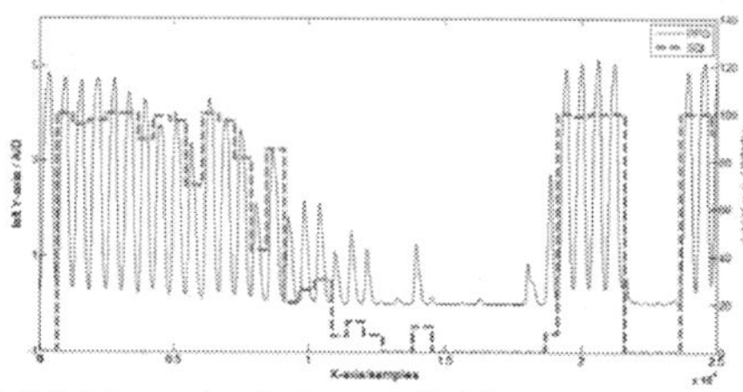

Fig.5 SQI Level of the PPG Signal with Asystole

5. Conclusion

It is proposed a new method for quality evaluation of PPG signals. Preprocessing is first implemented to PPG signals collected. Then, SSF method is applied to extract features thus realizing signal segments. Finally, the effective quality evaluation method is adopted to calculate the SQI value of each band. Experimental results show that the method can effectively calculate the SQI level of each band, so as to objectively reflect the variation trend of the signal, such as the occurrence of motion interferences and asystole, thus having great significance in clinical monitoring. On the other hand, since the cross-correlation processing on two complete heart beat bands should be implemented in the quality evaluation method and a period of training process is required to obtain the reference signal, the calculation amount has been increased, thus affecting the real-time performance of signal detection. To solve this problem, spline interpolation can be performed on the peak and valley points of the known standard signal to construct a standard reference signal, so as to reduce the time of updating the reference signal.

Above all, the method can effectively make signal segmentation, evaluate the signal quality of each band, and effectively identify abnormal wavebands caused by movement interferences, asystole, and sensor loss, thus acquiring the overall variation trend of the PPG signal. Therefore, the method has a good application prospect both in clinical care and in quality evaluation of periodic signals with distinct characteristics.

Acknowledgements

We acknowledge the support of key research projects of provincial natural science foundation of department of Education of Anhui. Research on Low Conductivity Fluid Flow Measurement Technology. KJ2018A0596.

References

[1] Zhang Genxuan, Shi Bo, Cao Yang. A New Method for Calculating Pulse Oxygen Saturation[J]. Journal of Biomedical Engineering, 2017, (01): 62-65.

[2] Edward D. Chan, Michael M, Chan, et al. Pulse oximetry: understanding its basic principles facilitates appreciation of its limitations. Respiratory Medicine (2013) 107, 789-799.

[3] Spurek P, Tabor J, Rola P, et al. ICA based on asymmetry[J]. Pattern Recognition, 2017, 67:230-244.

[4] Wang W T, Huang H C. Regularized Principal Component Analysis for Spatial Data[J]. Statistics, 2016.

[5] Tian He, Zhu Huanyan, Zhang Wei. Research on Multi-functional Fitness Detector Based on FFT

and Photoelectric Sensor[J]. Chinese Journal of Medical Instrumentation, 2013, 37(1): 23-26.
[6] Latour M, Piluso V, Rizzano G. Free from damage beam-to-column joints: Testing and design of DST connections with friction pads[J]. Engineering Structures, 2015, 85:219-233.
[7] Grill W M, Wongsarnpigoon A. Waveform shapes for treating neurological disorders optimized for energy efficiency: US, US9089708[P]. 2015.
[8] Cole S R, Voytek B. Brain Oscillations and the Importance of Waveform Shape[J]. Trends in Cognitive Sciences, 2017, 21(2):137.
[9] Wang Guowei, Mei Zhaomin, Zeng Ruili, et al. Extraction of gearbox early fault features based on segmenting with Features of a curve and multilevel fractional Fourier transformation filtering [J]. Journal of Vibration and Shock, 2016, 35(11):161-166.
[10] Li Q, Li P, Su Z, et al. Effect of a novel motion correction algorithm (SSF) on the image quality of coronary CTA with intermediate heart rates: Segment-based and vessel-based analyses[J]. European Journal of Radiology, 2014, 83(11):2024-32.
[11] Saeh I, Mustafa M W, Al-Geelani N A. New classifier design for static security evaluation using artificial intelligence techniques[J]. International Journal of Electrical & Computer Engineering, (2088), 2016.
[12] Zhang Mei, Wen Jinghua. Point Cloud Registration Method of Normalized Correlation Number and Iterative Recent Surface [J]. Computer Engineering, 2016, 42(10):271-276.
[13] Zhu T, Pimentel M A, Clifford G D, et al. Bayesian fusion of algorithms for the robust estimation of respiratory rate from the photoplethysmogram[C]// Engineering in Medicine and Biology Society. IEEE, 2015:6138-6141.

IOP Conf. Series: Materials Science and Engineering **711** (2020) 012101 doi:10.1088/1757-899X/711/1/012101

Simulation Study on Laser Damage of Iron Plate

Tianyu Wang[1,2,a], Jintian Bian[1,2,b] and Qing Ye[1,2,c*], Xin Li[1,2,d], Xiaoquan Sun[1,2,e]

[1] The State Key Laboratory of Pulsed Power Laser Technology of China, 460 Huangshan Road, Shushan District, Hefei City, Anhui Province, China

[2] Anhui Laboratory of Advanced Laser Technology, 460 Huangshan Road, Shushan District, Hefei City, Anhui Province, China

[a]714030068@qq.com, [b]bianjt@126.com, [c*]yeqing0518@sina.com, [d]lixinkiller@126.com, [e]sunxq@ustc.edu.cn

Abstract. Studying laser damage effects are of great significance in the field of industrial production. In order to evaluate the damage effect of laser on iron, the process of pulsed laser and continuous laser irradiate an iron plate was simulated by using COMSOL. The relationship between temperature and surface energy absorptivity of iron plate was discussed. The initial simulation parameters were set as follows: the pulsed laser wavelength is 1064nm, the pulse width is 10ns, the pulse energy is changed from 2mJ to 8mJ, and the average power of continuous laser is 225mW. The simulation results show that pulsed laser can rise the surface temperature of iron plate very quickly, and improve surface energy absorptivity of the plate. Thus using pulsed laser to radiate an iron plate before using continuous laser can improve damage efficiency of continuous laser on an iron plate.

1. Introduction

Laser damage is a very important part in instrument design, precision measurement and other industrial production activeties. Therefore, it is significant to know laser damage mechanism of materials.

During the past few years, many reserch has been put into effect, most of which, however, focused on only one type of laser to irradiate materials[1~7]. Other studies, which concentrate on using combination laser to irradiat targets, were carried out too[8~9]. Among these cases, high power pulsed laser were used to damage target materials, and continuous laser play the part of keeping temperature of targets. We didn't find anyone of them using continuous laser to damage materials while using pulsed laser to rise the surface temperature of targets. In this paper, damage effect of the special combination type has been identified.

Meanwhile, metallic materials are widely used in our daily life and industrial production. Besides all the metallic materials, iron is one of the most important one. In this paper, we establish a model of an iron plate to simulate the process of laser heating.

2. Model Established

Theoretical Model. The effect of laser damage includes heat effect and mechanical effect. The iron plate absorbs heat from laser. For lasers are non-uniform spatial distribution, heat absorbing is non-uniform spatial distribution too. Then the energy passes inner the plate, leading to thermal stress between different parts of it. When the volume of the plate is small, mechanical effect is not significant. In order to simplify the operation, mechanical effect is ignored, while heat effect is mainly discussed in this paper.

For metal materials have lower electronic-lattice power and skin tropism, there is a huge number of free electronics on the surface of them. Two-temperature equations are used to describe the radiate process [10]. The equations divest the whole heat transfer process into two steps. First of all, the free electronics on the surface absorb energy from laser. Then the energy is passed from electronics which are heated to cold lattices. Eq. 1, the classical one-dimensional two-temperature equations, are shown below.

$$\begin{cases} C_e \dfrac{\partial T_e}{\partial t} = \partial\left(K_e \dfrac{\partial T_e}{\partial x}\right) - G\left(T_e - T_l\right) + s\left(x,t\right) \\ C_l = \dfrac{\partial T_l}{\partial t} = G\left(T_e - T_l\right) \end{cases} \tag{1}$$

In the equations, "C_e" stands for heat capacity of electronics, and "C_l" stands for that of lattices. "G" is the energy exchange function between electronics and lattices. "s" is used to describe the heat absorbed from laser.

In practical research, if the power of laser is not too large or pulse width is not too short, heat passing between electronics and lattices can be ignored. During the reaction time of middle power and low power nanosecond laser, electronics and lattices can reach heat balance.

Surface absorptivity of iron is one of the most important parameters which influence the surface temperature of the plate under irradiated by the same laser. Ignoring other effects, energy that the plate absorb only depends on the power of laser and the reflex rate of the plate. In ideal vacuum environment, iron's one wavelength reflex rate can be calculated by Hagen-Rubens equation [11]:

$$R \approx 1 - \sqrt{\frac{c}{\sigma\lambda_0}} \tag{2}$$

"c" is light speed in vacuum. "σ" stands for DC conductivity of iron. "λ_0" is the wavelength of laser.

DC conductivity of iron changes with the surface temperature of the target. In ideal vacuum environment, it can be described by the Eq. 3 [11].

$$\sigma = \begin{cases} 1/\left(8.79\times10^{-13}\times T^2 + 8.54\times10^{-12}\times T + 1.41\times10^{-9}\right), T \le 1100K \\ 1/\left(1.62\times10^{-15}\times T^3 + 6.95\times10^{-12}\times T^2 + 1.02\times10^{-9}\times T - 3.86\times10^{-6}\right), T \ge 1100K \end{cases} \tag{3}$$

Substituting Ep. 1 into Ep. 2, surface absorbity of iron can be described by Eq. 4[11].

$$A = \begin{cases} 3.19T/10000, T \le 1100K \\ 0.3476, T \ge 1100K \end{cases} \tag{4}$$

Basic Parameters. Some basic parameters of iron are shown in Table 1.

Table 1 Parameters of Iron

Material	Density[kg/m^3]	Specific heat capacity[J/(kg·K)]	Melting point[K]	Latent heat of fusion[J/kg]	Heat conduction[W/(m·K)]
Fe	7860	460	1812	269550	82

The plate is put in ideal vacuum environment. The ambient temperature is 298K, the pressure is 1atm, and the initial surface energy absorptivity of the plate is 0.0951. Thermal convection effect is ignored in the simulation. The radius of light plot on the surface of the plate is 0.5mm.

Fig. 1 is the model of iron has been established.

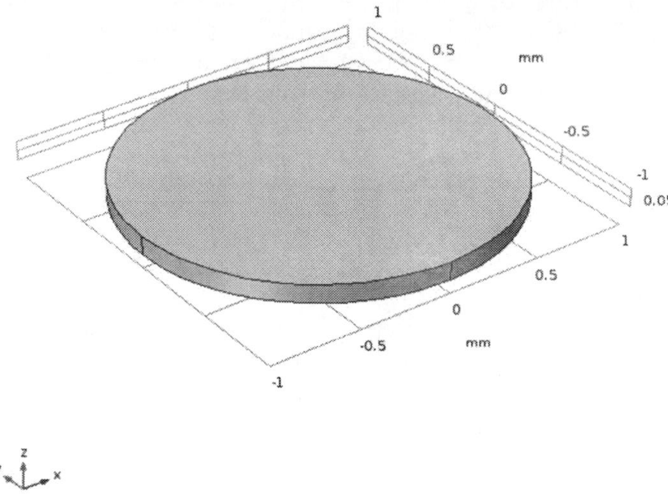

Fig.1 Plate Model

Lasers Model. In the ideal case, laser pumped from a Q-switch Nd:YAG laser is fundamental mode Gaussian beam. Both time and space distribution of power is in gauss distribution. Power distribution of pulsed laser in time after normalization is shown in Fig. 2.

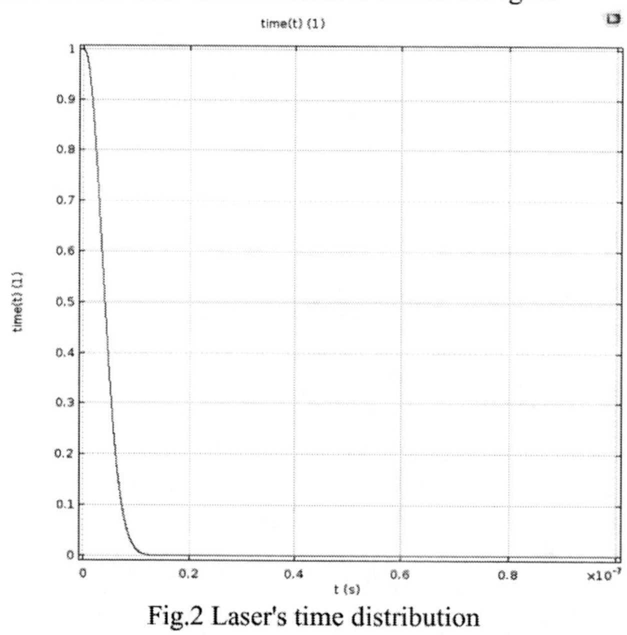

Fig.2 Laser's time distribution

3. Simulation Result

Pulsed Laser Ablate the Iron Plate. Pulsed laser, which energy of one pulse is 2mJ, 5mJ and 8mJ, irradiate the iron plate. Using one pulse to heat the plate, 3D temperature patterns of the three situations are shown from Fig. 3 to Fig. 5.

Fig.3 A 2mJ Pulse Heats the Plate

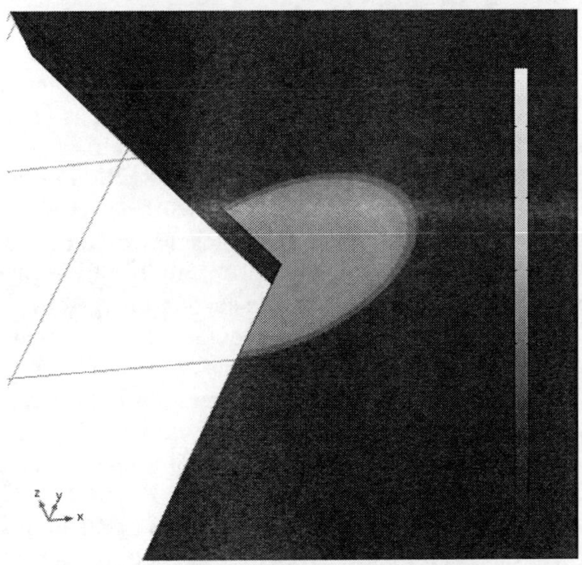

Fig.4 A 5mJ Pulse Heats the Plate

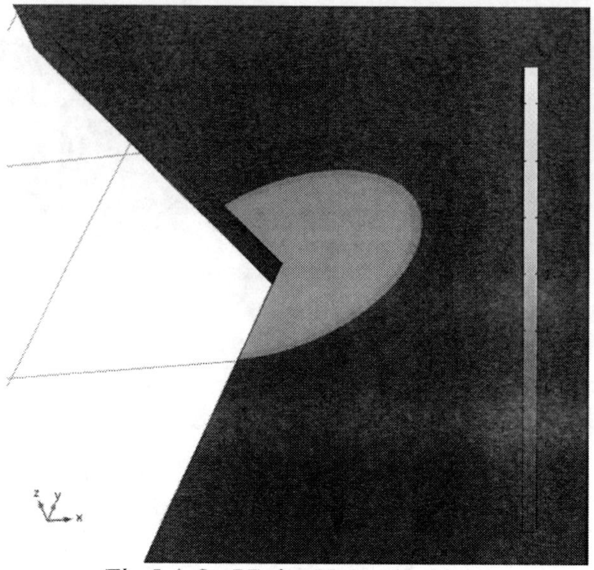

Fig.5 A 8mJ Pulse Heats the Plate

The surface temperature of the plot area arises quickly in all three situations, and nearly the same area has been heated. The higher energy of a pulse, the higher surface temperature of the heated area is.

When the energy of pulse is 2mJ, the highest temperature of the plate is 460K, the surface energy absorptivity of the plate rises to 0.1467. When the energy of pulse is 5mJ, the highest temperature of the plate is 1434.3K, the surface energy absorptivity of the plate rises to 0.3476. When the energy of pulse is 8mJ, the temperature of the plot area is more than 1812K. In other words, damage has taken place during the burning time. However, when the energy of pulse is 8mJ, the power density of the laser is $6.35\times10^{11}W/m^2$, lower than $3\times10^{11}W/cm^2$, no plasma would be produced.

As it is shown in Ep. 4 that the surface energy absorptivity of the plate doesn't change when the surface temperature is over 1100K. It is unable to rise surface energy absorptivity of the plate by rising the energy of laser under the condition, so it's unnecessary to use pulsed laser with too high energy. Thus the pulse with the energy of 5mJ is chosen to use as on part of combination laser.

Continuous Laser Ablate the Iron Plate. Different with pulsed laser, the reaction time of continuous laser is long. When volume of the iron plate isn't too big, all the plate would be heated during several seconds. As is shown in Fig. 5, while using a pulse whose energy is 5mJ to heat the plate, the temperature of the spot area can reach 1434.3K. In other words, the temperature rises by 1136.3K. Ignoring heat exchange effect, we can use Eq. 5 to calculate how much energy continuous laser need if the temperature rises to the same level with the pulsed laser.

$$P=\frac{c\rho v\Delta T}{\Delta t} \tag{5}$$

"c" is heat capacity of iron. "ρ" is iron's density. "V" stands for volume of the model plate.

Substituting values, when the reaction time is 20s, P=258.3mW. Continuous laser which average power is 225mW is chosen to irradiate the plate. The result is shown by Fig. 6. The highest temperature of the plate reaches 1496.7K.

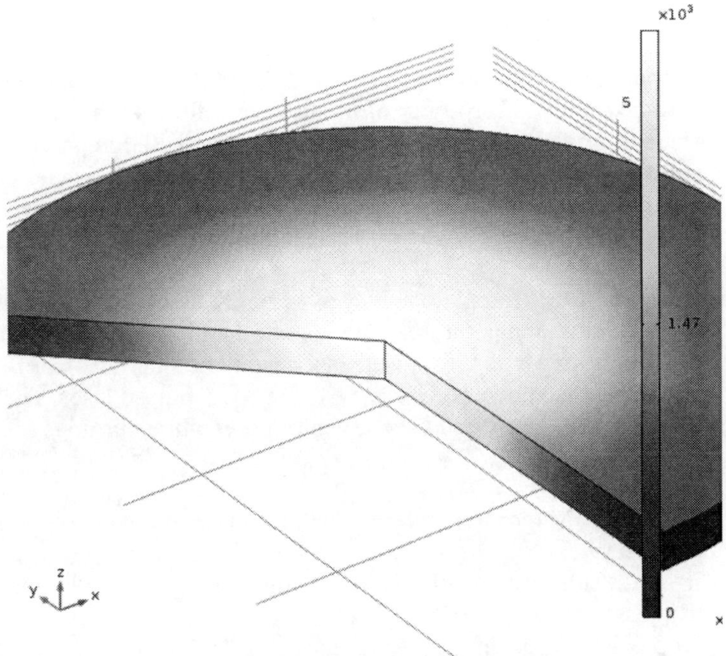

Fig.6 225mW Continuous Laser Heat the Plate

Combination Laser Ablate the Iron Plate. Pulsed laser with energy of 5mJ and continuous laser with average power is 225mW are chosen to irradiate the plate. The burning effect of the combination laser is shown in Fig. 6.

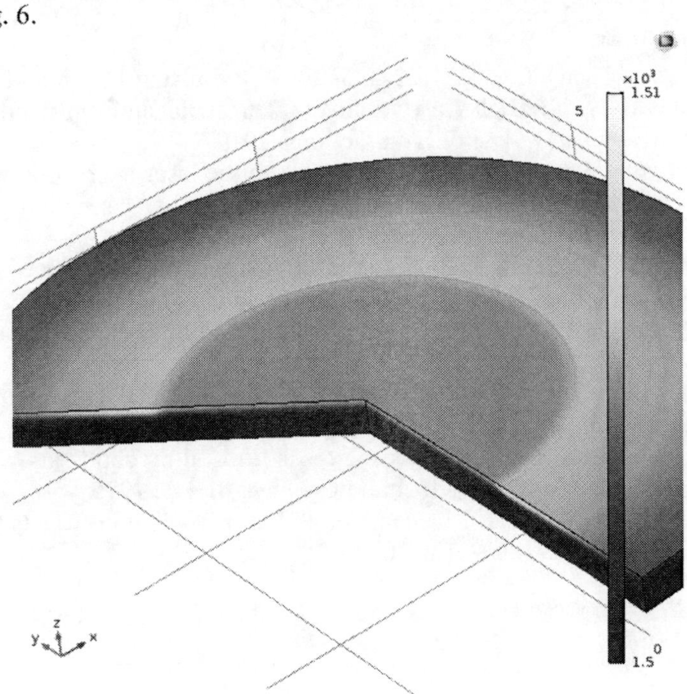

Fig.7 Combination Laser Heat the Plate

As is shown in Fig. 6, the average temperature achieves 1563.2K. It is higher than both that of only using the continuous laser to irradiate the plate and that of only using the pulsed laser. What's more, damage has happens at the center point after 20s burning. So using pulsed laser before burning can improve the ablation efficiency of the iron plate.

4. Summary

When the surface temperature of the iron plate is under 1100K, the higher surface temperature the plate is, the higher surface energy absorptivity of it. However, if the surface temperature of the plate is over 1100K, the surface energy absorptivity of it doesn't change with temperature. The surface energy absorptivity of the plate rises to 0.0976 from 0.0951 by using pulsed laser with energy of 5mJ and the pulse width of 10ns. If the energy of the pulse rises to5mJ, the surface energy absorptivity of the plate rises to 0.3476. However, if the energy of the pulse becomes 8mJ, the surface energy absorptivity of it doesn't go up anymore.

The highest temperature of the iron plate achieves 1496.7K by using continuous laser with average power is 225mW. The maximum temperature difference of the plate within 1K after burning.

The combination laser consists of the pulsed laser with the energy of 5mJ at the front and the continuous laser with the average power of 225mW follows the pulsed laser. The average temperature reaches 1563.2K, and damage has happened at the center point while burning.

References

[1] Allison S W, Gillies G T, Magnuson D W, et al. Pulsed laser damage to optical fibers.[J]. Applied Optics, 1985, 24(19):3140.
[2] Silipigni L, Torrisi L, Cutroneo M. Physical investigations on the radiation damage of graphene oxide by IR pulsed laser[C]// 2018.
[3] Xia Z, Wu Y, Kong F, et al. Using a cover layer to improve the damage resistance of gold-coated gratings induced by a picosecond pulsed laser[J]. Applied Surface Science, 2018, 436:362-366.
[4] Yuan D, Di W, Zhi W, et al. Study on the inversion of doped concentration induced by millisecond pulsed laser irradiation silicon-based avalanche photodiode[J]. Applied Optics, 2018, 57(5):1051-.
[5] Huang Y , Wu K , Cheng J , et al. Li2ZnGeS4: A Promising Diamond-like Infrared Nonlinear Optical Material with High Laser Damage Threshold and Outstanding Second-Harmonic Generation Response[J]. Dalton Transactions, 2019.
[6] During A, Bertussi B, Bonneau F, et al. Localized pulsed laser interaction with sub-micronic gold particles embedded in silica: a method for investigating laser damage initiation[J]. Optics Express, 2003, 11(7):824.
[7] Wang W , Baida Lü, Liu C , et al. The morphology of pulsed laser-induced damage on the surface of a copper film mirror[J]. Optics and Laser Technology, 2003, 35(4):303-307.
[8] Stonesifer F R, Randles P W, Nemes J A. Assessment of Single-Pulse Laser Damage to Structural Materials for Spacecraft,[J]. 1992.
[9] Zhang M , Nie J , Sun K , et al. Numerical analysis on thermal function of single crystal silicon irradiated by combined laser[J]. Infrared and Laser Engineering, 2018, 47(11).
[10] Electron emission from metal surfaces exposed to ultra short laser pulses. Anisimov S I,Kapeliovich B L,Perel‘man T L et al. Soy. Phys. JETP . 1975
[11] Sun Chengwei, Lu Qisheng, Fan Xiuzheng, et al. Laser Irradiation Effects [M]. Beijing: National Defense Industry Press, 2002. (in Chinese).

Simulation on Molten Pool Characteristics of Al2O3 Laser Selective Melting Based on Energy Input Model

Xiaobing Chen[1, 2], Quan Sun[1, 2], Quanping Sun[1, 2], Haibing Wu[1, 2] and Qian Xu[1, 2]

[1]Key Lab of Advanced Manufacturing Technology of Jiangsu Province, Huaiyin Institute of Technology, Huai'an, Jiangsu Province, 223003, China

[2]Faculty of Computer and Software Engineering, Huaiyin Institute of Technology, Huai'an, Jiangsu Province ,223003,China

Abstract. In order to obtain the formation rule of molten pool characteristics in SLM process of ceramics, an energy input model for SLM process of Al2O3 ceramics is established. The model is related to laser power P, scanning speed v, scanning spacing s and thickness h of powder layer. The influence of process parameters in the energy model on molten pool characteristics is analyzed by simulation of temperature field of multi-layer and multi-channel SLM process. The simulation results show that when the laser power P is 105W, the scanning speed v is 50mm/s, the initial temperature of powder is 330°C and the thickness of powder layer is 0.20mm, the molten pool formed is easy to form Al2O3 ceramic specimens.

1. Introduction

The process of selective laser melting (SLM) is very complicated. Foreign researchers have summarized up to 130 factors influencing the quality of SLM. The technological parameters that affect the quality of parts are: laser power, scanning speed, scanning distance, powder composition and particle size. R.organ [3] et al. discussed the relationship between energy input and density, and believed that energy input varies with laser power, scanning speed, scanning spacing and other process parameters, but ignored the influence of the thickness of the powder layer on energy input. A. semimchi [4] deeply studied the relationship between energy input and density, and believed that within a certain range of technological parameters, the density of molded parts increased with the increase of energy input. Molten pool convection is one of the important factors affecting SLM forming process. Due to the influence of flow on heat and mass transfer, molten pool convection drives bubbles to move in the molten pool, affects the morphology of the molten pool, and further affects the density of specimens [1-2]. Based on the establishment of SLM energy input model, the influence of energy model parameters on the formation of molten pool is analyzed by ANSYS software simulation.

2. SLM energy input model

SLM forming process mainly includes: material melting and solidification, molten pool morphology and flow, other physical and chemical changes of materials, etc. [5-6]. The input and output of energy in the molding process will directly lead to the change of the physical state of the powder. In the interaction between laser and powder materials, energy is transferred from laser to powder, which complies with the conservation of energy. Energy transfer can be divided into three parts: part of laser energy is reflected by the surface of powder, and dissipates to the air above by convection; The other part of laser energy is transferred to ceramic powder by heat conduction, making the powder melt. The rest of the laser energy is transferred to the material in the region near the heat source, heating up the surrounding powder and melting layer. Figure 1 is a model of energy transfer involved in SLM processing, in which laser input energy is Q1, energy transferred to powder layer and substrate is Q2, and energy escaping from powder in the form of radiation and convection is Q3. The red area represents the molten pool, and the red dotted line represents the gaussian distributed laser heat source.

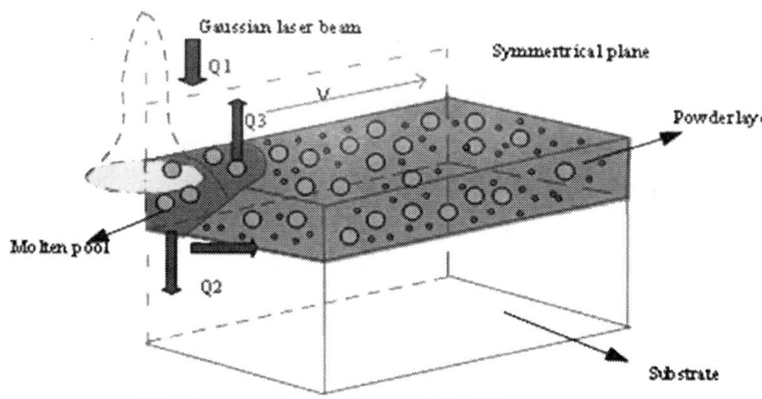

Fig. 1 energy transfer model of SLM process

The amount of energy input directly determines whether there is remelting between different melted channels and different powder layers, and thus affects the density of the molded parts. On the basis of Simchi energy input model, energy input is defined as a function of laser power, scanning speed, scanning interval and powder layer thickness. Its physical significance is the total energy input per unit volume of powder in unit time:

$$\psi = \frac{P}{vsh} \tag{1}$$

P is laser power, v is laser scanning speed, s is scanning interval, and h is powder layer thickness. These parameters are called energy model parameters.

3. Multi-layer and multi-channel SLM temperature field simulation

3.1 Heat Transfer Control Equation

Selective laser melting is a typical non-linear transient heat conduction process, which is described by the Control equation (Fourier equation)[7]:

$$\frac{\partial}{\partial x}\left(k_{xx}\frac{\partial T}{\partial x}\right)+\frac{\partial}{\partial y}\left(k_{yy}\frac{\partial T}{\partial y}\right)+\frac{\partial}{\partial z}\left(k_{zz}\frac{\partial T}{\partial z}\right)+Q=\rho c\frac{\partial T}{\partial t} \tag{2}$$

In formula ρ——material density (kg/m^3);

c—— specific heat capacity ($J/(kg\cdot K)$);

T—— powder temperature (K);

t—— laser scanning time (s);

Q—— heat source density (W/m^3);

k_{xx}, k_{yy}, k_{zz}—— heat conductivity ($W/(m\cdot K)$);

To solve the differential equation (2), it is necessary to determine the initial conditions and various boundary conditions of powder and substrate.Assuming that the temperature at any point of the powder layer is T_0, the powder layer at t=0:

$$T(x, y, z, t)\,|_{t=0} = T_0(x, y, z) \tag{3}$$

The boundary conditions refer to the heat exchange between the outer surface of the bed and the surrounding medium. The second, third and fourth type of boundary condition are used to simulate the temperature field[7].

3.2 Simulation analysis process

In order to simulate the temperature field in SLM forming process, a three-layer and three-channel scanning path is planned, as shown in Figure 2. The ANSYS software is used to establish the temperature field analysis model and carry out the finite element analysis. In ANSYS software, the movement of heat source is controlled by triple loop statement. The whole powder layer is divided into three layers, and there are three channels in each layer. Zig-zag scanning strategy is adopted, and the influence between the same layer and the adjacent layer is researched by analyzing the temperature at the midpoint of each channel.

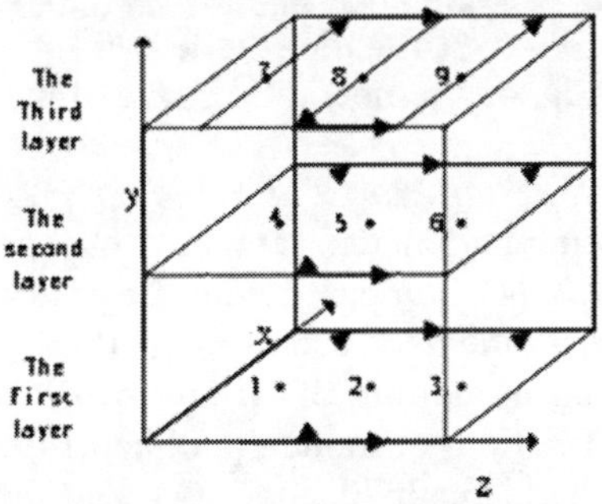

Fig. 2 the diagram of laser scan sequence

The whole model is divided into two parts: powder layer and substrate layer. The powder and substrate materials are the same, as shown in Figure 3. The size of powder layer is 5 mm×0.20 mm ×0.45 mm, cell size is 0.1 mm×0.07 mm×0.05 mm, and three layers are constructed. The laser radius is 0.15 mm, and substrate size is 5 mm×2 mm×2.5 mm. The displacement of each load step is 0.001 mm. Solid70 thermal analysis unit is used for mesh generation, and the life and death element method is used for simulation. The thermal physical parameters of alumina are shown in Table 1.

Fig. 3 the meshed solid model

Table 1. Thermal physical parameters of alumina

Temperapture /K	Thermal conductivity/ (W · m^{-1} · K^{-1})	Specific heat capacity/(J · kg^{-1} · K^{-1})
300	35	779
600	15.8	838
1000	7.85	1224
1700	5.54	1320
2200	5.5	1330

4. Characteristics of multilayer multichannel temperature pool

SLM simulation process parameters are: laser power P=105W, scanning speed v=50mm/s, powder layer thickness h=0.25mm, mark 1 ~ 9 in figure 2, corresponding node Numbers: 4875, 4876, 4877, 4904, 4905, 4906, 2670, 2669, 2668. It takes 16.66s to complete a single scan, and a total of 159 seconds to complete all the scans. The length, width and depth of the melted pool at node 2670 (point 7 in Fig.2) are 0.23mm, 0.15mm and 0.03mm respectively.

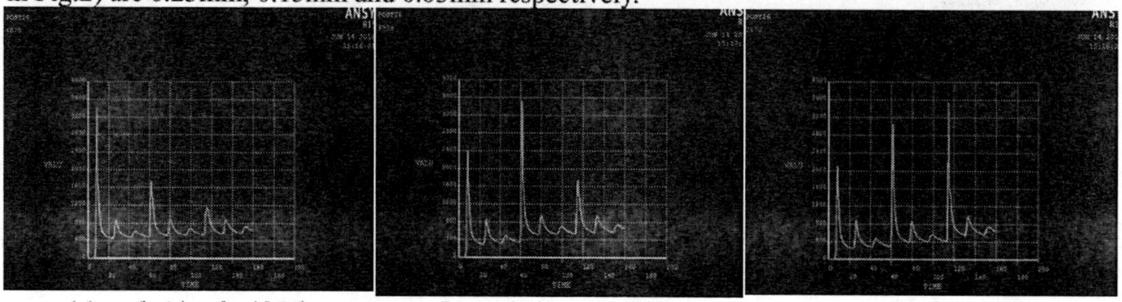

(a)mark 1(node 4875) (b)mark 4(node 4904) (c)ark 7(node 2670)

Fig. 4. temperature curve of the node

4.1. Influence of laser power on the characteristics of molten pool

When P=80W and other processing parameters remain unchanged, the top view of temperature cloud at 2670 node is shown in Fig. 5(a). (1) Compared with the power of 105W, the maximum temperature in the whole process decreases with the decrease of laser power, and the maximum temperature drops from 3246°C to 2273°C, resulting in melting of Al_2O_3 powder. (2) The length of the molten pool is 0.13mm, which decreases. (3) The width of molten pool is 0.14mm, which reduce slightly. Figure 5(b) is the left view of temperature field of 2670 node. (4) The molten pool depth is 0.025mm, which decreases a little. The powder layer is only slightly melted and most of the powder is at 1174°C.

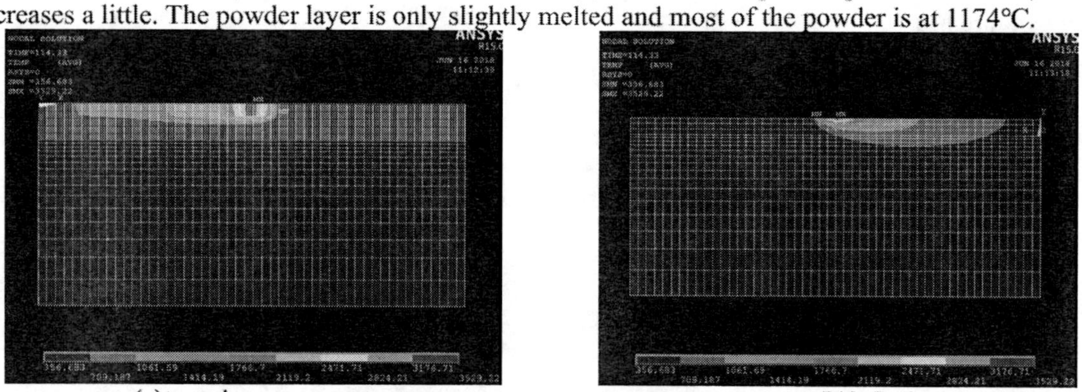

(a)top view (b)left view

Fig. 5 temperature field at NO.2670 node at 80

4.2. Influence of scanning speed on the characteristics of molten pool

When the speed is doubled, v=100mm/s, and other process parameters remain unchanged, the simulation results are shown in Fig. 6. (1) The melting pool width is 0.14mm, which reduced slightly. (2) The melting pool depth is 0.014mm, reducing 54%, and only a small part of the powder melted. Under the same material, the depth of molten pool is affected by the laser power and velocity, and velocity has a great influence.

IOP Conf. Series: Materials Science and Engineering **711** (2020) 012102 doi:10.1088/1757-899X/711/1/012102

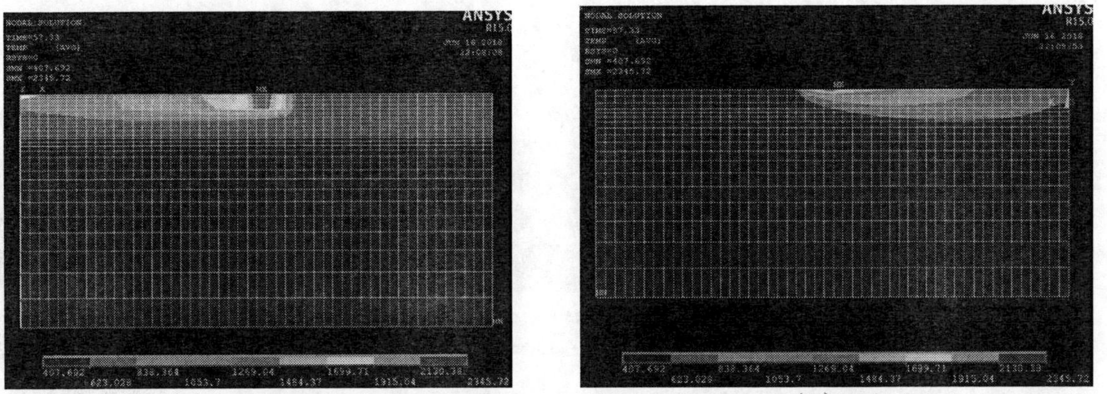

(a) top view (b) left view

Fig. 6 temperature field at NO.2670 node when speed doubles

4.3. influence of initial powder temperature on the characteristics of molten bath

P=80W, v=50mm/s, h=0.25mm, and the initial temperature of the powder is increased to 330°C. The simulation results are shown in Fig. 7. (1) Remelting begins between layers, as shown in figure 7(a). When the heat source is scanned to 2670 node, the temperature of node 4904 can still reach 2100°C. (2) There is no remelting between the channels. Taking node 4875 as an example, the temperature when scanning to the mark points 1, 2 and 3 is 2100°C, 800°C and 500°C, respectively. Obviously, the influence of the latter two melting channels on mark point 1 is not enough to reach the melting temperature. (3) As shown in Fig. 7(b)(c), the length and width of the molten pool change little, and the depth of the molten pool reaches 0.036mm. It can be seen that increasing the initial powder temperature has a great impact on the depth of the molten pool.

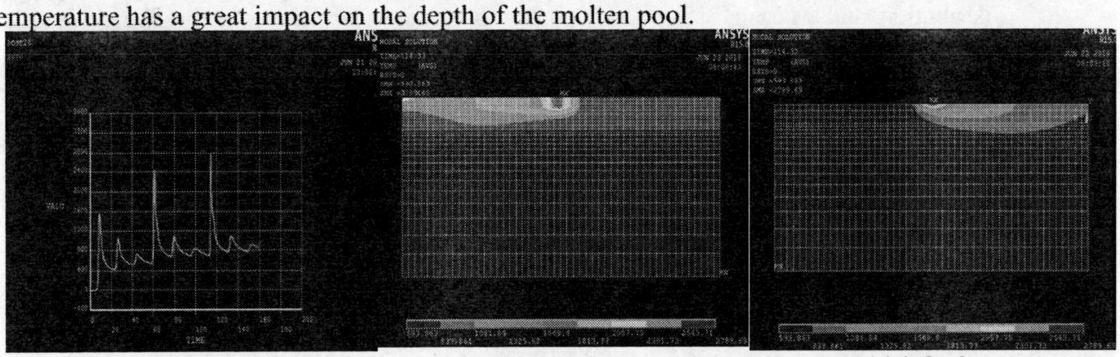

(a) temperature curve (b) top view (c) left view

Fig. 7 temperature field at NO.2670 node when initialization temperature increased

4.4. Influence of powder layer thickness on molten pool characteristics

P =105W, v=50mm/s, the initial temperature of the powder is 330°C, and the thickness of the powder layer is set to 0.20mm. The simulation results are shown in Fig. 8. (1) Remelting begins between layers, as shown in figure 8(a) below. When the heat source scans to the 2670 node, the temperature of node 4904 can still reach 3000°C, meeting the remelting condition. (2) There is no remelting between the channels, as shown in Fig. 8(b). The width of the molten pool is 0.15mm, which did not change. (3) As shown in Fig. 8(c), the molten pool depth reaches 0.068mm, and the powder layer thickness has a great influence on the molten pool depth.

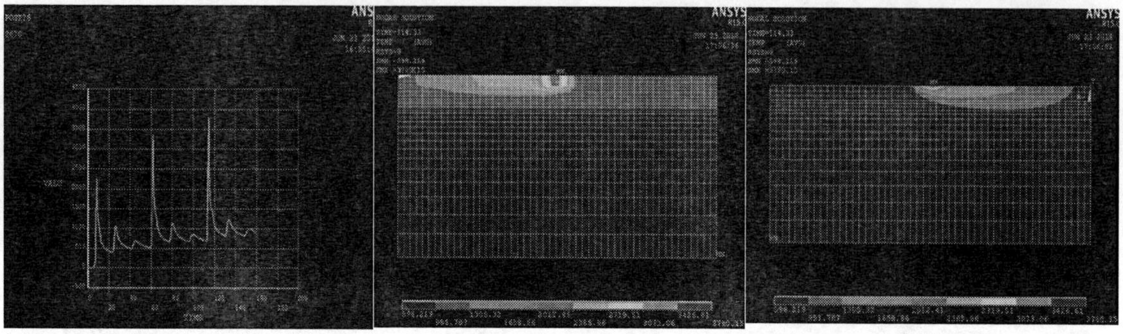

(a) temperature curve (b) top view (c) left view

Fig. 8 temperature field at NO.2670 node when powder layer thickness decreased

5. Conclusion

The powder thickness, initial temperature, scanning speed and laser power of the energy model have great influence on the depth of the molten pool. The influence of scanning speed and laser power on the length of molten pool is the second. These parameters have a little influence on the width of molten pool, among which the laser powder has greater influence on the width of molten pool. Appropriate energy model parameters are required to determine whether remelting occurs between different layers. When the simulation process parameters P=105W, v=50mm/s, the initial temperature of the powder is 330°C, and the thickness of the powder layer is 0.20mm, the characteristics of the molten pool that is easy to be formed can be obtained.

References

[1] Chen Q, Guillemot G, Gandin C A, et al. Numerical modelling of the impact of energy distribution and Marangoni surface tension on track shape in selective laser melting of ceramic material[J]. Additive Manufacturing, 2018:713-723.

[2] Panwisawas C, Qiu C L, Sovani Y, et al. On the role of thermal fluid dynamics into the evolution of porosity during selective laser melting[J]. Scripta Materialia, 2015, 105:14-17.

[3] Morgan R-H, Papworth A-J, Sutcliffe C, et al. High density net shape components by direct laser re-melting of single-phase powders[J]. Journal of Materials Science, 2002, 37(15):3093-3100.

[4] Simchi A, Petzoldt F, Pohl H. On the development of direct metal laser sintering for rapid tooling[J]. Journal of Materials Processing Tech, 2003, 141(3):319-328.

[5] Bertol L-S, Júnior W-K, Silva F-P-D, et al. Medical design: Direct metal laser sintering of Ti–6Al–4V[J]. Materials & Design, 2010, 31(8):3982-3988.

[6] Vandenbroucke B, Kruth J. Selective laser melting of biocompatible metals for rapid manufacturing of medical parts[J]. Rapid Prototyping Journal, 2013, 13(4):196-203.

[7] Duley W-W. Laser processing & analysis of materials.N.Y.Plenum Pr,1983.1-459

IOP Conf. Series: Materials Science and Engineering **711** (2020) 012103 doi:10.1088/1757-899X/711/1/012103

Properties of mode-locked ultrafast fiber lasers

Youwen Jia[1, a]

163 Xianlin Avenue, Nanjing, Jiangsu Province, PRC

161240028@smail.nju.edu.cn

Abstract. Lasers, including solid-state lasers, fiber lasers and ultrafast fiber laser, are critical to our daily lives. Different kinds of laser have their own advantages and disadvantages. For solid-state lasers, they can generate short duration, high energy pulses but are difficult to maintain and expensive. Fiber lasers, which are compact, low-price, high-efficiency, can provide a novel solution to these issues. Here, we illustrate the theory and main features of the most important ultrafast fiber laser designs: soliton laser, dispersion-management soliton, dissipative soliton and Mamyshev oscillator. At last, we focus primarily on the promising future of the ultra-fast fiber lasers.

1. Introduction

High-power ultrafast lasers are widely used in additive 3D printing, welding, micromachining, medical imaging etc [1-3]. In the past, the main workhorses were solid-state lasers as they have been carefully engineered for over 30 years. For instance, solid-state Ti:sapphire lasers offers a remarkable combination of short duration and high pulse energy, which completely changed the ultrafast science and industry since they were invented. However, these lasers suffer from large footprint, difficulty to manufacture and maintain, and high price.

With the increasing demand of ultrafast pulses, more-compact, user-friendly and economical laser sources are needed. Fiber lasers are invented to resolve the problems of solid-state lasers, which were listed above. The clear advantages of fiber laser solutions (compared to their solid-state counterparts) are: compact design, easy thermal management, minimal alignment, high spatial beam quality, and low cost. Consequently, fiber systems have already become a competent option for applications requiring continuous wave or long-pulse operation.

Nevertheless, the pulsed fiber lasers have their own challenges. In the pulsed operations, the benefits of fiber lasers come at the cost of tighter confinement of the light, which cause larger nonlinear phase accumulation that can rapidly degrade the pulse quality [4]. In order to cope with the problem, different types of mode-locked fiber lasers have been invented [5]. Furthermore, recent works had shown that much higher pulse energies and peak powers can be achieved in fiber lasers that operate at large normal dispersion. For instance, self-similar laser and dissipative soliton fiber lasers can provide over 10-nJ and less than 100 femtosecond pulse width, which are competitive to most of commercial solid-state lasers [6-8].

The aim of this paper is to present the properties of different ultrafast fiber lasers, as well as the physics governing the formation of stable pulses. The paper is organized as follows: stating the main features of the each laser, as well as its theoretical knowledge one by one, with the comparison of the specifications of the four lasers in the end.

2. Fiber lasers

2.1 Soliton Lasers:

In optics, soliton generally refers to an optical field that remains unchanged temporally or spatially during propagation. The soliton is classified as temporal solitons and spatial solitons. Its difference mainly comes from the formation mechanism: nonlinearity and dispersion for the temporal soliton, and nonlinearity and spatial diffraction for spatial soliton. In ultrafast laser field, soliton laser typically means temporal soliton. Soliton lasers are created by incorporating passive single-mode fiber and active gain fiber and keeping a large net cavity dispersion negative. This is accomplished by either exploiting anomalous dispersion fiber or dispersive components. The optical Kerr effect is one of the most common nonlinear effects in such fibers [4]. The generated pulse in a cavity is also determined by the dissipative effects of gain, loss, saturable absorption, and spectral filtering. Despite this, dissipative effects and gain remain only of secondary importance in soliton lasers when compared to as dispersion and Kerr nonlinearity.

Therefore, soliton laser in single mode fibers is described by the nonlinear Schrodinger equation as follows:

$$\frac{\partial A}{\partial z} = -i\frac{\beta_2}{2}\frac{\partial^2 A}{\partial t^2} + i\gamma|A|^2 A \quad (1)$$

where A is the slow varying envelope of the optical pulse, γ is the nonlinear parameter, β_2 is the second order dispersion coefficient which corresponds to the group velocity dispersion effect.

The fundamental bright soliton solution is given by:

$$A(z,t) = \sqrt{2\gamma}\eta exp\{-4i(\zeta^2 - \eta^2)z - 2i\zeta t + i\varphi\} sech\,(2\eta(t - t_0) + 8\eta\zeta z) \quad (2)$$

where η, ζ, φ and t_0 are all constants and β_2= -2.

Conventionally, soliton oscillators output approximately transform-limited pulses. With the laser action builds up from noise, the cavity's gain and loss elements periodically perturbed the pulse. Following each perturbation, the pulse adjusts itself to satisfy the soliton area theorem, at the same time shedding some energy as a dispersive wave. This resonant phenomenon generates narrow spectral sidebands, and restricts the pulse durations and energy [9,10].

In general, with its simplicity and lower requirement for saturable absorber, the soliton laser is still the most widely used laser in industry and laboratories.

2.2 Dispersion-Managed Soliton Lasers:

The dispersion-managed soliton laser, which is also called stretched pulses laser, is a soliton in a cavity-averaged sense. Fibers with different signs of dispersion are placed in different sections of the cavity. Pulse stretches and compresses temporally twice per roundtrip as it passes through each dispersion elements (positive and negative), reducing the averaged nonlinearity per roundtrip and permitting higher energy. Net cavity dispersions ranging from slightly anomalous to slightly normal are possible for stable output [11]. The shortest pulses are achieved near zero net dispersion [12].

Large temporal breathing reduces the accumulated nonlinearity compared to a static solitary pulse of the same energy, allowing more energetic pulses than a comparable soliton laser. The output coupler can be place at different places in the cavity, and the output pulses may be up-chirped or down-chirped.

The dispersion management soliton can be modeled by Master Equation:

$$\frac{1}{T_R}\frac{\partial}{\partial T}a = (g - l)a + \left(\frac{1}{\Omega_f^2} + jD_{net}\right)\frac{\partial^2}{\partial t^2}a + (\gamma_0 - j\delta_0)|A_0|^2\left(1 - \mu\frac{t^2}{\tau^2}\right)a \quad (3)$$

In the equation stated above, *(g-l)* means the net gain of the bandwidth and the modulation frequency, T_R means the round-trip time, A_0 means the amplifier when n=*0*, τ satisfies the equation $\tau^4 = \frac{2g}{(M\Omega_m^2\Omega_g^2)}$ (Ω_m is the modulation frequency), a and A is a Fourier-transform with the Fourier-transform pairs when A is the amplifier, T is the long term time variable, while the γ satisfies the equation $\gamma = \frac{s_0}{I_{sat}A_{eff}}$ (I_{sat} means the saturation intensity of the absorber, s_0 is the unsaturated loss, A_{eff} is the intensity multiplied by the effective area of the mode), D is the group velocity dispersion parameter, δ is the Kerr-coefficient whose value is $((2\pi/\lambda)m2L/A_{eff})$, Ω_f is the filter of bandwidth, μ is a coefficient which values variationally.

This equation has a group of Gaussian-pulse solutions. In the experiments, the measured pulses are also closer to a Gaussian shape compare to the soliton lasers. The output bandwidth can be greater than the gain medium bandwidth. The first dispersion management soliton laser was achieved at 1550 nm by inserting a new Erbium doped fiber with normal dispersion into the laser cavity. The net dispersion is close to zero and the pulse was stretched and compressed strongly (20-fold) in the cavity. At other wavelengths, such as 1030 nm, the anomalous components can also come from dispersion delay systems, which are similar to soliton lasers.

The typical performance for the stretched pulse cavity design is <100 fs de-chirped pulse duration with energies of ~2 nJ [11,12].

2.3 Dissipative Soliton Lasers:

In contrast to soliton lasers, dissipative soliton lasers are formed as a combination between the mutual nonlinear interactions among the normal self-phase modulation, cavity dispersion, effective gain bandwidth filtering and gain saturation, properties of which are determined by the Ginzburg-Landau equation [13].

Optical pulse evolution in the presence of Kerr nonlinearity, dispersion, dissipative and gain process can be modeled by the cubic-quintic Ginzburg-Landau equation:

$$U_z = gU + \left(\frac{1}{\Omega} - i\frac{D}{2}\right)U_{tt} + (\alpha + i\gamma)|U|^2U + \delta|U|^4U \quad (4)$$

where U is the electric field envelope, z represents the propagation coordinate, t is the retarded time, D is GVD, g is the net gain, Ω is relevant with the filter bandwidth, α is a cubic saturable absorber term, γ is the cubic refractive nonlinearity of the medium, and δ is the quantic saturable absorber term [14].

From the solution of this equation, one can find the spectral filter plays a significant role in the pulse shaping process. It contributes to both the dissipation and spectral filtering and the filter bandwidth strongly determines the output pulse shape, duration and energy [14]. The output pulse owns a sharp spectral boundary, which differs dissipative soliton from any other laser cavities. In spite of their sharp spectral features, dissipative solitons can usually be compressed to within 20% of the transform limit and still maintain a low secondary structure. Compared to conventional soliton lasers, dissipative soliton lasers are able to generate up to 20 nJ and 200 fs or ~10 nJ and 80 fs pulses. This is significantly greater than soliton and stretched pulse designs. Using photonics crystal fiber, the pulse energy can be scaled up to hundreds of nanojoules.

2.4 Mamyshev Oscillator:

The development of the above fiber lasers improved the output pulse energy. However, in order to be widely used outside the laboratory, the environmental stability is another big concern. Many research have been explored to obtain the environmental-stable design [15-17]. However, their performance is not as good as the traditional designs. This prevents the lasers from having even more importance in wider industrial applications.

One promising solution for this is the Mamyshev Oscillator. This oscillator includes two major parts: One is a laser cavity with two gain segments and two filters and the other is a starting arm that creates a fluctuating field that seeds the mode-locked state. This cavity design originates from the idea of signal regeneration in telecommunication field [18]. Laser based on reamplification and reshaping can be used to generate short pulses [19–23]. Self-similar pulses propagate in both cavity arms. And near linear phase is accumulated during the propagation. Self-phase-modulation (SPM) and two spectral filters shaped the pulse. The two spectral filters limit the continuous wave growing, which lead to larger nonlinear phase for the pulses. Therefore, the pulse energy is higher than all other laser designs. All the cavity is made by polarization maintaining fiber and it can tolerate strong environmental perturbation. Using single mode fiber, this is the first fiber laser that reach megawatt peak power. Energy scaling was demonstrated by Sidorenko et. al [23].

3. Conclusion

Table 1. Performance summary of fiber oscillators for different pulse evolutions

Pulse evolution	Nonlinear Phase	Typical Performance	Best Performance
Soliton	near 0	100 pJ, 300fs	less than 1nJ, 100fs
DM Soliton	0-π	1nJ, 100fs	up to 3nJ, down to 50fs
Dissipative Soliton	2π-10π	6nJ, 150fs	up to 20nJ, down to 70fs
Mamyshev oscillator	>60π(experiment) >140π(simulation)	/ /	50nJ, 40fs(experiment) >190nJ, <20fs(simulation)

With the development of the lasers, from soliton, DM soliton to dissipative soliton and Mamyshev oscillator, the nonlinear phase gradually increases, which cause both the typical performance and the best performance to improve.

In this article, we discussed the four types of soliton lasers---soliton lasers, dispersion management soliton lasers, dissipative soliton lasers and Mamyshev oscillator, all of them have their own advantages and disadvantages. We also emphasize the theory as well as demonstrate the performances of these four lasers. All these four lasers have already been applied and employed in the new materials, new further relevant researches. It can be anticipated that the fiber lasers and its techniques will be a powerful tool for the studies in the future [24].

References

[1] W. R. Zipfel, R. M. Williams, and W. W. Webb, "Nonlinear magic: multiphoton microscopy in the biosciences," Nat. Biotechnol. 21, 1369–1377 (2003).
[2] R. R. Gattass and E. Mazur, "Femtosecond laser micromachining in transparent materials," Nat. Photonics 2, 219–225 (2008).
[3] X. Liu, D. Du, and G. Mourou, "Laser ablation and micromachining with ultrashort laser pulses," IEEE J. Quantum Electron. 33, 1706-1716 (1997).
[4] G. P. Agrawal. Nonlinear Fiber Optics, Third Edition. Academic Press (2007).
[5] F. W. Wise, A. Chong, and W. H. Renninger, "High-energy femtosecond fiber lasers based on pulse propagation at normal dispersion," Laser Photonics Rev. 2, 58–73 (2008).
[6] F. O. Ilday, J. R. Buckley, W. G. Clark, and F. W. Wise, "Self-similar evolution of parabolic pulses in a laser," Phys. Rev. Lett. 92, 213902 (2004).
[7] A. Chong, J. Buckley, W. Renninger, and F. W. Wise, "All-normal dispersion femtosecond fiber laser," Opt. Express 14, 10095–10100 (2006).
[8] B. Nie, D. Pestov, F. W. Wise, and M. Dantus, "Generation of 42-fs and 10-nJ pulses from a fiber laser with self-similar evolution in the gain segment," Opt. Express 19, 12074–12080 (2011).
[9] S. M. J. Kelly, "Characteristic sideband instability of periodically amplified average soliton," Electron. Lett. 28, 806-807 (1992).
[10] J. P. Gordon. "Dispersive perturbations of solitons of the nonlinear Schrödinger equation." Journal of the Optical Society of America B 9, 1 (1992).
[11] K. Tamura, E. P. Ippen, H. A. Haus, and L. E. Nelson. "77-fs pulse generation from a stretched-pulse mode-locked all-fiber ring laser." Optics Letters 18, 13 (1993).
[12] D. Ma, Y. Cai, C. Zhou, W. Zong, L. Chen, and Z. Zhang, "37.4 fs pulse generation in an Er:fiber laser at a 225 MHz repetition rate," Opt. Lett. 35(17), 2858–2860 (2010).
[13] W. Renninger, A. Chong, and F. W. Wise, "Dissipative solitons in normal-dispersion fiber lasers," Phys. Rev. A 77(2), 023814 (2008).
[14] P. Grelu and N. Akhmediev. "Dissipative solitons for mode-locked lasers." Nature Photonics 6, 2 (2012).

[15] A. Chong, W. H. Renninger, and F. W. Wise, "Environmentally stable allnormal-dispersion femtosecond fiber laser," Opt. Lett. 33, 1071–1073 (2008).

[16] C. Aguergaray, N. G. R. Broderick, M. Erkintalo, J. S. Y. Chen, and V. Kruglov, "Mode-locked femtosecond all-normal all-PM Yb-doped fiber laser using a nonlinear amplifying loop mirror," Opt. Express 20, 10545–10551 (2012).

[17] C. Aguergaray, R. Hawker, A. F. J. Runge, M. Erkintalo, and N. G. R. Broderick, "120 fs, 4.2 nJ pulses from an all-normal-dispersion, polarization-maintaining, fiber laser," Appl. Phys. Lett. 103, 121111 (2013).

[18] P. V. Mamyshev, "All-optical data regeneration based on self-phase modulation effect," in 24th European Conference on Optical Communication, Madrid, Spain (IEEE, 1998), pp. 475–476.

[19] M. Piche, "Mode locking through nonlinear frequency broadening and spectral filtering," Proc. SPIE, 2041, 358–365 (1994).

[20] K. Sun, M. Rochette, and L. R. Chen, "Output characterization of a selfpulsating and aperiodic optical fiber source based on cascaded regeneration," Opt. Express 17, 10419–10432 (2009).

[21] T. North and M. Rochette, "Regenerative self-pulsating sources of large bandwidths," Opt. Lett. 39, 174–177 (2014).

[22] Z. Liu, Z. Ziegler, L. G. Wright, and F. W. Wise. "Megawatt peak power from a Mamyshev oscillator." Optica **4**, 649 (2017).

[23] P. Sidorenko, W. Fu, L. G. Wright, M. Olivier, and F. W. Wise. "Self-seeded, multi-megawatt, Mamyshev oscillator." Optics Letters **43**, 2672 (2018).

[24] A. Chong, L. G. Wright, and F. W. Wise, "Ultrafast fiber lasers based on self-similar pulse evolution: a review of current progress," Rep. Prog. Phys. 78, 113901 (2015).

Research on Temperature and Humidity Decoupling Control of Constant Temperature and Humidity Test Chamber

Liheng Wang[1,a], Zhifeng Zhu[1,b]

[1]Wuhan Institute of Technology, Wuhan 430205, China

[a]wlhhust@163.com,[b]zzfeng8@163.com

Abstract: The temperature and humidity system is a system commonly used in industry and is recognized as one of the most complex systems. Temperature and humidity are important parameters for improving production quality in industrial production. To carry out effective control, factors such as nonlinearity, large hysteresis and real-time change must be considered, and there is mutual coupling between temperature and humidity. Temperature change will lead to humidity. The change of humidity and the same change of temperature will also affect the temperature change. The positional PID control can achieve satisfactory results. In this paper, through the analysis of the temperature and humidity system model, the feedforward decoupling compensation method is adopted for the characteristics of the temperature and humidity control system, which provides an effective way for the modeling and control of the temperature and humidity system. The simulation results show that the feedforward decoupling controller designed in this paper can achieve satisfactory results under various experimental conditions.

1. Introduction

The constant temperature and humidity test chamber is widely used in product development testing and scientific research to provide a constant temperature and humidity test environment for the tested products.When controlling the test environment, the change of temperature will lead to the change of humidity. The change of humidity will also affect the change of temperature. The two parameters are coupled with each other, resulting in deterioration of control quality.Therefore, when controlling the parameters of the constant temperature and humidity chamber, decoupling control is required. In this paper, the decoupling control method is studied from the characteristics of system objects, and the effectiveness of the control algorithm is verified by simulation.

2. The composition and working principle of the constant temperature and humidity test chamber

The schematic diagram of the experimental device is shown in Figure 1. It mainly includes the box body, the constant temperature control system, the constant humidity control system, and the circulation control system[1].

1. Temperature control system mainly includes two parts: heating system and cooling system.

(1) Heating system: When the controller receives the heating command, it will output voltage to the solid state relay to make it suck. The voltage is applied to both ends of the heater to start heating up, and the circulation system starts to start, and the heat of the heater is brought to the box through the fan. In the body, the temperature rise process of the constant temperature and humidity test chamber is completed.The controller controls the solid state relay to output 100% power before the

IOP Conf. Series: Materials Science and Engineering 711 (2020) 012104 doi:10.1088/1757-899X/711/1/012104

temperature reaches the dead zone value set by the PID; when the temperature exceeds the set dead zone value, the controller performs on-off regulation of the solid state relay by outputting the PFM signal.

(2) Cooling system: When the controller receives the cooling command, the compressor, condenser and evaporator start to start. The low-temperature and low-pressure gas is injected into the compressor to turn it into a high-temperature and high-pressure gas, and the high-temperature and high-pressure gas is condensed into a liquid to discharge heat through the condenser, and the heat in the tank is discharged through the circulation fan, and then absorbed by the refrigerant in the evaporator. The heat turns the low-pressure liquid into a low-temperature low-pressure gas and finally returns to the compressor to achieve the cooling effect, and finally cools the constant temperature and humidity test chamber.

2. Humidity control system mainly includes two parts: humidification system and dehumidification system.

(1) Humidification system: When the controller receives the humidification command, because the heater of the constant temperature and humidity test chamber is equipped with a water tray, the water temperature can be raised by controlling the heating of the heater, and the water temperature rises along with the surface layer of the water tray. High, the water vapor pressure increases, and the difference between the water vapor pressure in the air in the tank increases, which increases the water vapor diffusion and the exchange of fluid.In this way, the superheat of water vapor is significantly reduced, and finally the constant temperature and humidity test chamber is humidified[2].

(2) Dehumidification system: When the controller receives the dehumidification command, the air is cooled to below the dew point temperature, so that the water vapor larger than the saturated moisture content is condensed and precipitated, so that the constant temperature and humidity test chamber is dehumidified[3].

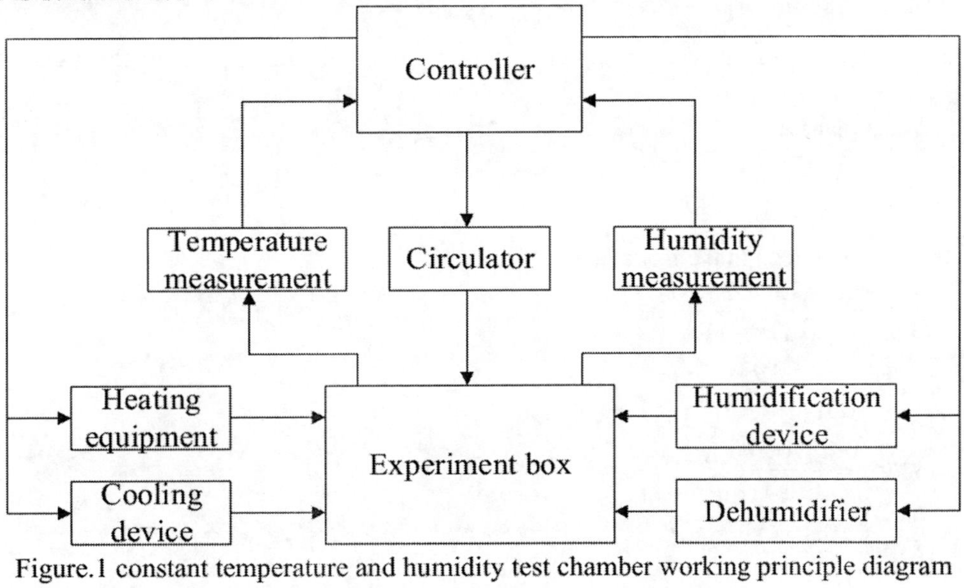

Figure.1 constant temperature and humidity test chamber working principle diagram

3. Establishment of test chamber temperature and humidity model

The controlled volume of the test box is: temperature, humidity, temperature controlled by heater and refrigerator; humidity is controlled by heating evaporation tray and air conditioning dehumidification. The control block diagram of the system is shown in Figure 2:

IOP Conf. Series: Materials Science and Engineering **711** (2020) 012104 doi:10.1088/1757-899X/711/1/012104

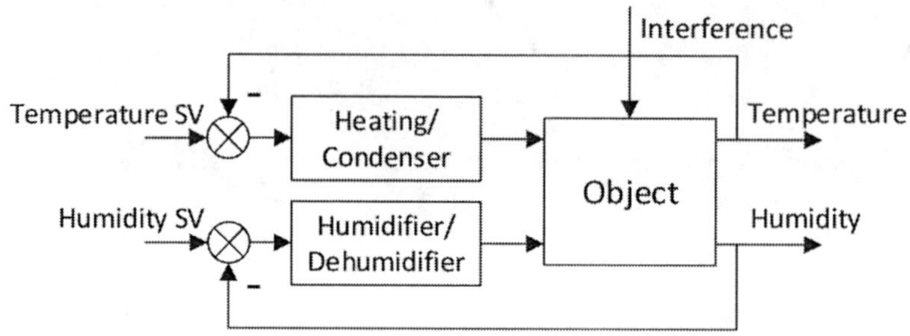

Figure.2 constant temperature and humidity control system block diagram

It can be seen from Fig. 2 that the constant temperature and humidity system is a dual input dual output system, and the control object is accompanied by external interference.

The constant temperature and humidity system is a relatively common complex system. It is a nonlinear real-time change system with strong coupling and large hysteresis[4].Using the mechanism modeling method, a complex coupled real-time nonlinear system model is obtained, which is not conducive to the study of decoupling control.In order to facilitate the control, the system identification method is adopted. Under the open loop, the system step excitation is given to obtain the temperature and humidity response curve. The system model is identified by using multiple sets of unit step curves to obtain the temperature response model of the following system:

$$T(s)=\frac{0.01813(152s+1)}{s(87s+1)}e^{-19s}\cdot C_t(s)+\frac{0.01109}{s(52s+1)}e^{-75s}\cdot C_h(s) \quad (1)$$

Humidity response model:

$$H(s)=-\frac{41.193}{408.09s+1}e^{-24s}\cdot C_t(s)+\frac{0.1575}{s(27.4s+1)}e^{-31s}\cdot C_h(s) \quad (2)$$

In the formula, $Y_1(s)$,is the temperature control amount, and $C_h(s)$ is the humidity control amount.

Write it as a matrix:

$$\begin{bmatrix} T \\ H \end{bmatrix}=\begin{bmatrix} G_{11} & G_{12} \\ G_{21} & G_{22} \end{bmatrix}\begin{bmatrix} C_t \\ C_h \end{bmatrix} \quad (3)$$

The transfer functions in the formula are:

$$G_{11}(s)=\frac{0.01813(152s+1)}{s(87s+1)}e^{-19s} \quad (4)$$

$$G_{21}(s)=-\frac{41.193}{408.09s+1}e^{-24s} \quad (5)$$

$$G_{12}(s)=\frac{0.01109}{s(52s+1)}e^{-75s} \quad (6)$$

$$G_{22}(s)=\frac{0.1575}{s(27.4s+1)}e^{-31s} \quad (7)$$

The control block diagram of Figure 2 can be changed to the following structure:

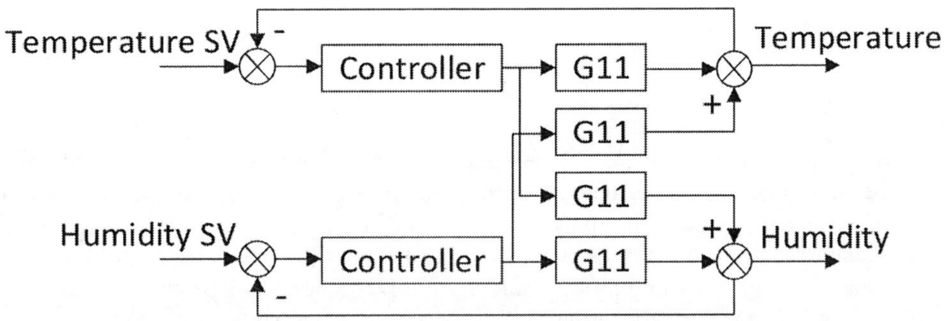

Figure.3 coupling of temperature and humidity control block diagram

4. Decoupling controller design

A system based on the strong coupling characteristics between the two in the temperature and humidity control process is also called a temperature and humidity coupling system.If such systems are not decoupled, each loop in the system will interfere with each other during operation, and it will be difficult to obtain an ideal control effect.Therefore, the decoupling of the temperature and humidity coupling system can be realized by decoupling methods such as feedforward compensation decoupling, state feedback decoupling, series compensation decoupling and static decoupling.The decoupler is placed in front of the control object or on the feedback channel in a compensated manner to attenuate or even eliminate the coupling between temperature and relative humidity[5].

From the transfer function of the temperature and humidity control system, their coupling relationship is that the influence of temperature on the relative humidity variable is much greater than the relative humidity on the temperature variable, and the relative humidity changes, the humidity changes much more slowly, and can be compensated completely[6]. Whether from theoretical analysis or practical experience, the relative humidity drops by 2%-3% for every one degree Celsius increase in temperature, and such coupling effects are negligible.Therefore, it is completely possible to use single-edge decoupling, temperature and humidity compensation, temperature and humidity variables can be controlled as a single variable[7].

In this paper, the feedforward compensation decoupling method based on the principle of feedforward compensation is selected.The feedforward compensation decoupling is to regard the coupling channel as the disturbance signal of the control object, so as to realize the decoupling of the disturbance signal. The feedforward compensation decoupling system diagram is shown in Figure 4:

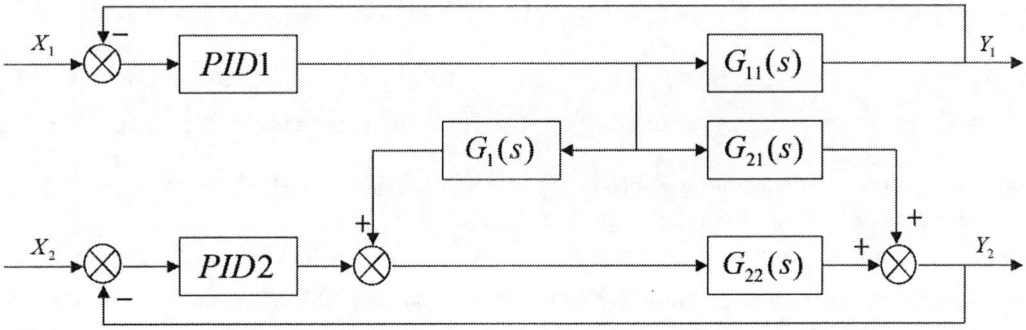

Figure.4 Feedforward compensation decoupling control system

According to the principle of invariance of feedforward control, we can know:

$$\frac{Y_2(s)}{V_1(s)} = G_{21}(s) + G_1(s)G_{22}(s) = 0 \tag{8}$$

From the above formula, the feedforward decoupler $G_1(s)$ can be determined as:

$$G_1(s) = -\frac{G_{21}(s)}{G_{22}(s)} \tag{9}$$

5. Simulation

According to the reference model of the temperature and humidity system, the coupled system is subjected to feedforward compensation decoupling. Under the step signal, the temperature jumps from 0 °C to 60 °C, and the humidity jumps from 0% to 85%.Curves 1, 2 are the set values of relative humidity and temperature, and curves 3 and 4 are the simulated values of relative humidity and temperature respectively. The simulation diagram before decoupling of temperature and humidity system and its simulation results are shown in Figure 5:

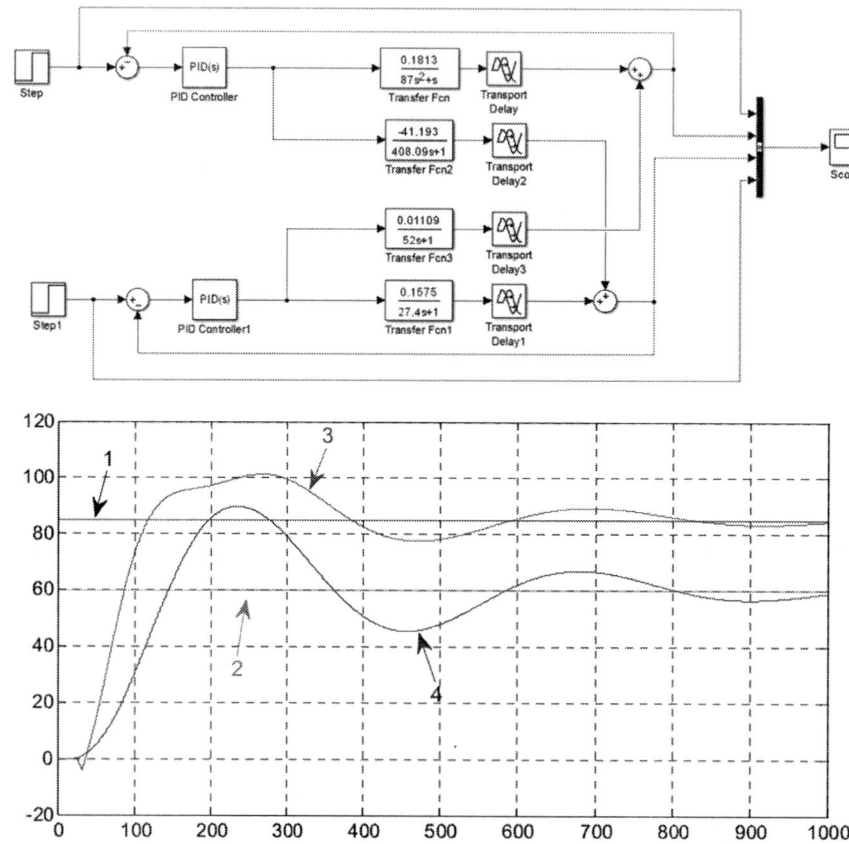

Figure.5 before the temperature and humidity system decoupling simulation diagram and its results

Below we analyze the decoupling effect of the coupling channel between temperature and humidity under the feedforward compensation decoupling:

(1) When the humidification heating amount is $U_2(s)=0$, the heating heating amount $U_1(s)$ affects the controlled quantity temperature $Y_1(s)$, the simulation result after decoupling is shown in curve 1 of Fig. 6(a); the control quantity $U_1(s)$ is the controlled quantity humidity The effect of $Y_2(s)$, the decoupling simulation results are shown in curve 2 of Figure 6 (a).

(2) When the heating heating amount is $U_1(s)=0$, the effect of humidifying heating quantity

$U_2(s)$ on the controlled quantity humidity $Y_2(s)$, the simulation result after decoupling is shown in curve 1 of Fig. 6(b); the influence of control quantity $U_2(s)$ on the controlled quantity temperature $Y_1(s)$ The decoupling simulation results are shown in curve 2 of Figure 6(b).

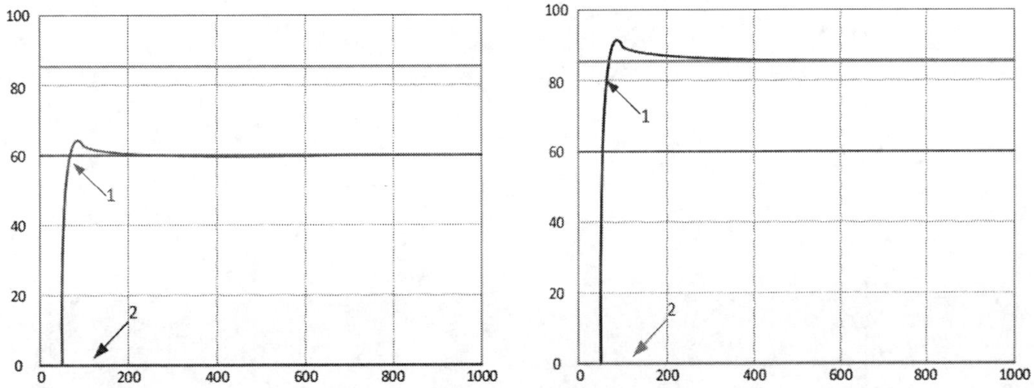

(a)Temperature versus humidity decoupling system simulation diagram

(b)Humidity-to-temperature decoupling system simulation diagram

Figure. 6 Simulation diagram of decoupling of temperature and humidity system

It can be seen from the simulation curves in the two graphs in Fig. 6 that the control amount $U_2(s)$ does not affect the controlled quantity $Y_1(s)$, and the control quantity $U_1(s)$ does not affect the controlled quantity $Y_2(s)$; the controlled quantity temperature $Y_1(s)$ is only affected by $U_1(s)$. The controlled humidity $Y_2(s)$ is also only affected by $U_2(s)$. that is, the feedforward compensation decoupling removes the correlation of the temperature and humidity coupling channel, making the system two independent, non-influenced single-entry systems.

By comparing the simulation results of the temperature and humidity system before and after decoupling, the control effects of temperature and relative humidity have been greatly improved. The feedforward compensation decoupling improves the robustness of the temperature and humidity control system and improves the performance of the temperature and humidity control system.

Therefore, the feedforward compensation decoupling can cancel the coupling relationship between temperature and humidity, eliminating the influence of temperature on relative humidity. By changing the different set values of the control object, the feedforward compensation decoupling method can still achieve the control. Decoupling of objects. It is thus verified that the feedforward compensation decoupling method can eliminate the coupling effect of temperature and relative humidity.

6. Experimental result

The temperature and humidity test chamber has a temperature of 60 ° C and a relative humidity of 85%. The results are shown in Figure 7. Curves 1, 2 are the relative humidity and temperature settings, and curves 3 and 4 are the relative humidity and temperature. value. According to China's measurement instrumentation error standard, the temperature is 1% error and the humidity error is 2%. It can be seen from the experimental results that very good control results can be obtained.

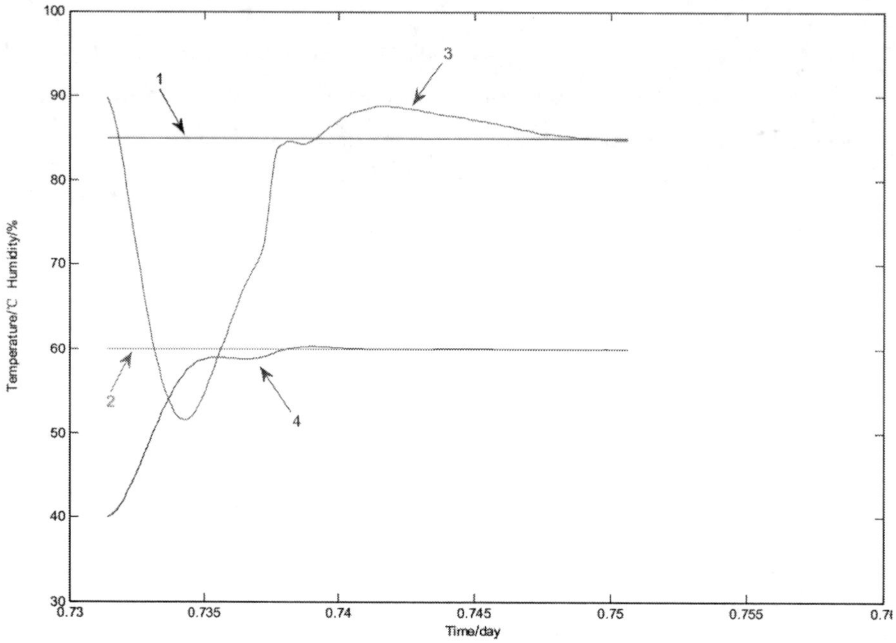

Figure.7 the actual temperature and humidity effect

7. Conclusion

The temperature and humidity control system of the constant temperature and humidity test chamber is a system with large hysteresis and strong coupling. After decoupling simulation results, it can be known that the positional PID algorithm can achieve good control effect. In the current control system control process, sometimes some oscillations occur. We need to further optimize the controller and optimize the parameters of the controller to reduce the oscillation and make the system more stable.

References

[1] Sun Bi. Application and research of fuzzy adaptive control in temperature and humidity system [D]. South China University of Technology, 2015.

[2] Peng Tong. Low temperature and high humidity stress achieved by shallow water dish humidification [J]. Environmental technology, 2010, 29 (04): 15-17.

[3] Hao Dongsheng, Liu Yuhui, Li Peng, Lian Sumei. Construction and use of constant temperature and humidity laboratory for cotton inspection [J]. Cotton Science, 2019, 41 (02): 36-40.

[4] Becker M, Oestreich D, Hasse H. Fuzzy control for temperature and humidity in refrigeration systems[C]. Proceedings of the Third IEEE Conference. IEEE, 1994:1607-1612

[5] Xu Jian, Fang Yanjun, Lin Shizhuang. Application of Fuzzy PID Control in Alternating Damp and Heat Control [J]. Microcomputer Information, 2008 (25): 14-15+68.

[6] Pan Chunyue. Design of temperature and humidity control system based on genetic algorithm and decoupling control [J]. Industrial instruments and automation devices, 2006 (03): 19-21.

[7] Zhang Panpan, Zhao Gang. Fuzzy PID-adaptive Smith control with temperature and humidity decoupling [J]. Chemical automation and instrumentation, 2012,39(10): 1260-1264.

IOP Conf. Series: Materials Science and Engineering **711** (2020) 012105 doi:10.1088/1757-899X/711/1/012105

Effects of Typical Emission Waveform on Laser Reflective Tomography Imaging

Xin-yuan Zhang[1,2,a], Yi-hua Hu[1,2,b], Nan-xiang Zhao[1,2,c] and Shi-long Xu[1,2,d]

[1]State Key Laboratory of Pulsed Power Laser Technology, National University of Defense Technology, Hefei 230037, China

[2]Key Laboratory of Electronic Restriction Technology Anhui, National University of Defense Technology, Hefei 230037, China

[a]565598675@qq.com, [b]skl_hyh@163.com, [c]southfly@163.com, [d]xushi1988@yeah.net

Abstract. Laser reflective tomography imaging is a new imaging method of long distance and high resolution. The echo could be expressed as the convolution of incident waveform and the reflectance projection distribution, so the deconvolution effect of the echo obtained by different emission waveforms is also different. In order to explore the reflectance projection distribution obtained by deconvolution of different emission waveforms under the same SNR, we simulated four kinds of emission pulses and from the comparison of the RMS of Euclidean distance between deconvolution results and reality results, we found that Gaussian pulse could be selected as the most suitable emission waveform when the SNR was less than 40, and when the SNR was larger, Sawtooth pulse was the best choice.

1.Introduction

Remote dark detection has gotten much interests following the development of deep space explorations. Traditional detection methods lack the ability to distinguish space debris targets with the magnitude of cm-level. Laser reflective tomography imaging(LRTI) is a new imaging method of long distance and high resolution. The space-based lidar is not restricted by geographical location or meteorological conditions, and the coverage of space domain and time domain is greatly increased. Also, it can realize long-range and high-precision detection throughout the day, and is an ideal means for space debris detection.

Developed from Computed Tomography(CT), LRTI was first proposed in 1988 by Parker, Kningt et al. from the Lincoln laboratory in the United States[1]. In 1989, the experimental system was modified to replace the photomultiplier with an EG&G streak camera by Knight F K et al[2]. Matson et al. have further studied the theory and application of reflection tomography imaging[3,4,5]. In addition to, they obtained reconstruction images of satellite from a ground-based lidar[6].

In image reconstruction, the modulation of the pulse waveform by the target is a typical convolution process, which can result in the reduction of the resolution of the reconstructed target image and the blurred image details. Generally, the method of deconvolution is used to obtain the reflectance projection distribution, and the effect of deconvolution of different emission waveforms is different. Moreover, the results of reflectance projection distribution obtained by deconvolution of different emission waveforms under the same Signal to Noise Ratio(SNR) are different. Therefore, in this research, we analysed the emission and reception regularly by establishing analytic model, and

IOP Conf. Series: Materials Science and Engineering **711** (2020) 012105 doi:10.1088/1757-899X/711/1/012105

compared the results of reflectance projection distribution obtained by deconvolution of several different emission waveforms. Finally, we found the most suitable emission waveform for LRTI, and proposed the requirements for waveform selection and SNR.

The rest of this paper is organized as follows: in Section 2, we briefly review the principle of laser reflective tomography. The model of laser reflection tomography is established in Section 3. The results of reflectance projection distribution obtained by deconvolution of different emission waveforms under the same SNR are displayed and analyzed in Section 4. Finally, the concluding remarks are made in Section 5.

2.Principle of laser reflective tomography imaging

The basic principles of LRTI are to illuminating the object of laser at multiple angles, collecting the echo of the target at multiple angles, and reconstruct the 2D cross-sectional image of the target according to the reflectance projection distribution. As shown in Fig. 1(a), the parallel laser beams irradiate a 2-D target, when the irradiation angle is ϕ, The reflectance projection distribution of the target at angle ϕ is defined as

$$p(r,\phi)=\int_{L_{r,s}} f(x,y)ds. \tag{1}$$

Where $L_{r,\phi}$ is a straight line perpendicular to the direction of light, with a function of $r=x\cos\phi+y\sin\phi$, $f(x,y)$ is the reflectance distribution of the target. An angle projection image is shown in Fig. 1(b), the parallel lines with arrows unfold in the laser beams irradiate region, and the distance of the projection depends on the depth of the region irradiate by the laser. It should be noted that the reflectance projection distribution is actually nonzero only on the surface, as laser beams can't penetrate the target. That is

$$f(x,y)=0,(x,y)\notin D. \tag{2}$$

where D is the set of points of the target surface. Therefore, the actual projection can be represented as

$$p(r,\phi)=\int_{L_{r,\phi}\cap D} f(x,y)ds. \tag{3}$$

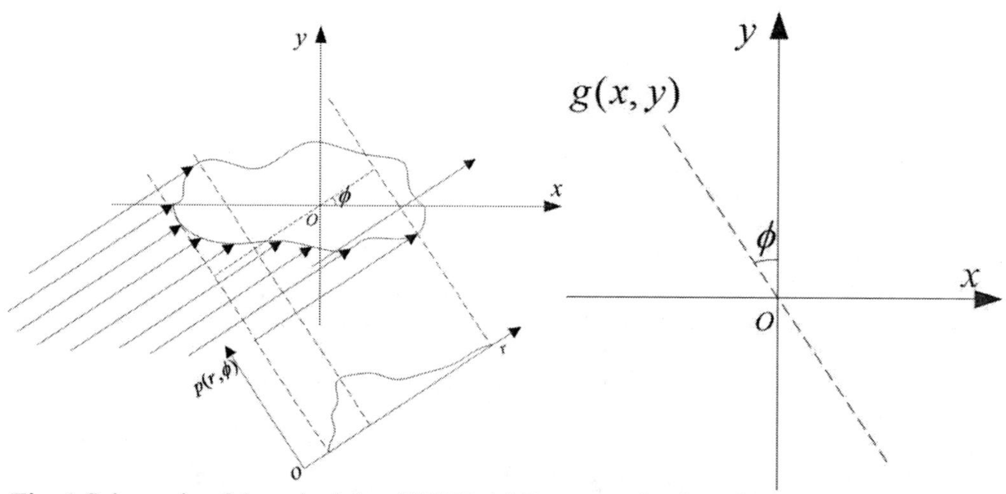

Fig. 1 Schematic of the principle of LRTI. (a) Target projection; (b) Data backprojection

The main idea of the direct back-projection method is smearing the scanned path("original path") of the measured projection data in angle to the pixel passed by the path. That is, the values of all the points passing through this coordinate point are accumulated or averaged after accumulated, which are taken as the values of the pixels in the reconstructed 2D image. However, this method can generate artifacts and decrease the resolution of image, so the general method of LRTI is Filtered Back projection (FBP) algorithm.

3.Model of LRTI

3.1 Echo convolution model. In laser reflective tomography imaging, the depth information is contained in the echo wave, from which the reflectance projection distribution can be extracted. Theoretically, the echo can be expressed as the convolution of the emission pulse and the reflectance projection distribution

$$w(r,\phi) = p(r,\phi) \otimes g + \xi(r,\phi) = \int_{-\infty}^{+\infty} p(s,\phi) \cdot g(r-s) + \xi(r,\phi). \tag{4}$$

Where $w(r,\phi)$ is the echo at angle ϕ, $p(r,\phi)$ is the reflectance projection distribution at angle ϕ. g is the expression of the emission pulse, and $\xi(r,\phi)$ is the additive noise. The additive white gaussian noise(AWGN) is one of the most basic models of noise and interference[8]. In the process of simulation, the random normal distribution function is firstly used to generate white noise, and then the average energy of signal and noise is calculated. Then given SNR, the variance value set by noise is calculated. Finally, white noise is formed according to the average energy of noise, and the actual echo is the superposition of convolution results and noise. In order to get the expression of reflectance projection distribution, some deconvolution methods should be applied in the solving process such as multi-frame iterative blind deconvolution method[9] and maximum likelihood method [10] .

3.2 Multi-angle echo model. The geometry of LRTI is shown in Fig. 2(a), and the target is situated in the origin of the coordinate system. The laser pulse source and avalanche photodiode(APD) detector are put together as a LRTI system to rotate around the target along the circle, or equivalently the system stays static and the target rotate around. the $\overrightarrow{ox}$ axis points to the system at the beginning when the angle between the $\overrightarrow{ox}$ axis and the sight of view of the system is $\phi = 0°$.

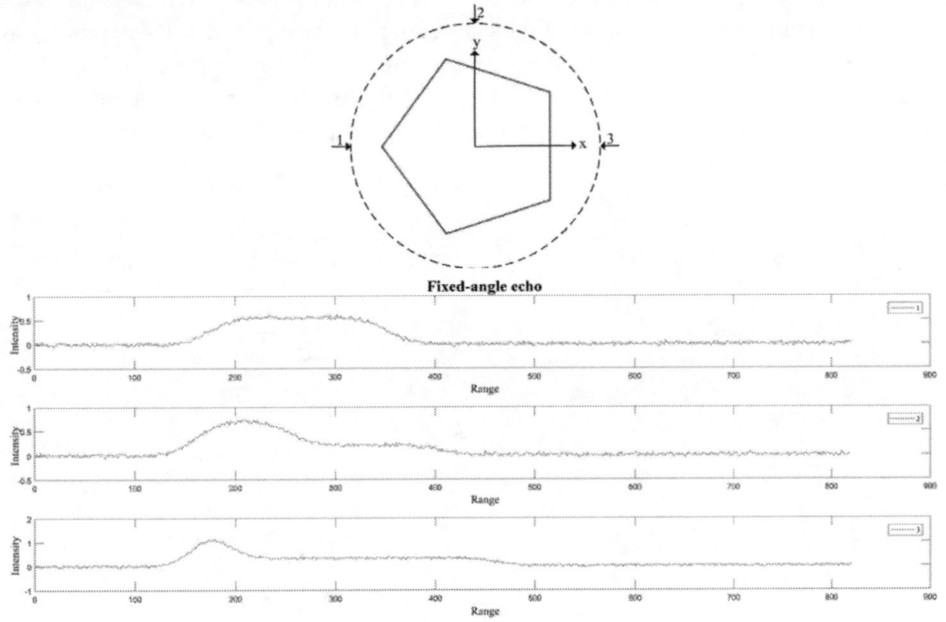

Fig.2 Schematic diagram of echo model.
(a) Diagram of the geometry of LRTI; (b) Echo from different angles

When the target rotates slowly along the $\overrightarrow{ox}$ axis of rotation, it is equivalent to the laser beams moving slowly on the target surface with a certain fixed track. If the target rotates at a constant angular speed, the echo received by the detector can be arranged at fixed projection angle intervals. The projection angle starts from $\phi = 0°$ with the $\overrightarrow{ox}$ axis and rotates clockwise to $\phi = 90°$, $\phi = 180°$, etc. The laser beams irradiates the target from three directions of 1-3 in the Fig. 2(a), and the echoes of the receiving system at these three specific angles are shown in Fig. 2(b).

4. Simulation results and analysis

In order to explore the reflectance projection distribution obtained by deconvolution of different emission waveforms under the same SNR. We simulated four kinds of emission pulses: Gaussian pulse, Triangular pulse, Square pulse, and Sawtooth pulse. Through the analysis of the reflectance projection distribution, we found the most suitable emission waveform for LRTI, and proposed the requirements for waveform selection and SNR. The schematic diagrams of different emission waveform are shown in Fig. 3.

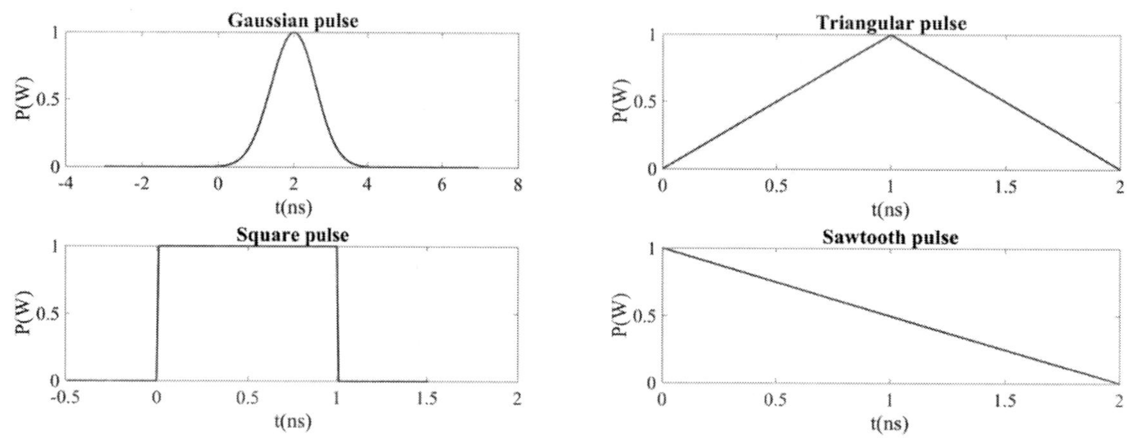

Fig. 3 Schematic diagrams of different emission waveform

Fig. 3 indicates that the full width at half maximum(FWHM) of Gaussian pulse is set as $FWHM = 1ns$, and the pulse amplitude of the Gaussian pulse is set as $A = 1W$. In order to control the invariant, we set the other pulses to have the same amplitude. Furthermore, we set the signal pulse energy of each pulse to be the same, correspond to the pulse width of the Square pulse is $width = 1ns$, and the FWHM of the Triangular pulse and the Sawtooth pulse is $FWHM = 1ns$.

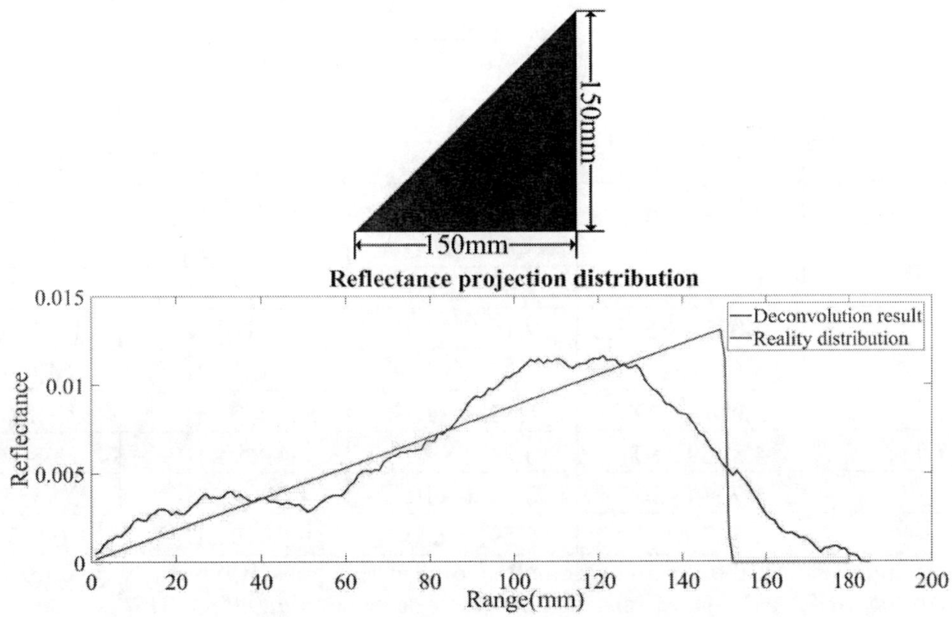

Fig. 4 Schematic of echo deconvolution effect. (a) Reflection projection distribution of target; (b) Comparison between the deconvolution result and the reality

According to the distance resolution equation of LRTI

$$R = \frac{c\tau}{2}. \tag{5}$$

Where c is the speed of light, τ is the FWHM of the Gaussian pulse. We set a target with the reflection projection distribution, with the depth range of 0.15m. The laser beams irradiate the target horizontally from the left, and the reflectance projection distribution is obtained by deconvolution of the echo of the target irradiated by the Gaussian pulse. As shown in Fig. 4(b), The red curve shows the reality reflectance projection distribution and the blue curve shows the deconvolved reflectance projection distribution. The noncausal wiener filter(NCWF) is a regularization inverse filter which has the advantage of simple calculation and efficiency[9], and this method is widely applied in LRTI. Therefore, in the process of deconvolution, we used NCWF to estimate the reflection projection distribution, and the SNR was set to 30. We believed that the deconvolution result at 150mm was significantly different from the reality result due to the Gaussian pulse had the certain FWHM, which could not be responded well by the change of reflectance with a large slope.

Euclidean distance is the most easily understood distance calculation method, which derives from the distance formula between two points in Euclidean space[11]. In order to denote the quality of reflectance projection distribution obtained by deconvolution more intuitively, it is vital to consider that the points of the curve are discrete. The evaluation standard is defined as the root mean square(RMS) of Euclidean distance in discrete form

$$\mathrm{R}_E = \sqrt{\frac{\sum_{i=1}^{N}\left(p_d(i) - p(i)\right)^2}{N}}. \tag{6}$$

Where N is the number of points on the curve, $p_d(i)$ is the deconvolution result and $p(i)$ is the reality. Due to the noise was random, we repeated each experiment 100 times to avoid accidental errors. The RMS of Euclidean distance of the reflectivity of the four pulses under a certain SNR are shown in Table 1.

Table 1 RMS of Euclidean distance of the reflectivity of the four pulses under a certain SNR

RMS of Euclidean distance				
SNR \ Waveform	Gaussian pulse	Triangular pulse	Square pulse	Sawtooth pulse
5	1.2085×10^{-4}	2.3×10^{-3}	2.99×10^{-2}	2.21×10^{-2}
10	8.1077×10^{-5}	1.6×10^{-3}	1.4×10^{-2}	1.01×10^{-2}
15	3.9680×10^{-5}	8.1366×10^{-4}	5.3×10^{-3}	3.7×10^{-3}
20	2.0363×10^{-5}	3.8510×10^{-4}	1.9×10^{-3}	1.3×10^{-3}
25	1.2225×10^{-5}	1.6976×10^{-4}	6.6105×10^{-4}	4.4802×10^{-4}
30	7.0386×10^{-6}	7.1087×10^{-5}	2.3315×10^{-4}	1.5927×10^{-4}
35	5.6199×10^{-6}	3.2597×10^{-5}	8.4432×10^{-5}	5.6264×10^{-5}
40	5.0818×10^{-6}	1.4189×10^{-5}	3.1977×10^{-5}	2.1218×10^{-5}
45	4.8939×10^{-6}	6.2475×10^{-6}	1.1919×10^{-5}	7.6871×10^{-6}
50	4.7214×10^{-6}	2.8580×10^{-6}	4.2292×10^{-6}	2.5534×10^{-6}
100	4.5896×10^{-6}	6.5968×10^{-7}	9.8819×10^{-8}	1.0365×10^{-10}

As shown in Table 1, the deconvolution effect of Gaussian pulse is obviously superior to others in the case of low SNR, and Square pulse has the worst deconvolution effect. This is different from the estimated results, as it is believed that the NCWF loses part of the high frequency signal in order to suppress the noise, and the echo bandwidth of Square pulse after modulation is relatively wide. In order to verify these analysis, the frequency domain analysis of echo without the noise was tested, and the result are shown in Fig. 5. Through the frequency domain analysis of the echo without noise, it is found that the convolution results of Square pulse and reflection projection distribution has more high frequency components, so NCWF deconvolution has the worst effect under the condition of low SNR, while Gaussian pulse has the least high frequency components, so the deconvolution effect is the best. That is, Gaussian pulse has strong anti-noise ability. In addition, it can be informed from Table 1 that the deconvolution effects of both of Sawtooth pulse and Square pulse are better than Gaussian pulse under the condition of high SNR, this might be because the pulse front of Sawtooth pulse and Square pulse are narrow, which can be well responded by the change of reflectance with a large slope. In general, in order to suppress noise, part of high frequency signals will be lost in the deconvolution process. Therefore, Gaussian pulse can be selected as the most suitable emission waveform when the SNR is less than 40, while, when the SNR is larger, Sawtooth pulse is the best choice.

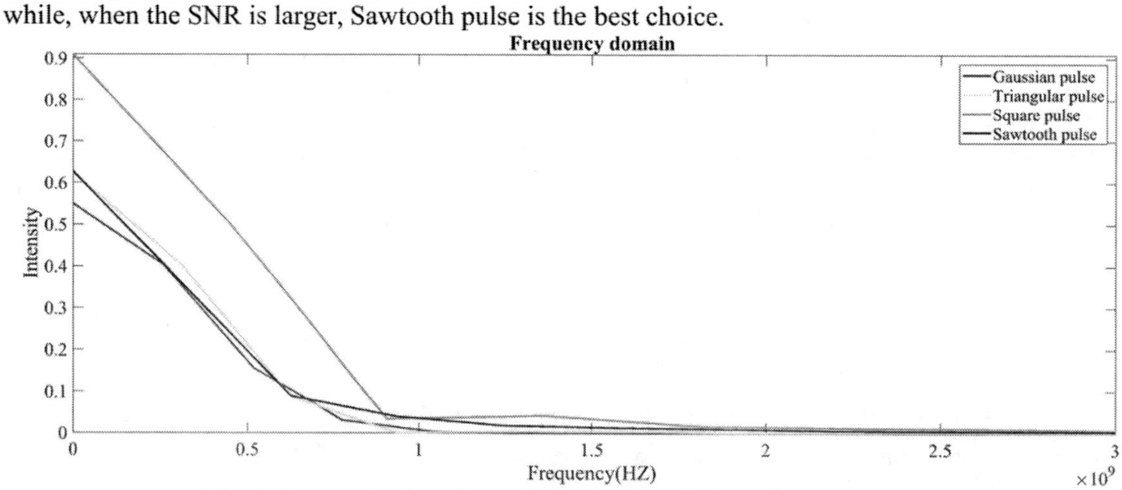

Fig. 5 Frequency domain analysis results of echo without the noise

5. Conclusion

We studied the basic principle of LRTI, discussed the deconvolution problem in echo processing, and used NCWF to get the reflection projection distribution from the deconvolution in echo. Moreover, in order to explore the reflectance projection distribution obtained by deconvolution of different emission waveforms under the same SNR, we simulated four kinds of typical emission pulses. Additionally, from the comparison of the RMS of Euclidean distance between deconvolution results and reality results, we found that Gaussian pulse could be selected as the most suitable emission waveform for LRTI when the SNR was less than 40, while, Sawtooth pulse was the best choice when the SNR was larger. However, in this research, only NCWF deconvolution was used, in the future reseach, subsequent simulations can be carried out based on the commonly used deconvolution methods of LRTI, through which the conclusion might be more universal and convincing.

Acknowledgements

This work was supported by the National Natural Science Foundation of China(Grant No.61871389) and the Research Plan Project of the National University of Defense Technology(Grant No.ZK18-01-02).

References

[1] Parker J K, Craig E B, Klick D I, et al. Applied Optics, 1988, 27(13):2642-2643.
[2] Knight F K, Klick D, Ryanhoward D P, et al. Applied Optics, 1989, 28(12):2196-2198.
[3] Matson, Charles L. Optical Engineering, 1995, 34(9):2811-2820.
[4] Matson C L, Boger J. Appl Opt, 1997, 36(14):3165-3173.
[5] Matson C L. Optics Communications, 1997, 137(4-6):343-358.
[6] Matson C L, Mosley D E. Applied Optics, 2001, 40(14):2290-2296.
[7] Lin F, Wang J C, Lei W H, et al. Optics Communications, 2017, 402:540-544.
[8] Jog V, Anantharam V. IEEE Transactions on Information Theory, 2015, 62(8):4413-4438.
[9] Lin F, Hu Y H, Wang J C, et al. Acta Optica Sinica, 2017, 37(9): 0911001-2.
[10] Wang J C, Zhou S W, Shi L, et al. Optics Communications, 2016, 359:177-183.
[11] Chen C, Zhuang Y, Xiao J, et al. Computer Animation & Virtual Worlds, 2009, 20(2-3):267-277.

IOP Conf. Series: Materials Science and Engineering **711** (2020) 012106 doi:10.1088/1757-899X/711/1/012106

Detection Performance of Single Pulse Photon Accumulation with Typical Emission Waveform

A-Hui Hou[1,2,a], Yi-Hua Hu[1,2,b] and Nan-Xiang Zhao[1,2,c]

[1]State Key Laboratory of Pulsed Power Laser Technology, National University of Defense Technology, Hefei 230037, China

[2]Key Laboratory of Electronic Restriction Technology Anhui, National University of Defense Technology, Hefei 230037, China

[a]hou_a_hui068@163.com, [b]skl_hyh@163.com, [c]southfly@163.com

Abstract. Based on the theory of photon statistics, the cumulative detection probability model of mono-pulse echo is established. The influences of rectangular, Gaussian and triangular pulses on the photon detection performance are simulated, and it is analyzed the detection performances of three waveforms with different pulse widths. The results showed that the detection performance of rectangular pulse was slightly better than that of triangular pulse, and the detection performance of the Gaussian pulse was the worst. The variation of the pulse width of the three waveforms within a certain range had no effect on the detection performance.

1. Introduction

Photon counting laser radar is widely used in space detection, remote sensing imaging, aerial reconnaissance and other fields with high sensitivity, high detection range and low system energy consumption. Photon counting laser radars often use Geiger-mode single photon detection. In order to accurately describe the detection performance of the system, two parameters of detection probability (PD) and false alarm probability (Pf) are generally selected. In 2003, S. Johnson et al. [1] first proposed the use of discrete intervals to study the detection performance of single-photon detectors, which laid the foundation for the study of the detection performance of subsequent single-photon detectors. In 2003, D.G. Fouche et al. [2] studied the expression of detection and false-alarm probability of single-photon detection Lidar by means of multi-pulse cumulative detection and mono-pulse direct detection. In 2005, H Markus [3] analyzed the detection and false alarm probability expressions of photon detection mode, and studied the detection performance of multi-pulse cumulative detection. In 2011, Ying Guo [4] studied the multi-pulse cumulative detection of photon counting three-dimensional imaging laser radar, showing the method of multi-cumulative detection distinguishes echo photon and dark count. In 2012, Lu Xu et al. [6] studied the single-pulse cumulative detection performance under long dead time, indicating that the detection probability was significantly improved when the cumulative detection reached a certain number of times. In 2013, Hanjun Luo [7] studied the mono-pulse cumulative detection theory of Q-switched pulses, and analyzed the influence of pulse width on the detection probability. In 2017, Hanjun L et al. [8] studied the influence of turbulence on the detection performance of single-photon detection Lidar by means of mono-pulse accumulation, and the results showed mono-pulse accumulation could improve the detection probability. It can be found the detection efficiency of single photon detection has always been a hot issue. Recently, research has been inclined to use multi-pulse cumulative detection to improve detection efficiency, and there are few studies on single pulse cumulative detection, especially based on the basic theoretical level of single-pulse cumulative detection.

In this paper, the photon number distribution of photon counting Lidar echo was studied. Based on the statistical distribution of echo photons, the principle of mono-pulse cumulative detection was analyzed. The single pulse cumulative detection performance of rectangular, Gaussian and triangular pulse echoes was compared. The effects of pulse width on detection performance under three waveforms were studied.

2. Single Pulse Photon Accumulation Detection Model

2.1 Signal Photon Echo Distribution

The emitted laser power is P_t, the distance between the laser radar and the target is R, and the target is Lambert. According to the Lidar equation, the wide pulse laser echo energy and photon numbers received within t_1-t_2 are[7]

$$N_{s(t_1-t_2)} \approx \eta_q \frac{E_{t_1-t_2}}{h\upsilon} = \rho\tau^2\eta_q\eta_l\eta_r \cos\theta \frac{A_t}{A_l} \bullet \frac{A_r}{\pi R^2} \bullet \frac{1}{h\upsilon} \bullet \int_{t_1}^{t_2} P_t dt \text{ ,and } E_{t_1-t_2} = \int_{t_1}^{t_2} P_r dt \quad (1)$$

Wherein, $N_{s(t_1-t_2)}$ is average signal photon number of detector response within t_1-t_2; $E_{t_1-t_2}$ is laser echo energy; η_q is quantum efficiency of detector; $h\upsilon$ is single photon energy, ρ is target emissivity; τ is the atmospheric transmittance; η_l and η_r are the optical efficiencies of transmitting and receiving system; θ is the angle between the optical axis and target; A_r is the effective receiving area; A_t and A_l are the projected areas of laser beam in cross-sectional and at the target.

The noise mainly includes the background noise and the dark count of the detector, which are independent of each other. The total noise photon number of the system is the sum of the two.

$$N_n = N_b + N_d \quad (2)$$

Wherein, N_d is the dark count of the detector, N_b is the background noise entering the receiving optical system within a unit time;

In order to improve the detection accuracy and working efficiency of the detector, this paper only measures within a certain period, which is called distance gate. Because the photon counting needs to consider the sampling time, the distance gate is divided into several time intervals[7] (also called time bin). The discussion and research below in this paper are based on the above settings.

2.2 Photon Count Detection Probability Model

In the Lidar detection system, it is generally believed that the detection probability is the probability of photons of the real target detected in the target time interval. In addition to this interval, the probability in other time intervals of detecting the number of photons caused by background or other noises is false alarm probability, also known as false alarm rate.

According to optical statistical theory, the photon number of noise and signal echo detected by the detector obeys Poisson distribution [2][10]. Based on the Poisson distribution, the probability of m event being detected within t_1-t_2 is

$$P(m;M) = \frac{1}{m!}\left[M(t_1,t_2)\right]^m \exp\left[-M(t_1,t_2)\right] \text{ ,and } M(t_1,t_2) = \int_{t_1}^{t_2} R_r dt \quad (3)$$

Wherein, $M(t_1,t_2)$ is the average number of primary photons within t_1-t_2 ; R_r is the rate function of the primary photoelectron.

The probabilities that no photons are excited and of generating a detectable pulse at a single time interval are respectively

$$P(0,M)=\exp\left[-M(t_1,t_2)\right];\ \overline{P}=1-\exp\left[-M(t_1,t_2)\right] \tag{4}$$

2.3 Single Pulse Cumulative Detection Probability Model

When the echo information occupies multiple time intervals, the single pulse cumulative detection model can obtain higher detection probability and lower false-alarm rate. The model is studied when the dead time is zero or short, that is, the influence of dead time is not considered. The key point of mono-pulse cumulative detection is how to stack multiple detection [6][8].

For photon counting Lidar target detection, the detection probability in $i-th$ interval is

$$\begin{aligned} P(i)&=\exp\left[-\int_0^{(i-1)\tau_{bin}} R_r dt\right]\bullet\left\{1-\exp\left[\int_{(i-1)\tau_{bin}}^{i\tau_{bin}} R_r dt\right]\right\} \\ &=\exp\left[-\sum_{j=1}^{i-1} N(j)\right]\bullet\left\{1-\exp\left[-N(i)\right]\right\} \end{aligned} \tag{5}$$

Wherein, τ_{bin} is the time interval; $N(i)$ is the detectable photon number in $i-th$ interval, i.e.

$N(i)=N_s(i)+N_n(i)$;and $N_s(i)=\eta_q \dfrac{E(i)}{h\upsilon}=\eta_q \dfrac{\int_{i\tau_{bin}}^{(i+1)\tau_{bin}} P_r dt}{h\upsilon}$, $N_n(i)=N_n \bullet \tau_{bin}=n_b$.

Combined with the above equation, we can get

$$P(i)=\begin{cases} \exp\left[-(i-1)n_b\right]\bullet\left[1-\exp(-n_b)\right] & 1\le i<T \\ \exp\left[-(i-1)n_b\right]\bullet\prod_{j=T}^{i-1} P\left[0,N_s(j)\right]\bullet\left\{1-P\left[0,N_s(i)\right]\right\} & T\le i\le T+r \\ \exp\left[-(i-1)n_b\right]\bullet\prod_{j=T}^{T+r+1} P\left[0,N_s(j)\right]\bullet\left[1-\exp(-n_b)\right] & T+r<i\le b \end{cases} \tag{6}$$

Wherein, $T=\tau_0/\tau_{bin}$, $r=\tau/\tau_{bin}$, $b=T_g/\tau_{bin}$, τ_0 is the time interval between the echo signal and the beginning of the distance gate, τ is the emitted laser pulse width, T_g is the distance gate width.

In order to reduce false-alarm rate, the threshold of the number of photons that can be detected in a single time interval is set as N_{th}, that is, when the number of input photons is lower than N_{th}, it is considered as noise and not accumulated. Therefore, the cumulative lower limit and upper limit are respectively set as $Th1$ and $Th2$.

Therefore, the detection probability and false-alarm probability of mono-pulse cumulative detection are respectively [11]

$$Pd=\sum_{i=Th1}^{Th2} P(i);\text{and } Pf=1-Pd-\exp\left(-\sum_{j=T+1}^{T+r} N_s(j)-N_n \bullet T_g\right) \tag{7}$$

Wherein, the second term of the formula (7) means the probability that the signal and noise are present at the same time, but the detector has no response.

IOP Conf. Series: Materials Science and Engineering **711** (2020) 012106 doi:10.1088/1757-899X/711/1/012106

3. Simulation of Detection Performance

3.1 Different Waveform Detection Performance

Studying the influence of waveforms on detection probability is helpful to determine what kind of modulation at the transmitter can make the best detection performance. Suppose the emission pulse occupies multiple time intervals; the dead time of the single-photon detector is ignored, that is, as long as there is a signal pulse exceeding the threshold, the single-photon detector will produce a response. Based on the single-pulse cumulative detection probability model, the detection and false alarm rates of three waveforms are calculated, and the best waveform is obtained by comparison.

Reference [12] points that the mono-pulse energy of laser pulse is the integral of its power function over time. Suppose the energy and peak power of three pulses are equal. The single pulse energy is $0.2\mu J$, the pulse width of the rectangular and triangular pulse is $10ns$, the half-height width of the Gaussian is $8ns$, the divergence angle is $1mrad$, $T_g = 1\mu s$, $\tau_{bin} = 1ns$, $\theta = 0$, $\tau = 1$, $\eta_l = 1$, $\lambda = 1064nm, \rho = 0.5, R = 100km, \eta_r = 0.75, \eta_q = 0.35, N_d = 200kHz$ .Required $P_f \le 5\%$, $P_D \ge 90\%$ the maximum allowable number of noise photons in time bin is calculated as 2×10^{-3}, so $N_{th} = 0.02$.

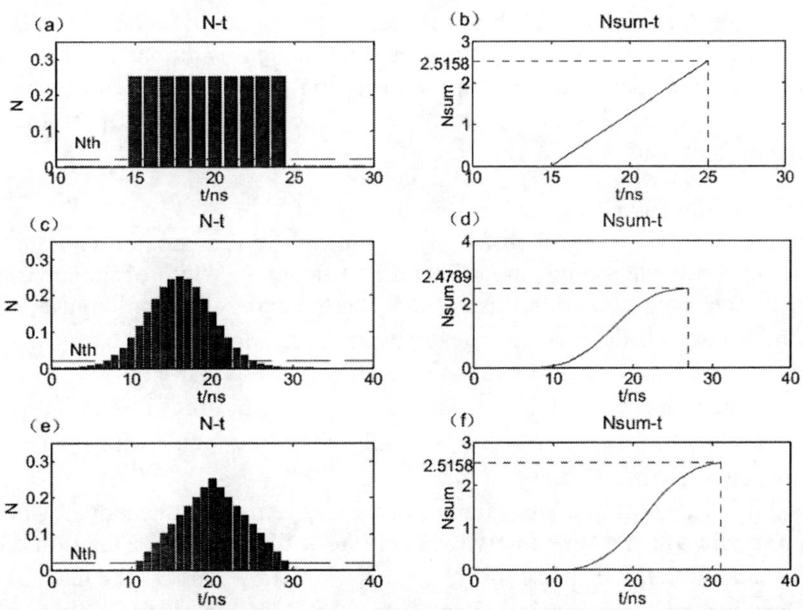

Fig. 1 Rectangular, Gaussian and triangular pulses transmission characteristics (a)(c)(e)The numbers of peak photons of rectangular, Gaussian and triangular pulse are 0.2518;(b)(d)(f) the cumulative-detection photons of rectangular, Gaussian and triangular pulse are 2.5158,0.4789,2.5158.

The simulation results are shown in Fig.1, (a) (c) (e) represents the photon number distribution received by the detector, which is the sum of signal and noise photons; (b) (d) (f) represents the cumulative number of photons that can cause response within the pulse width.

The detection and false alarm rate of the three waveforms were calculated as shown in Table 1.

Table 1 Detection probability and false alarm probability of three waveforms

waveform	Rectangular	Gaussian	Triangular
detection probability /%	91.78	91.37	91.60
false-alarm rate /%	1.60	1.77	1.78

pulse duration /ns	10	>20	20
cumulative-detection times	10	19	19
cumulative-detection photons	2.5158	2.4789	2.5158

Table 1 showed that the probabilities of single pulse detection of three waveforms were greater than 90%, and false-alarm rates were less than 2%. The detection probability was: rectangular > triangular > Gaussian; the false-alarm probability was: triangular > Gaussian > rectangular. The single pulse energy of the three waveforms is equal, that is, the maximum cumulative number of detected photons is equal. The cumulative photon number of the response caused by the pulse width of three waveforms was compared: rectangular = triangular > Gaussian, and the detection probability of the Gaussian pulse was the lowest, so we can get the conclusion: the probability of detection is related to the cumulative number of detected photons causing the response in the pulse. The durations of different waveform pulses are different, resulting in different cumulative times: Gaussian = triangular > rectangular, and found that the rectangular pulse had the lowest false-alarm probability. It is concluded as the cumulative time increases, the detection probability increases, and the noise is superimposed, which increases the false-alarm probability to some extent. The (a) (c) (e) revealed only few low-signals time interval in the Gaussian pulse were filtered out when the threshold was 0.02, and the cumulative detected photons of the three pulses were basically the same, besides the detection probabilities were similar. So we can get the conclusion: three kinds of pulse detection performances are no difference in nature under low threshold.

In summary: 1) the detection probability is related to the cumulative number of detected photons in response to the pulse; 2) the increase of cumulative times increases the false-alarm probability to some extent; 3) there is no essential difference in the detection performance of the three pulses when the threshold is low; 4) In general, rectangular pulse has the best detection performance, followed by triangular pulse and Gaussian pulse.

3.2 Different Pulse Width Detection Performance

In 3.1, the detection performances of different waveforms were compared and analyzed. In order to further compare the three waveforms, the influences of the pulse width of the three waveforms on the detection performance were studied in this section. Set the energy of single pulse with different pulse widths to be the same, the dark count rate to be $200kHz$, ignore the background noise, and the threshold to be the same as above. The simulation results are shown in Fig. 2.

The abscissa is the pulse width, and the value is $0-50ns$. In order to facilitate the analysis of the detection performance when the pulse width is smaller than time bin, during $0-1ns$ the step is $0.2ns$, and the remaining steps are $1ns$.

Fig.2 showed the detection and false-alarm probability remained unchanged when widths of three waveforms were smaller than a time interval. When the width was wider than time bin, the detection probability decreased with the increase of pulse width. Within $0-50ns$, the detection probabilities of rectangular, Gaussian and triangular pulse reduced 0.73%, 8.10%, 5.35%; the false alarm probabilities increased $0.74\%, 2.75\%, 2.28\%$. In order to analyze the changes of data in more detail, some other data are listed in Table 2.

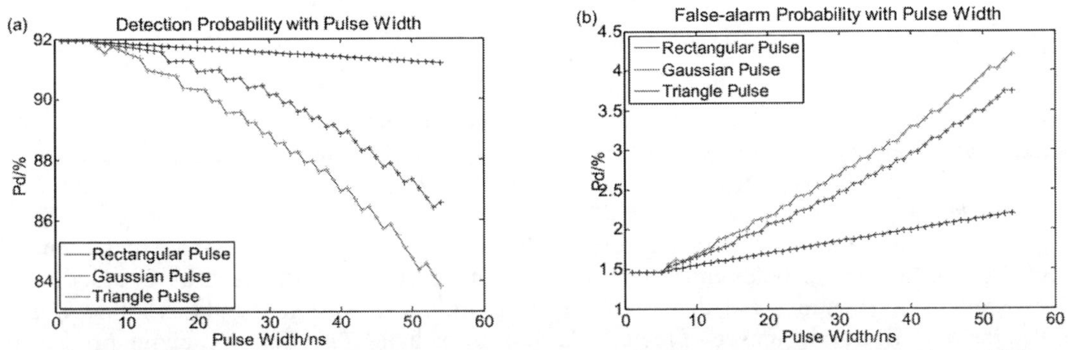

Fig. 2 Detection performances with rectangular, Gaussian and triangular Pulse Width
(a)(b) detection probability and false-alarm probability with pulse width of three waveforms

Table 2 Comparison of partial pulse width detection performance of rectangular, Gaussian and triangular pulse

waveform	Pulse width/ns	1	5	10	20	30	40	50
Rectangular	cumulative-detection times	1	5	10	20	30	40	50
	cumulative-detection photons	2.5158	2.5158	2.5158	2.5158	2.5158	2.5158	2.5158
Gaussian	cumulative-detection times	1	13	21	35	47	55	59
	cumulative-detection photons	2.5158	2.5001	2.4331	2.3207	2.2270	2.0959	1.9222
Triangular	cumulative-detection times	1	9	19	33	45	55	61
	cumulative-detection photons	2.5158	2.5158	2.5158	2.4403	2.3592	2.2705	2.1333

Table 2 showed the cumulative number of detections of three waveforms increased with the increase of pulse width; the cumulative number of detected photons of rectangular pulse was constant; that of Gaussian pulse decreased 0.5936; that of triangular pulse decreased 0.3824.

For the reason of analysis, the rectangular pulse width increases, but the number of photons in each time interval is still higher than the threshold, the cumulative number of detected photons of the rectangular pulse remains unchanged, so the detection probability of rectangular pulse remains basically unchanged; False alarm probability increases with the increase of detection probability and cumulative times. Therefore, it is concluded that different pulse widths of rectangular pulses have little effect on the detection performance. The Gaussian and triangular pulse widths increase, and the number of photons in each time interval gradually decrease. Particularly, the numbers of photons at both ends of the Gaussian and triangular pulse are lower than the threshold, so that the cumulative numbers of detected photons gradually reduce, the detection probabilities decreases. The cumulative-detection times increases and the false-alarms probabilities of Gaussian and triangular pulse increase. When the pulse width was $0-50ns$, the pulse width of Gaussian and triangular pulse had a great influence on the detection performance and cannot be ignored. However, when the Gaussian pulse width was $0-10ns$ and the triangular pulse width was $0-20ns$, the detection probability and the false-alarm probability of Gaussian and triangular pulse changed around 1%, and the influences of the pulse width on the detection performance can be ignored. Therefore, it is concluded that when the Gaussian pulse width is smaller than $10ns$ and the Triangle pulse width is smaller than $20ns$, the influences of different pulse widths on the detection performance are negligible. Comparing the three waveforms, the variation of Gaussian pulse width has the greatest impact on the detection performance, while the rectangular pulse width has no effect on the detection performance basically. The reason is analyzed: rectangular pulse has the highest utilization rate of energy, triangle pulse has the second, Gaussian pulse has the lowest. There are many low energy distributions at both ends of the triangle and Gaussian pulse, which cannot generate response. As the pulse width increases, the energy is more dispersed, the energy utilization rate is further lowered, and the detection performance is further worse.

In conclusion: 1) when the pulse width is smaller than time bin, the detection performances of different pulse widths are basically equal; 2) the Gaussian pulse width has the greatest influence on the detection performance of single-photon detector; 3) when pulse is less than $10ns$, the influence of different pulse widths of the three waveforms on detection performance can be ignored.

4. Conclusion

This paper mainly studies the basic theoretical level problems based on single pulse cumulative detection. The content includes the detection performance of different waveforms based on single pulse cumulative detection, and that of three waveforms with different pulse widths. The results showed the detection performance of rectangular pulse was better than that of triangular pulse, and the Gaussian pulse was the worst. However, due to the small difference in the number of photons detected by the three waveforms, the detection performances were not much different, and the detection probability was around 91.5% and false alarm probability was within 2%. The Gaussian width had the greatest influence on the detection performance, the detection probability reduced 8.1% within $0-50ns$, and the false-alarm probability increased 2.75%. The pulse width of the triangular pulse had a slightly smaller influence, the detection probability reduced 5.35% and the false-alarm probability increased 2.28%, and the influence of the pulse width of the rectangular pulse was negligible. When the pulse width is within $0-10ns$, the influence of the pulse width variation of the three waveforms on the detection performance is negligible.

Acknowledgements

This work was supported by the National Natural Science Foundation of China (Grant No.61871389) and the Research Plan Project of the National University of Defense Technology (Grant No.ZK18-01-02).

References

[1] Johnson S , Gatt P , Nichols T:Proceedings of SPIE - The International Society for Optical Engineering(2003).
[2] Fouche D G:Applied Optics (2003)
[3] Markus H:Applied Optics (2005).
[4] GUO Ying: University of Chinese Academy of Science (2011).
[5] HOU Libing: University of Chinese Academy of Science (2013)
[6] XU Wei, ZHANG Yu, ZHANG Yong, et al:Chinese Journal of Lasers(2012).
[7] Luo Hanjun:Huazhong University of Science and Technology(2013).
[8] Hanjun L , Zhengbiao O , Qiang L , et al.: (2017).
[9] Dai Yongjiang: Publishing House of Electronics Industry (2010).
[10] Johnson S , Gatt P , Nichols T: Proceedings of SPIE - The International Society for Optical Engineering (2003).
[11] Oh M S , Kong H J , Kim T H , et al.: Journal of the Optical Society of America A (2011).
[12] Zhou Bingzhen. : National Defense Industry Press (2014).

Preparation and Electrochemical Properties of $LiMnPO_4$ Cathode Material with Controllable Morphology

Zhongxu WANG[1, a *] and Tingting LI[2, b]

[1]Xidian University, Xi'an China

[2]China academy of aerospace standardization and product assurance, Beijing, China

[a]email: wzx7240@163.com, [b]email: rolineting@126.com

Abstract. The morphology of particles was a really vital parameter influencing the electrochemical properties of $LiMnPO_4$ based cathode materials. In this work, nano-sized $LiMnPO_4$ with various kinds of morphology were prepared by solvothermal method. Then the phase structure, micro morphology and electrochemical properties were deeply analyzed by using XRD, SEM, electrochemical workstation and battery test system. The results indicated that obtained samples with porous and spindle-shaped powders show the best electrochemical properties, contributing to the decrease of Li^+ diffusion coefficient and enhancement of wetting properties between the active powders and the electrolyte.

1. Introduction

Due to the advantages of large energy density, long cycling lives, big output power density and excellent safety properties, rechargeable lithium-ion batteries (named as LIBs) had been extensively used as the power and energy reserve devices for electric vehicles, networks and other industrial applications. Generally, LIBs consist of cathode, anode, separator and electrolyte. Cathode electrodes often considered to directly influence the electrochemical properties, prices and security risks of LIBs, and are considered the key part of the batteries. Accordingly, many efforts have been played on the improvement for cathode electrodes in LIBs [1].

Layered structure $LiCoO_2$, olivine type $LiFePO_4$ and spinel structure Li-rich oxides cathode materials have been widely used and fully researched. Among them, polyanionic $LiMnPO_4$ has been thought as an exciting negative electrode in the reason for their extremely structure stabilization coming from strongly combined phosphorus and oxygen covalent bond [2]. Besides, $LiMnPO_4$ shows the merits in aspect of little money spent, small environmental pollution and cheap raw materials for preparation. Moreover, high energy density (700 Wh/kg) and high theoretical capacity (170 Ah/kg) are also attracted many attentions of researchers [3-4]. Therefore, $LiMnPO_4$ had been considering as a newly developed positive electrode materials for LIBs. Nevertheless, the weakness of this materials is also obviously on the poor conductivity of electronics and slow diffusion of Li ion in the host materials, which influencing the further using of LIBs in many situations [5]. Many methods had been introduced to enhance the properties of $LiMnPO_4$.

Particle size was firstly found to exist the huge influence for the electrochemical properties of $LiMnPO_4$ based cathode powders. The impact of grain scale for the stabilization of micro crystal structure during charging process of lithium manganese phosphate was discussed by Kim et al. [6], and they confirmed that the structure change of lithium manganese phosphate-based cathode powders might change along with particle dimeter. It had been confirmed that nano sized particle size was helpful to raise significant infiltration between lithium manganese phosphate electrode and liquefied electrolytes, further to improving the power capabilities of the battery. The particles with little dimeter often thought to help to reduce the diffusion path of lithium ion at the surface of the grain, leaded to a bigger capability at bigger charge/discharge electric currents, resulting great power capabilities [7]. Nanocrystalline (100 nm) $LiMnPO_4$ cathode particles had been synthesized by Yeo .et al [8] using a hydrothermal method. The cathode nanocrystalline exhibited the largest specific capacities of 153 mAh per gram for the initial cycle and 146 mAh per gram during 110 cycles, respectively. However, if the grain size continuous decrease, it was found that it could decrease the tap density during the preparation process [9]. Therefore, it is very harmful to apply this material-based electrode in large application of LIBs.

Except size effect, the morphology of particles is as well the significant parameter affecting electrochemical properties of $LiMnPO_4$ cathode powders. The anisotropy during Li^+ disinsertion and insertion process in $LiMnPO_4$ makes it very sensitive to the morphology of particles. Choi et al. [10] synthesized nano sized lithium manganese phosphate plates with olivine type structure and the thickness was about 50 nm by a novel solid-state reaction. This prepared $LiMnPO_4$ active material expressed the capacity achieving up to 167 mAh per gram. Then highly fragmented plate-like $LiMnPO_4$ powders were prepared by Zhen et al. using hydrothermal method [11]. The thickness of this prepared plate was about 100 nm. Their good electrochemical properties were attributed to its big ratio of exposed lattice plane on (0 1 0) and also the slight thickness directing the [0 1 0] direction.

In this work, nano-sized $LiMnPO_4$ with various kinds of morphology was synthesized by solvothermal method. Different solvents and the temperature were used to change the powder morphology. Then the phase, micromorphology and electrochemical properties were analyzed by XRD, SEM, electrochemical workstation and battery test system.

2. Experimental section

2.1 Preparation

$LiMnPO_4$ cathode powder was synthesized by the solvothermal method. All used reagents were analytically pure and commercially available, obtained from Fuyu Chemical Company and applied as raw powders without any treatment. Lithium hydroxide, manganese carbonate and phosphoric acid were applied as the raw reagents with the suitable ratio of 3/1/1. To investigate the controllable morphology of the $LiMnPO_4$ architecture, parallel experiments were conducted by keeping all the parameters constant and only changing the kinds of solvothermal solvent.

Firstly, H_3PO_4 and $MnSO_4$ was thoroughly dissolved in deionized water to obtain the solution with 1 mol/L. Then LiOH water solution (3 mol/L) was introduced into the mix H_3PO_4 and $MnSO_4$ contained solution drop by drop with strongly stirring for about 0.5 h. Ammonia solution was introduced drop by drop to adjust the pH-value of the prepared solution to 7. Then the mixed solution was dumped into an autoclave with volume of 100 mL and treated at different temperature for twelve hours. After reaction, the resultant precipitations were collected and staved in baking oven at ninety degree for twenty-four hours. Then specimens with only water solution treated at 180 °C, with ethylene glycol (EG) at 180 °C, with ethylene glycol (EG) and citric acid (CA) mixture solution treat at 180 °C, 200 °C and with CA treated at 200 °C, were named as LMP, LMP-EG, LMP-EG-CA180, LMP-EG-CA200 and LPM-CA, respectively.

The prepared active samples were then blended with conductive agent (AC), PVDF in the weight ratio of 8/10/10 with NMP to obtain a cathode slurry by strongly grinding about four hours. Then prepared cathode slurry was printed on the surface of aluminum foil and heat-treated at 120 °C in a vacuum oven last for 24 hours with the aim to volatilize the NMP. Then, oven-dried active electrodes were cut into circular planchet with diameter about 30 mm. After that, the prepared electrodes were dried in the vacuum oven for 6 hours again in a vacuum oven, and then uses to assemble to the coin-type battery cells. During the batteries making process, metal type Li, Celgard 2400 membrane and 1 M lithium hexafluorophosphate solution were used as anode electrode, separator and liquefied electrolytes, respectively. Finally, the coin batteries (CR-2032) were fabricated in the glove compartment filled with Ar gas (Mikrouna, Super, China).

2.2 Characterization

Micro grain structures of the obtained cathode materials were carried out by X-ray diffraction (XRD, D8 Advance, Bruker, Germany) using Kα radiation of copper with the speed about 0.04 °/s over a 2θ range of 25-75 °. Scanning electron microscope (SEM, 7500F, Jeol, Japan) was applied to observe the surface appearance of the prepared active material.

Electrochemical impedance spectra (EIS) was carried out by the electrochemical workstation (CHI630E, Chinstruments, China) to discuss the resistance of the active material under the AC voltage at five millivolt and the test frequency range from 10^{-2} Hz to 0.1 MHz. The electrochemical properties of the prepared coin cells were tested by the battery measurement system (CT2001A, Land, China) under the voltage range of 2.3 to 4.4 V undergone different rates. In this work, 1C was defined as 170 mA/g. The cycling measurement was carried out at room temperature and the charge/discharge properties were tested under the similar rate ranging from 1/10 C to 5 C.

3. Results and discussion

Figure 1 displays the XRD patterns of lithium manganese phosphate series samples prepared with various solvents. Standard spectrum of $LiMnPO_4$ (JCPDS, No: 33-0804) is also shown in the bottom of the figure 1. It can be found that the obtained powders with olivine-type structure in the orthorhombic space were clearly observed. These results are well consistent with the reported works [12-13]. This means the obtained $LiMnPO_4$ powders are pure and does not exist any extra impurity peaks. On the other hand, the characteristic diffraction peaks are very sharp indicate that prepared powders with high crystallinity. According to the phase structure in figure 1, we can infer that solvothermal synthesis may synthetize olivine powder materials in single-phase with different solvent without calcination. The results also indicate that the kinds of solvent express little effects on the phase structure of $LiMnPO_4$ powders by using solvothermal synthesis method.

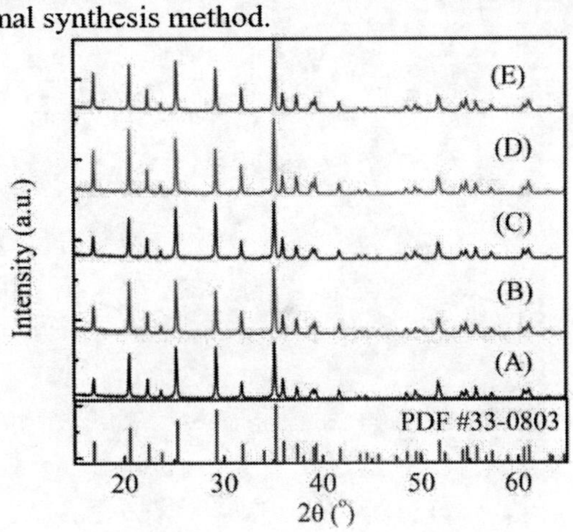

Figure 1. XRD patterns of prepared $LiMnPO_4$ cathode powders, LMP (A), LMP-EG (B), LMP-EG-CA180 (C), LMP-EG-CA200 (D) and LPM-CA (E), respectively.

Figure 2 is the SEM pictures of prepared $LiMnPO_4$ cathode powders. It can be found that the morphology is obviously different as the synthesis conditions changed. When the solvent is only deionized water, the prepared powders are rectangular shape and about 10 μm. And some nanoscale small particles are attached at the surface of big particle. These pellets which have inconsistency size

IOP Conf. Series: Materials Science and Engineering **711** (2020) 012107 doi:10.1088/1757-899X/711/1/012107

unevenly distributed, as shown in figure 2(A). Then, ethylene glycol (EG) was introduced into the solvent, the particle size becomes much small (500 nm) with uniform distribution, as shown in figure 2(B). This indicates that EG is helpful to decrease the particle by using solvothermal method.

When the mixture solution of ethylene glycol (EG) and citric acid (CA) was used as the solvothermal solvent, porous and spindle-shaped powders appeared. In addition, as can be seen in figure 2 (C) and (D), this morphology does not change as the treatment temperature increased from 180 to 200 °C. This porous structure may help to decrease the lithium ion diffusion length and enhance wetting properties between cathode powders and the liquefied electrolytes, further enhancing the electrochemical properties of the batteries as discussed in the following. However, if only citric acid (CA) is used in the solution, the powders are rod-shaped with the diameter is about 200 nm, as shown in figure 2(E).

Figure 2. SEM pictures of prepared cathode powders lithium manganese phosphate. (A) LMP, (B) LMP-EG, (C) LMP-EG-CA180, (D) LMP-EG-CA200 and (E) LPM-CA, respectively.

Figure 3 shows the rate capabilities of all samples. The batteries are firstly tested at 1/10 C, then charged/discharged at 1/2 C, 1 C, 2 C and 5 C, respectively, finally cycled at 1/10 C again. Obviously, electrode with LMP-EG-AC200 shows the great rate capacity, and in different rate the discharge capacity is 141.8, 122.5, 116.8, 107.9, and 92.6 mAh/g, respectively. The rate capabilities of samples LMP-EG-AC200 and LMP-EG-AC180 are almost the same in different rate. This may attribute to the similar morphology of these two samples. However, sample LMP expresses the worst rate performance. Then the cycle performances are tested in 1 C for 100 cycles, as shown in figure 3 (B). Among all samples, LMP-EG-AC200 expresses the highest and most outstanding discharge properties and still maintains to 94.6 mAh per gram. That is means, the loss of discharge capacity is about 19.5%. However, the discharge capacities for LMP, LMP-EG, LMP-EG-AC180, LMP-EG-AC200 and LMP-AC are 26.1, 62.6, 79.6, 94.7 and 47.2 mAh per gram after 100 cycling. Based on calculation in figure 3 (B), the retentions of capacity are 57.6%, 69.1%, 70.9%, 80.55 and 61.8%, respectively.

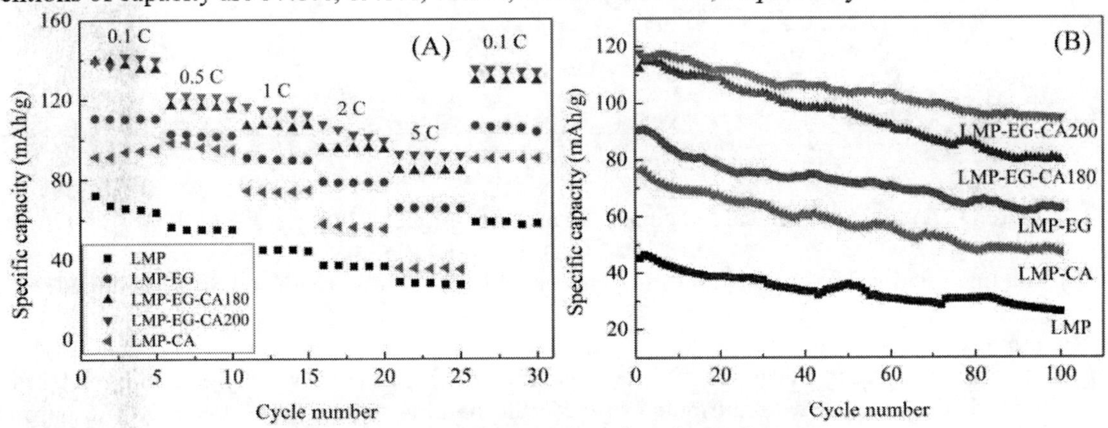

Figure 3. Rate capabilities (A) and cycle performances (B) of various samples with different morphology.

Electrochemical impedance spectra (EIS) tests were applied to analyze the electrochemical properties arising in active material/electrolyte interfaces. Also, in inside of the lithium manganese phosphate, the intercalation/de-intercalation behavior of lithium ion can be discussed by EIS. The EIS plots of these obtained cathode materials are shown in figure 4(A). It is obviously that the Nyquist curve contains two parts curves. Firstly, there is a quasi-semicircle at the frequency range from high to middle. In this range, the intercept at the real part axis in the high frequency area is considered as the ohmic resistance (Re) in liquefied electrolytes. Secondly, at the low-frequency range, there is a straight sloping line according to the Zw. The semicircles in the middle-frequency area are often considered as the charge transfer resistance [4, 14-16]. It can be found that the Re are close to each other (about 8-10 Ω), because of all the electrolyte of the batteries are the same. However, the Rct is obviously different. The sample LMP shows the largest Rct as high as 398 Ω, while the sample LMP-EG-CA200 shows the lowest Rct as 122 Ω. Moreover, the sample the Rct of samples LMP-EG-CA180 and LMP-EG-CA200 are almost the same. It is well known that the Rct is corresponds to the transfer transversion of active charge during the charge/discharge process of $LiMnPO_4$. The samples LMP-EG-CA180 and LMP-EG-CA200 with porous and spindle-shaped may help to enhance the wetting properties between the active powders and the electrolyte. Therefore, these samples show the low Rct.

On the other hand, based on the number of electron transport (n) and the concentration of lithium ions (C) during the electrochemical reactions in the active electrode, diffusion coefficient of Li^+ (D_{Li}) often obtained by the formula as shown in following [4, 12, 16],

$$D_{Li} = \frac{R^2T^2}{2A^2n^4F^4C^2\sigma^2} \qquad (1)$$

here, R, T, A and F are constant, representing the gas state constant, temperature, effective working area, and Faraday constant, respectively. Besides, σ is often considered with sloping line which can be calculated based on the plot slope of Zre vs $\omega^{-1/2}$, and also considered as the Warburg factor, as shown in figure 4(B). It can be inferred that the sample LMP- EG-CA200 shows the largest diffusion coefficient, while the smallest diffusion coefficient is obtained for sample LMP. As shown in figure 2, the sample LMP- EG-CA200 shows the porous and spindle-shaped morphology, helping to decrease the diffusion length of Li^+.

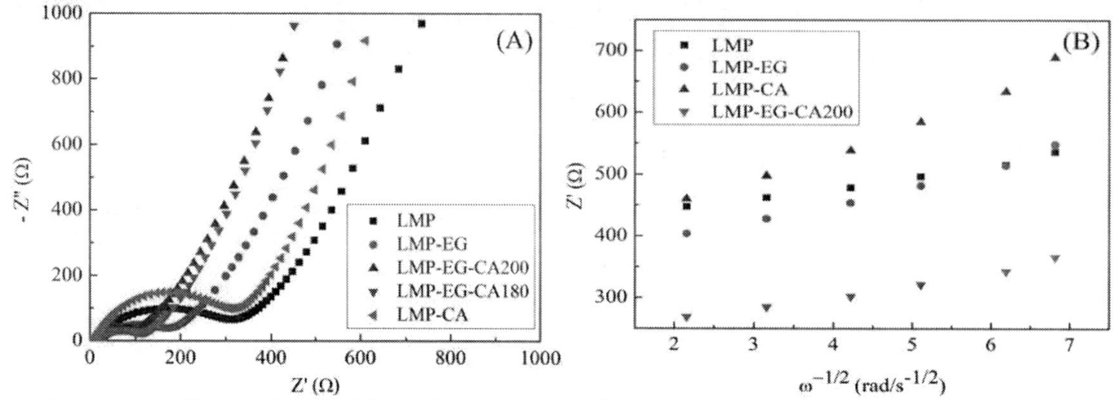

Figure. 4 (A) Electrochemical impedance spectra of samples LMP, LMP-EG, LMP-EG-CA200, LMP-EG-CA180 and LMP-CA.
(B) And line relationship of real part of impedance and $\omega^{-1/2}$ in low frequency of the obtained powders.

4.Summary

In this work, nano-sized $LiMnPO_4$ with various kinds of morphology were prepared by solvothermal method. All the samples were attributed to the orthorhombic olivine-type structure. The porous and spindle-shaped powders could be obtained by using ethylene glycol and citric acid mixture solution. the samples LMP-EG-AC200 expressed the highest discharge capacities as 141.8, 122.5, 116.8, 107.9 and 92.6 mAh/g at different rate of 1/10 C, 1/2 C, 1 C, 2 C and 5 C, respectively. These improvements were contributed to the well-developed porous and spindle-shape and decrease of Li^+ diffusion coefficient.

References

[1] Y. Hong, Z.L. Tang, W. Quan, Controllable synthesis of $LiMnPO_4$ nanocrystals: Morphology evolution and their size-dependent electrochemical properties, Ceram. Int. 42 (2016) 8769-8778.
[2] Y.M. Wang, F. Wang, X.J. Feng, Porous nest-like $LiMnPO_4$ microstructures assembled by nanosheets for lithium ion battery cathodes, J. Mater. Sci. - Mater. Electron. 29 (2018) 1426-1434.
[3] Y. Wan, Q.J. Zheng, D.M. Lin, Recent development of $LiMnPO_4$ as cathode materials of lithium-ion batteries, Acta. Chim. Sinica. 72 (2014) 537-551.
[4] S.L. Luo, Y. Sun, S. Bao, Synthesis of Er-doped $LiMnPO_4$/C by a sol-assisted hydrothermal process with superior rate capability, J. Electroanal. Chem. 832 (2019) 196-203.
[5] E.K. Redouan, A. Abdellah, L. Ali, Solution combustion synthesis of $LiMnPO_4$/C cathode material: The effect of four fuel sources on the electrochemical performances, Int. J. Hydrogen Energ. 44 (2019) 18272-18282.
[6] J. Kim, K.Y. Park, I.P. Park, J.K. Yoo, D.H. Seo, S.W. Kim, and K. Kang, The effect of particle size on phase stability of the delithiated Li_xMnPO_4, J. Electrochem. Soc. 159 (2012) 55-59.
[7] K. Zaghib, A. Guerfi, P. Hovington, A. Vijh, M. Trudeau, A. Mauger, J.B. Goodenough, C.M. Julien, Review and analysis of nanostructured olivine-based lithium recheargeable batteries: Status and trends, J. Power Sources. 232 (2013) 357-369.
[8] H.C. Dinh, S.M. Yongku Kang, I.H. Yeo, Large discharge capacities at high current rates for carbon-coated $LiMnPO_4$ nanocrystalline cathodes, J. Power Sources. 244 (2013) 189-195.

[9] Z.H. Chen, J.R. Dahn, Reducing carbon in $LiFePO_4$/ccomposite electrodes to maximize specific energy, volumetric energy, and tap density, J. Electrochem. Soc. 149 (2002) 1184-1189.
[10] D. Choi, D.H. Wang, I.T. Bae, J. Xiao, Z.M. Nie, W. Wang, V.V. Viswanathan, Y.J. Lee, J.G. Zhang, G.L. Graff, Z.G. Yang, J. Liu, $LiMnPO_4$ nanoplate grown via solid-state reaction in molten hydrocarbon for Li-ion battery cathode, Nano Lett. 10 (2010) 2799-2803.
[11] X.L. Pan, C.Y. Xu, D. Hong, H.T. Fang, L. Zhen, Hydrothermal synthesis of well-dispersed $LiMnPO_4$ plates for lithium ion batteries cathode, Electrochim. Acta. 87 (2013) 303-308
[12] Y.B. Cao, L. Xu, X.M. Xie, K.C. Mu, Y.Q. Xie, Z.C. Xue, G.R. Hu, K. Du, Z.D. Peng, Controllable synthesis of micronano-structured $LiMnPO_4$/C cathode with hierarchical spindle for lithium ion batteries, Ceram. Int. 45 (2019) 4886-4891.
[13] X.N. Fu, K. Chang, H.W. Tang, B. Li, Y.H. Li, Z.R. Chang, Environmentally compatible synthesis of $LiMnPO_4$/RGO using pure water system, Solid State Ion. 337 (2019) 115-121.
[14] Z. Wu, Y.F. Long, X.Y. Yan, J. Su, Y.X. Wen, Microwave heating synthesis of spindle-like $LiMnPO_4$/C in a deep eutectic solvent, Ceram. Int. 43 (2017) 6089-6093.
[15] L. Bao, Y.F. Chen, G. Xu, T. Yang, Z.G. Ji, Hydrothermal synthesis of monodispersed $LiMnPO_4$ (010) nanobelts and [001] nanorods and their applications in Li-ion batteries, Eur. J. Inor. Chem. 13 (2018) 1533-1539.
[16] L.Q. Kou, F.J. Chen, F. Tao, Y. Dong, L. Cheng, High rate capability and cycle performance of Ce-doped $LiMnPO_4$/C via an efficient solvothermal synthesis in water/diethylene glycol system, Electrochim. Acta. 173 (2015) 721-727.

IOP Conf. Series: Materials Science and Engineering **711** (2020) 012108 doi:10.1088/1757-899X/711/1/012108

Fourier single-pixel imaging based on edge attenuation technique

Yang-Di Hu[1,a], Zheng-Dong Cheng[1,b], Zhen-Yu Liang[1,c], Xiang-zhai[1,d], Zhe-Wang[2]

[1]National University of Defense Technology, State Key Laboratory of Pulsed Power Laser Technology, Hefei 230037, China

[2]92664 Troops, China

[a]huyangdi2014@163.com, [b]chengzdmaths@163.com, [c]liangzhenyueei@163.com, [d]njzhaixiang@gmail.com

Abstract: Fourier single-pixel imaging has been widely concerned with its real-time and high quality. However, *DMD* (Digital Micromirror Array) as an illumination source cannot produce a continuous gray level of the Fourier basis pattern. Therefore, the inevitable error is produced. This paper proposes a halftone technique based on edge attenuation, which can be applied to the Fourier basis pattern generation method on *DMD* without loss of resolution, and compared with the traditional methods, the invention solves the problem that the traditional methods has a single scanning direction and accumulated errors, finally; we improve the image quality.

1. Introduction

Single-pixel imaging is a new type of imaging technology that uses a bucket detector with no spatial resolution to receive light intensity for imaging. According to the spectral detection capability of the bucket detector, single-pixel imaging can even achieve invisible spectrum imaging such as infrared. Single-pixel imaging has attracted wide attention due to its low cost, high sensitivity and wide spectral characteristics, at the same time, it has the disadvantages of low imaging quality and slow speed. With the deepening of research, the research on single-pixel imaging is moving towards real-time and high-quality direction. Tao proposed the theory of compressed sensing, which broke the Nyquist sampling law and synchronizes data acquisition with data compression, avoiding the waste of resources caused by the traditional sampling and compression methods, and increased the rate. In 2006, RICE University applied the theory of compressed sensing to single-pixel imaging and successfully designed a single-pixel camera[1]. In 2014, Zhang proposed the Fourier single-pixel imaging theory and applied it successfully[2]. The binary grating pattern on the *DMD* was used as an illumination source to project on the object, and solved the Fourier spectral coefficient by the intensity value from the bucket detector, then made it through Fourier inverse operation, in 2017, Zhang used the spatial averaging effect to bind the pixels around the aimed pixel to sharing diffusion error[3], which improved the image quality, but the pixel binding caused the drop of spatial resolution, In 2019, Liang used Sierra-Lite error diffusion kernel to improve the accuracy of gray pattern based on *DMD* by improving the binary algorithm, and improved the quality of reconstruct images[4].

Images output devices are usually binary devices[5], compared to the grating image produced by the projector, the *DMD* can only output "0" and "1" by controlling the two mirror flip angles, but the high-speed mirror flip of the *DMD* provides the possibility of real-time measurement.

(a) (b)

Fig. 1. *DMD* output in binary mode (a) , (b)

As is shown in Fig. 1, the *DMD* control mirror flip angle output "0" "1", because the inevitably error caused in the binary illumination mode, therefore, there is plenty of room for improvement in this field by improving the projected grating binary pattern and making the pattern maintain better sinusoidal property to improve the measurement accuracy. To this end, this paper proposes an algorithm based on Liang's method, which improves the image quality by edge attenuation method[6], which compared with the traditional method for reducing the spatial resolution.

2. Principle

2.1. Four-step phase shift

In the 2D Fourier transformed domain, the image can be viewed as a weighted sum of 2D sinusoidal light patterns with different frequencies, so by solving each Fourier coefficient, the image can be reconstructed, each Fourier coefficient is calculated by measuring the intensity of light from four π/2 phase shifted grating lighting patterns.

$$P(\mathrm{x},\mathrm{y},\mathrm{f}_x,\mathrm{f}_y,\phi)=\mathrm{a}+\mathrm{b}\cdot\cos(2\pi\,\mathrm{f}_x\mathrm{x}+2\pi\,\mathrm{f}_y\mathrm{y}+\phi)\,. \tag{1}$$

P is a known illumination grating expression, a is the average light intensity, b is the contrast, and ϕ is the initial phase.

Record $R(\mathrm{x},\mathrm{y})$ as the reflectivity of the surface of the object, s is the illumination area, the grating pattern is projected onto the object, and the resulting reflection intensity I is:

$$I(\mathrm{f}_x,\mathrm{f}_y)=\iint_s R(\mathrm{x},\mathrm{y})\cdot\mathrm{P}(\mathrm{x},\mathrm{y},\mathrm{f}_x,\mathrm{f}_y,\phi)dxdy\,. \tag{2}$$

Due to the presence of ambient light, the detector response can be expressed as D_ϕ :

$$D_\phi(\mathrm{f}_x,\mathrm{f}_y)=\mathrm{D}_n+k\cdot I_\phi(\mathrm{f}_x,\mathrm{f}_y)\,. \tag{3}$$

D_n is noise from ambient light, and the k-factor is related to the size and position of the detector.

Four pairs of gratings with π/2 phase shift are sequentially projected, and the Fourier coefficient C is solved by the following formula:

$$C(\mathrm{f}_x,\mathrm{f}_y)=[\mathrm{D}_0(\mathrm{f}_x,\mathrm{f}_y)-\mathrm{D}_\pi(\mathrm{f}_x,\mathrm{f}_y)]+j[\mathrm{D}_{\pi/2}(\mathrm{f}_x,\mathrm{f}_y)-\mathrm{D}_{3\pi/2}(\mathrm{f}_x,\mathrm{f}_y)]\,. \tag{4}$$

In addition, with the sparsity of the natural image Fourier frequency, the image can be reconstructed with fewer samples.

2.2. Error diffusion method

Error diffusion is a popular algorithm for viewing continuous tone images on a binary display. It is known for its good reproduction grayscale and preservation ability[7], the technology was proposed by Floyd and Steinberg[8]. The core idea is to distribute the quantization error of the current pixel to the adjacent pixels with different weights, that is, multi-pixels share the excessive error caused by the single pixel judged by the threshold value, which can be represented by Fig. 2(a).

IOP Conf. Series: Materials Science and Engineering **711** (2020) 012108 doi:10.1088/1757-899X/711/1/012108

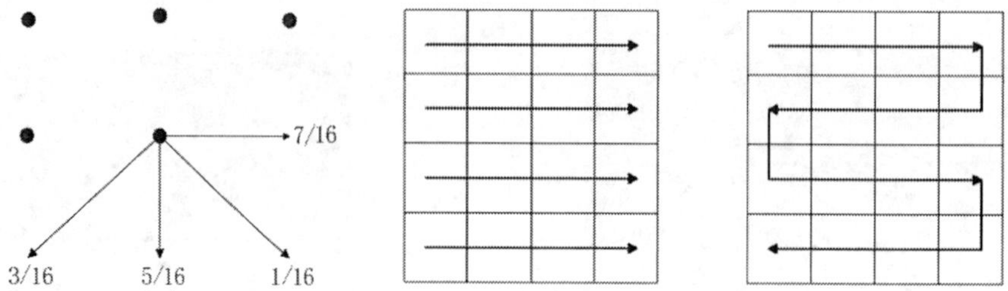

(a) (b) (c)

Fig. 2. Floyd-Steinberg error diffusion map (a) traditional scan path (b) "s" type scan path (c)

"•" stands for the pixels that have been processed. The error caused by the binarization of the current pixel is assigned to the right, bottom right, bottom, and bottom left with a weight of 7:1:5:3. This method delivers the error to the surrounding pixels and reduces the visual error, traditional Floyd-Steinberg error diffusion is a popular halftone method. As is shown in Fig.2 (b), the single scanning mode of this method is easy to cause one-way error accumulation, inspired by Liang, the error diffusion method of "S" scan path is proposed as shown in Fig. 2 (c). The "S" scanning path adopts a different direction depending on odd and even line scanning, which can control the error accumulation caused by the traditional method.

2.3. Error diffusion based on edge attenuation technique

As a popular halftone technology, error diffusion technology can perfectly restore the original grayscale image. However, due to the "poor" nature of the gray level, the image and the original image still have different degrees of deviation. Cheng found that the process of two-dimensional image error diffusion actually spreads the low frequency part of the image to the high frequency part in the frequency domain[9], that is, in the error diffusion process, with the low frequency conversion, the image gradually deviates from the original image, and the error diffusion affects the spectrum. Therefore, in the process of generating a sinusoidal grating, if the loss of low frequency energy can be effectively suppressed, the sinusoidal can be more faithfully retained. As is shown in Fig. 3, taking the lena128×128 standard diagram as an example, Fig. 3(a) can clearly see that the image spectrum is mainly concentrated in the low-frequency component, and after the error diffusion processing Fig. 3 (b) the high-frequency component is gradually increased.

(a) (b)

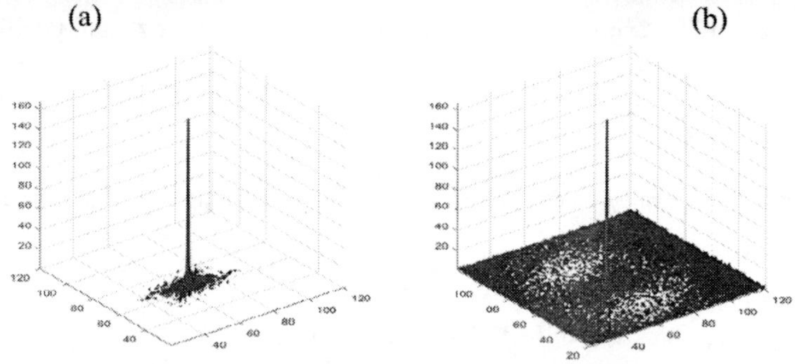

(c) (d)

Fig. 3 . Original image(a); Image after error diffusion (b); The normalized spatial spectrum of the original image of "a" (c); the normalized spatial spectrum of "b" (d)

In order to suppress the high frequency conversion in the error diffusion process, the edge attenuation technique is proposed. The core idea is to superimpose the image by itself before the error diffusion of the grating pattern and introduce the parameter L to control the sharpness of the image[10]. The high and low frequency distribution is fitted by controlling the sharpening degree. The flow chart is shown in Fig. 4:

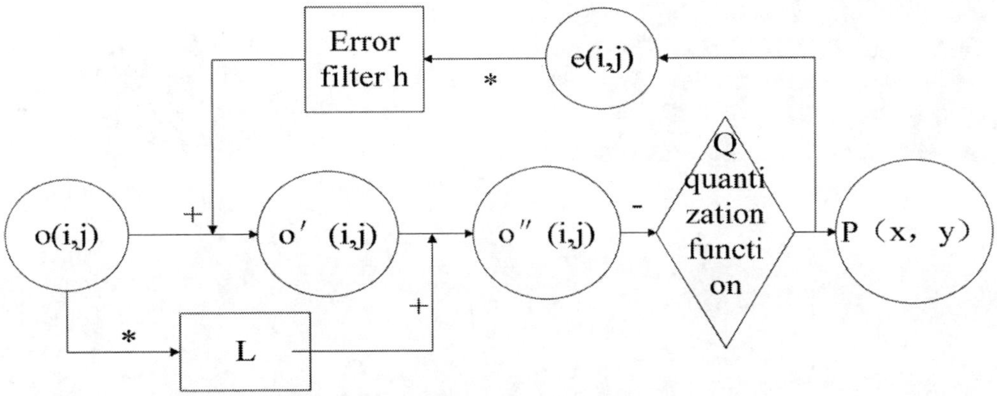

Fig. 4. Improved error diffusion flow chart

Where o(i,j) is the input signal, e(i,j) is the error generated by the previous pixel, h is the error filter, the distribution coefficient is controlled, Q is the quantization function, determining the current pixel is black or white. p(i,j) is the output signal, and the input signal o(i,j) is superimposed with L by itself after being diffused by the error of the previous pixel, and the output P(i ,j) is judged by the quantization function Q, Where, the error quantity e (i,j) is passed to the next pixel. The process can be expressed by the following mathematical formula:

$$o'(i,j) = o(i,j) + e(i,j) \times h \quad (5)$$

$$o''(i,j) = o'(i,j) + o(i,j) \times L \quad (6)$$

$$p(i,j) = Q(o''(i,j)) = \begin{cases} 1, o''(i,j) > 0.5 \\ 0, o''(i,j) \le 0.5 \end{cases} \quad (7)$$

$$e(i,j) = p(i,j) - o''(i,j) \quad (8)$$

It should be noted that when L is a positive value, the image information is reintroduced in the threshold to achieve the purpose of enhancing the edge of the image, and the high frequency component is increased in the frequency domain. When L is a negative value, it is actually the edge information is attenuation and appears as a process of increasing the low-frequency component in the frequency domain. However, it is not as small as L is as small as possible, because when the edge is sufficiently attenuation, numerous low frequency parts are concentrated on the fundamental frequency, and an increase in the lower-order harmonic part also causes image distortion, We traversed the parameter L in the range of -1 to 1 with the step of 0.1, and got the result that when L=-0.6 was the best parameter. Here, a 256×256 grayscale grating image is taken as an example:

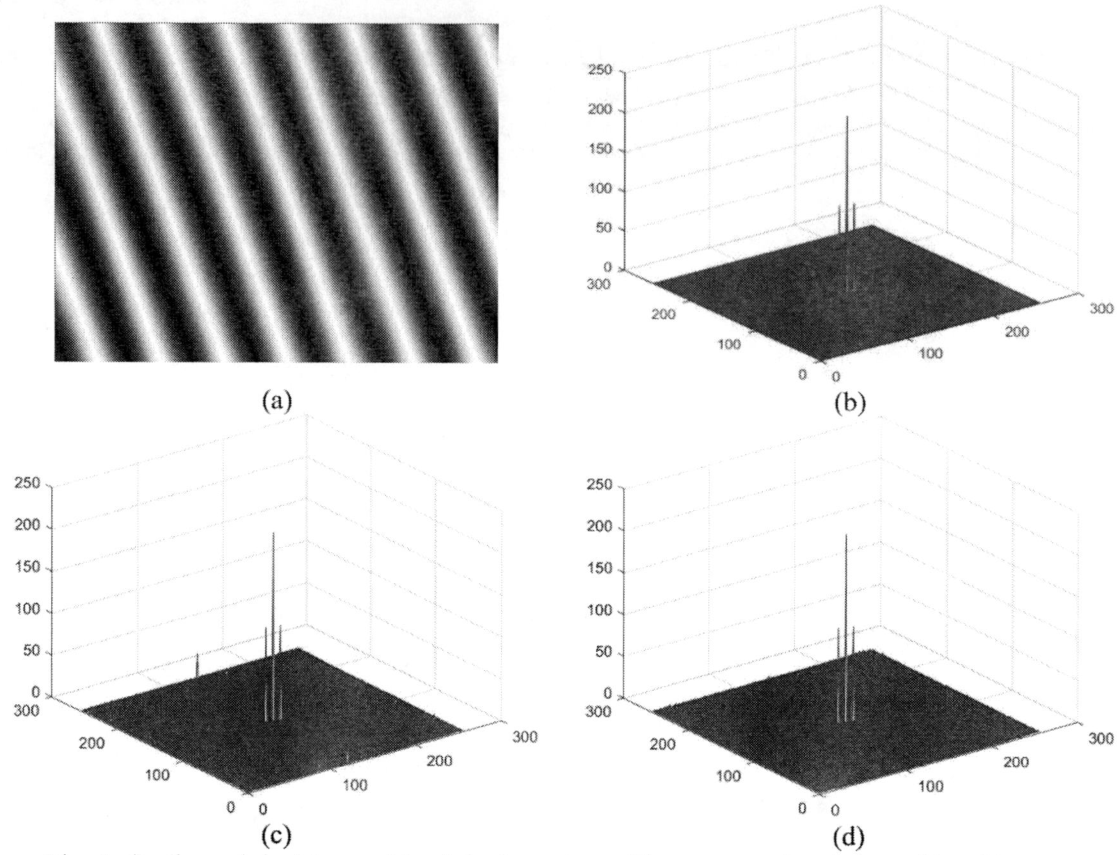

Fig. 5. Grating original image (a) original spectrum (b)spectrum when L is positive (c) spectrum when L is negative (d)

Fig. 5(a) shows the original image of the grating. The center of the spectrum is the DC component, which is represented by the illumination of the background light. The fundamental frequency component is conjugated symmetric, and the entire spectrum is clear and pure. (c) and (d) are positive and negative for L respectively. Compared with the original spectrum, we can clearly see from the graph that the high-frequency component away from the zero-frequency increases when L is positive, and the image details are enhanced on the image, and the edge is steep. The spectrum shows that the fundamental frequency component is attenuation, the sinusoidal contrast is reduced, the harmonic component is increased, the high-frequency component is suppressed when L is negative, and the low-frequency component is close to zero-frequency, which is smoother, more "pure" and closer to the original spectrum, so the edge attenuation technique is useful to maintain the better grayscale grating.

3. Experiment

In MATLAB program, we apply edge attenuation technique to error diffusion, using serpentine scanning type, compared with traditional Floyd-Steinberg method and Liang method, using Lena128×128 pixel image as the image to be tested, the sampling rates are 0.01, 0.05, 0.1. 0.2, with structural similarity (SSIM) and peak signal-to-noise ratio (PSNR) as image quality evaluation criteria. Table 1 gives the objective evaluation data of the imaging quality at each sampling rate, and Fig. 6 shows the comparison of the quality evaluation indicators under the different sampling rates of the three algorithms of the Lena image.

Table .1. Comparison of methods when the sampling rates are 0.01, 0.05 ,0.1 ,0.2

(a) Comparison of methods when the sampling rate is 0.01

algorithm	SSIM%	RMSE	PSNR/dB
Floyd-Steinberg method	42.72	0.1051	18.98
Liang method	42.89	0.1047	19.11
Edge attenuation method	43.14	0.1046	19.16

(b) Comparison of methods when the sampling rate is 0.05

algorithm	SSIM%	RMSE	PSNR/dB
Floyd-Steinberg method	65.65	0.0706	22.70
Liang method	66.45	0.0688	22.93
Edge attenuation method	66.79	0.0681	22.94

(c) Comparison of methods when the sampling rate is 0.1

algorithm	SSIM%	RMSE	PSNR/dB
Floyd-Steinberg method	74.58	0.0582	24.56
Liang method	75.77	0.0551	24.95
Edge attenuation method	76.36	0.0538	25.05

(d) Comparison of methods when the sampling rate is 0.2

algorithm	SSIM%	RMSE	PSNR/dB
Floyd-Steinberg method	82.62	0.0466	26.93
Liang method	83.60	0.0432	27.43
Edge attenuation method	84.04	0.0411	27.68

It can be seen from Table.1 that compared with the other two methods, the SSIM% and PSNR of the proposed method are improved, and the RMSE is also reduced, in order to further compare the quality in different sampling rates. We plot Fig. 6 as follows:

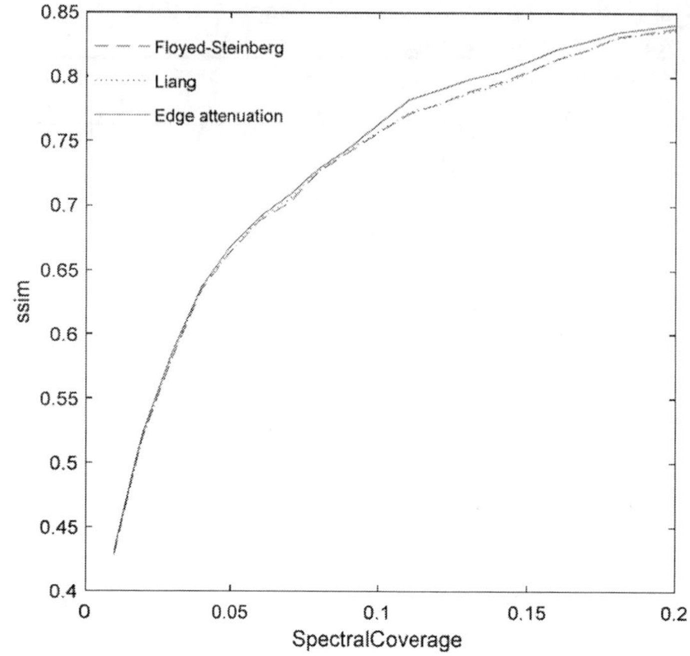

(a)SSIM of reconstructed images by each method

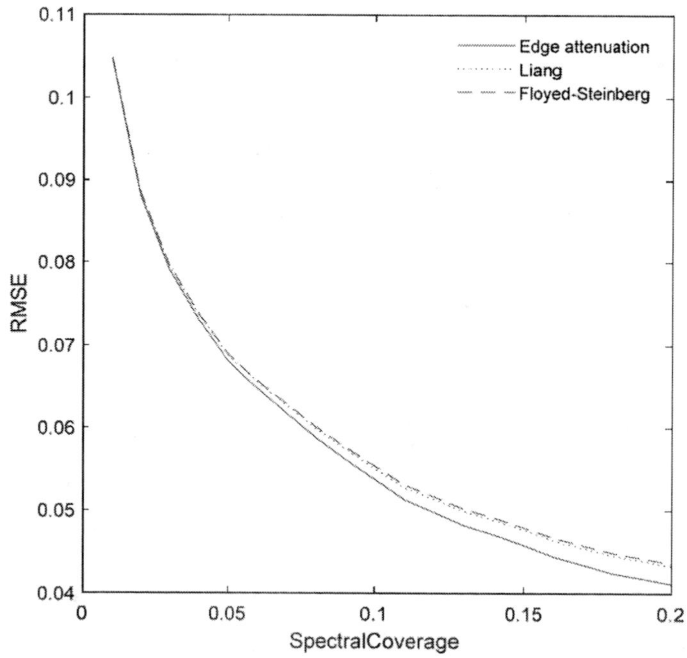

(b) RMSE of reconstructed images by each method

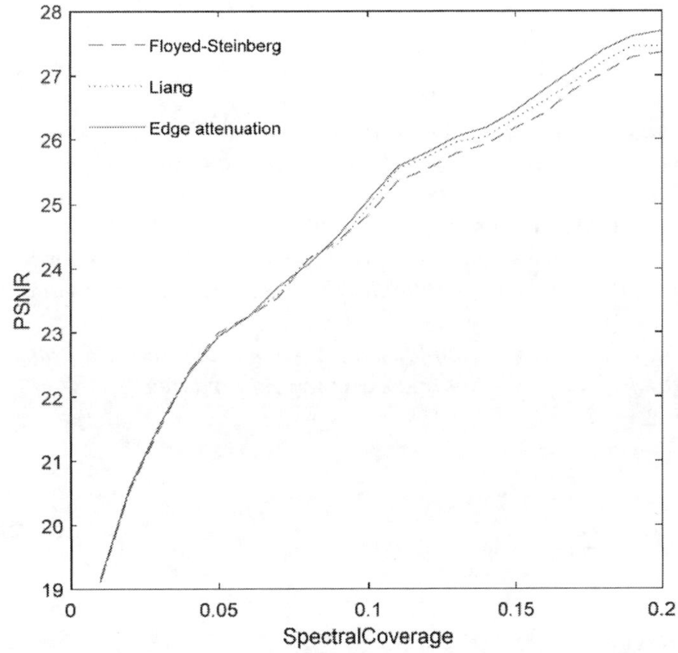

(c) PSNR of reconstructed images by each method

Fig. 6. Comparison of images reconstructed by each method

Fig. 6 shows the change of the evaluation index of the image quality reconstructed by different methods under the conditions of each sampling rate. "—" represents the data obtained from edge attenuation method,"---" represents the data obtained from Liang method,"…"represents the data obtained from Floyd-Steinberg method. It can be seen from Fig. 6 that the Liang method is similar to the traditional Floyd-Steinberg method, and the quality of reconstructed image controlled by the negative parameter is slightly higher than the two, the SSIM% and the PSNR are better than Liang method. The average increase is 0.52,0.08dB, and the RMSE is decreased by an average of 0.0013.

4. Conclusion

The error diffusion method based on edge attenuation technique can generate a higher quality of the sinusoidal grating. In this paper, we analyzed the frequency domain of the sinusoidal grating, and the negative parameter L was introduced to control the sharpness of a grating pattern to reduce the increments caused by error diffusion. The low-frequency components of the original grating were preserved to have better similarity with the original grating. Then, by traversing method, we found when L =-0.6 we could achieve the best effect, in addition, we could get better parameters by reducing the step size or selecting the appropriate filtering window[11]. Finally, we verified the rationality of the proposed method through simulation in MATLAB, we made a comparison of the traditional Floyd-Steinberg method and Liang method, the objective data of image quality evaluation shows that the proposed algorithm has an advantage than others.

References

[1] Takhar D, Laska J N, Wakin M B, et al. *A new compressive imaging camera architecture using optical-domain compression*[J]. Proc Computational Imaging IV,(2006)

[2] Zhang Z, Ma X, Zhong J. *Single-pixel imaging by means of Fourier spectrum acquisition*[J]. Nature Communications,(2014)

[3] Zhang Z, Wang X, Zheng G, et al. *Fast Fourier single-pixel imaging via binary illumination*[J]. Sci Rep,(2017)

[4] Liang Z-Y, Cheng Z-D, Liu Y-Y, et al. *Fast Fourier single-pixel imaging basisd on Sierra–Lite dithering algorithm*[J]. Chinese physics B,(2019)
[5] Yi yaohua, su hai. *Adaptive green noise halftone algorithm for optimal hysteresis coefficient [J].* Journal of wuhan university (information science edition),(2012)
[6] Eschbach R, Knox K T. *Error-diffusion algorithm with edge enhancement*[J]. J.opt.soc.am.a,(1991)
[7] Knuth D E. *Digital Halftones by Dot Diffusion*[J]. Acm Transactions on Graphics,(1987)
[8] Floyd R, Steinberg L. *An adaptive algorithm for spatial gray scale sid symposium*[J]. Sid Digest,(1975)
[9] Cheng ju, su xianyu. *Application research of binary error diffusion grating in phase measurement profilometry* [J]. Journal of huaiyin normal university (natural science edition),(2008)
[10] Algie, S. H. *Resolution and tonal continuity in bilevel printed picture quality*[J]. Computer Vision Graphics & Image Processing,(1983)
[11] Sun M J, Huang J Y, Penuelas J. *Suppressing the noise in binarized Fourier single-pixel imaging utilizing defocus blur*[J]. Optics and Lasers in Engineering,(2018)

IOP Conf. Series: Materials Science and Engineering **711** (2020) 012109 doi:10.1088/1757-899X/711/1/012109

Study on the effect of storage time on the quality of mutton of different brands by low-field nuclear magnetic resonance

Fang Yang[1], Qing Lin[1], Jiping Li[1,a] *, YuYan Huang[1] and Jianmei Xu[1,2,b] *

[1]College of Medical Informatics, Hainan Medical University, Haikou 571199, China.

[2]Modern Technology Education Centre, Hainan Medical University, Haikou 571199, China.

[a]hyyangfang@yeah.net (J.L.); [b]linqinglab@126.com (J.X.)

Abstract: Based on low-field magnetic resonance technology, the quality of Hainan Dong goat meat was studied at the molecular level. With the prolongation of storage time, the pH value of Hainan Dong goat meat showed an overall upward trend, and the T_{22} peak and T_{23} peak of transverse relaxation gradually shifted to the left. As the storage time prolonged, the transverse relaxation times of the peaks of samples T_{22} and T_{23} showed a decreasing trend. The T_{22} peak area of the water that is not easy to flow in the sample gradually decreases, while the peak area of the free water T_{23} gradually increases. The non-flowable water in Hainan East Goat meat was converted into free water. At the same time, the correlation between the transverse relaxation time and the pH value of the mutton quality index was analyzed, and the relationship between the quality of the mutton and the change of storage time was obtained. Finally, this paper compares Hainan Dong goat meat and Guangxi goat meat. The results show that the value of the hard-to-flow water T_{22} of Hainan Dong goat meat is small and the meat quality is tender. This provides data support for Hainan Dong goat meat to become the brand of lamb.

1. Introduction

Hainan East Goat was mainly produced in the Dongshan Mountains of Wanning, Hainan, so it is abbreviated as the East Goat by the local people. Hainan has abundant conditions of light, temperature, water and heat, and geothermal. Common forage can grow in Hainan for a long time. When sheep eat Baicao Fertilizer, they should choose forage with good growth, high yield, rich nutrition and suitable for tropical areas. Hainan East Goat is famous for its high quality mutton both inside and outside the island. Hainan East goat meat is tender and sweet, with high nutritional value, and has the effect of warming, benefiting qi and tonifying deficiency. Due to the demand of mutton market, Hainan mutton can not meet the dietary supply of local residents, so every day, foreign mutton flows into Hainan Island, mostly from Henan, Guangxi and other places. Due to the high content of placenta in Hainan East Goat, the price of Hainan East Goat meat is higher than that of foreign mutton. At the same time, Hainan East Goat was listed in the first list of protected breeds of livestock and poultry genetic resources by Hainan Provincial Department of Agriculture (2009).

Water is the most abundant chemical component in mutton. Many chemical and physical changes in mutton are related to water. These chemical and physical changes will affect the quality, texture and flavor of mutton [1-2]. At present, the traditional methods of food moisture detection generally use drying method, distillation method and Karl Fischer method [3]. In the process of detection, some proteins in Mutton may change physically or chemically, which destroys the integrity and composition structure of

IOP Conf. Series: Materials Science and Engineering **711** (2020) 012109 doi:10.1088/1757-899X/711/1/012109

the tested samples and can not reflect the physical state information of water in mutton. At the same time, other new technologies (infrared analysis technology [4]) to determine the moisture content of mutton only reflect part of the moisture information in the tested samples, and their application also has some limitations. In contrast, low-field NMR technology has the advantages of non-destructive and non-radiation detection, can effectively obtain the distribution information of water components in mutton, and has obvious non-destructive advantages [5-6]. Based on low-field magnetic resonance technology, this paper studied the quality change rule of Hainan East goat meat at molecular level, and detected the difference of water distribution and state between Hainan East goat meat and Guangxi goat meat, providing data support for Hainan East goat meat as a brand mutton.

2. Material and Analysis Method

The whole sheep carcass was purchased from the slaughterhouse and transported back to the laboratory in the incubator. The carcass was discharged at 0-4°C for 24 hours. The meat of the front and back legs, longissimus dorsi and other parts were taken out, and the connective tissue such as visible fat and fascia were removed.Meso MR23-060H-I Nuclear Magnetic Resonance Imaging Analyser: Suzhou Newmark Analytical Instruments Co., Ltd.JA1003 Electronic Balance: Shanghai Liangping Instruments and Instruments Co., Ltd.HH.S21 thermostat bath pot: Shanghai Boxun Industrial Co., Ltd.PH-5F flat pen pH meter: Shanghai Sanxin Instrument Factory.

Using CPMG (hard pulse echo) sequence to detect transverse relaxation pattern of meat products has great influence on the intensity and signal-to-noise ratio of sampled signal because of setting different sequence parameters. If the sampling frequency SW is too small, the effective information in the sample may be lost. The larger the SW, the shorter the sampling time, and the larger the sample frequency, the better the complete sample information can be obtained. The shorter the relaxation time of the sample, the larger the SW needed to be set. SW = 100KHz. With the increase of the number of echoes NECH, the transverse relaxation interval of CPMG echo curve becomes larger. Generally speaking, the number of echoes should be set large enough to attenuate the signal completely. However, if the number of echoes is too large, the determination time will be prolonged, the experimental detection efficiency will be reduced, and the signal-to-noise ratio of sampling signal will be reduced; if the number of echoes is too small, the effective information in the sample will be lost. In theory, the number of echoes is the minimum number of echoes to relax the sample. As shown in Figure 1, the signal occupies one third of the window in the image. Set NECH = 5000.

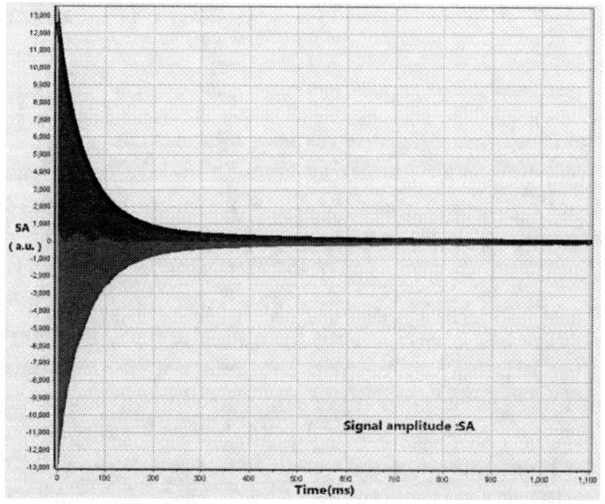

Fig.1 Sampling signal of mutton

Sampling point TD, in CPMG experiment, software automatically calculates TD according to the set SW and NECH values. The number of repeated sampling is usually 2 times of NS. The larger the NS, the higher the signal-to-noise ratio of the sampled signal, but the longer the scanning time. Set the number of repeated sampling NS = 8.The relaxation time inversion fitting software T-InvfitGeneral was used to

IOP Conf. Series: Materials Science and Engineering **711** (2020) 012109 doi:10.1088/1757-899X/711/1/012109

inverse the data. Origin 8.0 is used for data processing and drawing. The Pearson correlation coefficient was analyzed by SPSS24.0.

3. Results and analysis

3.1 Detection of Hainan East Goat Meat and Guangxi Goat Meat by Low Field Nuclear Magnetic Resonance

The quality of Hainan East goat meat and Guangxi goat meat in Fig. 2 and 3 are 20$\pm$0.02g. There are three peaks in the lateral relaxation curves of Hainan East goat meat and Guangxi goat meat. These three peaks represent bound water T_{21} (0-10ms), non-flowing water T_{22} (10-100ms) and free water T_{23} (100-1000ms), respectively. Among them, T_{21} represents the water molecular layer on which the polar groups on the surface of protein molecules are closely bound to water molecules; T_{22} indicates the water which is not easy to flow between myofibrils, myofibrils and membranes, and is easily affected by the changes of protein structure and charge; and T_{23} indicates the water which can flow freely in the extracellular space and is maintained by capillary force [8-9].

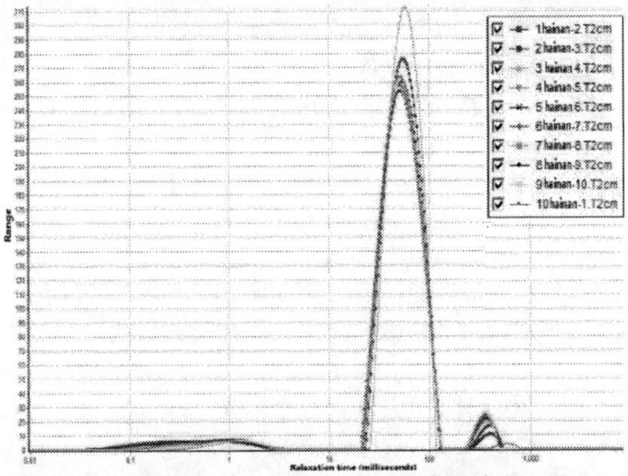

Fig.2 Lateral Relaxation Map of Hainan East Goat Meat at at Different Storage Time

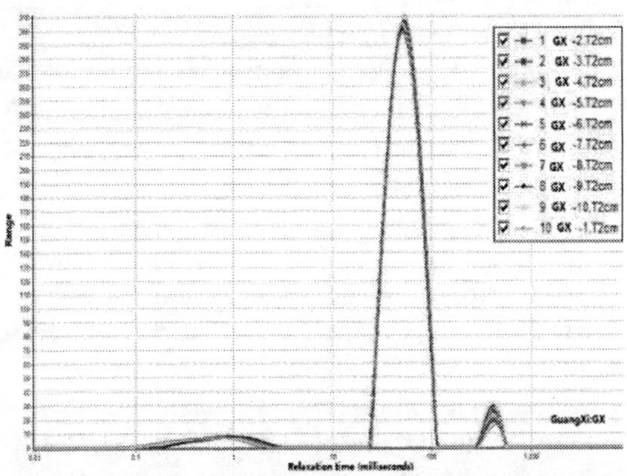

Fig.3 Transverse Relaxation Atlas of Guangxi Goat Meat at at Different Storage Time

It can be seen from the graph that the peak area of T_{21} of Hainan East Goat and Guangxi Goat has little change with the storage time. The peak area of T_{22} and T_{23} shifts to the left gradually [10]. The peak area of T_{22} decreases gradually, while the peak area of T_{23} increases gradually. With the prolongation of storage time, the muscle fibers in meat gradually become loose and compact, and can not absorb more water. There is a flow of water between myofibrils and membranes that is not easy to flow to the interstitial space outside the cell, and it changes into free water, which reduces the flow of water and increases the free water [11]. At the same time, with the prolongation of storage time, the total peak area of Hainan East goat meat and Guangxi goat meat gradually decreased, and the growth of microorganisms in meat consumed part of water, resulting in the decrease of the total moisture content of mutton.

Table 1. The variation of the peak transverse relaxation time of Guangxi goat meat with storage time.

Time(day)	1	2	3	4	5
T2（ms）	1.054	1.113	1.083	1.054	0.848
T2（ms）	59.391	56.242	54.731	53.260	53.260
T2（ms）	601.527	399.768	399.768	378.572	368.399
Time(day)	6	7	8	9	10
T2（ms）	0.895	0.971	0.971	0.920	1.276
T2（ms）	51.829	51.829	51.829	51.829	51.829
T2（ms）	368.399	368.399	358.499	358.499	358.499

IOP Conf. Series: Materials Science and Engineering **711** (2020) 012109 doi:10.1088/1757-899X/711/1/012109

From Table 1 and Table 2, it can be seen that the binding of macromolecular proteins to water in Hainan East goat meat and Guangxi goat meat is relatively close. With the extension of time, the transverse relaxation time of T_{21} peak shows irregular change and no significant change, which is consistent with the results of Magnolia et al. [12]. The transverse relaxation time of the peaks of T_{22} and T_{23} decreases with the storage time. This is because the content of free water and non-flowing water decreases and the number of hydrogen nuclei decreases, which shortens the time from excited state to equilibrium state, so the relaxation time of samples becomes shorter.

Table 2. The variation of the peak transverse relaxation time of Hainan East Goat meat with storage time.

Time(day)	1	2	3	4	5
T_2（ms）	0.848	0.803	0.803	0.895	0.825
T_2（ms）	53.2598	53.2598	51.8286	51.8286	51.8286
T_2（ms）	445.787	410.807	410.807	410.807	410.807
Time(day)	6	7	8	9	10
T_2（ms）	0.871	0.848	0.895	1.054	0.803
T_2（ms）	51.8286	50.4359	50.4359	50.4359	50.4359
T_2（ms）	410.807	410.807	410.807	399.768	399.768

At present, nuclear magnetic resonance (NMR) refers to the splitting of nuclear energy levels in a magnetic nucleus under the action of a static magnetic field. When the external radio frequency electromagnetic wave acts on the nuclear system under certain energy conditions, the atomic nucleus absorbs the incident electromagnetic wave energy and transits from low energy level to adjacent high energy level. When the RF pulse is turned off, the time for the hydrogen nucleus to recover from the excited state to the equilibrium state is related to the state of the hydrogen nucleus and its physical and chemical environment. Therefore, the transverse relaxation map can be used to reflect the properties, distribution and dynamic information of water in meat products. Water content in Mutton accounts for more than 75% of meat, and physical and chemical changes in mutton are related to water. By studying the water distribution and state change in mutton, we can know the influencing factors of mutton quality change. Figure 4 shows the transverse relaxation T_2 inversion results of Hainan East goat meat and Guangxi goat meat at the same time. It can be seen from the figure that the area of Hainan East goat meat T_{22} peak is larger than that of Guangxi goat meat T_{22} peak, indicating that Hainan East goat meat exists between myofibrils and membranes and is not easy to flow water. The transverse relaxation time of the peak value of T_{22} in Hainan East goat meat is smaller than that in Guangxi goat meat. The shorter the relaxation time, the tighter the combination of water and substrates, and the longer the relaxation time, the freer the water [13]. That is to say, Hainan Dongshan goat meat is not only not easy to flow more water, but also closely bound with myofibril, so the meat quality of Hainan Dongshan goat meat is tender. Therefore, Hainan East Goat has the characteristics of thin skin, tender meat, fat but not greasy.

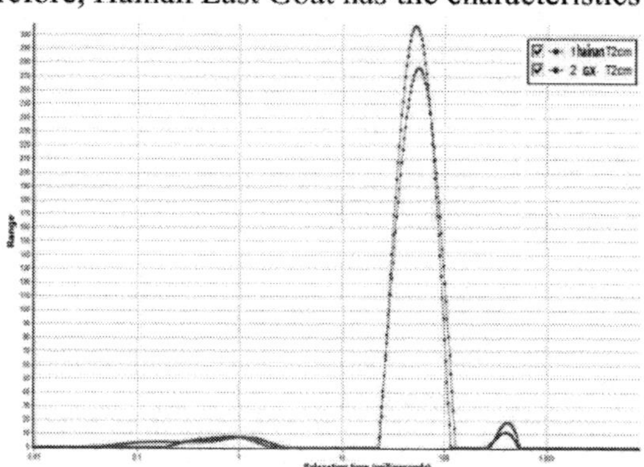

Fig. 4 Comparison of T_2 Parameters between Hainan East Goat Meat and Guangxi Goat Meat

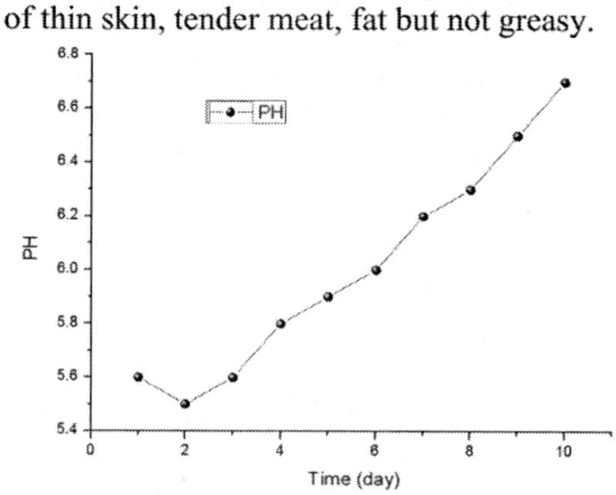

Fig. 5 Changes of pH value of Hainan Donggoat meat with time

3.2 Changes of PH Value of Hainan East Goat Meat with Storage Time

Hainan East Goat has excellent meat quality, uniform fat distribution, tender skin, thick meat, delicious and tasteless. At present, the pH value is an important index to detect the freshness of mutton. When the pH value drops to near the protein isoelectric point or denatures the protein, it will directly affect the meat quality [14]. The change of pH value of Hainan Donggoat meat with storage time is shown in Fig. 5. The change of mutton pH value in the first three days was not significant, because the physiological metabolism terminated after slaughter, aerobic respiration changed into anaerobic respiration, glycolysis occurred, and the accumulation of lactic acid in the final product decreased the pH value [15]. The pH value of mutton increased continuously from 4 days to 7 days, which may be due to the degradation of protein, glycogen and other macromolecule substances into polypeptides and amino acids under the decomposition of bacteria and enzymes, and the release of alkaline groups and other alkaline substances, thus increasing the pH value of mutton [16].

3.3 Preson correlation analysis between transverse relaxation parameters and pH value of Hainan East goat meat

Table 3 is the Pearson correlation data of transverse relaxation parameters and pH value of Hainan East Goat Meat. The Pearson correlation of T_{21}, T_{22}, T_{23} and pH value of quality index was analyzed. The results showed that there was a significant negative correlation between transverse relaxation time T_{22} and mutton pH value, and the correlation coefficient was -0.755. The correlation coefficient between transverse relaxation time T_{23} and mutton pH value was -0.522. By comparing the correlation analysis between T_{21}, T_{22}, T_{23} and pH value, it was found that the relaxation time of non-flowing water T_{22} was more significantly related to mutton quality. It can be concluded that the change of transverse relaxation time of mutton is closely related to the change of mutton quality. With the change of water content and distribution of mutton, the pH value of mutton changes accordingly, which affects the quality of mutton.

Table 3. Correlation analysis between transverse relaxation parameters and pH value

Variable	T_{21}	T_{22}	T_{23}	pH
T_{21}	1			
T_{22}	0.239	1		
T_{23}	0.151	0.901**	1	
pH	0.065	-0.755*	-0.522	1

Note: * $P < 0.05$, ** $P < 0.01$

4. Conclusion

In this paper, the effect of storage time on the quality of mutton of different brands was studied by low field nuclear magnetic resonance technology. The results showed that with the prolongation of refrigeration time, the peak value of T_{22} and T_{23}, the lateral relaxation time and the total peak area of Hainan East Goat meat showed a downward trend. It was not easy for flowing water to be converted into free water, which resulted in the decrease of the content of flowing water and the increase of the content of free water. Compared with Guangxi goat meat, Hainan East goat meat is not easy to flow more water, and closely bound with myofibril, so the meat quality of Hainan East goat meat is tender. According to the Pearson analysis of relaxation parameters and pH value of quality index, there was a significant correlation between relaxation time of T_{22} and pH value of Hainan Dongshan goat meat. Therefore, low-field nuclear magnetic resonance technology can fully reflect the changes of water content and distribution of meat, and characterize the changes of meat quality. In this paper, the difference of water distribution and its status between Hainan Donggoat meat and Guangxi goat meat was detected to provide data support for Hainan Donggoat meat as a brand mutton.

Acknowledgments

This work was financially supported by Hainan Provincial Natural Science Foundation of China (No. 818MS065) ,and Scientific Research Projects of Higher Education Institutions in Hainan Province (Project No. Hnjg2017-41).F.Yang and Q.Lin contributed equally to this work.J.M.Xu and Q.Lin participated in experimental design. F.Yang and J.P.Li performed the experiments. F.Yang and Y.Y.Huang contributed reagents/materials/analysis tools. J.P.Li and J.M.Xu are co-corresponding author. All authors discussed the results and commented on the manuscript.

References

[1] Møller SM,Gunvig A ,Bertram H C.Effect of starter culture and fermentation temperature on water mobility and distribution in fermented sausages and correlation to microbial safety studied by nuclear magnetic resonance relaxometry.Meat Science,2010,86(2):462-467.
[2] Pearce K L, Rosenvold K,Andersen H J,et al.Water distribution and mobility in meat during the conversion of muscle to meat and ageing and the impacts on fresh meat quality attributes-A review.Meat Science,2011,89(2):111-124.
[3] Wang Aiping,Gong Wei,et al.The Sample Pretreatment Methods for the Determination of Water with Karl Fischer Method.PTCA(PART B:CHEM.ANAL.),2016,52(3):369-372.
[4] Mitsumoto M,Maeda S,Mitsuhashi T,et al.Near-Infrared Spectroscopy Determination of Physical and Chemical Characteristics in Beef Cuts.Journal of Food Science,1991,56(6):1493-1496.
[5] Belton P S,Gil A M,Webb G A,et al.NMR relaxometry and MRI for food quality control:Application to dairy products and processes.SPECIAL PUBLICATION-ROYAL SOCIETY OF CHEMISTRY,2003,1(2):84-86.
[6] YANG He-hong,LI Pei-jun,et al.Application of low field nuclear magnetic resonance in meat science.Science and Technology of Food Industry,2012,13:400-405.
[7] WANG Zhiyong,WANG Xin,et al.Assessment of Fresh Pork Quality by LF-NMR:Parameter Optimization.Food Science,2015,36(12):177-184.
[8] Bertram H C, Andersen H J,Karlsson A H.Comparative study of low-field NMR relaxation measurements and two traditional methods in the determination of water holding capacity of pork. Meat Science,2001,57(2):125-132.
[9] Ma Tianlan, Wu Longguo,et al.Detection of Tan-sheep Meat Tenderness based on Low-field Nuclear Magnetic Resonance.Science and Technology of Food Industry,2017,38(2):69-74.
[10] Zhang LL,Li YY, Peng YK,et al. Determination of pork freshness attributes by hyperspectral imaging technique.Trans Chin Soc Agric Eng, 2012, 28(7): 254-259.
[11] Bertram H C,Karlsson A H,Andersen H J.The significance of cooling rate on water dynamics in porcine muscle from heterozygote carriers and non-carriers of the halothane gene-a low-field NMR relaxation study. Meat Science,2003,65(4):1281-1291.
[12] MA Tianlan,WANG Songlei,et al.Study on the Detection of Mutton Freshness by LF-NMR Relaxation Characteristics.Journal of Nuclear Agricultural Sciences,2017,31(6):1110 -1118.
[13] Chaland B,FRANÇOIS MARIETTE,Marchal P,et al.1H nuclear magnetic resonance relaxometric characterization of fat and water states in soft and hard cheese.Journal of Dairy Research, 2000, 67(4):609-618.
[14] Yin Jing Dong.Animal Muscle Biology and Meat Quality.BeiJing:China Agricultural University Press,2011.
[15] Offer, Gerald.Modelling of the formation of pale,soft and exudative meat:Effects of chilling regime and rate and extent of glycolysis.Meat Science,1991,30(2):157-184.
[16] Saito K,Ahhmed A M,Kawahara S,et al.Evaluation of the performance of osmotic dehydration sheets on freshness parameters in cold-stored beef biceps femoris muscle.Meat Science,2009,82(2):260-265.

Numerical Simulation of Flows Field Characteristics of High Altitude Liquid Rocket Exhaust Plume

YE Qing[1,a], LIU Zunyang[2,b *], LI Jie[3,c] and SU Rongyuan[3,d]

[1] State Key Laboratory of Pulsed Power Laser Technology, National University of Defense Technology, Hefei 230037, China

[2] College of Electronic Countermeasures, National University of Defense Technology, Hefei 230037, China

[3] College of Space Science, National University of Defense Technology, Changsha 410073, China

[a]yeqing0518@sina.com, [b]liukp2003@163.com, [c]lijie_gfkd@163.com, [d]1156372010@qq.com

Abstract. The simulation of flow field structures of high altitude liquid rocket exhaust plume is of great significance for infrared radiation caclution. Considering the flow field's symmetry construction, half of the 2-D zone was chosen as the calculation zone. In order to reduce the amount of calculation, we had made an equivalent reduction of the calculation zone. By using the direct simulation Monte Carlo (DSMC) method, a solver was constructed to simulate the flow field structure of the 120km altitude liquid rocket exhaust plume. The simulation results show that the flow field of the high altitude liquid rocket exhaust plume has the following three characteristics: a. the whole flow field indeed contains continuous flow and rarefied flow; b. rarefied gas effect plays a significant role to form a recirculation zones that surpass the nozzle; c. Non-equilibrium effect is remarkable in shock wave region and the boundary layers.

1. Introduction

The rocket engine plume is the product of the interaction between the rocket engine jet and the atmospheric environment. Different working conditions or atmospheric environment for rocket engine will result in different flow field structures of the exhaust plume, which in turn will display diversified infrared radiation characteristics.

As we all know, there are two types of rocket engines: solid and liquid. The flow field characteristics in high altitude are different due to the different rocket engines. The liquid rocket exhaust plume in high altitude contains continuous gas flow and rarefied gas flow without particles. The solid rocket exhaust plume is characterized by a highly rarefied two-phase flow composed of alumina particles whose flow field simulation is more complicated. Therefore, this study only focuses on the simulation of flow field characteristics of high altitude liquid rocket exhaust plume, which is the basis of infrared radiation simulation of liquid rocket exhaust plume.

In the past decades, many studies had been carried out on flow field simulation of rocket exhaust plume, most of which, however, were focused on relatively low to medium altitudes used by computational fluid dynamics (CFD) method [1]. The high altitude rocket exhaust plume, above 100 km, for which flow field structure may have different characteristics from that at the low or medium altitudes.

In recent years, some simulations have been done on the flow field of high altitude plumes. Candler et al.[2],Vitkin et al.[3] carried out the solution of Navier-Stokes equations for the gas simulation, which is not accurate for rarefied gas flow of the plume. Burt and Boyd conducted their studies using a DSMC-based simulation (Direct Simulation Monte Carlo) which allows the treatment of rarefied aspects of the plume[4]. However, DSMC method is too long for calculating continuous flow without necessary simplification. The domestic Cai Guojun and Xu Xiaoying team numerically simulated the high-altitude molecular flow field and independently developed the plume calculation software (PWS) using the DSMC method, but they did not take disequilibrium into account[5,6].

In order to make up for the limitations of domestic and foreign research, our study deals with simulation of flow field characteristics of high altitude plume in liquid rocket using a DSMC, with the aspects of rarefied and disequilibrium taken into account.

2. The direct simulation Monte Carlo solver

The direct simulation Monte Carlo (DSMC) method was first used by Bird (1994), which has gained acceptance as an accurate, robust and broadly applicable technique for modeling gas flows under disequilibrium conditions over several decades[7].

A solver is constructed to simulate the flow field structure of the high altitude plumes based on DSMC method as follow steps[8]. Frist, the flow field area is divided into several computational grids. A limited number of simulated molecules are used in each computational grid to replace a large number of real gas molecules. Second, the molecular motion is decoupled from the collision. Molecular motion is solved according to Newton's law and updated the grid in which the molecules are located based on the latest position. Third, a hard sphere collision model is established to descrip intermolecular collisons. During each time step, some fraction of the molecules in a grid collided with each other, and probabilistic techniques are used for calculations of individual collisions. All molecules are then moved through the grid according to assigned velocities. Finally, all the calculation grids are iterated several times, and the macroscopic physical parameter distribution of the flow field is statistically obtained.

As no assumptions are made about the shape of the gas velocity distribution, the DSMC method retains validity for all flow regimes between free molecular and continuum, so long as binary collisions are the dominant type of molecular interaction. The equilibrating effect of these collisions may be characterized by the Knudsen number (Kn), which is defined as the ratio of the gas mean free path to a macroscopic length scale based on boundary dimensions, flow structures or gradients in bulk gas properties[9].

3. Parameters of instance

The reference coordinate system is set first for the ease of analysis. We set the nozzle axis as the z-axis, the center of the nozzle as z=0, and the direction away from the nozzle as positive. The solid angle direction of the zenith angle is set to coincide with the positive direction of the z-axis.

The parameters of the flow field are set on the supposition that rockets fly at the altitude of 120 km as follows: considering the flow field's symmetry construction, half of the 2-D zone is chosen as the calculation zone shown in Fig. 1. The size of the calculation zone is 150 m×30 m, the radius of the throat and outlet of nozzle are 10 cm and 30 cm respectively. The boundary conditions definitions are shown in Table 1. Neglecting the unimportant species, the species calculated in this paper are shown in Table 2. Collision between gas phases is selected as a hard ball collision model.

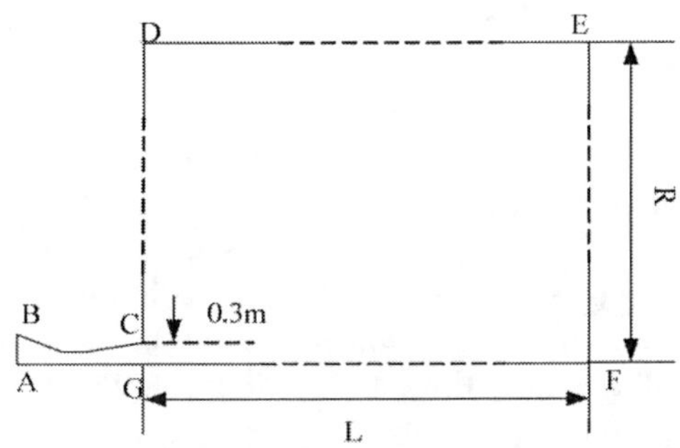

Figure 1. Calculating zone

Table 1. Boundary conditions definitions

zone	Boundary type	pressure (MPa)	Temperature (k)	Velocity (Ma)
AB	Pressure inlet	4.8	3550	——
CD	Pressure far field	0.1	500	0.2
DE	Pressure far field	0.1	500	0.2
EF	Pressure outlet	0.1	500	——

Table 2. Mole fractions of species in the combustion chamber and in the free stream

Species	Combustion chamber	Free flow
H_2O	0.5	0
CO_2	0.1	0
N_2	0.4	0

In order to reduce the computational complexity, it is considered that the nozzles are evenly distributed, and the nozzle parameters are calculated by the FLUNT software to obtain an average gas velocity of 2776.364 m/s, a temperature of 1918.450 K, a pressure of 0.916 atm, and a density of 0.011 kg/m^3. The time step is 1×10^{-7}s.

Although the mesh division as mentioned above is set in perfect condition, the calculation needs further simplifying, as the actual amount is still excessively large. The final calculation domain size is reduced by 5000 times for calculation. The grid is automatically generated by the program, 300 grids are arranged in the x direction, 100 grids are arranged in the y direction, and two sub-grids are arranged in the x and y directions in each grid. For the adaption of the DSMC's grid scale requirements (approximately one-third of the mean free path), a uniform grid is used in x-direction, and an exponential function relationship grid is used in y-direction. The final calculated grid is shown in Fig. 2.

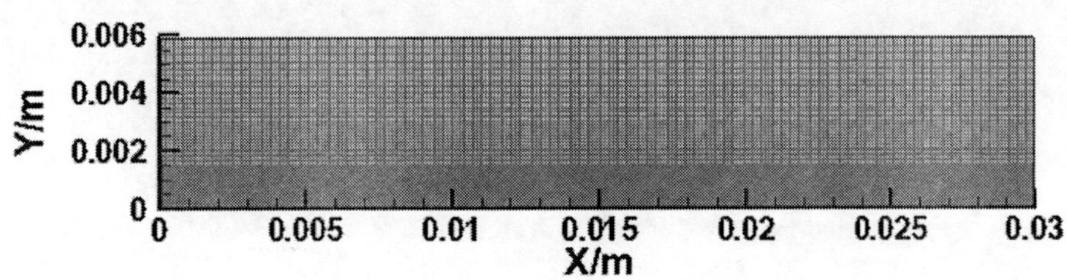

Figure 2. Scale-down calculating zone

IOP Conf. Series: Materials Science and Engineering **711** (2020) 012110 doi:10.1088/1757-899X/711/1/012110

4. Results and Discussions

Substituting the calculated initial conditions into the above calculation steps, the flow field calculation results are as follows: Fig. 3 presents the average density profiles of mixed gas. Fig. 4 presents the density contour map of mixed gas. Fig. 5 presents the speed cloud map in the X direction.

As shown in Fig. 3 and Fig. 4, since the external environment is a vacuum, the gas rapidly expands after exiting the nozzle, and no low-altitude Mach disk is formed. Moreover, the gas density is larger near the outlet of the nozzle, and the farther the gas is diffused away from the nozzle, the density gradually decreases. As can be seen from Fig. 5, there is a region where the x velocity is negative above the nozzle, indicatingthe existing of a significant recirculation zone. That is one of the characteristics which could not be found at the low flow field. The calculation results are consistent with the conclusions given in literature[10].

It can be seen in conjunction with Fig. 3, Fig. 4 and Fig. 5 that there is a region where the density gradient changes more intensely in the region close to the recirculation zone, and the contour changes are unstable. The reason for this phenomenon is that the rarefied gas effect of these regions is significant, and the number of simulated molecules is small, resulting in a large statistical fluctuation. In the future, it is necessary to increase the mesh density in these regions, and to translate the calculated boundary from the CD to the AB position(as shown in Fig. 1), which will lead to more accurate calculation results.

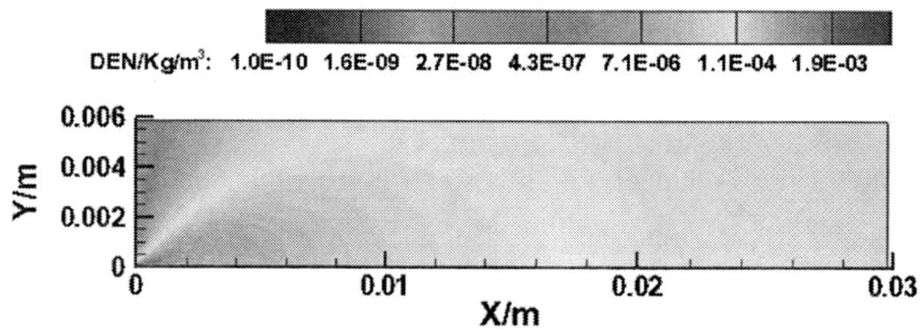

Figure 3. The average density profiles of mixed gas

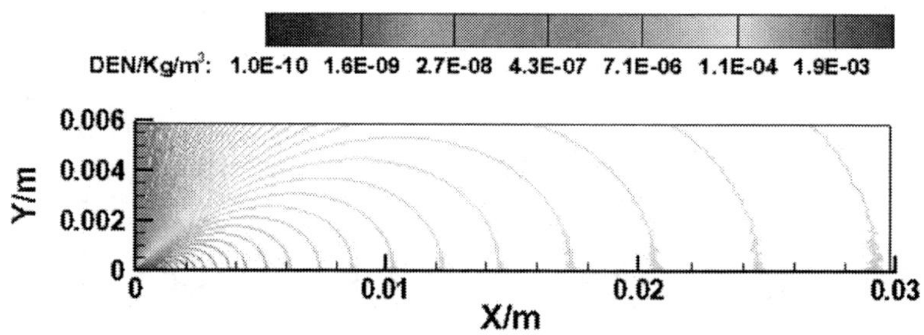

Figure 4. The density contour map of mixed gas

IOP Conf. Series: Materials Science and Engineering **711** (2020) 012110 doi:10.1088/1757-899X/711/1/012110

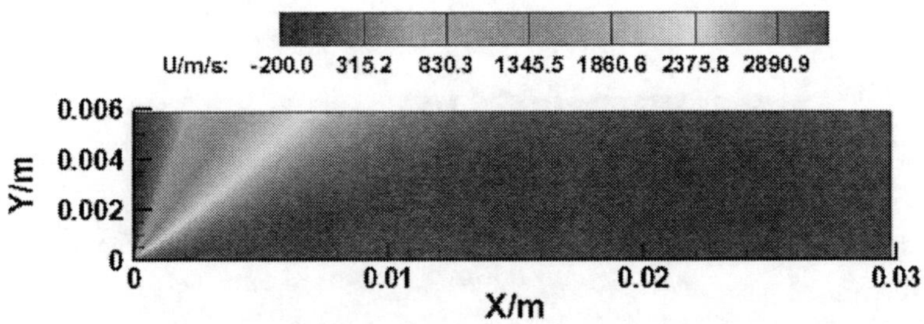

Figure 5. The speed cloud map in the X direction.

The temperature cloud diagram of the mixed gas is shown in Fig. 6, the translational temperature cloud diagram is shown in Fig. 7, and the rotation temperature cloud diagram is shown in Fig. 8. In rarefied gas, the translational temperature is a measure of the difference in molecular translational kinetic energy.

It can be seen from Fig. 7 that in the recirculation zone, the difference in the kinetic energy of the gas molecules in the grid is large, resulting in a high temperature distribution in the recirculation zone. The rotational temperature is proportional to the rotational energy and inversely proportional to the rotational freedom. It can be seen from Fig. 7 and Fig. 8 that the distribution trend of the translational temperature and the rotational temperature is basically the same, but the rotational temperature is higher than the translational temperature as a whole, which indicates that shock wave and the boundary layer have a non-equilibrium effect, and the rarefied gas effect is remarkable.

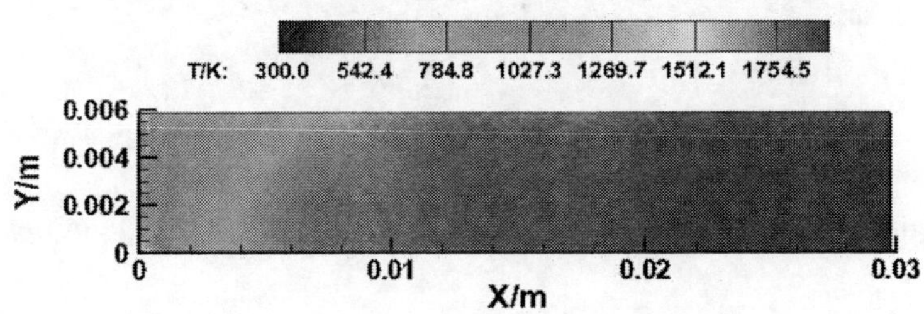

Figure 6. The temperature cloud diagram of the mixed gas

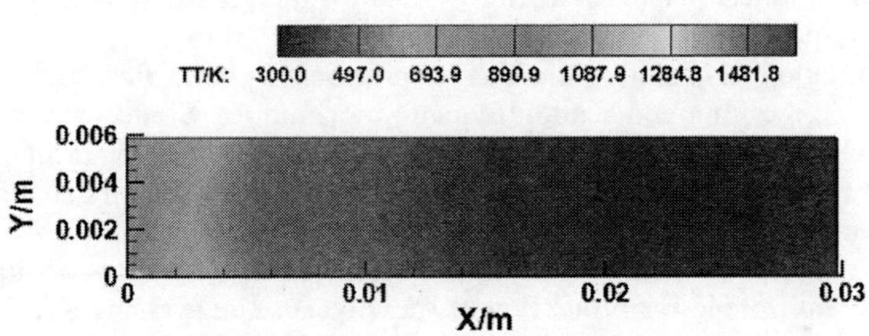

Figure 7. The translational temperature cloud diagram of the mixed gas

IOP Conf. Series: Materials Science and Engineering **711** (2020) 012110 doi:10.1088/1757-899X/711/1/012110

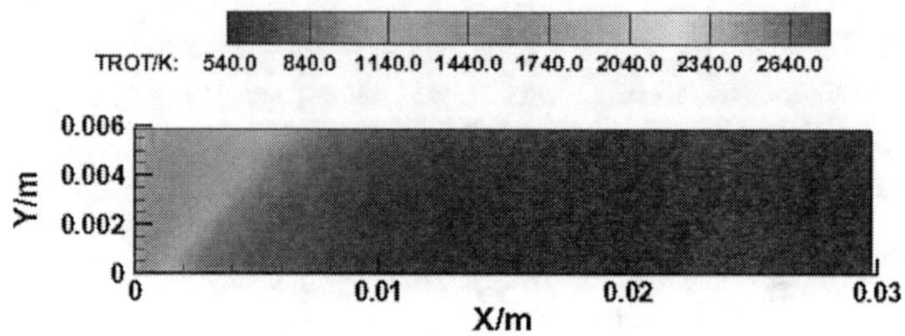

Figure 8. The rotation temperature cloud diagram of the mixed gas

The Knudsen number is usually used to divide the boundary between the rarefied flow and the continuous flow. The region where the Knudsen number surpassing 0.05 is generally considered to be a rarefied flow region. Therefore, we will enlarge the nozzle outlet, and the Knudsen number cloud map is shown in Figure 9. The red areas are rarefied flow, and the other color areas are continuous flow. The entire liquid rocket plume flow field is indeed a mixed flow field containing continuous flow and rarefied flow.

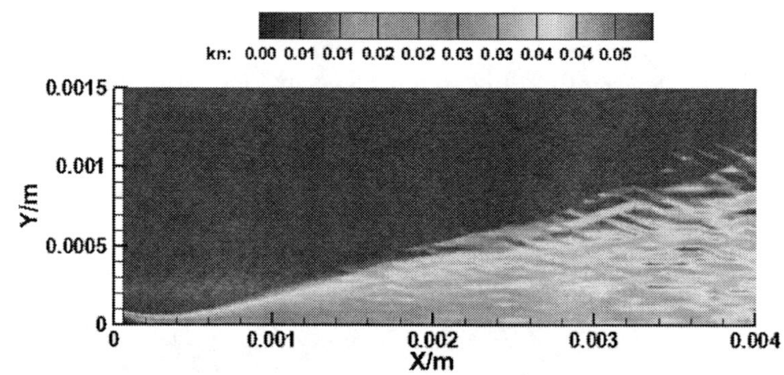

Figure 9. The Knudsen number cloud map near the nozzle outlet

5. Summary

The high altitude liquid rocket exhaust plume flow field contains continuous flow and rarefied flow, so the traditional CFD method is no longer applicable. The direct simulation DSMC method is needed. Existing studies are only limited on the calculation of the flow field characteristics of high altitude liquid rocket exhaust plume in partial rarefied flow, instead of the whole flow.

To remedy the limitation, this study simulates liquid rocket plume flow field at high altitude by DSMC through dynamic meshing and scaling reduction techniques. Significant rarefied gas effect in high altitude liquid rocket flow field and rapid gas-phase molecular expansion brings about recirculation zones that surpass the nozzle. The rotational temperature is higher than translational and rotational temperature, which indicates non-equilibrium effect is remarkable in shock wave region and the boundary layers. It is foreseeable that high altitude flow field characteristics of liquid rocket exhaust plume, which are different from low-altitude flow field, will cause huge changes in infrared radiation of high altitude liquid rocket exhaust plume.

Various rational approximation is employed in the above-mentioned algorithm indeed, such as the assumption that the parameters at the nozzle exit are evenly distributed and the adoption of shrink ratio thought to reduce the calculation amount on flow field. Further and deeper studies study will be carried out to obtain the infrared radiation data in the future on the basis of the accurate simulation of high altitude exhaust plume flow field as mentioned above. These data are worthwhile to provide relevant industrial sector for the detection and penetration of rockets.

Acknowledgements
This research was financially supported by Anhui Natural Science Foundation (Grant No. 1908085MF199).

References
[1] F.S. Simmons, Rocket Exhaust Plume Phenomenology, Aerospace Press, CA, El Segundo, 2000.
[2] G.V. Candler, D.A. Levin, R.J. Collins, P.W. Erdman, E. Zipf, C. Howlett, Theory of plume radiance from the bow shock ultraviolet 2 rocket flight, J. Thermophys. Heat Tran. 7 (4) (1993) 709–716.
[3] E. Vitkin, V. Karelin, A. Kirillov, A. Suprun, J.V. Khadyka, A physico-mathematical model of rocket exhaust plumes, Int. J. Heat Mass Tran. 40 (5) (1997) 1227–1241.
[4] J.M. Burt, I.D. Boyd, High altitude plume simulations for a solid propellant rocket, AIAA Journal 45 (12) (2007) 2872–2884.
[5] ZHANG Xiaoying，XIANG Hongjun，ZHU Dingqiang. Study on Plume Infrared Radiation of Solid Rocket in Ground Test and Flight Condition [J]. Infrared Technology ,2016, 38 (1) :81-87.
[6] ZHANG Xiaoying，XIANG Hongjun，ZHU Dingqiang. Study on Plume Infrared Radiation of Solid Rocket in Ground Test and Flight Condition [J]. Infrared Technology ,2016, 38 (1) :81-87.
[7] S. Dietrich and I. D. Boyd. Scalar and Parallel Optimized Implementation of the Direct Simulation Monte Carlo Method. Journal of Computational Physics, 1996, 126:328-342.
[8] M. A. Gallis, J. R. Torczynski, and D. J. Rader. An Approach for Simulating the Transport of Spherical Particles in a Rarefied Gas Flow via the Direct Simulation Monte Carlo Method. Physics of Fluids, 2001,13(11):3482-3492.
[9] G.A.Bird. Molecular gas dynamics and direct simulation of gas flow, Oxford Science publication Publications, Oxford, UK, 1994.
[10] T. Hyakutake and Kyoji Yamamoto. Numerical Simulation of Rarefied Plume Flow Exhausting from a Small Nozzle[C]. AIP Conference Proceedings, 2003,663(1):604-611.

IOP Conf. Series: Materials Science and Engineering **711** (2020) 012111 doi:10.1088/1757-899X/711/1/012111

Data Augmentation Based on Wasserstein Generative Adversarial Nets Under Few Samples

Yuchen Jiang[1, a], Bin Zhu[1, b], Qi Ma[1, c]

[1]State Key Laboratory of Pulsed Power Laser Technology, National University of Defense Technology, Hefei, Anhui, China

[a]jyc2647118@126.com, [b]zhubineei@163.com, [c]905303927@qq.com

Abstract. Aiming at the problem of low accuracy of image classification under the condition of few samples, an improved method based on Wasserstein Generative Adversarial Nets is proposed. The small data sets are augmented by generating target samples through minimax game between the generator and the discriminator. Firstly, label information is introduced into the network to save the cost of manual labeling for the generated images, and then classifier loss is introduced to further improve the quality of generated images, and the classification accuracy is improved after expanding the data set.

1. Introduction

In visual processing and image classification, the performance of convolutional neural network is superior, but the good performance of convolutional neural network depends on the support of large-scale image data sets. The size of training sets plays a vital role in the performance of CNN.

However, in the real world, due to the influence of natural factors and the limitation of data recording conditions, it is usually unrealistic to obtain large-scale labeled data sets, often with a small number of labeled data samples. In the field of data augmentation, Generative Adversarial Nets (GAN) [1] are widely used in various data augmentation problems because of its good generating ability. Yang Yinan et al. improved the original GAN in document [2] and expanded the harmonics of power system. Chen Wenbin et al. used Conditional generative adversarial nets [3] and integrated Gauss mixture model to enhance the data of dense fog weather situation in document [4]. In document [5], Xing Enxu et al. combined transfer learning with generative adversarial nets, trained in large sample Chinese character font dataset, and the weight obtained was transferred to small sample Chinese character font dataset to generate a large number of small sample Chinese character fonts.

After data augmentation, a commonly used evaluation index is used for classification tasks. The accuracy of classification is used to judge the quality of generated samples. Starting from the evaluation index of data augmentation, this paper distinguishes it from the generator which is not related to classifier in the literature mentioned above. The loss of classifier classification is added to the generator, and the discriminator and classifier jointly guide the generator to generate samples for classification.

2. Related Work

Generative Adversarial Nets. In 2014, Goodfellow et al. first proposed the concept of Generative Adversarial Nets, and set up a generator G and a discriminator D, in which the generator is responsible for generating data close to the real data, while the discriminator tries to distinguish the real data from the pseudo data created by the generator. The input of generator G is an arbitrary random variable. In his original paper, Goodfellow mapped the noise signal $p_z(z)$ as input variable to the data space $G(z,\theta_g)$ of the generator, and obtained the probability distribution p_g of the generator. Thus, the generator G is represented by the parameters θ_g in the multi-layer perceptron. The input of discriminator D is real data or generated data from generated model G. A multi-layer perceptron $D(x,\theta_d)$ is also defined and a scalar is output. D(x) denotes the probability that x comes from real data instead of generating data.

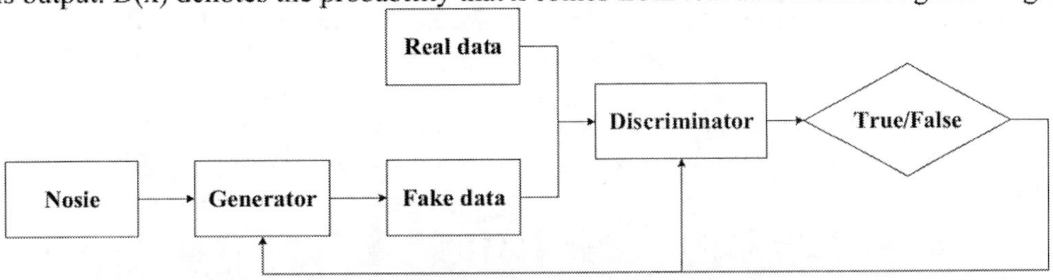

Fig. 1: Basic Structure Framework of GAN

In GAN, training D maximizes the probability of assigning correct labels to both real data and generated data, while training G minimizes log(1 − D(G(z))). Its optimization function is shown in equation (1):

$$\min_G \max_D V(D,G) = E_{x\sim p_{data}(x)}[\log D(x)] + E_{z\sim p_z(z)}[\log(1-D(G(z)))] \tag{1}$$

Updating discriminator D by reducing the gradient:

$$\nabla_{\theta_d} \frac{1}{m}\sum_{i=1}^{m}[\log D(x^{(i)}) + \log(1-D(G(z^{(i)})))] \tag{2}$$

Updating Generator G by reducing the gradient:

$$\nabla_{\theta_g} \frac{1}{m}\sum_{i=1}^{m}[1-\log(1-D(G(z^{(i)})))] \tag{3}$$

Finally, they reach Nash equilibrium, and the output of discriminator is 0.5. That is to say, discriminator D can not judge whether the input is from real data or from generated data. However, under the approximate optimal discriminator in the original GAN, the loss of the minimization generator is equivalent to minimizing the JS divergence between the real data distribution and the generated data distribution, which eventually leads to the gradient of the generator being zero and the gradient disappearing.

Wasserstein Generative Adversarial Nets. Martin Arjovsky et al. proposed Wasserstein Generative Adversarial Nets (WGAN) in document [6]. This model uses Wasserstein distance (also known as Earth-mover, EM distance) instead of Jensen-Shannon (JS) divergence to evaluate the distance between actual samples and generated samples. The advantage of EM distance over JS divergence is that it can measure the distance between two non-coincidence parts, and it further enhances the stability of training compared with the original GAN. Gulrajani et al. further improved WGAN in document [7], and Wasserstein Generative Adversarial Nets-Gradient Penalty (WGAN-GP) model was proposed. The weight clipping used in WGAN to ensure Lipschitz restriction was adjusted to gradient penalty.

3. Conditional Wasserstein Generative Adversarial Nets

This paper improves the WGAN-GP model. There are two main works: The first is to use the idea of conditional generative adversarial nets for reference, and add conditional information into the input of generator and discriminator in order to control the type of generated samples and reduce the cost of labeling generated samples manually. The second is to add a classifier and work together with the discriminator to generator to control the generation of samples.

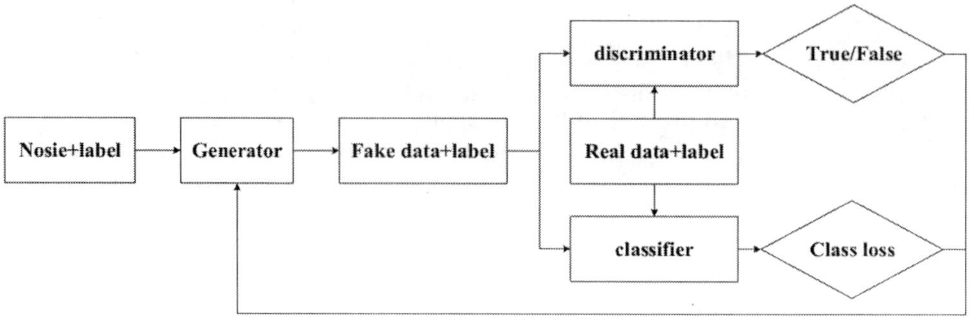

Fig. 2: Basic Structural Framework of Conditional WGAN-GP

The discriminator loss function adds label information on the basis of WGAN-GP, as shown in Equation (4):

$$L = D_w(\tilde{x}|y) - D_w(\mathrm{x}|y) + \lambda(\left\|\nabla_{\hat{x}} D_w(\hat{x}|y)\right\|_2 - 1)^2 \tag{4}$$

Where $\tilde{x}$ is the generated image of the generator, x is the real image, y is the label, and $\hat{x}$ is the interpolated image of the generated image and the real image. The value is shown in Equation (5):

$$\hat{x} = \varepsilon x + (1-\varepsilon)\tilde{x} \tag{5}$$

Two classifier losses are added to the generator loss function:

$$L = -D_w(\tilde{x}|y) - y\log(\mathrm{C}(\hat{x})) - \mathrm{ylog}(\mathrm{C}(\mathrm{x})) \tag{6}$$

Discriminator structure. The discriminator uses four layers of convolution. The size of convolution kernel is 3 and the step size is 2. The number of channels increases gradually. After the convolution layer, Batch normalization is used and activated by Leaky ReLU. Then dropout regularization is used. The structure diagram is as follows:

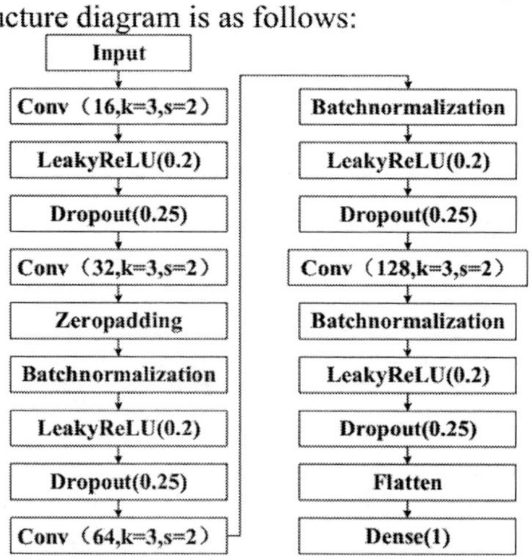

Fig. 3: Discriminator structure

Generator structure. In the generator, the input layer is Dense layer, using two Upsampling layers and three convolution layers. The size of convolution kernel is 4 and the step size is 1. The number of channels decreases gradually. After the convolution layer, Batch normalization is used and activated by ReLU. The activation function of the output layer is tanh. Its structure sketch is as follows:

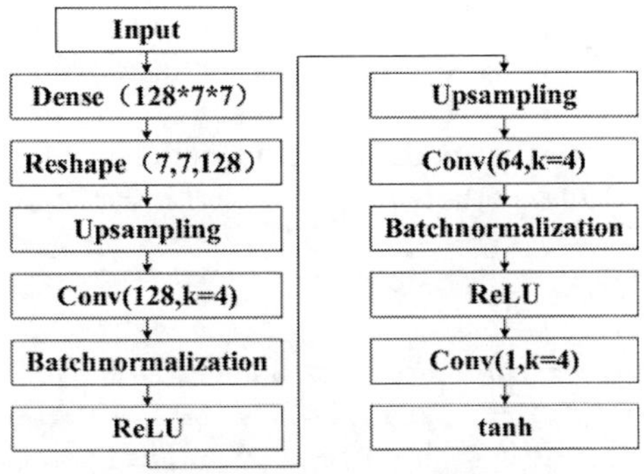

Fig. 4: Generator structure

Classifier structure. The classifier uses one layer of convolution, the size of convolution kernel is 5, the step size is 1, and the number of channels is 32. Then Maxpooling and Dropout regularization are used. And two Dense layers are used. The output activation function is Softmax. The structure diagram is as follows:

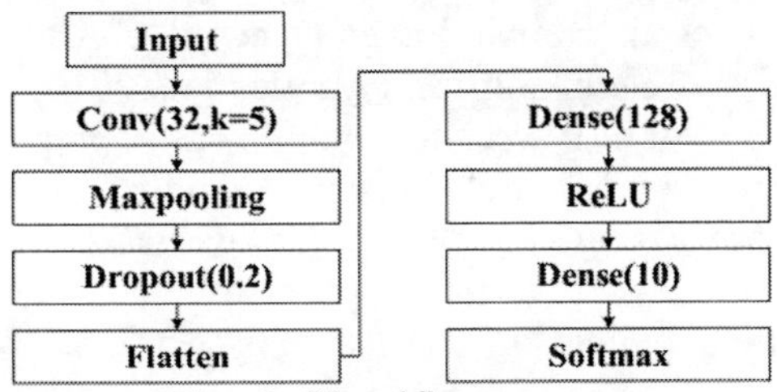

Fig. 5: Classifier structure

4. Experiments

The experimental data set used in this paper is a subset of the MNIST database of handwritten digits. The MNIST data set consists of 60,000 training samples and 10,000 test samples. Each sample is a 28*28 pixel gray handwritten digital image. In the experiment, a total of 500 pictures were randomly selected from 10 types of the original MNIST training set, and 200 pictures were randomly selected in the original MNIST test set, 100 as validation set and 100 as test set.

During the experiment, three data sets were selected as comparative experiments. The first training data set selected the subset of 500 pictures. The second training data set added augmented pictures to the subset, and the augmented model chose CWGAN-GP, but did not introduce classifier loss. 500 pictures after 20,000 epochs of training were selected to augment, a total of 1000 images. The third training data set also added augmented pictures to the subset, but the augmented model used CWGAN-GP combined with classifier loss. 500 images after 20,000 epochs of training are selected to augment the original data set, a total of 1000 images. The results of using the same classifier to classify the validation set and the test set are shown in the following table:

Table 1. Classification accuracy under different data sets

Data set	Val	Test
1	83%	82%
2	86%	85%
3	87%	88%

It can be seen that the proposed CWGAN-GP+Classifier model improves the accuracy of the classifier under the condition of fewer samples, and performs better than CWGAN-GP. It also shows that adding classification loss to the generative adversarial nets can improve the ability of sample generation.

5. Conclusion

Aiming at the problem of low classification accuracy under the condition of few samples, this paper improves the WGAN-GP model. Firstly, label information is added as a condition to reduce the cost of manual labeling after image generation. Then classifier loss is introduced into generator loss function. The proposed CWGAN-GP+Classifier model achieves better results in the subset of MNIST dataset than CWGAN-GP model, and has certain reference significance for the classification task of few samples.

References

[1] Goodfellow I, Pouget-Abadie J, Mirza M, et al. Generative adversarial nets[C].Advances in neural information processing systems. 2014: 2672-2680.
[2] Yang Yinan, Qi Linhai, et al. Research on Small Sample Data Generation Technology Based on Generative Adversarial Nets [J]. Power construction. 2019, 40(05):75-81.
[3] Mirza M, Osindero S. Conditional generative adversarial nets [J]. arXiv preprint arXiv:1411.1784, 2014.
[4] Chen Wenbin, Guan Zhengxiong, et al. Data Augmentation Method Based on Conditional Generative Adversarial Nets [J]. Computer application. 2018, 38(11):259-265.
[5] Xin Enxu, Wu Xiaoyong, et al. Double-Layer Generative Adversarial Nets Based on Transfer Learning [J]. Computer Engineering and Application, 2019,55(15):38-48.
[6] Arjovsky M, Chintala S, Bottou L. Wasserstein gan[J]. arXiv preprint arXiv:1701.07875, 2017.
[7] Gulrajani I, Ahmed F, Arjovsky M, et al. Improved training of wasserstein gans[C].Advances in neural information processing systems. 2017: 5767-5777.

IOP Conf. Series: Materials Science and Engineering **711** (2020) 012112 doi:10.1088/1757-899X/711/1/012112

Hot-wire Anemometer Based on Frosted Fiber Bragg Grating Coated with Silver Film

Jing Yang[1,a], Xuke Chen[1,b], Xinyong Dong[1,c]

[1]Institute of Optoelectronic Technology, China Jiliang University, Hangzhou, China

[a]15757164349@163.com, [b]517525328@qq.com, [c]xydong@cjlu.edu.cn

Abstract. Hot-wire anemometer is proposed based on a frosted fiber Bragg grating (FBG) coated with silver film assisted by a coupler of lumbar-enlarged optical fiber taper. The silver film adhered to the surface of frosted FBG absorbs pumping light, which is coupled into the fiber cladding by the lumbar-enlarged taper to generate heat. A strong diffuse reflection occurs on the surface of frosted FBG, which increases absorption of light by silver film. Center wavelength of frosted FBG changes with velocity of airflow according to a relationship. Experimental results reveal that sensitivity is 5.3 pm/mW and the measurement range from 0 to 20.36 m/s.

1. Introduction

Flow measurement is a vital parameter in all kinds of territory such as chemical engineering, physical detection, automobile manufacturing, medical equipment in various applications technology and so on[1, 2]. Optical fiber anemometers have caused extensive concern due to the superiority such as simple, sensitivity, strong to electromagnetic interference, long-distance detecting, low power consumption, electrically passive operation and integrated structures[3-5]. Flow sensors are mainly based on microelectronic and mechanical system technology, some of them are based on optical fiber device[5, 6]. However, they also have limitations such as corrosive environment, temperatures extremes and radiation environment under the relatively harsh working environment.

In view of the above deficiencies, thermal anemometry stands out and is used commonly in air flow measurement. In the late years, hot-wire anemometers optical sensor based on fiber Bragg grating (FBG) has been proposed as a new research focus[7-10]. It is realized that anemometer contains a pump laser, FBG coated with metal film and an optical coupler, which couples the pumping light from fiber core into cladding. The metal film absorbs the pump laser from cladding, temperature of metal film rises, which drifts the Bragg wavelength of FBG.

In this paper, we propose an optical fiber hot-wire anemometer based on a lumbar-enlarged optical fiber coupler combining with a frosted FBG. The coupler possesses not only high but also stable coupling efficiency so that most power of the pumping light can be coupled from fiber core into cladding. Fiber Bragg grating is ground by a grinder, silver film is adhered to the FBG. A strong diffuse reflection occurs on the surface of frosted FBG, and the sliver film absorbs more light and obtain a higher initial temperature. When air flows through anemometer, the temperature of anemometer drops and the Bragg wavelength drifts.

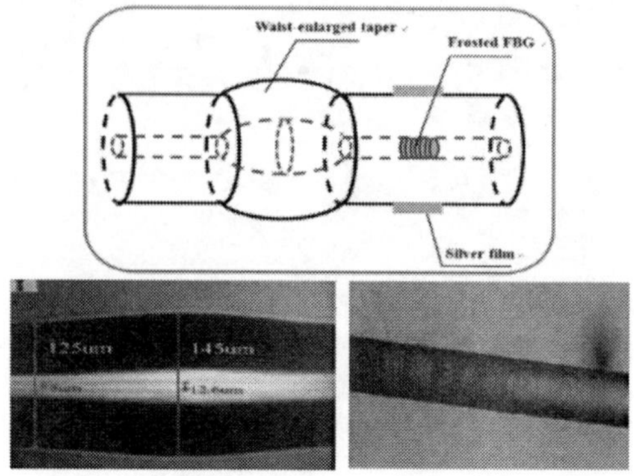

Fig. 1. Schematic diagram of optical fiber anemometer, with micro-graph view of the lumbar-enlarged fiber coupler and frosted FBG.

2. Sensor Fabrication and Principle

The proposed optical fiber anemometer is schematically presented in Fig. 1, which constants a lumbar-enlarged optical fiber taper and frosted FBG with silver film coated. The taper, with a maximal diameter of 145 μm, can couple pump laser light from fiber core into cladding. The silver film absorbs energy of the pump laser and heats the FBG. The FBG was written in a hydrogen-loaded single-mode fiber (Corning SMF-28) with a scanning 244-nm UV laser beam through the phase-mask method. Bragg wavelength was 1555.088 nm and 3-dB bandwidth of 0.2082 nm with a reflectivity of 98%. It should be noted that the taper was proximit to the FBG, because there was a high loss when the light propagates in the cladding. It was 3 mm in our sensor.

A lumbar-enlarged optical fiber taper was fabricated 1mm long by using a fusion splicer (Fujikura FSM-60s) under the auto-splicing mode with a much larger overlap distance of 150 μm and a large discharge time of 1500 ms. The core dimeter of SMF was 8 μm, cladding dimeter was 125 μm, cladding refractive index was 1.45. The inset of Fig. 1 shows a micrograph view of the coupler and frosted FBG. The dimeter of the fiber transforms to 145 μm from 125 μm. Insertion loss was ~11 dB, which meaned ~92% of incident light was coupled into cladding by the taper.

The FBG was ground by the grinder in order to keep the cladding rough, which makes the sliver absorb more light transmitted in the cladding. The illustration of Fig. 1 demonstrates a micrograph view of frosted FBG. After that, silver film was adhered to the surface of the FBG through vacuum evaporating method with a thickness of ~120 nm. To prevent oxidation, a quartz film with thickness of ~100 nm was deposited on the silver.

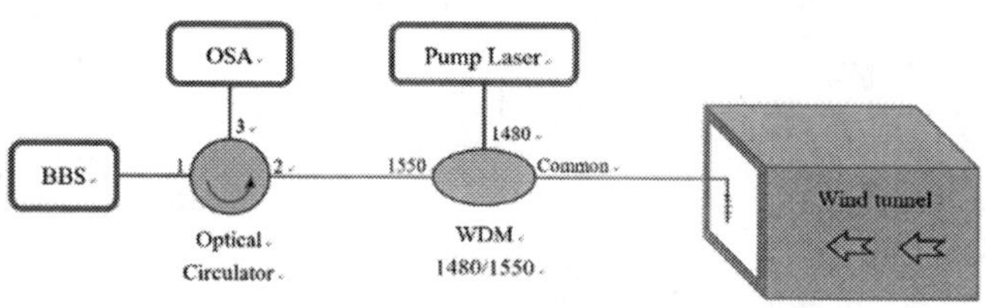

Fig. 2. Experimental system.

The silver film absorbs power of pumping light in order to generate heat, which makes temperature of the FBG rise. Result that a red shift of center wavelength of the FBG. When wind passes, part of heat is taken away, which makes the resonant wavelength of the FBG transfer to short wavelength. A dynamic thermal equilibrium based on the pumping light power and airflow velocity can be achieved.

The hot-wire anemometer is an instrument for measuring the velocity of flow by monitoring wavelength shift of the FBG based on the solution of the heat convection equation (King'S equation). When the airflow rate is greater than 0.5 m/s, forced convection plays a major role, free convection can be ignored. By monitoring and analyzing wavelength shift of the FBG, flow measurement can be achieved. The heat transfer of convective heat transfer is calculated by Newton cooling formula:

$$Q = h(t_w - t_f) = h\Delta t \quad (1)$$

When heat exchange area is S,

$$Q_S = Sh(t_w - t_f) = Sh\Delta t \quad (2)$$

here, Δt is average temperature difference of heat exchange area, h is convective heat transfer coefficient of surface, A and B are empirical calibration constant, ν is airflow rate.
Based on previous research,

$$Q_S = P\varphi a = (A + B\sqrt{v})\ \Delta t \quad (3)$$

here, P is input energy, φ is coupling coefficient of lumbar-enlarged fiber, a is absorption coefficient of the silver film to pump laser.

$$\lambda - \lambda_0 = kP\Delta t\lambda_0 \quad (4)$$

here, k is the FBG's temperature coefficient of sensitivity, λ_0 is the initial wavelength of the FBG. The relevance between Bragg wavelength of the FBG and the airflow rate can be showed as[11-13]

$$\lambda = \lambda_0\left(1 + \frac{kP\varphi a}{A+B\sqrt{v}}\right) \quad (5)$$

3. Experimental Results and Discussions

The experimental device is presented in Fig. 2. A wind tunnel was used to provide stable airflows, whose caliber was 60 cm × 60 cm. In experimenting, the airflow rate is varied from 0 to 20.36 m/s. The output laser power of the pump laser turned to 264 mW at 1480 nm in order to heat the anemometer through a 1480/1550 nm WDM coupler. The output wavelength of broadband source (BBS) range from 1550 nm to 1565 nm. An optical spectral analyzer (OSA) with precision of 20 pm interrogated Bragg wavelength of the FBG.

Fig. 3 (a) presents the reflection spectra of FBG anemometer with and without rough before and after laser pumping. It shows that unfrosted FBG has a center wavelength shift of 0.908 nm while frosted FBG has a red shift of 1.404 nm, due to pump laser heating. It also can be noticed that, compared with unfrosted FBG, the reflection spectral of the frosted FBG with and without pumping light has a distinguishable deformation, the reflectivity reduces slightly and the reflection profile changes. On the condition of continuous absorption by the sliver film, power of the pumping lighting and temperature are reducing gradually along the FBG, which is called chirp effect. Reduction in reflectivity of the FBG will intensify if the pump power increase furtherly. However, there is no obvious effect on measurements. Fig. 3 (b) revels wavelength shift of frosted FBG and unfrosted FBG against power of the pumping light. The wavelength shift is linear, the slope of frosted FBG anemometer is 5.3 pm/mW, which is 1.5 times as high as unfrosted FBG anemometer, which means frosted FBG absorbed more light of cladding than unfrosted one indeed.

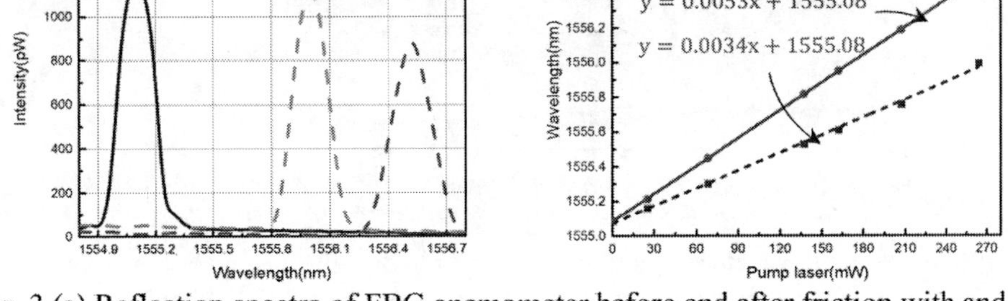

Fig. 3 (a) Reflection spectra of FBG anemometer before and after friction with and without laser pumping. (b) Wavelength shift of FBG before and after friction against power of the pumping light.

The measurement of airflow rate was handled based on the fixed pump laser power of 264 mW. Airflow measurement was developed on the condition of a fixed temperature of 25°C in case of the influences of environmental temperature. The airflow rate in the wind tunnel range from 0 to 20.36 m/s. Fig. 4 takes on the reflection spectra of frosted FBG anemometer under different airflow rate. It can be noticed that, center wavelength diverts gradually to short wavelength with the increase of the airflow velocity. It is also should be observed that with increase of the airflow velocity, the reflectivity is enlarged step-by-step and the spectral profile return to the start state. The reason why is that the chirping effect was eliminated gradually along when the sensor was cooled down by airflow. This wavelength shifts induced by the laser pumping is more than double than previous FBG anemometer coated with silver film and combined by a core-offset fusion splice, even if the pump laser power in this experiment is 76 mW lower[13]. Thus the absorption efficiency of silver film is improved significantly because of frosted FBG.

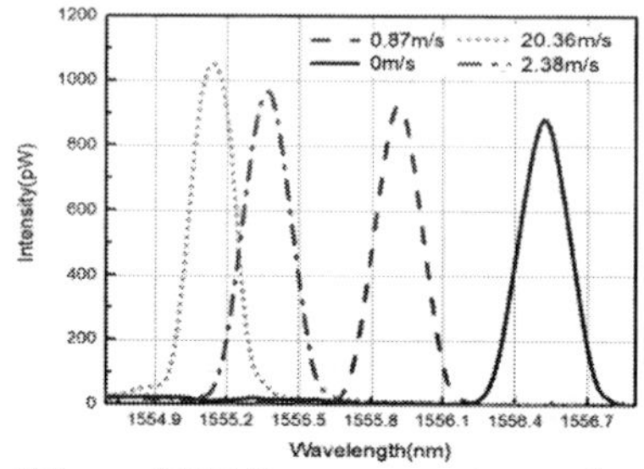

Fig. 4 Reflection spectra of frosted FBG anemometer under various airflow velocities.

From the demonstration of Fig. 5, it should be noted that reflectivity of the FBG returns slightly to the start one with increase of the airflow rate. However, as the sensor is poor, which is hard to stand a stronger airflow. Under condition of 10.9 pm/°C temperature sensitivity, we can calculate that the FBG is 83°C. The temperature of airflow at the rate of 20.36 m/s is 6 degrees lower than the frosted FBG anemometer.

Bring data into Eq. (4), a fitting curve with R-squared value of 0.998 can be achieved, as reveled in Fig. 5. Eq. 4 can be written as the following expression:

$$y = \frac{0.49}{0.35+\sqrt{x}} + 1555.08 \tag{5}$$

Based on the above formula, airflow rate is worked out from the measured center wavelength of the FBG sensor. The variant can be written as:

$$v = \left(\frac{0.49}{\lambda-1555.08} - 0.35\right)^2 \tag{6}$$

Noticed that Eq. (5) and Eq. (6) are workable only under aforementioned conditions and measurement process. Any variation of single condition, including but not limited to coupling coefficient of the lumbar-enlarged fiber taper, power of the pumping light, environmental (or airflow) temperature and initial center wavelength of the FBG, may change parameter of the equations.

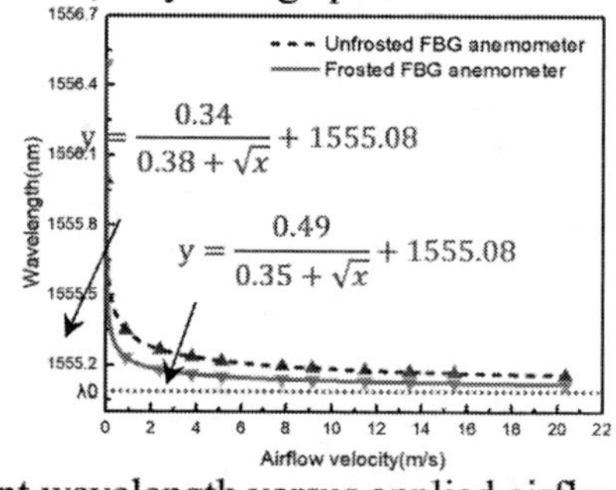

Fig. 5 Resonant wavelength versus applied airflow velocity.

It can be seen that the airflow velocity is almost invariable and the response becomes smaller and smaller when airflow rate is over 6 m/s in the demonstration of Fig. 5. What can't be ignored is that center wavelength of the FBG declines more quickly under a greater airflow rate. The sensitivity and resolution of FBG anemometer under different airflow velocity can be obtained from the analysis of Fig.

5, as shown in Table 1.

Table 1. The sensitivity and resolution of FBG anemometer under different airflow velocity.

Airflow velocity(m/s)		0.873	2.384	3.794	5.127
Resolution(pm/m/s)	unfrosted	398.06	38.39	17.59	10.65
	frosted	738.95	24.25	10.13	5.78

It can be clearly seen that the sensitivity of unfrosted FBG anemometer is up to 398.06 pm/(m/s) and the resolution can reach 0.05 m/s, when the airflow rate is lower than 6m/s. The response is almost linear when the airflow rate is between 1.0 m/s and 6.0 m/s, the sensitivity and the resolution is ~54.83 pm/(m/s) and 0.36 m/s, respectively. The highest sensitivity of frosted FBG anemometer is 738.95 pm/(m/s), the resolution is 0.027 m/s. The corresponding temperature change, under condition that temperature sensitivity of the FBG is 10.9 pm/°C, is ~128°C. The sensor is still 3°C higher than the airflow rate of 20.36 m/s.

4. Summary

A hot-wire anemometer is proposed based on a frosted fiber Bragg grating (FBG) coated with silver film assisted by a lumbar-enlarged optical fiber taper. A 1480 nm pumping light with a maximum output power of 264 mW was operated to heat the FBG through the optical fiber taper coupler. The sensitivity of the FBG anemometer without rough is up to 398.06 pm/(m/s) and the resolution can reach 0.05 m/s. When airflow velocity is between 1.0 m/s and 6.0 m/s, the sensitivity and the resolution are ~54.83 pm/(m/s) and 0.36 m/s respectively. The highest sensitivity of frosted FBG anemometer is 738.95 pm/(m/s), the resolution is 0.027 m/s. Frosted FBG anemometer increases the absorption of light conveyed in the cladding, which effectively improves the sensitivity. The optical fiber anemometer is dominant with the superiority of cost, manufacture, physical strength, structure, which enjoys good potential in all kinds of applications.

References

[1] A.Venugopal, AmitAgrawal, S.V.Prabhu. Review on vortex flowmeter-Designer perspective. Sens. Actuators A, Phys. 2011, 170(1): 8-23.

[2] S. Oda, M Anzai, S Uematsu, K. Watanabe. A silicon micromachined flow sensor using thermopiles for heat transfer measurements. Instrumentation & Measurement IEEE Transactions on. 2003, 52(4): 1155-1159.

[3] Charles Jewart, Ben McMillen, Sung Kwon Cho, et al. X-probe flow sensor using self-powered active fiber Bragg gratings. Sensors & Actuators A Physical. 2006, 127(1): 63-68.

[4] Gerald D Byrne, Stephen W James, Ralph P Tatam. A Bragg grating based fibre optic reference beam laser Doppler anemometer. Measurement Science & Technology. 2001, 12(7): 909-913.

[5] Ying-Li Hsiao, Chai-Ming Li, Tsai-Ching Chiang, et al. A novel airflow sensor based on a reflective tapered fiber interferometer, in 16th Opto-Electronics and Communications Conference (OECC). 2011, 709-710.

[6] Cheng-Ling Lee, Wan-Yu Hong, Hsin-Jung Hsieh. Air Gap Fiber Fabry–Pérot Interferometer for Highly Sensitive Micro-Airflow Sensing. IEEE Photonics Technology Letters. 2011, 23(13): 905-907.

[7] LJ Cashdollar, K. P. Chen. Fiber Bragg grating flow sensors powered by in-fiber light. IEEE Sensors Journal. 2005, 5(6): 1327-1331.

[8] Xinhuai Wang, Xinyong Dong, Yan Zhou. Hot-Wire Anemometer Based on Silver-Coated Fiber Bragg Grating Assisted by No-Core Fiber. IEEE Photonics Technology Letters. 2013, 25(24): 2458-2461.

[9] Caldas P, Jorge PA, Rego G, et al. Fiber optic hot-wire flowmeter based on a metallic coated hybrid long period grating/fiber Bragg grating structure. Applied Optics. 2011, 50(17): 2738-2743.

[10] Xinyong Dong, Yan Zhou, Wenjun Zhou, et al. Compact Anemometer Using Silver-Coated Fiber Bragg Grating. 2012, 4(5): 1381-1386.

[11] CG Lomas, MS Korman. Fundamentals of Hot Wire Anemometry. Cambridge University Press.

1990, 87(3): 1379.
[12] H. H. Bruun. Hot-Wire Anemometry Principles and Signal Analysis. MEASUREMENT SCIENCE and TECHNOLOGY. 1995, 7(10).
[13] Xinhuai Wang, Xinyong Dong, Yan Zhou, et al. Optical fiber anemometer using silver-coated fiber Bragg grating and taper. Sensors and Actuators A: Physical. 2014, 214: 230-233.

IOP Conf. Series: Materials Science and Engineering **711** (2020) 012113 doi:10.1088/1757-899X/711/1/012113

Influences of Feedstocks and Laser Remelting on Erosion Resistance of Al_2O_3-13wt.%TiO_2 Coatings Prepared by Plasma Spraying

Wang Dongsheng[1, a*] **and Zhou Xinghua**[2, b]

[1] College of Mechanical Engineering, Tongling University, Tongling, China, 244000

[2] Department of Students Affairs, Tongling University, Tongling 244000, China, 244000

[a]wangdongsheng@tlu.edu.cn, [b]zhouxh@tlu.edu.cn

Abstract. Conventional and nanostructured ceramic coatings were prepared with conventional and nanostructured aggregate Al_2O_3-13wt.%TiO_2 ceramic powders through plasma spraying and plasma spraying-laser remelting combined technique. Influences of feedstocks and laser remelting on erosion resistance of the prepared coatings were discussed, and the erosion failure mechanisms of four types of coatings were discussed. Results demonstrate that compared with coatings prepared by plasma spraying, the coating prepared by laser remelting has better erosion resistance. Given the equal conditions, nanostructured coating shows stronger erosion resistance than conventional coating. Conventional ceramic coating shows typical brittle erosion characteristics and nanostructured ceramic coating also shows brittle erosion characteristics, accompanied with certain degree of plastic erosion characteristics. Erosive wearing of plasma coating is mainly lamellar peeling, accompanied with certain degree of brittle ceramic particle crushing. Coating prepared by laser remelting present crack initiation and propagation on near surface, finally leading to crystal crushing and peeling of remelting layers.

1. Introduction

Erosive wear is a common form of wearing. With some inherent characteristics, ceramic materials are applied more and more as protective coatings in erosion environment. Plasma spraying is one of the most common technologies to prepare ceramic coatings in China and abroad [1-2]. However, inherent characteristics of plasma spraying determine that lamellar superposition structure of coating structures. Moreover, coating prepared by plasma spraying contains abundant pores and microcracks that restrict its application ranges and service life. Laser remelting is a laser surface strengthening technology that combines laser technology and thermal processing. It can eliminate lamellar structure and most pores of coating layers, thus enabling to form uniform and dense ceramic coatings. Coating prepared by laser remelting can protect workpieces and prolong service life of the workpiece [3-5]. Due to unique structure, nanomaterials have excellent performance beyond the capability of ordinary materials and provide favorable conditions to improve performance of ceramic coatings. Recent studies prove that nanostructural ceramic coating is significantly superior to traditional ceramic coating [6-8].

In this study, conventional and nanostructured ceramic coatings were prepared with conventional and nanostructured aggregate Al_2O_3-13wt.%TiO_2 ceramic powder through plasma spraying and plasma spraying-laser remelting combined technique. Influences of feedstocks and laser remelting on erosion resistance of coating were discussed. The erosion failure mechanisms of coatings were investigated.

2. Experimental method

The γ-TiAl-based alloy (20 mm×20 mm×8 mm) melted by the High-temperature Material Research Institute of Iron and Steel Research Institute was used as the base material. The transition layer MCrAlY used the NiCoCrAl superalloy powder dispersed by Y_2O_3. NiCoCrAl superalloy powder was made by Metal Material Institute of Beijing General Research Institute of Mining and Metallurgy and its grain size ranged -140~+325 meshes. Ordinary ceramic materials chose the ordinary commercial Al_2O_3-13 wt.%TiO_2 composite ceramic powder which was produced by Shenyang Ronghua, with a grain size of 15~45μm. Nanostructured aggregate powder applied the Nanox S2613P powder (Inframat, USA) prepared by spraying drying method. The nominal component of nanostructured aggregate powder was Al_2O_3-13 wt.%TiO_2 and the size range was 10~50 μm. The original nanoparticle was 30~80 nm. The corresponding powder morphology is shown in Fig. 1.

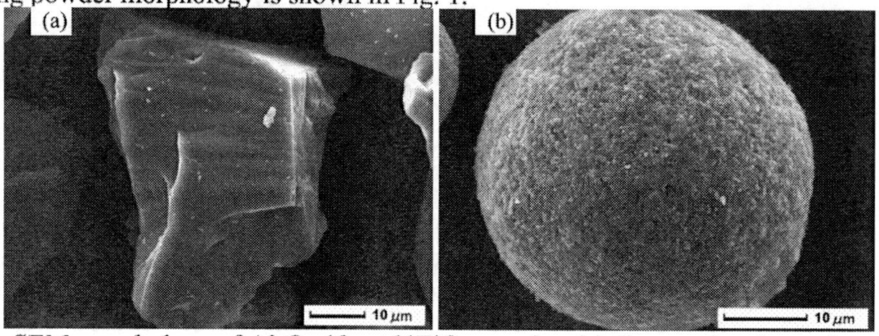

Fig. 1. SEM morphology of Al_2O_3-13 wt.%TiO_2 powder: (a) Conventional powder and (b) nanostructured aggregate powder.

The 3710 plasma spraying system made by Chemplex Company (USA) was applied for pre-manufacturing of conventional and nanostructured Al_2O_3-13 wt.%TiO_2 ceramic coating. The technological parameters of plasma spraying are listed in Table 1. Before coating, samples were preprocessed through grinding, deoiling and abrasive blasting. Laser remelting applied SLCF-X12×25 CO_2 laser machine. Technological parameters of laser remelting were introduced as follows. For conventional coating, the laser power, rectangular spot size, laser scanning direction, scanning rate and amount of overlap were 650 W, 5 mm×3 mm, along 3mm of the spot, 1200 mm·min^{-1} and 20%, respectively. For nano-coating, the laser power, rectangular spot size, laser scanning direction, scanning rate and amount of overlap were 500 W, 5×3 mm, along 3mm of the spot, 700 mm·min^{-1} and 20%, respectively.

Table 1 Technical parameters of plasma spraying

Process parameters	MCrAlY	Conventional Al_2O_3-13 wt.%TiO_2	Nanostructured Al_2O_3-13 wt.%TiO_2
Current [A]	710	850	870
Voltage [V]	42	42	42
Primary gas (Ar) [PSI]	65	45	45
Secondary gas (He) [PSI]	115	140	140
Powder carrier gas (Ar) [PSI]	45	45	45
Powder feed rate [r·min^{-1}]	2	3	3
Spray distance [mm]	110	100	100
Traverse speed [mm·s^{-1}]	100	100	100
Coating thickness [μm]	~100	~350	~350

Solid particle erosion test was accomplished on an erosion tester which was improved from a sand-blasting machine. The working principle is shown in Fig. 2(a). Compressed air was used as the momentum to drive abrasive particle sprayed onto surface of specimens, thus realizing microcutting and

IOP Conf. Series: Materials Science and Engineering **711** (2020) 012113 doi:10.1088/1757-899X/711/1/012113

impact of sample surface. The erosion test was carried out by using natural adamantine spar (Al_2O_3) as the abrasive particles (particle size=150~250 μm, prismatic, HV=2000~2300 kg·mm^{-2}, Fig. 2(b)) under room temperature. Airflow speed was adjusted by controlling the inlet pressure valve and erosion angle was controlled by adjusting the support through the self-made erosion clamp. Erosion parameters included erosion surface size (Φ17.5 mm), inner diameter of nozzle (6 mm), distance from nozzle to samples (150 mm), air pressure (0.6 MPa), erosion angle (30°, 60° and 90°), and erosion time 30+30+60+60+60 s. Samples were cleaned by ultrasonic treatment with acetone before and after erosion, and weight loss after drying was weighted by a FA1004 analytical balance with a weighting sensing of 10^{-4} g (all results chose mean of 3 samples). Morphology before and after erosion was observed under JSM-7100F (JEOL) field emission scanning electron microscope and erosion failure mechanism of ceramic coatings was discussed.

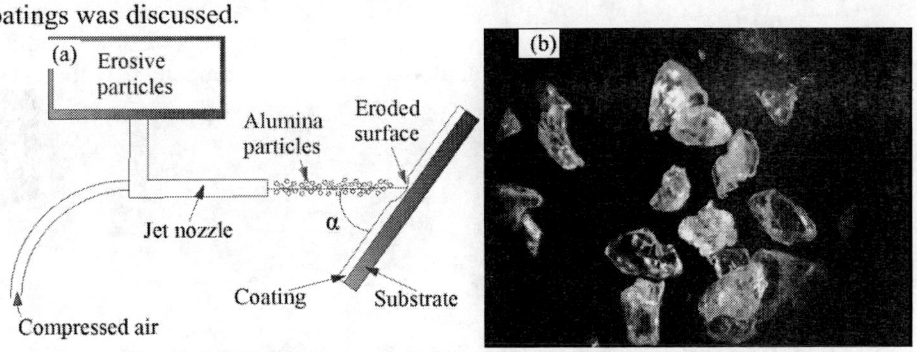

Fig. 2. (a) Erosion test and (b) pictures of abrasive particles.

3. Experimental results and discussions

Structure of ceramic coating. Morphologies of conventional and nanostructured Al_2O_3-13 wt.%TiO_2 ceramic coatings prepared by plasma spraying and laser remelting are shown in Fig. 3. Conventional ceramic coating based on plasma spraying shows typical lamellar superposition characteristics, while the nanostructured coating based on plasma spraying is composed of fully melted region and partially melted region of nanoparticles, thus resulting in a two-phase structure. After laser remelting, most internal defects of the original coating prepared by plasma spraying have been deleted and coating surface forms dense, uniform and small isometric crystal remelting zone. Compared with conventional ceramic coating prepared by laser remelting, the isometric crystal tissues in the remelting zone of nano-coating prepared by laser remelting is smaller and denser. Due to fast heating and cooling of laser remelting, some remained nanoparticles in some melting regions of plasma coating are still retained in the remelting zone. Microstructure and formation mechanism of conventional and nanostructured Al_2O_3-13 wt.%TiO_2 ceramic coatings which are prepared by plasma spraying and laser remelting have been introduced in Reference [9-11]. These are not introduced in the present study.

Erosion test results. Erosive wear test results of four Al_2O_3-13 wt.%TiO_2 ceramic coatings under different erosion angles are shown in Fig. 4. Generally, the nanostructured ceramic coating prepared by laser remelting shows the highest erosion resistance, followed by the nanostructured ceramic coating prepared by plasma spraying, conventional ceramic coating prepared by laser remelting and conventional ceramic coating prepared by plasma spraying. However, erosion-induced weight loss of coatings prepared by laser remelting increased significantly as the erosion time prolongs, especially the conventional ceramic coating prepared by laser remelting. Given 90° erosion angle, the erosion-induced weight loss of conventional ceramic coating prepared by laser remelting is the highest after 240s of erosion. Besides, it can be seen from Fig. 4(a), (b) and (c) that weight loss of all ceramic coatings is the lowest when the erosion angle is 30° and it reaches the highest when the erosion angle is 90°. This reflects that the erosion-induced weight loss is positively related with erosion angle and coatings represent typical erosion characteristics of brittle materials. Take the conventional ceramic coating prepared by plasma spraying for example. The total weight losses at 240s are 40.4 mg, 72.3 mg and 110.3 mg when the erosion angle is 30°, 60° and 90°, respectively.

Erosion results analysis of coatings. The erosive surface morphology of conventional Al_2O_3-13 wt.%TiO_2 coating prepared by plasma spraying is shown in Fig. 5(a). Clearly, coating surface after erosion presents evident peeling behaviors and micro-breakage to some extent. For the conventional ceramic coating prepared by plasma spraying, its structure is composed of embedding and superposition of lamellar particles and particles are mainly combined through mechanical riveting. Due to the weak binding force, the bonding interface of particles is easy to be damaged. In the process of erosive wear, cracks propagate from the uncombined interface to the combined interface along inter-lamella structure under repeated impact of erosion particles, thus making the upper lamellar tissues under direct impact peeling off continuously [12]. The erosive failure of conventional Al_2O_3-13 wt.%TiO_2 ceramic coating prepared by plasma spraying is shown in Fig. 6. Moreover, conventional ceramic has low tenacity and high brittleness, and erosive energy cannot be consumed by plastic deformation. Under the alternative compressive-tensile stresses produced by impact of erosive particles, the coating is easy to be broken and separated from the base significantly. Therefore, erosive wear of conventional ceramic coating prepared by plasma spraying is dominated by lamellar peeling, accompanied with certain degree of brittle ceramic grain breakage.

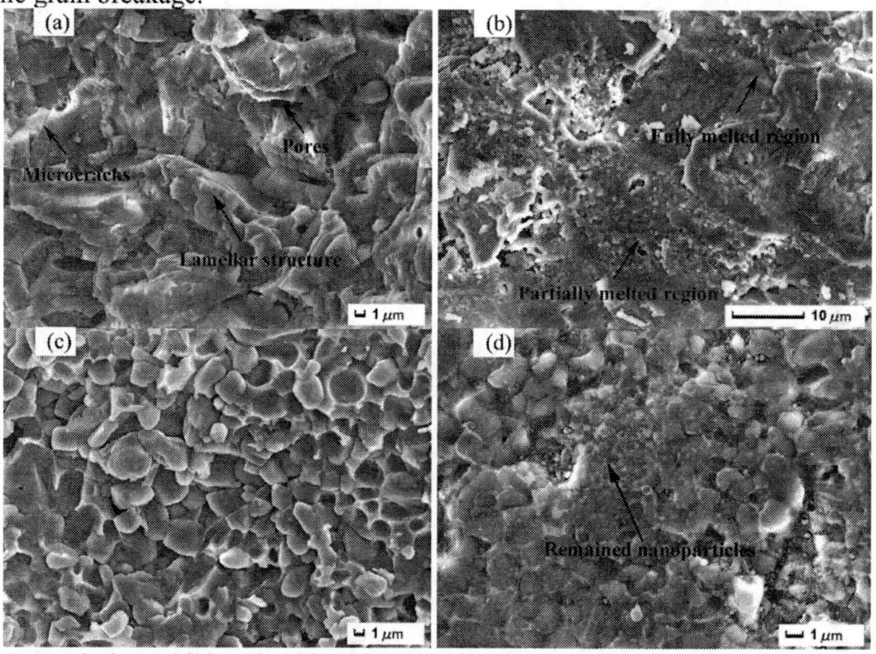

Fig. 3. SEM morphology Al_2O_3-13 wt.%TiO_2 ceramic coating: (a) conventional coating prepared by plasma spraying, (b) nano-coating prepared by plasma spraying, (c) conventional coating prepared by laser remelting and (d) nano-coating prepared by laser remelting.

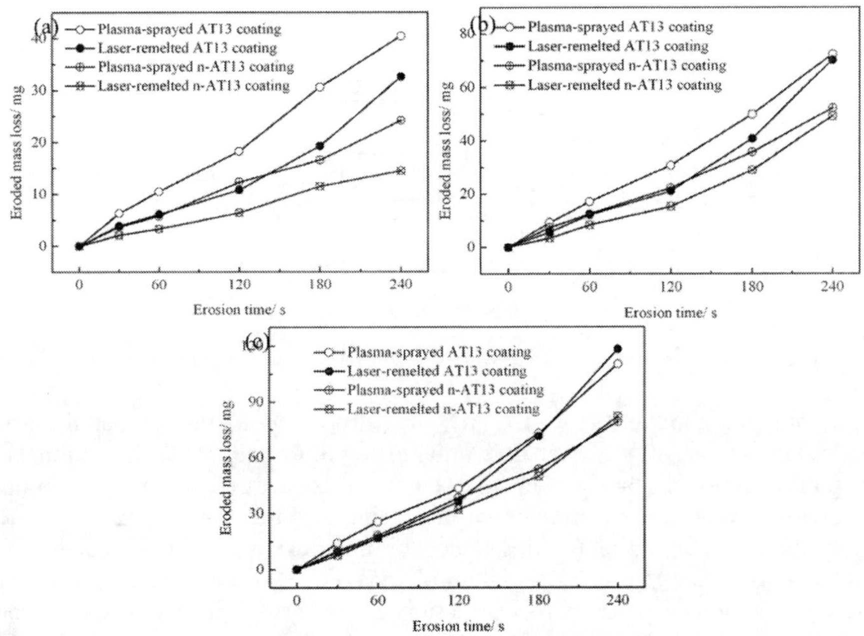

Fig. 4. Erosive wear test results of ceramic coatings under different erosion angles(α): (a)30°, (b)60° and (c)90°.

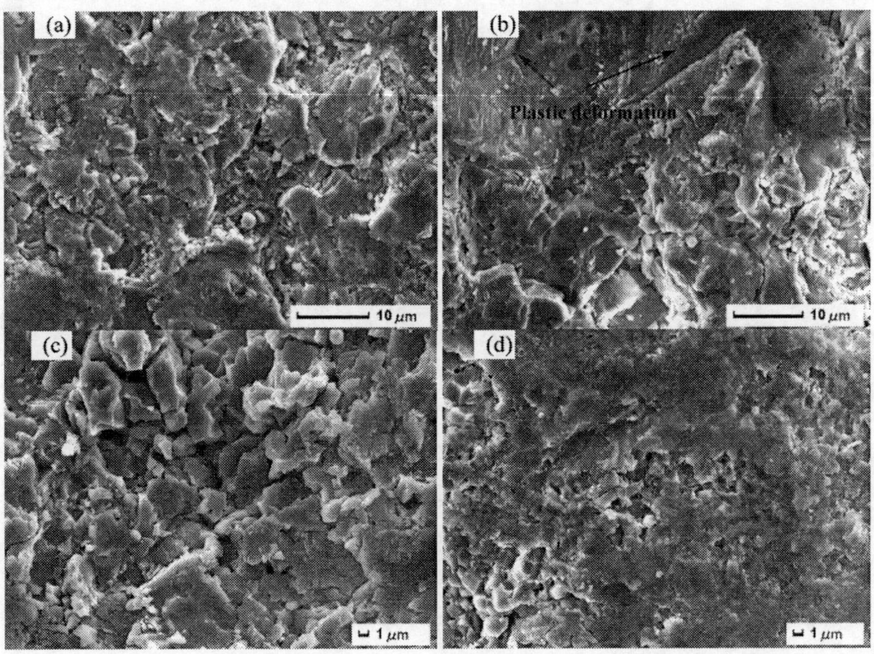

Fig. 5. Typical morphology of erosive surface of Al_2O_3-13 wt.%TiO_2 ceramic coating under 60°: (a) conventional coating prepared by plasma spraying, (b) nano-coating prepared by plasma spraying, (c) conventional coating prepared by laser remelting and (d) nano-coating prepared by laser remelting.

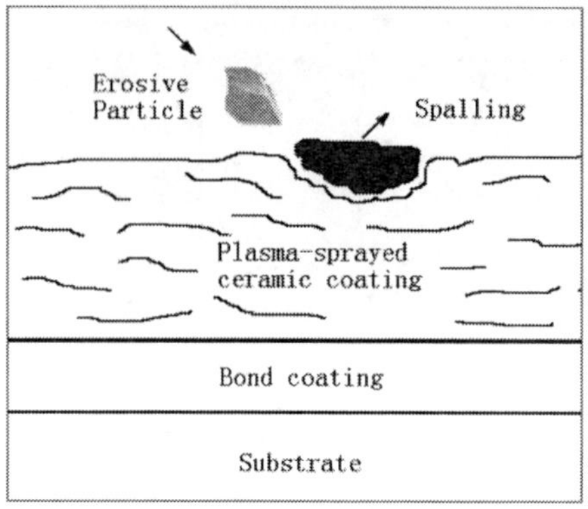

Fig. 6. Erosive failure of conventional Al_2O_3-13 wt.%TiO_2 coating prepared by plasma spraying.

SEM morphology of the Al_2O_3-13 wt.%TiO_2 nanostructured coating prepared by plasma spraying after erosion is shown in Fig. 5(b). Compared with erosion morphology of conventional ceramic coating prepared by plasma spraying, the erosion surface of nano-structured ceramic coating presents similar peeling and breaking features, but the degree of peeling and breaking features is lower. In addition, there's obvious plastic deformation (cutting traces) on the erosion surface, especially in partial melting region (micro-nanostructure region). This is not observed on erosion surface of conventional coating. Such plastic deformation is because nanostructure can improve plasticity and tenacity of ceramic coating. The erosive failure of Al_2O_3-13 wt.%TiO_2 nanostructured coating prepared by plasma spraying is shown in Fig. 7.

The nanostructured coating prepared by plasma spraying maintains a considerable proportion of nanostructure, which increases tenacity of ceramic coating [6,13]. Therefore, this is beneficial to improve erosion resistance of nanostructured coating. Moreover, erosion resistance of coating is related with bonding strength of coating. At plasma spraying onto the nanostructured aggregate powder, the powder surface is easy to be heated and melted due to the high specific surface and high activity, and powder surface is melted well. Therefore, nano particles develop great deformation after impacting onto the base or the deposited surface, and the pavement of nano particles is better than conventional powder [14]. Therefore, bonding of lamellar structure of nanostructured coating is better than that of the conventional coating prepared by plasma spraying, manifested by the high bonding strength. A good combination increases the resistance of nanostructured coating to cracks on the bonding interface, thus enabling to increase erosion resistance of nanostructured coating accordingly.

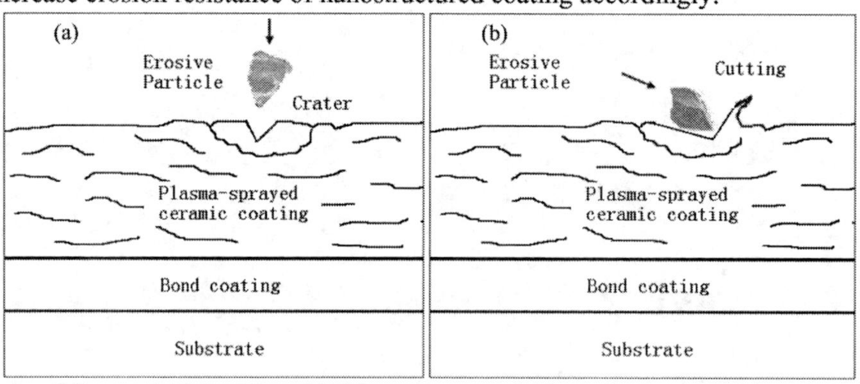

Fig. 7. Erosive failure of Al_2O_3-13 wt.%TiO_2 nanostructured coating prepared by plasma spraying.

Typical erosion morphology of conventional Al_2O_3-13 wt.%TiO_2 coating prepared by laser remelting is shown in Fig. 5(c). Compared with samples prepared by plasma spraying, samples prepared by laser remelting develop significantly smaller degree of peeling and breaking after erosion. However, there are still serious breakages in local regions. After laser remelting, ceramic coating prepared by plasma spraying develops remelting crystals on the uneven and porous surface layer, and the lamellar structure disappears. Instead, a dense remelting layer was formed. Due to high cohesion in coatings prepared by laser remelting, it is difficult to have peeling during the erosion, but cracks initiate and propagate on near surface, finally resulting in coating breakage. The erosion failure principle is shown in Fig. 8. During initial stage of erosion, the dense remelting layer with strong bonding and hardness shows high resistance to crack initiation and propagation, leading to the small erosion-induced weight loss and high erosion resistance [5]. However, ceramic materials which have poor impact resistance and low breaking tenacity are easy to produce cracks, pores and even peeling under dramatic heating and cooling conditions during laser remelting. Given repeated impacts of erosive particles, there's a large region with defects and deep macroscopic cracks on local remelting surface, which further causes peeling in a large area and high peeling thickness in some positions. Some peeling even reaches the interface between remelting layer and residual plasma spraying layer. This differs significantly from stepwise lamellar peeling mechanism of the coating prepared by plasma spraying after erosion. Therefore, the erosion-induced weight loss of coatings prepared by laser remelting is accelerated under erosion conditions for a long period.

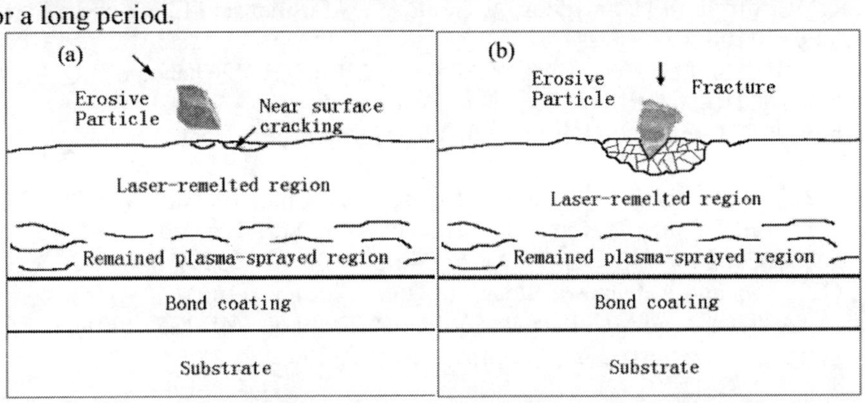

Fig. 8. Erosion failure of conventional Al_2O_3-13 wt.%TiO_2 coating prepared by laser remelting.

Erosion morphology of Al_2O_3-13 wt.%TiO_2 nanostructure coating prepared by laser remelting is shown in Fig. 5(d). Obviously, the coating develops the slightest peeling and breaking. This is because the nanostructured ceramic coating prepared by combining laser remelting and plasma spraying is equipped with not only surface hardness and high density of samples prepared by laser remelting, but also high tenacity of nanostructured coating. The tenacity and hardness are unified well, so that the coating is equipped with the strongest erosion resistance. Although erosion-induced weight loss of nanostructured ceramic coating prepared by combining laser remelting and plasma spraying is accelerating slightly, the erosion-induced weight loss is significantly smaller than that of conventional ceramic coating prepared by laser remelting. Therefore, nanostructured ceramic coating prepared by laser remelting has good erosion resistance in early erosion and after erosion for a long period.

4. Summary

Conventional and nanostructured ceramic coatings were prepared with conventional and nanostructured aggregate Al_2O_3-13wt.%TiO_2 ceramic powders through plasma spraying and plasma spraying-laser remelting combined technique. Influences of feedstocks and laser remelting on erosion resistance of the prepared coatings were discussed, and the erosion failure mechanisms of four types of coatings were discussed.

(1) Compared with conventional ceramic coating prepared by plasma spraying, nanostructured coating has stronger erosion resistance to some extent and the erosion resistance of nanostructured coating can be further improved after laser remelting. The nanostructured ceramic coating prepared by laser remelting presents the best erosion resistance.

(2) Conventional ceramic coating presents typical brittle erosion characteristics and nanostructured ceramic coating presents obvious brittle erosion characteristics, accompanied with certain plastic erosion characteristics.

(3) Erosive wear of coating prepared by plasma coating is dominated by lamellar peeling. Moreover, the brittle ceramic particles are crushed to some extent. However, coating prepared by laser remelting mainly develops crack initiation and propagation on the near surface, finally resulting in breaking and peeling of crystal grains on the remelting layer.

References

[1] K. Szymański, A. Hernas, G. Moskal and H. Myalska: Surf. Coat. Technol. Vol. 268 (2015), p. 153
[2] R.G. Song: Surf. Coat. Technol. Vol. 168 (2013), p.191
[3] Y.Wang, G.Darut, T. Poirier, J. Stella, H. Liao and M.P. Planche: J. Eur. Ceram. Soc. Vol. 37 (2017), p. 3623
[4] J.A.Wahab and M.J. Ghazali: Wear Vol. 432-433 (2019), p. 202937
[5] P.C. Tsai, J.H. Lee and C.L. Chang: Surf. Coat. Technol. Vol. 202 (2007), p. 719
[6] P. Carpio, A. Borrell, M.D. Salvador, A. Gómez, E. Martínez and E. Sánchez: Ceram. Int. Vol. 41 (2015), p. 4108
[7] Y. Wang, Y. Bai, K. Liu, J.W. Wang, Y.X. Kang, J.R. Li, H.Y. Chen and B.Q. Li: Appl. Surf. Sci. Vol. 363 (2016), p. 101
[8] A. Lbrahim, R.S. Lima, C.C. Berndt and B.R. Marple: Surf. Coat. Technol. Vol. 201 (2007), p. 7589
[9] D. Wang, Z. Tian, L. Shen, Z. Liu and Y. Huang: Surf. Coat. Technol. Vol. 203 (2009), p. 1298
[10] D. Wang, Z. Tian, L. Shen, Z. Liu and Y. Huang: Appl. Surf. Sci. Vol. 255 (2009), p. 4606
[11] D. Wang: IOP Conference Series: Mater. Sci. Eng. Vol. 585 (2019), p. 012021
[12] C.J. Li, G.J. Yang and A. Ohmori: Wear Vol. 260 (2006), p. 1166
[13] R.S. Lima, A. Kucuk and C.C. Berndt: Surf. Coat. Technol. Vol. 135 (2001), p. 166
[14] B. Liang and C. Ding: Surf. Coat. Technol. Vol. 197 (2005), p. 185

IOP Conf. Series: Materials Science and Engineering **711** (2020) 012114 doi:10.1088/1757-899X/711/1/012114

Study on Mnospectral Peak Identification of Hard Target Based on Brewster Angle

LIU Jia Xiang[1,a], LIU Jia ju[3,a], ZHU Zhi Zheng[1,2,a], FANG Yong Hua[1,a], Dong Xiao[4,5,a],Zhang Xin Yuan[4,5,a], Pan Ying[1,a]

[1]Anhui Institute of Optics and Fine Mechanics, Hefei, 230031, China

[2]University of Science and Technology of China, Hefei, 230026, China

[3]Anhui agricultural university, Hefei, 230036, China

[4]State Key Laboratory of Pulsed Power Laser Technology, National University of Defense Technology, Hefei, 230037, China

[5]Key Laboratory of Electronic Restriction Technology Anhui, National University of Defense Technology, Hefei 230037, China

[a]luckuis@163.com

Abstract. It is difficult to obtain the wide band spectrum of the target in the hard target detection with strong background weak signal by laser. A single spectral peak recognition method based on brewster angle is proposed. By combining the characteristics of pPolarized reflectivity and characteristics of typical target extinction coefficient, the brewster angle solution of hard target is obtained by extremum finding by the second derivative in the hard target detection model. The results show that the brewster angle of nonmetallic hard targets is correlated with inverse triangulation of double index. Therefore, the single spectral peak method can be used to identify the target material by means of the laser which covers the peak spectral segment of the target refractive index.

1.Introduction

In the field of weak signal detection under strong background noise, such as detection of dangerous road condition[1], determine internal quality attributes agricultural products[2], telemetry of dangerous goods at airport[3], space target detection[4], surface contamination detection[5] etc., laser was used for its high-directivity, good monochrome performance, strong power and so on. There are several methods of hard target detection: 1)LIBS（Laser-induced breakdown spectroscopy）[7,9],and this technique excites the material to the plasma state by laser and obtains the information of the composition and concentration of the element by analyzing the position and intensity of the spectrum line of the light emitted by the plasma. LIBS is widely used in the analysis of elements, but rarely used in the analysis of organic compounds with the same molecular structure of elements. 2)Raman spectroscopy[10,12], and this technique stimulates inelastic Raman scattering and identify the target by analyzing Raman characteristic spectrum by laser irradiation. The disadvantage of Raman spectroscopy is its susceptibility to fluorescence interference. The latest trend of Raman spectroscopy is ultraviolet band Raman spectroscopy, that is, hard target is detected in ultraviolet band to avoid fluorescence interference, but there are not many substances with Raman characteristics in the ultraviolet band, so it is not widely used.

IOP Conf. Series: Materials Science and Engineering **711** (2020) 012114 doi:10.1088/1757-899X/711/1/012114

3)Absorption spectrum technology [13,15], and this technique identifies the target by detecting the characteristic absorption spectrum of the target, typical technologies are TDLAS (Tunable Diode Laser Absorption Spectroscopy) and PAS (Photoacoustic Spectroscopy), etc. TDLAS detects target transmission by combining a narrow line width and high-power laser with a photoelectric detector. PAS detects target absorption based on photoacoustic absorption. After several years of development, the theory of absorption spectrum technology has been very mature, and a fairly complete spectrum library has been established. That's to say, absorption spectrum technology has a very wide range of applications. However, in the application of hard target detection, in order to identify the target material, absorption spectrum technology needs to obtain the target broadband spectrum by wide-band laser or multiple laser combination detection to accomplish the spectral identification. But too many lasers in detection system make the system bulky and costly. In order to solve this problem, a method of single-band spectrum hard target recognition based on featured brewster angle is presented.

2.Hard target detection model

As Fig 1 shows, hard target detection model can be described in three layers[5]. The first layer is radiance source, the temperature of radiance source is T_1,the equivalent blackbody radiance of radiance source is $B(T_1)$.The second layer is the surface of hard target, it can be regarded as contamination of hard target, the temperature of contamination is T_2,the equivalent blackbody radiance of contamination is $B(T_2)$,the reflectance of . contamination is R. The Third layer is the rest of hard target, it can be regarded as substrate. The temperature of substrate is T_3,the emittance of substrate is 1-R. The equivalent blackbody radiance of it is $B(T_3)$ .As the hard target in thermal equilibrium, $T_2 = T_3$,Then.

$$B(T_2) = B(T_3) \tag{1}$$

There two radiation can be got by detector. The first radiation is the reflected light from source shown int blue arrow in Fig 1,the value of it is $B(T_1) * R$.The second radiance is the substrate radiance shown in black arrow in Fig 1,the value of it is $B(T_2) * (1 - R)$.

$$L = B(T_2) * (1 - R) + B(T_1) * R \tag{2}$$

Merge similar terms in equations (2).

$$L = \left(B(T_2) - B(T_2)\right) * R) + B(T_2) \tag{3}$$

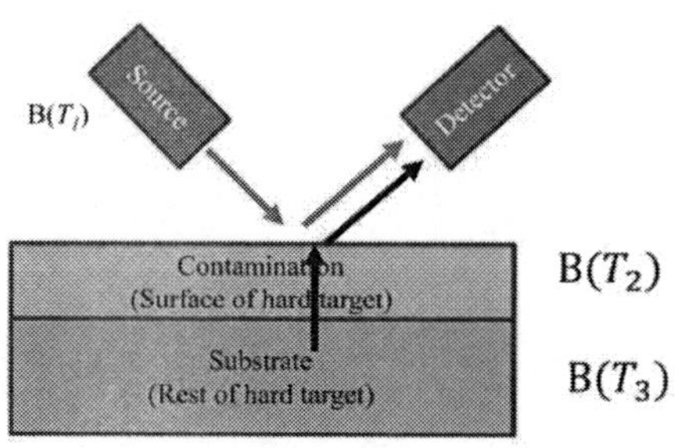

Fig 1 Hard target detection model

Equation (3) is the expression of hard target detection model, from which it can be seen that: The difference of radiation source and substrate's equivalent black body radiation brightness exists in the product term of hard target's reflection features. Although the difference does not include hard target's reflection features, it will amplify the detection signals. The greater the difference is, the higher the detection sensitivity is, and the harder the target's features are easier to obtain. When the equivalent radiation brightness of the external radiation source is much larger than the equivalent radiation brightness of the substrate, the influence of the substrate can be ignored, and the reflection characteristic information of hard target can be extracted directly from the signal.

3.Hard target detection model

Although the reflection feature of the hard target can be used to identify the hard target, in the practical application of hard target material identification, the reflection feature usually refers to the spectrum consisting of reflections from multiple bands of the hard target. It can be seen from the pPolarized reflection spectra in 800cm^{-1}-1300cm^{-1} in Fig 2 that DMMP (dimethyl methylphosphonate) and PDMS (polydimethylsiloxane)have obvious spectral characteristics and can distinguish the two substances by spectrum. But the two spectra intersect in the black circle, where they have the same reflectivity. In the end, it is difficult to identify the hard target directly by the reflected signal when using the laser with narrow band.

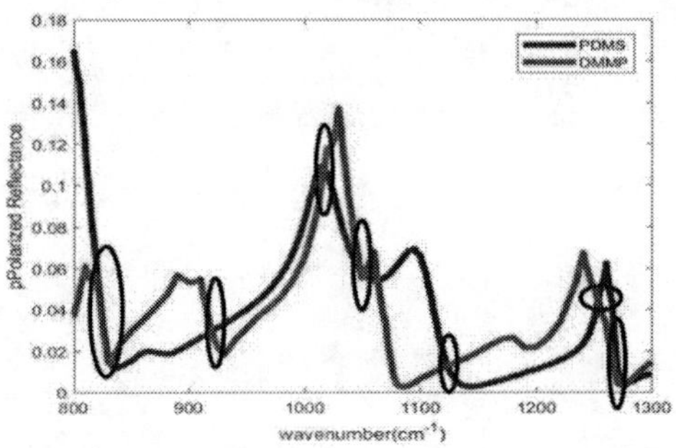

Fig 2 pPolarization reflection spectra of DMMP and PDMS. a) the overall shape of the two spectra is very different, and b) the two spectra intersect at the black circle

According to Fresnel formula, the sPolarized reflectivity of hard target R^s,pPolarized reflectivity of hard target R^p are:

$$R^s = \frac{(n-\cos\theta_1)^2+k^2}{(n+\cos\theta_1)^2+k^2} \tag{4}$$

$$R^p = \frac{(n-\frac{1}{\cos\theta_1})^2+k^2}{(n+\frac{1}{\cos\theta_1})^2+k^2} \tag{5}$$

Where n is refractive index of hard target, k is extinction coefficient. From equations (4) and (5), it can be seen that the polarization reflectance is a complex expression related to the incident Angle, extinction coefficient of hard target and refractive index, which takes great trouble to the process of target recognition. The refractive index and extinction coefficient are both parameters of hard target that reflect the characteristics of hard target materials. As shown in Fig 3, the extinction absorption coefficient spectrum have characteristic peaks at 1024cm^{-1}, 1095cm^{-1}, 1261cm^{-1}, while the characteristic peaks of the refractive index spectrum are peak at 1009cm^{-1} and peak at 1255cm^{-1}. The spectral features of the two are independent and in different shape. Therefore, the polarization reflectivity mixed with extinction coefficient and refractive index makes it difficult to identify hard target materials directly.

In fact, when the target is observed at different incident angles, the polarization reflectivity of the target will vary with the incident angle. According to equations (4)-(6), the relation between the polarization reflectivity of PDMS at 1000cm-1 and the incidence angle can be calculated, as shown in Fig 4.And PDMS' complex refractive index is shown in Fig 3.

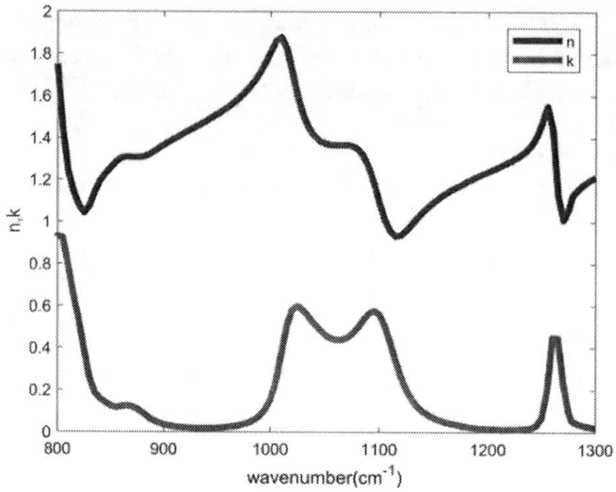

Fig 3[6] complex refrative index of PDMS

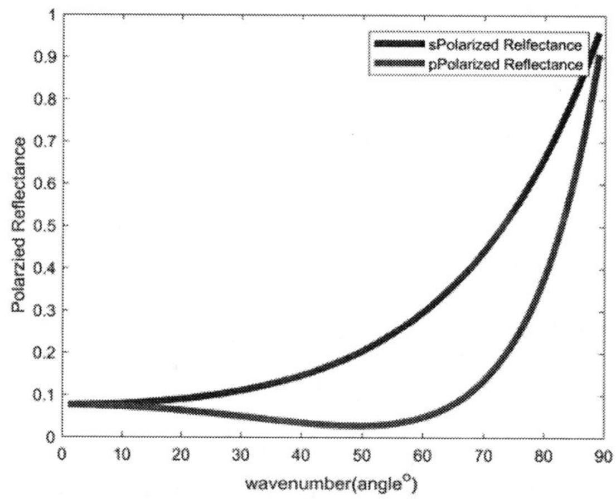

Fig 4 polarized reflectance varies with incidence angle

It can be seen from Fig 4 that the polarization reflectivity of p polarization decreases first and then increases with the increase of incidence angle, and there is an inflection point, which is the brewster angle of hard target in this band. When the hard target in the detection band is transparent, that is, the absorption coefficient is zero, the relationship between the brewster angle and the refractive index is shown in equation (6):

$$\tan(\theta_{brewster}) = n \tag{6}$$

When the target is opaque, that is, the absorption coefficient is not zero, the mathematical expression of brewster angle ($\theta_{brewster}$) in the detection band of hard target is more complicated, which needs to be obtained by the method of calculating the extreme value of the second derivative. The second derivative of equation (14) can be obtained:

$$\frac{2}{\Psi} - \frac{2((n-x)^2+k^2)^2}{\Psi^2} + \frac{2(2n+2x)^2((n-x)^2+k^2)^2}{\Psi^3} + \frac{2(2n-2x)(2n+2x)}{\Psi^2} \tag{7}$$

x and Ψ are listed in equations (8), (9).

$$x = \frac{1}{\cos\theta_1} \tag{8}$$

$$\Psi = (n+x)^2 + k^2 \tag{9}$$

IOP Conf. Series: Materials Science and Engineering **711** (2020) 012114 doi:10.1088/1757-899X/711/1/012114

The inflection point of the function is the point where the second derivative is zero. By extracting the molecule from equation (7) and establishing the equation, we can get

$$16k^2n^2 + 24k^2nx + 16n^4 + 24n^3x - 8nx^3 = 0 \tag{10}$$

To simplify the solving process, the common factor of equation (10) is extracted:

$$8n(2k^2n + 3k^2x + 2n^3 + 3n^2x - x^3) = 0 \tag{11}$$

Since the refractive index n is not zero, equation (11) is equivalent to equation (12).

$$2k^2n + 3k^2x + 2n^3 + 3n^2x - x^3 = 0 \tag{12}$$

Solution to equation (12) is obtained

$$\arccos\left(\frac{1}{2\sqrt{n^2+k^2}\cos\left(\frac{\arccos\left(\frac{n}{\sqrt{n^2+k^2}}\right)}{3}\right)}\right) \tag{13}$$

Equation (13) is the mathematical expression of brewster angle in the laser detection of hard targets. It can be seen that the mathematical expression is relatively complex, which is a trigonometric function and a mixed mathematical expression related to refractive index and extinction coefficient, and it is still difficult to find out the law. Therefore, it is necessary to combine the physical meaning to further analyze. In fact, in many occasions of hard target detection, such as road detection, agricultural products detection and explosives telemetry, the target surface is composed of non-metallic materials with tiny extinction coefficient (<1)[6]. Therefore, the square term of extinction coefficient k in equation (13) can be omitted and simplified to obtain.

$$\arccos\left(\frac{1}{2n}\right) \tag{14}$$

Equation (14) is the mathematical expression of the hard target brewster angle of non-metallic opaque materials, from which it can be concluded that the brewster angle has nothing to do with extinction coefficient, but is related to the refractive index of the target. Therefore, "pure" spectral information can be obtained by detecting the brewster angle of the hard target, and the target material can be identified according to the spectrum.

4. Discussion

According to equation (5), the incidence angle corresponding to the minimum value of polarization reflectance of p polarization direction of each wave in 800cm^{-1}-1300cm^{-1} is taken as the brewster angle, as shown in Fig 5. For convenience of comparison, the brewster angle calculated by equation (14) based on the hard target detection model is also given in the figure. It can be seen from the figure that the spectral shape is very close, and the positions of the characteristic peaks of the 1000cm^{-1} band and the 1250cm^{-1} band are almost identical. The angle deviation in figure (b) in Fig 5 is due to the influence of k value in the model. The successful verification of the hard target detection model makes it possible to refer to the characteristic peak spectrum of refractive index in practical detection, select the laser covering the band, and obtain the brewster angle for detecting the band by angle scanning, so to realize the material identification of hard target.

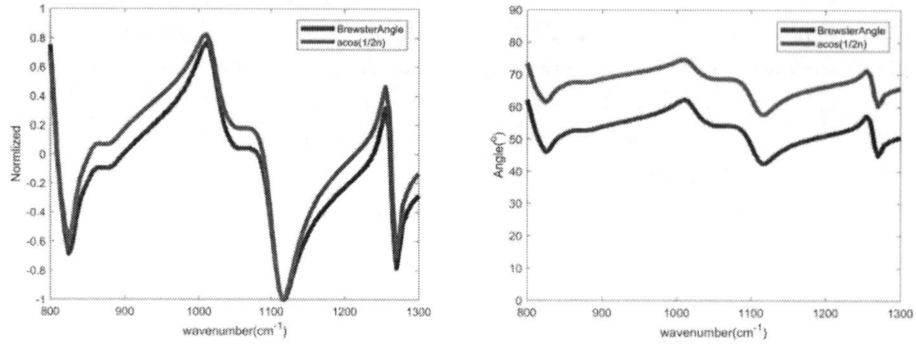

(a) (b)

Fig 5 Brewster angle of PDMS in 800cm^{-1}-1300cm^{-1},the blue line is the incidence angle corresponding to the minimum value of polarization reflectance of p polarization direction of each wave in 800cm^{-1}-1300cm^{-1}, and the red line is the brewster angle calculated by equation (14) based on the hard target detection model (b)the ordinate is the angle, (a)the normalized value of figure (b).

5.Summary

A new method based on brewster angle for material identification of hard target is proposed. It is found that the brewster angle of nonmetallic opaque hard target is similar to the transparent target, and has nothing to do with extinction coefficient, but is related to the refractive index of the target. In the practical application of hard target laser detection, the target material can be detected and identified by the laser at the specific spectral peak based on the characteristic peak spectrum segment of refractive index.

Acknowledgments

This work was supported by the National Natural Science Foundation of China (Grant No.61871389) and the Research Plan Project of the National University of Defense Technology (Grant No. ZK18-01-02).

References

[1] A. Piccardi, L. Colace:Sensors.Vol. 19(2019),p.1360
[2] S.H Hua, C. P. Chen, Pin Han: Applied Optics.Vol.22(2017), p.6235
[3] Y. Fu, H Liu, Q Hu, et al. Pin Han: Proc Of SPIE.Vol.10183(101830B),p. 101830B-1
[4] H. Tao, H. H. Ling, H. Y. Hua. Laser & Infrared.Vol.40(2010),p.685.
[5] J. X. Liu. Research on detection method of surface contamination based on infrared spectroscopy[D].2015
[6] Information on: https://refractiveindex.info/
[7] Y. Jia, N. J. Zhao, W. Q. Liu, et al: Chinese Journal of Lasers.Vol.45(2018), p,0611001.
[8] J. Moros, J. A. Lorenzo, J. J. Laserna: Anal Bioanal Chem.400(2011), p,3353.
[9] G. G. Dai, G. Q. Wang, H. C, H. C. Chen: Journal of China University of Metrology.Vol.4(28), p,430.
[10] S. Almavia, R. Chirico, M. Nuvoli, et al: Talanta.Vol.144(2015), p.420.
[11] R. Chirico, S. Almaviva, F. Colao, et al: Sensors.Vol.16(2016), p.s16010008.
[12] J. C. Carter, S. M. Angel, M. L. Snyder, et al: Applied Spectroscopy. Vol.6(59), p.759.
[13] C. G. Li, L. Dong, C. T. Zheng, et al: Sensors and Actuators B: Chemical.Vol.232(2016), p.188
[14] Y. Wang, Y. B. Wei, J. Chang, et al: IEEE Sensors Journal.Vol.15(17), p.4975.
[15] M. B. Frish, M. C. Laderer, C. J. Smith, et al. Proc of SPIE.Vol.1(9730), p.97300P-1.

A New Single Pixel Imaging System Based on Paired Rotary Diffraction Plates

Qi Yang[1,2], Zhihao Yuan[1], Shaowu Wang[1], Jianjun Lai[1,2,3*]

[1]Wuhan National Laboratory for Optoelectronics, Huazhong University of Science and Technology, Wuhan 430074, China

[2]Ezhou Institute of Industrial Technology, Huazhong University of Science and Technology, Ezhou 436000, China

[3]State Key Laboratory of applied optics, Changchun Institute of Optics, Fine Mechanics and Physics, Chinese Academy of Sciences, Changchun 130033, China

* Corresponding Author: Jianjun Lai

Email: jjlai@hust.edu.cn.

Abstract: We present a new single-pixel imaging system based on a paired rotary diffractive plates. Compared with tradition imaging system, our designed single-pixel imaging system contains simple structure and lower computational complexity. Theoretical simulation results show the single-pixel imaging system has the feasibility of thermal object imaging at wavelength λ=10.6μm, therefore which can be applied for SF_6 gas imaging.

1. Introduction

Traditional single-pixel imaging recognition is mainly used in ghost imaging [1]. In the 1980s, Klyshko proposed a ghost imaging scheme based on entanglement. Until 1995 Shi Yanhua et al firstly realized ghost imaging and ghost interference through entangled photon pairs. It was once believed that only entangled light could achieve ghost imaging. but in 2002 Bennink at Rochester University used pulsed laser to achieve the "ghost" imaging of the classical light source [2]. Soon afterwards Shi Yanhua et al realized the "ghost" imaging of the classical thermal light source, which formally opens the application door of "ghost" imaging in military and life.

One of the single-pixel imaging applications is as the cheap detector for gas imaging system. Traditionally gas imaging systems contain amount of infrared radiation detectors, which cause a greater cost [3]. Recently, a single pixel imaging camera for short wavlength infrared region was developed to detect methane gas leak [4]. A digital micromirror device (DMD) is usedto pattern the laser output with a sequence of Hadamard masks (256 mask patterns, corresponding to an image resolution of 16x16). However, DMD is not the idea space light modulator for long-wave infrared application due to diffraction effect [5]. In this paper, a new single-pixel detection system at wavelength of 10.6 um is proposed to construct a imaging system for SF_6 gas. Different from traditional imaging system based on a change of reference and sampling components, our single-pixel detection system is a rotated double-plate diffractive structure. Preliminary theoretical simulation has been done to demonstrate the feasibility of new principle for single-pixel imaging, the simulation results show that the calculating error can be limited less than 8%.

IOP Conf. Series: Materials Science and Engineering **711** (2020) 012115 doi:10.1088/1757-899X/711/1/012115

2. Theoretical model and structural design

2.1 Structural design and details

Figure 1 demonstrates the whole structure of single-pixel imaging system, including of light source information, optical platform, single-pixel detector, processor and controller. The optical platform is constructed by four parts. Part 1 and 2 are the same optical rectangular frames to fix two similar rotary platforms (part 3 and 4). Two platforms have four apertures in total signed as C1, C2, C3, C4, and which are placed at the same optical axis with the light source and single-pixel detector. Besides, two diffraction plates are placed in the apertures C2 and C4, which have different holes-distribution as shown in Fig. 3 and possess non-uniform diffraction characteristics,

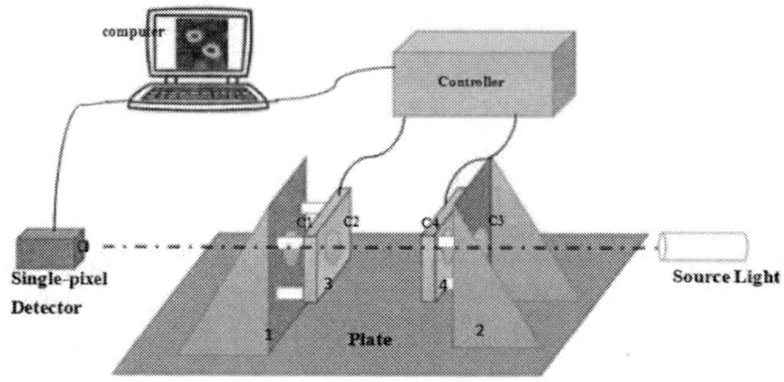

Fig. 1. Schematic diagram of the whole structure of single-pixel imaging system

In Fig. 2, we illustrate the framework of a paired rotary diffractive plates. The rotary plates are signed as T_1 and T_2, besides S and V represent light source and single pixel detector, respectively. When the light emitted by the source S passes through the plats T_1 and T_2 in turns, two different diffraction effect will occur, finally the transmitted light is detected by the single pixel detector V. According to the detected results and model algorithm, the original distribution information of light intensity can be obtained.

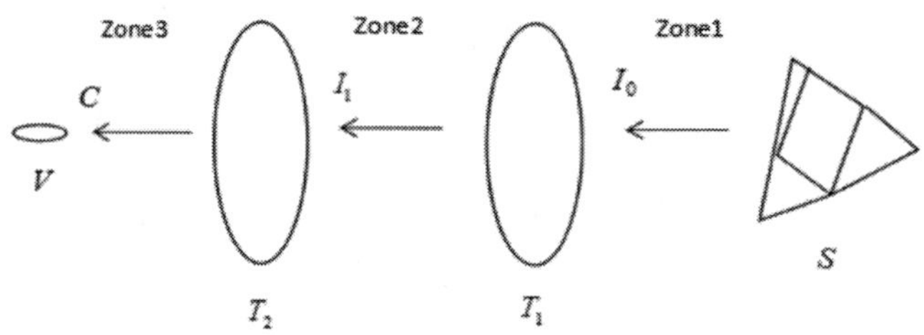

Fig. 2. Framework of the paired rotary diffractive plates

Figure 3 shows the detailed structure of those two plates. Point A is the centre of the plate and the centre is not punched, the remained five points are five punched holes to make light go through from the plates. For this model, we select following parameters: $D = 15\mu m$, $l = 165\mu m$, $f_0^1 = 2cm$, $f_1^2 = 1.55mm$. The weight is set as $\varphi_p = 0.6$. D is diameter of circular hole, l is distance between adjacent rings on T_1 and T_2, f_0^1 is distance between T_1 and T_2, f_1^2 is distance between T_2 and the single pixel detector V .

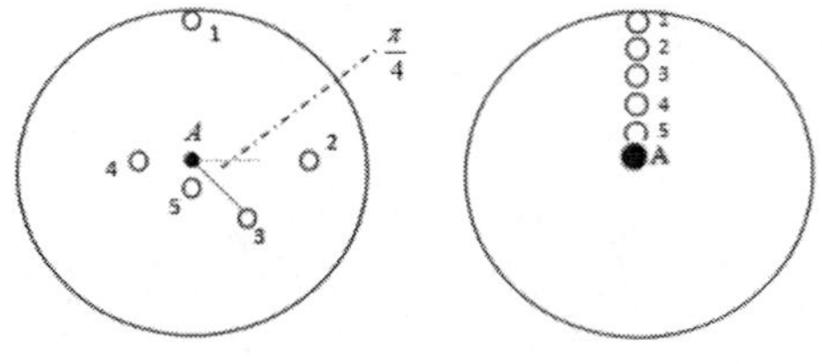

(a) T_1 structure (b) T_2 structure

Fig. 3. Schematic diagram of double-plate structure

2.2 Implementation Methods

When no.1 hole's original value has been calculated successfully, T_2 begins to rotate to align T_1's no.2 hole with T_2's no.2 hole so that no.2 hole's original light signal value of this fixed radius can be calculated. Repeat the above operation to obtain no.3,no.4 and no.5 hole's original values of this fixed radius. As 5 hole's original light signal values has been obtained, T_1 rotates and repeat the above operation to calculate the next radius' data.

2.3 Model Establishing

According to the theory of circular aperture diffraction, the light intensity I_p at point p of circular aperture diffraction at the display screen can be expressed as [6]:

$$I_p = I_0\left[\frac{J_1(2m)}{m}\right]^2, m = \frac{\pi D\rho}{2\lambda f} \tag{1}$$

Where I_0 is the intensity of incident light, D is the diameter of circular aperture, ρ is the distance from the point p at display screen to the maximum point of diffraction, and f is the distance from the circular aperture diffraction plane to the target display screen.

According to Fig. 4, light will be diffracted by plates T1 and T2, the final formula is as follows:

$$I_1(i) = \sum_j\left[\frac{J_1(2m_0^1(j))}{m_0^1(j)}\right]^2 I_0(j), m_0^1(j) = \frac{\pi D_0(j)\rho_0^i(j)}{2\lambda f_0^1} \tag{2}$$

$$C_k = \sum_i\left[\frac{J_1(2m_1^2(i))}{m_1^2(i)}\right]^2 I_1(i), m_1^2(i) = \frac{\pi D_0(i)\rho_1(i)}{2\lambda f_1^2} \tag{3}$$

Here, $I_1(i)$ is the intensity value of the i point on the T_2 plate after the diffraction of T_1, C_k is the intensity value of the light, which is from T_1 diffracted into the T_2 plate, secondly diffracted into the single-pixel detector from T_2 when measuring the five points of the no. k radius.

Through mathematical discretization and error analysis, light intensity can be restored by weight method.The diffraction effect of light passing through a circular aperture can be controlled by controlling parameters D、f and ρ, so that in the final detection value C_k, it can be known that one of the approximate proportions is the value of light diffraction from the aligned point, and by choosing the above parameters, the proportions which can let us get good results can reach more than 60%(this proportions' related value can be detected). The diffraction light is main generated by the alignment points but not the other non-alignment points. The diffraction caused by T1 and T2 corresponds to the weight σ_i and τ_p,respectively. When the point p is aligned, we can get:

$$I_0(p) = a_1(p)\cdot a_0(j)\cdot \tau_p\sigma_i C_k \tag{4}$$

IOP Conf. Series: Materials Science and Engineering **711** (2020) 012115 doi:10.1088/1757-899X/711/1/012115

Here $i = p$, $a_1(p) = \sum_j \left[\frac{J_1(2m_0^1(j))}{m_0^1(j)} \right]^2$, $a_0(i) = \left[\frac{J_1(2m_1^2(i))}{m_1^2(i)} \right]^2$ .The incident light intensity of point p can be restored. Through making $\varphi_p = \tau_p \sigma_p$, we can get the reduction value of the alignment point as follows:

$$I_0(p) = a_1(p) \cdot a_0(p) \cdot \varphi_p C_k \tag{5}$$

3. Results

Taking the simulation test of 10.6 um(Infrared absorption peak) infrared light source as an example, the parameters of ciculaare selected as follows: The intensity of incident light is set as $I_0 = (100, 30, 50, 1, 70)$ to present the distribution of the intensity. After 5 times calculation of one radius' holes, the detected value is $C = (164.8036, 49.4411, 82.4018, 1.6480, 115.3625)$, after calculating, the final data is $I_0^p = (101.0374, 30.3112, 49.8059, 1.0102, 70.7362)$.The error is near 1% through comparing I_0 and I_0^p , which reveals that the detected values are almost the same as the intensity of incident light, as shown in figure 4.

(a) Original data (b) calculated data

Fig. 4. Linear numerical simulation diagram

According to the linear idea, each linear point in the image is simulated and each linear restored value is spliced into an image at the end. The two-dimensional picture can be restored by combining the linear points. The restoration of the image is not a direct rotational restoration, but divided into several linear points, and finally the restored results are spliced into one image. In Fig. 5, we present the original and calculated pictures. The results show that the degree of restoration between the calculated image and the original image is high. But the original image points are too dense and large, so less than 8% of the error accumulation makes the restored image deviate in some details. In addition, when calculating the reduction by computer, the default value is used to reserve three digits behind the decimal point, and the digits behind are omitted which increases the reduction error.

(a) Original picture (b) calculated picture

Fig. 5. Two-Dimensional Image Simulation Chart of Subregional Linear Restoration

For an image, the rotation simulation test is carried out directly. The test method is to use the idea of rotation and polar coordinates to directly simulate and test a circular highlight map without subarea. In this simulation, considering that there are only five layers of perforated rings, the simulated image is a smaller five-layer thermal image, and the results after restoration are within the expected range,as shown in figure 6. However, careful observation reveals that there are slight differences in edge details between the restored image and the original image, the reason is similar with the above simulation.

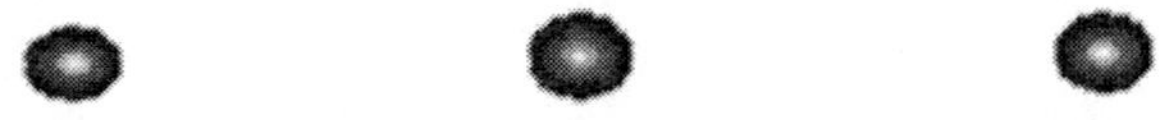

(a) Original picture (b) calculated gray picture (c) calculated hot map
Fig. 6. Two-Dimensional Graph of Rotating Model Restoration

4. Discussion and conclusions

A new single-pixel imaging model based on paired diffractive plates is established. A set of fixed parameters D、f、ρ and weight are selected, which have good effect on infrared image restoration at wavelength of 10.6 um, but there are shortcomings: the results are successful but the points' number are insufficient.

In conclusion, the double rotation of weights can solve the complicated problem of data acquisition in ghost imaging, and can be applied to SF_6 gas imaging sensing and recognition. Besides, the cost can be reduced significantly and the results are reasonable. But there are still further investigations to be made theoretically and experimentally.

Acknowledgement

This research was funded by Fundamental Research Funds for the Central Universities (HUST: 2016YXMS022) and also by the State Key Laboratory of applied optics, Changchun Institute of Optics, Fine Mechanics and Physics, Chinese Academy of Sciences.

Reference

[1] Liu H C, Zhang S. Computational ghost imaging of hot objects in long-wave infrared range. Applied Physics Letters, 2017, 111(3): 031110.
[2] Bennink R S, Bentley S J, Boyd R W, et al. Quantum and classical coincidence imaging. Physical review letters, 2004, 92(3): 033601.
[3] Gibson G M, Sun B, Edgar M P, et al. Real-time imaging of methane gas leaks using a single-pixel camera. Optics express, 2017, 25(4): 2998-3005.
[4] Han Q, Wang J, Zhang J, et al. Diffraction analysis for digital micromirror device scene projectors in the long-wave infrared. Optical Engineering, 2016, 55(8): 085105.
[5] Vandenrijt J F, Thizy C, Georges M P, et al. Long-wave infrared digital holography for the qualification of large space reflectors. International Conference on Space Optics—ICSO 2012. International Society for Optics and Photonics, 2017, 10564: 1056403.
[6] Dunham R W. Diffraction by a Circular Aperture. JOSA, 1964, 54(9): 1102-1105.

IOP Conf. Series: Materials Science and Engineering **711** (2020) 012116 doi:10.1088/1757-899X/711/1/012116

Numerical calculation and analysis of infrared radiation characteristics of stratospheric airship

FENG Yunsong[1,2,3,a] , **LU Yuan**[1,2,3, b], **JIN Wei**[1,2,3,c]

[1] Infrared and Low Temperature Plasma Key Laboratory of Anhui Province, Hefei 230037, China

[2] State Key Laboratory of Pulsed Power Laser Technology, Hefei 230037, China

[3]Po box 904, 460 Huangshan Road, Hefei city, Anhui Province, P.R. China

[a]fyseei@163.com, [b]luyuanmail@163.com, [c]kingvee@163.com

Abstract. In order to estimate the spatial distribution of infrared radiation of stratospheric airship, the geometric model of stratospheric airship was firstly constructed in this paper. Then, considering the influence of environmental radiation, the one-dimensional heat conduction differential equation of airship skin was established, the equation was solved by backward difference method, and the temperature distribution of airship skin was numerically calculated. On this basis, the spatial distribution of airship infrared radiation was further calculated. The results show that the infrared radiation intensity of the airship in the sunny face is higher than that in the shadow face, and the infrared radiation intensity of the airship in the 8~14μm is obviously higher than that in the 3-5μm.

1.Introduction

Stratospheric airships can achieve longer flight time, higher cost efficiency and lower energy consumption, so airships have attracted more and more research attention in recent years[1]. The infrared radiation characteristics of airship are studied by real time measurement, but the experimental data are relatively few due to the high cost and difficulty, and it is difficult to establish a reliable empirical model.

This paper mainly studies the infrared radiation characteristics of airship surface and obtains the infrared radiation model of airship surface by numerical calculation. To study the infrared radiation characteristics of airship surface, the key step is to calculate the temperature distribution of airship skin. According to the thermal properties of the surface material, the infrared radiation of the airship itself can be obtained. Then, combined with the influence of environmental factors on the infrared radiation characteristics of the airship skin, the infrared radiation reflected by the airship can be calculated.

2. Computational model

2.1 Airship geometry model

The shape of the airship is approximately an ellipsoid[2]. The three parameters a, b and c of the ellipsoid are the major axis, the middle axis and the short axis of the ellipsoid respectively, and b=c. From these three parameters, the equation of the ellipsoid in the spatial cartesian coordinate system can be expressed as $\frac{x^2}{a^2}+\frac{y^2}{b^2}+\frac{z^2}{c^2}=1$. Taking points (2a-1) along the long axis of the ellipsoid, divide the long axis of the ellipsoid into 2a equal parts, and cut the ellipsoid into 2a small rings with a width of 1, as shown in figure 1. Then, each ring is evenly spaced with 8 points, and each small ring is further divided, as shown in figure 2. The ellipsoid is divided into (2a-1) *8 facets.

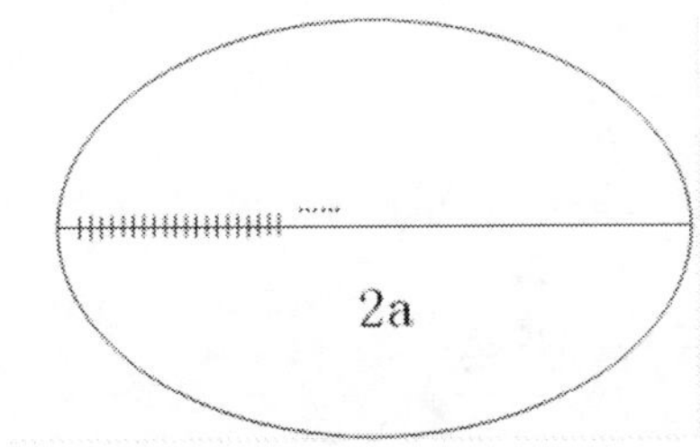

Fig. 1 division of the long axis

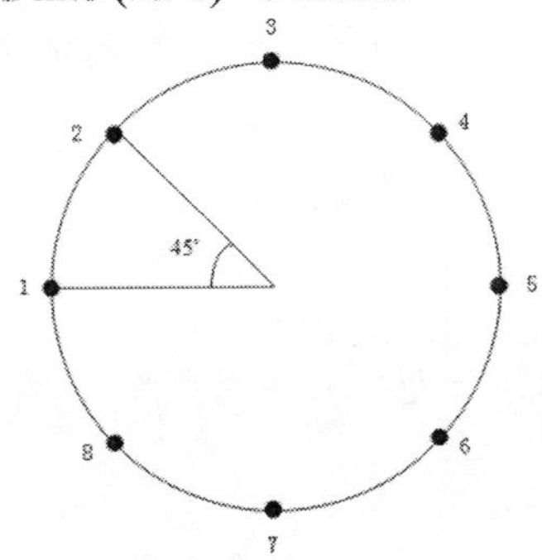

Fig. 2 division of the ring

2.2 Skin material properties

The surface of modern airships is mostly laminated composite materials[3], as shown in figure 3. The coating material used as the outer surface coating is generally polyvinyl chloride coating (PVC) or polytetrafluoroethylene coating (PTFE), among which polytetrafluoroethylene is the ideal coating for the stratospheric airship, and this coating can roughly meet the various material requirements of airships. Due to the low infrared emissivity of PTFE, it is of great significance for airship stealth in infrared[4].

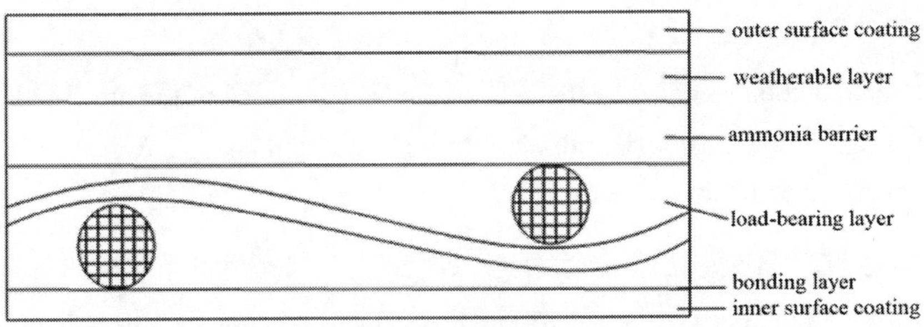

Fig.3 Typical laminated membrane structure

3. Airship skin radiation heat transfer factor

The energy of airship surface heat balance includes solar direct radiation Q_{sundir}, reflection of earth to solar radiation Q_{sunref}, scattering of atmosphere to solar radiation Q_{sundis}, earth radiation Q_{ground}, sky radiation Q_{skyref}, aerodynamic heating $Q_{outside}$, the skins outer surface thermal radiation Q_{self} and internal heat transfer.

3.1Skin self radiation

The radiation loss of the skin surface to the outer space can be obtained by the Stephen-Boltzmann law.

$$Q_{self}=\varepsilon_{skin}\sigma T_{skin}{}^4 \tag{1}$$

where ε_{skin} is the emissivity of the skin outer surface, σ is the Stephen-Boltzmann constant, T_{skin} is the temperature of the skin outer surface.

3.2 Solar radiation

The total illumination of the solar radiation received by the skin surface includes the direct radiation E_{sundir}, the reflection of the earth to the solar radiation E_{sunref} and the scattering of the atmosphere to the solar radiation E_{sundis}, so the total illumination of the solar radiation received by the skin is

$$E_{sun} = E_{sundir} + E_{sundis} + E_{sunref} \tag{2}$$

E_{sun} has to do with weather conditions, the sun's position, the height angle and the relative position of the irradiated surface.

So the solar radiation heat flux received by the target surface is

$$Q_{sun} = \alpha_{skin} E_{sun} \tag{3}$$

where α_{skin} is the absorption of the skin surface, and $\alpha_{skin} = \varepsilon_{skin}$.

3.3 Earth radiation

The radiation from the ground is E_{ground}, then the radiation heat flux that each surface of the airship receives from the ground is

$$Q_{ground} = \alpha_{skin} E_{ground} F_{ground-j} \tag{4}$$

where $F_{ground-j}$ is the angular coefficient of airship surfaces to ground radiation.

3.4 Sky radiation

The atmospheric radiation received by the skin surface is composed of the direct radiation E_{sky} from the atmosphere and the atmospheric radiation through the earth $\rho_{ground} E_{sky}$, then the radiation heat flux each surface of the airship receiving is

$$Q_{sky} = \alpha_{skin} (E_{sky} F_{sky-j} + \rho_{ground} E_{sky} F_{ground-j}) \tag{5}$$

where F_{sky-j} is the angular coefficient of airship surfaces to atmospheric radiation, ρ_{ground} is the reflection of the ground radiation.

3.5 Convective heat transfer

There is a temperature difference between the airship skin and the air, and convective heat transfer occurs. Assuming that the air temperature is T_a, the temperature of the (*i, j*) unit on the surface of the airship is $T_{i,j}$. The heat $Q_{outside}$ generated by the convective heat transfer between the surface of the airship and the atmosphere near the surface of the skin is

$$Q_{outside} = \hbar_a \left(T_a - T_{i,j}\right) \tag{6}$$

where $\hbar_a$ is the heat transfer coefficient of the airship surface skin and the nearby air, its value is 0.4.

IOP Conf. Series: Materials Science and Engineering **711** (2020) 012116 doi:10.1088/1757-899X/711/1/012116

4.Skin temperature field

4.1Heat conduction differential equation

When there is no inner heat source and the height and width of the plane partition is much larger than the thickness, and the basic equation of the airship skin temperature field can be simplified to one-dimensional heat conduction differential equation without inner heat source[5].

$$\rho c \frac{\partial T}{\partial \tau} = k \frac{\partial^2 T}{\partial x^2} \tag{7}$$

In the formula (7), c is the specific heat; ρ is density; τ is time; k is the thermal conductivity.

4.2 Establishment of boundary conditions

According to the analysis of the heat exchange between the skin surface and the environment, the outer boundary condition is obtained.

$$k \frac{\partial T}{\partial r}\Big|_{\text{outer}} = Q_{sun} + Q_{ground} + Q_{sky} + Q_{outside} - Q_{self} \tag{8}$$

Where r is the outer normal direction of the boundary outer edge, the inner wall (the inner boundary) temperature is generally considered to constant temperature.

$$T\big|_{\text{inner}} = T_0 \tag{9}$$

4.3Numerical calculation and results of skin temperature

4.3.1 Numerical calculation

The skin thickness is X, and it can be divided into n thin layers, then the thickness of a thin layer is $\Delta x = X/n$, if the calculation time $t = m\Delta\tau$, $m = 0,1,2,3\cdots$, then at the time t, the center temperature of the i layer can be expressed as $T(m,i)$, as shown in Figure 4.

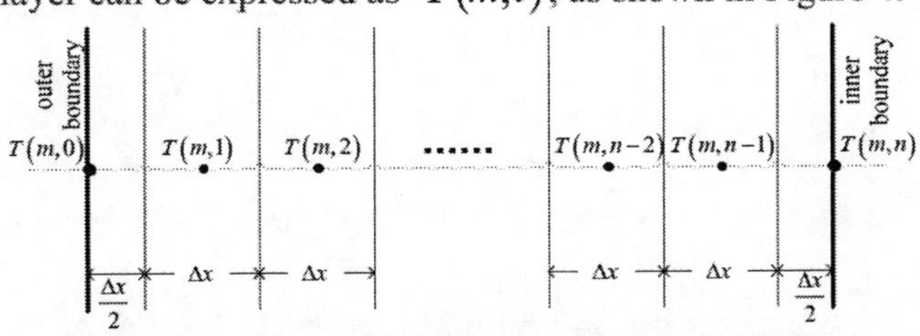

Fig.4 Partitions and nodes of skin heat conduction

Through using the backward difference scheme[6], the heat conduction differential equation (7) is dispersed to

$$\rho c \frac{T(m+1,i) - T(m,i)}{\Delta\tau} = k \frac{T(m+1,i+1) + T(m+1,i-1) - 2T(m+1,i)}{(\Delta x)^2} \tag{10}$$

To the boundary node, according to the energy balance the node equation can be expressed as

$$\rho c \frac{T(m+1,0) - T(m,0)}{\Delta\tau} \cdot \frac{\Delta x}{2} = k \frac{T(m+1,1) - T(m+1,0)}{\Delta x} + Q_{sun}^{m+1} + Q_{ground}^{m+1} + Q_{sky}^{m+1} + Q_{outside}^{m+1} - Q_{self}^{m+1} \tag{11}$$

For the initial condition, the skin temperature can be considered as a linear distribution along the thickness direction.

$$T(0,i)=T_2+i\frac{T_1-T_2}{n} \quad (i=0,1,2\cdots n) \tag{12}$$

where T_1 and T_2 is initial temperature of inner and outer surfaces respectively.

By using the implicit scheme, the differential equations of all nodes are solved iteratively, then the temperature $T(m,0)$ of the skin outer surfaces can be obtained at any time.

4.3.2 Numerical results

The above thermal differential equation can be solved by using MATLAB software[7]. The time is 9:00 am on July 1, 2018. The position of the airship is（N 30°， E 120°），and the flying height of the airship is 30km. Through calculation, the airship surface temperature is obtained, what is shown in the figure 5 and figure 6.

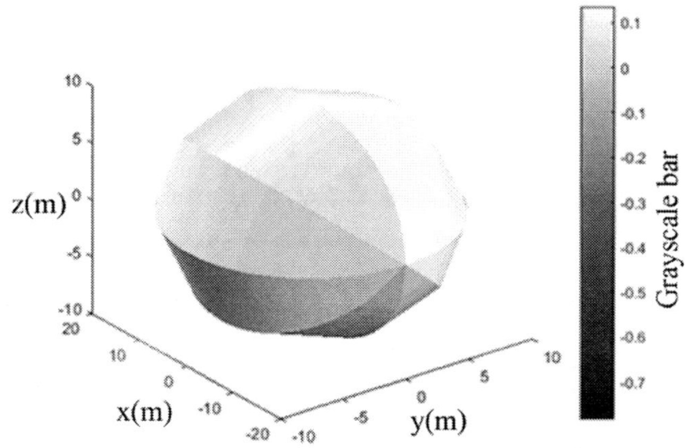

Fig.5 Grayscale map of airship surface temperature at 9:00 am

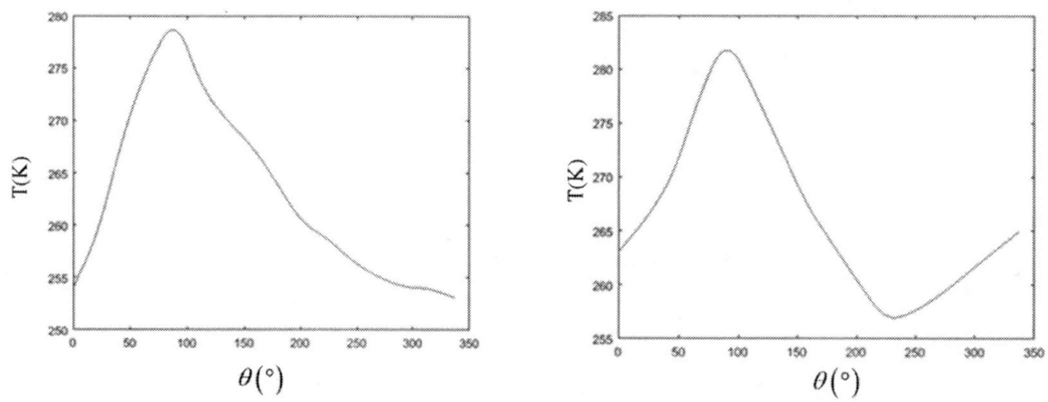

(a) xoy plane (b) xoz plane

Fig.6 Temperature curve of the airship surface at 9:00 am

5.Calculation results and analysis of infrared radiation

Usually the airship skin can be considered as a gray body, and the infrared radiation of the airship contains its own radiation and reflection of environmental radiation.

5.1 Infrared radiation calculation

In the daytime, the environment infrared radiation which is received by any face of the model is

$$E_{bg-j}^{\Delta\lambda} = E_{sun}^{\Delta\lambda} + E_{ground}^{\Delta\lambda} F_{ground-j} + E_{sky}^{\Delta\lambda} F_{sky-j} + \rho_{ground} E_{sky}^{\Delta\lambda} F_{ground-j} \tag{13}$$

Then the reflective infrared radiation of skin to environment radiation is

$$M_{ref-j}^{\Delta\lambda} = \rho_{skin} E_{bg-j}^{\Delta\lambda} \tag{14}$$

where the skin reflectivity $\rho_{skin} = 1 - \varepsilon_{skin}$.

Therefore, the infrared radiation intensity of any face in a certain direction θ due to reflecting the environment radiation can be expressed as

$$I_{ref-j}^{\Delta\lambda} = \frac{M_{ref-j}^{\Delta\lambda}}{\pi} \cos\theta \cdot A_j = \frac{\rho_{skin} \cdot E_{bg-j}^{\Delta\lambda} \cdot A_j \cdot \cos\theta}{\pi} \tag{15}$$

The self infrared radiation intensity of any face in a certain direction θ can be expressed as

$$I_{self-j}^{\Delta\lambda} = \frac{M_{self-j}^{\Delta\lambda}}{\pi} A_j \cos\theta \tag{16}$$

Therefore, the total infrared radiation intensity of any face in a certain direction θ is

$$I_{\theta-j}^{\Delta\lambda} = I_{ref-j}^{\Delta\lambda} + I_{self-j}^{\Delta\lambda} = \frac{\cos\theta}{\pi} \cdot A_j \cdot (\rho_{skin} \cdot E_{bg-j}^{\Delta\lambda} + M_{self-j}^{\Delta\lambda}) \tag{17}$$

The detection direction is $\vec{r}_\theta$, then with the change of θ [8], the intensity distribution of the entire airship skin in the XOY section is

$$I_\theta^{\Delta\lambda} = \sum_{j=(2a-1)*8} I_{\theta-j}^{\Delta\lambda} \cdot \max[\vec{r}_j \cdot \vec{r}_\theta, 0] \tag{18}$$

In the above expressions all of $\Delta\lambda$ represent $3 \sim 5\mu m$ or $8 \sim 14\mu m$.

5.2 Calculation results and analysis of infrared radiation

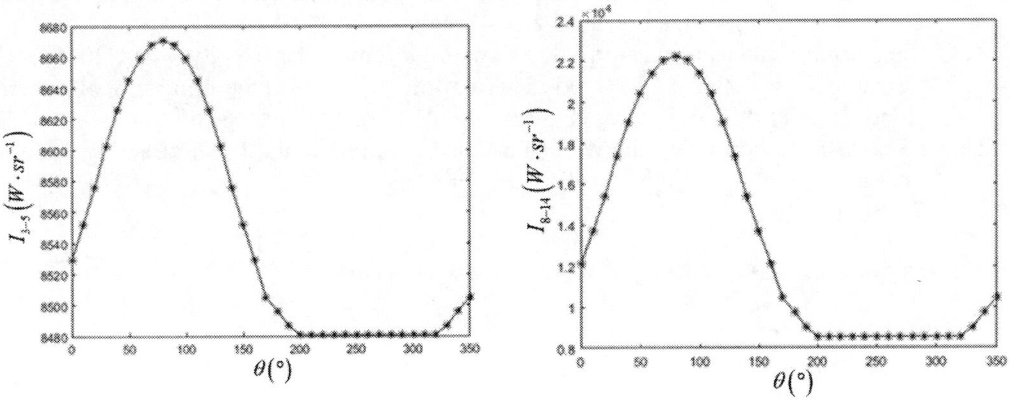

(a) 3-5um (b) 8-14um

Fig.7 Distribution of airship infrared radiation intensity of at 9:00 am

Figure 7 is the mid and far infrared radiation intensity distribution of the airship skin at 9:00 am on July 1, 2018 while the airship aviates. It can be seen from the figure 7, the maximum infrared radiation of the airship surface is 8670.8 $W \cdot sr^{-1}$ in the 3-5μm, and the minimum value is 8480.7 $W \cdot sr^{-1}$. the maximum value of infrared radiation in the 8-14μm is 22253 $W \cdot sr^{-1}$, and the minimum value is 8531.3 $W \cdot sr^{-1}$. The infrared radiation of the airship is mainly the infrared radiation that reflects the environment. In the infrared radiation of the reflective environment, the proportion of the infrared radiation received by the skin from the solar radiation is the largest, and the radiation of the airship skin is relatively small for the reflection of the ambient radiation. The impact on the infrared radiation intensity of the airship is relatively small. From the comparison of the two figures, we can see that the skin has more radiation in the 8-14μm than in the 3-5μm, but their spatial distribution is almost the same regardless of the infrared radiation in which band. The infrared radiation of the airship is the strongest in the 80° direction and the weakest in the 200° direction. The infrared radiation is mainly concentrated in the 0°-150° direction, that is, in the upper half of the airship.

6.Conclusion

The geometric model of stratospheric airship was firstly constructed in this paper. Then, considering the influence of environmental radiation, the one-dimensional heat conduction differential equation of airship skin was established, the equation was solved by backward difference method, and the temperature distribution of airship skin was numerically calculated. On this basis, the spatial distribution of airship infrared radiation was further calculated. The results show that the infrared radiation intensity of the airship in the sunny face is higher than that in the shadow face, and the infrared radiation intensity of the airship in the 8~14μm is obviously higher than that in the 3-5μm.

Acknowledgements

This research was financially supported by the National Nature Science Foundation of China(Grant No. 61405248).

References

[1] Fang Xiande, Wang Weizhi, Li Xiaojian. A Study of Thermal Simulation of Stratospheric Airships[J]. SPACECRAFT RECOVERY &REMOTE SENSING,2007(02):5-9.
[2] Shi Junhai, Zhu Xinjian, Sui Sheng. Modeling and Analysis of Solar Energy Receiving Surface of Stratospheric Airship[J]. Computer Simulation,2007(08):64-67.
[3] Li Hui, He Jingyu, Sun Na. Research on thermal Performance of a stratospheric airship at High-Altitude Station-Keeping Conditions[J]. Journal of Xi'an Aeronautical University,2016,34(01):7-12.
[4] Hong Shi, Shanshan Geng, Xiaohui Qian. Thermodynamics analysis of a stratospheric airship with hovering capability[J]. Applied Thermal Engineering,2019,146.
[5] W. Q. Tao. Numerical Heat Transfer[M]. Xi'an: Xi'an Traffic University Press, 2010,28-39.
[6] C. M. Cao, X. H. Zhang. Infrared Thermal-diagnosis of One-dimensional Unsteady Heat Conduction Problem[J]. Laser & Infrared, 2007, 37(11): 1198-1200.
[7] Zheng Kun-peng, Zhou Guo-yin, Wang Pu-kai, Lv Liang-dong. The Extracting Method of Tank Feature Vector from Infrared Images Based on MATLAB[J]. Infrared Technology,2010,32(11): 625-628.
[8] J. W. Lu, Q. Wang. Aircraft-skin Infrared Radiation Characteristics Modeling and Analysis[J]. Chinese Journal of Aeronautics, 2009,22(5): 493-497.

AUTHOR INDEX

Institute of Physics
Dirac House, Temple Back
Bristol BS1 6BE UK

ISSN: 1757-8981
ISBN 978-1-7138-2570-8